Discrete Mathematics for Computer Science

or

(A Bit of) The Math that Computer Scientists Need to Know

David Liben-Nowell
Professor of Computer Science
Carleton College

WILEY

VP AND EDITORIAL DIRECTOR	Laurie Rosatone
SENIOR DIRECTOR	Don Fowley
ACQUISITIONS EDITOR	Linda Ratts
EDITORIAL MANAGER	Gladys Soto
CONTENT MANAGEMENT DIRECTOR	Lisa Wojcik
CONTENT MANAGER	Nichole Urban
SENIOR CONTENT SPECIALIST	Nicole Repasky
PRODUCTION EDITOR	Rajeshkumar Nallusamy
PHOTO RESEARCHER	Billy Ray
COVER PHOTO CREDIT	© slobo/Getty Images, Inc.

This book was set in TeXGyrePagella 10/12 by SPi Global and printed and bound by Strategic Content Imaging.

This book is printed on acid free paper. ∞

Founded in 1807, John Wiley & Sons, Inc. has been a valued source of knowledge and understanding for more than 200 years, helping people around the world meet their needs and fulfill their aspirations. Our company is built on a foundation of principles that include responsibility to the communities we serve and where we live and work. In 2008, we launched a Corporate Citizenship Initiative, a global effort to address the environmental, social, economic, and ethical challenges we face in our business. Among the issues we are addressing are carbon impact, paper specifications and procurement, ethical conduct within our business and among our vendors, and community and charitable support. For more information, please visit our website: www.wiley.com/go/citizenship.

ISBN: 978-1-118-06553-2 (PBK)
ISBN: 978-1-119-07073-3 (EVALC)

Library of Congress Cataloging in Publication Data:

LCCN: 2017025007

The inside back cover will contain printing identification and country of origin if omitted from this page. In addition, if the ISBN on the back cover differs from the ISBN on this page, the one on the back cover is correct.

To MDSWM, with never-ending appreciation, and in loving memory of my grandfather, Jay Liben, who brought more joy, curiosity, and kvetching to this world than anyone else I know.

Contents

List of Computer Science Connections

Chapter 2: Basic Data Types

Chapter 3: Logic

Chapter 4: Proofs

Chapter 5: Mathematical Induction

Chapter 6: Analysis of Algorithms

Chapter 7: Number Theory

Acknowledgements

> Would thou hadst less deserved,
> That the proportion both of thanks and payment
> Might have been mine! only I have left to say,
> More is thy due than more than all can pay.
>
> William Shakespeare (1564–1616)
> The Scottish Play

To everyone who has helped, directly and indirectly, with everything over these last years—these words cannot adequately convey my thanks, but at least they're a start: *thank you!*

I owe special thanks to a very long list of generous and warm people—many more than I can mention here—for advice and kindness and support, both technical and emotional, as this book came into being. For those whom I haven't named by name, please know that it's only because I have gotten such great support from so many people, and I hope that you'll consider this sentence the promise that, when we next see each other, the first round's on me. While I'm leaving out the names of the many people who have helped make my life happy and fulfilling while I've been working on this book, I do want to give specific thanks to a few people:

I want to thank my colleagues—near and far, including many who are not just colleagues but also dear friends and beloved family members—for their wisdom and patience, for answering my endlessly annoying questions, and for conversations that led to examples or exercises or bug fixes or the very existence of this entire book (even if you didn't know that's what we were talking about at the time): Eric Alexander, Tanya Berger-Wolf, Kelly Connole, Amy Csizmar Dalal, Josh Davis, Roger Downs, Laura Effinger-Dean, Eric Egge, Adriana Estill, Andy Exley, Alex Freeman, Sherri Goings, Jack Goldfeather, Deanna Haunsperger, Pierre Hecker, David Huyck, Sue Jandro, Sarah Jansen, Iris Jastram, Jon Kleinberg, Carissa Knipe, Mark Krusemeyer, Jessica Leiman, Lynn Liben, Jadrian Miles, Dave Musicant, Gail Nelson, Rich Nowell, Layla Oesper, Jeff Ondich, Sam Patterson, Anna Rafferty, Alexa Sharp, Julia Strand, Mike Tie, Zach Weinersmith, Tom Wexler, Kevin Woods, Jed Yang, and Steve Zdancewic.

I also owe my appreciation to Don Fowley, Bryan Gambrel, Beth Golub, Jessy Moor, Anna Pham, Sondra Scott, and Gladys Soto at Wiley. Thanks to Judy Brody for relentless and efficient pursuit of permissions (from many different people and publishers) to use the quotes that appear as epigraphs throughout the book. And thanks as well to the many insightful reviewers of previous drafts of this material. So many times I got chapter reviews back and put them aside in a huff, only to come back to the reviewers' comments months later and realize that their suggestions were exactly right. (And, to be clear: blame me, not them, for the errors that I'm sure remain.)

I specifically want to thank Eric Alexander, Laura Biester, Josh Davis, Charlotte Foran, Jadrian Miles, Dave Musicant, Layla Oesper, Anna Rafferty, Jed Yang, and the Carleton CS 202 students from 2013–2017 for their willingness to work with early, and buggy, drafts of this book. And thanks to those and many other students at Carleton for their patience, and for sending their comments and suggestions for improvements—in particular: Hami Abdi, David Abel, Alexander Auyeung, Andrew Bacon, Kharmen Bharucha, John Blake, Caleb Braun, Macallan Brown, Adam Canady, Noah Carnahan, Yitong Chen, Jinny Cho, Leah Cole, Katja Collier, Lila Conlee, Eric Ewing, Greg Fournier, Andy Freeland, Emma Freeman, Samuel Greaves, Reilly Hallstrom, Jacob Hamalian, Sylvie Hauser, Jack Hessel, Joy Hill, Matt Javaly, Emily Johnston, Emily Kampa, Carlton Keedy, Henry Keiter, Jonathan Knudson, Julia Kroll, Brennan Kuo, Edward Kwiatkowski, Dimitri Lang, Tristan Leigh, Zach Levonian, Daniel Levy, Rhys Lindmark, Gordon Loery, David Long, Robert Lord, Inara Makhmudova, Elliot Mawby, Javier Moran Lemus, Sean Mullan, Micah Nacht, Justin Norden, Laurel Orr, Raven Pillmann, Josh Pitkofsky, Matthew Pruyne, Nikki Rhodes, Will Schifeling, Colby Seyferth, Alex Simonides, Oscar Smith, Kyung Song, Frederik Stensaeth, Patrick Stephen, Maximiliano Villarreal, Alex Voorhees, Allie Warren, Ben Wedin, Michael Wheatman, Jack Wines, Christopher Winter, and Andrew Yang.

This book would not have been possible without the support of Carleton College, not only for the direct support of this project, but also for providing a wonderfully engaging place to make my professional home. When I started at Carleton, my friends and family back east thought that moving to Minnesota (the frontier!) was nothing less than a sign that I had finally lost it, and I have to admit that I thought they had a point. But it's been a fabulous place to have landed, with great friends and colleagues and students—the kind who don't let you get away with anything, but in a good way.

Some of the late stages of the work on this book occurred while I was visiting the University of Cambridge. Thanks to Churchill College and the Computer Laboratory, and especially to Melissa Hines and Cecilia Mascolo, for their hospitality and support.

And my thanks to the somewhat less formal host institutions that have fueled this writing: Brick Oven Bakery, Cakewalk, Goodbye Blue Monday, Tandem Bagels, The Hideaway (Northfield, MN); Anodyne, Blue Moon, Bull Run, Caffetto, Common Roots, Espresso Royale, Isles Bun & Coffee, Keen Eye, Plan B, Precision Grind, Reverie, Spyhouse, Sebastian Joe's, The Beat, The Nicollet, The Purple Onion, Turtle Bread Company, Uncommon Grounds, Urban Bean (Minneapolis, MN); Ginkgo, Grand Central, Kopplin's (St. Paul, MN); Collegetown Bagels (Ithaca, NY); Slave to the Grind (Bronxville, NY); Bloc Eleven, Diesel Cafe (Somerville, MA); Lyndell's (Cambridge, MA); Tryst (Washington, DC); Hot Numbers, Espresso Library (Cambridge, England); and various Starbucks, Caribous, and Dunn Brothers.

And, last but certainly not least, my deepest gratitude to my friends and family for all your help and support while this project has consumed both hours and years. You know who you are, and I hope you also know how much I appreciate you. *Thank you!*

David Liben-Nowell
Northfield, MN
May 2017

PS: I would be delighted to receive any comments or suggestions from readers. Please don't hesitate to get in touch.

Credits

This book was typeset using LaTeX, and I produced all but a few figures from scratch using a combination of PSTricks and TikZ. The other figures are reprinted with permission from their copyright holders. The illustrations that open every chapter were drawn by Carissa Knipe (`http://carissaknipe.com`), who was a complete delight to work with—both on these illustrations and when she was a student at Carleton. I took the photograph of a house in Figure 2.48 myself. Figure 4.5 (the Therac-25 diagram) is reproduced from Nancy Leveson's book *Safeware: System Safety and Computers* with permission from Pearson Education. Figure 4.27 (a poem proving the undecidability of the Halting Problem) is reproduced with permission from Geoffrey K. Pullum. Figure 5.22 (triangulations of a rabbit) is reproduced from a paper by Tobias Isenberg, Knut Hartmann, and Henry König with permission from the Society for Modeling and Simulation International (SCS). Figure 11.15 (a map of some European train routes) is reproduced with permission from RGBAlpha/Getty Images.[1]

[1] Nancy Leveson. *Safeware: System Safety and Computers*. Pearson Education, Inc., New York, 1995; Tobias Isenberg, Knut Hartmann, and Henry König. Interest value driven adaptive subdivision. In *Simulation and Visualisation (SimVis)*, pages 139–149. SCS European Publishing House, 2003; and Geoffrey K. Pullum. Scooping the loop snooper: A proof that the halting problem is undecidable. *Mathematics Magazine*, 73(4):319–320, 2000. Used by permission of Geoffrey K. Pullum.

For their kind permission to use quotes that appear as epigraphs in sections throughout the book, thanks to:

Kurt Vonnegut, p. 102. Excerpt from *Hocus Pocus* by Kurt Vonnegut, copyright ©1990 by Kurt Vonnegut. Used by permission of G. P. Putnam's Sons, an imprint of Penguin Publishing Group, a division of Penguin Random House LLC. All rights reserved. Any third party use of this material, outside of this publication, is prohibited. Interested parties must apply directly to Penguin Random House LLC for permission.

Pablo Picasso, p. 203. ©2017 Estate of Pablo Picasso / Artists Rights Society (ARS), New York. Reprinted with permission.

Laurence J. Peter, p. 317. Reprinted with permission of the estate of Laurence J. Peter.

Carl Sagan, p. 331. From *Broca's Brain: Reflections on the Romance of Science*, ©1979 Carl Sagan. Reprinted with permission from Democritus Properties, LLC.

Peter De Vries, p. 349. Copyright ©1967 by Peter De Vries. Reprinted by permission of Curtis Brown, Ltd. All rights reserved.

Edna St. Vincent Millay, p. 521. Edna St. Vincent Millay, excerpt from a letter to Arthur Davidson Ficke (October 24, 1930) from *Letters of Edna St. Vincent Millay,* edited by Allan Ross Macdougall, ©1952 by Norma Millay Ellis. Reprinted with the permission of The Permissions Company, Inc., on behalf of Holly Peppe, Literary Executor, The Millay Society, www.millay.org.

George C. Marshall, p. 533. Reprinted with permission of the George C. Marshall Foundation.

Peter Drucker, p. 602. Reprinted with permission of the Drucker 1996 Literary Works Trust.

Bob Dylan, p. 603. Lyrics from Bob Dylan's "Don't Think Twice, It's All Right" (1963). Copyright ©1963 by Warner Bros. Inc.; renewed 1991 by Special Rider Music. All rights reserved. International copyright secured. Reprinted by permission.

Mario Andretti, p. 617. Printed with permission of Sports Management Network, Inc.

E. B. White, p. 631. E. B. White / *The New Yorker*; ©Conde Nast. The quote originally appeared in the *Notes and Comment* section of the July 3, 1943 issue of *The New Yorker,* "The 40s: The Story of a Decade." Reprinted with permission.

Charles de Gaulle, p. 647. © Editions Plon. Reprinted with permission.

W. H. Auden, p. 703. "Notes on the Comic" from *The Dyer's Hand and Other Essays* by W. H. Auden, copyright ©1948, 1950, 1952, 1953, 1954, 1956, 1957, 1958, 1960, 1962 by W. H. Auden. Used by permission of Random House, an imprint and division of Penguin Random House LLC. All rights reserved. Any third party use of this material, outside of this publication, is prohibited. Interested parties must apply directly to Penguin Random House LLC for permission.

Bill Watterson, p. 833. Quote from a *Calvin & Hobbes* cartoon; reprinted with permission from Universal Uclick.

Tom Lehrer, p. 926. Lyrics from "Poisoning Pigeons In The Park" reprinted with permission from Maelstrom Music/Tom Lehrer.

Dick Cavett, p. 1021. Reprinted with permission from Dick Cavett.

Tom Stoppard, p. 1108. Excerpts from *Rosencrantz and Guildenstern Are Dead*, copyright © 1967 by Tom Stoppard. Used by permission of Grove/Atlantic, Inc. Any third party use of this material, outside of this publication, is prohibited.

Marshall Dodge and Robert Bryan, p. 1129. From "Which Way to Millinocket?," *Bert and I* (1958). Reprinted with permission from Islandport Press, Inc.

1
On the Point of this Book

In which our heroes decide, possibly encouraged by a requirement for graduation, to set out to explore the world.

Why You Might Care

> Just because some of us can read and write and do a little math, that doesn't mean we deserve to conquer the Universe.
>
> Kurt Vonnegut (1922–2007)
> *Hocus Pocus* (1990)

This book is designed for an undergraduate student who has taken a computer science class or three—most likely, you are a sophomore or junior prospective or current computer science major taking your first non-programming-based CS class. If you are a student in this position, you may be wondering why you're taking this class (or why you *have* to take this class!). Computer science students taking a class like this one sometimes don't see why this material has anything to do with computer science—particularly if you enjoy CS because you enjoy programming.

I want to be clear: programming is awesome! I get lost in code all the time—let's not count the number of hours that I spent writing the code to draw the fractals in Figure 5.1 in LaTeX, for example. (LaTeX, the tool used to typeset this book, is the standard typesetting package for computer scientists, and it's actually also a full-fledged, if somewhat bizarre, programming language.)

But there's more to CS than programming. In fact, many seemingly unrelated problems rely on the same sorts of abstract thinking. It's not at all obvious that an optimizing compiler (a program that translates source code in a programming language like C into something directly executable by a computer) would have anything important in common with a program to play chess perfectly. But, in fact, they're both tasks that are best understood using *logic* (Chapter 3) as a central component of any solution. Similarly, filtering spam out of your inbox ("given a message m, should m be categorized as spam?") and doing speech recognition ("given an audio stream s of a person speaking in English, what is the best 'transcript' reflecting the words spoken in s?") are both best understood using *probability* (Chapter 10).

And these, of course, are just examples; there are many, many ways in which we can gain insight and efficiency by thinking more abstractly about the commonalities of interesting and important CS problems. That is the goal of this book: to introduce the kind of mathematical, formal thinking that will allow you to understand ideas that are shared among disparate applications of computer science—and to make it easier for you to make your own connections, and to extend CS in even more new directions.

How To Use This Book

> Read much, but not many Books.
>
> Benjamin Franklin (1706–1790)
> *Poor Richard's Almanack* (1738)

The brief version of the advice for how to use this book is: *it's your book; use it however you'd like.* (Will Shortz, the puzzle editor of *The New York Times*, gives the analogous advice about crossword puzzles when he's asked whether Googling for an answer is cheating.) But my experience is that students do best when they read actively, with scrap paper close by; most people end up with a deeper understanding of a problem by trying to solve it themselves *first*, before they look at the solution.

I've assumed throughout that you're comfortable with programming in at least one language, including familiarity with recursion. It doesn't much matter which particular programming language you know; we'll use features that are shared by almost all modern languages—things like conditionals, loops, functions, and recursion. You may or may not have had more than one programming-based CS course; many, but not all, institutions require Data Structures as a prerequisite for this material. There are times in the book when a data structures background may give you a deeper understanding (but the same is true in reverse if you study data structures after this material). There are similarly a handful of topics for which rudimentary calculus background is valuable. But knowing/remembering calculus will be specifically useful only a handful of times in this book; the mathematical prerequisite for this material is really algebra and "mathematical maturity," which basically means having some degree of comfort with the idea of a mathematical definition and with the manipulation of a mathematical expression. (The few places where calculus is helpful are explicitly marked.)

There are 10 chapters after this one in the book. Their dependencies are as shown at right. Aside from these dependencies, there are some occasional references to other chapters, but these references are light. If you've skipped Chapter 6—many instructors will choose not cover this material, as it is frequently included in a course on Algorithms instead of this one—then it will still be useful to have an informal sense of O, Ω, and Θ notation in the context of the worst-case running time of an algorithm. (You might skim Sections 6.1 and 6.6 before reading Chapters 7–11.)

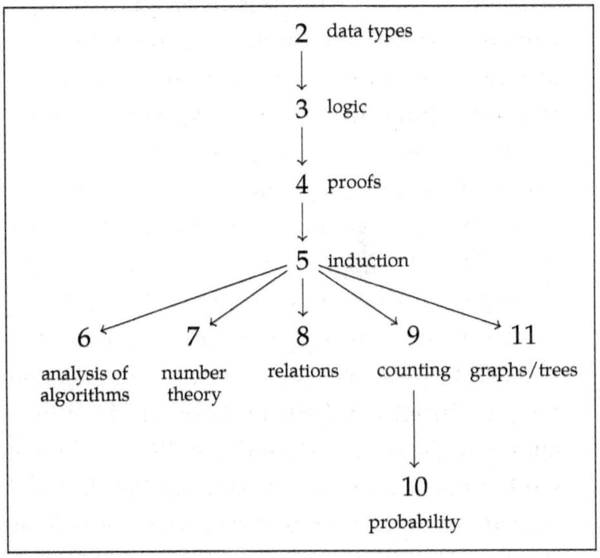

I've tried to include some helpful tips for problem solving in the margins throughout the book, along with a few warnings about common confusions and some notes on terminology/notation that may be helpful in keeping the words and symbols straight. There are also two kinds of extensions to the main material. The "Taking it Further" blocks give more technical details about the material under discussion—an alternate way of thinking about a definition, or a way that a concept is used in CS or a related field. You should read the "Taking it Further" blocks if—but only if!—you find them engaging. Each section also ends with one or more boxed-off "Computer Science Connections" that show how the core material can be used to solve a wide variety of (interesting, I hope!) CS applications. No matter how interesting the core technical material may be, I think that it is what we can *do* with it that makes it worth studying.

What This Book Is About

> All truths are easy to understand once they are discovered; the point is to discover them.
>
> Galileo Galilei (1564–1642)

This book focuses on *discrete* mathematics, in which the entities of interest are distinct and separate. Discrete mathematics contrasts with *continuous* mathematics, as in calculus, which addresses infinitesimally small objects, which cannot be separated. We'll use summations rather than integrals, and we'll generally be thinking about things more like the integers ("$1, 2, 3, \ldots$") than like the real numbers ("all numbers between π and 42"). Because this book is mostly focused on non-programming-based parts of computer science, in general the "output" that you produce when solving a problem will be something different from a program. Most typically, you will be asked to answer some question (quantitatively or qualitatively) and to justify that answer—that is, to *prove* your answer. (A *proof* is an ironclad, airtight argument that convinces its reader of your claim.) Remember that your task in solving a problem is to persuade your reader that your purported solution genuinely solves the problem. Above all, that means that your main task in writing is communication and persuasion.

Be careful; there are two different words that are pronounced identically:

discrete, adj.: individually separate and distinct.

discreet, adj.: careful and judicious in speech, especially to maintain privacy or avoid embarrassment.

You wouldn't read a book about discreet mathematics; instead, someone who trusts you might quietly share it while making sure no one was eavesdropping.

There are three very reasonable ways of thinking about this book.

View #1 is that this book is about the mathematical foundations of computation. This book is designed to give you a firm foundation in mathematical concepts that are crucial to computer science: sets and sequences and functions, logic, proofs, probability, number theory, graphs, and so forth.

View #2 is that this book is about practice. Essentially no particular example that we consider matters; what's crucial is for you to get exposure to and experience with formal reasoning. Learning specific facts about specific topics is less important than developing your ability to reason rigorously about formally defined structures.

View #3 is that this book is about applications of computer science: it's about error-correcting codes (how to represent data redundantly so that the original information is recoverable even in the face of data corruption); cryptography (how to communicate securely so that your information is understood by its intended recipient but not by anyone else); natural language processing (how to interpret the "meaning" of an English sentence spoken by a human using an automated customer service system); and so forth. But, because solutions to these problems rely fundamentally on sets and counting and number theory and logic, we have to understand basic abstract structures in order to understand the solutions to these applied problems.

In the end, of course, all three views are right: I hope that this book will help to introduce some of the foundational technical concepts and techniques of theoretical computer science, and I hope that it will also help demonstrate that these theoretical approaches have relevance and value in work throughout computer science—in topics both theoretical and applied. And I hope that it will be at least a little bit of fun.

Bon voyage!

2
Basic Data Types

In which our heroes equip themselves for the journey ahead, by taking on the basic provisions that they will need along the road.

2.1 *Why You Might Care*

It is a capital mistake to theorize before one has data.

Sir Arthur Conan Doyle (1859–1930),
A Scandal in Bohemia (1892)

This chapter will introduce concepts, terminology, and notation related to the most common data types that recur throughout this book, and throughout computer science. These basic entities—the Booleans (True and False), numbers (integers, rationals, and reals), sets, sequences, functions—are also the basic data types we use in modern programming languages. Essentially every common primitive data type in programs appears on this list: a Boolean, an integer (or an `int`), a real number (or a `float`), and a string (an ordered sequence of characters). Ordered sequences of other elements are usually called *arrays* or *lists*. If you've taken a course on data structures, you've probably worked on several implementations of sets that allow you to insert an element into an unordered collection and to test whether a particular object is a "member" of the collection. And functions that map a given input to a corresponding output are the basic building blocks of programs.

Virtually every interesting computer science application uses these basic data types extensively. *Cryptography*, which is devoted to the secure storage and transmission of information in such a way that a malicious third party cannot decipher that information, is typically based directly on integers, particularly large prime numbers. A ubiquitous task in *machine learning* is to "cluster" a set of entities into a collection of nonoverlapping subsets so that two entities in the same subset are similar and two entities in different subsets are dissimilar. In *information retrieval*, where we might seek to find the document from a large collection that is most relevant to a given query, it is common to represent each document by a vector (a sequence of numbers) based on the words used in the document, and to find the most relevant documents by identifying which ones "point in the same direction" as the query's vector. And functions are everywhere in CS, from data structures like hash tables to the routing that's done for every packet of information on the internet.

In this chapter, we'll describe these basic entities and some standard notation that's associated with them. Some closely related topics will appear later in the book, as well. Chapter 7, on number theory, will discuss some subtler properties of the integers, particularly divisibility and prime numbers. Chapter 8 will discuss relations, a generalization of functions. But, really, every chapter of this book is related to this chapter: our whole enterprise will involve building complex objects out of these simple ones (and, to be ready to understand the more complex objects, we have to understand the simple pieces first). And before we launch into the sea of applications, we need to establish some basic shared language. Much of the basic material in this chapter may be familiar, but regardless of whether you have seen it before, it is important and standard content with which it is important to be comfortable.

2.2 *Booleans, Numbers, and Arithmetic*

Everything you can imagine is real.

Pablo Picasso (1881–1973)

We start with the most basic types of data: *Boolean* values (True and False), *integers* $(\ldots, -2, -1, 0, 1, 2, \ldots)$, *rational numbers* (fractions with integers as numerators and denominators), and *real numbers* (including the integers and all the numbers in between them). The rest of this section will then introduce some basic numerical operations: absolute values and rounding, exponentiation and logarithms, summations and products. Figure 2.1 summarizes this section's notation and definitions.

2.2.1 *Booleans: True and False*

The most basic unit of data is the *bit*: a single piece of information, which either takes on the value 0 or the value 1. Every piece of stored data in a digital computer is stored as a sequence of bits. (See Section 2.4 for a formal definition of sequences.)

We'll view bits from several different perspectives: 1 and 0, on and off, yes and no, *True* and *False*. Bits viewed under the last of these perspectives have a special name, the *Booleans*:

Booleans are named after George Boole (1815–1864), a British mathematician, who was the first person to think about True as 1 and False as 0.

Definition 2.1 (Booleans)
A Boolean value *is either True or False.*

The Booleans are the central object of study of Chapter 3, on logic. In fact, they are in a sense the central object of study of this entire book: simply, we are interested in making true statements, with a proof to justify why the statement is true.

2.2.2 *Numbers: Integers, Reals, and Rationals*

The superficially unintuitive notation for the integers, the symbol $\mathbb{Z}$, is a stylized "Z" that was chosen because of the German word *Zahlen*, which means "numbers." The name *rationals* comes from the word *ratio*; the symbol $\mathbb{Q}$ comes from its synonym *quotient*. (Besides, the symbol $\mathbb{R}$ was already taken by the reals, so the rationals got stuck with their second choice.)

We'll often encounter a few common types of numbers—*integers*, *reals*, and *rationals*:

Definition 2.2 (Integers, Reals, and Rationals)

- *The* integers, *denoted by* $\mathbb{Z}$, *are those numbers with no fractional part: 0, the positive integers* $(1, 2, \ldots)$, *and the negative integers* $(-1, -2, -3, \ldots)$.
- *The* real numbers, *denoted by* $\mathbb{R}$, *are those numbers that can be (approximately) represented by decimal numbers; informally, the reals include all integers and all numbers "between" any two integers.*
- *The* rational numbers, *denoted by* $\mathbb{Q}$, *are those real numbers that can be represented as a ratio* $\frac{n}{m}$ *of two integers* n *and* m, *where* n *is called the* numerator *and* $m \neq 0$ *is called the* denominator. *A real number that is not rational is called an* irrational *number.*

Here are a few examples of each of these types of numbers:

Booleans	True and False
$\mathbb{Z}$	integers $(\ldots, -3, -2, -1, 0, 1, 2, 3, \ldots)$
$\mathbb{Q}$	rational numbers
$\mathbb{R}$	real numbers
$[a, b]$	those real numbers x where $a \leq x \leq b$
(a, b)	those real numbers x where $a < x < b$
$[a, b)$	those real numbers x where $a \leq x < b$
$(a, b]$	those real numbers x where $a < x \leq b$
$\lvert x \rvert$	absolute value of x: $\lvert x \rvert := -x$ if $x < 0$; $\lvert x \rvert := x$ if $x \geq 0$
$\lfloor x \rfloor$	floor of x: x rounded down to the nearest integer
$\lceil x \rceil$	ceiling of x: x rounded up to the nearest integer
b^n	b multiplied by itself n times
$b^{1/n}$, or $\sqrt[n]{b}$	a number y such that $y^n = b$ (where $y \geq 0$ if possible), if one exists
$b^{m/n}$	$(b^{1/n})^m$
$\log_b x$	logarithm: $\log_b x$ is the value y such that $b^y = x$, if one exists
$n \bmod k$	modulo: $n \bmod k :=$ the remainder when dividing n by k
$k \mid n$	k (evenly) divides n
$\sum$	summation: $\sum_{i=1}^{n} x_i := x_1 + x_2 + \cdots + x_n$
$\prod$	product: $\prod_{i=1}^{n} x_i := x_1 \cdot x_2 \cdot \cdots \cdot x_n$

Figure 2.1: Summary of the basic mathematical notation introduced in Section 2.2.

Example 2.1 (Integers, reals, and rationals)

The following are all examples of integers: $1, 42, 0$, and -17.

All of the following are real numbers: 1, 99.44, the ratio of the circumference of a circle to its diameter $\pi \approx 3.141592653\cdots$, and the so-called *golden ratio* $\phi = (1 + \sqrt{5})/2 \approx 1.61803\cdots$.

Examples of rational numbers include $\frac{3}{2}$, $\frac{9}{5}$, $\frac{16}{4}$, and $\frac{4}{1}$. (In Chapter 8, we'll talk about the familiar notion of the equivalence of two rational numbers like $\frac{1}{2}$ and $\frac{2}{4}$, or like $\frac{16}{4}$ and $\frac{4}{1}$, based on common divisors. See Example 8.36.) Of the example real numbers above, both 1 and 99.44 are rational numbers; we can write them as $\frac{1}{1}$ and $\frac{4972}{50}$, for example. Both π and ϕ are irrational.

Here are a few useful points relating these three types of numbers:

- All integers are rational numbers (with denominator equal to 1).
- All rational numbers are real numbers.
- But not all rational numbers are integers and not all real numbers are rational: for example, $\frac{3}{2}$ is not an integer, and $\sqrt{2}$ is not rational. (We'll prove that $\sqrt{2}$ is not rational in Example 4.21.)

Taking it further: Definition 2.2 specifies $\mathbb{Z}$, $\mathbb{Q}$, and $\mathbb{R}$ somewhat informally. To be completely rigorous, one can define the nonnegative integers as the smallest collection of numbers such that: (i) 0 is an integer; and (ii) if x is an integer, then $x + 1$ is also an integer. See Section 5.4.1. (Of course, for even this definition to make sense, we'd need to give a rigorous definition of the number zero and a rigorous definition of the operation of adding one.) With a proper definition of the integers, it's fairly easy to define the rationals as ratios of integers. But formally defining the real numbers is surprisingly challenging; it was a major enterprise of mathematics in the late 1800s, and is often the focus of a first course in analysis in an undergraduate mathematics curriculum.

Virtually every programming language supports both integers (usually known as `ints`) and real numbers (usually known as `floats`); see p. 217 for some discussion of the way that these basic numerical types are implemented in real computers. (Rational numbers are much less frequently implemented as basic data types in programming languages, though there are some exceptions, like Scheme.)

In addition to the basic symbols that we've introduced to represent the integers, the rationals, and the reals ($\mathbb{Z}$, $\mathbb{Q}$, and $\mathbb{R}$), we will also introduce special notation for some specific subsets of these numbers. We will write $\mathbb{Z}^{\geq 0}$ and $\mathbb{Z}^{\leq 0}$ to denote the nonnegative integers $(0, 1, 2, \ldots)$ and nonpositive integers $(0, -1, -2, \ldots)$, respectively. Generally, when we write $\mathbb{Z}$ with a superscripted condition, we mean all those integers for which the stated condition is true. For example, $\mathbb{Z}^{\neq 1}$ denotes all integers aside from 1. Similarly, we write $\mathbb{R}^{>0}$ to denote the positive real numbers (every real number $x > 0$). Other conditions in the superscript of $\mathbb{R}$ are analogous.

We'll also use standard notation for *intervals* of real numbers, denoting all real numbers between two specified values. There are two variants of this notation, which allow "between two specified values" to either *include* or *exclude* those specified values. We use round parentheses to mean "exclude the endpoint" and square brackets to mean "include the endpoint" when we denote a range:

- (a, b) denotes those real numbers x for which $a < x < b$.
- $[a, b]$ denotes those real numbers x for which $a \leq x \leq b$.
- $(a, b]$ denotes those real numbers x for which $a < x \leq b$.
- $[a, b)$ denotes those real numbers x for which $a \leq x < b$.

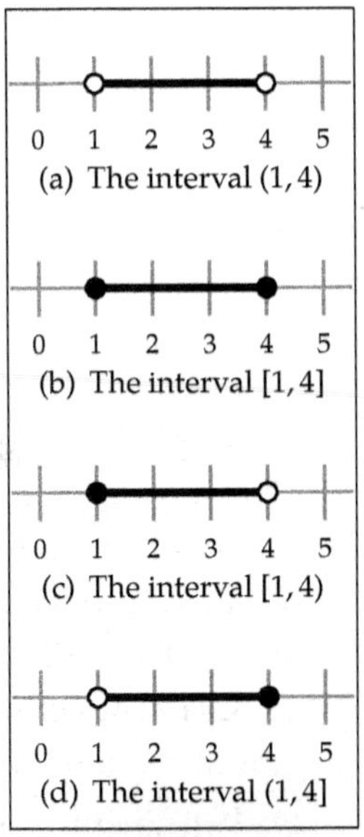

(a) The interval $(1, 4)$
(b) The interval $[1, 4]$
(c) The interval $[1, 4)$
(d) The interval $(1, 4]$

Figure 2.2: Number lines representing real numbers between 1 and 4, with 1 included in the range in (b, c), and 4 included in the range in (b, d).

Sometimes (a, b) and $[a, b]$ are, respectively, called the *open interval* and *closed interval* between a and b. These four types of intervals are also sometimes denoted via a *number line*, with open and closed circles denoting open and closed intervals; see Figure 2.2 for an example. For two real numbers x and y, we will use the standard notation "$x \approx y$" to denote that x is *approximately equal to* y. This notation is defined informally, because what counts as "close enough" to be approximately equal will depend heavily on context.

2.2.3 *Absolute Value, Floor, and Ceiling*

In the remaining subsections of Section 2.2, we will give definitions of some standard arithmetic operations that involve the numbers we just defined. We'll start in this subsection with three operations on a real number: absolute value, floor, and ceiling.

The *absolute value* of a real number x, written $|x|$, denotes how far x is from 0, disregarding the *sign* of x (that is, disregarding whether x is positive or negative):

Definition 2.3 (Absolute Value)
The absolute value *of a real number x is* $|x| := \begin{cases} x & \text{if } x \geq 0 \\ -x & \text{otherwise.} \end{cases}$

For example, $|42.42| = 42.42$ and $|-128| = 128$. (Definition 2.3 uses standard notation for defining "by cases": the value of $|x|$ is x when $x \geq 0$, and the value of $|x|$ is $-x$ otherwise—that is, when $x < 0$.)

For a real number x, we can consider x "rounded down" or "rounded up," which are called the *floor* and *ceiling* of x, respectively:

Definition 2.4 (Floor and ceiling)
The floor *of a real number x, written* $\lfloor x \rfloor$, *denotes the largest integer that is less than or equal to x. The* ceiling *of a real number x, written* $\lceil x \rceil$, *denotes the smallest integer that is greater than or equal to x.*

Note that Definition 2.4 defines the floor and ceiling of negative numbers, too; the definition doesn't care whether x is greater than or less than 0.

Here are a few examples of floor and ceiling:

Example 2.2 (Floor and ceiling)
We have $\lfloor \sqrt{2} \rfloor = \lfloor 1.4142 \cdots \rfloor = 1$, $\lfloor 2\pi \rfloor = \lfloor 6.28318 \cdots \rfloor = 6$, and $\lfloor 3 \rfloor = 3$. For ceilings, we have $\lceil \sqrt{2} \rceil = 2$, $\lceil 2\pi \rceil = 7$, and $\lceil 3 \rceil = 3$.

For negative numbers, $\lfloor -\sqrt{2} \rfloor = \lfloor -1.4142 \cdots \rfloor = -2$, and $\lceil -\sqrt{2} \rceil = -1$.

The number line may give an intuitive way to think about floor and ceiling: $\lfloor x \rfloor$ denotes the first integer that we encounter moving left in the number line starting at x; $\lceil x \rceil$ denotes the first integer that we encounter moving right from x. (And x itself counts for both definitions.) See Figure 2.3.

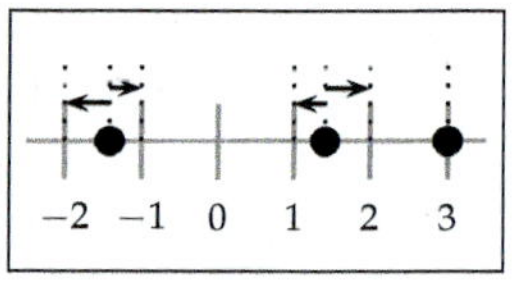

Figure 2.3: The floor and ceiling of $-\sqrt{2}$, $\sqrt{2}$, and 3.

2.2.4 *Exponentiation*

We next consider raising a number to an *exponent* or *power*.

Definition 2.5 (Raising a number to an integer power)
For a real number b and a nonnegative integer n, the number b^n denotes the result of multiplying b by itself n times:

$$b^0 := 1 \qquad \text{and, for } n \geq 1,\ \ b^n := \underbrace{b \cdot b \cdots b}_{n \text{ times}}.$$

The number b is called the base *and the integer n is called the* exponent.

For example, $2^0 = 1$, $2^2 = 2 \cdot 2 = 4$, $2^5 = 2 \cdot 2 \cdot 2 \cdot 2 \cdot 2 = 32$, and $5^2 = 5 \cdot 5 = 25$.

Note again that $b^0 = 1$ for *any* base b, including $b = 0$. (The case of 0^0 is tricky: one is tempted to say *both* "0 to the anything is 0" *and* "anything to the 0 is 1." But, of course, these two statements are inconsistent. For us, the latter trumps the former, and $0^0 = 1$, as in Definition 2.5.)

Raising a base to nonintegral exponents

Consider the expression b^x for an exponent $x > 0$ that is not an integer. (It's all too easy to have done this calculation by typing numbers into a calculator without actually thinking about what the expression actually means!) Here's the definition of $b^{m/n}$ when the exponent $\frac{m}{n}$ is a rational number:

Definition 2.6 (Raising a number to a positive rational power)
For any real number b and for any positive integers m and $n \neq 0$:

- *$b^{1/n}$ denotes the number y such that $y^n = b$. The value $b^{1/n}$ is called the* nth root of *b, and it can also be denoted by $\sqrt[n]{b}$. If there are two values y such that $y^n = b$, then by $b^{1/n}$ we mean the number $y \geq 0$ such that $y^n = b$. If there are no such values y, then we'll treat $b^{1/n}$ as undefined.*

- *$b^{m/n}$ denotes the* mth power *of $b^{1/n}$: that is, $b^{m/n} := (b^{1/n})^m$.*

Here are a few examples:

Example 2.3 (Some fractional exponents)

- $16^{1/2}$ is the value y such that $y^2 = 16$, so $16^{1/2} = 4$ (because $4^2 = 16$). Similarly, $16^{1/4} = 2$ because $2^4 = 16$.

- The value of $5^{1/2}$ is roughly 2.2360679774, because $2.2360679774^2 \approx 5$. (But note that this value of $5^{1/2}$ is only an approximation, because actually $2.2360679774^2 = 4.99999999955372691076 \neq 5$.)

- As the definition implies, there may be more than one y such that $y^n = b$. For example, consider $4^{1/2}$. We need a number y such that $y^2 = 4$—and either $y = 2$ or $y = -2$ satisfies this condition. By the definition, if there are positive and negative values of y satisfying the requirement, we choose the positive one. So $4^{1/2} = 2$.

- For $(-8)^{1/3}$, we need a value y such that $y^3 = -8$. No $y \geq 0$ satisfies this condition, but $y = -2$ does. Thus $(-8)^{1/3} = -2$.

- For $(-8)^{1/2}$, we need a value y such that $y^2 = -8$. No $y \geq 0$ satisfies this condition, and no $y \leq 0$ does either. Thus we will treat $(-8)^{1/2}$ as undefined.

Taking it further: Definition 2.6 presents difficulties if we try to compute, say, $\sqrt{-1}$: the definition tells us that we need to find a number y such that $y^2 = -1$. But $y^2 \geq 0$ if $y \leq 0$ *and* if $y \geq 0$, so no real number y satisfies the requirement $y^2 = -1$. To handle this situation, one can define the *imaginary numbers*, specifically by defining $\mathbf{i} := \sqrt{-1}$. (The name "real" to describe real numbers was chosen to contrast with the imaginary numbers.)

We will not be concerned with imaginary numbers in this book, although—perhaps surprisingly—there are some very natural computational problems in which imaginary numbers are fundamental parts of the best algorithms solving them, such as in signal processing and speech processing (transcribing English words from a raw audio stream) or even quickly multiplying large numbers together.

When we write $\sqrt{b}$ without explicitly indicating which root is intended, then we are talking about the *square root* of b. In other words, $\sqrt{b} := \sqrt[2]{b}$ denotes the y such that $y^2 = b$. An integer n is called a *perfect square* if $\sqrt{n}$ is an integer.

Definition 2.7 (Raising a number to a negative power)
When the exponent x is negative, then b^x is defined as $\frac{1}{b^{-x}}$.

For example, $2^{-4} = \frac{1}{2^4} = \frac{1}{16}$ and $25^{-3/2} = \frac{1}{25^{3/2}} = \frac{1}{(25^{1/2})^3} = \frac{1}{5^3} = \frac{1}{125}$.

For an irrational exponent x, the value of b^x is approximated arbitrarily closely by choosing a rational number $\frac{m}{n}$ sufficiently close to x and computing the value of $b^{m/n}$.

$$2^3 = 8$$
$$2^{31/10} = 8.5741\cdots$$
$$2^{314/100} = 8.8815\cdots$$
$$2^{3141/1000} = 8.8213\cdots$$
$$2^{31415/10000} = 8.8244\cdots$$
$$2^{314159/100000} = 8.8249\cdots$$
$$\vdots$$

Figure 2.4: Approximating 2^π.

Taking it further: A fully rigorous treatment of irrational powers requires a formal definition of the real numbers and an (ε, δ)-style proof as in calculus; we will omit the details as they are tangential to our purposes in this book. The basic idea is to choose a rational number m/n that approximates x to within a small error—for example, approximate r by the first k digits of its decimal expansion (which can be written as $m/10^k$)—and approximate b^x by $b^{m/n}$. For example, 2^π is approximated by the sequence shown in Figure 2.4; the value of 2^π is the limit of this sequence of approximations.

While essentially every modern programming language supports exponentiation—including positive, fractional, and negative powers—in some form, often in a separate math library, the actual behind-the-scenes computation is rather complicated. See p. 218 for some discussion of the underlying steps that are done to compute a quantity like $\sqrt{x}$.

Here are a few useful facts about exponentiation:

Theorem 2.1 (Properties of exponentials)
For any real numbers a and b, and for any rational numbers x and y:

$$b^0 = 1 \quad (2.1.1)$$
$$b^1 = b \quad (2.1.2)$$
$$b^{x+y} = b^x \cdot b^y \quad (2.1.3)$$
$$(b^x)^y = b^{xy} \quad (2.1.4)$$
$$(ab)^x = a^x \cdot b^x \quad (2.1.5)$$

These properties follow fairly straightforwardly from the definition of exponentiation. (The properties of Theorem 2.1 carry over to irrational exponents, though the proofs are less straightforward.)

2.2.5 *Logarithms*

The *logarithm* (or *log*) is the inverse operation to exponentiation: the value of an exponential b^y is the result of multiplying a number b by itself y times, while the value of a logarithm $\log_b x$ is the number of times we must multiply b by itself to get x.

Problem-solving tip: I have found many CS students scared, and scarred, by logs. The fear appears to me to result from students attempting to *memorize* facts about logs without trying to think about what they *mean*. Mentally translating between logs and exponentials can help make these properties more intuitive and can help make them make sense. Often the intuition of a property of exponentials is reasonably straightforward to grasp.

Definition 2.8 (Logarithm)
For a positive real number $b \neq 1$ and a real number $x > 0$, the logarithm base b of x, *written $\log_b x$, is the real number y such that $b^y = x$.*

Here are a few simple examples:

Example 2.4 (Some logs)
- The quantity $\log_3 81$ is the power to which we must raise 3 to get 81—and thus $\log_3 81 = 4$, because $3^4 = 3 \cdot 3 \cdot 3 \cdot 3 = 81$.
- Similarly, $\log_4 16 = 2$, because $4^2 = 16$.

- Because $2 = \sqrt{4} = 4^{1/2}$, we have $\log_4 2 = 0.5$.
- $128^0 = 1$, so $\log_{128} 1 = 0$.
- $2^{1.5849625} = 2.999999998 \approx 3$, so $\log_2 3 \approx 1.5849625$.

For any base b, note that $\log_b x$ does get larger as the value of x increases, but it gets larger very slowly. Figure 2.5 illustrates the slow rate of growth of $\log_{10} x$ as x grows.

For a real number $x \leq 0$ and any base b, the expression $\log_b x$ is undefined. For example, the value of $\log_2(-4)$ would be the number y such that $2^y = -4$—but 2^y can never be negative. Similarly, logarithms base 1 are undefined: $\log_1 2$ would be the number y such that $1^y = 2$—but $1^y = 1$ for every value of y.

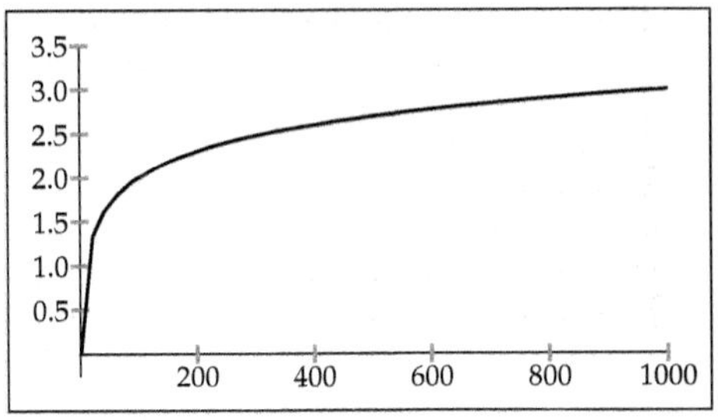

Figure 2.5: A graph of $\log_{10} x$.

Logarithms show up frequently in the analysis of data structures and algorithms, including a number that we will discuss in this book. Several facts about logarithms will be useful in these analyses, and are also useful in other settings. Here are a few:

Theorem 2.2 (Properties of logarithms)
For any real numbers $b > 1$, $c > 1$, $x > 0$, and $y > 0$, the following properties hold:

$$\log_b 1 = 0 \tag{2.2.1}$$

$$\log_b b = 1 \tag{2.2.2}$$

$$\log_b xy = \log_b x + \log_b y \qquad \textit{log of a product} \tag{2.2.3}$$

$$\log_b \tfrac{x}{y} = \log_b x - \log_b y \qquad \textit{log of a quotient} \tag{2.2.4}$$

$$\log_b x^y = y \log_b x \tag{2.2.5}$$

$$\log_b x = \frac{\log_c x}{\log_c b} \qquad \textit{“change of base” formula} \tag{2.2.6}$$

These properties generally follow directly from the analogous properties of exponentials in Theorem 2.1. You'll explore some properties of logarithms (including many of the properties from Theorem 2.2) in the exercises.

We will make use of one standard piece of notational shorthand: often the expression $\log x$ is written without an explicit base. When computer scientists write the expression $\log x$, we mean $\log_2 x$. One other base is commonly used in logarithms: the *natural logarithm* $\ln x$ denotes $\log_e x$, where $e \approx 2.718281828\cdots$ is defined from calculus as $e := \lim_{n\to\infty}(1 + \frac{1}{n})^n$.

Throughout this book (and throughout computer science), the assumed base of $\log x$ is 2. (Some computer scientists write $\lg x$ to denote $\log_2 x$; we'll simply write $\log x$.) But be aware that mathematicians or engineers may treat the default base to be e or 10.

2.2.6 *Moduli and Division*

So far, we've discussed multiplying numbers (repeatedly, to compute exponentials); in this subsection, we turn to the *division* of one number by another. When we consider dividing two integers—64 by 5, for example—there are several useful values to consider: regular-old division ($\frac{64}{5} = 12.8$), what's sometimes called *integer division* giving "the whole part" of the fraction ($\lfloor \frac{64}{5} \rfloor = 12$), and the *remainder* giving "the leftover part" of the fraction (the difference between 64 and $12 \cdot 5$, namely $64 - 60 = 4$).

We will return to these notions of division in great detail in Chapter 7, but we'll begin here with the formal definitions for the notions related to remainders:

Definition 2.9 (Modulus (remainder))
For any integers $k > 0$ and n, the integer $n \bmod k$ is the remainder when we divide n by k. Using the "floor" notation from Section 2.2.3, the value $n \bmod k$ is defined as $n \bmod k := n - k \cdot \lfloor \frac{n}{k} \rfloor$.

Here are examples of the value of a few integers mod 3:

Example 2.5 (Three values mod 3)

- $8 \bmod 3 = 2$, because 8 is 2 more than a multiple of 3, namely 6. (Or because $\left\lfloor \frac{8}{3} \right\rfloor = \lfloor 2.6666\cdots \rfloor = 2$, and $8 - 2 \cdot 3 = 8 - 6 = 2$.)
- $28 \bmod 3 = 1$, as $\left\lfloor \frac{28}{3} \right\rfloor = 9$, and $28 - 9 \cdot 3 = 28 - 27 = 1$.
- $48 \bmod 3 = 0$, because $\left\lfloor \frac{48}{3} \right\rfloor = \lfloor 16 \rfloor = 16$, and $48 - 16 \cdot 3 = 0$.

Taking it further: In many programming languages, the / operator performs integer division when its arguments are both integers, and performs "real" division when either argument is a floating point number. So the expression `64 / 5` will yield `12`, but `64.0 / 5` and `64 / 5.0` and `64.0 / 5.0` will all yield `12.8`. In this book, though, we will always mean "real" division when we write x/y or $\frac{x}{y}$.

The $n \bmod k$ operation is a standard one in programming languages—it's written as `n % k` in many languages, including Java, Python, and C/C++, for example.

In Definition 2.9, we allowed n to be a negative integer, which may stretch your intuition about remainders a bit. Here's an example of this case of the definition:

Example 2.6 (A negative integer mod 5)
We'll compute $-3 \bmod 5$ simply by following the definition of mod from Definition 2.9:

$$-3 \bmod 5 = (-3) - 5 \cdot \left\lfloor \frac{-3}{5} \right\rfloor = (-3) - 5 \cdot (-1) = (-3) + 5 = 2.$$

Viewed from an appropriate perspective, this calculation should actually be very intuitive: the value $r = n \bmod k$ gives the amount r by which n exceeds its closest multiple of k. (And -3 is 2 more than a multiple of 5, namely -5, so $-3 \bmod 5 = 2$.)

Notice that the value of $n \bmod k$ is always at least 0 and at most $k - 1$, for any n and any $k > 0$; the remainder when dividing by k can never be k or more. At one of these extreme points, when $\frac{n}{k}$ has zero remainder, then we say that k *(evenly) divides* n:

Definition 2.10 (Integer k (evenly) divides integer n)
For any integers $k > 0$ and n, we say that k divides n, written $k \mid n$, if $\frac{n}{k}$ is an integer. Notice that $k \mid n$ is equivalent to $n \bmod k = 0$.

Here's a simple example:

Example 2.7 (What 5 divides)
Because $5 \cdot \lfloor \frac{10}{5} \rfloor = 5 \cdot 2 = 10 = 10$, we know $5 \mid 10$. But $5 \cdot \lfloor \frac{9}{5} \rfloor = 5 \cdot 1 = 5 \neq 9$, so $5 \not\mid 9$.

By rearranging the floor-based definition from Definition 2.9 when $n \bmod k = 0$, we can see that the condition $k \mid n$ is also equivalent to the condition $k \cdot \lfloor \frac{n}{k} \rfloor = n$.

Some special numbers: evens, odds, primes, composites

A few special types of integers are defined in terms of their divisibility—specifically based on whether they are divisible by 2 (*evens* and *odds*), or whether they are divisible by any other integer except for 1 (*primes* and *composites*).

Definition 2.11 (Even, odd, and parity)
A nonnegative integer n is even *if $n \bmod 2 = 0$, and n is* odd *if $n \bmod 2 = 1$. The* parity *of n is its "oddness" or "evenness."*

For example, we have $17 \bmod 2 = 1$ and $42 \bmod 2 = 0$, so 17 is odd and 42 is even.

Taking it further: If we view 0 as False and 1 as True (see Section 2.2.1), then the value $n \bmod 2$ can be interpreted as a Boolean value. In fact, there's a deeper connection between arithmetic and the Booleans than might be readily apparent. The "exclusive or" of two Boolean values p and q (which we will encounter in Section 3.2.3) is denoted $p \oplus q$, and the expression $p \oplus q$ is true when one but not both of p and q is true. The exclusive or is sometimes referred to as the *parity function*, because $p + q$ is odd (viewing p and q as numerical values, 0 or 1) exactly when $p \oplus q$ is true (viewing p and q as Boolean values, False or True).

Definition 2.12 (Prime and composite numbers)
A positive integer $n > 1$ is prime *if the only positive integers that evenly divide n are 1 and n itself. A positive integer $n > 1$ is* composite *if it is not prime.*

Notice that the definition of prime numbers does not include 0 and 1, and neither does the definition of composite numbers: in other words, 0 and 1 are neither composite nor prime. Here are a few examples of prime and composite numbers:

Example 2.8 (Prime numbers)
Problem: Is 77 prime? What about 7?

Solution: 77 is not prime, because it is evenly divisible by 7. In other words, because $77 \bmod 7 = 0$ (and the integer 7 that evenly divides 77 is neither 1 nor 77 itself), 77 is composite.

On the other hand, 7 is prime. Convincing yourself that something *is* prime is harder than convincing yourself that something is *not* prime, but we can see it by trying all the possible divisors, namely every positive integer except 1 and 7: $7 \bmod 2 = 1$ and $7 \bmod 3 = 1$ and $7 \bmod 4 = 3$ and $7 \bmod 5 = 2$ and $7 \bmod 6 = 1$, and furthermore $7 \bmod d = 7$ for any $d \geq 8$. None of these remainders is zero, so 7 is prime.

Example 2.9 (Small primes and composites)
The first ten prime numbers are $2, 3, 5, 7, 11, 13, 17, 19, 23, 29$. The first ten composite numbers are $4, 6, 8, 9, 10, 12, 14, 15, 16, 18$.

Chapter 7 is devoted to the properties of modular arithmetic, prime numbers, and the like. These quantities have deep and important connections to cryptography, error-correcting codes, and other applications that we'll explore later.

2.2.7 *Summations and Products*

There is one final piece of notation related to numbers that we need to introduce: a simple way of expressing the *sum* or *product* of a collection of numbers. We'll start with the compact *summation notation* that allows us to express the result of adding many numbers:

Definition 2.13 (Summation notation)
Let $x_1, x_2, \ldots, x_n$ be a sequence of n numbers. We write $\sum_{i=1}^{n} x_i$ (usually read as "the sum for i equals 1 to n of x_i") to denote the sum of the x_is:

$$\sum_{i=1}^{n} x_i := x_1 + x_2 + \cdots + x_n.$$

The variable i is called the index of summation *or the* index variable.

Note that $\sum_{i=1}^{0} x_i = 0$: when you add nothing together, you end up with zero.

Here are a few very simple examples:

Example 2.10 (Some simple summations)
Let $a_1 = 2$, $a_2 = 4$, $a_3 = 8$, and $a_4 = 16$, and let $b_1 = 1$, $b_2 = 2$, $b_3 = 3$, and $b_4 = 4$. Then

$$\sum_{i=1}^{4} a_i = a_1 + a_2 + a_3 + a_4 = 2 + 4 + 8 + 16 = 30$$

$$\sum_{i=1}^{4} b_i = b_1 + b_2 + b_3 + b_4 = 1 + 2 + 3 + 4 = 10$$

We can interpret this summation notation as if it expressed a **for** loop, as shown in Figure 2.6. The **for** loop interpretation might help make the "empty sum" more intuitive: the value of $\sum_{i=1}^{0} x_i = 0$ is simply 0 because *result* is set to 0 in line 1, and it never changes, because $n = 0$ (and therefore line 3 is never executed).

```
1: result := 0
2: for i := 1, 2, ..., n
3:     result := result + x_i
4: return result
```

Figure 2.6: A **for** loop that returns the value of $\sum_{i=1}^{n} x_i$.

In general, instead of just adding x_i in the ith term of the sum, we can add any expression involving the index of summation. (We can also start the index of summation at a value other than 1: to denote the sum $x_j + x_{j+1} + \cdots + x_n$, we write $\sum_{i=j}^{n} x_i$.) Here are a few examples:

Example 2.11 (Some sums)
Let $a_1 = 2$, $a_2 = 4$, $a_3 = 8$, and $a_4 = 16$. Then

$$\begin{aligned}
\textstyle\sum_{i=1}^{4} a_i &= 2+4+8+16 &= 30\\
\textstyle\sum_{i=1}^{4} (a_i+1) &= (2+1)+(4+1)+(8+1)+(16+1) &= 34\\
\textstyle\sum_{i=1}^{4} i &= 1+2+3+4 &= 10
\end{aligned}$$

Example 2.12 (Some more sums)
Problem: As above, let $a_1 = 2$, $a_2 = 4$, $a_3 = 8$, and $a_4 = 16$. What are the values of the following expressions?

1. $\sum_{i=1}^{4} i^2$ 2. $\sum_{i=2}^{4} i^2$ 3. $\sum_{i=1}^{4} (a_i + i^2)$ 4. $\sum_{i=1}^{4} 5$

Solution: Here are the values of these sums:

1. $\sum_{i=1}^{4} i^2 = 1^2+2^2+3^2+4^2 = 30$
2. $\sum_{i=2}^{4} i^2 = 2^2+3^2+4^2 = 29$
3. $\sum_{i=1}^{4} (a_i + i^2) = (2+1^2)+(4+2^2)+(8+3^2)+(16+4^2) = 60$
4. $\sum_{i=1}^{4} 5 = 5+5+5+5 = 20$

Two special types of summations arise frequently enough to have special names. A *geometric series* is $\sum_{i=1}^{n} \alpha^i$ for some real number α; an *arithmetic series* is $\sum_{i=1}^{n} i \cdot \alpha$ for a real number α. See Section 5.2.2 for more on these types of summations.

We will very occasionally consider an *infinite* sequence of numbers $x_1, x_2, \ldots, x_i, \ldots$; we may write $\sum_{i=1}^{\infty} x_i$ to denote the infinite sum of these numbers.

Example 2.13 (An infinite sum)
Define $x_i := 1/2^i$, so that $x_1 = 1/2$, $x_2 = 1/4$, $x_3 = 1/8$, and so forth. We can write $\sum_{i=1}^{\infty} x_i$ to denote $1/2 + 1/4 + 1/8 + 1/16 + \cdots$. The value of this summation is 1: each term takes the sum halfway closer to 1.

While the **for** loop in Figure 2.6 would run forever if we tried to apply it to an infinite summation, the idea remains precisely the same: we successively add the value of each term to the *result* variable. (We will discuss this type of infinite sum in detail in Section 5.2.2, too.)

Reindexing summations

Just as in a **for** loop, the "name" of the index variable in a summation doesn't matter, as long as it's used consistently. For example, both $\sum_{i=1}^{5} a_i$ and $\sum_{j=1}^{5} a_j$ denote the value of $a_1 + a_2 + a_3 + a_4 + a_5$.

We can also rewrite a summation by *reindexing* it (also known as using a *change of index* or a *change of variable*), by adjusting both the limits of the sum (lower and upper) and what's being summed while ensuring that, overall, exactly the same things are being added together.

Example 2.14 (Shifting by two)
The sums $\sum_{i=3}^{n} i$ and $\sum_{j=1}^{n-2}(j+2)$ are equal, because both express $3+4+5+\cdots+n$. (We have applied the substitution $j := i-2$ to get from the first summation to the second.)

Example 2.15 (Counting backward)
The following two summations have the same value:

$$\sum_{i=0}^{n}(n-i) \qquad \text{and} \qquad \sum_{j=0}^{n} j.$$

We can produce one from the other by substituting $j := n-i$, so that $i = 0, 1, \ldots, n$ corresponds to $j = n-0, n-1, \ldots, n-n$ (or, more simply, to $j = n, n-1, \ldots, 0$).

Reindexing can be surprisingly helpful when we're confronted by ungainly summations; doing so can often turn the given summation into something more familiar.

NESTED SUMS

We can sum any expression that depends on the index variable—including summations. These summations are called *double summations* or, more generally, *nested summations*. Just as with nested loops in programs, the key is to read "from the inside out" in simplifying a summation. Here are two examples:

Example 2.16 (A double sum)
Let's compute $\sum_{i=1}^{6}\left[\sum_{j=1}^{i} 5\right]$.

Observe that, for any fixed value of $i \geq 0$, the value of $\sum_{j=1}^{i} 5$ is just $5i$, because we are summing i different copies of the number 5. Therefore

$$\sum_{i=1}^{6}\left[\sum_{j=1}^{i} 5\right] = \sum_{i=1}^{6} 5i = 5+10+15+20+25+30 = 105.$$

Example 2.17 (A slightly more complicated double sum)
<u>*Problem:*</u> What is $\sum_{i=1}^{6}\left[\sum_{j=1}^{i} j\right]$?

<u>*Solution:*</u> Observe that the inner sum ($\sum_{j=1}^{i} j$) has the following value, for each $1 \leq i \leq 6$:

- $\sum_{j=1}^{1} j = 1$
- $\sum_{j=1}^{2} j = 1+2 = 3$
- $\sum_{j=1}^{3} j = 1+2+3 = 6$
- $\sum_{j=1}^{4} j = 1+2+3+4 = 10$
- $\sum_{j=1}^{5} j = 1+2+3+4+5 = 15$
- $\sum_{j=1}^{6} j = 1+2+3+4+5+6 = 21$

Thus $\sum_{i=1}^{6}\sum_{j=1}^{i} j = 1+3+6+10+15+21 = 56.$

When you're programming and need to write two nested loops, it sometimes ends up being easier to write the loops with one variable in the outer loop rather than the other variable. Similarly, it may turn out to be easier to think about a nested sum by *reversing the summation*—that is, swapping which variable is the "outer" summation and which is the "inner." If we have any sequence $a_{i,j}$ of numbers indexed by two variables i and j, then $\sum_{i=1}^{n}\sum_{j=1}^{n} a_{i,j}$ and $\sum_{j=1}^{n}\sum_{i=1}^{n} a_{i,j}$ have precisely the same value.

	1	2	3	4
1	7	5	6	5
2	5	5	1	7
3	3	5	8	3

(a) A small table with some arbitrarily chosen numbers.

	$j=1$	2	3	4	5	6	7	8
$i=1$	−1	−1	−2	−2	−3	−3	−4	−4
2	1	1	2	2	3	3	4	4
3	−1	−1	−2	−2	−3	−3	−4	−4
4	1	1	2	2	3	3	4	4
5	−1	−1	−2	−2	−3	−3	−4	−4
6	1	1	2	2	3	3	4	4
7	−1	−1	−2	−2	−3	−3	−4	−4
8	1	1	2	2	3	3	4	4

(b) The terms of $\sum_{i=1}^{n}\sum_{j=1}^{n}\left((-1)^i \cdot \lceil \frac{j}{2} \rceil\right)$, for $n = 8$.

Figure 2.7: Two tables whose elements we'll sum "row-wise" and "column-wise."

Here are two examples of reversing the order of a double summation, for the tables shown in Figure 2.7:

Example 2.18 (A simple sum)

Consider the table in Figure 2.7(a). Write $a_{i,j}$ to denote the element in the ith row and jth column of the table. Then the sum of elements in the table is, by summing the row-sums,

$$\sum_{i=1}^{3}\left[\sum_{j=1}^{4} a_{i,j}\right] = \sum_{i=1}^{3} \text{the sum of elements in row } i \qquad = 23 + 18 + 19 \qquad = 60.$$

And, by summing the column-sums, the sum of elements in the table is also

$$\sum_{j=1}^{4}\left[\sum_{i=1}^{3} a_{i,j}\right] = \sum_{j=1}^{4} \text{the sum of elements in column } j \qquad = 15 + 15 + 15 + 15 \qquad = 60.$$

Example 2.19 (A double sum, reversed)

<u>*Problem:*</u> Let $n = 8$. What is the value of the following sum?

$$\sum_{i=1}^{n}\sum_{j=1}^{n}\left[(-1)^i \cdot \left\lceil \frac{j}{2} \right\rceil\right]$$

<u>*Solution:*</u> We are computing the sum of all the values contained in the table in Figure 2.7(b). The *hard* way to add up all of these values is by computing the row sums, and then adding them all up. (The given equation expresses this hard way.) The *easier* way is reverse the summation, and to instead compute

$$\sum_{j=1}^{n}\sum_{i=1}^{n}\left[(-1)^i \cdot \left\lceil \frac{j}{2} \right\rceil\right].$$

For any value of j, observe that $\sum_{i=1}^{n}(-1)^i \cdot \lceil \frac{j}{2} \rceil$ is actually zero! (This value is just $(\lceil \frac{j}{2} \rceil)\frac{n}{2} + (-\lceil \frac{j}{2} \rceil)\frac{n}{2}$.) In other words, every column sum in the table is zero. Thus the value of the entire summation is $\sum_{j=1}^{n} 0$, which is just 0.

Problem-solving tip: When you're looking at a complicated double summation, try reversing it; it may be much easier to analyze the other way around.

Note that computing the sum from Example 2.19 when $n = 100$ or $n = 100{,}000$ remains just as easy if we use the column-based approach: as long as n is an even number, every column sum is 0, and thus the entire summation is 0. (The row-based approach is ever-more painful to use as n gets large.)

Here's one more example—another view of the double sum $\sum_{i=1}^{6}\sum_{j=1}^{i} j$ from Example 2.17—where reversing the summation makes the calculation simpler:

Example 2.20 (A double sum, redone)
The value of $\sum_{i=1}^{6}\sum_{j=1}^{i} j$ is the sum of all the numbers in the table in Figure 2.8. We solved Example 2.17 by first computing $\sum_{j=1}^{i} j$, which is the sum of the numbers in the ith row. We then summed these values over the six different values of i to get 56.

Alternatively, we can compute the desired sum by looking at *columns* instead of *rows*. The sum of the table's elements is also $\sum_{j=1}^{6}\left[\sum_{i=j}^{6} j\right]$, where $\sum_{i=j}^{6} j$ is the sum of the numbers in the jth *column*. Because there are a total of $(7-j)$ terms in $\sum_{i=j}^{6} j$, the sum of the numbers in the jth column is precisely $j \cdot (7-j)$. (For example, the 4th column's sum is $4 \cdot (7-4) = 4 \cdot 3 = 12$.) Thus the overall summation can be written as

$$\sum_{i=1}^{6}\sum_{j=1}^{i} j = \sum_{j=1}^{6}\left[j \cdot (7-j)\right] = (1\cdot 6)+(2\cdot 5)+(3\cdot 4)+(4\cdot 3)+(5\cdot 2)+(6\cdot 1)$$

$$= 6+10+12+12+10+6 = 56.$$

	1	2	3	4	5	6
1	1					
2	1	2				
3	1	2	3			
4	1	2	3	4		
5	1	2	3	4	5	
6	1	2	3	4	5	6

Figure 2.8: The terms of $\sum_{i=1}^{6}\sum_{j=1}^{i} j$. We seek the sum of all entries in the table.

Products

The $\sum$ notation allows us to express repeated *addition* of a sequence of numbers; there is analogous notation to represent repeated *multiplication* of numbers, too:

The summation and product notation have a secret mnemonic to help you remember what each means: "Σ" is the Greek letter Sigma, which starts with the same letter as the word *sum*. And "Π" is the Greek letter Pi, which starts with the same letter as the word *product*.

Definition 2.14 (Product notation)
Let $x_1, x_2, \ldots, x_n$ be a sequence of n numbers. We write $\prod_{i=1}^{n} x_i$ (usually read as "the product for i equals 1 to n of x_i") to denote the product of the x_is:

$$\prod_{i=1}^{n} x_i := x_1 \cdot x_2 \cdot \;\cdots\; \cdot x_n.$$

There are direct analogues between the notions regarding $\sum$ and corresponding notions for $\prod$: the **for** loop interpretation (Figure 2.9), infinite products, reindexing, and nested products. One slight difference worthy of note: the value of $\prod_{i=1}^{0} x_i$ is 1; when we multiply by nothing, we're multiplying by one.

```
result := 1
for i := 1, 2, ..., n
    result := result · x_i
return result
```

Figure 2.9: A **for** loop that returns the value of $\prod_{i=1}^{n} x_i$.

Example 2.21 (Some products)
Here are a few simple products:

$$\begin{aligned}
\textstyle\prod_{i=1}^{4} i &= 1\cdot 2\cdot 3\cdot 4 &&= 24\\
\textstyle\prod_{i=0}^{4} i &= 0\cdot 1\cdot 2\cdot 3\cdot 4 &&= 0\\
\textstyle\prod_{i=1}^{4} i^2 &= 1^2\cdot 2^2\cdot 3^2\cdot 4^2 &&= 576\\
\textstyle\prod_{i=1}^{4} 5 &= 5\cdot 5\cdot 5\cdot 5 &&= 625
\end{aligned}$$

Computer Science Connections

Integers and `int`s, Reals and `float`s

Every modern programming language has types that correspond to the integers and the real numbers, often called something like `int` (short for "integer") and `float` (short for *floating-point number*; more about this name and the floating point representation is below).

In most programming languages, though, these types differ from $\mathbb{Z}$ and $\mathbb{R}$ in important ways. Every piece of data stored on a computer is stored as a sequence of bits, and typically the bit sequence storing a number has some fixed length. For example, an `int` stored using 7 bits can range from 0000000 (the number 0 represented in binary) to 1111111 (the number $2^7 - 1 = 127$ represented in binary). Typically, the first bit in an `int`'s representation is reserved as the *sign bit* (set to True for a negative number and False for a positive number), and the remaining bits store the value of the number. (See Figure 2.10.) Thus there's a bound on the largest `int`, depending on the number of bits used to represent `int`s in a particular programming language: 32,767 in Pascal ($= 2^{15} - 1$, using 16 bits per `int`: 1 sign bit and 15 data bits), and 2,147,483,647 in Java ($= 2^{31} - 1$; 32 bits, of which 1 is a sign bit). Similar constraints apply to the set of real numbers representable as a `float`.

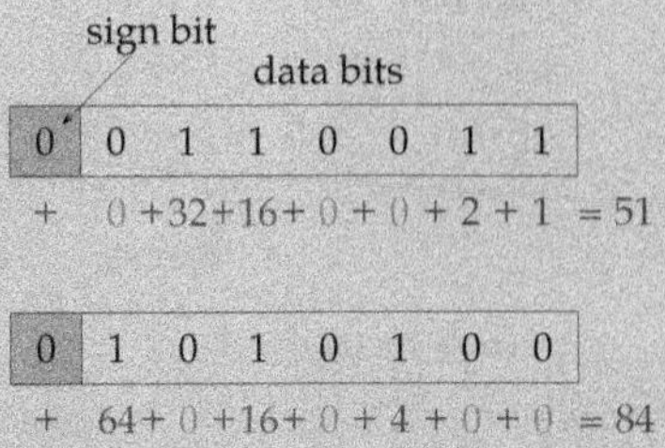

Figure 2.10: The integers 51 and 84, represented in binary as 8-bit signed integers.

A crucial point about $\mathbb{Z}$ and $\mathbb{R}$ is that they are *infinite*: there is no smallest integer, there's no biggest real number, and there isn't even a biggest real number that is smaller than 1. In almost every programming language, however, there is a smallest `int`, a biggest `float`, and a biggest `float` that's smaller than 1: after all, there are only finitely many possible `float`s (perhaps 2^{64} different values), and one of these 2^{64} values is the smallest `float`.

The finite nature of these programming language data types can cause some subtle bugs in programs. There are issues related to *integer overflow* if we try to store "too large" an integer: for example, when we compute 32767 + 1 in Pascal, the result is −32768. And there are bugs related to *underflow* if we try to store "too small" a floating-point number: for example, if we compute $(0.0000000001)^{33}$ in Python, the result is 0.0. (But $(0.0000000001)^{32}$ is, correctly, 10^{-320}.) Similarly, there are also rounding errors implicit in floating point representations of numbers: because there are only finitely many different `float`s, the infinitely many real numbers cannot all be stored exactly. For example, when I type `0.0006 - 0.0004 == 0.0002` into a Python interpreter, I get `False` as output. (That's because, according to Python, `0.0006 - 0.0004` is `0.00019999999999999993`, not `0.0002`.)

You can learn more about the details of how numerical values are stored on computers in a course on computer architecture. In addition to the floating-point standard, other interesting details include *2's complement* storage of integers, which allows a single representation of positive and negative integers so that addition "just works" the same way, even with a sign bit. You can learn more about this material in a good computer architecture textbook, such as

The name *float* originates with a clever idea that's used to mitigate (though not solve) the issues above: we allow the decimal point to "float" in the representation of different numbers. Consider decimal numbers like

$$x = 0.000000000000000000000000000000000000000000000000001$$
$$y = 1929192919291929192919291929192919291929192919291929.5.$$

If, say, we represent these numbers using a total of 64 bits, most of the 64 bits representing x are devoted to the part after decimal point, whereas most of the 64 bits representing y are devoted to the part before the decimal point.[1]

[1] David A. Patterson and John L. Hennessy. *Computer Organization and Design: the Hardware/Software Interface*. Morgan Kaufmann, 4th edition, 2008.

Computer Science Connections

Computing Square Roots, and Not Computing Square Roots

Programs can make use of numerical operations in surprisingly complex ways. Many programmers just happily use these numerical operations without thinking about how they're implemented—but a little knowledge of what's happening behind the scenes can actually help speed up our programs. Computer hardware can directly and efficiently execute basic arithmetic operations like addition and multiplication and division, but more complex operations may require many of these basic operations.

Consider the task of computing $\sqrt{x}$, given an input value x, for example. The basic idea is to use some kind of *iterative improvement* algorithm: we start with a guess y_0 of the value of $\sqrt{x}$, and then update our guess to a new guess y_1 (by observing in some way whether y_0 was too big or too small). We continue to improve our guess until we've reached a value y such that y^2 is "close enough" to x. (We can specify the *tolerance* of the algorithm—that is, how close counts as "close enough.")

A simple implementation of this idea is called *Heron's method*, named after the 1st-century Greek mathematician Heron of Alexandria and shown in Figure 2.11. It relies on the nonobvious fact that the average of y and $\frac{x}{y}$ is closer to $\sqrt{x}$ than y was. (Unless y is exactly equal to $\sqrt{x}$, of course; in that case, the new guess is identical to the old guess: the average of $\sqrt{x}$ and $\frac{x}{\sqrt{x}}$ is still $\sqrt{x}$.) Almost two millennia later, Isaac Newton developed a general technique for computing values of numerical expressions involving exponentials, among other things. This technique, known as *Newton's method*, involves calculus—specifically, using derivatives to figure out how far to move from a current guess y_i in making the next guess y_{i+1}. Like Heron's method, Newton's method is an example of a technique in *scientific computing*, the subfield of computer science devoted to efficient computation of numerical values, often for the purposes of simulating a complex system.[2]

Input: A positive real number x.
Output: A real number y such that $y^2 \approx x$.

1: Let y_0 be arbitrary, and let $i := 0$.
2: **while** $(y_i)^2$ is too far from x:
3: let $y_{i+1} := \frac{y_i + \frac{x}{y_i}}{2}$ and $i := i + 1$
4: **return** y_i

For example, here's the computation of the square root of $x = 42$, using $\frac{x}{2}$ as the initial guess:

i	y_i
0	21
1	11.5
2	7.576086956 ···
3	6.559922961 ···
4	6.481218587 ···
5	6.480740716 ···
6	6.480740698 ···

Figure 2.11: Heron's method for computing square roots, and an example.

Many interesting questions and techniques are used in scientific computing; one outstanding, and classic, reference for some of this material is the book

[2] William Press, Saul Teukolsky, William Vetterling, and Brian Flannery. *Numerical Recipes: The Art of Scientific Computing*. Cambridge University Press, 3rd edition, 2007.

Work in scientific computing has improved the efficiency of numerical computation. But even better is to be aware of the fact that operations like square roots require significant computation "under the hood," and to avoid them when possible. To take one particular example, consider applying a *blur filter* to an image: replace each pixel p by the average of all pixels within a radius-r circle centered at p in the original image. To compute the blurred version of a particular pixel p, we might look at every pixel q within $\pm r$ rows or columns and compute whether p and q are within distance r. (See Figure 2.12.) There are two natural ways to compute whether the two pixels p and q are within distance r:

1. the "obvious" way: test whether $\sqrt{(p_x + q_x)^2 + (p_y + q_y)^2} \leq r$.
2. the "other" way: test whether $(p_x + q_x)^2 + (p_y + q_y)^2 \leq r^2$.

While there is no important mathematical difference between these two formulas (we've simply squared both sides in the "other" way), there *is* a computational difference. Because square roots are expensive to compute, it turns out that in my Python implementation of a blur filter, using the "other" way was about 12% faster than using the "obvious" way.

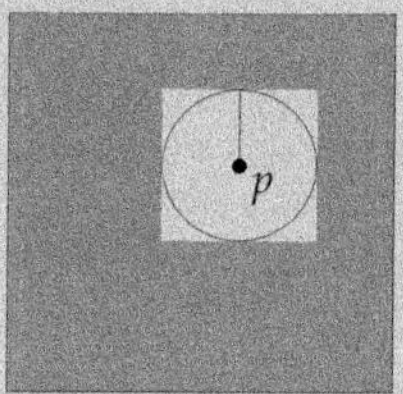

Figure 2.12: Implementing a blur filter. We wish to average all pixels within the circle to compute the new pixel p.

2.2.8 Exercises

What are the smallest and largest integers that are ...

2.1 ... in the interval $(111, 202)$?

2.2 ... in the interval $[111, 202)$?

2.3 ... in the interval $(17, 42)$ but not in the interval $(39, 99]$?

2.4 ... in the interval $[17, 42]$ but not in the interval $[39, 99)$?

Explain your answers to the following questions.

2.5 If x and y are integers, is $x + y$ necessarily an integer?

2.6 If x and y are rational numbers, is $x + y$ necessarily rational?

2.7 If x and y are irrational numbers, is $x + y$ necessarily irrational?

What is the value of each of the following expressions?

2.8 $\lfloor 2.5 \rfloor + \lceil 3.75 \rceil$

2.9 $\lfloor 3.14159 \rfloor \cdot \lceil 0.87853 \rceil$

2.10 $(\lfloor 3.14159 \rfloor)^{\lceil 3.14159 \rceil}$

2.11 Most programming languages provide two different functions called *floor* and *truncate* to trim real numbers to integers. In these languages, `floor`(x) is defined exactly as we defined $\lfloor x \rfloor$, and `trunc`(x) is defined to simply delete any digits that appear after the decimal point in writing x. So `trunc`(3.14159) = 3.14159 = 3. Explain why programming languages have both `floor` and `trunc`—that is, explain under what circumstances `floor`(x) and `trunc`(x) give different values.

Using floor, ceiling, and standard arithmetic notation, give an expression for a real number x ...

2.12 ... rounded to the nearest integer. ("Round up" for a number that's exactly between two integers—for example, 7.5 rounds to 8.)

2.13 ... rounded to the nearest 0.1.

2.14 ... rounded to the nearest 10^{-k}, for an arbitrary number k of digits after the decimal point.

2.15 ... truncated to k digits after the decimal point—that is, leaving off the $(k+1)$st digit and beyond. (For example, 3.1415926 truncated with 3 digits is 3.141, and truncated with 4 digits is 3.1415.)

> **Taking it further:** Many programming languages provide a facility for displaying formatted output, particularly numbers, in the style of Example 2.15. For example, `printf("%.3f", x)` says to "print (formatted)" the value of x with only 3 digits after the decimal point. (The "f" of "printf" stands for formatted; the "f" of `"%.3f"` stands for float.) This style of `printf` command appears in many languages: C, Java, Python, and others.

2.16 For what value(s) of x in the interval $[2, 3]$ is $x - \frac{\lfloor x \rfloor + \lceil x \rceil}{2}$ the largest?

2.17 For what value(s) of x in the interval $[2, 3]$ is $x - \frac{\lfloor x \rfloor + \lceil x \rceil}{2}$ the smallest?

Let x be a real number. Rewrite each of the following as simply as possible:

2.18 $\lfloor \lfloor x \rfloor \rfloor$

2.19 $\lceil \lceil x \rceil \rceil$

2.20 $\lfloor \lceil x \rceil \rfloor$

2.21 $\lceil \lfloor x \rfloor \rceil$

2.22 Are $|\lfloor x \rfloor|$ and $\lfloor |x| \rfloor$ always equal? Explain.

2.23 Are $1 + \lfloor x \rfloor$ and $\lfloor 1 + x \rfloor$ always equal? Explain.

2.24 Are $\lfloor x \rfloor + \lfloor y \rfloor$ and $\lfloor x + y \rfloor$ always equal? Explain.

2.25 Let x be a real number. Describe (in English) what $1 + \lfloor x \rfloor - \lceil x \rceil$ represents. Explain.

2.26 In performing a binary search for x in a sorted n-element array $A[1 \ldots n]$ (see Figure 6.17(a)), the first thing we do is to compare the value of x and the value of $A\left[\lfloor \frac{1+n}{2} \rfloor\right]$. Assume that all elements of A are distinct. How many elements of A are *less than* $A\left[\lfloor \frac{1+n}{2} \rfloor\right]$? How many are *greater*? Write your answers as simply as possible.

2.27 Which is bigger, 3^{10} or 10^3?

What is the value of each of the following expressions?

2.28 4^8

2.29 $(1/4)^8$

2.30 $(-4)^8$

2.31 $(-4)^9$

2.32 $256^{1/4}$

2.33 $8^{1/4}$

2.34 $8^{3/4}$

2.35 $(-9)^{1/4}$

What is the value of each of the following expressions?

2.36 $\log_2 8$

2.37 $\log_2(1/8)$

2.38 $\log_8 2$

2.39 $\log_{1/8} 2$

2.40 Which is bigger, $\log_{10} 17$ or $\log_{17} 10$?

Each of the following statements are general properties of logarithms (from Theorem 2.2), for any real numbers $b, c > 1$ and $x, y > 0$. Using the definition of logarithms and the properties of exponentials from Theorem 2.1, justify each of these properties.

2.41 $\log_b 1 = 0$

2.42 $\log_b b = 1$

2.43 $\log_b x^y = y \log_b x$

2.44 $\log_b xy = \log_b x + \log_b y$

2.45 $\log_b x = \frac{\log_c x}{\log_c b}$

Using the properties from Theorem 2.2 that you just proved, and the fact that $\log_b x = \log_b y$ exactly when $x = y$ (for any base $b > 1$), justify the following additional properties of logarithms:

2.46 For any real numbers $b > 1$ and $x > 0$, we have that $b^{[\log_b x]} = x$.

2.47 For any real numbers $b > 1$ and $a, n > 0$, we have that $n^{[\log_b a]} = a^{[\log_b n]}$.

2.48 Prove (2.2.4) from Theorem 2.2: for any $b > 1$ and $x, y > 0$, we have that $\log_b \frac{x}{y} = \log_b x - \log_b y$.

2.49 Using notation defined in this chapter, define the "hyperceiling" $\overline{\lceil n \rceil}$ of a positive integer n, where $\overline{\lceil n \rceil}$ is the smallest exact power of two that is greater than or equal to n. (That is, $\overline{\lceil n \rceil}$ denotes the smallest value of 2^k where $2^k \geq n$ and k is a nonnegative integer.)

2.50 Similar to the last exercise: when writing down an integer n on paper using standard decimal notation, we need enough columns for all the digits of n (and perhaps one additional column for a "$-$" if $n < 0$). Write down an expression indicating how many columns we need to represent n. *(Hint: use the case notation introduced in Definition 2.3, and be sure that your expression is well defined—that is, it doesn't "generate any errors"—for all integers n.)*

What are the values of the following expressions?

2.51 202 mod 2

2.52 202 mod 3

2.53 202 mod 10

2.54 -202 mod 10

2.55 17 mod 42

2.56 42 mod 17

2.57 17 mod 17

2.58 -42 mod 17

2.59 -42 mod 42

2.60 Observe the Python behavior of the `%` operator (the Python notation for mod) that's shown in Figure 2.13. The first two lines (3 mod 5 = 3 and -3 mod 5 = 2) are completely consistent with the definition that we gave for mod (Definition 2.9), including its use for n mod k when n is negative (as in Example 2.6). But we haven't defined what n mod k means for $k < 0$. Propose a formal definition of `%` in Python that's consistent with Figure 2.13.

```
>>> 3 % 5
3
>>> -3 % 5
2
>>> 3 % -5
-2
>>> -3 % -5
-3
```

Figure 2.13: Python's implementation of `%` ("mod"). (The value of the expression written after >>> is shown on the next line.)

What is the smallest positive integer n that has the following characteristics?

2.61 n mod 2 = 0, n mod 3 = 0, and n mod 5 = 0

2.62 n mod 2 = 1, n mod 3 = 1, and n mod 5 = 1

2.63 n mod 2 = 0, n mod 3 = 1, and n mod 5 = 0

2.64 n mod 3 = 2, n mod 5 = 3, and n mod 7 = 5

2.65 n mod 2 = 1, n mod 3 = 2, n mod 5 = 3, and n mod 7 = 4

2.66 *(programming required)* Write a program to determine whether a given positive integer n is prime by testing all possible divisors between 2 and $n - 1$. Use your program to find all prime numbers less than 202.

2.67 *(programming required)* A *perfect number* is a positive integer n that has the following property: n is equal to the sum of all positive integers $k < n$ that evenly divide n. For example, 6 is a perfect number, because 1, 2, and 3 are the positive integers less than 6 that evenly divide 6—and $6 = 1 + 2 + 3$. Write a program that finds the four smallest perfect numbers.

2.68 *(programming required)* Write a program to find all integers between 1 and 1000 that are evenly divisible by *exactly three* different integers.

Compute the values of the following summations and products.

2.69 $\sum_{i=1}^{6} 6$

2.70 $\sum_{i=1}^{6} i^2$

2.71 $\sum_{i=1}^{6} 2^{2i}$

2.72 $\sum_{i=1}^{6} i \cdot 2^i$

2.73 $\sum_{i=1}^{6} (i + 2^i)$

2.74 $\prod_{i=1}^{6} 6$

2.75 $\prod_{i=1}^{6} i^2$

2.76 $\prod_{i=1}^{6} 2^{2i}$

2.77 $\prod_{i=1}^{6} i \cdot 2^i$

2.78 $\prod_{i=1}^{6} (i + 2^i)$

Compute the values of the following nested summations.

2.79 $\sum_{i=1}^{6} \sum_{j=1}^{6} (i \cdot j)$

2.80 $\sum_{i=1}^{6} \sum_{j=i}^{6} (i \cdot j)$

2.81 $\sum_{i=1}^{6} \sum_{j=1}^{i} (i \cdot j)$

2.82 $\sum_{i=1}^{8} \sum_{j=i}^{8} i$

2.83 $\sum_{i=1}^{8} \sum_{j=i}^{8} j$

2.84 $\sum_{i=1}^{8} \sum_{j=i}^{8} (i + j)$

2.85 $\sum_{i=1}^{4} \sum_{j=i}^{4} (j^i)$

2.3 *Sets: Unordered Collections*

History is a set of lies agreed upon.

Napoleon Bonaparte (1769–1821)

Section 2.2 introduced the primitive types of objects that we'll use throughout the book. We turn now to *collections* of objects, analogous to lists and arrays in programming languages. We start in this section with *sets*, in which objects are collected without respect to order or repetition. (Section 2.4 will address *sequences*, which are collections of objects in which order and repetition *do* matter.) The definitions and notation related to sets are summarized in Figure 2.14.

Definition 2.15 (Sets)
A set *is an unordered collection of objects.*

Here are a few simple examples:

Sets are typically denoted by uppercase letters (generically $S, T, U, A, B, \ldots$), often by a mnemonic letter: S for a set of students, D for a set of documents, etc. As we saw, the common sets from mathematics defined in Section 2.2.2 are often written using a "blackboard bold" font: $\mathbb{Z}$, $\mathbb{R}$, and $\mathbb{Q}$.

Example 2.22 (Some sets)
Here are three sets: the set of bits $\{0, 1\}$, the set of prime numbers $\{2, 3, 5, 7, 11, \ldots\}$, and the set of basic arithmetic operators $\{+, -, \cdot, /\}$. (We've written these sets using standard notation by listing the objects in the set between curly braces { and }.)

Set membership—that is, the question *is the object x one of the objects in the collection S?*, for a particular object x and a particular set S—is the central notion for sets:

Definition 2.16 (Set membership)
For a set S and an object x, the expression $x \in S$ is true when x is one of the objects contained in the set S. When $x \in S$, we say that x is an element *or* member *of S or, more simply, that x is in S.*

The expression $x \notin S$ is the negation of the expression $x \in S$: that is, $x \notin S$ is true whenever x is not an element of S (and thus whenever $x \in S$ is false).

Example 2.23 (Some set memberships)
The integer 0 is an element of the set of bits, and + is in the set of basic arithmetic operators. But 1 is not an element of the set of prime numbers, and 8 is not in the set of bits.

A second key concept about a set is its *cardinality*, or *size*:

Definition 2.17 (Set cardinality)
The cardinality *of a set S, denoted by $|S|$, is the number of distinct elements in S.*

set membership	$x \in S$	x is one of the elements of S
cardinality	$\|S\|$	the number of distinct elements in the set S
set enumeration	$\{x_1, x_2, \ldots, x_k\}$	the set containing elements $x_1, x_2, \ldots, x_k$
set abstraction	$\{x \in U : P(x)\}$	the set containing all $x \in U$ for which $P(x)$ is true; U is the "universe" of candidate elements
empty set	$\{\}$ or $\varnothing$	the set containing no elements
complement	$\sim S := \{x \in U : x \notin S\}$	the set of all elements in the universe U that aren't in S; U may be left implicit if it's obvious from context
union	$S \cup T := \{x : x \in S \text{ or } x \in T\}$	the set of all elements in either S or T (or both)
intersection	$S \cap T := \{x : x \in S \text{ and } x \in T\}$	the set of all elements in both S and T
set difference	$S - T := \{x : x \in S \text{ and } x \notin T\}$	the set of all elements in S but not in T
set equality	$S = T$	every $x \in S$ is also in T, and every $x \in T$ is also in S
subset	$S \subseteq T$	every $x \in S$ is also in T
proper subset	$S \subset T$	$S \subseteq T$ but $S \neq T$
superset	$S \supseteq T$	every $x \in T$ is also in S
proper superset	$S \supset T$	$S \supseteq T$ but $S \neq T$
power set	$\mathscr{P}(S)$	the set of all subsets of S

Figure 2.14: A summary of set notation.

Example 2.24 (Some set sizes)

The cardinality of the set of bits is 2, because there are two distinct elements of that set (namely 0 and 1).

The cardinality of the set S of prime numbers between 10 and 20 is $|S| = 4$: the four elements of S are 11, 13, 17, and 19.

Chapter 9 is devoted entirely to the apparently trivial problem of *counting*—given a (possibly convoluted) description of a set S, find $|S|$—which turns out to have some interesting and useful applications, and isn't as easy as it seems.

Taking it further: In this book, we will be concerned almost exclusively with the cardinality of *finite* sets, but one can also ask questions about the cardinality of sets like $\mathbb{Z}$ or $\mathbb{R}$ that contain an infinite number of distinct elements. For example, it's possible to prove that $|\mathbb{Z}| = |\mathbb{Z}^{\geq 0}|$, which is a pretty amazing result: *there are as many nonnegative integers as there are integers!* (And that's true despite the fact that every nonnegative integer *is* an integer!) But it's also possible to prove that $|\mathbb{Z}| \neq |\mathbb{R}|$: *... but there are more real numbers than integers!* More amazingly, one can use similar ideas to prove that there are fewer computer programs than there are problems to solve, and that therefore there are some problems that are not solved by any computer program. This idea is the central focus of the study of *computability* and *uncomputability*. See Section 4.4.4 and the discussion on p. 937.

2.3.1 *Building Sets from Scratch*

There are two standard ways to specify a set "from scratch": by simply listing each of the elements of the set, or by defining the set as the collection of objects for which a particular logical condition is true.

Set definition via exhaustive enumeration

A set can be specified using an exhaustive listing its elements—that is, by writing a complete list of its elements inside the curly braces { and }. Here are a few examples:

Example 2.25 (Some exhaustively enumerated sets)

- The set of even prime numbers is $\{2\}$.

- The set of prime numbers between 10 and 20 is $\{11, 13, 17, 19\}$.
- The set of 2-digit perfect squares is $\{81, 64, 25, 16, 36, 49\}$.
- The set of bits is $\{0, 1\}$.
- The set of Turing Award winners between 1984 and 1987 inclusive is {Niklaus Wirth, Richard Karp, John Hopcroft, Robert Tarjan, John Cocke}.

Taking it further: The Turing Award is the most prestigious award given in computer science—the "Nobel Prize of CS," it's sometimes called. Niklaus Wirth developed a number of programming languages, including Pascal. Richard Karp made major contributions to the study of computational complexity, in particular with respect to the understanding of NP-Completeness. John Hopcroft and Robert Tarjan made massive early contributions in designing and analyzing algorithms and data structures for problems. John Cocke was a leader in compilers and computer architecture and is often credited with inventing the RISC architecture, which changed the way that computer chips and their corresponding instruction sets were designed.

Recall that a set is an *unordered* collection, and thus the order in which the elements are listed doesn't matter when specifying a set via exhaustive enumeration. Any repetition in the listed elements is also unimportant. For example:

Example 2.26 (The same set, three ways)
The set $\{2+2,\ 2\cdot 2,\ 2/2,\ 2-2\}$ is precisely identical to the set $\{0, 1, 4\}$, both of which are precisely identical to $\{4, 0, 1\}$. Also note that $|\{2+2,\ 2\cdot 2,\ 2/2,\ 2-2\}| = 3$; despite there being four entries in the list of elements, there are only three *distinct* objects in the set.

It's important to remember that the integer 2 and the set $\{2\}$ are two entirely different kinds of things. For example, note that $2 \in \{2\}$, but that $\{2\} \notin \{2\}$; the lone element in $\{2\}$ is *the number two*, not *the set containing the number two*.

Set definition via set abstraction

Instead of explicitly listing all of a set's elements, we can also define a set in terms of a condition that is true for the elements of the set and that's false for every object that is not an element of the set. Defining a set this way uses *set abstraction* notation:

Definition 2.18 (Set Abstraction)
Let U be a set of possible elements, called the universe. *Let $P(x)$ be a condition (also called a* predicate*) that, for every $x \in U$, is either true or false. Then*

$$\{x \in U : P(x)\}$$

denotes the set of all objects $x \in U$ for which $P(x)$ is true.

The colon in the notation for set abstraction is read as "such that," so the set in Definition 2.18 would be read "the set of all x in U such that P of x."

That is, for any candidate element $y \in U$, the element y is in the set $\{x \in U : P(x)\}$ when $P(y) = \text{True}$, and $y \notin \{x \in U : P(x)\}$ when $P(y) = \text{False}$. (A fully proper version of Definition 2.18 requires *functions*, described in Section 2.5.)

Example 2.27 (Most of Example 2.25, redone)

- The set of even prime numbers is $\{x \in \mathbb{Z}^{>1} : x \text{ is prime and } x \text{ is even}\}$.
- The set of 2-digit perfect squares is $\{n \in \mathbb{Z} : \sqrt{n} \in \mathbb{Z} \text{ and } 10 \leq n \leq 99\}$.
- The set of bits is $\{b \in \mathbb{Z} : b^2 = b\}$.

For this set abstraction notation to meaningfully define a set S, we must specify the universe U of candidates from which the elements of S are drawn. We will permit ourselves to be sloppy in our notation, and when the universe U is clear from context we will allow ourselves the liberty of writing $\{x : P(x)\}$ instead of $\{x \in U : P(x)\}$.

Taking it further: The notational sloppiness of omitting the universe in set abstraction will be a convenience for us, and it will not cause us any trouble—but it turns out that one must be careful! In certain strange scenarios when defining sets, there are subtle but troubling paradoxes that arise if we allow the universe to be anything at all. The key problem can be seen in *Russell's paradox*, named after the British philosopher/mathematician Bertrand Russell; Russell's discovery of this paradox revealed an inconsistency in the commonly accepted foundations of mathematics in the early 20th century.

Here is a brief sketch of Russell's Paradox. Let X denote the set of all sets that do not contain themselves: that is, let $X := \{S : S \notin S\}$. For example, $\{2\} \in X$ because $\{2\} \notin \{2\}$, and $\mathbb{R} \in X$ because $\mathbb{R}$ is not a real number, so $\mathbb{R} \notin \mathbb{R}$. On the other hand, if we let T^* denote the set of all sets, then $T^* \notin X$: because T^* is a set, and T^* contains all sets, then $T^* \in T^*$ and therefore $T^* \notin X$.

Here's the problem: is $X \in X$? Suppose that $X \in X$: then $X \in \{S : S \notin S\}$ by the definition of X, and thus $X \notin X$. But suppose that $X \notin X$; then, by the definition of X, we have $X \in X$. So if $X \in X$ then $X \notin X$, and if $X \notin X$ then $X \in X$—but that's absurd!

One standard way to escape this paradox is to say that the set X cannot be defined—because, to be able to define a set using set abstraction, we need to start from a defined universe of candidate elements. (And the set T^* cannot be defined either.) The *Liar's Paradox*, dating back about 3000 years, is a similar paradox: is "this sentence is false" true (nope!) or false (nope!)? In both Russell's Paradox and the Liar's Paradox, the fundamental issue relates to *self-reference*; many other mind-twisting paradoxes are generated through self-reference, too.[3]

For more on these and other paradoxes, see

[3] R. M. Sainsbury. *Paradoxes*. Cambridge University Press, 3rd edition, 2009.

Definition 2.18 lets us write $\{x \in U : P(x)\}$ to denote the set containing exactly those elements x of U for which $P(x)$ is True. We will extend this notation to allow ourselves to write more complicated expressions to the left of the colon, as in the following example:

Example 2.28 (2-digit perfect squares, again)

We can write the set of 2-digit perfect squares as $\{x^2 : x \in \mathbb{Z} \text{ and } 10 \leq x^2 \leq 99\}$ or as $\{x^2 : x \in \{4,5,6,7,8,9\}\} = \{4^2, 5^2, 6^2, 7^2, 8^2, 9^2\}$.

To properly define this extended form of the set-abstraction notation, we again need the idea of *functions*, which are defined in Section 2.5.1. See Definition 2.47 for a proper definition of this extended notation.

Taking it further: Almost all modern programming languages support the use of *lists* to store a collection of objects. While these lists store ordered collections, there are some very close parallels between these lists and sets. In fact, the ways we've described building sets have very close connections to ideas in certain programming languages like Scheme and Python; see p. 233 for some discussion.

The empty set

One particularly useful set—despite its simplicity—is the *empty set*, also sometimes called the *null set*:

Definition 2.19 (The empty set ∅)
The empty set, *denoted* $\{\}$ *or* $\varnothing$, *is the set that contains no elements.*

The definition of the empty set as $\{\}$ *is* an exhaustive listing of all of the elements of the set—though, because there aren't any elements, there are no elements in the list.

Alternatively, we could have used the set abstraction notation to define the empty set, as $\varnothing := \{x : \text{False}\}$. This definition may seem initially confusing, but it's in fact a direct application of Definition 2.18: the condition P for this set is $P(x) = \text{False}$ (that is: for every object x, the value of $P(x)$ is False), and we've defined $\varnothing$ to contain every object y such that $P(y) = \text{True}$. But there *isn't* any object y such that $P(y) = \text{True}$—because $P(y)$ is always false—and thus there's no $y \in \{x : P(x)\}$.

Notice that, because there are zero elements in $\varnothing$, its cardinality is zero: in other words, $|\varnothing| = 0$. One other special type of set is defined based on its cardinality; a *singleton set* is a set S that contains exactly one element—that is, a set S such that $|S| = 1$.

2.3.2 *Building Sets from Other Sets*

There are a number of ways to create new sets from two given sets A and B. We will define these operations formally, but it is sometimes more intuitive to look at a more visual representation of sets called a *Venn diagram*, which are drawings that represent sets as circular "blobs" that contain points (elements), enclosed in a rectangle that denotes the universe.

Venn diagrams are named after the 19th-century British logician/philosopher John Venn.

Example 2.29 (Venn diagram of odds and primes)
Let $U := \{1, 2, \ldots, 10\}$. Let $P := \{2, 3, 5, 7\}$ denote the set of primes in U, and let $O := \{1, 3, 5, 7, 9\}$ denote the set of odd numbers in U.

A Venn diagram illustrating these sets is shown in Figure 2.15: 3, 5, and 7 are elements of both P and O; 2 is in P but not O; 1 and 9 are in O but not P; and 4, 6, and 8 are in neither P nor O.

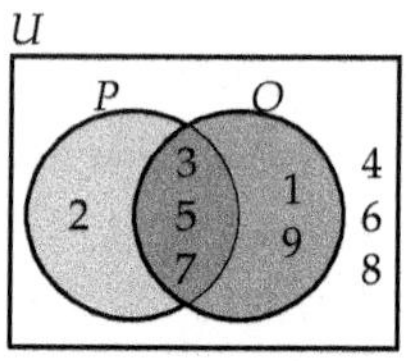

Figure 2.15: A Venn diagram for the set O of odd numbers and the set P of prime numbers between 1 and 9.

We will now define four standard ways of building a new set in terms of one or two existing sets: *complement, union, intersection,* and *set difference.*

Definition 2.20 (Set complement)
The complement *of a set A with respect to the universe U, written $\sim A$ (or sometimes $\overline{A}$), is the set of all elements* not *contained within A. Formally, $\sim A := \{x \in U : x \notin A\}$. (When the universe is obvious from context, we will leave it implicit.)*

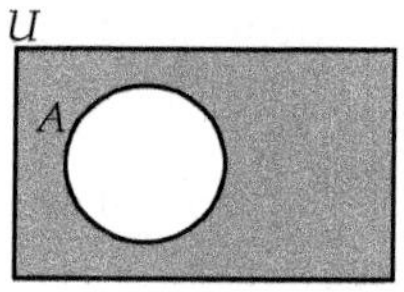

Figure 2.16: The complement of a set A. The shaded region represents the set $\sim A$ with respect to the universe U.

Figure 2.16 shows a Venn diagram illustrating the complement of A.

For example, if the universe is $\{1, 2, \ldots, 10\}$, then $\sim \{1, 2, 3\} = \{4, 5, 6, 7, 8, 9, 10\}$ and $\sim \{3, 4, 5, 6\} = \{1, 2, 7, 8, 9, 10\}$.

Definition 2.21 (Set union)
The union *of two sets A and B, denoted* $A \cup B$*, is the set of all elements in* either A or B *(or both). Formally,* $A \cup B := \{x : x \in A \text{ or } x \in B\}$. *Analogously to summation and product notation (*$\sum$ *and* $\prod$*), we will sometimes write* $\bigcup_{i=1}^{n} S_i$ *to denote* $S_1 \cup S_2 \cup \cdots \cup S_n$.

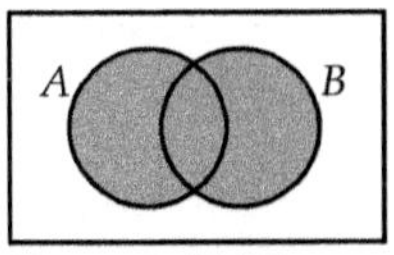

Figure 2.17: The union $A \cup B$ of two sets A and B.

Figure 2.17 shows a Venn diagram illustrating the union of A and B.

For example, $\{1,2,3\} \cup \{3,4,5,6\} = \{1,2,3,4,5,6\}$.

Definition 2.22 (Set intersection)
The intersection *of two sets A and B, denoted* $A \cap B$*, is the set of all elements in* both A and B. *Formally,* $A \cap B := \{x : x \in A \text{ and } x \in B\}$. *We will sometimes write* $\bigcap_{i=1}^{n} S_i$ *to denote* $S_1 \cap S_2 \cap \cdots \cap S_n$.

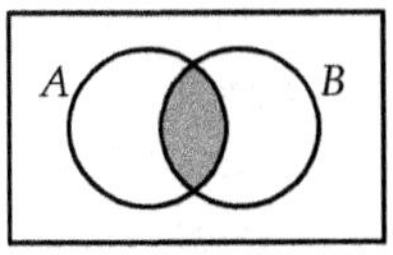

Figure 2.18: The intersection $A \cap B$ of sets A and B.

Figure 2.18 shows a Venn diagram illustrating $A \cap B$.

For example, $\{1,2,3\} \cap \{3,4,5,6\} = \{3\}$.

Definition 2.23 (Set difference)
The difference *of two sets A and B, denoted* $A - B$*, is the set of all elements contained* in the set A but not in the set B. *Formally,* $A - B := \{x : x \in A \text{ and } x \notin B\}$. *(Some people write* $A \setminus B$ *instead of* $A - B$ *to denote set difference.)*

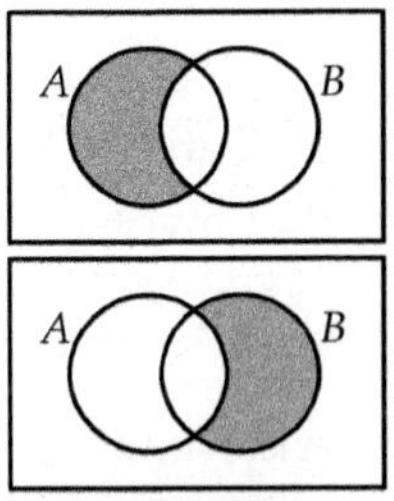

Figure 2.19: The difference of two sets A and B. The shaded region in the first panel represents the set $A - B$, and the shaded region in the second panel represents $B - A$.

Figure 2.19 shows a Venn diagram illustrating the set difference of A and B. Note that $A - B$ and $B - A$ are different sets; both are illustrated in Figure 2.19. For example, $\{1,2,3\} - \{3,4,5,6\} = \{1,2\}$ and $\{3,4,5,6\} - \{1,2,3\} = \{4,5,6\}$.

In more complicated expressions that use more than one of these set operators, the $\sim$ operator "binds tightest"—that is, in an expression like ${\sim}S \cup T$, we mean $({\sim}S) \cup T$ and not ${\sim}(S \cup T)$. We use parentheses to specify the order of operations among $\cap$, $\cup$, and $-$. Here's a slightly more complicated example that combines set operations:

Example 2.30 (Combining odds and primes)

<u>*Problem:*</u> As in Example 2.29, define $U := \{1,2,\ldots,10\}$, the set $P := \{2,3,5,7\}$ of primes in U, and the set $O := \{1,3,5,7,9\}$ of odd numbers in U. What are the following sets?

1. $P \cap {\sim}O$
2. ${\sim}(P \cup O)$
3. ${\sim}P - {\sim}O$

<u>*Solution:*</u> For each part, we simply plug in the definitions:

1. The set $P \cap {\sim}O$ is the set of all prime numbers that are also not odd.

$$\begin{aligned} P \cap {\sim}O &= \{2,3,5,7\} \cap {\sim}\{1,3,5,7,9\} \\ &= \{2,3,5,7\} \cap \{2,4,6,8,10\} \\ &= \{2\}. \end{aligned}$$

2. The set $\sim(P \cup O)$ consists of everything that is not an element of $P \cup O$—that is, $\sim(P \cup O)$ contains only nonprime even numbers.

$$\begin{aligned} \sim(P \cup O) &= \sim(\{2,3,5,7\} \cup \{1,3,5,7,9\}) \\ &= \sim \{1,2,3,5,7,9\} \\ &= \{4,6,8,10\}. \end{aligned}$$

3. The set $\sim P - \sim O$ consists of all elements of $\sim P$ except those that are elements of $\sim O$—in other words, all nonprime numbers that aren't nonodd, or, more simply stated, all nonprime odd numbers:

$$\begin{aligned} \sim P - \sim O &= \sim \{2,3,5,7\} - \sim \{1,3,5,7,9\} \\ &= \{1,4,6,8,9,10\} - \{2,4,6,8,10\} \\ &= \{1,9\}. \end{aligned}$$

Of course, we can also combine more than two sets in expressions using these set operators—for example, $A \cup B \cup C$ denotes the set $\{x : x \in A \text{ or } x \in B \text{ or } x \in C\}$. We can use Venn diagrams to visualize set operations that involve more than two sets; see Figure 2.20 for a few examples.

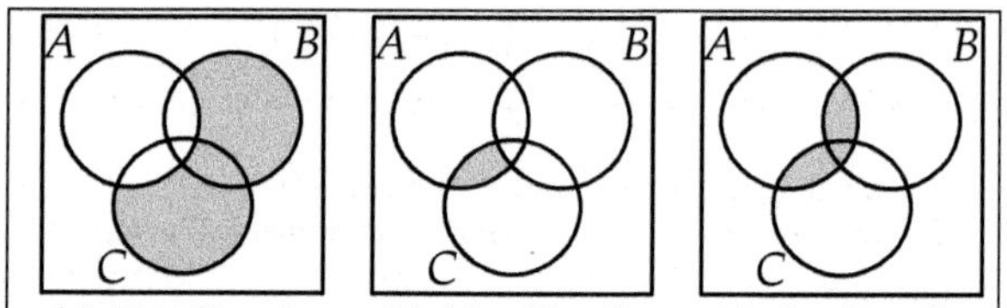

(a) $(B \cup C) - A$ (b) $(A - B) \cap C$ (c) $A \cap (B \cup C)$

Figure 2.20: Some three-set Venn diagrams.

ARITHMETIC OPERATIONS ON SETS

We'll end this subsection with a few pieces of notation that allow us to perform mathematical operations on the elements of a set. In Section 2.2.7, we introduced summation and product notation, so that we could write

$$\sum_{i=1}^{n} x_i \quad \text{and} \quad \prod_{i=1}^{n} x_i$$

to represent $x_1 + x_2 + \cdots + x_n$ and $x_1 \cdot x_2 \cdot \cdots \cdot x_n$. We will also sometimes wish to represent the sum or product of the elements of a particular set (instead of a sequence of values like $x_1, x_2, \ldots, x_n$). It will also sometimes be handy to refer to the smallest or largest element in a set.

Definition 2.24 (Sum, product, minimum, and maximum of a set)
Let S be a set. Then the expressions

$$\sum_{x \in S} x, \quad \prod_{x \in S} x, \quad \min_{x \in S} x, \quad \textit{and} \quad \max_{x \in S} x$$

respectively denote the sum of the elements of S, the product of the elements of S, the smallest element in S, and the largest element in S.

For example, for the set $S := \{1,2,4,8\}$, we have that the sum of the elements of S is

$\sum_{x \in S} x = 15$; the product of the elements of S is $\prod_{x \in S} x = 64$; the minimum of S is $\min_{x \in S} x = 1$; and the maximum of S is $\max_{x \in S} x = 8$.

2.3.3 *Comparing Sets*

In the same way that two numbers x and y can be compared (we can ask questions like: does $x = y$? is $x \leq y$? is $x \geq y$?), we can also compare two sets A and B. Here, we will define the analogous notions of comparison for sets. We'll begin by defining what it means for two sets to be equal:

Definition 2.25 (Set equality)
Two sets A and B are equal, *denoted $A = B$, if A and B have exactly the same elements. (In other words, sets A and B are not equal if there's an element $x \in A$ but $x \notin B$, or if there's an element $y \in B$ but $y \notin A$.)*

This definition formalizes the idea that order and repetition don't matter in sets: for example, the sets $\{4,4\}$ and $\{4\}$ are equal because there is no element $x \in \{4,4\}$ where $x \notin \{4\}$ and there is no element $y \in \{4\}$ where $y \notin \{4,4\}$. This definition also implies that the empty set is unique: any set containing no elements is identical to $\varnothing$.

Taking it further: Definition 2.25 is sometimes called the *axiom of extensionality*. (All of mathematics, including a completely rigorous definition of the integers and all of arithmetic, can be built up from a small number of axioms about sets, including this one.) The point is that the only way to compare two sets is by their "externally observable" properties. For example, the following two sets are *exactly* the same set: $\{x : x > 10$ is an even prime number$\}$, and $\{y : y$ is a country with a 128-letter name$\}$. (Namely, both of these sets are $\varnothing$.)

The other common type of comparison between two sets A and B is the *subset* relationship, which expresses that every element of A is also an element of B:

Definition 2.26 (Subset)
A set A is a subset *of a set B, written $A \subseteq B$, if every $x \in A$ is also an element of B. (In other words, $A \subseteq B$ is equivalent to $A - B = \{\}$.)*

For example, $\{1,3,5\} \subseteq \{1,2,3,4,5\}$, because $1 \in \{1,2,3,4,5\}$ and $3 \in \{1,2,3,4,5\}$ and $5 \in \{1,2,3,4,5\}$.

Notice that $\{\} \subseteq S$ for *any* set S: it's impossible for there to be an $x \in \{\}$ that satisfies $x \notin S$, because there is no element $x \in \{\}$ in the first place—and if there's no $x \in \{\}$ at all, then there's certainly no $x \in \{\}$ such that $x \notin S$.

Definition 2.27 (Proper subset)
A set A is a proper subset *of a set B, written $A \subset B$, if $A \subseteq B$ and $A \neq B$. In other words, $A \subset B$ whenever $A \subseteq B$ but $B \not\subseteq A$.*

For example, let $A := \{1,2,3\}$. Then $A \subseteq \{1,2,3,4\}$ and $A \subseteq \{1,2,3\}$ and $A \subset \{1,2,3,4\}$, but A is not a proper subset of $\{1,2,3\}$.

When $A \subset B$ or $A \subseteq B$, we refer to A as the (possibly proper) subset of B; we can also call B the (possibly proper) *superset* of A:

Definition 2.28 (Superset and proper superset)
Let A be a set. A set B is a superset *of A, written $B \supseteq A$, if $A \subseteq B$. The set B is a* proper superset *of A, written $B \supset A$, if $A \subset B$.*

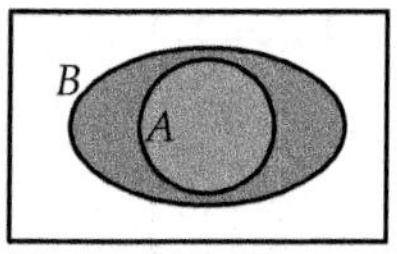

Figure 2.21: Two sets satisfying $A \subseteq B$ and, equivalently, $B \supseteq A$. The sets satisfy $A \subset B$ (and $B \supset A$) if there's at least one element in the darker shaded region, and they satisfy $A = B$ if there's no element in that region.

Figure 2.21 illustrates subsets, proper subsets, supersets, and proper supersets. Here's an example involving these relationships:

Example 2.31 (Subsets and supersets)
Problem: Let $A := \{3,4,5\}$ and $B := \{4,5,6\}$. Identify a set C satisfying the following conditions, or state that the requirement is impossible to achieve and explain why.

1. $A \subseteq C$ and $C \supseteq B$
2. $A \supseteq C$ and $C \subseteq B$
3. $A \supseteq C$ and $C \supseteq B$

Solution: The first two conditions are achievable, but the third isn't.

1. Let $C := \{3,4,5,6\}$; both A and B are (proper) subsets of this set.
2. We can choose $C := \{4,5\}$, because $\{4,5\} \subseteq A$ and $\{4,5\} \subseteq B$.
3. It's impossible to satisfy $\{3,4,5\} \supseteq C$ and $C \supseteq \{4,5,6\}$ simultaneously. If $6 \in C$ then we don't have $\{3,4,5\} \supseteq C$, but if $6 \notin C$ we don't have $C \supseteq \{4,5,6\}$. We can't have $6 \in C$ and we can't have $6 \notin C$, so we're stuck with an impossibility.

We'll end the section with one last piece of terminology. Two sets A and B are called *disjoint* if they have no elements in common:

Definition 2.29 (Disjoint sets)
Two sets A and B are disjoint *if there is no $x \in A$ where $x \in B$—in other words, if $A \cap B = \{\}$.*

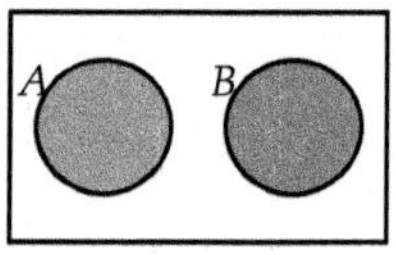

Figure 2.22: Disjoint sets A and B.

For example, the sets $\{1,2,3\}$ and $\{4,5,6\}$ are disjoint because $\{1,2,3\} \cap \{4,5,6\} = \{\}$, but the sets $\{2,3,5,7\}$ and $\{2,4,6,8\}$ are not disjoint because 2 is an element of both. See Figure 2.22 for a diagram of two disjoint sets.

2.3.4 *Sets of Sets*

Just as we can have a list of lists in a programming language like Scheme or Java, we can also consider a set that has sets as its elements. (After all, sets are just collections of objects, and one kind of object that can be collected is a set itself.)

Example 2.32 (Set of sets of numbers)
The set $A := \{\mathbb{Z}, \mathbb{R}, \mathbb{Q}\}$ of the sets defined in Section 2.2.2 is itself a set. This set has cardinality $|A| = 3$, because A has three distinct elements—namely $\mathbb{Z}$ and $\mathbb{R}$ and $\mathbb{Q}$. (Of course, all three of these elements of A are themselves sets, and each of these three elements of A has infinite cardinality.)

Example 2.33 (A set of smaller sets)
Consider the set $B := \{\{\}, \{1,2,3\}\}$. Note that $|B| = 2$: B has two elements, namely $\{\}$ and $\{1,2,3\}$. Therefore $\{\} \in B$ because $\{\}$ is one of the two elements of B. However $1 \notin B$, because 1 is not one of the two elements of B—that is, $1 \neq \{\}$ and $1 \neq \{1,2,3\}$—although 1 *is* an element of one of the two elements of B.

There are two important types of sets of sets that we will define in the remainder of this section, both derived from a base set S.

Partitions

The first interesting use of a set of sets is to form a *partition* of S into a set of disjoint subsets whose union is precisely S.

Definition 2.30 (Partition)
A partition *of a set S is a set $\{A_1, A_2, \ldots, A_k\}$ of nonempty sets $A_1, A_2, \ldots, A_k$, for some $k \geq 1$, such that:*

- *$A_1 \cup A_2 \cup \cdots \cup A_k = S$; and*
- *for any distinct $i, j \in \{1, \ldots, k\}$, the sets A_i and A_j are disjoint.*

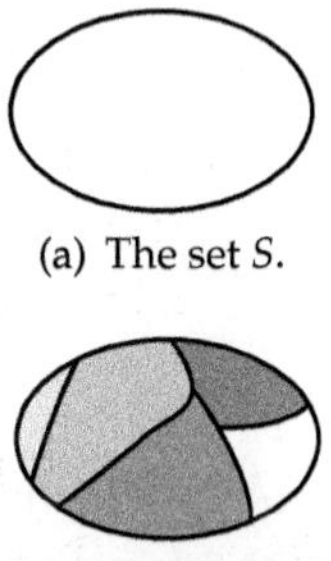
(a) The set S.
(b) S partitioned into 5 subsets.
Figure 2.23: A visualization of partitioning a set S into disjoint nonempty subsets whose union equals S itself.

A useful way of thinking about a partition of a set S is that we've divided S up into several (nonoverlapping) subcategories. See Figure 2.23 for an illustration of a partition of a set S. Here's an example of one set partitioned many different ways:

Example 2.34 (Several partitions of the same set)
Consider the set $S := \{1,2,3,4,5,6,7,8,9,10\}$. Here are some different ways to partition S:

$\{\{1,3,5,7,9\}, \{2,4,6,8,10\}\}$	(evens and odds)
$\{\{1,2,3,4,5,6,7,8,9\}, \{10\}\}$	(one- and two-digit numbers)
$\{\{1,4,7,10\}, \{2,5,8\}, \{3,6,9\}\}$	(x mod 3 = 0 and = 1 and = 2)
$\{\{1\}, \{2\}, \{3\}, \{4\}, \{5\}, \{6\}, \{7\}, \{8\}, \{9\}, \{10\}\}$	(all separate)
$\{\{1,2,3,4,5,6,7,8,9,10\}\}$	(all together)

In each case, each of the 10 numbers from S is in one, and only one, of the listed sets (and no elements not in S appear in any of the listed sets).

It's worth noting that the last two ways of partitioning S in Example 2.34 genuinely *are* partitions. For the partition $\{\{1\}, \{2\}, \{3\}, \{4\}, \{5\}, \{6\}, \{7\}, \{8\}, \{9\}, \{10\}\}$, we have $k = 10$ different disjoint sets whose union is precisely S. For the partition $\{\{1,2,3,4,5,6,7,8,9,10\}\}$, we have $k = 1$: there's only one "subcategory" in the partitioning, and every $x \in S$ is indeed contained in one (the only one!) of these "subcategories." (And no two distinct subcategories overlap, because there aren't even two distinct subcategories at all!)

Taking it further: One way to helpfully organize a massive set S of data—for example, students or restaurants or web pages—is to partition S into small *clusters*. The idea is that two elements in the same cluster will be "similar," and two entities in different clusters will be "dissimilar." (So students might be clustered by their majors or dorms; restaurants might be clustered by their cuisine or geography; and web pages might be clustered based on the set of words that appear in them.) For more about clustering, see the discussion on p. 234.

Power sets

Our second important type of a set of sets is the *power set* of a set S, which is the set of all subsets of S:

Definition 2.31 (Power set)
The power set *of a set S, written $\mathscr{P}(S)$, denotes the set of all subsets of S: that is, a set A is an element of $\mathscr{P}(A)$ precisely if $A \subseteq S$. In other words, $\mathscr{P}(S) := \{A : A \subseteq S\}$.*

The power set of S is also occasionally denoted by 2^S, in part because—as we'll see in Chapter 9—$|\mathscr{P}(S)|$ is $2^{|S|}$. The name "power set" also comes from this fact: the cardinality of $\mathscr{P}(S)$ is 2 to the power of $|S|$.

Here are some simple examples, and one example that's a bit more complicated:

Example 2.35 (Some small power sets)
Here are the power sets of $\{0\}$, $\{0,1\}$, and $\{0,1,2\}$:

$$\mathscr{P}(\{0\}) = \{\{\},\{0\}\}$$
$$\mathscr{P}(\{0,1\}) = \{\{\},\{0\},\{1\},\{0,1\}\}$$
$$\mathscr{P}(\{0,1,2\}) = \{\{\},\{0\},\{1\},\{2\},\{0,1\},\{0,2\},\{1,2\},\{0,1,2\}\}$$

A quick check for the second of these examples: there are four elements in $\mathscr{P}(\{0,1\})$: the empty set, two singleton sets $\{0\}$ and $\{1\}$, and the two-element set $\{0,1\}$ itself, because $\{0,1\} \subseteq \{0,1\}$ is a subset of itself.

Example 2.36 ($\mathscr{P}(\mathscr{P}(\{0,1\}))$)
The power set of the power set of $\{0,1\}$ is

$$\begin{aligned}
&\mathscr{P}(\mathscr{P}(\{0,1\})) \\
&= \mathscr{P}(\{\{\},\{0\},\{1\},\{0,1\}\}) \\
&= \left\{\begin{array}{ll}
\{\}, & \textit{1 set with 0 elements} \\
\{\{\}\},\{\{0\}\},\{\{1\}\},\{\{0,1\}\}, & \textit{4 sets with 1 element} \\
\{\{\},\{0\}\},\{\{\},\{1\}\},\{\{\},\{0,1\}\}, & \textit{6 sets with 2 elements} \\
\quad \{\{0\},\{1\}\},\{\{0\},\{0,1\}\},\{\{1\},\{0,1\}\}, & \\
\{\{0\},\{1\},\{0,1\}\},\{\{\},\{1\},\{0,1\}\}, & \textit{4 sets with 3 elements} \\
\quad \{\{\},\{0\},\{0,1\}\},\{\{\},\{0\},\{1\}\}, & \\
\{\{\},\{0\},\{1\},\{0,1\}\} & \textit{1 set with 4 elements}
\end{array}\right\}.
\end{aligned}$$

Computer Science Connections

Set Building in Languages

Programming languages like Python, Scheme, or ML make heavy use of lists and also allow higher-order functions (functions that take other functions as parameters); if you have experience programming in these languages, the set-construction notions from Section 2.3.1 may seem familiar. These mechanisms for building sets in mathematical notation closely parallel built-in functionality for building *lists* in programs in these languages:

- build a list from scratch by writing out its elements.
- build a list from an existing list using the function `filter`, which takes two parameters (a list `U`, corresponding to the universe, and a function `P`) and returns a new list containing all $x \in$ `U` for which `P`(x) is true.
- build a list from an existing list using the function `map`, which takes two parameters (a list `U` and a function `f`) and returns a new list containing `f`(x) for every element x of `U`.

Unlike sets, the map function can cause repetitions in the stored list: `map(square,L)` where `L` contains both 2 and -2 will lead to 4 being present twice. (Some languages, including Python, also have syntax for *sets* instead of *lists*, creating an unordered, duplicate-free collection of elements.)

Python has `filter` and `map` built in; some versions of Scheme have `filter` and `map` either built in or in a standard library. In Python, there's even an explicit *list comprehension* syntax to create a list without using `filter` or `map`, which even more closely parallels the set-abstraction notation from Definitions 2.18 and 2.47. Here are some examples:

In set notation:	*In Python:*	*In Scheme:*

$L = \{1,2,4,8,16\}$
$M = \{x \in L : x < 10\}$
$N = \{x \in L : x \text{ is even}\}$
$O = \{x^2 : x \in L\}$
$P = \{x^2 : x \in L \text{ and } x \text{ is even}\}$
$Q = \{x \in L : \text{False}\}$
	`def even(x):     return x % 2 == 0`
	`def square(x):   return x**2`
	`def false(x):    return False`
	
	`L = [1,2,4,8,16]`
	`M = [x for x in L if x < 10]`
	`N = filter(even, L)`
	`O = map(square, L)`
	`P = [square(x) for x in L if even(x)]`
	`Q = [x for x in L if false(x)]`
		`(define even?`
		`   (lambda (x) (= (modulo x 2) 0)))`
		`(define square (lambda (x) (* x x)))`
		`(define false? (lambda (x) #f))`
		
		`(define L (list 1 2 4 8 16))`
		`;;; no simple Scheme is analogous to M in Python`
		`(define N (filter even? L))`
		`(define O (map square L))`
		`(define P (map square (filter even? L)))`
		`(define Q (filter false? L))`

$L = \{1,2,4,8,16\}$
$M = \{1,2,4,8\}$
$N = \{2,4,8,16\}$
$O = \{1,4,16,64,256\}$
$P = \{4,16,64,256\}$
$Q = \{\}$
	`>>> L`
	`[1, 2, 4, 8, 16]`
	`>>> M`
	`[1, 2, 4, 8]`
	`>>> N`
	`[2, 4, 8, 16]`
	`>>> O`
	`[1, 4, 16, 64, 256]`
	`>>> P`
	`[4, 16, 64, 256]`
	`>>> Q`
	`[]`
		`> L`
		`(1 2 4 8 16)`
		
		`> N`
		`(2 4 8 16)`
		`> O`
		`(1 4 16 64 256)`
		`> P`
		`(4 16 64 256)`
		`> Q`
		`()`

While the technical details are a bit different, the basic idea underlying `map` forms half of a programming model called *MapReduce* that's become increasingly popular for processing very large datasets.[4] MapReduce is a distributed-computing framework that processes data using two user-specified functions: a "map" function that's applied to every element of the dataset, and a "reduce" function that collects together the outputs of the map function. Implementations of MapReduce allow these computations to occur in parallel, on a cluster of machines, vastly speeding processing time.

[4] Jeffrey Dean and Sanjay Ghemawat. MapReduce: simplified data processing on large clusters. *Communications of the ACM*, 51(1):107–113, 2008.

Computer Science Connections

Clustering

Partitioning a set is a task that arises frequently in various applications, usually with a goal like *clustering* a large collection of data points. The goal is that elements placed into the same cluster should be "very similar," and elements in different clusters should be "not very similar."[5] Why might we want to perform clustering on a data set? For example, we might try to cluster a set N of news articles into "topics" $C_1, C_2, \ldots, C_k$, where any two articles x, y that are both in the same cluster C_i are similar (say, with respect to the words contained within them), but if $x \in C_i$ and $y \in C_{j \neq i}$ then x and y are not very similar. Or we might try to cluster the people in a social network into *communities*, so that a person in community c has a large fraction of her friends who are also in community c. Understanding these clusters—and understanding what properties of a data point "cause" it to be in one cluster rather than another—can help reveal the structure of a large data set, and can also be useful in building a system to react to new data. Or we might want to use clusters for *anomaly detection*: given a large data set—for example, of user behavior on a computer system, or the trajectory of a car on a highway—we might be able to identify those data points that do not seem to be part of a normal pattern. These data points may be the result of suspicious behavior that's worth further investigation (or that might trigger a warning to the driver of the car that he or she has strayed from a lane).

You can read more about clustering, and clustering algorithms, in a data-mining book like
[5] Jure Leskovec, Anand Rajaraman, and Jeff Ullman. *Mining of Massive Datasets*. Cambridge University Press, 2nd edition, 2014.

Here's one (vastly simplified) example application for clustering: *speech processing*. Software systems that interact with users as they speak in natural language—that is, as they talk in English—have developed with rapidly increasing quality over the last decade. *Speech recognition*—taking an audio input, and identifying what English word is being spoken from the acoustic properties of the audio signal—turns out to be a very challenging problem. Figure 2.24 illustrates some of the reasons for the difficulty, showing a *spectrogram* generated by the Praat software tool.[6] In a spectrogram, the x-axis is time, and the y-axis is frequency; a darkly shaded frequency f at time t shows that the speech at time t had an intense component at frequency f. But we can partition a *training set* of many speakers saying a collection of common words into subsets based on which word was spoken, and then use the average acoustic properties of the utterances to guess which word was spoken. Figure 2.25 shows the frequencies of the two lowest *formants*—frequencies of very high intensity—in the utterances of a half-dozen college students pronouncing the words *bat* and *beat*. First, the formants' frequencies are shown unclustered; second, they are shown partitioned by the pronounced word. The *centroid* of each cluster (the center of mass of the points) can serve as a prototypical version of each word's acoustics.

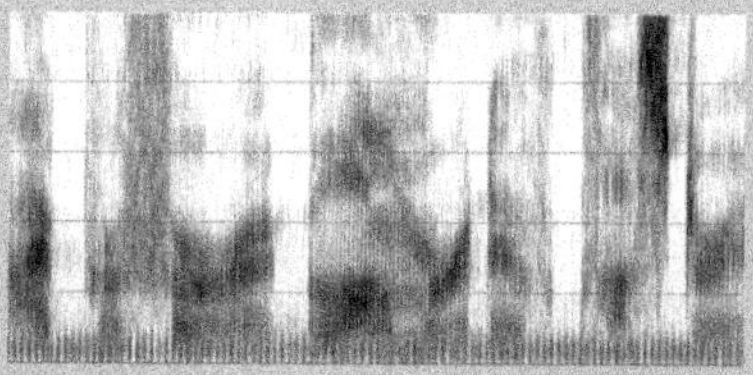

Figure 2.24: A *spectrogram* generated by Praat of me pronouncing the sentence "I prefer agglomerative clustering." There are essentially no acoustic correlates to the divisions between words, which is one reason speech recognition is so difficult.

[6] Paul Boersma and David Weenink. Praat: doing phonetics by computer. `http://www.praat.org`, 2012. Version 5.3.22.

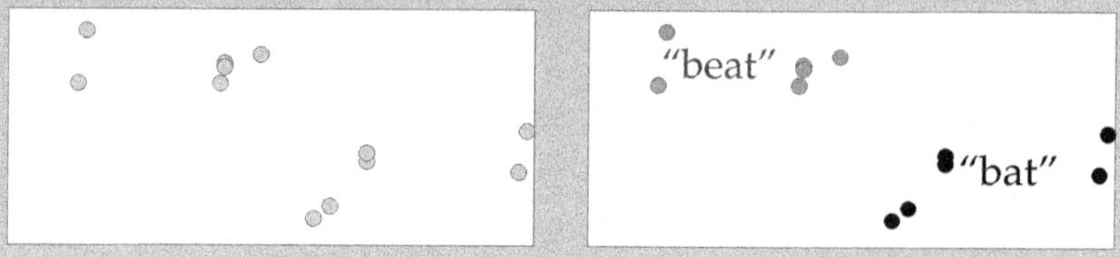

Figure 2.25: The frequencies of the first two formants in utterances by six speakers saying the words *beat* and *bat*.

2.3.5 Exercises

Let $H := \{0,1,2,3,4,5,6,7,8,9,\texttt{a},\texttt{b},\texttt{c},\texttt{d},\texttt{e},\texttt{f}\}$ denote the set of hexadecimal digits.

2.86 Is $6 \in H$?

2.87 Is $\texttt{h} \in H$?

2.88 Is $\texttt{a70e} \in H$?

2.89 What is $|H|$?

Let $S := \{0+0,\ 0+1,\ 1+0,\ 1+1,\ 0\cdot 0,\ 0\cdot 1,\ 1\cdot 0,\ 1\cdot 1\}$ be the set of results of adding any two bits together or multiplying any two bits together.

2.90 Which of 0, 1, 2, and 3 are elements of S?

2.91 What is $|S|$?

Let $T := \{n \in \mathbb{Z} : 0 \leq n \leq 20$ and $n \bmod 2 = n \bmod 3\}$. Let $H := \{0,1,2,3,4,5,6,7,8,9,\texttt{a},\texttt{b},\texttt{c},\texttt{d},\texttt{e},\texttt{f}\}$ and $S := \{0+0,\ 0+1,\ 1+0,\ 1+1,\ 0\cdot 0,\ 0\cdot 1,\ 1\cdot 0,\ 1\cdot 1\}$, as in the previous blocks of exercises.

2.92 Identify at least one element of H that is not an element of T.

2.93 Identify at least one element of T that is not an element of H.

2.94 Identify at least one element of T that is not an element of S.

2.95 Identify at least one element of S that is not an element of T.

2.96 What is $|T|$?

Rewrite the following sets by exhaustively listing their elements:

2.97 $\{n \in \mathbb{Z} : 0 \leq n \leq 20 \text{ and } n \bmod 5 = n \bmod 7\}$

2.98 $\{n \in \mathbb{Z} : 10 \leq n \leq 30 \text{ and } n \bmod 5 = n \bmod 7\}$

Let $A := \{1,3,4,5,7,8,9\}$ and let $B := \{0,4,5,9\}$. What are the following sets?

2.99 $A \cap B$

2.100 $A \cup B$

2.101 $A - B$

2.102 $B - A$

Assume the universe is the set $U := \{0,1,2,\ldots,9\}$. Define $C := \{0,3,6,9\}$, and let $A := \{1,3,4,5,7,8,9\}$ and $B := \{0,4,5,9\}$ as before. What are the following sets?

2.103 ${\sim}B$

2.104 $A \cup {\sim}C$

2.105 ${\sim}C - {\sim}B$

2.106 $C - {\sim}C$

2.107 ${\sim}(C - {\sim}A)$

2.108 In general, $A - B$ and $B - A$ do *not* denote the same set. (See Figure 2.26.) But your friends Evan and Yasmin wander by and tell you the following. Let E denote the set of CS homework questions that Evan has not yet solved. Let Y denote the set of CS homework questions that Yasmin has not yet solved. Evan and Yasmin claim that $E - Y = Y - E$. Is this possible? If so, under what circumstances? If not, why not? Justify your answer.

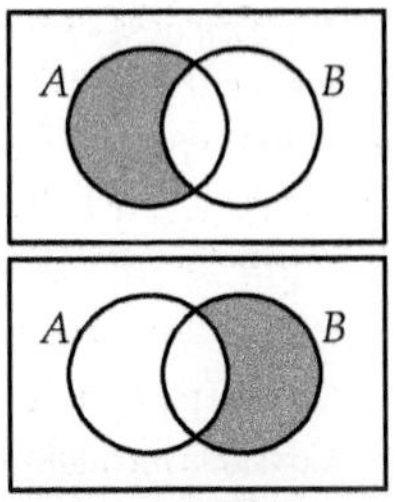

Figure 2.26: In general, the sets $A - B$ and $B - A$ are different.

Let D and E be arbitrary sets. For each set given below, indicate which of the following statements is true:

- *the given set* must *be a subset of D (for every choice of D and E);*
- *the given set* may *be a subset of D (for certain choices of D and E); or*
- *the given set* cannot *be a subset of D (for any choice of D and E).*

If you answer "must" or "cannot," justify your answer (1–2 sentences). If you answer "may," identify an example D_1, E_1 for which the given set is a subset of D_1, and an example D_2, E_2 for which the given set is not a subset of D_2.

2.109 $D \cup E$

2.110 $D \cap E$

2.111 $D - E$

2.112 $E - D$

2.113 ${\sim}D$

Let $F := \{1,2,4,8\}$, let $G := \{1,3,9\}$, and let $H := \{0,5,6,7\}$. Let $U := \{0,1,2,\ldots,9\}$ be the universe. Which of the following pairs of sets are disjoint?

2.114 F and G

2.115 G and ${\sim}F$

2.116 $F \cap G$ and H

2.117 H and ${\sim}H$

Let S and T be two sets, with $n = |S|$ and $m = |T|$. For each of the following sets, state the smallest cardinality that the given set can have. Give examples of the minimum-sized sets for each part. (You should give a family *of examples—that is, describe a smallest-possible set for* any *values of n and m.)*

2.118 $S \cup T$

2.119 $S \cap T$

2.120 $S - T$

Repeat the last three exercises for the largest *set: for two sets S and T with $n = |S|$ and $m = |T|$, state the largest cardinality that the given set can have. Give a family of examples of the largest-possible sets for each part.*

2.121 $S \cup T$

2.122 $S \cap T$

2.123 $S - T$

In a variety of CS applications, it's useful to be able to compute the similarity *of two sets A and B. (More about one of these applications, collaborative filtering, below.) There are a number of different ideas of how to measure set similarity, all based on the intuition that the larger $|A \cap B|$ is, the more similar the sets A and B are. Here are two basic measures of set similarity that are sometimes used:*

- the cardinality measure: *the similarity of A and B is* $|A \cap B|$.
- the Jaccard coefficient:[7] *the similarity of A and B is* $\frac{|A \cap B|}{|A \cup B|}$.

The Jaccard coefficient is named after the Swiss botanist Paul Jaccard, from around the turn of the 20th century, who was interested in how similar or different the distributions of various plants were in different regions.

[7] P. Jaccard. Distribution de la flore alpine dans le bassin des dranses et dans quelques régions voisines. *Bulletin de la Société Vaudoise des Sciences Naturelles*, 37:241–272, 1901.

2.124 Let $A :=$ {chocolate, hazelnut, cheese}; $B :=$ {chocolate, cheese, cardamom, cherries}; and $C :=$ {chocolate}. Compute the similarities of each pair of these sets using the cardinality measure.

2.125 Repeat the previous exercise for the Jaccard coefficient.

Suppose we have a collection of sets $A_1, A_2, \ldots, A_n$. Consider the following claim:

> **Claim:** *Suppose that the set A_v is the most similar set to the set A_u in this collection (aside from A_u itself). Then A_u is necessarily the set that is most similar to A_v (aside from A_v itself).*

2.126 Decide whether you think this claim is true for the cardinality measure of set similarity, and justify your answer. (That is, argue why it must be true, or give an example showing that it's false.)

2.127 Repeat the previous exercise for the Jaccard coefficient.

> **Taking it further:** A *collaborative filtering system*, or *recommender system*, seeks to suggest new products to a user u on the basis of the similarity of u's past behavior to the past behavior of other users in the system. Collaborative filtering systems are mainstays of many popular commercial online sites (like Amazon or Netflix, for example). One common approach to collaborative filtering is the following. Let U denote the set of users of the system, and for each user $u \in U$, define the set S_u of products that u has purchased. To make a product recommendation to a user $u \in U$:
>
> (i) Identify the user $v \in U - \{u\}$ such that S_v is the set "most similar" to S_u.
> (ii) Recommend the products in $S_v - S_u$ to user u (if any exist).
>
> This approach is called *nearest-neighbor collaborative filtering,* because the v found in step (i) is the other person closest to u. The measure of set similarity used in step (i) is all that's left to decide, and either cardinality or the Jaccard coefficient are reasonable choices. The idea behind the Jaccard coefficient is that the *fraction* of agreement matters more than the *total amount* of agreement: a {Cat's Cradle, Catch 22} purchaser is more similar to a {Slaughterhouse Five, Cat's Cradle} purchaser than someone who bought *every* book Amazon sells.

For each of the following claims, decide whether you think the statement is true for all sets of integers A, B, C. If it's true for every A, B, C, then explain why. (A Venn diagram may be helpful.) If it's not true for every A, B, C, then provide an example for which it does not hold.

2.128 $A \cap B = \sim(\sim A \cup \sim B)$

2.129 $A \cup B = \sim(\sim A \cap \sim B)$

2.130 $(A - B) \cup (B - C) = (A \cup B) - C$

2.131 $(B - A) \cap (C - A) = (B \cap C) - A$

2.132 List all of the different ways to partition the set $\{1, 2, 3\}$.

Consider the table of distances shown in Figure 2.27 for a set P = {Alice, ..., Frank} of people. Suppose we partition P into subsets $S_1, \ldots, S_k$. Define the intracluster distance *as the largest distance between two people who are in the same cluster:*

$$\max_i \left[\max_{x,y \in S_i} \text{distance between } x \text{ and } y \right].$$

Define the intercluster distance *as the smallest distance between two people who are in different clusters:*

$$\min_{i,j \neq i} \left[\min_{x \in S_i, y \in S_j} \text{distance between } x \text{ and } y \right].$$

	Alice	*Bob*	*Charlie*	*David*	*Eve*	*Frank*
Alice	0.0	1.7	1.2	0.8	7.2	2.9
Bob	1.7	0.0	4.3	1.1	4.3	3.4
Charlie	1.2	4.3	0.0	7.8	5.2	1.3
David	0.8	1.1	7.8	0.0	2.1	1.9
Eve	7.2	4.3	5.2	2.1	0.0	1.9
Frank	2.9	3.4	1.3	1.9	1.9	0.0

Figure 2.27: Some distances between people.

In each of the following questions, partition P into ...

2.133 ... 3 or fewer subsets so that the intracluster distance is ≤ 2.0.

2.134 ... subsets $S_1, \ldots, S_k$ so the intracluster distance is as small as possible. (You choose k.)

2.135 ... subsets $S_1, \ldots, S_k$ so the intercluster distance is as large as possible. (Again, you choose k.)

2.136 Define $S := \{1, 2, \ldots, 100\}$. Let $W := \{x \in S : x \bmod 2 = 0\}$, $H := \{x \in S : x \bmod 3 = 0\}$, and $O := S - H - W$. Is $\{W, H, O\}$ a partition of S?

What is the power set of each of the following sets?

2.137 $\{1, a\}$

2.138 $\{1\}$

2.139 $\{\}$

2.140 $\mathscr{P}(1)$

2.4 *Sequences, Vectors, and Matrices: Ordered Collections*

> Watch out for the fellow who talks about putting things in order! Putting things in order always means getting other people under your control.
>
> Denis Diderot (1713–1784)
> *Supplément au voyage de Bougainville* (1796)

In Section 2.3, we introduced sets—collections of objects in which the order of those objects doesn't matter. In many circumstances, though, order *does* matter: if a Java method takes two parameters, then swapping the order of those parameters will usually change what the method does; if there's an interesting site at longitude x and latitude y, then showing up at longitude y and latitude x won't do. In this section, we turn to *ordered* collections of objects, called *sequences*. A summary of the notation related to sequences is given in Figure 2.29.

Definition 2.32 (Sequence, list, and tuple)
A sequence*—also known as a* list *or* tuple*—is an ordered collection of objects, typically called* components *or* entries. *When the number of objects in the collection is* 2, 3, 4, *or* n, *the sequence is called an* (ordered) pair, triple, quadruple, *or,* n-tuple, *respectively.*

We'll write a sequence inside angle brackets, as in $\langle$Northfield, Minnesota$\rangle$ or $\langle 0, 1 \rangle$. (Some people use parentheses instead of angle brackets, as in $(128, 128, 0)$ instead of $\langle 128, 128, 0 \rangle$.) For two sets A and B, we frequently will refer to the set of ordered pairs whose two elements, in order, come from A and B:

Definition 2.33 (Cartesian product)
The Cartesian product *of two sets A and B, denoted $A \times B$, is the set*

$$A \times B = \{\langle a, b \rangle : a \in A \text{ and } b \in B\}$$

containing all ordered pairs where the first component comes from A and the second from B.

The Cartesian product is named after René Descartes, the 17th-century French philosopher/ mathematician. (The English adjectival form uses only the *cartes* part of his last name Des*cartes*.)

For example, $\{0, 1\} \times \{2, 3\}$ is the set $\{\langle 0, 2 \rangle, \langle 0, 3 \rangle, \langle 1, 2 \rangle, \langle 1, 3 \rangle\}$. We can also view any particular cell in a 2-dimensional grid—like a cell in a spreadsheet, or a square on a chess board—as a sequence:

Example 2.37 (Chess positions)
A chess board is an 8-by-8 grid. Chess players use what's called "Algebraic notation" to refer to the columns (which they call *files*) using the letters a through h, and they refer to the rows (which they call *ranks*) using the numbers 1 through 8. (See Figure 2.28.)

Thus the square containing the white queen ♕ is $\langle \mathsf{d}, 1 \rangle$; the full set of squares of the chess board is $\{\mathsf{a}, \mathsf{b}, \mathsf{c}, \mathsf{d}, \mathsf{e}, \mathsf{f}, \mathsf{g}, \mathsf{h}\} \times \{1, 2, 3, 4, 5, 6, 7, 8\}$; and the squares containing knights—the ♘ pieces (both white and black)—are $\{\langle \mathsf{b}, 1 \rangle, \langle \mathsf{g}, 1 \rangle, \langle \mathsf{b}, 8 \rangle, \langle \mathsf{g}, 8 \rangle\}$. The set of squares with knights could also be written as $\{\mathsf{b}, \mathsf{g}\} \times \{1, 8\}$.

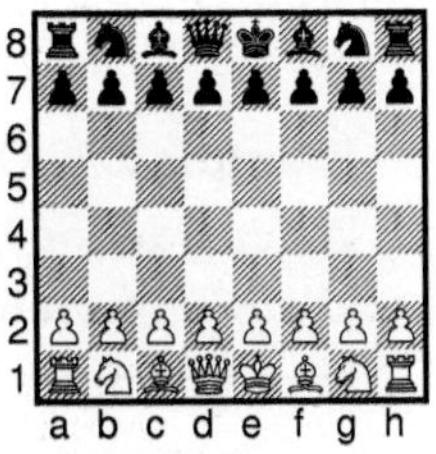

Figure 2.28: The squares of a chess board, written using Algebraic notation.

sequence/ordered tuple	$\langle a_1, a_2, \ldots, a_n \rangle$
Cartesian product	$A \times B := \{\langle a, b\rangle : a \in A \text{ and } b \in B\}$
the set of all n-element sequences of S	$S^n := S \times S \times \cdots \times S$ (n times)
vector	$x \in \mathbb{R}^n$
vector length, for $x \in \mathbb{R}^n$	$\|x\| := \sqrt{\sum_{i=1}^n x_i^2}$
vector addition, for vectors $x, y \in \mathbb{R}^n$	$x + y := \langle x_1 + y_1, x_2 + y_2, \ldots, x_n + y_n\rangle$
scalar product, for $a \in \mathbb{R}$ and $x \in \mathbb{R}^n$	$ax := \langle a \cdot x_1, a \cdot x_2, \ldots, a \cdot x_n\rangle$
dot product, for vectors $x, y \in \mathbb{R}^n$	$x \bullet y := \sum_{i=1}^n x_i \cdot y_i$
matrix	$M \in \mathbb{R}^{n \times m}$
identity matrix	a matrix $I \in \mathbb{R}^{n \times n}$ where $I = \begin{bmatrix} 1 & 0 & \ldots & 0 \\ 0 & 1 & \ldots & 0 \\ \vdots & \vdots & \ddots & \vdots \\ 0 & 0 & \ldots & 1 \end{bmatrix}$
scalar multiplication, for $\alpha \in \mathbb{R}$ and $M \in \mathbb{R}^{n \times m}$	a matrix $N \in \mathbb{R}^{n \times m}$ where $N_{i,j} := \alpha \cdot M_{i,j}$
matrix addition, for $M, M' \in \mathbb{R}^{n \times m}$	a matrix $N \in \mathbb{R}^{n \times m}$ where $N_{i,j} := M_{i,j} + M'_{i,j}$
matrix multiplication, for $A \in \mathbb{R}^{n \times m}$ and $B \in \mathbb{R}^{m \times p}$	a matrix $M \in \mathbb{R}^{n \times p}$ where $M_{i,j} = \sum_{k=1}^m A_{i,k} B_{k,j}$
matrix inverse, for $M \in \mathbb{R}^{n \times n}$	a matrix $M^{-1} \in \mathbb{R}^{n \times n}$ where $MM^{-1} = I$ (if any such M^{-1} exists)

Figure 2.29: A summary of notation for sequences, vectors, and matrices.

Here's another example, about color representation on computers:

Example 2.38 (RGB color values)
The *RGB color space* represents colors as ordered triples, where each component is an element of $\{0, 1, \ldots, 255\}$. RGB stands for *red–green–blue;* the three components of a color c, respectively, represent how red, how green, and how blue the color c is. Formally, a color c is an element of $\{0, 1, \ldots, 255\} \times \{0, 1, \ldots, 255\} \times \{0, 1, \ldots, 255\}$.

The order of these components matters; for example, the color $\langle 0, 0, 255\rangle$ is pure blue, while the color $\langle 255, 0, 0\rangle$ is pure red. See Figure 2.30 for a few examples.

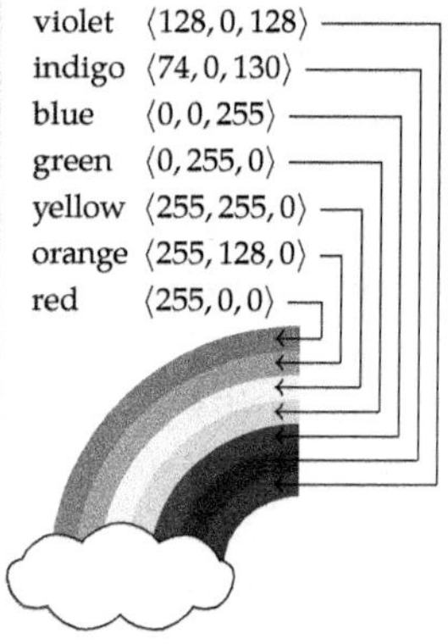

Figure 2.30: A few RGB values of colors.

Taking it further: An annoying pedantic point: we are being sloppy with notation in Example 2.38; we only defined the Cartesian product for two sets, so when we write $S \times S \times S$ we "must" mean either $S \times (S \times S)$ or $(S \times S) \times S$. We're going to ignore this issue, and simply write statements like $\langle 0, 1, 1\rangle \in \{0, 1\} \times \{0, 1\} \times \{0, 1\}$—even though we *ought* to instead be writing statements like $\langle 0, \langle 1, 1\rangle\rangle \in \{0, 1\} \times (\{0, 1\} \times \{0, 1\})$. (A similar shorthand shows up in programming languages like Scheme, where pairing—"cons"ing—a single element `3` with a list `(2 1)` yields the three-element list `(3 2 1)`, rather than the two-element pair `(3 . (2 1))`, where the second element is a two-element list.)

Beyond the "obvious" sequences like Examples 2.37 and 2.38, we've also already seen some definitions that don't seem to involve sequences, but implicitly *are* about ordered tuples of values. One example is the rational numbers (see Section 2.2.2):

Example 2.39 (Rational numbers as sequences)
We can define the *rational numbers* (also known as *fractions*) as the set $\mathbb{Q} := \mathbb{Z} \times \mathbb{Z}^{>0}$. Under this view, a rational number would be represented as a pair $\langle n, d\rangle \in \mathbb{Z} \times \mathbb{Z}^{>0}$, with a numerator n and a denominator d.

For example, the fractions $\frac{1}{2}$ and $\frac{202}{808}$ would be represented as $\langle 1, 2\rangle$ and $\langle 202, 808\rangle$, respectively. (To flesh out the details of this representation, we also have to consider reducing fractions to lowest terms, to establish the equivalence of fractions like $\langle 2, 4\rangle$ and $\langle 1, 2\rangle$. In Example 8.36, we'll formalize this equivalence.)

We will often consider sequences of elements that are all drawn from the same set, and there is special notation for such a sequence:

Definition 2.34 (Sequences of elements from the same set)
For a set S and a positive integer n, we write S^n to denote

$$S^n := \underbrace{S \times S \times \ldots \times S}_{n \text{ times}}.$$

Thus S^n denotes the set of all sequences of length n where each component of the sequence is an element the set S. For example, the RGB values from Example 2.38 are elements of $\{0, 1, \ldots, 255\}^3$, and $\{0, 1\}^3$ denotes the set

$$\{\langle 0,0,0\rangle, \langle 0,0,1\rangle, \langle 0,1,0\rangle, \langle 0,1,1\rangle, \langle 1,0,0\rangle, \langle 1,0,1\rangle, \langle 1,1,0\rangle, \langle 1,1,1\rangle\}.$$

This notation also lets us write $\mathbb{R} \times \mathbb{R}$, called the *Cartesian plane*, as $\mathbb{R}^2$—the way you might have written it in a high school algebra class. (See Figure 2.31.)

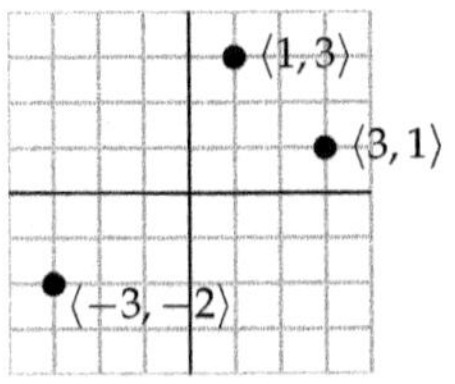

Figure 2.31: Three points in $\mathbb{R}^2$. The first component represents the x-axis (horizontal) position; the second component represents the y-axis (vertical) position.

Taking it further: René Descartes, the namesake of the Cartesian product and the Cartesian plane, was a major contributor in mathematics, particularly geometry. But Descartes is probably most famous as a philosopher, for the *cogito ergo sum* ("I think therefore I am") argument, in which Descartes—after adopting a highly skeptical view about all claims, even apparently obviously true ones—attempts to argue that he himself must exist.

In certain contexts, sequences of elements from the same set (as in Definition 2.34) are called *strings*. For a set Σ, called an *alphabet*, a *string over* Σ is an element of Σ^n for some nonnegative integer n. (In other words, a string is any element of $\bigcup_{n \in \mathbb{Z}^{\geq 0}} \Sigma^n$.) The *length* of a string $x \in \Sigma^n$ is n. For example, the set of 5-letter words in English is a subset of $\{\texttt{A}, \texttt{B}, \ldots, \texttt{Z}\}^5$. We allow strings to have length zero: for any alphabet Σ, there is only one sequence of elements from Σ of length 0, called the *empty string*; it's denoted by ε, and for any alphabet Σ, we have $\Sigma^0 := \{\varepsilon\}$. When writing strings, it is customary to omit the punctuation (angle brackets and commas), so we write $\texttt{ABRACADABRA} \in \{\texttt{A}, \texttt{B}, \ldots, \texttt{Z}\}^{11}$ and $11010011 \in \{0, 1\}^8$.

2.4.1 *Vectors*

As we've already seen, we can create sequences of many types of things: we can view sequences of letters as strings (like $\texttt{ABRACADABRA} \in \{\texttt{A}, \texttt{B}, \ldots, \texttt{Z}\}^{11}$), or sequences of three integers between 0 and 255 as colors (like $\langle 119, 136, 153\rangle \in \{0, 1, \ldots, 255\}^3$, officially called "light slate gray"). Perhaps the most pervasive type of sequence, though, is a sequence of real numbers, called a *vector*.

Taking it further: Vectors are used in a tremendous variety of computational contexts: computer graphics (representing the line-of-sight from the viewer's eye to an object in a scene), machine learning (a *feature vector* describing which characteristics a particular object has, which can be used in trying to classify that object as satisfying a condition or failing to satisfy a condition), among many others. The discussion on p. 248 describes the *vector-space model* for representing a document d as a vector whose components correspond to the number of times each word appears in d.

Vectors and matrices (the topics of this and the next subsection) are the main focus of a math course in linear algebra. In these subsections, we're only mentioning a few highlights of vectors and matrices; you can find much more in any good textbook on linear algebra.

Definition 2.35 (Vector)
A vector *(or* n-vector*)* x *is a sequence* $x \in \mathbb{R}^n$*, for some positive integer* n*. For a vector* $x \in \mathbb{R}^n$ *and for any index* $i \in \{1, 2, \ldots, n\}$*, we write* x_i *to denote the* i*th component of* x*.*

A warning for C or Java or Python (or ...) programmers: notice that our vectors' components are indexed starting at one, not zero. For a vector $x \in \mathbb{R}^n$, the expression x_i is meaningless unless $i \in \{1, 2, \ldots, n\}$. In particular, the expression x_0 doesn't mean anything.

For example, $\langle 0, 1 \rangle$, $\langle 1, 0 \rangle$, and $\langle \frac{1}{\sqrt{2}}, \frac{1}{\sqrt{2}} \rangle$ are all vectors in $\mathbb{R}^2$. For the vector $x := \langle 1/2, \sqrt{3}/2 \rangle$, we have $x_1 = 1/2$ and $x_2 = \sqrt{3}/2$.

Vectors are sometimes contrasted with *scalars*, which are just numbers: that is, a scalar is an element of $\mathbb{R}$. Vectors are also sometimes written in square brackets, so we may see an n-vector x written as $x = [x_1, x_2, \ldots, x_n]$. We may encounter vectors in which the components are a restricted kind of number—for example, integers or bits. Elements of $\{0, 1\}^n$ are often called *bit vectors* or *bitstrings*.

Here's an example of using vectors to compute distances between points:

Example 2.40 (Train stations in Manhattan)

Problem: Let's (very roughly!) represent a location in Manhattan as a vector—specifically, as a point $\langle x, y \rangle \in \mathbb{R}^2$ representing the intersection of xth Avenue and yth Street. Define the *walking distance* between points p and q in Manhattan as $|p_1 - q_1| + |p_2 - q_2|$: the number of east–west blocks between p and q *plus* the number of north–south blocks between p and q. (Note that walking distance is different from the straight-line distance between the points!)

1. The two major train stations in Manhattan are Penn Station, located at $s := \langle 8, 33 \rangle$, and Grand Central Station, located at $g := \langle 4, 42 \rangle$. What's the walking distance between Penn Station and Grand Central?
2. Describe the set of all points that are closer (in walking distance) to Penn Station than to Grand Central.

Solution: 1. The distance between $s = \langle 8, 33 \rangle$ and $g = \langle 4, 42 \rangle$ is $|s_1 - g_1| + |s_2 - g_2| = |8 - 4| + |33 - 42| = 4 + 9 = 13$.

2. Let's compute some points that are equidistant to the two stations. (Those points are on the boundary of the region of points closer to g and the region of points closer to s.) For example, a point $\langle 4, y \rangle$ has distances $|42 - y|$ and $4 + |y - 33|$ to the stations; these distances are both equal to 6.5 when $y = 35.5$.

More generally, let's think about a point whose x-coordinate falls between 4 and 8. For any offset $0 \leq \delta \leq 4$, the distance between the point $\langle 4 + \delta, y \rangle$ and the two stations are $\delta + |42 - y|$ and $4 - \delta + |y - 33|$. These two values are both equal to 6.5 when $y = 35.5 + \delta$. (For example, when $\delta = 4$, then $y = 39.5$.) Thus the points $\langle 4 + 0, 35.5 + 0 \rangle = \langle 4, 35.5 \rangle$ and $\langle 4 + 4, 35.5 + 4 \rangle = \langle 8, 39.5 \rangle$ are both equidistant to s and g, as are all points on the line segment between them. (See Figure 2.32.)

The remaining cases of the analysis—figuring out which points with x-coordinate less than 4 or greater than 8 are closer to s or g (the regions marked with "?" in Figure 2.32)—are left to you in Exercises 2.184 and 2.185.

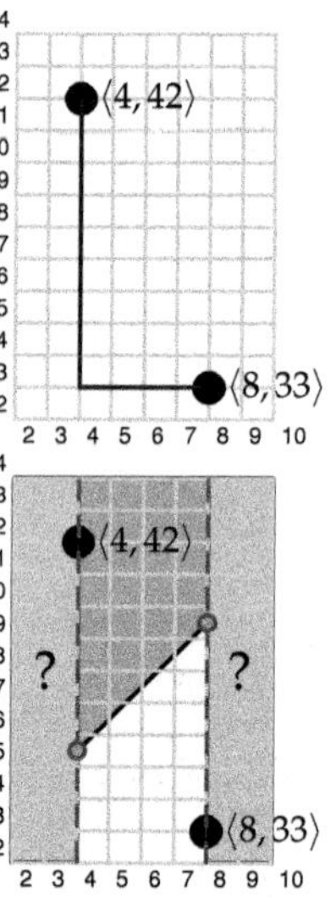

Figure 2.32: Illustrations of Manhattan train stations. In the second panel, the dark shaded points are closer (in walking distance) to $\langle 4, 42 \rangle$ than to $\langle 8, 33 \rangle$. The white shaded points are closer to $\langle 8, 33 \rangle$ than to $\langle 4, 42 \rangle$.

Taking it further: The measure of walking distance between points that we used in Example 2.40 is used surprisingly commonly in computer science applications—and, appropriately enough, it's actually named after Manhattan. The *Manhattan distance* between two points $p, q \in \mathbb{R}^n$ is defined as $\sum_{i=1}^{n} |p_i - q_i|$. (We're summing the number of "blocks" of difference in each of the n dimensions; we take the absolute value of the difference in each component because we care about the difference in each dimension rather than which point has the higher value in that component.)

Here's one more useful definition about vectors:

Definition 2.36 (Vector length)
The length *of a vector $x \in \mathbb{R}^n$ is defined as $\|x\| := \sqrt{\sum_{i=1}^{n} (x_i)^2}$.*

For example, $\|\langle 2,8 \rangle\| = \sqrt{2^2 + 8^2} = \sqrt{4+64} = \sqrt{68} \approx 8.246$. If we draw a vector $x \in \mathbb{R}^2$ in the Cartesian plane, then $\|x\|$ denotes the length of the line from $\langle 0,0 \rangle$ to x. (See Figure 2.33.) A vector $x \in \mathbb{R}^n$ is called a *unit vector* if $\|x\| = 1$.

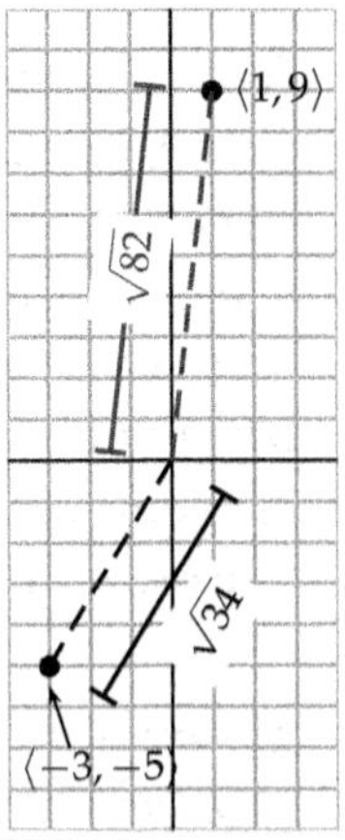

Figure 2.33: Two vector lengths: $\|\langle 1,9 \rangle\|$ is $\sqrt{1+81} = \sqrt{82}$, and $\|\langle -3,-5 \rangle\|$ is $\sqrt{9+25} = \sqrt{34}$.

VECTOR ARITHMETIC

We will now define basic arithmetic for vectors: *vector addition*, which is performed component-wise (adding the corresponding elements of the two vectors), and two forms of multiplication—one for multiplying a vector by a scalar (also component-wise) and one for multiplying two vectors together. We'll start with addition:

Definition 2.37 (Vector addition)
The sum *of two vectors $x, y \in \mathbb{R}^n$, written $x + y$, is a vector $z \in \mathbb{R}^n$, where for every index $i \in \{1, 2, \ldots, n\}$ we have $z_i := x_i + y_i$. (Note that the sum of two vectors with different sizes is meaningless.)*

For example, $\langle 1.1, 2.2, 3.3 \rangle + \langle 2, 0, 2 \rangle = \langle 3.1, 2.2, 5.3 \rangle$.

The first type of multiplication for vectors is *scalar multiplication*, when we multiply a vector by a real number. As with vector addition, scalar multiplication acts on each component independently, by rescaling each component by the same factor:

Definition 2.38 (Scalar product)
Given a vector $x \in \mathbb{R}^n$ and a real number $\alpha \in \mathbb{R}$, the scalar product *αx is a vector $z \in \mathbb{R}^n$, where $z_i := \alpha x_i$ for every index $i \in \{1, 2, \ldots, n\}$.*

For example, we have $3 \cdot \langle 1,2,3 \rangle = \langle 3,6,9 \rangle$. Similarly $-1.5 \cdot \langle 1, -1 \rangle = \langle -1.5, 1.5 \rangle$ and $0 \cdot \langle 1,2,3,5,8 \rangle = \langle 0,0,0,0,0 \rangle$.

The second type of vector multiplication, the *dot product*, takes two vectors as input and multiplies them together to produce a single scalar as output:

As with vector addition, the dimensions of the vectors in a dot product have to match up: if $x \in \mathbb{R}^n$ and $y \in \mathbb{R}^m$ are vectors where $n \neq m$, then $x \bullet y$ is meaningless.

Definition 2.39 (Dot product)
Given two vectors $x, y \in \mathbb{R}^n$, the dot product *of x and y, denoted $x \bullet y$, is given by summing the products of the corresponding components:*

$$x \bullet y = \sum_{i=1}^{n} x_i \cdot y_i.$$

For example, $\langle 1,2,3\rangle \bullet \langle 4,5,6\rangle = 1\cdot 4 + 2\cdot 5 + 3\cdot 6 = 4 + 10 + 18 = 32$.

Intuitively, the dot product of two vectors measures the extent to which they point in the "same direction." Here's an example with a few unit vectors:

Example 2.41 (Dot products of unit vectors)
Consider the unit vectors $a := \langle 0,1\rangle$, $b := \langle 1,0\rangle$, $c := \langle 1/\sqrt{2}, 1/\sqrt{2}\rangle$, and $d := \langle 0,-1\rangle$. (See Figure 2.34.) Here is the dot product of c with each of these vectors:

$$\begin{aligned} & c \bullet a \\ &= c_1 \cdot a_1 + c_2 \cdot a_2 \\ &= \tfrac{1}{\sqrt{2}} \cdot 0 + \tfrac{1}{\sqrt{2}} \cdot 1 \\ &= \tfrac{1}{\sqrt{2}}. \end{aligned}$$

$$\begin{aligned} & c \bullet b \\ &= c_1 \cdot b_1 + c_2 \cdot b_2 \\ &= \tfrac{1}{\sqrt{2}} \cdot 1 + \tfrac{1}{\sqrt{2}} \cdot 0 \\ &= \tfrac{1}{\sqrt{2}}. \end{aligned}$$

$$\begin{aligned} & c \bullet c \\ &= c_1 \cdot c_1 + c_2 \cdot c_2 \\ &= \tfrac{1}{\sqrt{2}} \cdot \tfrac{1}{\sqrt{2}} + \tfrac{1}{\sqrt{2}} \cdot \tfrac{1}{\sqrt{2}} \\ &= \tfrac{1}{2} + \tfrac{1}{2} = 1. \end{aligned}$$

$$\begin{aligned} & c \bullet d \\ &= c_1 \cdot d_1 + c_2 \cdot d_2 \\ &= \tfrac{1}{\sqrt{2}} \cdot 0 + \tfrac{1}{\sqrt{2}} \cdot -1 \\ &= -\tfrac{1}{\sqrt{2}}. \end{aligned}$$

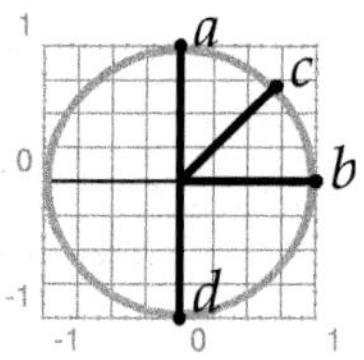

Figure 2.34: Four unit vectors.

Here are two examples using dot products for simple applications:

Example 2.42 (Common classes)
Let $C := \langle \text{CS1}, \text{CS2}, \ldots, \text{CS8}\rangle$ denote the list of all courses offered by a (somewhat narrowly focused) university. For a particular student, let the bit vector u represent the courses taken by that student, so that $u_i := 1$ if the student has taken course c_i (and $u_i := 0$ otherwise). For example, a student who's taken only CS1 and CS8 would be represented by $x := \langle 1,0,0,0,0,0,0,1\rangle$, and a student who's taken everything except CS3 would be represented by $y := \langle 1,1,0,1,1,1,1,1\rangle$.

The dot product of two student vectors represents the number of common courses that they've taken. For example, the number of common classes taken by x and y is

$$\begin{aligned} x \bullet y = \sum_{i=1}^{8} x_i y_i &= 1\cdot 1 + 0\cdot 1 + 0\cdot 0 + 0\cdot 1 + 0\cdot 1 + 0\cdot 1 + 0\cdot 1 + 1\cdot 1 \\ &= 1 + 0 + 0 + 0 + 0 + 0 + 0 + 1 \qquad = 2. \end{aligned}$$

Specifically, the two common courses taken by x and y are CS1 and CS8.

Example 2.43 (GPAs)
Let $g \in \mathbb{R}^n$ be an n-vector where g_i denotes the grade (measured on the grade point scale) that you got in the ith class that you've taken in your college career. Let $c \in \mathbb{R}^n$ be an n-vector where c_i denotes the number of credit hours for the ith class you took in your college career. Then your grade point average (GPA) is given by $\frac{g \bullet c}{\sum_{i=1}^{n} c_i}$.

For example, suppose your school gives grade points on the scale 4.0 = A, 3.7 = A-, 3.3 = B+, 3.0 = B, etc. Suppose you took CS 111 (6 credits), CS 201 (6 credits), and Mbira Lessons (4 credits), and got grades of B+, A-, and B, respectively. Then $g = \langle 3.3, 3.7, 3.0\rangle$ and $c = \langle 6,6,4\rangle$, and your GPA is given by

$$\frac{g \bullet c}{\sum_{i=1}^{3} c_i} = \frac{3.3\cdot 6 + 3.7\cdot 6 + 3.0\cdot 4}{6+6+4} = \frac{19.8 + 22.2 + 12.0}{16} = \frac{54}{16} = 3.375.$$

2.4.2 *Matrices*

If a vector is analogous to an array of numbers, then a *matrix* is analogous to a two-dimensional array of numbers:

$$\begin{bmatrix} M_{1,1} & M_{1,2} & \cdots & M_{1,m} \\ M_{2,1} & M_{2,2} & \cdots & M_{2,m} \\ \vdots & \vdots & \ddots & \vdots \\ M_{n,1} & M_{n,2} & \cdots & M_{n,m} \end{bmatrix}$$

Figure 2.35: A matrix M.

Definition 2.40 (Matrix)
An n-by-m matrix M *is a two-dimensional table of real numbers containing* n rows *and* m columns. *The* $\langle i,j\rangle$*th entry of the matrix appears in the* i*th row and* j*th column, and we denote that entry by* $M_{i,j}$*, as shown in Figure 2.35. Such a matrix* M *is an element of* $\mathbb{R}^{n\times m}$*, and we refer to* M *as having* size *or* dimension n-by-m.

Here are a few very small example matrices:

The plural of matrix is *matrices* (which rhymes with the word "cheese").

Example 2.44 (Three matrices)
Here are three matrices. (The $\langle 2,1\rangle$st entry is circled in each.)

$$A = \begin{bmatrix} 3 & 1 & 4 \\ \textcircled{9} & 7 & 2 \end{bmatrix} \qquad B = \begin{bmatrix} 5 & 3 \\ \textcircled{4} & 8 \\ 6 & 9 \end{bmatrix} \qquad I = \begin{bmatrix} 1 & 0 & 0 \\ \textcircled{0} & 1 & 0 \\ 0 & 0 & 1 \end{bmatrix}.$$

In these examples, A is a 2-by-3 matrix, B is a 3-by-2 matrix, and I is a 3-by-3 matrix.

One can think of a two-dimensional array in a programming language as a one-dimensional array *of one-dimensional arrays*. Similarly, if you prefer, you can think of an n-by-m matrix as a sequence of n vectors, all of which are elements of $\mathbb{R}^m$. This view of an n-by-m matrix is as an element of $(\mathbb{R}^n)^m$. One simple application of matrices is as an easy way to represent images:

$$\begin{bmatrix}
1&1&1&1&1&1&1&1&1&1&1&1&1&1&1&1&1&1&1&1\\
1&1&1&1&1&1&1&1&1&1&1&1&1&1&0&1&1&1&1&1\\
1&1&1&1&1&1&1&1&1&1&1&1&1&1&0&0&1&1&1&1\\
1&1&1&1&1&1&1&1&1&1&1&1&1&1&0&0&0&1&1&1\\
0&0&0&0&0&0&0&0&0&0&0&0&0&0&0&0&0&0&1&1\\
1&1&1&1&1&1&1&1&1&1&1&1&1&1&0&0&0&1&1&1\\
1&1&1&1&1&1&1&1&1&1&1&1&1&1&0&0&1&1&1&1\\
1&1&1&1&1&1&1&1&1&1&1&1&1&1&0&1&1&1&1&1\\
1&1&1&1&1&1&1&1&1&1&1&1&1&1&1&1&1&1&1&1
\end{bmatrix}$$

(a) A matrix.

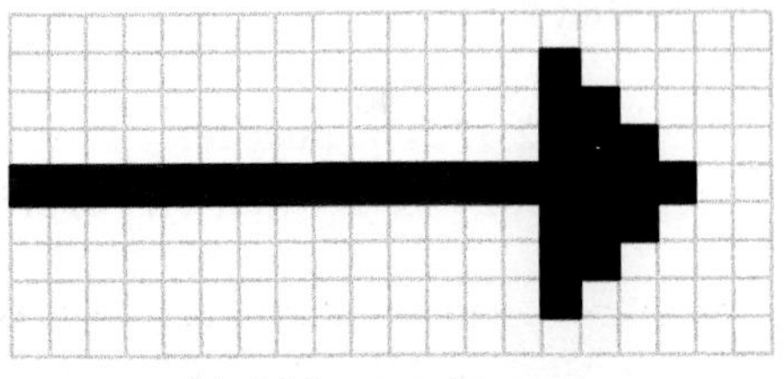

(b) A bitmapped image.

Figure 2.36: A matrix representing a black-and-white bitmapped image, and the image.

Example 2.45 (Bitmaps)
A black-and-white image can be represented as a matrix with all entries in $\{0,1\}$: each 1 entry represents white in the corresponding pixel; each 0 represents black. For example, the matrix in Figure 2.36(a) could represent the image in Figure 2.36(b).

<u>**Taking it further:**</u> The picture shown in Figure 2.36 is a simple black-and-white image, but we can use the same basic structure for grayscale or color images. Instead of just an integer in $\{0,1\}$ as each entry in the matrix, a grayscale pixel could be represented using a real number in $[0,1]$—or, more practically, a number in $\{\frac{0}{255}, \frac{1}{255}, \ldots, \frac{255}{255}\}$. For color images, each entry would be an RGB triple (see Example 2.38).

These matrix-based representations of an image are often called *bitmaps*. Bitmaps are highly inefficient ways of storing images; most computer graphics file formats use much cleverer (and more space-efficient) representations.

Here are few other examples of the pervasive applications of matrices in computer science. A *term–document matrix* can be used to represent a collection of documents: the entry $M_{d,k}$ of the matrix M stores the number of times that keyword k appears in document d. An *adjacency matrix* (see Chapter 11) can represent the page-to-page hyperlinks of the web in a matrix M, where $M_{i,j} = 1$ if web page i has a hyperlink to web page j (and $M_{i,j} = 0$ otherwise). A *rotation matrix* can be used in computer graphics to re-render a scene from a different perspective; see p. 249 for some discussion.

A matrix $M \in \mathbb{R}^{m \times n}$ is called *square* if $m = n$. For a square matrix $M \in \mathbb{R}^{n \times n}$, we may say that the size of M is n (rather than saying that its size is n-by-n). A square matrix M is called *symmetric* if, for all indices $i, j \in \{1, 2, \ldots, n\}$, we have $M_{i,j} = M_{j,i}$. The *main diagonal* of a square matrix $M \in \mathbb{R}^{n \times n}$ is the sequence consisting of the entries $M_{i,i}$ for $i = 1, 2, \ldots, n$. For example:

$$\begin{bmatrix} \textcircled{1} & 2 & 3 \\ 4 & \textcircled{5} & 6 \\ 7 & 8 & \textcircled{9} \end{bmatrix}$$

Figure 2.37: A matrix M with the entries of the main diagonal circled.

Example 2.46 (Main diagonal)
Consider the 3-by-3 square matrix M shown in Figure 2.37. The main diagonal of M is $\langle M_{1,1}, M_{2,2}, M_{3,3} \rangle = \langle 1, 5, 9 \rangle$.

One special square matrix that will arise frequently is the *identity matrix*, which has ones on the main diagonal and zeros everywhere else (see Figure 2.38):

$$\begin{bmatrix} 1 & 0 & \cdots & 0 \\ 0 & 1 & \cdots & 0 \\ \vdots & \vdots & \ddots & \vdots \\ 0 & 0 & \cdots & 1 \end{bmatrix}$$

Figure 2.38: The identity matrix I.

Definition 2.41 (Identity matrix)
The n-by-n identity matrix *is the matrix $I \in \mathbb{R}^{n \times n}$ whose entries satisfy*

$$I_{i,j} = \begin{cases} 1 & \text{if } i = j \\ 0 & \text{if } i \neq j. \end{cases}$$

Note that there is a different n-by-n identity matrix for every $n \geq 1$:

Example 2.47 (The smallest identity matrices)
Here are the identity matrices of size up to 5:

$$\begin{bmatrix} 1 \end{bmatrix} \quad \begin{bmatrix} 1 & 0 \\ 0 & 1 \end{bmatrix} \quad \begin{bmatrix} 1 & 0 & 0 \\ 0 & 1 & 0 \\ 0 & 0 & 1 \end{bmatrix} \quad \begin{bmatrix} 1 & 0 & 0 & 0 \\ 0 & 1 & 0 & 0 \\ 0 & 0 & 1 & 0 \\ 0 & 0 & 0 & 1 \end{bmatrix} \quad \begin{bmatrix} 1 & 0 & 0 & 0 & 0 \\ 0 & 1 & 0 & 0 & 0 \\ 0 & 0 & 1 & 0 & 0 \\ 0 & 0 & 0 & 1 & 0 \\ 0 & 0 & 0 & 0 & 1 \end{bmatrix}$$

As with vectors, we will need to define the basic arithmetic operations of addition and multiplication for matrices. Just as with vectors, adding two n-by-m matrices or multiplying a matrix by a scalar is done component by component.

Definition 2.42 (Matrix addition and scalar multiplication)
Given two matrices $M, M' \in \mathbb{R}^{n \times m}$ and a real number $\alpha \in \mathbb{R}$:

- *The product αM is a matrix $N \in \mathbb{R}^{n \times m}$ where $N_{i,j} := \alpha M_{i,j}$ for all indices $i \in \{1, 2, \ldots, n\}$ and $j \in \{1, 2, \ldots, m\}$.*

- *The sum* $M + M'$ *is a matrix* $N \in \mathbb{R}^{n \times m}$ *where* $N_{i,j} := M_{i,j} + M'_{i,j}$ *for all indices* $i \in \{1, 2, \ldots, n\}$ *and* $j \in \{1, 2, \ldots, m\}$.

Again, just as with vectors, adding two matrices that are not the same size is meaningless. Here are some small examples:

Example 2.48 (Simple matrix arithmetic)
Consider the following matrices:

$$A := \begin{bmatrix} 0 & 2 & 2 \\ 2 & 0 & 2 \\ 2 & 2 & 0 \end{bmatrix} \qquad B := \begin{bmatrix} 1 & 2 & 3 \\ 0 & 0 & 6 \\ 0 & 0 & 4 \end{bmatrix} \qquad I := \begin{bmatrix} 1 & 0 & 0 \\ 0 & 1 & 0 \\ 0 & 0 & 1 \end{bmatrix}$$

Then we have:

$$A + B = \begin{bmatrix} 1 & 4 & 5 \\ 2 & 0 & 8 \\ 2 & 2 & 4 \end{bmatrix} \qquad 4B = \begin{bmatrix} 4 & 8 & 12 \\ 0 & 0 & 24 \\ 0 & 0 & 16 \end{bmatrix}$$

$$A + 3I = \begin{bmatrix} 3 & 2 & 2 \\ 2 & 3 & 2 \\ 2 & 2 & 3 \end{bmatrix} \qquad A - 3I = \begin{bmatrix} -3 & 2 & 2 \\ 2 & -3 & 2 \\ 2 & 2 & -3 \end{bmatrix}$$

Matrix multiplication

Multiplying matrices is a bit more complicated than the other vector/matrix operations that we've seen so far. The product of two matrices is a *matrix*, rather than a single number: the entry in the ith row and jth column of AB is derived from the ith row of A and the j column of B. More precisely:

Definition 2.43 (Matrix multiplication)
The product AB *of two matrices* $A \in \mathbb{R}^{n \times m}$ *and* $B \in \mathbb{R}^{m \times p}$ *is an* n*-by-*p *matrix* $M \in \mathbb{R}^{n \times p}$ *whose entries are, for any* $i \in \{1, 2, \ldots n\}$ *and* $j \in \{1, 2, \ldots, p\}$,

$$M_{i,j} := \sum_{k=1}^{m} A_{i,k} B_{k,j}.$$

As usual, if the dimensions of the matrices A and B don't match—if the number of columns in A is different from the number of rows in B—then AB is undefined.

Example 2.49 (Multiplying some small matrices)
Let's compute the product of a sample 2-by-3 matrix and a 3-by-2 matrix:

$$\begin{bmatrix} 1 & 2 & 3 \\ 4 & 5 & 6 \end{bmatrix} \cdot \begin{bmatrix} 7 & 8 \\ 1 & 3 \\ 9 & 0 \end{bmatrix}$$

Note that, by definition, the result will be a 2-by-2 matrix. Let's compute its entries:

$$\begin{bmatrix} 1 & 2 & 3 \\ 4 & 5 & 6 \end{bmatrix} \cdot \begin{bmatrix} 7 & 8 \\ 1 & 3 \\ 9 & 0 \end{bmatrix} = \begin{bmatrix} 1\cdot 7+2\cdot 1+3\cdot 9 & 1\cdot 8+2\cdot 3+3\cdot 0 \\ 4\cdot 7+5\cdot 1+6\cdot 9 & 4\cdot 8+5\cdot 3+6\cdot 0 \end{bmatrix}$$

$$= \begin{bmatrix} 7+2+27 & 8+6+0 \\ 28+5+54 & 32+15+0 \end{bmatrix}$$

$$= \begin{bmatrix} 36 & 14 \\ 87 & 47 \end{bmatrix}.$$

For example, the 14 in ⟨row #1, column #2⟩ of the result was calculated by successively multiplying the first matrix's first row $\langle 1,2,3\rangle$ by the second matrix's second column $\langle 8,3,0\rangle$. Alternatively, here's a visual representation of this multiplication:

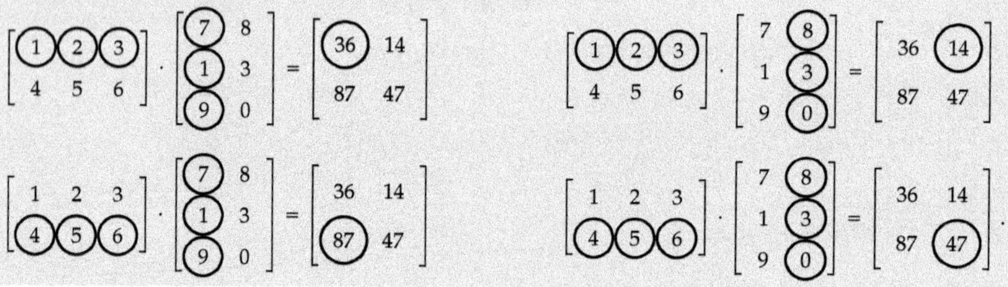

Problem-solving tip: To help keep matrix multiplication straight, it may be helpful to compute the $\langle i,j\rangle$th entry of AB by simultaneously tracing the ith row of A with the index finger of your left hand, and the jth column of B with the index finger of your right hand. Multiply the two numbers that you're pointing at, and add the result to a running tally; when you've traced the whole row/column, the running tally is $(AB)_{i,j}$.

More compactly, we could write matrix multiplication using the dot product from Definition 2.39: for two matrices $A \in \mathbb{R}^{n\times m}$ and $B \in \mathbb{R}^{m\times p}$, the $\langle i,j\rangle$th entry of AB is the value of $A_{i,(1\ldots m)} \bullet B_{(1\ldots m),j}$.

Be careful: matrix multiplication is not commutative—that is, for matrices A and B, the values AB and BA are generally different! (This asymmetry is unlike numerical multiplication: for $x, y \in \mathbb{R}$, it is always the case that $xy = yx$.) In fact, because the number of columns of A must match the number of rows of B for AB to even be meaningful, it's possible for BA to be meaningless or a different size from AB.

Example 2.50 (Multiplying the other way around)

If we multiply the matrices from Example 2.49 in the other order, we get

$$\begin{bmatrix} 7 & 8 \\ 1 & 3 \\ 9 & 0 \end{bmatrix} \cdot \begin{bmatrix} 1 & 2 & 3 \\ 4 & 5 & 6 \end{bmatrix} = \begin{bmatrix} 39 & 54 & 69 \\ 13 & 17 & 21 \\ 9 & 18 & 27 \end{bmatrix}$$

This matrix differs from the result in Example 2.49—it's not even the same size!

You'll show in the exercises that, for any n-by-m matrix A, the result of multiplying A by the identity matrix I yields A itself: that is, $AI = A$. You'll also explore the *inverse* of a matrix A: that is, the matrix A^{-1} such that $AA^{-1} = I$ (if any such A^{-1} exists).

Here's another example of using matrices, and matrix multiplication, to combine different types of information:

Example 2.51 (Programming language knowledge)

Problem: Let A be an n-by-m matrix where $A_{i,j} = 1$ if student i has taken class j (and $A_{i,j} = 0$ otherwise). Let B be an m-by-p matrix where $B_{j,k} = 1$ if class j uses programming language k (and $B_{j,k} = 0$ otherwise). What does the matrix AB represent?

Solution: First, note that the resulting matrix AB has n rows and p columns; that is, its size is (number of students)-by-(number of languages). For a student i and a programming language k, we have by definition that

$$\begin{aligned}(AB)_{i,k} &= \sum_{j=1}^{m} A_{i,j}B_{j,k} \\ &= \sum_{j=1}^{m} \left[\begin{cases} 1 & \text{if student } i \text{ took class } j \textit{ and } j \text{ uses language } k \\ 0 & \text{otherwise} \end{cases} \right]\end{aligned}$$

because $0 \cdot 0 = 0 \cdot 1 = 1 \cdot 0 = 0$, so the only terms of the sum that are 1 occur when both $A_{i,j}$ ("student i took class j?") and $B_{j,k}$ ("class j uses language k?") are true (that is, 1). Thus $(AB)_{i,k}$ denotes the number of classes that use language k that student i took.

Example 2.52 (A concrete example of Example 2.51)

Concretely, consider these 3 students, 5 courses, and 7 programming languages:

$A :=$

	intro	*data structures*	*org/arch*	*prog langs*	*theory of comp*
Alice	0	1	1	1	1
Bob	1	1	0	1	0
Charlie	1	0	0	0	1

$B :=$

	Perl	*Python*	*C*	*Java*	*Assembly*	*C++*	*Scheme*
intro	0	1	0	0	0	0	0
data struct	0	1	0	1	0	0	0
org/arch	0	0	1	0	1	0	0
prog lang	0	1	1	1	1	1	1
theory of comp	0	0	0	0	0	0	0

.

For these matrices, we have

$AB =$

	Perl	*Python*	*C*	*Java*	*Assembly*	*C++*	*Scheme*
Alice	0	2	2	2	2	1	1
Bob	0	3	1	2	1	1	1
Charlie	0	1	0	0	0	0	0

.

(For example, the Alice/C cell is computed by $\langle 0,1,1,1,1 \rangle \bullet \langle 0,0,1,1,0 \rangle$—the dot product of Alice's row of A with C's column of B—which has the value

$$0 \cdot 0 + 1 \cdot 0 + 1 \cdot 1 + 1 \cdot 1 + 1 \cdot 0 = 2.$$

This entry reflects the fact that Alice has taken two classes that use C: organization/architecture and programming languages.)

Computer Science Connections

The Vector Space Model

Here's a classic application of vectors, taken from *information retrieval*, the subfield of computer science devoted to searching for information relevant to a given query in large datasets. We start with a large *corpus* of documents—for example, transcripts of all email messages that you've sent in your entire life. (The word *corpus* comes from the Latin for "body"; it simply means a body of texts.) Tasks involving the corpus might include *clustering* the documents into subcollections ("which of my email messages are spam?"), or finding the stored documents most similar to a given query ("find me the 10 emails most relevant to 'good restaurants in Chicago' in my archives").

The *vector space model* is a standard approach to representing text documents for the purposes of information retrieval. We choose a list of n *terms* that might appear in a document. We then represent a document d as an n-vector x of integers, where x_i is the number of times that the ith term appears in the document d. See Figure 2.39 for an example.

Because documents that are about similar topics tend to contain similar vocabulary, we can judge the similarity of documents d and d' based on "how similar" their corresponding vectors x and x' are:

- A first stab at measuring similarity between x and x' is to compute the dot product $x \bullet x'$; this approach counts the number of times any word in d appears in d'. (And if a word appears twice in d, then each appearance in d' counts twice for the dot product.)
- This first approach has an issue in that it favors longer documents: a document that lists all the words in the dictionary would correspond to a vector $[1, 1, 1, 1, 1, \ldots]$—which would therefore have a large dot product with all documents in the corpus. To compensate for the fact that longer documents have more words, we *normalize* these vectors so that they have the same length, by using $x / \|x\|$ and $x' / \|x'\|$ to represent the documents. It turns out that the dot product of the normalized vectors computes the cosine of the *angle* between these representations of the documents.
- This second approach suffers from counting common occurrences of the word *the* and the word *normalize* as equally indicative of the similarity of documents. Information retrieval systems apply different weights to different terms in measuring similarity; one common approach is called *term frequency–inverse document frequency (TFIDF)*, which downweights terms that appear in many documents in the corpus.

It's worth noting that real information retrieval systems are usually quite a lot more complicated than we've discussed so far. For example, a document that talks about *sofas* would be judged to be completely unrelated to a document that talks about *couches*, which seems like a naïve judgement. Handling synonyms requires a more complicated approach, often based around analyzing the *term–document matrix* that simultaneously represents the entire corpus. (For example, if documents that discuss *sofas* use very similar *other* words to documents that discuss *couches*—like *change* and *cushion* and *nap*—then we might be able to infer something about *sofas* and *couches*.)[8]

d_1	*Three is one of the loneliest numbers.*
d_2	*A one and a two and a one, two, three.*
d_3	*One, two, buckle my shoe.*

↓

d_1	$[1, 0, 1]$
d_2	$[2, 2, 1]$
d_3	$[1, 1, 0]$

(a) Three documents translated into vectors using the keywords *'one'*, *'two'*, and *'three'*.

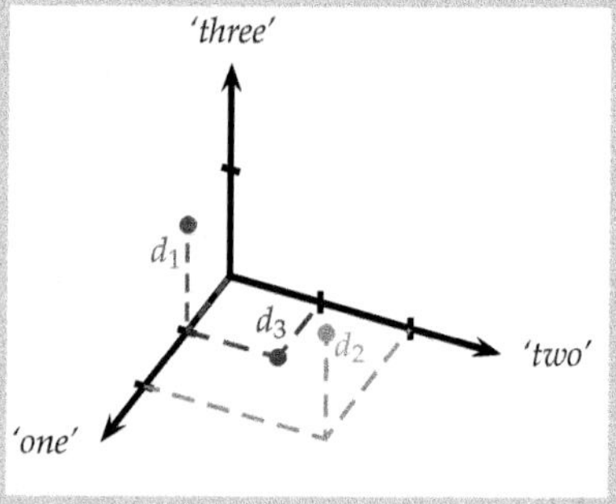

(b) A plot of the three documents in $\mathbb{R}^3$

Figure 2.39: An example from the vector-space model.

For much more on information retrieval, see the excellent text

[8] Christopher D. Manning, Prabhakar Raghavan, and Hinrich Schütze. *Introduction to Information Retrieval*. Cambridge University Press, 2008.

Computer Science Connections

Rotation Matrices

When an image is *rendered* (drawn) using computer graphics, we typically proceed by transforming a 3-dimensional representation of a *scene*, a model of the world, into a 2-dimensional *image* fit for a screen. The scene is typically represented by a collection of points in $\mathbb{R}^3$, each defining a vertex of a polygon. The *camera* (the eye from which the scene is viewed) is another point in $\mathbb{R}^3$, with an orientation describing the direction of view. We then *project* the polygons' points into $\mathbb{R}^2$. This computation is done using matrix multiplications, by taking into account the position and direction of view of the camera, and the position of the given point. While a full account of this rendering algorithm isn't *too* difficult, we'll stick with a simpler problem that still includes the interesting matrix computations.[9] We'll instead consider the *rotation* of a set of points in $\mathbb{R}^2$ by an angle θ. (The full-scale problem requires thinking about the angle of view with two parameters, akin to "azimuth" and "elevation" in orienteering: the direction θ in the horizontal plane and the angle φ away from a straight horizontal view.) Suppose that we have a scene that consists of a collection of points in $\mathbb{R}^2$. As an example, Figure 2.40 shows a collection of hand-collected points in $\mathbb{R}^2$ that represent the borders of the state of Nevada.

You can learn more about way that the full-scale computer graphics algorithms work in a textbook like

[9] John F. Hughes, Andries van Dam, Morgan McGuire, David F. Sklar, James D. Foley, Steven K. Feiner, and Kurt Akeley. *Computer Graphics: Principles and Practice*. Addison-Wesley, 3rd edition, 2013.

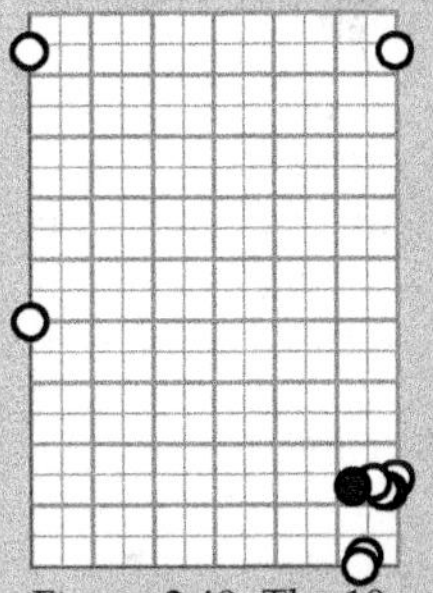

Figure 2.40: The 10 points in $\mathbb{R}^2$ representing the boundaries of Nevada.

Suppose that we wish to rotate a point $\langle x, y \rangle$ by an angle θ around the point $\langle 0, 0 \rangle$. You should be able to convince yourself with a drawing that we can rotate a point $\langle x, 0 \rangle$ around the point $\langle 0, 0 \rangle$ by moving it to $\langle x \cos\theta, x \sin\theta \rangle$. More generally, the point $\langle x, y \rangle$ becomes the point $\langle x\cos\theta - y\sin\theta, x\sin\theta + y\cos\theta \rangle$ when it's rotated.

Suppose we wish to rotate the points $\langle x_1, y_1 \rangle, \ldots, \langle x_n, y_n \rangle$ by angle θ. Write a matrix with the ith column corresponding to the ith point, and perform matrix multiplication as follows:

$$\begin{bmatrix} \cos\theta & -\sin\theta \\ \sin\theta & \cos\theta \end{bmatrix} \begin{bmatrix} x_1 & x_2 & \cdots & x_n \\ y_1 & y_2 & \cdots & y_n \end{bmatrix} = \begin{bmatrix} x_1\cos\theta - y_1\sin\theta & x_2\cos\theta - y_2\sin\theta & \cdots & x_n\cos\theta - y_n\sin\theta \\ x_1\sin\theta + y_1\cos\theta & x_2\sin\theta + y_2\cos\theta & \cdots & x_n\sin\theta + y_n\cos\theta \end{bmatrix}$$

(The matrix $R = \begin{bmatrix} \cos\theta & -\sin\theta \\ \sin\theta & \cos\theta \end{bmatrix}$ is called a *rotation matrix*.)

The result is that we have rotated an entire collection of points—arranged in the 2-by-n matrix M—by multiplying M by this rotation matrix. In other words, RM is a 2-by-n matrix of the rotated points. See Figure 2.41.

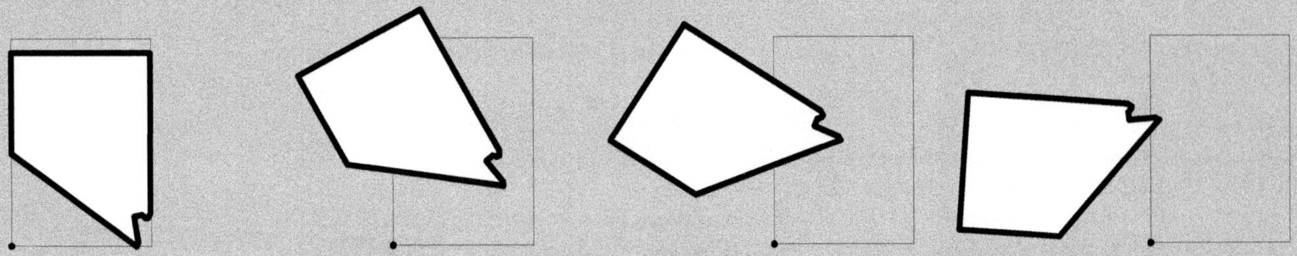

Figure 2.41: Nevada, as above and rotated by three different angles.

2.4.3 *Exercises*

2.141 What is $\{1,2,3\} \times \{1,4,16\}$?

2.142 What is $\{1,4,16\} \times \{1,2,3\}$?

2.143 What is $\{1\} \times \{1\} \times \{1\}$?

2.144 What is $\{1,2\} \times \{2,3\} \times \{1,4,16\}$?

2.145 Suppose $A \times B = \{\langle 1,1\rangle, \langle 2,1\rangle\}$. What are A and B?

Let $S := \{1,2,3,4,5,6,7,8\}$, and let T be an unknown set. From the following, what can you conclude about T? Be as precise as possible: if you can list the elements of T exhaustively, do so; if you can't, identify any elements that you can conclude must be (or must not be) in T.

2.146 $|S \times T| = 16$ and $\langle 1,2\rangle, \langle 3,4\rangle \in S \times T$

2.147 $S \times T = \varnothing$

2.148 $(S \times T) \cap (T \times S) = \{\langle 3,3\rangle\}$

2.149 $S \times T = T \times S$

Recall that Algebraic notation denotes the squares of the chess board as $\{\mathtt{a},\mathtt{b},\mathtt{c},\mathtt{d},\mathtt{e},\mathtt{f},\mathtt{g},\mathtt{h}\} \times \{1,2,3,4,5,6,7,8\}$, as in Figure 2.42. For each of the following questions, identify sets S and T such that the set of cells containing the designated pieces can be described as $S \times T$.

2.150 the white rooks (♖)

2.151 the bishops (♝, white or black)

2.152 the pawns (♙, white or black)

2.153 no pieces at all

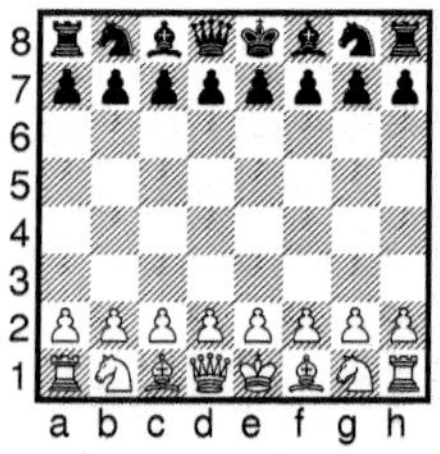

Figure 2.42: The squares of a chess board, written using Algebraic notation.

Write out the elements of the following sets.

2.154 $\{0,1,2\}^3$

2.155 $\{\mathtt{A},\mathtt{B}\} \times \{\mathtt{C},\mathtt{D}\}^2 \times \{\mathtt{E}\}$

2.156 $\bigcup_{i=1}^{3} \{0,1\}^i$

Let $\Sigma := \{\mathtt{A},\mathtt{B},\ldots,\mathtt{Z}\}$ denote the English alphabet. Using notation from this chapter, give an expression that denotes each of the following sets. It may be useful to recall that Σ^k denotes the set of strings consisting of a sequence of k elements from Σ, so Σ^0 contains the unique string of length 0 (called the empty string, *and typically denoted by ε—or by `""` in most programming languages).*

2.157 The set of 8-letter strings.

2.158 The set of 5-letter strings that do not contain any vowels $\{\mathtt{A},\mathtt{E},\mathtt{I},\mathtt{O},\mathtt{U}\}$.

2.159 The set of 6-letter strings that do not contain more than one vowel. (So `GRITTY`, `QWERTY`, and `BRRRRR` are fine; but `EEEEEE`, `THREAT`, `STRENGTHS`, and `A` are not.)

2.160 The set of 6-letter strings that contain at most one type of vowel—multiple uses of the same vowel are fine, but no two different vowels can appear. (So `BANANA`, `RHYTHM`, and `BOOBOO` are fine; `ESCAPE` and `STRAIN` are not.)

Recall that the length of a vector $x \in \mathbb{R}^n$ is given by $\|x\| = \sqrt{\sum_{i=1}^{n} x_i^2}$. Considering the vectors $a := \langle 1,3\rangle$, $b := \langle 2,-2\rangle$, $c := \langle 4,0\rangle$, and $d := \langle -3,-1\rangle$, state the values of each of the following:

2.161 $\|a\|$

2.162 $\|b\|$

2.163 $\|c\|$

2.164 $a+b$

2.165 $3d$

2.166 $2a+c-3b$

2.167 $\|a\|+\|c\|$ and $\|a+c\|$

2.168 $\|a\|+\|b\|$ and $\|a+b\|$

2.169 $3\|d\|$ and $\|3d\|$

2.170 Explain why, for an arbitrary vector $x \in \mathbb{R}^n$ and an arbitrary scalar $a \in \mathbb{R}$, $\|ax\| = a\|x\|$.

2.171 For any two vectors $x, y \in \mathbb{R}^n$, we have $\|x\| + \|y\| \geq \|x+y\|$. Under precisely what circumstances do we have $\|x\| + \|y\| = \|x+y\|$ for $x, y \in \mathbb{R}^n$? Explain briefly.

Still considering the same vectors $a := \langle 1,3\rangle$, $b := \langle 2,-2\rangle$, $c := \langle 4,0\rangle$, and $d := \langle -3,-1\rangle$, what are the following?

2.172 $a \bullet b$

2.173 $a \bullet d$

2.174 $c \bullet c$

Recall that the Manhattan distance *between vectors $x, y \in \mathbb{R}^n$ is defined as $\sum_{i=1}^{n} |x_i - y_i|$. The* Euclidean distance *between two vectors $x, y \in \mathbb{R}^n$ is $\sqrt{\sum_{i=1}^{n}(x_i - y_i)^2}$. What is the Manhattan/Euclidean distances between the following pairs of vectors?*

2.175 a and b

2.176 a and d

2.177 b and c

Suppose that the Manhattan distance between two vectors $x, y \in \mathbb{R}^2$ is 1. Justify your answers:

2.178 What's the largest possible Euclidean distance between x and y?

2.179 What's the smallest possible Euclidean distance between x and y?

2.180 What's the smallest possible Euclidean distance between x and y if $x, y \in \mathbb{R}^n$ (not just $n = 2$)?

Consider Figure 2.43, and sketch the following sets:

2.181 $\{x \in \mathbb{R}^2 :$ the Euclidean distance between x and $\langle 0,0\rangle$ is at most $2\}$.

2.182 $\{x \in \mathbb{R}^2 :$ the Manhattan distance between x and $\langle 0,0\rangle$ is at most $2\}$.

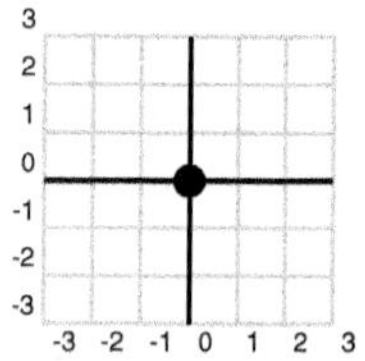

Figure 2.43: The plane.

In Example 2.40, we considered two train stations located at points $s := \langle 8,33 \rangle$ and $g := \langle 4,42 \rangle$. (See Figure 2.44(a).) In that example, we showed that, for an offset $\delta \in [0,4]$, the Manhattan distance between the point $\langle 4+\delta, y \rangle$ and s is smaller than the Manhattan distance between the point $\langle 4+\delta, y \rangle$ and g when $y < 35.5+\delta$.

2.183 Show that the point $\langle 16,40 \rangle$ is closer to one station under Manhattan distance, and to the other under Euclidean distance.

Let $\delta \geq 0$. Under Manhattan distance, describe the values of y for which the following point is closer to s than to g:

2.184 $\langle 8+\delta, y \rangle$

2.185 $\langle 4-\delta, y \rangle$

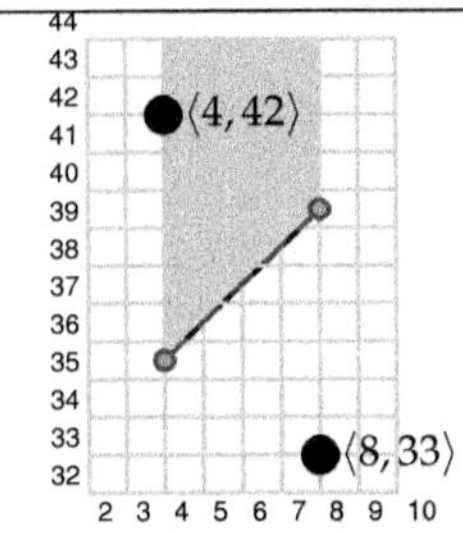

(a) The unscaled version.

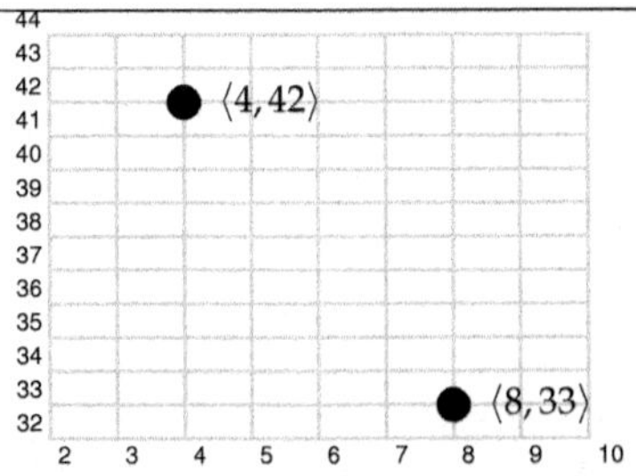

(b) The scaled version.

Figure 2.44: Manhattan train stations.

2.186 In the real-world island of Manhattan, the east–west blocks are roughly twice the length of the north–south blocks. As such, the more accurate picture of distances in the city is shown in Figure 2.44(b). Assuming it takes 1.5 minutes to walk a north–south (up–down) block and 3 minutes to walk an east–west (left–right) block, give a formula for the walking distance between $\langle x,y \rangle$ and Penn Station, at $s := \langle 8,33 \rangle$.

A Voronoi diagram—*named after the 20th-century Russian mathematician Georgy Voronoy—is a decomposition of the plane $\mathbb{R}^2$ into regions based on a given set S of points. The region "belonging" to a point $x \in S$ is $\{y \in \mathbb{R}^2 : d(x,y) \leq \min_{z \in S} d(z,y)\}$, where $d(\cdot,\cdot)$ denotes Euclidean distance—in other words, the region "belonging" to point x is that portion of the plane that's closer to x than any other point in S.*

2.187 Compute the Voronoi diagram of the set of points $\{\langle 0,0 \rangle, \langle 4,5 \rangle, \langle 3,1 \rangle\}$. That is, compute:

- the set of points $y \in \mathbb{R}^2$ that are closer to $\langle 0,0 \rangle$ than $\langle 4,5 \rangle$ or $\langle 3,1 \rangle$ under Euclidean distance;
- the set of points $y \in \mathbb{R}^2$ that are closer to $\langle 4,5 \rangle$ than $\langle 0,0 \rangle$ or $\langle 3,1 \rangle$ under Euclidean distance; and
- the set of points $y \in \mathbb{R}^2$ that are closer to $\langle 3,1 \rangle$ than $\langle 0,0 \rangle$ or $\langle 4,5 \rangle$ under Euclidean distance.

2.188 Compute the Voronoi diagram of the set of points $\{\langle 2,2 \rangle, \langle 8,1 \rangle, \langle 5,8 \rangle\}$.

2.189 Compute the Voronoi diagram of the set of points $\{\langle 0,7 \rangle, \langle 3,3 \rangle, \langle 8,1 \rangle\}$.

2.190 *(programming required)* Write a program that takes three points as input and produces a representation of the Voronoi diagram of those three points as output.

Taking it further: Voronoi diagrams are used frequently in computational geometry, among other areas of computer science. (For example, a coffee-shop chain might like to build a mobile app that is able to quickly answer the question *What store is closest to me right now?* for any customer at any time. Voronoi diagrams can allow precomputation of these answers.)

Given any set S of n points, it's reasonably straightforward to compute (an inefficient representation of) the Voronoi diagram of those points by computing the line that's equidistant between each pair of points, as you saw in the last few exercises. But there are cleverer ways of computing Voronoi diagrams more efficiently; see a good textbook on computational geometry for more.[10]

[10] Mark de Berg, Marc van Kreveld, Mark Overmars, and Otfried Schwarzkopf. *Computational Geometry*. Springer-Verlag, 2nd edition, 2000.

Consider the following matrix:

$$M = \begin{bmatrix} 3 & 9 & 2 \\ 0 & 9 & 8 \\ 6 & 2 & 0 \\ 7 & 5 & 5 \\ 7 & 2 & 4 \\ 1 & 6 & 7 \end{bmatrix}$$

2.191 What size is M?

2.192 What is $M_{3,1}$?

2.193 List every $\langle i,j \rangle$ such that $M_{i,j} = 7$.

2.194 What is 3M?

Considering the following matrices, what are the values of the given expressions (if they're defined)?

$$A = \begin{bmatrix} 0 & 8 & 0 \\ 9 & 6 & 0 \\ 2 & 3 & 3 \end{bmatrix} \quad B = \begin{bmatrix} 5 & 8 \\ 7 & 5 \\ 3 & 2 \end{bmatrix} \quad C = \begin{bmatrix} 7 & 2 & 7 \\ 3 & 5 & 6 \\ 1 & 2 & 5 \end{bmatrix} \quad D = \begin{bmatrix} 3 & 1 \\ 0 & 8 \end{bmatrix} \quad E = \begin{bmatrix} 8 & 4 \\ 3 & 2 \end{bmatrix} \quad F = \begin{bmatrix} 1 & 2 & 9 \\ 5 & 4 & 0 \end{bmatrix}$$

(If the given quantity is undefined, say so—and say why.)

2.195 $A+C$

2.196 $B+F$

2.197 $D+E$

2.198 $A+A$

2.199 $-2D$

2.200 $0.5F$

2.201 AB

2.202 AC

2.203 AF

2.204 BC

2.205 DE

2.206 ED

Consider the matrices

$$A = \begin{bmatrix} 1 & 0 & 0 \\ 1 & 0 & 0 \\ 1 & 1 & 0 \end{bmatrix} \quad \text{and} \quad B = \begin{bmatrix} 0 & 0 & 0 \\ 0 & 1 & 0 \\ 1 & 1 & 1 \end{bmatrix}.$$

2.207 What is $0.25A + 0.75B$?

2.208 What is $0.5A + 0.5B$?

2.209 Identify two *other* matrices C and D with the same average—that is, such that $\{A, B\} \neq \{C, D\}$ but $0.5A + 0.5B = 0.5C + 0.5D$.

2.210 *(programming required)* A common computer graphics effect in the spirit of the last few exercises is *morphing* one image into another—that is, slowly changing the first image into the second. There are sophisticated techniques for this task, but a simple form can be achieved just by averaging. Given two n-by-m images represented by matrices A and B—say grayscale images, with each entry in $[0, 1]$—we can produce a "weighted average" of the images as $\lambda A + (1 - \lambda)B$, for a parameter $\lambda \in [0, 1]$. See Figure 2.45.

Write a program, in a programming language of your choice, that takes three inputs—an image A, an image B, and a weight $\lambda \in [0, 1]$—and produces a new image $\lambda A + (1 - \lambda)B$. (You'll need to research an image-processing library to use in your program.)

Figure 2.45: Clubs to hearts (0%, 20%, 40%, 60%, 80%, and 100%).

2.211 Let A be an m-by-n matrix. Let I be the n-by-n identity matrix. Explain why the matrix AI is identical to the matrix A.

If M is an n-by-n matrix, then the product of M with itself is also an n-by-n matrix. We write matrix powers in the normal way that we defined powers of integers (or of the Cartesian product of sets): $M^k = M \cdot M \cdots M$, multiplied k times. (M^0 is the n-by-n identity matrix I.) What are the following? (Hint: $M^{2k} = (M^k)^2$.)

2.212 $\begin{bmatrix} 2 & 3 \\ 1 & 1 \end{bmatrix}^3$

2.213 $\begin{bmatrix} 1 & 1 \\ 1 & 0 \end{bmatrix}^2$

2.214 $\begin{bmatrix} 1 & 1 \\ 1 & 0 \end{bmatrix}^4$

2.215 $\begin{bmatrix} 1 & 1 \\ 1 & 0 \end{bmatrix}^9$

> **Taking it further:** *The* Fibonacci numbers *are defined recursively as the sequence* $f_1 := 1, f_2 := 1$, *and* $f_n := f_{n-1} + f_{n-2}$ *for* $n \geq 3$. *The first several Fibonacci numbers are* $1, 1, 2, 3, 5, 8, 13, \ldots$. *As we'll see in Exercises 5.56 and 6.99, there's a very fast algorithm to compute the nth Fibonacci number based on computing the nth power of the matrix from Exercises 2.213–2.215.*

Let A by an n-by-n matrix. The inverse *of A, denoted A^{-1}, is also an n-by-n matrix, with the property that $AA^{-1} = I$. There's a general algorithm that one can develop to invert matrices, but in the next few exercises you'll calculate inverses of some small matrices by hand.*

2.216 Note that $\begin{bmatrix} 1 & 1 \\ 2 & 1 \end{bmatrix} \cdot \begin{bmatrix} x & y \\ z & w \end{bmatrix} = \begin{bmatrix} x+z & y+w \\ 2x+z & 2y+w \end{bmatrix}$. Thus $\begin{bmatrix} 1 & 1 \\ 2 & 1 \end{bmatrix}^{-1}$ is the matrix $\begin{bmatrix} x & y \\ z & w \end{bmatrix}$, where the following four conditions hold: $x + z = 1$ and $y + w = 0$ and $2x + z = 0$ and $2y + w = 1$. Find the values of x, y, w, and z that satisfy these four conditions.

Using the same approach as the last exercise, find the inverse of the following matrices:

2.217 $\begin{bmatrix} 1 & 2 \\ 3 & 4 \end{bmatrix}$

2.218 $\begin{bmatrix} 0 & 1 \\ 1 & 0 \end{bmatrix}$

2.219 $\begin{bmatrix} 1 & 0 \\ 0 & 1 \end{bmatrix}$

2.220 Not all matrices have inverses—for example, $\begin{bmatrix} 1 & 1 \\ 1 & 1 \end{bmatrix}$ doesn't have an inverse. Explain why not.

An error-correcting code *(see Section 4.2) is a method for redundantly encoding information so that the information can still be retrieved even in the face of some errors in transmission/storage. The* Hamming code *is a particular error-correcting code for 4-bit chunks of information. The Hamming code can be described using matrix multiplication: given a* message $m \in \{0, 1\}^4$, *we* encode m *as* $mG \bmod 2$, *where*

$$G = \begin{bmatrix} 1 & 0 & 0 & 0 & 0 & 1 & 1 \\ 0 & 1 & 0 & 0 & 1 & 0 & 1 \\ 0 & 0 & 1 & 0 & 1 & 1 & 0 \\ 0 & 0 & 0 & 1 & 1 & 1 & 1 \end{bmatrix}.$$

*(Here you should interpret the "*mod 2*" as describing an operation to* each element *of the output vector.) For example,* $[1, 1, 1, 1] \cdot G = [1, 1, 1, 1, 3, 3, 3]$, *so we'd encode* $[1, 1, 1, 1]$ *as* $[1, 1, 1, 1, 3, 3, 3] \bmod 2 = [1, 1, 1, 1, 1, 1, 1]$. *What is the Hamming code encoding of the following messages?*

2.221 $[0, 0, 0, 0]$

2.222 $[0, 1, 1, 0]$

2.223 $[1, 0, 0, 1]$

2.5 *Functions*

> There is no passion like that of a functionary for his function.
>
> Georges Clemenceau (1841–1929)

A *function* transforms an input value into an output value; that is, a function f takes an *argument* or *parameter* x, and *returns* a value $f(x)$. Functions are familiar from both algebra and from programming. In algebra, we frequently encounter mathematical functions like $f(x) = x + 6$, which means that, for example, we have $f(3) = 9$ and $f(4) = 10$. In programming, we often write or invoke functions that use an algorithm to transform an input into an output, like a function **sort**—so that **sort**($\langle 3, 1, 4, 1, 5, 9 \rangle$) = $\langle 1, 1, 3, 4, 5, 9 \rangle$, for example.

In this section, we will give formal definitions of functions and of some terminology related to functions, and also discuss a few special types of functions. (Functions themselves are a special case of *relations*, and we will revisit the definition of functions in Chapter 8 when we discuss relations.)

2.5.1 *Basic Definitions*

We start with the definition of a function itself:

Definition 2.44 (Function)
Let A and B be sets. A function f from A to B, *written $f : A \to B$, assigns to each input value $a \in A$ a unique output value $b \in B$; the unique value b assigned to a is denoted by $f(a)$. We sometimes say that f* maps *a* to *$f(a)$.*

Note that A and B are allowed to be the same set; for example, a function might have inputs and outputs that are both elements of $\mathbb{Z}$.

Here are two simple examples. First, we define a function *not* for Boolean inputs that maps True to False, and False to True:

Example 2.53 (Not function)
The function *not* : {True, False} $\to$ {True, False} can be defined with the table in Figure 2.46. Given an input x, we find the output value *not*(x) by locating x in the first column of the table and reading the value in that row's second column. Thus *not*(True) = False and *not*(False) = True.

x	*not*(x)
True	False
False	True

Figure 2.46: The function *not*.

As another simple example, we can also define a function *square* that returns its input multiplied by itself:

Example 2.54 (Square function)
The function *square* : $\mathbb{R} \to \mathbb{R}$ can be defined as *square*(x) := x^2: for any input $x \in \mathbb{R}$, the output is the real number x^2. Thus, for example, *square*(8) = 64, because the function *square* assigns the output $8^2 = 64$ to the input 8.

Note, too, that a function $f : A \to B$ might have a set A of inputs that are *pairs*; for example, the function that takes two numbers and returns their average is the function $average : \mathbb{R} \times \mathbb{R} \to \mathbb{R}$, where $average(\langle x, y \rangle) := (x + y)/2$. (We interpret $\mathbb{R} \times \mathbb{R} \to \mathbb{R}$ as $(\mathbb{R} \times \mathbb{R}) \to \mathbb{R}$.) When there is no danger of confusion, we drop the angle brackets and simply write, for example, $average(3, 2)$ instead of $average(\langle 3, 2 \rangle)$.

As we've already seen in Examples 2.53 and 2.54, the rule by which a function assigns an output to a given input can be specified either symbolically—typically via an algebraic expression—or exhaustively, by giving a table describing the input/output relationship. The table-based definition only makes sense when the set of possible inputs is *finite*; otherwise the table would have to be infinitely large. (And it's only *practical* to define a function with a table if the set of possible inputs is pretty small!)

Here's an example of specifying the same function in two different ways, once symbolically and once using a table:

Example 2.55 (Doubling function)
Let's define the function *double* that doubles its input value, for any input in $\{0, 1, \ldots, 7\}$. (That is, we are defining a function $double : \{0, 1, \ldots, 7\} \to \mathbb{Z}$.)

We can write *double* symbolically by defining

$$double(x) := 2 \cdot x.$$

To define *double* using a table, we specify the output corresponding to every one of the 8 possible inputs, as shown in Figure 2.47.

x	$double(x)$
0	0
1	2
2	4
3	6
4	8
5	10
6	12
7	14

Figure 2.47: The *double* function, specified using a table.

The functions that we've discussed so far are all fairly simple, but even simple functions can have some valuable applications. Here's an example of another simple function that can be used in compressing images so that they take up less space:

Example 2.56 (Reducing the colorspace of an image)
The pixels in a grayscale image are all elements of $\{0, 1, \ldots, 255\}$. To reduce the space requirements for a large image, we can consider a form of *lossy compression* (that is, compression that loses some amount of data) by replacing each pixel with one chosen from a smaller list of candidate colors. That is, instead of having 256 different shades of gray, we might have 128 or 64 or even fewer shades.

Define $quantize : \{0, 1, \ldots, 255\} \to \{0, 1, \ldots, 255\}$ as follows:

$$quantize(n) := \begin{cases} 26 & \text{if } 0 \leq n \leq 51 \\ 78 & \text{if } 52 \leq n \leq 103 \\ 130 & \text{if } 104 \leq n \leq 155 \\ 182 & \text{if } 156 \leq n \leq 207 \\ 234 & \text{if } 208 \leq n \leq 255. \end{cases}$$

We can apply *quantize* to every pixel in a grayscale image, and then use a much smaller number of bits per pixel in storing the resulting image. See Figure 2.48 for an example.

(a) The function *quantize*.

(b) An image of a house.

(c) The same image, compressed to use only 5 shades of gray using the *quantize* function.

Figure 2.48: A visual representation of the color-mapping function (each input color in the left column is assigned the corresponding color in the right column), applied to an example image. In PNG format, the file for the second image takes up less than 14% of the space consumed by the first image.

Taking it further: A *byte* is a sequence of 8 bits. Using 8 bits, we can represent the numbers from 00000000 to 11111111—that is, from 0 to 255. Thus a pixel with $\{0, 1, \ldots, 255\}$ as possible grayscale values in an image requires one byte of storage for each pixel. If we don't do something cleverer, a moderately sized 2048-by-1536 image (the size of an iPad) requires over 3 megabytes even if it's grayscale. (A color image requires three times that amount of space.) Techniques similar to the compression function from Example 2.56 are used in a variety of CS applications—including, for example, in automatic speech recognition, where each sample from a sound stream is stored using one of only, say, 256 different possible values instead of a floating-point number, which requires much more space.

Domain and codomain

The *domain* and *codomain* of a function are its sets of possible inputs and outputs:

Definition 2.45 (Domain/codomain)
For a function $f : A \to B$, the set A is called the domain *of the function $f : A \to B$, and the set B is called the* codomain *of the function $f : A \to B$.*

Let's identify the domain and codomain from the previous examples of this section:

Example 2.57 (Some domains and codomains)
For the functions from Examples 2.53–2.56, we have:

function	domain	codomain
not (Example 2.53)	$\{\text{True}, \text{False}\}$	$\{\text{True}, \text{False}\}$
square (Example 2.54)	$\mathbb{R}$	$\mathbb{R}$
double (Example 2.55)	$\{0, 1, \ldots, 7\}$	$\mathbb{Z}$
quantize (Example 2.56)	$\{0, 1, \ldots, 255\}$	$\{0, 1, \ldots, 255\}$

Note that for three of these functions, the domain and codomain are actually the same set; for the function $double : \{0, 1, \ldots, 7\} \to \mathbb{Z}$, they're different.

When the domain and codomain are clear from context (or they are unimportant for the purposes of a discussion), then they may be left unwritten.

Taking it further: This possibility of implicitly representing the domain and codomain of a function is also present in code. Some programming languages (like Java) require the programmer to explicitly write out the types of the inputs and outputs of a function; in some (like Python), the input and output types are left implicit. In Java, for example, one would write an `isPrime` function with the explicit declaration that the input is an integer (`int`) and the output is a Boolean (`boolean`). In Python, one would write the function without any explicit type information.

```
boolean isPrime(int n) {
  /* code to check primality of n */
}
```

```
def isPrime(n):
  # code to check primality of n
```

But regardless of whether they're written out or left implicit, these functions *do* have a domain (the set of valid inputs) and a codomain (the set of possible outputs).

Range/Image

For a function $f : A \to B$, the set A (the domain) is the set of all possible inputs, and the set B (the codomain) is the set of all possible outputs. But not all of the possible outputs are necessarily actually *achieved*: in other words, there may be an element $b \in B$ for which there's no $a \in A$ with $f(a) = b$. For example, we defined $square : \mathbb{R} \to \mathbb{R}$ in Example 2.54, but there is no real number x such that $square(x) = -1$. The *range* or *image* defines the set of actually achieved outputs:

Definition 2.46 (Range/image)
The range *or* image *of a function $f : A \to B$ is the set of all $b \in B$ such that $f(a) = b$ for some $a \in A$. Using the notation of Section 2.3, the range of f is the set*

$$\{y \in B : \text{there exists at least one } x \in A \text{ such that } f(x) = y\}.$$

We'll start with the four functions defined earlier in this section:

Example 2.58 (Some ranges)
For the functions from Examples 2.53–2.56, we have:

function	range
not (Example 2.53)	$\{\text{True}, \text{False}\}$
square (Example 2.54)	$\mathbb{R}^{\geq 0}$
double (Example 2.55)	$\{0, 2, 4, 6, 8, 10, 12, 14\}$
quantize (Example 2.56)	$\{26, 78, 130, 182, 234\}$

For *not*, *double*, and *quantize*, the range is easy to determine: it's precisely the set of values that appear in the "output" column of the table defining the function.

For *square*, it's clear that the range includes no negative numbers, because there's no $y \in \mathbb{R}$ such that $y^2 < 0$. In fact, the range of *square* is precisely $\mathbb{R}^{\geq 0}$: for any $x \in \mathbb{R}^{\geq 0}$, there's an input to *square* that produces x as output—specifically $\sqrt{x}$.

Here's another example, for a slightly more complex function:

Example 2.59 (The smallest divisor function)

Problem: Define a function $sd : \mathbb{Z}^{\geq 2} \to \mathbb{Z}^{\geq 2}$ as follows. Given an input $n \in \mathbb{Z}^{\geq 2}$, the value of $sd(n)$ is the *smallest integer $k \geq 2$ that evenly divides n*. For example:

- $sd(2) = 2$ (because $2 \mid 2$);
- $sd(3) = 3$ (because $3 \mid 3$ but $2 \nmid 3$);
- $sd(4) = 2$ (because $2 \mid 4$); and
- $sd(121) = 11$ (because $11 \mid 121$ but $2 \nmid 121, 3 \nmid 121, \ldots, 10 \nmid 121$).

What are the domain, codomain, and range of sd?

Solution: The domain and codomain of sd are easy to determine: they are both $\mathbb{Z}^{\geq 2}$. Any integer $n \geq 2$ is a valid input to sd, and we defined the function sd as producing an integer $k \geq 2$ as its output. (The domain and codomain are simply written in the function's definition, before and after the arrow in $sd : \mathbb{Z}^{\geq 2} \to \mathbb{Z}^{\geq 2}$.) The range is a bit harder to see, but it turns out to be the set P of all prime numbers. Let's argue that P is the range of sd by showing that (i) every prime number $p \in P$ is in the range of sd, and (ii) every number p in the range of P is a prime number.

(i) Let $p \in \mathbb{Z}^{\geq 2}$ be any prime number. Then $sd(p) = p$: by the definition of primality, the only integers than evenly divide p are 1 and p itself (and $1 \geq 2$ isn't true!). Therefore every prime number p is in the range of sd, because there's an input to sd such that the output is p.

(ii) Let p be any number in the range of sd—that is, suppose $sd(n) = p$ for some n. We will argue that p must be prime. Imagine that p were instead composite—that is, there is an integer k satisfying $2 \leq k < p$ that evenly divides p. But then $sd(n) = p$ is impossible: if p evenly divides n, then k *also* evenly divides n, and $k < p$, so k would be a smaller divisor of n. (For example, if n were evenly divisible by the composite number 15, then n would *also* be evenly divisible by 3 and 5—two factors of 15—so $sd(n) \neq 15$.) Therefore every number in the range of sd is prime.

Putting together the facts from (i) and (ii), we conclude that the range of sd is precisely the set of all prime numbers.

Problem-solving tip: Example 2.59 illustrates a useful general technique if we wish to show that two sets A and B are equal. One nice way to establish that $A = B$ is to show that $A \subseteq B$ and $B \subseteq A$. That's what we did to establish the range of sd in Example 2.59:

- define P as the set of all prime numbers.
- define R as the range of sd.

We showed in (i) that every element of P is in R (that is, $P \subseteq R$); and in (ii) that every element of R is in P (that is, $R \subseteq P$). Together these facts establish that $R = P$.

We will also introduce a minor extension to the set-abstraction notation from Section 2.3.1 that's related to the range of a function. (We used this notation informally in Example 2.28.) Consider a function $f : A \to B$ and a set $U \subseteq A$. We denote by $\{f(x) : x \in U\}$ the set of all output values of the function f when it's applied to the elements $x \in U$:

Definition 2.47 (Set abstraction using functions)

For a function $f : A \to B$ and a set $U \subseteq A$, we write $\{f(x) : x \in U\}$ as shorthand for the set $\{b \in B :$ there exists some $u \in U$ for which $f(u) = b\}$.

Remember that order and repetition of elements in a set don't matter, which means that the set $\{f(x) : x \in A\}$ is precisely the range of the function $f : A \to B$.

A visual representation of functions

The table-based and symbolic representations of functions that we've discussed fully represent a function, but sometimes a more visual representation of a function is clearer. Consider a function $f : A \to B$. We can give a picture representing f by putting the elements of A into one column, the elements of B into a second column, and drawing an arrow from each $a \in A$ to the value of $f(a) \in B$. Notice that the definition of a function guarantees that *every element in the first column has one and only one arrow going from it to the second column*: if $f : A \to B$ is a function, then every $a \in A$ is assigned a unique output $f(a) \in B$. Here's a simple example:

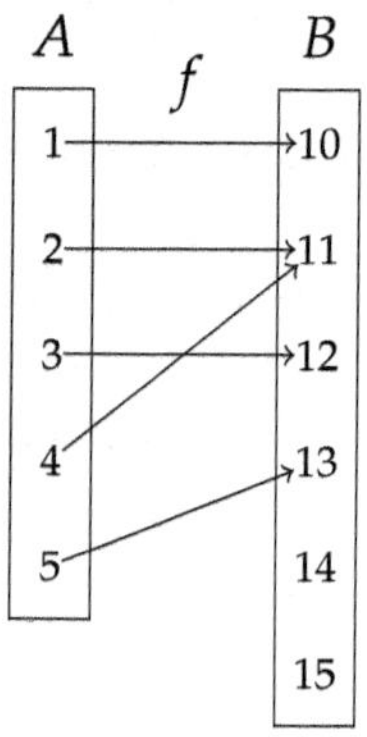

Figure 2.49: A picture of a function $f : A \to B$, where $A = \{1, \ldots, 5\}$ and $B = \{10, \ldots, 15\}$.

Example 2.60 (A picture of a function)
Figure 2.49 displays a function $f : \{1, \ldots, 5\} \to \{10, \ldots, 15\}$, where $f(1) = 10$ and $f(2) = f(4) = 11$ and $f(3) = 12$ and $f(5) = 13$.

We can read the domain, codomain, and range directly from this picture: the domain is the set of elements in the first column; the codomain is the set of elements in the second column; and the range is the set of elements in the second column *for which there is at least one incoming arrow.* For instance, the range of f from Example 2.60 is $\{10, 11, 12, 13\}$. (There are no arrows pointing to 14 or 15, so these two numbers are in the codomain but not the range of f.)

Function composition

Suppose we have two functions $f : A \to B$ and $g : B \to C$. Given an input $a \in A$, we can find $f(a) \in B$, and then apply g to map $f(a)$ to an element of C, namely $g(f(a)) \in C$. This successive application of f and g defines a new function, called the *composition* of f and g, whose domain is A and whose codomain is C:

Definition 2.48 (Function composition)
For two functions $f : A \to B$ and $g : B \to C$, the function $g \circ f : A \to C$ maps an element $a \in A$ to $g(f(a)) \in C$. The function $g \circ f$ is called the composition of f and g.

Notice a slight oddity of the notation: $g \circ f$ applies the function f *first* and the function g *second*, even though g is written first.

Here's an example of the functions that result from composing two simple functions in four different ways:

Example 2.61 (Function composition, four ways)
Let $f : \mathbb{R} \to \mathbb{R}$ and $g : \mathbb{R} \to \mathbb{R}$ be defined by $f(x) := 2x + 1$ and $g(x) := x^2$.

1. The function $g \circ f$, given an input x, produces output
$$g(f(x)) = g(2x + 1) = (2x + 1)^2 = 4x^2 + 4x + 1.$$
2. The function $f \circ g$ maps x to $f(g(x)) = f(x^2) = 2x^2 + 1$.
3. The function $g \circ g$ maps x to $g(g(x)) = g(x^2) = (x^2)^2 = x^4$.
4. The function $f \circ f$ maps x to $f(f(x)) = f(2x + 1) = 2(2x + 1) + 1 = 4x + 3$.

As with many function-related concepts, the visual representation of functions gives a nice way of thinking about function composition: the function $g \circ f$ corresponds to the "short-circuiting" of the pictures of the functions f and g. Here is a small example of this visualization:

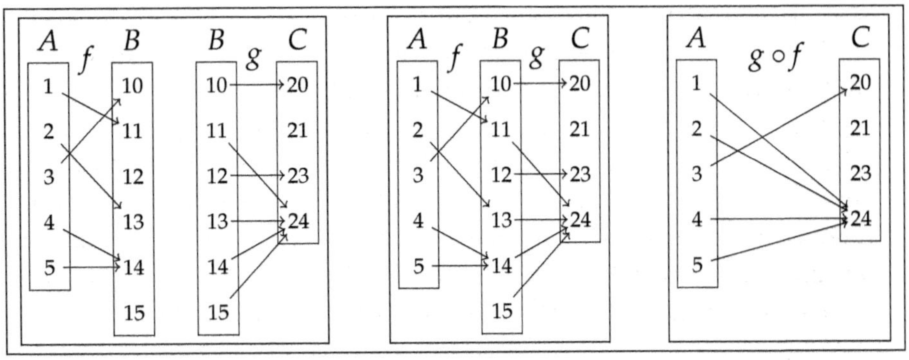

Figure 2.50: A picture of functions $f : A \to B$ and $g : B \to C$, first separately and then pasted together. The third panel shows $g \circ f$, based on successively following two arrows from the second panel.

Example 2.62 (Function composition, by picture)
Figure 2.50 shows functions $f : A \to B$ and $g : B \to C$. Their composition $g \circ f$ is given by following *two* arrows in the diagram. For example, the value of $(g \circ f)(1)$ is $g(f(1))$, which is $g(11)$ because $f(1) = 11$. And $g(11) = 24$ because of g's arrow from 11 to 24.

2.5.2 *Onto and One-to-One Functions*

We now turn to two special categories of functions—*onto* and *one-to-one* functions—that are distinguished by how many different input values (always at least one? never more than one?) are mapped to each output value.

ONTO FUNCTIONS

A function $f : A \to B$ is *onto* if every *possible* output in B is, in fact, an *actual* output:

Definition 2.49 (Onto functions)
A function $f : A \to B$ is called onto *if, for every $b \in B$, there exists at least one $a \in A$ for which $f(a) = b$. An onto function is also sometimes called a* surjective *function.*

Alternatively, using the terminology of Section 2.5.1, a function f is onto if f's codomain equals f's range. As an example, here are two of our previous functions, one of which is onto and one of which isn't:

Example 2.63 (An onto function)
The function *not* : {True, False} $\to$ {True, False} is onto: there's an input value that produces True (namely False), and there's an input value that produces False (namely True). Every element of the codomain is "hit" by *not*, so the function is onto.

Example 2.64 (A non-onto function)
The function *quantize* : $\{0, 1, \ldots, 255\} \to \{0, 1, \ldots, 255\}$ from Example 2.56 is not onto. Recall that the only output values achieved were $\{26, 78, 130, 182, 234\}$. For example,

then, there is no value of x for which $quantize(x) = 42$. Thus 42 is not in the range of *quantize*, and therefore this function is not onto.

Here is a collection of a few more examples, where we'll try to construct onto and non-onto functions meeting a certain description:

Example 2.65 (Sample onto/non-onto functions)

Problem: Let $A := \{0, 1, 2\}$ and $B := \{3, 4\}$. Give an example of a function that satisfies the following descriptions; if there's no such function, explain why it's impossible.

1. an onto function $f : A \to B$.
2. a function $g : A \to B$ that is *not* onto.
3. an onto function $h : B \to A$.

Solution: The first two are possible, but the third is not:

1. Define $f(0) := 3$, $f(1) := 4$, and $f(2) := 4$.
2. Define $g(0) := 3$, $g(1) := 3$, and $g(2) := 3$.
3. Impossible! A function h whose domain is $\{3, 4\}$ only has two output values, namely $h(3)$ and $h(4)$. For a function whose codomain is $\{0, 1, 2\}$ to be onto, we need three different output values to be achieved. These two conditions cannot be simultaneously satisfied, so there is no onto function from B to A.

It may be easier to think about onto functions using the visual representation that we just introduced: a function f is onto if *there's at least one arrow pointing at every element in the second column*. Figure 2.51 illustrates the functions from Example 2.65.1 and Example 2.65.2; the fact that f is onto and g is not onto is immediately visible.

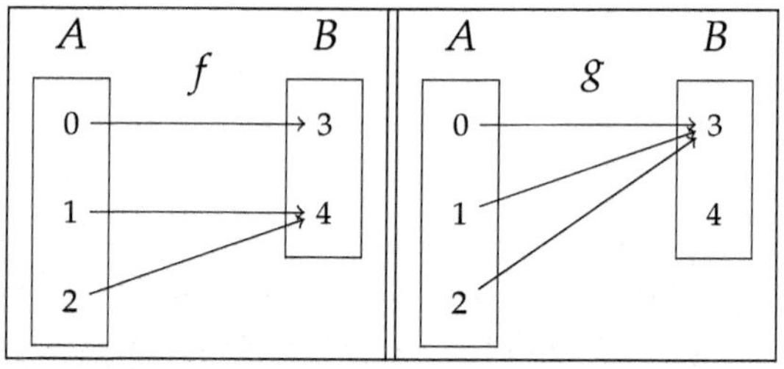

Figure 2.51: An onto function $f : \{0, 1, 2\} \to \{3, 4\}$ and a non-onto function $g : \{0, 1, 2\} \to \{3, 4\}$.

One-to-one functions

An onto function $f : A \to B$ guarantees that every element $b \in B$ is "hit at least once" by f—that is, that $b = f(a)$ for at least one $a \in A$. A *one-to-one function* $f : A \to B$ guarantees that every element $b \in B$ is "hit *at most once*" by f:

Definition 2.50 (One-to-one functions)

A function $f : A \to B$ is called one-to-one *if, for any $b \in B$, there is at most one $a \in A$ such that $f(a) = b$. A one-to-one function is also sometimes called an* injective *function.*

(Terminologically, a one-to-one function sits in contrast to a *many-to-one* function, in which many different input values map to the same output value. Thinking about what a many-to-one function would mean may help to make the name "one-to-one" more intuitive.)

Taking it further: One of the many places that functions are used in computer science is in designing the data structure known as a *hash table*, discussed on p. 267. The idea is that we will store a piece of data called x in a location $h(x)$, for some function h called a *hash function*. We want to choose h to ensure that this function is "not-too-many-to-one" so that no location has to store too much information.

As an example, we'll consider two of our previous functions, *double* and *quantize*, and evaluate whether they are one-to-one:

Example 2.66 (A one-to-one function)
The function $double : \{0, 1, \ldots, 7\} \rightarrow \mathbb{Z}$, defined in Example 2.55, is one-to-one. By examining the table of outputs for the function (reproduced in Figure 2.52), we see that no number appears more than once in the second column. Because every element of the codomain is "hit" by *double* at most once, the function is one-to-one.

x	$double(x)$
0	0
1	2
2	4
3	6
4	8
5	10
6	12
7	14

Figure 2.52: The *double* function.

Observe that $double : \{0, 1, \ldots, 7\} \rightarrow \mathbb{Z}$ is not onto, because there are elements of the codomain that are "hit" zero times—but it is one-to-one, because no element of the codomain is hit twice. Here's an example of a function that is not one-to-one:

Example 2.67 (A non–one-to-one function)
The function $quantize : \{0, 1, \ldots, 255\} \rightarrow \{0, 1, \ldots, 255\}$ from Example 2.56 is not one-to-one. Recall that $quantize(42) = 26$ and $quantize(17) = 26$. Thus 26 is the output for two or more distinct inputs, and therefore this function is not one-to-one.

As with the definition of onto, it may be easier to think about one-to-one functions using our visual two-column representation: a function f is one-to-one if *there's at most one arrow pointing at every element in the second column.* Here are two simple examples using this visual perspective: the function f in Figure 2.53 is one-to-one, because no element of B has multiple incoming arrows. But the function g is not one-to-one, because $4 \in B$ has two incoming arrows.

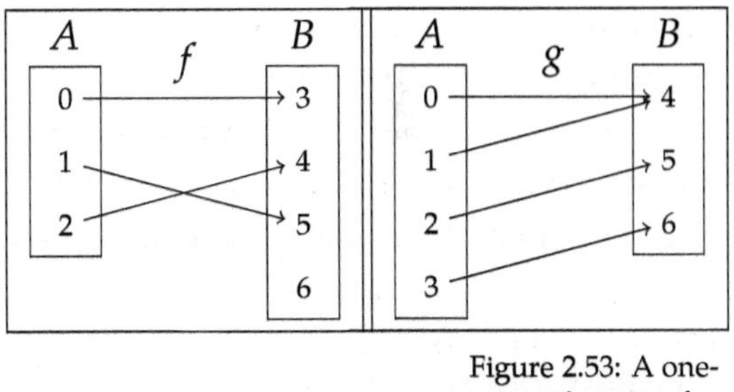

Figure 2.53: A one-to-one function f and a non–one-to-one function g.

ONE-TO-ONE AND ONTO FUNCTIONS

One way of restating the definitions of onto and one-to-one functions is as follows. Let $f : A \rightarrow B$ be a function. Then

- f is *onto* if, for every $b \in B$, we have $|\{a \in A : f(a) = b\}| \geq 1$.
- f is *one-to-one* if, for every $b \in B$, we have $|\{a \in A : f(a) = b\}| \leq 1$.

Therefore a function $f : A \rightarrow B$ that is *both* one-to-one *and* onto guarantees that $|\{a \in A : f(a) = b\}| = 1$— that is, for any $b \in B$, there is *exactly* one element $a \in A$ so that $f(a) = b$. (There is at most one such a because f is one-to-one, and at least one such a because f is onto.) A function with both of these properties is called a *bijection*:

Definition 2.51 (Bijection)
A function $f : A \to B$ is called a bijection *iff f is one-to-one and onto—that is, if $|\{a \in A : f(a) = b\}| = 1$ for every $b \in B$.*

Here are two examples of bijections:

Example 2.68 (Two bijections)
The function *not* : {True, False} → {True, False} from Example 2.53 and the function $f : \mathbb{R} \to \mathbb{R}$ defined by $f(x) := x - 1$ are both bijections.

For *not*, there's exactly one input value whose output is True, namely False; and there's exactly one input value whose output is False, namely True.

Similarly, for f, for every $b \in \mathbb{R}$, there is exactly one a such that $f(a) = b$: specifically, the value $a = b + 1$.

If $f : A \to B$ is a bijection, then every input in A is assigned by f to a unique value in B. We can define a new function, denoted f^{-1}, that reverses this assignment—given $b \in B$, the function $f^{-1}(b)$ identifies the $a \in A$ to which b was assigned by f. This function f^{-1} called the *inverse* of f:

Definition 2.52 (Function inverses)
Let f be a bijection. Then $f^{-1} : B \to A$ is a function called the inverse *of f, where $f^{-1}(b) = a$ whenever $f(a) = b$.*

Here is an example of finding inverses of a few functions:

Example 2.69 (Three inverses)
Problem: What is the inverse of each of the following functions?

1. $f : \mathbb{R} \to \mathbb{R}$, where $f(x) = \frac{x}{2}$.
2. *square* : $\mathbb{R}^{\geq 0} \to \mathbb{R}^{\geq 0}$, where $square(x) = x^2$.
3. *not* : {True, False} → {True, False}.

Solution: 1. We can find the function f^{-1}, the inverse of f, by solving the equation $y = \frac{x}{2}$ for x. We see that $2y = x$. Thus the function $f^{-1} : \mathbb{R} \to \mathbb{R}$ is given by $f^{-1}(y) = 2y$. For any real number $x \in \mathbb{R}$, we have that $f(x) = \frac{x}{2}$ and $f^{-1}(\frac{x}{2}) = x$. (For example, $f(3) = 1.5$ and $f^{-1}(1.5) = 3$.)

2. Notice that *square* : $\mathbb{R}^{\geq 0} \to \mathbb{R}^{\geq 0}$ *is* a bijection—otherwise this problem wouldn't be solvable!—because the domain and the codomain are both the equal to the set of nonnegative real numbers. (For example, $3^2 = 9$ and $(-3)^2 = 9$; if we had allowed both negative and positive inputs, then *square* would not have been one-to-one. And there's no $x \in \mathbb{R}$ such that $x^2 = -9$; if we had allowed negative outputs, then *square* would not have been onto.) The inverse of *square* is the function $square^{-1}(y) = \sqrt{y}$.

3. Note that *not*(*not*(True)) = *not*(False) = True and *not*(*not*(False)) = *not*(True) = False. Thus the inverse of the function *not* is the function *not* itself!

If $f : A \to B$ is a bijection, then, for any $a \in A$, observe that applying f^{-1} to $f(a)$ gives a back as output: that is, $f^{-1}(f(a)) = a$. In other words, the function $f^{-1} \circ f$ is the *identity function*, defined by $id : A \to A$ where $id(a) := a$.

A bijection $f : A \to B$ has exactly one arrow coming into every element in the second column, and by definition it also has exactly one arrow leaving every element in the first column. The inverse of f is precisely the function that results from reversing the direction of each arrow. (The fact that every right-hand column element has exactly one incoming arrow under f is precisely what guarantees that reversing the direction of each arrow still results in the arrow diagram of a *function*.)

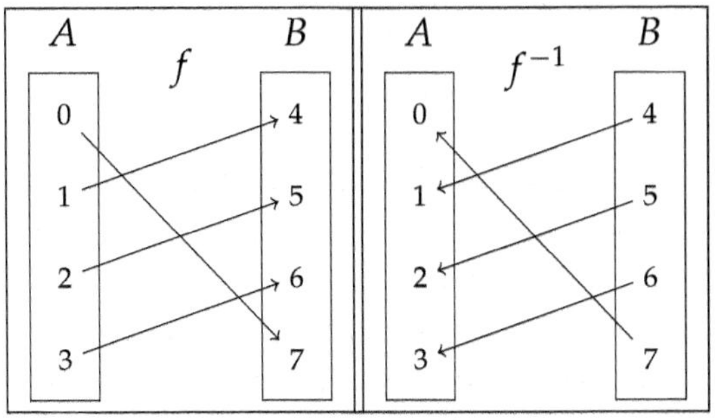

Figure 2.54: A bijection $f : \{0,1,2,3\} \to \{4,5,6,7\}$ and its inverse $f^{-1} : \{4,5,6,7\} \to \{0,1,2,3\}$.

Figure 2.54 shows an example of a bijection and its inverse illustrated in this manner. This picture-based approach should help to illustrate why a function that is not onto or that is not one-to-one fails to have an inverse. If $f : A \to B$ is not onto, then there exists some element $b^* \in B$ that's never the value of f, so $f^{-1}(b^*)$ would be undefined. On the other hand, if f is not one-to-one, then there exists $b^\dagger$ such that $f(a) = b^\dagger$ and $f(a') = b^\dagger$ for $a \neq a'$; thus $f^{-1}(b^\dagger)$ would have to be *both* a and a', which is forbidden by the definition of a function.

2.5.3 *Polynomials*

We'll turn now to *polynomials*, a special type of function whose input and output are both real numbers, and where $f(x)$ is the sum of powers of x:

Definition 2.53 (Polynomial)
A polynomial *is a function* $f : \mathbb{R} \to \mathbb{R}$ *of the form*

$$f(x) = a_0 + a_1x + a_2x^2 + \cdots + a_kx^k$$

where each $a_i \in \mathbb{R}$ *and* $a_k \neq 0$, *for some* $k \in \mathbb{Z}^{\geq 0}$. *(More compactly, we can write this function as* $f(x) = \sum_{i=0}^{k} a_ix^i$.*)*

The real numbers $a_0, a_1, \ldots, a_k$ *are called the* coefficients *of the polynomial, and the values* $a_0, a_1x, a_2x^2, \ldots, a_kx^k$ *being added together are called the* terms *of the polynomial.*

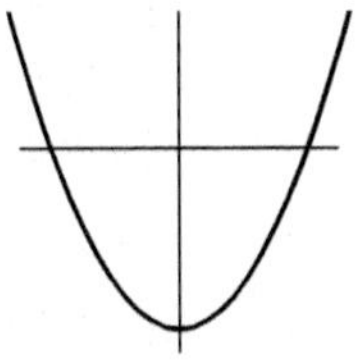

Figure 2.55: A graph of the polynomial $h(x) = x^2 - 2$.

Here are a few examples:

Example 2.70 (Some polynomials)
Here are a few polynomials: $f(x) = 7x$, $g(x) = x^{202} - 201x^{111}$, and $h(x) = x^2 - 2$. The function h is graphed in Figure 2.55—in other words, for every $x \in \mathbb{R}$, the point $\langle x, h(x) \rangle$ is drawn.

There are two additional definitions related to polynomials that will be useful. The first is the *degree* of the polynomial $p(x)$, which is the highest power of x in p's terms:

Definition 2.54 (Degree)
The degree of a polynomial $f(x) = \sum_{i=0}^{k} a_i x^i$ *is the largest index* i *such that* $a_i \neq 0$*—that is, the highest power of* x *with a nonzero coefficient.*

Here are a few examples:

Example 2.71 (Some degrees)
For the polynomials $f(x) = x + x^3$ and $g(x) = x^9$, the degree of f is 3 and the degree of g is 9. For the polynomial $p(x)$ with $a_0 = 1$, $a_1 = 3$, and $a_2 = 0$, the degree of p is 1, because $p(x) = 1 + 3x + 0x^2 = 1 + 3x$.

Some more examples of polynomials with small degrees (namely 0, 1, 2, 3, and 4) are shown in Figure 2.56.

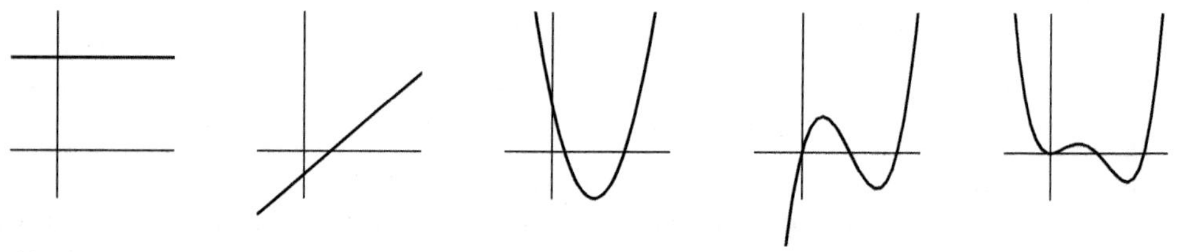

(a) Degree 0. (b) Degree 1. (c) Degree 2. (d) Degree 3. (e) Degree 4.

Figure 2.56: Graphs of some polynomials of degree 0, 1, 2, 3, and 4.

The second useful notion about a polynomial $p(x)$ is a *root*, which is a value of x where the graph of p crosses the x axis:

Definition 2.55 (Roots)
The roots of a polynomial $p(x)$ *are the values in the set* $\{x \in \mathbb{R} : p(x) = 0\}$.

Here are a few simple examples:

Example 2.72 (Some roots)
The roots of the polynomial $f(x) = x + x^2$ are 0 and -1. For the polynomial $g(x) = x^9$, the only root is 0.

A useful general theorem relates the number of different roots for a polynomial to its degree: a polynomial p with degree k has at most k different values of x for which $p(x) = 0$ (unless p is *always* equal to 0):

Theorem 2.3 ((Nonzero) polynomials of degree k have at most k roots)
Let $p(x)$ *be a polynomial of degree at most* k*. Then* p *has at most* k *roots* unless $p(x)$ is zero for every value $x \in \mathbb{R}$.

When $p(x)$ is zero for every value $x \in \mathbb{R}$, we sometimes write $p(x) \equiv 0$ and say that p is *identically zero*.

We won't give a formal proof of Theorem 2.3, but here's one way to convince yourself of the basic idea. Think about how many times a polynomial of degree k can "change direction" from increasing to decreasing or from decreasing to increasing.

Observe that a polynomial p must change directions between any two roots. (Draw a picture!) A polynomial of degree 0 never changes direction, so it's either always zero or never zero. A polynomial $p(x)$ of degree $d \geq 1$ can change directions only at a point where its slope is precisely equal to zero—that is, a point x where the derivative p' of p satisfies $p'(x) = 0$. Using calculus, we can show that the derivative of a polynomial of degree $d \geq 1$ is a polynomial of degree $d - 1$. The idea of a *proof by mathematical induction* is to combine the above intuition to prove the theorem.

Taking it further: Here's some more detailed intuition of how to prove Theorem 2.3 using a proof by mathematical induction; see Chapter 5 for much more detail on this form of proof.

Think first about a degree-zero polynomial—that is, a constant function $p(x) = a$. The theorem is clear for this case: either $a = 0$ (in which case $p(x) \equiv 0$); or $a \neq 0$, in which case $p(x) \neq 0$ for any x. (See Figure 2.56(a).)

Now think about a degree-1 polynomial—that is, $p(x) = ax + b$ for $a \neq 0$. The derivative of p is a constant function—namely $p'(x) = a \neq 0$. Imagine what it would mean for p to have two roots: as we move from smaller x to larger x, at some point r we cross the x-axis, say from $p(r - \varepsilon) < 0$ to $p(r + \varepsilon) > 0$. (See Figure 2.56(b).) In order to find another root larger than r, the function p would have to change from increasing to decreasing—in other words, there would have to be a point at which $p'(x) = 0$. But we just argued that a degree-zero polynomial like $p'(x)$ that is not identically zero is never zero. So we can't find another root.

Now think about a degree-2 polynomial—that is, $p(x) = ax^2 + bx + c$ for $a \neq 0$. After a root, p will have to change direction to head back toward the x-axis. That is, between any two roots of p, there must be a point where the derivative of p is zero: that is, there is a root of the degree-one polynomial $p'(x) = 2ax + b$ between any two roots of p. But p' has at most one root, as we just argued, so p has at most two roots.

And so forth! We can apply the same argument for degree 3, then degree 4, and so on, up to any degree k. (See Chapter 5.)

2.5.4 *Algorithms*

While functions are a valuable mathematical abstraction, computer scientists are fundamentally interested in *computing* things. So, in addition to the type of functions that we've discussed so far in this section, we will also often talk about mapping an input x to a corresponding output $f(x)$ in the way that a computer program would, by computing the value of $f(x)$ using an *algorithm*:

Definition 2.56 (Algorithm)
An algorithm *is step-by-step procedure to transform an input into an output.*

In other words, an algorithm is function—but specified as a sequence of simple operations, of the type that could be written as a program in your favorite programming language; in fact, these step-by-step procedures are even *called* functions in many programming languages. (It's probably worth noting that it's unusual for a book like this one to introduce algorithms in the context of functions. But, because the point of an algorithm really *is* to transform inputs into outputs, it can be helpful to think of an algorithm as a description a function f that specifies *how* to calculate the output $f(x)$ from a given input x, instead of simply describing *what* the value $f(x)$ is.)

We will write algorithms in *pseudocode*, rather than in any particular programming language. In other words, we will specify the steps of the algorithm in a style that is neither Python nor Java nor English, but something in between; it's written in a style that "looks" like a program, but is designed to communicate the steps to a human reader, rather than to a computer executing the code. We will aim to write pseudocode that can be interpreted straightforwardly by a reader who has used any modern programming language; we will always try to avoid getting bogged down in detailed syntax, and instead emphasize trying to communicate algorithms clearly. Translating the pseudocode for an algorithm into any programming language should be straightforward.

We will make use of the standard elements of any programming language in our pseudocode: conditionals ("if"), loops ("for" and "while"), function definitions and function calls (including recursive function calls), and functions returning values. We will use the symbol ":=" to denote assignment and the symbol "=" to denote equality testing, so that $x := 3$ sets the value of x to be 3, and $x = 3$ is True (if x is 3) or False (if x is not 3). We assume a basic familiarity with these basic programming constructs throughout the book.

Our notation of := for assignment and = for equality testing is borrowed from the programming language Pascal. In a lot of other programming languages, like C and Java and Python, assignment is expressed using = and equality testing is expressed using ==.

We will spend significant energy later in the book on proving algorithms correct (Chapters 4 and 5)—that is, showing that an algorithm computes the correct output for any given input—and on analyzing the efficiency of algorithms (Chapter 6). But here is one simple example to get us started:

findMaxIndex(L):
Input: A list L with $n \geq 1$ elements $L[1], \ldots, L[n]$.
Output: An index i such that $L[i]$ is the maximum value in L.

```
maxIndex := 1
for i := 2 to n:
    if L[i] > L[maxIndex] then
        maxIndex := i
return maxIndex
```

Figure 2.57: An algorithm to find the index of the maximum element of a list.

Example 2.73 (Max finder)
An algorithm to find the index of the maximum element of a list is shown in Figure 2.57. (More properly, this algorithm finds the index of the *first* maximum element.)

Computer Science Connections

Hash Tables and Hash Functions

Consider the following scenario: we have a set S of elements that we must store, each of which is chosen from a *universe* U of all possible elements. We need to be able to answer the question "is x in S?" quickly. (We might also have data associated with each $x \in S$, and seek to find the associated data rather than just determining membership.) Furthermore, the set S might change over time, either by insertion of a new element or deletion of an existing element. How might we efficiently organize the data to support these operations?

A *hash table*, one of the most frequently used data structures in computer science, is designed to store a set like S, as follows:

- we define a table $T[1 \ldots n]$.
- we choose a *hash function* $h : U \to \{1, \ldots, n\}$.
- each element $x \in S$ is stored in the cell $T[h(x)]$.

There are several different choices about how to handle *collisions*, when we try to store two different elements in the same cell, but for simplicity let's assume that we store them all in that cell, in a list. For example, see the hash function and hash table in Figure 2.58:

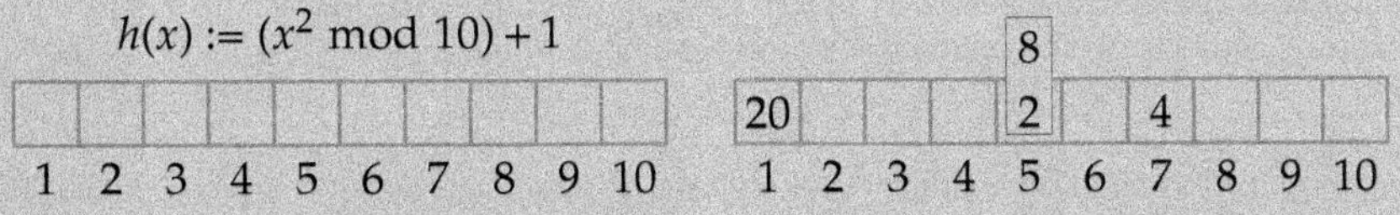

(a) A hash table with hash function h.

(b) The table, filled with 4, 2, 8, and 20.

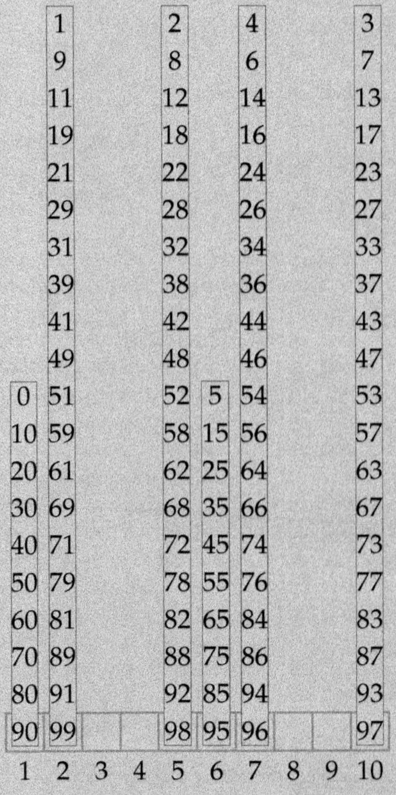

(c) The table filled with $\{0, 1, \ldots, 99\}$.

Figure 2.58: A hash table, empty and filled. If we're asked to store 4 and 2 and 20 and 8, they would go into cells $h(4) = (16 \bmod 10) + 1 = 7$ and $h(2) = 5$ and $h(20) = 1$ and $h(8) = 5$. Panel (c) shows every element from the universe $\{0, 1, \ldots, 99\}$; the fact that the number of elements per cell is so variable means that this hash function does a poor job of spreading out its inputs across the table.

To insert a value x into the table, we merely need to compute $h(x)$ and place the value into the list in the cell $T[h(x)]$. Answering the question "is x stored in the table?" is similar; we compute $h(x)$ and look through whatever entries are stored in that list. As a result, the performance of this data structure is almost entirely dependent on how many collisions are generated—that is, how long the lists are in the cells of the table.

A "good" hash function $h : U \to \{1, \ldots, n\}$ is one that distributes the possible values of U as evenly as possible across the n different cells. The more evenly the function spreads out U across the cells of the table, the smaller the typical length of the list in a cell, and therefore the more efficiently the program would run. (Figure 2.58(c) says that the above hash function is not a very good one.) Programming languages like Python and Java have built-in implementations of hash tables, and they use some mildly complex iterative arithmetic operations in their hash functions. But designing a good hash function for whatever kind of data you end up storing can be the difference between a slow implementation and a blazingly fast one.

Incidentally, there are two other concerns with efficiency: first, the hash function must be able to be computed quickly, and there's also some cleverness in choosing the size of the table and in deciding when to *rehash* everything in the table into a bigger table if the lists get too long (on average).

2.5.5 Exercises

Consider the function $f : \{0,1,\ldots 7\} \to \{0,1,\ldots 7\}$ *defined by* $f(x) := (x^2+3) \bmod 8$.

2.224 What is $f(3)$?

2.225 What is $f(7)$?

2.226 For what x is $f(x) = 3$?

2.227 Redefine f using a table.

$$\textbf{quantize}(n) := \begin{cases} 26 & \text{if } 0 \le n \le 51 \\ 78 & \text{if } 52 \le n \le 103 \\ 130 & \text{if } 104 \le n \le 155 \\ 182 & \text{if } 156 \le n \le 207 \\ 234 & \text{if } 208 \le n \le 255 \end{cases}$$

Figure 2.59: The function from Example 2.56.

2.228 In Example 2.56, we introduced a function **quantize** for compressing a grayscale image to use only five different shades of gray. (See Figure 2.59 for a reminder of the function.) Using basic arithmetic notation (including $\lfloor\ \rfloor$ and/or $\lceil\ \rceil$ if appropriate), redefine **quantize** without using cases.

Let's generalize the quantization idea from the previous exercise to be a two-argument *function, so that* **quantize**(n,k) *takes an input color* $n \in \{0,1,\ldots,255\}$ *and a number* k *of "quanta." (We insist that* $1 \le k \le 256$.*) In other words,* k *is the number of different equally spaced output values, and the input color* n *is translated to the closest of these* k *values. (The ranges associated with the quanta are only* approximately *equal because of issues of integrality: for example, in the* $k = 5$ *case from Figure 2.59, the first four quanta correspond to 52 different colors; the last quantum corresponds to only* $256 - 52 \cdot 4 = 48$ *different colors.)*

2.229 What are the domain and range of **quantize**(n,k)?

2.230 Repeat Exercise 2.228 for **quantize**(n,k). You should ensure that **quantize**$(n,5)$ yields the function from Figure 2.59. *(Hint: first determine how big a range of colors should be mapped to a particular quantum, rounding the size up. Then figure out which quantum the given input n corresponds to.)*

2.231 A function $f : A \to B$ is said to be c-to-1 if, for every output value $b \in B$, there are exactly c different values $a \in A$ such that $f(a) = b$. (These functions are useful in counting; see the Division Rule in Theorem 9.11.) For what values of k is it *possible* to define a c-to-1 (for some integer c) quantizing function that transforms into $\{0,1,\ldots,255\}$ into a set of k quanta?

2.232 *(programming required)* Implement quantization for image files, in a programming language of your choice. Specifically, implement **quantize**(n,k), and apply it to every pixel of a given image. (You'll need to research an image-processing library to use in your program.)

Many of the pieces of basic numerical notation that we've introduced can be thought of as functions. For each of the following, state the domain *and* range *of the given function.*

2.233 $f(x) = |x|$

2.234 $f(x) = \lfloor x \rfloor$

2.235 $f(x) = 2^x$

2.236 $f(x) = \log_2 x$

2.237 $f(x) = x \bmod 2$

2.238 $f(x) = 2 \bmod x$

2.239 $f(x,y) = x \bmod y$

2.240 $f(x) = 2 \mid x$

2.241 $f(x) = \|x\|$

2.242 $f(\theta) = \langle \cos\theta, \sin\theta \rangle$

2.243 Let $T = \{1,\ldots,12\} \times \{0,1,\ldots,59\}$ denote the set of numbers that can be displayed on a digital clock in twelve-hour mode. Define a function *add* : $T \times \mathbb{Z}^{\geq 0} \to T$ so that *add*(t,x) denotes the time that's x minutes later than t. Do so using only standard symbols from arithmetic.

Define the functions $f(x) := x \bmod 10$, $g(x) := x+3$, *and* $h(x) := 2x$. *What are the following? (That is, rewrite the definition of the given function using a single algebraic expression. For example, the function* $g \circ g$ *is given by the definition* $(g \circ g)(x) = g(g(x)) = x+6$.*)*

2.244 $f \circ f$

2.245 $h \circ h$

2.246 $f \circ g$

2.247 $g \circ h$

2.248 $h \circ g$

2.249 $f \circ h$

2.250 $f \circ g \circ h$

Let $f(x) := 3x+1$ *and let* $g(x) := 2x$. *Identify a function* h *such that ...*

2.251 ... $g \circ h$ and f are identical.

2.252 ... $h \circ g$ and f are identical.

Which of the following functions $f : \{0,1,2,3\} \to \{0,1,2,3\}$ *are onto?*

2.253 $f(x) = x$

2.254 $f(x) = x^2 \bmod 4$

2.255 $f(x) = x^2 - x \bmod 4$

2.256 $f(0) = 3, f(1) = 2, f(2) = 1, f(3) = 0$

2.257 $f(0) = 1, f(1) = 2, f(2) = 1, f(3) = 2$

Which of the following functions $f : \{0,1,2,3\} \to \{0,1,\ldots,7\}$ *are one-to-one?*

2.258 $f(x) = x^2 \bmod 8$

2.259 $f(x) = x^3 \bmod 8$

2.260 $f(x) = (x^3 - x) \bmod 8$

2.261 $f(x) = (x^3 + 2x) \bmod 8$

2.262 $f(0) = 3, f(1) = 1, f(2) = 4, f(3) = 1$

A heap *is a data structure that is used to represent a collection of items, each of which has an associated priority. (See p. 529.) A heap can be represented as a complete binary tree—a binary tree with no "holes" as you read in left-to-right, top-to-bottom order—but a heap can also be stored more efficiently as an array, in which the elements are stored in that same left-to-right and top-to-bottom order. (See Figure 2.60.) To do so, we define three functions that allow us to compute the index of the* parent *of a node; the index of the* left *child of a node; and the index of the* right *child of a node. (For example, the parent of the node labeled 8 in Figure 2.60 is labeled 9, the left child of the node labeled 8 is labeled 3, and the right child is labeled 5.) Here are the functions: given an index i into the array, we define*

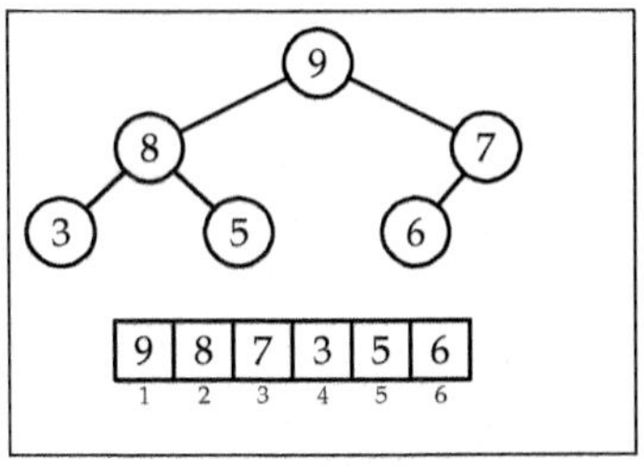

Figure 2.60: A maximum heap, as a tree and as an array.

$$\textbf{parent}(i) := \left\lfloor \frac{i}{2} \right\rfloor \qquad \textbf{left}(i) := 2i \qquad \textbf{right}(i) := 2i + 1.$$

For example, the node labeled 8 *has index* 2 *in the array, and* **parent**(2) = 1 *(the index of the node labeled* 9*);* **left**(2) = 4 *(the index of the node labeled* 3*); and* **right**(2) = 5 *(the index of the node labeled* 5*).*

2.263 Suppose that we have a heap stored as an array $A[1 \ldots n]$. State the domain and range of the function **parent**. Is parent one-to-one?

2.264 State the domain and range of **left** and **right** for the heap as stored in $A[1 \ldots n]$. Are **left** and **right** one-to-one?

Give both a mathematical description and an English-language description of the meanings of the following heap-related functions. Assume for the purposes of these questions that the array A is infinite (that is, don't worry about the possibility of encountering an i such that **left**(i) *or* **right**(i) *is undefined).*

2.265 **parent** ∘ **left**

2.266 **parent** ∘ **right**

2.267 **left** ∘ **parent**

2.268 **right** ∘ **parent**

What are the inverses of the following functions?

2.269 $f : \mathbb{R} \to \mathbb{R}$, where $f(x) = 3x + 1$.

2.270 $g : \mathbb{R}^{\geq 0} \to \mathbb{R}^{\geq 0}$, where $g(x) = x^3$.

2.271 $h : \mathbb{R}^{\geq 0} \to \mathbb{R}^{\geq 1}$, where $h(x) = 3^x$.

2.272 Why doesn't the function $f : \{0, \ldots, 23\} \to \{0, \ldots, 11\}$ where $f(n) = n \bmod 12$ have an inverse?

What are the degrees of the following polynomials?

2.273 $p(x) = 3x^3 + 2x^2 + x + 0$

2.274 $p(x) = 9x^3$

2.275 $p(x) = 4x^4 + x^2 - (2x)^2$

Suppose that p and q are polynomials, both with degree 7*. What are the smallest and largest possible degrees of the following polynomials?*

2.276 $f(x) = p(x) + q(x)$

2.277 $f(x) = p(x) \cdot q(x)$

2.278 $f(x) = p(q(x))$

Give an example of a polynomial p of degree 2 *such that ...*

2.279 ... p has exactly 0 roots.

2.280 ... p has exactly 1 root.

2.281 ... p has exactly 2 roots.

2.282 The *median* of a list L of n numbers is the number in the "middle" of L in sorted order. Describe an algorithm to find the median of a list L. (Don't worry about efficiency.) You may find it useful to make use of the algorithm in Figure 2.57.

2.6 *Chapter at a Glance*

Booleans, Numbers, and Arithmetic

A *Boolean value* is True or False. The *integers* $\mathbb{Z}$ are $\{\ldots, -3, -2, -1, 0, 1, 2, 3, \ldots\}$. The *real* numbers $\mathbb{R}$ are the integers and all numbers in between. The *closed interval* $[a, b]$ consists of all real numbers x where $a \leq x \leq b$; the *open interval* (a, b) excludes a and b. The *rational* numbers $\mathbb{Q}$ are those numbers that can be represented as a/b for integers a and $b \neq 0$. Here is some useful notation involving numbers:

- *exponentiation*: b^k is $b \cdot b \cdot \cdots \cdot b$, where b is multiplied k times;
- *logarithms*: $\log_b x$ is the number y such that $b^y = x$;
- *absolute value*: $|x|$ is x for $x \geq 0$, and $|x| = -x$ for $x < 0$;
- *floor* and *ceiling*: $\lfloor x \rfloor$ is the largest integer $n \leq x$; $\lceil x \rceil$ is the smallest integer $n \geq x$;
- *modulus*: $n \bmod k$ is the remainder when n is divided by k.

If $n \bmod d = 0$, then d is a *factor* of n or *evenly divides* n, written $d \mid n$. If $2 \mid n$ for a positive integer n, then n is *even* ("has even *parity*"); otherwise n is *odd*. An integer $n \geq 2$ is *prime* if it has no positive integer factors other than 1 and n; otherwise n is *composite*. (Note that 0 and 1 are neither prime nor composite.)

For a collection of numbers $x_1, x_2, \ldots, x_n$, their sum $x_1 + x_2 + \cdots + x_n$ is written formally as $\sum_{i=1}^{n} x_i$, and their product $x_1 \cdot x_2 \cdot \cdots \cdot x_n$ is written $\prod_{i=1}^{n} x_i$.

Sets: Unordered Collections

A *set* is an unordered collection of objects called *elements*. A set can be specified by listing its elements inside braces, as $\{x_1, x_2, \ldots, x_n\}$. A set can also be denoted by $\{x : P(x)\}$, which contains all objects x such that $P(x)$ is true. The set of possible values x that are considered is the *universe* U, which is sometimes left implicit.

Standard sets include the *empty set* $\{\}$ (also written $\varnothing$), which contains no elements; the *integers* $\mathbb{Z}$; the *real numbers* $\mathbb{R}$; and the *booleans* $\{\text{True}, \text{False}\}$. We write $\mathbb{Z}^{\geq 0} = \{0, 1, 2, \ldots\}$ and $\mathbb{Z}^{<0} = \{-1, -2, \ldots\}$, etc. For a set A and an object x, the expression $x \in A$ ("x is in A") is true whenever x is in the set A. (So $y \in \{x : P(x)\}$ whenever $P(y) = \text{True}$, and $y \in \{x_1, x_2 \ldots, x_n\}$ whenever $x_i = y$ for some i.) The *cardinality* of a set A, written $|A|$, is the number of distinct elements in A.

Given two sets A and B, the *union* of A and B is $A \cup B = \{x : x \in A \text{ or } x \in B\}$. The *intersection* of A and B is $A \cap B = \{x : x \in A \text{ and } x \in B\}$. The *set difference* of A and B is $A - B = \{x : x \in A \text{ and } x \notin B\}$. The *complement* of a set A is $\sim\!A = U - A = \{x : x \in U \text{ and } x \notin A\}$, where U is the universe.

A *subset* of a set B is a set A such that every element of A is also an element of B; this relationship is denoted by $A \subseteq B$. If A is a subset of B, then B is a *superset* of A, written $B \supseteq A$. A *proper subset* of B is a set A that is a subset of B but $A \neq B$, written $A \subset B$. Such a set B is a *proper superset* of A, written $B \supset A$. Two sets A and B are *disjoint* if $A \cap B = \varnothing$. A *partition* of a set S is a collection of sets $A_1, A_2, \ldots, A_k$, where $A_1 \cup A_2 \cup \cdots \cup A_k = S$ and, for any distinct i and j, the sets A_i and A_j are disjoint.

The *power set* of a set A, written $\mathscr{P}(A)$, is the set of all subsets of A.

Sequences, Vectors, and Matrices: Ordered Collections

A *sequence* (or *tuple, (ordered) pair, triple, quadruple, ..., n-tuple, ...*) is an ordered collection of objects called *components* or *entries*, written inside angle brackets. The set $A \times B = \{\langle a,b\rangle : a \in A \text{ and } b \in B\}$ is the *Cartesian product* of sets A and B; the set $A \times B$ contains all pairs where the first component comes from A and the second from B. For a set S and a number $n \geq 0$, the set S^n denotes the n-fold Cartesian product of S with itself: $S^n = S \times S \times \ldots \times S$, where S occurs n times in this product.

A *vector* (or *n-vector*) is an element of $\mathbb{R}^n$, for some positive integer $n \geq 2$. (An element of $\mathbb{R}^1 = \mathbb{R}$ is called a *scalar*.) A *bit vector* is an element of $\{0,1\}^n$. Vectors are sometimes written in square brackets: $x = [x_1, x_2, \ldots, x_n]$. For a vector x, write x_i to denote the ith component of x. (But x_i is meaningless unless $i \in \{1, 2, \ldots, n\}$.) The *size* or *dimensionality* of $x \in \mathbb{R}^n$ is n.

For a vector $x \in \mathbb{R}^n$ and a real number $\alpha \in \mathbb{R}$, the *scalar product* αx is a vector where $(\alpha x)_i = \alpha x_i$. For two vectors $x, y \in \mathbb{R}^n$, the sum of x and y is a vector $x + y$, where $(x+y)_i = x_i + y_i$. The *dot product* of two vectors $x, y \in \mathbb{R}^n$ is $x \bullet y = \sum_{i=1}^{n} x_i y_i$. Both $x + y$ and $x \bullet y$ are meaningless unless x and y have the same dimensionality.

An *n-by-m matrix* M is an element of $(\mathbb{R}^n)^m$, which is also sometimes written $\mathbb{R}^{n \times m}$. Such a matrix M has n *rows* and m *columns*, as in Figure 2.61. A matrix $M \in \mathbb{R}^{n \times m}$ is *square* if $n = m$. For a size n, the *identity matrix* is $I \in \mathbb{R}^{n \times n}$ has ones on the main diagonal (the entries $I_{i,i} = 1$) and zeros everywhere else.

$$M = \begin{bmatrix} M_{1,1} & M_{1,2} & \cdots & M_{1,m} \\ M_{2,1} & M_{2,2} & \cdots & M_{2,m} \\ \vdots & \vdots & \ddots & \vdots \\ M_{n,1} & M_{n,2} & \cdots & M_{n,m} \end{bmatrix}$$

Figure 2.61: A matrix.

Given a matrix $M \in \mathbb{R}^{n \times m}$ and a real number $\alpha \in \mathbb{R}$, the matrix αM is specified by $(\alpha M)_{i,j} = \alpha M_{i,j}$. Given two matrices $M, M' \in \mathbb{R}^{n \times m}$, the matrix $M + M'$ is specified by $(M + M')_{i,j} = M_{i,j} + M'_{i,j}$. (The sum $M + M'$ is meaningless if M and M' have different dimensions.) The product of two matrices $A \in \mathbb{R}^{n \times m}$ and $B \in \mathbb{R}^{m \times p}$ is a matrix $AB \in \mathbb{R}^{n \times p}$ whose components are given by $(AB)_{i,j} = \sum_{k=1}^{m} A_{i,k} B_{k,j}$. (More compactly, $(AB)_{i,j} = A_{i,(1\ldots m)} \bullet B_{(1\ldots m),j}$.) If the number of rows in A is different from the number of columns in B then AB is meaningless. The *inverse* of M is a matrix M^{-1} such that $MM^{-1} = I$ (if any such matrix M^{-1} exists).

Functions

A *function* $f : A \to B$ maps every element $a \in A$ to some element $f(a) \in B$. The *domain* of f is A and the *codomain* is B. The *image* or *range* of f is $\{f(x) : x \in A\}$, the set of elements of the codomain "hit" by some element of A according to f.

The *composition* of a function $f : A \to B$ and $g : B \to C$ is written $g \circ f : A \to C$, and $(g \circ f)(x) = g(f(x))$. A function $f : A \to B$ is *one-to-one* or *injective* if $f(x) = f(y)$ implies that $x = y$. The function f is *onto* or *surjective* if the image is equal to the codomain. If $f : A \to B$ is one-to-one and onto, it is *bijective*. For a bijection $f : A \to B$, the function $f^{-1} : B \to A$ is the *inverse* of f, where $f^{-1}(b) = a$ when $f(a) = b$.

A *polynomial* $p : \mathbb{R} \to \mathbb{R}$ is a function $p(x) = a_0 + a_1 x + \cdots + a_k x^k$, where each $a_i \in \mathbb{R}$ is a *coefficient*. The *degree* of p is k. The *roots* of p are $\{x : p(x) = 0\}$. A polynomial of degree k that is not always zero has at most k different roots.

An *algorithm* is a step-by-step procedure that transforms an input into an output.

Key Terms and Results

Key Terms

BOOLEANS, NUMBERS, ARITHMETIC

- booleans, integers, reals, rationals
- open intervals, closed intervals
- absolute value $|x|$, floor $\lfloor x \rfloor$, ceiling $\lceil x \rceil$
- exponentiation, logarithms
- modulus, remainder, divides
- even, odd, prime, parity
- summation $\sum$, product $\prod$
- nested summations, nested products

SETS

- set, element, membership, cardinality
- exhaustive enumeration
- set abstraction, universe
- the empty set $\varnothing = \{\}$
- Venn diagram
- complement $\sim$, union $\cup$, intersection $\cap$
- set difference $-$
- (proper) subset, (proper) superset
- disjoint sets
- partitions
- power set

SEQUENCES, VECTORS, MATRICES

- sequence, list, ordered pair, n-tuple
- Cartesian product
- vector, dot product
- matrix, identity matrix
- matrix multiplication
- matrix inverse

FUNCTIONS

- domain, codomain, image/range
- function composition
- one-to-one, onto functions
- bijection, inverse
- polynomial, degree, roots
- algorithm

Key Results

BOOLEANS, NUMBERS, AND ARITHMETIC

1. The value of b^n is $b \cdot b \cdots b$, multiplied together n times. If $n < 0$, then $b^n = 1/(b^{-n})$. For rational exponents, $b^{1/m}$ is the number x such that $x^m = b$, and $b^{n/m} = (b^{1/m})^n$.
2. For a positive real number $b \neq 1$ and a real number $x > 0$, the quantity $\log_b x$ (the log base b of x) is the real number y such that $b^y = x$.
3. Consider integers $k > 0$ and n. Then $k \mid n$ ("k divides n") if $\frac{n}{k}$ is an integer—or, equivalently, if $n \bmod k = 0$.
4. As long as the terms being added remain unchanged, we can reindex a summation (for example, shifting the variable over which the sum is taken, or reversing the order of nested sums) without affecting the total value of the sum. The same is true for products.

SETS: UNORDERED COLLECTIONS

1. A set can be specified using exhaustive enumeration (a list of its elements), or by abstraction (a condition describing when an object is an element of the set).
2. Two sets S and T are equal if every element of S is an element of T and every element of T is an element of S.

SEQUENCES, VECTORS, AND MATRICES

1. For vectors $x, y \in \mathbb{R}^n$, the *dot product* of x and y is $x \bullet y = \sum_{i=1}^{n} x_i y_i$.
2. The product AB of two matrices $A \in \mathbb{R}^{n \times m}$ and $B \in \mathbb{R}^{m \times p}$ is an n-by-p matrix $M \in \mathbb{R}^{n \times p}$ whose components are given by $M_{i,j} = \sum_{k=1}^{m} A_{i,k} B_{k,j}$.

FUNCTIONS

1. A one-to-one and onto function $f : A \to B$ has an inverse function $f^{-1} : B \to A$, where $f(a) = b$ precisely when $f^{-1}(b) = a$.
2. A polynomial of degree k that is not always zero has at most k different roots.

3

Logic

In which our heroes move carefully through the marsh, making sure that each step follows safely from the one before it.

3.1 *Why You Might Care*

How fondly dost thou reason!

William Shakespeare (1564–1616)
The Comedy of Errors

Logic is the study of truth and falsity, of theorem and proof, of valid reasoning in any context. In this chapter, we focus on *formal logic*, in which it is the "form" of the argument that matters, rather than the "content." This chapter will introduce the two major types of formal logic:

- *propositional logic* (Sections 3.2 and 3.3), in which we will study the truth and falsity of statements, how to construct logical statements from basic logical operators (like **and** and **or**), and how to reason about those statements.
- *predicate logic* (Sections 3.4 and 3.5), which gives us a framework to write logical statements of the form "every x ..." or "there's some x such that ...."

One of our main goals in this chapter will be to define a precise, formal, and unambiguous language to express reasoning—in which writer and reader agree on what each word means.

Logic is the foundation of all of computer science; it's the reasoning that you use when you write the condition of an **if** statement or when you design a circuit to add two 32-bit integers or when you design a program to beat a grandmaster at chess. Because logic is the study of valid reasoning, any endeavor in which one wishes to state and justify claims rigorously—such as that of this book—must at its core rely on logic. Every condition that you write in a loop is a logical statement. When you sit down to write binary search in Python, it is through a (perhaps tacit) use of logical reasoning that you ensure that your code works properly for any input. When you use a search engine to look for web pages on the topic "beatles and not john or paul or george or ringo" you've implicitly used logical reasoning to select this particular query. Solving a Sudoku puzzle is nothing more and nothing less than following logical constraints to their conclusion. The central component of a natural language processing (NLP) system is to take an utterance by a human user that's made in a "natural" language like English and "understand" what it means—and understanding what a sentence means is essentially the same task as understanding the circumstances under which the sentence is true, and thus is a question of logic.

And these are just a handful of examples; for a computer scientist, logic is the basis of the discipline. Indeed, the processor of a computer is built up from almost unthinkably simple logical components: wires and physical implementations of logical operations like **and**, **or**, and **not**. Our main goal in this chapter will be to introduce the basic constructs of logic. But along the way, we will encounter applications of logic to natural language processing, circuits, programming languages, optimizing compilers, and building artificially intelligent systems to play chess and other games.

3.2 *An Introduction to Propositional Logic*

> Everyone wishes to have truth on his side, but not everyone wishes to be on the side of truth.
>
> Richard Whately (1787–1863)

A *proposition* is a statement that is either true or false—*In December 2012, Facebook had over one billion users* or *Java is a programming language that uses indentation to denote block structure*, for example. *Propositional logic* is the study of propositions, including how to formulate statements as propositions, how to evaluate whether a proposition is true or false, and how to manipulate propositions. The goal of this section is to introduce propositions—including related terminology, standard notation, and some techniques for reasoning about propositions.

3.2.1 *Propositions and Truth Values*

We'll begin, briefly, with propositions themselves:

Definition 3.1 (Propositions and Truth Values)
A proposition *is a statement that is either true or false. For a particular proposition p, the* truth value *of p is its truth or falsity.*

A proposition is also sometimes called a *Boolean expression* or a *Boolean formula*. (See Section 2.2.1.) A proposition is written in English as a declarative sentence, the kind of sentence that usually ends with a period. (Questions and demands—like *Did you try binary search?* or *Use quicksort!*—aren't the kinds of things that are true or false, and so they're not propositions.) Here are a few examples:

Example 3.1 (Some sample propositions)
The following statements are all propositions:

1. $2+2=4$.
2. 33 is a prime number.
3. Barack Obama is the 44th person to be president of the United States.
4. Every even integer greater than 2 can be written as the sum of two prime numbers.

(The last of these propositions is called *Goldbach's conjecture*; it's more complicated than the other propositions in this example, and we'll return to it in Section 3.4.)

Let's determine the above propositions' truth values:

Example 3.2 (Determining truth values)
Problem: What are the truth values of the propositions from Example 3.1?

Solution: These propositions' truth values are

1. **True.** It really is the case that $2 + 2$ equals 4.
2. **False.** The integer 33 is not a prime number because $33 = 3 \cdot 11$. (Prime numbers are evenly divisible only by 1 and themselves; 33 is evenly divisible by 3 and 11.)
3. **False.** Although Barack Obama is called president #44, Grover Cleveland was president #22 *and* #24. So Barack Obama is actually the *43rd* person to be president of the United States, not the 44th.
4. **Unknown (!).** Goldbach's conjecture was first made in 1742, but has thus far resisted proof—or disproof! It's easy to check that particular small even integers can be written as the sum of two prime numbers; for example, $4 = 2 + 2$, $6 = 3 + 3$, $8 = 3 + 5$, $10 = 3 + 7$, and so on. But is it true for *all* even integers greater than 2? We simply don't know! Many even integers have been tested, and no violation has been found in any of these tests. But, as far as we know, the next even integer we test *can't* be written as the sum of two primes. See Example 3.47 and Exercises 3.178–3.181.

Before we move on from Example 3.2, there's an important point to make about statements that have an unknown truth value. Even though we don't *know* the truth value of Goldbach's conjecture, it is still a proposition and thus it *has* a truth value. That is, Goldbach's conjecture is indeed either true or false; it's just that we don't know which it is. (Like the proposition *The person currently sitting next to you is wearing clean underwear:* it has a truth value, you just don't know what truth value it has.)

Taking it further: Goldbach's conjecture stands in contrast to declarative sentences whose truth is ill-defined—for example, *This book is boring* and *Logic is fun*. Whether these claims are true or false depends on the (imprecise) definitions of words like *boring* and *fun*. We're going to de-emphasize subtle "shades of truth" questions of this form throughout the book, but see p. 314 for some discussion, including the role of ambiguity in software systems that interact with humans via English language input and output.

There is also a potentially interesting philosophical puzzle that's hiding in questions about the truth values of natural-language utterances. Here's a silly (but obviously true) statement: *The sentence "snow is white" is true if and only if snow is white.* (Of course!) This claim becomes a bit less trivial if the embedded proposition is stated in a different language—Spanish or Dutch, say: *The sentence "La nieve es blanca" is true if and only if snow is white*; or *The sentence "Sneeuw is wit" is true if and only if snow is white.* But there's a troubling paradox lurking here. Surely we would like to believe that the English sentence x and the French translation of the English sentence x have the same truth value. For example, *Snow is white* and *La neige est blanche* surely are both true, or they're both false. (And, in fact, it's the former.) But this belief leads to a problem with certain self-referential sentences: for example, *This sentence starts with a 'T'* is true, but *Cette phrase commence par un 'T'* is, surely, false.[1]

For more on paradoxes and puzzles of translation, see

[1] Douglas Hofstadter. *Le Ton Beau de Marot: In Praise of the Music of Language.* Basic Books, 1998; and R. M. Sainsbury. *Paradoxes.* Cambridge University Press, 3rd edition, 2009.

3.2.2 *Atomic and Compound Propositions*

We will distinguish between two types of propositions, those that cannot be broken down into conceptually simpler pieces and those that can be:

Definition 3.2 (Atomic and compound propositions)
An atomic proposition *is a proposition that is conceptually indivisible. A* compound proposition *is a proposition that is built up out of conceptually simpler propositions.*

Here's a simple example of the difference:

Example 3.3 (Atomic and compound propositions)
The University of Minnesota's mascot is the Badger is an atomic proposition, because it is not conceptually divisible into any simpler claim.

The University of Washington's mascot is the Duck or the University of Oregon's mascot is the Duck is a compound proposition, because it is conceptually divisible into two simpler claims—namely *The University of Washington's mascot is the Duck* and *The University of Oregon's mascot is the Duck*.

Atomic propositions are also sometimes called *Boolean variables*; see Section 2.2.1. A compound proposition that contains Boolean variables $p_1, \ldots, p_k$ is sometimes called a *Boolean expression* or *Boolean formula over* $p_1, \ldots, p_k$.

Example 3.4 (Password validity as a compound proposition)
A certain small college sends the following instructions to its users when they are required to change their password:

> Your password is valid only if it is at least 8 characters long, you have not previously used it as your password, and it contains at least three different types of characters (lowercase letters, uppercase letters, digits, non-alphanumeric characters).

This compound proposition involves seven different atomic propositions:

- p: the password is valid
- q: the password is at least 8 characters long
- r: the password has been used previously by you
- s: the password contains lowercase letters
- t: the password contains uppercase letters
- u: the password contains digits
- v: the password contains non-alphanumeric characters

The form of the compound proposition is "p, only if q and not r and at-least-three-of $\{s, t, u, v\}$ are true." (Later we'll see how to write this compound proposition in standard logical notation; see Example 3.15.)

3.2.3 *Logical Connectives*

Logical connectives are the glue that creates the more complicated compound propositions from simpler propositions. Here are definitions of our first three of these logical connectives—*not*, *and*, and *or*:

Definition 3.3 (Negation (not): $\neg$)
The proposition $\neg p$ *("not p," called the* negation *of the proposition p) is true when the proposition p is false, and is false when p is true.*

Definition 3.4 (Conjunction (and): $\wedge$)
The proposition $p \wedge q$ ("p and q," the conjunction *of the propositions p and q) is true when both of the propositions p and q are true, and is false when one or both of p or q is false.*

Definition 3.5 (Disjunction (or): $\vee$)
The proposition $p \vee q$ ("p or q," the disjunction *of the propositions p and q) is true when one or both of the propositions p or q is true, and is false when both p and q are false.*

The prefix *con-* means "together" and *dis-* means "apart." (*Junct* means "join.") The *con*junction $p \wedge q$ is true when p and q are true together; the *dis*junction $p \vee q$ is true when p is true "apart from" q, or the other way around.

In the conjunction $p \wedge q$, the propositions p and q are called *conjuncts*; in $p \vee q$, they are called *disjuncts*. Here's a simple example:

Example 3.5 (Some simple compound propositions)
Let p denote the proposition *Ohio State's mascot is the Buckeye* and let q denote the proposition *Michigan's mascot is the Wolverine*. Then:

- $\neg q$ denotes the proposition *Michigan's mascot is not the Wolverine.*
- $p \wedge q$ denotes the proposition *Ohio State's mascot is the Buckeye, and Michigan's mascot is the Wolverine.*
- $p \vee q$ denotes the proposition *Ohio State's mascot is the Buckeye, or Michigan's mascot is the Wolverine.*

To help keep the symbols straight, it may be helpful to notice that the symbol $\wedge$ is the angular version of the symbol $\cap$ (intersection), while the symbol $\vee$ is the angular version of the symbol $\cup$ (union). The set $S \cap T$ is the set of all elements contained in S *and* T; the set $S \cup T$ is the set of all elements contained in S *or* T.

Here's an example of translating some English statements that express compound propositions into standard logical notation:

Example 3.6 (From English statements to compound propositions)
Problem: Translate each of the following statements into logical notation. (Name the atomic propositions using appropriate Boolean variables.)

1. Carissa is majoring in computer science and studio art.
2. Either Dave took a formal logic class, or he is a quick learner.
3. Eli broke his hand and didn't take the test as scheduled.
4. Fred knows Python or he has programmed in both C and Java.

Solution: Let's first name the atomic propositions within these English statements:

p = Carissa is majoring in computer science.	t = Eli broke his hand.
q = Carissa is majoring in studio art.	u = Eli took the test as scheduled.
r = Dave took a formal logic class.	v = Fred knows Python.
s = Dave is a quick learner.	w = Fred has programmed in C.
	x = Fred has programmed in Java.

We can now translate the four given statements as: (1) $p \wedge q$; (2) $r \vee s$; (3) $t \wedge \neg u$; and (4) $v \vee (w \wedge x)$.

Implication (if/then)

Another important logical connective is $\Rightarrow$, which denotes *implication*. It expresses a familiar idea from everyday life, though one that's not quite captured by a single

English word. Consider the sentence *If you scratch my back, then I'll scratch yours.* It's easiest to think of this sentence as a promise: I've promised that I'll scratch your back *as long as you scratch mine.* I haven't promised anything about what I'll do if you fail to scratch my back—I can abstain from back scratching, or I might generously scratch your back anyway, but I haven't *guaranteed* anything. (You'd justifiably call me a liar if you scratched my back and I failed to scratch yours in return.) This kind of promise is expressed as an *implication* in propositional logic:

One initially confusing aspect of logical implication is that the word "implies" seems to hint at something about causation—but $p \Rightarrow q$ doesn't actually say anything about p *causing* q, only that p being true *implies that* q is true (or, in other words, p being true *lets us conclude that* q is true).

Definition 3.6 (Implication: $\Rightarrow$)
The proposition $p \Rightarrow q$ is true when the truth of p implies the truth of q. In other words, $p \Rightarrow q$ is true unless p is true and q is false.

In the implication $p \Rightarrow q$, the proposition p is called the *antecedent* or the *hypothesis*, and the proposition q is called the *consequent* or the *conclusion*.

Here are a few examples of statements involving implication:

Example 3.7 (Some implications)
The following propositions are all true:

- $1+1=2$ implies that $2+3=5$. ("True implies True" is true.)
- $2+3=4$ implies that $2+2=4$. ("False implies True" is true.)
- $2+3=4$ implies that $2+3=6$. ("False implies False" is true.)

But the following proposition is false:

- $2+2=4$ implies that $2+1=5$. ("True implies False" is false.)

This last proposition is false because $2+2=4$ is true, but $2+1=5$ is false.

There are many different ways to express the proposition $p \Rightarrow q$ in English, including all of those in Figure 3.1.

"p implies q"	"q, if p"
"if p, then q"	"q is necessary for p"
"p only if q"	"p is sufficient for q"
"q whenever p"	

Figure 3.1: Some ways of expressing $p \Rightarrow q$ in English.

Here is an example of the same implication being stated in English in many different ways:

Example 3.8 (Expressing implications in English)
According to United States law, people who can legally vote must be American citizens, and they must also satisfy some other various conditions that vary from state to state (for example, registering in advance or not being a felon). Thus the following compound proposition is true:

$$\text{you are a legal U.S. voter} \Rightarrow \text{you are an American citizen.}$$

All of the following sentences express this proposition in English:

If you are a legal U.S. voter, then you are an American citizen.
You being a legal U.S. voter implies that you are an American citizen.
You are a legal U.S. voter only if you are an American citizen.

You are an American citizen if you are a legal U.S. voter.
You are an American citizen whenever you are a legal U.S. voter.
You being an American citizen is necessary for you to be a legal U.S. voter.
You being a legal U.S. voter is sufficient for you to be an American citizen.

Most of these sentences are reasonably natural ways to express the stated implication, though the last phrasing seems awkward. But it's easier to understand if we slightly rephrase it as "You being a legal U.S. voter *is sufficient for one to conclude that you are* an American citizen."

Here's another example of restating implications:

Example 3.9 (More implications in English)
Consider the proposition

$$\underbrace{\textit{The nondisclosure agreement is valid}}_{p}\ \textit{only if}\ \underbrace{\textit{you signed it}}_{q}.$$

(This statement is *different* from "if you signed, then the agreement is valid": for example, the agreement might not be valid because you're legally a minor and thus not legally allowed to sign away rights.) We can restate $p \Rightarrow q$ as "if p then q":

If the nondisclosure agreement is valid, then you signed it.

We can also restate this implication equivalently—and perhaps more intuitively—using the so-called contrapositive $\neg q \Rightarrow \neg p$ (see Example 3.21):

The nondisclosure agreement is invalid if you didn't sign it.

"Exclusive or" and "if and only if"

The four logical connectives that we have defined so far ($\neg$, $\vee$, $\wedge$, and $\Rightarrow$) are the ones that are most frequently used, but we'll define two other common connectives too. The first is *exclusive or*:

The connective $\oplus$ is usually pronounced like "ex ore" (a former significant other + some rock with high precious-metal content).

Definition 3.7 (Exclusive or: $\oplus$)
The proposition $p \oplus q$ ("p exclusive or q" or, more briefly, "p xor q") is true when one of the propositions p or q is true, but not both. Thus $p \oplus q$ is false when both p and q are true, and when both p and q are false.

When we want to emphasize the distinction between $\vee$ and $\oplus$, we refer to $\vee$ as *inclusive or*. This terminology highlights the fact that $p \vee q$ *includes* the possibility that both p and q are true, while $p \oplus q$ *excludes* that possibility. Unfortunately, the word "or" in English can mean either inclusive or exclusive or, depending on the context in which it's being used. When you see the word "or," you'll have to think carefully about which meaning is intended.

Here's an example of distinguishing inclusive and exclusive or:

Example 3.10 (Inclusive versus exclusive or in English)

Problem: Translate these statements from a cover letter for a job into logical notation:

> You may contact me by email or by phone. I am available for an on-site day-long interview on October 8th in Minneapolis or Hong Kong.

Use the following Boolean variables:

p = you may contact me by phone
q = you may contact me by email
r = I am physically available for an interview in Minneapolis
s = I am physically available for an interview in Hong Kong

Solution: The "or" in "email or phone" is *inclusive*, because you could receive both an email and a call. However, the "or" in "Minneapolis or Hong Kong" is *exclusive*, because it's not physically possible to be simultaneously present in Minneapolis and Hong Kong. Thus a correct translation of these statements is $(p \vee q) \wedge (r \oplus s)$.

We are now ready to define our last logical connective:

Sometimes you'll see $\Leftrightarrow$ abbreviated in sentences as "iff" as shorthand for "if and only if." We'll avoid this notation in this book, but you should understand it if you see it elsewhere.

Definition 3.8 (If and only if: $\Leftrightarrow$)
The proposition $p \Leftrightarrow q$ ("p if and only if q") is true when the propositions p or q have the same truth value (both p and q are true, or both p and q are false), and false otherwise.

The reason that $\Leftrightarrow$ is read as "if and only if" is that $p \Leftrightarrow q$ means the same thing as the compound proposition $(p \Rightarrow q) \wedge (q \Rightarrow p)$. (We'll prove this equivalence in Example 3.23.) Furthermore, the propositions $p \Rightarrow q$ and $q \Rightarrow p$ can be rendered, respectively, as "p only if q" and "p, if q." Thus $p \Leftrightarrow q$ expresses "p if q, and p only if q"—or, more compactly, "p if and only if q." (The connective $\Leftrightarrow$ is also sometimes called the *biconditional*, because an implication can also be called a *conditional*.)

Unfortunately, just like with "or," the word "if" is ambiguous in English. Sometimes "if" is used to express an implication, and sometimes it's used to express an if-and-only-if definition. When you see the word "if" in a sentence, you'll need to think carefully about whether it means $\Rightarrow$ or $\Leftrightarrow$. Here's an example:

Example 3.11 ("If" versus "if and only if" in English)

Problem: Think of a number between 10 and 1,000,000. Let

p := your number is prime.
q := your number is even.
r := your number is evenly divisible by an integer other than 1 and itself.

Now translate the following two sentences into logical notation:

1. If the number you're thinking of is even, then it isn't prime.
2. The number you're thinking of isn't prime if it's evenly divisible by an integer other than 1 and itself.

Solution: The "if" in (1) is an implication, and the "if" in (2) is "if and only if." A correct translation of these sentences is (1) $q \Rightarrow \neg p$; and (2) $\neg p \Leftrightarrow r$.

3.2.4 *Combining Logical Connectives*

The six standard logical connectives that we've defined so far ($\neg$, $\wedge$, $\vee$, $\Rightarrow$, $\oplus$, and $\Leftrightarrow$) are summarized in Figure 3.2. The logical connective $\neg$ is a *unary operator*, because it builds a compound proposition from a single simpler proposition. The other five connectives are *binary* operators, which build a compound proposition from two simpler propositions. (We'll encounter the full list of binary logical connectives later; see Exercises 4.66–4.71.)

negation	$\neg p$	"not p"	*highest precedence*
conjunction	$p \wedge q$	"p and q"	↓
disjunction	$p \vee q$	"p or q"	
exclusive or	$p \oplus q$	"p xor q"	
implication	$p \Rightarrow q$	"if p, then q" or "p implies q"	
if and only if	$p \Leftrightarrow q$	"p if and only if q"	*lowest precedence*

Figure 3.2: Summary of notation for propositional logic.

Taking it further: The unary-vs.-binary categorization of logical connectives based on how many "arguments" they accept also occurs in other contexts—for example, arithmetic and programming. In arithmetic, for example, one might distinguish between "unary minus" and "binary minus": the former denotes negation, as in -3; the latter subtraction, as in $2 - 3$.

In programming languages, the number of arguments that a function takes is called its *arity*. (The arity of `length` is one; the arity of `equals` is two.) You will sometimes encounter *variable arity* functions that can take a different number of arguments each time they're invoked. Common examples include the print functions in many languages—C's `printf` and Python's `print`, for example, can take any number of arguments—or arithmetic in prefix languages like Scheme, where you can write an expression like `(+ 1 2 3 4)` to denote $1+2+3+4$ ($= 10$).

Order of operations

A full description of the syntax of a programming language always includes a table of the *precedence* of operators, arranged from "binds the tightest" (highest precedence) to "binds the loosest" (lowest precedence). These precedence rules tell us when we have to include parentheses in an expression to make it mean what we want it to mean, and when the parentheses are optional. In the same way, we'll adopt some standard conventions regarding the precedence of our logical connectives:

- Negation ($\neg$) binds the tightest.
- After negation, there is a three-way tie among $\wedge$, $\vee$, and $\oplus$. (We'll always use parentheses in propositions containing more than one of these three operators, just as we should in programs.)
- The trifecta ($\wedge$, $\vee$, and $\oplus$) is followed by $\Rightarrow$.
- $\Rightarrow$ is followed finally by $\Leftrightarrow$.

The word "precedence" (*pre* before, *cede* go) means "what comes first," so precedence rules tell us the order of which the operators "get to go." For example, consider a proposition like $p \wedge q \Rightarrow r$. If $\wedge$ "goes first," the proposition is $(p \wedge q) \Rightarrow r$; if $\Rightarrow$ "goes first," it means $p \wedge (q \Rightarrow r)$. Figure 3.2 says that the former is the correct interpretation.

The horizontal lines in Figure 3.2 separate the logical connectives by their precedence, so that operators closer to the top of the table have higher precedence. For example:

Example 3.12 (Precedence of logical connectives)
The propositions $p \vee \neg q$ and $p \vee q \Rightarrow \neg r \Leftrightarrow p$ mean, respectively,

$$p \vee (\neg q) \qquad \text{and} \qquad \Big((p \vee q) \Rightarrow (\neg r)\Big) \Leftrightarrow p,$$

which we can see by simply applying the relevant precedence rules ("$\neg$ goes first, then $\vee$, then $\Rightarrow$, then $\Leftrightarrow$").

Taking it further: The precedence rules that we've described here match the precedence rules in most programming languages. In Java, for example, the condition `!p && q`—that's "not p and q" in Java syntax—will be interpreted as `(!p) && q`, because not/$\neg$/`!` binds tighter than and/$\wedge$/`&&`.

The precedence rules for operators tell us the order in which two different operators are applied in an expression. For a sequence of applications of the *same* binary operator, we'll use the convention that the operator *associates to the left*. For example, $p \wedge q \wedge r$ will mean $(p \wedge q) \wedge r$ and not $p \wedge (q \wedge r)$.

Example 3.13 (Precedence of logical connectives)

Problem: Fully parenthesize each of the following propositions. (In other words, add parentheses around each operator without changing the meaning.)

1. $p \vee q \Leftrightarrow p$
2. $p \oplus p \oplus q \oplus q$
3. $\neg p \Leftrightarrow p \Leftrightarrow \neg(p \Leftrightarrow p)$
4. $p \wedge \neg q \Rightarrow r \Leftrightarrow s$
5. $p \Rightarrow q \Rightarrow r \wedge s$

Solution: Using the precedence rules from Figure 3.2 and left associativity, we get:

1. $(p \vee q) \Leftrightarrow p$
2. $((p \oplus p) \oplus q) \oplus q$
3. $((\neg p) \Leftrightarrow p) \Leftrightarrow (\neg(p \Leftrightarrow p))$
4. $((p \wedge (\neg q)) \Rightarrow r) \Leftrightarrow s$
5. $(p \Rightarrow q) \Rightarrow (r \wedge s)$

The choice that logical operators associate to the left (instead of associating to the right) won't matter for most of the logical connectives anyway. For example, the propositions $(p \wedge q) \wedge r$ and $p \wedge (q \wedge r)$ are true under exactly the same circumstances, as we'll see shortly. In fact, of the binary operators $\{\wedge, \vee, \oplus, \Rightarrow, \Leftrightarrow\}$, the only one for which the order of application matters is implication. See Exercises 3.45–3.47.

Writing tip: Because the order of application does matter for implication, it's considered good style to include the optional parentheses so that it's clear what you mean.

3.2.5 *Truth Tables*

In Section 3.2.3, we described the logical connectives $\neg$, $\wedge$, $\vee$, $\Rightarrow$, $\oplus$, and $\Leftrightarrow$, but we can more systematically define these connectives by using a *truth table* that collects the value yielded by the logical connective under every *truth assignment*.

Definition 3.9 (Truth assignment)

A truth assignment *for a proposition over variables* $p_1, p_2, \ldots, p_k$ *is a function that assigns a truth value to each* p_i.

For example, the function f where $f(p) = \text{T}$ and $f(q) = \text{F}$ is a truth assignment for the proposition $p \vee \neg q$. (Each "T" abbreviates a truth value of true; each "F" abbreviates a truth value of false.)

For any particular proposition and for any particular truth assignment f for that proposition, we can *evaluate* the proposition under f to figure out the truth value of the entire proposition. In the previous example, the proposition $p \vee \neg q$ is true under the truth assignment with $p = \text{T}$ and $q = \text{F}$ (because $T \vee \neg\text{F}$ is $\text{T} \vee \text{T}$, which is true). A *truth table* displays a proposition's truth value (evaluated in the way we just described) under all truth assignments:

Definition 3.10 (Truth table)
A truth table *for a proposition lists, for each possible truth assignment for that proposition (with one truth assignment per row in the table), the truth value of the entire proposition.*

p	q	$p \wedge q$
T	T	T
T	F	F
F	T	F
F	F	F

Figure 3.3: The truth table for $\wedge$.

For example, the truth table that defines $\wedge$ is shown in Figure 3.3. A few words about this truth table are in order:

- Columns #1 and #2 correspond to the atomic propositions p and q. There is a row in the table corresponding to each possible truth assignment for $p \wedge q$—that is, for every pair of truth values for p and q. (So there are four rows: TT, TF, FT, and FF.)
- The third column corresponds to the compound proposition $p \wedge q$, and it has a T only in the first row. That is, the truth value of $p \wedge q$ is false unless both p and q are true—just as Definition 3.4 said.

The truth tables for the six basic logical connectives (negation, conjunction, disjunction, exclusive or, implication, and "if and only if") are shown in Figure 3.4. It's worth paying special attention to the column for $p \Rightarrow q$: the *only* truth assignment under which $p \Rightarrow q$ is false is when p is true and q is false. *False implies anything! Anything implies true!* For example, both of the following are true propositions:

p	$\neg p$
T	F
F	T

p	q	$p \wedge q$	$p \vee q$	$p \Rightarrow q$	$p \oplus q$	$p \Leftrightarrow q$
T	T	T	T	T	F	T
T	F	F	T	F	T	F
F	T	F	T	T	T	F
F	F	F	F	T	F	T

Figure 3.4: Truth tables for the basic logical connectives.

If $2 + 3 = 4$, *then you will eat tofu for dinner.* (if false, then anything)
If you are your own mother, then $2 + 3 = 5$. (if anything, then true)

To emphasize the point, observe that the first statement is true *even if* you would never eat tofu if it were the last so-called food on earth; the hypothesis "$2 + 3 = 4$" of the proposition wasn't true, so the truth of the proposition doesn't depend on what your dinner plans are.

For more complicated compound propositions, we can fill in a truth table by repeatedly applying the rules in Figure 3.4. For example, to find the truth table for $(p \Rightarrow q) \wedge (q \vee p)$, we compute the truth tables for $p \Rightarrow q$ and $q \vee p$, and put a "T" in the $(p \Rightarrow q) \wedge (q \vee p)$ column for precisely those rows in which the truth tables for $p \Rightarrow q$ and $q \vee p$ both had "T"s. Here's a simple example, and a somewhat more complicated one:

Example 3.14 (A small truth table)
Here is a truth table for the proposition $(p \wedge q) \Rightarrow \neg q$:

p	q	$p \wedge q$	$\neg q$	$(p \wedge q) \Rightarrow \neg q$
T	T	T	F	F
T	F	F	T	T
F	T	F	F	T
F	F	F	T	T

This truth table shows that the given proposition $(p \wedge q) \Rightarrow \neg q$ is true precisely when at least one of p and q is false.

Example 3.15 (Three (or more) of four, formalized)
In Example 3.4 (on the validity of passwords), we had a sentence of the form

"p, only if q and not r and at-least-three-of $\{s, t, u, v\}$ are true."

Let's translate this sentence into propositional logic. The tricky part will be translating "at least three of $\{s, t, u, v\}$ are true." There are many solutions, but one relatively simple way to do it is to explicitly write out four cases, one corresponding to allowing a different one of the four variables $\{s, t, u, v\}$ to be false:

$$(s \wedge t \wedge u) \vee (s \wedge t \wedge v) \vee (s \wedge u \wedge v) \vee (t \wedge u \wedge v)$$

We can verify that we've gotten this proposition right with a (big!) truth table, shown in Figure 3.5. Indeed, the five rows in which the last column has a "T" are exactly the five rows in which there are three or four "T"s in the columns for s, t, u, and v.

To finish the translation, recall that "x only if y" means $x \Rightarrow y$, so the given sentence can be translated as $p \Rightarrow q \wedge \neg r \wedge$ (the proposition above)—that is,

$$p \Rightarrow q \wedge \neg r \wedge \Big((s \wedge t \wedge u) \vee (s \wedge t \wedge v) \vee (s \wedge u \wedge v) \vee (t \wedge u \wedge v)\Big).$$

Taking it further: It's worth pondering why there are five different rows of the truth table in Figure 3.5 in which the last column is true: there are four different truth assignments corresponding to exactly three of $\{s, t, u, v\}$ being true (stu, suv, stv, tuv), and there is one truth assignment corresponding to all four being true ($stuv$). In Chapter 9, on counting, we'll re-encounter this style of question. (And, actually, precisely the same reasoning as in this example will allow us to prove something interesting about error-correcting codes—see Section 4.2.5.)

Figure 3.5: A truth table for Example 3.15.

s	t	u	v	$s \wedge t \wedge u$	$s \wedge t \wedge v$	$s \wedge u \wedge v$	$t \wedge u \wedge v$	$(s \wedge t \wedge u)$ $\vee (s \wedge t \wedge v)$ $\vee (s \wedge u \wedge v)$ $\vee (t \wedge u \wedge v)$
T	T	T	T	T	T	T	T	T
T	T	T	F	T	F	F	F	T
T	T	F	T	F	T	F	F	T
T	T	F	F	F	F	F	F	F
T	F	T	T	F	F	T	F	T
T	F	T	F	F	F	F	F	F
T	F	F	T	F	F	F	F	F
T	F	F	F	F	F	F	F	F
F	T	T	T	F	F	F	T	T
F	T	T	F	F	F	F	F	F
F	T	F	T	F	F	F	F	F
F	T	F	F	F	F	F	F	F
F	F	T	T	F	F	F	F	F
F	F	T	F	F	F	F	F	F
F	F	F	T	F	F	F	F	F
F	F	F	F	F	F	F	F	F

Computer Science Connections

Natural Language Processing, Ambiguity, and Truth

Our main interest in this book is in developing (and understanding) precise and unambiguous language to express mathematical notions; in this chapter specifically, we're thinking about the truth values of completely precise statements. But thinking about the truth of ambiguous or ill-defined terms is absolutely crucial to any computational system that's designed to interact with users via natural language. (A *natural language* is one like English or French or Xhosa; these languages contrast with *artificial languages* like Java or Python or, arguably, Esperanto or Klingon.)

Natural language processing (NLP) (or the roughly similar *computational linguistics*) is the subfield of computer science that lies at the discipline's interface with linguistics.[2] In NLP, we work to develop software systems that can interact with users in a natural language. A necessary step in an NLP system is to take an utterance made by the human user and "understand it." ("Understanding what a sentence means" is more or less the same as "understanding the circumstances under which it is true"—which is fundamentally a question of logic.)

For more, you can look for a textbook on NLP like

[2] Daniel Jurafsky and James H. Martin. *Speech and Language Processing: An Introduction to Natural Language Processing, Computational Linguistics, and Speech Recognition.* Pearson Prentice Hall, 2nd edition, 2008.

One major reason that NLP is hard is that there is a tremendous amount of ambiguity in natural-language utterances. We can have *lexical ambiguity*, in which two different words are spelled identically but have two different meanings; we have to determine which word is meant in a sentence. Or there's *syntactic ambiguity*, in which a sentence's structure can be interpreted very differently. (See Figure 3.6.) But there are also subtleties about when a statement is true, even if the meaning of each word and the sentence's structure are clear.

A: *Do you prefer coffee or tea?*
B: *Do you prefer cream or sugar?*
C: *We ate cake with walnuts.*
D: *We ate cake with forks.*

Figure 3.6: Examples of *lexical* (A and B) and *syntactic ambiguity* (C and D). The *or* of A/B can be either inclusive or exclusive; simply answering "yes" is a reasonable response to question B, but a bizarre one to question A. The *with* of C/D can either attach to the *cake* or the *eating*; the sentences' structures are consistent with using walnuts as an eating utensil in C, or the cake containing forks as an ingredient in D.

Consider, for example, designing and implementing a conversational system designed to assist with travel planning. (Many airlines or train companies have such systems.) Such a system might engage in a dialogue like the one in Figure 3.7 with a human user. There's no hard-and-fast rule for what other flights should count as "slightly later" and "too much more expensive." This conversational system has to be able to decide the truth of statements like *Delta #2931 is slightly later than Delta #1927* and *Delta #2931 isn't too much more expensive than Delta #1927*, even though the "truth" of these statements depends on heavy use of conversational context and pragmatic reasoning. Of course, even though one cannot unambiguously determine whether these sentences are true or false, they're the kind of statement made continually in natural language. So systems that process natural language must deal with this issue with great frequency.

User: *I want to fly from MSP to BOS on 28 December.*

System: *Delta #1927 is a nonstop flight from MSP to BOS on Delta Airlines for $472 that leaves at 8:45am.*

User: *Is there a slightly later flight that isn't too much more expensive?*

Figure 3.7: A sample dialogue. Suppose that Delta #2931 is a second nonstop flight from MSP to BOS that leaves at 10:33am and costs $529.

One approach for handling these statements whose truth value is ambiguous is called *fuzzy logic*, in which each proposition has a truth value that is a real number between 0 and 1. (So *10:33a is slightly later than 8:45a* is "more true" than *12:19p is slightly later than 8:45a*—so the former might have a truth value of 0.74, while the latter might have a truth value of 0.34. But *7:30a is slightly later than 8:45a* would have a truth value of 0.00, as 7:30a is unambiguously *not* slightly later than 8:45a.)

3.2.6 Exercises

What are the truth values of the following propositions?

3.1 $2^2 + 3^2 = 4^2$

3.2 The number 202 is written 11010010 in binary.

3.3 After executing the C code fragment in Figure 3.8 (shown at right), the variable x has the value 1.

```
int x = 202;
while (x > 2) {
   x = x / 2;
}
```

Figure 3.8: Snippet of C code. Note that x/2 denotes integer division; for example, 7/2 = 3.

Consider the following atomic propositions:

p:	x + y *is valid Python*	u:	x *is a numeric value*
q:	x * y *is valid Python*	v:	y *is a numeric value*
r:	x ** y *is valid Python*	w:	x *is a list*
s:	x * y *is a list*	z:	y *is a list*
t:	x + y *is a list*		

*Using these atomic propositions, translate the following (true!) statements about legal Python programs into logical notation. (Note that these statements do not come close to fully characterizing the set of valid Python statements, for several reasons: first, they're about particular variables—*x *and* y*—rather than about generic variables. And, second, they omit some important common-sense facts—for example, it's not simultaneously possible to be both a list and a numeric value. That is, for example, we have $\neg v \vee \neg z$.)*

3.4 x ** y is valid Python if and only if x and y are both numeric values.

3.5 x + y is valid Python if and only if x and y are both numeric values, or they're both lists.

3.6 x * y is valid Python if and only if x and y are both numeric values, or if one of x and y is a list and the other is numeric.

3.7 x * y is a list if x * y is valid Python and x and y are not both numeric values.

3.8 if x + y is a list, then x * y is not a list.

3.9 x + y and x * y are both valid Python only if x is not a list.

3.10 True story: a 29-year-old friend of mine who does not have an advance care directive was asked the following question on a form at a doctor's office. What should she answer?

If you're over 55 years old, do you have an advance care directive? Circle one: YES NO

In Example 3.15, we constructed a proposition corresponding to "at least three of $\{s, t, u, v\}$ are true." Generalize this construction by building a proposition ...

3.11 ... expressing "at least 3 of $\{p_1, \ldots, p_n\}$ are true."

3.12 ... expressing "at least $n - 1$ of $\{p_1, \ldots, p_n\}$ are true."

The identity *of a binary operator $\diamond$ is a value i such that, for any x, the expressions $\{x, x \diamond i, i \diamond x\}$ are all equivalent. The* zero *of $\diamond$ is a value z such that, for any x, the expressions $\{z, x \diamond z, z \diamond x\}$ are all equivalent. For an example from arithmetic, the identity of $+$ is 0, because $x + 0 = 0 + x = x$ for any number x. And the zero of multiplication is 0, because $x \cdot 0 = 0 \cdot x = 0$ for any number x. For each of the following, identify the identity or zero of the given logical operator. Justify your answer. Some operators do not have an identity or a zero; if the given operator fails to have the stated identity/zero, explain why it doesn't exist.*

3.13 What is the identity of $\vee$?

3.14 What is the identity of $\wedge$?

3.15 What is the identity of $\Leftrightarrow$?

3.16 What is the identity of $\oplus$?

3.17 What is the zero of $\vee$?

3.18 What is the zero of $\wedge$?

3.19 What is the zero of $\Leftrightarrow$?

3.20 What is the zero of $\oplus$?

Because $\Rightarrow$ is not commutative (that is, because $p \Rightarrow q$ and $q \Rightarrow p$ mean different things), it is not too surprising that $\Rightarrow$ has neither an identity nor a zero. But there are a pair of related definitions that apply to this type of operator:

3.21 The *left identity* of a binary operator $\diamond$ is a value i_ℓ such that, for any x, the expressions x and $i_\ell \diamond x$ are equivalent. The *right identity* of $\diamond$ is a value i_r such that, for any x, the expressions x and $x \diamond i_r$ are equivalent. (Again, some operators may not have left or right identities.) What are the left and right identities of $\Rightarrow$ (if they exist)?

3.22 The *left zero* of a binary operator $\diamond$ is a value z_ℓ such that, for any x, the expressions z_ℓ and $z_\ell \diamond x$ are equivalent; similarly, the *right zero* is a value z_r such that, for any x, the expressions z_r and $x \diamond z_r$ are equivalent. (Again, some operators may not have left or right zeros.) What are the left and right zeros for $\Rightarrow$ (if they exist)?

In many programming languages, the Boolean values True and False are actually stored as the numerical values 1 and 0, respectively. In Python, for example, both `0 == False` *and* `1 == True` *are True. Thus, despite appearances, we can add or subtract or multiply Boolean values! Furthermore, in many languages (including Python), anything that is not False (in other words, anything other than 0) is considered True for the purposes of conditionals. For example, in many programming languages, including Python, code like* `if 2 print "yes" else print "no"` *will print "yes."*

Suppose that x and y are two Boolean variables in a programming language, like Python, where True *and* False *are 1 and 0, respectively—that is, the values of x and y are both 0 or 1. Each of the following code snippets includes a conditional statement based on an arithmetic expression using x and y. For each, rewrite the given condition using the standard notation of propositional logic.*

3.23 `if x * y ...`

3.24 `if x + y ...`

3.25 `if 1 - x ...`

3.26 `if (x * (1 - y)) + ((1 - x) * y) ...`

We can use the common programming language features described in in the previous block of exercises to give a simple programming solution to Exercises 3.11–3.12. Assume that $\{p_1, \ldots, p_n\}$ *are all Boolean variables in Python—that is, their values are all 0 or 1. Write a Python conditional expressing the condition that ...*

3.27 ... at least 3 of $\{p_1, \ldots, p_n\}$ are true.

3.28 ... at least $n - 1$ of $\{p_1, \ldots, p_n\}$ are true.

In addition to purely logical operations, computer circuitry has to be built to do simple arithmetic very quickly. Here you'll explore some pieces of using propositional logic and binary representation of integers to express arithmetic operations. (It's straightforward to convert your answers into circuits.)

Consider a number $x \in \{0, \ldots, 15\}$ *represented as a 4-bit binary number, as shown in Figure 3.9. Denote by* x_0 *the least-significant bit of x, by* x_1 *the next bit, and so forth. For example, for the number* $x = 12$ *(written 1100 in binary) would have* $x_0 = 0$, $x_1 = 0$, $x_2 = 1$, *and* $x_3 = 1$*). For each of the following conditions, give a proposition over the Boolean variables* $\{x_0, x_1, x_2, x_3\}$ *that expresses the stated condition. (Think of 0 as false and 1 as true.)*

3.29 x is greater than or equal to 8.

3.30 x is evenly divisible by 4.

3.31 x is evenly divisible by 5. *(Hint: use a truth table, and then build a proposition from the table.)*

3.32 x is an exact power of two.

x_3	x_2	x_1	x_0	
0	0	1	1	
0 +	0 +	2 +	1	= 3

x_3	x_2	x_1	x_0	
1	1	0	0	
8 +	4 +	0 +	0	= 12

Figure 3.9: Representing $x \in \{0, \ldots, 15\}$ using 4-bits.

3.33 Suppose that we have *two* 4-bit input integers x and y, represented as in Exercises 3.29–3.32. Give a proposition over $\{x_0, x_1, x_2, x_3, y_0, y_1, y_2, y_3\}$ that expresses the condition that $x = y$.

3.34 Given two 4-bit integers x and y as in the previous exercise, give a proposition over the Boolean variables $\{x_0, x_1, x_2, x_3, y_0, y_1, y_2, y_3\}$ that expresses the condition that $x \leq y$.

3.35 Suppose that we have a 4-bit input integer x, represented by four Boolean variables $\{x_0, x_1, x_2, x_3\}$ as in Exercises 3.29–3.32. Let y be the integer $x + 1$, represented again as a 4-bit value $\{y_0, y_1, y_2, y_3\}$. (For the purposes of this question, treat $15 + 1 = 0$—that is, we're really defining $y = (x + 1) \bmod 16$.) For example, for $x = 11$ (which is 1011 in binary), we have that $y = 12$ (which is 1100 in binary). For each $i \in \{0, 1, 2, 3\}$, give a proposition over the Boolean variables $\{x_0, x_1, x_2, x_3\}$ that expresses the value of y_i.

The remaining problems in this section ask you to build a program to compute various facts about a given proposition φ*. To make your life as easy as possible, you should consider a simple representation of* φ*, based on representing any compound proposition as a list. In such a list, the first element will be the logical connective, and the remaining elements will be the subpropositions. For example, the proposition* $p \Rightarrow (\neg q)$ *will be represented as*

```
["implies", ["or", "p", "r"], ["not", "q"]]
```

Now, using this representation of propositions, write a program, in a programming language of your choice, to accomplish the following operations:

We'll occasionally use lowercase Greek letters, particularly φ ("phi") or ψ ("psi"), to denote not-necessarily-atomic propositions.

3.36 *(programming required)* Given a proposition φ, compute the set of all atomic propositions contained within φ. The following recursive formulation may be helpful:

$$\textbf{variables}(p) := \{p\} \qquad \textbf{variables}(\neg\varphi) := \textbf{variables}(\varphi)$$

$$\textbf{variables}(\varphi \diamond \psi) := \textbf{variables}(\varphi) \cup \textbf{variables}(\psi) \qquad \text{for any connective } \diamond \in \{\wedge, \vee, \Rightarrow, \Leftrightarrow, \oplus, \ldots\}$$

3.37 *(programming required)* Given a proposition φ and a truth assignment for each variable in φ, evaluate whether φ is true or false under this truth assignment.

3.38 *(programming required)* Given a proposition φ, compute the set of all truth assignments for the variables in φ that make φ true. (One good approach: use your solution to Exercise 3.36 to compute all the variables in φ, then build the full list of truth assignments for those variables, and then evaluate φ under each of these truth assignments using your solution to Exercise 3.37.)

3.3 *Propositional Logic: Some Extensions*

Against logic there is no armor like ignorance.

Laurence J. Peter (1919–1990)

With the definitions from Section 3.2 in hand, we turn to a few extensions: some special types of propositions, and some special ways of representing propositions.

3.3.1 *Tautology and Satisfiability*

Several important types of propositions are defined in terms of their truth tables: those that are always true (*tautologies*), sometimes true (*satisfiable* propositions), or never true (*unsatisfiable* propositions). We will explore each of these types in turn.

TAUTOLOGIES

We'll start by considering propositions that are always true:

Etymologically, the word *tautology* comes from *taut* "same" (*to* + *auto*) + *logy* "word." Another meaning for the word "tautology" (in real life, not just in logic) is the unnecessary repetition of an idea: "a canine dog." (The etymology and the secondary street meaning are not totally removed from the usage in logic.)

Definition 3.11 (Tautology)
A proposition is a tautology *if it is true under every truth assignment.*

One reason that tautologies are important is that we can use them to reason about logical statements, which can be particularly valuable when we're trying to prove a claim.

Examples 3.16 and 3.17 illustrate two important tautologies. The first of these tautologies is the proposition $p \vee \neg p$, which is called the *law of the excluded middle*: for any proposition p, either p is true or p is false; there is nothing "in between."

Example 3.16 (Law of the Excluded Middle)
Here is the truth table for the proposition $p \vee \neg p$:

p	$\neg p$	$p \vee \neg p$
T	F	T
F	T	T

The third column is filled with "T"s, so $p \vee \neg p$ is a tautology.

The second tautology is the proposition $p \wedge (p \Rightarrow q) \Rightarrow q$, called *modus ponens*: if we know both that (a) p is true and that (b) the truth of p implies the truth of q, then we can conclude that q is true.

Modus ponens rhymes with "goad us phone-ins"; literally, it means "the mood that affirms" in Latin.

Example 3.17 (Modus Ponens)
Here is the truth table for $p \wedge (p \Rightarrow q) \Rightarrow q$ (with a few extra columns of "scratch work," for each of the constituent pieces of the desired final proposition):

p	q	$p \Rightarrow q$	$p \wedge (p \Rightarrow q)$	$p \wedge (p \Rightarrow q) \Rightarrow q$
T	T	T	T	T
T	F	F	F	T
F	T	T	F	T
F	F	T	F	T

There are only "T"s in the last column of this truth table, which establishes that modus ponens is a tautology.

Figure 3.10 contains a number of tautologies that you may find interesting and occasionally helpful. (Exercises 3.60–3.72 ask you to build truth tables to verify that these propositions really are tautologies.)

One terminological note from Figure 3.10: *modus tollens* is the proposition $(p \Rightarrow q) \wedge \neg q \Rightarrow \neg p$, and it's the counterpoint to modus ponens: if we know both that (a) the truth of p implies the truth of q and that (b) q is not true, then we can conclude that p cannot be true either. (Modus tollens means "the mood that denies" in Latin.)

$(p \Rightarrow q) \wedge p \Rightarrow q$	Modus Ponens
$(p \Rightarrow q) \wedge \neg q \Rightarrow \neg p$	Modus Tollens
$p \vee \neg p$	Law of the Excluded Middle
$p \Leftrightarrow \neg\neg p$	Double Negation
$p \Leftrightarrow p$	
$p \Rightarrow p \vee q$	
$p \wedge q \Rightarrow p$	
$(p \vee q) \wedge \neg p \Rightarrow q$	
$(p \Rightarrow q) \wedge (\neg p \Rightarrow q) \Rightarrow q$	
$(p \Rightarrow q) \wedge (q \Rightarrow r) \Rightarrow (p \Rightarrow r)$	
$(p \Rightarrow q) \wedge (p \Rightarrow r) \Leftrightarrow p \Rightarrow q \wedge r$	
$(p \Rightarrow q) \vee (p \Rightarrow r) \Leftrightarrow p \Rightarrow q \vee r$	
$p \wedge (q \vee r) \Leftrightarrow (p \wedge q) \vee (p \wedge r)$	
$p \Rightarrow (q \Rightarrow r) \Leftrightarrow p \wedge q \Rightarrow r$	

Figure 3.10: Some tautologies.

Satisfiable and unsatisfiable propositions

We now turn to propositions that are sometimes true, and those propositions that are never true:

Definition 3.12 (Satisfiable propositions)
A proposition is satisfiable *if it is true under at least one truth assignment.*

If f is a truth assignment under which a proposition is true, then we say that the proposition is *satisfied by* f.

Definition 3.13 (Unsatisfiable propositions/contradictions)
A proposition is unsatisfiable *if it is not satisfiable. Such a proposition is also called a* contradiction.

Thus a proposition is satisfiable if it is true under at least one truth assignment, and unsatisfiable if it is false under every truth assignment. (And it's a tautology if it is true under every truth assignment.) Here are some examples:

Example 3.18 (Contradiction of $p \Leftrightarrow q$ and $p \oplus q$)
Here is the truth table for $(p \Leftrightarrow q) \wedge (p \oplus q)$:

p	q	$p \Leftrightarrow q$	$p \oplus q$	$(p \Leftrightarrow q) \wedge (p \oplus q)$
T	T	T	F	F
T	F	F	T	F
F	T	F	T	F
F	F	T	F	F

Because the column of the truth table corresponding to the given proposition has no "T"s in it, the proposition $(p \Leftrightarrow q) \wedge (p \oplus q)$ is unsatisfiable.

Though it might not have been immediately apparent when they were defined, the logical connectives $\oplus$ and $\Leftrightarrow$ demand precisely opposite things of their arguments: the proposition $p \oplus q$ is true when p and q have *different* truth values, while $p \Leftrightarrow q$ is true when p and q have the *same* truth values. Because p and q cannot simultaneously have the same and different truth values, the conjunction $(p \Leftrightarrow q) \wedge (p \oplus q)$ is a contradiction.

Example 3.19 (Demanding satisfaction)
Problem: Is the proposition $p \vee q \Rightarrow \neg p \wedge \neg q$ satisfiable?

Solution: We'll answer the question by building a truth table for the given proposition:

p	q	$p \vee q$	$\neg p$	$\neg q$	$\neg p \wedge \neg q$	$p \vee q \Rightarrow \neg p \wedge \neg q$
T	T	T	F	F	F	F
T	F	T	F	T	F	F
F	T	T	T	F	F	F
F	F	F	T	T	T	T

Because there is at least one "T" in the last column in the truth table, the proposition is satisfiable. Specifically, this proposition is satisfied by the truth assignment p = False, q = False. (Under this truth assignment, the hypothesis $p \vee q$ is false; because false implies anything, the entire implication is true.)

Let φ be *any* proposition. Then φ is a tautology exactly when $\neg\varphi$ is unsatisfiable: φ is a tautology when the truth table for φ is all "T"s, which happens exactly when the truth table for $\neg\varphi$ is all "F"s. And that's precisely the definition of $\neg\varphi$ being unsatisfiable!

As we said in Section 3.2.6, we occasionally denote generic propositions by lowercase Greek letters, particularly φ ("phi") or ψ ("psi").

Taking it further: While satisfiability seems like a pretty precise technical definition that wouldn't matter all that much, the *satisfiability problem*—given a proposition φ, determine whether φ is satisfiable—turns out to be at the heart of the biggest open question in computer science today. If you figure out how to solve the satisfiability problem efficiently (or prove that it's impossible to solve efficiently), then you'll be the most famous computer scientist of the century. See the discussion on p. 326.

3.3.2 *Logical Equivalence*

We'll now turn to a special type of *pairs* of propositions. When two propositions "mean the same thing" (that is, they are true under precisely the same circumstances), they are called *logically equivalent*:

Definition 3.14 (Logical equivalence)
Two propositions φ and ψ are logically equivalent, *written $\varphi \equiv \psi$, if they have exactly identical truth tables (in other words, their truth values are the same under every truth assignment).*

To state it differently: propositions φ and ψ are logically equivalent whenever $\varphi \Leftrightarrow \psi$ is a tautology. Here's a simple example of logical equivalence:

Example 3.20 ($\neg(p \wedge q) \equiv (p \wedge q) \Rightarrow \neg q$)
In Example 3.14, we found that $(p \wedge q) \Rightarrow \neg q$ is true except when p and q are both true. Thus $\neg(p \wedge q)$ is logically equivalent to $(p \wedge q) \Rightarrow \neg q$, as this truth table shows:

p	q	$(p \wedge q) \Rightarrow \neg q$	$\neg(p \wedge q)$
T	T	F	F
T	F	T	T
F	T	T	T
F	F	T	T

Writing tip: Now that we have a reasonable amount of experience in writing truth tables, we will permit ourselves to skip columns when they're both obvious and not central to the point of a particular example. When you're writing *anything*—whether as a food critic or a Shakespeare scholar or a computer scientist—you should always think about the intended audience, and how much detail is appropriate for them.

Implication, converse, contrapositive, inverse, and mutual implication

We'll now turn to an important question of logical equivalence that involves the proposition $p \Rightarrow q$ and three other implications derived from it:

Definition 3.15 (Converse, Contrapositive, and Inverse)
Consider an implication $p \Rightarrow q$. Then:

- *The* converse *of $p \Rightarrow q$ is the proposition $q \Rightarrow p$.*
- *The* contrapositive *of $p \Rightarrow q$ is the proposition $\neg q \Rightarrow \neg p$.*
- *The* inverse *of $p \Rightarrow q$ is the proposition $\neg p \Rightarrow \neg q$.*

These three new implications derived from the original implication $p \Rightarrow q$—particularly the converse and the contrapositive—will arise frequently. Let's compare the three new implications to the original in light of logical equivalence:

p	q	*proposition* $p \Rightarrow q$	*converse* $q \Rightarrow p$	*contrapositive* $\neg q \Rightarrow \neg p$	*inverse* $\neg p \Rightarrow \neg q$
T	T	T	T	T	T
T	F	F	T	F	T
F	T	T	F	T	F
F	F	T	T	T	T

Figure 3.11: The truth table for an implication and its contrapositive, converse, and inverse.

Example 3.21 (Implications, contrapositives, converses, inverses)
<u>*Problem:*</u> Consider the implication $p \Rightarrow q$. Which of the converse, contrapositive, and inverse of $p \Rightarrow q$ are logically equivalent to the original proposition $p \Rightarrow q$?

<u>*Solution:*</u> To answer this question, let's build the truth table; see Figure 3.11. Thus the proposition $p \Rightarrow q$ is logically equivalent to its contrapositive $\neg q \Rightarrow \neg p$, but *not* to its inverse or its converse.

Here's a real-world example to make these results more intuitive:

Thanks to Jeff Ondich for Example 3.22.

Example 3.22 (Contrapositives, converses, and inverses)
Consider the following (true!) proposition, of the form $p \Rightarrow q$:

$$\underbrace{\textit{If you were President of the U.S. in 2006}}_{p}\textit{, then } \underbrace{\textit{your name is George}}_{q}.$$

The contrapositive of this proposition is $\neg q \Rightarrow \neg p$, which is also true:

If your name isn't George, then you weren't President of the U.S. in 2006.

But the converse $q \Rightarrow p$ and the inverse $\neg p \Rightarrow \neg q$ are both blatantly false:

If your name is George, then you were President of the U.S. in 2006.
If you weren't President of the U.S. in 2006, then your name isn't George.

Consider, for example, George Clooney, Saint George, George Lucas, and Curious George—all named George, and none the President in 2006.

For emphasis, let's summarize the results from Example 3.21. Any implication $p \Rightarrow q$ is logically equivalent to its contrapositive $\neg q \Rightarrow \neg p$, but it is *not* logically equivalent to its converse $q \Rightarrow p$ or its inverse $\neg p \Rightarrow \neg q$. You might notice, though, that the inverse of $p \Rightarrow q$ is the contrapositive of the converse of $p \Rightarrow q$ (!), so the inverse and the converse *are* logically equivalent to each other.

Here's another example of the concepts of tautology and satisfiability, as they relate to implications and converses:

Example 3.23 (Mutual implication)

Problem: Consider the conjunction of the implication $p \Rightarrow q$ and its converse: in other words, consider $(p \Rightarrow q) \wedge (q \Rightarrow p)$. Is this proposition a tautology? Satisfiable? Unsatisfiable? Is there a simpler proposition to which it's logically equivalent?

Solution: We can answer this question with a truth table:

p	q	$p \Rightarrow q$	$q \Rightarrow p$	$(p \Rightarrow q) \wedge (q \Rightarrow p)$
T	T	T	T	T
T	F	F	T	F
F	T	T	F	F
F	F	T	T	T

Because there is a "T" in its column, $(p \Rightarrow q) \wedge (q \Rightarrow p)$ *is* satisfiable (and thus isn't a contradiction). But that column does contain an "F" as well, and therefore $(p \Rightarrow q) \wedge (q \Rightarrow p)$ is *not* a tautology.

Notice that the truth table for $(p \Rightarrow q) \wedge (q \Rightarrow p)$ is identical to the truth table for $p \Leftrightarrow q$. (See Figure 3.4.) Thus $p \Leftrightarrow q$ and $(p \Rightarrow q) \wedge (q \Rightarrow p)$ are logically equivalent. (And $\Leftrightarrow$ is called *mutual implication* for this reason: p and q imply each other.)

Some other logically equivalent statements

Figure 3.12 contains a large collection of logical equivalences. These equivalences may use some unfamiliar terminology, which we'll define here. Informally, an operator is *commutative* if the order of its arguments doesn't matter; an operator is *associative* if the way we parenthesize successive applications doesn't matter; and an operator is *idempotent* if applying it to the same argument twice gives that argument back. (In addition to these definitions, there are two other frequently discussed concepts: the *identity* and the *zero* of the operator; logical equivalences involving identities and zeros were left to you, in Exercises 3.13–3.22.) For each equivalence in Figure 3.12, it's worth

Latin: *idem* "same" + *potent* "strength."

Commutativity	$p \vee q \equiv q \vee p$
	$p \wedge q \equiv q \wedge p$
	$p \oplus q \equiv q \oplus p$
	$p \Leftrightarrow q \equiv q \Leftrightarrow p$
Associativity	$p \vee (q \vee r) \equiv (p \vee q) \vee r$
	$p \wedge (q \wedge r) \equiv (p \wedge q) \wedge r$
	$p \oplus (q \oplus r) \equiv (p \oplus q) \oplus r$
	$p \Leftrightarrow (q \Leftrightarrow r) \equiv (p \Leftrightarrow q) \Leftrightarrow r$
Idempotence	$p \vee p \equiv p$
	$p \wedge p \equiv p$

Distribution of $\wedge$ over $\vee$	$p \wedge (q \vee r) \equiv (p \wedge q) \vee (p \wedge r)$
Distribution of $\vee$ over $\wedge$	$p \vee (q \wedge r) \equiv (p \vee q) \wedge (p \vee r)$
Contrapositive	$p \Rightarrow q \equiv \neg q \Rightarrow \neg p$
	$p \Rightarrow q \equiv \neg p \vee q$
	$p \Rightarrow (q \Rightarrow r) \equiv p \wedge q \Rightarrow r$
	$p \Leftrightarrow q \equiv \neg p \Leftrightarrow \neg q$
Mutual Implication	$(p \Rightarrow q) \wedge (q \Rightarrow p) \equiv p \Leftrightarrow q$
De Morgan's Laws	$\neg(p \wedge q) \equiv \neg p \vee \neg q$
	$\neg(p \vee q) \equiv \neg p \wedge \neg q$

Figure 3.12: Some logically equivalent propositions.

taking a few minutes to think about why the two propositions are logically equivalent. See also Exercises 3.73–3.82.

De Morgan's Laws are named after Augustus De Morgan, a 19th-century British mathematician.

Taking it further: There are at least two ways in which the types of logical equivalences shown in Figure 3.12 play an important role in programming. (See the discussion on p. 327.) First, most modern languages have a feature called *short-circuit evaluation* of logical expressions—they evaluate conjunctions and disjunctions from left to right, and stop as soon as the truth value of the logical expression is known—and programmers can exploit this feature to make their code cleaner or more efficient. Second, in compiled languages, an optimizing compiler can make use of logical equivalences to simplify the machine code that ends up being executed.

3.3.3 *Representing Propositions: Circuits and Normal Forms*

Now that we've established the core concepts of propositional logic, we'll turn to some bigger and more applied questions. We'll spend the rest of this section exploring two specific ways of representing propositions: *circuits*, the wires and connections from which physical computers are built; and two *normal forms*, in which the structure of propositions is restricted in a particular way.

The approach we're taking with normal forms is a commonly used idea to make reasoning about some language L easier: we define a *subset* S of L, with two goals: (1) any statement in L is equivalent to some statement in S; and (2) S is "simple" in some way. Then we can consider any statement from the "full" language L, which we can then "translate" into a simple-but-equivalent statement of S. Defining this subset and its accompanying translation will make it easier to accomplish some task for *all* expressions in L, while still making it easy to write statements clearly.

Taking it further: The idea of translating all propositions into a particular form has a natural analogue in designing and implementing programming languages. For example, every **for** loop can be expressed as a **while** loop instead, but it would be very annoying to program in a language that doesn't have **for** loops. A nice compromise is to allow **for** loops, but behind the scenes to translate each **for** loop into a **while** loop. This compromise makes the language easier for the "user" programmer to use (**for** loops exist!) *and* also makes the job of the programmer of the compiler/interpreter easier (she can worry exclusively about implementing and optimizing **while** loops!).

In programming languages, this translation is captured by the notion of *syntactic sugar*. (The phrase is meant to suggest that the addition of **for** to the language is a bonus for the programmer—"sugar on top," maybe—that adds to the syntax of the language.) The programming language Scheme is perhaps the pinnacle of syntactic sugar; the core language is almost unbelievably simple. Here's one illustration: `(and x y)` (Scheme for "$x \wedge y$") is syntactic sugar for `(if x y #f)` (that's "if x then y else false"). So a Scheme programmer can use `and`, but there's no "real" `and` that has to be handled by the interpreter.

Circuits

We'll introduce the idea of circuits by using the proposition $(p \wedge \neg q) \vee (\neg p \wedge q)$ as an example. (Note, by the way, that this proposition is logically equivalent to $p \oplus q$.)

Observe that the stated proposition is a disjunction of two smaller propositions, $p \wedge \neg q$ and $\neg p \wedge q$. Similarly, $p \wedge \neg q$ is a conjunction of two even simpler propositions, namely p and $\neg q$. A representation of a proposition called a *tree* continues to break down every compound proposition embedded within it. (We'll talk about trees in detail in Chapter 11.) The tree for $(p \wedge \neg q) \vee (\neg p \wedge q)$ is shown in Figure 3.13. The tree-based view isn't much of a change from our usual notation $(p \wedge \neg q) \vee (\neg p \wedge q)$; all we've done is use the parentheses and order-of-operation rules to organize the logical connectives. But this representation is closely related to a very important way of viewing logical propositions: *circuits.*

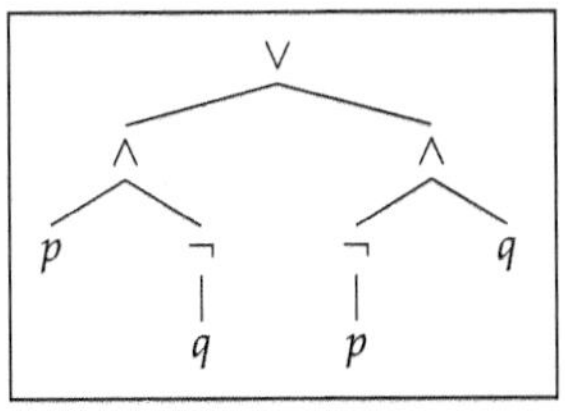

Figure 3.13: A tree-based view of $(p \wedge \neg q) \vee (\neg p \wedge q)$.

Figure 3.14 shows the same proposition redrawn as a collection of *wires* and *gates*. Wires carry a truth value from one physical location to another; gates are physical implementations of logical connectives. We can think of truth values "flowing in" as inputs to the left side of each gate, and a truth value "flowing out" as output from the right side of the gate. (The only substantive difference between Figures 3.13 and 3.14—aside from which way is up—is whether the two p inputs come from the same wire, and likewise whether the two q inputs do.)

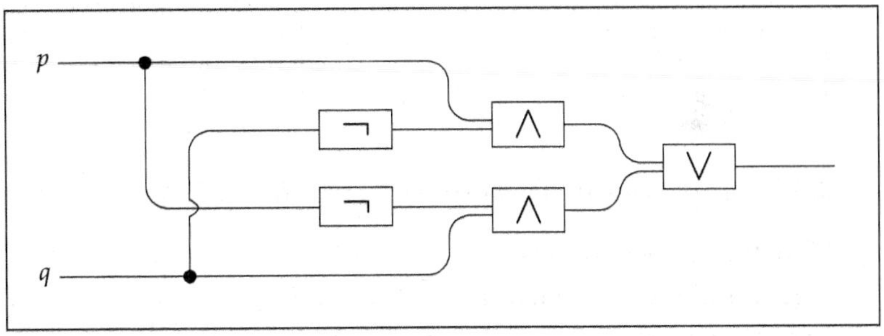

Figure 3.14: A circuit-based view.

Example 3.24 (Using and and not for or)

Problem: Build a circuit for $p \vee q$ using only $\wedge$ and $\neg$ gates.

Solution: We'll use one of De Morgan's Laws, which says that $p \vee q \equiv \neg(\neg p \wedge \neg q)$:

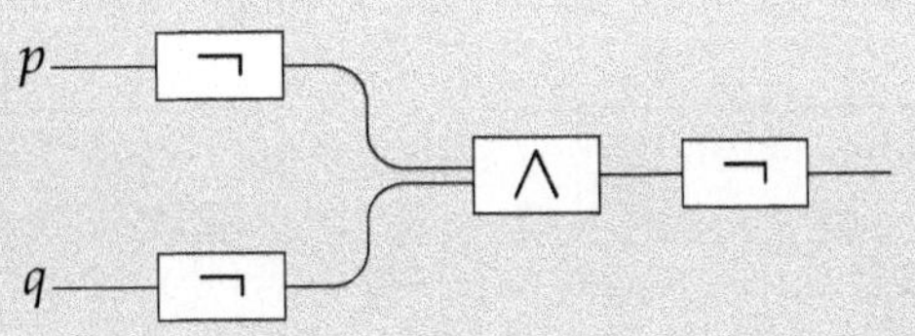

This basic idea—of replacing one logical connective by another one (or by multiple other ones)—is a crucial part of the construction of computers themselves; we'll return to this idea in Section 4.4.1.

Conjunctive and Disjunctive Normal Forms

In the rest of this section, we'll consider a way to simplify propositions: *conjunctive* and *disjunctive normal forms*, which constrain propositions to have a particular format. To define these restricted types of propositions, we need a basic definition: a *literal* is a Boolean variable (a.k.a. an atomic proposition) or the negation of a Boolean variable. (So p and $\neg p$ are both literals.)

Definition 3.16 (Conjunctive normal form)
A proposition is in conjunctive normal form (CNF) *if it is the conjunction of one or more* clauses, *where each* clause *is the disjunction of one or more literals.*

Definition 3.17 (Disjunctive normal form)
A proposition is in disjunctive normal form (DNF) *if it is the disjunction of one or more* clauses, *where each* clause *is the conjunction of one or more literals.*

Less formally, a proposition in conjunctive normal form is "the and of a bunch of ors," and a proposition in disjunctive normal form is "the or of a bunch of ands."

Taking it further: In computer architecture and digital electronics, people usually refer to a proposition in CNF as being a *product of sums*, and a proposition in DNF as being a *sum of products*. (There is a deep way of thinking about formal logic based on $\wedge$ as multiplication, $\vee$ as addition, 0 as False, and 1 as True; see Exercises 3.23–3.26.)

Here is a simple example of both CNF and DNF:

Example 3.25 (Simple propositions in CNF and DNF)
The proposition $(\neg p \vee q \vee r) \wedge (\neg q \vee \neg r) \wedge (r)$ is in conjunctive normal form. It has three clauses: $\neg p \vee q \vee r$ and $\neg q \vee \neg r$ and r.

The proposition $(\neg p \wedge q \wedge r) \vee (\neg q \wedge \neg r) \vee (r)$ is in disjunctive normal form, again with three clauses: $\neg p \wedge q \wedge r$ and $\neg q \wedge \neg r$ and r.

While conjunctive and disjunctive normal forms seem like heavy restrictions on the format of propositions, it turns out that *every* proposition is logically equivalent to a CNF proposition and to a DNF proposition:

Theorem 3.1 (All propositions are expressible in CNF)
For any proposition φ, there is a proposition φ_{cnf} over the same Boolean variables and in conjunctive normal form such that $\varphi \equiv \varphi_{cnf}$.

Theorem 3.2 (All propositions are expressible in DNF)
For any proposition φ, there is a proposition ψ_{dnf} over the same Boolean variables and in disjunctive normal form such that $\varphi \equiv \psi_{dnf}$.

These two theorems are perhaps the first results that we've encountered that are unexpected, or at least unintuitive. There's no particular reason for it to be clear that they're true—let alone how we might prove them. But we can, and we will: we'll prove both theorems in Section 4.4.1 and again in Section 5.4.3, after we've introduced some relevant proof techniques. But, for now, here are a few examples of translating propositions into DNF/CNF.

Problem-solving tip: A good strategy when you're trying to prove a not-at-all-obvious claim is to test out some small examples, and then try to start to figure a general pattern.

Example 3.26 (Translating basic connectives into DNF)

Problem: Give propositions in disjunctive normal form that are logically equivalent to each of the following:

1. $p \vee q$
2. $p \wedge q$
3. $p \Rightarrow q$
4. $p \Leftrightarrow q$

Solution: 1 & 2. These questions are boring: both propositions are already in DNF, with 2 clauses (p and q) and 1 clause ($p \wedge q$), respectively.

3. Figure 3.12 tells us that $p \Rightarrow q \equiv \neg p \vee q$, and $\neg p \vee q$ is in DNF.

4. The proposition $p \Leftrightarrow q$ is true when p and q are either both true or both false, and false otherwise. So we can rewrite $p \Leftrightarrow q$ as $(p \wedge q) \vee (\neg p \wedge \neg q)$. We can check that we've gotten this proposition right with a truth table:

p	q	$p \wedge q$	$\neg p \wedge \neg q$	$(p \wedge q) \vee (\neg p \wedge \neg q)$	$p \Leftrightarrow q$
T	T	T	F	T	T
T	F	F	F	F	F
F	T	F	F	F	F
F	F	F	T	T	T

And here's the task of translating basic logical connectives into CNF:

Example 3.27 (Translating basic connectives into CNF)

Problem: Give propositions in conjunctive normal form that are logically equivalent to each of the following:

1. $p \Rightarrow q$
2. $p \Leftrightarrow q$
3. $p \oplus q$

(Note that, as with DNF, both $p \vee q$ and $p \wedge q$ are already in CNF.)

Solution: 1. As above, we know that $p \Rightarrow q \equiv \neg p \vee q$, and $\neg p \vee q$ is also in CNF.

2. We can rewrite $p \Leftrightarrow q$ as follows:

$$\begin{aligned} p \Leftrightarrow q &\equiv (p \Rightarrow q) \wedge (q \Rightarrow p) && \textit{mutual implication (Example 3.23)} \\ &\equiv (\neg p \vee q) \wedge (\neg q \vee p) && x \Rightarrow y \equiv \neg x \vee y \textit{ (Figure 3.12), used twice} \end{aligned}$$

The proposition $(\neg p \vee q) \wedge (\neg q \vee p)$ is in CNF.

3. Because $p \oplus q$ is true as long as one of $\{p, q\}$ is true and one of $\{p, q\}$ is false, it's easy to verify via truth table that $p \oplus q \equiv (p \vee q) \wedge (\neg p \vee \neg q)$, which is in CNF.

We've only given some examples of converting a (simple) proposition into a new proposition, logically equivalent to the original, that's in either CNF or DNF. We will figure out how to generalize this technique to *any* proposition in Section 4.4.1.

Computer Science Connections

Computational Complexity, Satisfiability, and $1,000,000

Complexity theory is the subfield of computer science devoted to understanding the resources—time and memory, usually—necessary to solve particular problems. It's the subject of a great deal of fascinating current research in theoretical computer science.[3] Here is a central problem of complexity theory, the *satisfiability problem:*

Given: A Boolean formula φ over variables $p_1, p_2, \ldots, p_n$.
Output: Is φ satisfiable?

The satisfiability problem is pretty simple to solve. In fact, we've implicitly described an algorithm for this problem already:

- construct the truth table for the n-variable proposition φ; and
- check to see whether there are any "T"s in φ's column of the table.

But this algorithm is not very fast, because the truth table for φ has lots and lots of rows—2^n rows, to be precise. (We've already seen this for $n = 1$, for negation, and $n = 2$, for all the binary connectives, with $2^1 = 2$ and $2^2 = 4$ rows each; in Chapter 9, we'll address this counting issue formally.) And then even a moderate value of n means that this algorithm will not terminate in your lifetime; 2^{300} exceeds the number of particles in the known universe.

So, it's clear that there is an algorithm that solves the SAT problem. What's not clear is whether there is a substantially more efficient algorithm to solve the SAT problem. It's so unclear, in fact, that nobody knows the answer, and this question is one of the biggest open problems in computer science and mathematics today. (Arguably, it's *the* biggest.) The Clay Mathematics Institute will even give a $1,000,000 prize to anyone who solves it.

Why is this problem so important? The reason is that, in a precise technical sense, SAT is *just as hard* as a slew of other problems that have a plethora of unspeakably useful applications: the traveling salesman problem, protein folding, optimally packing the trunk of a car with suitcases. This slew is a class of computational problems known as NP ("nondeterministic polynomial time"), for which it is easy to "verify" correct answers. In the context of SAT, that means that whenever you've got a satisfiable proposition φ, it's very easy for you to (efficiently) convince me that φ is satisfiable. Here's how: you'll simply tell me a truth assignment under which φ evaluates to true. And I can make sure that you didn't try to fool me by plugging and chugging: I substitute your truth assignment in for every variable, and then I make sure that the final truth value of φ is indeed True.

One of the most important results in theoretical computer science in the 20th century—that's saying something for a field that was founded in the 20th century!—is the *Cook–Levin Theorem:*[4] *if one can solve SAT efficiently, then one can solve* any *problem in* NP *efficiently*. The major open question is what's known as the *P-versus-NP question*. A problem that's in P is easy to solve from scratch. A problem that's in NP is easy to verify (in the way described above). So the question is: does P = NP? Is verifying an answer to a problem no easier than solving the problem from scratch? (It seems intuitively "clear" that the answer is no—but nobody has been able to prove it!)

You can read more about complexity theory in general, and the P-versus-NP question addressed here in particular, in most books on algorithms or the theory of computing. Some excellent places to read more are:

[3] Thomas H. Cormen, Charles E. Leiserson, Ronald L. Rivest, and Clifford Stein. *Introduction to Algorithms*. MIT Press, 3rd edition, 2009; Jon Kleinberg and Éva Tardos. *Algorithm Design*. Addison–Wesley, 2006; and Michael Sipser. *Introduction to the Theory of Computation*. Course Technology, 3rd edition, 2012.

[4] Stephen Cook. The complexity of theorem proving procedures. In *Proceedings of the Third Annual ACM Symposium on Theory of Computing*, pages 151–158, 1971; and Leonid Levin. Universal search problems. *Problems of Information Transmission*, 9(3):265–266, 1973. In Russian.

Computer Science Connections

Short-Circuit Evaluation, Optimization, and Modern Compilers

The logical equivalences in Figure 3.12 may seem far removed from "real" programming, but logical equivalences are actually central in modern programming. Here are two ways in which they play an important role:

Short-circuit evaluation: In most modern programming languages, a logical expression involving **and**s and **or**s will only be evaluated until the truth value of the expression can be determined. For an example in Java, see Figure 3.15. Like most modern languages, Java evaluates an $\wedge$ expression from left to right and stops as soon as it finds a false conjunct. Similarly, Java evaluates an $\vee$ expression from left to right and stops as soon as it finds a true disjunct, because True $\vee$ *anything* $\equiv$ True. This style of evaluation is called *short-circuit evaluation.*

Two slick ways in which programmers can take advantage of short-circuit evaluation are shown in Figure 3.16.

- Lines 1–4 use short-circuit evaluation to avoid deeply nested **if** statements to handle exceptional cases. When $x = 0$, evaluating the second disjunct would cause a divide-by-zero error—but the second disjunct isn't evaluated when $x = 0$ because the first disjunct was true!
- Lines 6–9 use short-circuit evaluation to make code faster. If the second conjunct typically takes much longer to evaluate (or if it is much more frequently true) than the first conjunct, then careful ordering of conjuncts avoids a long and usually fruitless computation.

Compile-time optimization: For a program written in a compiled language like C, the source code is translated into machine-readable form by the *compiler*. But this translation is not verbatim; instead, the compiler streamlines your code (when it can!) to make it run faster.

One of the simplest types of compiler optimizations is *constant folding*: if some of the values in an arithmetic or logical expression are constants—known to the compiler at "compile time," and thus unchanged at "run time"—then the compiler can "fold" those constants together. Using the rules of logical or arithmetic equivalence broadens the types of code that can be folded in this way. For example, in C, when you write an assignment statement like `y = x + 2 + 3`, most compilers will translate it into `y = x + 5`. But what about `z = 7 * x * 8`? A modern compiler *will* optimize it into `z = x * 56`, using the commutativity of multiplication. Because the compiler can reorder the multiplicands without affecting the value, and this reordering allows the 7 and 8 to be folded into 56, the compiler does the reordering and the folding.

An example using logical equivalences is shown in Figure 3.17. Because $p \vee \neg p$ is a tautology—the law of the excluded middle—*no matter what the value of p*, the "then" clause is executed, not the "else" clause. Thus the compiler doesn't even have to waste time checking whether p is true or false, and this optimization can be applied.

```
if (2 > 3 && x + y < 9) {
  ...
} else {
  ...
}
```

Figure 3.15: A snippet of Java code. In Java, && denotes $\wedge$ and || denotes $\vee$. The second conjunct of the `if` condition will actually never be evaluated, because `2 > 3` is false, and False $\wedge$ *anything* $\equiv$ False.

```
1 if (x == 0
2     || (x-1) / x > 0.5) {
3   ...
4 }
5
6 if (simpleOrOftenFalse(x)
7     && complexOrOftenTrue(x)) {
8   ...
9 }
```

Figure 3.16: Two handy ways to rely on short-circuit evaluation.

```
if (p || !p) {  /* "p or not p" */
    x = 51;
} else {
    x = 63;
}
```

```
x = 51;
```

Figure 3.17: Two snippets of C code. When this code is compiled on a modern optimizing compiler (gcc 4.3.4, with optimization turned on), the machine code that is produced is *exactly* identical for both snippets.

3.3.4 Exercises

The operators $\wedge$ and $\vee$ are idempotent (see Figure 3.12)—that is, $p \wedge p \equiv p \vee p \equiv p$. But $\Rightarrow$, $\oplus$, and $\Leftrightarrow$ are not idempotent. Simplify—that is, give as-simple-as-possible propositions that are logically equivalent to—the following:

3.39 $p \Rightarrow p$

3.40 $p \oplus p$

3.41 $p \Leftrightarrow p$

Consider the proposition $p \Rightarrow \neg p \Rightarrow p \Rightarrow q$. Add parentheses to this proposition so that the resulting proposition ...

3.42 ... is logically equivalent to True (that is, the result is a tautology).

3.43 ... is logically equivalent to q.

3.44 Give as simple as possible a proposition logically equivalent to the (unparenthesized) original.

Unlike the binary connectives $\{\wedge, \vee, \oplus, \Leftrightarrow\}$, implication is not associative. In other words, $p \Rightarrow (q \Rightarrow r)$ and $(p \Rightarrow q) \Rightarrow r$ are not *logically equivalent. The next few exercises explore the non-associativity of $\Rightarrow$.*

3.45 Prove that implication is not associative by giving a truth assignment in which $p \Rightarrow (q \Rightarrow r)$ and $(p \Rightarrow q) \Rightarrow r$ have different truth values.

3.46 Consider the propositions $p \Rightarrow (q \Rightarrow q)$ and $(p \Rightarrow q) \Rightarrow q$. One of these is a tautology; one of them is not. Which is which? Prove your answer.

3.47 Consider the propositions $p \Rightarrow (p \Rightarrow q)$ and $(p \Rightarrow p) \Rightarrow q$. Is either one a tautology? Satisfiable? Unsatisfiable? What is the simplest proposition to which each is logically equivalent?

On an exam, I once asked students to write a proposition logically equivalent to $p \oplus q$ using only the logical connectives $\Rightarrow$, $\neg$, and $\wedge$. Here are some of the students' answers. Which ones are right?

3.48 $\neg(p \wedge q) \Rightarrow (\neg p \wedge \neg q)$

3.49 $(p \Rightarrow \neg q) \wedge (q \Rightarrow \neg p)$

3.50 $(\neg p \Rightarrow q) \wedge \neg(p \wedge q)$

3.51 $\neg\left[(p \wedge \neg q \Rightarrow \neg p \wedge q) \wedge (\neg p \wedge q \Rightarrow p \wedge \neg q)\right]$

3.52 Write a proposition logically equivalent to $p \oplus q$ using only the logical connectives $\Rightarrow$, $\neg$, and $\vee$.

The following code uses nested conditionals, or compound propositions as conditions. Simplify each as much as possible. (For example, if $p \Rightarrow q$, it's a waste of time to test whether q holds in a block where p is known to be true.)

3.53

```
if (x > 20
      or (x <= 20 and y < 0))
then foo(x,y)
else bar(x,y)
```

3.54

```
if (y >= 0
      or y <= x
      or (x - y) * y >= 0)
  then foo(x,y)
  else bar(x,y)
```

3.55

```
if (x % 12 == 0):
  then if not (x % 4 == 0):
           then foo(x)
           else bar(x)
  else if (x == 17):
           then baz(x)
           else quz(x)
```

(Note that `x % k == 0` is true when $x \bmod k = 0$, also known as when $k \mid x$.)

Simplify the following propositions as much as possible.

3.56 $(\neg p \Rightarrow q) \wedge (q \wedge p \Rightarrow \neg p)$

3.57 $(p \Rightarrow \neg p) \Rightarrow ((q \Rightarrow (p \Rightarrow p)) \Rightarrow p)$

3.58 $(p \Rightarrow p) \Rightarrow (\neg p \Rightarrow \neg p) \wedge q$

3.59 Is the following claim true or false? Prove your answer.

Claim: Every proposition over the single variable p is either logically equivalent to p or it is logically equivalent to $\neg p$.

Show using truth tables that these propositions from Figure 3.10 are tautologies:

3.60 $(p \Rightarrow q) \wedge \neg q \Rightarrow \neg p$ (Modus Tollens)

3.61 $p \Rightarrow p \vee q$

3.62 $p \wedge q \Rightarrow p$

3.63 $(p \vee q) \wedge \neg p \Rightarrow q$

3.64 $(p \Rightarrow q) \wedge (\neg p \Rightarrow q) \Rightarrow q$

3.65 $(p \Rightarrow q) \wedge (q \Rightarrow r) \Rightarrow (p \Rightarrow r)$

3.66 $(p \Rightarrow q) \wedge (p \Rightarrow r) \Leftrightarrow p \Rightarrow q \wedge r$

3.67 $(p \Rightarrow q) \vee (p \Rightarrow r) \Leftrightarrow p \Rightarrow q \vee r$

3.68 $p \wedge (q \vee r) \Leftrightarrow (p \wedge q) \vee (p \wedge r)$

3.69 $p \Rightarrow (q \Rightarrow r) \Leftrightarrow p \wedge q \Rightarrow r$

Show that the following propositions are tautologies:

3.70 $p \vee (p \wedge q) \Leftrightarrow p$

3.71 $p \wedge (p \vee q) \Leftrightarrow p$

3.72 $p \oplus q \Rightarrow p \vee q$

Prove De Morgan's Laws:

3.73 $\neg(p \wedge q) \equiv \neg p \vee \neg q$

3.74 $\neg(p \vee q) \equiv \neg p \wedge \neg q$

Show the following logical equivalences regarding associativity using truth tables:

3.75 $p \vee (q \vee r) \equiv (p \vee q) \vee r$

3.76 $p \wedge (q \wedge r) \equiv (p \wedge q) \wedge r$

3.77 $p \oplus (q \oplus r) \equiv (p \oplus q) \oplus r$

3.78 $p \Leftrightarrow (q \Leftrightarrow r) \equiv (p \Leftrightarrow q) \Leftrightarrow r$

Show using truth tables that the following logical equivalences hold:

3.79 $p \Rightarrow q \equiv \neg p \vee q$

3.80 $p \Rightarrow (q \Rightarrow r) \equiv p \wedge q \Rightarrow r$

3.81 $p \Leftrightarrow q \equiv \neg p \Leftrightarrow \neg q$

3.82 $\neg(p \Rightarrow q) \equiv p \wedge \neg q$

3.83 On p. 327, we discussed the use of tautologies in optimizing compilers. In particular, these compilers will perform the following optimization, transforming the first block of code into the second:

```
if (p || !p) {  /* "p or not p" */
   x = 51;
} else {
   x = 63;
}
```

```
x = 51;
```

The compiler performs this transformation because $p \vee \neg p$ is a tautology—no matter what the truth value of p, the proposition $p \vee \neg p$ is true. But there *are* situations in which this code translation actually changes the behavior of the program, *if* `p` *can be an arbitrary expression* (rather than just a Boolean variable)! Describe such a situation. *(Hint: why do (some) people watch auto racing?)*

The unknown circuit in Figure 3.18 takes three inputs $\{p, q, r\}$, and either turns on a light bulb (output of the circuit = true) or leaves it off (output = false). For each of the following, draw a circuit—using at most three $\wedge$, $\vee$, and $\neg$ gates—that is consistent with the listed behavior. The light's status is unknown for unlisted inputs. (If multiple circuits are consistent with the given behavior, draw any one them.)

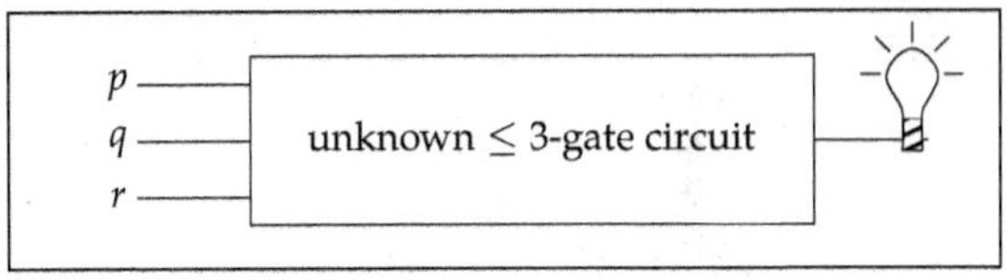

Figure 3.18: A circuit with at most 3 gates.

3.84 The light is on when the true inputs are $\{q\}$ or $\{r\}$. The light is off when the true inputs are $\{p\}$ or $\{p, q\}$ or $\{p, q, r\}$.

3.85 The light is on when the true inputs are $\{p, q\}$ or $\{p, r\}$. The light is off when the true inputs are $\{p\}$ or $\{q\}$ or $\{r\}$.

3.86 The light is off when the true inputs are $\{p\}$ or $\{q\}$ or $\{r\}$ or $\{p, q, r\}$.

3.87 The light is off when the true inputs are $\{p, q\}$ or $\{p, r\}$ or $\{q, r\}$ or $\{p, q, r\}$.

3.88 Consider a simplified class of circuits like those from Exercises 3.84–3.87: there are *two* inputs $\{p, q\}$ and at most *two* gates, each of which is $\wedge$, $\vee$, or $\neg$. There are a total of $2^4 = 16$ distinct propositions over inputs $\{p, q\}$: four different input configurations, each of which can turn the light on or leave it off. Which, if any, of these 16 propositions *cannot* be expressed using up to two $\{\wedge, \vee, \neg\}$ gates?

3.89 *(programming required)* Consider the class of circuits from Exercises 3.84–3.87: inputs $\{p, q, r\}$, and at most three gates chosen from $\{\wedge, \vee, \neg\}$. There are a total of $2^8 = 256$ distinct propositions over inputs $\{p, q, r\}$: eight different input configurations, each of which can turn the light on or leave it off. Write a program to determine how many of these 256 propositions can be represented by a circuit of this type. (If you design it well, your program will let you check your answers to Exercises 3.84–3.88.)

3.90 Consider a set $S = \{p, q, r, s, t\}$ of Boolean variables. Let $\varphi = p \oplus q \oplus r \oplus s \oplus t$. Describe *briefly* the conditions under which φ is true. Use English and, if appropriate, standard (nonlogical) mathematical notation. *(Hint: look at the symbol $\oplus$ itself. What's $p + q + r + s + t$, treating true as 1 and false as 0 as in Exercises 3.23–3.26?)*

3.91 *Dithering* is a technique for converting grayscale images to black-and-white images (for printed media like newspapers). The classic dithering algorithm proceeds as follows. For every pixel in the image, going from top to bottom ("north to south"), and from left to right ("west to east"):

- "Round" the current pixel to black or white. (If it's closer to black, make it black; if it's closer to white, make it white.)
- This alteration to the current pixel has created "rounding error" x (in other words, we have added $x > 0$ "whiteness units" by making it white, or $x < 0$ "whiteness units" by making it black). We compensate for this adding a total of $-x$ "whiteness units," distributed among the neighboring pixels to the "east" (add $-7x/16$ to the eastern neighboring pixel) "southwest" ($-3x/16$), "south" ($-5x/16$) and "southeast" ($-x/16$). If any of these neighboring pixels don't exist (because the current pixel is on the border of the image), simply ignore the corresponding fraction of $-x$ (and don't add it anywhere).

I assigned a dithering exercise in an introductory CS class, and I got, more or less, the code in Figure 3.19 from one student. This code is correct, but it is very repetitious. Reorganize this code so that it's not so repetitive. In particular, rewrite lines 7–63 ensuring that each "distribute the error" line (9, 11, 12, and 13) appears *only once* if your solution.

```
1  for y = 1 ... height:
2    for x = 1 ... width:
3      if P[x,y] is more white than black:
4        error = "white" - P[x,y]
5        P[x,y] = "white"
6
7        if x > 1:
8          if x < width and not (y < height):
9            add 7/16 · error to P[x+1,y]      (E)
10         else if x < width and y < height:
11           add 5/16 · error to P[x,y+1]      (S)
12           add 3/16 · error to P[x+1,y+1]    (SE)
13           add 1/16 · error to P[x-1,y+1]    (SW)
14           add 7/16 · error to P[x+1,y]      (E)
15         else if y < height
16                and not (x < width):
17           add 5/16 · error to P[x,y+1]      (S)
18           add 1/16 · error to P[x-1,y+1]    (SW)
19         else:
20           do nothing
21       else:
22         if x < width and not (y < height):
23           add 7/16 · error to P[x+1,y]      (E)
24         else if x < width and y < height:
25           add 5/16 · error to P[x,y+1]      (S)
26           add 3/16 · error to P[x+1,y+1]    (SE)
27           add 7/16 · error to P[x+1,y]      (E)
28         else if y < height
29                and not (x < width):
30           add 5/16 · error to P[x,y+1]      (S)
31         else:
32           do nothing
33
34     else:  # P[x,y] is closer to "black"
35       error = "black" - P[x,y]
36       P[x,y] = "black"
37
38       if x > 1:
39         if x < width and not (y < height):
40           add 7/16 · error to P[x+1,y]      (E)
41         else if x < width and y < height:
42           add 5/16 · error to P[x,y+1]      (S)
43           add 3/16 · error to P[x+1,y+1]    (SE)
44           add 1/16 · error to P[x-1,y+1]    (SW)
45           add 7/16 · error to P[x+1,y]      (E)
46         else if y < height
47                and not (x < width):
48           add 5/16 · error to P[x,y+1]      (S)
49           add 1/16 · error to P[x-1,y+1]    (SW)
50         else:
51           do nothing
52       else:
53         if x < width and not (y < height):
54           add 7/16 · error to P[x+1,y]      (E)
55         else if x < width and y < height:
56           add 5/16 · error to P[x,y+1]      (S)
57           add 3/16 · error to P[x+1,y+1]    (SE)
58           add 7/16 · error to P[x+1,y]      (E)
59         else if y < height
60                and not (x < width):
61           add 5/16 · error to P[x,y+1]      (S)
62         else:
63           do nothing
```

Figure 3.19: Some dithering code.

Recall Definition 3.16: a proposition φ is in conjunctive normal form (CNF) if φ is the conjunction of one or more clauses, where each clause is the disjunction of one or more literals, and where a literal is an atomic proposition or its negation. Further, recall Definition 3.17: φ is in disjunctive normal form (DNF) if φ is the disjunction of one or more clauses, where each clause is the conjunction of one or more literals.

Give a proposition in disjunctive normal form that's logically equivalent to ...

3.92 $\neg(p \wedge q) \Rightarrow r$

3.93 $p \wedge (q \vee r) \Rightarrow (q \wedge r)$

3.94 $p \vee \neg(q \Leftrightarrow p \wedge r)$

3.95 $p \oplus (\neg p \Rightarrow (q \Rightarrow r) \wedge \neg r)$

Give a proposition in conjunctive normal form that's logically equivalent to ...

3.96 $\neg(p \wedge q) \Rightarrow r$

3.97 $p \wedge (q \Rightarrow (r \Rightarrow q \oplus r))$

3.98 $(p \Rightarrow q) \Rightarrow (q \Rightarrow r \wedge p)$

3.99 $p \Leftrightarrow (q \vee r \vee \neg p)$

A CNF proposition φ is in 3CNF *if each clause contains* exactly three *distinct literals. (Note that p and $\neg p$ are distinct literals.) In terms of the number of clauses, what's the smallest 3CNF formula ...*

3.100 ... that's a tautology?

3.101 ... that's not satisfiable?

Consider the set of 3CNF propositions over the variables $\{p, q, r\}$ for which no clause appears more than once. (Exercises 3.102–3.104 turn out to be boring without the restriction of no repeated clauses; we could repeat the same clause as many times as we please: $(p \vee q \vee r) \wedge (p \vee q \vee r) \wedge (p \vee q \vee r) \cdots$.) Two clauses that contain precisely the same literals (in any order) do not count as distinct. (But recall that a single clause can *contain a variable in both negated and unnegated form.) In terms of the number of clauses, what's the largest 3-variable distinct-clause 3CNF proposition ...*

3.102 ... at all (with no further restrictions)?

3.103 ... that's a tautology?

3.104 ... that's satisfiable?

A proposition φ is in 3DNF *if it is the disjunction of one or more clauses, each of which is the conjunction of exactly three distinct literals. In terms of the number of clauses, what's the smallest 3DNF formula ...*

3.105 ... that's a tautology?

3.106 ... that's not satisfiable?

3.4 *An Introduction to Predicate Logic*

> But the fact that some geniuses were laughed at does not imply that all who are laughed at are geniuses. They laughed at Columbus, they laughed at Fulton, they laughed at the Wright brothers. But they also laughed at Bozo the Clown.
>
> Carl Sagan (1934–1996)
> *Broca's Brain: Reflections on the Romance of Science* (1979)

Propositional logic, which we have been discussing thus far, gives us formal notation to encode Boolean expressions. But these expressions are relatively simple, a sort of "unstructured programming" style of logic. *Predicate logic* is a more general type of logic that allows us to write function-like logical expressions called *predicates*, and to express a broader range of notions than in propositional logic.

3.4.1 *Predicates*

Informally, a predicate is a property that a particular entity might or might not have; for example, *being a vowel* is a property that some letters do have (A, E, ...) and some letters do not have (B, C, ...). A predicate isn't the kind of thing that's true or false, so predicates are different from propositions; rather, a predicate is like a "proposition with blanks" waiting to be filled in. For example:

Example 3.28 (Some predicates)
- "The integer ___ is prime."
- "The string ___ is a palindrome."
- "The person ___ costarred in a movie with Kevin Bacon."
- "The string ___ is alphabetically after the string ___."
- "The integer ___ evenly divides the integer ___."

Once the blanks of a predicate are filled in, the resulting expression is a proposition. Here are some examples of propositions—some true, some false—derived from the predicates in Example 3.28:

Example 3.29 (Some propositions derived from Example 3.28)
- "The integer 57 is prime."
- "The string TENET is a palindrome."
- "The person Sean Connery costarred in a movie with Kevin Bacon."
- "The string PYTHON is alphabetically after the string PYTHAGOREAN."
- "The integer 17 evenly divides the integer 42."

We can now give a formal definition of predicates:

Definition 3.18 (Predicate)
A predicate P is a Boolean-valued function—that is, P is a function $P : U \to \{\text{True}, \text{False}\}$ for a set U. The set U is called the universe *or the* domain of discourse, *and we say that P is a predicate* over U.

When the universe U is clear from context, we will allow ourselves to be sloppy with notation by leaving U implicit.

Although we didn't use the name at the time, we've already encountered predicates, in Chapter 2. Definition 2.18 introduced the notation $\{x \in U : P(x)\}$ to denote the set of those objects $x \in U$ for which P is true. The set abstraction notation "selects" the elements of U for which the predicate P is true.

Example 3.30 (Some example predicates)
Here are a few more sample predicates based on arithmetic:

1. *isPrime*(n): the positive integer n is a prime number.
2. *isPowerOf*(n, k): the integer n is an exact power of k: $n = k^i$ for some $i \in \mathbb{Z}^{\geq 0}$.
3. *onlyPowersOfTwo*(S): every element of the set S is a power of two.
4. $Q(n, a, b)$: positive integer n satisfies $n = a + b$, and integers a and b are both prime.
5. *sumOfTwoPrimes*(n): positive integer n is equal to the sum of two prime numbers.

(To reiterate Definition 3.18, the *isPrime* predicate, for example, is a function *isPrime* : $\mathbb{Z}^{>0} \to \{\text{True}, \text{False}\}$.)

Deriving propositions from predicates

Again, by plugging particular values into the predicates from Example 3.30, we get propositions, each of which has a truth value:

Example 3.31 (Propositions derived from predicates)
Using the predicates in Example 3.30, let's figure out the truth values of the propositions *isPrime*(261), *isPrime*(262), $Q(8, 3, 5)$, and $Q(9, 3, 6)$. For each, we'll simply plug the given arguments into the definition of the predicate and figure out the truth value of the resulting proposition.

- A little arithmetic shows that $261 = 3 \cdot 87$; thus *isPrime*(261) = False.
- Similarly, we have $262 = 2 \cdot 131$, so *isPrime*(262) = False.
- To compute the truth value of $Q(8, 3, 5)$, we simply plug $n = 8$, $a = 3$, and $b = 5$ into the definition of $Q(n, a, b)$. The proposition $Q(8, 3, 5)$ requires that *the positive integer 8 satisfies $8 = 3 + 5$, and the integers 3 and 5 are both prime.* All of the requirements are met, so $Q(8, 3, 5) = \text{True}$.
- On the other hand, $Q(9, 3, 6) = \text{False}$ because $Q(9, 3, 6)$ requires that $9 = 3 + 6$, *and* that the integers 3 and 6 are both prime. But 6 isn't prime.

Just like the propositional logical connectives, each predicate takes a fixed number of arguments. So a predicate might be *unary* (taking one argument, like the predicate *isPrime*); or *binary* (taking two arguments, like *isPowerOf*); or *ternary* (taking three arguments, like Q from Example 3.30); and so forth. Here are a few more examples:

Example 3.32 (More propositions derived from predicates)
Problem: Using the predicates in Example 3.30, find the truth values of these propositions:

1. *sumOfTwoPrimes*(17) and *sumOfTwoPrimes*(34)
2. *isPowerOf*(16, 2) and *isPowerOf*(2, 16)
3. *onlyPowersOfTwo*($\{1, 2, 8, 128\}$)

Solution: As before, we just plug the given arguments into the definition:

1. *sumOfTwoPrimes*(17) = False: the only way to get an odd number n by adding two prime numbers is for one of those prime numbers to be 2—but $17 - 2 = 15$, and 15 isn't prime. But *sumOfTwoPrimes*(34) = True, because $34 = 17 + 17$, and 17 is prime. (And the other 17 is prime, too.)
2. *isPowerOf*(16, 2) = True because $2^4 = 16$ (and the exponent 4 is an integer), but *isPowerOf*(2, 16) = False because $16^{1/4} = 2$ (and $1/4$ is not an integer).
3. *onlyPowersOfTwo*($\{1, 2, 8, 128\}$) = True because every element of $\{1, 2, 8, 128\}$ is a power of two: $\{1, 2, 8, 128\} = \{2^0, 2^1, 2^3, 2^7\}$.

These brief examples may already be enough to begin to give you a sense of the power of logical abstraction that predicates grant us: we can now consider the same logical "condition" applied to two different "arguments." In a sense, propositional logic is like programming without functions; letting ourselves use predicates allows us to write two related propositions using related notation, and to reason simultaneously about multiple propositions—just like writing a function in Java allows you to think simultaneously about the same function applied to different arguments.

Taking it further: Predicates give a convenient way of representing the state of play of multiplayer games like Tic-Tac-Toe, checkers, and chess. The basic idea is to define a predicate $P(B)$ that expresses "Player 1 will win from board position B if both players play optimally." For more on this idea, and on the application of logic (both predicate and propositional) to playing these kinds of games, see the discussion on p. 344.

3.4.2 *Quantifiers*

We've seen that we can form a proposition from a predicate by applying that predicate to a particular argument. But we can also form a proposition from a predicate using *quantifiers*, which allow us to formalize statements like *every Java program contains at least four* **for** *loops* (false!) or *there is a proposition that cannot be expressed using only the connectives* $\wedge$ *and* $\vee$ (true! See Exercise 4.71).

These types of statements are expressed by the two standard quantifiers, the *universal* ("every") and *existential* ("some") quantifiers (see Figure 3.20):

$\forall x \in S : P(x)$	"for all" (universal quantifier)	true if $P(x)$ is true for *every* $x \in S$.
$\exists x \in S : P(x)$	"there exists" (existential quantifier)	true if $P(x)$ is true for *at least one* $x \in S$.

Figure 3.20: Summary of notation for predicate logic.

Definition 3.19 (Universal quantifier ("for all"): $\forall$)
Let P be a predicate over the universe S. The proposition $\forall x \in S : P(x)$ ("for all x in S, P(x)") is true if, for every possible $x \in S$, P(x) is true.

The *for all* notation is $\forall$, an upside-down 'A' as in "all"; the *exists* notation is $\exists$, a backward 'E' as in "exists." (Annoyingly, they had to be flipped in different directions: a backward 'A' is still an 'A,' and an upside-down 'E' is still an 'E.')

Definition 3.20 (Existential quantifier ("there exists"): $\exists$)
Let P be a predicate over the universe S. The proposition $\exists x \in S : P(x)$ ("there exists an x in S such that P(x)") is true if, for at least one possible $x \in S$, we have that P(x) is true.

Here's an example of two simple numerical propositions using these quantifiers:

Example 3.33 (Simple propositions using quantifiers)
Problem: What are the truth values of the following two propositions?

1. $\forall n \in \mathbb{Z}^{\geq 2} : isPrime(n)$
2. $\exists n \in \mathbb{Z}^{\geq 2} : isPrime(n)$

Solution: 1. **False.** This proposition says "every integer $n \geq 2$ is prime." This statement is false because, for example, the integer 32 is greater than or equal to 2 and is not prime.

2. **True.** The proposition says "there exists an integer $n \geq 2$ that is prime." This statement is true because, for example, the integer 31 (which is greater than or equal to 2) *is* prime.

In addition, we can make precise many intuitive statements using quantifiers. For example, we can use quantifiers to formalize the predicates from Example 3.30. (See Figure 3.21 for a reminder.)

isPrime(n): $n \in \mathbb{Z}^{>0}$ is a prime number. *isPowerOf*(n,k): $n \in \mathbb{Z}$ is an exact power of k.	*onlyPowersOfTwo*(S): every element of S is a power of two. $Q(n,a,b)$: $n \in \mathbb{Z}^{>0}$ satisfies $n = a + b$, and $a, b \in \mathbb{Z}$ are both prime.	*sumOfTwoPrimes*(n): $n \in \mathbb{Z}^{>0}$ is equal to the sum of two prime numbers.

Figure 3.21: Reminder of the predicates from Example 3.30.

Example 3.34 (Some example predicates, formalized)
isPrime(n): An integer $n \in \mathbb{Z}^{>0}$ is prime if and only if $n \geq 2$ and the only integers that evenly divide n are 1 and n itself. Thus we are really expressing a condition on every candidate divisor d: either $d \in \{1, n\}$, or d doesn't evenly divide n. Using the "divides" notation from Definition 2.10, we can formalize *isPrime*(n) as

$$n \geq 2 \wedge \left[\forall d \in \mathbb{Z}^{\geq 1} : (d \mid n \Rightarrow d = 1 \vee d = n)\right].$$

isPowerOf(n,k): We can formalize this predicate as $\exists i \in \mathbb{Z}^{\geq 0} : n = k^i$.

onlyPowersOfTwo(S): Because *isPowerOf*(n, 2) expresses the condition that n is a power of two, we can formalize this predicate as $\forall x \in S : \mathit{isPowerOf}(x, 2)$.

$Q(n, a, b)$: Formalizing Q actually doesn't require a quantifier at all; we can simply write $Q(n, a, b)$ as $(n = a + b) \land \mathit{isPrime}(a) \land \mathit{isPrime}(b)$.

sumOfTwoPrimes(n): This predicate requires that *there exist* prime numbers a and b that sum to n. Given our definition of Q, we can write *sumOfTwoPrimes*(n) as

$$\exists \langle a, b \rangle \in \mathbb{Z} \times \mathbb{Z} : Q(n, a, b).$$

("There exists a pair of integers $\langle a, b \rangle$ such that $Q(n, a, b)$.") Or we could write *sumOfTwoPrimes*(n) as $\exists a \in \mathbb{Z} : [\exists b \in \mathbb{Z} : Q(n, a, b)]$, by *nesting* one quantifier within the other. (See Section 3.5.)

Here's one further example, regarding the *prefix* relationship between two strings:

Example 3.35 (Prefixes, formalized)
A binary string $x \in \{0,1\}^k$ is a *prefix* of the binary string $y \in \{0,1\}^n$, for $n \geq k$, if y is x with some extra bits added on at the end. For example, `01` and `0110` are both prefixes of `01101010`, but `1` is not a prefix of `01101010`. If we write $|x|$ and $|y|$ to denote the length of x and y, respectively, then we can formalize *isPrefixOf*(x, y) as

$$|x| \leq |y| \quad \land \quad \Big[\forall i \in \{i \in \mathbb{Z} : 1 \leq i \leq |x|\} \; : \; x_i = y_i\Big].$$

In other words, y must be no shorter than x, and the first $|x|$ characters of y must equal their corresponding characters in x.

Quantifiers as loops

One useful way of thinking about these quantifiers is by analogy to loops in programming. If we ever encounter an $x \in S$ for which $\neg P(x)$ = True, then we immediately know that $\forall x \in S : P(x)$ is false. Similarly, any $x \in S$ for which $Q(x)$ = True is enough to demonstrate that $\exists x \in S : Q(x)$ is true. But if we "loop through" all candidate values of x and fail to encounter an x with $\neg P(x)$ or $Q(x)$, we know that $\forall x \in S : P(x)$ is true or $\exists x \in S : Q(x)$ is false. By this analogy, we might think of the two standard quantifiers as executing the programs in Figure 3.22(a) for $\forall$, and Figure 3.22(b) for $\exists$.

```
for x in S:
    if not P(x) then
        return False
return True
```

(a) A loop corresponding to $\forall x \in S : P(x)$.

```
for x in S:
    if Q(x) then
        return True
return False
```

(b) A loop corresponding to $\exists x \in S : Q(x)$.

Figure 3.22: Two **for** loops that return the value of $\forall x \in S : P(x)$ and $\exists x \in S : Q(x)$.

Another intuitive and useful way to think about these quantifiers is as a supersized version of $\land$ and $\lor$:

$$\forall x \in \{x_1, x_2, \ldots, x_n\} : P(x) \equiv P(x_1) \land P(x_2) \land \cdots \land P(x_n)$$
$$\exists x \in \{x_1, x_2, \ldots, x_n\} : P(x) \equiv P(x_1) \lor P(x_2) \lor \cdots \lor P(x_n)$$

The first of these propositions is true only if *every one* of the $P(x_i)$ terms is true; the second is true if *at least one* of the $P(x_i)$ terms is true.

There is one way in which these analogies are loose, though: just as for $\sum$ (summation) and $\prod$ (product) notation (from Section 2.2.7), the loop analogy only makes sense when the domain of discourse is finite! The Figure 3.22(a) "program" for a true proposition $\forall x \in \mathbb{Z} : P(x)$ would have to complete an infinite number of iterations before returning True. But the intuition may still be helpful.

Precedence and parenthesization

As in propositional logic, we'll adopt standard conventions regarding order of operations so that we don't overdose on parentheses. We treat the quantifiers $\forall$ and $\exists$ as binding tighter than the propositional logical connectives. Thus

$$\forall x \in S : P(x) \;\Rightarrow\; \exists y \in S : P(y)$$

will be understood to mean

$$\Big[\forall x \in S : P(x)\Big] \;\Rightarrow\; \Big[\exists y \in S : P(y)\Big].$$

To express the other reading (which involves nested quantifiers; see Section 3.5), we can use parentheses explicitly, by writing $\forall x \in S : \big[P(x) \Rightarrow \exists y \in S : P(y)\big]$.

Free and bound variables

Consider the variables x and y in the expressions

$$3 \mid x \qquad \text{and} \qquad \forall y \in \mathbb{Z} : 3 \mid y.$$

Understanding the first of these expressions requires knowledge of what x means, whereas the second is a self-contained statement that can be understood without any outside knowledge. The variable x is called a *free* or *unbound variable*: its value is not fixed by the expression. In contrast, the variable y is a *bound variable*: its value is defined within the expression itself. We say that the quantifier *binds* the variable y, and the *scope* or *body* of the quantifier is the part of the expression in which it has bound y. (We've encountered bound variables before; they arise whenever a variable name is assigned a value within an expression. For example, the variable i is bound in the arithmetic expression $\sum_{i=1}^{10} i^2$, as is the variable n in $\{n \in \mathbb{Z} : |n| \leq |n^2|\}$.)

A single expression can contain both free and bound variables: for example, the expression $\exists y \in \mathbb{Z}^{\geq 0} : x \geq y$ contains a bound variable y and a free variable x. Here's another example:

Example 3.36 (Free and bound variables)

Problem: Which variables are free in the following expression?

$$\Big[\forall x \in \mathbb{Z} : x^2 \geq y\Big] \wedge \Big[\forall z \in \mathbb{Z} : y = z \vee z^y = 1\Big]$$

Solution: The variable y doesn't appear as the variable bound by either of the quantifiers in this expression, so y is a free variable. Both x and z are bound by the universal quantifiers. (Incidentally, this expression is true if and only if $y = 0$.)

To test whether a particular variable x is free or bound in an expression, we can (consistently) replace x by a different name in that expression. If the meaning stays the same, then x is bound; if the meaning changes, then x is free. For example:

Example 3.37 (Testing for free and bound variables)
Consider the following pairs of propositions:

$$\exists x \in S : x > 251 \quad \text{and} \quad \exists y \in S : y > 251 \qquad \text{(A)}$$

$$x \geq 42x \quad \text{and} \quad y \geq 42y \qquad \text{(B)}$$

The expressions in (A) express precisely the same condition, namely: *some element of* S *is greater than* 251. Thus, the variables x and y in these two expressions are *bound*.

But the expressions in (B) mean different things, in the sense that we can construct a context in which these two statements have different truth values (for example, $x = 3$ and $y = -2$). The first expression states a condition on the value of x, and the latter states a condition on the value of y. So x is a free variable in "$x \geq 42x$."

Taking it further: The free-versus-bound-variable distinction is also something that may be familiar from programming, at least in some programming languages. There are some interesting issues in the design and implementation of programming languages that center on how free variables in a function definition, for example, get their values. See the discussion on p. 345.

An expression of predicate logic that contains no free variables is called *fully quantified*. For expressions that are not fully quantified, we adopt a standard convention that any unbound variables in a stated claim are *implicitly* universally quantified. For example, consider these claims:

Claim A: If $x \geq 1$, then $x^2 \leq x^3$.
Claim B: For all $x \in \mathbb{R}$, if $x \geq 1$, then $x^2 \leq x^3$.

When we write a (true) claim like Claim A, we will implicitly interpret it to mean Claim B. (Note that Claim B also explicitly notes $\mathbb{R}$ as the domain of discourse, which was left implicit in Claim A.)

3.4.3 *Theorem and Proof in Predicate Logic*

Recall that a *tautology* is a proposition that is always true—in other words, it is true no matter what each Boolean variable p in the proposition "means" (that is, whether p is true or false). In this section, we will be interested in the corresponding notion of always-true statements of predicate logic, which are called *theorems*. A statement of predicate logic is "always true" when it's true no matter what its predicates mean. (Formally, the "meaning" of a predicate P is the set of elements of the universe U for which the predicate is true—that is, $\{x \in U : P(x)\}$.)

Definition 3.21 (Theorems in predicate logic)
A fully quantified expression of predicate logic is a theorem *if and only if it is true for every possible meaning of each of its predicates.*

Analogously, two fully quantified expressions are *logically equivalent* if, for every possible meaning of their predicates, the two expressions have the same truth values.

We'll begin with a simple example of a theorem and a nontheorem:

Example 3.38 (A theorem of predicate logic)

Let S be any set. The following claim is true *regardless of what the predicate P denotes:*

$$\forall x \in S : \Big[P(x) \vee \neg P(x) \Big].$$

Indeed, this claim simply says that every $x \in S$ either makes $P(x)$ true or $P(x)$ false. And that assertion is true if the predicate $P(x)$ is "$x \geq 42$" or "x has red hair" or "x prefers programming in Python to playing Parcheesi"—indeed, it's true for any predicate P.

Example 3.39 (A nontheorem)

Let's show that the following proposition is not a theorem:

$$\Big[\forall x \in S : P(x) \Big] \vee \Big[\forall x \in S : \neg P(x) \Big].$$

A theorem must be true regardless of P's meaning, so we can establish that this proposition isn't a theorem by giving an example predicate that makes it false. Here's one: let P be *isPrime* (where S is $\mathbb{Z}$). Observe that $\forall x \in \mathbb{Z} : isPrime(x)$ is false because $isPrime(4) = \text{False}$; and $\forall x \in \mathbb{Z} : \neg isPrime(x)$ is false because $\neg isPrime(5) = \text{False}$. Thus the given proposition is false when P is *isPrime*, and so it is not a theorem.

Note the crucial difference between Example 3.38, which states that *every element of S either makes P true or makes P false,* and Example 3.39, which states that *either every element of S makes P true, or every element of S makes P false.* (Intuitively, it's the difference between "Every letter is either a vowel or a consonant" and "Every letter is a vowel or every letter is a consonant." The former is true; the latter is false.)

Example 3.39 establishes that the proposition $[\forall x \in S : P(x)] \vee [\forall x \in S : \neg P(x)]$ isn't true for *every* meaning of the predicate P, but it may be true for *some* meanings. For example, if $P(x)$ is the predicate $x^2 \geq 0$ and S is the set $\mathbb{R}$, then this disjunction is true (because $\forall x \in \mathbb{R} : x^2 \geq 0$ is true).

The challenge of proofs in predicate logic

The remainder of this section states some theorems of predicate logic, along with an initial discussion of how we might prove that they're theorems. (A *proof* of a statement is simply a convincing argument that the statement is a theorem.) Much of the rest of the book will be devoted to developing and writing proofs of theorems like these, and Chapter 4 will be devoted exclusively to some techniques and strategies for proofs. (This section will preview some of the ideas we'll see there.) Some theorems of predicate logic are summarized in Figure 3.23; we'll prove a few of them here, and you'll return to some of the others in the exercises.

$\forall x \in S : \big[P(x) \vee \neg P(x)\big]$	
$\neg\big[\forall x \in S : P(x)\big] \Leftrightarrow \big[\exists x \in S : \neg P(x)\big]$	De Morgan's Laws (quantified form)
$\neg\big[\exists x \in S : P(x)\big] \Leftrightarrow \big[\forall x \in S : \neg P(x)\big]$	
$\big[\forall x \in S : P(x)\big] \Rightarrow \big[\exists x \in S : P(x)\big]$	*if the set S is nonempty*
$\forall x \in \varnothing : P(x)$	Vacuous quantification
$\neg\exists x \in \varnothing : P(x)$	
$\exists x \in S : \big[P(x) \vee Q(x)\big] \Leftrightarrow \big[\exists x \in S : P(x)\big] \vee \big[\exists x \in S : Q(x)\big]$	
$\forall x \in S : \big[P(x) \wedge Q(x)\big] \Leftrightarrow \big[\forall x \in S : P(x)\big] \wedge \big[\forall x \in S : Q(x)\big]$	
$\exists x \in S : \big[P(x) \wedge Q(x)\big] \Rightarrow \big[\exists x \in S : P(x)\big] \wedge \big[\exists x \in S : Q(x)\big]$	
$\forall x \in S : \big[P(x) \vee Q(x)\big] \Leftarrow \big[\forall x \in S : P(x)\big] \vee \big[\forall x \in S : Q(x)\big]$	
$\big[\forall x \in S : P(x) \Rightarrow Q(x)\big] \wedge \big[\forall x \in S : P(x)\big] \Rightarrow \big[\forall x \in S : Q(x)\big]$	
$\big[\forall x \in \{y \in S : P(y)\} : Q(x)\big] \Leftrightarrow \big[\forall x \in S : P(x) \Rightarrow Q(x)\big]$	
$\big[\exists x \in \{y \in S : P(y)\} : Q(x)\big] \Leftrightarrow \big[\exists x \in S : P(x) \wedge Q(x)\big]$	
$\varphi \wedge \big[\exists x \in S : P(x)\big] \Leftrightarrow \big[\exists x \in S : \varphi \wedge P(x)\big]$	*if x does not appear as a free variable in φ*
$\varphi \vee \big[\forall x \in S : P(x)\big] \Leftrightarrow \big[\forall x \in S : \varphi \vee P(x)\big]$	*if x does not appear as a free variable in φ*

Figure 3.23: A few theorems involving quantification.

While predicate logic allows us to express claims that we couldn't state without quantifiers, that extra expressiveness comes with a cost! For a quantifier-free proposition (like all propositions in Sections 3.2–3.3), there is a straightforward—if tedious—algorithm to decide whether a given proposition is a tautology: first, build a truth table for the proposition; and, second, check to make sure that the proposition is true in every row. It turns out that the analogous question for predicate logic is much more difficult—in fact, *impossible* to solve in general: there's no algorithm that's guaranteed to figure out whether a given fully quantified expression is a theorem! Demonstrating that a statement in predicate logic is a theorem will require you to *think* in a way that demonstrating that a statement in propositional logic is a tautology did not.

Taking it further: See the discussion on p. 346 for more about the fact that there's no algorithm guaranteed to determine whether a given proposition is a theorem. The absence of such an algorithm sounds like bad news; it means that proving predicate-logic statements is harder, because you can't just plug-and-chug into a simple algorithm to figure out whether a given statement is actually always true. But this fact is also precisely the reason that creativity plays a crucial role in proofs and in theoretical computer science more generally—and why, arguably, proving things can be fun! (For me, this difference is exactly why I find Sudoku less interesting than crossword puzzles: when there's no algorithm to solve a problem, we have to embrace the creative challenge in attacking it.)

3.4.4 *A Few Examples of Theorems and Proofs*

In the rest of this section, we will see a few further theorems of predicate logic, with proofs. As we've said, there's no formulaic approach to prove these theorems; we'll need to employ a variety of strategies in this endeavor.

NEGATING QUANTIFIERS: A FIRST EXAMPLE

Suppose that your egomaniacal, overconfident partner from Intro CS wanders into the lab and says *For any array A that you give me, partner, my implementation of insertion sort correctly sorts A.* You know, though, that your partner is wrong. (You spot a bug in his egomaniacal code.) What would that mean? Well, you might reply, gently but firmly: *There's an array A for which your implementation of insertion sort does not correctly sort A.* The equivalence that you're using is a theorem of predicate logic:

Example 3.40 (Negating universal quantifiers)
Let's prove the equivalence you're using to debunk your partner's claim:

$$\neg\big[\forall x \in S : P(x)\big] \Leftrightarrow \big[\exists x \in S : \neg P(x)\big].$$

Perhaps the easiest way to view this claim is as a quantified version of the tautology $\neg(p \wedge q) \Leftrightarrow \neg p \vee \neg q$, which was one of De Morgan's Laws from propositional logic. If we think of $\forall x \in S : P(x)$ as $P(x_1) \wedge P(x_2) \wedge P(x_3) \wedge \cdots$, then

$$\begin{aligned}\neg\big[\forall x \in S : P(x)\big] &\approx \neg\big[P(x_1) \wedge P(x_2) \wedge P(x_3) \wedge \cdots\big]\\ &\equiv \big[\neg P(x_1) \vee \neg P(x_2) \vee \neg P(x_3) \vee \cdots\big]\\ &\approx \exists x \in S : \neg P(x),\end{aligned}$$

where the second line follows by the propositional version of De Morgan's Laws. There is something slightly more subtle to our claim because the set S might be infinite, but the idea is identical. If there's an $a \in S$ such that $P(a) =$ False, then $\exists x \in S : \neg P(x)$ is true (because a is an example) and $\forall x \in S : P(x)$ is false (because a is a counterexample). And if every $a \in S$ has $P(a) =$ True, then $\exists x \in S : \neg P(x)$ is false and $\forall x \in S : P(x)$ is true.

The analogous claim for the negation of $\exists x \in S : P(x)$ is also a theorem:

Example 3.41 (Negating existential quantifiers)
Let's prove that this claim is a theorem, too:

$$\neg\big[\exists x \in S : P(x)\big] \Leftrightarrow \big[\forall x \in S : \neg P(x)\big].$$

To see that this claim is true for an arbitrary predicate P, we start with the claim from Example 3.40, but using the predicate $Q(x) := \neg P(x)$. (Note that Q is also a predicate—so Example 3.40 holds for Q too!) Thus we know that

$$\neg\big[\forall x \in S : Q(x)\big] \Leftrightarrow \big[\exists x \in S : \neg Q(x)\big],$$

and, because $p \Leftrightarrow q \equiv \neg p \Leftrightarrow \neg q$, we therefore also know that

$$\big[\forall x \in S : Q(x)\big] \Leftrightarrow \neg\big[\exists x \in S : \neg Q(x)\big].$$

But $Q(x)$ is just $\neg P(x)$ and $\neg Q(x)$ is just $P(x)$, by definition of Q, and so we have

$$\left[\forall x \in S : \neg P(x)\right] \Leftrightarrow \neg\left[\exists x \in S : P(x)\right].$$

Thus we've now shown that the desired claim is true for any predicate P, so it is a theorem.

All implies some: a proof of an implication

The entirety of Chapter 4 is devoted to proofs and proof techniques; there's lots more there about how to approach proving or disproving new claims. But here we'll preview a particularly useful proof strategy for proving an implication, and use it to establish another theorem of predicate logic. Here's the method of proof:

Definition 3.22 (Proof by assuming the antecedent)
Suppose that we must prove an implication $\varphi \Rightarrow \psi$. Because the only way for $\varphi \Rightarrow \psi$ to fail to be true is for φ to be true and ψ to be false, to prove that the implication $\varphi \Rightarrow \psi$ is always true, we will rule out the one scenario in which it wouldn't be. Specifically, we assume that φ is true, and then prove that ψ must be true too, under this assumption.

(Recall from the truth table of $\Rightarrow$ that the only way for the implication $\varphi \Rightarrow \psi$ to be false is when φ is true but ψ is false. Also recall that the proposition φ is called the *antecedent* of the implication $\varphi \Rightarrow \psi$; hence this proof technique is called *assuming the antecedent*.) Here are two examples of proofs that use this technique, one from propositional logic and one from arithmetic:

- Let's prove that $p \Rightarrow p \vee q$ is a tautology: we assume that the antecedent p is true, and we must prove that the consequent $p \vee q$ is true too. But that's obvious, because p is true (by our assumption), and True $\vee\, q \equiv$ True.
- Let's prove that *if x is a perfect square, then $4x$ is a perfect square*: assume that x is a perfect square, that is, assume that $x = k^2$ for an integer k. Then $4x = 4k^2 = (2k)^2$ is a perfect square too, because $2k$ is also an integer.

Finally, here's a theorem of predicate logic that we can prove using this technique:

Example 3.42 (If everybody's doing it, then somebody's doing it)
Consider the following proposition, for an arbitrary nonempty set S:

$$\left[\forall x \in S : P(x)\right] \Rightarrow \left[\exists x \in S : P(x)\right].$$

We'll prove this claim by assuming the antecedent. Specifically, we assume $\forall x \in S : P(x)$, and we need to prove that $\exists x \in S : P(x)$.

Because the set S is nonempty, we know that there's at least one element $a \in S$. By our assumption, we know that $P(a)$ is true. But because $P(a)$ is true, then it's immediately apparent that $\exists x \in S : P(x)$ is true too—because we can just pick $x := a$.

Problem-solving tip: When you're facing a statement that contains a lot of mathematical notation, try to understand it by rephrasing it as an English sentence. Restating the assertion from Example 3.42 in English makes it pretty obvious that it's true: *if everyone in S satisfies P—and there's actually someone in S—then of course someone in S satisfies P!*

Vacuous quantification

Consider the proposition *All even prime numbers greater than* 12 *have a* 3 *as their last digit*. Write P to denote the set of all even prime numbers greater than 12; formalized, then, this claim can be written as $\forall n \in P : n \bmod 10 = 3$. Is this claim true or false? It has to be true! The point is that P actually contains no elements (there *are* no even prime numbers other than 2, because an even number is by definition divisible by 2). Thus this claim says: for every $n \in \varnothing$, something-or-other is true of n. But there is no n in $\varnothing$, so the claim has to be true! The general statement of the theorem is

$$\forall x \in \varnothing : P(x).$$

Quantification over the empty set is called *vacuous quantification*; this proposition is said to be *vacuously true*.

Here's another way to see that $\forall x \in \varnothing : P(x)$ is a theorem, using the De Morgan–like view of quantification. The negation of $\forall x \in \varnothing : P(x)$ is $\exists x \in \varnothing : \neg P(x)$, but there never exists *any* element $x \in \varnothing$, let alone an element $x \in \varnothing$ such that $\neg P(x)$. Thus $\exists x \in \varnothing : \neg P(x)$ is false, and therefore its negation $\neg \exists x \in \varnothing : \neg P(x)$, which is equivalent to $\forall x \in \varnothing : P(x)$, is true.

Disjunctions and quantifiers

Here's one last example, where we'll figure out when the "or" of two quantified statements can be expressed as one single quantified statement:

Example 3.43 (Disjunctions and quantifiers)
Consider the following two propositions, for an arbitrary set S:

$$\forall x \in S : \Big[P(x) \lor Q(x)\Big] \;\Leftrightarrow\; \Big[\forall x \in S : P(x)\Big] \lor \Big[\forall x \in S : Q(x)\Big] \qquad \text{(A)}$$

$$\exists x \in S : \Big[P(x) \lor Q(x)\Big] \;\Leftrightarrow\; \Big[\exists x \in S : P(x)\Big] \lor \Big[\exists x \in S : Q(x)\Big] \qquad \text{(B)}$$

Problem: Is either (A) or (B) a theorem? Prove your answers.

Problem-solving tip: In thinking about a question like whether (A) from Example 3.43 is a theorem, it's often useful to get intuition by plugging in a few sample values for S, P, and Q.

Solution: Claim (B) is a theorem. To prove it, we'll show that the left-hand side implies the right-hand side, and vice versa. (That is, we're proving $p \Leftrightarrow q$ by proving both $p \Rightarrow q$ and $q \Rightarrow p$, which is a legitimate proof because $p \Leftrightarrow q \equiv (p \Rightarrow q) \land (q \Rightarrow p)$.) Both proofs will use the technique of assuming the antecedent.

- First, suppose that $\exists x \in S : [P(x) \lor Q(x)]$ is true. Then there is some particular $x^* \in S$ for which either $P(x^*)$ or $Q(x^*)$. But in either case, we're done: if $P(x^*)$ then $\exists x \in S : P(x)$—in particular, x^* satisfies the condition; if $Q(x^*)$ then $\exists x \in S : Q(x)$.
- Conversely, suppose that $[\exists x \in S : P(x)] \lor [\exists x \in S : Q(x)]$ is true. Thus either there's an $x^* \in S$ such that $P(x^*)$ or an $x^* \in S$ such that $Q(x^*)$. That x^* suffices to make the left-hand side true.

On the other hand, (A) is not a theorem, for much the same reason as in Example 3.39. (In fact, if $Q(x) := \neg P(x)$, then Examples 3.38 and 3.39 precisely show that (A) is not a theorem.) The set $\mathbb{Z}$ and the predicates *isOdd* and *isEven* make (A) false: the left-hand side is true ("all integers are either even or odd") but the right-hand side is false ("either (i) all integers are even, or (ii) all integers are odd").

Although (A) from this example is not a theorem, one direction of it is; we'll prove this implication as another example:

Example 3.44 (Disjunction, quantifiers, and one-way implications)
The $\Leftarrow$ direction of (A) from Example 3.43 is a theorem:

$$\forall x \in S : \Big[P(x) \vee Q(x)\Big] \quad \Leftarrow \quad \Big[\forall x \in S : P(x)\Big] \vee \Big[\forall x \in S : Q(x)\Big].$$

To convince yourself of this claim, observe that if $P(x)$ is true for an arbitrary $x \in S$, then it's certainly true that $P(x) \vee Q(x)$ is true for an arbitrary $x \in S$ too. And if $Q(x)$ is true for every $x \in S$, then, similarly, $P(x) \vee Q(x)$ is true for every $x \in S$.

To prove this claim, we assume the antecedent $[\forall x \in S : P(x)] \vee [\forall x \in S : Q(x)]$. Thus either $[\forall x \in S : P(x)]$ or $[\forall x \in S : Q(x)]$, and, in either case, we've argued that $P(x) \vee Q(x)$ is true for all $x \in S$.

You'll have a chance to consider a number of other theorems of predicate logic in the exercises, including the $\wedge$-analogy to Examples 3.43–3.44 (in Exercises 3.130–3.131).

COMPUTER SCIENCE CONNECTIONS

GAME TREES, LOGIC, AND WINNING TIC-TAC(-TOE)

In 1997, Deep Blue, a chess-playing program developed by IBM,[5] beat the chess Grandmaster Garry Kasparov in a six-match series. This event was a turning point in the public perception of computation and artificial intelligence; it was the first time that a computer had outperformed the best humans at something that most people tended to identify as a "human endeavor." Ten years later, a research group developed a program called Chinook, a perfect checkers-playing system: from any game position arising in its games, Chinook chooses *the* best possible legal move.[6]

[5] Murray Campbell, A. Joseph Hoane Jr., and Feng-hsiung Hsu. Deep Blue. *Artificial Intelligence*, 134:57–83, 2002.

[6] Jonathan Schaeffer, Neil Burch, Yngvi Bjornsson, Akihiro Kishimoto, Martin Muller, Rob Lake, Paul Lu, and Steve Sutphen. Checkers is solved. *Science*, 317(5844):1518–1522, 14 September 2007.

While chess and checkers are very complicated games, the basic ideas of playing them—ideas based on logic—are shared with simpler games. Consider *Tic-Tac*, a 2-by-2 version of Tic-Tac-Toe. Two players, O and X, make alternate moves, starting with O; a player wins by occupying a complete row or column. Diagonals don't count, and if the board is filled without O or X winning, then the game is a draw. Note that—unless O is tremendously dull—O will win the game, but we will use a *game tree* (Figure 3.24), which represents all possible moves, to systematize this reasoning.

Thanks to Jon Kleinberg for suggesting this game.

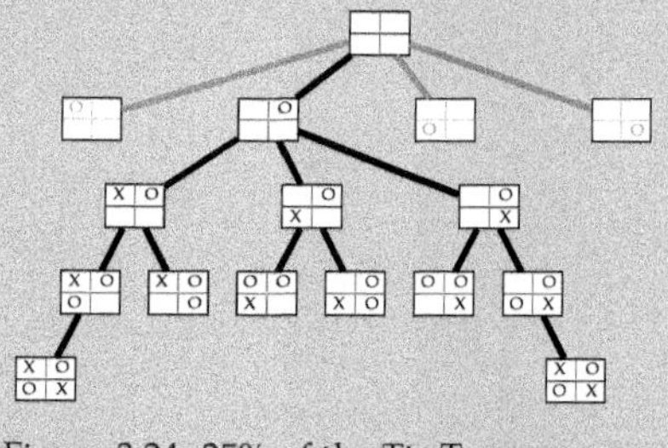

Figure 3.24: 25% of the Tic-Tac game tree. (The missing 75% is rotated, but otherwise identical.)

Here's the basic idea. Define $P(B)$ to be the predicate

$$P(B) := \text{"Player O wins under optimal play starting from board } B\text{."}$$

For example, $P(\begin{array}{c|c} X & \\ \hline O & O \end{array}) = \text{True}$ because O has already won; and $P(\begin{array}{c|c} O & X \\ \hline X & O \end{array}) = \text{False}$ because it's a draw. The answer to the question "does O win Tic-Tac if both players play optimally?" is the truth value of $P(\begin{array}{c|c} \; & \; \\ \hline \; & \; \end{array})$. If it's O's turn in board B, then $P(B)$ is true if and only if *there exists* a possible move for O leading to a board B' in which $P(B')$; if it's X's turn, then $P(B)$ is true if and only if *every* possible move made by X leads to a board B' in which $P(B')$. So

$$P(\begin{array}{c|c} \; & O \\ \hline \; & \; \end{array}) = P(\begin{array}{c|c} X & O \\ \hline \; & \; \end{array}) \wedge P(\begin{array}{c|c} \; & O \\ \hline X & \; \end{array}) \wedge P(\begin{array}{c|c} \; & O \\ \hline \; & X \end{array})$$
$$\text{and } P(\begin{array}{c|c} \; & \; \\ \hline \; & \; \end{array}) = P(\begin{array}{c|c} O & \; \\ \hline \; & \; \end{array}) \vee P(\begin{array}{c|c} \; & O \\ \hline \; & \; \end{array}) \vee P(\begin{array}{c|c} \; & \; \\ \hline O & \; \end{array}) \vee P(\begin{array}{c|c} \; & \; \\ \hline \; & O \end{array}).$$

The game tree, labeled appropriately, is shown in Figure 3.25. If we view the truth values from the leaves as "bubbling up" from the bottom of the tree, then a board B gets assigned the truth value True if and only if Player O can guarantee a win from the board B.

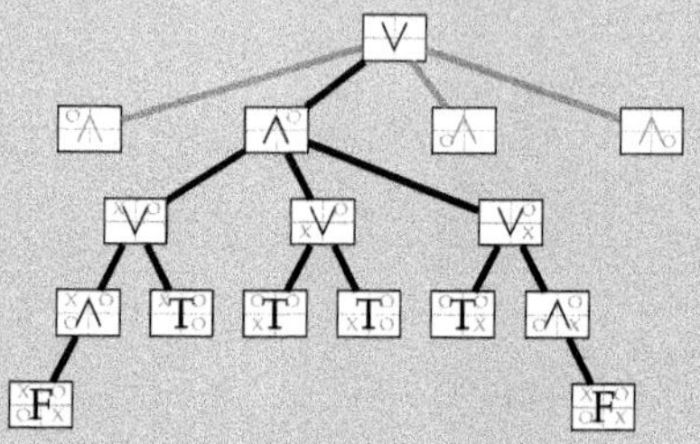

Figure 3.25: The game tree, with each win for O labeled by T, each loss/draw by F, $\vee$ if it's Player O's turn, and $\wedge$ if it's Player X's turn.

Some serious complications arise in writing a program to play more complicated games like checkers or chess. Here are just a few of the issues that one must confront in building a system like Deep Blue or Chinook:[7]

- There are $\approx$ 500,000,000,000,000,000,000 different checkers positions—and $\approx 10^{40}$ chess positions!—so we can't afford to represent them all. (Luckily, we can choose moves so most positions are never reached.)
- Approximately one bit per trillion is written incorrectly *merely in copying data on current hard disk technologies.* So a program constructing a massive structure like the checkers game tree must "check its work."
- For a game as big as chess, we can't afford to compute all the way to the bottom of the tree; instead, we *estimate* the quality of each position after computing a handful of layers deep in the game tree.

For more on game trees and algorithms for exploring large search spaces, see a good artificial intelligence (AI) text like

[7] Stuart Russell and Peter Norvig. *Artificial Intelligence: A Modern Approach.* Prentice Hall, 3rd edition, 2009.

Computer Science Connections

Nonlocal Variables and Lexical vs. Dynamic Scoping

In a function f written in a programming language—say, C or Python—we can use several different types of variables that store values:

- *local variables*, whose values are defined completely within the body of f;
- *parameters*, inputs to f whose value is specified when f is invoked;
- *nonlocal variables*, which get their value from other contexts. The most common type of these "other" variables is a *global variable*, which persists throughout the execution of the entire program.

```
int addB(int a) {
  return a + b;
}
```

```
def addB(a):
  return a + b
```

Figure 3.26: A function addB written in C and analogous function addB written in Python. Here addB takes one (integer) parameter a, accesses a nonlocal variable b, and returns a + b.

For an example function (written in C and Python as illustrative examples) that uses both a parameter and a nonlocal variable, see Figure 3.26. In the body of this function, the variable a is a *bound* variable; specifically, it is bound when the function is invoked with an actual parameter. But the variable b is *unbound*. (Just as with a quantified expression, an unbound variable is one for which the meaning of the function could change if we replaced that variable with a different name. If we changed the a to an x in both lines 1 and 2, then the function would behave identically, but if we changed the b to a y, then the function would behave differently.)

In this function, the variable b has to somehow get a value from somewhere if we are going to be able to invoke the function addB without causing an error. Often b will be a global variable, but it is also possible in Python or C (with appropriate compiler settings) to *nest* function definitions—just as quantifiers can be nested. (See Section 3.5.)

For more about lexical versus dynamic scope, and other related issues, see a textbook on programming languages. (One of the other interesting issues is that there are actually multiple paradigms for passing parameters to a function; we're discussing *call-by-value* parameter passing, which probably is the most common.) Some good books on programming languages include

One fundamental issue in the design and implementation in programming languages is illustrated in Figure 3.27.[8] Suppose x is an unbound variable in the definition of a function f. Generally, programming languages either use *lexical scope*, where x's value is found by looking "outward" where f is *defined*; or *dynamic scope*, where x's value is found by looking where f is *called*. Almost all modern programming languages use lexical scope, though *macros* in C and other languages use dynamic scope. While we're generally used to lexical scope and therefore it feels more intuitive, there are some circumstances in which macros can be tremendously useful and convenient.

[8] Michael L. Scott. *Programming Language Pragmatics*. Morgan Kaufmann Publishers, 3rd edition, 2009; and Kenneth C. Louden and Kenneth A. Lambert. *Programming Languages: Principles and Practices*. Course Technology, 3rd edition, 2011.

```
int b = 17;

int addB(int a) { return a + b; }
    /* a FUNCTION in C finds values for unbound */
    /* variables in the *defining* environment */

int test() {
  int b = 128;
  return addB(3);
}

test(3);        /* returns 20 */
```

```
int b = 17;

#define addB(a)     a + b
    /* a MACRO in C finds values for unbound */
    /* variables in the *calling* environment */

int test() {
  int b = 128;
  return addB(3);
}

test(3);        /* returns 131 */
```

Figure 3.27: Two C snippets defining addB, where the nonlocal variable b gets its value from different places.

Computer Science Connections

Gödel's Incompleteness Theorem

Given a fully quantified proposition φ, is φ a theorem? This apparently simple question drove the development of some of the most profound and mind-numbing results of the last hundred years. In the early 20th century, there was great interest in the "formalist program," advanced especially by the German mathematician David Hilbert. The formalist approach aimed to turn all of mathematical reasoning into a machine: one could feed in a mathematical statement φ as input, turn a hypothetical crank, and the machine would spit out a proof or disproof of φ as output. But this program was shattered by two closely related results—two of the greatest intellectual achievements of the 20th century.

The first blow to the formalist program was the proof by Kurt Gödel, in 1931, of what became known as *Gödel's Incompleteness Theorem*. Gödel's incompleteness theorem is based on the following two important and desirable properties of logical systems:

- A logical system is *consistent* if only true statements can be proven. (In other words, if there is a proof of φ in the system, then φ is true.)
- A logical system is *complete* if every true statement can be proven. (In other words, if φ is true, then there is a proof of φ in the system.)

Gödel's Incompleteness Theorem is the following troubling result:

Theorem 3.3 (Gödel's (First) Incompleteness Theorem)
Any sufficiently powerful logical system is either inconsistent or incomplete.

(Here "sufficiently powerful" just means "capable of expressing multiplication"; predicate logic as described here is certainly "sufficiently powerful.")

If the system is inconsistent, then there is a false statement φ that can be proven (which means that anything can be proven, as false implies anything!). And if the system is incomplete, then there is a true statement φ that cannot be proven. Gödel's proof proceeds by constructing a self-referential logical expression φ that means "φ is not provable." (So if φ is true, then the system is incomplete; and if φ is false, then the system is inconsistent.)

The second strike against the formalist program was the proof of the *undecidability of the halting problem*, shown independently by Alan Turing and Alonzo Church in the 1930s. We can think of the halting problem as asking the following question: given a function f written in Python and an input x, does running $f(x)$ get stuck in an infinite loop? (Or does it eventually terminate?) The *undecidability* of this problem means that *there is no algorithm that solves the halting problem.* A corollary of this result is that our problem—given a fully quantified proposition φ, is φ a theorem?—is also undecidable. We'll discuss uncomputability in more detail in Chapter 4.

Undecidability, incompleteness, and their profound consequences are the focus of a number of excellent textbooks on the theory of computation[9]—and also Douglas Hofstadter's fascinating masterpiece *Gödel, Escher, Bach,*[10] which is all-but-required reading for computer scientists.

See, for example:
[9] Dexter Kozen. *Automata and Computability*. Springer, 1997; and Michael Sipser. *Introduction to the Theory of Computation*. Course Technology, 3rd edition, 2012.

[10] Douglas Hofstadter. *Gödel, Escher, Bach: An Eternal Golden Braid*. Vintage, 1980.

3.4.5 Exercises

Figure 3.28 lists some well-known programming languages, with some characteristics. Using these characteristics, define a predicate that's true for each of the following lists of languages, and false for every other language in the table. For example, the predicate $P(x)$ = "x has strong typing and x is not functional" makes P(Pascal) and P(Java) true, and makes $P(x)$ false for every $x \in \{C, C{+}{+}, \LaTeX, ML, Perl, Scheme\}$.

3.107 Java

3.108 ML, Perl

3.109 Pascal, Scheme, Perl

3.110 LaTeX, Java, C++, Perl

3.111 C, Pascal, ML, C++, LaTeX, Scheme, Perl

	paradigm	typing	scope
C	imperative	weak	lexical
C++	object-oriented	weak	lexical
Java	object-oriented	strong	lexical
LaTeX	scripting	weak	dynamic
ML	functional	strong	lexical
Pascal	imperative	strong	lexical
Perl	scripting	weak	either
Scheme	functional	weak	either

Figure 3.28: Some programming languages.

Examples 3.4 and 3.15 construct a proposition corresponding to "the password contains at least three of four character types (digits, lowercase letters, uppercase letters, other)." In that example, we took "the password contains at least one digit" (and its analogues for the other character types) as an atomic proposition. But we could give a lower-level characterization of valid passwords. Let isDigit, isLower, *and* isUpper *be predicates that are true of single characters of the appropriate type. Use standard arithmetic notation and these predicates to formalize the following conditions on a password $x = \langle x_1, \ldots, x_n \rangle$, where x_i is the ith character in the password:*

3.112 x is at least 8 characters long.

3.113 x contains at least one lowercase letter.

3.114 x contains at least one non-alphanumeric character. (Remember that *isDigit*, *isLower*, and *isUpper* are the only predicates available!)

3.115 *(Inspired by a letter to the editor in* The New Yorker *by Alexander George from 24 December 2007.)* Steve Martin, the great comedian, reports in *Born Standing Up: A Comic's Life* that, inspired by Lewis Carroll, he started closing his shows with the following line.[11] (It got big laughs.)

> *I'm not going home tonight; I'm going to Bananaland, a place where only two things are true, only two things: One, all chairs are green; and two, no chairs are green.*

Steve Martin describes the joke as a contradiction—but, in fact, these two true things are not contradictory! Describe how it is possible for both "all chairs in Bananaland are green" and "no chairs in Bananaland are green" to be simultaneously true.

[11] Steve Martin. *Born Standing Up: A Comic's Life.* Simon & Schuster, 2008.

As a rough approximation, we can think of a database *as a two-dimensional table, where rows correspond to individual entities, and columns correspond to fields (data about those entities). A database* query *defines a predicate $Q(x)$ that consists of tests of the values from various columns, joined by the basic logical connectives. The database system then returns a list of rows/entities for which the predicate is true. We can think of this type of database access as involving predicates: in response to query Q, the system returns the list of all rows x for which $Q(x)$ is true.*

See Figure 3.29 for an example; here, to find a list of all students with grade point averages over 3.4 who have taken at least one CS course if and only if they're from Hawaii, we could query $\text{GPA}(x) \geq 3.4 \wedge (\text{CS?}(x) = \textit{yes} \Leftrightarrow \text{home}(x) = \textit{Hawaii})$. *For this database, this query would return Charlie (and not Alice, Bob, or Dave).*

Each of the following predicates $Q(x)$ uses tests on particular columns in x's row. For each, give a logically equivalent predicate in which each column's name appears at most once. *You may also use the symbols $\{\text{True}, \text{False}, \wedge, \vee, \neg, \Rightarrow\}$ as many times as you please. Use a truth table to prove that your answer is logically equivalent to the given predicate.*

name	GPA	CS?	home	...
Alice	4.0	yes	Alaska	...
Bob	3.14	yes	Bermuda	...
Charlie	3.54	no	California	...
Dave	3.8	yes	Delaware	...
		⋮		

Figure 3.29: A sample database.

3.116 $[age(x) < 18] \vee (\neg[age(x) < 18] \wedge [gpa(x) \geq 3.0])$

3.117 $cs(x) \Rightarrow \neg(hawaii(x) \Rightarrow (hawaii(x) \wedge cs(x)))$

3.118 $(hasMajor(x) \wedge \neg junior(x) \wedge oncampus(x)) \vee (hasMajor(x) \wedge \neg junior(x) \wedge \neg oncampus(x))$
$\vee (hasMajor(x) \wedge junior(x) \wedge \neg oncampus(x))$

3.119 Following the last few exercises, you might begin to think that any query can be rewritten without duplication. Can it? Consider a unary predicate that is built up from the predicates $P(x)$ and $Q(x)$ and the propositional symbols $\{\text{True}, \text{False}, \wedge, \vee, \neg, \Rightarrow\}$. Decide whether the following claim is true or false, and prove your answer:

Claim: Every such predicate is logically equivalent to a predicate that uses only the following symbols: (i) $\{\text{True}, \text{False}, \wedge, \vee, \neg, \Rightarrow\}$, all of which can be used as many times as you please; and (ii) the predicates $\{P(x), Q(x)\}$, which can appear *only one time each.*

Modern web search engines allow users to specify Boolean conditions in their queries. For example, "social OR networks" will return only web pages containing either the word "social" or the word "networks." You can view a query as a predicate Q; the search engine returns (in some order) the list of all pages p for which Q(p) is true. Consider the following queries:

A: "java AND program AND NOT computer"
B: "(computer OR algorithm) AND java"
C: "java AND NOT (computer OR algorithm OR program)"

Give an example of a web page—or a sentence—that would be returned . . .

3.120 . . . by query A but not by B or C.

3.121 . . . by query B but not by A or C.

3.122 . . . by query C but not by A or B.

3.123 Prove or disprove: $\forall n \in \mathbb{Z} : \mathit{isPrime}(n) \Rightarrow \frac{n}{2} \notin \mathbb{Z}$.

3.124 Translate this Groucho Marx quote into logical notation: *It isn't necessary to have relatives in Kansas City in order to be unhappy.* Let $P(x)$ be "x has relatives in Kansas City" and $Q(x)$ be "x is unhappy," and view the statement as implicitly making a claim that a particular kind of person exists.

Write an English sentence that expresses the logical negation of each given sentence. (Don't just say "It is not the case that ..."; give a genuine negation.) Some of the given sentences are ambiguous in their meaning; if so, describe all of the interpretations of the sentence that you can find, then choose one and give its negation.

3.125 Every entry in the array A is positive.

3.126 Every decent programming language denotes block structure with parentheses or braces.

3.127 There exists an odd number that is evenly divisible by a different odd number.

3.128 There is a point in Minnesota that is farther than ten miles from a lake.

3.129 Every sorting algorithm takes at least $n \log n$ steps on some n-element input array.

In Examples 3.43 and 3.44, we proved that

$$\exists x \in S : \Big[P(x) \vee Q(x)\Big] \Leftrightarrow \Big[\exists x \in S : P(x)\Big] \vee \Big[\exists x \in S : Q(x)\Big]$$
$$\forall x \in S : \Big[P(x) \vee Q(x)\Big] \Leftarrow \Big[\forall x \in S : P(x)\Big] \vee \Big[\forall x \in S : Q(x)\Big]$$

are theorems. Argue that the following $\wedge$*-analogies to these statements are also theorems:*

3.130 $\exists x \in S : \Big[P(x) \wedge Q(x)\Big] \Rightarrow \Big[\exists x \in S : P(x)\Big] \wedge \Big[\exists x \in S : Q(x)\Big]$

3.131 $\forall x \in S : \Big[P(x) \wedge Q(x)\Big] \Leftrightarrow \Big[\forall x \in S : P(x)\Big] \wedge \Big[\forall x \in S : Q(x)\Big]$

Explain why the following are theorems of predicate logic:

3.132 $\Big[\forall x \in S : P(x) \Rightarrow Q(x)\Big] \wedge \Big[\forall x \in S : P(x)\Big] \Rightarrow \Big[\forall x \in S : Q(x)\Big]$

3.133 $\Big[\forall x \in \{y \in S : P(y)\} : Q(x)\Big] \Leftrightarrow \Big[\forall x \in S : P(x) \Rightarrow Q(x)\Big]$

3.134 $\Big[\exists x \in \{y \in S : P(y)\} : Q(x)\Big] \Leftrightarrow \Big[\exists x \in S : P(x) \wedge Q(x)\Big]$

Explain why the following propositions are theorems of predicate logic, assuming that x does not appear as a free variable in the expression φ *(and assuming that S is nonempty):*

3.135 $\varphi \Leftrightarrow \Big[\forall x \in S : \varphi\Big]$

3.136 $\varphi \vee \Big[\forall x \in S : P(x)\Big] \Leftrightarrow \Big[\forall x \in S : \varphi \vee P(x)\Big]$

3.137 $\varphi \wedge \Big[\exists x \in S : P(x)\Big] \Leftrightarrow \Big[\exists x \in S : \varphi \wedge P(x)\Big]$

3.138 $\Big(\varphi \Rightarrow \Big[\exists x \in S : P(x)\Big]\Big) \Leftrightarrow \Big[\exists x \in S : \varphi \Rightarrow P(x)\Big]$

3.139 $\Big(\Big[\exists x \in S : P(x)\Big] \Rightarrow \varphi\Big) \Leftrightarrow \Big[\forall x \in S : P(x) \Rightarrow \varphi\Big]$

3.140 Give an example of a predicate P, a nonempty set S, and an expression φ containing x as a free variable such that the proposition from Exercise 3.136 is false. Because x has to get its meaning from somewhere, we will imagine a universal quantifier for x wrapped around the entire expression. Specifically, give an example of P, φ, and S for which

$$\forall x \in S : \Big[\varphi \vee [\forall x \in S : P(x)]\Big] \text{ is not logically equivalent to } \forall x \in S : \Big[[\forall x \in S : \varphi \vee P(x)]\Big].$$

3.5 *Predicate Logic: Nested Quantifiers*

Everybody hates me because I'm so universally liked.

Peter De Vries (1910–1993)

Just as we can place one loop inside another in a program, we can place one quantified statement inside another in predicate logic. In fact, the most interesting quantified statements almost always involve more than one quantifier. (For example: *during every semester, there's a computer science class that every student on campus can take.*) In formal notation, such a statement typically involves *nested quantifiers*—that is, multiple quantifiers in which one quantifier appears inside the scope of another.

We've encountered statements involving nested quantification before, although so far we've discussed them using English rather than mathematical notation. The definition of a partition of a set (Definition 2.30) or of an onto function (Definition 2.49) are two examples. (To make the latter definition's quantifiers more explicit: an onto function $f : A \to B$ is one where, for every element of B, there's an element of A such that $f(a) = b$: that is, $\forall b \in B : \left[\exists a \in A : f(a) = b\right]$.) Here are two other examples:

Example 3.45 (No unmatched elements in an array)

Let's express the condition that every element of an array $A[1 \ldots n]$ is a "double"—that is, appears at least twice in A. (For example, the array [3,2,1,1,4,4,2,3,1] satisfies this condition.) This condition requires that, for every index i, there exists another index j such that $A[i] = A[j]$. We can express the requirement as follows:

$$\forall i \in \{1,2,\ldots,n\} : \left[\exists j \in \{1,2,\ldots,n\} : i \neq j \wedge A[i] = A[j]\right].$$

Example 3.46 (Alphabetically later)

Let's formalize the predicate "The string __ is alphabetically after the string __" from Example 3.28. For two letters $a, b \in \{\texttt{A},\texttt{B},\ldots,\texttt{Z}\}$, write $a < b$ if a is earlier in the alphabet than b; we'll use this ordering on *letters* to define an ordering on *strings*. Let x and y be strings over $\{\texttt{A},\texttt{B},\ldots,\texttt{Z}\}$. There are two ways for x to be alphabetically later than y:

- y is a (proper) prefix of x. (See Example 3.35.) For example, FORTRAN is after FORT.
- x and y share an initial prefix of identical letters, and the first i for which $x_i \neq y_i$ has x_i later in the alphabet than y_i. For example, PASTOR comes after PASCAL.

Formally, then, $x \in \{\texttt{A},\texttt{B},\ldots,\texttt{Z}\}^n$ is alphabetically after $y \in \{\texttt{A},\texttt{B},\ldots,\texttt{Z}\}^m$ if

$$\left[m < n \wedge [\forall j \in \{1,2\ldots,m\} : x_j = y_j]\right] \qquad \textit{y is a proper prefix of x} \ldots$$
$$\vee \left[\exists i \in \{1,\ldots,\min(n,m)\} : x_i > y_i \wedge [\forall j \in \{1,2\ldots,i-1\} : x_i = y_i]\right]$$
$$\ldots \textit{or } x_{1,\ldots,i-1} = y_{1,\ldots,i-1} \textit{ and } x_i > y_i.$$

"Sorting alphabetically" is usually called *lexicographic ordering* in computer science. This ordering reflects the way that words are listed in the dictionary (also known as the *lexicon*).

Here is one more example of a statement that we've already seen—Goldbach's conjecture—that implicitly involves nested quantifiers; we'll formalize it in predicate logic. (Part of the point of this example is to illustrate how complex even some apparently simple concepts are; there's a good deal of complexity hidden in words like "even" and "prime," which at this point seem pretty intuitive!)

Writing tip: Just as with nested loops in programs, the deeper the nesting of quantifiers, the harder an expression is for a reader to follow. Using well-chosen predicates (like *isPrime*, for example) in a logical statement can make it much easier to read—just like using well-chosen (and well-named) functions makes your software easier to read!

Example 3.47 (Goldbach's Conjecture)

Problem: Recall Goldbach's conjecture, from Example 3.1:

> Every even integer greater than 2 can be written as the sum of two prime numbers.

Formalize this proposition using nested quantifiers.

Solution: Using the *sumOfTwoPrimes* predicate from Example 3.34, we can write this statement as either of the following:

$$\forall n \in \{n \in \mathbb{Z} : n > 2 \land 2 \mid n\} : \mathit{sumOfTwoPrimes}(n) \quad \text{(A)}$$

$$\forall n \in \mathbb{Z} : \left[n > 2 \land 2 \mid n \Rightarrow \mathit{sumOfTwoPrimes}(n) \right] \quad \text{(B)}$$

In (B), we quantify over *all* integers, but the implication $n > 2 \land 2 \mid n \Rightarrow \mathit{sumOfTwoPrimes}(n)$ is trivially true for an integer n that's not even or not greater than 2, because false implies anything! Thus the only instantiations of the quantifier in which the implication has any "meat" is for even integers greater than 2. As such, these two formulations are equivalent. (See Exercise 3.133.) Expanding the definition of *sumOfTwoPrimes*(n) from Example 3.34, we can also rewrite (B) as

$$\left[\begin{array}{l} \forall n \in \mathbb{Z} : n > 2 \land 2 \mid n \Rightarrow \\ \quad \exists p \in \mathbb{Z} : \exists q \in \mathbb{Z} : \left[\mathit{isPrime}(p) \land \mathit{isPrime}(q) \land n = p + q \right] \end{array} \right] \quad \text{(C)}$$

We've also already seen that the predicate *isPrime* implicitly contains quantifiers too ("for all potential divisors d, it is not the case that d evenly divides p")—and, for that matter, so does the "evenly divides" predicate $\mid$. In Exercises 3.178, 3.179, and 3.180, you'll show how to rewrite Goldbach's Conjecture in a few different ways, including using yet further layers of nested quantifiers.

3.5.1 *Order of Quantification*

In expressions that involve nested quantifiers, the order of the quantifiers matters! As a frivolous example, take the title of the 1947 hit song "Everybody Loves Somebody" (sung by Dean Martin). There are two plausible interpretations of the title:

$$\forall x : \exists y : x \text{ loves } y \qquad \text{and} \qquad \exists y : \forall x : x \text{ loves } y.$$

The former is the more natural reading; it says that every person x has someone that he or she loves, but each different x can love a different person. (As in: "every child loves his or her mother.") The latter says that there is one single person loved by *every* x. (As in: "Everybody loves Raymond.") These claims are different!

Taking it further: Disambiguating the order of quantification in English sentences is one of the most daunting challenges in natural language processing (NLP) systems. (See p. 314.) Compare *Every student received a diploma* and *Every student heard a commencement address*: there are, surely, many diplomas and only one address, but building a software system that understands that fact is tremendously challenging! There are many other vexing types of ambiguity in NLP systems, too. A classic example of ambiguity in natural language is the sentence *I saw the man with the telescope*. Is the man holding a telescope? Or did I use one to see him? Human listeners are able to use pragmatic knowledge about the world to disambiguate, but doing so properly in an NLP system is very difficult.

Figure 3.30 shows a visual representation of the importance of this order of quantification. Compare Figure 3.30(d) and Figure 3.30(f), for example: $\forall r : \exists c : P(r,c)$ is true if every row has at least one column with a filled cell in it, whereas $\exists c : \forall r : P(r,c)$ requires that there be a *single* column so that every row has that column's cell filled. The proposition $\exists c : \forall r : P(r,c)$ is *not* true in Figure 3.30(d)—though the proposition $\forall r : \exists c : P(r,c)$ is true in *both* Figure 3.30(d) and Figure 3.30(f).

Here's a mathematical example that illustrates the difference even more precisely.

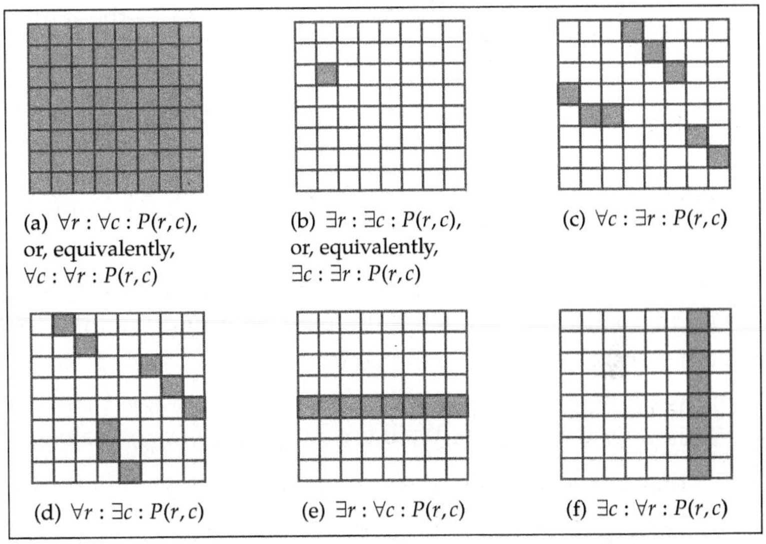

(a) $\forall r : \forall c : P(r,c)$, or, equivalently, $\forall c : \forall r : P(r,c)$

(b) $\exists r : \exists c : P(r,c)$, or, equivalently, $\exists c : \exists r : P(r,c)$

(c) $\forall c : \exists r : P(r,c)$

(d) $\forall r : \exists c : P(r,c)$

(e) $\exists r : \forall c : P(r,c)$

(f) $\exists c : \forall r : P(r,c)$

Figure 3.30: An illustration of order of quantification. Let r index a *row* of the grid, and let c index a *column*. If $P(r,c)$ is true in each filled cell, then the corresponding proposition is true.

Example 3.48 (The largest real number)

Problem: One of the following propositions is true; the other is false. Which is which?

$$\exists y \in \mathbb{R} : \forall x \in \mathbb{R} : x < y \qquad \text{(A)}$$

$$\forall x \in \mathbb{R} : \exists y \in \mathbb{R} : x < y \qquad \text{(B)}$$

Solution: Translating these propositions into English helps resolve this question. (A) says that there is a real number y for which the following property holds: every real number is less than y. ("There is a largest real number.") But there isn't a largest real number! So (A) is false. (If someone tells you that y^* satisfies $\forall x \in \mathbb{R} : x < y^*$, then you can convince him he's wrong by choosing $x = y^* + 1$.) On the other hand, (B) says that, for every real number x, there is a real number greater than x. And that's true: for any $x \in \mathbb{R}$, the number $x + 1$ is greater than x.

In fact, (B) is nearly the negation of (A). (Before you read through the derivation, can you figure out why we had to say "nearly" in the last sentence?)

$$
\begin{aligned}
&\neg[\exists y \in \mathbb{R} : \forall x \in \mathbb{R} : x < y] && \textit{negation of (A)} \\
&\equiv \forall y \in \mathbb{R} : \neg[\forall x \in \mathbb{R} : x < y] && \textit{De Morgan's Laws (quantified form)} \\
&\equiv \forall y \in \mathbb{R} : \exists x \in \mathbb{R} : \neg(x < y) && \textit{De Morgan's Laws (quantified form)} \\
&\equiv \forall y \in \mathbb{R} : \exists x \in \mathbb{R} : y \leq x && \neg(x < y) \Leftrightarrow y \leq x \\
&\equiv \forall x \in \mathbb{R} : \exists y \in \mathbb{R} : x \leq y. && \textit{renaming the bound variables}
\end{aligned}
$$

So (B) and the negation of (A) are almost—but not quite—identical: the latter has a $\leq$ where the former has a $<$. But both (B) and $\neg$(A) are theorems!

Although the order of quantifiers does matter when universal and existential quantifiers both appear in a proposition, the order of consecutive universal quantifiers doesn't matter, nor does the order of consecutive existential quantifiers. (Using our previously defined terminology—see Figure 3.12—these quantifiers are *commutative*.) Thus the following statements are theorems of predicate logic:

$$\forall x \in S : \forall y \in T : P(x,y) \;\Leftrightarrow\; \forall y \in T : \forall x \in S : P(x,y) \qquad (*)$$

$$\exists x \in S : \exists y \in T : P(x,y) \;\Leftrightarrow\; \exists y \in T : \exists x \in S : P(x,y) \qquad (**)$$

The point is simply that the left- and right-hand sides of $(*)$ are both true if and only if $P(x,y)$ is true for every pair $\langle x,y \rangle \in S \times T$, and the left- and right-hand sides of $(**)$ are both true if and only if $P(x,y)$ holds for at least one pair $\langle x,y \rangle \in S \times T$. See Figure 3.30(a) and Figure 3.30(b): both sides of $(*)$ require that all the cells be filled and both sides of $(**)$ require that at least one cell be filled. Because of these equivalences, as notational shorthand we'll sometimes write $\forall x, y \in S : P(x,y)$ instead of writing $\forall x \in S : \forall y \in S : P(x,y)$. We'll use $\exists x, y \in S : P(x,y)$ analogously.

NESTED QUANTIFICATION AND NESTED LOOPS

Just as it can be helpful to think of a quantifier in terms of a corresponding loop, it can be helpful to think of nested quantifiers in terms of nested loops. And a useful way to think about the importance of the order of quantification is through the way in which changing the order of nested loops changes what they compute. In Exercises 3.191–3.196, you'll get a chance to do some translations between quantified statements and nested loops.

Here's one example about how thinking about the nested-loop analogy for nested quantifiers can be helpful. Imagine writing a nested loop to examine every element of a 2-dimensional array. As long as iterations don't depend on each other, it doesn't matter whether we proceed through the array in *row-major order* ("for each row, look at all columns' entries") or *column-major order* ("for each column, look at all rows' entries"). Figure 3.31 illustrates a loop-based view of the logical equivalence expressed by $(**)$, above: both code segments always have the same return value. (The graphical view is that both check every cell of the "grid" of possible inputs to A, as in Figure 3.30(b), just in different orders.)

```
for j = 1 to m:
    for i = 1 to n:
        if A[i, j] then
            return True
return False
```

```
for i = 1 to n:
    for j = 1 to m:
        if A[i, j] then
            return True
return False
```

Figure 3.31: Two nested **for** loops that return the value of $\exists i : \exists j : A[i,j] \equiv \exists j : \exists i : A[i,j]$, by looping in row- or column-major orders.

3.5.2 *Negating Nested Quantifiers*

Recall the rules for negating quantifiers found earlier in the chapter:

$$\neg \forall x \in S : P(x) \;\Leftrightarrow\; \exists x \in S : \neg P(x)$$

$$\neg \exists x \in S : P(x) \;\Leftrightarrow\; \forall x \in S : \neg P(x)$$

Informally, these theorems say that "*everybody is P* is false" is equivalent to "*somebody isn't P*"; and, similarly, "*somebody is P* is false" is equivalent to "*everybody isn't P*."

Here we will consider negating a sequence of *nested* quantifiers. Negating nested quantifiers proceeds in precisely the same way as negating a single quantifier, just acting on one quantifier at a time. (We already saw this idea in Example 3.48, where we repeatedly applied these quantified versions of De Morgan's Laws to a sequence of nested quantifiers.) For example:

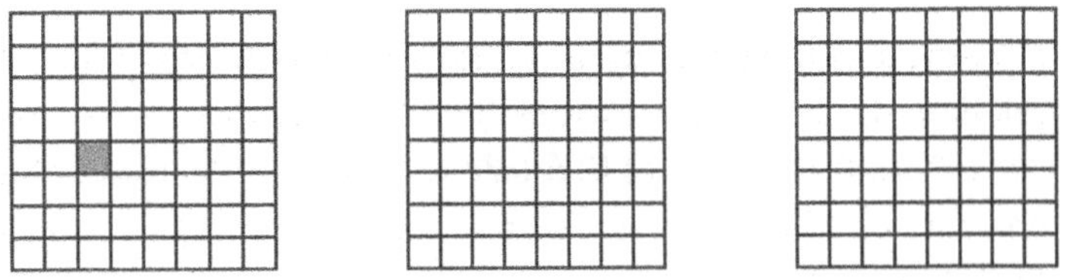

(a) $\exists r : \exists c : P(r,c)$ (b) $\neg(\exists r : \exists c : P(r,c))$ (c) $\forall r : \forall c : \neg P(r,c))$

Figure 3.32: Negating nested quantifiers: what it means for (a) a filled cell to exist; (b) it not to be the case that a filled cell exists; and (c) that every cell is unfilled.

Example 3.49 (No cell is filled ≡ every cell is empty)
Observe that $\exists r : \exists c : P(r,c)$ is true if any r and c makes $P(r,c)$ true—that is, visually, that any cell in the grid in Figure 3.32(a) is filled. For $\exists r : \exists c : P(r,c)$ to be false (Figure 3.32(b)), then we need:

$$\neg(\exists r : \exists c : P(r,c)) \equiv \forall r : \neg(\exists c : P(r,c)) \equiv \forall r : \forall c : \neg P(r,c).$$

That is, *$P(r,c)$ is false for every r and c*—that is, visually, every cell in the grid is unfilled (Figure 3.32(c)). Similarly,

$$\neg\exists r : \forall c : P(r,c) \equiv \forall r : \neg\forall c : P(r,c) \equiv \forall r : \exists c : \neg P(r,c).$$

Thus $\neg\exists r : \forall c : P(r,c)$ expresses that *it's not the case that there's a row with all columns filled*; using the above equivalence, we can rephrase the condition as *every row has at least one unfilled column.*

Example 3.50 (Triple negations)
Here's an example of negating a sequence of triply nested quantifiers:

$$\begin{aligned}\neg\exists x : \forall y : \exists z : P(x,y,z) &\equiv \forall x : \neg\forall y : \exists z : P(x,y,z)\\ &\equiv \forall x : \exists y : \neg\exists z : P(x,y,z)\\ &\equiv \forall x : \exists y : \forall z : \neg P(x,y,z).\end{aligned}$$

Here's a last example, which requires translation from English into logical notation:

Example 3.51 (Negating nested quantifiers)
Problem: Negate the following sentence:

> *For every iPhone user, there's an iPhone app that every one of that user's iPhone-using friends has downloaded.*

Solution: First, let's reason about how the given statement would be false: there would be some iPhone user—we'll call him Steve—such that, for every iPhone app, Steve has a friend who didn't download that app.

Write U and A for the sets of iPhone users and apps, respectively. In (pseudo)logical notation, the original claim looks like

$$\forall u \in U : \exists a \in A : \forall v \in U : \big[(u, v \text{ friends}) \Rightarrow (v \text{ downloaded } a)\big].$$

To negate this statement, we apply the quantified De Morgan's laws, once per quantifier:

$$\begin{aligned} &\neg\forall u \in U : \exists a \in A : \forall v \in U : [(u, v \text{ friends}) \Rightarrow (v \text{ downloaded } a)] \\ &\equiv \exists u \in U : \neg\exists a \in A : \forall v \in U : [(u, v \text{ friends}) \Rightarrow (v \text{ downloaded } a)] \\ &\equiv \exists u \in U : \forall a \in A : \neg\forall v \in U : [(u, v \text{ friends}) \Rightarrow (v \text{ downloaded } a)] \\ &\equiv \exists u \in U : \forall a \in A : \exists v \in U : \neg[(u, v \text{ friends}) \Rightarrow (v \text{ downloaded } a)]. \end{aligned}$$

Using $\neg(p \Rightarrow q) \equiv p \wedge \neg q$ (Exercise 3.82), we can further write this expression as:

$$\equiv \exists u \in U : \forall a \in A : \exists v \in U : [(u, v \text{ friends}) \wedge \neg(v \text{ downloaded } a)].$$

This last proposition, translated into English, matches the informal description above as to why the original claim would be false: *there's some person such that, for every app, that person has a friend who hasn't downloaded that app.*

3.5.3 *Two New Ways of Considering Nested Quantifiers*

We'll close this section with two different but useful ways to think about nested quantification. As a running example, consider the following (true!) proposition

$$\forall x \in \mathbb{Z} : \exists y \in \mathbb{Z} : x = y + 1, \qquad (\dagger)$$

which says that the number that's one less than every integer is an integer too. We'll discuss two ways of thinking about propositions like (†) with nested quantifiers: as a "game with a demon" in which you battle against an all-knowing demon to try to make the innermost quantifier's body true;[12] and as a single quantifier whose body is a predicate, but a predicate that just happens to be expressed using quantifiers.

Thanks to Dexter Kozen for teaching me this way of thinking of nested quantifiers. See:

[12] Dexter Kozen. *Automata and Computability.* Springer, 1997.

Nested quantifiers as demon games

One way to think about any proposition involving nested quantifiers is as a "game" played between you and a demon. Here are the rules of the game:

- Your goal is to make the innermost statement—$x = y + 1$ for our running example (†)—turn out to be true; the demon's goal is to make that statement false.
- Every "for all" quantifier in the expression is a choice that the demon makes; every "there exists" quantifier in the expression is a choice that you get to make. (That

is, in the expression $\forall a \in S : \cdots$, the demon chooses a particular value of $a \in S$, and the game continues in the "$\cdots$" part of the expression. And in the expression $\exists b \in S : \cdots$, you choose a particular value of $b \in S$, and, again, the game continues in the "$\cdots$" part.)

- Your choices and the demon's choices are made in the left-to-right order (from the outside in) of the quantifiers.
- You win the game—in other words, the proposition in question is true—if, no matter how cleverly the demon plays, you make the innermost statement true.

Here are two examples of viewing quantified statement as demon games, one for a true statement and one for a false statement:

Example 3.52 (Showing that (†) is true)
We'll use a "game with the demon" to argue that $\forall x \in \mathbb{Z} : \exists y \in \mathbb{Z} : x = y + 1$ is true.

1. The outermost quantifier is $\forall$, so the demon picks a value for $x \in \mathbb{Z}$.
2. Now you get to pick a value $y \in \mathbb{Z}$. A good choice for you is $y := x - 1$.
3. Because you chose $y = x - 1$, indeed $x = y + 1$. You win!

(For example, if the demon picks 16, you pick 15. If the demon picks -19, you pick -20. And so forth.) No matter what the demon picks, your strategy will make you win—and therefore (†) is true!

By contrast, consider (†) with the order of quantification reversed:

Example 3.53 (A losing demon game)
Consider playing a demon game for the proposition

$$\exists y \in \mathbb{Z} : \forall x \in \mathbb{Z} : x = y + 1.$$

Unfortunately, the $\exists$ is first, which means that you have to make the first move. But when you pick a number y, the demon *then* gets to pick an x—and there are an infinitude of x values that the demon can choose so that $x \neq y + 1$. (You pick 42? The demon picks 666. You pick 17? The demon picks 666. You pick 665? The demon picks 616.) Therefore you can't guarantee that you win the game, so we haven't established this claim.

By the way, you *could* win a demon game to prove the negation of the claim in Example 3.53:

$$\neg(\text{the claim from Example 3.53}) \equiv \forall y \in \mathbb{Z} : \exists x \in \mathbb{Z} : x \neq y + 1.$$

First, the demon picks some unknown $y \in \mathbb{Z}$. Then you have to pick an $x \in \mathbb{Z}$ such that $x \neq y + 1$—but that's easy: for any y the demon picks, you pick $x = y$. You win!

NESTED QUANTIFIERS AS SINGLE QUANTIFIERS

In our running example—$\forall x \in \mathbb{Z} : \underline{\exists y \in \mathbb{Z} : x = y + 1}$—what kind of thing is the underlined piece of the expression? It can't be a proposition, because x is a free variable in it. But once we plug in a value for x, the expression becomes true or false. In other words, the expression $\exists y \in \mathbb{Z} : x - 1 = y$ is itself a (unary) predicate: once we are given a value of x, we can compute the truth value of the expression. Similarly, the expression $x - 1 = y$ is also a predicate—but a binary predicate, taking both an x and y as arguments. Let's define two predicates:

- $P(x, y)$ denotes the predicate $x - 1 = y$.
- $hasIntPredecessor(x)$ denotes the predicate $\exists y \in \mathbb{Z} : x - 1 = y$.

Using this notation, we can write (†) as

$$\begin{aligned} \forall x \in \mathbb{Z} : \overbrace{\exists y \in \mathbb{Z} : \overbrace{x - 1 = y}^{P(x,y)}}^{hasIntPredecessor(x)} &\equiv \forall x \in \mathbb{Z} : \exists y \in \mathbb{Z} : P(x, y) \\ &\equiv \forall x \in \mathbb{Z} : hasIntPredecessor(x). \qquad (\ddagger) \end{aligned}$$

One implication of this view is that negating nested quantifiers is really just the same as negating non-nested quantifiers. For example:

Example 3.54 (Negating nested quantifiers)
We can view the negation of (†), as written in (‡), as follows:

$$\begin{aligned} \neg(\dagger) &\equiv \neg \forall x \in \mathbb{Z} : hasIntPredecessor(x) \\ &\equiv \exists x \in \mathbb{Z} : \neg hasIntPredecessor(x). \end{aligned}$$

And, re-expanding the definition of *hasIntPredecessor* and again applying the quantified De Morgan's Law, we have that

$$\begin{aligned} \neg hasIntPredecessor(x) &\equiv \neg \exists y \in \mathbb{Z} : P(x, y) \\ &\equiv \forall y \in \mathbb{Z} : \neg P(x, y) \\ &\equiv \forall y \in \mathbb{Z} : x - 1 \neq y. \end{aligned}$$

Together, these two negations show

$$\begin{aligned} \neg(\dagger) &\equiv \exists x \in \mathbb{Z} : \neg hasIntPredecessor(x) \\ &\equiv \exists x \in \mathbb{Z} : \forall y \in \mathbb{Z} : \neg P(x, y) \\ &\equiv \exists x \in \mathbb{Z} : \forall y \in \mathbb{Z} : x - 1 \neq y. \end{aligned}$$

Taking it further: This view of nested quantifiers as a single quantifier whose body just happens to express its condition using quantifiers has a close analogy with writing a particular kind of function in a programming language. If we look at a two-argument function in the right light, we can see it as a function that takes one argument *and returns a function that takes one argument*. This approach is called *Currying*; see p. 357 for some discussion.

Computer Science Connections

Currying

Consider a binary predicate $P(x,y)$, as used in a quantified expression like $\forall y : \forall x : P(x,y)$. As we discussed, we can think of this expression as first plugging in a value for y, which then yields a unary predicate $\forall x : P(x,y)$ which then takes the argument x.

There's an interesting parallel between this view of nested quantifiers and a way of writing functions in some programming languages. For concreteness, let's think about a very simple function that takes two arguments and returns their sum. Figure 3.33 shows implementations of this function in three different programming languages: ML, Python, and Scheme. A few notes about syntax:

- For ML: `fun` is a keyword that says we're defining a function; `sum` is the name of it; `a b` is the list of arguments; and that function is defined to return the value of `a + b`.
- For Scheme: `(lambda args body)` denotes the function that takes arguments `args` and returns the value of the function body `body`. Applying the function `f` to arguments `arg1`, `arg2`, ..., `argN` is expressed as `(f arg1 arg2 ... argN)`. For example, `(+ 1 2)` has the value 3.

We can then use the function `sum` to actually add numbers; see Figure 3.34.

But now suppose that we wanted to write a new function that takes one argument and adds 3 to it. Can we make use of the `sum` function to do so? (The analogy to predicates is that taking a two-argument predicate and applying it to one argument gives one-argument predicate; here we're trying to take a two-argument function in a programming language and apply it to one argument to yield a one-argument function.) It turns out that creating the "add 3" function using `sum` is very easy in ML: we simply apply `sum` to one argument, and the result is a function that "still wants" one more argument. See Figure 3.35.

A function like `sum` in ML, which takes its multiple arguments "one at a time," is said to be *Curried*—in honor of Haskell Curry, a 20th-century American logician. (The programming language Haskell is also named in his honor.) Thinking about Curried functions is a classical topic in the study of the structure of programming languages.[13] While writing Curried functions is almost automatic in ML, one can also write Curried functions in other programming languages, too. Examples of a Curried version of `sum` in Python and Scheme are in Figure 3.36; it's even possible to write Curried functions in C or Java, though it's much less natural than in ML/Python/Scheme.

```
fun sum a b = a + b;                (* ML *)

def sum(a,b):                       # Python
  return a + b

(define sum                         ; Scheme
  (lambda (a b)
    (+ a b)))
```

Figure 3.33: Implementations of sum$(a,b) = a + b$ in three languages.

```
sum 2 3;            (* returns 5 *)
sum 99 12;          (* returns 111 *)
```

```
sum(2,3)            # returns 5
sum(99,12)          # returns 111
```

```
(sum 2 3)           ; returns 5
(sum 99 12)         ; returns 111
```

Figure 3.34: Using sum in all three languages.

```
(* define a "value" add3 as sum
   applied to 3, making add3 a
   1-argument function *)
val add3 = sum 3;

add3 0;             (* returns 3 *)
add3 108;           (* returns 111 *)
add3 199;           (* returns 202 *)
```

Figure 3.35: Applying `sum` to one of two arguments in ML, and then applying the resulting function to a second argument.

For more, see the classic text

[13] Harold Abelson and Gerald Jay Sussman with Julie Sussman. *Structure and Interpretation of Computer Programs*. MIT Press/McGraw-Hill, 2nd edition, 1996.

```
def sum(a):
  def sumA(b):
     return a + b
  return sumA

sum(3)(2)           # returns 5
add3 = sum(3)
add3(2)             # returns 5
```

```
(define sum                 ;; define sum as
  (lambda (a)               ;; the function taking argument a and returning
    (lambda (b) (+ a b))))  ;;   [the function taking argument b and returning a+b]

((sum 3) 2)                 ;; returns 5
(define add3 (sum 3))
(add3 2)                    ;; returns 5
```

Figure 3.36: Python/Scheme Currying.

3.5.4 Exercises

Let F denote the set of all functions $f : \mathbb{R} \to \mathbb{R}$ taking real numbers as input and producing real numbers as output. (For one example, $\mathsf{plusone}(x) = x + 1$ *is a function* $\mathsf{plusone} : \mathbb{R} \to \mathbb{R}$*, so* $\mathsf{plusone} \in F$*.) Are the following propositions true or false? Justify your answers, including a description of the function(s) if they exist.*

3.141 $\forall c \in \mathbb{R} : \left[\exists f \in F : f(0) = c\right]$

3.142 $\exists f \in F : \left[\forall c \in \mathbb{R} : f(0) = c\right]$

3.143 $\forall c \in \mathbb{R} : \left[\exists f \in F : f(c) = 0\right]$

3.144 $\exists f \in F : \left[\forall c \in \mathbb{R} : f(c) = 0\right]$

Under many operating systems, users can schedule a task to be run at a specified time in the future. In Unix-like operating systems, this type of scheduled job is called a cron job*. (For example, a backup might run nightly at 2:00am, and a scratch drive might be emptied out weekly on Friday night at 11:50pm.)*

Greek: *chron-* "time."

Let $T = \{1, 2, \ldots, t_{\max}\}$ be a set of times (measured in minutes, let's say), and let J be a set of jobs. Let scheduledAt *be a predicate so that* scheduledAt(j, t) *is true if and only if job j is scheduled at time t. (Assume that jobs do not last more than one minute.) Formalize the following conditions using only standard quantifiers, arithmetic operators, logical connectives, and the* scheduledAt *predicate.*

3.145 There is never more than one job scheduled at the same time.

3.146 Every job is scheduled at least once.

3.147 Job A is never run twice within two minutes.

3.148 Job B is run at least three times.

3.149 Job C is run at most twice.

3.150 Job D is run sometime after the last time that Job E is run.

3.151 Job F is run at least once between consecutive executions of Job G.

3.152 Job H is run at most once between consecutive executions of Job I.

Let $P[1 \ldots n, 1 \ldots m]$ be a 2-dimensional array of the pixels of a black-and-white image: for every x and y, the value of $P[x, y] = 0$ if the $\langle x, y \rangle$th pixel is black, and $P[x, y] = 1$ if it's white. Translate these statements into predicate logic:

3.153 Every pixel in the image is black.

3.154 There is at least one white pixel.

3.155 Every row has at least one white pixel.

3.156 There are never two consecutive white pixels in the same column.

A standard American crossword puzzle is a 15-by-15 grid, which can be represented as a two-dimensional 15-by-15 array G, where $G[i, j] =$ True *if and only if the cell in the ith row and jth column is "open" (a.k.a. unfilled, a.k.a. not a black square). Maximal contiguous horizontal or vertical sequences of two or more open squares are called* words*. For any $i \leq 0$, $i > 15$, $j \leq 0$, or $j > 15$, treat $G[i, j] =$* False.

Taking it further: *The assumption that the $\langle i, j \rangle$th cell of G is False except when $1 \leq i, j \leq 15$ can be re-expressed as us pretending that our real grid is surrounded by black squares. In CS, this style of structure is called a* sentinel*, wherein we introduce boundary values to avoid having to write out verbose special cases.*

There are certain customs that G must obey to be a standard American puzzle. (See Figure 3.37, for example.) Rewrite the informally stated conditions that follow as fully formal definitions.

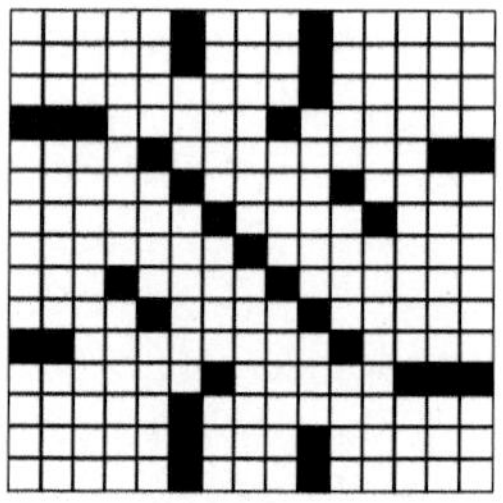

Figure 3.37: A sample American crossword puzzle.

3.157 *no unchecked letters:* every open cell appears in both a down word and an across word.

3.158 *no two-letter words:* every word has length at least 3.

3.159 *rotational symmetry:* if the entire grid is rotated by 180°, then the rotated grid is identical to the original grid.

3.160 *overall interlock:* for any two open squares, there is a path of open squares that connects the first to the second. (That is, we can get from *here* to *there* through words.) Your answer should formally define a predicate $P(i, j, x, y)$ that is true exactly when there exists is a path from $\langle i, j \rangle$ to $\langle x, y \rangle$: "there exists a sequence of open squares starting with $\langle i, j \rangle$ such that ...".)

3.161 Definition 2.30 defines a *partition* of a set S as a set $\{A_1, A_2, \ldots, A_k\}$ of sets such that (i) $A_1, A_2, \ldots, A_k$ are all nonempty; (ii) $A_1 \cup A_2 \cup \cdots \cup A_k = S$; and (iii) for any distinct $i, j \in \{1, \ldots, k\}$, the sets A_i and A_j are disjoint. Formalize this definition using nested quantifiers and basic set notation.

3.162 Consider the "maximum" problem: given an array of numbers, return the maximum element of that array. Complete the formal specification for this problem by finishing the specification under "output":

Input: An array $A[1 \ldots n]$, where each $A[i] \in \mathbb{Z}$.
Output: An integer $x \in \mathbb{Z}$ such that ...

Let $T = \{1,\ldots,12\} \times \{0,1,\ldots,59\}$ denote the set of numbers that can be displayed on a digital clock in twelve-hour mode. (A clock actually displays a colon between the two numbers.) We can think of a clock as a function $c : T \to T$, so that when the real time is $t \in T$, then the clock displays the time $c(t)$. (For example, if fastby7 *runs seven minutes fast, then* fastby7(12:00) = 12:07.)

For several of these questions, it may be helpful to make use of the function $\text{add} : T \times \mathbb{Z}^{\geq 0} \to T$ *so that* $\text{add}(t,x)$ *denotes the time that's x minutes later than t. See Exercise 2.243.*

Formalize each of the following predicates using only the standard quantifiers and equality symbols.

3.163 A clock is *right* if it always displays the correct time. Formalize the predicate *right*.

3.164 A clock *keeps time* if there's some fixed offset by which it is always off from being right. (For example, *fastby7* above correctly keeps time.) Formalize the predicate *keepsTime*.

3.165 A clock is *close enough* if it always displays a time that's within two minutes of the correct time. Formalize the predicate *closeEnough*.

3.166 A clock is *broken* if there's some fixed time that it always displays, regardless of the real time. Formalize the predicate *broken*.

3.167 "Even a broken clock is right twice a day," they say. (They mean: "even a broken clock displays the correct time at least once per T.") Formalize the adage and prove it true.

A classic topic of study for computational biologists is genomic distance measures: *given two genomes, we'd like to report a single number that represents how different those two genomes are. These distance computations are useful in, for example, reconstructing the evolutionary tree of a collection of species.*

Consider two genomes A and B of bacterium. Let's label the n genes that appear in A's chromosome, in order, as $\pi_A = 1,2,\ldots,n$. The same genes appear in a different order in B—say, in the order $\pi_B = r_1, r_2, \ldots r_n$. A particular model of genomic distance will define a specific way in which this list of numbers can mutate; the question at hand is to find the minimum-length sequence of these mutations that are necessary to explain the difference between the orders π_A and π_B. One type of biologically motivated mutation is the prefix reversal—*in which some prefix of π_B is reversed, as in $\langle \underline{3,2,1},4,5\rangle \to \langle \underline{1,2,3},4,5\rangle$. It turns out that this model is exactly the* pancake-flipping problem, *the subject of the lone academic paper with Bill Gates as an author.*[14] *(See Figure 3.38.)*

[14] W. H. Gates and C. H. Papadimitriou. Bounds for sorting by prefix reversals. *Discrete Mathematics*, 27:47–57, 1979.

(a) Two pancake-flipping instances. Given a stack of pancakes, with radii labeled from top to bottom, we must sort the pile by radius. We sort with a sequence of *flips*: turn the top k pancakes upside down, for some k, and replace them (inverted) on top of the remaining pancakes. The left instance is $\langle 4,3,2,1,5\rangle$; the right is $\langle 5,4,3,1,2\rangle$. They require 1 and 2 flips, respectively, to solve (as shown).

(b) A biological view. Think of a chromosome as a sequence of genes. If, in the course of cell activity, one end of the chromosome comes in contact with a point in the middle of the chromosome, a loop forms. If the loop untangles itself "the other way around," the effect is to reverse the order of the genes in that loop. This transformation effects a prefix reversal on those genes. Here 123456789*abc* becomes 987654321*abc*.

Figure 3.38: The pancake-flipping problem, and its biological significance.

Suppose that you are given a stack of pancake radii $r_1, r_2, \ldots, r_n$, arranged from top to bottom, where $\{r_1, r_2, \ldots, r_n\} = \{1,2,\ldots,n\}$ (but not necessarily in order). Write down a fully quantified logical expression that expresses the condition that ...

3.168 ... the given pancakes are sorted.

3.169 ... the given pancakes can be sorted with exactly one flip (see Figure 3.38).

3.170 ... the given pancakes can be sorted with exactly two flips. *(Hint: writing a program to verify that your indices aren't off by one is a very good idea!)*

Let P be a set of people, and let T be a set of times. Let friends(x,y) *be a predicate denoting that $x \in P$ and $y \in P$ are friends. Let* bought(x,t) *be a predicate denoting that $x \in P$ bought an iPad at time $t \in T$.*

3.171 Formalize this statement in predicate logic: "Everyone who bought an iPad has a friend who bought one previously."

3.172 Is the claim from Exercise 3.171 true (in the real world)? Justify your answer.

In programming, an assertion *is a logical statement that announces ("asserts") a condition φ that the programmer believes to be true. For example, a programmer who is about to access the 202nd element of an array A might assert that* length(A) $\geq$ 202 *before accessing this element. When an executing program in languages like C and Java reaches an* `assert` *statement, the program aborts if the condition in the statement isn't true.*
For the following, give a nonempty *input array A* that would cause the stated assertion from Figure 3.39 to fail *(that is, for the asserted condition to be false).*

3.173 `foo`

3.174 `bar`

3.175 `baz`

```
foo(A[1...n]):
   last = 0
   for index = 1 ... n-1:
      if A[index] > A[index+1]:
         last = index
   assert(last >= 1 and last <= n-1)
   swap A[last], A[last+1]
```

```
bar(A[1...n]):
   total = A[1]
   i = 1
   for i = 2 ... n-1:
      if A[i+1] > A[i]:
         total = total + A[i]
         assert(total > A[1])
   return total
```

```
baz(A[1...n]):
   for start = 1 ... n-1:
      min = start
      for i = start+1 ... n:
         assert(start == 1
                   or A[i] > A[start-1])
         if A[min] > A[i]:
            min = i
         swap A[start], A[min]
```

Figure 3.39: Some functions using `assert` statements.

Taking it further: Using assertions can be an extremely valuable way of documenting and debugging programs, particularly because liberally including assertions will allow the revelation of unexpected data values much earlier in the execution of a program. And these languages have a global toggle that allows the testing of assertions to be turned off, so once the programmer is satisfied that the program is working properly, she doesn't have to worry about any running-time overhead for these checks.

While the quantifiers $\forall$ and $\exists$ are by far the most common, there are some other quantifiers that are sometimes used. For each of the following quantifiers, write an expression that is logically equivalent to the given statement that uses only the quantifiers $\forall$ and $\exists$; standard propositional logic notation ($\wedge, \neg, \vee, \Rightarrow$); standard equality/inequality notation ($=, \geq, \leq, <, >$); and the predicate P in the question.

3.176 Write an equivalent expression to $\exists! x \in \mathbb{Z} : P(x)$ ("there exists a unique $x \in \mathbb{Z}$ such that $P(x)$"), which is true when there is one and only one value of x in the set $\mathbb{Z}$ such that $P(x)$ is true.

3.177 Write an equivalent expression to $\exists_\infty x \in \mathbb{Z} : P(x)$ ("there exist infinitely many $x \in \mathbb{Z}$ such that $P(x)$"), which is true when there are infinitely many different values of $x \in \mathbb{Z}$ such that $P(x)$ is true.

Here are two formulations of Goldbach's conjecture (see Example 3.47):

$$\forall n \in \mathbb{Z} : \left[\; n > 2 \wedge 2 \mid n \Rightarrow \quad \left(\exists p \in \mathbb{Z} : \exists q \in \mathbb{Z} : \left[\text{isPrime}(p) \wedge \text{isPrime}(q) \wedge n = p + q\right]\right) \;\right]$$
$$\forall n \in \mathbb{Z} : \exists p \in \mathbb{Z} : \exists q \in \mathbb{Z} : \left[\; n \leq 2 \vee 2 \nmid n \vee \left[\text{isPrime}(p) \wedge \text{isPrime}(q) \wedge n = p + q\right] \;\right].$$

3.178 Prove that these two formulations of Goldbach's conjecture are logically equivalent.

3.179 Rewrite Goldbach's conjecture without using *isPrime*—that is, using only quantifiers, the $\mid$ predicate, and standard arithmetic ($+, \cdot, \geq$, etc.).

3.180 Even the $\mid$ predicate implicitly involves a quantifier: $p \mid q$ is equivalent to $\exists k \in \mathbb{Z} : p \cdot k = q$. Rewrite Goldbach's conjecture without using the $\mid$ predicate either—that is, use only quantifiers and standard arithmetic symbols ($+, \cdot, \geq$, etc.).

3.181 *(programming required)* As we discussed, the truth value of Goldbach's conjecture is currently unknown. As of April 2012, the conjecture has been verified for all even integers from 4 up to 4×10^{18}, through a massive distributed computation effort led by Tomás Oliveira e Silva. Write a program to test Goldbach's conjecture, in a programming language of your choice, for even integers up to 10,000.

Most real-world English utterances are ambiguous*—that is, there are multiple possible interpretations of the given sentence. A particularly common type of ambiguity involves* order of quantification. *For each of the following English sentences, find as many different logical readings based on order of quantification as you can. Write down those interpretations using pseudological notation, and also write a sentence that expresses each meaning unambiguously.*

3.182 A computer crashes every day.

3.183 Every prime number except 2 is divisible by an odd integer greater than 1.

3.184 Every student takes a class every term.

3.185 Every submitted program failed on a case submitted by a student.

3.186 You should have found two different logical interpretations in Exercise 3.183. One of these interpretations is a theorem, and one of them is not. Decide which is which, and prove your answers.

Let S be an arbitrary nonempty set and let P be an arbitrary binary predicate. Decide whether the following statements are always true (for any P and S), or whether they can be false. Prove your answers.

3.187 $\left[\exists y \in S : \forall x \in S : P(x,y)\right] \Rightarrow \left[\forall x \in S : \exists y \in S : P(x,y)\right]$

3.188 $\left[\forall x \in S : \exists y \in S : P(x,y)\right] \Rightarrow \left[\exists y \in S : \forall x \in S : P(x,y)\right]$

Consider any unary predicate P(x) over a nonempty set S. It turns out that both of the following propositions are theorems of propositional logic. Prove them both.

3.189 $\forall x \in S : \left[P(x) \Rightarrow (\exists y \in S : P(y))\right]$

3.190 $\exists x \in S : \left[P(x) \Rightarrow (\forall y \in S : P(y))\right]$

The following blocks of code use nested loops to compute some fact about a predicate P. For each, write a fully quantified statement of predicate logic whose truth value matches the value returned by the given code. (Assume that S is a finite universe.)

3.191

```
for x in S:
   for y in S:
      flag = False
      if P(x) or P(y):
         flag = True
   if flag:
      return True
return False
```

3.192

```
for x in S:
   flag = False
   for y in S:
      if not P(x,y):
         flag = True
   if flag:
      return True
return False
```

3.193

```
for x in S:
   flag = True
   for y in S:
      if not P(x,y):
         flag = False
   if flag:
      return True
return False
```

3.194

```
for x in S:
   flag = False
   for y in S:
      if not P(x,y):
         flag = True
   if not flag:
      return False
return True
```

3.195

```
for x in S:
   for y in S:
      if P(x,y):
         return False
return True
```

3.196

```
flag = False
for x in S:
   for y in S:
      if P(x,y):
         flag = True
return flag
```

3.197 As we've discussed, there is no algorithm that can decide whether a given fully quantified proposition φ is a theorem of predicate logic. But there are several specific types of fully quantified propositions for which we *can* decide whether a given statement is a theorem. Here you'll show that, when quantification is only over a *finite* set, it is possible to give an algorithm to determine whether φ is a theorem. Suppose that you are given a fully quantified proposition φ, where the domain for every quantifier is a finite set—say $S = \{0, 1\}$. Describe an algorithm that is guaranteed to figure out whether φ is a theorem.

3.6 *Chapter at a Glance*

Propositional Logic

A *proposition* is the kind of thing that is either true or false. An *atomic proposition* (or *Boolean variable*) is a conceptually indivisible proposition. A *compound proposition* (or *Boolean formula*) is one built up using a *logical connective* and one or more simpler propositions. The most common logical connectives are the ones shown in Figure 3.40. A proposition that contains the atomic propositions $p_1, \ldots, p_k$ is sometimes called a *Boolean formula over* $p_1, \ldots, p_k$ or a *Boolean expression over* $p_1, \ldots, p_k$.

negation		$\neg p$	"not p"
disjunction	(inclusive: "p, q, or both")	$p \vee q$	"p or q"
conjunction		$p \wedge q$	"p and q"
implication		$p \Rightarrow q$	"if p, then q" or "p implies q"
equivalence		$p \Leftrightarrow q$	"p if and only if q"
exclusive or	("p or q, but not both")	$p \oplus q$	"p xor q"

Figure 3.40: Logical connectives.

The *truth value* of a proposition is its truth or falsity. (The truth value of a Boolean formula over $p_1, \ldots, p_k$ is determined only by the truth values of each of $p_1, \ldots, p_k$.) Each logical connective is defined by how the truth value of the compound proposition formed using that connective relates to the truth values of the constituent propositions. A *truth table* defines a connective by listing, for each possible assignment of truth values for the constituent propositions, the truth value of the entire compound proposition. See Figure 3.41. Observe that the proposition $p \Rightarrow q$ is true if, whenever p is true, q is too. So the only situation in which $p \Rightarrow q$ is false is when p is true and q is false. False implies anything! Anything implies true!

Consider a Boolean formula over variables $p_1, \ldots, p_k$. A *truth assignment* is a setting to true or false for each variable. (So a truth assignment corresponds to a row of the truth table for the proposition.) A truth assignment *satisfies* the proposition if, when the values from the truth assignment are plugged in, the proposition is true. A Boolean formula is a *tautology* if *every* truth assignment satisfies it; it's *satisfiable* if *some* truth assignment satisfies it; and it's *unsatisfiable* or a *contradiction* if no truth assignment does. Two Boolean propositions are *logically equivalent* if they're satisfied by exactly the same truth assignments (that is, they have identical truth tables).

p	$\neg p$
T	F
F	T

p	q	$p \wedge q$	$p \vee q$
T	T	T	T
T	F	F	T
F	T	F	T
F	F	F	F

p	q	$p \Rightarrow q$
T	T	T
T	F	F
F	T	T
F	F	T

p	q	$p \oplus q$	$p \Leftrightarrow q$
T	T	F	T
T	F	T	F
F	T	T	F
F	F	F	T

Figure 3.41: Truth tables for the basic logical connectives.

Consider an implication $p \Rightarrow q$. The *antecedent* or *hypothesis* of the implication is p; the *consequent* or *conclusion* of the implication is q. The *converse* of the implication $p \Rightarrow q$ is the implication $q \Rightarrow p$. The *contrapositive* is the implication $\neg q \Rightarrow \neg p$. Any implication is logically equivalent to its contrapositive. But an implication is *not* logically equivalent to its converse!

A *literal* is a Boolean variable or the negation of a Boolean variable. A proposition is in *conjunctive normal form (CNF)* if it is the conjunction (and) of a collection of clauses, where a clause is a disjunction (or) of a collection of literals. A proposition is in *disjunctive normal form (DNF)* if it is the disjunction of a collection of clauses, where a clause is a conjunction of a collection of literals. Every proposition is logically equivalent to a proposition that is in CNF, and to another that is in DNF.

Predicate Logic

A *predicate* is a statement containing some number of variables that has a truth value once values are plugged in for those variables. (Alternatively, a *predicate* is a Boolean-valued function.) Once particular values for these variables are plugged in, the resulting expression is a proposition. A proposition can also be formed from a predicate through *quantifiers:*

- The *universal quantifier* $\forall$ ("for all"): the proposition $\forall x \in U : P(x)$ is true if, for every $x \in U$, we have that $P(x)$ is true.
- The *existential quantifier* $\exists$ ("there exists"): the proposition $\exists x \in U : P(x)$ is true if, for at least one $x \in U$, we have that $P(x)$ is true.

The set U is called the *universe* or *domain of discourse*. When the universe is clear from context, it may be omitted from the notation.

In the expression $[\forall x : \underline{\quad}]$ or $[\exists x : \underline{\quad}]$, the *scope* or *body* of the quantifier is the underlined blank, and the variable x is *bound* by the quantifier. A *free* or *unbound* variable is one that is not bound by any quantifier. A *fully quantified* expression is one with no free variables.

A *theorem* of predicate logic is a fully quantified expression that is true for all possible meanings of the predicates in it. Two expressions are *logically equivalent* if they are true under precisely the same set of meanings for their predicates. (Alternatively, two expressions φ and ψ are *logically equivalent* if $\varphi \Leftrightarrow \psi$ is a theorem.) Two useful theorems of predicate logic are De Morgan's laws: $\neg\forall x \in S : P(x) \Leftrightarrow \exists x \in S : \neg P(x)$ and $\neg\exists x \in S : P(x) \Leftrightarrow \forall x \in S : \neg P(x)$.

There is no general algorithm that can test whether any given expression is a theorem. If we wish to prove that an implication $\varphi \Rightarrow \psi$ is an theorem, we can do so with a *proof by assuming the antecedent:* to prove that the implication $\varphi \Rightarrow \psi$ is always true, we will rule out the one scenario in which it wouldn't be; specifically, we *assume* that φ is true, and then *prove* that ψ must be true too, under this assumption.

A *vacuously quantified* statement is one in which the domain of discourse is the empty set. The vacuous universal quantification $\forall x \in \varnothing : P(x)$ is a theorem; the vacuous existential quantification $\exists x \in \varnothing : P(x)$ is always false.

Quantifiers are *nested* if one quantifier is inside the scope of another quantifier. Nested quantifiers work in precisely the same way as single quantifiers, applied in sequence. A proposition involving nested quantifier like $\forall x \in S : \exists y \in T : R(x, y)$ is true if, for every choice of x, there is some choice of y (which can depend on the choice of x) for which $R(x, y)$ is true. Order of quantification matters in general; the expressions $\forall x : \exists y : R(x, y)$ and $\exists y : \forall x : R(x, y)$ are *not* logically equivalent.

Key Terms and Results

Key Terms

PROPOSITIONAL LOGIC

- proposition
- truth value
- atomic and compound propositions
- logical connectives:
  - negation ($\neg$)
  - conjunction ($\wedge$)
  - disjunction ($\vee$)
  - implication ($\Rightarrow$)
  - exclusive or ($\oplus$)
  - if and only if ($\Leftrightarrow$)
- truth assignments and truth tables
- tautology
- satisfiability/unsatisfiability
- logical equivalence
- antecedent and consequent
- converse, contrapositive, and inverse
- conjunctive normal form (CNF)
- disjunctive normal form (DNF)

PREDICATE LOGIC

- predicate
- quantifiers:
  - universal quantifier ($\forall$)
  - existential quantifier ($\exists$)
- free and bound variables
- fully quantified expression
- theorems of predicate logic
- logical equivalence in predicate logic
- proof by assuming the antecedent
- vacuous quantification
- nested quantifiers

Key Results

PROPOSITIONAL LOGIC

1. We can build a truth table for any proposition by repeatedly applying the definitions of each of the logical connectives, as shown in Figure 3.4.
2. Two propositions φ and ψ are logically equivalent if and only if $\varphi \Leftrightarrow \psi$ is a tautology.
3. An implication $p \Rightarrow q$ is logically equivalent to its contrapositive $\neg q \Rightarrow \neg p$, but not to its converse $q \Rightarrow p$.
4. There are many important propositional tautologies and logical equivalences, some of which are shown in Figures 3.10 and 3.12.
5. We can show that propositions are logically equivalent by showing that every row of their truth tables are the same.
6. Every proposition is logically equivalent to one that is in disjunctive normal form (DNF) and to one that is in conjunctive normal form (CNF).

PREDICATE LOGIC

1. We can build a proposition from a predicate $P(x)$ by plugging in a particular value for x, or by quantifying over x as in $\forall x : P(x)$ or $\exists x : P(x)$.
2. Unlike with propositional logic, there is no algorithm that is guaranteed to determine whether a given fully quantified predicate-logic expression is a theorem.
3. There are many important predicate-logic theorems, some of which are shown in Figure 3.23.
4. The statements $\neg\forall x : P(x)$ and $\exists x : \neg P(x)$ are logically equivalent. So are $\neg\exists x : P(x)$ and $\forall x : \neg P(x)$.
5. We can think of nested quantifiers as a sequence of single quantifiers, or as "games with a demon."

4

Proofs

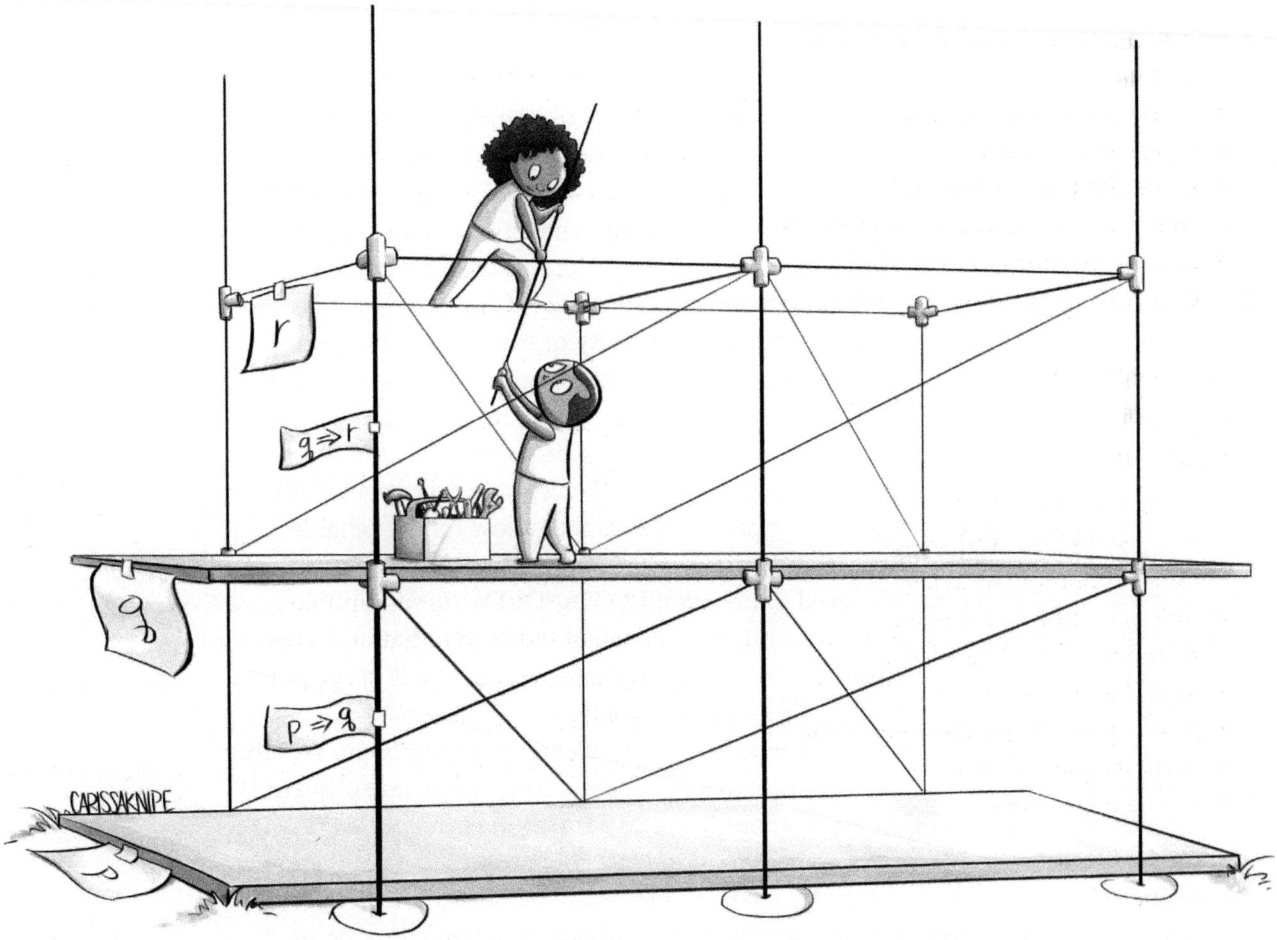

In which our heroes build ironclad scaffolding to support their claims, thereby making them impervious to any perils they might encounter.

4.1 *Why You Might Care*

By far the best proof is experience.

Sir Francis Bacon (1561–1626)

A *proof* is a convincing argument that establishes a particular claim as fact. That claim might be something explicitly computational: *Bubble Sort performs fewer comparisons than Merge Sort when the input array is already sorted*, for example. Or the claim might be noncomputational, at least superficially: a property of an operating system, a structural fact about the minimum-length sequence of flips to sort pancakes, the impossibility of designing a voting system with a certain set of properties.

Generally speaking, our goal—in this chapter, in this book—is to establish new facts. And that's precisely the point of a proof: to derive a new fact from old facts, while persuading the reader that the new fact is, indeed, a fact. (For example, we can derive a new fact using Modus Ponens: if we know both p and $p \Rightarrow q$, then we can conclude that q is a fact, too.) In Section 4.3, the technical meat of this chapter, we will develop a toolbox of techniques to use in proofs, and some strategies for choosing among these techniques. (In Section 4.5, we'll also catalogue some common types of mistakes in purported proofs, so that you can avoid them—and recognize bogus proofs when others attempt them.) We'll illustrate these proof techniques throughout Section 4.3 with a hefty collection of examples about arithmetic.

While the proof techniques themselves are the "point" of this chapter, in many cases the *fact* that we're proving is at least as interesting as the *proof of that fact*. Throughout our tour of proof techniques, we'll encounter a variety of examples of (fingers crossed!) interesting facts: about propositional logic, including the fact that we need only one logical connective ("nand") to express every proposition; about geometry (the Pythagorean theorem); about prime numbers; and about *uncomputability* (there are problems that cannot be solved by any computer!). We begin in Section 4.2 with an extended exploration of *error-correcting codes*, systems that allow for the reliable transmission and storage of information even in environments that corrupt data as it's stored/transmitted/received/retrieved. (For example, CDs/DVDs are susceptible to scratches, and deep-space satellites' transmissions are susceptible to radiation.) This section will merely scratch the surface of error-correcting codes, but it will serve as a nice introduction to error-correcting codes—and to proofs.

Why are proofs useful in computer science? First, proofs help prevent bugs. Whether or not she writes down in full detail a proof that her code is correct, a good software developer is always reasoning carefully about whether a function performs the task it's supposed to perform, or whether a particular optimization continues to meet the given specification. For a theoretical computer scientist, proofs are bread and butter: proofs of correctness for novel algorithms, or proofs of the hardness of solving a particular problem. For both theoretically and practically oriented computer scientists, a proof often yields great insight that can avoid a brute force solution, improve the efficiency of the code, or unearth some structural property of a problem that reveals that the problem doesn't even need to be solved in the first place.

4.2 *An Extended Application with Proofs: Error-Correcting Codes*

> Irrationally held truths may be more harmful than reasoned errors.
>
> Thomas H. Huxley (1825–1895)

This section introduces *error-correcting codes,* a way of encoding data so that it can be transmitted correctly even in the face of (a limited number of) errors in transmission. These codes are used widely—for example, on DVDs/CDs and in file transfer protocols—and they're interesting to study on their own. But, despite appearances, they are not the point of this section! Rather, they're mostly an excuse to introduce a technical topic with some interesting (and nonobvious) results—and to persuade you of a few of those results. In other words, this section is really about proofs.

Error-detecting and error-correcting codes: the basic idea

Visa and Mastercard use 16-digit numbers for their credit and debit cards, but it turns out that there are only 10^{15} valid credit-card numbers: a number is valid only if a particular arithmetic calculation on the digits—more or less, adding up the digits and taking the result modulo 10—always turns out to be zero. (See Exercises 4.1–4.5 for details of the calculation.) Or, to describe this fact in another way: if you get a (mildly gullible) friend to read you any 15 digits of his or her credit-card number, you can figure out the 16th digit. Less creepily, this system means that there's an *error-detection* mechanism built into credit-card numbers: if any one digit in your number is mistranscribed, then a very simple algorithm can reject that incorrect card number as invalid (because the calculation above will yield an answer other than zero).

In this section, we'll explore encoding schemes with this sort of error-handling capability. Suppose that you have some binary data that you wish to transmit to a friend across an imperfect channel—that is, one that (due to cosmic rays, hardware failures, or whatever) occasionally mistransmits a 0 as a 1, or vice versa. (When we refer to an *error* in a bitstring x, what we mean is a "substitution error," where some single bit in x is flipped.) The fundamental idea will be to add redundancy to the transmitted data; if there is enough redundancy relative to the number of errors, then enough correct information will be transmitted to allow the receiver to reconstruct the original message. We'll explore both *error-detecting codes* that are able to recognize *whether* an error has occurred (at least, as long as there aren't too many errors) and *error-correcting codes* that can *fix* a small number of errors. To reiterate the above, though: although we're focusing on error-correcting and error-detecting codes in this section, the fundamental purpose of this section is to introduce proof techniques. Along the way, we'll see some interesting results about error-correcting codes, but the takeaway message is really about the methods that we'll use to prove those results.

> **Taking it further:** Aside from credit-card numbers, other examples of error-detecting or error-correcting codes include *checksums* on a transferred file—we might break a large file we wish to transmit into 32-bit blocks, transmit those blocks individually, and transmit as a final 32-bit block the XOR of all previously transmitted blocks—as a way to check that the file was transmitted properly. Error-correcting codes are also used in storing data on media (hard disks and CDs/DVDs, for example) so that one can reconstruct stored data even in the face of hardware errors (or scratches on the disc).

The idea of error detection appears in other contexts, too. UPC ("universal product code") bar codes on products in supermarkets use error checking similar to that in credit-card numbers. There are error-detection aspects in DNA. And "the buddy system" from elementary school field trips detects any one "deletion error" among the group (though two "deletions" may evade detection of the system).

4.2.1 *A Formal Introduction*

Imagine a *sender* who wishes to transmit a *message* $m \in \{0,1\}^k$ to a *receiver.* A *code* $\mathcal{C}$ is a subset of $\{0,1\}^n$, listing the set of legal *codewords*; each k-bit *message* m is encoded as an n-bit codeword $c \in \mathcal{C}$. The codeword is then transmitted to the receiver, but it may be corrupted during transmission. The recipient of the (possibly corrupted) n-bit string c' decodes c' into a new message $m' \in \{0,1\}^k$. The goal is that, so long as the corruption is limited, the decoded message is identical to the original message—in other words, that $m = m'$ as long as $c' \approx c$. (We'll make the meaning of "$\approx$" precise soon.) Figure 4.1 shows a schematic of the process.

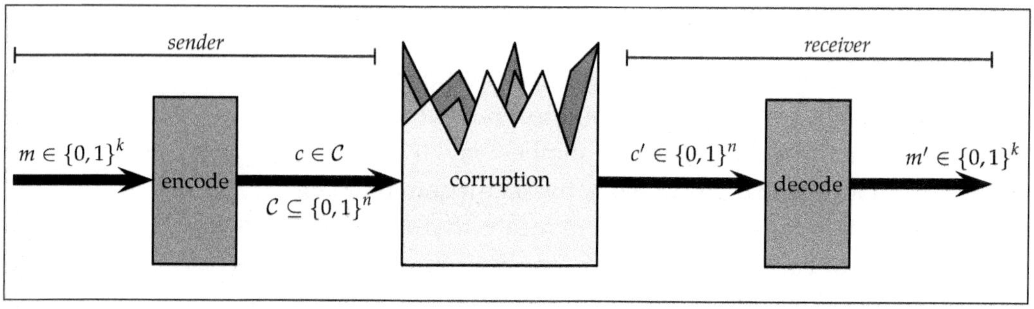

Figure 4.1: A schematic view of error-correcting codes. The goal is that, as long as there isn't *too* much corruption, the received message m' is identical to the sent message m.

(For an error-*detecting* code, the receiver still receives the bitstring c', but determines *whether the originally transmitted codeword was corrupted* instead of determining *which codeword was originally transmitted,* as in an error-correcting code.)

MEASURING THE DISTANCE BETWEEN BITSTRINGS

Before we get to codes themselves, we need a way of quantifying how similar or different two bitstrings are:

The Hamming distance is named after Richard Hamming, a 20th-century American mathematician/computer scientist who was the third winner of the Turing Award.

Definition 4.1 (Hamming distance)
Let $x, y \in \{0,1\}^n$ be two n-bit strings. The Hamming distance *between x and y, denoted by $\Delta(x,y)$, is the number of positions in which x and y differ. In other words,*

$$\Delta(x,y) := \left|\left\{i \in \{1,2,\ldots,n\} : x_i \neq y_i\right\}\right|.$$

(Hamming distance is undefined if x and y don't have the same length.)

For example, $\Delta(011, 101) = 2$ because 011 and 101 differ in bit positions #1 and #2, and $\Delta(0011, 0111) = 1$ because 0011 and 0111 differ in bit #2. Similarly, $\Delta(0000, 1111) = 4$ because all four bits differ, and $\Delta(10101, 10101) = 0$ because all five bits match.

In Exercise 4.6, you'll show that the Hamming distance is a *metric,* which means that it satisfies the following properties, for all bitstrings $x, y, z \in \{0,1\}^n$:

- *"reflexivity"*: $\Delta(x, y) = 0$ if and only if $x = y$;
- *"symmetry"*: $\Delta(x, y) = \Delta(y, x)$; and
- *"the triangle inequality"*: $\Delta(x, y) \leq \Delta(x, z) + \Delta(z, y)$. (See Figure 4.2.)

Informally, the fact that Δ is a metric means that it generally matches your intuitions about geometric (Euclidean) distance.

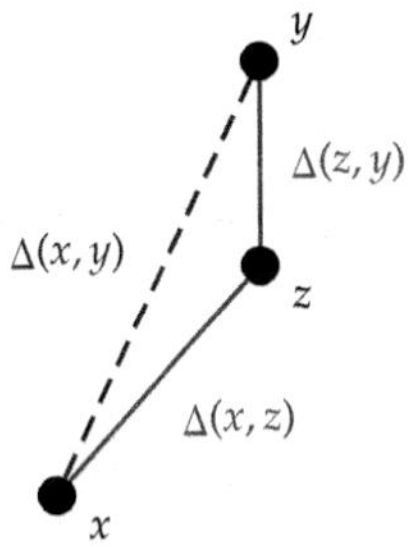

Figure 4.2: The triangle inequality. The distance from x to y isn't decreased by "stopping off" at z along the way.

ERROR-DETECTING AND ERROR-CORRECTING CODES

Definition 4.2 (Codes, messages, and codewords)
A code is a set $C \subseteq \{0,1\}^n$, where $|C| = 2^k$ for some integer $1 \leq k \leq n$. Any element of $\{0,1\}^k$ is called a message, *and the elements of C are called* codewords.

(It might seem a bit strange to require that the number of codewords in C be a precise power of two—but doing so is convenient, as it allows us to consider all k-bit strings as the set of possible messages, for $k := \log_2 |C|$.) Here's an example of a code:

Example 4.1 (A small code)
The set $C := \{000000, 101010, 000111, 100001\}$ is a code. Because $|C| = 4 = 2^2$, there are four messages, namely the four elements of $\{0,1\}^2 = \{00, 01, 10, 11\}$. And because $C \subseteq \{0,1\}^6$, the codewords—the four elements of the set C—are elements of $\{0,1\}^6$.

We can think of a code as being defined by a pair of operations:

- *encoding*: given a message $m \in \{0,1\}^k$, which codeword in C should we transmit? (We'd break up a longer message into a sequence of k-bit message chunks.)
- *decoding*: from a received (possibly corrupted) bitstring $c' \in \{0,1\}^n$, what message should we infer was sent? (Or, if we trying to *detect* errors rather than correct them: from a received bitstring $c' \in \{0,1\}^n$, do we say that an error occurred, or not?)

For the moment, we'll consider a generic (and slow) way of encoding and decoding. Given C, we build a table mapping messages to codewords, by matching up the ith-largest message with the ith-largest codeword (with both the messages from $\{0,1\}^k$ and the codewords in C sorted in numerical order):

- We encode a message m by the codeword in row m of the table.
- We detect an error in a received bitstring c' by reporting "no error" if c' appears in the table, and reporting "error" if c' does not appear in the table.
- We decode a received bitstring c' by identifying the codeword $c \in C$ *that's closest to c'*, measured by Hamming distance. We decode c' as the message in row c of the table. (If there's a tie, we choose one of the tied-for-closest codewords arbitrarily.)

message	codeword
00	000000
01	000111
10	100001
11	101010

Figure 4.3: The message/codeword table for the code from Example 4.1.

Example 4.2 (Encoding and decoding with a small code)
Recall the code $\{000000, 101010, 000111, 100001\}$ from Example 4.1. Sorting the four codewords (and the messages from $\{0,1\}^2$), we get the table in Figure 4.3.

For example, we encode the message 10 as the codeword 100001.

If we receive the bitstring 111110, we report "error" because 111110 is not in C.

To decode the received bitstring 111110, we see that $\Delta(111110, 000000) = 5$, $\Delta(111110, 000111) = 4$, $\Delta(111110, 100001) = 5$, and $\Delta(111110, 101010) = 2$. The last of these distances is smallest, so we would decode 111110 as the message 11 (corresponding to codeword 101010).

The danger in error detection is that we're sent a codeword $c \in C$ that's corrupted into a bitstring c', but we report "no error" because $c' \in C$. (Note that we're never wrong when we report "error.") The danger in error correction is that we report another codeword $c'' \in C$ because c' is closer to c'' than it is to c. (As we'll see soon, these dangers are really about Hamming distance *between codewords*: we might make a mistake if two codewords in C are too close together, relative to the number of errors.) Here are the precise definitions of error-detecting and error-correcting codes:

Definition 4.3 (Error-detecting and error-correcting codes)
Let $C \subseteq \{0,1\}^n$ be a code, and let $\ell \geq 1$ be any integer.

We say that C can detect *ℓ errors if, for any codeword $c \in C$ and for any sequence of up to ℓ errors applied to c, we can correctly report "error" or "no error."*

The code C can correct *ℓ errors if, for any codeword $c \in C$ and for any sequence of up to ℓ errors applied to c, we can correctly identify that c was the original codeword.*

Here's an example, for our small example code:

Example 4.3 (Error detection and correction in a small code)
Recall $C = \{000000, 101010, 000111, 100001\}$ from Example 4.1. Figure 4.4 shows every bitstring $x \in \{0,1\}^6$, and the Hamming distance between x and each codeword in C.

There are 24 single-bit errors that can happen to codewords in C: there are 4 choices of codeword, and, for each, 6 different one-bit errors that can occur:

no errors:	000000	101010	000111	100001
one error:	100000	001010	100111	000001
	010000	111010	010111	110001
	001000	100010	001111	101001
	000100	101110	000011	100101
	000010	101000	000101	100011
	000001	101011	000110	100000

This code can detect one error, because the 24 bitstrings below the line are all different from the 4 bitstrings above the line; we can correctly report whether the bitstring in question is a codeword (no errors) or one of the 24 non-codewords (one error). Or, to state this fact in a different way: the four starred lines of Figure 4.4 corresponding to uncorrupted codewords are not within one error of any other codeword. On the other hand, C *cannot* detect two errors. If we receive the bitstring 000000, we can't distinguish whether the original codeword was 000000 (and no errors occurred) or whether the original codeword was 100001 (and two errors occurred, in 000000). (Receiving the bitstring 100001 creates the same problem.)

c'	$\Delta(c', 000000)$	$\Delta(c', 000111)$	$\Delta(c', 100001)$	$\Delta(c', 101010)$
000000*	0	3	2	3
000001†	1	2	1	4
000010	1	2	3	2
000011	2	1	2	3
000100	1	2	3	4
000101	2	1	2	5
000110	2	1	4	3
000111*	3	0	3	4
001000	1	4	3	2
001001	2	3	2	3
001010	2	3	4	1
001011	3	2	3	2
001100	2	3	4	3
001101	3	2	3	4
001110	3	2	5	2
001111	4	1	4	3
010000	1	4	3	4
010001	2	3	2	5
010010	2	3	4	3
010011	3	2	3	4
010100	2	3	4	5
010101	3	2	3	6
010110	3	2	5	4
010111	4	1	4	5
011000	2	5	4	3
011001	3	4	3	4
011010	3	4	5	2
011011	4	3	4	3
011100	3	4	5	4
011101	4	3	4	5
011110	4	3	6	3
011111	5	2	5	4
100000†	1	4	1	2
100001*	2	3	0	3
100010	2	3	2	1
100011	3	2	1	2
100100	2	3	2	3
100101	3	2	1	4
100110	3	2	3	2
100111	4	1	2	3
101000	2	5	2	1
101001	3	4	1	2
101010*	3	4	3	0
101011	4	3	2	1
101100	3	4	3	2
101101	4	3	2	3
101110	4	3	4	1
101111	5	2	3	2
110000	2	5	2	3
110001	3	4	1	4
110010	3	4	3	2
110011	4	3	2	3
110100	3	4	3	4
110101	4	3	2	5
110110	4	3	4	3
110111	5	2	3	4
111000	3	6	3	2
111001	4	5	2	3
111010	4	5	4	1
111011	5	4	3	2
111100	4	5	4	3
111101	5	4	3	4
111110	5	4	5	2
111111	6	3	4	3

Figure 4.4: The Hamming distance of every 6-bit string to all codewords from Example 4.1.

The code C also cannot *correct* even one error. Consider the bitstring 100000. We cannot distinguish (i) the original codeword was 000000 (and one error occurred) from (ii) the original codeword was 100001 (and one error occurred). Or, to state this fact differently: the two lines of Figure 4.4 marked with † are only one error away from two *different* codewords. (That is, 100000 appears *twice* in the list of 24 bitstrings below the line.)

4.2.2 *Distance and Rate*

Our goal with error-correcting codes is to ensure that the decoded message m' is identical to the original message m, as long as there aren't too many errors in the transmission. At a high level, we will achieve this goal by ensuring that the codewords in our code are all "very different" from each other. If every pair of distinct codewords c_1 and c_2 are far apart (in Hamming distance), then the closest codeword c to the received transmission c' will correspond to the original message, even if "a few" errors occur. (We'll quantify "very" and "a few" soon.)

Intuitively, this desire suggests adding a lot of redundancy to our codewords, by making them more redundant. But we must balance this desire for robustness against another desire that pulls in the opposite direction: we'd like to transmit a small number of bits (so that the number of "wasted" non-data bits is small). There's a seeming trade-off between these two measures of the quality of a code: increasing error tolerance suggests making the codewords longer (so there's room for them to differ more); increasing efficiency suggests making the codewords shorter (so there are fewer wasted bits). Let's formally define both of these measures of code quality:

Definition 4.4 (Minimum distance)
The minimum distance *of a code C is the smallest Hamming distance between two distinct codewords of C: that is, the minimum distance of C is* $\min\{\Delta(x, y) : x, y \in C \text{ and } x \neq y\}$.

(Quiz question: if we hadn't restricted the minimum in this definition to be only over pairs such that $x \neq y$, what would the minimum distance have been?)

Definition 4.5 (Rate)
The rate *of a code C is the ratio between message length and codeword length. That is, if C is a code where $|C| = 2^k$ and $C \subseteq \{0,1\}^n$, then the rate of C is the ratio $\frac{k}{n}$.*

Let's compute the rate and minimum distance for our running example:

	000000	000111	100001	101010
000000	0	3	2	3
000111	3	0	3	4
100001	2	3	0	3
101010	3	4	3	0

Figure 4.5: The Hamming distance between codewords of C from Example 4.1.

Example 4.4 (Distance and rate in a small code)
Recall the code $C = \{000000, 101010, 000111, 100001\}$ from Example 4.1.

The minimum distance of C is 2, because $\Delta(000000, 100001) = 2$. You can check Figure 4.4 (or see Figure 4.5) to see that no other pair of codewords is closer.

The rate of C is $\frac{2}{6}$, because $|C| = 4 = |\{0,1\}^2|$, and the codewords have length 6.

Relating minimum distance and error detection/correction

We have now defined enough of the concepts that we can state a first nontrivial theorem, which characterizes the error-detecting and error-correcting capabilities of a code C in terms of the minimum distance of C. Here is the statement:

Theorem 4.1 (Relationship of minimum distance to detecting/correcting errors) *Let $t \geq 0$ be any integer. If the minimum distance of a code C is $2t + 1$, then C can detect $2t$ errors and correct t errors.*

Problem-solving tip: Step #1 in proving any claim is to understand what it's saying! (You can't persuade someone of something you don't understand.) One good way to start to do so is by plugging particular values into the statement.

We're now going to try to prove Theorem 4.1—that is, we're going to try to generate a convincing argument that this statement is true. As with any statement that you try to prove, our first task is to *understand* what exactly the claim is saying. In this case, the theorem makes a statement about a generic nonnegative integer t and a generic code C. Plugging in particular values for t can help make the claim clearer:

- If the minimum distance of a code C is 9—that is, the minimum distance is $2t + 1$ for $t = 4$—then the claim says C can detect $2t = 2 \cdot 4 = 8$ errors and correct $t = 4$ errors.
- Suppose the minimum distance of C is 7. Writing $7 = 2t + 1$ for $t = 3$, the claim states that C can detect 6 errors and correct 3 errors.
- If the minimum distance of C is 5, then C can detect 4 errors and correct 2 errors.
- If the minimum distance of C is 3, then C can detect 2 errors and correct 1 error.
- If the minimum distance of C is 1, then C can detect 0 errors and correct 0 errors.

Now that we have a better sense of what the theorem says, let's prove it:

Problem-solving tip: Draw a picture to help you clarify/understand the statement you're trying to prove.

Proof of Theorem 4.1. First we'll prove the error-detection condition. We must argue for the following claim: if a code C has minimum distance $2t + 1$, then C can detect $2t$ errors. In other words, for an arbitrary codeword $c \in C$ and an arbitrary received bitstring c' with $\Delta(c, c') \leq 2t$, our error-detection algorithm must be correct. (If $\Delta(c, c') > 2t$, then we're not obliged to correctly state that an error occurred, because we're only arguing that we can detect $2t$ errors.) Recall that our error-detection algorithm reports "no error" if $c' \in C$, and it reports "error" if $c' \notin C$. Thus:

- If $\Delta(c, c') = 0$, then no error occurred (because the received bitstring matches the transmitted one). In this case, our error-detection algorithm correctly reports "no error"—because $c' \in C$ (because $c' = c$, and c was a codeword).
- On the other hand, suppose $1 \leq \Delta(c, c') \leq 2t$—so an error occurred. The only way that we'd fail to detect the error is if the received bitstring c' is itself *another* codeword. But this situation can't happen, by the definition of minimum distance: for any codeword $c \in C$, the set $\{c' : \Delta(c, c') \leq 2t\}$ *cannot* contain any elements of C—otherwise the minimum distance of C would be $2t$ or smaller.

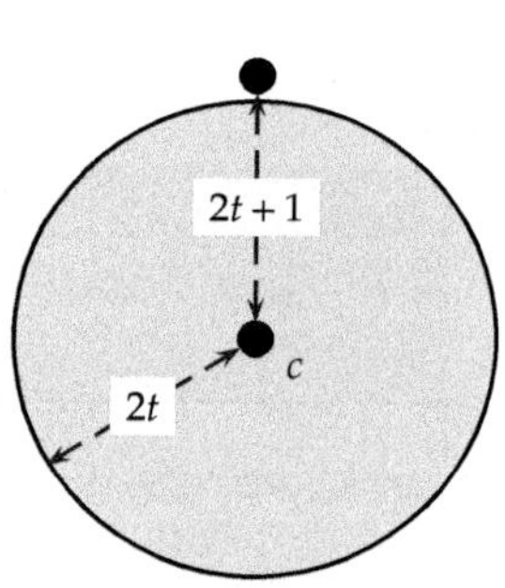

Figure 4.6: If the minimum distance is $2t + 1$, no codewords are within distance $2t$ of each other.

It may be helpful to think about this proof via Figure 4.6.

For the error-correction condition, suppose that $x \in C$ is the transmitted codeword, and the received bitstring c' satisfies $\Delta(x, c') \leq t$. We have to persuade ourselves that x is the codeword closest to c' in Hamming distance. Let $y \in C - \{x\}$

be any other codeword. We'll start from the triangle inequality, which tells us that $\Delta(x,y) \leq \Delta(x,c') + \Delta(c',y)$ and therefore that $\Delta(c',y) \geq \Delta(x,y) - \Delta(x,c')$, and prove that c' is closer to x than it is to y:

$$
\begin{aligned}
\Delta(c',y) &\geq \Delta(x,y) - \Delta(x,c') && \textit{triangle inequality} \\
&\geq (2t+1) - \Delta(x,c') && \Delta(x,y) \geq 2t+1 \textit{ by definition of minimum distance} \\
&\geq (2t+1) - t && \Delta(x,c') \leq t \textit{ by assumption} \\
&= t+1 \\
&> t \\
&\geq \Delta(x,c'). && \Delta(x,c') \leq t \textit{ by assumption}
\end{aligned}
$$

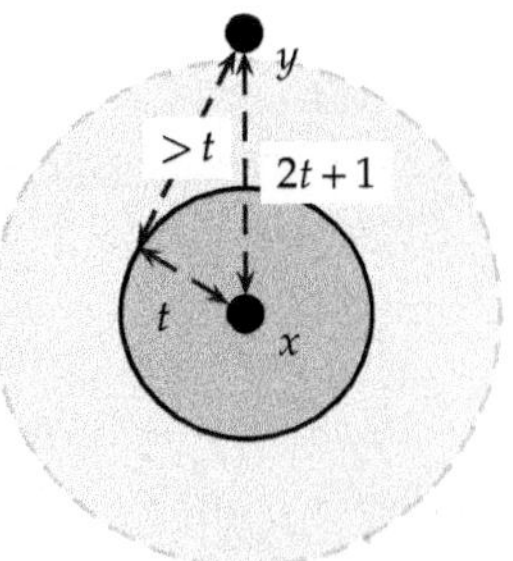

Figure 4.7: If the minimum distance is $2t+1$, a bitstring within distance t of one codeword is more than t away from every other codeword.

This chain of inequalities shows c' is closer to x than it is to y. (Pedantically speaking, we're also relying on the *symmetry* of Hamming distance here: $\Delta(c',y) = \Delta(y,c')$. Again, see Exercise 4.6.) Because y was a generic codeword in $C - \{x\}$, we can conclude that the original codeword x is the one closest to c'. (See Figure 4.7.) □

Before we move on from the theorem, let's reflect a little bit on the proof. (We'll concentrate on the error-correction half.) The most complicated part was unwinding the definitions in the theorem statement, in particular of "C has minimum distance $2t+1$" and "C can correct t errors." Eventually, we had to argue for the claim

for every $x \in C$, $y \in C - \{x\}$, and $c' \in \{0,1\}^n$: if $\Delta(x,c') \leq t$ then $\Delta(x,c') < \Delta(y,c')$.

(In other words, if c' is within t errors of x, then c' is closer to x than to any other codeword.) In the end, we were able to state the proof as a relatively simple sequence of inequalities. After proving a theorem, it's also worth briefly reflecting on what the theorem does *not* say. Theorem 4.1, for example, only addresses codes with a minimum distance that's an odd number. You'll be asked to consider the error-correcting and error-detecting properties of a code C with an even minimum distance in Exercise 4.13. We also didn't show that we couldn't do better: Theorem 4.1 says that a code C with minimum distance $2t+1$ *can* correct t errors, but the theorem doesn't say that C *can't* correct $t+1$ (or more) errors. (But, in fact, it can't; see Exercise 4.12.)

Problem-solving tip: When you're trying to prove a claim of the form $p \Rightarrow q$, try to massage p to look as much like q as possible. A good first step in doing so is to expand out the definitions of the premises, and then try to see what additional facts you can infer.

Outline of the remainder of the section

Intuitively, rate and minimum distance are measures of the inherent tension in an error-correcting code. A code that has a higher distance means that we are more robust to errors: the farther apart codewords are, the more corruption can occur before we're unable to reconstruct the original message. A code that has a higher rate means that we are "wasting" fewer bits in providing this robustness: the larger the rate, the more our codeword contains "data" rather than "redundancy." In the rest of this section, we're going to prove several more theorems about error-correcting codes, exploring the trade-off between rate and distance. (But it's also worth noting that it's not a strict trade-off: sometimes we can improve in one measure without costing ourselves in the other!) And, as we go, we'll continue to try to reflect on the proof techniques that we use to establish these claims.

It is customary to mark the end of one's proofs typographically; here, we're using a traditional box symbol: □. Other people may write "QED," short for the Latin phrase *quod erat demonstrandum* ("that which was to be demonstrated").

Here are the three main theorems that we'll prove in the rest of this section:

Theorem 4.2 (Good news)
There exists a code with 4-bit messages, minimum distance 3, and rate $\frac{1}{3}$.

Theorem 4.3 (Better news)
There exists a code with 4-bit messages, minimum distance 3, and rate $\frac{4}{7}$.

Theorem 4.4 (Bad news)
There does not exist a code with 4-bit messages, minimum distance 3, and any rate strictly better than $\frac{4}{7}$.

Notice that the first two of these results say that a code with particular properties exists, while the third result says that it's impossible to create a code with a different set of properties. Also notice that Theorem 4.3 is an improvement on Theorem 4.2: we've made the rate better (higher) without making the minimum distance worse. (When we can, we'll prove more general versions of these theorems, too, not limited to 4-bit messages with minimum distance 3.)

We'll prove Theorem 4.2 and Theorem 4.3 "by construction"—specifically, by building a code with the desired parameters. But, because Theorem 4.4 says that a code with certain properties fails to exist, we'll prove the result with a *proof by contradiction:* we assume that a code with 4-bit messages with distance 3 and rate strictly better than $\frac{4}{7}$ *does exist,* and reasoning logically from that assumption, we will derive a false statement (a contradiction). Because $p \Rightarrow \text{False} \equiv \neg p$, we can conclude that the assumption must have been false, and no such code can exist.

4.2.3 *Repetition Codes*

Intuitively, a good error-correcting code will amplify even a small difference between two different messages—a single differing bit—into a larger difference between the corresponding codewords. Perhaps the most obvious implementation of this idea is simply to encode a message m by repeating the bits of m several times. This idea gives rise to a simple error-correcting code, called the *repetition code.* (Actually, there are many different versions of the repetition code, depending on how many times we repeat m in the codeword.) Here's the basic definition:

Definition 4.6 (Repetition code)
Let $\ell \in \mathbb{Z}^{\geq 2}$. The REPETITION$_\ell$ code *for k-bit messages consists of the codewords*

$$\Big\{\underbrace{m\,m\,\cdots\,m}_{\ell \text{ times}} : m \in \{0,1\}^k\Big\}.$$

That is, the codeword corresponding to a message $m \in \{0,1\}^k$ is the ℓ-fold repetition of the message m, so each codeword is an element of $\{0,1\}^{k\ell}$.

Here are some small examples of encoding/decoding using repetition codes:

Example 4.5 (Some codewords for the repetition code)
If we encode the message 00111 using the REPETITION3 code, we get the codeword 00111 00111 00111. If we encode the same message using the REPETITION5 code, we get the codeword 00111 00111 00111 00111 00111.

For an example of decoding, suppose that we receive the (possibly corrupted) bitstring $c' = 0010\ 0110\ 0010$ under the REPETITION3 code. We detect that an error occurred: c' is not a codeword, because the only codewords are 12-bit strings where all three 4-bit thirds are identical. For error correction, note that the closest codeword to c' is 0010 0$\underline{0}$10 0010, so we decode c' as corresponding to the message 0010.

m	c
0000	0000 0000 0000
0001	0001 0001 0001
0010	0010 0010 0010
0011	0011 0011 0011
0100	0100 0100 0100
0101	0101 0101 0101
0110	0110 0110 0110
0111	0111 0111 0111
1000	1000 1000 1000
1001	1001 1001 1001
1010	1010 1010 1010
1011	1011 1011 1011
1100	1100 1100 1100
1101	1101 1101 1101
1110	1110 1110 1110
1111	1111 1111 1111

Figure 4.8: The REPETITION3 code for 4-bit messages.

The message/codeword table for the REPETITION3 code for 4-bit messages is shown in Figure 4.8. The distance and rate properties of the repetition code are relatively easy to see (from the definition or from this style of table):

Lemma 4.5 (Distance and rate of the repetition code)
The REPETITION$_\ell$ code has rate $\frac{1}{\ell}$ and minimum distance ℓ.

Proof. Recall that the rate of a code is the ratio $\frac{k}{n}$, where k is the length of the messages and n is the length of the codewords. A k-bit message is encoded as a $(k\ell)$-bit codeword (ℓ repetitions of k bits), and so the rate of this code is $\frac{k}{k\ell} = \frac{1}{\ell}$.

For the minimum distance, consider any two distinct messages $m, m' \in \{0,1\}^k$ with $m' \neq m$. We know that m and m' must differ in at least one bit position, say bit position i. (Otherwise $m = m'$.) But if $m_i \neq m'_i$, then

$$\text{the codeword corresponding to } m = m\ \ m\ \ \cdots\ m \text{ and}$$
$$\text{the codeword corresponding to } m' = \underbrace{m'\ \ m'\ \ \cdots\ m'}_{\ell \text{ times}}$$

differ in at least one bit in each of the ℓ "blocks" (in the ith position of the block)—for a total of at least ℓ differences. Furthermore, the REPETITION$_\ell$ encodings of the messages $000\cdots0$ and $100\cdots0$ differ in only ℓ places (the first bit of each "block"). Thus the minimum distance of the REPETITION$_\ell$ code is exactly ℓ. □

Lemma 4.5 says that the REPETITION3 code on 4-bit messages (see Figure 4.8) has minimum distance 3 and rate $\frac{1}{3}$. Thus we've proven Theorem 4.2: we had to show that a code with these parameters exists, and we did so *by explicitly building such a code.* This proof is an example of a "proof by construction": to show that an object with a particular property exists, we've explicitly built an object with that property.

Problem-solving tip: When you're trying to prove a claim of the form $\exists x : P(x)$, try using a proof by construction first. (There are other ways to prove an existential claim, but this approach is great when it's possible.)

It's also worth noticing that we started out by describing a *generic* way to do encoding and decoding for error-correcting codes in Section 4.2: after we build the table (like the one in Figure 4.8), we encode a message by finding the corresponding codeword in the table, and we decode a bitstring c' by looking at every codeword and identifying the one closest to c'. For particular codes, we may be to give a *much* more efficient algorithm—and, indeed, we can do so for repetition codes. See Exercise 4.21.

4.2.4 *Hamming Codes*

When we're encoding 4-bit messages, the Repetition3 code achieves minimum distance 3 with 12-bit codewords. (So its rate is $\frac{1}{3}$.) But it turns out that we can do better by defining another, cleverer code: the *Hamming code*[1] maintains the same minimum distance, while improving the rate from $\frac{1}{3}$ to $\frac{4}{7}$.

The basic idea of the Hamming code is to use an extra bit that, like the 16th digit of a credit card number, redundantly reports a value computed from the previous components of the message. Concretely, we could tack a single bit b onto the message m, where b reports the *parity* of m—that is, whether there are an even or odd number of bits set to 1 in m. If a single error occurs in the message, then b would be inconsistent with the message m, and we'd detect that error. (See Exercise 4.19.) In fact, for the Hamming code, we'll use several *different* parity bits, corresponding to different subsets of the bits of m.

The Hamming code, like the Hamming distance, is named after Richard Hamming, who invented this code in 1950. (He was frustrated that programs he started running on Friday nights often failed over the weekend because of a single bit error in memory.)

[1] R. W. Hamming. Error detecting and error correcting codes. *The Bell System Technical Journal*, XXIX(2):147–160, April 1950.

Definition 4.7 (Parity function)
The parity *of a sequence* $\langle a_1, a_2, \ldots, a_k \rangle$ *of bits is denoted either* $\text{parity}(a_1, a_2, \ldots, a_k)$ *or* $a_1 \oplus a_2 \oplus \cdots \oplus a_k$*, and its value is*

$$a_1 \oplus a_2 \oplus \cdots \oplus a_k := \begin{cases} 1 & \text{if there are an odd number of } i \text{ such that } a_i = 1 \\ 0 & \text{if there are an even number of } i \text{ such that } a_i = 1. \end{cases}$$

(We could also have defined this function as $\text{parity}(a_1, \ldots, a_k) := [\sum_{i=1}^{k} a_i] \bmod 2$.*)*

The parity of a and b can be denoted as $a \oplus b$, because if you think of $a, b \in \{0,1\}$, where True $=$ 1 and False $=$ 0, then *parity*(a,b) is the XOR of a and b.

Hamming's insight was that it's possible to achieve good error-correction properties by using three different parity bits, corresponding to different subsets of the message bits. It's easiest to think of this code in terms of its encoding algorithm:

Definition 4.8 (Hamming code)
The Hamming code *is defined via the following encoding function. We will encode a 4-bit message* $\langle a, b, c, d \rangle$ *as the following 7-bit codeword:*

$$\langle \underbrace{a, b, c, d,}_{\text{message bits}} \; \underbrace{b \oplus c \oplus d, \; a \oplus c \oplus d, \; a \oplus b \oplus d}_{\text{parity bits}} \rangle.$$

Applying this encoding to every 4-bit message yields the table of messages and their corresponding codewords shown in Figure 4.9; here are a few examples in detail:

m	c
0000	0000000
0001	0001111
0010	0010110
0011	0011001
0100	0100101
0101	0101010
0110	0110011
0111	0111100
1000	1000011
1001	1001100
1010	1010101
1011	1011010
1100	1100110
1101	1101001
1110	1110000
1111	1111111

Figure 4.9: The Hamming code for 4-bit messages.

Example 4.6 (Sample Hamming code encodings)

message	codeword
a, b, c, d	$a, b, c, d, (b \oplus c \oplus d), (a \oplus c \oplus d), (a \oplus b \oplus d)$
$0,0,0,0$	$0,0,0,0,(0 \oplus 0 \oplus 0),(0 \oplus 0 \oplus 0),(0 \oplus 0 \oplus 0) = 0000000$
$1,0,0,0$	$1,0,0,0,(0 \oplus 0 \oplus 0),(1 \oplus 0 \oplus 0),(1 \oplus 0 \oplus 0) = 1000011$
$1,1,1,0$	$1,1,1,0,(1 \oplus 1 \oplus 0),(1 \oplus 1 \oplus 0),(1 \oplus 1 \oplus 0) = 1110000.$

(We could have described encoding for the Hamming code using matrix multiplication instead; see Exercises 2.221–2.223.)

Before we analyze the rate and minimum distance of the Hamming code, let's start to develop some intuition by looking at a few received (possibly corrupted) codewords. (We'll also begin to work out an efficient decoding algorithm as we go.)

Example 4.7 (Some Hamming code decoding problems)

Problem: You receive the following (possibly corrupted) Hamming code codewords. Find the original message, assuming at most one error occurred in transmission.

1. 0000010
2. 1000000
3. 1011010
4. 1110111

Solution: 1. We've received message bits 0000 and parity bits 010. Everything in the received codeword is consistent with the message being $m = 0000$, except for the second parity bit. So we infer that the second parity bit was corrupted, the transmitted codeword was 0000000, and the message was 0000.

Could there have been a one-bit error in message bits instead? No: these parity bits are consistent only with a message $\langle a, b, c, d \rangle$ with $a \neq b$ (because the first two received parity bits differ), and therefore with $d = 1$ (because $a \neq b$ implies that $a \oplus b \oplus d = 1 \oplus d = \neg d$, and the third parity bit $a \oplus b \oplus d$ is 0). But 10?1 and 01?1 are both at least two errors away from the received message 0000.

2. We've received message bits 1000 and parity bits 000. If the message bits were uncorrupted, then the correct parity bits would have been 011. But then we would have to have suffered *two* transmission errors in the parity bits, and we're assuming that at most one error occurred. Thus the error is in the message bits; the original message is 0000, and the first bit of the message was corrupted.

3. The parity bits for the message 1011 are indeed 010, so 1011010 is itself a legal codeword for the message 1011, and no errors occurred at all.

4. These received bits are consistent with the message 1111 with parity bits 111, where the fourth bit of the message was flipped.

Recall that, for a message a, b, c, d, the bits of the uncorrupted codeword are:

1. a
2. b
3. c
4. d
5. $b \oplus c \oplus d$
6. $a \oplus c \oplus d$
7. $a \oplus b \oplus d$

From this example, the basic approach to decoding the Hamming code should start to coalesce. Briefly, we compute what the parity bits *should have been,* supposing that the received message bits (the first four bits of the received codeword) are correct; comparing the computed parity bits to the received parity bits allows us to deduce which, if any, of the transmitted bits were erroneous. (More on efficient decoding later.) Why does this approach to decoding work? (And, relatedly, why were the parity bits of the Hamming code chosen the way that they were?) Here are two critical properties in the Hamming code's parity bits:

- *every message bit appears in at least two parity bits.* Thus any error in a received parity bit is distinguishable from an error in a received message bit: an erroneous message bit will cause at least two parity bits to look wrong; an erroneous parity bit will cause only that one parity bit to look wrong.

- *no two message bits appear in precisely the same set of parity bits.* Thus any error in a received message bit has a different "signature" of wrong-looking parity bits: an error in bit a affects parity bits #2 and #3; b affects parity bits #1 and #3; c affects #1 and #2; and d affects all three parity bits. Because all four of these signatures are different, we can distinguish *which* message bit was corrupted based on which set of two or more parity bits look wrong.

RATE AND MINIMUM DISTANCE OF THE HAMMING CODE

Let's use the intuition that we've developed so far to establish the rate and minimum distance for the Hamming code:

> **Lemma 4.6 (Distance and rate of the Hamming code)**
> *The Hamming code has rate $\frac{4}{7}$ and minimum distance* 3.

Proof. The rate is straightforward to compute: we have 4-bit messages and 7-bit codewords, so the rate is $\frac{4}{7}$ by definition.

There are several ways to convince yourself that the minimum distance is 3—perhaps the simplest way (though certainly the most tedious) is to compute the Hamming distance between each pair of codewords in Figure 4.9. (There are only 16 codewords, so we just have to check that all $(16 \cdot 15)/2 = 120$ pairs of distinct codewords have Hamming distance at least three.) You'll write a program to verify this claim in Exercise 4.24. But here's a different argument.

Consider any two distinct messages $m \in \{0,1\}^4$ and $m' \in \{0,1\}^4$. We must establish that the codewords c and c' associated with m and m' satisfy $\Delta(c,c') \geq 3$. We'll argue for this fact by looking at three separate cases, depending on $\Delta(m,m')$:

Case I: $\Delta(m,m') \geq 3$. Then we're done immediately: the message bits of c and c' differ in at least three positions (even without looking at the parity bits).

Case II: $\Delta(m,m') = 2$. Then at least one of the three parity bits contains one of the bit positions where $m_i \neq m_i'$ but not the other. (This fact follows from the second crucial property above, that no two message bits appear in precisely the same set of parity bits.) Therefore this parity bit differs in c and c'. Thus there are two message bits and at least one parity bit that differ, so $\Delta(c,c') \geq 3$.

Case III: $\Delta(m,m') = 1$. Then at least two of the three parity bits contain the bit position where $m_i \neq m_i'$. (This fact follows from the first crucial property above, that every message bit appears in at least two parity bits.) Thus there are at least two parity bits and one message bit that differ, and $\Delta(c,c') \geq 3$.

Note that $\Delta(m,m')$ must be $1, 2$, or ≥ 3—it can't be zero because $m \neq m'$—so, no matter what $\Delta(m,m')$, we've established that $\Delta(c,c') \geq 3$.

Because, for the codewords corresponding to messages 0000 and 1110, we have $\Delta(0000000, 1110000) = 3$, the minimum distance is in fact exactly equal to three. □

Lemma 4.6 says that the Hamming code encodes 4-bit messages with minimum distance 3 and rate $\frac{4}{7}$; thus we've proven Theorem 4.3. Let's again reflect a little on the proof. Our proof of the minimum distance in Lemma 4.6 was a *proof by cases:* we divided pairs of codewords into three different categories (differing in $1, 2$, or ≥ 3 bits), and then used three different arguments to show that the corresponding codewords differed in ≥ 3 places. So we showed that the desired distance property was true in all three cases—and, crucially, that one of the cases applies for every pair of codewords.

Problem-solving tip: If you discover that a proposition seems true "for different reasons" in different circumstances (and those circumstances seem to cover all possible scenarios!), then a proof by cases may be a good strategy to employ.

Although we're mostly omitting any discussion of the efficiency of encoding and decoding, it's worth a brief mention here. (The speed of these algorithms is a big deal for error-correcting codes used in practice!) The algorithm for decoding under the Hamming code is suggested by Figure 4.10: we calculate what the parity bits would have been if the received message bits were uncorrupted, and identify which received parity bits don't match those calculated parity bits. Figure 4.10 tells us what inference to draw from each constellation of mismatched parity bits.

parity bit #1: $b \oplus c \oplus d$	parity bit #2: $a \oplus c \oplus d$	parity bit #3: $a \oplus b \oplus d$	location of error
			no error!
✗			parity #1
	✗		parity #2
		✗	parity #3
✗	✗		bit c
✗		✗	bit b
	✗	✗	bit a
✗	✗	✗	bit d

Figure 4.10: Decoding the Hamming code. We conclude that the stated error occurred if the received parity bits and those calculated from the received message bits mismatch in the listed places.

Why does this decoding algorithm allow us to correct any single error? First, a low-level answer: the Hamming code has a minimum distance of $3 = 2 \cdot 1 + 1$, so Lemma 4.1 tells us that we can correct up to one error. So we know that a decoding scheme is possible. At a higher level, the reason that this decoding procedure works properly is that there are eight possible "≤ 1 error" corruptions of a codeword x—namely one 0-error string (x itself) and seven 1-error strings (one corresponding to an error in each of the seven bit positions of x)—and furthermore there are eight different subsets of the three parity bits that can be "wrong." The Hamming code works by carefully selecting the parity bits in a way that each of these eight bitstrings corresponds to a different one of the eight parity-bit subsets. In Exercises 4.25–4.28, you'll explore longer versions of the Hamming code (with longer messages and more parity bits) with the same relationship.

Taking it further: As we've said, our attention here is mostly on the proofs and the proof techniques that we've used to establish the claims in this section, rather than on error-correcting codes themselves. But see p. 418 for an introduction to *Reed–Solomon codes,* the basis of the error-correcting codes used in CDs/DVDs (among other applications).

4.2.5 Upper Bounds on Rates

In the last two sections, we've constructed two different codes, both for 4-bit messages with minimum distance 3: the repetition code (rate $\frac{4}{12}$) and the Hamming code (rate $\frac{4}{7}$). Because the message lengths and minimum distances match, and because higher rates are better, the Hamming code is better. Here we'll consider whether we can improve the rate further, while still encoding 4-bit messages with minimum distance 3. (In other words, can we make the codewords shorter than 7 bits?) The answer turns out to be "no"—and we'll prove that it's impossible.

"Balls" around codewords

We'll start by thinking about "balls" around codewords in a general code. (The *ball of radius r around $x \in \{0,1\}^n$* is the set $\{x' : \Delta(x, x') \leq r\}$—that is, the set of all points that are within Hamming distance r of x.) Here's a first observation:

Lemma 4.7 (The size of a ball of radius 1 in $\{0,1\}^n$)
Let $x \in \{0,1\}^n$, and define $X := \{x' \in \{0,1\}^n : \Delta(x,x') \leq 1\}$. Then $|X| = n+1$.

Proof. The bitstring x itself is an element of X, as are all bitstrings x' that differ from x in exactly one position. There are n such strings x': one that is x with the first bit flipped, one that is x with the second bit flipped; ...; and one that is x with the nth bit flipped. Thus there are $1+n$ total bitstrings in X. □

Here's a second useful fact about these balls: in a code $\mathcal{C}$, the balls around codewords (of radius related to the minimum distance of $\mathcal{C}$) cannot overlap.

Lemma 4.8 (Balls around codewords are disjoint)
Let $\mathcal{C} \subseteq \{0,1\}^n$ be a code with minimum distance $2t+1$. For distinct codewords $x, y \in \mathcal{C}$, the sets $\{x' \in \{0,1\}^n : \Delta(x,x') \leq t\}$ and $\{y' \in \{0,1\}^n : \Delta(y,y') \leq t\}$ are disjoint.

Proof. Suppose not: that is, suppose that the sets $X := \{x' \in \{0,1\}^n : \Delta(x,x') \leq t\}$ and $Y := \{y' \in \{0,1\}^n : \Delta(y,y') \leq t\}$ are *not* disjoint. We will derive a contradiction from this assumption—that is, a statement that can't possibly be true. Thus we'll have proven that $X \cap Y \neq \varnothing \Rightarrow$ False, which allows us to conclude that $X \cap Y = \varnothing$, because $\neg p \Rightarrow \text{False} \equiv p$. That is, we're using a *proof by contradiction.*

To start again from the beginning: suppose that X and Y are not disjoint. That is, suppose that there is some bitstring $z \in \{0,1\}^n$ such that $z \in X$ and $z \in Y$. In other words, by definition of X and Y, there is a bitstring $z \in \{0,1\}^n$ such that $\Delta(x,z) \leq t$ *and* $\Delta(y,z) \leq t$. But if $\Delta(x,z) \leq t$ and $\Delta(y,z) \leq t$, then, by the triangle inequality, we know

$$\Delta(x,y) \leq \Delta(x,z) + \Delta(z,y) \leq t + t = 2t.$$

Therefore $\Delta(x,y) \leq 2t$—but then we have two distinct codewords $x, y \in \mathcal{C}$ with $\Delta(x,y) \leq 2$. This condition contradicts the assumption that the minimum distance of $\mathcal{C}$ is $2t+1$. (See Figure 4.11.) □

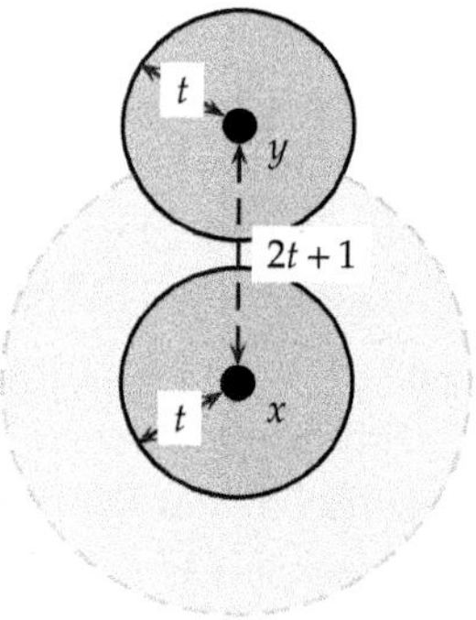

Figure 4.11: If the minimum distance is $2t+1$, the "balls" of radius t around each codeword are disjoint.

We could have used Lemma 4.8 to establish the error-correction part of Theorem 4.1—a bitstring corrupted by $\leq t$ errors from a codeword c is closer to c than to any other codeword—but here we'll use it, plus Lemma 4.7, to establish a upper bound on the rate of codes. But, first, let's pause to look at a similar argument in a different (but presumably more familiar) domain: normal Euclidean geometry.

Problem-solving tip: When you're facing a problem in a less familiar domain, try to find an analogous problem in a different, more familiar setting to help gain intuition.

In a *circle-packing* problem, we are given an enclosing shape, and we're asked to place ("pack") as many nonoverlapping unit circles (of radius 1) into that shape as possible. (*Sphere packing*—what grocers have to do with oranges—is the 3-dimensional analogue.) How many unit circles can we fit into a 6-by-6 square, for example? (See Figure 4.12.) Here's an argument that it's at most 11: a unit circle has area $\pi \cdot 1^2 = \pi$, and the 6-by-6 square has area 36; thus we certainly can't fit more than $\frac{36}{\pi} \approx 11.459$ nonoverlapping circles into the square. There isn't *room* for 12. (In fact, we can't even fit 10, because the circles won't nestle together without wasting space "in between." Thus, in this case we'd say that the area-based bound is *loose*.)

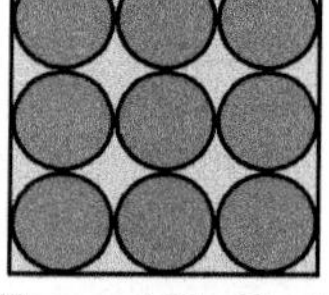

Figure 4.12: Circles packed in a square.

Using packing arguments to derive bounds on error-correcting codes

Now, let's return to error-correcting codes, and use the circle-packing intuition (and the last two lemmas) to prove a bound on the number of n-bit codewords that can "fit" into $\{0,1\}^n$ with minimum distance 3:

Lemma 4.9 (The "sphere-packing bound": distance-3 version)
Let $C \subseteq \{0,1\}^n$ be a code with minimum distance three. Then $|C| \leq 2^n/(n+1)$.

Proof. For each $x \in C$, let $S_x := \{x' \in \{0,1\}^n : \Delta(x', x) \leq 1\}$ be the ball of radius 1 around x. Lemma 4.7 says that $|S_x| = n+1$ for each x. Further, Lemma 4.8 says that every element of $\{0,1\}^n$ is in at most one S_x because the balls are disjoint. Therefore,

$$|\{x' \in \{0,1\}^n : x' \text{ is in one of the } S_x \text{ balls}\}| = \sum_{x \in C} |S_x| = \sum_{x \in C} (n+1) = |C| \cdot (n+1).$$

Also observe that every element of any S_x is an n-bit string. There are only 2^n different n-bit strings, so therefore

$$|\{x' \in \{0,1\}^n : x' \text{ is in one of the } S_x \text{ balls}\}| \leq 2^n.$$

Putting together these two facts, we see that $|C| \cdot (n+1) \leq 2^n$. Solving for $|C|$ yields the desired relationship: $|C| \leq \frac{2^n}{n+1}$. ☐

Corollary 4.10 (The Hamming code is optimal)
Any code with messages of length 4 and minimum distance 3 has codewords of length ≥ 7. (Thus the Hamming code has the best possible rate among all such codes.)

Proof. By Lemma 4.9, we know that $|C| \leq 2^n/(n+1)$. With 4-bit messages we have $|C| = 16$, so we know that $16 \leq 2^n/(n+1)$, or, equivalently, that $2^n \geq 16(n+1)$. And $2^7 = 16(7+1)$, while for any $n < 7$ this inequality does not hold. ☐

Corollary 4.10 implies Theorem 4.4, so we've now proven the three claims that we set out to establish. Before we close, though, we'll mention a few extensions. Lemma 4.8 was general, for any code with an odd minimum distance. But Lemma 4.7 was specifically about codes with minimum distance 3. To generalize the latter lemma, we'd need techniques from *counting* (see Chapter 9, specifically Section 9.4.)

Another interesting question: when is the bound from Lemma 4.9 exactly achievable? If we have k-bit messages, n-bit codewords, and minimum distance 3, then Lemma 4.9 says that $2^k \leq 2^n/(n+1)$, or, taking logs, that $k \leq n - \log_2(n+1)$. Because k has to be an integer, this bound is exactly achievable only when $n+1$ is an exact power of two. (For example, if $n = 9$, this bound requires us to have $2^k \leq 2^9/10 = 512/10 = 51.2$. In other words, we need $k \leq \log_2 51.2 \approx 5.678$. But, because $k \in \mathbb{Z}$, in fact we need $k \leq 5$. That means that this bound is *not* exactly achievable for $n = 9$.) However, it's possible to give a version of the Hamming code for $n = 15$ and $k = 7$ with minimum distance 3, as you'll show in Exercise 4.26. (In fact, there's a version of the Hamming code for any $n = 2^\ell - 1$; see Exercise 4.28.)

Computer Science Connections

Reed–Solomon Codes

The error-correcting codes that are used in CDs and DVDs are a bit more complicated than Repetition or Hamming codes, but they perform better. We'll leave out a lot of the details, but here is a brief sketch of how they work. These codes are called *Reed–Solomon codes*, and they're based on polynomials and modular arithmetic. First, we're going to go beyond bits, to a larger "alphabet" of characters in our messages and codewords: instead of encoding messages from $\{0,1\}^k$, we're going to encode messages from $\{0,1,\ldots,q\}^k$, for some integer q. Here's the basic idea: given a message $m = \langle m_1, m_2, \ldots, m_k \rangle$, we will define a polynomial $p_m(x)$ as follows, with the *coefficients of the polynomial corresponding to the characters of the message*:

$$p_m(x) := \sum_{i=1}^{k} m_i x^i.$$

To encode the message m, we will evaluate the polynomial for several values of x: $encode(m) := \langle p_m(1), p_m(2), \ldots, p_m(n) \rangle$. See Figure 4.13 for an example.

Suppose that we use a k-character message and an n-character output. It's easy enough to compute that the rate is $\frac{k}{n}$. But what about the minimum distance? Consider two distinct messages m and m'. Note that p_m and $p_{m'}$ are both polynomials of degree at most k. Therefore $f(x) := p_m(x) - p_{m'}(x)$ is a polynomial of degree at most k, too—and $f(x) \not\equiv 0$, because $m \neq m'$. Notice that $\{x : f(x) = 0\} = \{x : p_m(x) = p_{m'}(x)\}$. And $|\{x : f(x) = 0\}| \leq k$, by Lemma 2.3 ("degree-$k$ polynomials have at most k roots"). Therefore $|\{x : f(x) = 0\} \cap \{1, 2, \ldots, n\}| \leq k$: there are at most k values x for which $p_m(x) = p_{m'}(x)$. We encoded m and m' by evaluating p_m and $p_{m'}$ on n different inputs, so there are at least $n - k$ inputs on which these two polynomials *disagree*. Thus the minimum distance is at least $n - k$. For example, if we pick $n = 2k$, then we achieve rate $\frac{1}{2}$ and minimum distance k.

How might we decode Reed–Solomon codes? Efficient decoding algorithms rely on some results from linear algebra, but the basic idea is to find the degree-k polynomial that goes through as many of the given points as possible. As a simple example, suppose you're looking for a 2-character message (that is, something encoded as a quadratic), and you receive the codeword $\langle 2, 6, 12, 13, 30, 42 \rangle$. What was the original message? Plot the codeword and see! See Figure 4.14: all but one of the components of the received codeword is consistent with the polynomial $p_m(x) = x + x^2$, so you can decode this codeword as the message $\langle 1, 1 \rangle$.

We've left out several important details of actual Reed-Solomon codes here. One is that our computation of the rate was misleading: we only counted the number of slots, rather than the "size" of those slots. (Figure 4.13 shows that the numbers can get pretty big!) In real Reed–Solomon codes, every value is stored *modulo a prime*. See p. 731 for discussion of how (and why) this fix works. There's also a clever trick used in the physical layout of the encoded information on a CD/DVD: the bits for a particular codeword are spread out over the disc, so that a single physical scratch doesn't cause errors all to occur in the same codeword.

Reed–Solomon codes are named after Irving Reed and Gustave Solomon, 20th-century American mathematicians who invented them in 1960.

Consider the message $m = \langle 1,3,2 \rangle$. Then $p_m(x) = x + 3x^2 + 2x^3$. If we choose $n = 6$, then the encoding of this message will be

$$\begin{aligned}
&\langle 1(1) + 3(1)^2 + 2(1)^3, \\
&\ 1(2) + 3(2)^2 + 2(2)^3, \\
&\ 1(3) + 3(3)^2 + 2(3)^3, \\
&\ 1(4) + 3(4)^2 + 2(4)^3, \\
&\ 1(5) + 3(5)^2 + 2(5)^3, \\
&\ 1(6) + 3(6)^2 + 2(6)^3 \rangle \\
=&\langle 6, 30, 84, 180, 330, 546 \rangle.
\end{aligned}$$

Alternatively, consider the message $m' = \langle 3,0,3 \rangle$. Then $p_{m'}(x) = 3x + 3x^3$. Again for $n = 6$, the encoding of m' is

$$\begin{aligned}
&\langle 3(1) + 3(1)^3, \\
&\ 3(2) + 3(2)^3, \\
&\ 3(3) + 3(3)^3, \\
&\ 3(4) + 3(4)^3, \\
&\ 3(5) + 3(5)^3, \\
&\ 3(6) + 3(6)^3 \rangle \\
=&\langle 6, 30, 90, 204, 390, 666 \rangle.
\end{aligned}$$

Figure 4.13: An example Reed–Solomon encoding.

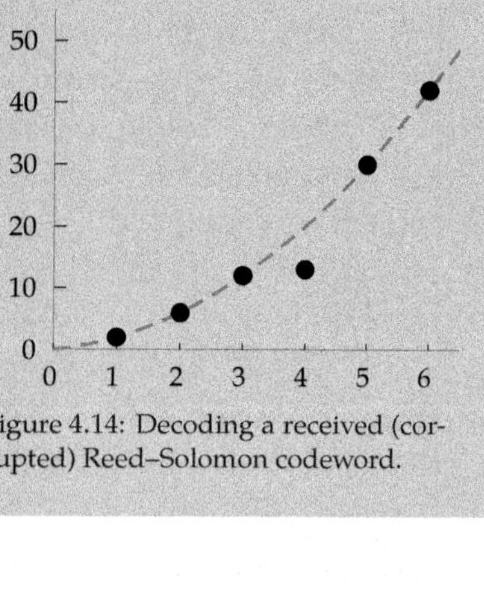

Figure 4.14: Decoding a received (corrupted) Reed–Solomon codeword.

4.2.6 Exercises

```
cc-check(n):
Input: a 16-digit credit-card number n ∈ {0,1,...,9}^16
1: sum := 0
2: for i = 1,2,...,16:
3:     if i is odd then
4:         d_i := 2 · n_i
5:     else
6:         d_i := n_i
7:     Increase sum by the ones' and tens' digits of d_i. (That is, sum := sum + (d_i mod 10) + ⌊d_i/10⌋.)
8: return True if sum mod 10 = 0, and False otherwise.
```

Figure 4.15: An algorithm for testing the validity of credit-card numbers.

The algorithm for testing whether a given credit-card number is valid is shown in Figure 4.15. Here's an example of the calculation that **cc-check**(4471 8329 $\cdots$) *performs:*

(original number)	4	4	7	1	8	3	2	9...
(odd-indexed digits doubled)	8	4	14	1	16	3	4	9...
(digits summed)	4 +	8 +	1+4 +	1	+1+6 +	3 +	4 +	9...

(Try executing **cc-check** *from Figure 4.15 on a few credit-card numbers, to make sure that you've understood the algorithm correctly.) This code can detect any one substitution error, because*

$$0, 2, 4, 6, 8, 1 = 1 + 0, 3 = 1 + 2, 5 = 1 + 4, 7 = 1 + 6, 9 = 1 + 8$$

are all distinct (so, even in odd-indexed digits, changing the digit changes the overall value of sum*).*

4.1 *(programming required)* Implement **cc-check** in a programming language of your choice. Extend your implementation so that, if it's given any 16-digit credit/debit-card number with a single digit replaced by a "?", it computes and outputs the correct missing digit.

4.2 Suppose that we modified **cc-check** so that, instead of *adding the ones digit and (if it exists) the tens digit* to *sum* in Line 7 of the algorithm, we instead simply added the ones digit. (That is, replace Line 7 by $sum := sum + d_i$.) Does this modified code still allow us to detect any single substitution error?

4.3 Suppose that we modified **cc-check** so that, instead of *doubling* odd-indexed digits in Line 4 of the algorithm, we instead *tripled* the odd-indexed digits. (That is, replace Line 4 by $d_i := 3 \cdot n_i$.) Does this modified code still allow us to detect any single substitution error?

4.4 What if we replace Line 4 by $d_i := 5 \cdot n_i$?

4.5 There are simpler schemes that can detect a single substitution error than the one in **cc-check**: for example, we could simply ensure that the sum of all the digits themselves (undoubled) is divisible by 10. (Just skip the doubling step.) The credit-card encoding system includes the more complicated doubling step to help it detect a different type of error, called a *transposition error*, where two adjacent digits are recorded in reverse order. (If two digits are swapped, then the "wrong" digit is multiplied by two, and so this kind of error might be detectable.) Does **cc-check** detect every possible transposition error?

A metric space *consists of a set X and a function* $d : X \times X \to \mathbb{R}^{\geq 0}$, *called a* distance function, *where d obeys the following three properties:*

- reflexivity: *for any x and y in X, we have* $d(x,x) = 0$, *and* $d(x,y) \neq 0$ *whenever* $x \neq y$.
- symmetry: *for any* $x, y \in X$, *we have* $d(x,y) = d(y,x)$.
- triangle inequality: *for any* $x, y, z \in X$, *we have* $d(x,y) \leq d(x,z) + d(z,y)$.

When it satisfies all three conditions, we call the function d a metric.

4.6 In this section, we've been measuring the distance between bitstrings using the Hamming distance, which is a function $\Delta : \{0,1\}^n \times \{0,1\}^n \to \mathbb{Z}^{\geq 0}$, denoting the number of positions in which x and y differ. Prove that Δ is a metric. *(Hint: think about one bit at a time.)*

The next few exercises propose a different distance function $d : \{0,1\}^n \times \{0,1\}^n \to \mathbb{Z}^{\geq 0}$. For each, decide whether you think the given function d is a metric or not, and prove your answer. (In other words, prove that d satisfies reflexivity, symmetry, and the triangle inequality; or prove that d fails to satisfy one or more of these properties.)

4.7 For $x, y \in \{0,1\}^n$, define $d(x,y)$ as the smallest $i \in \{0,1,\ldots,n\}$ such that $x_{i+1,\ldots,n} = y_{i+1,\ldots,n}$. For example, $d(01000, 10101) = 5$ and $d(01\underline{000}, 10\underline{100}) = 3$ and $d(01\underline{000}, 10\underline{000}) = 2$ and $d(1\underline{1010}, 0\underline{1010}) = 1$. (This function measures how far into x and y we must go before the remaining parts match; we could also define $d(x,y)$ as the largest $i \in \{0,1,\ldots,n\}$ such that $x_i \neq y_i$, where we treat $x_0 \neq y_0$.) Is d a metric?

4.8 For $x, y \in \{0,1\}^n$, define $d(x,y)$ as the length of the longest consecutive run of differing bits in corresponding positions of x and y—that is, $d(x,y) := \max\{j - i : \text{for all } k = i, i+1, \ldots, j \text{ we have } x_k \neq y_k\}$. For example, $d(\underline{010}00, \underline{101}01) = 3$ and $d(0\underline{010}0, 0\underline{101}0) = 3$ and $d(\underline{01}000, \underline{10}000) = 2$ and $d(\underline{1}10\underline{1}0, \underline{0}10\underline{0}0) = 1$. Is d a metric?

4.9 For $x, y \in \{0,1\}^n$, define $d(x,y)$ as the difference in the number of ones that appears in the two bitstrings—that is, $d(x,y) := \big|\, |\{i : x_i = 1\}| - |\{i : y_i = 1\}| \,\big|$. (The vertical bars here are a little confusing: the bars around $|\{i : x_i = 1\}|$ and $|\{i : y_i = 1\}|$ denote set cardinality, while the outer vertical bars denote absolute value.) For example, $d(01000, 10101) = |1 - 3| = 2$ and $d(01000, 10100) = |1 - 2| = 1$ and $d(01000, 10000) = |1 - 1| = 0$ and $d(11010, 01010) = |2 - 2| = 0$. Is d a metric?

4.10 The distance version of the *Sørensen index* (a.k.a. the *Dice coefficient*) defines the distance based on the fraction of ones in x or y that are in the same positions. Specifically,

$$d(x,y) := 1 - \frac{2\sum_i x_i \cdot y_i}{\sum_i x_i + y_i}.$$

For example, $d(01000, 10101) = 1 - \frac{2 \cdot 0}{1+3} = 1 - \frac{0}{4} = 1$ and $d(00\underline{1}00, 01\underline{1}10) = 1 - \frac{2 \cdot 1}{1+3} = 1 - \frac{2}{4} = 1/2$ and $d(0\underline{1}000, 1\underline{1}000) = 1 - \frac{2 \cdot 1}{1+2} = 1 - \frac{2}{3} = 1/3$ and $d(1\underline{1}0\underline{1}0, 0\underline{1}0\underline{1}0) = 1 - \frac{2 \cdot 2}{3+2} = 1 - \frac{2}{5} = 3/5$. Is d a metric?

The Sørensen/Dice measure is named after independent work by two ecologists from the 1940s, the Danish botanist Thorvald Sørensen and the American mammalogist Lee Raymond Dice.

4.11 For $x, y \in \{0,1\}^n$, define $d(x,y)$ as the difference in the numbers that are represented by the two strings in binary. Writing this function formally is probably less helpful (particularly because the higher powers of 2 have lower indices), but here it is: $d(x,y) := \left|\sum_{i=1}^{n} x_i \cdot 2^{n-i} - \sum_{i=1}^{n} y_i 2^{n-i}\right|$. For example, $d(01000, 10101) = |8 - 21| = 13$ and $d(01000, 10100) = |8 - 20| = 12$ and $d(01000, 10000) = |8 - 16| = 8$ and $d(11010, 01010) = |26 - 10| = 16$. Is d a metric?

4.12 Show that we can't improve on the parameters in Theorem 4.1: for any integer $t \geq 0$, prove that a code with minimum distance $2t+1$ cannot correct $t+1$ or detect $2t+1$ errors.

4.13 Theorem 4.1 describes the error-detecting and error-correcting properties for a code whose minimum distance is any odd integer. This exercise asks you to give the analogous analysis for a code whose minimum distance is any even integer. Let $t \geq 1$ be any integer, and let C be a code with minimum distance $2t$. Determine how many errors C can detect and correct, and prove your answers.

Let $c \in \{0,1\}^n$ be a codeword. Until now, we've mostly talked about substitution errors, *in which a single bit of c is flipped from 0 to 1, or from 1 to 0. The next few exercises explore two other types of errors.*

An erasure error *occurs when a bit of c isn't successfully transmitted, but the recipient is informed that the transmission of the corresponding bit wasn't successful. We can view an erasure error as replacing a bit c_i from c with a '?' (as in Exercise 4.1, for credit-card numbers). Thus, unlike a substitution error, the recipient* knows *which bit was erased. (So a codeword* 1100110 *might become* 1?0011? *after two erasure errors.) When codeword $c \in \{0,1\}^n$ is sent, the receiver gets a corrupted codeword $c' \in \{0,1,?\}^n$ and where all unerased bits were transmitted correctly (that is, if $c'_i \in \{0,1\}$, then $c'_i = c_i$).*

A deletion error *is like a "silent erasure" error: a bit fails to be transmitted, but there's no indication to the recipient as to where the deletion occurred. (So a codeword* 1100110 *might become* 10011 *after two deletion errors.)*

4.14 Let C be a code that can *detect* t substitution errors. Prove that C can *correct* t erasure errors.

4.15 Let C be a code that can correct t deletion errors. Prove that C can correct t erasure errors.

4.16 Give an example of a code that *can* correct one erasure error, but *can't* correct one deletion error.

Consider the following codes. For each, determine the rate and minimum distance of this code. How many errors can it detect/correct?

4.17 the "code" where all n-bit strings are codewords. (That is, $C := \{0,1\}^n$.)

4.18 the *trivial code*, defined as $C := \{0^n, 1^n\}$.

4.19 the *parity-check code*, defined as follows: the codewords are all n-bit strings with an even number of bits set to 1.

4.20 Let's extend the idea of the parity-check code, from the previous exercise, as an add-on to any existing code with odd minimum distance.

Let $C \subseteq \{0,1\}^n$ be a code with minimum distance $2t+1$, for some integer $t \geq 0$. Consider a new code C', in which we augment every codeword of C by adding a *parity bit*, which is zero if the number of ones in the original codeword is even and one if the number is odd, as follows:

$$C' := \left\{ \langle x_1, x_2, \ldots, x_n, (\textstyle\sum_{i=1}^{n} x_i) \bmod 2 \rangle : x \in C \right\}.$$

Then the minimum distance of C' is $2t+2$. *(Hint: consider two distinct codewords $x, y \in C$. You have to argue that the corresponding codewords $x', y' \in C$ have Hamming distance $2t+2$ or more. Use two different cases, depending on the value of $\Delta(x,y)$.)*

4.21 Show that we can correctly decode the REPETITION$_\ell$ code as follows: given a bitstring c', for each bit position i, we take the majority vote of the ℓ blocks' ith bit in c', breaking ties arbitrarily. (In other words, prove that this algorithm actually gives the codeword that's closest to c'.)

In some error-correcting codes, for certain errors, we may be able to correct more errors than Theorem 4.1 suggests: that is, the minimum distance is $2t+1$, but we can correct certain sequences of $> t$ errors. We've already seen that we can't successfully correct every *such sequence of errors, but we can successfully handle* some *sequences of errors using the standard algorithm for error correction (returning the closest codeword).*

4.22 The REPETITION$_3$ code with 4-bit messages is only guaranteed to correct 1 error. What's the largest number of errors that can possibly be corrected successfully by this code? Explain your answer.

4.23 In the Hamming code, we *never* correct more than 1 error successfully. Prove why not.

4.24 *(programming required)* Write a program, in a programming language of your choice, to verify that any two codewords in the Hamming code differ in at least three bit positions.

Let's find the "next" Hamming code, with 7-bit messages and 11-bit codewords and a minimum distance of 3. We'll use the same style of codeword as in Definition 4.8: the first 7 bits of the codeword will simply be the message, and the next 4 bits will be parity bits (each for some subset of the message bits).

4.25 To achieve minimum distance 3, it will suffice to have parity bits with the following properties:

(a) each bit of the original message appears in at least two parity bits.
(b) no two bits of the original message appear in exactly the same set of parity bits.

Prove that these conditions are sufficient. That is, prove that any set of parity bits that satisfy conditions (a) and (b) ensure that the resulting code has minimum distance 3.

4.26 Define 4 parity bits for 11-bit messages that satisfy conditions (a) and (b) from Exercise 4.25.

4.27 Define 5 parity bits for 26-bit messages that satisfy conditions (a) and (b) from Exercise 4.25.

4.28 Let $\ell \in \mathbb{Z}^{>0}$, and let $n := 2^\ell - 1$. Prove that a code with n-bit codewords, minimum distance 3, and messages of length $n - \ell$ is achievable. *(Hint: look at all ℓ-bit bitstrings; use the bits to identify which message bits are part of which parity bits.)*

4.29 You have come into possession of 8 bottles of "poison," except, you've learned, 7 are fake poison and only 1 is really poisonous. Your master plan to take over the world requires you to identify the poison *by tomorrow.* Luckily, as an evil genius, you have a small collection of very expensive rats, which you can use for testing. You can give samples from bottles to multiple rats simultaneously (a rat can receive a mixture of samples from more than one bottle), and then wait for a day to see which ones die. Obviously you can identify the real poison with 8 rats (one bottle each), or even with 7 (one bottle each, one unused bottle; if all rats survive then the leftover bottle is the poison). But how many rats do you *need* to identify the poison? (Make the number as small as possible.)

Let $c \in \{0,1\}^{23}$. A handy fact (which you'll show in Exercise 9.132, after we've developed the necessary tools for counting to figure out this quantity): the number of 23-bit strings c' with $\Delta(c, c') \leq 3$ is exactly $2048 = 2^{11} = 2^{23-12}$. This fact means that (according to a generalization of Lemma 4.9) it might be possible to achieve the following code parameters:

- *12-bit messages;*
- *23-bit codewords; and*
- *minimum distance 7.*

In fact, these parameters are achievable—and a code that achieves these parameters is surprisingly simple to construct. The Golay code *is an error-correcting code that can be constructed by the following so-called "greedy" algorithm in Figure 4.16. (The loop should consider the strings x in* lexicographic order*: first $00\cdots00$, then $00\cdots01$, then $00\cdots10$, going all the way up to $11\cdots11$. Notice that therefore the all-zero vector will be added to S in the first iteration of the* **while** *loop; a hundred and twenty-seven iterations later, 00000000000000001111111 will be the second element added to S, and so forth.)*

```
1: S := ∅
2: for x ∈ {0,1}^23 (in numerical order):
3:     if Δ(x,y) ≥ 7 for all y ∈ S then
4:         add x to S
5: return S.
```

Figure 4.16: The "greedy algorithm" for generating the Golay code.

4.30 *(programming required)* Write a program, in a language of your choice (but see the warning below), that implements the algorithm in Figure 4.16, and outputs the list of the $2^{12} = 4096$ different 23-bit codewords of the Golay code in a file, one per line.

Implementation hint: suppose you represent the set S as an array, appending each element that passes the test in Line 3 to the end of the array. When you add a bitstring x to S, the very next thing you do is to consider adding $x + 1$ to S. Implementing Line 3 by starting at the x-end of the array will make your code *much* faster than if you start at the 00000000000000000000000-end of the array. Think about why!

Implementation warning: this algorithm is not very efficient! We're doing 2^{23} iterations, each of which might involve checking the Hamming distance of as many as 2^{12} pairs of strings. On a mildly aging laptop, my Python solution took about ten minutes to complete; if you ignore the implementation hint from the previous paragraph, it took 80 minutes. (I also implemented a solution in C; it took about 10 seconds following the hint, and 100 seconds not following the hint.)

The Golay code is named after Marcel Golay, a Swiss researcher who discovered them in 1949, just before Hamming discovered what would later be called the Hamming code. A slight variant of the Golay code was used by NASA around 1980 to communicate with the Voyager spacecraft as they traveled to Saturn and Jupiter.

4.31 You and six other friends are imprisoned by an evil genius, in a room filled with eight bubbling bottles marked as "poison." (Though, really, seven of them look perfectly safe to you.) The evil genius, though, admires skill with bitstrings and computation, and offers you all a deal.

You and your friends will each have a red or blue hat placed on your heads randomly. (Each hat has a 50% chance of being red and 50% chance of being blue, independent of all other hats' colors.) Each person can each see all hats except his or her own. After a brief moment to look at each others' hats, all of you must simultaneously say one of three things: RED, BLUE, or PASS. The evil genius will release all of you from your imprisonment if:

- everyone who says RED or BLUE correctly identifies their hat color; and
- at least one person says a color (that is, not everybody says PASS).

You may collaborate on a strategy before the hats are placed on your heads, but once the hat is in place, no communication is allowed.

An example strategy: all 7 of you pick a random color and say it. (You succeed with probability $(1/2)^7 = 1/128 \approx 0.0078$.) Another example: you number yourselves $1, 2, \ldots, 7$, and person #7 picks a random color and says it; everyone else passes. (You succeed with probability $1/2$.)

Can you succeed with probability better than $1/2$? If so, how?

4.32 In Section 4.2.5, we proved an upper bound for the rate of a code with a particular minimum distance, based on the volume of "spheres" around each codeword. There are other bounds that we can prove, with different justifications.

Suppose that we have a code $C \subseteq \{0,1\}^n$ with $|C| = 2^k$ and minimum distance d. Prove the *Singleton bound,* which states that $k \leq n - d + 1$. *(Hint: what happens if we delete the first $d - 1$ bits from each codeword?)*

Confusingly, the Singleton bound is named after Richard Singleton, a 20th-century American computer scientist; it has nothing to do with singleton sets (sets containing only one element).

4.3 *Proofs and Proof Techniques*

> Arguments are to be avoided; they are always vulgar and often convincing.
>
> Oscar Wilde (1854–1900)

In Section 4.2, we saw a number of claims about error-correcting codes—and, more importantly, proofs that those claims were true. These proofs used several different styles of argument: proofs that involved straightforward reasoning by starting from the relevant definitions; proofs that used "case-based" reasoning; and proofs "by contradiction" that argued that x must be true because something impossible would happen if x were false. Indeed, whenever you face a claim that you need to prove, a variety of different strategies (including these strategies from Section 4.2) are possible approaches for you to employ. This section is devoted to outlining these and some other common proof strategies. We'll first catalogue these techniques in Section 4.3.1, and then, in Section 4.3.2, we'll reflect briefly on the strategies and how to choose among them—and also reflect on the *writing* part of writing proofs.

What is a proof?

This chapter is devoted to techniques for proving claims—but before we explore proof techniques, let's spend a few words discussing what a proof actually *is:*

> **Definition 4.9 (Proof)**
> *A* proof *of a proposition is a convincing argument that the proposition is true.*

Definition 4.9 says that a proof is a "convincing argument," but it doesn't say *to whom* the argument should be convincing. The answer is: *to your reader.* This definition may be frustrating, but the point is that a proof is a piece of writing, and—just like with fiction or a persuasive essay—you must write *for your audience.*

> **Taking it further:** Different audiences will have very different expectations for what counts as "convincing." A formal logician might not find an argument convincing unless she saw every last step, no matter how allegedly obvious or apparently trivial. An instructor of early-to-mid-level computer science class might be convinced by a proof written in paragraph form that omits some simple steps, like those that invoke the commutativity of addition, for example. A professional CS researcher reading a publication in conference proceedings would expect "elementary" calculus to be omitted.
>
> Some of the debates over what counts as convincing to an audience—in other words, what counts as a "proof"—were surprisingly controversial, particularly as computer scientists began to consider claims that had previously been the exclusive province of mathematicians. See the discussion on p. 437 of the *Four-Color Theorem*, which triggered many of these discussions in earnest.

To give an example of writing for different audiences, we'll give several proofs of the same result. Here's a claim regarding divisibility and factorials. (Recall that $n!$, pronounced "n factorial," is defined as $n! := n \cdot (n-1) \cdot (n-2) \cdots 1$.) Before reading further, spend a minute trying to convince yourself why (†) is true:

> Let n be a positive integer and let k be any integer satisfying $2 \leq k \leq n$.
> Then $n! + 1$ is not evenly divisible by k. (†)

We'll prove Claim (†) three times, using three different levels of detail:

Example 4.8 (Factorials: Proof I)

Proof (heavy detail). By the definition of factorial, we have that $n! = \prod_{i=1}^{n} i$, which can be rewritten as $n! = \left[\prod_{i=1}^{k-1} i\right] \cdot k \cdot \left[\prod_{i=k+1}^{n} i\right]$. Let $m = \left[\prod_{i=1}^{k-1} i\right] \cdot \left[\prod_{i=k+1}^{n} i\right]$. Thus we have that $n! = k \cdot m$ and $m \in \mathbb{Z}$, because the product of any finite set of integers is also an integer.

Observe that $n! + 1 = mk + 1$. We claim that there is no integer ℓ such that $k\ell = n! + 1$. First, there is no $\ell \leq m$ such that $k\ell = n! + 1$, because $k\ell \leq km = n! < n! + 1$. Second, there is no $\ell \geq m + 1$ such that $k\ell = n! + 1$, because $k \geq 2$ implies that $k\ell \geq k(m+1) = n! + k > n! + 1$. Because there is no such integer $\ell \leq m$ and no such integer $\ell > m$, the claim follows. □

Example 4.9 (Factorials: Proof II)

Proof (medium detail). Define $m = n!/k$, so that $n! = mk$ and $n! + 1 = mk + 1$. Because k is an integer between 2 and n, the definition of factorial implies that m is an integer. But because $k \geq 2$, we know $mk < mk + 1 < (m+1)k$. Thus $mk + 1$ is not evenly divisible by k, because this quantity is strictly between two consecutive integral multiples of k, namely $m \cdot k$ and $(m+1) \cdot k$. □

Example 4.10 (Factorials: Proof III)

Proof (light detail). Note that k evenly divides $n!$. The next integer evenly divisible by k is $n! + k$. But $k \geq 2$, so $n! < n! + 1 < n! + k$. The claim follows immediately. □

Which of the three proofs from Examples 4.8, 4.9, and 4.10 is best? *It depends!* The right level of detail depends on your intended reader. A typical reader of this book would probably be happiest with the medium-detail proof from Example 4.9, but it is up to you to tailor your proof to your desired reader.

Writing tip: As you study the material in this book, you will frequently be given a claim and asked to prove it. To complete this task well, you must think about the question of *for whom* you are writing your proof. A reasonable guideline is that your audience for your proofs is a classmate or a fellow reader of this book who has read and understood everything up to the point of the claim that you're proving, but hasn't thought about this particular claim at all.

Taking it further: It turns out that one can encode literally all of mathematics using a handful of set-theoretic axioms, and a lot of patience. It's possible to write down everything in this book in ultraformal set-theoretic notation, which serves the purpose of making arguments 100% airtight. But the high-level computer science content can be hard to see in that style of proof. If you've ever programmed in assembly language before, there's a close analogy: you can express every program that you've ever written in extremely low-level machine code, or you can write it in a high-level language like C or Java or Python or Scheme (and, one hopes, make the algorithm much more understandable for the reader). We'll prove a lot of facts in this book, but at the Python-like level of proof. Someone could "compile" our proofs down into the low-level set-theoretic language—but we won't bother. (Lest you underestimate the difficulty of this task: a proof that $2 + 2 = 4$ would require hundreds of steps in this low-level proof!)

There are subfields of computer science ("formal methods" or "formal verification," or "automated theorem proving") that take this ultrarigorous approach: start from a list of axioms, and a list of inference rules, and a desired theorem, and derive the theorem by applying the inference rules. When it is absolutely life-or-death critical that the proof be 100% verified, then these approaches tend to be used: in verifying protocols in distributed computing, or in verifying certain crucial components of a processor, for example.

4.3.1 *Proof Techniques*

We will describe three general strategies for proofs:

- *direct proof:* we prove a statement φ by repeatedly inferring new facts from known facts to eventually conclude φ. (Sometimes we'll divide our work into separate cases and give different proofs in each case. And if φ is of the form $p \Rightarrow q$, we'll generally assume p and then try to infer q under that assumption.)

- *proof by contrapositive:* when the statement that we're trying to prove is an implication $p \Rightarrow q$, we can instead prove $\neg q \Rightarrow \neg p$—the *contrapositive* of the original claim. The contrapositive is logically equivalent to the original implication, so once we've proven $\neg q \Rightarrow \neg p$, we can also conclude $p \Rightarrow q$.

- *proof by contradiction:* we prove a statement φ by repeatedly *assuming* $\neg\varphi$, and proving something impossible—that is, proving $\neg\varphi \Rightarrow$ False. Because $\neg\varphi$ therefore cannot be true, we can conclude that φ must be true.

"When you have eliminated the impossible, whatever remains, however improbable, must be the truth."
— Sir Arthur Conan Doyle (1859–1930), *The Sign of the Four* (1890).

We'll give some additional examples of each proof technique as we go, proving some purely arithmetic claims to illustrate the strategy.

Almost every claim that we'll prove here—or that you'll ever need to prove—will be a universally quantified statement, of the form $\forall x \in S : P(x)$. (Often the quantification will not be explicit: we view any unquantified variable in a statement as being implicitly universally quantified.) To prove a claim of the form $\forall x \in S : P(x)$, we usually proceed by considering a generic element $x \in S$, and then proving that $P(x)$ holds. (Considering a "generic" element means that we make no further assumptions about x, other than assuming that $x \in S$.) Because this proof establishes that an arbitrary $x \in S$ makes $P(x)$ true, we can conclude that $\forall x \in S : P(x)$.

Direct proofs

The simplest type of proof for a statement φ is a derivation of φ from known facts. This type of argument is called a *direct proof*:

Definition 4.10 (Direct Proof)
A direct proof *of a proposition* φ *starts from known facts and implications, and repeatedly applies logical deduction to derive new facts, eventually leading to the conclusion* φ.

Most of the proofs in Section 4.2 were direct proofs. Here's another, simpler example:

Example 4.11 (Divisibility by 4)
Let's prove the correctness of a simple test of whether a given integer is divisible by 4:

Claim: Any positive integer n is divisible by 4 if and only if its last two digits are themselves divisible by 4. (That is, n is divisible by 4 if and only if n's last two digits are in $\{00, 04, 08, \ldots, 92, 96\}$.)

Proof. Let $d_k, d_{k-1}, \ldots, d_1, d_0$ denote the digits of n, reading from left to right, so that

$$n = d_0 + 10d_1 + 100d_2 + 1000d_3 + \cdots + 10^k d_k,$$

or, dividing both sides by 4,

$$n/4 = (d_0 + 10d_1)/4 + 25d_2 + 250d_3 + \cdots + 25 \cdot 10^{k-2} d_k. \qquad (*)$$

The integer n is a divisible by 4 if and only if $n/4$ is an integer, which because of $(*)$ occurs if and only if the right-hand side of $(*)$ is an integer. And that's true if and only if $(d_0 + 10d_1)/4$ is an integer, because all other terms in the right-hand side of $(*)$ are integers. Therefore $4 \mid n$ if and only if $4 \mid (d_0 + 10d_1)$. The last two digits of n are precisely $d_0 + 10d_1$, so the claim follows. □

Note that this argument considers a generic positive integer n, and establishes the result for that generic n. The proof relies on two previously known facts: (1) an integer n is divisible by 4 if and only if $n/4$ is an integer; and (2) for an integer a, we have that $x + a$ is an integer if and only if x is an integer. The argument itself uses these two basic facts to derive the desired claim.

Let's give another example, this time for an implication. The proof strategy of *assuming the antecedent*, discussed in Definition 3.22 in Section 3.4.3, is a form of direct proof. To prove an implication of the form $\varphi \Rightarrow \psi$, we *assume* the antecedent φ and then prove ψ under this assumption. This proof establishes $\varphi \Rightarrow \psi$ because the only way for the implication to be false is when φ is true but ψ is false, but the proof shows that ψ is true whenever φ is true. Here's an example of this type of direct proof, for a basic fact about rational numbers. (Recall that a number x is *rational* if and only if there exist integers n and $d \neq 0$ such that $x = \frac{n}{d}$.)

Example 4.12 (The product of rational numbers is rational)
Claim: If x and y are rational numbers, then so is xy.
Proof. Assume the antecedent—that is, assume that x and y are rational. By the definition of rationality, then, there exist integers n_x, n_y, $d_x \neq 0$, and $d_y \neq 0$ such that $x = \frac{n_x}{d_x}$ and $y = \frac{n_y}{d_y}$. Therefore

$$xy = \frac{n_x}{d_x} \cdot \frac{n_y}{d_y} = \frac{n_x n_y}{d_x d_y}.$$

Both $n_x n_y$ and $d_x d_y$ are integers, because the product of any two integers is also an integer. And $d_x d_y \neq 0$ because both $d_x \neq 0$ and $d_y \neq 0$. Thus xy is a rational number, by the definition of rationality. □

Proof by cases

Sometimes we'll be asked to prove a statement of the form $\forall x \in S : P(x)$ that indeed seems true for every $x \in S$—but the "reason" that $P(x)$ is true seems to be different for different "kinds" of elements x. For example, Lemma 4.6 argued that the Hamming distance between two Hamming-code codewords was at least three, based on three different arguments based on whether the corresponding messages differed in 1, 2, or ≥ 3 positions. This proof was an example of a *proof by cases:*

Definition 4.11 (Proof by cases)
To give a proof by cases *of a proposition φ, we identify a set of* cases *and then prove two different types of facts: (1) "in every case, φ holds"; and (2) one of the cases has to hold.*

(Proofs by cases need not be direct proofs, but plenty of them are.) Here are two simple examples of proofs by cases:

Example 4.13 (Certain squares)
Claim: Let n be any integer. Then $n \cdot (n+1)^2$ is even.
Proof. We'll give a proof by cases, based on the parity of n:

- If n is even, then any multiple of n is also even, so we're done.
- If n is odd, then $n+1$ must be even. Thus any multiple of $n+1$ is also even, so we're done again.

Because the integer n must be either even or odd, and the quantity $n \cdot (n+1)^2$ is an even number in either case, the claim follows. □

Example 4.14 (An easy fact about absolute values)
Claim: Let $x \in \mathbb{R}$. Then $-|x| \leq x \leq |x|$.
Proof. Observe that $x \geq 0$ or $x \leq 0$. In both cases, we'll show the desired inequality:

- For the case that $x \geq 0$, we know $-x \leq 0 \leq x$. By the definition of absolute value, we have $|x| = x$ and $-|x| = -x$. Thus $-|x| = -x \leq 0 \leq x = |x|$.
- For the case that $x < 0$, we know $x \leq 0 \leq -x$. By the definition of absolute value, we have $|x| = -x$ and $-|x| = x$. Thus $-|x| = x \leq 0 \leq -x = |x|$. □

Note that a proof by cases is only valid if the cases are *exhaustive*—that is, if every situation falls into one of the cases. (If, for example, you try to prove $\forall x \in \mathbb{R} : P(x)$ with the cases $x > 0$ and $x < 0$, you've left out $x = 0$—and your proof isn't valid!) But the cases do not need to be mutually exclusive (that is, they're allowed to overlap), as long as the cases really do cover all the possibilities; in Example 4.14, we handled the $x = 0$ case in *both* cases $x \geq 0$ and $x \leq 0$. If all possible values of x are covered by *at least* one case, and the claim is true in every case, then the proof is valid.

Here's another slightly more complex example, where we'll prove the triangle inequality for the absolute value function. (See Figure 4.2.)

The phrase "without loss of generality" indicates that we won't explicitly write out all the cases in the proof, because the omitted ones are virtually identical to the ones that we *are* writing out. It allows you to avoid cut-and-paste-and-search-and-replace arguments for two very similar cases.

Example 4.15 (Triangle inequality for absolute values)
Claim: Let $x, y, z \in \mathbb{R}$. Then $|x - y| \leq |x - z| + |y - z|$.
Proof. Without loss of generality, assume that $x \leq y$. (If $y \leq x$, then we simply swap the names of x and y, and nothing changes in the claim.)

Because we're assuming $x \leq y$, we must show that $|x - z| + |y - z| \geq |x - y| = y - x$. We'll consider three cases: $z \leq x$, or $x \leq z \leq y$, or $y \leq z$. See Figure 4.17.

Case I: $z \leq x$. Then

$$\begin{aligned} |x - z| + |y - z| &\geq |y - z| && |x - z| \geq 0 \text{ by the definition of absolute value.} \\ &= y - z && x \leq y \text{ by assumption and } z \leq x \text{ in Case I, so } z \leq y \text{ too.} \\ &\geq y - x. && z \leq x \text{ in Case I, so } -z \geq -x. \end{aligned}$$

Case II: $x \leq z \leq y$. Then

$$\begin{aligned} |x - z| + |y - z| &= (z - x) + |y - z| && \text{definition of absolute value and } x \leq z \text{ in Case II.} \\ &= (z - x) + (y - z) && \text{definition of absolute value and } z \leq y \text{ in Case II.} \\ &= y - x. && \text{algebra/rearranging terms.} \end{aligned}$$

Case III: $y \leq z$. Then

$$\begin{aligned} |x - z| + |y - z| &\geq |x - z| && |y - z| \geq 0 \text{ by the definition of absolute value.} \\ &= z - x && x \leq y \text{ by assumption and } y \leq z \text{ in Case III, so } x \leq z \text{ too.} \\ &\geq y - x. && z \geq y \text{ in Case III.} \end{aligned}$$

In all three cases, we've shown that $|x - z| + |y - z| \geq y - x$, so the claim follows. □

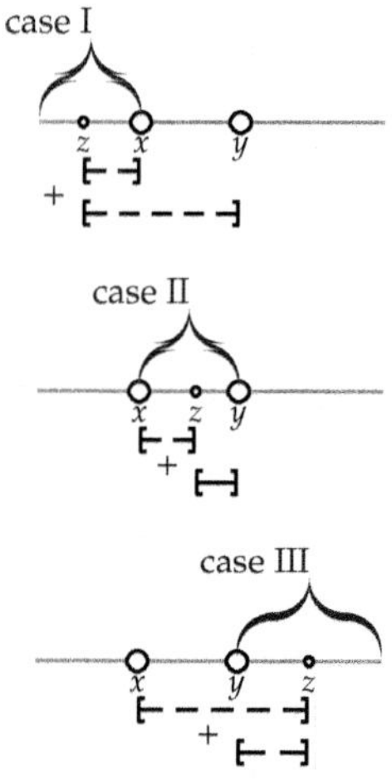

Figure 4.17: The three cases for Example 4.15: z can fall to the left of x, between x and y, or to the right of y. In each case, we argue that the sum of the lengths of the dashed lines is at least $y - x$.

Notice the creative demand if you choose to develop a proof by cases: you have to choose which cases to use! The proposition itself does not necessarily make obvious an appropriate choice of which different cases to use.

Proof by contrapositive

When we seek to prove a claim φ, it suffices to instead prove any proposition that is logically equivalent to φ. (For example, a proof by cases with two cases q and $\neg q$ corresponds to the logical equivalence $p \equiv (q \Rightarrow p) \wedge (\neg q \Rightarrow p)$.) A valid proof of any logically equivalent proposition can be used to prove that φ is true, but a few logical equivalences turn out to be particularly useful. A *proof by contrapositive* is a very common proof technique that relies on this principle:

Definition 4.12 (Proof by contrapositive)
To give a proof by contrapositive *of an implication $\varphi \Rightarrow \psi$, we instead give a proof of the implication $\neg\psi \Rightarrow \neg\varphi$.*

Recall from Section 3.4.3 that an implication $p \Rightarrow q$ is logically equivalent to its *contrapositive* $\neg q \Rightarrow \neg p$. (An implication is true unless its antecedent is true and its conclusion is false, so $\neg q \Rightarrow \neg p$ is true unless $\neg q$ is true and $\neg p$ is false, which is precisely when $p \Rightarrow q$ is false.) Here are two simple examples of proofs using the contrapositive, one about absolute values and one about rational numbers:

Example 4.16 (The sum of the absolute values vs. the absolute value of the sum)
Claim: If $|x| + |y| \neq |x + y|$, then $xy < 0$.
Proof. We'll prove the contrapositive:

$$\text{If } xy \geq 0\text{, then } |x| + |y| = |x + y|. \qquad (*)$$

To prove (∗), assume the antecedent; that is, assume that $xy \geq 0$. We must prove $|x| + |y| = |x + y|$. Because $xy \geq 0$, there are two cases: either both $x \geq 0$ and $y \geq 0$, or both $x \leq 0$ and $y \leq 0$.

Case I: $x \geq 0$ and $y \geq 0$. Then $|x| + |y| = x + y$, by the definition of absolute value. And $|x + y| = x + y$ too, because $x \geq 0$ and $y \geq 0$ implies that $x + y \geq 0$ as well.

Case II: $x \leq 0$ and $y \leq 0$. Then $|x| + |y| = -x + -y$, by the definition of absolute value. And $|x + y| = -(x + y) = -x + -y$ too, because $x \leq 0$ and $y \leq 0$ implies that $x + y \leq 0$ as well. □

Writing tip: Help your reader figure out what's going on! If you're going to use a proof by contrapositive, *say you're using a proof by contrapositive!* Don't leave 'em guessing. This tip applies for all proof techniques: your job is to convince your reader, so be kind and informative to your reader.

Example 4.17 (Irrational quotients have an irrational numerator or denominator)
Claim: Let $y \neq 0$. If x/y is irrational, then either x is irrational or y is irrational.
Proof. We will prove the contrapositive:

$$\text{If } x \text{ is rational and } y \text{ is rational, then } x/y \text{ is rational.} \qquad (\dagger)$$

(Note that, by De Morgan's Laws, $\neg$ (x is irrational or y is irrational) is equivalent to x being rational and y being rational.)

To prove (†), assume the antecedent—that is, assume that x is rational and y is rational. By definition, then, there exist four integers n_x, n_y, $d_x \neq 0$, and $d_y \neq 0$ such that $x = \frac{n_x}{d_x}$ and $y = \frac{n_y}{d_y}$. Thus $\frac{x}{y} = \frac{n_x d_y}{d_x n_y}$. (By the assumption that $y \neq 0$, we know that $n_y \neq 0$, and thus $d_x n_y \neq 0$.) Both the numerator and denominator are integers, so $\frac{x}{y}$ is rational. □

Of course, you can always reuse previous results in any proof—and Example 4.12 is particularly useful for the claim in Example 4.17. Here's a second, shorter proof:

Example 4.18 (Irrational quotients, Version B)
Claim: Let $y \neq 0$. If x/y is irrational, then either x is irrational or y is irrational.
Proof. We prove the contrapositive. Assume that x and y are rational. By definition, then, $y = \frac{n}{d}$ for some integers n and $d \neq 0$. Therefore $\frac{1}{y} = \frac{d}{n}$ is rational too. (By the assumption that $y \neq 0$, we know that $n \neq 0$.) But $\frac{x}{y} = x \cdot \frac{1}{y}$, and both x and $\frac{1}{y}$ are rational. Therefore Example 4.12 implies that $\frac{x}{y}$ is rational too. □

Here's one more example of a proof that uses the contrapositive. When proving an "if and only if" statement $\varphi \Leftrightarrow \psi$, we can instead give proofs of both $\varphi \Rightarrow \psi$ and $\psi \Rightarrow \varphi$, because $\varphi \Leftrightarrow \psi$ and $(\varphi \Rightarrow \psi) \wedge (\psi \Rightarrow \varphi)$ are logically equivalent. This type of proof is sometimes called a *proof by mutual implication.* (We can also prove $\varphi \Leftrightarrow \psi$

by giving a chain of logically equivalent statements that transform φ into ψ, but it is often easier to prove one direction at a time.) Here's an example of a proof by mutual implication, which also uses the contrapositive to prove one of the directions:

Example 4.19 (Even integers (and only even integers) have even squares)
Claim: Let n be any integer. Then n is even if and only if n^2 is even.
Proof. We proceed by mutual implication.

First, we will show that if n is even, then n^2 is even too. Assume that n is even. Then, by definition, there exists an integer k such that $n = 2k$. Therefore $n^2 = (2k)^2 = 4k^2 = 2 \cdot (2k^2)$. Thus n^2 is even too, because there exists an integer ℓ such that $n^2 = 2\ell$. (Namely, $\ell = 2k^2$.)

Second, we will show the converse: if n^2 is even, then n is even. We will instead prove the contrapositive: if n is not even, then n^2 is not even. Assume that n is not even. Then n is odd, and there exists an integer k such that $n = 2k + 1$. Therefore $n^2 = (2k+1)^2 = 4k^2 + 4k + 1 = 2(2k^2 + 2k) + 1$. Thus n^2 is odd too, because there exists an integer ℓ such that $n^2 = 2\ell + 1$. (Namely, $\ell = 2k^2 + 2k$.) □

PROOFS BY CONTRADICTION

The proof techniques that we've described so far establish a claim φ by arguing that φ *must be true.* Here, we'll look at the other side of the coin, and prove φ has to be true by proving that φ *cannot be false.* This approach is called a *proof by contradiction:* we prove that something impossible must happen if φ is false (that is, we prove $\neg\varphi \Rightarrow \text{False}$); thus the assumption $\neg\varphi$ led us to an absurd conclusion, and we must reject the assumption $\neg\varphi$ and instead conclude its negation φ:

A proof by contradiction is also called *reductio ad absurdum* (Latin: "reduction to an absurdity").

Definition 4.13 (Proof by contradiction)
To prove φ using a proof by contradiction, we assume *the negation of φ and derive a* contradiction; *that is, we assume $\neg\varphi$ and prove False.*

(This proof technique is based on the logical equivalence of φ and the proposition $\neg\varphi \Rightarrow \text{False}$.) We used a proof by contradiction in Lemma 4.8: to show that two particular sets X and Y were disjoint, we assumed that there *was* an element $z \in X \cap Y$ (that is, we assumed that X and Y were *not* disjoint), and we showed that this assumption led to a violation of the assumptions in the definitions of X and Y. Here's another simple example:

As my grandfather always used to say: "If the conclusion is obviously false, reexamine the premises."
—Jay Liben (1913–2006)

Example 4.20 ($15x + 111y = 55057$ for integers x and y?)
Claim: Suppose $15x + 111y = 55057$, for two real numbers x and y. Then either x or y (or both) is not an integer.
Proof. Suppose not: that is, suppose that x and y are integers with $15x + 111y = 55057$. But $15x + 111y = 3 \cdot (5x + 37y)$, so $\frac{55057}{3} = 5x + 37y$. But then $\frac{55057}{3}$ must therefore be an integer, because $5x + 37y$ is—but $\frac{55057}{3} = 18352.333\cdots \notin \mathbb{Z}$. Therefore the

assumption that both $x \in \mathbb{Z}$ and $y \in \mathbb{Z}$ was false, and at least one of x and y must be nonintegral. □

Here is another example of a proof by contradiction, for a classical result showing that there are numbers that aren't rational:

Writing tip: It's always a good idea to help your reader with "signposts" in your writing. In a proof by contradiction, announce at the outset that you're assuming $\neg\varphi$ for the purposes of deriving a contradiction; when you reach a contradiction, *say* that you've reached a contradiction, and declare that therefore the assumption $\neg\varphi$ was false, and φ is true.

Example 4.21 (The irrationality of $\sqrt{2}$)
Claim: $\sqrt{2}$ is not rational.
Proof. We proceed by contradiction.

Assume that $\sqrt{2}$ is rational. Therefore, by the definition of rationality, there exist integers n and $d \neq 0$ such that $n/d = \sqrt{2}$, where n and d are in lowest terms (that is, where n and d have no common divisors).

Squaring both sides yields that $n^2/d^2 = 2$, and therefore that $n^2 = 2d^2$. Because $2d^2$ is even, we know that n^2 is even. Therefore, by Example 4.19 ("n is even if and only if n^2 is even") we have that n is itself even.

Because n is even, there exists an integer k such that $n = 2k$, which implies that $n^2 = 4k^2$. Thus $n^2 = 4k^2$ and $n^2 = 2d^2$, so $2d^2 = 4k^2$ and $d^2 = 2k^2$. Hence d^2 is even, and—again using Example 4.19—we have that d is even.

But now we have a contradiction: we assumed that n/d was in lowest terms, but we have now shown that n and d are both even! Thus the original assumption that $\sqrt{2}$ was rational was false, and we can conclude that $\sqrt{2}$ is irrational. □

Note again the structure of this proof: *suppose that* $\sqrt{2}$ is rational; *therefore* we can write $\sqrt{2} = n/k$ where n and k have no common divisors, and (a few steps later) *therefore* n and k are both even. Because n and k cannot *both* have no common divisors *and* also both be even, we've derived an absurdity. The only way we could have gotten to this absurdity is via our assumption that $\sqrt{2}$ was rational—so we conclude that this assumption must have been false, and therefore $\sqrt{2}$ is irrational.

Note that, when you're trying to prove an implication $\varphi \Rightarrow \psi$, a proof by contrapositive has some similarity to a proof by contradiction:

- in a proof by contrapositive, we prove $\neg\psi \Rightarrow \neg\varphi$, by assuming $\neg\psi$ and proving $\neg\varphi$.
- in a proof by contradiction, we prove False under the assumption $\neg(\varphi \Rightarrow \psi)$—that is, under the assumption that $\varphi \wedge \neg\psi$. (Note that there's an extra creative demand here: you have to figure out which contradiction to derive—something that's not generally made immediately clear by the given claim.)

Proofs by contrapositive are generally preferred over proofs by contradiction when a proof by contrapositive is possible. A proof by contradiction can be hard to follow because we're asking the reader to temporarily accept an assumption that we'll later show to be false, and there can be a mental strain in keeping track of what's been assumed and what was previously known. (Notice that the claim in Example 4.21 wasn't an implication, so a proof by contrapositive wasn't an option. The proofs of Lemma 4.8 and Example 4.20, though, could have been rephrased as proofs by contrapositive.)

PROOFS BY CONSTRUCTION AND DISPROOFS BY COUNTEREXAMPLE

So far we've concentrated on proofs of universally quantified statements, where you are asked to show that some property holds for all elements of a given set. (Every example proof in this section, except the two proofs by contradiction about the irrationality of $\sqrt{2}$ and the infinitude of primes, were proofs of a "for all" statement—and, actually, even those two claims could have been phrased as universal quantifications. For example, we could have phrased Example 4.21 as the following claim: for all integers n and d, we have $n \neq d \cdot \sqrt{2}$.) Sometimes you'll confront a universally quantified statement that's false, though. The easiest way to prove that $\forall x \in S : P(x)$ is false is using a *disproof by counterexample:*

Definition 4.14 (Disproof by counterexample)
A counterexample *to a claim $\forall x \in S : P(x)$ is a particular element $y \in S$ such that $P(y)$ is false. A* disproof by counterexample *of $\neg\forall x \in S : P(x)$ is such a counterexample $y \in S$, together with a proof that $P(y)$ is false.*

Finding a counterexample for a claim requires creativity: you have to think about why a claim might not be true, and then try to construct an example that embodies that reason. Here is a simple example:

Problem-solving tip: One way you might try to identify counterexample to a claim is by writing a program: write a loop that tries a bunch of examples; if you ever find one for which the claim is false, then you've found a counterexample. Just because you haven't found a counterexample with your program doesn't mean that there isn't one—unless you've tried *all* the elements of S—but if you *do* find a counterexample, it's still a counterexample no matter how you found it!

Example 4.22 (Unique sums of squares)
Claim: Let n be a positive integer such that $n = a^2 + b^2$ for positive integers a and b. Then n cannot be expressed as the sum of the squares of two positive integers except a and b. (Alternatively, this claim could be written more tersely as: *No positive integer is expressible in two different ways as the sum of two perfect squares.)*

The claim is false, and we will prove that it is false by counterexample. We can start trying some examples. One easy class of potential counterexamples is $a^2 + 1$ for an integer a. $1^2 + 1^2 = 2$ can't be expressed a different way. What about 5? 10? 17? 26? 37? 50? 65? 82? By testing these examples, we find that 65 is a counterexample to the claim. Observe that $1^2 + 8^2 = 1 + 64 = 65$, and $4^2 + 7^2 = 16 + 49 = 65$. Another counterexample is 50, as $50 = 5^2 + 5^2 = 1^2 + 7^2$.

What about when you're asked to prove an existential claim $\exists x : P(x)$? One approach is to prove the claim by contradiction: you assume $\forall x : \neg P(x)$, and then derive some contradiction. This type of proof is called *nonconstructive*: you have proven that an object with a certain property must exist, but you haven't actually described a particular object with that property. In contrast, a *proof by construction* actually identifies a specific object that has the desired property:

Definition 4.15 (Proof by construction)
A constructive proof *or* proof by construction *for a claim $\exists x \in S : P(x)$ actually builds an object satisfying the property P: first, we identify a particular element $y \in S$; and, second, we prove $P(y)$.*

For example, here's a simple claim that we'll prove twice, once nonconstructively and once constructively:

Example 4.23 (The last two digits of some squares)
Claim: There exist distinct integers $x, y \in \{1901, 1902, \ldots, 2014\}$ such that the last two digits of x^2 and y^2 are the same. (In other words, $x^2 \bmod 100 = y^2 \bmod 100$.)
Nonconstructive. There are 114 different numbers in the set $\{1901, 1902, \ldots, 2014\}$. There are only 100 different possible values for the last two digits of numbers. Thus, because there are 114 elements assigned to only 100 categories, there must be some category that contains more than one element. □

Constructive. Let $x = 1986$ and $y = 1964$. Both numbers' squares have 96 as their last two digits: $1986^2 = 3{,}944{,}196$ and $1964^2 = 3{,}857{,}296$. □

It's generally preferable to give a constructive proof when you can. A constructive proof is sometimes harder to develop than a nonconstructive proof, though: it may require more insight about the kind of object that can satisfy a given property, and more creativity in figuring out how to actually construct that object.

Taking it further: A constructive proof of a claim is generally more satisfying for the reader than a nonconstructive proof. A proof by contradiction may leave a reader unsettled—okay, the claim is true, but what can we do with that?—while a constructive proof may be useful in designing an algorithm, or it may suggest further possible claims to try to prove. (There's even a school of thought in logic called *constructivism* that doesn't count a proof by contradiction as a proof!)

4.3.2 *Some Brief Thoughts about Proof Strategy*

So far in this section, we've concentrated on developing a toolbox of proof techniques. But when you're confronted with a new claim and asked to prove it, you face a difficult task in figuring out which approach to take. (It's even harder if you're asked to formulate a claim and *then* prove it!) As we discussed in Chapter 3, there's no formulaic approach that's guaranteed to work—you must be creative, open-minded, persistent. You will have to accept that you will explore approaches that end up being dead ends. This section will give a few brief pointers about proof strategy—some things to try when you're just starting to attack a new problem. We'll start with some concrete advice in the form of a three-step plan, largely inspired by an outstanding book by George Pólya.[2] (I highly recommend Pólya as further reading!)

[2] George Pólya. *How to Solve It.* Doubleday, Garden City, NY, 1957.

1. *Understand what you're trying to do.* Read the statement that you're trying to prove. Reread it. What are the assumptions? What is the desired conclusion? (That is, what are you trying to prove under the given assumptions?) Remind yourself of any unfamiliar notation or terminology. Pick a simple example and make sure the alleged theorem holds for your example. (If not, either you've misunderstood something or the claim is false.) Reread the statement again.

   If you're not given a specific claim—for example, you're asked to prove or disprove a given statement, or if you're asked for the "best possible" solution to a problem—then it's harder but even more important to understand what you're trying to do. Play around with some examples to generate a sense of what might be plausibly true. Then try to form a conjecture based on these examples or the intuition that you've developed.

2. *Do it.* Now that you have an understanding of the statement that you're trying to prove, it's time to actually prove it. You might start by trying to think about slightly different problems to help grant yourself insight about this one. Are there results that you already know that "look similar" to this one? Can you solve a more general problem? Make the premises look as much like the conclusion as possible. Expand out the definitions; write down what you know and what you have to derive, in primitive terms. Can you derive some facts from the given hypotheses? Are there easier-to-prove statements that would suffice to prove the desired conclusion?

   Look for a special case: add assumptions until the problem is easy, and then see if you can remove the extra assumptions. Restate the problem. Restate it again. Make analogies to problems that you've already solved. Could those related problems be directly valuable? Or could you use a similar technique to what you used in that setting? Try to use a direct proof first; if you're finding it difficult to construct a direct proof of an implication, try working on the contrapositive instead. If both of these approaches fail, try a proof by contradiction. When you have a candidate plan of attack, try to execute it. If there's a picture that will help clarify the steps in your plan, draw it. Sketch out the "big" steps that you'd need to make the whole proof work. Make sure they fit together. Then crank through the details of each big step. Do the algebra. Check the algebra. If it all works out, great! If not, go back and try again. Where did things go off the rails, and can you fix them?

   Think about how to present your proof; then actually write it. Note that what you did in *figuring out* how to prove the result might or might not be the best way to *present* the proof.

Problem-solving tip: If you're totally stuck in attempting to prove a statement true, switch to trying to prove it false. If you succeed, you're done—or, by figuring out why you're struggling to construct a counterexample, you may figure out how to prove that the statement is true.

3. *Think about what you've done.* Check to make sure your proof is reasonable. Did you actually use all the assumptions? (If you didn't, do you believe the stronger claim that has the smaller set of assumptions?) Look over all the steps of your proof. Turn your internal skepticism dial to its maximum, and reread what you just wrote. Ask yourself *Why?* as you think through each step. Don't let yourself get away with anything.

   After you're satisfied that your proof is correct, work to improve it. Can you strengthen the result by making the conclusion stronger or the assumptions weaker? Can you make the proof constructive? Simplify the argument as much as you can. Are there unnecessary steps? Are there unnecessarily complex steps? Are there subclaims that would be better as separate lemmas?

Problem-solving tip: Check your work! If your claim says something about a general n, test it for $n = 1$. Compare your answer to a plot, or the output of a quick program.

It's important to be willing to move back and forth among these steps. You'll try to prove a claim φ, and then you'll discover a counterexample to φ—so you go back and modify the claim to a new claim φ' and try to prove φ' instead. You'll formulate a draft of a proof of φ' but discover a bug when you check your work while reflecting on the proof. You'll go back to proving φ', fix the bug, and discover a new proof that's bugfree. You'll think about your proof and realize that it didn't use all the assumptions of φ', so you'll formulate a stronger claim φ'' and then go through the proof of φ'' and reflect again about the proof.

Taking it further: One of the most famous—and prolific!—mathematicians of modern times was Paul Erdős (1913–1996), a Hungarian mathematician who wrote literally thousands of papers over his career, on a huge range of topics. Erdős used to talk about a mythical "Book" of proofs, containing the perfect proof of every theorem (the clearest, the most elegant—the best!). See p. 438 for some more discussion of The Book, and of Paul Erdős himself.

4.3.3 *Some Brief Thoughts about Writing Good Proofs*

When you're writing a proof, it's important to remember that you are *writing*. Proofs, like novels or persuasive essays, form a particular genre of writing. Treat writing a proof with the same care and attention that you would give to writing an essay.

Make your argument self-contained; include definitions of all variables and all nonstandard notation. State all assumptions, and explain your notation. Choose your notation and terminology carefully; name your variables well. Here's an example.

Writing tip: Draft. Write. Edit. *Rewrite.*

Example 4.24 (Pythagorean Theorem, stated poorly)
Theorem: $a^2 + b^2 = c^2$.

This formulation is a *terrible* way of phrasing the theorem: the reader has no idea what a, b, and c *are*, or even that the theorem has anything whatsoever to do with geometry. (The Pythagorean Theorem, from geometry, states that the square of the hypotenuse of a right triangle is equal to the sum of the squares of its legs.) Here's a much better statement of the Pythagorean Theorem:

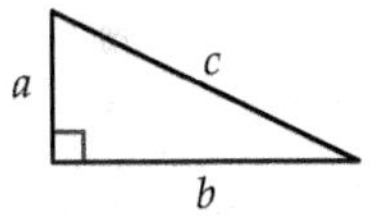

Figure 4.18: A right triangle.

Example 4.25 (Pythagorean Theorem, stated well)
Theorem: Let a and b denote the lengths of the legs of a right triangle, and let c denote the length of its hypotenuse. Then $a^2 + b^2 = c^2$.

If you are worried that your audience has forgotten the geometric terminology from this statement, then you might add the following clarification:

> As reminder from geometry, a *right triangle* is a 3-sided polygon with one 90° angle, called a *right angle*. The two sides adjacent to the right angle are called *legs* and the third side is called the *hypotenuse*. Figure 4.18 shows an example of a right triangle. Here the legs are labeled a and b, and the hypotenuse is labeled c. As is customary, the right angle is marked with the special square-shaped symbol $\square$.

Thanks to Josh Davis for suggesting Examples 4.24 and 4.25.

Because the "standard" phrasing of the Pythagorean Theorem—which you might have heard in high school—calls the length of the legs a and b and the length of the hypotenuse c, we use the standard variable names. Calling the leg lengths θ and ϕ and the hypotenuse r would be hard on the reader; conventionally in geometry θ and ϕ are angles, while r is a radius. *Whenever you can, make life as easy as possible for your reader.*

(By the way, we'll prove the Pythagorean Theorem in Example 4.14, and you'll prove it again in Exercise 4.75.)

Above all, remember that your primary goal in writing is communication. Just as when you are programming, it is possible to write two solutions to a problem that both "work," but which differ tremendously in readability. Document! Comment your code; explain *why* this statement follows from previous statements. Make your proofs—and your code!—a pleasure to read.

Writing tip: In writing a proof, keep your reader informed about the status of every sentence. And make sure that everything you write *is* a sentence. For example, every sentence contains a verb. (Note that a symbol like "=" is read as "is equal to" and *is* a verb.) Is the sentence an assumption? A goal? A conclusion? Annotate your sentences with signaling words and phrases to make it clear what each statement is doing. For example, introduce statements that follow logically from previous statements with words like *hence*, *thus*, *so*, *therefore*, and *then*.

COMPUTER SCIENCE CONNECTIONS

ARE MASSIVE COMPUTER-GENERATED PROOFS PROOFS?

As we've said, what we mean by a "proof" is an argument that convinces the audience that the claim is true. What, then, is the status of the so-called proof of the claim *Checkers is a draw when both players play optimally*? The "proof" of this claim that we discussed on p. 344 hinged on showing that the software system Chinook can never lose at checkers—which was established via massive computation to perform a large-scale search of the checkers game tree.[3] Is that "proof" convincing? Can such a proof *ever* be convincing? It's clear that a human reader cannot accommodate the 5×10^{20} checkers board positions in his or her brain, so it's not convincing in the sense that a reader would be able to verify every step of the argument. But, on the other hand, a reader could potentially be convinced that Chinook's *code* is correct, even if the *output* is too big for a reader to find convincing.

The philosophical question about whether a large-scale computer-generated proof "counts" actually as a proof first arose in the late 1970s, when the *Four-Color Theorem* was first proven(?).[4] Here is the theorem:

> Any "map" of contiguous geometric regions can be colored using four colors so that no two adjacent regions share the same color.

Two quick notes: first, *adjacent* means sharing a positive-length border; two regions meeting at a point don't need different colors. Second, the requirement of regions being *contiguous* means the map can't require two disconnected regions (like the Lower 48 States and Alaska) to get the same color.

The computational proof of four-color theorem given by Appel and Haken proceeds as follows. Appel and Haken first identified a set of 1476 different map configurations and proved (in the traditional way, by giving a convincing argument) that, if the four-color theorem were false, it would fail on one of these 1476 configurations. They then wrote a computer program that showed how to color each one of these 1476 configurations using only four colors. The theorem follows ("if there were a counterexample at all, there'd be a counterexample in one of the 1476 cases—and there are no counterexamples in the 1476 cases").

A great deal of controversy followed the publication of Appel and Haken's work. Some mathematicians felt strongly that a proof that's too massive for a human to understand is not a proof at all. Others were happy to accept the proof, particularly because the four-colorability question had been posed, and remained unresolved, for centuries. Computer scientists, by our nature, tend to be more accepting of computational proof than mathematicians—but there are still plenty of interesting questions to ponder. For example, as we discussed on p. 344, some errors in the *execution* of the code that generates Chinook's proof are known to have occurred, simply because hardware errors happen at a high enough rate that they will arise in a computation of this size. Thus bit-level corruption may have occurred, without 100% correction, in Chinook's proof that checkers is a draw under optimal play. So is Chinook's "proof" really a proof? (Of course, there are also plenty of human-generated purported proofs that contain errors!)

[3] Jonathan Schaeffer, Neil Burch, Yngvi Bjornsson, Akihiro Kishimoto, Martin Muller, Rob Lake, Paul Lu, and Steve Sutphen. Checkers is solved. *Science*, 317(5844):1518–1522, 14 September 2007.

[4] Kenneth Appel and Wolfgang Haken. Solution of the four color map problem. *Scientific American*, 237(4):108–121, October 1977.

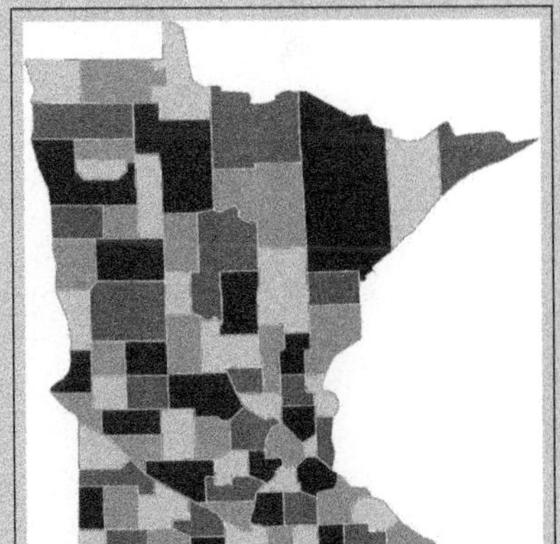

Figure 4.19: A four-colored map of the 87 counties in Minnesota.

Computer Science Connections

Paul Erdős, "The Book," and Erdős Numbers

After you've completed a proof of a claim—and after you've celebrated completing it—you should think again about the problem. In programming, there are often many fundamentally different algorithms to solve a particular problem; in proofs, there are often many fundamentally different ways of proving a particular theorem. And, just as in programming, some approaches will be more elegant, more clear, or more efficient than others.

Paul Erdős, a prolific and world-famous mathematician who published approximately 1500 papers before his death in 1996 (including papers on math, physics, and computer science), used to talk about "The Book" of proofs. "The Book" contains the ideal proof of each theorem—the most elegant, insightful, and beautiful proof. (If you believe in God, then The Book contains God's proofs.) There's even a non-metaphorical book called *Proofs from The Book* that collects some of the most elegant known proofs of some theorems.[5] Proving a theorem is great, but giving a beautiful proof is even better. Strive for the "book proof" of every theorem.

[5] Martin Aigner and Günter Ziegler. *Proofs from The Book*. Springer, 4th edition, 2009.

Erdős was one of the most respected mathematicians of his time—and one of the most eccentric, too. (He forswore most material possessions, and instead traveled the world, crashing in the guest rooms of his research collaborators for months at time.) Because of Erdős's prolific publication record and his great respect from the research community, a measure of a certain type of fame for researchers has sprung up around him. A researcher's *Erdős number* is 1 if she has coauthored a published paper with Erdős; it's 2 if she has coauthored a published paper with someone with an Erdős number of one; and so forth. For example, Bill Gates has an Erdős number of 4: he wrote a paper on the pancake-flipping problem with Christos Papadimitriou, who has coauthored a paper with someone (Xiao Tie Deng) who wrote a paper with someone (Pavol Hell) who wrote a paper with Paul Erdős.

If you're more of a movie person than a peripatetic mathematician person, then you may be more familiar with a very similar notion from the entertainment world, the so-called *Bacon game*. The goal here is to connect a given actor to Kevin Bacon via the shortest possible chain of intermediaries, where two actors are linked if they have appeared together in a movie.

The Erdős Number Project, maintained at `http://www.oakland.edu/enp` by Jerry Grossman of Oakland University, is a good place to look for more information. You can see more about the Bacon game at the Oracle of Bacon, at `http://oracleofbacon.org`.

It is a source of great pride for researchers to have small Erdős numbers. And, although Erdős numbers themselves are really nothing more than a nerdy source of amusement, the ideas underlying them are fundamental in *graph theory*, the subject of Chapter 11. A closely related topic is the *small-world phenomenon*, also known as "six degrees of separation," the principle that almost any two people are likely to be connected by a short chain of intermediate friends. The "six degrees of separation" phrase came from an important early paper by the social psychologist Stanley Milgram;[6] it has spawned a massive amount of recent research by computer scientists, who have begun working to analyze questions about human behavior that have only become visible in the "Facebook era" in which it is now possible to study collective decision making on an massive scale.

[6] Stanley Milgram. The small world problem. *Psychology Today*, 1:61–67, May 1967.

4.3.4 Exercises

Prove the following claims about divisibility.

4.33 The binary representation of any odd integer ends with a 1.

4.34 A positive integer n is divisible by 5 if and only if its last digit is 0 or 5.

4.35 Let k be any positive integer. Then any positive integer n is divisible by 2^k if and only if its last k digits are divisible by 2^k. (This exercise is a generalization of Example 4.11.)

Prove the following claims about rationality.

4.36 If x and y are rational numbers, then $x - y$ is also rational.

4.37 If x and y are rational numbers and $y \neq 0$, then $\frac{x}{y}$ is also rational.

4.38 One of the following statements is true and one is false:

- If xy and x are both rational, then y is too.
- If $x - y$ and x are both rational, then y is too.

Decide which statement is true and which is false, and give proof/disproof of both.

4.39 Let n be any integer. Prove by cases that $n^3 - n$ is evenly divisible by 3.

4.40 Let n be any integer. Prove by cases that $n^2 + 1$ is *not* evenly divisible by 3.

4.41 Prove that $|x| + |y| \geq |x + y|$ for any real numbers x and y.

4.42 Prove that $|x| - |y| \leq |x - y|$ for any real numbers x and y.

4.43 Prove that the product of the absolute values of x and y is equal to the absolute value of their product—that is, prove that $|x| \cdot |y| = |x \cdot y|$ for any real numbers x and y.

4.44 Suppose that $x, y \in \mathbb{R}$ satisfy $|x| \leq |y|$. Prove that $\frac{|x+y|}{2} \leq |y|$.

4.45 Let A and B be sets. Prove that $A \times B = B \times A$ if and only if $A = \varnothing$ or $B = \varnothing$ or $A = B$. Prove the result by mutual implication, where the proof of the $\Leftarrow$ direction proceeds by contrapositive.

Let $x \geq 0$ and $y \geq 0$ be arbitrary real numbers. The arithmetic mean *of x and y is $(x+y)/2$, their average. The* geometric mean *of x and y is $\sqrt{xy}$.*

4.46 First, a warm-up exercise: prove that $x^2 \geq 0$ for any real number x. *(Hint: yes, it's easy.)*

4.47 Prove the *Arithmetic Mean–Geometric Mean inequality*: for $x, y \in \mathbb{R}^{\geq 0}$, we have $\sqrt{xy} \leq (x+y)/2$. *(Hint: $(x-y)^2 \geq 0$ by Exercise 4.46. Use algebraic manipulation to make this inequality look like the desired one.)*

4.48 Prove that the arithmetic mean and geometric mean of x and y are equal if and only if $x = y$.

In Chapter 2, when we defined square roots, we introduced Heron's method, *a first-century algorithm to compute $\sqrt{x}$ given x. See p. 218, or Figure 4.20 for a reminder. Here you'll prove two properties that help establish why this algorithm correctly computes square roots:*

4.49 Assume that $y_0 \geq \sqrt{x}$. Prove that, for every $i \geq 1$, we have $y_i \geq \sqrt{x}$. In other words, prove that if $y \geq \sqrt{x}$ then $(y + \frac{x}{y})/2 \geq \sqrt{x}$ too.

4.50 Suppose that $y > \sqrt{x}$. Prove that $\frac{x}{y}$ is closer to $\sqrt{x}$ than y is—that is, prove that $|\frac{x}{y} - \sqrt{x}| < |y - \sqrt{x}|$. *(Hint: show that $|y - \sqrt{x}| - |\sqrt{x} - \frac{x}{y}| > 0$.)*

Now, using this result and Exercise 4.44, prove that y_{i+1} as computed in Heron's Method is closer to $\sqrt{x}$ than y_i, as long as $y_i > \sqrt{x}$.

```
Input: A positive real number x
Output: A real number y where y^2 ≈ x

  Let y_0 be arbitrary, and let i := 0.
  while (y_i)^2 is too far away from x
     let y_{i+1} := (y_i + x/y_i)/2, and let i := i + 1.
  return y_i
```

Figure 4.20: A reminder of Heron's method for computing square roots.

The second property that you just proved (Exercise 4.50) shows that Heron's method improves its estimate of $\sqrt{x}$ in every iteration. (We haven't *shown "how much" improvement Heron's method achieves in an iteration, or even that this algorithm is converging to the correct answer—let alone quickly!—but, in fact, it is.)*

Prove the following claims using a proof by contrapositive.

4.51 Let $n \in \mathbb{Z}^{\geq 0}$. If $n \bmod 4 \in \{2, 3\}$, then n is not a perfect square.

4.52 Let n and m be integers. If nm is not evenly divisible by 3, then neither n nor m is evenly divisible by 3. (In fact, the converse is true too, but you don't have to prove it.)

4.53 Let $n \in \mathbb{Z}^{\geq 0}$. If $2n^4 + n + 5$ is odd, then n is even.

Prove the following claims using a proof by mutual implication, using a proof by contrapositive for one direction.

4.54 Let n be any integer. Then n^3 is even if and only if n is even.

4.55 Let n be any integer. Then n is divisible by 3 if and only if n^2 is divisible by 3.

Prove the following claims using a proof by contradiction.

4.56 Let x, y be positive real numbers. If $x^2 - y^2 = 1$, then x or y (or both) is not an integer.

4.57 Suppose $12x + 3y = 254$, for real numbers x and y. Then either x or y (or both) is not an integer.

4.58 Adapt Example 4.21 to prove that $\sqrt[3]{2} = 2^{1/3}$ is irrational. (You may find Exercise 4.54 helpful.)

4.59 Adapt Example 4.21 to prove that $\sqrt{3}$ is irrational. (You may find Exercise 4.55 helpful.)

4.60 Consider an array $A[1 \ldots n]$. A value x is called a *strict majority element of A* if strictly more than half of the elements in A are equal to x—in other words, if

$$\Big| \{i \in \{1, 2, \ldots, n\} : A[i] = x\} \Big| > \frac{n}{2}.$$

Give a proof by contradiction that every array has at most one strict majority element.

In Example 4.12, Exercise 4.36, and Exercise 4.37, we proved that if x and y are both rational, then so are all three of xy, $x - y$, and $\frac{x}{y}$. The converse of each of these three statements is false. ***Disprove*** *the following claims by giving counterexamples:*

4.61 If xy is rational, then x and y are rational.

4.62 If $x - y$ is rational, then x and y are rational.

4.63 If $\frac{x}{y}$ is rational, then x and y are rational.

4.64 In Example 4.22, we disproved the following claim by giving a counterexample:

Claim 1: No positive integer is expressible in two different ways as the sum of two perfect squares.

Let's consider a related claim that is not disproved by our counterexamples from Example 4.22:

Claim 2: No positive integer is expressible in *three* different ways as the sum of two perfect squares.

Disprove Claim 2 by giving a counterexample.

4.65 Leonhard Euler, an 18th-century Swiss mathematician to whom the idea of an abstract formal model of networks (graphs; see Chapter 11) is due, made the observation that the polynomial

$$f(n) = n^2 + n + 41$$

yields a prime number when it's evaluated for many small integers n: for example, $f(0) = 41$ and $f(1) = 43$ and $f(2) = 47$ and $f(3) = 53$, and so forth. Prove or disprove the following claim: *the function $f(n)$ yields a prime for every nonnegative integer n.*

4.4 *Some Examples of Proofs*

> Few things are harder to put up with than the annoyance of a good example.
>
> Mark Twain (1835–1910)
> *Pudd'nhead Wilson* (1894)

We've now catalogued a variety of proof techniques, discussed some strategies for proving novel statements, and described some ideas about presenting proofs well. Section 4.3 illustrated some proof techniques with a few simple examples each, entirely about numbers and arithmetic. In this section, we'll give a few "bigger"—and perhaps more interesting!—examples of theorems and proofs.

4.4.1 *A Proof about Propositional Logic: Conjunctive/Disjunctive Normal Form*

We'll start with a result about propositional logic, namely showing that any proposition is logically equivalent to another proposition that has a "simpler" structure. Recall the definitions of *conjunctive* and *disjunctive normal form*:

Definition 4.16 (Reminder: Conjunctive/Disjunctive Normal Form)
In propositional logic, a literal *is a Boolean variable or its negation (like p or $\neg p$).*

A proposition φ is in conjunctive normal form (CNF) *if φ is the conjunction of one or more* clauses, *where each* clause *is the disjunction of one or more literals.*

A proposition φ is in disjunctive normal form (DNF) *if φ is the disjunction of one or more* clauses, *where each* clause *is the conjunction of one or more literals.*

Here are two small examples of CNF and DNF:

$$(\neg p \vee q \vee \neg r) \wedge (\neg q \vee r) \qquad \text{(conjunctive normal form)}$$

$$(\neg p \wedge \neg q \wedge r) \vee (\neg q \wedge \neg r \vee s) \vee (r). \qquad \text{(disjunctive normal form)}$$

Back in Chapter 3, we claimed that every proposition is logically equivalent to one in CNF and one in DNF, but we didn't prove it. Here we will.

First, though, let's recall an example from Chapter 3 and brainstorm a bit about how to generalize that result into the desired theorem. In Example 3.26, we converted $p \Leftrightarrow q$ into DNF as the logically equivalent proposition $(p \wedge q) \vee (\neg p \wedge \neg q)$. Note that this expression has two clauses $p \wedge q$ and $\neg p \wedge \neg q$, *each of which is true in one and only one row of the truth table.* And our full proposition $(p \wedge q) \vee (\neg p \wedge \neg q)$ is true in precisely two rows of the truth table. (See Figure 4.21.)

p	q	$p \Leftrightarrow q$	$p \wedge q$	$\neg p \wedge \neg q$
T	T	T	T	F
T	F	F	F	F
F	T	F	F	F
F	F	T	F	T

Figure 4.21: Truth table for $p \Leftrightarrow q$ and the clauses for converting it to DNF.

Can we make this idea general? Yes! For an arbitrary proposition φ, and for any particular row of the truth table for φ, we can construct a clause that's true in that row and only in that row. We can then build a DNF proposition that's logically equivalent to φ by "or"ing together each of the clauses corresponding to the rows in which φ is true. And then we're done!

(Well, we're almost *done! There is one subtle bug in the proof sketch in the previous paragraph—can you find it? We'll fix the issue in the last paragraph of the proof below.)*

Theorem 4.11 (All propositions are expressible in DNF (Theorem 3.2))
For any proposition φ, there exists a proposition ψ_{dnf} in disjunctive normal form such that $\varphi \equiv \psi_{dnf}$.

Proof. Let φ be an arbitrary proposition, say over the Boolean variables $p_1, \ldots, p_k$.

For any particular truth assignment ρ for the variables $p_1, \ldots, p_k$, we'll construct a conjunction c_ρ that's true under ρ and false under all other truth assignments. Let $x_1, x_2, \ldots, x_\ell$ be the variables assigned true by ρ, and $y_1, y_2, \ldots, y_{k-\ell}$ be the variables assigned false by ρ. Then the clause

$$c_\rho := x_1 \wedge x_2 \wedge \cdots \wedge x_\ell \wedge \neg y_1 \wedge \neg y_2 \wedge \cdots \wedge \neg y_{k-\ell}$$

is true under ρ, and c_ρ is false under every other truth assignment.

We can now construct a DNF proposition ψ_{dnf} that is logically equivalent to φ by "or"ing together the clause c_ρ for each truth assignment ρ that makes φ true. Build the truth table for φ, and let S_φ denote the set of truth assignments for $p_1, \ldots, p_k$ under which φ is true. If the truth assignments in S_φ are $\{\rho_1, \rho_2, \ldots, \rho_m\}$, then define

$$\psi_{\text{dnf}} := c_{\rho_1} \vee c_{\rho_2} \vee \cdots \vee c_{\rho_m}. \qquad (*)$$

It's easy to see that ψ_{dnf} is true under every truth assignment ρ under which φ was true (because the clause c_ρ is true under ρ). And, for a truth assignment ρ under which φ was false, every disjunct in ψ_{dnf} evaluates to false, so the entire disjunction is false under such a ρ, too. Thus $\varphi \equiv \psi_{\text{dnf}}$.

There's one thing we have to be careful about: what happens if $S_\varphi = \varnothing$—that is, if φ is unsatisfiable? (This issue is the minor bug we mentioned before the theorem statement.) The construction in $(*)$ doesn't work, but it's easy to handle this case too: we simply choose an unsatisfiable DNF proposition like $p \wedge \neg p$ as ψ_{dnf}. □

Problem-solving tip: Be on the lookout for special cases (like an unsatisfiable φ in Theorem 4.11), and see whether you can handle them separately from the argument for the "typical" case.

Note that, although we didn't phrase it as such from the beginning, our proof of Theorem 4.11 was actually a proof by cases, with two cases corresponding to φ being unsatisfiable and φ being satisfiable.

As an illustration, let's use the construction from Theorem 4.11 to transform an example proposition into DNF:

Example 4.26 (Converting $p \Rightarrow (q \wedge r)$ to DNF)
Problem: Find a proposition in DNF logically equivalent to $p \Rightarrow (q \wedge r)$.

Solution: To convert $p \Rightarrow (q \wedge r)$ to DNF, we start from the truth table, and then "or" together the propositions corresponding to each row that's marked with as True:

p	q	r	$q \wedge r$	$p \Rightarrow (q \wedge r)$	
T	T	T	T	T	$p \wedge q \wedge r$
T	T	F	F	F	$p \wedge q \wedge \neg r$
T	F	T	F	F	$p \wedge \neg q \wedge r$
T	F	F	F	F	$p \wedge \neg q \wedge \neg r$
F	T	T	T	T	$\neg p \wedge q \wedge r$
F	T	F	F	T	$\neg p \wedge q \wedge \neg r$
F	F	T	F	T	$\neg p \wedge \neg q \wedge r$
F	F	F	F	T	$\neg p \wedge \neg q \wedge \neg r$

Our DNF proposition will therefore have five clauses, one for each of the five truth assignments under which this implication is true:

$$\underbrace{(p \wedge q \wedge r)}_{\text{TTT}} \vee \underbrace{(\neg p \wedge q \wedge r)}_{\text{FTT}} \vee \underbrace{(\neg p \wedge q \wedge \neg r)}_{\text{FTF}} \vee \underbrace{(\neg p \wedge \neg q \wedge r)}_{\text{FFT}} \vee \underbrace{(\neg p \wedge \neg q \wedge \neg r)}_{\text{FFF}}.$$

Conjunctive normal form

Now that we've proven that we can translate any proposition into disjunctive normal form (the "or of ands"), we'll turn our attention to conjunctive normal form (the "and of ors").

Theorem 4.12 (All propositions are expressible in CNF)
For any proposition φ, there exists a proposition φ_{cnf} in conjunctive normal form such that $\varphi \equiv \varphi_{cnf}$.

Though it's not initially obvious, Theorem 4.12 actually turns out to be easy to prove by making use of the DNF result. The crucial idea—and, once again, it's an idea that requires some genuine creativity to come up with!—is that it's fairly simple to turn the *negation* of a DNF proposition into a CNF proposition. So, to build a CNF proposition logically equivalent to φ, we'll construct a DNF proposition that is logically equivalent to $\neg\varphi$; we can then negate that DNF proposition and use De Morgan's Laws to convert the resulting proposition into CNF. Here are the details:

Problem-solving tip: Try being lazy first! Think about whether there's a way to use a previously established result to make the current problem easier.

Proof. If φ is a tautology, the task is easy; just define $\varphi_{\text{cnf}} = p \vee \neg p$.

Otherwise, φ is a nontautology, say over the variables $p_1, \ldots, p_k$. Using Theorem 4.11, we can construct a DNF proposition ψ that is logically equivalent to $\neg\varphi$. (Note that, using our construction from Theorem 4.11, the proposition ψ will have k literals in every clause, because $\neg\varphi$ is satisfiable.) Thus the form of ψ will be

$$\psi = (c_1^1 \wedge \cdots \wedge c_k^1) \vee (c_1^2 \wedge \cdots \wedge c_k^2) \vee \cdots \vee (c_1^m \wedge \cdots \wedge c_k^m)$$

for some $m \geq 1$, where each c_i^j is a literal. Recall that $\psi \equiv \neg\varphi$, so we also know that $\neg\psi \equiv \varphi$. Let's negate ψ:

$$\begin{aligned}
\neg\psi &= \neg\left[(c_1^1 \wedge \cdots \wedge c_k^1) \vee (c_1^2 \wedge \cdots \wedge c_k^2) \vee \cdots \vee (c_1^m \wedge \cdots \wedge c_k^m)\right] \\
&\equiv \neg(c_1^1 \wedge \cdots \wedge c_k^1) \wedge \neg(c_1^2 \wedge \cdots \wedge c_k^2) \wedge \cdots \wedge \neg(c_1^m \wedge \cdots \wedge c_k^m) \\
&\qquad \textit{De Morgan's Law: } \neg(p \vee q) \equiv \neg p \wedge \neg q \\
&\equiv (\neg c_1^1 \vee \cdots \vee \neg c_k^1) \wedge (\neg c_1^2 \vee \cdots \vee \neg c_k^2) \cdots \wedge (\neg c_1^m \vee \cdots \vee \neg c_k^m). \\
&\qquad \textit{De Morgan's Law: } \neg(p \wedge q) \equiv \neg p \vee \neg q, \textit{ applied once per clause}
\end{aligned}$$

But this expression is in CNF once we remove any doubly negated literals—that is, we replace any occurrences of $\neg\neg p$ by p instead. Thus we've constructed a proposition in conjunctive normal form that's logically equivalent to $\neg\psi \equiv \varphi$. □

As an illustration of this construction, let's convert $p \Rightarrow (q \wedge r)$—which we converted to DNF in Example 4.26—to conjunctive normal form too:

Example 4.27 (Converting $p \Rightarrow (q \wedge r)$ to CNF)

In Example 4.26, we converted the proposition $\varphi = p \Rightarrow (q \wedge r)$ into DNF. Here we'll convert it into CNF, using Theorem 4.12. Again, we start from the truth table for $\neg\varphi$:

p	q	r	$q \wedge r$	φ: $p \Rightarrow (q \wedge r)$	$\neg\varphi$: $\neg(p \Rightarrow (q \wedge r))$	
T	T	T	T	T	F	$p \wedge q \wedge r$
T	T	F	F	F	T	$p \wedge q \wedge \neg r$
T	F	T	F	F	T	$p \wedge \neg q \wedge r$
T	F	F	F	F	T	$p \wedge \neg q \wedge \neg r$
F	T	T	T	T	F	$\neg p \wedge q \wedge r$
F	T	F	F	T	F	$\neg p \wedge q \wedge \neg r$
F	F	T	F	T	F	$\neg p \wedge \neg q \wedge r$
F	F	F	F	T	F	$\neg p \wedge \neg q \wedge \neg r$

We first construct a DNF proposition equivalent to $\neg\varphi$. This proposition has three clauses, one for each of the truth assignments under which $\neg\varphi$ is true (and φ is false):

$$\neg\varphi \equiv \underbrace{(p \wedge q \wedge \neg r)}_{\text{TTF}} \vee \underbrace{(p \wedge \neg q \wedge r)}_{\text{TFT}} \vee \underbrace{(p \wedge \neg q \wedge \neg r)}_{\text{TFF}}$$

We negate this proposition and use De Morgan's Laws to push around the negations:

$$\begin{aligned}
\varphi &\equiv \neg\left[(p \wedge q \wedge \neg r) \vee (p \wedge \neg q \wedge r) \vee (p \wedge \neg q \wedge \neg r)\right] \\
&\equiv \neg(p \wedge q \wedge \neg r) \wedge \neg(p \wedge \neg q \wedge r) \wedge \neg(p \wedge \neg q \wedge \neg r) && \textit{De Morgan} \\
&\equiv (\neg p \vee \neg q \vee \neg\neg r) \wedge (\neg p \vee \neg\neg q \vee \neg r) \wedge (\neg p \vee \neg\neg q \vee \neg\neg r) && \textit{De Morgan} \\
&\equiv (\neg p \vee \neg q \vee r) \wedge (\neg p \vee q \vee \neg r) \wedge (\neg p \vee q \vee r). && \textit{Double Negation}
\end{aligned}$$

So $(\neg p \vee \neg q \vee r) \wedge (\neg p \vee q \vee \neg r) \wedge (\neg p \vee q \vee r)$ is a CNF proposition that's logically equivalent to $p \Rightarrow (q \wedge r)$. We can verify via truth table that this proposition is indeed logically equivalent to $p \Rightarrow (q \wedge r)$.

One last comment about these proofs: it's worth emphasizing again that there's genuine creativity required in proving these theorems. Through the strategies from Section 4.3.2 and through practice, you can get better at having the kinds of creative ideas that lead to proofs—but that doesn't mean that these results should have been "obvious" to you in advance. It takes a real moment of insight to see how to use the truth table to develop the DNF proposition to prove Theorem 4.11, or how to use the DNF formula of the negation to prove Theorem 4.12.

Taking it further: Theorems 4.11 and 4.12 said that "a proposition ψ (of a particular form) exists for every φ"—but our proofs actually described an algorithm to *build* ψ from φ. (That's a more computational way to approach a question: a statement like "such-and-such exists!" is the kind of thing more typically proven by mathematicians, and "a such-and-such can be found with this algorithm!" is a claim more typical of computer scientists.) Our algorithms in Theorems 4.11 and 4.12 aren't very efficient, unfortunately; they require 2^k steps just to build the truth table for a k-variable proposition. We'll give a (sometimes, and somewhat) more efficient algorithm in Chapter 5 (see Section 5.4.3) that operates directly on the form of the proposition ("syntax") rather than on using the truth table ("semantics").

SOME OTHER RESULTS ABOUT PROPOSITIONAL LOGIC

In the exercises, you'll be asked to prove a large collection of other facts about propositional logic. We'll highlight one of them, which is similar in spirit to the theorems about DNF and CNF: you'll show that any proposition φ is logically equivalent to a simpler proposition that uses only one kind of logical connective, called "nand." For reasons of physics, building the physical circuitry for the logical connective *nand*—as in "not and," where p *nand* q means $\neg(p \wedge q)$—is much simpler than other logical connectives. (The physical reasons relate specifically to the way that *transistors*—the most basic building blocks for digital circuits—work.) The truth table for nand—also known as the *Sheffer stroke* $|$—appears in Figure 4.22.

The Sheffer stroke $|$ is named after the early-20th-century logician Henry Sheffer.

It turns out that every (*every!*) logical connective can be expressed in terms of $|$. In other words, if you have enough nand gates, then you will be able to build *any logical circuit* that you want. Here is a theorem that formally states this result:

p	q	$p \mid q$	$p \downarrow q$
T	T	F	F
T	F	T	F
F	T	T	F
F	F	T	T

Figure 4.22: The truth table for nand (also known as the Sheffer stroke $|$), and nor (also known as Peirce's arrow $\downarrow$).

Theorem 4.13 (All propositions are expressible using only $|$)
For any Boolean formula φ over $p_1, \ldots, p_k$, there exists a proposition $\psi_{\text{nand-only}}$ such that (i) $\varphi \equiv \psi_{\text{nand-only}}$, and (ii) $\psi_{\text{nand-only}}$ contains only $p_1, \ldots, p_k$ and the logical connective $|$.

The theorem follows from Exercise 4.69, where you'll show that every logical connective can be expressed in terms of $|$. (To give a fully rigorous proof, we will need to use mathematical induction, the subject of Chapter 5. Mathematical induction will essentially allow us to apply the results of Exercise 4.69 recursively to translate an arbitrary proposition φ into $\psi_{\text{nand-only}}$.)

Taking it further: Indeed, real circuits are typically built exclusively out of nand gates, using logical equivalences to construct and/or/not gates from a small number of nand gates. Although it may be initially implausible if this is the first time that you've heard it, the processor of a physical computer is essentially nothing more than a giant circuit built out of nand gates and wires. With some thought, you can build a circuit that takes two integers (represented in binary, as a 64-bit sequence) and computes their sum. Similarly, but more thought-provokingly, you can build a circuit that takes an *instruction* (add these numbers; compare those numbers; save this thing in memory; load the other thing from memory) and performs the requested action. That circuit is a computer!

Peirce's arrow is named after the 18th-century logician Charles Peirce. Its truth table is also shown in Figure 4.22.

Incidentally, all of the logical connectives can also be defined in terms of the logical connective known as *Peirce's arrow* $\downarrow$ and also known as *nor*, as in "not or." (You'll prove the analogous result to Theorem 4.13 for Peirce's arrow in Exercise 4.70.)

4.4.2 *The Pythagorean Theorem*

Example 4.24 presented the Pythagorean Theorem, which you probably once saw in a long-ago geometry class: the square of the length of hypotenuse of a right triangle equals the sum of the squares of the lengths of the legs. Let's prove it. In brainstorming about this theorem, here's an idea that turns out to be helpful. Because the statement of Pythagorean theorem involves side lengths raised to the second power ("squared"), we might be able to think about the problem using geometric squares, appropriately configured. Here's a proof that proceeds using this geometric idea:

The original formulation of the Pythagorean Theorem is attributed to Pythagoras, a Greek mathematician/philosopher who lived around 500 BCE.

Theorem 4.14 (The Pythagorean Theorem)
Let a and b denote the lengths of the legs of a right triangle, and let c denote the length of its hypotenuse. Then $a^2 + b^2 = c^2$.

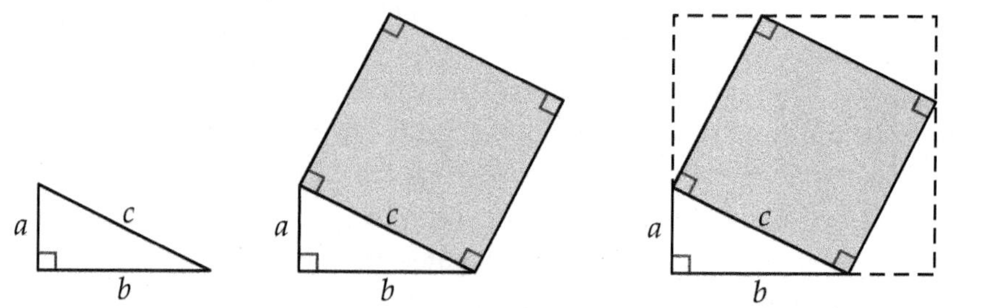

(a) The right triangle. (b) ... with an added square. (c) ... and three added triangles.

Figure 4.23: Illustrations for the proof of the Pythagorean Theorem, Theorem 4.14.

Proof. Starting with the given right triangle in Figure 4.23(a), draw a square with side length c, where one side of the square coincides with the hypotenuse of the given triangle, as in Figure 4.23(b). Now draw three new triangles, each identical to the first. Place these three new triangles symmetrically around the square that we just drew, so that each side of the square coincides with the hypotenuse of one of the four triangles, as in Figure 4.23(c). Each of these four triangles has leg lengths a and b and hypotenuse c. Including both the c-by-c square and the four triangles, the resulting figure is a square with side length $a + b$.

To complete the proof, we will account for the area of Figure 4.23(c) in two different ways. First, because a square with side length x has area x^2, we have that

$$\text{area of the enclosing square} = (a + b)^2 = a^2 + 2ab + b^2.$$

Second, this enclosing square can be decomposed into a c-by-c square and four identical right triangles with leg lengths a and b. Because the area of a right triangle with leg lengths x and y is $xy/2$, we also have that

$$\begin{aligned}\text{area of the enclosing square} &= 4 \cdot (\text{area of one triangle}) + c^2\\ &= 4 \cdot \frac{1}{2}ab + c^2\\ &= 2ab + c^2.\end{aligned}$$

But the area of the enclosing square is the same regardless of whether we count it all together, or in its five disjoint pieces. Therefore $a^2 + 2ab + b^2 = 2ab + c^2$. The theorem follows by subtracting $2ab$ from both sides. □

There are *many* proofs of the Pythagorean theorem—in fact, hundreds! There is a classic proof attributed to Euclid (see p. 447), and many subsequent and different proof approaches followed over the millennia. There's even a book that collects over 350 different proofs of the result![7] There's an important lesson to draw from the many proofs of this theorem: *there's more than one way to do it.* Just as there are usually many fundamentally different algorithms for the same problem (think about sorting, for example), there are usually many fundamentally different techniques that can prove the same theorem. Keep an open mind; there is absolutely no shame in proving a result using a different approach than the "standard" way!

[7] Elisha Scott Loomis. *The Pythagorean Proposition.* National Council of Teachers of Mathematics, June 1968.

"There's more than one way to do it" is also the motto of the programming language Perl.

4.4.3 *Prime Numbers*

We'll return to arithmetic for our next set of examples, a pair of proofs about the prime numbers. Recall that a positive integer $n \geq 2$ is *prime* if and only if the only positive integers that divide n evenly are 1 and n itself. Also recall that a positive integer $n \geq 2$ that is not prime is called *composite*. (That is, the integer n is composite if and only if there exists a positive integer $k \notin \{1, n\}$ such that k divides n evenly.)

We'll start with another example of a proof by contradiction:

> **Theorem 4.15 (An infinitude of primes)**
> *There are infinitely many prime numbers.*

A similar proof to the one for Theorem 4.15 dates back around 2300 years. It's due to Euclid, the ancient Greek mathematician after whom Euclidean geometry—and the Euclidean algorithm (see Section 7.2.4)—is named.

Proof. We proceed by contradiction.

Suppose, for the purposes of deriving a contradiction, that there are only finitely many primes. This assumption means that there is a largest prime number, which we will call p. Consider the integer $p!$, the factorial of this largest prime p. Let's consider two separate cases: either $p! + 1$ is prime, or $p! + 1$ is not prime.

- If $p! + 1$ is prime, then we have a contradiction of the assumption that p is the largest prime, because $p! + 1 > p$ is also prime.
- If $p! + 1$ is not prime, then by definition it is evenly divisible by some integer k satisfying $2 \leq k \leq p!$. But we proved in Example 4.8 that $p! + 1$ is not evenly divisible by any integer between 2 and p, inclusive. Thus the smallest integer k that evenly divides $p! + 1$ must exceed p. Further, this integer k must be prime—otherwise some $2 \leq k' < k$ divides k and therefore divides $p! + 1$, but k was the smallest divisor of $p! + 1$. Thus $k > p$ is prime, and again we have a contradiction of the assumption that p is the largest prime.

In either case, we have a contradiction! Thus the original assumption—there are only finitely many prime numbers—is false, and so there are infinitely many primes. □

We'll now turn to another result about prime numbers, relating to the *primality testing* problem: you are given a positive integer n, and you have to determine whether n is prime. The definition of primality says that n is composite if there's an integer $k \in \mathbb{Z} - \{1, n\}$ such that $k \mid n$, but it should be easy to see that n is composite if and only if there's an integer $k \in \{2, 3, \ldots, n-1\}$ such that $k \mid n$. (That is, the largest possible divisor of n is $n - 1$.) But we can do better, strengthening this result by shrinking the largest candidate value of k:

> **Theorem 4.16 (A composite number n has a factor $\leq \sqrt{n}$)**
> *A positive integer $n \geq 2$ is evenly divisible by some other integer $k \in \{2, 3, \ldots, \lceil \sqrt{n} \rceil\}$ if and only if n is composite.*

Proof. We'll proceed by mutual implication.

The forward direction is easy: if there's some integer $k \in \{2, 3, \ldots, \lceil\sqrt{n}\rceil\}$ with $k \neq n$ such that k evenly divides n, then by definition n is composite. (That integer k satisfies $k \mid n$ and $k \notin \{1, n\}$.)

For the other direction, assume that the integer $n \geq 2$ is composite. By definition of composite, there exists a positive integer $k \notin \{1, n\}$ such that $n \bmod k = 0$—that is, there exist positive integers $k \notin \{1, n\}$ and d such that $dk = n$, so $d \mid n$ and $k \mid n$. We must have that $d \neq 1$ (otherwise $dk = 1 \cdot k = k = n$, but $k < n$) and $d \neq n$ (otherwise $dk = nk > n$, but $dk = n$). Thus there exist positive integers $d, k \notin \{1, n\}$ such that $dk = n$. But if both $d > \sqrt{n}$ and $k > \sqrt{n}$, then $dk > \sqrt{n} \cdot \sqrt{n} = n$, which contradicts the fact that $dk = n$. Thus either $d \leq \sqrt{n}$ or $k \leq \sqrt{n}$. □

Taking it further: Generating large prime numbers (and testing the primality of large numbers) is a crucial step in many modern cryptographic systems. See the discussion on p. 454 for some discussion of algorithms for testing primality suggested by these proofs, and a bit about the role that they play in modern cryptographic systems.

4.4.4 *Uncomputability*

We'll close this section with one of the most important results in computer science, dating from the early 20th century: *there are problems that cannot be solved by computers.* At that time, great thinkers were pondering some of the most fundamental questions that can be asked in CS. What is a computer? What is computation? What is a program? What tasks can be solved by computers/programs? One of the deepest and most mind-numbing results of this time was a proof, developed independently by Alan Turing and by Alonzo Church, that there are *uncomputable* problems. That is, there is a problem P for which it's possible to give a completely formal description of the right answer—but it's not possible to write a program that solves P.

Here, we'll prove this theorem. Specifically, we'll describe the *halting problem,* and prove that it's uncomputable. (Informally, the halting problem is: *given a function p written in Python and an input x, does p get stuck in an infinite loop when it's run on x?*) The result is a great example of a proof by contradiction, where we will exploit the abyss of self-reference to produce the contradiction.

Problems

Before we address the computability of the halting problem, we have to define precisely what we mean by a "problem" and "computable." A *problem* is the kind of task that we wish to solve with a computer program. We will focus on yes–no problems, called *decision problems:*

Definition 4.17 (Problem)
A problem *is a description of a set of valid inputs, and a specification of the corresponding output for each them. A* decision problem *is one where the output is either "yes" or "no."*

(In other words, a decision problem is specified by a description of a set of possible inputs, along with a description of those inputs for which the correct answer is "yes.") We've already encountered several decision problems:

Example 4.28 (Some sample decision problems)

- PRIMALITY: the set of possible inputs is the set of positive integers; the set of "yes" inputs is the set of prime numbers. (The "no" inputs are 1 and the composites.)
- SATISFIABILITY: any propositional-logic proposition φ is a valid input, and φ is a "yes" input if and only if φ is satisfiable.

An *instance* of a problem is a valid input for that problem. (An invalid input is one that isn't the right "kind of thing" for that problem.) We will refer to an instance x of a problem P as a *yes-instance* if the correct output is "yes," and as a *no-instance* if the correct output is "no." For example, 17 or 18 are both instances of PRIMALITY; 17 is a yes-instance, while 18 is a no-instance; $p \vee \neg p$ is an invalid input.

COMPUTABILITY

Problems are the things that we'll be interested in solving via computer programs. Informally, problems that can be solved by computer are called *computable* and those that cannot be solved by computer are called *uncomputable*. It'll be easiest to think of computability in terms of your favorite programming language, whatever it may be. For the sake of concreteness, we'll pretend it's Python, though any language would do.

Taking it further: The original definition of computability given by Alan Turing used an abstract device called a *Turing machine*; a programming language is called *Turing complete* if it can solve any problem that can be solved by a Turing machine. Every non-toy programming language is Turing complete: Java, C, C++, Python, Ruby, Perl, Haskell, BASIC, Fortran, Assembly Language, whatever.

Formally, we'll define computability in terms of the existence of an algorithm, which we will think of as a function written in Python:

Definition 4.18 (Computability)
A decision problem P is computable *if there exists a Python function $\mathcal{A}$ that solves P. That is, P is computable if there exists a Python function $\mathcal{A}$ such that, on any input x:*

(i) $\mathcal{A}$ terminates when run on x.
(ii) $\mathcal{A}(x)$ returns true if and only if x is a yes-instance of P.

Notice that we insist that the Python function $\mathcal{A}$ must actually terminate on any input x: it's not allowed to run forever. Furthermore, running $\mathcal{A}(x)$ returns True if x is a yes-instance of P and running $\mathcal{A}(x)$ returns False if x is a no-instance of P.

The decision problems from Example 4.28 are both computable:

Example 4.29 (Computability of some sample decision problems)

- PRIMALITY is computable: both **isPrime** and **isPrimeBetter** (p. 454) are algorithms that could be implemented as a Python function that (i) terminates when run on any positive integer, and (ii) returns True on input n if and only if n is prime.

- SATISFIABILITY is computable, too: as we discussed in Section 3.3.1, we can exhaustively try all truth assignments for φ, checking whether any of them satisfies φ. This algorithm is slow—if φ has n variables, there are 2^n different truth assignments—but it is guaranteed to terminate for any input φ, and correctly decides whether φ is satisfiable.

PROGRAMS THAT TAKE SOURCE CODE AS INPUT

The inputs to the problems or programs that we've talked about so far have been integers (for PRIMALITY) or Boolean formulas (for SATISFIABILITY). Of course, other input types like rational numbers or lists are possible, too. *Programs that take programs as input* are a particularly important category.

Taking it further: Although you might not have thought about them in these terms, you've frequently encountered programs that take programs as input. For example, in any introductory CS class, you've seen one frequently: the Python interpreter `python`, the Java compiler `javac`, and the C compiler `gcc` all take programs (written in Python or Java or C, respectively) as input.

It's easy to think up some decision problems where the input is a Python program. Here's one, about commenting code. (For example, it's not hard to imagine an Intro CS instructor setting up an automated grading system for programs that gives an automatic zero to any submitted assignment that contains no comments.)

```
def commentedTester(sourceCode):
  for character in sourceCode:
     if character == "#"
         and isn't inside quotes:
       return True
  return False
```

```
def absoluteValue(n):
   if n > 0:
      return n
   else:
      return -1 * n
```

```
def isEven(n):
   # % is Python's mod operator
   if n % 2 == 0:
      return True  # n is even
   else:
      return False # n is odd
```

Figure 4.24: Python source code for three functions.

Example 4.30 (The COMMENTED decision problem)
Define the decision problem COMMENTED as follows:

Input: the Python source code s for a function
Output: "yes" if s contains at least one comment; "no" otherwise.

In Python, a comment starts with # and goes until the end of the line, so as long as a # appears somewhere in the source code s—and not inside quotation marks—then s is a yes-instance of COMMENTED; otherwise s is a no-instance.

The COMMENTED problem is computable: testing whether s is a yes-instance can be done by looking at the characters of s one by one, and testing to see whether any one of those characters starts a comment. A Python program `commentedTester` that solves COMMENTED is shown in Figure 4.24. (The details of testing whether `character` is inside quotes are omitted from the source code, but otherwise the code for `commentedTester` is valid, runnable Python code.)

Consider running `commentedTester` on the other instances shown Figure 4.24. Observe that `absoluteValue` is a no-instance of COMMENTED, because it doesn't contain the comment character # at all, and `isEven` is a yes-instance of COMMENTED, because it contains three comments. As desired, if we ran `commentedTester` on these two pieces of source code, the output would be `False` and `True`, respectively.

Example 4.30 showed that the decision problem COMMENTED is computable by giving a Python function `commentedTester` that solves COMMENTED. Because we can run `commentedTester` on any piece of Python source code we please, let's do something a little bizarre: let's run `commentedTester` on the source code for `commentedTester` itself (!). There weren't any comments in `commentedTester`—the only # in the code is inside quotes—so the source code of `commentedTester` is a no-instance of COMMENTED. Put a different way, if s_{ct} denotes the source code of `commentedTester`, then running s_{ct} on s_{ct} returns `False`. This idea of taking some source code s and running s *on s itself* will be essential in the rest of this section.

```
def commentedTester(sourceCode):
  for character in sourceCode:
     if character == "#"
         and isn't inside quotes:
        return True
  return False
```

Figure 4.25: A reminder of the Python source code for `commentedTester`.

THE HALTING PROBLEM

The key decision problem that we'll consider is the *halting problem*:

Definition 4.19 (The Halting Problem)
Define the decision problem HALTINGPROBLEM *as follows:*

Input: a pair $\langle s, x \rangle$, where s is the source code of a syntactically valid Python function that takes one argument, and x is any value;
Output: "yes" if s terminates when run on input x; "no" otherwise.

That is, $\langle s, x \rangle$ is a yes-instance of HALTINGPROBLEM *if $s(x)$ terminates (doesn't get stuck in an infinite loop), and it's a no-instance if $s(x)$ does get stuck in an infinite loop.*

We can now use the idea of running a function with itself as input to show that the Halting Problem is uncomputable, by contradiction:

Theorem 4.17 (Uncomputability of the Halting Problem)
HALTINGPROBLEM *is uncomputable.*

Proof. We give a proof by contradiction. Suppose for the sake of contradiction that the Halting Problem is computable—that is, assume

There's a Python function $\mathcal{A}_{\text{halting}}$ solving the Halting Problem. (1)

(In other words, for the Python source code s of a one-argument function, and any value x, running $\mathcal{A}_{\text{halting}}(s, x)$ always terminates, and returns `True` if and only if running s on x does not result in an infinite loop.)

Now consider the Python function `makeSelfSafe` in Figure 4.26. The function `makeSelfSafe` takes as input the Python source code s of a one-argument function, tests whether running s on s itself is "safe" (does not cause an infinite loop), and if it's safe then it runs s on s. We claim that `makeSelfSafe` never gets stuck in an infinite loop:

```
makeSelfSafe(s):   # the input s is the Python source
                   # code of a one-argument function.
   safe = Ahalting(s,s)
   if safe:
      run s on input s
   return True
```

Figure 4.26: The Python code for `makeSelfSafe`.

For any Python source code s, `makeSelfSafe`(s) terminates. (2)

To see that (2) is true, observe that Step 1 of the algorithm always terminates, by assumption (1). Step 2 of the algorithm ensures that s is called on input s if and only if $\mathcal{A}_{\text{halting}}(s, s)$ said that s terminates when run on s. And, by assumption, $\mathcal{A}_{\text{halting}}$ is always correct. Thus s is run on input s *only if* s terminates when run on input s. So Step 2 of the algorithm always terminates. And Step 3 of the algorithm doesn't do anything except return, so it terminates immediately. Thus (2) follows.

Write s_{mss} to denote the Python source code of `makeSelfSafe`. Because s_{mss} is itself Python source code, Fact (2) implies that

$$\texttt{makeSelfSafe}(s_{\text{mss}}) \text{ terminates.} \tag{3}$$

In other words, running s_{mss} on s_{mss} terminates. Thus, by the assumption (1) that $\mathcal{A}_{\text{halting}}$ is correct, we can conclude that

$$\mathcal{A}_{\text{halting}}(s_{\text{mss}}, s_{\text{mss}}) \text{ returns true.} \tag{4}$$

But now consider what happens when we run `makeSelfSafe` on its own source code—that is, when we compute `makeSelfSafe`(s_{mss}). Observe that `safe` is set to true in Step 1 of the algorithm, by Fact (4). Thus Step 2 calls `makeSelfSafe`(s_{mss}) recursively! But therefore `makeSelfSafe`(s_{mss}) calls `makeSelfSafe`(s_{mss}), which calls `makeSelfSafe`(s_{mss}), and so on, *ad infinitum*. In other words,

$$\texttt{makeSelfSafe}(s_{\text{mss}}) \text{ does not terminate.} \tag{5}$$

But (3) and (5) are contradictory! Thus the only assumption that we made, namely (1), was false. Therefore there does not exist a correct always-terminating algorithm for the Halting Problem. That is, the Halting Problem is uncomputable. □

To summarize Theorem 4.17: we showed that the assumption of the existence of an algorithm for the halting problem leads to a contradiction, and therefore we conclude that such an algorithm cannot exist. The contradiction is, at its heart, about self-reference—an algorithmic version of the Liar's Paradox: *This sentence is false.*

Taking it further: *Computability theory* is the study of what problems can and cannot be solved by computers. Computability was a primary focus of theoretical computer science from the 1930s through roughly the 1970s. (After that time, the focus of theoretical computer scientists began to shift to *complexity theory*, which addresses the question of what problems can and cannot be solved *efficiently* by computers.) You can read more about the halting problem in any textbook on computability theory, and in Douglas Hofstadter's amazing book *Gödel, Escher, Bach.*[8] For extra amusement, you can even find a full proof of Theorem 4.17 in poem form, in Figure 4.27. And see p. 455 for a discussion of some practically relevant problems that are also uncomputable.

[8] Dexter Kozen. *Automata and Computability*. Springer, 1997; Michael Sipser. *Introduction to the Theory of Computation*. Course Technology, 3rd edition, 2012; and Douglas Hofstadter. *Gödel, Escher, Bach: An Eternal Golden Braid*. Vintage, 1980.

Scooping the Loop Snooper: A proof that the Halting Problem is undecidable
Geoffrey K. Pullum

No general procedure for bug checks will do.
Now, I won't just assert that, I'll prove it to you.
I will prove that although you might work till you drop,
you cannot tell if computation will stop.

For imagine we have a procedure called *P*
that for specified input permits you to see
whether specified source code, with all of its faults,
defines a routine that eventually halts.

You feed in your program, with suitable data,
and *P* gets to work, and a little while later
(in finite compute time) correctly infers
whether infinite looping behavior occurs.

If there will be no looping, then *P* prints out 'Good.'
That means work on this input will halt, as it should.
But if it detects an unstoppable loop,
then *P* reports 'Bad!'—which means you're in the soup.

Well, the truth is that *P* cannot possibly be,
because if you wrote it and gave it to me,
I could use it to set up a logical bind
that would shatter your reason and scramble your mind.

Here's the trick that I'll use—and it's simple to do.
I'll define a procedure, which I will call *Q*,
that will use *P*'s predictions of halting success
to stir up a terrible logical mess.

For a specified program, say *A*, one supplies,
the first step of this program called *Q* I devise
is to find out from *P* what's the right thing to say
of the looping behavior of *A* run on *A*.

If *P*'s answer is 'Bad!', *Q* will suddenly stop.
But otherwise, *Q* will go back to the top,
and start off again, looping endlessly back,
till the universe dies and turns frozen and black.

And this program called *Q* wouldn't stay on the shelf;
I would ask it to forecast its run on *itself.*
When it reads its own source code, just what will it do?
What's the looping behavior of *Q* run on *Q*?

If *P* warns of infinite loops, *Q* will quit;
yet *P* is supposed to speak truly of it!
And if *Q*'s going to quit, then *P* should say 'Good.'
Which makes *Q* start to loop! (*P* denied that it would.)

No matter how *P* might perform, *Q* will scoop it:
Q uses *P*'s output to make *P* look stupid.
Whatever *P* says, it cannot predict *Q*:
P is right when it's wrong, and is false when it's true!

I've created a paradox, neat as can be—
and simply by using your putative *P*.
When you posited *P* you stepped into a snare;
Your assumption has led you right into my lair.

So where can this argument possibly go?
I don't have to tell you; I'm sure you must know.
A *reductio:* There cannot possibly be
a procedure that acts like the mythical *P*.

You can never find general mechanical means
for predicting the acts of computing machines;
it's something that cannot be done. So we users
must find our own bugs. Our computers are losers!

Figure 4.27: A proof of Theorem 4.17, in poetic form, from

[9] Geoffrey K. Pullum. Scooping the loop snooper: A proof that the halting problem is undecidable. *Mathematics Magazine*, 73(4):319–320, 2000. Used by permission of Geoffrey K. Pullum.

Computer Science Connections

Cryptography and the Generation of Prime Numbers

As we'll see in Section 7.5, prime numbers are used extensively in cryptography. The *RSA cryptosystem*—named after the first letters of its inventors' last names[10]—uses as a primary step the generation of two large prime numbers, perhaps ≈128-bit integers.

The primary reason that prime numbers are useful in cryptography is an asymmetry in the apparent difficulty of two directions of a problem. If you are given two (big) prime numbers p and q, then computing their product pq is easy. But if you are given a number n that is guaranteed to be the product of two prime numbers, finding those two numbers—*factoring n*—appears to be much harder. For example, if you're told that $n = 504{,}761$, it will probably take you a long time to figure out that $n = 251 \cdot 2011$. But if you're told that $p = 251$ and $q = 2011$, then you should be able to calculate $pq = 504{,}761$ in just a few seconds.

A crucial step in RSA, then, is the generation of large prime numbers. This step can be accomplished by choosing a random integer of the appropriate size and then testing whether that number is prime. (We keep retrying until the random number turns out to be prime.)

A little consideration of the definition of primality implies that we can test whether an integer n is prime using the algorithm in Figure 4.28, which tests all candidate divisors between 2 and $n-1$. This algorithm requires us to do roughly n divisibility checks (actually, to be precise, $n-2$ divisibility checks). Using Theorem 4.16, the algorithm can be improved to do only about $\sqrt{n}$ divisibility checks, as Figure 4.29.

We can test these two algorithms empirically. A Python implementation using $n-1$ calls to **isPrime** to find all primes in the integers $\{2, \ldots, n\}$ took about three minutes for $n = 65{,}536$ on a 2010-era laptop. For the same n, **isPrimeBetter** took about a second. This difference is a nice example of the way in which theoretical, proof-based techniques can improve actual widely used algorithms.

In part because of its importance to cryptography, there has been significant work on algorithms for primality testing over recent decades—improving far beyond the roughly $\sqrt{n}$ division tests of **isPrimeBetter**. In general, an efficient algorithm for a number n should require a number of steps proportional to $\log n$ rather than proportional to n or even $\sqrt{n}$. (For example, when you add two 10-digit numbers by hand, you want to do about 10 operations, rather than about 1,000,000,000 operations.) Thus **isPrimeBetter** is still not as efficient as we'd like.

There are some very efficient *randomized* algorithms for primality testing which are actually used in real cryptosystems, including the *Miller-Rabin test*.[11] This randomized algorithm performs a (randomly chosen) test that all prime numbers pass and most composite numbers fail; repeating with many different randomly chosen tests decreases the probability of getting a wrong answer to an arbitrarily small number. (See p. 742.) And more recently, three researchers gave the first theoretically efficient algorithm for primality testing that's not randomized.[12]

[10] R. L. Rivest, A. Shamir, and L. Adleman. A method for obtaining digital signatures and public-key cryptosystems. *Communications of the ACM*, 21:120–126, February 1978.

```
isPrime(n):
  k := 2
  while k < n:
      if n is evenly divisible by k then
          return False
      k := k + 1
  return True
```

Figure 4.28: Slow primality testing.

```
isPrimeBetter(n):
  k := 2
  while k ≤ ⌈√n⌉:
      if n is evenly divisible by k and n ≠ k then
          return False
      k := k + 1
  return True
```

Figure 4.29: Faster primality testing. (We could further save roughly another factor of two by checking only $k = 2$ and odd $k \geq 3$.)

[11] Gary L. Miller. Riemann's hypothesis and tests for primality. *Journal of Computer and System Sciences*, 13(3):300–317, 1976; and Michael O. Rabin. Probabilistic algorithm for testing primality. *Journal of Number Theory*, 12(1):128–138, 1980.

[12] Manindra Agrawal, Neeraj Kayal, and Nitin Saxena. Primes is in P. *Annals of Mathematics*, 160:781–793, 2004.

Computer Science Connections

Other Uncomputable Problems (That You Might Care About)

The Halting Problem may seem like a purely abstract problem, and therefore one that doesn't matter in the real world—sure, it'd be nice to have an infinite-loop detector in your Python interpreter or Java compiler, but would it just be a vaguely helpful feature for students in Intro CS classes but nobody else? The answer is a resounding no: while the Halting Problem itself may seem obscure, there are many uncomputable problems that, if solved, would vastly improve operating systems or compilers. But they're uncomputable, and therefore the desired improvements cannot be made.

Here's one example. Modern operating systems use *virtual memory* for their applications. The physical computer has a limited amount of physical memory—say, eight gigabytes of RAM—that applications can use. But the operating system "pretends" that it has a much larger amount of memory, so that the word processor, web browser, Java compiler, and solitaire game can *each* act as though they had even more than eight gigabytes of memory that they don't have to share. Memory (both virtual and real) is divided into chunks of a fixed size, called *pages*. The operating system stores pages that are actively in use in physical memory (RAM), and relegates some of the not-currently-used pages to the hard drive. At every point in time, the operating system's *paging system* decides which pages to leave in physical memory, and which pages to "eject" to the hard drive. (This idea is the same as what you do when you're cooking several dishes in a kitchen with limited counter space: you have to relegate some of the not-currently-being-prepared ingredients to the fridge. And at every moment you have to decide which ingredients to leave on the counter, and which to "eject" to the fridge.) See Figure 4.30.

Here's a problem that a paging system would love to solve: given a page p of memory that an application has used, will that application ever access the contents of p again? Let's call this problem willBeUsedAgain. When the paging system needs to eject a page, ideally it would eject a page that's a no-instance of willBeUsedAgain, because it will never have to bring that page back into physical memory. (When you're out of counter space, you would of course prefer to put away some ingredient that you're done using.)

Unfortunately for operating system designers, willBeUsedAgain is uncomputable. There's a very quick proof, based on the uncomputability of the Halting Problem. Consider the algorithm:

1. run the Python function f on the input x.
2. if $f(x)$ terminates, then access some memory from page p.

This algorithm accesses page p if and only if $\langle f, x \rangle$ is a yes-instance of the Halting Problem.

Therefore *if we could give an algorithm to solve the* willBeUsedAgain *problem, then we could give an algorithm to solve the Halting Problem.* But we already know that we can't give an algorithm to solve the Halting Problem. If $p \Rightarrow q$ and $\neg q$, then we can conclude $\neg p$; therefore willBeUsedAgain is uncomputable.

(a) Initial configuration, with pages #1,2,6 in memory, and remaining pages on disk.

(b) Program requests data on page #2. It's in memory, so it's just fetched; nothing else happens.

(c) Program requests data on page #4. It's on disk, so it's fetched and replaces some page in RAM—say, #1.

(d) Program requests data on page #1. It's on disk, so it's fetched and replaces some page in RAM—say, #6.

Figure 4.30: A sample sequence of memory fetches in a paged memory system.

4.4.5 Exercises

Figure 4.31 shows the truth tables for all 16 different binary logical operators, with each column named if it's a logical operator that we've already seen:

p	q	[1] True	[2] $p \vee q$	[3] $p \Leftarrow q$	[4] p	[5] $p \Rightarrow q$	[6] q	[7] $p \Leftrightarrow q$	[8] $p \wedge q$	[9] $p \mid q$	[10] $p \oplus q$	[11] $\neg q$	[12]	[13] $\neg p$	[14]	[15] $p \downarrow q$	[16] False
T	T	T	T	T	T	T	T	T	T	F	F	F	F	F	F	F	F
T	F	T	T	T	T	F	F	F	F	T	T	T	T	F	F	F	F
F	T	T	T	F	F	T	T	F	F	T	T	F	F	T	T	F	F
F	F	T	F	T	F	T	F	T	F	T	F	T	F	T	F	T	F

Figure 4.31: The full set of binary logical operators.

A set S of binary operators is said to be universal *if every binary logical operation can be expressed using some combination of the operators in S. Formally, a set S is universal if, for every Boolean expression φ over variables $p_1, \ldots, p_k$, there exists a Boolean expression ψ that is logically equivalent to φ where ψ uses only the variables $p_1, \ldots, p_k$ and the logical connectives in S.*

4.66 Prove that the set $\{\vee, \wedge, \Rightarrow, \neg\}$ is universal. *(Hint: To do so, you need to show that, for each column* [1] *through* [16] *of Figure 4.31, you can build a Boolean expression φ_i over the variables p and q that uses only the operators $\{\vee, \wedge, \Rightarrow, \neg\}$, and such that φ_i is logically equivalent to p* [*i*] *q.)*

4.67 Prove that the set $\{\vee, \wedge, \neg\}$ is universal. *(Hint: once you've done Exercise 4.66, all you have to do is show that you can express $\Rightarrow$ using $\{\vee, \wedge, \neg\}$.)*

4.68 Prove that $\{\vee, \neg\}$ and $\{\wedge, \neg\}$ are both universal.

4.69 Prove that the set $\{|\}$—the set containing just the Sheffer stroke, that is, *nand*—is universal.

4.70 Prove that the singleton set $\{\downarrow\}$ is universal.

4.71 Prove that the set $\{\wedge, \vee\}$ is *not* universal. *(Hint: what happens under the all-true truth assignment?)*

4.72 Let φ be a fully quantified proposition of predicate logic. Prove that φ is logically equivalent to a fully quantified proposition ψ in which *all quantifiers are at the outermost level of ψ*. In other words, the proposition ψ must be of the form

$$\forall\!/\!\exists\, x_1 : \forall\!/\!\exists\, x_2 : \cdots \forall\!/\!\exists\, x_k : P(x_1, x_2, \ldots, x_k),$$

where each $\forall\!/\!\exists$ is either a universal or existential quantifier. (The transformation that you performed in Exercise 3.178 put Goldbach's Conjecture in this special form.) *(Hint: you might find the results from Exercises 4.66–4.71 helpful. Using these results, you can assume that φ has a very particular form.)*

4.73 Prove that, for any integer $n \geq 1$, there is an n-variable logical proposition φ in conjunctive normal form such that the truth-table translation to DNF (from Theorem 4.11) yields an DNF proposition with exponentially more clauses than φ has.

4.74 Prove that the area of a right triangle with legs x and y is $xy/2$.

4.75 Use Figure 4.32(a) as an outline to give a different proof of the Pythagorean theorem.

4.76 Exercise 4.47 asked you to prove (via algebra) the *Arithmetic Mean–Geometric Mean inequality*: for $x, y \in \mathbb{R}^{\geq 0}$, we have $\sqrt{xy} \leq (x+y)/2$. Here you'll reprove the result geometrically. Suppose that $x \geq y$, and draw two circles of radius x and y tangent to each other, and tangent to a horizontal line. See Figure 4.32(b). Considering the right triangle shown in that diagram, and using the Pythagorean theorem and the fact that the hypotenuse is the longest side of a right triangle, prove the result again.

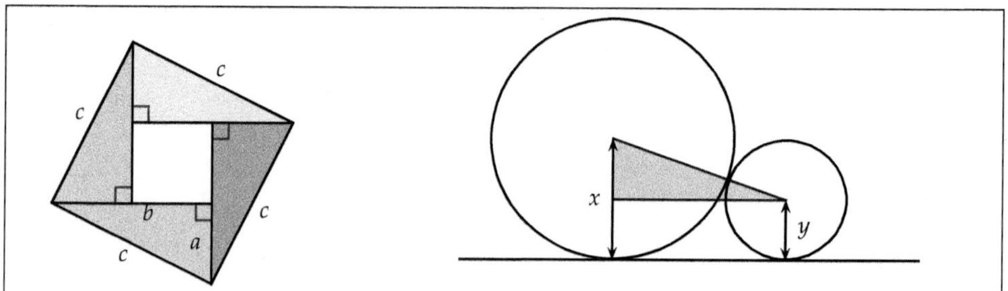

(a) Another way to prove the Pythagorean Theorem.

(b) Using the Pythagorean Theorem for the Arithmetic Mean/Geometric Mean inequality.

Figure 4.32: More on the Pythagorean Theorem.

Let $x, y \in \mathbb{R}^2$ be two points in the plane. As usual, denote their coordinates by x_1 and x_2, and y_1 and y_2, respectively. The Euclidean distance *between these points is the length of the line that connects them:* $\sqrt{(x_1 - y_1)^2 + (x_2 - y_2)^2}$. *The* Manhattan distance *between them is* $|x_1 - y_1| + |x_2 - y_2|$*: the number of blocks that you would have to walk "over" plus the number that you'd have to walk "up" to get from one point to the other. Denote these distances by* $d_{euclidean}$ *and* $d_{manhattan}$.

4.77 Prove that $d_{\text{euclidean}}(x, y) \leq d_{\text{manhattan}}(x, y)$ for any two points x, y.

4.78 Prove that there exists a constant a such that both

- $d_{\text{manhattan}}(x, y) \leq a \cdot d_{\text{euclidean}}(x, y)$ for all points x and y; and
- there exist points x^*, y^* such that $d_{\text{manhattan}}(x^*, y^*) = a \cdot d_{\text{euclidean}}(x^*, y^*)$

A positive integer n is called a perfect number *if it is equal to the sum of all positive integer factors $1 \leq k < n$ of n. For example, the number 14 is not perfect: the numbers less than 14 that evenly divide 14 are $\{1, 2, 7\}$, but* $1 + 2 + 7 = 10 \neq 14$.

4.79 Prove that at least one perfect number exists.

4.80 Prove that, for any prime integer p, the positive integer p^2 is not a perfect number.

4.81 Let $n \geq 10$ be any positive integer. Prove that the set $\{n, n+1, \ldots, n+5\}$ contains at most two prime numbers.

4.82 Let n be any positive integer. Prove or disprove: any set of ten consecutive positive integers $\{n, n+1, \ldots, n+9\}$ contains at least one prime number.

4.83 *(Thanks to the NPR radio show Car Talk, from which I learned this exercise.)* Imagine a junior high school, with 100 lockers, numbered 1 through 100. All lockers are initially closed. There are 100 students, each of whom—brimming with teenage angst—systematically goes through the lockers and slams some of them shut and yanks some of them open. Specifically, in round $i := 1, 2, \ldots, 100$, student #i changes the state of every ith locker: if the door is open, then it's slammed shut; if the door is closed, then it's opened. (So student #1 opens them all, student #2 closes all the even-numbered lockers, etc.) Which lockers are open after this whole process is over? Prove your answer.

4.84 We proved the following claim in Theorem 4.16: *A positive integer $n \geq 2$ is evenly divisible by some other integer $k \in \{2, 3, \ldots, \lceil\sqrt{n}\rceil\}$ if and only if n is composite.* If we delete the word "other," this claim becomes false. Prove that this modified claim is false.

4.85 Prove that the unmodified claim (retaining the word "other") remains true if the bounds on k are changed from $k \in \{2, 3, \ldots, \lceil\sqrt{n}\rceil\}$ to $k \in \{\lceil\sqrt{n}\rceil, \ldots, n-1\}$.

4.86 Prove that the bound *cannot* be changed from $k \in \{2, 3, \ldots, \lceil\sqrt{n}\rceil\}$ to $k \in \{\lfloor\sqrt{n}/2\rfloor, \ldots, \lfloor 3\sqrt{n}/2\rfloor\}$. That is, prove that the following claim is false: *A positive integer $n \geq 2$ is evenly divisible by some other integer $k \in \{\lfloor\sqrt{n}/2\rfloor, \ldots, \lfloor 3\sqrt{n}/2\rfloor\}$ if and only if n is composite.*

4.87 Let n be any positive integer, and let p_n denote the smallest prime number that evenly divides n. Prove that there are infinite number of integers n such that $p_n \geq \sqrt{n}$. (This fact establishes that we cannot change the bound in the aforementioned theorem to anything smaller than $\sqrt{n}$.)

4.5 *Common Errors in Proofs*

Mistakes were made.

Ron Ziegler (1939–2003), press secretary for President Richard Nixon during Watergate

We've now spent considerable time establishing a catalogue of proof techniques that you can use to prove theorems, along with some examples of these techniques in action. We'll close this chapter with a brief overview of some common *flaws* in proofs, so that you can avoid them in your own work (and be on the lookout for them in the work of others). Recall that a proof consists of a sequence of logical inferences, deriving new facts from assumptions or previously established facts. A *valid* inference is one whose conclusion is always true as long as the facts that it relies on were true. (That is, a valid step never creates a false statement from true ones.) An *invalid* inference is one in which the conclusion can be false *even if the premises are all true.* An invalid argument can also be called a *logical fallacy,* a *fallacious argument,* or just a *fallacy.* In a correct proof, of course, every step is valid. Here are a few examples of a single logical inference, some of which might be fallacious:

Example 4.31 (Some (valid and invalid) logical inferences)

Problem: Here are several inferences. In each case, there are two premises, and a conclusion that is claimed to follow logically from those premises. Which of these inferences are valid, and which are fallacies?

1. *Premises:* (a) All software is buggy. (b) Windows is a piece of software.
   Conclusion: Therefore, Windows is buggy.
2. *Premises:* (a) All people are annoying sometimes. (b) Mark Zuckerberg is a person.
   Conclusion: Therefore, Mark Zuckerberg is annoying sometimes.
3. *Premises:* (a) If you handed in an exam without your name on it, then you got a zero. (b) You handed in an exam without your name on it.
   Conclusion: Therefore, you got a zero.
4. *Premises:* (a) If you handed in an exam without your name on it, then you got a zero. (b) You handed in an exam with your name on it.
   Conclusion: Therefore, you didn't get a zero.

Solution: We abstract away from buggy software and annoying people by rewriting these arguments in purely logical form:

Problem-solving tip: To make the logical structure of an argument clearer, consider an abstract form of the argument in which you use variables to name the atomic propositions.

1. Assume $a \in S$ and assume $\forall x \in S : P(x)$. Conclude $P(a)$.
2. Assume $a \in S$ and assume $\forall x \in S : P(x)$. Conclude $P(a)$.
3. Assume $p \Rightarrow q$ and assume p. Conclude q.
4. Assume $p \Rightarrow q$ and assume $\neg p$. Conclude $\neg q$.

In this format, we see first that (1) and (2) are actually the same logical argument (with different meanings for the symbols), and they're both valid. Argument (3) is precisely an invocation of Modus Ponens (see Chapter 3), and it's valid. But (4) is a fallacy: the fact that $p \Rightarrow q$ and $\neg p$ is consistent with either q or $\neg q$, so in particular when p = False and q = True the premises are true but the conclusion is false.

Each of these examples purports to convince its reader of its conclusion, *under the assumption* that the premises are true. Valid arguments will convince any (reasonable) reader that their conclusion follows from their premises. Fallacious arguments are buggy; a vigilant reader will not accept the conclusion of a fallacious argument even if she accepts the premises.

Taking it further: A useful way to think about validity and fallacy is as follows. An argument with premises $p_1, p_2, \ldots, p_k$ and conclusion c is valid if and only if $p_1 \wedge p_2 \wedge \cdots \wedge p_k \Rightarrow c$ is a theorem. If there is a circumstance in which $p_1 \wedge p_2 \wedge \cdots \wedge p_k \Rightarrow c$ is false—in other words, where the premises $p_1 \wedge p_2 \wedge \cdots \wedge p_k$ are all true but the conclusion c is false—then the argument is fallacious.

Some of the most famous disasters in the history of computer science have come from some bugs that arose because of an erroneous understanding of some property of a system—and a lack of valid proof of correctness for the system. These bugs have been costly, with both lives and many dollars lost. See p. 464 for a few highlights/lowlights.

Your main job in proofs is simple: avoid fallacies! But that can be harder than it sounds. The remainder of this section is devoted to a few types of common mistakes in proofs—that is, some common types of fallacies.

A BROKEN PROOF

The most common mistake in a purported proof is simple but insidious: a single statement is alleged to follow logically from previous statements, but it doesn't. Here's a somewhat subtle example:

Example 4.32 (What's wrong with this logic?)

Problem: Find the error in this purported proof, and give a counterexample to the claim.

False Theorem: Let $F_n = \{k \in \mathbb{Z}^{\geq 1} : k \mid n\}$ denote the factors of an integer $n \geq 2$. Then $|F_n|$ is even.

Proof. Let $F_{\text{small}} \subseteq F$ be the set of factors of n that are less than $\sqrt{n}$. Let $F_{\text{big}} \subseteq F$ be the set of factors of n that are greater than $\sqrt{n}$. Observe that every $d \in F_{\text{small}}$ has a unique entry n/d corresponding to it in F_{big}. Therefore $|F_{\text{small}}| = |F_{\text{big}}|$. Let $k = |F_{\text{small}}| = |F_{\text{big}}|$. Note that k is an integer. Thus F_n contains precisely k elements less than $\sqrt{n}$ and k elements greater than $\sqrt{n}$, and so $|F_n| = 2k$, which is an even number. □

Solution: The problem comes right at the end of the proof:

> *Thus F_n contains precisely k elements less than $\sqrt{n}$ and k elements greater than $\sqrt{n}$, and so $|F_n| = 2k$.*

The problem is that this statement discounts the possibility that $\sqrt{n}$ itself might be in F. For an integer n that's a perfect square, we have that $\sqrt{n} \in F$, and therefore $|F| = 2k + 1$. For example, the integer 9 is a counterexample, because $F_9 = \{1, 3, 9\}$ and $|F_9| = 3$.

Problem-solving tip: The kind of mistake in Example 4.32, in which there's a single step that doesn't follow from the previous step, can sometimes be difficult to sniff out. But it's the kind of bug that you can spot by simply being überskeptical of everything that's written in a purported proof.

But while an error of this form—one step in the proof that doesn't actually follow from the previously established facts—may be the most common type of bug in a proof, there are some other, more structural errors that can arise. Most of these structural errors come from errors of propositional logic—namely by proving a new proposition that's *not* in fact logically equivalent to the given proposition. Here are a few of these types of flawed reasoning.

FALLACY: PROVING TRUE

We are considering a claim φ. We proceed as follows: we assume φ, and (correctly) prove True under that assumption. (Usually, for some reason, the "proof" writer puts a little check mark in their alleged proof at this point: ✓.) What can we conclude about φ? The answer is: *absolutely nothing!* The reason: we've proven that $\varphi \Rightarrow$ True, but *anything implies true.* (Both True $\Rightarrow$ True and False $\Rightarrow$ True are true implications.) Here's a classical example of a bogus proof that uses this fallacious reasoning:

Example 4.33 (What's wrong with this logic?)
Problem: Find the error in this purported proof.

False Theorem: $1 = 0$.
Proof. Suppose that $1 = 0$. Then:

$$\begin{aligned} & & 1 &= 0 \\ &\text{therefore, multiplying both sides by 0} & 0 \cdot 1 &= 0 \cdot 0 \\ &\text{and therefore,} & 0 &= 0. \checkmark \end{aligned}$$

And, indeed, $0 = 0$.
Thus the assumption that $1 = 0$ was correct, and the theorem follows. □

Solution: We have merely shown that $(1 = 0) \Rightarrow (0 = 0)$, which does not say anything about the truth or falsity of $1 = 0$; anything implies true.

Writing tip: When you're trying to prove that two quantities a and b are equal, it's generally preferable to manipulate a until it equals b, rather than "meeting in the middle" by manipulating both sides of the equation until you reach a line in which the two sides are equal. The "manipulate a until it equals b" style of argument makes it clear to the reader that you are proving $a = b$ rather than proving $(a = b) \Rightarrow$ True.

FALLACY: AFFIRMING THE CONSEQUENT

We are considering a claim φ. We prove (correctly) that $\varphi \Rightarrow \psi$, and we prove (correctly) that ψ. We then conclude φ. (Recall that ψ is the *consequent* of the implication $\varphi \Rightarrow \psi$, and we have "affirmed" it by proving ψ.) This "proof" is wrong because it confuses necessary and sufficient conditions: when we prove $\varphi \Rightarrow \psi$, we've shown that *one way* for ψ to be true is for φ to be true. But there might be other reasons that φ is true! Here's an example of a fallacious argument that uses this bogus logic:

Example 4.34 (What's wrong with this logic?)
Problem: Find the error in this argument:

Premises: (1) If it's raining, then the computer burning will be postponed.
(2) The computer burning was postponed.
Conclusion: Therefore, it's raining.

Solution: This fallacious argument is an example of affirming the consequent. The first premise here merely says that the computer burning will be postponed if it rains; it does not say that rain is the only reason that the burning could be postponed. There may be many other reasons why the burning might be delayed: for example, the inability to find a match, the sudden vigilance of the health and safety office, or a last-minute stay of execution by the owner of the computer.

FALLACY: DENYING THE HYPOTHESIS

Denying the hypothesis is a closely related fallacy to affirming the consequent: we prove (correctly) that $\psi \Rightarrow \varphi$, and we prove (correctly) that $\neg\psi$; we then (fallaciously) conclude $\neg\varphi$. This logic is buggy for essentially the same reason as affirming the consequent. (In fact, denying the hypothesis is the contrapositive of affirming the consequent—and therefore a fallacy too, because it's logically equivalent to a fallacy.) The implication $\psi \Rightarrow \varphi$ means that one way of φ being true is for ψ to be true, but it does *not* mean that there is no other way for φ to be true. Here's an example of a fallacious argument of this type:

Example 4.35 (What's wrong with this logic?)

Problem: Find the error in this argument:

Premises: (1) If you have resolved the P-versus-NP question, then you are famous.
(2) You have not resolved the P-versus-NP question.

Conclusion: Therefore, you are not famous.

Solution: This fallacious argument is an example of denying the hypothesis. The first premise says that one way to be famous is to resolve the P-versus-NP question (see p. 326 for a brief description of this problem), but it does not say that resolving the P-versus-NP question is the only way to be famous. For example, you could be famous by being the President of the United States or by founding Google.

FALLACY: FALSE DICHOTOMY

A *false dichotomy* or *false dilemma* is a fallacious argument in which two nonexhaustive alternatives are presented as exhaustive (without acknowledgement that there are any unmentioned alternatives).

Example 4.36 (False Dichotomy)

The flawed step in Example 4.32 can be interpreted as a false dilemma: implicitly, that proof relied on the assertion that if k evenly divides n, then

$$k \in F_{\text{small}} = \{\text{factors of } n \text{ that are less than } \sqrt{n}\} \text{ or}$$
$$k \in F_{\text{big}} = \{\text{factors of } n \text{ that are greater than } \sqrt{n}\}.$$

But of course the third unmentioned possibility is that $k = \sqrt{n}$.

(The classical false dichotomy, often found in political rhetoric, is "either you're with us or you're against us": actually, you might be neutral on the issue, and therefore neither "with" nor "against" us!)

FALLACY: BEGGING THE QUESTION

We wish to prove a proposition φ. A purported proof of φ that *begs the question* is one that assumes φ along the way. That is, the "proof" assumes precisely the thing that it purports to prove, and thus actually proves $\varphi \Rightarrow \varphi$. Although this type of fallacious reasoning sounds ridiculous, the assumption of the desired result can be very subtle; you must be vigilant to catch this type of error. Here's an example of a fallacious argument of this kind:

Example 4.37 (What's wrong with this logic?)

Problem: Find the error in this proof:

False Theorem: Let n be a positive integer such that $n + n^2$ is even. Then n is odd.

Proof. Assume the antecedent—that is, assume that $n + n^2$ is even. Let k be the integer such that $n = 2k + 1$. Then

$$\begin{aligned} n + n^2 &= 2k + 1 + (2k + 1)^2 \\ &= 2k + 1 + 4k^2 + 4k + 1 \\ &= 4k^2 + 6k + 2 \\ &= 2 \cdot (2k^2 + 3k + 1), \end{aligned}$$

which is even because it is equal to 2 times an integer. But $n^2 = (2k + 1)^2 = 4k^2 + 4k + 1$ is odd (because $4k^2$ and $4k$ are both even). Therefore

$$n = \underbrace{n + n^2}_{\text{even by the above argument}} - \underbrace{n^2}_{\text{odd by the above argument}}.$$

An even number less an odd number is an odd number, which implies that n must be odd too. □

Solution: The problem comes very early in the "proof," in the sentence

Let k be the integer such that $n = 2k + 1$.

But this statement implicitly assumes that n is an odd integer; an integer k such that $n = 2k + 1$ exists *only if* n is odd. So the proof begs the question: it assumes that n is odd, and—after some algebraic shenanigans—concludes that n is odd.

Problem-solving tip: Even without identifying the specific bug in Example 4.37, we could notice that there's something fishy by doing the post-proof plausibility check to make sure that all premises were actually used. The "proof" states that it is assuming the antecedent, but we actually *derived* the fact that $n + n^2$ is even. So we never used that assumption in the "proof." (In fact, $n + n^2$ is even for *any* positive integer n.) But, because we didn't use the assumption, the same proof works just as well without it as an assumption, so we could use the same "proof" to establish this claim instead:

Patently False Theorem: *Let n be a positive integer. Then n is odd.*

Given that this new claim is obviously false, there *must* be a bug in the proof. The only challenge is to find that bug.

OTHER FALLACIES

We have discussed a reasonably large collection of logical fallacies into which some less-than-careful or less-than-scrupulous proof writers may fall. But there are many other types of flaws in arguments that more typically arise in informal contexts; these are the kinds of flawed arguments that are—sadly—often used in politics. (Some of them have analogues in more mathematical settings, too.) Here are a few examples of other types of fallacies that you may encounter in "real-world" arguments:

- *Confusing correlation and causation.* Phenomena A and B are said to be *(positively) correlated* if they occur together more often than their individual frequencies would predict. (See Chapter 10.) But just because A and B are correlated does *not* mean that one *causes* the other! For example, the user population of Facebook is much younger than is the population at large. We could say, correctly, that *Being young is correlated with using Facebook.* But *Using Facebook makes you young* is an obviously absurd conclusion. (Some correlation-versus-causation mistakes are subtler; your reaction to *Being young makes you use Facebook* is probably less virulent, but it is equally unsupported by the facts that we've cited here.) Always be wary when attempting to infer causal relationships!

- *Ad hominem attacks.* An *ad hominem* attack ignores the logical argument and speaks to the arguer: *Bob doesn't know the difference between contrapositive and converse, and he says that n is prime. So n must be composite.*

Latin *ad hominem*: "to the man."

- *Equivocation* or *shifting language.* This type of argument relies on changes in the meanings of the words/variables in an argument. This shift can be grammatical: *Time waits for no man,* and *no man is an island*; therefore, *time waits for an island.* Or it can be in the semantics of a particular word: 1024 *is a prime example of an exact power of two,* and *prime numbers are evenly divisible only by* 1 *and themselves*; therefore, 1024 *is not divisible by* 4. A similar type of fallacy can also occur when a variable in a proof is introduced to mean two different things.

Taking it further: This listing is just a brief outline of some of the many invalid techniques of persuasion/propaganda; a much more extensive and thorough list is maintained by Gary Curtis at `http://www.fallacyfiles.org/`. You might also be interested in books that catalogue fallacious techniques of argument.[13]

For example,
[13] Madsen Pirie. *How to Win Every Argument: The Use and Abuse of Logic.* Continuum, 2007.

It is always your job to be vigilant—both when reading proofs written by others, and in developing your own proofs—to avoid fallacious reasoning.

COMPUTER SCIENCE CONNECTIONS

THE COST OF MISSING PROOFS: SOME FAMOUS BUGS IN CS

There's an apocryphal story that the first use of the word "bug" to refer to a flaw in a computer system was in the 1940s: Grace Hopper, a rear admiral in the US Navy and a pioneer in early programming, found a moth (a literal, physical moth) jamming a piece of computer equipment and causing a malfunction. (The story is true, but the *Oxford English Dictionary* reports uses of "bug" to refer to a technological fault dating back to Thomas Edison in the late 1800s.) But there are many other stories of bugs that are both more important and more true. When a computer system "almost" works—when there's no proof that it works correctly in all circumstances—there can be grave repercussions, in dollars and lives lost. Here are a few of the most famous, and most costly, bugs in history:[14]

For a list of one person's view of the ten worst bugs in history, including these three and some other sordid tales, see:
[14] Simpson Garfinkel. History's worst software bugs. *Wired Magazine*, 2005.

The Pentium division bug: In 1994, Thomas Nicely, at the time a math professor, discovered a hardware bug in Intel's new Pentium chip that caused incorrect results when some floating-point numbers were divided by certain other floating-point numbers. The flaw resulted from a lookup table for the division operation that was missing a handful of entries. Although the range of numbers that were incorrectly divided was limited, the resulting brouhaha led to a full Pentium recall and about $500 million in losses for Intel.[15]

For more information on these bugs and their aftermath, see:
[15] Ivars Peterson. MathTrek: Pentium bug revisited. *MAA Online*, May 1997.

The Ariane 5 rocket: The European Space Agency's rocket, carrying a $400,000,000 payload of satellites, exploded 40 seconds into its first flight, in 1996. The rocket had engaged its self-destruct system, which was correctly triggered when it strayed from its intended trajectory. But the altered trajectory was caused by a sequence of errors, including an *integer overflow* error: the rocket's velocity was too big to fit into the 16-bit variable that was being used to store it.[16] (An Ariane 5 rocket was much faster than the Ariane 4 rockets for which the code was originally developed.) Embarrassingly, the overflow caused a subsystem to output a diagnostic error code that was interpreted as navigation data. More embarrassingly still, this entire subsystem played no role in navigation after liftoff, and would have caused no harm if it were just turned off.

[16] J. L. Lions. Ariane 5 flight 501 failure report: Report by the enquiry board, 1996.

The Therac-25: The Therac-25 was a medical device in use in the mid-1980s that treated tumors with a focused beam of radiation. The device fired a concentrated X-ray beam of extremely high dosage into a diffuser that would reduce the beam's intensity to the desired levels before it was directed at the patient. But it turned out that a particularly fast touch-typing operator could cause the high-intensity beam to be fired without the diffuser in place: hitting `enter` at the precise moment that an internal variable reset to zero caused the undiffused beam to be fired. (This kind of bug is called a *race condition*, in which the output of a system depends crucially on the precise timing of events like operator input.) At least five patients were killed by radiation overdoses.[17]

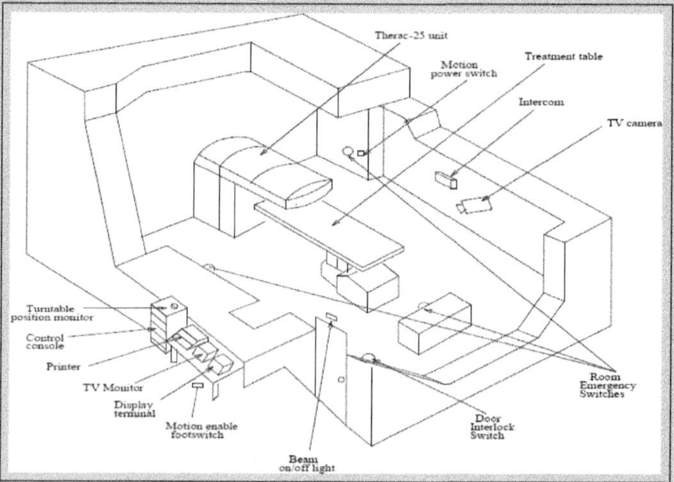

Figure 4.33: Image of the Therac-25. Reprinted with permission from
[17] Nancy Leveson. *Safeware: System Safety and Computers*. Pearson Education, Inc., New York, 1995.

4.5.1 Exercises

Identify whether the following arguments are valid or fallacious. Justify your answers.

4.88 *Premises:* (a) Every programming language that uses garbage collection is slow; and (b) C does not use garbage collection.
Conclusion: Therefore, C is slow.

4.89 *Premises:* (a) If a piece of software is written well, then it was built with its user in mind; and (b) The Firefox web browser is a piece of software that was written with its user in mind.
Conclusion: Therefore, the Firefox web browser is written well.

4.90 *Premises:* (a) If a processor overheats while multiplying, then it overheats while computing square roots; and (b) The xMax processor does not overheat while computing square roots.
Conclusion: Therefore, the xMax processor does not overheat while multiplying.

4.91 *Premises:* (a) Every data structure is either slow at insertions or lookups; and (b) The data structure called the *Hackmatack tree* is slow at insertions.
Conclusion: Therefore, the Hackmatack tree is slow at lookups.

4.92 *Premises:* (a) Every web server has an IP address; and (b) `www.cia.gov` is a web server.
Conclusion: Therefore, `www.cia.gov` has an IP address.

4.93 *Premises:* (a) If a computer system is hacked, then there was user error or the system had a design flaw; and (b) A computer at NASA was hacked; and (c) That computer did not have a design flaw.
Conclusion: Therefore, there was user error.

In the next several problems, you will be presented with a false claim and a bogus proof of that false claim. For each, you'll be asked to (a) identify the precise error in the proof, and (b) give a counterexample to the claim. (Note that saying why the claim is false does not address (a) in the slightest—it would be possible to give a bogus proof a true claim!)

False Claim #1: *Let n be a positive integer and let $p, q \in \mathbb{Z}^{\geq 2}$, where p and q are prime. If n is evenly divisible by both p and q, then n is also evenly divisible by pq.* (FC-1)

Bogus proof of (FC-1). Because $p \mid n$, there exists a positive integer k such that $n = pk$. Thus, by assumption, we know that $q \mid pk$. Because p and q are both prime, we know that p does not evenly divide q, and thus the only way that $q \mid pk$ can hold is if $q \mid k$. Hence $k = q\ell$ for some positive integer ℓ, and thus $n = pk = pq\ell$. Therefore $pq \mid n$. □

4.94 State precisely what's wrong with the proof of (FC-1).
4.95 Give a counterexample to (FC-1).

False Claim #2: 721 *is prime.* (FC-2)

Bogus proof of (FC-2). In Example 4.8, we proved that $n! + 1$ is not evenly divisible by any k satisfying $2 \leq k \leq n$. Observe that $6! = 720$. Therefore, $721 = 6! + 1$ isn't evenly divisible by any integer between 2 and 720 inclusive, and therefore 721 is prime. □

4.96 State precisely what's wrong with the proof of (FC-2).
4.97 *Without using a calculator,* disprove (FC-2).
4.98 *Without using a calculator,* find an integer n such that $n! + 1$ *is* prime.

False Claim #3: $\sqrt{2}/4$ *and* $8/\sqrt{2}$ *are both rational.* (FC-3)

Bogus proof of (FC-3). In Example 4.12, we proved that if x and y are rational then xy is rational too. Here, let $x = \sqrt{2}/4$ and $y = 8/\sqrt{2}$. Then $xy = \frac{\sqrt{2}}{4} \cdot \frac{8}{\sqrt{2}} = \frac{8\sqrt{2}}{4\sqrt{2}} = 2$. So $xy = 2$ is rational, and x and y are too. □

4.99 State precisely what's wrong with the proof of (FC-3).
4.100 Prove that $8/\sqrt{2}$ isn't rational.

False Claim #4: *Let n be any integer. Then $12 \mid n$ if and only if $12 \mid n^2$.* (FC-4)

Bogus proof of (FC-4), similar to Example 4.19. We proceed by mutual implication.

- First, assume that $12 \mid n$. Then, by definition, there exists an integer k such that $n = 12k$. Therefore $n^2 = (12k)^2 = 12 \cdot (12k^2)$. Thus $12 \mid n^2$ too.
- Second, we must show the converse: if $12 \mid n^2$, then $12 \mid n$. We prove the contrapositive. Assume that $12 \not\mid n$. Then there exist integers k and $r \in \{1, \ldots, 11\}$ such that $n = 12k + r$. Therefore $n^2 = (12k + r)^2 = 144k^2 + 24kr + r^2 = 12(12k^2 + 2kr) + r^2$. Because $r < 12$, adding r^2 to a multiple of 12 does not result in another multiple of 12. Thus $12 \not\mid n^2$. □

4.101 State precisely what's wrong with the proof of (FC-4).
4.102 Disprove (FC-4).

False Claim #5: *$\sqrt{4}$ is irrational.* (FC-5)

Bogus proof of (FC-5). We'll follow the same outline as Example 4.21. Our proof is by contradiction.

Assume that $\sqrt{4}$ is rational. Therefore, there exist integers n and $d \neq 0$ such that $n/d = \sqrt{4}$, where n and d have no common divisors.

Squaring both sides yields that $n^2/d^2 = 4$, and therefore that $n^2 = 4d^2$. Because $4d^2$ is divisible by 4, we know that n^2 is divisible by 4. Therefore, by the same logic as in Example 4.19, we have that n is itself divisible by 4.

Because n is divisible by 4, there exists an integer k such that $n = 4k$, which implies that $n^2 = 16k^2$. Thus $n^2 = 16k^2$ and $n^2 = 4d^2$, so $d^2 = 4k^2$. Hence d^2 is divisible by four.

But now we have a contradiction: we assumed that n/d was in lowest terms, but we have now shown that n^2 and d^2 are both divisible by 4, and therefore both n and d must be even! Thus the original assumption was false, and $\sqrt{4}$ is irrational. □

4.103 State precisely what's wrong with the proof of (FC-5).

False Claim #6: $3 \leq 2$. (FC-6)

Bogus proof of (FC-6). Let x and y be arbitrary nonnegative numbers. Because $y \geq 0$ implies $-y \leq y$, we can add x to both sides of this inequality to get

$$x - y \leq x + y. \tag{1}$$

Similarly, adding $y - 3x$ to both sides of $-x \leq x$ yields

$$y - 4x \leq y - 2x. \tag{2}$$

Observe that whenever $a \leq b$ and $c \leq d$, we know that $ac \leq bd$. So we can combine (1) and (2) to get

$$(x - y)(y - 4x) \leq (x + y)(y - 2x). \tag{3}$$

Multiplying out and then combining like terms, we have

$$xy - 4x^2 - y^2 + 4xy \leq xy - 2x^2 + y^2 - 2xy, \text{ and} \tag{4}$$
$$6xy \leq 2x^2 + 2y^2. \tag{5}$$

This calculation was valid for any $x, y \geq 0$. For $x = y = \sqrt{1/2}$, we have $xy = x^2 = y^2 = (\sqrt{1/2})^2 = 1/2$. Plugging into (5), we have

$$(6/2) \leq (2/2) + (2/2). \tag{6}$$

In other words, we have $3 \leq 2$. □

4.104 State precisely what's wrong with the proof of (FC-6).

Computer vision *is the subfield of computer science devoted to developing algorithms that can "understand" images. For example, some security systems use facial recognition software to decide whether to grant access to a particular person. We desire to maximize the probability that the vision algorithm we choose gets the answer right—that is, grants access to the person if and only if that person is authorized to enter.*

Suppose that we have two algorithms, $\mathcal{A}$ and $\mathcal{B}$, that we have employed on two different cameras in a test run. Suppose that algorithm $\mathcal{A}$ is deployed on Camera I. It makes the correct decision on 75% of the CS majors at Camera I and 60% of philosophy majors at Camera I. (That is, when a CS major arrives at Camera I, algorithm $\mathcal{A}$ correctly decides whether to grant her access 75% of the time.) Algorithm $\mathcal{B}$, deployed at Camera II, makes the correct decision on 70% of CS majors and 50% of philosophy majors. The following claim seems obvious, because Algorithm $\mathcal{A}$ performed better for both philosophy majors and CS majors:

Claim: *Algorithm $\mathcal{A}$ is right a higher fraction of the time (overall, combining both majors) than Algorithm $\mathcal{B}$.*

But the claim is false, as you'll show!

4.105 The falsehood of this claim (for example, in the scenario illustrated by the next exercise) is called *Simpson's Paradox* because the behavior is so counterintuitive. State precisely where the following argument goes wrong:

> *Observe that Algorithm $\mathcal{A}$ had a better success probability with CS majors, and also had a better success probability with philosophy majors. Therefore Algorithm $\mathcal{A}$ was right a higher fraction of the time (in total, for both philosophy majors and CS majors) than Algorithm $\mathcal{B}$.*

4.106 Suppose that there were 100 CS majors and 100 philosophy majors who went by Camera I. Suppose that 1000 CS majors and 100 philosophy majors went by Camera II. Calculate the success rate for Algorithm $\mathcal{A}$ at Camera I, over all people. Do the same for Algorithm $\mathcal{B}$ at Camera II.

4.107 Here is an obviously false theorem, together with a (nonobviously) bogus proof. Identify precisely the flaw in the argument and explain where the proof fails.

False Theorem: $1 = 0$.

Proof. Consider the four shapes in Figure 4.34(a), and the two arrangements thereof in Figure 4.34(b). (See below.)

The area of the triangle in the first configuration is $13 \cdot 5/2 = 65/2$, as it forms a right triangle with height 5 and base 13. But the second configuration also forms a right triangle with height 5 and base 13 as well, and therefore it too has area 65/2. But the second configuration has one unfilled square in the triangle, and thus we have

$$\begin{aligned} 0 &= \frac{65}{2} - \frac{65}{2} \\ &= \text{area of the second bounding triangle} - \text{area of the first bounding triangle} \\ &= (1 + \text{area of four constituent shapes}) - (\text{area of four constituent shapes}) \\ &= 1. \end{aligned}$$

Thus $0 = 1$. □

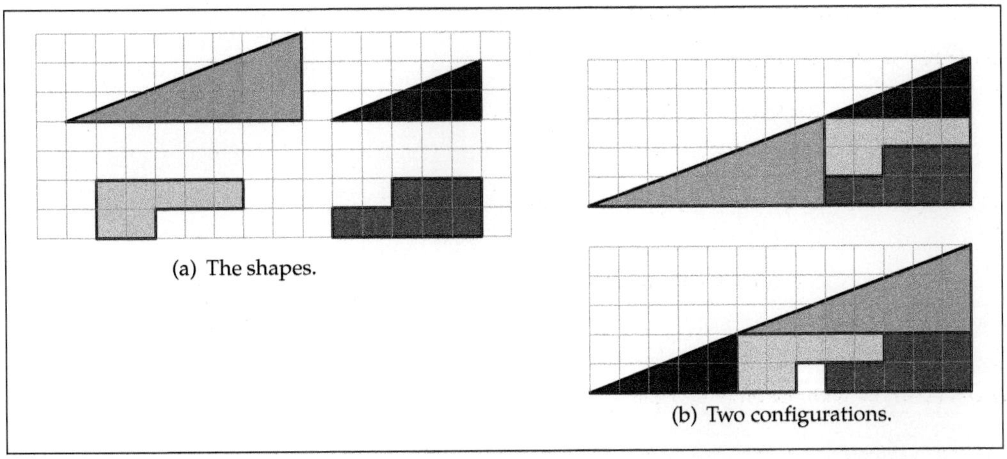

(a) The shapes.

(b) Two configurations.

Figure 4.34: Some shapes and their arrangements, for Exercise 4.107.

The following two statements are theorems from geometry that you may recall from high school:

- *the angles of a triangle sum to precisely* 180°.
- *if the three angles of triangle T_1 are precisely equal to the three angles of T_2, then T_1 and T_2 are* similar, *and their sides are in the same ratios. (That is, if the side lengths of T_1 are a, b, c and the side lengths of T_2 are x, y, z, then $a/x = b/y = c/z$.)*

These statements are theorems, but they're used in the following utterly bogus "proof" of the Pythagorean Theorem (actually one that was published, in 1896!).

4.108 State precisely what's wrong with the following purported proof of the Pythagorean Theorem.

Proof. Consider an arbitrary right triangle. Let the two legs and hypotenuse, respectively, have length a, b, and c, and let the angles between the legs and the hypotenuse be given by θ and $\phi = 90° - \theta$. (See Figure 4.35(a).) Draw a line perpendicular to the hypotenuse to the opposite vertex, dividing the interior of the triangle into two separate sections, which are shaded with different colors in Figure 4.35(b). Observe that the unlabeled angle within the smaller shaded interior triangle must be $\phi = 90° - \theta$, because the other angles of the smaller shaded interior triangle are (just like for the enclosing triangle) 90° and θ. Similarly, the unlabeled angle within the larger shaded interior triangle must be θ. Therefore we have three similar triangles, all with angles 90°, θ, and ϕ. Call the lengths of the previously unnamed sides x, y, and z as in Figure 4.35(c). Now we can assemble our known facts. By assumption,

$$a^2 = x^2 + y^2, \qquad b^2 = x^2 + z^2, \text{ and} \qquad (y+z)^2 = a^2 + b^2,$$

which we can combine to yield

$$(y+z)^2 = 2x^2 + y^2 + z^2. \tag{1}$$

Expanding $(y+z)^2 = y^2 + 2yz + z^2$ and subtracting common terms from both sides, we have

$$2yz = 2x^2, \tag{2}$$

which, dividing both sides by two, yields

$$yz = x^2. \tag{3}$$

But (3) is immediate: we know that

$$x/y = z/x \tag{4}$$

because the two shaded triangles are similar, and therefore the two triangles have the same ratio of the length of the hypotenuse to the length of the longer leg. Multiplying both sides of (4) by xy gives us $x^2 = yz$, as desired. □

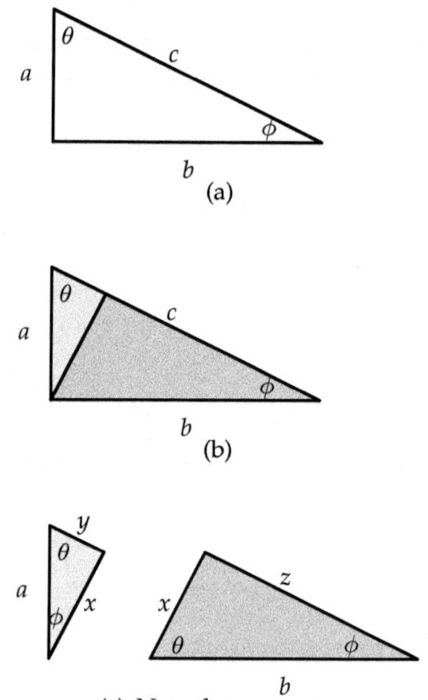

(c) Note that $c = y + z$.

Figure 4.35: Diagrams for Exercise 4.108.

4.6 *Chapter at a Glance*

Error-Correcting Codes

Although the main purpose of this section was to introduce proofs, here's a brief summary of the results about error-correcting and error-detecting codes, too.

A *code* is a set $C \subseteq \{0,1\}^n$, where $|C| = 2^k$ for some integer $1 \leq k \leq n$. A *message* is an element of $\{0,1\}^k$; the elements of C are called *codewords*. Consider any codeword $c \in C$ and for any sequence of up to ℓ errors applied to c to produce c'. The code C can *detect* $\ell \geq 0$ errors if we can always correctly report "error" or "no error," and can *correct* ℓ errors if we can always correctly identify that c was the original codeword.

The *Hamming distance* between strings $x, y \in \{0,1\}^n$, denoted $\Delta(x,y)$, is the number of positions i in which $x_i \neq y_i$. The *minimum distance* of a code C is the smallest Hamming distance between two distinct codewords of C. The *rate* of a code with k-bit messages and n-bit codewords is k/n. If the minimum distance of a code C is $2t + 1$ for an integer t, then C can detect $2t$ errors and correct t errors.

The REPETITION$_\ell$ *code* creates codewords via the ℓ-fold repetition of the message. This code has rate $1/\ell$ and minimum distance ℓ. The *Hamming code* creates 7-bit codewords from 4-bit messages by adding three different parity bits to the message. This code has rate $4/7$ and minimum distance 3. Any code with messages of length 4 and minimum distance 3 has codewords of length ≥ 7. (Thus the Hamming code has the best possible rate among all such codes.) We can prove this result via a "sphere-packing" argument and a proof by contradiction.

Proofs and Proof Techniques

A *proof* of a claim φ is a convincing argument that φ is true. (A proof should be written with its audience in mind.) A variety of useful proof techniques can be employed to prove a given claim φ:

- *direct proof:* we prove φ by repeatedly inferring new facts from known facts to eventually conclude φ. (Sometimes we divide a proof into multiple cases, or "assume the antecedent," where we prove $p \Rightarrow q$ by assuming p and deriving q.)

You may also prove φ by proving a claim logically equivalent to φ:

- *proof by contrapositive:* to prove $p \Rightarrow q$, we instead prove $\neg q \Rightarrow \neg p$.

- *proof by contradiction* (or *reductio ad absurdum*): to prove φ, we instead prove that $\neg\varphi \Rightarrow$ False—that is, we prove that $\neg\varphi$ leads to an absurdity.

We say that $y \in S$ with $\neg P(y)$ is a *counterexample* to the claim $\forall x \in S : P(x)$. A *proof by construction* of the claim $\exists x \in S : P(x)$ proceeds by constructing a particular $y \in S$ and proving that $P(y)$. A *nonconstructive proof* establishes $\exists x \in S : P(x)$ without giving an explicit $y \in S$ for which $P(x)$—for example, by proving $\exists x \in S : P(x)$ by contradiction.

The process of developing a proof requires persistence, open-mindedness, and creativity. Here's a helpful three-step plan to use when developing a new proof: (1)

understand what you're trying to do (checking definitions and small examples); (2) do it (by trying the proof techniques catalogued here, and thinking about analogies from similar problems that you've solved previously); and (3) think about what you've done (reflecting on and trying to improve your proof). Remember that writing a proof is a form of writing! Be kind to your reader.

Some Examples of Proofs

We can use these proof techniques to establish a wide variety of facts—about arithmetic, propositional logic, geometry, prime numbers, and computability. For more extensive examples, see Section 4.4. We'll highlight one result: there are problems that we can formally define, but that cannot be solved by any computer program; these problems (including the *Halting Problem*) are called *uncomputable.*

Common Errors in Proofs

A *valid* inference is one whose conclusion is always true as long as the facts that it relies on were true. An *invalid* inference is one in which the conclusion can be false *even if the premises are all true.* An invalid, or fallacious, argument can also be called a *logical fallacy* or just a *fallacy.* In a correct proof, of course, every step is valid.

Perhaps the most common error in a proof is simply asserting that a fact φ follows from previously established facts, when actually φ is not implied by those facts. Other common types of fallacious reasoning are structural errors that involve purporting to prove a statement φ, but instead proving a statement that is not logically equivalent to φ. (For example, the *fallacy of proving true*: a "proof" of φ that assumes φ and proves True. But $\varphi \Rightarrow$ True is true regardless of the truth of φ, so this purported proof proves nothing.) Be vigilant; do not let anyone—yourself or others!—get away with fallacious reasoning.

Key Terms and Results

Key Terms

ERROR-CORRECTING CODES

- Hamming distance
- code, message, codeword
- error-detecting/correcting code
- minimum distance, rate
- repetition code
- Hamming code

PROOFS AND PROOF TECHNIQUES

- proof
- proof techniques:
  - direct proof
  - proof by contrapositive
  - proof by contradiction
- counterexample
- constructive/nonconstructive proof

SOME EXAMPLES OF PROOFS

- conjunctive/disjunctive normal form
- uncomputability
- the Halting Problem

VALID AND FALLACIOUS ARGUMENTS

- valid argument
- fallacious/invalid argument; fallacy
- fallacy: proving true
- fallacy: affirming the consequent
- fallacy: denying the hypothesis
- fallacy: false dichotomy
- fallacy: begging the question

Key Results

ERROR-CORRECTING CODES

1. If the minimum distance of a code $\mathcal{C}$ is $2t + 1$ for an integer $t \geq 0$, then $\mathcal{C}$ can detect $2t$ errors and correct t errors.
2. For 4-bit messages and minimum distance 3, there exist codes with rate $\frac{1}{3}$ (such as the REPETITION3 code) and with rate $\frac{4}{7}$ (such as the Hamming code), but not with rate better than $\frac{4}{7}$.

PROOFS AND PROOF TECHNIQUES

1. You can prove a claim φ with a direct proof, or by instead proving a different claim that is logically equivalent to φ. Examples include proofs by contrapositive and proofs by contradiction.
2. A useful three-step process for developing proofs is: (1) understand what you're trying to do; (2) do it; and (3) think about what you've done. All three steps are important, and doing each will help with the other steps.
3. Writing a proof is a form of writing.

SOME EXAMPLES OF PROOFS

1. All logical propositions are equivalent to propositions in conjunctive/disjunctive normal form, or using only *nand*.
2. There are infinitely many prime numbers.
3. There are problems that can be specified completely formally that are uncomputable (that is, cannot be solved by any computer program). The Halting Problem is one example.

VALID AND FALLACIOUS ARGUMENTS

1. There are many common mistakes in proofs that are centered on several types of fallacious reasoning. These fallacies are essentially all the result of purporting to prove a statement φ by instead proving a statement ψ, where ψ fails to be logically equivalent to φ.

5
Mathematical Induction

In which our heroes wistfully dream about having dreams about dreaming about a very simple and pleasant world in which no one sleeps at all.

5.1 *Why You Might Care*

> Each problem that I solved became a rule which served afterwards to solve other problems.
>
> René Descartes (1596–1650)

Recursion is a powerful technique in computer science. If we can express a solution to problem X in terms of solutions to smaller instances of the same problem X—and we can solve X directly for the "smallest" inputs—then we can solve X for all inputs. There are many examples. We can sort an n-element array A by sorting the left half of A and the right half of A and merging the results together; 1-element arrays are trivially sorted. (That's *merge sort.*) We can build an efficient data structure for storing and searching a set of keys by selecting one of those keys k, and building two such data structures for keys $< k$ and for keys $> k$; to search for a key x, we compare x to k and search for x in the appropriate substructure. And a trivial empty data structure can store an empty set of keys. (That's a *binary search tree.*) And many other things are best understood recursively: factorials, the Fibonacci numbers, fractals (see Figure 5.1), and finding the median element of an unsorted array, for example.

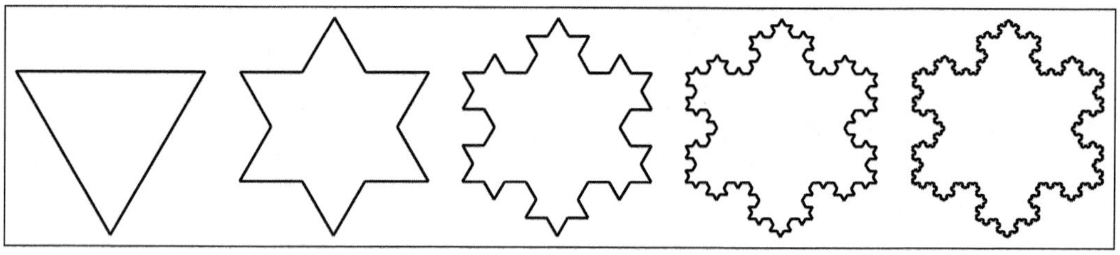

Figure 5.1: The Von Koch Snowflake fractal, shown at levels $\{0, 1, 2, 3, 4\}$. A level-ℓ snowflake consists of three level-ℓ lines. A level-0 line is ——; a level-ℓ line consists of four level-$(\ell - 1)$ lines arranged in the shape _/_.

Mathematical induction is a technique for proofs that is directly analogous to recursion: to prove that $P(n)$ holds for all nonnegative integers n, we prove that $P(0)$ is true, and we prove that for an arbitrary $n \geq 1$, if $P(n-1)$ is true, then $P(n)$ is true too. The proof of $P(0)$ is called the *base case*, and the proof that $P(n-1) \Rightarrow P(n)$ is called the *inductive case.* In the same way that a recursive solution to a problem relies on solutions to a smaller instance of the same problem, an inductive proof of a claim relies on proofs of a smaller instance of the same claim.

A full understanding of recursion depends on a thorough understanding of mathematical induction. And many other applications of mathematical induction will arise throughout the book: analyzing the running time of algorithms, counting the number of bitstrings that have a particular form, and many others.

In this chapter, we will introduce mathematical induction, including a few variations and extensions of this proof technique. We will start with the "vanilla" form of proofs by mathematical induction (Section 5.2). We will then introduce *strong induction* (Section 5.3), a form of proof by induction in which the proof of $P(n)$ in the inductive case may rely on the truth of all of $P(0), P(1), \ldots,$ and $P(n-1)$ instead of just on $P(n-1)$. Finally, we will turn to *structural induction* (Section 5.4), a form of inductive proof that operates directly on recursively defined structures like linked lists, binary trees, or well-formed formulas of propositional logic.

5.2 *Proofs by Mathematical Induction*

> So if you find nothing in the corridors open the doors, if you find nothing behind these doors there are more floors, and if you find nothing up there, don't worry, just leap up another flight of stairs. As long as you don't stop climbing, the stairs won't end, under your climbing feet they will go on growing upwards.
>
> Franz Kafka (1883–1924)
> *Fürsprecher (Advocates)* (c. 1922)

5.2.1 *An Overview of Proofs by Mathematical Induction*

The *principle of mathematical induction* says the following: to prove that a statement $P(n)$ is true for all nonnegative integers n, we can prove that P "starts being true" (the *base case*) and that P "never stops being true" (the *inductive case*). Formally, a proof by mathematical induction proceeds as follows:

Definition 5.1 (Proof by mathematical induction)
Suppose that we want to prove that $P(n)$ holds for all $n \in \mathbb{Z}^{\geq 0}$. To give a proof by mathematical induction *of $\forall n \in \mathbb{Z}^{\geq 0} : P(n)$, we prove the following:*

1. *the* base case: *prove $P(0)$.*
2. *the* inductive case: *for every $n \geq 1$, prove $P(n-1) \Rightarrow P(n)$.*

When we've proven both the base case and the inductive case as in Definition 5.1, we have established that $P(n)$ holds for all $n \in \mathbb{Z}^{\geq 0}$. Here's an example to illustrate how the base case and inductive case combine to establish this fact:

Example 5.1 (Proving $P(5)$ from a base case and inductive case)
Problem: Suppose we've proven both the base case ($P(0)$) and the inductive case ($P(n-1) \Rightarrow P(n)$, for any $n \geq 1$) as in Definition 5.1. Why do these two facts establish that $P(n)$ holds for all $n \in \mathbb{Z}^{\geq 0}$? For example, why do they establish $P(5)$?

Solution: Here is a proof of $P(5)$, using the base case once and the inductive case five times. (At each stage we make use of *modus ponens*—which, as a reminder, states that from $p \Rightarrow q$ and p, we can conclude q.)

We know	$P(0)$	*base case*	(5.1)
and we know	$P(0) \Rightarrow P(1)$	*inductive case, with $n = 1$*	(5.2)
and thus we can conclude	$P(1)$.	*(5.1), (5.2), and modus ponens*	(5.3)
We know	$P(1) \Rightarrow P(2)$	*inductive case, with $n = 2$*	(5.4)
and thus we can conclude	$P(2)$.	*(5.3), (5.4), and modus ponens*	(5.5)
We know	$P(2) \Rightarrow P(3)$	*inductive case, with $n = 3$*	(5.6)
and thus we can conclude	$P(3)$.	*(5.5), (5.6), and modus ponens*	(5.7)

We know	$P(3) \Rightarrow P(4)$	*inductive case, with n = 4*	(5.8)
and thus we can conclude	$P(4)$.	*(5.7), (5.8), and modus ponens*	(5.9)
We know	$P(4) \Rightarrow P(5)$	*inductive case, with n = 5*	(5.10)
and thus we can conclude	$P(5)$.	*(5.9), (5.10), and modus ponens*	(5.11)

This sequence of inferences established that $P(5)$ is true. We can use the same technique to prove that $P(n)$ holds for an arbitrary integer $n \geq 0$, using the base case once and the inductive case n times.

The principle of mathematical induction is as simple as in Example 5.1—we apply the base case to get started, and then repeatedly apply the inductive case to conclude $P(n)$ for any larger n—but there are several analogies that can help to make proofs by mathematical induction more intuitive; see Figure 5.2.

Dominoes falling: We have an infinitely long line of dominoes, numbered $0, 1, 2, \ldots, n, \ldots$. To convince someone that the nth domino falls over, you can convince them that

- the 0th domino falls over, and
- whenever one domino falls over, the next domino falls over too.

(One domino falls, and they keep on falling. Thus, for any $n \geq 0$, the nth domino falls.)

Climbing a ladder: We have a ladder with rungs numbered $0, 1, 2, \ldots, n, \ldots$. To convince someone that a climber climbing the ladder reaches the nth rung, you can convince them that

- the climber steps onto rung #0.
- if the climber steps onto one rung, then she also steps onto the next rung.

(The climber starts to climb, and the climber never stops climbing. Thus, for any $n \geq 0$, the climber reaches the nth rung.)

Whispering down the alley: We have an infinitely long line of people, with the people numbered $0, 1, 2, \ldots, n, \ldots$. To argue that everyone in the line learns a secret, we can argue that

- person #0 learns the secret.
- if person #n learns the secret, then she tells person #$(n+1)$ the secret.

(The person at the front of the line learns the secret, and everyone who learns it tells the secret to the next person in line. Thus, for any $n \geq 0$, the nth person learns the secret.)

Falling into the depths of despair: Consider the Pit of Infinite Despair, which is filled with nothing but despair and goes infinitely far down beneath the surface of the earth. (The Pit does not respect physics.) Suppose that:

- the Evil Villain is pushed into the pit (that is, She is in the Pit zero meters below the surface).
- if someone is in the Pit at a depth of n meters beneath the surface, then She falls to depth $n+1$ meters beneath the surface.

(The Villain starts to fall, and if the Villain has fallen to a certain depth then She falls another meter further. Thus, for any $n \geq 0$, the Evil Villain eventually reaches depth n in the Pit.)

Figure 5.2: Some analogies to make mathematical induction more intuitive.

Taking it further: "Mathematical induction" is somewhat unfortunately named because its name collides with a distinction made by philosophers between two types of reasoning. *Deductive* reasoning is the use of logic (particularly rules of inference) to reach conclusions—what computer scientists would call a *proof.* A proof by mathematical induction is an example of deductive reasoning. For a philosopher, though, *inductive reasoning* is the type of reasoning that draws conclusions from empirical observations. If you've seen a few hundred ravens in your life, and every one that you've seen is black, then you might conclude *All ravens are black.* Of course, it might turn out that your conclusion is false, because you haven't happened upon any of the albino ravens that exist in the world; hence what philosophers call inductive reasoning leads to conclusions that may turn out to be false.

A FIRST EXAMPLE: SUMMING POWERS OF TWO

Let's use mathematical induction to prove a simple arithmetic property:

Theorem 5.1 (A formula for the sum of powers of two)
For any nonnegative integer n, we have

$$\sum_{i=0}^{n} 2^i = 2^{n+1} - 1.$$

As a plausibility check, let's test the given formula for some small values of n:

$$\begin{array}{lll} n = 1: & 2^0 + 2^1 = 1 + 2 = 3 & 2^2 - 1 = 3 \\ n = 2: & 2^0 + 2^1 + 2^2 = 1 + 2 + 4 = 7 & 2^3 - 1 = 7 \\ n = 3: & 2^0 + 2^1 + 2^2 + 2^3 = 1 + 2 + 4 + 8 = 15 & 2^4 - 1 = 15 \end{array}$$

Problem-solving tip: Do this kind of plausibility check, and test out a claim for small values of n before you try to prove it. Often the process of testing small examples either reveals a misunderstanding of the claim *or* helps you see why the claim is true in general.

These small examples all check out, so it's reasonable to try to prove the claim. Here is our first example of a proof by induction:

Example 5.2 (A proof of Theorem 5.1)
Let $P(n)$ denote the property

$$\sum_{i=0}^{n} 2^i = 2^{n+1} - 1.$$

We'll prove that $\forall n \in \mathbb{Z}^{\geq 0} : P(n)$ by induction on n.

base case ($n = 0$): We must prove $P(0)$. That is, we must prove $\sum_{i=0}^{0} 2^i = 2^{0+1} - 1$. But this fact is easy to prove, because both sides are equal to 1: $\sum_{i=0}^{0} 2^i = 2^0 = 1$, and $2^{0+1} - 1 = 2 - 1 = 1$.

inductive case ($n \geq 1$): We must prove that $P(n-1) \Rightarrow P(n)$, for an arbitrary integer $n \geq 1$. We prove this implication by assuming the antecedent—namely, we assume $P(n-1)$ and prove $P(n)$. The assumption $P(n-1)$ is

$$\sum_{i=0}^{n-1} 2^i = 2^{(n-1)+1} - 1. \qquad (*)$$

We can now prove $P(n)$—under the assumption $(*)$—by showing that the left-hand and right-hand sides of $P(n)$ are equal:

$$\begin{aligned} \sum_{i=0}^{n} 2^i &= \left[\sum_{i=0}^{n-1} 2^i\right] + 2^n && \text{by the definition of summations} \\ &= \left[2^{(n-1)+1} - 1\right] + 2^n && \text{by } (*)\text{, a.k.a. by the assumption that } P(n-1) \\ &= 2^n - 1 + 2^n && \text{by algebraic manipulation} \\ &= 2 \cdot 2^n - 1 \\ &= 2^{n+1} - 1. \end{aligned}$$

We've thus shown that $\sum_{i=0}^{n} 2^i = 2^{n+1} - 1$—in other words, we've proven $P(n)$.

We've proven the base case $P(0)$ and the inductive case $P(n-1) \Rightarrow P(n)$, so by the principle of mathematical induction we have shown that $P(n)$ holds for all $n \in \mathbb{Z}^{\geq 0}$.

Taking it further: In case the inductive proof doesn't feel 100% natural, here's another way to make the result from Example 5.2 intuitive: think about binary representations of numbers. Written in binary, the number $\sum_{i=0}^{n} 2^i$ will look like $11 \cdots 111$, with $n+1$ ones. What happens when we add 1 to, say, 11111111 (= 255)? It's a colossal sequence of carrying (as $1+1=0$, carrying the 1 to the next place):

$$\begin{array}{rcccccccc} & {\scriptstyle 1} & {\scriptstyle 1} & {\scriptstyle 1} & {\scriptstyle 1} & {\scriptstyle 1} & {\scriptstyle 1} & {\scriptstyle 1} & \\ & 1 & 1 & 1 & 1 & 1 & 1 & 1 & 1 \\ + & 0 & 0 & 0 & 0 & 0 & 0 & 0 & 1 \\ \hline 1 & 0 & 0 & 0 & 0 & 0 & 0 & 0 & 0. \end{array}$$

In other words, $2^{n+1} - 1$ is written in binary as a sequence of $n+1$ ones—that is, $2^{n+1} - 1 = \sum_{i=0}^{n} 2^i$.

Example 5.2 follows the standard outline of a proof by mathematical induction. We will *always* prove the inductive case $P(n-1) \Rightarrow P(n)$ by assuming the antecedent $P(n-1)$ and proving $P(n)$. The assumed antecedent $P(n-1)$ in the inductive case of the proof is called the *inductive hypothesis*.

You may see "inductive hypothesis" abbreviated as *IH*.

A SECOND EXAMPLE, AND A TEMPLATE FOR PROOFS BY INDUCTION

Here's another proof by induction, with the parts of the proof carefully labeled:

Example 5.3 (Summing powers of −1)

Claim: For any integer $n \geq 0$, we have that $\sum_{i=0}^{n} (-1)^i = \begin{cases} 1 & \text{if } n \text{ is even} \\ 0 & \text{if } n \text{ is odd.} \end{cases}$

Proof.

Step #1: *Clearly state the claim to be proven. Clearly state that the proof will be by induction, and clearly state the variable upon which induction will be performed.*

Let $P(n)$ denote the property

$$\sum_{i=0}^{n} (-1)^i = \begin{cases} 1 & \text{if } n \text{ is even} \\ 0 & \text{if } n \text{ is odd.} \end{cases}$$

We'll prove that $\forall n \in \mathbb{Z}^{\geq 0} : P(n)$ by induction on n.

Warning! $P(n)$ denotes a *proposition*—that is, $P(n)$ is either true or false. (We're proving that, in fact, it's true for every n.) Despite its apparent temptation to people new to inductive proofs, it is nonsensical to treat $P(n)$ as a number.

Step #2: *State and prove the base case.*

base case ($n = 0$): We must prove $P(0)$. But $\sum_{i=0}^{0} (-1)^i = (-1)^0 = 1$, and 0 is even.

Step #3: *State and prove the inductive case. Within the statement and proof of the inductive case . . .*

. . . **Step #3a:** *state the inductive hypothesis.*

inductive case ($n \geq 1$): We assume the inductive hypothesis $P(n-1)$, namely

$$\sum_{i=0}^{n-1} (-1)^i = \begin{cases} 1 & \text{if } n-1 \text{ is even} \\ 0 & \text{if } n-1 \text{ is odd.} \end{cases}$$

> ... **Step #3b:** *state what we need to prove.*

We must prove $P(n)$.

> ... **Step #3c:** *prove it,* making use of the inductive hypothesis *and stating where it was used.*

$$\sum_{i=0}^{n}(-1)^i = \left[\sum_{i=0}^{n-1}(-1)^i\right] + (-1)^n \qquad \textit{definition of summations}$$

$$= \begin{cases} 1+(-1)^n & \text{if } n-1 \text{ is even} \\ 0+(-1)^n & \text{if } n-1 \text{ is odd.} \end{cases} \qquad \textit{inductive hypothesis}$$

$$= \begin{cases} 1+(-1)^n & \text{if } n \text{ is odd} \\ 0+(-1)^n & \text{if } n \text{ is even.} \end{cases} \qquad n \text{ is odd} \Leftrightarrow n-1 \text{ is even}$$

$$= \begin{cases} 1+-1 & \text{if } n \text{ is odd} \\ 0+1 & \text{if } n \text{ is even.} \end{cases} \qquad (-1)^n = \pm 1, \text{ depending on whether } n \text{ is even; see Exercise 5.3.}$$

$$= \begin{cases} 0 & \text{if } n \text{ is odd} \\ 1 & \text{if } n \text{ is even.} \end{cases}$$

Thus we have proven $P(n)$, and the theorem follows. □

Writing tip: In the inductive case of a proof of an equality—like Example 5.3—start from the left-hand side of the equality and manipulate it until you derive the right-hand side of the equality *exactly*. If you work from both sides simultaneously, you're at risk of the fallacy of proving true—or at least the appearance of that fallacy!

We can treat the labeled pieces of Example 5.3 as a checklist for writing proofs by induction. You should ensure that when you write an inductive proof, you include each of these steps. These steps are summarized in Figure 5.3.

> Checklist for a proof by mathematical induction:
>
> 1. A clear statement of the claim to be proven—that is, a clear definition of the property $P(n)$ that will be proven true for all $n \geq 0$—and a statement that the proof is by induction, including specifically identifying the variable n upon which induction is being performed. (Some claims involve multiple variables, and it can be confusing if you aren't clear about which is the variable upon which you are performing induction.)
> 2. A statement and proof of the base case—that is, a proof of $P(0)$.
> 3. A statement and proof of the inductive case—that is, a proof of $P(n-1) \Rightarrow P(n)$, for a generic value of $n \geq 1$. The proof of the inductive case should include all of the following:
>
>   (a) a statement of the inductive hypothesis $P(n-1)$.
>   (b) a statement of the claim $P(n)$ that needs to be proven.
>   (c) a proof of $P(n)$, which at some point makes use of the assumed inductive hypothesis.

Figure 5.3: A checklist of the steps required for a proof by mathematical induction.

THE SUM OF THE FIRST n INTEGERS

We'll do another simple example of an inductive proof of an arithmetic property, by showing that the sum of the integers between 0 and n is $\frac{n(n+1)}{2}$. (For example, for $n = 4$ we have $0+1+2+3+4 = 10 = \frac{4(4+1)}{2}$.) Here's a proof:

Example 5.4 (Sum of the first n integers)
Problem: Show that $0+1+\cdots+n$ is $\frac{n(n+1)}{2}$, for any integer $n \geq 0$.

Solution: First, we must phrase this problem in terms of a property $P(n)$ that we'll prove true for every $n \geq 0$. For a particular integer n, let $P(n)$ denote the claim that

$$\sum_{i=0}^{n} i = \frac{n(n+1)}{2}.$$

We will prove that $P(n)$ holds for all integers $n \geq 0$ by induction on n.

base case ($n = 0$): Note that $\sum_{i=1}^{0} i = 0$ and $\frac{0(0+1)}{2} = 0$ too. Thus $P(0)$ follows.

inductive case ($n \geq 1$): Assume the inductive hypothesis $P(n-1)$, namely

$$\sum_{i=0}^{n-1} i = \frac{(n-1)((n-1)+1)}{2}.$$

We must prove $P(n)$—that is, we must prove that $\sum_{i=0}^{n} i = \frac{n(n+1)}{2}$. Here is the proof:

$$\begin{aligned}
\sum_{i=0}^{n} i &= \left[\sum_{i=0}^{n-1} i\right] + n && \textit{definition of summations} \\
&= \frac{(n-1)((n-1)+1)}{2} + n && \textit{inductive hypothesis} \\
&= \frac{(n-1)n + 2n}{2} && \textit{putting terms over common denominator} \\
&= \frac{n(n-1+2)}{2} && \textit{factoring} \\
&= \frac{n(n+1)}{2}.
\end{aligned}$$

Thus we've shown $P(n)$ assuming $P(n-1)$, which completes the proof.

Problem-solving tip: Your first task in giving a proof by induction is to identify the property $P(n)$ that you'll prove true for every integer $n \geq 0$. Sometimes the property is given to you more or less directly and sometimes you'll have to formulate it yourself, but in any case you need to identify the precise property you're going to prove before you can prove it!

Taking it further: While the summation that we analyzed in Example 5.4 may seem like a purely arithmetic example, it also has direct applications in CS—particularly in the *analysis of algorithms*. Chapter 6 is devoted to this topic, and there's much more there, but here's a brief preview.

A basic step in analyzing an algorithm is counting how many steps that algorithm takes, for an input of arbitrary size. One particular example is Insertion Sort, which sorts an n-element array by repeatedly ensuring that the first k elements of the array are in sorted order (by swapping the kth element backward until it's in position). The total number of swaps that are done in the kth iteration can be as high as $k-1$—so the total number of swaps can be as high as $\sum_{k=1}^{n} k - 1 = \sum_{i=0}^{n-1} i$. Thus Example 5.4 tells us that Insertion Sort can require as many as $n(n-1)/2$ swaps.

Generating a conjecture: segments in a fractal

In the inductive proofs that we've seen thus far, we were given a problem statement that described exactly what property we needed to prove. Solving these problems "just" requires proving the base case and the inductive case—which may or may not be *easy*, but at least we know what we're trying to prove! In other problems, though, you may also have to first figure out what you're going to prove, and *then* prove it. Obviously this task is generally harder. Here's one example of such a proof, about the Von Koch snowflake fractal from Figure 5.1:

Example 5.5 (Vertices in a Von Koch Line)

Problem: A Von Koch line of level 0 is a straight line segment; a Von Koch line of level $\ell \geq 1$ consists of four Von Koch lines of level $(\ell - 1)$, arranged in the shape ⌃. (See Figure 5.4.) Conjecture a formula for the number of *vertices* (that is, the number of segment endpoints) in a Von Koch line of level ℓ. Prove your formula by induction.

Figure 5.4: Von Koch lines of level $0, 1, \ldots, 5$. (A Von Koch snowflake consists of three Von Koch lines, all of the same level, arranged in a triangle; see Figure 5.1.)

Solution: Our first task is to formulate a conjecture for the number of vertices in a Von Koch line of level ℓ. Let's start with a few small examples, based on Figure 5.4:

- a level-0 line has 2 endpoints (and 1 segment).
- a level-1 line has 5 endpoints (and 4 segments): the two at the far left and far right, plus the three in the start, middle, and end of the "bump" in the center.
- a level-2 line—after some tedious counting in the picture in Figure 5.4—turns out to have 17 endpoints (and 16 segments).

There are a few ways to think about this pattern. Here's one that turns out to be helpful: a level-ℓ line contains 4 lines of level $(\ell - 1)$, so it contains 16 lines of level $(\ell - 2)$. And thus, expanding it all the way out, the level-ℓ line contains 4^ℓ lines of level 0. The number of endpoints that we observe is $2 = 4^0 + 1$, then $5 = 4^1 + 1$, then $17 = 4^2 + 1$. (Why the "+1?" Each segment starts where the previous segment ended—so there is one more endpoint than segment, because of the last segment's second endpoint.)

So it looks like there are $4^\ell + 1$ endpoints in a Von Koch line of level ℓ. Let's turn this observation into a formal claim, with an inductive proof:

Claim: For any $\ell \geq 0$, a Von Koch line of level ℓ has $4^\ell + 1$ endpoints.

Proof. Let $P(\ell)$ denote the claim that a Von Koch line of level ℓ has $4^\ell + 1$ endpoints. We'll prove that $P(\ell)$ holds for all integers $\ell \geq 0$ by induction on ℓ.

base case ($\ell = 0$): We must prove $P(0)$. By definition, a Von Koch line of level 0 is a single line segment, which has 2 endpoints. Indeed, $4^0 + 1 = 1 + 1 = 2$.

inductive case ($\ell \geq 1$): We assume the inductive hypothesis, namely $P(\ell - 1)$, and we must prove $P(\ell)$. The key observation is that a Von Koch line of level ℓ consists of four Von Koch lines of level $(\ell - 1)$—and the last endpoint of line #1 is identical to the first endpoint of line #2; the last endpoint of #2 is the first of #3, and the last endpoint of #3 is the first of #4. Therefore there are three endpoints that are shared among the four lines of level $(\ell - 1)$. Thus:

the number of endpoints in a Von Koch line of level ℓ

$$= 4 \cdot \Big[\text{the number of endpoints in a Von Koch line of level } (\ell - 1)\Big] - 3$$

by the definition of a Von Koch line, and by the above discussion

$$= 4 \cdot \left[4^{\ell-1} + 1\right] - 3 \qquad \textit{by the inductive hypothesis}$$

$$= 4^\ell + 4 - 3 \qquad \textit{multiplying through}$$

$$= 4^\ell + 1. \qquad \textit{algebra}$$

Thus $P(\ell)$ follows, completing the proof. □

A NOTE AND TWO VARIATIONS ON THE INDUCTIVE TEMPLATE

The basic idea of induction is simple: the reason that $P(n)$ holds is that $P(n-1)$ held, and the reason that $P(n-1)$ held is that $P(n-2)$ held—and so forth, until eventually the proof finally rests on $P(0)$, the base case. A proof by induction can sometimes look superficially like it's circular reasoning—that we're assuming precisely the thing that we're trying to prove. *But it's not!* In the inductive case, we're assuming $P(n-1)$ and proving $P(n)$—we are *not* assuming $P(n)$ and proving $P(n)$.

Taking it further: The superficial appearance of circularity in a proof by induction is equivalent to the superficial appearance that a recursive function in a program will run forever. (A recursive function f *will* run forever if calling f on n results in f calling itself on n again! That's the same circularity that would happen if we assumed $P(n)$ and proved $P(n)$.) The correspondence between these aspects of induction and recursion should be no surprise; induction and recursion are essentially the same thing. In fact, it's not too hard to write a recursive function that "implements" an inductive proof by outputting a step-by-step argument establishing $P(n)$ for an arbitrary n, as in Example 5.1.

Warning! If you do not use the inductive hypothesis $P(n-1)$ in the proof of $P(n)$, then something is wrong—or, at least, your proof is not actually a proof by induction!

Our proofs so far have shown $\forall n \in \mathbb{Z}^{\geq 0} : P(n)$ by proving $P(0)$ as a base case. If we instead want to prove $\forall n \in \mathbb{Z}^{\geq k} : P(n)$ for some integer k, we can prove $P(k)$ as the base case, and then prove the inductive case $P(n-1) \Rightarrow P(n)$ for all $n \geq k+1$.

Another variation in writing inductive proofs relates to the statement of the inductive case. We've proven $P(0)$ and $P(n-1) \Rightarrow P(n)$ for arbitrary $n \geq 1$. Some writers prefer to prove $P(0)$ and $P(n) \Rightarrow P(n+1)$ for arbitrary $n \geq 0$. The difference is merely a reindexing, not a substantive difference: it's just a matter of whether one thinks of induction as "the nth domino falls because the $(n-1)$st domino fell into it" or as "the nth domino falls and therefore knocks over the $(n+1)$st domino."

In the remainder of this section, we'll give some more examples of proofs by mathematical induction, following the template of Figure 5.3. While the examples that we've used so far have almost all related to summations, the same style of inductive proof can be used for a wide variety of claims. We'll encounter many inductive proofs throughout the book, and you'll find inductive proofs ubiquitous throughout computer science. We'll start with some more summation-based proofs, and then move on to inductive proofs of some other types of statements.

5.2.2 *Some Numerical Examples: Geometric, Arithmetic, and Harmonic Series*

We'll now introduce three types of summations that arise frequently in computer science: *geometric* sequences $(1, 2, 4, 8, 16, \ldots)$; *arithmetic* sequences $(2, 4, 6, 8, 10, \ldots)$; and the *harmonic* sequence $(1, \frac{1}{2}, \frac{1}{3}, \frac{1}{4}, \frac{1}{5}, \ldots)$. Summations involving all of these types of sequences can be analyzed inductively, and we'll address all three of them here and in the exercises. (The statements we'll prove are both useful facts to know about geometric/arithmetic/harmonic sequences, and good practice with induction.)

GEOMETRIC SERIES

Definition 5.2 (Geometric sequences and series)
A geometric sequence *is a sequence of numbers where each number is generated by multiplying the previous entry by a fixed ratio* $\alpha \in \mathbb{R}$*, starting from an initial value* x_0.

(Thus the sequence is $\langle x_0, x_0 \cdot \alpha, x_0 \cdot \alpha^2, x_0 \cdot \alpha^3, \ldots \rangle$.) A geometric series *or* geometric sum *is $\sum_{i=0}^{n} x_0 \alpha^i$.*

Examples include $\langle 2, 4, 8, 16, 32, \ldots \rangle$; or $\langle 1, \frac{1}{3}, \frac{1}{9}, \frac{1}{27}, \ldots \rangle$; or $\langle 1, 1, 1, 1, 1, \ldots \rangle$.

It turns out that there is a relatively simple formula expressing the sum of the first n terms of a geometric sequence:

Theorem 5.2 (Analysis of geometric series)
Let $\alpha \in \mathbb{R}$ where $\alpha \neq 1$, and let $n \in \mathbb{Z}^{\geq 0}$. Then

$$\sum_{i=0}^{n} \alpha^i = \frac{\alpha^{n+1} - 1}{\alpha - 1}.$$

(If $\alpha = 1$, then $\sum_{i=0}^{n} \alpha^i = n + 1$.)

(For simplicity, we stated Theorem 5.2 without reference to x_0. Because we can pull a constant multiplicative factor out of a summation, we can use the theorem to conclude that $\sum_{i=0}^{n} x_0 \alpha^i = x_0 \cdot \sum_{i=0}^{n} \alpha^i = x_0 \cdot \frac{\alpha^{n+1}-1}{\alpha-1}$.)

We will be able to prove Theorem 5.2 using a proof by mathematical induction:

Example 5.6 (Geometric series)
Proof of Theorem 5.2. Consider a fixed real number α with $\alpha \neq 1$, and let $P(n)$ denote the property that

$$\sum_{i=0}^{n} \alpha^i = \frac{\alpha^{n+1} - 1}{\alpha - 1}.$$

We'll prove that $P(n)$ holds for all integers $n \geq 0$ by induction on n.

base case ($n = 0$): Note that $\sum_{i=0}^{0} \alpha^i = \alpha^0$ and $\frac{\alpha^{0+1}-1}{\alpha-1}$ both equal 1. Thus $P(0)$ holds.

inductive case ($n \geq 1$): We assume the inductive hypothesis $P(n-1)$, namely

$$\sum_{i=0}^{n-1} \alpha^i = \frac{\alpha^n - 1}{\alpha - 1},$$

and we must prove $P(n)$. Here is the proof:

$$\begin{aligned}
\sum_{i=0}^{n} \alpha^i &= \alpha^n + \sum_{i=0}^{n-1} \alpha^i && \textit{definition of summation} \\
&= \alpha^n + \frac{\alpha^n - 1}{\alpha - 1} && \textit{inductive hypothesis} \\
&= \frac{\alpha^n(\alpha - 1) + \alpha^n - 1}{\alpha - 1} && \textit{putting the fractions over a common denominator} \\
&= \frac{\alpha^{n+1} - \alpha^n + \alpha^n - 1}{\alpha - 1} && \textit{multiplying out} \\
&= \frac{\alpha^{n+1} - 1}{\alpha - 1}. && \textit{simplifying}
\end{aligned}$$

Thus $P(n)$ holds, and the theorem follows. □

Problem-solving tip: The inductive cases of many inductive proofs follow the same pattern: first, we use some kind of structural definition to "pull apart" the statement about n into something kind of statement about $n - 1$ (plus some "leftover" other stuff), then apply the inductive hypothesis to simplify the $n - 1$ part. We then manipulate the result of using the inductive hypothesis plus the leftovers to get the desired equation.

Notice that Examples 5.2 and 5.3 were both special cases of Theorem 5.2. For the former, Theorem 5.2 tells us that $\sum_{i=0}^{n} 2^i = \frac{2^{n+1}-1}{2-1} = 2^{n+1} - 1$; for the latter, this theorem tells us that

$$\sum_{i=0}^{n}(-1)^i = \frac{(-1)^{n+1}-1}{-1-1} = \frac{1-(-1)^{n+1}}{2} = \begin{cases} \frac{1-(-1)}{2} = 1 & \text{if } n \text{ is even} \\ \frac{1-1}{2} = 0 & \text{if } n \text{ is odd.} \end{cases}$$

A corollary of Theorem 5.2 addressing *infinite* geometric sums will turn out to be useful later, so we'll state it now. (You can skip over the proof if you don't know calculus, or if you haven't thought about calculus recently.)

Corollary 5.3

Let $\alpha \in \mathbb{R}$ where $0 \leq \alpha < 1$, and define $f(n) = \sum_{i=0}^{n} \alpha^i$. Then:

1. *$\sum_{i=0}^{\infty} \alpha^i = \frac{1}{1-\alpha}$, and*
2. *For all $n \geq 0$, we have $1 \leq f(n) \leq \frac{1}{1-\alpha}$.*

Proof. The proof of (1) requires calculus. Theorem 5.2 says that $f(n) = \frac{\alpha^{n+1}-1}{\alpha-1}$, and we take the limit as $n \to \infty$. Because $\alpha < 1$, we have that $\lim_{n\to\infty} \alpha^{n+1} = 0$. Thus as $n \to \infty$ the numerator $\alpha^{n+1} - 1$ tends to -1, and the entire ratio tends to $1/(1-\alpha)$.

For (2), observe that $\sum_{i=0}^{n} \alpha^i$ is definitely greater than or equal to $\sum_{i=0}^{0} \alpha^i$ (because $\alpha \geq 0$ and so the latter results by eliminating n nonnegative terms from the former). Similarly, $\sum_{i=0}^{n} \alpha^i$ is definitely less than or equal to $\sum_{i=0}^{\infty} \alpha^i$. Thus:

$$f(n) = \sum_{i=0}^{n} \alpha^i \geq \sum_{i=0}^{0} \alpha^i = \alpha^0 = 1$$
$$f(n) = \sum_{i=0}^{n} \alpha^i \leq \sum_{i=0}^{\infty} \alpha^i = \frac{1}{1-\alpha}.$$ □

Arithmetic series

Definition 5.3 (Arithmetic sequences and series)

An arithmetic sequence *is a sequence of numbers where each number is generated by adding a fixed step-size $\alpha \in \mathbb{R}$ to the previous number in the sequence. The first entry in the sequence is some initial value $x_0 \in \mathbb{R}$. (Thus the sequence is $\langle x_0, x_0 + \alpha, x_0 + 2\alpha, x_0 + 3\alpha, \ldots \rangle$.)* An arithmetic series *or* sum *is $\sum_{i=0}^{n}(x_0 + i\alpha)$.*

Examples include $\langle 2, 4, 6, 8, 10, \ldots \rangle$; or $\langle 1, \frac{1}{3}, -\frac{1}{3}, -1, -\frac{5}{3}, \ldots \rangle$; or $\langle 1, 1, 1, 1, 1, \ldots \rangle$. You'll prove a general formula for an arithmetic sum in the exercises.

Harmonic series

Definition 5.4 (Harmonic series)

A harmonic series *is the sum of a sequence of numbers whose kth number is $\frac{1}{k}$. The nth* harmonic number *is defined by $H_n := \sum_{k=1}^{n} \frac{1}{k}$.*

Thus, for example, we have $H_1 = 1$, $H_2 = 1 + \frac{1}{2} = 1.5$, $H_3 = 1 + \frac{1}{2} + \frac{1}{3} \approx 1.8333$, and $H_4 = 1 + \frac{1}{2} + \frac{1}{3} + \frac{1}{4} \approx 2.0833$.

Giving a precise equation for the value of H_n requires a bit more work, but we can very easily prove upper and lower bounds on H_n by induction. (If you've had calculus, then there's a simple way for you to approximate the value of H_n, as

$$H_n = \sum_{x=1}^{n} \frac{1}{x} \approx \int_{x=1}^{n} \frac{1}{x}\, dx = \ln n.$$

But we'll do a calculus-free version here.) We will be able to prove the following, which captures the value of H_n to within a factor of 2, at least when n is a power of 2:

The name "harmonic" comes from music: when a note at frequency f is played, *overtones* of that note—other high-intensity frequencies—can be heard at frequencies $2f, 3f, 4f, \ldots$. The *wavelengths* of the corresponding sound waves are $\frac{1}{f}, \frac{1}{2f}, \frac{1}{3f}, \frac{1}{4f}, \ldots$.

Theorem 5.4 (Bounds on the (2^k)th harmonic number)
For any integer $k \geq 0$, we have $k+1 \geq H_{2^k} \geq \frac{k}{2}+1$.

We'll prove half of Theorem 5.4 (namely $k+1 \geq H_{2^k}$) by induction in Example 5.7, leaving the other half to the exercises. We will also leave to the exercises a proof of upper and lower bounds for H_n when n is not an exact power of 2.

Example 5.7 (Inductive proof that $k+1 \geq H_{2^k}$)
Proof. Let $P(k)$ denote the property that $k+1 \geq H_{2^k}$. We'll use induction on k to prove that $P(k)$ holds for all integers $k \geq 0$.

base case ($k = 0$): We have that $H_{2^k} = H_{2^0} = H_1 = 1$, and $k+1 = 0+1 = 1$ as well. Therefore $H_{2^k} = 1 = k+1$.

inductive case ($k \geq 1$): Let $k \geq 1$ be an arbitrary integer. We must prove $P(k)$—that is, we must prove that $k+1 \geq H_{2^k}$. To do so, we assume the inductive hypothesis $P(k-1)$, namely that $k \geq H_{2^{k-1}}$. Consider H_{2^k}:

$$\begin{aligned}
H_{2^k} &= \sum_{i=1}^{2^k} \frac{1}{i} && \textit{definition of the harmonic numbers}\\
&= \left[\sum_{i=1}^{2^{k-1}} \frac{1}{i}\right] + \left[\sum_{i=2^{k-1}+1}^{2^k} \frac{1}{i}\right] && \textit{splitting the summation into parts}\\
&= H_{2^{k-1}} + \left[\sum_{i=2^{k-1}+1}^{2^k} \frac{1}{i}\right] && \textit{definition of the harmonic numbers, again}\\
&\leq H_{2^{k-1}} + \left[\sum_{i=2^{k-1}+1}^{2^k} \frac{1}{2^{k-1}}\right] && \textit{every term in the summation } \textstyle\sum_{i=2^{k-1}+1}^{2^k} \frac{1}{i} \textit{ is smaller than } \frac{1}{2^{k-1}}\\
&\leq H_{2^{k-1}} + 2^{k-1} \cdot \frac{1}{2^{k-1}} && \textit{there are } 2^{k-1} \textit{ terms in the summation}\\
&= H_{2^{k-1}} + 1 && \tfrac{1}{x} \cdot x = 1 \textit{ for any } x \neq 0\\
&\leq k+1. && \textit{inductive hypothesis}
\end{aligned}$$

Thus we've proven that $H_{2^k} \leq k+1$—that is, we've proven $P(k)$. This proof completes the inductive case, and the theorem follows. □

The proof in Example 5.7 is perhaps the first time in this chapter in which we needed some serious insight and creativity to establish the inductive case. The structure of a proof by induction is rigid—we must prove a base case $P(0)$; we must prove an inductive case $P(n-1) \Rightarrow P(n)$—but that doesn't make the entire proof totally formulaic. (The proof of the inductive case must use the inductive hypothesis at some point, so its statement gives you a little guidance for the kinds of manipulations to try.) Just as with all the other proof techniques that we explored in Chapter 4, a proof by induction can require you to *think*—and all of strategies that we discussed in Chapter 4 may be helpful to deploy.

5.2.3 *Some More Examples*

We'll close this section with a few more examples of proofs by mathematical induction, but we'll focus on things other than analyzing summations. Some of these examples are still about arithmetic properties, but they should at least hint at the breadth of possible statements that we might be able to prove by induction.

n	2^n	n^2
0	1	0
1	2	1
2	4	4
3	8	9
4	16	16
5	32	25
6	64	36
7	128	49

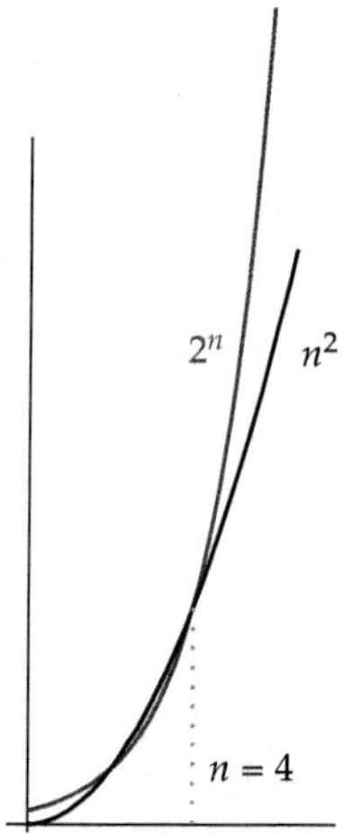

Figure 5.5: Small values of 2^n and n^2, and a plot of the functions.

COMPARING ALGORITHMS: WHICH IS FASTER?

Suppose that we have two different candidate algorithms that solve a problem related to a set S with n elements—a *brute-force algorithm* that tries all 2^n possible subsets of S, and a second algorithm that computes the solution by looking at only n^2 subsets of S. Which would be faster to use? It turns out that the latter algorithm is faster, and we can prove this fact (with a small caveat for small n) by induction:

Example 5.8 (2^n vs. n^2)
We'd like to prove that $2^n \geq n^2$ for all integers $n \geq 0$—but it turns out not to be true! (See Figure 5.5.) Indeed, $2^3 < 3^2$. But the relationship appears to begin to hold starting at $n = 4$. Let's prove it, by induction:

Claim: For all integers $n \geq 4$, we have $2^n \geq n^2$.

Proof. Let $P(n)$ denote the property $2^n \geq n^2$. We'll use induction on n to prove that $P(n)$ holds for all $n \geq 4$.

base case ($n = 4$): For $n = 4$, we have $2^n = 16 = n^2$, so the inequality $P(4)$ holds.

inductive case ($n \geq 5$): Assume the inductive hypothesis $P(n-1)$—that is, assume $2^{n-1} \geq (n-1)^2$. We must prove $P(n)$. For $n \geq 4$, note that $n^2 \geq 4n$ (by multiplying both sides of the inequality $n \geq 4$ by n). Thus $n^2 - 4n \geq 0$, and so

$$\begin{aligned}
2^n &= 2 \cdot (2^{n-1}) && \textit{definition of exponentiation} \\
&\geq 2 \cdot (n-1)^2 && \textit{inductive hypothesis} \\
&= 2n^2 - 4n + 2 && \textit{multiplying out} \\
&= n^2 + (n^2 - 4n) + 2 && \textit{rearranging} \\
&\geq n^2 + 0 + 2 && \textit{by the above discussion, we have } n^2 - 4n \geq 0 \\
&> n^2.
\end{aligned}$$

Thus we have shown $2^n > n^2$, which completes the proof of the inductive case. The claim follows. □

Taking it further: In analyzing the efficiency of algorithms, we will frequently have to do the type of comparison that we just completed, to compare the amount of time consumed by one algorithm versus another. Chapter 6 discusses this type of comparison in much greater detail, but here's one example of this sort.

Let X be a sequence. A *subsequence* of X results from selecting some of the entries in X—for example, `TURING` is a subsequence of `OUTSOURCING`. For two sequences X and Y, a *common subsequence* is a subsequence of both X and Y. The *longest common subsequence* of X and Y is, naturally, the common subsequence of X and Y that's longest. (For example, `TURING` is the longest common subsequence of `DISTURBINGLY` and `OUTSOURCING`.)

Given two sequences X and Y of length n, we can find the longest common subsequence fairly easily by testing *every possible subsequence of* X to see whether it's also a subsequence of Y. This brute-force solution takes requires testing 2^n subsequences of X. But there's a cleverer approach to solving this problem using an algorithmic design technique called *dynamic programming* (see p. 959 or a textbook on algorithms) that avoids redoing the same computation—here, testing the same sequence of letters to see if it appears in Y—more than once. The dynamic programming algorithm for longest common subsequence requires only about n^2 steps.

Proving algorithms correct: factorial

We just gave an example of using a proof by induction to analyze the efficiency of an algorithm, but we can also use mathematical induction to prove the *correctness* of a recursive algorithm. (That is, we'd like to show that a recursive algorithm always returns the desired output.) Here's a simple example, for the natural recursive algorithm to compute factorials (see Figure 5.6):

```
fact(n):
  if n = 1 then
      return 1
  else
      return n · fact(n − 1)
```

Figure 5.6: Pseudocode for factorial: given $n \in \mathbb{Z}^{\geq 1}$, we wish to compute the value of $n!$.

Example 5.9 (Factorial)

Consider the recursive algorithm **fact** in Figure 5.6. For a positive integer n, let $P(n)$ denote the property that $\textbf{fact}(n) = n!$. We'll prove by induction on n that, indeed, $P(n)$ holds for all integers $n \geq 1$.

base case ($n = 1$): Observe that **fact**(1) returns 1 immediately. And $1! = 1$ by definition. Thus $P(1)$ holds.

inductive case ($n \geq 2$): We assume the inductive hypothesis $P(n-1)$, namely that $\textbf{fact}(n-1)$ returns $(n-1)!$. We want to prove that $\textbf{fact}(n)$ returns $n!$. But this claim is easy to see:

$$\begin{aligned}\textbf{fact}(n) &= n \cdot \textbf{fact}(n-1) && \textit{by inspection of the algorithm}\\ &= n \cdot (n-1)! && \textit{by the inductive hypothesis}\\ &= n! && \textit{by definition of !}\end{aligned}$$

Therefore the claim holds by induction.

In fact, induction and recursion are basically the same thing: recursion "works" by leveraging a solution to a smaller instance of a problem to solve a larger instance of the same problem; a proof by induction "works" by leveraging a proof of a smaller instance of a claim to prove a larger instance of the same claim. (Actually, one common use of induction is to analyze the efficiency of a recursive algorithm. We'll discuss this type of analysis in great depth in Section 6.4.)

Taking it further: While induction is much more closely related to recursive algorithms than nonrecursive algorithms, we can also prove the correctness of an iterative algorithm using induction. The basic idea is to consider a statement, called a *loop invariant*, about the correct behavior of a loop; we can prove inductively that a loop invariant starts out true and stays true throughout the execution of the algorithm. See the discussion on p. 517.

Divisibility

We'll close this section with one more numerical example, about divisibility:

Writing tip: Example 5.10 illustrates why it is crucial to state clearly the variable upon which induction is being performed. This statement involves two variables, k and n, but we're performing induction on only one of them!

Example 5.10 ($k^n - 1$ is evenly divisible by $k - 1$)

Claim: For any $n \geq 0$ and $k \geq 2$, we have that $k^n - 1$ is evenly divisible by $k - 1$.

(For example, $7^n - 1$ is always divisible by 6, as in $7 - 1, 49 - 1$, and $343 - 1$. And $k^2 - 1$ is always divisible by $k - 1$; in fact, factoring $k^2 - 1$ yields $k^2 - 1 = (k-1)(k+1)$.)

Proof. We'll proceed by induction on n. That is, let $P(n)$ denote the claim

For all integers $k \geq 2$, we have that $k^n - 1$ is evenly divisible by $k - 1$.

We will prove that $P(n)$ holds for all integers $n \geq 0$ by induction on n.

base case ($n = 0$): For any k, we have $k^n - 1 = k^0 - 1 = 1 - 1 = 0$. And 0 is evenly divisible by any positive integer, including $k - 1$. Thus $P(0)$ holds.

inductive case ($n \geq 1$): We assume the inductive hypothesis $P(n-1)$, and we need to prove $P(n)$. Let $k \geq 2$ be an arbitrary integer. Then:

$$\begin{aligned} k^n - 1 &= k^n - k + k - 1 && \textit{antisimplification: } x = x + k - k. \\ &= k \cdot (k^{n-1} - 1) + k - 1 && \textit{factoring} \end{aligned}$$

By the inductive hypothesis, $k^{n-1} - 1$ is evenly divisible by $k - 1$. In other words, by the definition of divisibility, there exists a nonnegative integer a such that $a \cdot (k-1) = k^{n-1} - 1$. Therefore

$$\begin{aligned} k^n - 1 &= k \cdot a \cdot (k-1) + k - 1 \\ &= (k-1) \cdot (k \cdot a + 1). \end{aligned}$$

Because $k \cdot a + 1$ is a nonnegative integer, $(k-1) \cdot (k \cdot a + 1)$ is by definition evenly divisible by $k - 1$. Thus $k^n - 1 = (k-1) \cdot (k \cdot a + 1)$ is evenly divisible by $k - 1$. Our k was arbitrary, so $P(n)$ follows. □

Problem-solving tip: In inductive proofs, try to massage the expression in question into something—*anything!*—that matches the form of the inductive hypothesis. Here, the "antisimplification" step is obviously true but seems completely bizarre. Why did we do it? Our only hope in the inductive case is to somehow make use of the inductive hypothesis. Here, the inductive hypothesis tells us something about $k^{n-1} - 1$—so a good strategy is to transform $k^n - 1$ into an expression involving $k^{n-1} - 1$, plus some leftover stuff.

COMPUTER SCIENCE CONNECTIONS

LOOP INVARIANTS

In Example 5.9, we saw how to use a proof by induction to establish that a recursive algorithm correctly solves a particular problem. But proving the correctness of *iterative* algorithms seems different. An approach—pioneered in the 1960s by Robert Floyd and C. A. R. Hoare[1]—is based on *loop invariants*, and can be used to analyze nonrecursive algorithms. A *loop invariant* for a loop L is logical property P such that (i) P is true before L is first executed; and (ii) if P is true at the beginning of an iteration of L, then P is true after that iteration of L. The parallels to induction are clear; property (i) is the base case, and property (ii) is the inductive case. Together, they ensure that P is always true, and in particular P is true when the loop terminates.

[1] Robert W. Floyd. Assigning meanings to programs. In *Proceedings of Symposia in Applied Mathematics XIX*, American Mathematical Society, pages 19–32, 1967; and C. A. R. Hoare. An axiomatic basis for computer programming. *Communications of the ACM*, 12(10):576–585, October 1969.

Here's an example of a sketch of a proof of correctness of Insertion Sort (Figure 5.7) using loop invariants. (Many proofs using loop invariants would proceed with more formal detail.) We claim that the property

$$P(k) := A[1 \ldots k+1] \textit{ is sorted after completing } k \textit{ iterations of the outer } \textbf{while} \textit{ loop}$$

is true for all $k \geq 0$. (That is, P is a loop invariant for the outer **while** loop.)

Proof (sketch). For the base case ($k = 0$), we've completed zero iterations—that is, we have only executed line 1. But $A[1 \ldots k+1]$ is then vacuously sorted, because it contains only the lone element $A[1]$.

For the inductive case ($k \geq 1$), we assume the inductive hypothesis $P(k-1)$—that is, $A[1 \ldots k]$ was sorted before the kth iteration. The kth iteration of the loop executed lines 2–7, so we must show that the execution of these lines extended the sorted segment $A[1 \ldots k]$ to $A[1 \ldots k+1]$. A formal proof of this claim would use another loop invariant, like

$$Q(j) := \textit{both } A[1 \ldots j-1] \textit{ and } A[j \ldots i] \textit{ are sorted, and } A[j-1] < A[j+1]$$

but for this proof sketch we'll be satisfied by concluding the desired conclusion by inspection of the algorithm's code. □

```
insertionSort(A[1...n]):
1: i := 2
2: while i ≤ n:
3:     j := i
4:     while j > 1 and A[j] > A[j − 1]:
5:         swap A[j] and A[j − 1]
6:         j := j − 1
7:     i := i + 1
```

Figure 5.7: Insertion Sort.

Because $P(n-1)$ is true (after $n-1$ iterations of the loop), we know that $A[1 \ldots (n-1)+1] = A[1 \ldots n]$ is sorted, as desired.

Loop invariants can also be extremely valuable as part of the development of programs. For example, many people end up struggling to correctly write binary search—but by writing down loop invariants before actually writing the code, it's actually easy. If we think about the property

$$\textit{if } x \textit{ is in } A, \textit{ then } x \textit{ is one of } A[lo, \ldots, hi]$$

as a loop invariant as we write the program, binary search becomes much easier to get right. Many programming languages allow programmers to use *assertions* to state logical conditions that they believe to always be true at a particular point in the code. A simple `assert(P)` statement can help a programmer identify bugs earlier in the development process and avoid a great deal of debugging trauma later.

```
binarySearch(A[1...n], x):
// output: is x in the sorted array A?
lo := 1
hi := n
while lo ≤ hi:
    middle := ⌊(lo+hi)/2⌋
    if A[middle] = x then
        return True
    else if A[middle] > x then
        hi := middle − 1
    else
        lo := middle + 1
return False
```

Figure 5.8: Binary Search.

5.2.4 *Exercises*

Prove that the following claims hold for all integers $n \geq 0$, by induction on n:

5.1 $\sum_{i=0}^{n} i^2 = \frac{n(n+1)(2n+1)}{6}$

5.2 $\sum_{i=0}^{n} i^3 = \frac{n^4+2n^3+n^2}{4}$

5.3 $(-1)^n = \begin{cases} 1 & \text{if } n \text{ is even} \\ -1 & \text{if } n \text{ is odd} \end{cases}$

5.4 $\sum_{i=1}^{n} \frac{1}{i(i+1)} = \frac{n}{n+1}$

5.5 $\sum_{i=1}^{n} \frac{2}{i(i+2)} = \frac{3}{2} - \frac{1}{n+1} - \frac{1}{n+2}$

5.6 $\sum_{i=1}^{n} i \cdot (i!) = (n+1)! - 1$

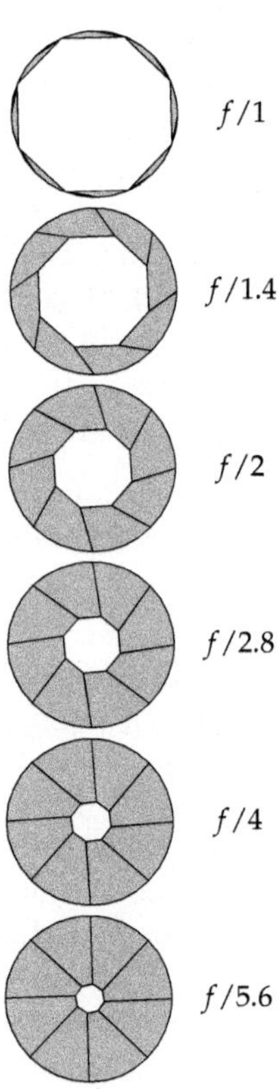

Figure 5.9: A particular lens of a camera, shown at several different f-stops. These configurations are only an approximation—the real blades are shaped somewhat differently than is shown here.

5.7 In a typical optical camera lens, the light that enters the lens (through the opening called the *aperture*) is controlled by a collection of movable blades that can be adjusted inward to narrow the area through which light can pass. (There are two effects of narrowing this opening: first, the amount of light entering the lens is reduced, darkening the resulting image; and, second, the *depth of field*—the range of distances from the lens at which objects are in focus in the image—increases.) Although some lenses allow continuous adjustment to their openings, many have a sequence of so-called *stops:* discrete steps by which the aperture narrows. (See Figure 5.9.) These steps are called *f-stops* (the "f" is short for "focal"), and they are denoted with some unusual notation that you'll unwind in this exercise. The "fastest" f-stop for a lens measures the ratio of two numbers: the focal length of the lens divided by the diameter of the aperture of the lens. (For example, you might use a lens that's 50mm long and that has a 25mm diameter, which yields an f-stop of $50\text{mm}/25\text{mm} = 2$.) One can also "stop down" a lens from this fastest setting by adjusting the blades to shrink the diameter of the aperture, as described above. (For example, for the 50mm-long lens with a 25mm diameter, you might reduce the diameter to 12.5mm, which yields an f-stop of $50\text{mm}/12.5\text{mm} = 4$.)

Consider a camera lens with a 50mm focal length, and let $d_0 := 50\text{mm}$ denote the diameter of the lens's aperture diameter. "Stopping down" the lens by one step causes the lens's aperture diameter to shrink by a factor of $\frac{1}{\sqrt{2}}$—that is, the next-smaller aperture diameter for a diameter d_i is defined as

$$d_{i+1} := \frac{d_i}{\sqrt{2}}, \text{ for any } i \geq 0.$$

Give a closed-form expression for d_n—that is, give a nonrecursive numerical expression whose value is equal to d_n (where your expression involves only real numbers and the variable n). Prove your answer correct by induction on n. Also give a closed-form expression for two further quantities:

- the "light-gathering" area (that is, the area of the aperture) of the lens when its diameter is set to d_n.
- the f-stop f_n of the lens when its diameter is set to d_n.

(Using your formula for f_n, can you explain the f-stop names from Figure 5.9?)

5.8 What is the sum of the first n odd positive integers? First, formulate a conjecture by trying a few examples (for example, what's $1+3$, for $n=2$? What's $1+3+5$, for $n=3$? What's $1+3+5+7$, for $n=4$?). Then prove your answer by induction.

5.9 What is the sum of the first n even positive integers? Prove your answer by induction.

5.10 Let $\alpha \in \mathbb{R}$ and let $n \in \mathbb{Z}^{\geq 0}$, and consider the arithmetic sequence $\langle x_0, x_0+\alpha, x_0+2\alpha, \ldots \rangle$. (Recall that each entry in an arithmetic sequence is a fixed amount more than the previous entry. Three examples are $\langle 1,3,5,7,9,\ldots \rangle$, with $x_0 = 1$ and $\alpha = 2$; $\langle 25,20,15,10,\ldots \rangle$, with $x_0 = 25$ and $\alpha = -5$; and $\langle 5,5,5,5,5,\ldots \rangle$, with $x_0 = 5$ and $\alpha = 0$.) An *arithmetic sum* or *arithmetic series* is the sum of an arithmetic sequence. For the arithmetic sequence $\langle x_0, x_0+\alpha, x_0+2\alpha, \ldots \rangle$, formulate and prove correct by induction a formula expressing the value of the arithmetic series

$$\sum_{i=0}^{n} (x_0 + i\alpha).$$

(Hint: note that $\sum_{i=0}^{n} i\alpha = \alpha \sum_{i=0}^{n} i = \frac{\alpha n(n+1)}{2}$, by Example 5.4.)

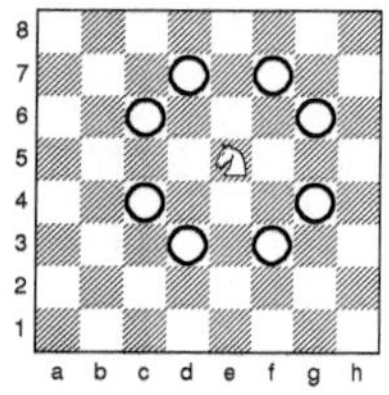

Figure 5.10: A chess board. The knight can move to any of the marked positions.

5.11 In chess, a knight at position $\langle r, c \rangle$ can move in an L-shaped pattern to any of eight positions: moving over one row and up/down two columns ($\langle r \pm 1, c \pm 2 \rangle$), or two rows over and one column up/down ($\langle r \pm 2, c \pm 1 \rangle$). (See Figure 5.10.) A *knight's walk* is a sequence of legal moves, starting from a square of your choice, that visits *every* square of the board. Prove by induction that there exists a knight's walk for any n-by-n chessboard for any $n \geq 4$. (A *knight's tour* is a knight's walk that visits every square *only* once. It turns out that knight's tours exist for all even $n \geq 6$, but you don't need to prove this fact.)

5.12 *(programming required)* In a programming language of your choice, implement your proof from Exercise 5.11 as a recursive algorithm that *computes* a knight's walk in an n-by-n chessboard.

5.13 In chess, a rook at position $\langle r, c \rangle$ can move in a straight line either horizontally or vertically (to $\langle r \pm x, c \rangle$ or $\langle r, c \pm x \rangle$, for any integer x). (See Figure 5.11.) A *rook's tour* is a sequence of legal moves, starting from a square of your choice, that visits *every* square of the board *once and only once.* Prove by induction that there exists a rook's tour for any n-by-n chessboard for any $n \geq 1$.

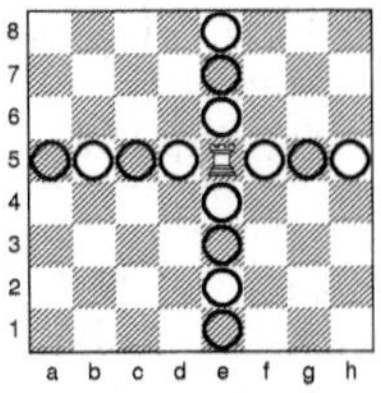

Figure 5.11: A rook can move to any of the positions marked with a circle.

Figure 5.12 shows three different fractals. One is the Von Koch snowflake (Figure 5.12(a)), which we've already seen: a Von Koch line of size s and level 0 is just a straight line segment; a Von Koch line of size s and level ℓ consists of four Von Koch lines of size $(s/3)$ and level $(\ell - 1)$ arranged in the shape *; a Von Koch snowflake of size s and level ℓ consists of a triangle of three Von Koch lines of size s and level ℓ.*

The other two fractals in Figure 5.12 are new. Figure 5.12(b) shows the Sierpinski triangle*: a Sierpinski triangle of level 0 and size s is an equilateral triangle of side length s; a Sierpinski triangle of level $(\ell + 1)$ is three Sierpinski triangles of level ℓ and side length $s/2$ arranged in a triangle. Figure 5.12(c) shows a related fractal called the* Sierpinski carpet, *recursively formed from 8 smaller Sierpinski carpets (arranged in a 3-by-3 grid with a hole in the middle); the base case is just a filled square.*

The Von Koch snowflake is named after Helge von Koch, a 19th/20th-century Swedish mathematician; the Sierpinski triangle/carpet are named after Wacław Sierpiński, a 20th-century Polish mathematician.

Suppose that we draw each of these fractals at level ℓ and with size 1. What is the perimeter of each of these fractals? (By "perimeter," we mean the total length of all boundaries separating regions inside the figure from regions outside—which includes, for example, the boundary of the "hole" in the Sierpinski carpet. For the Sierpinski fractals as drawn here, the perimeter is precisely the length of lines separating colored-in from uncolored-in regions.*) In each case, conjecture a formula and prove your answer correct by induction.*

5.14 Von Koch snowflake **5.15** Sierpinski triangle **5.16** Sierpinski carpet

Draw each of these fractals at level ℓ and with size 1. What is the enclosed area of each of these fractals? (Again, for the Sierpinski fractals as drawn here, the enclosed area is precisely the area of the colored-in regions.)

5.17 Von Koch snowflake **5.18** Sierpinski triangle **5.19** Sierpinski carpet

In the last few exercises, you computed the fractals' perimeter/area at level ℓ. But what if we continued the fractal-expansion process forever? What are the area and perimeter of an infinite-level *fractal? (Hint: use Corollary 5.3.)*

5.20 Von Koch snowflake **5.21** Sierpinski triangle **5.22** Sierpinski carpet

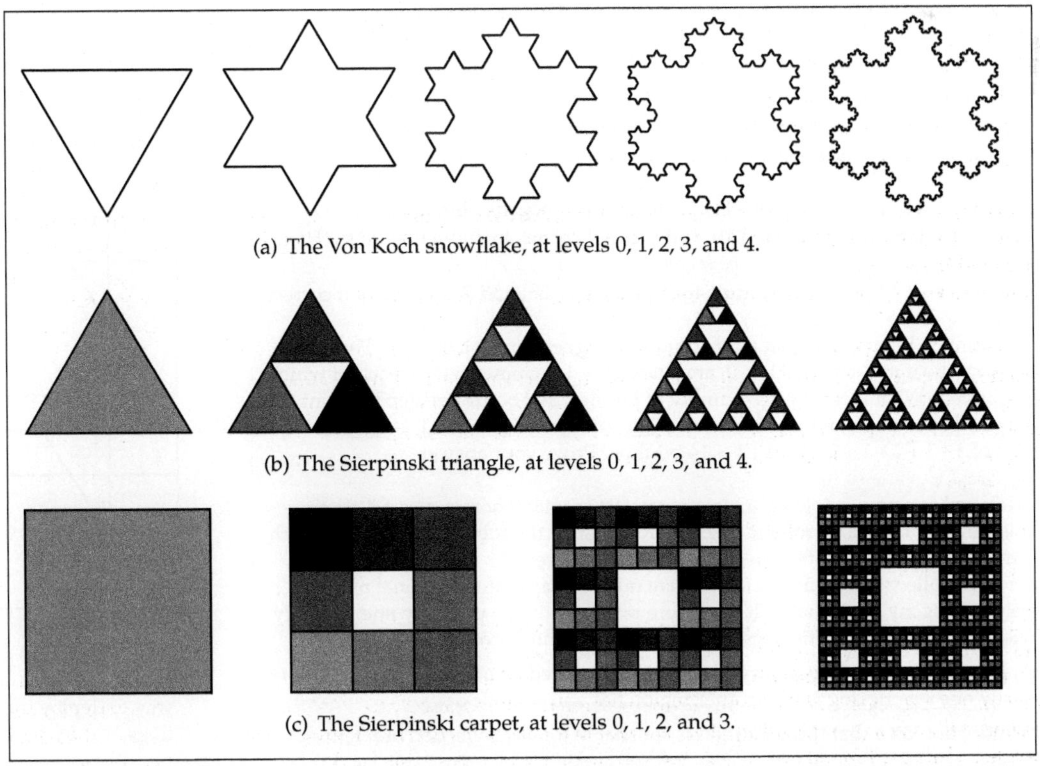

(a) The Von Koch snowflake, at levels 0, 1, 2, 3, and 4.

(b) The Sierpinski triangle, at levels 0, 1, 2, 3, and 4.

(c) The Sierpinski carpet, at levels 0, 1, 2, and 3.

Figure 5.12: Three fractals: the Von Koch snowflake, the Sierpinski triangle, and the Sierpinski carpet.

5.23 *(programming required)* Write a recursive function `sierpinskiTriangle`(*level*, *length*, x, y), in a language of your choice, to draw a Sierpinski triangle of side length *length* at level *level* with bottom-left coordinate $\langle x, y \rangle$. (You'll need to use some kind of graphics package with line-drawing capability.)

Write your function so that—in addition to drawing the fractal—it returns both the *total length* and *total area* of the triangles that it draws. Use your function to verify some small cases of Exercises 5.15 and 5.18.

5.24 *(programming required)* Write a recursive function `sierpinskiCarpet`(*level*, *length*, x, y), in a programming language of your choice, to draw a Sierpinski carpet. (See Exercise 5.23 for the meaning of the parameters.) Write your function so that—in addition to drawing the fractal—it also returns the *area* of the boxes that it encloses. Use your function to verify some small cases of your answer to Exercise 5.19.

5.25 An *n-by-n magic square* is an n-by-n grid into which the numbers $1, 2, \ldots, n^2$ are placed, once each. The "magic" is that each *row*, *column*, and *diagonal* must be filled with numbers that have the same sum. For example, a 3-by-3 magic square is shown in Figure 5.13. Conjecture and prove a formula for what the sum of each row/column/diagonal must be in an n-by-n magic square.

4	9	2
3	5	7
8	1	6

Figure 5.13: A Magic Square.

Recall from Section 5.2.2 the harmonic numbers, *where* $H_n := \sum_{i=1}^{n} \frac{1}{i}$ *is the sum of the reciprocals of the first n positive integers. Further recall Theorem 5.4, which states that* $k + 1 \geq H_{2^k} \geq \frac{k}{2} + 1$ *for any integer* $k \geq 0$.

5.26 In Example 5.7, we proved that $k + 1 \geq H_{2^k}$. Using the same type of reasoning as in the example, complete the proof of Theorem 5.4: show by induction that $H_{2^k} \geq \frac{k}{2} + 1$ for any integer $k \geq 0$.

5.27 Generalize Theorem 5.4 to numbers that aren't necessarily exact powers of 2. Specifically, prove that $\log n + 2 \geq H_n \geq (\log n - 1)/2 + 1$ for any real number $n \geq 1$. *(Hint: use Theorem 5.4.)*

5.28 Prove *Bernoulli's inequality:* let $x \geq -1$ be an arbitrary real number. Prove by induction on n that $(1 + x)^n \geq 1 + nx$ for any positive integer n.

Prove that the following inequalities $f(n) \leq g(n)$ hold "for sufficiently large n." That is, identify an integer k and then prove (by induction on n) that $f(n) \leq g(n)$ for all integers $n \geq k$.

5.29 $2^n \leq n!$

5.30 $b^n \leq n!$, for an arbitrary integer $b \geq 1$

5.31 $3n \leq n^2$

5.32 $n^3 \leq 2^n$

5.33 Prove that, for any nonnegative integer n, the algorithm **odd?**(n) returns True if and only if n is odd. (See Figure 5.14.)

5.34 Prove that the algorithm **sum**(n, m) returns $\sum_{i=n}^{m} i$ (again see Figure 5.14) for any $m \geq n$. *(Hint: perform induction on the value of $m - n$.)*

5.35 Describe how your proof from Exercise 5.34 would change if Line 4 from the **sum** algorithm in Figure 5.14 were changed to return $m + \textbf{sum}(n, m-1)$ instead of $n + \textbf{sum}(n+1, m)$.

```
odd?(n):
1: if n = 0 then
2:     return False
3: else
4:     return not odd?(n − 1)
```

```
sum(n, m):
1: if n = m then
2:     return m
3: else
4:     return n + sum(n + 1, m)
```

Figure 5.14: Two algorithms.

5.36 Prove by induction on n that $8^n - 3^n$ is divisible by 5 for any nonnegative integer n.

5.37 Conjecture a formula for the value of $9^n \bmod 10$, and prove it correct by induction on n. *(Hint: try computing $9^n \bmod 10$ for a few small values of n to generate your conjecture.)*

5.38 As in the previous exercise, conjecture a formula for the value of $2^n \bmod 7$, and prove it correct.

5.39 Suppose that we count, in binary, using an n-bit counter that goes from 0 to $2^n - 1$. There are 2^n different steps along the way: the initial step of $00\cdots0$, and then $2^n - 1$ increment steps, each of which causes at least one bit to be flipped. What is the *average* number of bit flips that occur per step? (Count the first step as changing all n bits.) For example, for $n = 3$, we have $\underline{000} \to 00\underline{1} \to 0\underline{10} \to 01\underline{1} \to \underline{100} \to 10\underline{1} \to 1\underline{10} \to 11\underline{1}$, which has a total of $3 + 1 + 2 + 1 + 3 + 1 + 2 + 1 = 14$ bit flips. Prove your answer.

5.40 To protect my backyard from my neighbor, a biology professor who is sometimes a little overfriendly, I have acquired a large army of vicious robotic dogs. Unfortunately the robotic dogs in this batch are very jealous, and they must be separated by fences—in fact, they can't even *face* each other directly through a fence. So I have built a collection of n fences to separate my backyard into polygonal regions, where each fence completely crosses my yard (that is, it goes from property line to property line, possibly crossing other fences). I wish to deploy my robotic dogs to satisfy the following property:

> For *any* two polygonal regions that share a boundary (that is, are separated by a fence segment), one of the two regions has exactly one robotic dog and the other region has zero robotic dogs.

(See Figure 5.15.) Prove by induction on n that this condition is satisfiable for *any* collection of n fences.

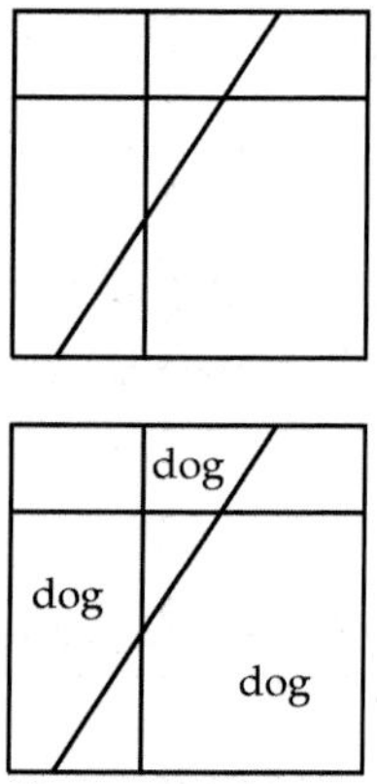

Figure 5.15: A configuration of fences, and a valid way to deploy my dogs.

5.3 *Strong Induction*

It's not true that life is one damn thing after another; it is one damn thing over and over.

Edna St. Vincent Millay (1892–1950)

In the proofs by induction in Section 5.2, we established the claim $\forall n \in \mathbb{Z}^{\geq 0} : P(n)$ by proving $P(0)$ [the base case] and proving that $P(n-1) \Rightarrow P(n)$ [the inductive case]. But let's think again about what happens in an inductive proof, as we build up facts about $P(n)$ for ever-increasing values of n. (Glance at Example 5.1 again.)

1. We prove $P(0)$.
2. We prove $P(0) \Rightarrow P(1)$, so we conclude $P(1)$, using Fact #1.

Now we wish to prove $P(2)$. In a proof by induction like those from Section 5.2, we'd proceed as follows:

3. We prove $P(1) \Rightarrow P(2)$, so we conclude $P(2)$, using Fact #2.

In a *proof by strong induction*, we allow ourselves to make use of more assumptions: namely, we know that $P(1)$ *and* $P(0)$ when we're trying to prove $P(2)$. (By way of contrast, we'll refer to proofs like those from Section 5.2 as using *weak* induction.) In a proof by strong induction, we proceed as follows instead:

3′. We prove $P(0) \wedge P(1) \Rightarrow P(2)$, so we conclude $P(2)$, using Fact #1 and Fact #2.

In a proof by strong induction, in the inductive case we prove $P(n)$ by assuming n different inductive hypotheses: $P(0), P(1), P(2), \ldots,$ and $P(n-1)$. Or, less formally: in the inductive case of a proof by weak induction, we show that *if P "was true last time" then it's still true this time;* in the inductive case of a proof by strong induction, we show that *if P "has been true up until now" then it's still true this time.*

5.3.1 *A Definition and a First Example*

Here is the formal definition of a proof by strong induction:

> **Definition 5.5 (Proof by strong induction)**
> *Suppose that we want to prove that $P(n)$ holds for all $n \in \mathbb{Z}^{\geq 0}$. To give a* proof by strong induction *of $\forall n \in \mathbb{Z}^{\geq 0} : P(n)$, we prove the following:*
>
> 1. *the* base case: *prove $P(0)$.*
> 2. *the* inductive case: *for every $n \geq 1$, prove $[P(0) \wedge P(1) \wedge \cdots \wedge P(n-1)] \Rightarrow P(n)$.*

Generally speaking, using strong induction makes sense when the "reason for" $P(n)$ is that $P(k)$ is true for more than one index $k \leq n-1$, or that $P(k)$ is true for some index $k \leq n-2$. (For weak induction, the "reason for" $P(n)$ is that $P(n-1)$ is true.)

Strong induction makes the inductive case easier to prove than weak induction, because the claim that we need to show—that is, $P(n)$—is the same, but we get to use more assumptions in strong induction: in strong induction, we've assumed all of $P(0) \wedge P(1) \wedge \ldots \wedge P(n-1)$; in weak induction, we've assumed only $P(n-1)$. We can always ignore those extra assumptions, so it's never harder to prove something by strong induction than with weak induction. (Strong induction is actually equivalent to weak induction; anything that can be proven with one can also be proven with the other. See Exercises 5.75–5.76.)

Writing tip: While anything that can be proven using weak induction can also be proven using strong induction, you should still use the tool that's best suited to the job—generally, the one that makes the argument easiest to understand.

A FIRST EXAMPLE: A SIMPLE ALGORITHM FOR PARITY

In the rest of this section, we'll give several examples of proofs by strong induction. We'll start here with a proof of correctness for a blazingly simple algorithm that computes the parity of a positive integer. (Recall that the *parity* of n is the "evenness" or "oddness" of n.) See Figure 5.16 for the **parity** algorithm.

We've already used (weak) induction to prove the correctness of recursive algorithms that, given an input of size n, call themselves on an input of size $n-1$. (That's how we proved the correctness of the factorial algorithm **fact** from Example 5.9.) But for recursive algorithms that call themselves on smaller inputs but not necessarily of size $n-1$, like **parity**, we can use strong induction to prove their correctness.

```
parity(n):                    // assume that n ≥ 0 is an integer.
1: if n ≤ 1 then
2:     return n
3: else
4:     return parity(n − 2)
```

Figure 5.16: A simple parity algorithm.

Example 5.11 (The correctness of parity)
Claim: For any nonnegative integer $n \geq 0$,

$$\textbf{parity}(n) = n \bmod 2.$$

Proof. Write $P(n)$ to denote the property that $\textbf{parity}(n) = n \bmod 2$. We proceed by strong induction on n to show that $P(n)$ holds for all $n \geq 0$:

base cases ($n = 0$ and $n = 1$): By inspection of the algorithm, $\textbf{parity}(0)$ returns 0 in Line 2, and, indeed, $0 \bmod 2 = 0$. Similarly, we have $\textbf{parity}(1) = 1$, and $1 \bmod 2 = 1$ too. Thus $P(0)$ and $P(1)$ hold.

inductive case ($n \geq 2$): Assume the inductive hypothesis $P(0) \wedge P(1) \wedge \cdots \wedge P(n-1)$. Namely, assume that

$$\text{for any integer } 0 \leq k < n, \text{ we have } \textbf{parity}(k) = k \bmod 2.$$

We must prove $P(n)$—that is, we must prove $\textbf{parity}(n) = n \bmod 2$:

$$\begin{aligned} \textbf{parity}(n) &= \textbf{parity}(n-2) && \textit{by inspection (specifically because } n \geq 2 \textit{ and by Line 4)} \\ &= (n-2) \bmod 2 && \textit{by the inductive hypothesis } P(n-2) \\ &= n \bmod 2, \end{aligned}$$

where $(n-2) \bmod 2 = n \bmod 2$ by Definition 2.9. (Note that the inductive hypothesis applies for $k := n-2$ because $n \geq 2$ and therefore $0 \leq n-2 < n$.) □

There are two things to note about the proof in Example 5.11. First, using strong induction instead of weak induction made sense because the inductive case relied on $P(n-2)$ to prove $P(n)$; we did *not* show $P(n-1) \Rightarrow P(n)$. Second, we needed two base cases: the "reason" that $P(1)$ holds is *not* that $P(-1)$ was true. (In fact, $P(-1)$ is false—**parity**(-1) isn't equal to 1! Think about why.) The inductive case of the proof in Example 5.11 does not correctly apply for $n = 1$, and therefore we had to handle that case separately.

5.3.2 *Some Further Examples of Strong Induction*

We'll continue this section with several more examples of proofs by strong induction. We'll first turn to a proof about *prime factorization* of integers, and then look at one geometric and one algorithmic claim.

Prime factorization

Recall that an integer $n \geq 2$ is called *prime* if the only positive integers that evenly divide it are 1 and n itself. It's a basic fact about numbers that any positive integer can be uniquely expressed as the product of primes:

The prime factorization theorem is also sometimes called the *Fundamental Theorem of Arithmetic.*

Theorem 5.5 (Prime Factorization Theorem)
Let $n \in \mathbb{Z}^{\geq 1}$ be a positive integer. Then there exist $k \geq 0$ prime numbers $p_1, p_2, \ldots, p_k$ such that $n = \prod_{i=1}^{k} p_i$. Furthermore, up to reordering, the primes $p_1, p_2, \ldots, p_k$ are unique.

While proving the *uniqueness* requires a bit more work (see Section 7.3.3), we can give a proof using strong induction to show that a prime factorization *exists.*

Example 5.12 (Prime factorization)
Let $P(n)$ denote the first part of Theorem 5.5, namely the claim

$$\text{there exist } k \geq 0 \text{ prime numbers } p_1, p_2, \ldots, p_k \text{ such that } n = \prod_{i=1}^{k} p_i.$$

We will prove that $P(n)$ holds for any integer $n \geq 1$, by strong induction on n.

base case ($n = 1$): Recall that the product of zero multiplicands is 1. (See Section 2.2.7.) Thus we can write n as the product of *zero* prime numbers. Thus $P(1)$ holds.

inductive case ($n \geq 2$): We assume the inductive hypothesis—namely, we assume that $P(n')$ holds for any positive integer n' where $1 \leq n' \leq n-1$. We must prove $P(n)$. There are two cases:

- If n is prime, then there's nothing to do: define $p_1 := n$, and we're done immediately. (We've written n as the product of 1 prime number.)
- If n is not prime, then by definition n can be written as the product $n = a \cdot b$, for positive integers a and b satisfying $2 \leq a \leq n-1$ and $2 \leq b \leq n-1$. (The definition of (non)primality says that $n = a \cdot b$ for $a \notin \{1, n\}$; it should be easy to

convince yourself that neither a nor b can be smaller than 2 or larger than $n - 1$.) By the inductive hypotheses $P(a)$ and $P(b)$, we have

$$a = q_1 \cdot q_2 \cdot \cdots \cdot q_\ell \qquad \text{and} \qquad b = r_1 \cdot r_2 \cdot \cdots \cdot r_m \qquad (*)$$

for prime numbers $q_1, \ldots, q_\ell$ and $r_1, \ldots, r_m$. By $(*)$ and the fact that $n = a \cdot b$,

$$n = q_1 \cdot q_2 \cdot \cdots \cdot q_\ell \cdot r_1 \cdot r_2 \cdot \cdots \cdot r_m.$$

Because each q_i and r_i is prime, we have now written n as the product of $\ell + m$ prime numbers, and $P(n)$ holds. The theorem follows.

Taking it further: As with any inductive proof, it may be useful to view the proof from Example 5.12 as a recursive algorithm, as shown in Figure 5.17. (Notice that there's some magic in the "algorithm," in the sense that Line 7 doesn't tell us *how* to find the values of a and b—but we do know that such values exist, by definition.) We can think of the inductive case of an inductive proof as "making a recursive call" to a proof for a smaller input.

```
primeFactor(n):
1: if n = 1 then
2:     return ⟨⟩                                             or "P(1) is true!"
3: else
4:     if n is prime then
5:         return ⟨n⟩                                        or "P(n) is true!"
6:     else
7:         find factors a, b where 2 ≤ a ≤ n − 1 and 2 ≤ b ≤ n − 1 such that n = a · b.
8:         ⟨q_1, ..., q_k⟩ := primeFactor(a)
9:         ⟨r_1, ..., r_m⟩ := primeFactor(b)
10:        return ⟨q_1, ..., q_k, r_1, ..., r_m⟩    or "P(n) is true, because P(a) ∧ P(b)!"
```

Figure 5.17: The proof of Example 5.12, interpreted as a recursive algorithm.

For example, **primeFactor**(2) returns $\langle 2 \rangle$ and **primeFactor**(5) returns $\langle 5 \rangle$, because both 2 and 5 are prime. For another example, the result of **primeFactor**(10) is $\langle 2, 5 \rangle$, because 10 is not prime, but we can write $10 = 2 \cdot 5$ and **primeFactor**(2) returns $\langle 2 \rangle$ and **primeFactor**(5) returns $\langle 5 \rangle$. The result of **primeFactor**(70) could be $\langle 7, 2, 5 \rangle$, because 70 is not prime, but we can write $70 = 7 \cdot 10$ and **primeFactor**(7) returns $\langle 7 \rangle$ and **primeFactor**(10) returns $\langle 2, 5 \rangle$. Or **primeFactor**(70) could be $\langle 7, 5, 2 \rangle$ because $70 = 35 \cdot 2$, and **primeFactor**(35) returns $\langle 7, 5 \rangle$ and **primeFactor**(2) returns $\langle 2 \rangle$. (Which ordering of the values is the output depends on the magic of Line 7. The second part of Theorem 5.5, about the uniqueness of the prime factorization, says that it is only the ordering of these numbers that depends on the magic; the numbers themselves must the same.)

Triangulating a polygon

We'll now turn to a proof by strong induction about a geometric question, instead of a numerical one. A *convex polygon* is, informally, the points "inside" a set of n vertices: imagine stretching a giant rubber band around n points in the plane; the polygon is defined as the set of all points contained inside the rubber band. See Figure 5.18 for an example. Here we will show that an arbitrary convex polygon can be decomposed into a collection of nonoverlapping triangles.

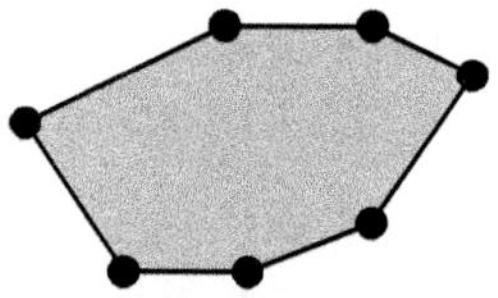

Figure 5.18: A polygon. The dots are called *vertices*; the lines connecting them are the *sides*; and the shaded region (excluding the boundary) is the *interior*.

Example 5.13 (Decomposing a polygon into triangles)

Problem: Prove the following claim:

Claim: Any convex polygon P with $k \geq 3$ vertices can be decomposed into a set of $k - 2$ triangles whose interiors do not overlap.

(For an example, and an outline of a possible proof, see Figure 5.19.)

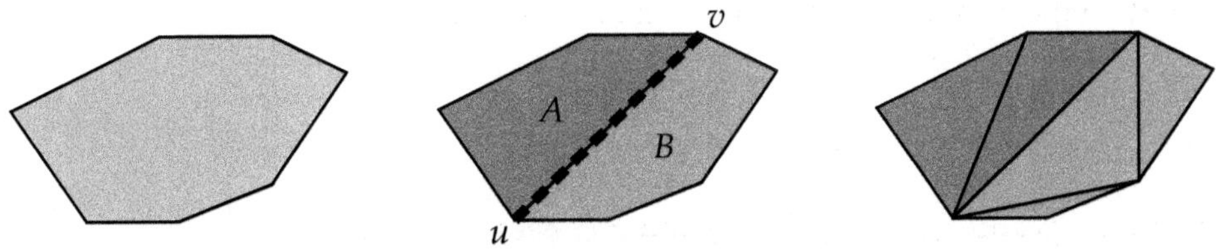

(a) The original polygon P.

(b) Two vertices u, v of P, and P divided into A and B (above and below the $\langle u, v \rangle$ line).

(c) The subpolygons A and B divided into triangles, using the inductive hypothesis.

Figure 5.19: An example of the recursive decomposition of a polygon into interior-disjoint triangles.

Solution: Let $Q(k)$ denote the claim that any k-vertex polygon can be decomposed into a set of $k - 2$ interior-disjoint triangles. We'll give a proof by strong induction on k that $Q(k)$ holds for all $k \geq 3$. (Note that strong induction isn't strictly necessary to prove this claim; we could give an alternative proof using weak induction.)

base case ($k = 3$): There's nothing to do: any 3-vertex polygon P is itself a triangle, so the collection $\{P\}$ is a set of $k - 2 = 1$ triangles whose interiors do not intersect (vacuously, because there is only one triangle). Thus $Q(3)$ holds.

inductive case ($k \geq 4$): We assume the inductive hypothesis: any convex polygon with $3 \leq \ell < k$ vertices can be decomposed into a set of $\ell - 2$ interior-disjoint triangles. (That is, we assume $Q(3), Q(4), \ldots, Q(k-1)$.) We must prove $Q(k)$.

Let P be an arbitrary k-vertex polygon. Let u and v be any two nonadjacent vertices of P. (Because $k \geq 4$, such a pair exists.) Define A as the "above the $\langle u, v \rangle$ line" piece of P and B as the "below the $\langle u, v \rangle$ line" piece of P. Notice that $P = A \cup B$, both A and B are convex, and the interiors of A and B are disjoint. Let ℓ be the number of vertices in A. Observe that $\ell \geq 3$ and $\ell < k$ because u and v are nonadjacent. Also observe that B contains precisely $k - \ell + 2$ vertices. (The "+2" is because vertices u and v appear in both A and B.) Note that both $3 \leq \ell \leq k - 1$ and $3 \leq k - \ell + 2 \leq k - 1$, so we can apply the inductive hypothesis to both ℓ and $k - \ell + 2$.

Therefore, by the inductive hypothesis $Q(\ell)$, the polygon A is decomposable into a set S of $\ell - 2$ interior-disjoint triangles. Again by the inductive hypothesis $Q(k - \ell + 2)$, the polygon B is decomposable into a set T of $k - \ell + 2 - 2 = k - \ell$ interior-disjoint triangles. Furthermore because A and B are interior disjoint, the triangles of $S \cup T$ all have disjoint interiors. Thus P itself can be decomposed into the union of these two sets of triangles, yielding a total of $\ell - 2 + k - \ell = k - 2$ interior-disjoint triangles.

We've shown both $Q(3)$ and $Q(3) \land \cdots \land Q(k-1) \Rightarrow Q(k)$ for any $k \geq 4$, which completes the proof by strong induction.

Taking it further: The style of *triangulation* from Example 5.13 has particularly important implications in computer graphics, in which we seek to render representations of complicated real-world scenes using computational techniques. In many computer graphics applications, complex surfaces are decomposed into small triangular regions, which are then rendered individually. See p. 528 for more discussion.

```
quickSort(A[1...n]):
if n ≤ 1 then
    return A
else
    choose pivot ∈ {1,...,n}, somehow.
    L := ⟨⟩
    R := ⟨⟩
    for i ∈ {1,...,n} with i ≠ pivot:
        if A[i] < A[pivot] then
            append A[i] to L
        else
            append A[i] to R
    L := quickSort(L)
    R := quickSort(R)
    return L + ⟨A[pivot]⟩ + R
```

(a) The pseudocode.

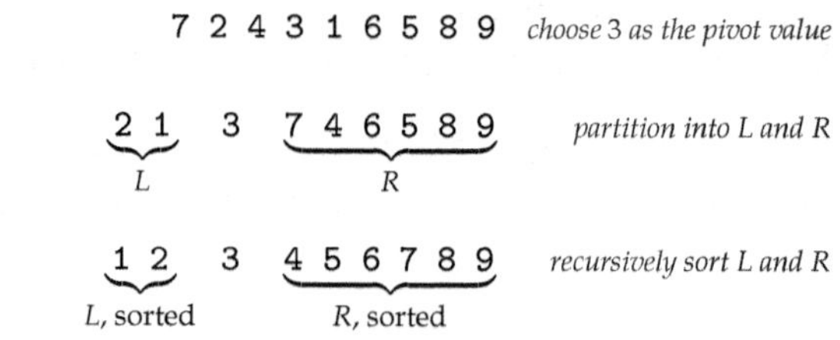

(b) An example of quick sort. Starting from an array 724316589, we (through whatever mechanism) choose 3 as the pivot value, divide the array into the elements < 3 and those > 3, and recursively sort those two pieces.

Figure 5.20: Quick Sort: pseudocode, and an example.

Proving algorithms correct: Quick Sort

We've now seen a proof of correctness by strong induction for a simple recursive algorithm (for parity), and proofs of somewhat more complicated non-algorithmic properties. Here we'll prove the correctness of a somewhat more complicated algorithm—the recursive sorting algorithm called *Quick Sort*—again using strong induction.

The idea of the Quick Sort algorithm is to select a *pivot* value x from an input array A; we then partition the elements of A into those less than x (which we then sort recursively), then x itself, and finally the elements of A greater than x (which we again sort recursively). We also need a base case: an input array with fewer than 2 elements is already sorted. (See Figure 5.20(a) for the algorithm.) For example, suppose we wish to sort all 43 U.S. Presidents by birthday. (Grover Cleveland will appear only once.) Barack Obama's birthday is August 4th. If we choose him as the pivot, then Quick Sort would first divide all the other presidents into two lists, of those with pre–August 4th and post–August 4th birthdays,

Even without two Grover Cleveland entries in the array, the simplifying assumption that we're making about distinct elements actually doesn't apply for the U.S. Presidents: James Polk and Warren Harding were both born on November 2nd. (Think about how you'd modify the proof that follows to handle duplicates.)

$$\textit{before}[1\ldots 23] := \langle \text{George Washington [February 22nd]}, \ldots, \text{George W. Bush [July 6th]} \rangle$$
$$\textit{after}[1\ldots 19] := \langle \text{John Adams [October 30th]}, \ldots, \text{Bill Clinton [August 19th]} \rangle,$$

and then recursively sort *before* and *after*. Then the final sorted list will be

before in sorted order	Barack Obama	*after* in sorted order
prez[1], ..., *prez*[23],	*prez*[24],	*prez*[25], ..., *prez*[43]

(See Figure 5.20(b) for another example of Quick Sort.)

While the efficiency of Quick Sort depends crucially on *how* we choose the pivot value (see Chapter 6), the correctness of the algorithm holds regardless of that choice. For simplicity, we will prove that Quick Sort correctly sorts its input under the assumption that all the elements of the input array A are distinct. (The more general case, in which there may be duplicate elements, is conceptually no harder, but is a bit more tedious.) It is easy to see by inspection of the algorithm that **quickSort**(A) returns a reordering of the input array A—that is, Quick Sort neither deletes or inserts elements. Thus the real work is to prove that Quick Sort returns a sorted array:

Example 5.14 (Correctness of Quick Sort)

Claim: For any array A with distinct elements, **quickSort**(A) returns a sorted array.

Proof. Let $P(n)$ denote the claim that **quickSort**($A[1 \ldots n]$) returns a sorted array for any n-element array A with distinct elements. We'll prove $P(n)$ for every $n \geq 0$, by strong induction on n.

base cases ($n = 0$ and $n = 1$): Both $P(0)$ and $P(1)$ are trivial: any array of length 0 or 1 is sorted.

inductive case ($n \geq 2$): We assume the inductive hypothesis $P(0), \ldots, P(n-1)$: for any array $B[1 \ldots k]$ with distinct elements and length $k < n$, **quickSort**(B) returns a sorted array. We must prove $P(n)$. Let $A[1 \ldots n]$ be an arbitrary array with distinct elements. Let $pivot \in \{1, \ldots, n\}$ be arbitrary. We must prove that x appears before y in **quickSort**(A) if and only if $x < y$. We proceed by cases, based on the relationship between x, y, and $A[pivot]$. (See Figure 5.21.)

Case 1: $x = A[pivot]$. The elements appearing after x in **quickSort**(A) are precisely the elements of R. And R is exactly the set of elements greater than x. Thus x appears before y if and only if y appears in R, which occurs if and only if $x < y$.

Case 2: $y = A[pivot]$. Analogously to Case 1, x appears before y if and only if x appears in L, which occurs if and only if $x < y$.

Case 3: $x < A[pivot]$ and $y < A[pivot]$. Then both x and y appear in L. Because $A[pivot]$ does *not* appear in L, we know that L contains at most $n - 1$ elements, all of which are distinct because they're a subset of the distinct elements of A. Thus the inductive hypothesis $P(|L|)$ says that x appears before y in **quickSort**(L) if and only if $x < y$. And x appears before y in **quickSort**(A) if and only if x appears before y in **quickSort**(L).

Case 4: $x > A[pivot]$ and $y > A[pivot]$. Then both x and y appear in R. An analogous argument to Case 3 shows that x appears before y if and only if $x < y$.

Case 5: $x < A[pivot]$ and $y > A[pivot]$. It is immediate both that x appears before y (because x is in L and y is in R) and that $x < y$.

Case 6: $x > A[pivot]$ and $y < A[pivot]$. It is immediately apparent that x does not appear before y and that $x \not< y$.

In all six cases, we have established that $x < y$ if and only if x appears before y in the output array; furthermore, the cases are exhaustive. The claim follows. □

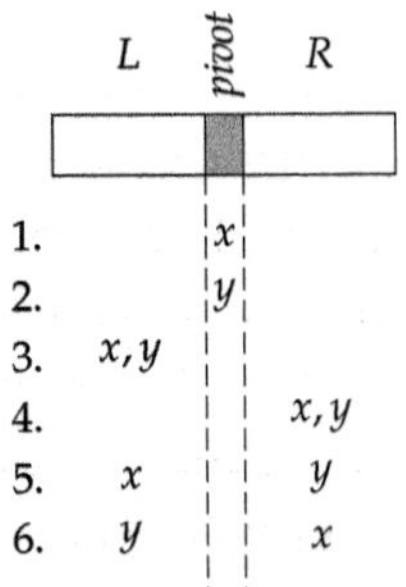

Figure 5.21: The cases of the proof in Example 5.14.

Taking it further: In addition to proofs of correctness for algorithms, like the one for **quickSort** that we just gave, strong induction is crucial in analyzing the efficiency of recursive algorithms; we'll see many examples in Section 6.4. And strong induction can also be fruitfully applied to understanding (and designing!) data structures—for example, see p. 529 for a discussion of *maximum heaps*.

COMPUTER SCIENCE CONNECTIONS

TRIANGULATION, COMPUTER GRAPHICS, AND 3D SURFACES

Here is a typical problem in computer graphics: we are given a three-dimensional *scene* consisting of a collection of objects of various shapes and sizes, and we must render a two-dimensional *image* that is a visual display of the scene. (Computer graphics uses a lot of matrix computation to facilitate the *projection* of a 3-dimensional shape onto a 2-dimensional surface.)

A typical approach—to simplify and speed the algorithms for displaying these scenes—approximates the three-dimensional shapes of the objects in the scene using triangles instead of arbitrary shapes. Triangles are the easiest shape to process computationally: the "real" triangle in the scene can be specified completely by three 3-dimensional points corresponding to the vertices; and the rendered shape in the image is still a triangle specified completely by 2-dimensional points corresponding to the vertices' projections onto the image. Specialized hardware called a *graphics processing unit (GPU)* makes these computations extremely fast on many modern computers.

When rendering a scene, we might compute a single color c that best represents the color of a given triangle in the real scene, and then display a solid c-colored (projected) triangle in the image. We can approximate any three-dimensional shape arbitrarily well using a collection of triangles, and we can *refine* a triangulation by dividing splitting one triangle into two pieces, and then properly rendering each constituent triangle:

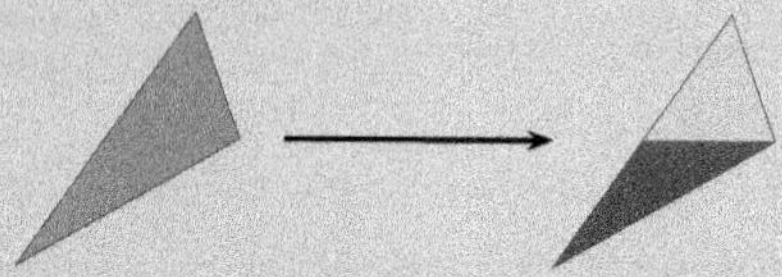

Note that there are many different ways to subdivide a given triangle into two separate triangles. Which subdivision we pick might depend on the geometry of the scene; for example, we might try to make the subtriangles roughly similar in size, or maximally different in color.

The larger the number of triangles we use, the better the match between the real 3-dimensional shape and the triangulated approximation. But, of course, the more triangles we use, the more computation must be done (and the slower the rendering will be). By identifying particularly important triangles—for example, those whose colors vary particularly widely, or those at a particularly steep angle to their neighbors, or those whose angles to the viewer are particularly extreme—we can selectively refine "the most important parts" of the triangulation to produce higher quality images.[2] (See Figure 5.22.)

Figure 5.22: Three strategies for refining a triangulation of a rabbit. Reprinted, with permission, from:

[2] Tobias Isenberg, Knut Hartmann, and Henry König. Interest value driven adaptive subdivision. In *Simulation and Visualisation (SimVis)*, pages 139–149. SCS European Publishing House, 2003.

Computer Science Connections

Max Heaps

When we design data structures to support particular operations, it is often the case that we wish to maintain some properties in the way that the data are stored. Here's one example, for an implementation of *priority queues*, that we'll establish using a proof by mathematical induction. A priority queue is a data structure that stores a set of jobs, each of which has a *priority*; we wish to be able to insert new jobs (with specified priorities) and identify/extract the existing job with the highest priority.

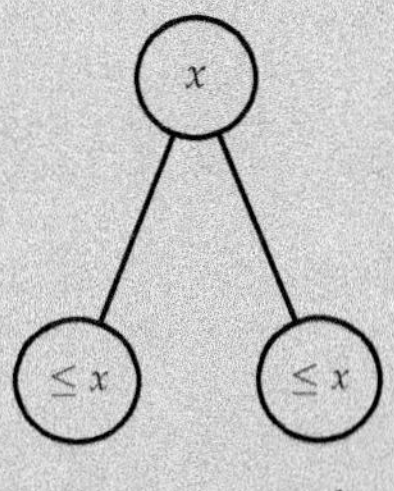

Figure 5.23: The maximum heap property. For a node with value x, the children must have values $\leq x$.

A *maximum heap* is one way of implementing priority queues. A maximum heap is a binary tree—see Section 5.4 or Chapter 11—in which every node stores a job with an associated priority. Every node in the tree satisfies the *(maximum) heap property* (see Figure 5.23): the priority of node u must be greater than or equal to the priorities of each of u's children. (A heap must also satisfy another property, being "nearly complete"—intuitively, a heap has no "missing nodes" except in the bottommost layer; this "nearly complete" property is what guarantees that heaps implement priority queues very efficiently.) An example of a heap is shown in Figure 5.24.

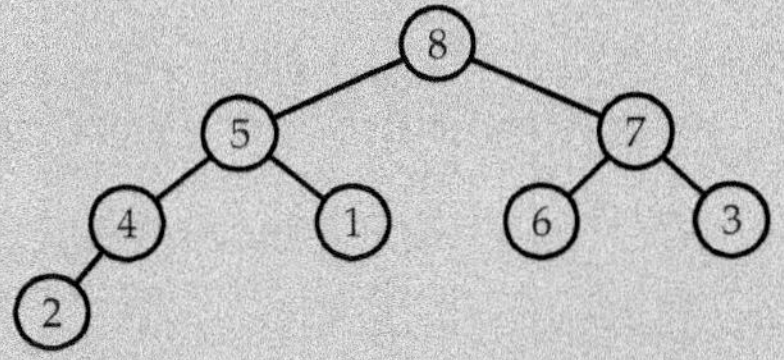

Figure 5.24: A maximum heap.

It's easy to check that the topmost node (the *root*) of the maximum heap in Figure 5.24 has the highest priority. Heaps are designed so that the root of the tree contains the node with the highest priority—but this claim requires proof. Here is a proof by induction:

Claim: In any binary tree in which every node satisfies the maximum heap property, the node with the highest priority is the root.

Proof. We'll proceed by strong induction on the number of layers of nodes in the tree. (This proof is an example of a situation in which it's not immediately clear upon what quantity to perform induction, but once we've chosen the quantity well, the proof itself is fairly easy.) Let $P(\ell)$ denote the claim

> In any tree containing ℓ layers of nodes, in which every node satisfies the maximum heap property, the node with the highest priority is the root of the tree.

We will prove that $P(\ell)$ holds for all $\ell \geq 1$ by strong induction on ℓ.

base case ($\ell = 1$): The tree has only one level—that is, the root *is* the only node in the tree. Thus, vacuously, the root has the highest priority, because there are no other nodes.

inductive case ($\ell \geq 2$): We assume the inductive hypothesis $P(1), \ldots, P(\ell - 1)$. Let x be the priority of the root of the tree. If the root has only one child, say with priority a, then by the inductive hypothesis every element y beneath a satisfies $y \leq a$. (There are at most $\ell - 1$ layers in the tree beneath a, so the inductive hypothesis applies.) By the heap property, we know $a \leq x$, and thus every element y satisfies $y \leq x$. If the root has a second child, say with priority b, then by the inductive hypothesis every element z beneath b satisfies $z \leq b$. (There are at most $\ell - 1$ layers in the tree beneath b, so the inductive hypothesis applies again.) Again, by the heap property, we have $b \leq x$, so every element z satisfies $z \leq x$. □

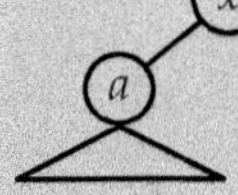

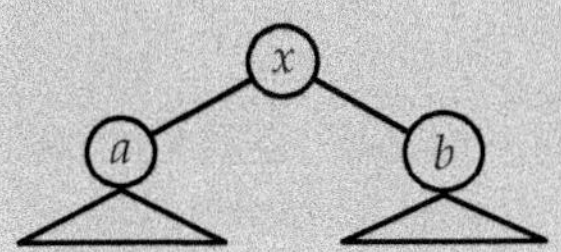

5.3.3 *Exercises*

5.41 In Example 5.11, we showed the correctness of the **parity** function (see Figure 5.25)—that is, for any $n \geq 0$, we have that **parity**$(n) = n \bmod 2$. Prove by strong induction on n that the *depth* of the recursion (that is, the total number of calls to **parity** made) for **parity**(n) is $1 + \lfloor n/2 \rfloor$.

5.42 Consider the algorithm in Figure 5.25, which finds the binary representation of a given integer $n \geq 0$. For example, **toBinary**$(13) = \langle 1, 1, 0, 1 \rangle$, and $1 \cdot 2^3 + 1 \cdot 2^2 + 0 \cdot 2^1 + 1 \cdot 2^0 = 8 + 4 + 0 + 1 = 13$.

Prove the correctness of **toBinary** by strong induction—that is, prove that for any $n \geq 0$, we have $\sum_{i=0}^{k} b_i 2^i = n$ where **toBinary**$(n) = \langle b_k, \ldots, b_0 \rangle$.

```
parity(n):                          // assume that n ≥ 0 is an integer.
  if n ≤ 1 then
     return n
  else
     return parity(n − 2)
```

```
toBinary(n):                        // assume that n ≥ 0 is an integer.
  if n ≤ 1 then
     return ⟨n⟩
  else
     ⟨b_k, ..., b_0⟩ := toBinary(⌊n/2⌋)
     x := parity(n)
     return ⟨b_k, ..., b_0, x⟩
```

Figure 5.25: A reminder of the parity algorithm (from Figure 5.16), and an algorithm to convert an integer to binary.

Your proof of the correctness of **toBinary**(n) *establishes that any nonnegative integer can be represented in binary. Now you'll show that this binary representation is unique—or, at least, unique up to leading zeros. (For example, we can represent 7 in binary as* 111 *or* 0111 *or* 00111*, but only* 111 *has no leading zeros.)*

5.43 Prove that every nonnegative integer n that can be represented as a k-bit string is *uniquely* represented as a k-bit bitstring. In other words, prove the following claim, for any integer $k \geq 1$:

> Let $a := \langle a_k, a_{k-1}, \ldots, a_0 \rangle$ and $b := \langle b_k, b_{k-1}, \ldots, b_0 \rangle$ be two k-bit sequences.
> If $\sum_{i=0}^{k} a_i 2^i = \sum_{i=0}^{k} b_i 2^i$, then for all $i \in \{k, k-1, \ldots, 0\}$ we have $a_i = b_i$.

Your proof should be by (weak) induction on k.

In Chapter 7, we'll talk in a great deal more detail about modular arithmetic, and we'll discuss a more general algorithm for converting from one base to another on *p. 714**. In Chapter 7, we'll do most of the computation iteratively; here you'll fill in a few pieces recursively.*

5.44 Generalize the **parity**(n) algorithm to **remainder**(n, k) to recursively compute the number $r \in \{0, 1, \ldots, k-1\}$ such that **remainder**$(n, k) = n \bmod k$. Assume that $k \geq 1$ and $n \geq 0$ are both integers, and follow the same algorithmic outline as in Figure 5.25. Prove your algorithm correct using strong induction on n.

5.45 Generalize the **toBinary**(n) algorithm to **baseConvert**(n, k) to recursively convert the integer n to base k. Assume that $k \geq 2$ and $n \geq 0$ are both integers, and follow the same algorithmic outline as in Figure 5.25. Prove using strong induction on n that if **baseConvert**$(n, k) = \langle b_\ell, b_{\ell-1}, \ldots, b_0 \rangle$ with each $b_i \in \{0, 1, \ldots, k-1\}$, then $n = \sum_{i=0}^{\ell} k^i b_i$.

5.46 Prove by strong induction on n that, for every integer $n \geq 4$, it is possible to make n dollars using only two- and five-dollar bills. (That is, prove that any integer $n \geq 4$ can be written as $n = 2a + 5b$ for some integer $a \geq 0$ and some integer $b \geq 0$.)

5.47 Consider a sport in which teams can score two types of goals, worth either 3 points or 7 points. For example, Team Vikings might (theoretically speaking) score 32 points by accumulating, in succession, 3, 7, 3, 7, 3, 3, 3, and 3 points. Find the smallest possible n_0 such that, for any $n \geq n_0$, a team can score exactly n points in a game. Prove your answer correct by strong induction.

5.48 You are sitting around the table with a crony you're in cahoots with. You and the crony decide to play the following silly game. (The two of you run a store called the Cis-Patriarchal Pet Shop that sells nothing but vicious robotic dogs. The loser of the game has to clean up the yard where the dogs roam—not a pleasant chore—so the stakes are high.) We start with $n \in \mathbb{Z}^{\geq 1}$ stolen credit cards on a table. The two players take turns removing cards from the table. In a single turn, a player can choose to remove either one or two cards. A player wins by taking the last credit card. (See Figure 5.26.)

Prove (by strong induction on n) that if n is divisible by three, then the second player to move can guarantee a win, and if n is not divisible by three, then the first player to move can guarantee a win.

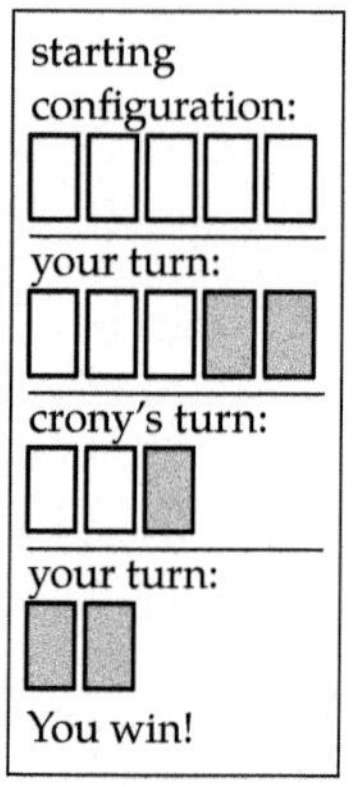

Figure 5.26: You start with $n = 5$ cards on the table, and you make the first move. You win because you took the last card.

Consider the following modifications of the game from Exercise 5.48. The two players start with n cards on the table, as before. Determine who wins the modified game: conjecture a condition on n that describes precisely when the first player can guarantee a win under the stated modification, and prove your answer.

5.49 Let $k \geq 2$ be any integer. As in the original game, the player who takes the last card wins—but each player is now allowed to take *any number of cards between* 1 *and* k in any single move.

5.50 As in the original game, players can take only 1 or 2 cards per turn—but the player who is forced to take the last card *loses* (instead of winning by managing to take the last card).

Define the Fibonacci numbers *by the sequence* $f_1 = 1, f_2 = 1$, *and* $f_n = f_{n-1} + f_{n-2}$ *for* $n \geq 3$. *Thus the first several Fibonacci numbers are* $1, 1, 2, 3, 5, 8, 13, 21, 34, 55, \ldots$. *(We'll see a lot more about the Fibonacci numbers in Section 6.4.) Prove each of the following statements by induction (weak or strong, as appropriate) on* n:

The Fibonacci numbers are named after Leonardo of Pisa (also sometimes known as Leonardo Bonacci or just as Fibonacci), a 13th-century Italian mathematician.

5.51 $f_n \bmod 2 = 0$ if and only if $n \bmod 3 = 0$. (That is, every third Fibonacci number is even.)

5.52 $f_n \bmod 3 = 0$ if and only if $n \bmod 4 = 0$.

5.53 $\sum_{i=1}^{n} f_i = f_{n+2} - 1$

5.54 $\sum_{i=1}^{n} (f_i)^2 = f_n \cdot f_{n+1}$

5.55 Prove *Cassini's identity*: $f_{n-1} \cdot f_{n+1} - (f_n)^2 = (-1)^n$ for any $n \geq 2$.

5.56 For a k-by-k matrix M, the matrix M^n is also k-by-k, and its value is the result of the n-fold multiplication of M by itself: $MM \cdots M$. Or we can define matrix exponentiation recursively: $M^0 := I$ (the k-by-k identity matrix), and $M^{n+1} := M \cdot M^n$. With this definition in mind, prove the following identity:

$$\begin{bmatrix} 1 & 1 \\ 1 & 0 \end{bmatrix}^{n-1} \cdot \begin{bmatrix} 1 \\ 0 \end{bmatrix} = \begin{bmatrix} f_n \\ f_{n-1} \end{bmatrix} \text{ for any } n \geq 2.$$

You may use the *associativity* of matrix multiplication in your answer: for any matrices A, B, and C of the appropriate dimensions, we have $A(BC) = (AB)C$.

Define the Lucas numbers *as* $L_1 = 1, L_2 = 3$, *and* $L_n = L_{n-1} + L_{n-2}$ *for* $n \geq 3$. *(The Fibonacci numbers are a much more famous cousin of the Lucas numbers; the Lucas numbers follow the same recursive definition as the Fibonacci numbers, but start from a different pair of base cases.) Prove the following facts about the Lucas numbers, by induction (weak or strong, as appropriate) on* n:

The Lucas numbers and Jacobsthal numbers are named after Édouard Lucas, a 19th-century French mathematician, and Ernst Jacobsthal, a 20th-century German mathematician, respectively.

5.57 $L_n = f_n + 2f_{n-1}$

5.58 $f_n = \dfrac{L_{n-1} + L_{n+1}}{5}$

5.59 $(L_n)^2 = 5(f_n)^2 + 4(-1)^n$

(Hint: for Exercise 5.59, you may need to conjecture a second property relating Lucas and Fibonacci numbers to complete the proof of the given property $P(n)$*—specifically, try to formulate a property* $Q(n)$ *relating* $L_n L_{n-1}$ *and* $f_n f_{n-1}$, *and prove* $P(n) \wedge Q(n)$ *with a single proof by strong induction.)*

Define the Jacobsthal numbers *as* $J_1 = 1, J_2 = 1$, *and* $J_n = J_{n-1} + 2J_{n-2}$ *for* $n \geq 3$. *(Thus the Jacobsthal numbers are a more distant relative of the Fibonacci numbers: they have the same base case, but a different recursive definition.) Prove the following facts about the Jacobsthal numbers by induction (weak or strong, as appropriate) on* n:

5.60 $J_n = 2J_{n-1} + (-1)^{n-1}$, for all $n \geq 2$.

5.61 $J_n = \dfrac{2^n - (-1)^n}{3}$

5.62 $J_n = 2^{n-1} - J_{n-1}$, for all $n \geq 2$.

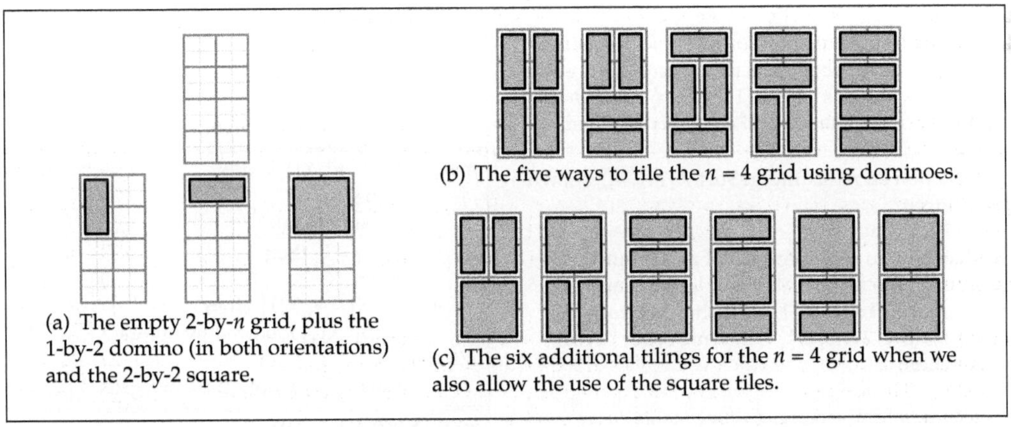

(a) The empty 2-by-n grid, plus the 1-by-2 domino (in both orientations) and the 2-by-2 square.

(b) The five ways to tile the $n = 4$ grid using dominoes.

(c) The six additional tilings for the $n = 4$ grid when we also allow the use of the square tiles.

Figure 5.27: A tiling problem, using 1-by-2 *dominoes* and 2-by-2 *squares*.

The next two problems are previews of Chapter 9, where we'll talk about how to count the size of sets (often, sets that are described in somewhat complicated ways). You should be able to attack these problems without the detailed results from Chapter 9, but feel free to glance ahead to Section 9.2 if you'd like.

5.63 You are given a 2-by-n grid that you must tile, using either 1-by-2 *dominoes* or 2-by-2 *squares*. The dominoes can be arranged either vertically or horizontally. (See Figure 5.27.) Prove by strong induction on n that the number of different ways of tiling the 2-by-n grid is precisely J_{n+1}. (Be careful: it's easy to accidentally count some configurations twice—for example, make sure that you count only once the tiling of a 2-by-3 grid that uses three horizontal dominoes.)

5.64 Suppose that you run out of squares, so you can now only use dominoes for tiling. (See Figure 5.27(b).) How does your answer to the last exercise change? How many different tilings of a 2-by-n grid are there now? Prove your answer.

The Fibonacci word fractal *defines a sequence of bitstrings using a similar recursive description to the Fibonacci numbers. Here's the definition:*

$$s_1 := 1 \qquad s_2 := 0 \qquad \text{for } n \geq 3,\ s_n := \underbrace{s_{n-1} \circ s_{n-2}}_{\text{the concatenation of } s_{n-1} \text{ and } s_{n-2}}.$$

For example, we have $s_3 = s_2 \circ s_1 = 01$ *and* $s_4 = s_3 \circ s_2 = 010$ *and* $s_5 = s_4 \circ s_3 = 01001$ *and* $s_6 = s_5 \circ s_4 = 01001010$. *It turns out that if we delete the last two bits from* s_n*, the resulting string is a palindrome (reading the same back-to-front and front-to-back). Here you'll prove a few slightly simpler properties, using strong induction on* n*:*

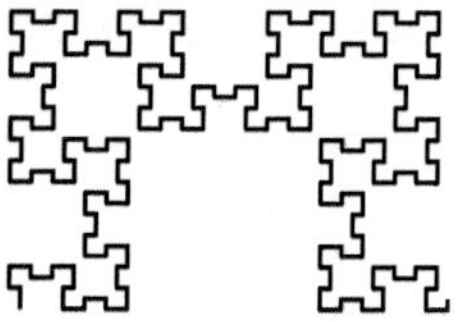

Figure 5.28: The Fibonacci word fractal s_{14}, visualized as in Exercise 5.68.

5.65 The number of bits in s_n is precisely f_n (the nth Fibonacci number).

5.66 The string s_n does not contain two consecutive 1s or three consecutive 0s.

5.67 Let $\#0(x)$ and $\#1(x)$ denote the number of 0s and 1s in a bitstring x, respectively. Show that, for all $n \geq 3$, the quantity $\#0(s_n) - \#1(s_n)$ is a Fibonacci number.

5.68 *(programming required)* The reason that s_n is called the "Fibonacci word *fractal*" is that it's possible to visualize these "words" (strings) as a geometric fractal by interpreting 0s and 1s as "turn" and "go straight," respectively. Specifically, here's the algorithm: start pointing east. For the ith symbol in s_n, for $i = 1, 2, \ldots, |s_n|$: if the symbol is 1 then do not turn; if the symbol is a 0 and i is even, turn 90° to the right; and if the symbol is a 0 and i is odd, turn 90° to the left. In any case, proceed in your current direction by one unit. (See Figure 5.28.) Write a program to draw a bitstring using these rules; then implement the recursive definition of the Fibonacci word fractal and "draw" the strings $s_1, s_2, \ldots, s_{16}$. (For efficiency's sake, you may want to compute s_n with a loop instead of recursively; see Figure 6.41 in Chapter 6 for some ideas.)

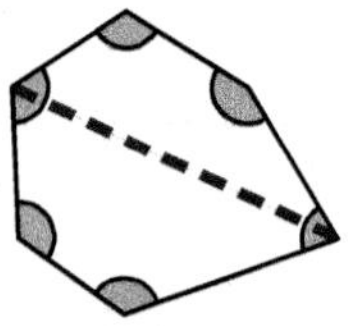

Figure 5.29: The interior angles and a diagonal for a polygon.

5.69 The sum of the interior angles of any triangle is 180°. Now, using this fact and induction, prove that any polygon with $k \geq 3$ vertices has interior angles that sum to $180k - 360$ degrees. (See Figure 5.29.)

5.70 A *diagonal* of a polygon is a line that connects two non-adjacent vertices. (See Figure 5.29.) How many diagonals are there in a triangle? A quadrilateral? A pentagon? Formulate a conjecture for the number $d(k)$ of diagonals in a k-gon, and prove your formula correct by induction. *(Hint: consider lopping off a triangle from the polygon.)*

5.71 Prove that the recursive binary search algorithm shown in Figure 5.30 is correct. That is, prove that the following condition is true, by strong induction on n: *For any sorted array* $A[1 \ldots n]$, **binarySearch**(A, x) *returns true if and only if* $x \in A$.

5.72 Prove by weak induction on the quantity $(n + m)$ that the **merge** algorithm in Figure 5.30 satisfies the following property for any $n \geq 0$ and $m \geq 0$: given any two sorted arrays $X[1 \ldots n]$ and $Y[1 \ldots m]$ as input, the output of **merge**(X, Y) is a sorted array containing all elements of X and all elements of Y.

5.73 Prove by strong induction on n that **mergeSort**$(A[1 \ldots n])$, shown in Figure 5.30, indeed sorts its input.

5.74 Give a recursive algorithm to compute a list of all permutations of a given set S. (That is, compute a list of all possible orderings of the elements of S. For example, **permutations**$(\{1,2,3\})$ should return $\{\langle 1,2,3\rangle, \langle 1,3,2\rangle, \langle 2,1,3\rangle, \langle 2,3,1\rangle, \langle 3,1,2\rangle, \langle 3,2,1\rangle\}$, in some order.) Prove your algorithm correct by induction.

Prove that weak induction, as defined in Section 5.2, and strong induction are equivalent. (Hint: in one of these two exercises, you will have to use a different predicate than P.)

5.75 Suppose that you've written a proof of $\forall n \in \mathbb{Z}^{\geq 0} : P(n)$ by weak induction. I'm in an evil mood, and I declare that you aren't allowed to prove anything by weak induction. Explain how to adapt your weak-induction proof to prove $\forall n \in \mathbb{Z}^{\geq 0} : P(n)$ using strong induction.

5.76 Now suppose that, obeying my new Draconian rules, you have written a proof of $\forall n \in \mathbb{Z}^{\geq 0} : P(n)$ by *strong* induction. In a doubly evil mood, I tell you that now you can only use weak induction to prove things. Explain how to adapt your strong-induction proof to prove $\forall n \in \mathbb{Z}^{\geq 0} : P(n)$ using weak induction.

```
binarySearch(A[1...n], x):
if n ≤ 0 then
    return False
middle := ⌊(1+n)/2⌋
if A[middle] = x then
    return True
else if A[middle] > x then
    return binarySearch(A[1...middle − 1], x)
else
    return binarySearch(A[middle + 1...n], x)
```

```
merge(X[1...n], Y[1...m]):
if n = 0 then
    return Y
else if m = 0 then
    return X
else if X[1] < Y[1] then
    return X[1] followed by merge(X[2...n], Y)
else
    return Y[1] followed by merge(X, Y[2...m])
```

```
mergeSort(A[1...n]):
if n = 1 then
    return A
else
    L := mergeSort(A[1...⌊n/2⌋])
    R := mergeSort(A[⌊n/2⌋ + 1...n])
    return merge(L, R)
```

Figure 5.30: Binary Search, Merge, and Merge Sort, recursively.

5.4 *Recursively Defined Structures and Structural Induction*

When a thing is done, it's done. Don't look back. Look forward to your next objective.

George C. Marshall (1880–1959)

In the proofs that we have written so far in this chapter, we have performed induction on an *integer*: the number that's the input to an algorithm, the number of vertices of a polygon, the number of elements in an array. In this section, we will address proofs about *recursively defined structures*, instead of about integers, using a version of induction called *structural induction* that proceeds over the defined structure itself, rather than just using numbers.

5.4.1 *Recursively Defined Structures*

A recursively defined structure, just like a recursive algorithm, is a structure defined in terms of one or more *base cases* and one or more *inductive cases*. Any data type that can be understood as either a trivial instance of the type or as being built up from a smaller instance (or smaller instances) of that type can be expressed in this way. For example, basic data structures like a *linked list* and a *binary tree* can be defined recursively. So too can well-formed sentences of a formal language—languages like Python, or propositional logic—among many other examples. In this section, we'll give recursive definitions for some of these examples.

LINKED LISTS

A *linked list* is a commonly used data structure in which we store a sequence of elements (just like the sequences from Section 2.4). The reasons that linked lists are useful are best left to a data structures course, but here is a brief synopsis of what a linked list actually is. Each element in the list, called a *node*, stores a data value and a "pointer" to the rest of the list. A special value, often called `null`, represents the empty list; the last node in the list stores this value as its pointer to represent that there are no further elements in the list. See Figure 5.31 for an example. (The slashed line in Figure 5.31 represents the `null` value.) Here is a recursive definition of a linked list:

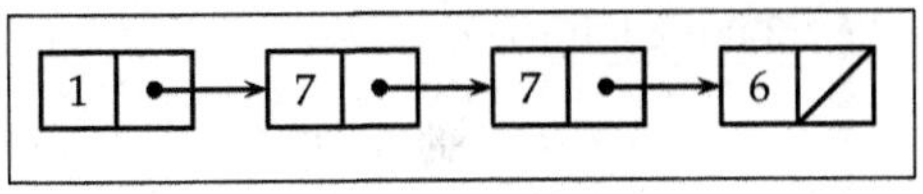

Figure 5.31: An example linked list.

Example 5.15 (Linked list)
A *linked list* is either:

1. $\langle\rangle$, known as the *empty list*; or
2. $\langle x, L\rangle$, where x is an arbitrary element and L is a linked list.

For example, Figure 5.31 shows the linked list that consists of 1 followed by the linked list containing 7, 7, and 6 (which is a linked list consisting of 7 followed by a linked list containing 7 and 6, which is a linked list consisting of 7 followed by the linked list containing 6, which is ...). That is, Figure 5.31 shows the linked list $\langle 1, \langle 7, \langle 7, \langle 6, \langle\rangle\rangle\rangle\rangle\rangle$.

Binary trees

We can also recursively define a *binary tree* (see Section 11.4.2). Again, deferring the discussion of why binary trees are useful to a course on data structures, here is a quick summary of what they are. Like a linked list, a binary tree is a collection of nodes that store data values and "pointers" to other nodes. Unlike a linked list, a node in a binary tree stores *two* pointers to other nodes (or `null`, representing an empty binary tree). These two pointers are to the *left child* and *right child* of the node. The *root* node is the one at the very top of the tree. See Figure 5.32 for an example; here the root node stores the value 1, and has a left child (the binary tree with root 3) and a right child (the binary tree with root 2). Here is a recursive definition:

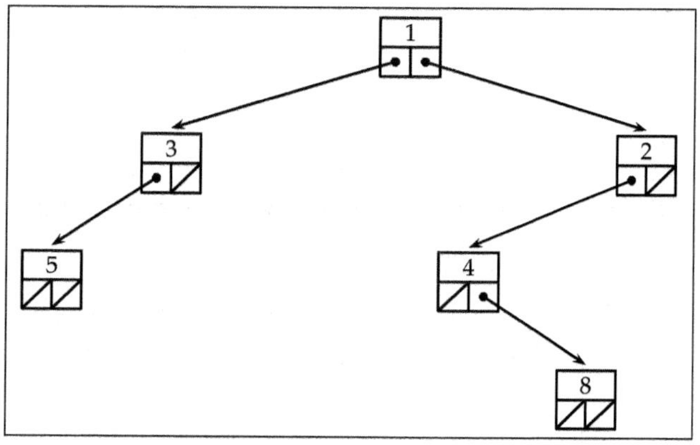

Figure 5.32: An example binary tree.

Example 5.16 (Binary trees)
A *binary tree* is either:

1. the empty tree, denoted by `null`; or
2. a *root node* x, a *left subtree* T_ℓ, and a *right subtree* T_r, where x is an arbitrary value and T_ℓ and T_r are both binary trees.

Taking it further: In many programming languages, we can explicitly define data types that echo these recursive definitions, where the base case is a trivial instance of the data structure (often `nil` or `None` or `null`). In C, for example, we can define a binary tree with integer-valued nodes as:

```
struct binaryTree {
  int root;
  struct binaryTree *leftSubtree;
  struct binaryTree *rightSubtree;
}
```

The base case—an empty binary tree—is `NULL`; the inductive case—a binary tree with a root node—has a value stored as its root, and then two binary trees (possibly empty) as its left and right subtrees. (In C, the symbol $*$ means that we're storing a *reference*, or *pointer*, to the subtrees, rather than the subtrees themselves, in the data structure.)

Define the *leaves* of a binary tree T to be those nodes contained in T whose left subtree and right subtree are both `null`. Define the *internal nodes* of T to be all nodes that are not leaves. In Figure 5.32, for example, the leaves are the nodes 5 and 8, and the internal nodes are $\{1, 2, 3, 4\}$.

Taking it further: Binary trees with certain additional properties turn out to be very useful ways of organizing data for efficient access. For example, a *binary search tree* is a binary tree in which each node stores a "key," and the tree is organized so that, for any node u, the key at node u is larger than all the keys in u's left subtree and smaller than all the keys in u's right subtree. (For example, we might store the email address of a student as a key; the tree is then organized alphabetically.) Another special type of a binary search tree is a *heap*, in which each node's key is larger than all the keys in its subtrees. These two data structures are very useful in making certain common operations very efficient; see p. 529 (for heaps) and p. 1160 (for binary search trees) for more discussion.

Sentences in a language

In addition to data structures, we can also define *sentences* in a language using a recursive definition—for example, arithmetic expressions of the type that are understood by a simple calculator; or propositions (as in Chapter 3's propositional logic):

Example 5.17 (Arithmetic expressions)
An *arithmetic expression* is any of the following:

1. any integer n;
2. $-E$, where E is an arithmetic expression; or
3. $E \odot F$, where E and F are arithmetic expressions and $\odot \in \{+, -, \cdot, /\}$ is an *operator*.

Example 5.18 (Sentences of propositional logic)
A *sentence of propositional logic* (also known as a *well-formed formula*, or *wff*) over the propositional variables X is one of the following:

1. x, for some $x \in X$;
2. $\neg P$, where P is a wff over X; or
3. $P \vee Q$, $P \wedge Q$, or $P \Rightarrow Q$, where P and Q are wffs over X.

We implicitly used the recursive definition of logical propositions from Example 5.18 throughout Chapter 3, but using this recursive definition explicitly allows us to express a number of concepts more concisely. For example, consider a truth assignment $f : X \to \{\text{True}, \text{False}\}$ that assigns True or False to each variable in X. Then the truth value of a proposition over X under the truth assignment f can be defined recursively for each case of the definition:

- the truth value of $x \in X$ under f is $f(x)$;
- the truth value of $\neg P$ under f is True if the truth value of P under f is False, and the truth value of $\neg P$ under f is False if the truth value of P under f is True;
- and so forth.

Taking it further: Linguists interested in syntax spend a lot of energy constructing recursive definitions (like those in Examples 5.17 and 5.18) of grammatical sentences of English. But one can also give a recursive definition for non-natural languages: in fact, another structure that can be defined recursively is *the grammar of a programming language itself*. As such, this type of recursive approach to defining (and processing) a grammar plays a key role not just in linguistics but also in computer science. See the discussion on p. 543 for more.

5.4.2 *Structural Induction*

The recursively defined structures from Section 5.4.1 are particularly amenable to inductive proofs. For example, recall from Example 5.16 that a binary tree is one of the following: (1) the empty tree, denoted by `null`; or (2) a *root node* x, a *left subtree* T_ℓ, and a *right subtree* T_r, where T_ℓ and T_r are both binary trees. To prove that some property P is true of all binary trees T, we can use (strong) induction on the number n of applications of rule #2 from the definition. Here is an example of such a proof:

Example 5.19 (Internal nodes vs. leaves in binary trees)
Recall that a *leaf* in a binary tree is a node whose left and right subtrees are both empty; an *internal node* is any non-leaf node. Write *leaves*(T) and *internals*(T) to denote the number of leaves and internal nodes in a binary tree T, respectively.

Claim: In any binary tree T, we have $\mathit{leaves}(T) \leq \mathit{internals}(T) + 1$.

An (abbreviated) reminder of the recursive definition of a binary tree:
Rule #1: `null` is a binary tree;
Rule #2: if T_ℓ and T_r are binary trees, then $\langle x, T_\ell, T_r \rangle$ is a binary tree.

Proof. We proceed by strong induction on the number of applications of rule #2 used to generate T. Specifically, let $P(n)$ denote the property that $\mathit{leaves}(T) \leq \mathit{internals}(T) + 1$ holds *for any binary tree T generated by n applications of rule* #2; we'll prove that $P(n)$ holds for all $n \geq 0$, which establishes the claim.

base case ($n = 0$): The only binary tree generated with 0 applications of rule #2 is the empty tree `null`. Indeed, *leaves*(`null`) = *internals*(`null`) = 0, and $0 \leq 0 + 1$.

inductive case ($n \geq 1$): Assume the inductive hypothesis $P(0) \wedge P(1) \wedge \cdots \wedge P(n-1)$: for any binary tree B generated using $k < n$ applications of rule #2, we have $\mathit{leaves}(B) \leq \mathit{internals}(B) + 1$. We must prove $P(n)$.

We'll handle the case $n = 1$ separately. (See Figure 5.33(a).) The only way to make a binary tree T using one application of rule #2 is to use rule #1 for both of T's subtrees, so T must contain only one node (which is itself a leaf). Then T contains 1 leaf and 0 internal nodes, and indeed $1 \leq 0 + 1$.

Otherwise $n \geq 2$. (See Figure 5.33(b).) Observe that the tree T must have been generated by (a) generating a left subtree T_ℓ using some number ℓ of applications of rule #2; (b) generating a right subtree T_r using some number r of applications of rule #2; and then (c) applying rule #2 to a root node x, T_ℓ, and T_r to produce T. Therefore $r + \ell + 1 = n$, and therefore $r < n$ and $\ell < n$. Ergo, we can apply the inductive hypothesis to both T_ℓ and T_r, and thus

$$\mathit{leaves}(T_\ell) \leq \mathit{internals}(T_\ell) + 1 \qquad (1)$$
$$\mathit{leaves}(T_r) \leq \mathit{internals}(T_r) + 1. \qquad (2)$$

Also observe that, because $r + \ell + 1 = n \geq 2$, either $T_r \neq$ `null` or $T_\ell \neq$ `null`, or both. Thus the leaves of T are the leaves of T_ℓ and T_r, and internal nodes of T are the internal nodes of T_ℓ and T_r plus the root x (which cannot be a leaf because at least one of T_ℓ and T_r is not empty). Therefore

$$\mathit{leaves}(T) = \mathit{leaves}(T_\ell) + \mathit{leaves}(T_r) \qquad (3)$$
$$\mathit{internals}(T) = \mathit{internals}(T_\ell) + \mathit{internals}(T_r) + 1. \qquad (4)$$

Putting together these facts, we have

$$\begin{aligned} \mathit{leaves}(T) &= \mathit{leaves}(T_\ell) + \mathit{leaves}(T_r) && \text{by (3)} \\ &\leq \mathit{internals}(T_\ell) + 1 + \mathit{internals}(T_r) + 1 && \text{by (1) and (2)} \\ &= \mathit{internals}(T) + 1. && \text{by (4)} \end{aligned}$$

Thus $P(n)$ holds, which completes the proof. □

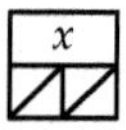

(a) The only binary tree produced by 1 application of rule #2 has one node, which is a leaf.

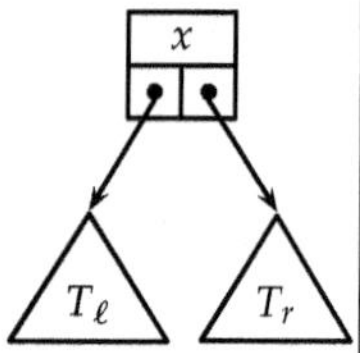

(b) If T was produced by ≥ 2 applications of rule #2, then at least one of T_ℓ and T_r is not `null`, and the leaves of T are precisely the leaves of T_ℓ plus the leaves of T_r.

Figure 5.33: Illustrations of the inductive case for Example 5.19.

Structural induction: the idea

The proof in Example 5.19 is perfectly legitimate, but there is another approach that we can use for recursively defined structures, called *structural induction*. The basic idea is to perform induction *on the structure of an object itself* rather than on some integer: instead of a case for $n = 0$ and a case for $n \geq 1$, in a proof by structural induction our cases correspond directly to the cases of the recursive structural definition.

For structural induction to make sense, we must impose some restrictions on the recursive definition. Specifically, the set of structures defined must be *well ordered*, which intuitively ensures that every invocation of the inductive case of the definition "makes progress" toward the base case(s) of the definition. (More precisely, a set of objects is well ordered if there's a "least" element among any collection of those objects.) For the type of recursive definitions that we're considering—where there are base cases in the definition, and all instances of the structure are produced by a finite-length sequence of applications of the inductive rules in the definition—structural induction is a valid technique to prove facts about the recursively defined structure.

Taking it further: More formally, a set S of structures is *well ordered* if there exists a "smaller than" relationship $\prec$ between elements of S such that, for any nonempty $T \subseteq S$, there exists a *minimal element* m in T—that is, there exists $m \in T$ such that no $x \in T$ satisfies $x \prec m$. (There might be more than one least element in T.) For example, the set $\mathbb{Z}^{\geq 0}$ is well ordered, using the normal $\leq$ relationship. However, the set $\mathbb{R}$ is not well ordered: for example, the set $\{x \in \mathbb{R} : x > 2\}$ has no smallest element using $\leq$. But the set of binary trees *is* well ordered; the relation $\prec$ is "is a subtree of."

One can prove that a set S is well ordered if and only if a proof by mathematical induction is valid on a set S (where the base cases are the minimal elements of S, and to prove $P(x)$ we assume the inductive hypotheses $P(y)$ for any $y \prec x$).

Proofs by structural induction

Here is the formal definition of a proof by structural induction:

Definition 5.6 (Proof by structural induction)
Suppose that we want to prove that $P(x)$ holds for every $x \in S$, where S is the (well-ordered) set of structures generated by a recursive definition, and P is some property. To give a proof by structural induction *of $\forall x \in S : P(x)$, we prove the following:*

1. Base cases: *for every x defined by a base case in the definition of S, prove $P(x)$.*
2. Inductive cases: *for every x defined in terms of $y_1, y_2, \ldots, y_k \in S$ by an inductive case in the definition of S, prove that $P(y_1) \wedge P(y_2) \wedge \cdots \wedge P(y_k) \Rightarrow P(x)$.*

In a proof by structural induction, we can view both base cases and inductive cases in the same light: each case assumes that the recursively constructed subpieces of a structure x satisfy the stated property, and we prove that x itself also satisfies the property. For a base case, the point is just that there *are no* recursively constructed pieces, so we actually are not making any assumption.

Notice that a proof by structural induction is identical in form to a proof by strong induction *on the number of applications of the inductive-case rules used to generate the object.* For example, we can immediately rephrase the proof in Example 5.19 to use structural induction instead. While the structure of the proof is identical, structural induction can streamline the proof and make it easier to read:

Example 5.20 (Internal nodes vs. leaves in binary trees, take II)
Claim: In any binary tree T, we have $leaves(T) \leq internals(T) + 1$.
Proof. Let $P(T)$ denote the property that $leaves(T) \leq internals(T) + 1$ for a binary tree T. We proceed by structural induction on the form of T.

base case (T = `null`): Then $leaves(T) = internals(T) = 0$, and indeed $0 \leq 0 + 1$.

inductive case (T has root x, left subtree T_ℓ, and right subtree T_r): We assume the inductive hypotheses $P(T_\ell)$ and $P(T_r)$, namely

$$leaves(T_\ell) \leq internals(T_\ell) + 1 \quad (1)$$
$$leaves(T_r) \leq internals(T_r) + 1. \quad (2)$$

- If x is itself a leaf, then $T_\ell = T_r =$ `null`, and therefore $leaves(T) = 1$ and $internals(T) = 0$, and indeed $1 \leq 0 + 1$.
- Otherwise x is not a leaf, and either $T_r \neq$ `null` or $T_\ell \neq$ `null`, or both. Thus the leaves of T are the leaves of T_ℓ and T_r, and internal nodes of T are the internal nodes of T_ℓ and T_r plus the root x. Therefore

$$leaves(T) = leaves(T_\ell) + leaves(T_r) \quad (3)$$
$$internals(T) = internals(T_\ell) + internals(T_r) + 1. \quad (4)$$

Putting together these facts, we have

$$\begin{aligned} leaves(T) &= leaves(T_\ell) + leaves(T_r) && \text{by (3)} \\ &\leq internals(T_\ell) + 1 + internals(T_r) + 1 && \text{by (1) and (2)} \\ &= internals(T) + 1. && \text{by (4)} \end{aligned}$$

Thus $P(n)$ holds, which completes the proof. □

5.4.3 *Some More Examples of Structural Induction: Propositional Logic*

We'll finish this section with two more proofs by structural induction, about propositional logic—using Example 5.18's recursive definition.

PROPOSITIONAL LOGIC USING ONLY $\neg$ AND $\wedge$

First, we'll give a formal proof using structural induction of the claim that any propositional logic statement can be expressed using $\neg$ and $\wedge$ as the only logical connectives. (See Exercise 4.68.)

Example 5.21 (All of propositional logic using $\neg$ and $\wedge$)
Claim: For any logical proposition φ using the connectives $\{\neg, \wedge, \vee, \Rightarrow\}$, there exists a proposition using only $\{\neg, \wedge\}$ that is logically equivalent to φ.

Proof. For a logical proposition φ, let $A(\varphi)$ denote the property that there exists a $\{\neg, \wedge\}$-only proposition logically equivalent to φ. We'll prove by structural induction on φ that $A(\varphi)$ holds for any well-formed formula φ (see Example 5.18):

base case: φ is a variable, say $\varphi = x$. The proposition x uses no connectives—and thus is vacuously $\{\neg, \wedge\}$-only—and is obviously logically equivalent to itself. Thus $A(x)$ follows.

inductive case I: φ is a negation, say $\varphi = \neg P$. We assume the inductive hypothesis $A(P)$. We must prove $A(\neg P)$. By the inductive hypothesis, there is a $\{\neg, \wedge\}$-only proposition Q such that $Q \equiv P$. Consider the proposition $\neg Q$. Because $Q \equiv P$, we have that $\neg Q \equiv \neg P$, and $\neg Q$ contains only the connectives $\{\neg, \wedge\}$. Thus $\neg Q$ is a $\{\neg, \wedge\}$-only proposition logically equivalent to $\neg P$. Thus $A(\neg P)$ follows.

inductive case II: φ is a conjunction, disjunction, or implication, say $\varphi = P_1 \wedge P_2$, $\varphi = P_1 \vee P_2$, or $\varphi = P_1 \Rightarrow P_2$. We assume the inductive hypotheses $A(P_1)$ and $A(P_2)$—that is, we assume there are $\{\neg, \wedge\}$-only propositions Q_1 and Q_2 with $Q_1 \equiv P_1$ and $Q_2 \equiv P_2$. We must prove $A(P_1 \wedge P_2)$, $A(P_1 \vee P_2)$, and $A(P_1 \Rightarrow P_2)$. Consider the propositions $Q_1 \wedge Q_2$, $\neg(\neg Q_1 \wedge \neg Q_2)$, and $\neg(Q_1 \wedge \neg Q_2)$. By De Morgan's Law, and the facts that $x \Rightarrow y \equiv \neg(x \wedge \neg y)$, $P_1 \equiv Q_1$, and $P_2 \equiv Q_2$:

$$\begin{aligned} Q_1 \wedge Q_2 &\equiv Q_1 \wedge Q_2 &&\equiv P_1 \wedge P_2 \\ \neg(\neg Q_1 \wedge \neg Q_2) &\equiv Q_1 \vee Q_2 &&\equiv P_1 \vee P_2 \\ \neg(Q_1 \wedge \neg Q_2) &\equiv Q_1 \Rightarrow Q_2 &&\equiv P_1 \Rightarrow P_2 \end{aligned}$$

Because Q_1 and Q_2 are $\{\neg, \wedge\}$-only, our three propositions are $\{\neg, \wedge\}$-only as well; therefore $A(P_1 \wedge P_2)$, $A(P_1 \vee P_2)$, and $A(P_1 \Rightarrow P_2)$ follow.

We've shown that $A(\varphi)$ holds for any proposition φ, so the claim follows. □

Taking it further: In the programming language ML, among others, a programmer can use both recursive definitions *and* a form of recursion that mimics structural induction. For example, we can give a simple implementation of the recursive definition of a well-formed formula from Example 5.18: a well-formed formula is a variable, or the negation of a well-formed formula, or the conjunction of a pair of well-formed formulas (`wff * wff`), or . . ..) In ML, we can also write a function that mimics the structure of the proof in Example 5.21, using ML's capability of *pattern matching* function arguments. See Figure 5.34 for both the recursive definition of the `wff` datatype and the recursive function `simplify`, which takes an arbitrary `wff` as input, and produces a `wff` that uses only `And` and `Not` as output.

```
datatype wff = Variable of string
             | Not of wff
             | And of (wff * wff)
             | Or of (wff * wff)
             | Implies of (wff * wff);


fun simplify (Variable var)        =  Variable var
  | simplify (Not P)               =  Not(simplify P)
  | simplify (And (P1, P2))        =  And(simplify P1, simplify P2)
  | simplify (Or (P1, P2))         =  Not(And(Not(simplify P1), Not(simplify P2)))
  | simplify (Implies (P1, P2)) =  Not(And(simplify P1, Not(simplify P2)));
```

Figure 5.34: Well-formed formulas in ML.

Conjunctive and Disjunctive Normal Forms

Here is another example of a proof by structural induction based on propositional logic, to establish Theorems 3.1 and 3.2, that any proposition is logically equivalent to one that's in conjunctive or disjunctive normal form.

(Recall that a proposition φ is in *conjunctive normal form (CNF)* if φ is the conjunction of one or more *clauses*, where each clause is the disjunction of one or more literals. A *literal* is a Boolean variable or the negation of a Boolean variable. A proposition φ is in *disjunctive normal form (DNF)* if φ is the disjunction of one or more *clauses*, where each clause is the conjunction of one or more literals.)

Theorem 5.6 (CNF/DNF suffice)
Let φ be a Boolean formula that uses the connectives $\{\wedge, \vee, \neg, \Rightarrow\}$. Then:

1. *there exists φ_{dnf} in disjunctive normal form so that φ and φ_{dnf} are logically equivalent.*
2. *there exists φ_{cnf} in conjunctive normal form so that φ and φ_{cnf} are logically equivalent.*

Problem-solving tip: Suppose we want to prove $\forall x : P(x)$ by induction. Here's a problem-solving strategy that's highly counterintuitive: it is sometimes easier to prove a *stronger* statement $\forall x : P(x) \wedge Q(x)$. It seems bizarre that trying to prove *more* than what we want is easier—but the advantage arises because the inductive hypothesis is a more powerful assumption! For example, I don't know how to prove that any proposition φ can be expressed in DNF (Theorem 5.6.1) by induction! But I do know how to prove that any proposition φ can be expressed in *both DNF and CNF* by induction, as is done in Example 5.22.

Perhaps bizarrely, it will turn out to be easier to prove that any proposition is logically equivalent to *both* one in CNF *and* one in DNF than to prove either claim on its own. So we will prove both parts of the theorem simultaneously, by structural induction.

We'll make use of some handy notation in this proof: analogous to summation and product notation, we write $\bigwedge_{i=1}^{n} p_i$ to denote $p_1 \wedge p_2 \wedge \cdots \wedge p_n$, and similarly $\bigvee_{i=1}^{n} p_i$ means $p_1 \vee p_2 \vee \cdots \vee p_n$. Here is the proof:

Example 5.22 (Conjunctive/disjunctive normal form)
Proof. We start by simplifying the task: we use Example 5.21 to ensure that φ contains only the connectives $\{\neg, \wedge\}$. Let $C(\varphi)$ and $D(\varphi)$, respectively, denote the property that φ is logically equivalent to a CNF proposition and a DNF proposition, respectively. We now proceed by structural induction on the form of φ—which now can only be a variable, negation, or conjunction—to show that $C(\varphi) \wedge D(\varphi)$ holds for any proposition φ.

base case: φ is a variable, say $\varphi = x$. We're done immediately; a single variable is actually in both CNF and DNF. We simply choose $\varphi_{\text{dnf}} = \varphi_{\text{cnf}} = x$. Thus $C(x)$ and $D(x)$ follow immediately.

inductive case I: φ is a negation, say $\varphi = \neg P$. We assume the inductive hypothesis $C(P) \wedge D(P)$—that is, we assume that there are propositions P_{cnf} and P_{dnf} such that $P \equiv P_{\text{cnf}} \equiv P_{\text{dnf}}$, where P_{cnf} is in CNF and P_{dnf} is in DNF. We must show $C(\neg P)$ and $D(\neg P)$.

We'll first show $D(\neg P)$—that is, that $\neg P$ can be rewritten in DNF. By the definition of conjunctive normal form, we know that the proposition P_{cnf} is of the form $P_{\text{cnf}} = \bigwedge_{i=1}^{n} c_i$, where c_i is a clause of the form $c_i = \bigvee_{j=1}^{m_i} c_i^j$, where c_i^j is a variable or its negation. Therefore we have

$$\neg P \equiv \neg P_{\text{cnf}} \equiv \neg \left[\bigwedge_{i=1}^{n} \left(\bigvee_{j=1}^{m_i} c_j^i \right) \right] \qquad \textit{inductive hypothesis } C(P) \textit{ and definition of CNF}$$

$$\equiv \left[\bigvee_{i=1}^{n} \neg \left(\bigvee_{j=1}^{m_i} c_j^i \right) \right] \qquad \textit{De Morgan's Law}$$

$$\equiv \bigvee_{i=1}^{n} \left(\bigwedge_{j=1}^{m_i} \neg c_j^i \right) \qquad \textit{De Morgan's Law}$$

Once we delete double negations (that is, if $c_i^j = \neg x$, then we write $\neg c_i^j$ as x rather than as $\neg\neg x$), this last proposition is in DNF, so $D(\neg P)$ follows.

The construction to show $C(\neg P)$—that is, to give an CNF proposition logically equivalent to $\neg P$—is strictly analogous; the only change to the argument is that we start from P_{dnf} instead of P_{cnf}.

inductive case II: φ is a conjunction, say $P \wedge Q$. We assume the inductive hypotheses $C(P) \wedge D(P)$ and $C(Q) \wedge D(Q)$—that is, we assume that there are CNF propositions P_{cnf} and Q_{cnf} and DNF propositions P_{dnf} and Q_{dnf} such that $P \equiv P_{\text{cnf}} \equiv P_{\text{dnf}}$ and $Q \equiv Q_{\text{cnf}} \equiv Q_{\text{dnf}}$. We must show $C(P \wedge Q)$ and $D(P \wedge Q)$.

- The argument for $C(P \wedge Q)$ is the easier of the two: we have propositions P_{cnf} and Q_{cnf} in CNF where $P_{\text{cnf}} \equiv P$ and $Q_{\text{cnf}} \equiv Q$. Thus $P \wedge Q \equiv P_{\text{cnf}} \wedge Q_{\text{cnf}}$—and the conjunction of two CNF formulas is itself in CNF. So $C(P \wedge Q)$ follows.
- We have to work a little harder to prove $D(P \wedge Q)$. Recall that, by the inductive hypothesis, there are propositions P_{dnf} and Q_{dnf} in DNF, where $P \equiv P_{\text{dnf}}$ and $Q \equiv Q_{\text{dnf}}$. By the definition of DNF, these propositions have the form $P_{\text{dnf}} = \bigvee_{i=1}^{n} c_i$ and $Q_{\text{dnf}} = \bigvee_{j=1}^{m} d_j$, where every c_i and d_j is a clause that is a conjunction of literals. Therefore

$$P \wedge Q \equiv P_{\text{dnf}} \wedge Q \equiv \left(\bigvee_{i=1}^{n} c_i \right) \wedge Q \qquad \textit{inductive hypothesis } D(P) \textit{ and definition of DNF}$$

$$\equiv \bigvee_{i=1}^{n} (c_i \wedge Q) \qquad \textit{distributivity of } \vee \textit{ over } \wedge$$

$$\equiv \bigvee_{i=1}^{n} \left(c_i \wedge \bigvee_{i=j}^{m} d_j \right) \qquad \textit{inductive hypothesis } D(Q) \textit{ and definition of DNF}$$

$$\equiv \bigvee_{i=1}^{n} \bigvee_{j=1}^{m} (c_i \wedge d_j) \,. \qquad \textit{distributivity of } \vee \textit{ over } \wedge$$

Because every c_i and d_j is a conjunction of literals, $c_i \wedge d_j$ is too, and thus this last proposition is in DNF! So $D(P \wedge Q)$ follows—as does the theorem. □

The construction for a conjunction $P \wedge Q$ in Theorem 5.22 is a little tricky, so let's illustrate it with a small example:

Example 5.23 (An example of the construction from Example 5.22)
Suppose that we are trying to transform a proposition $\varphi \wedge \psi$ into DNF. Suppose that we have (recursively) computed $\varphi_{\text{dnf}} = (p \wedge t) \vee q$ and $\psi_{\text{dnf}} = r \vee (s \wedge t)$. Then the construction from Example 5.22 lets us construct a proposition equivalent to $\varphi \wedge \psi$ as:

$$\begin{aligned}
\varphi \wedge \psi \equiv \varphi_{\text{dnf}} \wedge \psi_{\text{dnf}} &\equiv \Big[\underbrace{(p \wedge t)}_{c_1} \vee \underbrace{(q)}_{c_2}\Big] \wedge \Big[\underbrace{(r)}_{d_1} \vee \underbrace{(s \wedge t)}_{d_2}\Big] \\
&\equiv \Big[\underbrace{(p \wedge t)}_{c_1} \wedge \underbrace{[(r) \vee (s \wedge t)]}_{d_1 \vee d_2}\Big] \vee \Big[\underbrace{(q)}_{c_2} \wedge \underbrace{[(r) \vee (s \wedge t)]}_{d_1 \vee d_2}\Big] \\
&\equiv \Big[\underbrace{(p \wedge t \wedge r)}_{c_1 \wedge d_1} \vee \underbrace{(p \wedge t \wedge s \wedge t)}_{c_1 \wedge d_2}\Big] \vee \Big[\underbrace{(q \wedge r)}_{c_2 \wedge d_1} \vee \underbrace{(q \wedge s \wedge t)}_{c_2 \wedge d_2}\Big].
\end{aligned}$$

Then the construction yields

$$(p \wedge t \wedge r) \vee (p \wedge t \wedge s \wedge t) \vee (q \wedge r) \vee (q \wedge s \wedge t)$$

as the DNF proposition equivalent to $\varphi \wedge \psi$.

5.4.4 *The Integers, Recursively Defined*

Before we end the section, we'll close our discussion of recursively defined structures and structural induction with one more potentially interesting observation. Although the basic form of induction in Section 5.2 appears fairly different, that basic form of induction can actually be seen as structural induction, too. The key is to view the nonnegative integers $\mathbb{Z}^{\geq 0}$ as defined recursively:

Definition 5.7 (Nonnegative integers, recursively defined)
A nonnegative integer *is either:*

1. zero, *denoted by* 0*; or*
2. *the* successor *of a nonnegative integer, denoted by* s(*x*) *for a nonnegative integer x.*

Under this definition, a proof of $\forall n \in \mathbb{Z}^{\geq 0} : P(n)$ by structural induction and a proof of $\forall n \in \mathbb{Z}^{\geq 0} : P(n)$ by weak induction are identical:

- they have precisely the same base case: prove $P(0)$; and
- they have precisely the same inductive case: prove $P(n) \Rightarrow P(s(n))$—or, in other words, prove that $P(n) \Rightarrow P(n+1)$.

COMPUTER SCIENCE CONNECTIONS

GRAMMARS, PARSING, AND AMBIGUITY

In *interpreters* and *compilers*—systems that translate input source code written in a programming language like Python, Java, or C into a machine-executable format—a key initial step is to *parse* the input into a format that represents its structure. (A similar step occurs in systems designed to perform natural language processing.) The structured representation of such an expression is called a *parse tree*, in which the leaves of the tree correspond to the base cases of the recursive structural definition, and the internal nodes correspond to the inductive cases of the definition. We can then use the parse tree for whatever purpose we desire: evaluating arithmetic expressions, simplifying propositional logic, or any other manipulation. (See Figure 5.35.)

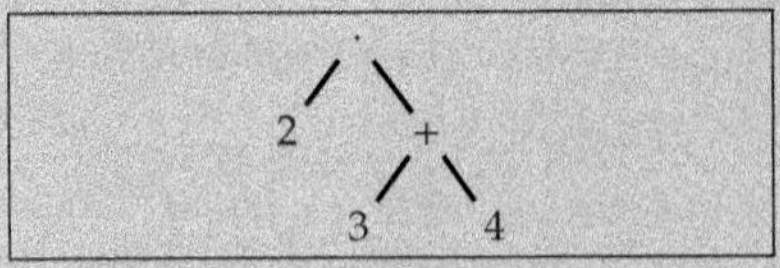

Figure 5.35: A parse tree for the arithmetic expression $2 \cdot (3 + 4)$.

In this setting, a recursively defined structure is written as a *context-free grammar (CFG)*. A grammar consists of a set of *rules* that can be used to generate a particular example of this defined structure. We'll take the definition of propositions over the variables $\{p, q, r\}$ (Example 5.18) as a running example. Here is a CFG for propositions, following that definition precisely. (Here "$\rightarrow$" means "can be rewritten as" and "$|$" means "or.")

This type of grammar is called *context free* because the rules defined by the grammar can be used any time—that is, without regard to the context in which the symbol on the left-hand side of the rule appears.

$S \rightarrow p \mid q \mid r$	S can be a propositional variable ...
$\mid \neg S$	... or the negation of a proposition ...
$\mid S \vee S \mid S \wedge S \mid S \Rightarrow S$	... or the $\wedge/\vee/\Rightarrow$ of two propositions.

An expression φ is a valid proposition over the variables $\{p, q, r\}$ if and only if φ can be generated by a finite-length sequence of applications of the rewriting rules in the grammar. For example, $\neg p \vee p$ is a valid proposition over $\{p, q, r\}$, because we can generate it as follows:

$$S \rightarrow S \vee S \rightarrow S \vee p \rightarrow \neg S \vee p \rightarrow \neg p \vee p.$$

(We used the rule $S \rightarrow p$ twice, the rule $S \rightarrow \neg S$ once, and the rule $S \rightarrow S \vee S$ once.) The parse tree corresponding to this sequence of rule applications is shown in Figure 5.36(a).

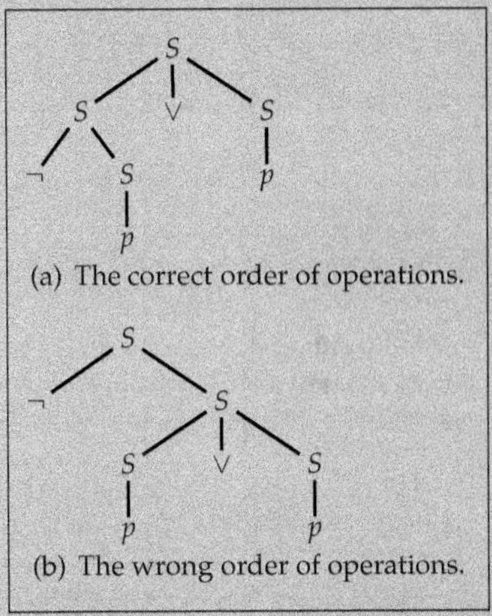

(a) The correct order of operations.

(b) The wrong order of operations.

Figure 5.36: Two parse trees for $\neg p \vee p$.

A complication that arises with the grammar given above is that it is *ambiguous*: the same proposition can be produced using a fundamentally different sequence of rule applications, which gives rise to a different parse tree, shown in Figure 5.36(b):

$$S \rightarrow \neg S \rightarrow \neg S \vee S \rightarrow \neg p \vee S \rightarrow \neg p \vee p.$$

The parse tree in Figure 5.36(b) corresponds to $\neg(p \vee p)$ instead of $(\neg p) \vee p$, which is the correct "order of operations" because $\neg$ binds tighter than $\vee$.

It's bad news if the grammar of a programming language is ambiguous, because certain valid code is then "allowed" to be interpreted in more than one way. (The classic example is the attachment of `else` clauses: in code like `if P then if Q then X else Y`, when should `Y` be executed? When `P` is true and `Q` is false? Or when `P` is false?) Thus programming language designers develop unambiguous grammars that reflect the desired behavior.[3]

More on context-free grammars and parsing, and their relationship to compilers and interpreters, can be found in books like

[3] Alfred V. Aho, Monica S. Lam, Ravi Sethi, and Jeffrey D. Ullman. *Compilers: Principles, Techniques, and Tools*. Prentice Hall, 2nd edition, 2006; Dexter Kozen. *Automata and Computability*. Springer, 1997; and Michael Sipser. *Introduction to the Theory of Computation*. Course Technology, 3rd edition, 2012.

5.4.5 *Exercises*

5.77 Let L be a linked list (as defined in Example 5.15). Prove by structural induction on L that **length**(L) returns the number of elements contained in L. (See Figure 5.37 for the algorithm.)

5.78 Let L be a linked list containing integers. Prove by structural induction on L that **sum**(L) returns the sum of the numbers contained in L. (See Figure 5.37 for the algorithm.)

```
length(L):                    // assume L is a linked list.
  if L = ⟨⟩ then
      return 0
  else if L = ⟨x, L'⟩ then
      return 1 + length(L')
```

```
sum(L):     // assume L is a linked list containing integers.
  if L = ⟨⟩ then
      return 0
  else if L = ⟨x, L'⟩ then
      return x + sum(L')
```

Figure 5.37: Two algorithms on linked lists.

5.79 In Example 5.15, we gave a recursive definition of a linked list. Here's a variant of that definition, where we insist that the elements be in increasing order. Define a *nonempty sorted list* as one of the following:

1. $\langle x, \langle\rangle\rangle$; or
2. $\langle x, \langle y, L\rangle\rangle$ where $x \leq y$ and $\langle y, L\rangle$ is a nonempty sorted list.

Prove by structural induction that in a nonempty sorted list $\langle x, L\rangle$, every element z in L satisfies $z \geq x$.

A string of balanced parentheses (with a close parenthesis that matches every open parenthesis, and appears to its right) is one of the following:

1. *the empty string (consisting of zero characters);*
2. *a string* `[ S ]` *where S is a string of balanced parentheses; or*
3. *a string S_1S_2 where S_1 and S_2 are both strings of balanced parentheses.*

For example, `[[]][]` *is a string of balanced parentheses, using Rule 3 on* `[[]]` *and* `[]`*. (Note that* `[]` *is a string of balanced parentheses using Rule 2 on the empty string (Rule 1), and therefore* `[[]]` *is by using Rule 2 on* `[]`*.)*

5.80 Prove by structural induction that every string of balanced parentheses according to this definition has exactly the same number of open parentheses as close parentheses.

5.81 Prove by structural induction that any prefix of a string of balanced parentheses according to this definition has at least as many open parentheses as it does close parentheses.

5.82 Recall from Definition 5.16 that we defined a *binary tree* as

1. an empty tree, denoted by `null`; or
2. a *root node* x, a *left subtree* T_ℓ, and a *right subtree* T_r, where x is an arbitrary value and T_ℓ and T_r are both binary trees.

Recall further that a *leaf* of a binary tree T is a node in T whose left subtree and right subtree are both `null`. Prove by structural induction that the algorithm **countLeaves**(T) in Figure 5.38 returns the number of leaves in a binary tree T.

```
countLeaves(T):
  if T = null then
      return 0
  else
      T_L, T_R := the left and right subtrees of T
      if T_L = T_R = null then
          return 1
      else
          return countLeaves(T_L) + countLeaves(T_R)
```

Figure 5.38: An algorithm to count leaves in a binary tree.

5.83 Recall that a *binary search tree (BST)* is a binary tree in which each node stores a "key," and, for any node u, the key at node u is larger than all keys in u's left subtree and smaller than all the keys in u's right subtree. (See p. 1160.) That is, a *BST* is either:

1. an empty tree, denoted by `null`; or
2. a *root node* x, a *left subtree* T_ℓ where all elements are less than x, and a *right subtree* T_r, where all elements are greater than x, and T_ℓ and T_r are both BSTs.

Prove that the smallest element in a nonempty BST is the bottommost leftmost node—that is, prove that

$$\text{the smallest element in a BST with root } x \text{ and left subtree } T_\ell = \begin{cases} x & \text{if } T_\ell = \texttt{null} \\ \text{the smallest element in } T_\ell & \text{if } T_\ell \neq \texttt{null}. \end{cases}$$

A heap is a binary tree where each node stores a priority, *and in which every node satisfies the* heap property: *the priority of a node u must be greater than or equal to the priorities of the roots of both of u's subtrees. (The restriction only applies for a subtree that is not* `null`*.)*

5.84 Give a recursive definition of a heap.

5.85 Prove by structural induction that every heap is empty, or that no element of the heap is larger than its root node. (That is, the root is a maximum element.)

5.86 Prove by structural induction that every heap is empty, or it has a leaf u such that u is no larger than any node in the heap. (That is, the leaf u is a minimum element.)

A 2–3 tree is a data structure, similar in spirit to a binary search tree (see Exercise 5.83)—or, more precisely, a balanced *form of BST, which is guaranteed to support fast operations like insertions, lookups, and deletions. The name "2–3 tree" comes from the fact that each internal node in the tree must have precisely 2 or 3 children; no node has a single child. Furthermore, all leaves in a 2–3 tree must be at the same "level" of the tree.*

5.87 Formally, a 2–3 tree of height h is one of the following:

1. a single node (in which case $h = 0$, and the node is called a *leaf*); or
2. a node with 2 subtrees, both of which are 2–3 trees of height $h - 1$; or
3. a node with 3 subtrees, all three of which are 2–3 trees of height $h - 1$.

Prove by structural induction that a 2–3 tree of height h has at least 2^h leaves and at most 3^h leaves. (Therefore a 2–3 tree that contains n leaf nodes has height between $\log_3 n$ and $\log_2 n$.)

5.88 A 2–3–4 tree is a similar data structure to a 2–3 tree, except that a tree can be a single node or a node with 2, 3, or 4 subtrees. Give a formal recursive definition of a 2–3–4 tree, and prove that a 2–3–4 tree of height h has at least 2^h leaves and at most 4^h leaves.

$(a + b) + c = a + (b + c)$	*Associativity of Addition*
$a + b = b + a$	*Commutativity of Addition*
$a + 0 = 0 + a = a$	*Additive Identity*
$(a \cdot b) \cdot c = a \cdot (b \cdot c)$	*Associativity of Multiplication*
$a \cdot b = b \cdot a$	*Commutativity of Multiplication*
$a \cdot 1 = 1 \cdot a = a$	*Multiplicative Identity*
$a \cdot 0 = 0 \cdot a = 0$	*Multiplicative Zero*

Figure 5.39: A few elementary-school facts about addition and multiplication.

The next few exercises give recursive definitions of some familiar arithmetic operations which are usually defined nonrecursively. In each, you're asked to prove a familiar property by structural induction. Think carefully when you choose the quantity upon which to perform induction, and don't skip any steps in your proof! You may use the elementary-school facts about addition and multiplication from Figure 5.39 in your proofs:

5.89 Let's define an *even number* as either (i) 0, or (ii) $2 + k$, where k is an even number. Prove by structural induction that the sum of any two even numbers is an even number.

5.90 Let's define a *power of two* as either (i) 1, or (ii) $2 \cdot k$, where k is a power of two. Prove by structural induction that the product of any two powers of two is itself a power of two.

5.91 Let $a_1, a_2, \ldots, a_k$ all be even numbers, for an arbitrary integer $k \geq 0$. Prove that $\left[\sum_{i=1}^{k} a_i\right]$ is also an even number. *(Hint: use weak induction and Exercise 5.89.)*

In Chapter 2, we defined b^n (for a base $b \in \mathbb{R}$ and an exponent $n \in \mathbb{Z}^{\geq 0}$) as denoting the result of multiplying b by itself n times (Definition 2.5). As an alternative to that definition of exponentiation, we could instead give a recursive *definition with integer exponents: $b^0 := 1$ and $b^{n+1} := b \cdot b^n$, for any nonnegative integer n.*

5.92 Using the associativity/commutativity/identity/zero properties in Figure 5.39, prove by induction that $b^m b^n = b^{m+n}$ for any integers $n \geq 0$ and $m \geq 0$. Don't skip any steps.

5.93 Using the facts in Figure 5.39 and Exercise 5.92, prove by induction that $(b^m)^n = b^{mn}$ for any integers $n \geq 0$ and $m \geq 0$. Again, don't skip any steps.

Recall Example 5.18, in which we defined a well-formed formula (a "wff") of propositional logic as a variable; the negation ($\neg$) of a wff; or the conjunction/disjunction/implication ($\wedge$, $\vee$, and $\Rightarrow$) of two wffs. Assuming we allow the corresponding new connective in the following exercises as part of a wff, give a proof using structural induction (see Example 5.21 for an example) that any wff is logically equivalent to one using only ...

5.94 Sheffer stroke $|$, where $p \mid q \equiv \neg(p \wedge q)$

5.95 Peirce's arrow $\downarrow$, where $p \downarrow q \equiv \neg(p \vee q)$

(programming required) *In the programming language ML (see Figure 5.34 for more), write a program to translate an arbitrary statement of propositional logic into a logically equivalent statement that has the following special form. (In other words, implement the proof of Exercises 5.94 and 5.95 as a recursive function.)*

5.96 $|$ is the only logical connective

5.97 $\downarrow$ is the only logical connective

5.98 Call a logical proposition *truth-preserving* if the proposition is true under the all-true truth assignment. That is, a proposition is truth-preserving if and only if the first row of its truth table is True.) Prove the following claim by structural induction on the form of the proposition:

Any logical proposition that uses only the logical connectives $\vee$ and $\wedge$ is truth-preserving.

(A solution to this exercise yields a rigorous solution to Exercise 4.71—there are propositions that cannot be expressed using only $\wedge$ and $\vee$. Explain.)

5.99 A *palindrome* is a string that reads the same front-to-back as it does back-to-front—for example, `RACECAR` or (ignoring spaces/punctuation) `A MAN, A PLAN, A CANAL--PANAMA!` or 10011001. Give a recursive definition of the set of palindromic bitstrings.

5.100 Let #0(s) and #1(s) denote the number of 0s and 1s in a bitstring s, respectively. Using your recursive definition from the previous exercise, prove by structural induction that, for any palindromic bitstring s, the value of $[\#0(s)] \cdot [\#1(s)]$ is an even number.

5.5 *Chapter at a Glance*

Proofs by Mathematical Induction

Suppose that we want to prove that a property $P(n)$ holds for all $n \in \mathbb{Z}^{\geq 0}$. To give a *proof by mathematical induction* of the claim $\forall n \in \mathbb{Z}^{\geq 0} : P(n)$, we prove the *base case* $P(0)$, and we prove the *inductive case*: for every $n \geq 1$, we have $P(n-1) \Rightarrow P(n)$.

When writing an inductive proof of the claim $\forall n \in \mathbb{Z}^{\geq 0} : P(n)$, include each of the following steps:

1. A clear statement of the claim to be proven—that is, a clear definition of the property $P(n)$ that will be proven true for all $n \geq 0$—and a statement that the proof is by induction, including specifically identifying the variable n upon which induction is being performed. (Some claims involve multiple variables, and it can be confusing if you aren't clear about which is the variable upon which you are performing induction.)

2. A statement and proof of the base case—that is, a proof of $P(0)$.

3. A statement and proof of the inductive case—that is, a proof of $P(n-1) \Rightarrow P(n)$, for a generic value of $n \geq 1$. The proof of the inductive case should include all of the following:

   (a) a statement of the inductive hypothesis $P(n-1)$.
   (b) a statement of the claim $P(n)$ that needs to be proven.
   (c) a proof of $P(n)$, which at some point makes use of the assumed inductive hypothesis $P(n-1)$.

We can use a proof by mathematical induction on arithmetic properties, like a formula for the sum of the nonnegative integers up to n—that is, $\sum_{i=0}^{n} i = \frac{n(n+1)}{2}$ for any integer $n \geq 0$—or a formula for a geometric series:

$$\text{if } \alpha \in \mathbb{R} \text{ where } \alpha \neq 1, \text{ and } n \in \mathbb{Z}^{\geq 0}, \text{ then } \sum_{i=0}^{n} \alpha^i = \frac{\alpha^{n+1} - 1}{\alpha - 1}.$$

(If $\alpha = 1$, then $\sum_{i=0}^{n} \alpha^i = n + 1$.) We can also use proofs by mathematical induction to prove the correctness of algorithms, particularly recursive algorithms.

Strong Induction

Suppose that we want to prove that $P(n)$ holds for all $n \in \mathbb{Z}^{\geq 0}$. To give a *proof by strong induction* of $\forall n \in \mathbb{Z}^{\geq 0} : P(n)$, we prove the *base case* $P(0)$, and we prove the *inductive case*: for every $n \geq 1$, we have $[P(0) \wedge P(1) \ldots \wedge P(n-1)] \Rightarrow P(n)$. Strong induction is actually completely equivalent to weak induction; anything that can be proven with one can also be proven with the other.

Generally speaking, using strong induction makes sense when the "reason" that $P(n)$ is true is that $P(k)$ is true for more than one value of $k < n$ (or a single value of $k < n$ with $k \neq n-1$). (For weak induction, the reason that $P(n)$ is true is just $P(n-1)$.) We can use strong induction to prove many claims, including part of the

Prime Factorization Theorem: if $n \in \mathbb{Z}^{\geq 1}$ is a positive integer, then there exist $k \geq 0$ prime numbers $p_1, p_2, \ldots, p_k$ such that $n = \prod_{i=1}^{k} p_i$.

Recursively Defined Structures and Structural Induction

A recursively defined structure, just like a recursive algorithm, is a structure defined in terms of one or more *base cases* and one or more *inductive cases*. Any data type that can be understood as either a trivial instance of the type or as being built up from a smaller instance (or smaller instances) of that type can be expressed in this way. The set of structures defined is *well ordered* if, intuitively, every invocation of the inductive case of the definition "makes progress" toward the base case(s) of the definition (and, more formally, that every nonempty subset of those structures has a "least" element).

Suppose that we want to prove that $P(x)$ holds for every $x \in S$, where S is the (well-ordered) set of structures generated by a recursive definition. To give a *proof by structural induction* of $\forall x \in S : P(x)$, we prove the following:

1. *Base cases*: for every x defined by a base case in the definition of S, prove $P(x)$.
2. *Inductive cases*: for every x defined in terms of $y_1, y_2, \ldots, y_k \in S$ by an inductive case in the definition of S, prove that $P(y_1) \wedge P(y_2) \ldots \wedge P(y_k) \Rightarrow P(x)$.

The form of a proof by structural induction that $\forall x \in S : P(x)$ for a well-ordered set of structures S is identical to the form of a proof using strong induction. Specifically, the proof by structural induction looks like a proof by strong induction of the claim $\forall n \in \mathbb{Z}^{\geq 0} : Q(n)$, where $Q(n)$ denotes the property "for any structure $x \in S$ that is generated using n applications of the inductive-case rules in the definition of S, we have $P(x)$."

Key Terms and Results

Key Terms

PROOFS BY MATHEMATICAL INDUCTION

- proof by mathematical induction
- base case
- inductive case
- inductive hypothesis
- geometric series
- arithmetic series
- harmonic series

STRONG INDUCTION

- strong induction
- prime factorization

RECURSIVELY DEFINED STRUCTURES AND STRUCTURAL INDUCTION

- recursively defined structures
- structural induction
- well-ordered set

Key Results

PROOFS BY MATHEMATICAL INDUCTION

1. Suppose that we want to prove that $P(n)$ holds for all $n \in \mathbb{Z}^{\geq 0}$. To give a *proof by mathematical induction* of $\forall n \in \mathbb{Z}^{\geq 0} : P(n)$, we prove the following:
   (a) the *base case* $P(0)$.
   (b) the *inductive case*: for every $n \geq 1$, we have $P(n-1) \Rightarrow P(n)$.
2. For any integer $n \geq 0$, we have $1 + 2 + \ldots + n = \frac{n(n+1)}{2}$.
3. Let $\alpha \in \mathbb{R}$ where $\alpha \neq 1$, and let $n \in \mathbb{Z}^{\geq 0}$. Then

$$\sum_{i=0}^{n} \alpha^i = \frac{\alpha^{n+1} - 1}{\alpha - 1}.$$

   (If $\alpha = 1$, then $\sum_{i=0}^{n} \alpha^i = n + 1$.)

STRONG INDUCTION

1. Suppose that we want to prove that $P(n)$ holds for all $n \in \mathbb{Z}^{\geq 0}$. To give a *proof by strong induction* of $\forall n \in \mathbb{Z}^{\geq 0} : P(n)$, we prove the following:
   (a) the *base case* $P(0)$.
   (b) the *inductive case*: for every $n \geq 1$, we have $[P(0) \wedge P(1) \ldots \wedge P(n-1)] \Rightarrow P(n)$.
2. The prime factorization theorem: let $n \in \mathbb{Z}^{\geq 1}$ be a positive integer. Then there exist $k \geq 0$ prime numbers $p_1, p_2, \ldots, p_k$ such that $n = \prod_{i=1}^{k} p_i$. Furthermore, up to reordering, the prime numbers $p_1, p_2, \ldots, p_k$ are unique.

RECURSIVELY DEFINED STRUCTURES AND STRUCTURAL INDUCTION

1. To give a *proof by structural induction* of $\forall x \in S : P(x)$, we prove the following:
   (a) the *base cases*: for every x defined by a base case in the definition of S, we have that $P(x)$.
   (b) the *inductive cases*: for every x defined in terms of $y_1, y_2, \ldots, y_k \in S$ by an inductive case in the definition of S, we have that $P(y_1) \wedge P(y_2) \ldots \wedge P(y_k) \Rightarrow P(x)$.

6
Analysis of Algorithms

In which our heroes stay beyond the reach of danger, by calculating precise bounds on how quickly they must move to stay safe.

6.1 *Why You Might Care*

> There is nothing so useless as doing efficiently that which should not be done at all.
>
> Peter Drucker (1909–2005)

Computer scientists are speed demons. When we are confronted by a computational problem that we need to solve, we want to solve that problem as quickly as possible. That "need for speed" has driven much of the advancement in computation over the last fifty years. We discover faster ways of solving important problems: developing data structures that support apparently instantaneous search of billions of web pages or hundreds of millions of users on a social networking site; or discovering new, faster algorithms that solve practical problems—such as finding shorter routes for delivery drivers or encrypting packets to be sent over the Internet. (Of course, the advances over the last fifty years have also been driven by improvements in computer hardware that ensure that *everything* we do computationally is faster!)

This chapter will introduce *asymptotic analysis*, the most common way in which computer scientists compare the speed of two possible solutions to the same problem. The basic idea is to think about the *rate of growth* of the running time of an algorithm—how much slower does the algorithm get if we double the size of the input?—in doing this analysis. We will think about "big" inputs to analyze the relative performance of the two algorithms, focusing on the long-run behavior instead of any small-input-size special cases for which one algorithm happens to perform exceptionally well. For the CS speed demon, asymptotic analysis is the speedometer. (Sometimes, instead of time, we measure the amount of space/memory or power/energy that an algorithm consumes.)

To take one example of why this kind of analysis of running time matters, consider sorting an n-element array A. One approach is to use brute force: try all $n!$ different permutations of A, and select the one permutation whose elements are in ascending order. Sorting algorithms like Selection Sort, Insertion Sort, or Bubble Sort require $\approx c \cdot n^2$ operations, for some constant c, to sort A. You may also have seen Merge Sort, which requires $\approx c \cdot n \log n$ operations. (We'll review these sorting algorithms in Section 6.3.) Figure 6.1 shows the number of operations required by these algorithms ($n!$, n^2, and $n \log n$). Given that some estimates say that the Earth will be swallowed by the sun in merely a few billion years,[1] there is plenty of reason to care about the differences in these running times. Asymptotic analysis is the first-cut approximation to making sure that our algorithms are fast enough—and that they will finish running while we're still around to view the output.

[1] David Appell. The sun will eventually engulf Earth—maybe. *Scientific American*, September 2008.

	$n = 10$	$n = 100$	$n = 1000$	$n = 10{,}000$	maximum n solvable in one minute on a machine that completes 1,000,000,000 operations per second
$n \log n$	33	664	9966	132,877	1.94×10^9
n^2	100	10,000	1,000,000	100,000,000	244,949
$n!$	3,628,800	9.333×10^{157}	4.029×10^{2567}	$2.846 \times 10^{35,659}$	13

Figure 6.1: The number of operations required for several algorithms with different running times, on several input sizes.

6.2 *Asymptotics*

I ain't sayin' you treated me unkind
You could have done better but I don't mind
You just kinda wasted my precious time
But don't think twice, it's all right

Bob Dylan (b. 1941)
"Don't Think Twice, It's All Right" (1963)

Generally speaking, we will be interested in the behavior of algorithms *ignoring constant factors*. There are two different senses in which we ignore constants. First, we will ignore constant multiplicative factors; for our purposes, the function $f(n)$ and the function $g(n) = 2 \cdot f(n)$ "grow at the same rate." (Exercises 6.1–6.4 discuss why we might evaluate efficiency of algorithms in this way.) Second, we will be interested in the long-run behavior of our algorithms, so we won't be concerned by any small input values for which the algorithm performs particularly quickly or slowly.

Example 6.1 (All of these things are quite the same)
The following functions all grow at the same rate:

$$f(n) = 3 \cdot n^2$$
$$g(n) = 0.01 \cdot n^2$$
$$h(n) = \begin{cases} 202 & \text{if } 0 < n < 100 \\ n^7 & \text{if } 100 \leq n < 1000 \\ 1776 \cdot n^2 & \text{otherwise.} \end{cases}$$

The functions f and g differ by a multiplicative factor. For $n \geq 1000$, the function h also differs by a constant multiplier from f and g; therefore for large enough n it too grows at the same rate as f and g.

This type of analysis is called *asymptotic analysis*.

asymptotic (Greek): *a* "without" + *symptotos* "falling together."

Taking it further: In mathematics, the *asymptote* of a function $f(n)$ is a line that $f(n)$ approaches as n gets very large. (Formally, this value is $\lim_{n\to\infty} f(n)$.) For example, the function $f(x) = \frac{1}{x}$ has an asymptote at 0: as x gets larger and larger, $f(x)$ gets closer and closer to 0. (Mathematicians also consider asymptotes where a function approaches, but does not reach, some particular value as the input approaches some point; for example, $\tan(\theta)$ has an asymptote of ∞ as $\theta \to \pi/2$ and $f(x) = -x/(x-2)$ has an asymptote of $-\infty$ as $x \to 2$ from below.) The asymptotic behavior of a function is similarly motivated: we're thinking about the growth rate of the function as n gets very large.

Consider two functions $f : \mathbb{R}^{\geq 0} \to \mathbb{R}^{\geq 0}$ and $g : \mathbb{R}^{\geq 0} \to \mathbb{R}^{\geq 0}$. (We will be interested in functions whose domain and range are both nonnegative because we're primarily thinking about functions that describe the number of steps of a particular algorithm on an input of a particular size, and neither input size nor number of computational steps executed can be negative.) The key concept of asymptotic analysis will be a definition of the *growth rates* of the functions f and g, and how those growth rates compare: that is, what it means to say that f grows faster (or, really, no slower) than g; or that f grows at the same rate as g; or that f grows slower (or no faster) than g.

6.2.1 *Big O*

Consider two functions f and g. To reiterate, our goal is to compare the rates at which these functions grow. We'll start by defining what it means for the function $f(n)$ to grow no faster than $g(n)$, written $f(n) = O(g(n))$.

The "=" in "$f(n) = O(g(n))$" is odd notation, but it's also very standard. This expression means *$f(n)$ has the property of being $O(g(n))$* and not *$f(n)$ is identical to $O(g(n))$*.

Taking it further: Philosophers sometimes distinguish between *the "is" of identity* and *the "is" of predication*. In a sentence like *Barbara Liskov is the 2008 Turing Award winner*, we are asserting that *Barbara Liskov* and *the 2008 Turing Award Winner* actually refer to the same thing—that is, they are identical. In a sentence like *Barbara Liskov is tall*, we are asserting that Barbara Liskov (the entity to which *Barbara Liskov* refers) has the property of being tall—that is, the predicate *x is tall* is true of Barbara Liskov. One should interpret the "=" in $f(n) = O(g(n))$ as an "is of predication."

One reasonably accurate way to distinguish these two uses of *is* is by considering what happens if you reverse the order of the sentence: *The 2008 Turing Award Winner is Barbara Liskov* is still a (true) well-formed sentence, but *Tall is Barbara Liskov* sounds very strange. Similarly, for an "is of identity" in a mathematical context, we can say either $x^2 - 1 = (x+1)(x-1)$ or $(x+1)(x-1) = x^2 - 1$. But, while "$f(n) = O(g(n))$" is a well-formed statement, it is nonsensical to say "$O(g(n)) = f(n)$."

Here is the formal definition:

O is pronounced "big oh."

Definition 6.1 ("Big O")
Consider two functions $f : \mathbb{R}^{\geq 0} \to \mathbb{R}^{\geq 0}$ *and* $g : \mathbb{R}^{\geq 0} \to \mathbb{R}^{\geq 0}$. *We say that* f grows no faster than g *if there exist constants* $c > 0$ *and* $n_0 \geq 0$ *such that*

$$\forall n \geq n_0 : f(n) \leq c \cdot g(n).$$

In this case, we write "$f(n)$ is $O(g(n))$" or "$f(n) = O(g(n))$."

The intuition of the definition is that $f(n) = O(g(n))$ if, for large enough n, we have $f(n) \leq constant \cdot g(n)$. Figure 6.2 shows five different functions $f : \mathbb{R}^{\geq 0} \to \mathbb{R}^{\geq 0}$ that all satisfy $f(n) = O(n)$. (In the figure, the value of x is "large enough" once x is outside of the gray box, and the multiplicative constant is equal to 3 in each subplot. For a function like $f(x) = 4x$, we'd show that $f(n) = O(n)$ by choosing some $c \geq 4$ as the multiplicative constant.)

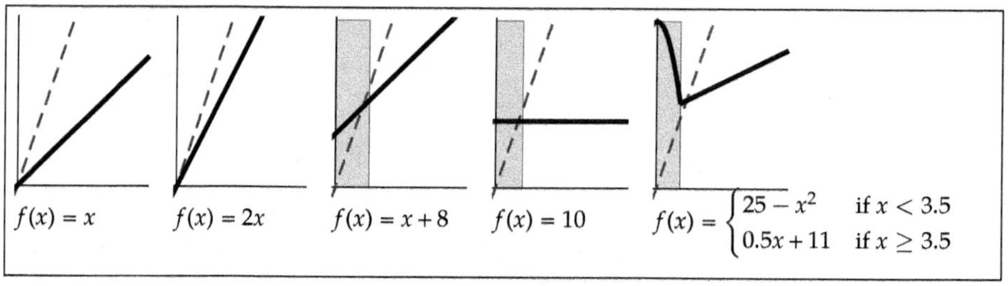

Figure 6.2: Five functions that are all $O(n)$. For any x beyond the gray box, we have $f(x) \leq 3x$.

More quantitatively, here are two simple examples of functions that are $O(n^2)$:

Example 6.2 (A square function)
Problem: Prove that the function $f(n) = 3n^2 + 2$ is $O(n^2)$.

Solution: To prove that $f(n) = 3n^2 + 2$ satisfies $f(n) = O(n^2)$, we must identify constants $c > 0$ and $n_0 \geq 0$ such that $\forall n \geq n_0 : 3n^2 + 2 \leq c \cdot n^2$. Let's select $c = 5$ and $n_0 = 1$. For all $n \geq 1$, observe that $2n^2 \geq 2$. Therefore, for all $n \geq 1$, we have

$$f(n) = 3n^2 + 2 \leq 3n^2 + 2n^2 = 5n^2 = c \cdot n^2.$$

Example 6.3 (Another square function)
Problem: Prove that the function $g(n) = 4n$ is also $O(n^2)$.

Solution: We wish to show that $4n \leq c \cdot n^2$ for all $n \geq n_0$, for constants $c > 0$ and $n_0 \geq 0$ that we get to choose. The two functions $g(n)$ and $q(n) := n^2$ are shown in Figure 6.3. Because the functions cross (with no constant multiplier), we can pick $c = 1$. Observe that $4n \leq n^2$ if and only if $n^2 - 4n = n(n-4) \geq 0$—that is, for $n \leq 0$ or $n \geq 4$. Thus $c = 1$ and $n_0 = 4$ suffice.

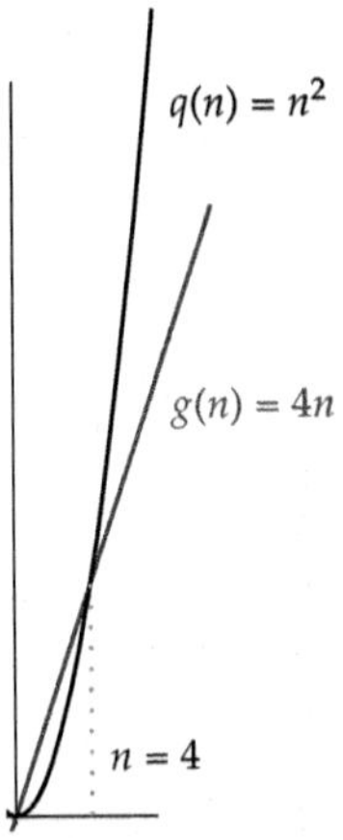

Figure 6.3: A plot of $g(n) = 4n$ and $q(n) = n^2$.

Note that, when $f(n) = O(g(n))$, there are *many* choices of c and n_0 that satisfy the definition. For example, we could have chosen $c = 4$ and $n_0 = 1$ in Example 6.3. (See Exercise 6.15.)

Example 6.4 (One nonsquare)
Problem: Prove that the function $h(n) = n^3$ is *not* $O(n^2)$.

Solution: To show that $h(n) = n^3$ is *not* $O(n^2)$, we need to argue that, for *all* constants n_0 and c, there exists an $n \geq n_0$ such that $h(n) > c \cdot n^2$—that is, that $n^3 > c \cdot n^2$.

Fix a purported n_0 and c. Let $n := \max(n_0, c+1)$. Then $n > c$ by our definition of n, so, by multiplying both sides of $n > c$ by the nonnegative quantity n^2, we have $n^3 = n \cdot n^2 > c \cdot n^2$. But we also have that $n \geq n_0$ by our definition of n, and thus we have identified an $n \geq n_0$ such that $n^3 > c \cdot n^2$.

Because n_0 and c were generic, we have shown that no such constants can exist, and therefore that $h(n) = n^3$ is *not* $O(n^2)$.

Some properties of $O(\cdot)$

Now that we've seen a few specific examples, let's turn to some more general results. There are many useful properties of $O(\cdot)$ that will come in handy later; we'll start here with a few of these properties, together with a proof of one. (The other proofs are left to you in Exercises 6.18–6.20.)

Lemma 6.1 (Asymptotic equivalence of max and sum)
We have $f(n) = O(g(n) + h(n))$ if and only if $f(n) = O(\max(g(n), h(n)))$.

Proof. We proceed by mutual implication. For the forward direction, suppose $f(n) = O(g(n) + h(n))$. Then by definition there exist constants $c > 0$ and $n_0 \geq 0$ such that

$$\text{for all } n \geq n_0 \qquad f(n) \leq c \cdot [g(n) + h(n)]. \tag{1}$$

For any $a, b \in \mathbb{R}$, we know that $a \leq \max(a, b)$ and $b \leq \max(a, b)$, so (1) implies

$$\begin{aligned} \text{for all } n \geq n_0 \qquad f(n) &\leq c \cdot [\max(g(n), h(n)) + \max(g(n), h(n))] \\ &= 2c \max(g(n), h(n)). \qquad (2) \end{aligned}$$

But (2) *is* the definition of $f(n) = O(\max(g(n), h(n)))$, using constants $n_0' = n_0$ and $c' = 2c$.

Conversely, suppose $f(n) = O(\max(g(n), h(n)))$. Then there exist constants $c > 0$ and $n_0 \geq 0$ such that

$$\text{for all } n \geq n_0 \qquad f(n) \leq c \cdot \max(g(n), h(n)). \tag{3}$$

For any $a, b \in \mathbb{R}^{\geq 0}$ we know $\max(a, b) \leq \max(a, b) + \min(a, b) = a + b$; thus (3) implies

$$\text{for all } n \geq n_0 \qquad f(n) \leq c \cdot [g(n) + h(n)]. \tag{4}$$

Thus (4) implies that $f(n) = O(g(n) + h(n))$, using the same constants, $n_0' = n_0$ and $c' = c$. □

Problem-solving tip: Don't force yourself to prove more than you have to! For example, when proving that an asymptotic relationship like $f(n) = O(g(n))$ holds, all we need to do is identify *some* pair of constants c, n_0 that satisfy Definition 6.1. Don't work too hard! Choose whatever c or n_0 makes your life easiest, even if they're much bigger than necessary. For asymptotic purposes, we care that the constants c and n_0 *exist*, but we *don't* care how big they are.

Lemma 6.2 (Transitivity of $O(\cdot)$)
If $f(n) = O(g(n))$ and $g(n) = O(h(n))$, then $f(n) = O(h(n))$.

Lemma 6.3 (Addition and multiplication preserve $O(\cdot)$-ness)
If $f(n) = O(h_1(n))$ and $g(n) = O(h_2(n))$, then:

- $f(n) + g(n) = O(h_1(n) + h_2(n))$.
- $f(n) \cdot g(n) = O(h_1(n) \cdot h_2(n))$.

Asymptotics of polynomials

So far, we've discussed properties of $O(\cdot)$ that are general with respect to the form of the functions in question. But because we're typically concerned with $O(\cdot)$ in the context of the running time of algorithms—and we are generally interested in algorithms that are efficient—we'll be particularly interested in the asymptotics of polynomials. The most salient point about the growth of a polynomial $p(n)$ is that $p(n)$'s asymptotic behavior is determined by the degree of $p(n)$—that is, the polynomial $p(n) = a_0 + a_1 n + a_2 n^2 + \cdots + a_k n^k$ behaves like n^k, asymptotically:

Lemma 6.4 (Asymptotics of polynomials)
Let $p(n) = \sum_{i=0}^{k} a_i n^i$ be a polynomial. Then $p(n) = O(n^k)$.

(If $a_k > 0$, then indeed $p(n) = O(n^k)$, and it is not possible to improve this bound—that is, in the notation of Section 6.2.2, we have that $p(n) = \Theta(n^k)$.)

The proof of Lemma 6.4 is deferred to Exercise 6.21, but we have already seen the intuition in previous examples: every term $a_i n^i$ satisfies $a_i n^i \leq |a_i| \cdot n^k$, for any $n \geq 1$.

Asymptotics of logarithms and exponentials

We will also often encounter logarithms and exponential functions, so it's worth identifying a few of their asymptotic properties. Again, we'll prove one of these properties as an example, and leave proofs of many of the remaining properties to the exercises. The first pair of properties is that logarithmic functions grow more slowly than polynomials, which grow more slowly than exponential functions:

Lemma 6.5 ($\log n$ grows slower than $n^{0.0000001}$)
Let $\varepsilon > 0$ be an arbitrary constant, and let $f(n) = \log n$. Then $f(n) = O(n^{\varepsilon})$.

Lemma 6.6 ($n^{1000000}$ grows slower than 1.0000001^n)
Let $b > 1$ and $k \geq 0$ be arbitrary constants, and let $p(n) = \sum_{i=0}^{k} a_i n^i$ be any polynomial. Then $p(n) = O(b^n)$.

The second pair of properties is that two logarithmic functions $\log_a n$ and $\log_b n$ grow at the same rate (for any bases $a > 1$ and $b > 1$) but that two exponential functions a^n and b^n do not (for any bases a and $b \neq a$):

Lemma 6.7 (The base of a logarithm doesn't matter, asymptotically)
Let $b > 1$ and $k > 0$ be arbitrary constants. Then $f(n) = \log_b(n^k)$ is $O(\log n)$.

Proof of Lemma 6.7. Using standard facts about logarithms, we have that

$$\begin{aligned} \log_b(n^k) &= k \cdot \log_b(n) && \text{(2.2.5): } \log_b x^y = y \log_b x \\ &= k \cdot \frac{\log n}{\log b}. && \text{change of base formula (2.2.6): } \log_b x = \frac{\log_c x}{\log_c b} \end{aligned}$$

Thus, for any $n \geq 1$, we have that $f(n) = \frac{k}{\log b} \cdot \log n$. Thus $f(n) = O(\log n)$ using the constants $n_0 = 1$ and $c = \frac{k}{\log b}$. □

Lemma 6.8 (The base of an exponential *does* matter, asymptotically)
Let $b \geq 1$ and $c \geq 1$ be arbitrary constants. Then $f(n) = b^n$ is $O(c^n)$ if and only if $b \leq c$.

Lemma 6.7 is the reason that, for example, binary search's running time is described as $O(\log n)$ rather than as $O(\log_2 n)$, without any concern for writing the "2": the base of the logarithm is inconsequential asymptotically, so $O(\log_{\sqrt{2}} n)$ and $O(\log_2 n)$ and $O(\ln n)$ all mean exactly the same thing. In contrast, for exponential functions, the base of the exponent *does* affect the asymptotic behavior: Lemma 6.8 says that, for example, the functions $f(n) = 2^n$ and $g(n) = (\sqrt{2})^n$ do *not* grow at the same rate. (See Exercises 6.25–6.28.)

Taking it further: Generally, exponential growth is a problem for computer scientists. Many computational problems that are important and useful to solve seem to require searching a very large space of possible answers: for example, testing the satisfiability of an n-variable logical proposition seems to require looking at about 2^n different truth assignments, and factoring an n-digit number seems to require looking at about 10^n different candidate divisors. The fact that exponential functions grow so quickly is exactly why we do not have algorithms that are practical for even moderately large instances of these problems.

But one of the most famous exponentially growing functions actually *helps* us to solve problems: the amount of computational power available to a "standard" user of a computer has been growing exponentially for decades: about every 18 months, the processing power of a standard computer has roughly doubled. This trend—dubbed *Moore's Law,* after Gordon Moore, the co-founder of Intel—is discussed on p. 613.

6.2.2 *Other Asymptotic Relationships: Ω, Θ, ω, and o*

There are several basic asymptotic notions (with accompanying notation), based around two core ideas (see Figure 6.4):

$f(n)$ grows no faster than $g(n)$: In other words, ignoring small inputs, for all n we have that $f(n) \leq$ constant $\cdot\, g(n)$. This relationship is expressed by the $O(\cdot)$ notation: $f(n) = O(g(n))$. We can also say that g is an *asymptotic upper bound* for f: if we plot n against $f(n)$ and $g(n)$, then $g(n)$ will be "above" $f(n)$ for large inputs.

$f(n)$ grows no slower than $g(n)$: The opposite relationship, in which g is an *asymptotic lower bound* on f, is expressed by $\Omega(\cdot)$ notation. Again, ignoring small inputs, $f(n) = \Omega(g(n))$ if for all n we have that $f(n) \geq$ constant $\cdot\, g(n)$. (Notice that the inequality swapped directions from the definition of $O(\cdot)$.)

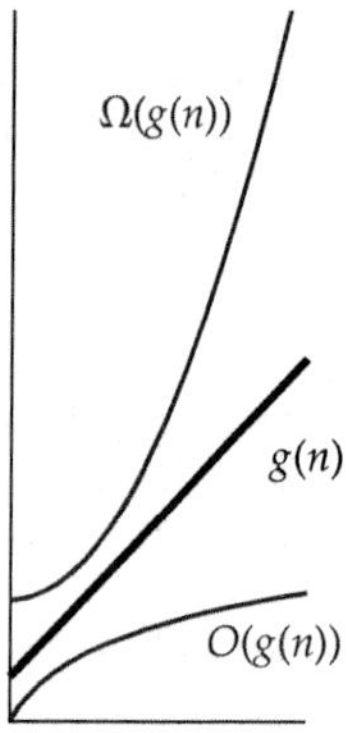

Figure 6.4: A function $g(n)$, a function that's $\Omega(g)$ (grows no slower than g), and a function that's $O(g)$ (grows no faster than g).

FORMAL DEFINITIONS

Here are the formal definitions of four other relationships based on these notions:

Definition 6.2 ("Big Omega")
A function f grows no slower than g, *written $f(n) = \Omega(g(n))$, if there exist constants $d > 0$ and $n_0 \geq 0$ such that $\forall n \geq n_0 : f(n) \geq d \cdot g(n)$.*

Ω is the Greek letter Omega written in upper case; ω is the same Greek letter written in lower case.

The two fundamental asymptotic relationships, $O(\cdot)$ and $\Omega(\cdot)$, are dual notions; they are related by the property that $f(n) = O(g(n))$ if and only if $g(n) = \Omega(f(n))$. (The proof is left as Exercise 6.30.)

There are three other pieces of asymptotic notation, corresponding to the situations in which $f(n)$ is both $O(g)$ and $\Omega(g)$, or $O(g)$ but not $\Omega(g)$, or $\Omega(g)$ but not $O(g)$:

Definition 6.3 ("Big Theta")
A function f grows at the same rate as g, *written $f(n) = \Theta(g(n))$, if $f(n) = O(g(n))$* and *$f(n) = \Omega(g(n))$.*

Definition 6.4 ("Little o")
A function f grows (strictly) slower than g, *written $f(n) = o(g(n))$, if $f(n) = O(g(n))$ but $f(n) \neq \Omega(g(n))$.*

Definition 6.5 ("Little omega")
A function f grows (strictly) faster than g, *written $f(n) = \omega(g(n))$, if $f(n) = \Omega(g(n))$ but $f(n) \neq O(g(n))$.*

This notation is summarized, in two different ways, in Figure 6.5.

	if $f(n) = O(g(n)) \ldots$	if $f(n) \neq O(g(n)) \ldots$
$\ldots$ and $f(n) = \Omega(g(n)) \ldots$	$\ldots$ then $f(n) = \Theta(g(n))$	$\ldots$ then $f(n) = \omega(g(n))$
$\ldots$ and $f(n) \neq \Omega(g(n)) \ldots$	$\ldots$ then $f(n) = o(g(n))$	—

	$\exists c > 0, n_0 \geq 0$ such that $\forall n \geq n_0 : f(n) \leq c \cdot g(n)$	$\exists d > 0, n_0 \geq 0$ such that $\forall n \geq n_0 : f(n) \geq d \cdot g(n)$	
$f(n) = O(g(n))$	*yes*	*don't care*	f grows no faster than g
$f(n) = \Omega(g(n))$	*don't care*	*yes*	f grows no slower than g
$f(n) = \Theta(g(n))$	*yes*	*yes*	f grows at the same rate as g
$f(n) = o(g(n))$	*yes*	*no*	f grows strictly slower than g
$f(n) = \omega(g(n))$	*no*	*yes*	f grows strictly faster than g

Figure 6.5: Summary of notation for asymptotic notation, in two different ways.

Example 6.5 ($f =$ ___(n))

Problem: Let $f(n) = 3n^2 + 1$. Is $f(n) = O(n)$? $\Omega(n)$? $\Theta(n)$? $o(n)$? $\omega(n)$? Prove your answers.

Solution: Once we determine whether $f(n) = O(n)$ and whether $f(n) = \Omega(n)$, we can answer all parts of the question using Figure 6.5(a).

- $f(n) = \Omega(n)$. For $n \geq 1$, we have $n \leq n^2 \leq 3n^2 + 1 = f(n)$. Thus selecting $d = 1$ and $n_0 = 1$ satisfies Definition 6.2.
- $f(n) \neq O(n)$. Let $c > 0$ be arbitrary. For any $n \geq \frac{c}{3}$, we have $3n^2 + 1 > 3n^2 \geq c \cdot n$. Therefore, for any $n_0 > 0$, there exists an $n \geq n_0$ such that $f(n) > c \cdot n$. (Namely, for $n = \max(n_0, c/3)$, we have $n \geq n_0$ and $f(n) > c \cdot n$.)

  Thus, every constant $c > 0$ fails to satisfy the requirements of Definition 6.1, and therefore $f(n) \neq O(n)$.

Assembling $f(n) = \Omega(n)$ and $f(n) \neq O(n)$ with Figure 6.5(a), we can also conclude that $f(n) = \omega(n)$, $f(n) \neq \Theta(n)$, and $f(n) \neq o(n)$.

Taking it further: We've given definitions of $O(\cdot)$, $\Omega(\cdot)$, $\Theta(\cdot)$, $o(\cdot)$, and $\omega(\cdot)$ that are based on nested quantifiers: there exists a multiplicative constant such that, for all sufficiently large n, ... For those with a more calculus-based mindset, we could also give an equivalent definition in terms of limits:

- $f(n) = O(g(n))$ if $\lim_{n\to\infty} f(n)/g(n)$ is finite;
- $f(n) = \Omega(g(n))$ if $\lim_{n\to\infty} f(n)/g(n)$ is nonzero;
- $f(n) = \Theta(g(n))$ if $\lim_{n\to\infty} f(n)/g(n)$ is finite and nonzero;
- $f(n) = o(g(n))$ if $\lim_{n\to\infty} f(n)/g(n) = 0$; and
- $f(n) = \omega(g(n))$ if $\lim_{n\to\infty} f(n)/g(n) = \infty$.

For the function $f(n) = 3n^2 + 1$ in Example 6.5, for example, observe that $\lim_{n\to\infty} \frac{f(n)}{n} = \infty$. Thus $f(n) = \Omega(n)$ and $f(n) = \omega(n)$, but none of the other asymptotic relationships holds.

A (POSSIBLY COUNTERINTUITIVE) EXAMPLE

Intuitively, the asymptotic symbols O, Ω, Θ, o, and ω correspond to the numerical comparison symbols $\leq$, $\geq$, $=$, $<$, and $>$—but the correspondence isn't perfect, as we'll see in this example:

Example 6.6 (Finding functions, to spec)

Problem: Fill in each blank in the following table with an example of a function f that satisfies the stated conditions.

	$f(n) = O(n^2) \ldots$	$f(n) \neq O(n^2) \ldots$
$\ldots$ and $f(n) = \Omega(n^2)$		
$\ldots$ and $f(n) \neq \Omega(n^2)$		

Solution: Three of these cells are easy to complete:

- $f(n) = n^2$ is $\Theta(n^2)$—that is, it satisfies both $O(n^2)$ and $\Omega(n^2)$;
- $f(n) = n$ is $o(n^2)$—that is, it satisfies $O(n^2)$ but not $\Omega(n^2)$; and
- $f(n) = n^3$ is $\omega(n^2)$—that is, it satisfies $\Omega(n^2)$ but not $O(n^2)$.

The lower-right cell—a function $f(n)$ that is *neither* $O(n^2)$ nor $\Omega(n^2)$—appears more challenging. For $f(n) \neq O(g(n))$, we need a function f such that, for any constants $c > 0$ and $n_0 \geq 0$, there exists $\overline{n} \geq n_0$ such that $f(\overline{n}) > c\overline{n}^2$. Similarly, for $f(n) \neq \Omega(n^2)$, we need, for any constants $d > 0$ and $n_0 \geq 0$, there to exist $\underline{n} \geq n_0$ such that $f(\underline{n}) < d\underline{n}^2$. How can we simultaneously achieve these conditions? Here's one way: we'll define the function f in a *piecewise* manner, so that for, say, even values of n the function grows faster than n^2, and for odd values it grows slower:

$$f(n) = \begin{cases} n^3 & \text{if } n \text{ is even} \\ n & \text{if } n \text{ is odd} \end{cases} = n^{2+(-1)^n}.$$

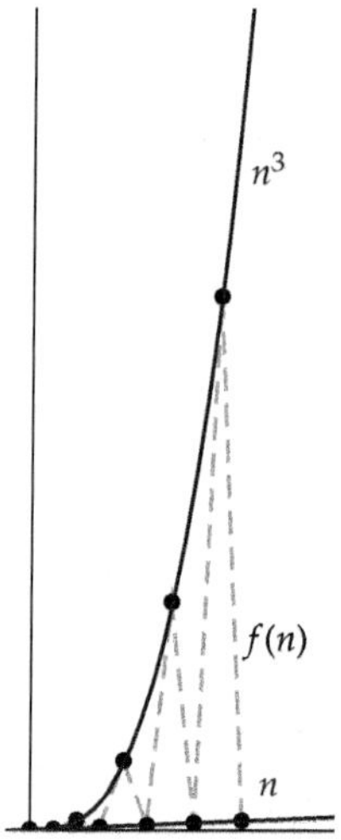

Figure 6.6: A plot of $f(n)$ from Example 6.6, where $f(n) = n$ when n is even, and $f(n) = n^3$ when n is odd. This function is neither $O(n^2)$ nor $\Omega(n^2)$.

(See Figure 6.6 for a plot of this function.)

Let's argue formally that $f(n) \neq O(n^2)$. Let $c > 0$ and $n_0 \geq 0$ be arbitrary. Let $\overline{n}$ be the smallest even number strictly greater than $\max(c, n_0)$. Then $f(\overline{n}) = \overline{n}^3$ and $\overline{n}^3 > c \cdot \overline{n}^2$ because we chose $\overline{n} > c$. But we just argued that, for arbitrary $c > 0$ and $n_0 \geq 0$, it is not the case that $\forall n \geq n_0 : f(n) \leq cn^2$. Thus $f(n) \neq O(n^2)$.

Together with the proof that $f(n) \neq \Omega(n^2)$, which is left to you as Exercise 6.44, the above arguments allow us to fill in the required table:

	$f(n) = O(n^2)$	$f(n) \neq O(n^2)$
$f(n) = \Omega(n^2)$	$f(n) = n^2$	$f(n) = n^3$
$f(n) \neq \Omega(n^2)$	$f(n) = n$	$f(n) = \begin{cases} n^3 & \text{if } n \text{ is even} \\ n & \text{if } n \text{ is odd.} \end{cases}$

Problem-solving tip: When you're confronted with a problem with seemingly contradictory constraints, as in the bottom-right cell of the table in Example 6.6, very carefully write down what the constraints require. This process can help you see why the constraints aren't actually contradictory.

SOME PROPERTIES OF Ω, Θ, o, AND ω

Many of the properties of $O(\cdot)$ also hold for the other four asymptotic notions; for example, all five of $O(\cdot)$, $\Omega(\cdot)$, $\Theta(\cdot)$, $o(\cdot)$, and $\omega(\cdot)$ obey transitivity, and several obey reflexivity. See Exercises 6.45–6.53.

One of the subtlest aspects of asymptotic notation is the fact that two functions can be *incomparable* with respect to their rates of growth: we can identify two functions f and g such that none of the asymptotic relationships holds. (That is, $f \neq O(g)$, $f \neq \Omega(g)$,

$f \neq \Theta(g), f \neq o(g)$, and $f \neq \omega(g)$.)

Let a and b be real numbers. The two inequalities $a \leq b$ and $b \leq a$ can be true and false in different combinations:

- When $a \leq b$ and $b \leq a$, then $a = b$.
- When $a \leq b$ and $b \not\leq a$, then $a < b$.
- When $a \not\leq b$ and $b \leq a$, then $a > b$.
- (It is not possible to have both $a \not\leq b$ and $b \not\leq a$.)

Intuitively, the relationship $f(n) = O(g(n))$ means (approximately!) that

$$\text{"the growth rate of } f \;\leq\; \text{the growth rate of } g.\text{"} \qquad \text{(A)}$$

And, again, intuitively, $f(n) = \Omega(g(n))$ means (approximately)

$$\text{"the growth rate of } f \;\geq\; \text{the growth rate of } g.\text{"} \qquad \text{(B)}$$

So Definitions 6.3, 6.4, and 6.5 correspond to these three combinations: (A) and (B) is Θ; (A) but not (B) is o; and (B) but not (A) is ω. But be careful! For $a, b \in \mathbb{R}$, it's true that either $a \leq b$ or $a \geq b$ must be true. But it's possible for *both of the inequalities* (A) *and* (B) *to be false!* The functions $g(n) = n^2$ and the function $f(n)$ from Example 6.6 that equals either n^3 or n depending on the parity of n are an example of a pair of functions for which *neither* (A) nor (B) is satisfied.

> **Taking it further:** The real numbers satisfy the mathematical property of *trichotomy* (Greek: "division into three parts"): for $a, b \in \mathbb{R}$, exactly one of $\{a < b, a = b, a > b\}$ holds. Functions compared asymptotically do not obey trichotomy: for two functions f and g, it's possible for *none* of $\{f = o(g), f = \Theta(g), f = \omega(g)\}$ to hold.

Before we begin to apply asymptotic notation to the analysis of algorithms, we'll close this section with a few notes about the use (and abuse) of asymptotic notation.

Using asymptotics in arithmetic expressions

It is often convenient to use asymptotic notation in arithmetic expressions. We permit ourselves to write something like $O(n \log n) + O(n^3) = O(n^3)$, which intuitively means that, given functions that grow no faster than $n \log n$ and n^3, their sum grows no faster than n^3 too. When asymptotic notation like $O(n^2)$ appears on the left-hand side of an equality, we interpret it to mean an arbitrary unnamed function that grows no faster than n^2. For example, making $\log n$ calls to an algorithm whose running time is $O(n)$ requires $\log n \cdot O(n) = O(n \log n)$ time.

Using asymptotics with multiple variables

It will also occasionally turn out to be convenient to be able to write asymptotic expressions that depend on more than one variable. Giving a precise technical definition of multivariate asymptotic notation is a bit subtle, but the intuition precisely matches the univariate definitions we've already given. We'll use the notation $g(n, m) = O(f(n, m))$ to mean "for all sufficiently large n and m, there exists a constant c such that $g(n, m) \leq c \cdot f(n, m)$." For example, the function $f(n, m) = n^2 + 3m - 5$ satisfies $f(n, m) = O(n^2 + m)$.

A common mistake and some meaningless language

There is a widespread—and incorrect—sloppy use of asymptotic notation: it is unfortunately common for people to use $O(\cdot)$ when they mean $\Theta(\cdot)$. You will sometimes encounter claims like:

$$\text{"I prefer } f \text{ to } g\text{, because } f(n) = O(n^2) \text{ and } g(n) = O(n^3)\text{."} \tag{1}$$

But this statement doesn't make sense: $O(\cdot)$ defines only an upper bound, so either of f or g might grow more slowly than the other! Saying (1) is like saying

"Alice is richer than Bob,
because Alice has at most \$1,000,000,000 and Bob has at most \$1,000,000." (2)

(Alice *might* be richer than Bob, but perhaps they both have twenty bucks each, or perhaps Bob has \$1,000,000 and Alice has nothing.) Use $O(\cdot)$ when you mean $O(\cdot)$, and to use $\Theta(\cdot)$ when you mean $\Theta(\cdot)$—and be aware that others may use $O(\cdot)$ improperly. (And, gently, correct them if they're doing so.)

There's a related imprecise use of asymptotics that leads to statements that don't mean anything. For example, consider statements like "$f(n)$ is at least $O(n^3)$" or "$f(n)$ is at most $\Omega(n^2)$." These sentences have no meaning: they say "$f(n)$ grows at least as fast as at most as fast as n^3" and "$f(n)$ grows at most as fast as at least as fast as n^2." (?!?) Be careful: use upper bounds as upper bounds, and use lower bounds as lower bounds! Again, by analogy, consider the sentences

Thanks to Tom Wexler for suggesting (5).

"My weight is more than $\leq$ 100 kilograms" (3)

or "I am shorter than some person who is taller than 4 feet tall." (4)

or "You could save up to 50% or more!" (5)

None of these sentences says anything!

Computer Science Connections

Moore's Law

In 1965, Gordon Moore, one of the co-founders of Intel, published an article making a basic prediction—and it's been reinterpreted many times—that processing power would double roughly once every 18–24 months.[2] (It's been debated and revised over time, by, for example, interpreting "processing power" as the number of transistors—the most basic element of a processor, out of which logic gates like AND, OR, and NOT are built—rather than what we can actually compute.) This prediction later came to be known as *Moore's Law*—it's not a real "law" like Ohm's Law or the Law of Large Numbers, of course, but rather simply a prediction. That said, it's proven to be a remarkably robust prediction: for something like 40 to 50 years, it has proven to be a consistent guide to the massive increase in processing power for a typical computer user over the last decades. (See Figure 6.7.)

[2] Gordon E. Moore. Cramming more components onto integrated circuits. *Electronics*, 38(8), April 1965.

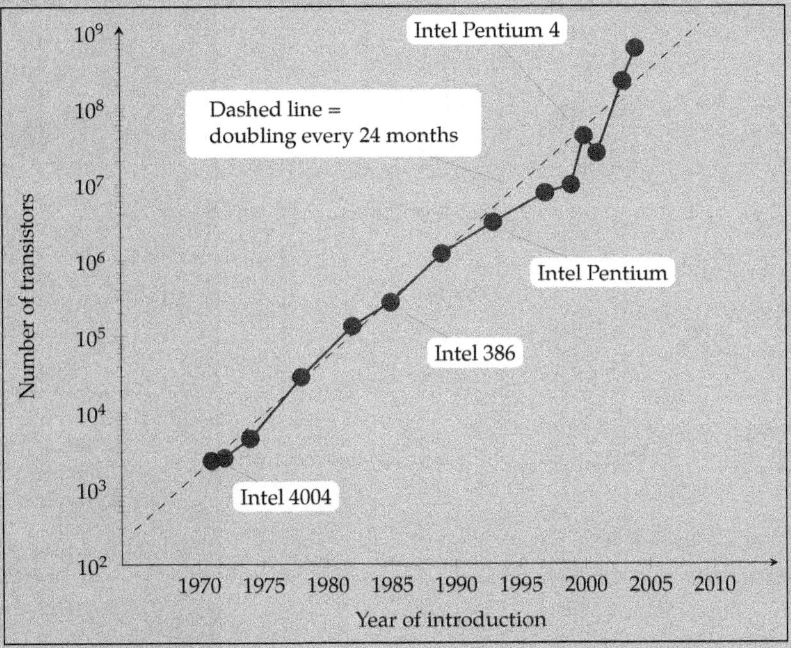

Figure 6.7: A plot of the number of transistors per processor, for about 15 Intel brand processors introduced over the last 50 years. (Data are from an Intel press release celebrating the 40th anniversary of the original publication of Moore's Law.) The dashed line indicates the rate of growth we'd see if the number of transistors per processor doubled every two years (starting with the Intel 4004 in 1971).

Claims that "Moore's Law is just about to end!" have been made for many decades—we're beginning to run up against physical limits in the size of transistors!—and yet Moore's Law has still proven to be remarkably accurate over time. Its imminent demise is still predicted today, and yet it's still a pretty good model of computing power.[3] One probable reason that Moore's Law has held for as long as it has is a little bizarre: the repeated publicity surrounding Moore's Law! Because chip manufacturing companies "know" that the public generally expects processors to have twice as many transistors in two years, these companies may actually be setting research-and-development targets based on meeting Moore's Law. (Just as in a physical system, we cannot observe a phenomenon without changing it!)

[3] Gordon E. Moore. No exponential is forever: but "forever" can be delayed! In *International Solid-State Circuits Conference*, 2003.

6.2.3 *Exercises*

Part of the motivation for asymptotic analysis was that algorithms are typically analyzed ignoring constant factors. Ignoring constant factors in analyzing an algorithm may seem strange: if algorithm $\mathcal{A}$ runs twice as fast as $\mathcal{B}$, then $\mathcal{A}$ is way faster! But the reason we care more about asymptotic running time is that even an improvement by a factor of 2 *is quickly swamped by an asymptotic improvement for even slightly larger inputs. Here are a few examples:*

6.1 Suppose that linear search can find an element in a sorted list of n elements in n steps on a particular machine. Binary search (perhaps not implemented especially efficiently) requires $100 \log n$ steps. For what values of $n \geq 2$ is linear search faster?

Alice implements Merge Sort so, on a particular machine, it requires exactly $\lceil 8n \log n \rceil$ steps to sort n elements. Bob implements Heap Sort so it requires exactly $\lceil 5n \log n \rceil$ steps to sort n elements. Charlie implements Selection Sort so it requires exactly $2n^2$ steps to sort n elements. Suppose that Alice can sort 1000 *elements in* 1 *minute.*

6.2 How many elements can Bob sort in a minute? How many can Charlie sort in a minute?

6.3 What is the largest value of n that Charlie can sort faster than Alice?

6.4 Charlie, devastated by the news from the last exercise, buys a computer that's twice the speed of Alice's. What is the largest value of n that Charlie can sort faster than Alice now?

Let $f(n) = 9n + 3$ and let $g(n) = 3n^3 - n^2$. (See the first plot in Figure 6.8.)

6.5 Prove that $f(n) = O(n)$.

6.6 Prove that $f(n) = O(n^2)$.

6.7 Prove that $f(n) = O(g(n))$.

6.8 Prove that $g(n) = O(n^3)$.

6.9 Prove that $g(n) = O(n^4)$.

6.10 Prove that $g(n)$ is not $O(n^2)$.

6.11 Prove that $g(n)$ is not $O(n^{3-\varepsilon})$, for any $\varepsilon > 0$.

Prove that the following functions are all $O(n^2)$. (See the second plot in Figure 6.8.)

6.12 $f(n) = 7n$

6.13 $g(n) = 3n^2 + \sin n$

6.14 $h(n) = 202$

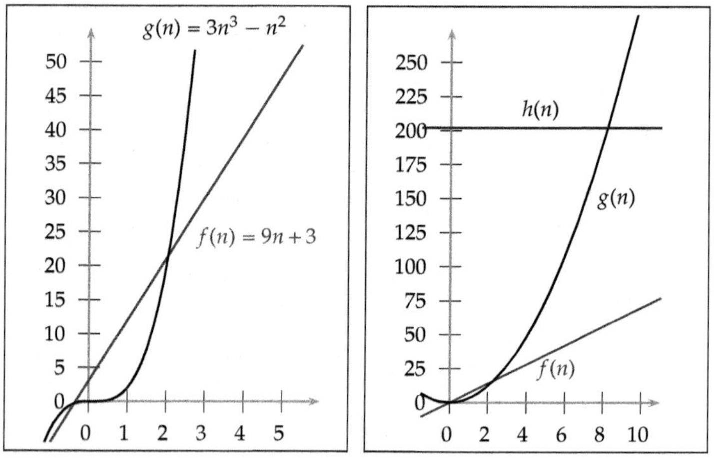

Figure 6.8: Two sets of functions, for Exercises 6.5–6.11 and 6.12–6.14.

The next few exercises ask you to explore the definition of $O(\cdot)$ in a little more detail.

6.15 Suppose $f(n) = O(g(n))$. Explain why there are infinitely many choices of c *and* infinitely many choices of n_0 that satisfy the definition of $O(\cdot)$.

Consider two functions $f, g : \mathbb{Z}^{\geq 0} \to \mathbb{Z}^{\geq 0}$. We defined $O(\cdot)$ notation as follows:

- *$f(n) = O(g(n))$ if there exist constants $c > 0$ and $n_0 \geq 0$ such that $\forall n \geq n_0 : f(n) \leq c \cdot g(n)$.*

It turns out that both c and n_0 are necessary to the definition. Define the following two pieces of alternative asymptotic notation, leaving out c (using $c = 1$) and n_0 (using $n_0 = 1$) from the definition:

- *$f(n) = P(g(n))$ if there exists a constant $n_0 \geq 0$ such that $\forall n \geq n_0 : f(n) \leq g(n)$.*
- *$f(n) = Q(g(n))$ if there exists a constant $c > 0$ such that $\forall n \geq 1 : f(n) \leq c \cdot g(n)$.*

Prove that $P(\cdot)$ and $Q(\cdot)$ are both different from $O(\cdot)$—that is, we can't just use either of the new definitions without changing what we meant. Specifically, prove that there exist functions f and g such that ...

6.16 ... either (i) $f = O(g)$ but $f \neq P(g)$, or (ii) $f \neq O(g)$ but $f = P(g)$.

6.17 ... either (i) $f = O(g)$ but $f \neq Q(g)$, or (ii) $f \neq O(g)$ but $f = Q(g)$.

The next several exercises ask you to prove some of properties of $O(\cdot)$ that we stated without proof earlier in the section. (For a model of a proof of this type of property, see Lemma 6.1 and its proof in this section.)

6.18 Prove Lemma 6.2, the transitivity of $O(\cdot)$: if $f(n) = O(g(n))$ and $g(n) = O(h(n))$, then $f(n) = O(h(n))$.

Prove Lemma 6.3: if $f(n) = O(h_1(n))$ and $g(n) = O(h_2(n))$, then ...

6.19 ... prove that $f(n) + g(n) = O(h_1(n) + h_2(n))$.

6.20 ... prove that $f(n) \cdot g(n) = O(h_1(n) \cdot h_2(n))$.

6.21 Prove Lemma 6.4: if $p(n) = \sum_{i=0}^{k} a_i n^i$ is a polynomial, then $p(n) = O(n^k)$.

6.22 Prove that the bound from the previous exercise cannot be improved. That is, prove that for $p(n) = \sum_{i=0}^{k} a_i n^i$ with $a_k > 0$, then $p(n)$ is not $O(n^{k-\varepsilon})$ for any $\varepsilon < k$.

Lemmas 6.5 and 6.6 state that all logarithmic functions grow slower than all polynomial functions, which grow slower than all exponential functions. (For example, $\log n = O(n^{0.000001})$ and $n^{1000000} = O(1.000001^n)$.) While fully general proofs are more calculus-intensive than we want to be in this book, here are a few simple results to prove:

6.23 Prove that Lemma 6.5 implies that any polylogarithmic function $f(n) = \log^k(n)$ satisfies $f(n) = O(n^\varepsilon)$ for any $\varepsilon > 0$ and any integer $k \geq 0$. (A *polylogarithmic* function is one that's a polynomial where the terms are powers of $\log n$ instead of powers of n—hence a poly(nomial of the)log function.)

6.24 Prove the special case of Lemma 6.5 for $\varepsilon = 1$: that is, prove that $\log n = O(n)$. Specifically, do so by proving that $\log n \leq n$ for all integers $n \geq 1$, using strong induction.

The next three exercises explore whether the asymptotic properties of two functions f and g "transfer over" to the functions $\log f$ and $\log g$. Specifically, consider two functions $f : \mathbb{Z}^{\geq 0} \to \mathbb{Z}^{\geq 1}$ and $g : \mathbb{Z}^{\geq 0} \to \mathbb{Z}^{\geq 1}$. (Note: the outputs of f and g are always positive, so that $\log(f(n))$ and $\log(g(n))$ are well defined.)

6.25 Assume that, for all n, we have $f(n) \geq n$ and $g(n) \geq n$. Furthermore assume that $f(n) = O(g(n))$. Prove that the function $\ell(n) := \log(f(n))$ satisfies $\ell(n) = O(\log(g(n)))$.

6.26 Prove that the converse of Exercise 6.25 is *not* true: identify functions $f(n)$ and $g(n)$ where $f(n) \geq n$ and $g(n) \geq n$ such that $\log(f(n)) = O(\log(g(n)))$ but $f(n) \neq O(g(n))$. *(Hint: what's $\log n^2$?)*

6.27 Prove that the assumption that $f(n) \geq n$ and $g(n) \geq n$ from Exercise 6.25 was necessary: identify functions $f : \mathbb{Z}^{\geq 0} \to \mathbb{Z}^{\geq 1}$ and $g : \mathbb{Z}^{\geq 0} \to \mathbb{Z}^{\geq 1}$ where $f(n) = O(g(n))$ but $\ell(n) \neq O(\log(g(n)))$ for the function $\ell(n) := \log(f(n))$.

6.28 For a real number $b \geq 1$, define the function $f(n) := b^n$. Prove Lemma 6.8: we have that $f(n) = O(c^n)$ if and only if $b \leq c$.

6.29 Something "going viral" online—a video, a joke, a hashtag, an app—can be reasonably modeled as a form of exponential growth: if each person who "adopts" the entity on a particular day causes two others to adopt that entity the next day, then 1 adopter on day #0 means 2 new ones on day #1 (for a total of 3), and 4 new ones on day #2 (for a total of 7), etc. Here we might call 2 the *spreading rate,* the number of people "infected" by each new adopter.

Let $b \in \mathbb{Z}^{\geq 1}$ be a spreading rate. Define $f(n) := \sum_{i=1}^{n} b^i$ to be the number of people who have adopted by day #n. Is $f(n) = O(b^n)$? Prove your answer.

6.30 Prove that $f(n) = O(g(n))$ if and only if $g(n) = \Omega(f(n))$.

Consider the function $f(n) := n + \frac{1}{n}$. (See Figure 6.9.) Because $f(0)$ is undefined and the output $f(n)$ is not an integer for any integer $n \geq 2$, treat f as a function from $\mathbb{Z}^{\geq 1}$ to $\mathbb{R}$. Prove all of your answers to the following questions:

6.31 Is $f(n) = O(1)$? $\Omega(1)$? $\Theta(1)$? $o(1)$? $\omega(1)$?

6.32 Is $f(n) = O(n)$? $\Omega(n)$? $\Theta(n)$? $o(n)$? $\omega(n)$?

6.33 Is $f(n) = O(n^2)$? $\Omega(n^2)$? $\Theta(n^2)$? $o(n^2)$? $\omega(n^2)$?

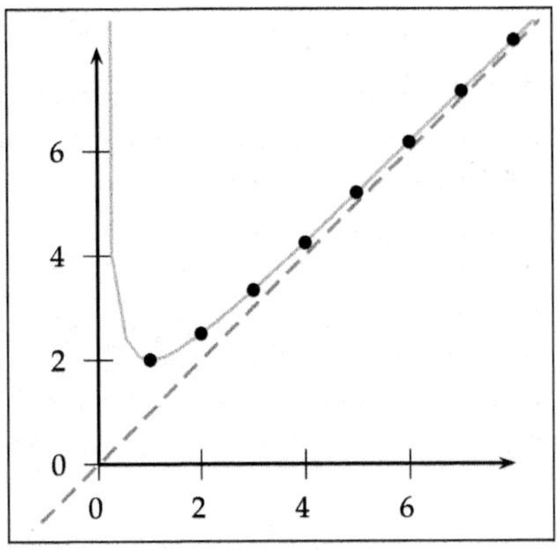

Figure 6.9: The function $f(n) = n + \frac{1}{n}$.

For an integer $n \geq 0$, let $k(n)$ denote the nonnegative integer such that $2^{k(n)} \leq n < 2^{k(n)+1}$. That is, $2^{k(n)}$ takes n and "rounds down" to a power of two: for example, $2^{k(4)} = 2^2 = 4$ and $2^{k(5)} = 2^2 = 4$ and $2^{k(202)} = 2^7 = 128$ and $2^{k(55{,}057)} = 2^{15} = 32{,}768$.

6.34 Prove that $2^{k(n)}$ and $2^{k(n)+1}$ are both $\Theta(n)$.

6.35 Prove that $k(n) = \Theta(\log n)$.

6.36 Let $b \geq 1$ be an arbitrary constant. Let $k_b(n)$ denote the nonnegative integer such that $b^{k_b(n)} \leq n < b^{k_b(n)+1}$. Prove that $k_b(n) = \Theta(\log n)$ *for any constant value $b > 1$.*

6.37 In Chapter 11, we'll talk about graphs and the "density" of graphs. If $f(n)$ denotes the number of edges in an n-node graph (we'll define those terms later!), then a graph is called *sparse* if $f(n) = O(n)$ and a graph is called *dense* if $f(n) = \Theta(n^2)$. Prove that there exists a function $f : \mathbb{Z}^{\geq 0} \to \mathbb{Z}^{\geq 0}$ satisfying $0 \leq f(n) \leq n^2$ such that neither $f(n) = \Theta(n^2)$ nor $f(n) = O(n)$.

6.38 Prove or disprove: the all-zero function $f(n) = 0$ is the *only* function that is $\Theta(0)$.

6.39 Give an example of a function $f(n)$ such that $f(n) = \Theta(f(n)^2)$.

6.40 Let $k \in \mathbb{Z}^{\geq 0}$ be any constant. Prove that $n^k = o(n!)$.

6.41 Let $f : \mathbb{Z}^{\geq 0} \to \mathbb{Z}^{\geq 0}$ be an arbitrary function. Define the function $g(n) = f(n) + 1$. Prove that $g(n) = O(f(n))$ *if and only if* $f(n) = \Omega(1)$.

6.42 Fill in each blank in the following table with an example of a function f that satisfies the stated conditions, or argue that it's impossible to satisfy both conditions:

$f(n)$ is ...	$o(n^2)$	$\neq o(n^2)$
... and $\omega(n^2)$		
... and $\neq \omega(n^2)$		

6.43 Let f and g be arbitrary functions. Prove that *at most one* of the three properties $f(n) = o(g(n))$ and $f(n) = \Theta(g(n))$ and $f(n) = \omega(g(n))$ can hold.

6.44 Complete the proof in Example 6.6: prove that $f(n) \neq \Omega(n^2)$, where $f(n)$ is the function

$$f(n) = \begin{cases} n^3 & \text{if } n \text{ is even} \\ n & \text{if } n \text{ is odd.} \end{cases}$$

Many of the properties of $O(\cdot)$ also hold for the other four asymptotic notions. Prove the following transitivity properties for arbitrary functions f, g, and h:

6.45 If $f(n) = \Omega(g(n))$ and $g(n) = \Omega(h(n))$, then $f(n) = \Omega(h(n))$.

6.46 If $f(n) = \Theta(g(n))$ and $g(n) = \Theta(h(n))$, then $f(n) = \Theta(h(n))$.

6.47 If $f(n) = o(g(n))$ and $g(n) = o(h(n))$, then $f(n) = o(h(n))$.

For each of the following purported properties related to symmetry, decide whether you think the statement is true or false, and—in either case—prove your answer.

6.48 Prove or disprove: if $f(n) = \Omega(g(n))$, then $g(n) = \Omega(f(n))$.

6.49 Prove or disprove: if $f(n) = \Theta(g(n))$, then $g(n) = \Theta(f(n))$.

6.50 Prove or disprove: if $f(n) = \omega(g(n))$, then $g(n) = \omega(f(n))$.

Do the same for the following purported properties related to reflexivity:

6.51 Prove or disprove: $f(n) = O(f(n))$.

6.52 Prove or disprove: $f(n) = \Omega(f(n))$.

6.53 Prove or disprove: $f(n) = \omega(f(n))$.

6.54 Consider the false claim (FC-6.1) below, and the bogus proof that follows. Where, precisely, does the proof of (FC-6.1) go wrong?

False Claim: The function $f(n) = n^2$ satisfies $f(n) = O(n)$. (FC-6.1)

Bogus proof of (FC-6.1). We proceed by induction on n:

base case ($n = 1$): Then $n^2 = 1$. Thus $f(1) = O(n)$ because $1 \leq n$ for all $n \geq 1$. (Choose $c = 1$ and $n_0 = 1$.)

inductive case ($n \geq 2$): Assume the inductive hypothesis—namely, assume that $(n-1)^2 = O(n)$. We must show that $n^2 = O(n)$. Here is the proof:

$$\begin{aligned} n^2 &= (n-1)^2 + 2n - 1 && \textit{by factoring} \\ &= O(n) + 2n - 1 && \textit{by the inductive hypothesis} \\ &= O(n) + O(n) && \textit{by definition of } O(\cdot) \textit{ and Lemma 6.3} \\ &= O(n). \end{aligned}$$

□

6.3 *Asymptotic Analysis of Algorithms*

> If everything seems under control, you're just not going fast enough.
>
> Mario Andretti (b. 1940)

The main reason that computer scientists are interested in asymptotic analysis is for its application to the *analysis of algorithms*. When, for example, we compare different algorithms that solve the same problem—say, Merge Sort, Selection Sort, and Insertion Sort—we want to be able to give a meaningful answer to the question *which algorithm is the fastest?* (And different inputs may trigger different behaviors in the algorithms under consideration: when the input array is sorted, for example, Insertion Sort is faster than Merge Sort and Selection Sort; when the input is very far from sorted, Merge Sort is fastest. But typically we still would like to identify a single answer to the question of which algorithm is the fastest.)

When evaluating the running time of an algorithm, we generally follow asymptotic principles. Specifically, we will generally ignore constants in the same two ways that $O(\cdot)$ and its asymptotic siblings do:

• First, we don't care much about what happens for small inputs: there might be small special-case inputs for which an algorithm is particularly fast, but this fast performance on a few special inputs doesn't mean that the algorithm is fast in general. For example, consider the algorithm for primality testing in Figure 6.10. Despite its speed on a few special cases ($n < 100$), we wouldn't consider **isPrime-tunedForDoubleDigits** a faster algorithm for primality testing *in general* than **isPrime**. We seek *general* answers to the question *which algorithm is faster?*, which leads us to pay little heed to special cases.

```
isPrime-tunedForDoubleDigits(n):
  if n ∈ {2,3,5,7,11,13,17,19,23,29,31,37,
      41,43,47,53,59,61,67,71,73,79,83,89,97} then
      return True
  else if n ≤ 100 then
      return False
  else
      return isPrime(n), from Figure 4.28.
```

Figure 6.10: A trivially faster algorithm for testing primality.

• Second, we typically evaluate the running time of an algorithm not by measuring elapsed time on the "wall clock," but rather by counting the number of steps that the algorithm takes to complete. (How long a program takes on your laptop, in terms of the wall clock, is affected by all sorts of things unrelated to the algorithm, like whether your virus checker is running while the algorithm executes.) We will generally ignore multiplicative constants in counting the number of steps consumed by an algorithm. One reason is so that we can give a machine-independent answer to the *which algorithm is faster?* question; how much is accomplished by one instruction on an Intel processor may be different from one instruction on an AMD processor, and ignoring constants allows us to compare algorithms in a way that doesn't depend on grungy details about the particular machine.

Definition 6.6 (Running time of an algorithm on a particular input)
Consider an algorithm $\mathcal{A}$ and an input x. The running time *of algorithm $\mathcal{A}$ on input x is the number of primitive steps that $\mathcal{A}$ takes when it's run on input x.*

For example, we can consider the running time of the algorithm **binarySearch** on the input $x = \langle [2,3,5,7,11,13,17,19,23,29,31], 4 \rangle$. The precise number of primitive steps in this execution depends on the particular machine on which the algorithm is being run, but it involves successively comparing 4 to 13, then 5, then 2, and finally 3.

Taking it further: Definition 6.6 is intentionally vague about what a "primitive step" is, but it's probably easiest to think of a single machine instruction as a primitive step. That single machine instruction might add or compare two numbers, increment a counter, return a value, etc. Different hardware systems might have different granularity in their "primitive steps"—perhaps a Mac desktop can "do more" in one machine instruction than an iPhone can do—but, as we just indicated, we'll look to analyze algorithms independently of this detail.

We typically evaluate an algorithm's efficiency by counting asymptotically of the number of primitive steps used by an algorithm's execution, rather than by using a stopwatch to measure how long the algorithm actually takes to run on a particular input on a particular machine. One reason is that it's very difficult to properly measure this type of performance; see p. 627 for some discussion about why.

In certain applications, particularly those in *scientific computing* (the subfield of CS devoted to processing and analyzing real-valued data, where we have to be concerned with issues like accumulated rounding errors in long calculations), it is typical to use a variation on asymptotic analysis. Calculations on integers are substantially cheaper than those involving floating point values; thus in this field one typically doesn't bother counting integer operations, and instead we only track floating point operations, or *flops*. Because flops are substantially more expensive, often we'll keep track of the constant on the leading (highest-degree) term—for example, an algorithm might require $\frac{3}{2}n^2 + O(n \log n)$ flops or $2n^2 + O(n)$ flops. (We'd choose the former.)

6.3.1 *Worst-Case Analysis*

We will generally evaluate the efficiency of an algorithm $\mathcal{A}$ by thinking about its performance as the input gets large: what happens to the number of steps consumed by $\mathcal{A}$ as a function of the input size n? Furthermore, we generally assume the worst: when we ask about the running time of an algorithm $\mathcal{A}$ on an input of size n, we are interested in the running time of $\mathcal{A}$ *on the input of size n for which* $\mathcal{A}$ *is the slowest*.

Definition 6.7 (Worst-case running time of an algorithm)
The worst-case running time of an algorithm $\mathcal{A}$ *is*

$$T_{\mathcal{A}}(n) = \max_{x:|x|=n} \left[\text{the number of primitive steps used by } \mathcal{A} \text{ on input } x \right].$$

We will be interested in the asymptotic behavior of the function $T_{\mathcal{A}}(n)$.

When we perform *worst-case analysis* of an algorithm—analyzing the asymptotic behavior of the function $T_{\mathcal{A}}(n)$—we seek to understand the rate at which the running time of the algorithm increases as the input size increases. Because a primary goal of algorithmic analysis is to provide a *guarantee* on the running time of an algorithm, we will be pessimistic, and think about how quickly $\mathcal{A}$ performs on the input of size n that's the worst for algorithm $\mathcal{A}$.

Taking it further: Occasionally we will perform *average-case analysis* instead of worst-case analysis: we will compute the *expected* (average) performance of algorithm $\mathcal{A}$ for inputs drawn from an appropriate distribution. It can be difficult to decide on an appropriate distribution, but sometimes this approach makes more sense than being purely pessimistic. See Section 6.3.2.

It's also worth noting that using asymptotic, worst-case analysis can sometimes be misleading. There are occasions in which an algorithm's performance in practice is very poor despite a "good" asymptotic running time—for example, because the multiplicative constant suppressed by the $O(\cdot)$ is massive. (And conversely: sometimes an algorithm that's asymptotically slow in the worst case might perform very well on problem instances that actually show up in real applications.) Asymptotics capture the high-level performance of an algorithm, but constants matter too!

Figure 6.11 shows a sampling of worst-case running times for a number of the algorithms you may have encountered earlier in this book or in previous CS classes. In the rest of this section, we'll prove some of these results as examples.

worst-case running time	sample algorithm(s)
$\Theta(1)$	push/pop in a stack
$\Theta(\log n)$	binary search
$\Theta(\sqrt{n})$	**isPrimeBetter** (p. 454)
$\Theta(n)$	linear search, **isPrime**
$\Theta(n \log n)$	merge sort
$\Theta(n^2)$	selection sort, insertion sort, bubble sort
$\Theta(n^3)$	naïve matrix multiplication
$\Theta(2^n)$	brute-force satisfiability algorithm

Figure 6.11: The running time of some sample algorithms.

Some examples: sorting algorithms

We'll now turn to a few examples of worst-case analysis of several different sorting and searching algorithms. We'll start with three sorting algorithms, illustrated in Figure 6.13:

- *Selection Sort:* repeatedly find the minimum element in the unsorted portion of A; then swap that minimum element into the first slot of the unsorted segment of A.
- *Insertion Sort:* maintain a sorted prefix of A (initially consisting only of the first element); repeatedly expand the sorted prefix by one element, by continuing to swap the first unsorted element backward in the array until it's in place.
- *Bubble Sort:* make n left-to-right passes through A; in each pass, swap each pair of adjacent elements that are out of order.

We'll start our analysis with Selection Sort, whose pseudocode is shown in Figure 6.12. (The pseudocode for the other algorithms will accompany their analysis.)

```
selectionSort(A[1...n]):
1: for i := 1 to n:
2:     minIndex := i
3:     for j := i+1 to n:
4:         if A[j] < A[minIndex] then
5:             minIndex := j
6:     swap A[i] and A[minIndex]
```

Figure 6.12: Selection Sort.

Example 6.7 (Selection Sort)

Problem: What is the worst-case running time of Selection Sort?

Solution: The outer **for** loop's body (lines 2–6) is executed n times, once each for $i = 1 \ldots n$. We complete the body of the inner **for** loop (lines 4–5) a total of $n - i$ times in iteration i. Thus the total number of times that we execute lines 4–5 is

$$\sum_{i=1}^{n} n - i = n^2 - \sum_{i=1}^{n} i = n^2 - \frac{n(n+1)}{2} = \frac{n^2 - n}{2},$$

where $\sum_{i=1}^{n} i = \frac{n(n+1)}{2}$ by Lemma 5.4.

Notice that the only variation in the running time of Selection Sort based on the particular input array $A[1 \ldots n]$ is in line 5; the number of times that *minIndex* is reassigned can vary from as low as 0 to as high as $n - i$. The remainder of the algorithm behaves precisely identically regardless of the input array values.

Thus, for some constants $c_1 > 0$ and $c_2 > 0$ the total number of primitive steps used by the algorithm is $c_1 n + c_2 n^2$ (for lines 1, 2, 3, 4, and 6), plus some number x of executions of line 5, where $0 \leq x \leq \sum_{i=1}^{n} n - i \leq n^2$, each of which takes a constant c_3 number of steps. Thus the total running time is between $c_1 n + c_2 n^2$ and $c_1 n + (c_2 + c_3) n^2$. The asymptotic worst-case running time of Selection Sort is therefore $\Theta(n^2)$.

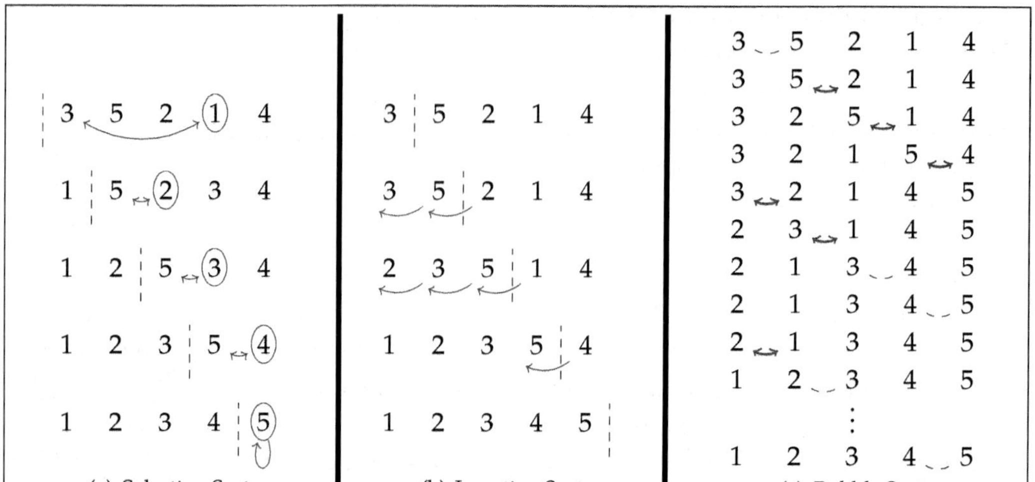

(a) Selection Sort (b) Insertion Sort (c) Bubble Sort

Figure 6.13: Three sorting algorithms applied to the list 3, 5, 2, 1, 4. Selection Sort repeatedly finds the minimum element in the unsorted segment and swaps it into place. Insertion Sort repeatedly extends a sorted prefix by swapping the next element backward into position. Bubble Sort repeatedly compares adjacent elements and swaps them if they're out of order.

We are generally interested in the asymptotic performance of algorithms, so the particular values of the constants c_1, c_2, and c_3 from Example 6.7, which reflect the number of primitive steps corresponding to each line of the pseudocode in Figure 6.12, are irrelevant to our final answer. (One exception is that we may sometimes try to count exactly the number of *comparisons* between elements of A, or *swaps* of elements of A; see Exercises 6.55–6.63.)

We'll now turn to our second sorting algorithm, Insertion Sort (Figure 6.14). Insertion Sort proceeds by maintaining a sorted prefix of the given array (initially the sorted prefix consists only of the first element); it then repeatedly expands the sorted prefix one element at a time, by continuing to swap the first unsorted element backward.

```
insertionSort(A[1...n]):
1: for i := 2 to n:
2:     j := i
3:     while j > 1 and A[j] < A[j − 1]:
4:         swap A[j] and A[j − 1]
5:         j := j − 1
```

Figure 6.14: Insertion Sort.

Example 6.8 (Insertion Sort)

Insertion Sort is more sensitive to the structure of its input than Selection Sort: if A is in sorted order, then the **while** loop in lines 3–5 terminates immediately (because the test $A[j] > A[j-1]$ fails); whereas if the input array is in *reverse* sorted order, then the **while** loop in lines 3–5 completes $i-1$ iterations. In fact, the reverse-sorted array is the worst-case input for Insertion Sort: there can be as many as $i-1$ iterations of the **while** loop, and there cannot be more than $i-1$ iterations. If the **while** loop goes through $i-1$ iterations, then the total amount of work done is

$$\begin{aligned}\sum_{i=1}^{n} c + (i-1)d &= (c-d)n + \sum_{i=1}^{n} id \\ &= (c-d)n + d \cdot \tfrac{n(n+1)}{2} \\ &= (c - \tfrac{d}{2})n + \tfrac{d}{2}n^2,\end{aligned}$$

where c and d are constants corresponding to the work of lines 1–2 and 3–5, respectively. This function is $\Theta(n^2)$, so Insertion Sort's worst-case running time is $\Theta(n^2)$.

Finally, we will analyze a third sorting algorithm: Bubble Sort (Figure 6.15), which makes n left-to-right passes through the array; in each pass, adjacent elements that are out of order are swapped. Bubble Sort is a very simple sorting algorithm to analyze. (But, in practice, it is also a comparatively slow sorting algorithm to run!)

```
bubbleSort(A[1...n]):
1: for i := 1 to n:
2:     for j := 1 to n − i:
3:         if A[j] > A[j+1] then
4:             swap A[j] and A[j+1]
```

Figure 6.15: Bubble Sort.

Example 6.9 (Bubble Sort)

Bubble Sort simply repeatedly compares $A[j]$ and $A[j+1]$ (swapping the two elements if necessary) for many different values of j. Every time the body of the inner loop, Lines 3–4, is executed, the algorithm does a constant amount of work: exactly one comparison and either zero or one swaps. Thus there are two constants $c > 0$ and $d > 0$ such that any particular execution of Lines 3–4 takes an amount of time t satisfying $c \leq t \leq d$. Therefore the total running time of Bubble Sort is somewhere between $\sum_{i=1}^{n} \sum_{j=1}^{n-i} c$ and $\sum_{i=1}^{n} \sum_{j=1}^{n-i} d$. The summation $\sum_{i=1}^{n} n - i$ is $\Theta(n^2)$, precisely as we analyzed in Example 6.7, and thus Bubble Sort's running time is $\Omega(cn^2) = \Omega(n^2)$ and $O(dn^2) = O(n^2)$. Therefore Bubble Sort is $\Theta(n^2)$.

Problem-solving tip: Precisely speaking, the number of primitive steps required to execute, for example, Lines 3–4 of Bubble Sort varies based on whether a swap has to occur. In Example 6.9, we carried through the analysis considering two different constants representing this difference. But, more simply, we could say that Lines 3–4 of Bubble Sort take $\Theta(1)$ time, without caring about the particular constants. You can use this simpler approach to streamline arguments like the one in Example 6.9.

Before we close, we'll mention one more sorting algorithm, Merge Sort, which proceeds recursively by splitting the input array in half, recursively sorting each half, and then "merging" the sorted subarrays into a single sorted array. But we will defer the analysis of Merge Sort to Section 6.4: to analyze recursive algorithms like Merge Sort, we will use *recurrence relations* which represent *the algorithm's running time itself* as a recursive function.

Some more examples: search algorithms

We will now turn to some examples of search algorithms, which determine whether a particular value x appears in an array A. We'll start with Linear Search (see Figure 6.16), which simply walks through the (possibly unsorted) array A and successively compares each element to the sought value x.

Unless otherwise specified (and we will rarely specify otherwise), we are interested in the worst-case behavior of algorithms. *This concern with worst-case behavior includes lower bounds!* Here's an example of the analysis of an algorithm that suffers from this confusion:

```
linearSearch(A[1...n], x):
Input: an array A[1...n] and an element x
Output: is x in the (possibly unsorted) array A?
1: for i := 1 to n:
2:     if A[i] = x then
3:         return True
4: return False
```

Figure 6.16: Linear Search.

Example 6.10 (Linear Search, unsatisfactorily analyzed)

Problem: What is incomplete or incorrect in the following analysis of the worst-case running time of Linear Search?

> The running time of Linear Search is obviously $O(n)$: we at most iterate over every element of the array, performing a constant number of operations per element. And it's obviously $\Omega(1)$: no matter what the inputs A and x are, the algorithm certainly at least does one operation (setting $i := 1$ in line 1), even if it immediately returns because $A[1] = x$.

Solution: The analysis is correct, but it gives a looser lower bound than can be shown: specifically, the running time of Linear-Search is $\Omega(n)$, and not just $\Omega(1)$. If we call **linearSearch**$(A, 42)$ for an array $A[1 \ldots n]$ that does not contain the number 42, then the total number of steps required by the algorithm will be at least n, because every element of A is compared to 42. Performing n comparisons takes $\Omega(n)$ time.

Taking it further: When we're analyzing an algorithm $\mathcal{A}$'s running time, we can generally prove several different lower and upper bounds for $\mathcal{A}$. For example, we might be able to prove that the running time is $\Omega(1)$, $\Omega(\log n)$, $\Omega(n)$, $O(n^2)$, and $O(n^3)$. The bound $\Omega(1)$ is a *loose bound*, because it is superseded by the bound $\Omega(\log n)$. (That is, if $f(n) = \Omega(\log n)$ then $f(n) = \Omega(1)$.) Similarly, $O(n^3)$ is a loose bound, because it is implied by $O(n^2)$.

We seek asymptotic bounds that are as tight as possible—so we always want to prove $f(n) = \Omega(g(n))$ and $f(n) = O(h(n))$ for the fastest-growing function g and slowest-growing function h that we can. If $g = h$, then we have proven a *tight bound*, or, equivalently, that $f(n) = \Theta(g(n))$. Sometimes there are algorithms for which we don't know a tight bound; we can prove $\Omega(n)$ and $O(n^2)$, but the algorithm might be $\Theta(n)$ or $\Theta(n^2)$ or $\Theta(n \log n \log \log \log n)$ or whatever. In general, we want to give upper and lower bounds that are as close together as possible.

Here is a terser writeup of the analysis of Linear Search:

Example 6.11 (Linear Search)
The worst case for Linear Search is an array $A[1 \ldots n]$ that doesn't contain the element x. In this case, the algorithm compares x to all n elements of A, taking $\Theta(n)$ time.

Binary Search (see Figure 6.17(a)) is another search algorithm for locating a value x in an array $A[1 \ldots n]$, if the array is sorted. It proceeds by defining a range of the array in which x would be found if it is present, and then repeatedly halving the size of that range by comparing x to the middle entry in that range. Let's analyze the running time of Binary Search.

```
binarySearch(A[1...n], x):
Input: a sorted array A[1...n]; an element x
Output: is x in the (sorted) array A?
1: lo := 1
2: hi := n
3: while lo ≤ hi:
4:     middle := ⌊(lo+hi)/2⌋
5:     if A[middle] = x then
6:         return True
7:     else if A[middle] > x then
8:         hi := middle − 1
9:     else
10:        lo := middle + 1
11: return False
```

(a) The code.

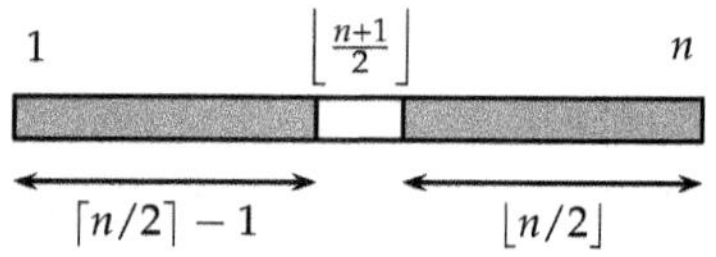

When $lo = 1$ and $hi = n$, then $middle = \lfloor (n+1)/2 \rfloor$. Because $\lfloor (n+1)/2 \rfloor = \lceil n/2 \rceil$, there are $\lceil n/2 \rceil - 1$ elements before *middle* and $\lfloor n/2 \rfloor$ elements after *middle*.

(b) An illustration of the split.

Figure 6.17: Binary Search.

Example 6.12 (Binary Search)
The intuition is fairly straightforward. In every iteration of the **while** loop in lines 3–10, we halve the range of elements under consideration—that is, $|\{i : lo \leq i \leq hi\}|$. We can halve a set of size n only $\log_2 n$ times before there's only one element left, and therefore we have at most $1 + \log_2 n$ iterations of the **while** loop. Each of those iterations takes a constant amount of time, and therefore the total running time is $O(\log n)$.

To translate this intuition into a more formal proof, suppose that the range of elements under consideration at the beginning of an iteration of the **while** loop is $A[lo, \ldots, hi]$, which contains $k = hi - lo + 1$ elements. There are $\lceil k/2 \rceil - 1$ elements in $A[lo, \ldots, middle - 1]$ and $\lfloor k/2 \rfloor$ elements in $A[middle + 1, \ldots, hi]$. Then, after comparing x to $A[middle]$, one of three things happens:

- we find that $x = A[middle]$, and the algorithm terminates.
- we find that $x < A[middle]$, and we continue on a range of the array that contains $\lceil k/2 \rceil - 1 \leq k/2$ elements.
- we find that $x > A[middle]$, and we continue on a range of the array that contains $\lfloor k/2 \rfloor \leq k/2$ elements.

In any of the three cases, we have at most $k/2$ elements under consideration in the next iteration of the loop. (See Figure 6.17(b).)

Initially, the number of elements under consideration has size n. Therefore after i iterations, there are at most $n/2^i$ elements left under consideration. (This claim can be proven by induction.) Therefore, after at most $\log_2 n$ iterations, there is only one element left under consideration. Once the range contains only one element, we complete at most one more iteration of the **while** loop. Thus the total number of iterations is at most $1 + \log_2 n$. Each iteration takes a constant number of steps, and thus the total running time is $O(\log n)$.

Notice that analyzing the running time of any single iteration of the **while** loop in the algorithm was easy; the challenge in determining the running time of **binarySearch** lies in figuring out how many iterations occur.

Here we have only shown an upper bound on the running time of Binary Search; in Example 6.26, we'll prove that, in fact, Binary Search takes $\Theta(\log n)$ time. (Just as for Linear Search, the worst-case input for Binary Search is an n-element array that does not contain the sought value x; in this case, we complete all logarithmically many iterations of the loops, and the running time is therefore $\Omega(\log n)$ too.)

6.3.2 *Some Other Types of Analysis*

So far we have focused on asymptotically analyzing the worst-case running time of algorithms. While this type of analysis is the one most commonly used in the analysis of algorithms, there are other interesting types of questions that we can ask about algorithms. We'll sketch two of them in this section: instead of being completely pessimistic about the particular input that we get, we might instead consider either the *best* possible case or the "average" case.

Best-case analysis of running time

Best-case running time simply replaces the "max" from Definition 6.7 with a "min":

Definition 6.8 (Best-case running time of an algorithm)
The best-case running time of an algorithm $\mathcal{A}$ *on an input of size n is*

$$T_{\mathcal{A}}^{best}(n) = \min_{x:|x|=n} \left[\textit{the number of primitive steps used by } \mathcal{A} \textit{ on input } x\right].$$

"Optimism, n. The doctrine or belief that everything is beautiful, including what is ugly."
— Ambrose Bierce (1842–≈1913), *The Devil's Dictionary* (1911)

Best-case analysis is rarely used; knowing that an algorithm *might* be fast (on inputs for which it is particularly well tuned) doesn't help much in drawing generalizable conclusions about its performance (on the input that it's actually called on).

Average-case analysis of running time

The "average" running time of an algorithm $\mathcal{A}$ is subtler to state formally, because "average" means that we have to have a notion of which values are more or less likely to be chosen as inputs. (For example, consider sorting. In many settings, an already-sorted array is the most common input type to the sorting algorithm; the programmer just wanted to "make sure" that the input was sorted, even though he might have been pretty confident that it already was.) The simplest way to do average-case analysis is to consider inputs that are chosen *uniformly at random* from the space of all possible inputs. For example, for sorting algorithms, we would consider each of the $n!$ different orderings of $\{1, 2, \ldots, n\}$ to be equally likely inputs of size n.

Definition 6.9 (Average-case running time of an algorithm)
Let X denote the set of all possible inputs to an algorithm $\mathcal{A}$. *The* average-case running time of an algorithm $\mathcal{A}$ for a uniformly chosen input *of size n is*

$$T_{\mathcal{A}}^{avg}(n) = \frac{1}{|\{y \in X : |y| = n\}|} \cdot \sum_{x \in X:|x|=n} \left[\textit{number of primitive steps used by } \mathcal{A} \textit{ on } x\right].$$

Taking it further: Let ρ_n be a probability distribution over $\{x \in X : |x| = n\}$—that is, let ρ_n be a function such that $\rho_n(x)$ denotes the fraction of the time that a size-n input to $\mathcal{A}$ is x. Definition 6.9 considers the uniform distribution, where $\rho_n(x) = 1/|\{x \in X : |x| = n\}|$.

The average-case running time of $\mathcal{A}$ on inputs of size n is the *expected running time* of $\mathcal{A}$ for an input x of size n chosen according to the probability distribution ρ_n. We will explore both probability distributions and expectation in detail in Chapter 10, which is devoted to probability. (If someone refers to the average case of an algorithm without specifying the probability distribution ρ, then they probably mean that ρ is the uniform distribution, as in Definition 6.9.)

We will still consider the asymptotic behavior of the best-case and average-case running times, for the same reasons that we are generally interested in the asymptotic behavior in the worst case.

Best- and average-case analysis of sorting algorithms

We'll close this section with the best- and average-case analyses of our three sorting algorithms. (See Figure 6.18 for a reminder of the algorithms.)

```
insertionSort(A[1...n]):
1: for i := 2 to n:
2:     j := i
3:     while j > 1 and A[j] < A[j − 1]:
4:         swap A[j] and A[j − 1]
5:         j := j − 1
```

```
selectionSort(A[1...n]):
1: for i := 1 to n:
2:     minIndex := i
3:     for j := i + 1 to n:
4:         if A[j] < A[minIndex] then
5:             minIndex := j
6:     swap A[i] and A[minIndex]
```

```
bubbleSort(A[1...n]):
1: for i := 1 to n:
2:     for j := 1 to n − i:
3:         if A[j] > A[j + 1] then
4:             swap A[j] and A[j + 1]
```

Figure 6.18: A reminder of the sorting algorithms.

Example 6.13 (Insertion Sort, best- and average-case)

In Example 6.8, we showed that the worst-case running time of Insertion Sort is $\Theta(n^2)$. Let's analyze the best- and average-case running times of Insertion Sort.

The best-case running time for Insertion Sort is much faster: if the input array is already in sorted order, the **while** loop that swaps each $A[i]$ into place (lines 3–5) terminates immediately without doing any swaps, because $A[i] > A[i-1]$. Each iteration of the **for** loop therefore takes $\Theta(1)$ time, so the total running time is $\Theta(n)$.

We will defer a fully formal analysis of the average-case running time of Insertion Sort to Chapter 10 (see Example 10.45), but here is an informal analysis. Consider iteration #i of the **for** loop of Insertion Sort. When that iteration starts, the first $i-1$ elements of A—that is, $A[1,\ldots,i-1]$—are in sorted order. The next element $A[i]$ has an equal chance of falling into any one of the i "slots" in the sorted $A[1,\ldots,i-1]$: before $A[1]$, between $A[1]$ and $A[2]$, ..., between $A[i-2]$ and $A[i-1]$, and after $A[i-1]$. On average, then, we complete $i/2$ swaps in the ith iteration of the **for** loop. Thus the total average running time will be $\sum_{i=1}^{n-1} i/2 = n(n-1)/4$, which is $\Theta(n^2)$.

While we will typically use formal mathematical analysis to address the best- and average-case performance of algorithms (as in Example 6.13), sometimes the kind of empirical analysis discussed above—where we measure an algorithm's performance by running it on an actual computer on an actual input and measuring how much time elapses before the algorithm terminates—can also be useful. Figure 6.19 shows the elapsed time on an aging laptop during executions of Insertion, Selection, and Bubble Sorts on sorted arrays, reverse-sorted arrays, and a randomly shuffled array.

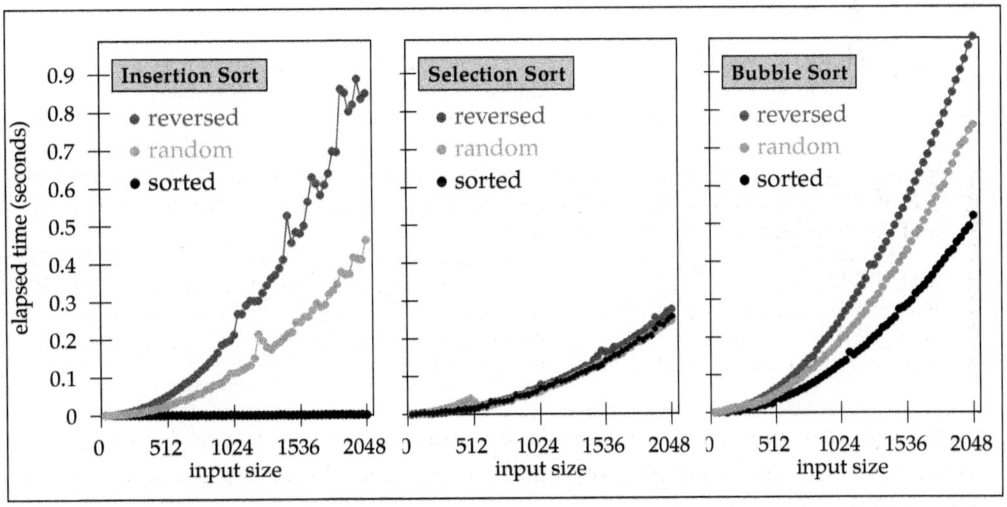

Figure 6.19: The elapsed-time running time for Insertion, Selection, and Bubble Sorts.

Figure 6.19(a) confirms the formal analysis from Example 6.13: Insertion Sort's worst case is about twice as slow as its average case, and both are $\Theta(n^2)$; the best case of Insertion Sort is virtually invisible along the x-axis. On the other hand, Figure 6.19(b) suggests that Selection Sort's performance does not seem to depend very much on the structure of its input. Let's analyze this algorithm formally:

Example 6.14 (Selection Sort, best- and average-case)
In Selection Sort (see Figure 6.18), the only effect of the input array's structure is the number of times that line 5 is executed. (That's why the reverse-sorted input tends to perform ever-so-slightly worse in Figure 6.19(b).) Thus the best- and average-case running time of Selection Sort is $\Theta(n^2)$, just like the worst-case running time established in Example 6.7.

Figure 6.19(c) suggests that Bubble Sort's performance varies only by a constant factor; indeed, the worst-, average-, and best-case running times are all $\Theta(n^2)$:

Example 6.15 (Bubble Sort, best- and average-case)
Again, the only difference in running time based on the structure of the input array is in how many times line 4 is executed—that is, how many swaps occur. (The number of swaps ranges between 0 for a sorted array and $n(n-1)/2$ for a reverse-sorted array.) But line 3 is executed $\Theta(n^2)$ times in any case, and $\Theta(n^2) + 0$ and $\Theta(n^2) + n^2$ are both $\Theta(n^2)$.

More careful examination of Bubble Sort shows that we can improve the algorithm's best-case performance without affecting the worst- and average-case performance asymptotically; see Exercise 6.65.

Taking it further: The tools from this chapter can be used to analyze the consumption of any resource by an algorithm. So far, the only resource that we have considered is *time*: how many primitive steps are used by the algorithm on an particular input? The other resource whose consumption is most commonly analyzed is the *space* used by the algorithm—that is, the amount of memory used by the algorithm. As with time, we almost always consider the worst-case space use of the algorithm. See the discussion on p. 628 for more on the subfield of CS called *computational complexity*, which seeks to understand the resources required to solve any particular problem.

While time and space are the resources most frequently analyzed by complexity theorists, there are other resources that are interesting to track, too. For example, *randomized algorithms* "flip coins" as they run—that is, they make decisions about how to continue based on a randomly generated bit. Generating a truly random bit is expensive, and so we can view randomness itself as a resource, and try to minimize the number of random bits used. And, particularly in mobile processors, *power consumption*—and therefore the amount of battery life consumed, and the amount of heat generated—may be a more limiting resource than time or space. Thus energy can also be viewed as a resource that an algorithm might consume.[4]

For some of the research from an architecture perspective on power-aware computing, see

[4] Stefanos Kaxiras and Margaret Martonosi. *Computer Architecture Techniques for Power-Efficiency*. Morgan Claypool, 2008.

Computer Science Connections

Multitasking, Garbage Collection, and Wall Clocks

One reason that we typically measure the running time of algorithms by counting (asymptotically) the number of primitive operations consumed by the algorithm on (worst-case) inputs is that measuring running time by so-called *wall-clock time* can be difficult to interpret—and potentially misleading.

All modern operating systems (everything that's been widely deployed for several decades: Windows, MacOS, Linux, iOS, Android, ...) are *multitasking* operating systems. That is, the user is typically running many applications simultaneously—perhaps an application to play music, a web browser, a programming environment, a word processor, a virus checker, and that sorting program that you wrote for your CS class. While it appears to the user that these applications are all running simultaneously, the operating system is actually pulling off a trick. There's typically only one processor (or maybe two or four, in increasingly used multicore machines), and the operating system uses *time-sharing* to allow each running application to have a "turn" using the processor. (When it's the next application's turn, the operating system *swaps out* one application, and *swaps in* the next one that gets a slice of time on the processor.) If there were more processes running when you ran Merge Sort than when you ran Bubble Sort, then the elapsed time for Merge Sort could look worse than it should.

Many operating systems can report the total amount of processor time that a particular process consumed, so we can avoid the multitasking concern—but even within a single process, total processor time consumed can be misleading. While a program in Python or Java, for example, is running, periodically the *garbage collector* runs to reclaim "garbage" memory (previously allocated memory that won't be used again) for future use. When the garbage collector runs, the code that you were executing stops running.

Figure 6.20 shows the elapsed time while running four sorting algorithms, written in Python, executed on sorted inputs $[1, 2, \ldots, n]$, reverse sorted inputs $[n, n-1, \ldots, 1]$, and a randomly permuted n-element array. The "spikiness" of the elapsed times within the second panel may be because I launched a large presentation-editing application while the Insertion Sort test was running on inputs in descending sorted order, or because the garbage collector happened to start running during those trials.

Even putting aside the difficulty of measuring running times accurately, there's another fundamental issue that we must address: we have to decide *on what* inputs to run the algorithms. The three panels of Figure 6.20 show why this choice can be significant. When the input is in sorted order, Insertion Sort is the best algorithm (in fact, it's barely visually distinguishable from the x-axis!). When the input is in reverse sorted order, Insertion Sort is terrible, and Merge Sort is the fastest. When the input is randomized, Insertion Sort is somewhere in the middle, and Merge Sort is again the fastest. Selection Sort is essentially unaffected by which type of input we consider.

The fact that we get such different pictures from the three different input types says that we have to decide which input to consider. (Typically we choose *the worst-case input for the particular algorithm*, as we've discussed.)

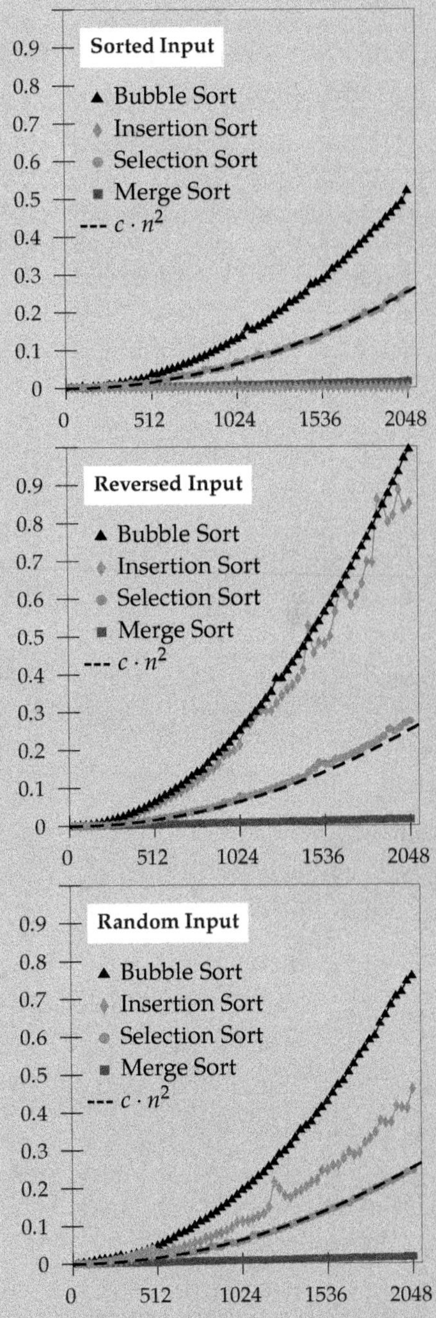

Figure 6.20: The wall-clock running time of four sorting algorithms on three different types of input. For n-element inputs of each type, the plot shows the number of seconds elapsed for the given sorting algorithms. The function $f(n) = 0.00000006 \cdot n^2$ is shown in each panel for comparison.

Computer Science Connections

Time, Space, and Complexity

Computational complexity is the subfield of computer science devoted to the study of the resources required to solve computational problems. Computational complexity is the domain of the most important open question in all of computer science, the P-versus-NP problem. That problem is described elsewhere in this book (see p. 326), but here we'll describe some of the basic entities that are studied by complexity theorists.

A *complexity class* is a set of problems that can be solved using a given constraint on resources consumed. Those resources are most typically the *time* or *space* used by an algorithm that solves the problem. For example, the complexity class EXPTIME includes precisely those problems solvable in exponential time—that is, $O(2^{n^k})$ time for some constant integer k.

One of the most important complexity classes is P, which denotes the set of all problems Π for which there is a polynomial-time algorithm $\mathcal{A}$ that solves Π. In other words,

$\Pi \in \mathsf{P} \Leftrightarrow$ there exists an algorithm $\mathcal{A}$ and an integer $k \in \mathbb{Z}^{\geq 0}$ such that $\mathcal{A}$ solves Π and the worst-case running time of $\mathcal{A}$ on an input of size n is $O(n^k)$.

Although the practical efficiency of an algorithm that runs in time $\Theta(n^{1000})$ is highly suspect, it has turned out that essentially any (non-contrived) problem that has been shown to be in P has actually also had a reasonably efficient algorithm—almost always $O(n^5)$ or better. As a result, one might think of the entire subfield of CS devoted to algorithms as really being devoted to understanding what problems can be solved in polynomial time. (Of course, improving the exponent of the polynomial is always a goal!)

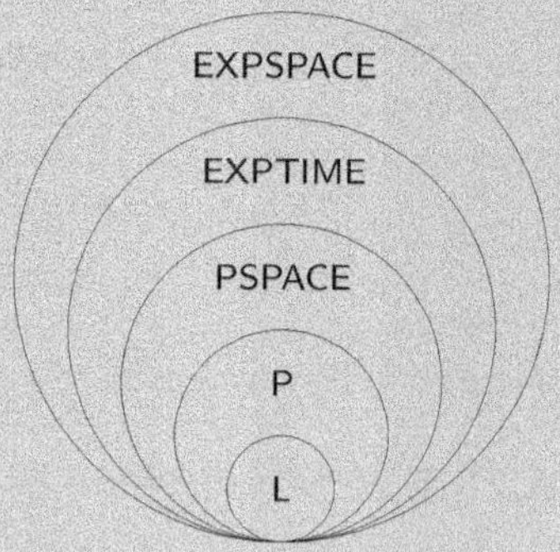

Figure 6.21: A few complexity classes, and their relationships.

Other commonly studied complexity classes are defined in terms of the space (memory) that they use:

- PSPACE: problems solvable using a polynomial amount of space;
- L: problems solvable using $O(\log n)$ space (beyond the input itself); and
- EXPSPACE: problems solvable in exponential space.

While a great deal of effort has been devoted to complexity theory over the last half century, surprisingly little is known about how much time or space is actually required to solve problems—including some very important problems! It is reasonably easy to prove the relationships among the complexity classes shown in Figure 6.21, namely

$$\mathsf{L} \subseteq \mathsf{P} \subseteq \mathsf{PSPACE} \subseteq \mathsf{EXPTIME} \subseteq \mathsf{EXPSPACE}.$$

Although the proofs are trickier, it has also been known since the 1960s that $\mathsf{P} \neq \mathsf{EXPTIME}$ (using the "time hierarchy theorem"), and that both $\mathsf{L} \neq \mathsf{PSPACE}$ and $\mathsf{PSPACE} \neq \mathsf{EXPSPACE}$ (using the "space hierarchy theorem"). But that's just about all that we know about the relationship among these complexity classes! For example, for all we know $\mathsf{L} = \mathsf{P}$ or $\mathsf{P} = \mathsf{PSPACE}$—but not both, because we *do* know that $\mathsf{L} \neq \mathsf{PSPACE}$. These foundational complexity-theoretic questions remain open—awaiting the insights of a new generation of computer scientists![5]

For more, see any good textbook on computational complexity (also known as complexity theory). For example,

[5] Michael Sipser. *Introduction to the Theory of Computation*. Course Technology, 3rd edition, 2012; and Christos H. Papadimitriou. *Computational Complexity*. Addison Wesley, 1994.

6.3.3 *Exercises*

A comparison-based *sorting algorithm reorders its input array* $A[1 \ldots n]$ *with two fundamental operations:*

- *the* comparison *of a pair of elements (to determine which one is bigger); and*
- *the* swap *of a pair of elements (to exchange their positions in the array).*

See Figure 6.22 for another reminder of three comparison-based sorting algorithms: Selection, Insertion, and Bubble Sorts. For each of the following problems, give an exact *answer (not an asymptotic one), and prove your answer. For the worst-case input array of size n, how many comparisons are done by these algorithms?*

6.55 **selectionSort**

6.56 **insertionSort**

6.57 **bubbleSort**

We'll now turn to counting swaps. In these exercises, you should count as a "swap" the exchange of an element $A[i]$ *with itself. (So if* $i = minIndex$ *in Line 6 of* **selectionSort**, *Line 6 still counts as performing as swap.) For the worst-case input array of size n, how many swaps are done by these algorithms?*

6.58 **selectionSort**

6.59 **insertionSort**

6.60 **bubbleSort**

```
selectionSort(A[1...n]):
1: for i := 1 to n:
2:     minIndex := i
3:     for j := i+1 to n:
4:         if A[j] < A[minIndex] then
5:             minIndex := j
6:     swap A[i] and A[minIndex]
```

```
insertionSort(A[1...n]):
for i := 2 to n:
    j := i
    while j > 1 and A[j] < A[j − 1]:
        swap A[j] and A[j − 1]
        j := j − 1
```

```
bubbleSort(A[1...n]):
for i := 1 to n:
    for j := 1 to n − i:
        if A[j] > A[j+1] then
            swap A[j] and A[j+1]
```

Figure 6.22: Another reminder of the sorting algorithms.

Repeat the previous exercises for the best-case input: *that is, for the input array* $A[1 \ldots n]$ *on which the given algorithm performs the best, how many comparisons/swaps does the algorithm do? (If the best-case array for swaps is different from the best-case array for comparisons, say so and explain why, and analyze the number of comparisons/swaps in the two different "best" arrays.) In the best case, how many comparisons and how many swaps are done by these algorithms?*

6.61 **selectionSort**

6.62 **insertionSort**

6.63 **bubbleSort**

Two variations of the basic **bubbleSort** *algorithm are shown in Figure 6.23. In the next few exercises, you'll explore whether they're asymptotic improvements.*

6.64 What's the worst-case running time of **early-stopping-bubbleSort**?

6.65 Show that the *best-case* running time of **early-stopping-bubbleSort** is asymptotically better than the best-case running time of **bubbleSort**.

6.66 Show that the running time of **forward-backward-bubbleSort** on a reverse-sorted array $A[1 \ldots n]$ is $\Theta(n)$. (The reverse-sorted input is the worst case for both **bubbleSort** and **early-stopping-bubbleSort**.)

Prove that the worst-case running time of **forward-backward-bubbleSort** *is ...*

6.67 $\ldots O(n^2)$.

6.68 $\ldots \Omega(n^2)$ (despite the apparent improvement!). To prove this claim, explicitly describe an array $A[1 \ldots n]$ for which **early-stopping-bubbleSort** performs poorly—that is, in $\Omega(n^2)$ time—on both A and the reverse of A.

6.69 *(programming required)* Implement the three versions of Bubble Sort (including the two in Figure 6.23) in a programming language of your choice.

6.70 *(programming required)* Modify your implementations from Exercise 6.69 to count the number of swaps and comparisons each algorithm performs. Then run all three algorithms on each of the $8! = 40{,}320$ different orderings of the elements $\{1, 2, \ldots, 8\}$. How do the algorithms' performances compare, on average?

```
early-stopping-bubbleSort(A[1...n]):
1: for i := 1 to n:
2:     swapped := False
3:     for j := 1 to n − i:
4:         if A[j] > A[j+1] then
5:             swap A[j] and A[j+1]
6:             swapped := True
7:     if swapped = False then
8:         return A
```

```
forward-backward-bubbleSort(A[1...n]):
Construct R[1...n], the reverse of A, where
  R[i] := A[n − i + 1] for each i.
for i := 1 to n:
    Run one iteration of lines 2–8 of
      early-stopping-bubbleSort on A.
    Run one iteration of lines 2–8 of
      early-stopping-bubbleSort on R.
    if either A or R is now sorted then
        return whichever is sorted
```

Figure 6.23: Bubble Sort, improved.

In Chapter 9, we will meet a sorting algorithm called Counting Sort *that sorts an array* $A[1 \ldots n]$ where each $A[i] \in \{1, 2, \ldots, k\}$ *as follows: for each possible value* $x \in \{1, 2, \ldots, k\}$, *we walk through A to compute* $c_x := |\{i : A[i] = x\}|$. *(We can compute all k values of* $c_1, \ldots, c_k$ *in a single pass through A.) The output array consists of* c_1 *copies of* 1*, followed by* c_2 *copies of* 2*, and so forth, ending with* c_k *copies of k. (See Figure 6.24.) Counting sort is particularly good when k is small.*

```
countingSort(A[1...n]):
                                  // assume each A[i] ∈ {1,2,...,k}
for v := 1 to k:
    count[v] := 0
for i := 1 to n:
    count[A[i]] := count[A[i]] + 1
i := 1
for v := 1 to k:
    for t := 1 to count[v]:
        A[i] := v
        i := i + 1
```

Figure 6.24: Counting Sort.

6.71 In terms of n, what is the worst-case running time of **countingSort** on an input array of n letters from the alphabet (so $k = 26$, and n is arbitrary)?

6.72 *(programming required)* Implement Counting Sort and one of the $\Theta(n^2)$-time sorting algorithms from this section. Collect some data to determine, on a particular computer, for what values of k you'd generally prefer Counting Sort over the $\Theta(n^2)$-time algorithm when $n = 4096 = 2^{12}$ elements are each chosen uniformly at random from the set $\{1, 2, \ldots, k\}$.

6.73 *Radix Sort* is a sorting algorithm based on Counting Sort that proceeds by repeatedly applying Counting Sort to the ith-most significant bit in the input integers, for increasing i. Do some online research to learn more about Radix Sort, then write pseudocode for Radix Sort and compare its running time (in terms of n and k) to Counting Sort.

In Example 5.14, we proved the correctness of Quick Sort, *a recursive sorting algorithm (see Figure 6.25 for a reminder, or Figure 5.20(a) for more detail). The basic idea is to choose a* pivot *element of the input array A, then* partition *A into those elements smaller than the pivot and those elements larger than the pivot. We can then recursively sort the two "halves" and paste them together, around the pivot, to produce a sorted version of A. The algorithm performs very well if the two "halves" are genuinely about half the size of A; it performs very poorly if one "half" contains almost all the elements of A. The running time of the algorithm therefore hinges on how we select the pivot, in Line 4. (A very good choice of pivot is actually a* random *element of A, but here we'll think only about deterministic rules for choosing a pivot.)*

```
quickSort(A[1...n]):
1: if n ≤ 1 then
2:     return A
3: else
4:     Choose pivotIndex ∈ {1,...,n}, somehow.
5:     Let less (those elements smaller than A[pivotIndex]),
       same and greater be empty arrays.
6:     for i := 1 to n:
7:         compare A[i] to A[pivotIndex], and append A[i] to
           the appropriate array less, same, or greater.
8:     return quickSort(less) + same + quickSort(greater).
```

Figure 6.25: A high-level reminder of Quick Sort.

6.74 Suppose that we always choose *pivotIndex* := 1. (That is, the first element of the array is the pivot value.) Describe (for an arbitrary n) an input array $A[1 \ldots n]$ that causes **quickSort** under this pivot rule to make either *less* or *greater* empty.

6.75 Argue that, for the array you found in Exercise 6.74, the running time of Quick Sort is $\Theta(n^2)$.

6.76 Suppose that we always choose *pivotIndex* := $\lfloor n/2 \rfloor$. (That is, the middle element of the array is the pivot value.) What input array $A[1 \ldots n]$ causes worst-case performance (that is, one of the two sides of the partition—*less* or *greater*—is empty) for this pivot rule?

6.77 A fairly commonly used pivot rule is called the *Median of Three* rule: we choose *pivotIndex* $\in \{1, \lfloor n/2 \rfloor, n\}$ so that $A[pivotIndex]$ is the median of the three values $A[1]$, $A[\lfloor n/2 \rfloor]$, and $A[n]$. Argue that there is still an input array of size n that results in $\Omega(n^2)$ running time for Quick Sort.

6.78 Earlier we described a linear-search algorithm that looks for an element x in an array $A[1 \ldots n]$ by comparing x to $A[i]$ for each $i = 1, 2, \ldots n$. (See Figure 6.16.) But if A is sorted, we can determine that x is not in A earlier, as shown in Figure 6.26: once we've passed where x "should" be, we know that it's not in A. (Our original version omitted lines 4–5.) What is the worst-case running time of the early-stopping version of linear search?

6.79 Consider the algorithm in Figure 6.26 for counting the number of times the letter Z appears in a given string s. What is the worst-case running time of this algorithm on an input string of length n? Assume that testing whether Z is in s (line 2) and removing a letter from s (line 4) both take $c \cdot |s|$ time, for some constant c.

```
early-stopping-linearSearch(A[1...n], x):
1: for i := 1 to n:
2:     if A[i] = x then
3:         return True
4:     else if A[i] < x then
5:         return False
6: return False
```

```
countZ(s):
1: z := 0
2: while there exists i such that s_i = Z:
3:     z := z + 1
4:     remove s_i from s
       (that is, set s := s_1 ... s_{i-1} s_{i+1} ... s_n)
5: return z
```

Figure 6.26: Linear Search and counting ZZZs.

6.4 *Recurrence Relations: Analyzing Recursive Algorithms*

> Democracy is the recurrent suspicion that more than half of the people are right more than half the time.
>
> E. B. White (1899–1985)

The nonrecursive algorithms in Section 6.3 could be analyzed by simple counting and manipulation of summations. First we figured out the number of iterations of each loop, and then figured out how long each iteration takes. By summing this work over the iterations and simplifying the summation, we were able to compute the running time of the algorithm. Determining the running time of a recursive algorithm is harder. Instead of merely containing loops that can be analyzed as above, the algorithm's running time on an input of size n depends on the same algorithm's running time for inputs of size smaller than n.

We'll use the classical recursive sorting algorithm Merge Sort (Figure 6.27) as an example. Merge Sort sorts an array by recursively sorting the first half, recursively sorting the second half, and finally "merging" the resulting sorted lists. (On an input array of size 1, Merge Sort just returns the array as is.) You'll argue in Exercise 6.100 that merging two $\frac{n}{2}$-element arrays takes $\Theta(n)$ time, but what does that mean for the overall running time of Merge Sort? We can think about Merge Sort's running time by drawing a picture of all of the work that is done in its execution, in the form of a *recursion tree*:

```
mergeSort(A[1...n]):
  if n = 1 then
      return A
  else
      L := mergeSort(A[1 ... ⌊n/2⌋])
      R := mergeSort(A[⌊n/2⌋ + 1 ... n])
      return merge(L, R)
```

Figure 6.27: Merge Sort. The **merge** function takes two sorted arrays and combines them into a single sorted array. (See Exercise 5.72 or 6.100.)

Definition 6.10 (Recursion tree)
The recursion tree *for a recursive algorithm $\mathcal{A}$ is a tree that shows all of the recursive calls spawned by a call to $\mathcal{A}$ on an input of size n. Each node in the tree is annotated with the amount of work, aside from any recursive calls, done by that call.*

Figure 6.28 shows the recursion tree for Merge Sort. For ease, we will assume that n is an exact power of 2. We denote by $c \cdot n$ the amount of time needed to process an n-element array *aside from the recursive calls*—that is, the time to split and merge.

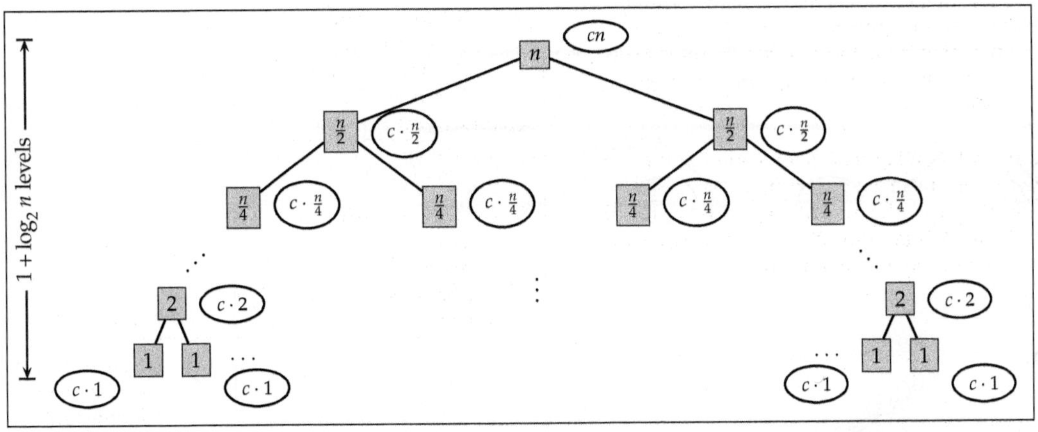

Figure 6.28: The recursion tree for Merge Sort. The size of the input itself is shown in the shaded square node; the $\Theta(n)$ amount of time required for splitting and merging an n-element input is shown in the oval adjacent to that node, as $c \cdot n$.

There are many different ways to analyze the total amount of work done by Merge Sort on an n-element input array, but one of the easiest is to use the recursion tree:

Example 6.16 (Analyzing Merge Sort via recursion tree)

Problem: How quickly does Merge Sort run on an n-element input array? (Assume that n is a power of two.)

Solution: The total amount of work done by Merge Sort is precisely the sum of the circled values contained in the tree. (At the root, by definition the total work aside from the recursive calls is $c \cdot n$; inductively, the work done in the recursive calls is the sum of the circled values in the left and right subtrees.)

The easiest way to sum up the work in the tree is to sum "row-wise." (See Figure 6.29.) The first "row" of the tree (one call on an input of size n) generates cn work. The second row (two calls on inputs of size $n/2$) generates $2 \cdot (cn/2) = cn$ work. The third row (four calls on inputs of size $n/4$) generates $4 \cdot (cn/4) = cn$ work. In general, row #k of the tree contains 2^{k-1} calls on inputs of size $n/2^{k-1}$, and generates $2^{k-1} \cdot c \cdot n/2^{k-1} = cn$ work—that is, the work at the kth level of the tree is cn, independent of the value of k.

There are $1 + \log_2 n$ rows in the tree, and so the total work in this tree is

$$\sum_{k=1}^{1+\log_2 n} 2^{k-1} \cdot c \cdot \frac{n}{2^{k-1}} = \sum_{k=1}^{1+\log_2 n} cn$$

$$= cn(1 + \log_2 n)$$

and thus is $\Theta(n \log n)$ in total.

Taking it further: Here's a different argument as to why Merge Sort requires $\Theta(n \log n)$ time: *every* element of the input array is merged once in an array of size 1, once in an array of size 2, once in an array of size 4, once in an array of size 8, etc. So each element is merged $\log_2 n$ times, so thus the total work is $\Theta(n \cdot \log_2 n)$.

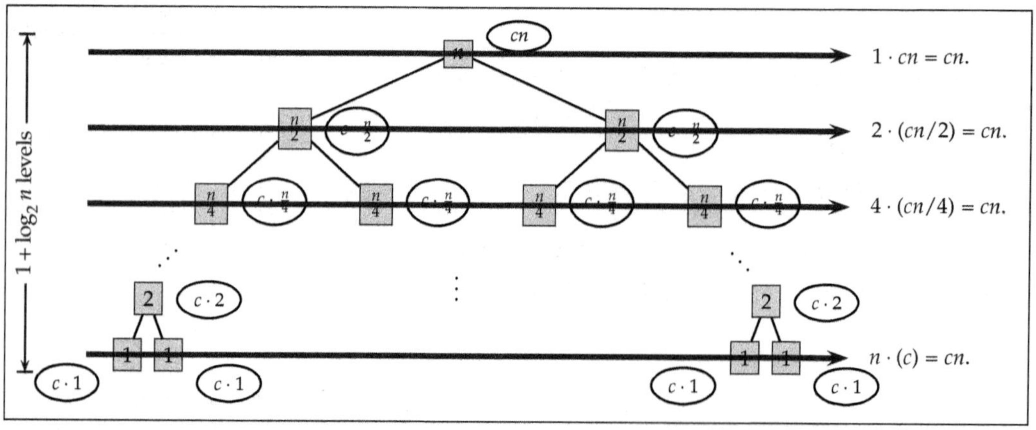

Figure 6.29: The row-wise sum of the tree in Figure 6.28.

6.4.1 *Recurrence Relations*

Recursion trees are an excellent way to gain intuition about the running time of a recursive algorithm, and to analyze it. We now turn to another way of thinking about recursion trees, which suggests a rigorous (and in many ways easier to use) approach to analyzing recursive algorithms: the *recurrence relation*. Because at least one of the steps in a recursive algorithm $\mathcal{A}$ is to call $\mathcal{A}$ on a smaller input, the running time of $\mathcal{A}$ on an input of size n depends on $\mathcal{A}$'s running time for inputs of size smaller than n. We will therefore express $\mathcal{A}$'s running time recursively, too:

Definition 6.11 (Recurrence relation)
A recurrence relation *(sometimes simply called a* recurrence*) is a function $T(n)$ that is defined (for some n) in terms of the values of $T(k)$ for input values $k < n$.*

A recurrence relation is called a recurrence relation because T *recurs* ("occurs again") on the right-hand side of the equation. That's the same reason that recursion is called recursion.

Here's a first example, about compounding interest in a bank account:

Example 6.17 (Compound interest)
Suppose that, in year #0, Alice puts \$1000 in a bank account that pays 2% annual compound interest. Writing $A(n)$ to denote the balance of Alice's account in year #n, we have

$$A(0) = 1000 \qquad\qquad A(n) = 1.02 \cdot A(n-1).$$

If Bob opens a bank account with the same interest rate, and deposits \$10 into the account each year (starting in year #0), then Bob's balance is given by the recurrence

$$B(0) = 10 \qquad\qquad B(n) = 1.02 \cdot B(n-1) + 10.$$

In computer science, the most common type of recurrence relation that we'll encounter is one where $T(n)$ denotes the worst-case number of steps taken by a particular recursive algorithm on an input of size n. Here are a few examples:

```
fact(n):
1: if n = 1 then
2:     return 1
3: else
4:     return n · fact(n − 1)
```

Figure 6.30: A recursive algorithm for factorial.

Example 6.18 (Factorial)
Let $T(n)$ denote the worst-case running time of **fact** (Figure 6.30). Then:

$$T(1) = d$$
$$T(n) = T(n-1) + c$$

where c is a constant denoting the work of the comparison–conditional–multiplication–return, and d is a constant denoting the work of the comparison–conditional–return.

Example 6.19 (Merge Sort)
Let $T(n)$ denote the worst-case running time of Merge Sort (Figure 6.27) on an input array containing n elements. Then, for a constant c, we have:

$$T(1) = c$$
$$T(n) = T(\lfloor \tfrac{n}{2} \rfloor) + T(\lceil \tfrac{n}{2} \rceil) + cn.$$

Just as for nonrecursive algorithms, we will generally be interested in the asymptotic running times of these recursive algorithms, so we will usually not fret about the particular values of the constants in recurrences. We will often abuse notation and use a single constant to represent different $\Theta(1)$-time operations, for example.

In Example 6.19, for instance, we are being sloppy in our recurrence, using a single variable c to represent two different values. The use of one constant to have two different meanings (plus the '=' sign) is an abuse of notation, but when we care about asymptotic values, this abuse doesn't matter. We will even sometimes write 1 to stand for this constant. (See Exercise 6.126.)

```
binarySearch(A[1...n], x):
if n ≤ 0 then
    return False
middle := ⌊(1+n)/2⌋
if A[middle] = x then
    return True
else if A[middle] > x then
    return binarySearch(A[1...middle − 1], x)
else
    return binarySearch(A[middle + 1...n], x)
```

Figure 6.31: Binary Search, recursively.

Here's another recurrence relation, for the recursive version of Binary Search:

Example 6.20 (Binary Search)
Let $T(n)$ denote the worst-case running time of the recursive **binarySearch** (Figure 6.31) on an n-element array. Then:

$$T(0) = c$$
$$T(n) = \begin{cases} T(\frac{n}{2}) + c & \text{if } n \text{ is even} \\ T(\frac{n-1}{2}) + c & \text{if } n \text{ is odd.} \end{cases}$$

Although our interest in recurrence relations will be almost exclusively about the running times of recursive algorithms, there are other interesting recurrence relations, too. The most famous of these is the recurrence for the *Fibonacci numbers* (which will turn out to have some interesting CS applications, too):

Example 6.21 (Fibonacci numbers)
The Fibonacci numbers are defined by

$$f_1 = 1$$
$$f_2 = 1$$
$$f_n = f_{n-1} + f_{n-2} \qquad \text{for } n \geq 3$$

The first several Fibonacci numbers are $1, 1, 2, 3, 5, 8, 13, 21, 34, 55, 89 \ldots$.

6.4.2 *Solving Recurrences: Induction*

When we *solve* a recurrence relation, we find a closed-form (that is, nonrecursive) equivalent expression. Because recurrence relations are recursively defined quantities, induction is the easiest way to prove that a conjectured solution is correct. (The hard part is figuring out what solution to conjecture, as we'll see.)

In the remainder of this section, we will solve all of the recurrences from Section 6.4.1—starting with Alice and Bob and their bank accounts:

Example 6.22 (Compound interest)
Recall the recurrences from Example 6.17:

$$A(0) = 1000 \qquad A(n) = 1.02 \cdot A(n-1) \qquad \textit{(Alice)}$$
$$B(0) = 10 \qquad B(n) = 1.02 \cdot B(n-1) + 10. \qquad \textit{(Bob)}$$

The recurrence for Alice is the easier of the two to solve: we can prove relatively straightforwardly by induction that $A(n) = 1000 \cdot 1.02^n$ for any $n \geq 1$.

For Bob, the analysis is a little trickier. Here's some intuition: at time n, Bob has had \$10 sitting in his account since year #0 (earning interest for n years); \$10 in his account since year #1 (earning interest for $n-1$ years); etc. A \$10 deposit that has accumulated interest for i years has, as with Alice, grown to $10 \cdot 1.02^i$. Thus the total amount of money in Bob's account in year #n will be

$$\sum_{i=0}^{n}\left[10 \cdot 1.02^i\right] = 10 \cdot \left[\sum_{i=0}^{n} 1.02^i\right] = 10 \cdot \frac{1.02^{n+1}-1}{1.02-1} = 510 \cdot 1.02^n - 500$$

where the second equality follows from Theorem 5.2 (the analysis of a geometric series). Let's prove the property that $B(n) = 510 \cdot 1.02^n - 500$, by induction on n:

base case ($n = 0$): Then $B(0) = 10$, and indeed $510 \cdot 1.02^0 - 500 = 510 - 500 = 10$.

inductive case ($n \geq 1$): We assume the inductive hypothesis $B(n-1) = 510 \cdot 1.02^{n-1} - 500$; we must show that $B(n) = 510 \cdot 1.02^n - 500$. Then:

$$\begin{aligned} B(n) &= 1.02 \cdot B(n-1) + 10 && \textit{definition of } B(n) \\ &= 1.02 \cdot \left[510 \cdot 1.02^{n-1} - 500\right] + 10 && \textit{inductive hypothesis} \\ &= 1.02 \cdot 510 \cdot 1.02^{n-1} - 1.02 \cdot 500 + 10 && \textit{multiplying through} \\ &= 510 \cdot 1.02^n - 510 + 10 && \textit{simplifying} \\ &= 510 \cdot 1.02^n - 500, \end{aligned}$$

precisely as desired.

Taking it further: As Example 6.22 suggests, some familiar kinds of summations like arithmetic and geometric series can be expressed using recurrence relations. Other familiar summations can also be expressed using recurrence relations; for example, the sum of the first n integers is given by the recurrence $T(1) = 1$ and $T(n) = T(n-1) + n$. (See Section 5.2 for some closed-form solutions.)

Factorial

One good way to generate a conjecture that we then prove correct by induction is by "iterating" the recurrence: expand out a few layers of the recursion to see what the values of $T(n)$ are for a few small values of n. We'll illustrate this technique with the simplest recurrence from the last section, for the recursive factorial function.

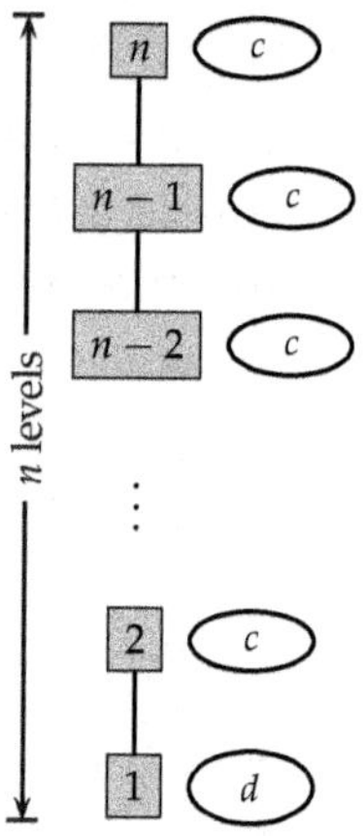

Figure 6.32: The (agonizingly simple) recursion tree for **fact**.

Example 6.23 (Factorial)

<u>Problem:</u> Recall the recurrence from Example 6.18:

$$T(1) = d \qquad\qquad T(n) = T(n-1) + c.$$

Give an exact closed-form (nonrecursive) solution for $T(n)$.

<u>Solution:</u> See Figure 6.32 for the recursion tree, which may help give some intuition. Let's iterate the recurrence a few times:

- $T(1) = d$
- $T(2) = c + T(1) = c + d$
- $T(3) = c + T(2) = 2c + d$
- $T(4) = c + T(3) = 3c + d.$

From these small values, we conjecture that $T(n) = (n-1)c + d$.

Let's prove this conjecture correct by induction. For the base case ($n = 1$), we have $T(1) = d$ by definition of the recurrence, which is $0 \cdot c + d$, as desired. For the inductive case, assume the inductive hypothesis $T(n-1) = (n-2)c + d$. We want to show that $T(n) = (n-1)c + d$. Here's the proof:

$$\begin{aligned} T(n) &= T(n-1) + c && \textit{by definition of the recurrence} \\ &= (n-2)c + d + c && \textit{by the inductive hypothesis} \\ &= (n-1)c + d. && \textit{by algebraic manipulation} \end{aligned}$$

Thus $T(n) = (n-1)c + d$.

Problem-solving tip: Try iterating a recurrence to generate its first few values. Once we have a few values, we can often conjecture a general solution (which we then prove correct via induction).

Merge Sort

Recall the Merge Sort recurrence, where $T(n) = T(\lceil \frac{n}{2} \rceil) + T(\lfloor \frac{n}{2} \rfloor) + cn$ and $T(1) = c$. It will be easier to address the case in which n is an exact power of 2 first (so that the floors and ceilings don't complicate the picture), so we'll start with that case first, and generalize later:

Example 6.24 (Merge Sort, for powers of 2)

<u>Problem:</u> Recall the Merge Sort recurrence from Example 6.19:

$$T(1) = c \qquad\qquad T(n) = T(\lceil \tfrac{n}{2} \rceil) + T(\lfloor \tfrac{n}{2} \rfloor) + cn.$$

For convenience, assume that n is an exact power of two. Give an exact closed-form (nonrecursive) solution for $T(n)$.

Solution: Because n is an exact power of two, we can write $n = 2^k$ for some $k \in \mathbb{Z}^{\geq 0}$. (Note that for $n = 2^k$ we have $\lceil \frac{n}{2} \rceil = \lfloor \frac{n}{2} \rfloor = \frac{n}{2} = 2^{k-1}$.) Define $R(k) = T(2^k)$; then $R(0) = T(1) = c$ and $R(k) = T(2^k) = 2 \cdot T(2^{k-1}) + c \cdot 2^k = 2 \cdot R(k-1) + c \cdot 2^k$, so we can instead solve the recurrence

$$R(0) = c \qquad R(k) = 2 \cdot R(k-1) + c \cdot 2^k.$$

Iterating R a few times, we see

- $R(0) = c$
- $R(1) = c \cdot 2^1 + 2 \cdot R(0) = 4c$
- $R(2) = c \cdot 2^2 + 2 \cdot R(1) = 12c$
- $R(3) = c \cdot 2^3 + 2 \cdot R(2) = 32c$

We conjecture

$$R(k) = (1+k)2^k \cdot c \qquad (*)$$

(How might we get to this conjecture? The pattern from iterating R matches it. Alternatively, looking at the recursion tree might help: there are $k+1$ levels of the tree, and there are 2^{k-i} copies of $2^i \cdot c$ work in the ith row of the tree—so that's $(k+1)2^{k-i}2^i c = (k+1)2^k c$. Or, we'd expect a solution that's the product of $\approx k$ and $\approx 2^k$ so that we get $T(n) \approx n \log n$. And if we check the $k = 0$ case—$R(0) = 1$—it looks like we'd better multiply by $k+1$ rather than k.)

Let's prove $(*)$, by induction on k. In the base case, $R(0) = c$ and indeed we have that $(1+0)2^0 \cdot c = 1 \cdot 1 \cdot c$. In the inductive case, we have

$$\begin{aligned} R(k) &= 2R(k-1) + c \cdot 2^k && \text{by definition of the recurrence} \\ &= 2(1+k-1)2^{k-1} \cdot c + c \cdot 2^k && \text{by the inductive hypothesis} \\ &= 2k \cdot 2^{k-1} \cdot c + 2^k \cdot c \\ &= (k+1)2^k \cdot c. \end{aligned}$$

Thus $R(k) = (k+1)2^k \cdot c$, completing the inductive case—and the proof of $(*)$.

Because we defined $R(k) = T(2^k)$, we can conclude that $T(n) = R(\log_2 n)$, by substituting. Thus $T(n) = (1 + \log_2 n) \cdot 2^{\log_2 n} \cdot c = n(1 + \log_2 n) \cdot c$.

Problem-solving tip: A useful technique for solving recurrences is to do a *variable substitution.* If you can express the recurrence in terms of a different variable and solve the new recurrence easily, you can then substitute back into the original recurrence to solve it. Transforming an unfamiliar recurrence into a familiar one will make life easy!

Thinking only about powers of two in Example 6.24 made our life simpler, but it leaves a hole in the analysis: what is the running time of Merge Sort when the input array's length is *not* precisely a power of two? The more general analysis is actually simple, given the result we just derived:

Example 6.25 (Merge Sort, for general n)
Problem: Solve the Merge Sort recurrence (asymptotically), for any integer $n \geq 1$:

$$T(1) = c \qquad T(n) = T(\lceil \tfrac{n}{2} \rceil) + T(\lfloor \tfrac{n}{2} \rfloor) + cn.$$

Solution: We'll use the fact that $T(n) \geq T(n')$ if $n \geq n'$—that is, T is *monotonic*. (See Exercise 6.101.) So let k be the nonnegative integer such that $2^k \leq n < 2^{k+1}$. Then

$$\begin{aligned} T(n) &\geq T(2^k) && \textit{monotonicity} \\ &= ((\log_2 2^k) + 1)2^k \cdot c && \textit{Example 6.24} \\ &> (\log_2 \tfrac{n}{2} + 1) \cdot \tfrac{n}{2} \cdot c. && \textit{definition of } k\textit{: we have } \tfrac{n}{2} < 2^k \end{aligned}$$

Thus we know $T(n) = \Omega(n \log n)$. Similarly,

$$\begin{aligned} T(n) &< T(2^{k+1}) && \textit{monotonicity} \\ &= ((\log_2 2^{k+1}) + 1)2^{k+1} \cdot c && \textit{Example 6.24} \\ &\leq (\log_2 2n + 1) \cdot 2n \cdot c. && \textit{definition of } k\textit{: we have } 2n \geq 2^{k+1} \end{aligned}$$

Thus $T(n) = O(n \log n)$. Combining these facts yields that $T(n) = \Theta(n \log n)$.

BINARY SEARCH

There is a very simple intuitive argument for why Binary Search takes logarithmic time, which we used in Example 6.12:

> In the worst case, when the sought item x isn't in the array, we repeatedly compare x to the middle of the valid range of the array, and halve the size of that valid range. We can halve an n-element range exactly $\log_2 n$ times, and thus the running time of Binary Search is logarithmic.

While this intuitive argument is plausible, there's a subtle but nontrivial issue: the so-called "halving" in this description isn't actually *exactly* halving. If there are n elements in the valid range, then after comparing x to the middle element of the range, we will end up with a valid range of size either $\frac{n}{2}$ or $\frac{n-1}{2}$, depending on the parity of n—not *exactly* $\frac{n}{2}$. (We *have* already shown that Binary Search's worst-case running time is $O(\log n)$, in Example 6.12, because if there are n elements in the valid range, then after so-called halving we end up with a valid range of size *at most* $\frac{n}{2}$. The issue here is that we have not ruled out the possibility that the running time might be *faster* than $\Theta(\log n)$, because we've "better-than-halved" at every stage.)

We can resolve this issue by rigorously analyzing the correct recurrence relation—and we can prove that the running time *is* in fact $\Theta(\log n)$.

Example 6.26 (Binary Search)
Problem: Solve the Binary Search recurrence:

$$T(0) = 1 \qquad T(n) = \begin{cases} T(\frac{n}{2}) + 1 & \text{if } n \text{ is even} \\ T(\frac{n-1}{2}) + 1 & \text{if } n \text{ is odd.} \end{cases}$$

(Note that we've changed the additive constants to 1 instead of c; changing it back to c would only have the effect of multiplying the entire solution by c.)

Solution: We conjecture that $T(n) = \lfloor \log_2 n \rfloor + 2$ for all $n \geq 1$. We'll prove the conjecture correct by strong induction on n.

For the base case ($n = 1$), we have $T(1) = T(0) + 1 = 1 + 1 = 2$ by definition of the recurrence, and indeed $2 = \lfloor 0 \rfloor + 2 = \lfloor \log_2 1 \rfloor + 2$.

For the inductive case ($n \geq 2$), assume the inductive hypothesis, that $T(k) = \lfloor \log_2 k \rfloor + 2$ for any $k < n$. We'll proceed in two cases:

- If n is even:

$$\begin{aligned}
T(n) &= T(\tfrac{n}{2}) + 1 && \text{by definition of the recurrence}\\
&= \lfloor \log_2(\tfrac{n}{2}) \rfloor + 2 + 1 && \text{by the inductive hypothesis}\\
&= \lfloor (\log_2 n) - 1 \rfloor + 3 && \text{because } \log(\tfrac{a}{b}) = \log a - \log b \text{, and } \log_2 2 = 1\\
&= \lfloor \log_2 n \rfloor + 2. && \text{because } \lfloor x+1 \rfloor = \lfloor x \rfloor + 1
\end{aligned}$$

- If n is odd:

$$\begin{aligned}
T(n) &= T(\tfrac{n-1}{2}) + 1 && \text{by definition of the recurrence}\\
&= \lfloor \log_2(\tfrac{n-1}{2}) \rfloor + 2 + 1 && \text{by the inductive hypothesis}\\
&= \lfloor \log_2(n-1) \rfloor + 2 && \text{by the same manipulations as in the even case}\\
&= \lfloor \log_2 n \rfloor + 2. && \text{because } \lfloor \log_2(n-1) \rfloor = \lfloor \log_2 n \rfloor \text{ for any odd integer } n > 1
\end{aligned}$$

Because we've shown that $T(n) = \lfloor \log_2 n \rfloor + 2$ in either case, we've proven the claim. Therefore $T(n) = \Theta(\log n)$.

Problem-solving tip: When solving a new recurrence, we can try to generate conjectures (to prove correct via induction) by iterating the recurrence, drawing out the recursion tree, or by straight-up guessing a solution (or recognizing a similar pattern to previously seen recurrences). To generate my conjecture for Example 6.26, I actually wrote a program that implemented the recurrence. I ran the program for $n \in \{1, 2, \ldots, 1000\}$ and printed out the smallest integer n for which $T(n) = 1$, then the smallest for which $T(n) = 2$, etc. (See Figure 6.33.) The conjecture followed from the observation that the breakpoints all happened at $n = 2^k - 1$ for an integer k.

As a general matter, the appearance of floors and ceilings inside a recurrence won't matter to the asymptotic running time, nor will small additive adjustments inside the recursive term. For example, $T(n) = T(\lceil \frac{n}{2} \rceil) + 1$ and $T(n) = T(\lfloor \frac{n}{2} \rfloor - 2) + 1$ both have $T(n) = \Theta(\log n)$ solutions. Intuitively, floors and ceilings don't change this type of recurrence because they don't affect the total depth of the recursion tree by more than a $\Theta(1)$ number of calls, and a $\Theta(1)$ difference in depth is asymptotically irrelevant. Typically, understanding the running time for the "pure" version of the recurrence will

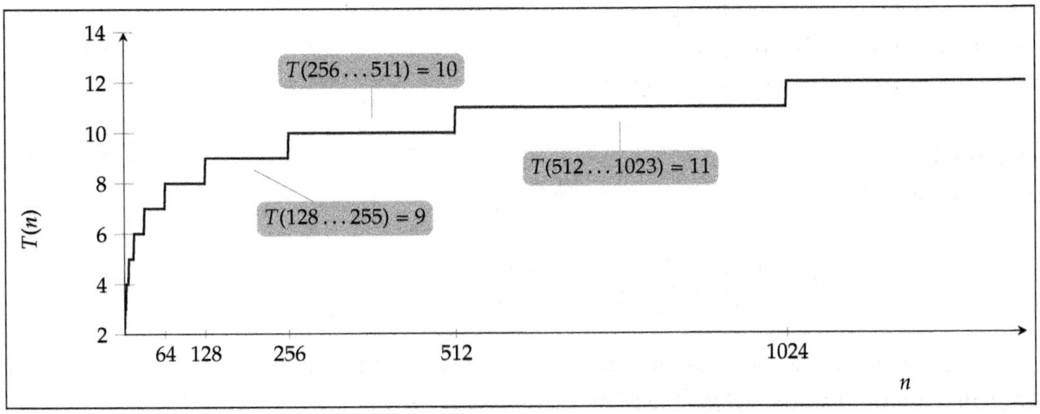

Figure 6.33: A plot of n versus $T(n)$ for the binary search recurrence.

give a correct understanding of the more complicated version. As such, we'll often be sloppy in our notation, and write $T(n) = T(\frac{n}{2}) + 1$ when we really mean $T(\lfloor \frac{n}{2} \rfloor)$ or $T(\lceil \frac{n}{2} \rceil)$. (This abuse of notation is fairly common.)

Taking it further: There's a general theorem called the *"sloppiness" theorem*, which states conditions under which it is safe to ignore floors and ceilings in recurrence relations. (As long as we actually prove inductively that our conjectured solution to a recurrence relation is correct, it's always fine in generating conjectures.) As a rough guideline, as long as $T(n)$ is monotonic ($n \leq n' \Rightarrow T(n) \leq T(n')$) and doesn't grow too quickly ($T(n)$ is $O(n^k)$ for some constant k), then this "sloppiness" is fine. The details of the theorem, and its precise assumptions, are presented in many algorithms textbooks.

6.4.3 *The Fibonacci Numbers*

We'll close with another example of a recurrence relation—the Fibonacci recurrence—that we will analyze using induction. But this time we will solve the recurrence exactly (that is, nonasymptotically):

Example 6.27 (The Fibonacci Numbers)
Problem: Recall the *Fibonacci numbers*, defined by the recurrence

$$f_1 = 1 \qquad f_2 = 2 \qquad f_n = f_{n-1} + f_{n-2}.$$

Prove that f_n grows exponentially: that is, prove that there exist $a \in \mathbb{R}^{>0}$ and $r \in \mathbb{R}^{>1}$ such that $f_n \geq ar^n$.

Brainstorming: Let's start in the middle: suppose that we've somehow magically figured out values of a and r to make the base cases ($n \in \{1, 2\}$) work, and we're in the middle of an inductive proof. (There are two base cases because $f_2 \neq f_1 + f_0$; f_0 isn't even defined!) We'd be able to prove this:

$$f_n = f_{n-1} + f_{n-2} \geq ar^{n-1} + ar^{n-2} = ar^{n-2}(r+1). \qquad \textit{inductive hypothesis/algebra}$$

But what we *want* to prove is $f_n \geq ar^n$. So we'd be done if only $r + 1 = r^2$—that is, if $r^2 - r - 1 = 0$. But we get to pick the value of r (!). Using the quadratic formula, we find that there are two solutions to this equation, which we'll name ϕ and $\hat{\phi}$:

$$\phi = \frac{1+\sqrt{5}}{2} \qquad \hat{\phi} = \frac{1-\sqrt{5}}{2}.$$

Let's use $r = \phi$. To get the base cases to work, we would need to have $f_1 = 1 \geq a\phi$ and $f_2 = 1 \geq a\phi^2 = a(1+\phi)$. Because $1 + \phi > \phi$, the latter is the harder one to achieve. To ensure that $a(1+\phi) \leq 1$, we must have

$$a \leq \frac{1}{1+\phi} = \frac{1}{1+\frac{1+\sqrt{5}}{2}} = \frac{2}{3+\sqrt{5}}.$$

Figure 6.34: Some brainstorming for Example 6.27.

Problem-solving tip: Sometimes starting in the middle of a proof helps! You still need to go back and connect the dots, but imagining that you've gotten somewhere may help you figure out how to get there.

Example 6.27 (The Fibonacci Numbers, continued)
Solution: Based on the brainstorming in Figure 6.34 (which identifies a value ϕ such that $\phi + 1 = \phi^2$ and a corresponding value for a), we'll prove the following claim:

Claim: $f_n \geq \frac{2}{3+\sqrt{5}} \cdot \phi^n$, where $\phi = \frac{1+\sqrt{5}}{2}$.

Proof (by strong induction on n). There are two base cases:

- For $n = 1$, we have $\frac{2}{3+\sqrt{5}} \cdot \phi^1 = \frac{2}{3+\sqrt{5}} \cdot \frac{1+\sqrt{5}}{2} = \frac{1+\sqrt{5}}{3+\sqrt{5}} < 1 = f_1$.
- For $n = 2$: we have

$$\begin{aligned} \frac{2}{3+\sqrt{5}} \cdot \phi^2 &= \frac{2}{3+\sqrt{5}} \cdot (1 + \phi) && \textit{we chose } \phi \textit{ so that } \phi + 1 = \phi^2 \\ &= \frac{2}{3+\sqrt{5}} \cdot \frac{3+\sqrt{5}}{2} = 1 = f_2. \end{aligned}$$

For the inductive case ($n \geq 3$), we assume the inductive hypothesis, namely that $f_k \geq \frac{2}{3+\sqrt{5}} \cdot \phi^k$ for $1 \leq k \leq n-1$. Then:

$$\begin{aligned} f_n &= f_{n-1} + f_{n-2} && \textit{definition of the Fibonaccis} \\ &\geq \frac{2}{3+\sqrt{5}} \cdot \phi^{n-1} + \frac{2}{3+\sqrt{5}} \cdot \phi^{n-2} && \textit{inductive hypothesis, twice} \\ &= \frac{2}{3+\sqrt{5}} \cdot \phi^{n-2} \cdot (\phi + 1) && \textit{factoring} \\ &= \frac{2}{3+\sqrt{5}} \cdot \phi^{n-2} \cdot \phi^2 && \textit{we chose } \phi \textit{ so that } \phi + 1 = \phi^2 \\ &= \frac{2}{3+\sqrt{5}} \cdot \phi^n. \end{aligned}$$

Therefore the claim follows by induction. □

Taking it further: The value $\phi = \frac{1+\sqrt{5}}{2} \approx 1.61803\cdots$ is called *the golden ratio.* It has a number of interesting characteristics, including both remarkable mathematical and aesthetic properties. For example, a rectangle whose side lengths are in the ratio ϕ-to-1 can be divided into a square and a rectangle whose side lengths are in the ratio 1-to-ϕ. That's because, for these rectangles to have the same ratios, we need $\frac{\phi}{1} = \frac{1}{\phi-1}$—that is, we need $\phi(\phi - 1) = 1$, which means $\phi^2 - \phi = 1$. (See Figure 6.35.) The golden ratio, it has been argued, describes proportions in famous works of art ranging from the Acropolis to Leonardo da Vinci's drawings.

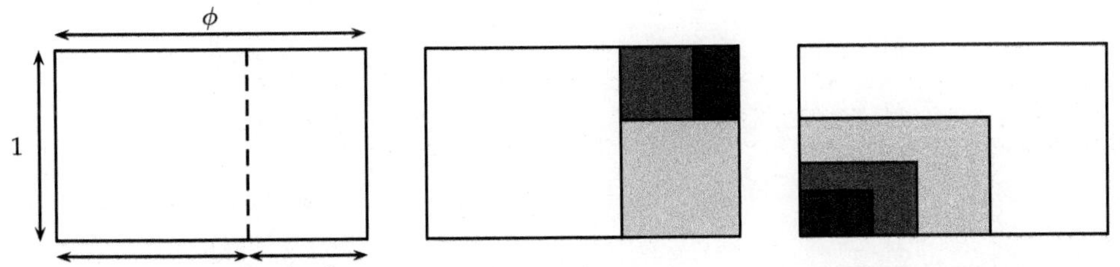

(a) A rectangle with sides in ratio ϕ-to-1, with a 1-by-1 square inscribed.

(b) Repeatedly inscribing a square in the "leftover" rectangle.

(c) The same rectangles, rotated and shifted to share a lower-left corner.

Figure 6.35: Some golden rectangles.

A closed-form formula for the Fibonaccis

While Example 6.27 establishes a lower bound on the Fibonacci numbers—in asymptotic notation, it proves that $f_n = \Omega(\phi^n)$—we have not yet established a closed-form solution for the nth Fibonacci number. Here's a solution that does so, based on the following ideas. The trick will be to make use of $\hat{\phi}$. The inductive case would go through perfectly, just as in Example 6.27, if we tried to prove $f_n = a\phi^n + b\hat{\phi}^n$, for constants a and b. But what about the base cases? For f_1, we would need $1 = a\phi + b\hat{\phi}$; for f_2, we would need $1 = a\phi^2 + b(\hat{\phi}^2) = a(1+\phi) + b(1+\hat{\phi})$. That's two linear equations with two unknowns, and some algebra will reveal that $a = \frac{1}{\sqrt{5}}$ and $b = \frac{-1}{\sqrt{5}}$ solves these equations. Let's use these ideas to give a closed-form solution for the Fibonaccis, and a proof:

Example 6.28 (A closed-form solution for the Fibonaccis)

Problem: Prove the following claim:

Claim: $f_n = \frac{\phi^n - \hat{\phi}^n}{\sqrt{5}}$, where $\phi = \frac{1+\sqrt{5}}{2}$ and $\hat{\phi} = \frac{1-\sqrt{5}}{2}$.

Solution: *Proof (by strong induction on n).* For the base cases ($n = 1$ and $n = 2$):

- For $n = 1$, we have

$$\begin{aligned} \frac{\phi^1 - \hat{\phi}^1}{\sqrt{5}} &= \frac{\frac{1+\sqrt{5}}{2} - \frac{1-\sqrt{5}}{2}}{\sqrt{5}} && \textit{definition of } \phi \textit{ and } \hat{\phi} \\ &= \frac{\frac{2\sqrt{5}}{2}}{\sqrt{5}} && \textit{algebra} \\ &= 1 \\ &= f_1. \end{aligned}$$

- For $n = 2$, we have that

$$\begin{aligned} \frac{\phi^2 - \hat{\phi}^2}{\sqrt{5}} &= \frac{1+\phi-(1+\hat{\phi})}{\sqrt{5}} && \phi^2 = 1+\phi \textit{ and } \hat{\phi}^2 = 1+\hat{\phi} \\ &= 1 && \textit{by the previous case} \\ &= f_2. \end{aligned}$$

For the inductive case ($n \geq 3$), we assume the inductive hypothesis: for any $k < n$, we have $f_k = \frac{\phi^k - \hat{\phi}^k}{\sqrt{5}}$. Then:

$$\begin{aligned} f_n &= f_{n-1} + f_{n-2} && \textit{definition of the Fibonaccis} \\ &= \frac{\phi^{n-1} - \hat{\phi}^{n-1}}{\sqrt{5}} + \frac{\phi^{n-2} - \hat{\phi}^{n-2}}{\sqrt{5}} && \textit{inductive hypothesis} \\ &= \frac{\phi^{n-2}(\phi+1) - \hat{\phi}^{n-2}(\hat{\phi}+1)}{\sqrt{5}} && \textit{factoring} \\ &= \frac{\phi^{n-2}\phi^2 - \hat{\phi}^{n-2}\hat{\phi}^2}{\sqrt{5}} && \phi+1 = \phi^2 \textit{ and } \hat{\phi}+1 = \hat{\phi}^2 \\ &= \frac{\phi^n - \hat{\phi}^n}{\sqrt{5}}. \end{aligned}$$

□

Taking it further: The Fibonacci numbers show up all over the place in nature—and in computation. One computational application in which they're relevant is in the design and analysis of a data structure called an *AVL tree,* a form of binary search tree that guarantees that the tree supports all its operations efficiently. See the discussion on p. 643.

Computer Science Connections

AVL Trees

A *binary search tree* is a data structure that allows us to store a dynamic set of elements, supporting Insert, Delete, and Find operations. (We'll discuss binary search trees themselves in Chapter 11.) A binary search tree consists of a *root node* at the top; each node u can have zero, one, or two *children* directly attached beneath u. (A node with no children is called a *leaf*.)

The *height* of a node in a tree is the number of levels of nodes beneath it. (Again, see Chapter 11 for more.) A single node has height 1; a node with one or two children that are leaves has height 2; etc. (We think of a nonexistent tree has having height 0.)

An *AVL tree* is a special type of binary search tree that ensures that the tree is "balanced" and therefore supports its operations very efficiently.[6] The whole point of a balanced binary search tree is that the height of the tree is supposed to be "small," because the cost of almost every operation on binary search trees is proportional to the height of the tree. (The height of the tree is the height of the root.)

AVL trees were developed by two Russian computer scientists in 1962:

[6] A. Adelson-Velskii and E. M. Landis. An algorithm for the organization of information. *Proceedings of the USSR Academy of Sciences*, 146:263–266, 1962.

Since then, a number of other schemes for maintaining *balanced binary search trees* have been developed, most prominently red–black trees.

An *AVL tree* is a binary search tree in which, for any node u, the height of u's left child and the height of u's right child can only differ by one. Alternatively, we can define AVL trees recursively:

Definition 6.12 (AVL trees)
Any empty tree (consisting of zero nodes) is an AVL tree of height 0.

A tree of height $h \geq 1$ *is an AVL tree if*

(i) the subtrees rooted at the two children of the root are both AVL trees; and
(ii) the heights of the root's children are either both $h-1$, *or one is* $h-1$ *and the other is* $h-2$.

In other words, for any node u in an AVL tree, the height h_ℓ of u's left subtree and the height h_r of u's right subtree must satisfy $|h_\ell - h_r| \leq 1$.

A few examples of AVL trees are shown in Figure 6.36. If you studied AVL trees before, you were probably told "AVL trees have logarithmic height." Here, we'll prove it.

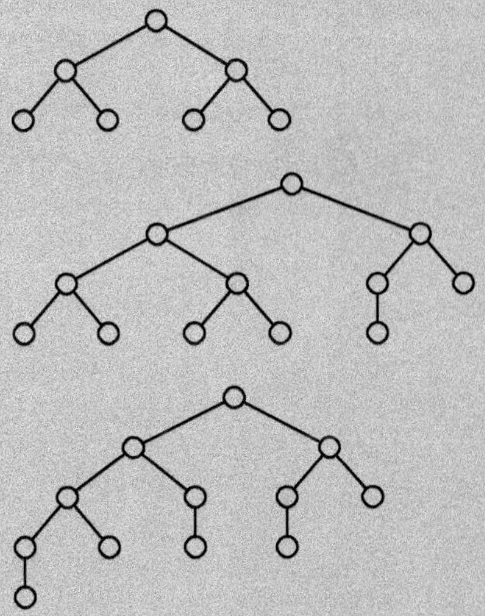

Figure 6.36: Three AVL trees. Take any node u in any of the three trees; one can verify that the number of layers beneath u's left child and u's right child differ by at most one.

An upper bound

Consider an AVL tree T of height h. After a little contemplation, it should be clear that T will contain the maximum possible number of nodes (out of all AVL trees of height h) when both of the children of T's root node have height $h-1$, and furthermore that both subtrees of the root have as many nodes as an AVL tree of height $h-1$ can have.

Let $M(h)$ denote the maximum number of nodes that can appear in an AVL tree of height h. There can be only one node in a height 1 tree, so $M(1) = 1$. For $h \geq 2$, the discussion in the previous paragraph shows that

$$M(h) = \underbrace{M(h-1)}_{\text{the left subtree}} + \underbrace{M(h-1)}_{\text{the right subtree}} + \underbrace{1}_{\text{the root node}} . \qquad (*)$$

Computer Science Connections

AVL Trees, continued

Claim: $M(h) = 2^h - 1$.

Proof. The proof is straightforward by induction. For the base case ($h = 1$), we have $M(h) = 1$ by definition, and $2^1 - 1 = 2 - 1 = 1$. For the inductive case, we have $M(h) = 2M(h) + 1 = 2 \cdot 2(2^{h-1} - 1) + 1$ by (∗) and the inductive hypothesis. Simplifying yields $M(h) = 2^h - 2 + 1 = 2^h - 1$. □

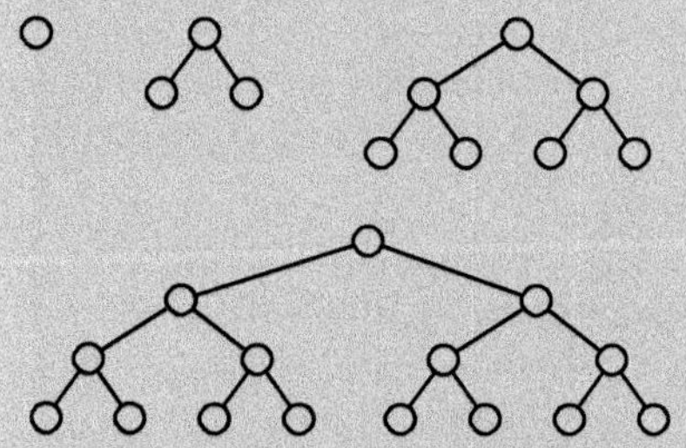

Figure 6.37: The fullest-possible AVL trees of height $h \in \{1, 2, 3, 4\}$, respectively containing $1 = 2^1 - 1$, $3 = 2^2 - 1$, $7 = 2^3 - 1$, and $15 = 2^4 - 1$ nodes.

A lower bound

Let's now analyze the other direction: what is the *fewest* nodes that can appear in an AVL of height h? (We can transform this analysis into one that finds the largest possible height of an AVL tree with n nodes.)

Define $N(h)$ as the minimum number of nodes in an AVL tree of height h. As before, any height 1 tree has one node, so $N(1) = 1$. It's also immediate that $N(2) = 2$. It's easy to see that the minimum number of nodes in an AVL tree is achieved when the root has one child of height $h - 1$ and one child of height $h - 2$—and furthermore when the root's subtrees contain as few nodes as legally possible. That is,

$$N(h) = \underbrace{N(h-1)}_{\text{the left subtree}} + \underbrace{N(h-2)}_{\text{the right subtree}} + \underbrace{1}_{\text{the root node}} . \qquad (\dagger)$$

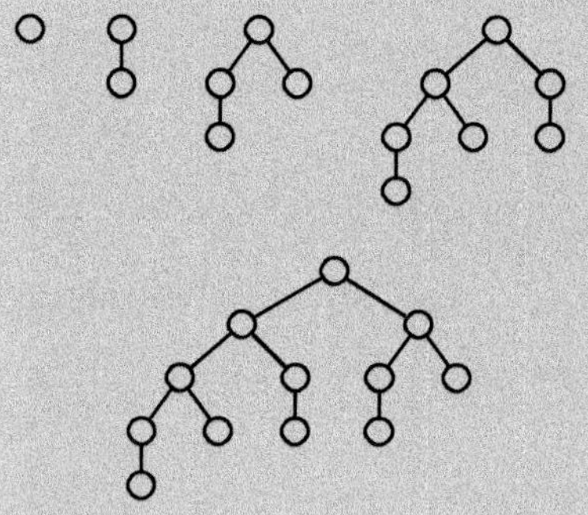

Figure 6.38: The emptiest-possible AVL trees of height $h \in \{1, 2, 3, 4, 5\}$, which contain 1, 2, 4, 7, and 12 nodes.

Observe that $N(h) = 1 + N(h-1) + N(h-2) \geq 1 + 2 \cdot N(h-2)$ because $N(h-1) \geq N(h-2)$. Therefore $N(h) \geq 2^{h/2} - 1$.

We can do better, though, with a bit more work. Define $P(h) = 1 + N(h)$. Adding one to both sides of (†), in this new notation, we have that $P(h) = P(h-1) + P(h-2)$. (This recurrence should look familiar: it's the same recurrence as for the Fibonacci numbers!) Because $P(1) = 1 + N(1) = 2 = f_3$ and $P(2) = 1 + N(2) = 3 = f_4$, we can prove inductively that $P(h) = f_{h+2}$.

Claim: $N(h) \geq \phi^h - 1$.

Proof. Using the definition of P, the proof in Example 6.27, and the fact that $\frac{1}{\phi^2} = \frac{2}{3+\sqrt{5}}$, we have

$$N(h) = P(h) - 1 = f_{h+2} - 1 \geq \frac{2}{3+\sqrt{5}} \cdot \phi^{h+2} - 1 = \phi^h - 1. \qquad □$$

Putting it all together

The analysis above will let us prove the following theorem:

> **Theorem 6.9**
> *The height h of any n-node AVL tree satisfies* $\log_\phi(n+1) \geq h \geq \log_2(n+1)$.

Proof. By the first claim above, we have $2^h - 1 = M(h) \geq n$. Thus $2^h \geq n + 1$, and—taking logs of both sides—we have $h \geq \log_2(n+1)$.

By the second claim above, we have $\phi^h - 1 = N(h) \leq n$. Thus $\phi^h \leq n + 1$, and—taking $\log_\phi$ of both sides—we have $h \leq \log_\phi(n+1)$. □

By changing log bases, we have

$$\begin{aligned} \log_\phi(x) &= \log_2(x) / \log_2(\phi) \\ &\approx \log_2(x)/0.69424\cdots \\ &\approx 1.4404 \cdot \log_2(x) \end{aligned}$$

Thus this theorem says that an n-node AVL tree has height between $\log_2(n+1)$ and $1.44 \log_2(n+1)$. In fact, there are AVL trees whose height is as large as $1.44 \log_2(n+1)$, so this analysis is tight.

6.4.4 Exercises

A quadtree *is a data structure typically used to store a collection of n points in* $\mathbb{R}^2$. *The basic idea is to start with a bounding box that includes all n points, and then subdivide, into four equal-sized subregions, any region that contains more than a designated number k of points. (For simplicity, we will subdivide any region with more than k = 1 point.) The* height *of a quadtree is the number of levels of the deepest subdivision of the tree. Figure 6.39 shows an example of the regions and the corresponding tree. (Figure 6.39's quadtree contains 17 regions, and its height is 4. A region's children are its subregions, clockwise from the upper left.)*

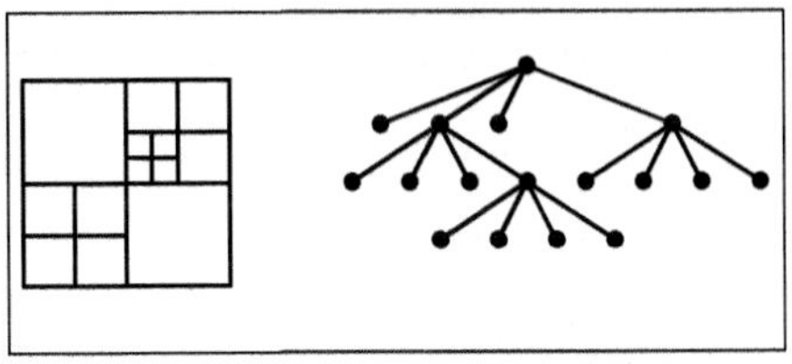

Figure 6.39: The decomposition of the plane to build a quadtree.

6.80 Let $R(h)$ denote the *largest number of regions* that a quadtree of height h can contain. Write a recurrence relation for $R(h)$.

6.81 Let $S(h)$ denote the *smallest number of regions* that a quadtree of height h can contain. Write a recurrence relation for $S(h)$.

6.82 It turns out that most efficient division of n points in a quadtree occurs when each subregion contains precisely $n/4$ points. Let $T(n)$ denote the *smallest number of regions* that a quadtree with n points can contain. Using the above assertion without proof, write a recurrence relation for $T(n)$.

For the recursive algorithms shown in Figure 6.40, write down a recurrence relation expressing their running time. (Assume that selecting a subarray takes $\Theta(1)$ *time.)*

6.83 **foo**

6.84 **bar**

6.85 **baz**

Using your recurrences, prove by induction that each algorithm requires $O(n)$ *time:*

6.86 **foo**

6.87 **bar** (for ease, you may assume n is a power of 2)

6.88 **baz**

Still considering the recursive algorithms shown in Figure 6.40:

6.89 What problem do the algorithms **foo**, **bar**, and **baz** solve?

Consider the following ternary search *algorithm, a variation on binary search. Suppose you have a sorted array* $A[1 \ldots n]$ *and you're searching for a particular value x in it. If* $n \leq 2$*, just check whether x is one of the one or two entries in A. Otherwise, compare x to* $A[n/3]$ *and* $A[2n/3]$*, and do the following:*

- *if* $x = A[\lfloor \frac{n}{3} \rfloor]$ *or* $x = A[\lceil \frac{2n}{3} \rceil]$*, return true.*
- *if* $x < A[\lfloor \frac{n}{3} \rfloor]$*, recursively search* $A[1 \ldots \lfloor \frac{n}{3} \rfloor - 1]$*.*
- *if* $A[\lfloor \frac{n}{3} \rfloor] < x < A[\lceil \frac{2n}{3} \rceil]$*, recursively search* $A[\lfloor \frac{n}{3} \rfloor + 1 \ldots \lceil \frac{2n}{3} \rceil - 1]$*.*
- *if* $x > A[\lceil \frac{2n}{3} \rceil]$*, recursively search* $A[\lceil \frac{2n}{3} \rceil + 1 \ldots n]$*.*

6.90 Analyze the asymptotic worst-case running time of ternary search. Prove your answer correct using induction. For convenience, you may assume that n is a power of three.

6.91 Does ternary search perform better or worse than binary search? Here you should count the *exact* number of comparisons that each algorithm performs—don't give an asymptotic answer.

6.92 Consider a simplified (and thus slightly erroneous) version of the recurrence for Binary Search: $T(n) = T(n/2) + c$ and $T(1) = c$. (This recurrence ignores the off-by-one complications.) Prove that $T(n) = c(1 + \log n)$ when n is a power of two by induction.

```
foo(A[1...n]):
  if n = 0 then
      return 0
  else if A[1] < 0 then
      return 1 + foo(A[2...n])
  else
      return foo(A[2...n])

bar(A[1...n]):
  if n = 0 or (n = 1 and A[1] ≥ 0) then
      return 0
  else if n = 1 and A[1] < 0 then
      return 1
  else
      count := 0
      count := count + bar(A[1...⌊n/2⌋])
      count := count + bar(A[⌊n/2⌋ + 1...n])
      return count

baz(A[1...n]):
  if n = 0 or (n = 1 and A[1] ≥ 0) then
      return 0
  else if n = 1 and A[1] < 0 then
      return 1
  else
      count := 0
      count := count + baz(A[1...⌊n/4⌋])
      count := count + baz(A[⌊n/4⌋ + 1...⌊3n/4⌋])
      count := count + baz(A[⌊3n/4⌋ + 1...n])
      return count
```

Figure 6.40: Three recursive algorithms.

The next two exercises ask you to analyze **quickSort***, discussed in Example 5.14 and Exercises 6.74–6.77.*

6.93 Consider the recurrence relation from Exercise 6.77, based on the "Median of Three" pivoting rule for **quickSort**, namely $T(1) = T(2) = 1$ and $T(n) = T(n-2) + cn$. Prove that $T(n) = \Theta(n^2)$.

6.94 Generalize your argument from the previous exercise to show that the recurrence

$$T(n) = \begin{cases} 1 & \text{if } n \leq k \\ T(n-k) + n & \text{otherwise} \end{cases}$$

has solution $T(n) = \Theta(n^2)$ *for any integer* $k \geq 1$.

```
fibNaive(n):
  if n = 0 or n = 1 then
      return 1
  else
      return fibNaive(n − 1) +
      fibNaive(n − 2)
```

```
fibMatrix(n):
  Compute (using repeated
  squaring)
      [x; y] := [1 1; 1 0]^n · [1; 1].
  return x
```

```
fibMedium(n):
  ⟨f_n, f_{n−1}⟩ := helper(n)
  return f_n

helper(n):
  if n = 0 then
      return ⟨1, undefined⟩
  else if n = 1 then
      return ⟨1, 1⟩
  else
      ⟨f_{n−1}, f_{n−2}⟩ := helper(n − 1)
      return ⟨f_{n−1} + f_{n−2}, f_{n−1}⟩
```

```
fibClever(n):
  return (exp(φ, n) − exp(φ̂, n)) / √5

exp(b, n):
  if n = 0 then
      return 1
  else
      s := exp(b, ⌊n/2⌋)
      if n is odd then
          return b · s · s
      else
          return s · s
```

Figure 6.41: Four algorithms for the Fibonaccis. The values ϕ and $\hat{\phi}$ satisfy $f_n = \frac{\phi^n - \hat{\phi}^n}{\sqrt{5}}$; see Example 6.28.

Recall that the Fibonacci numbers are defined by the recurrence $f_1 = f_2 = 1$ and $f_n = f_{n-1} + f_{n-2}$. The next several exercises refer to this recurrence and the algorithms for computing the Fibonacci numbers in Figure 6.41.

6.95 First, a warmup unrelated to the algorithms in Figure 6.41: prove by induction that $f_n \leq 2^n$.

6.96 Prove that **fibNaive**$(n - k)$ appears a total of f_{k+1} times in the call tree for **fibNaive**(n).

6.97 Write down and solve a recurrence for the running time of **helper** (and therefore **fibMedium**).

6.98 Write down and solve a recurrence for the running time of **exp** (and therefore **fibClever**).

6.99 The reference to "repeated squaring" in **fibMatrix** is precisely the same as the idea of **exp**. Implement **fibMatrix** using this idea in a programming language of your choice. (See Exercise 5.56.)

6.100 Recall from Chapter 5 (or see Figure 6.42) an algorithm that *merges* two sorted arrays into a single sorted array. Give a recurrence relation $T(n)$ describing the running time of **merge** on two input arrays with a total of n elements, and prove that $T(n) = \Theta(n)$.

6.101 Consider the recurrence for the running time of **mergeSort** (again, see Figure 6.42):

$$T(1) = c \qquad \text{and} \qquad T(n) = T(\lceil n/2 \rceil) + T(\lfloor n/2 \rfloor) + cn.$$

Prove that $T(n) \leq T(n')$ if $n \leq n'$—that is, T is monotonic.

```
merge(X[1...n], Y[1...m]):
  if n = 0 then
      return Y
  else if m = 0 then
      return X
  else if X[1] < Y[1] then
      return X[1] followed by merge(X[2...n], Y)
  else
      return Y[1] followed by merge(X, Y[2...m])
```

Figure 6.42: The "merging" of two sorted arrays.

6.102 Here is a recurrence relation for the number of *comparisons* done by **mergeSort** on an input array of size n (once again, see Figure 6.42):

$$C(1) = 0 \qquad \text{and} \qquad C(n) = 2C(n/2) + n - 1.$$

(For ease, we'll assume that n is a power of two.) Explain the recurrence relation, and then prove that $C(n) = n \log n - n + 1$ by induction.

The next few exercises refer to the algorithms in Figure 6.43, both which solve the same problem.

6.103 Give and solve (using induction) a recurrence relation for the running time of **f**.

6.104 Give a recurrence relation for **g**, and use it to prove that **g**(n) runs in $O(\log^2 n)$ time.

6.105 Describe the set of input values n that cause the worst-case behavior for **g**(n).

6.106 What problem do **f** and **g** solve? Prove your answer.

```
f(n):
  if n ≤ 1 then
      return n
  else
      return f(n − 2)
```

```
g(n):
  if n ≤ 1 then
      return n
  else
      x := 1
      while n ≥ 2x:
          x := 2 · x
      return g(n − x)
```

Figure 6.43: Two algorithms.

Two copies of an out-of-print book were listed online by Seller A and Seller B. Their prices were over \$1,000,000 each—and the next day, both prices were over \$2,000,000, and they kept going up. By watching the prices over several days, it became clear that the two sellers were using algorithms to set their prices in response to each other.

Let a_n and b_n be the prices on day n by Seller A and Seller B, respectively. The prices were set by two (badly conceived) algorithms such that $a_n = \alpha \cdot b_{n-1}$ and $b_n = \beta \cdot a_n$ where $\alpha = 0.9983$ and $\beta = 1.27059$.

6.107 Suppose that $b_0 = 1$. Find the closed form solution for a_n and b_n. Prove your answer.

6.108 State a necessary and sufficient condition on α, β, and b_0 such that $a_n = \Theta(1)$ and $b_n = \Theta(1)$.

Exercises 6.107–6.108 are based on a story from Michael Eisen's blog post "Amazon's \$23,698,655.93 book about flies."

6.5 *Recurrence Relations: The Master Method*

> In order to become the master, the politician poses as the servant.
>
> Charles de Gaulle (1890–1970)

In the remainder of this section, we'll turn to a more formulaic method, called the *Master Method,* of solving recurrence relations that have a certain form: in analyzing algorithms, we will frequently encounter recurrences that look like

$$T(n) = aT\left(\tfrac{n}{b}\right) + c \cdot n^k,$$

for four constants $a \geq 1, b > 1, c > 0$, and $k \geq 0$.

Why do these recurrences come up frequently? Consider a recursive algorithm that has the following structure: if the input is small—say, $n = 1$—then we compute the solution directly; otherwise, to solve an instance of size n:

- we make a different recursive calls on inputs of size $\frac{n}{b}$; and
- to construct the smaller instances and then to reconstruct the solution to the given instance from the recursive solutions, we spend $\Theta(n^k)$ time.

(These algorithms are usually called *divide-and-conquer algorithms:* they "divide" their input into a pieces, and then recursively "conquer" those subproblems.) To be precise, the recurrence often has ceilings and floors as part of its recursive calls, but for now assume that n is exact power of b, so that the floors and ceilings don't matter.

Here are a few examples of recursive algorithms with recurrences of this form:

Example 6.29 (Binary Search)
We spend $c = \Theta(1)$ time to compare the sought element to the middle of the range; we then make one recursive call to search for the element in the appropriate half of the array. If n is an exact power of two, then the recurrence is

$$T(n) = T(\tfrac{n}{2}) + c.$$

(So $a = 1$, $b = 2$, and $k = 0$, because $c = c \cdot 1 = c \cdot n^0$.)

Example 6.30 (Merge Sort)
We spend $\Theta(1)$ time to divide the array in half. We make two recursive calls on the left and right subarrays, and then spend $\Theta(n)$ time to merge the resulting sorted subarrays into a single sorted array. If n is an exact power of two, then the recurrence is

$$T(n) = 2T(\tfrac{n}{2}) + c \cdot n.$$

(So $a = 2$, $b = 2$, and $k = 1$.)

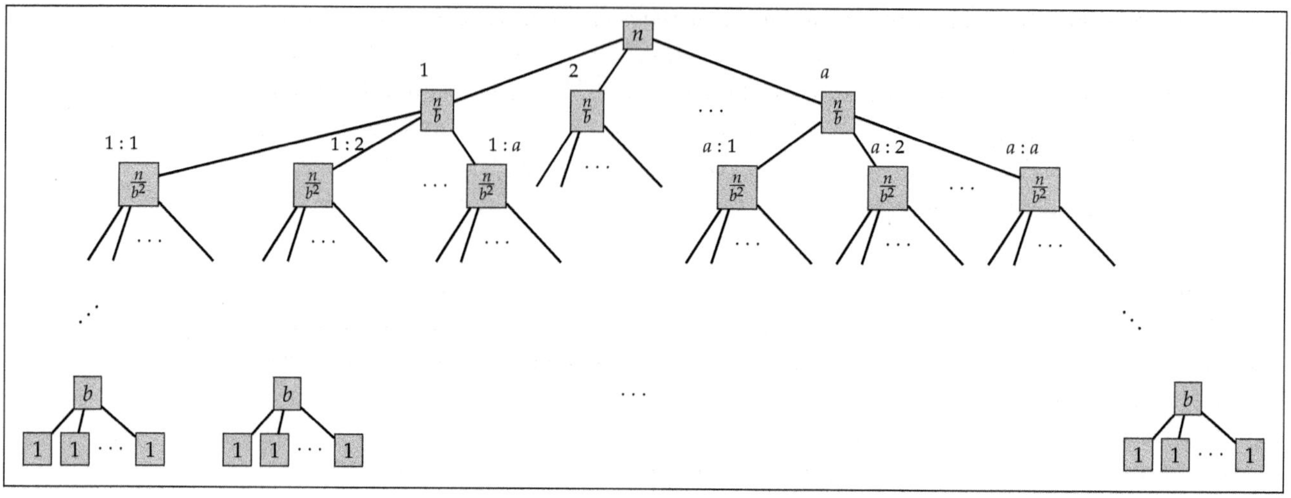

Figure 6.44: The recursion tree for a recurrence relation $T(n) = aT(\frac{n}{b}) + c \cdot n^k$, of the master method's form. Assume that n is an exact power of b.

6.5.1 *The Master Method: Some Intuition*

The *Master Method* is a technique that allows us to solve any recurrence relation of the form $T(n) = aT(\frac{n}{b}) + c \cdot n^k$ very easily. The Master Method is based on examining the recursion tree for this recurrence (see Figure 6.44), and the *Master Theorem* (Theorem 6.10) that describes the total amount of work represented by this tree.

Here's the intuition for the Master Method. Let's think about the ith level of the recursion tree (again, see Figure 6.44)—in other words, the work done by the recursive calls that are i levels beneath the root of the recursion tree. Observe the following:

There are a^i different calls at level i. There is $1 = a^0$ call at the 0th level, then $a = a^1$ calls at 1st level, then a^2 calls at the 2nd level, and so forth.

Each of the the calls at the ith level operates on an input of size $\frac{n}{b^i}$. The input size is $\frac{n}{1} = n$ at the 0th level, then $\frac{n}{b}$ at the 1st level, then $\frac{n}{b^2}$ at the 2nd, and so forth.

Thus the total amount of work in the ith level of the tree is $a^i \cdot c \cdot (\frac{n}{b^i})^k$. Or, simplifying, the total work at this level is $cn^k \cdot (\frac{a}{b^k})^i$.

Thus the total amount of work contained within the entire tree is

$$\sum_i \left[cn^k \cdot \left(\frac{a}{b^k}\right)^i \right] = cn^k \cdot \sum_i \left[\left(\frac{a}{b^k}\right)^i \right]. \qquad (*)$$

(We'll worry about the bounds on the summation later.)

Note that $(*)$ expresses the total work in the recursion tree as a geometric sum $\sum_i r^i$, in which the ratio between terms is given by $r := \frac{a}{b^k}$. (See Section 5.2.2.) As with any geometric sum, the critical question is how the ratio compares to 1: if $r < 1$, then the terms of the sum are getting smaller and smaller as i increases; if $r > 1$, then the terms of the sum are getting bigger and bigger as i increases. (And if $r = 1$, then each term is simply equal to 1.)

The Master Theorem has three cases, each of which corresponds to one of these three natural cases for the summation in $(*)$: its terms *increase exponentially* with i, its terms *decrease exponentially* with i, or its terms are *constant* with respect to i. In these cases, respectively, almost all of the work is done at the leaves of the tree; almost all of the work is done at the root of the tree; or the work is spread evenly across the levels of the tree. (Here "almost all the work" means "a constant fraction of the work," which means that the total work in the tree is asymptotically equivalent to the work done solely at the root or at the leaves.)

A trio of examples

Before we prove the general theorem, we'll solve a few recurrences that illustrate the cases of the Master Method, and then we'll prove the result in general. The three example recurrences are

$$T(n) = 2T(\tfrac{n}{2}) + 1$$
$$T(n) = 2T(\tfrac{n}{2}) + n$$
$$\text{and } T(n) = 2T(\tfrac{n}{2}) + n^2,$$

all with $T(1) = 1$. Figure 6.45 shows the recursion trees for these recurrences.

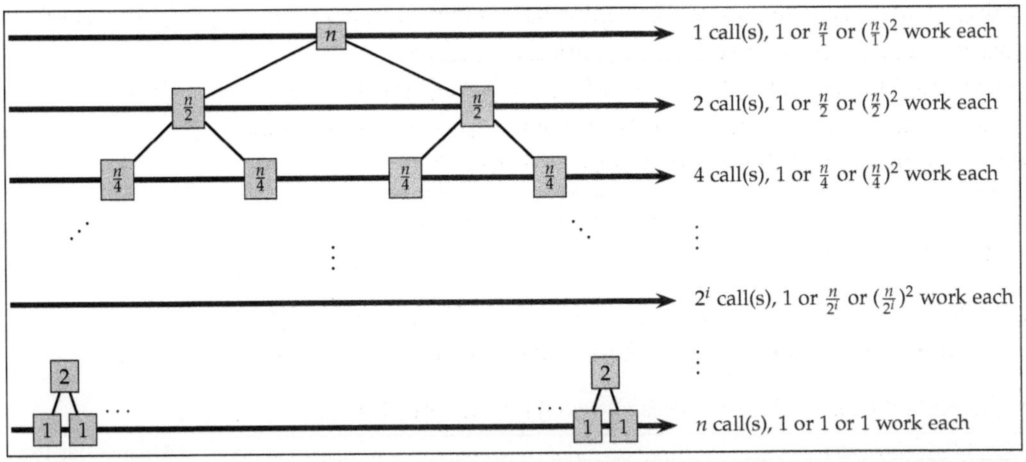

Figure 6.45: The recursion trees for three different recurrences: $T(n) = 2T(\frac{n}{2}) + f(n)$, for $f(n) \in \{1, n, n^2\}$. The annotation in each row of the tree shows both the number of calls at that level of the tree, plus the additional work done by each call at that level.

In each of these recurrences, we divide the input by two at every level of the recursion. Thus, the total depth of the recursion tree is $\log_2 n$. (Assume n is an exact power of two.) In the recursion tree for any one of these recurrences, consider the ith level of the tree beneath the root. (The root of the recursion tree has depth 0.) We have divided n by 2 a total of i times, and thus the input size at that level is $\frac{n}{2^i}$. Furthermore, there are 2^i different calls at the ith level of the tree.

Solving the three recurrences

To solve each recurrence, we will sum the total amount of work generated at each level of the tree. The recursion trees for each of these three recurrences are shown in Figures 6.46, 6.47, and 6.48.

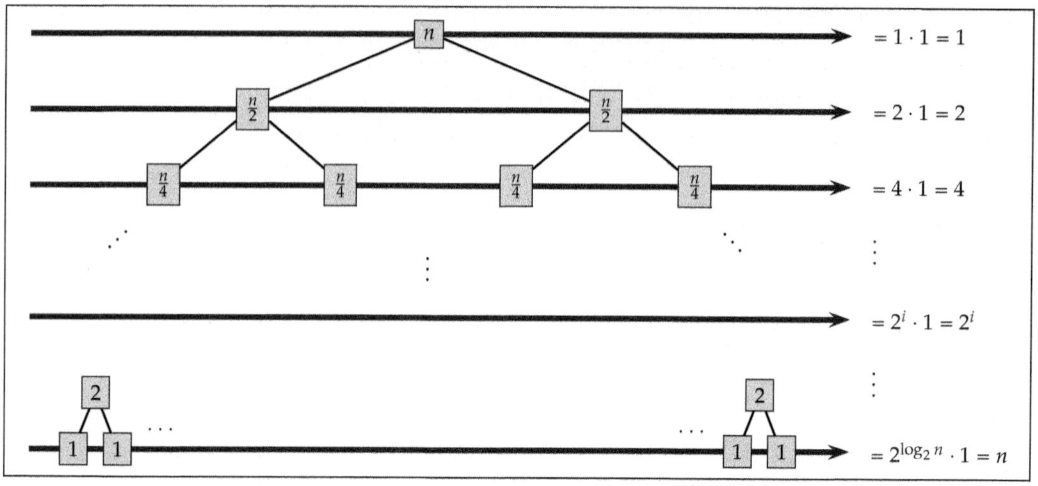

Figure 6.46: The recursion tree for $T(n) = 2T(\frac{n}{2}) + 1$, with the "row-wise" sums of work. The work at each level is twice the work at the level above it; thus the work is increasing exponentially at each level of the tree.

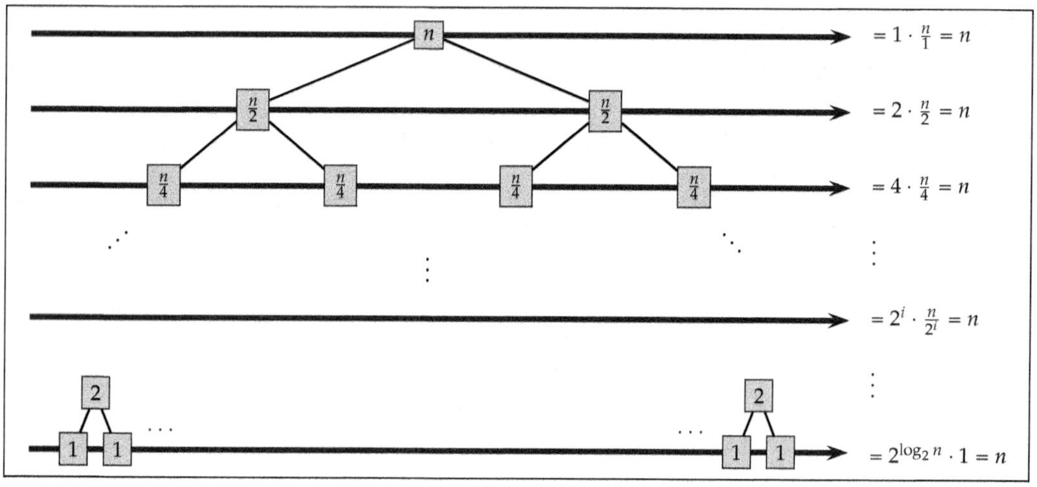

Figure 6.47: The recursion tree for $T(n) = 2T(\frac{n}{2}) + n$. The work at each level is exactly n; thus the work is constant across the levels of the tree.

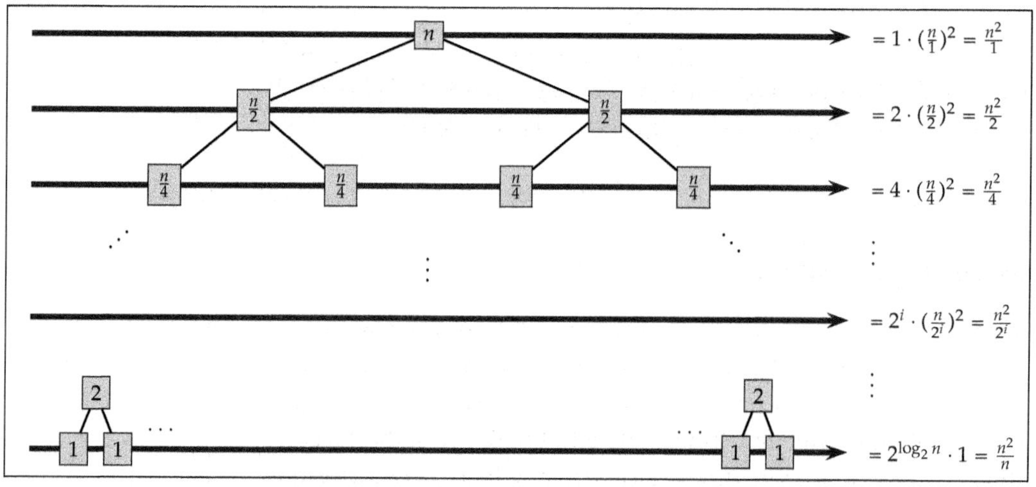

Figure 6.48: The recursion tree for $T(n) = 2T(\frac{n}{2}) + n^2$. The work at each level is half of the work at the level above it; thus the work is decreasing exponentially at each level of the tree.

Example 6.31 (Solving $T(n) = 2T(\frac{n}{2}) + 1$)
Figure 6.46 shows the recursion tree for this recurrence. There are 2^i different calls at the ith level, each of which is on an input of size $\frac{n}{2^i}$—and we do 1 unit of work for each of these 2^i calls. Thus the total amount of work at level i is 2^i. The total amount of work in the entire tree is therefore

$$T(n) = \sum_{i=0}^{\log_2 n} 2^i = \frac{2^{1+\log_2 n} - 1}{2-1} = 2 \cdot 2^{\log_2 n} = 2n$$

by Theorem 5.2. And, indeed, $T(n) = \Theta(n)$.

Example 6.32 (Solving $T(n) = 2T(\frac{n}{2}) + n$)
Figure 6.47 shows the recursion tree. There are 2^i calls at the ith level of the recursion tree, on inputs of size $\frac{n}{2^i}$. We do $\frac{n}{2^i}$ units of work at each call, so the total work at the ith level is $2^i \cdot (\frac{n}{2^i}) = n$. Note that the amount of work at level i is independent of the level i. The total amount of work in the tree is therefore

$$T(n) = \sum_{i=0}^{\log_2 n} \underbrace{n}_{\text{work at level } \#i} = n \cdot \sum_{i=0}^{\log_2 n} 1 = n(1 + \log_2 n) = \Theta(n \log n).$$

Example 6.33 (Solving $T(n) = 2T(\frac{n}{2}) + n^2$)
Figure 6.48 shows the recursion tree. There are 2^i calls at the ith level of the tree, and we do $(\frac{n}{2^i})^2$ work at each call at this level. Thus the work represented by the ith row of the recursion tree is $(\frac{n}{2^i})^2 \cdot 2^i = \frac{n^2}{2^i}$. The total amount of work in the tree is therefore

$$T(n) = \sum_{i=0}^{\log_2 n} (\tfrac{1}{2})^i n^2 = n^2 \cdot \sum_{i=0}^{\log_2 n} (\tfrac{1}{2})^i.$$

Notice that $\sum_{i=0}^{\log_2 n} (\frac{1}{2})^i = 1 + \frac{1}{2} + \frac{1}{4} + \cdots + \frac{1}{2^{\log_2 n}}$, which is certainly at least 1. But, by the fact that $1 + \frac{1}{2} + \frac{1}{4} + \ldots + \frac{1}{2^\ell} < 2$ (see Theorem 5.2), we also know $\sum_{i=0}^{\log_2 n} (\frac{1}{2})^i \leq 2$. Therefore $n^2 \leq T(n) \leq 2n^2$, which allows us to conclude that $T(n) = \Theta(n^2)$.

6.5.2 *The Master Method: The Formal Statement and a Proof*

Examples 6.31, 6.32, and 6.33 were designed to build the necessary intuition about the three different cases of the master method: work increases exponentially across levels of the recursion tree; work stays constant across levels; or work decreases exponentially across levels. Precisely the same intuition will yield the proof of the Master Theorem. Here is the formal statement of the Master Theorem, which generalizes the idea of these examples to all recurrences of the form $T(n) = aT(\frac{n}{b}) + cn^k$:

Theorem 6.10 (Master Theorem)
Consider the recurrence

$$T(1) = c$$
$$T(n) = a \cdot T(n/b) + c \cdot n^k$$

for constants $a \geq 1$, $b > 1$, $c > 0$, *and* $k \geq 0$. *Then:*

Case (i), "the leaves dominate": *if* $b^k < a$, *then* $T(n) = \Theta(n^{\log_b(a)})$.

Case (ii), "all levels are equal": *if* $b^k = a$, *then* $T(n) = \Theta(n^k \cdot \log n)$.

Case (iii), "the root dominates": *if* $b^k > a$, *then* $T(n) = \Theta(n^k)$.

(As we discussed previously, we are abusing notation by using c to denote two different constants in this theorem statement. Again, as you'll prove in Exercise 6.126, the recurrence $T(1) = d$ with a constant $d > 0$ possibly different than c has precisely the same asymptotic solution.)

Proving the theorem

While the Master Theorem holds even when the input n is not an exact power of b—we just have to fix the recurrence by adding floors or ceilings so that it still makes sense—we will prove the result for exact powers of b only.[7] We will show that the total amount work contained in the recursion tree is

$$T(n) = cn^k \cdot \sum_{i=0}^{\log_b n} \left(\frac{a}{b^k}\right)^i. \qquad (\dagger)$$

A full proof of the Master Theorem, including for the case when n is not an exact power of b, can be found in
[7] Thomas H. Cormen, Charles E. Leisersen, Ronald L. Rivest, and Clifford Stein. *Introduction to Algorithms*. MIT Press, 3rd edition, 2009.

As before, the formula (†) should make intuitive the fact that $a = b^k$ (that is, $\frac{a}{b^k} = 1$) is the critical value. The value of $\frac{a}{b^k}$ corresponds to whether the work at each level of the tree is increasing ($\frac{a}{b^k} > 1$), steady ($\frac{a}{b^k} = 1$), or decreasing ($\frac{a}{b^k} < 1$). The summation in (†) is a geometric sum, and as we saw in Chapter 5 geometric sums behave fundamentally differently based on whether their ratio is less than, equal to, or greater than one.

Proof of Theorem 6.10 (for n an exact power of b). For all three cases, we begin by examining the recursion tree (Figure 6.44). Summing the total amount of work in the tree "row-wise," we see that there are a^i nodes at the ith level of the tree (where, again, the root is at level zero), each of which corresponds to an input of size n/b^i and therefore contributes $c \cdot (n/b^i)^k$ work to the total. The tree continues until the inputs are of size 1—that is, until $n/b^i = 1$, or when $i = \log_b n$. Thus the total amount of work in the tree is

$$T(n) = \sum_{i=0}^{\log_b n} a^i \cdot c \cdot \left(\frac{n}{b^i}\right)^k = cn^k \sum_{i=0}^{\log_b n} \left(\frac{a}{b^k}\right)^i.$$

(See the note at the end of this proof for another justification for this summation, or see Exercise 6.127.) We'll examine this summation in each of the three cases, depending on the value of $\frac{a}{b^k}$—and we'll handle the cases in order of ease, rather than in numerical order:

Case (ii): If $a = b^k$, then (†) says that

$$T(n) = cn^k \sum_{i=0}^{\log_b n} \left(\frac{a}{b^k}\right)^i = cn^k \sum_{i=0}^{\log_b n} 1 = cn^k(1 + \log_b n).$$

Thus the total work is $\Theta(n^k \log n)$.

Case (iii): If $a < b^k$, then (†) is a geometric sum whose ratio is strictly less than 1. Corollary 5.3 states that any geometric sum whose ratio is strictly between 0 and 1 is $\Theta(1)$. (Namely, the summation $\sum_{i=0}^{\log_b n} (\frac{a}{b^k})^i$ is lower-bounded by 1 and upper-bounded by $\frac{1}{1-a/b^k}$, both of which are positive constants when $a < b^k$.) Therefore:

$$\begin{aligned} T(n) &= cn^k \sum_{i=0}^{\log_b n} \left(\frac{a}{b^k}\right)^i \\ &= cn^k \cdot \Theta(1). && \textit{by Corollary 5.3} \end{aligned}$$

Therefore the total work is $\Theta(n^k)$.

Case (i): If $a > b^k$, then (†) is a geometric sum whose ratio is strictly larger than one. But we can make this summation look more like Case (iii), using a little algebraic manipulation. Notice that, for any $\alpha \neq 0$, we can rewrite $\sum_{i=0}^{m} \alpha^i$ as follows:

$$\sum_{i=0}^{m} \alpha^i = \alpha^m \cdot \sum_{i=0}^{m} \alpha^{i-m} = \alpha^m \cdot \sum_{i=0}^{m} \left(\frac{1}{\alpha}\right)^{m-i} = \alpha^m \cdot \sum_{j=0}^{m} \left(\frac{1}{\alpha}\right)^j \qquad (‡)$$

where the last equality follows by reindexing the summation (so that we set $j = m - i$). Applying this manipulation to (†), we have

$$\begin{aligned} T(n) &= cn^k \sum_{i=0}^{\log_b n} \left(\frac{a}{b^k}\right)^i && \textit{by (†)} \\ &= cn^k \cdot \left(\frac{a}{b^k}\right)^{\log_b n} \cdot \sum_{j=0}^{\log_b n} \left(\frac{b^k}{a}\right)^j && \textit{by (‡)} \\ &= n^k \cdot \left(\frac{a}{b^k}\right)^{\log_b n} \cdot \Theta(1) && \textit{Corollary 5.3, because } \tfrac{b^k}{a} < 1. \\ &= n^k \cdot \frac{a^{\log_b n}}{(b^k)^{\log_b n}} \cdot \Theta(1) \\ &= n^k \cdot \frac{a^{\log_b n}}{n^k} \cdot \Theta(1) && (b^k)^{\log_b n} = b^{k \log_b n} = b^{\log_b n^k} = n^k \\ &= a^{\log_b n} \cdot \Theta(1). \end{aligned}$$

Therefore the total work is $\Theta(a^{\log_b n})$. And $a^{\log_b n} = n^{\log_b a}$, which we can verify by log manipulations:

$$a^{\log_b n} = b^{\log_b [a^{\log_b n}]} = b^{[\log_b n] \cdot [\log_b a]} = b^{[\log_b a] \cdot [\log_b n]} = b^{\log_b [n^{\log_b a}]} = n^{\log_b a}.$$

Therefore the total work in this case is $\Theta(a^{\log_b n}) = \Theta(n^{\log_b a})$. □

Taking it further: Another way to make the formula (†)—which was the entire basis of the Master Theorem—a little more intuitive is to consider iterating the recurrence a few times:

$$\begin{aligned}
T(n) &= cn^k + a\cdot T(\tfrac{n}{b}) && = \sum_{i=0}^{0} ca^i\left(\tfrac{n}{b^i}\right)^k + aT\left(\tfrac{n}{b}\right)\\
&= cn^k + a\left[c\left(\tfrac{n}{b}\right)^k + aT(\tfrac{n}{b^2})\right]\\
&= cn^k + ac\left(\tfrac{n}{b}\right)^k + a^2T(\tfrac{n}{b^2}) && = \sum_{i=0}^{1} ca^i\left(\tfrac{n}{b^i}\right)^k + a^2T\left(\tfrac{n}{b^2}\right)\\
&= cn^k + ac\left(\tfrac{n}{b}\right)^k + a^2\left[c\left(\tfrac{n}{b^2}\right)^k + aT(\tfrac{n}{b^3})\right]\\
&= cn^k + ac\left(\tfrac{n}{b}\right)^k + a^2c\left(\tfrac{n}{b^2}\right)^k + a^3T(\tfrac{n}{b^3}) && = \sum_{i=0}^{2} ca^i\left(\tfrac{n}{b^i}\right)^k + a^3T\left(\tfrac{n}{b^3}\right).
\end{aligned}$$

At every iteration, we generate another term of the form $ca^i(n/b^i)^k$. Eventually n/b^i will equal 1—specifically when $i = \log_b n$—and the recursion will terminate. By iterating the recurrence $\log_b n$ times, we would get to

$$T(n) = \sum_{i=0}^{(\log_b n)-1} ca^i\left(\frac{n}{b^i}\right)^k + a^{\log_b n}T\left(\frac{n}{b^{\log_b n}}\right). \qquad (6.10.1)$$

Because $T(n/b^{\log_b n}) = T(1) = c = 1^k c = (n/b^{\log_b n})^k c$, from (6.10.1) we can conclude

$$T(n) = \sum_{i=0}^{(\log_b n)-1} ca^i\left(\frac{n}{b^i}\right)^k + a^{\log_b n}(n/b^{\log_b n})^k c = \sum_{i=0}^{\log_b n} ca^i\left(\frac{n}{b^i}\right)^k,$$

which is precisely the summation (†).

The Master Method: a few examples

We'll conclude with a few easy examples using the Master Method, reproducing the recursion-tree analysis of Examples 6.31, 6.32, and 6.33:

Example 6.34 (Solving $T(n) = 2T(n/2) + \{1, n, n^2\}$)

Recall the recurrences

$$T(n) = 2T(\tfrac{n}{2}) + 1 \qquad (1)$$
$$T(n) = 2T(\tfrac{n}{2}) + n \qquad (2)$$
$$T(n) = 2T(\tfrac{n}{2}) + n^2, \qquad (3)$$

all with $T(1) = 1$.

For (1), we have $a = 2, b = 2, c = 1$, and $k = 0$; because $b^k = 2^0 = 1 < 2 = a$, case (i) of the Master Method says that $T(n) = \Theta(n^{\log_2 2}) = \Theta(n)$.

For (2), we have $a = 2, b = 2, c = 1$, and $k = 1$; because $b^k = 2^1 = 2 = a$, case (ii) of the Master Method says that $T(n) = \Theta(n^1 \log n) = \Theta(n \log n)$.

For (3), we have $a = 2, b = 2, c = 1$, and $k = 2$; because $b^k = 2^2 = 4 > 2 = a$, case (iii) of the Master Method says that $T(n) = \Theta(n^2)$.

Taking it further: Although we've mostly presented "algorithmic design" and "algorithmic analysis" as two separate phases, in fact there's interplay between these pieces. See p. 655 for a discussion of a particular computational problem—matrix multiplication—and algorithms for it, including a straightforward but slow algorithm and another that (with inspiration from the Master Method) improves upon that slow algorithm.

Computer Science Connections

Divide-and-Conquer Algorithms and Matrix Multiplication

Matrix multiplication (see Definition 2.43) is a fundamental operation with wide-ranging applications throughout CS: in computer graphics, in data mining, and in social-network analysis, just to name a few. Often the matrices in question are quite large—perhaps a matrix of hyperlinks among thousands or millions of web pages, for example. Thus asymptotic improvements to matrix multiplication algorithms have potential practical importance, too. For simplicity, we'll concentrate on multiplying square (n-by-n) matrices. The obvious algorithm for matrix multiplication simply follows the definition: separately for each of the n^2 entries in the output matrix, perform the $\Theta(n)$ multiplications/additions to compute the entry. (See Figure 6.49.) But, in the spirit of this section, what might we be able to do with a recursive algorithm?

```
matmult(M ∈ ℝ^{n×n}, N ∈ ℝ^{n×n}):
for i = 1,2,... n:
    for j = 1,2,...,n:
        P_{i,j} := 0
        for k = 1,2,...,n:
            P_{i,j} := P_{i,j} + M_{i,k} N_{k,j}
return P
```

Figure 6.49: The naïve algorithm for matrix multiplication for n-by-n matrices. For matrices $M \in \mathbb{R}^{n\times n}$ and $N \in \mathbb{R}^{n\times n}$, the product is a matrix $P \in \mathbb{R}^{n\times n}$ where $P_{i,j} := \sum_{k=1}^{n} M_{i,k}N_{k,j}$.

There is indeed a nice way to think about matrix multiplication recursively. To multiply two n-by-n matrices M and N, divide M and N each into four quarters, which we can label $M^{11}, M^{12}, \ldots$, as follows:

$$M = \begin{bmatrix} M^{11} & M^{12} \\ M^{21} & M^{22} \end{bmatrix}, \ N = \begin{bmatrix} N^{11} & N^{12} \\ N^{21} & N^{22} \end{bmatrix}.$$

Each of these quarters $M^{11}, M^{12}, \ldots$ is an $\frac{n}{2}$-by-$\frac{n}{2}$ matrix. It turns out that

$$MN = \begin{bmatrix} (MN)^{11} & (MN)^{12} \\ (MN)^{21} & (MN)^{22} \end{bmatrix} = \begin{bmatrix} M^{11}N^{11} + M^{12}N^{21} & M^{11}N^{12} + M^{12}N^{22} \\ M^{21}N^{11} + M^{22}N^{21} & M^{21}N^{12} + M^{22}N^{22} \end{bmatrix}.$$

This fact suggests a recursive, divide-and-conquer algorithm for multiplying matrices, with the recurrence $T(n) = 8T(\frac{n}{2}) + n^2$. (It takes $c \cdot n^2$ time to combine the result of the recursive calls.) By the Master Method ($a = 8, b = 2, k = 2$; case (i)), we have $T(n) = \Theta(n^{\log_2(8)}) = \Theta(n^3)$—so not an improvement over Figure 6.49 at all!

But, in a major algorithmic breakthrough, in 1969 Volker Strassen found a way to use *seven* recursive calls instead of *eight*. (See Figure 6.50.) This change makes the recurrence $T(n) = 7T(\frac{n}{2}) + n^2$; now the Master Method ($a = 7, b = 2, k = 2$; still case (i)), says that $T(n) = \Theta(n^{\log_2 7}) = \Theta(n^{2.8073\cdots})$—a nice improvement! (For example, $1000^{\log_2 7}$ is only about 25% of 1000^3.)

Compute these values recursively:

$$\begin{aligned}
A &:= (M^{11} + M^{22})(N^{11} + N^{22}) \\
B &:= (M^{21} + M^{22})N^{11} \\
C &:= M^{11}(N^{12} - N^{22}) \\
D &:= M^{22}(N^{21} - N^{11}) \\
E &:= (M^{11} + M^{12})N^{22} \\
F &:= (M^{21} - M^{11})(N^{11} + N^{12}) \\
G &:= (M^{12} - M^{22})(N^{21} + N^{22}).
\end{aligned}$$

Then compute MN as

$$\begin{bmatrix} A + D - E + G & C + E \\ B + D & A - B + C + F \end{bmatrix}.$$

Figure 6.50: The multiplications for Strassen's Algorithm. After we compute $A, B, \ldots, G$ recursively, we then add/subtract the results as indicated. (This addition/subtraction takes $c \cdot n^2$ time.)

Once the Master Method–style recurrence is in mind, one can investigate other Strassen-like algorithms (making fewer recursive calls, and combining them more cleverly). In 1978, Victor Pan gave a further running-time improvement using this style of algorithm—though more complicatedly!—using a total of 143,640 recursive calls on inputs of size $\frac{n}{70}$ (!), plus $\Theta(n^2)$ additional work. Using the Master Method, that algorithm yields a running time of $\Theta(n^{\log_{70} 143{,}640}) = \Theta(n^{2.7951\cdots})$. Algorithms continued to improve for several years, culminating in 1990 with an $\Theta(n^{2.3754\cdots})$-time algorithm due to Don Coppersmith and Shmuel Winograd. That algorithm was the best known for two decades, but in the last few years some new researchers with new insights have come along, and the exponent is now down to 2.373. For whatever it's worth, many people think that there might be an $\Theta(n^2)$ algorithm for multiplying n-by-n matrices—but no one has found it yet![8]

For more about matrix multiplication and the recent algorithmic improvements, see the following survey paper by Virginia Vassilevska Williams, one of the researchers responsible for the reinvigorated progress in improving this exponent:

[8] Virginia Vassilevska Williams. An overview of the recent progress on matrix multiplication. *ACM SIGACT News*, 43(4), December 2012.

6.5.3 Exercises

The following recurrence relations follow the form of the Master Method. Solve each.

6.109	$T(n) = 4T(n/3) + n^2$	**6.117**	$T(n) = 2T(n/2) + n^2$
6.110	$T(n) = 3T(n/4) + n^2$	**6.118**	$T(n) = 2T(n/2) + n$
6.111	$T(n) = 2T(n/3) + n^4$	**6.119**	$T(n) = 2T(n/4) + n^2$
6.112	$T(n) = 3T(n/3) + n$	**6.120**	$T(n) = 2T(n/4) + n$
6.113	$T(n) = 16T(n/4) + n^2$	**6.121**	$T(n) = 4T(n/2) + n^2$
6.114	$T(n) = 2T(n/4) + 1$	**6.122**	$T(n) = 4T(n/2) + n$
6.115	$T(n) = 4T(n/2) + 1$	**6.123**	$T(n) = 4T(n/4) + n^2$
6.116	$T(n) = 3T(n/3) + 1$	**6.124**	$T(n) = 4T(n/4) + n$

6.125 Solve the recurrence $T(1) = 1$ and $T(n) = 1 + 4T(n/4)$ (see Exercise 6.82, regarding the number of regions defined by quadtrees), using the Master Method.

6.126 Prove that the recurrences $T(n) = aT(\frac{n}{b}) + c \cdot n^k$ and $T(1) = d$ and $S(n) = aS(\frac{n}{b}) + n^k$ and $S(1) = 1$ have the same asymptotic solution, for any constants $a \geq 1, b > 1, c > 0, d > 0$, and $k \geq 0$.

6.127 Consider the Master Method recurrence $T(n) = aT(\frac{n}{b}) + n^k$ and $T(1) = 1$. Using induction, prove the summation (†) from the proof of the Master Theorem: prove that

$$T(n) = n^k \cdot \sum_{i=0}^{\log_b n} \left(\frac{a}{b^k}\right)^i$$

for any n that's an exact power of b.

6.128 The Master Method does not apply for the recurrence $T(n) = 2T(\frac{n}{2}) + n \log n$, but the same idea—considering the summation of all the work in the recursion tree—will still work. Prove that $T(n) = \Theta(n \log^2 n)$ by analyzing the summation analogous to (†).

Each of the following problems gives a brief description of an algorithm for an interesting problem in computer science. (Sometimes the recurrence relation is explicitly written; sometimes it's up to you to write down the recurrence.) For each, state the recurrence (if it's missing) and give a Θ-bound on the running time. If the Master Method applies, you may use it. If not, give a proof by induction.

6.129 The *Towers of Hanoi* is a classic puzzle, as follows. There are three posts (the "towers"); post A starts with n concentric discs stacked from top-to-bottom in order of decreasing radius. We must move all the discs to post B, never placing a disc of larger radius on top of a disc of smaller radius. The easiest way to solve this puzzle is with recursion: (i) recursively move the top $n - 1$ discs from A to C; (ii) move the nth disc from A to B; and (iii) recursively move the $n - 1$ discs from C to B. The total number of moves made satisfies $T(n) = 2T(n - 1) + 1$ and $T(1) = 1$. Prove that $T(n) = 2^n - 1$.

6.130 Suppose we are given a sorted array $A[1 \ldots n]$, and we wish to determine where in A the element x belongs—that is, the index i such that $A[i - 1] < x \leq A[i]$. (Binary Search solves this problem.) Here's a sketch of an algorithm **rootSearch** to solve this problem:

- if n is small (say, less than 100), find the index by brute force. Otherwise:
- define $mileposts := A[\sqrt{n}], A[2\sqrt{n}], A[3\sqrt{n}], \ldots, A[n]$ to be a list of every $(\sqrt{n})$th element of A.
- recursively, find $post :=$ **rootSearch**$(mileposts, x)$.
- return **rootSearch**$(A[(post - 1)\sqrt{n}, \ldots, post\sqrt{n}], x)$.

(Note that **rootSearch** makes *two* recursive calls.) Find a recurrence relation for the running time of this algorithm, and solve it.

6.131 A *van Emde Boas tree* is a recursive data structure (with somewhat similar inspiration to the previous exercise) that allows us to insert, delete, and look up *keys* drawn from a set $U = \{1, 2, \ldots, u\}$ quickly. (It solves the same problem that binary search trees solve, but our running time will be in terms of the size of the universe U rather than in terms of the number of keys stored.) A van Emde Boas tree achieves a running time given by $T(n) = T(\sqrt{n}) + 1$ and $T(1) = 1$. Solve this recurrence. *(Hint: define $R(k) := T(2^k)$. Solving $R(k)$ is easy!)*

6.6 *Chapter at a Glance*

Asymptotics

Asymptotic analysis considers the rate of growth of functions, ignoring multiplicative constant factors and concentrating on the long-run behavior of the function on large inputs.

Consider two functions $f : \mathbb{R}^{\geq 0} \to \mathbb{R}^{\geq 0}$ and $g : \mathbb{R}^{\geq 0} \to \mathbb{R}^{\geq 0}$. Then $f(n) = O(g(n))$ ("f grows no faster than g") if there exist $c > 0$ and $n_0 \geq 0$ such that $f(n) \leq c \cdot g(n)$ for all $n \geq n_0$. Some useful properties of $O(\cdot)$:

- $f(n) = O(g(n) + h(n))$ if and only if $f(n) = O(\max(g(n), h(n)))$.
- if $f(n) = O(g(n))$ and $g(n) = O(h(n))$, then $f(n) = O(h(n))$.
- if $f(n) = O(h_1(n))$ and $g(n) = O(h_2(n))$, then $f(n) + g(n) = O(h_1(n) + h_2(n))$ and $f(n) \cdot g(n) = O(h_1(n) \cdot h_2(n))$.
- a polynomial $p(n) = a_k n^k + \cdots a_1 n + a_0$ satisfies $p(n) = O(n^k)$.
- $\log n = O(n^\varepsilon)$ for any $\varepsilon > 0$.
- for any base b and exponent k, we have $\log_b(n^k) = O(\log n)$.
- for constants $b, c \geq 1$, we have $b^n = O(c^n)$ if and only if $b \leq c$.

There are several other forms of asymptotic notation, to capture other relationships between functions. A function f *grows no slower than* g, written $f(n) = \Omega(g(n))$, if there exist constants $d > 0$ and $n_0 \geq 0$ such that $\forall n \geq n_0 : f(n) \geq d \cdot g(n)$. Two functions f and g satisfy $f(n) = O(g(n))$ if and only if $g(n) = \Omega(f(n))$.

A function f *grows at the same rate as* g, written $f(n) = \Theta(g(n))$, if $f(n) = O(g(n))$ *and* $f(n) = \Omega(g(n))$; it *grows (strictly) slower than* g, written $f(n) = o(g(n))$, if $f(n) = O(g(n))$ but $f(n) \neq \Omega(g(n))$; and it *grows (strictly) faster than* g, written $f(n) = \omega(g(n))$, if $f(n) = \Omega(g(n))$ but $f(n) \neq O(g(n))$. Many of the properties of O have analogous properties for Ω, Θ, o, and ω. One possibly surprising point is that there are functions that are *incomparable*: there are functions f and g such that *neither* $f(n) = O(g(n))$ *nor* $f(n) = \Omega(g(n))$.

Asymptotic Analysis of Algorithms

Our main interest in asymptotics is in the *analysis of algorithms*, so that we can make statements about which of two algorithms that solve the same problem is faster. The *running time* of an algorithm is a count of the number of primitive steps that the algorithm takes to complete on a particular input. (Think of one machine instruction as a primitive step.)

We generally evaluate the efficiency of an algorithm $\mathcal{A}$ using *worst-case analysis*: as a function of n, how many primitive steps does $\mathcal{A}$ take *on the input of size n for which $\mathcal{A}$ is the slowest*. (A primary goal of algorithmic analysis is to provide a guarantee on the running time of an algorithm, so we will be pessimistic.) We can also analyze the *space* used by an algorithm, in the same way. Sometimes we will instead consider *average-case running time* of an algorithm $\mathcal{A}$, which computes the running time of $\mathcal{A}$, averaged over all inputs of size n. Almost never will we consider an algorithm's running time on the input of size n for which $\mathcal{A}$ is the fastest (known as *best-case analysis*); this type of analysis is rarely used.

Recurrence Relations: Analyzing Recursive Algorithms

Typically, for nonrecursive algorithms, we compute the running time by inspecting the algorithm and writing down a summation corresponding to the operations done in each iteration of each loop, summed over the iterations, and then simplifying. For recursive algorithms, we typically record the work using a *recurrence relation* that expresses the (worst-case) running time on inputs of size n in terms of the (worst-case) running time on inputs of size less than n. (For small inputs, the running time is a constant—say, $T(1) = c$.) For example, ignoring floors and ceilings, $T(1) = c$ and $T(n) = 2T(\frac{n}{2}) + cn$ is the recurrence relation for Merge Sort. (Almost always, we can safely ignore floors and ceilings.)

A *solution* to a recurrence relation is a closed-form (nonrecursive) expression for $T(n)$. Recurrence relations can be solved by conjecturing a solution and proving that conjecture correct by induction.

A recurrence relation can be represented using a *recursion tree,* where each node is annotated with the work that is performed there, aside from the recursive calls. Recurrence relations can also be solved by summing up all of the work contained within the recursion tree.

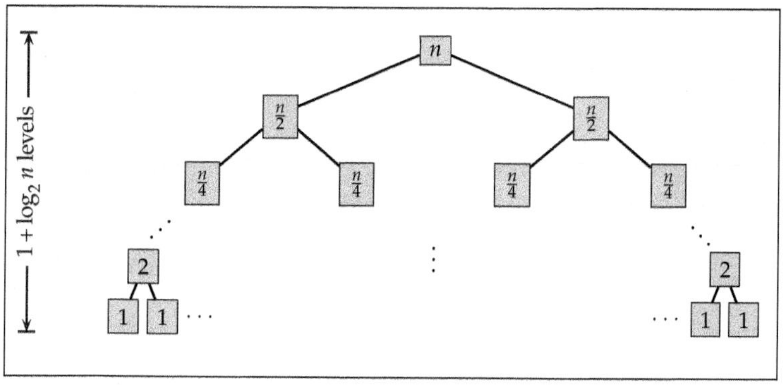

Recurrence Relations: The Master Method

A particularly common type of recurrence relation is one of the form

$$T(n) = aT(\tfrac{n}{b}) + c \cdot n^k,$$

for constants $a \geq 1, b > 1, c > 0$, and $k \geq 0$. This type of recurrence arises in divide-and-conquer algorithms that solve an instance of size n by making a different recursive calls on inputs of size $\frac{n}{b}$, and reconstructing the solution to the given instance in $\Theta(n^k)$ time. The *Master Theorem* states that the solution to any such recurrence relation is given by:

1. if $b^k < a$, then $T(n) = \Theta(n^{\log_b(a)})$. *"The leaves dominate."*
2. if $b^k = a$, then $T(n) = \Theta(n^k \cdot \log n)$. *"All levels are equal."*
3. if $b^k > a$, then $T(n) = \Theta(n^k)$. *"The root dominates."*

The proof follows by building the recursion tree, and summing the work at each level of the tree; the cases correspond to whether the work increases exponentially, decreases exponentially, or stays constant across levels of the tree.

Key Terms and Results

Key Terms

ASYMPTOTICS

- asymptotic analysis
- O (big oh)
- Ω (big omega)
- Θ (big theta)
- ω (little omega)
- o (little oh)

ANALYSIS OF ALGORITHMS

- running time
- worst-case analysis
- average-case analysis
- best-case analysis

RECURRENCE RELATIONS

- recurrence relation
- recursion tree
- iterating a recurrence

MASTER METHOD

- Master Theorem
- "the leaves dominate"
- "all levels are equal"
- "the root dominates"

Key Results

ASYMPTOTICS

1. Some sample useful properties of $O(\cdot)$:
   - $f(n) = O(g(n) + h(n)) \Leftrightarrow f(n) = O(\max(g(n), h(n)))$.
   - $O(\cdot)$ is transitive.
   - any degree-k polynomial satisfies $p(n) = O(n^k)$.
   - $\log n = O(n^\varepsilon)$ for any $\varepsilon > 0$.
   - if $f(n) = O(g(n))$ then $\log f(n) = O(\log g(n))$.
   - for any b and k, we have $\log_b(n^k) = O(\log n)$.
   - for constants $b, c \geq 1$, we have $b^n = O(c^n) \Leftrightarrow b \leq c$.
2. Two functions f and g satisfy $f(n) = O(g(n))$ if and only if $g(n) = \Omega(f(n))$.
3. There are pairs of functions f and g such that neither $f(n) = O(g(n))$ nor $f(n) = \Omega(g(n))$.

ANALYSIS OF ALGORITHMS

1. We generally evaluate the efficiency of an algorithm $\mathcal{A}$ using worst-case analysis: what happens (asymptotically) to the number of steps consumed by $\mathcal{A}$ as function of the input size n *on the input of size n for which $\mathcal{A}$ is the slowest*?
2. Typically we can analyze the running time of a nonrecursive algorithm by simple counting and manipulation of summations.

RECURRENCE RELATIONS

1. The running time of a recursive algorithm can be expressed using a recurrence relation, which can be solved by figuring out a conjecture of a closed-form formula for the relation, and then verifying by induction.

MASTER METHOD

1. Recurrence relations of the form $T(n) = aT(\frac{n}{b}) + cn^k$ (and $T(1) = c$) can be solved using the Master Method:

   Case 1: if $b^k < a$, then $T(n) = \Theta(n^{\log_b(a)})$.
   Case 2: if $b^k = a$, then $T(n) = \Theta(n^k \cdot \log n)$.
   Case 3: if $b^k > a$, then $T(n) = \Theta(n^k)$.

7
Number Theory

In which, after becoming separated, our heroes arrange a place to meet, by sending messages that stay secret even as snooping spies listen in.

7.1 *Why You Might Care*

> When you can measure what you are speaking about, and express it in numbers, you know something about it; but when you cannot express it in numbers, your knowledge is of a meager and unsatisfactory kind.
>
> Sir William Thomson, Lord Kelvin (1824–1907)

A chapter about numbers (particularly when it's so far along in this book!) probably seems a little bizarre—after all, what is there to say about numbers that you didn't figure out by elementary school?!? But, more so than any other chapter of the book, the technical material in this chapter leads directly to a single absolutely crucial (and ubiquitous!) modern application of computer science: *cryptography*, which deals with protocols to allow multiple parties to communicate securely, even in the presence of eavesdropping adversaries (or worse!). Cryptographic systems are used throughout our daily lives—both in the security layers that connect us as users to servers (for example, in banking online or in registering for courses at a college), and in the backend systems that, we hope, protect our data even when we aren't interacting with it.

cryptography (Greek): *kryptos* "concealed/secret" + *graph* "writing."

Our goal in this chapter will be to build up the technical machinery necessary to define and understand the *RSA cryptosystem*, one of the most commonly used cryptographic systems today. (RSA is named after the initials of its three discoverers, Rivest, Shamir, and Adleman.) By the end of the chapter, in Section 7.5, we'll be able to give a full treatment of RSA, along with sketched outlines of a few other important ideas from cryptography. (Later in the book, in Chapter 9, we'll also encounter the historical code*breaking* work of Alan Turing and colleagues, which deciphered the German encryption in World War II—a major part of the allied victory. See p. 960.)

To get there, we'll need to develop some concepts and tools from *number theory*. ("Number theory" is just a slightly fancy name for "arithmetic on integers.") Our focus will be on *modular arithmetic*: that is, the numbers on which we'll be doing arithmetic will be a set of integers $\{0, 1, 2, \ldots, n-1\}$, where—like on a clock—the numbers "wrap around" from $n-1$ back to 0. In other words, we'll interpret numerical expressions *modulo n*, always considering each expression via its remainder when we divide by n. We begin in Section 7.2 with formal definitions of modular arithmetic, and the adaptation of some basic ideas from elementary-school arithmetic to this new setting. We'll then turn in Section 7.3 to *primality* (when a number has no divisors other than 1 and itself) and *relative primality* (when two numbers have no common divisors other than 1). Modular arithmetic begins to diverge more substantially when we start to think about division: there's no integer that's one fifth of 3 ... but, on a clock where we treat 12:00 as 0, there *is* an integer that's a fifth of 3—namely 5, because $5+5+5$ is 3 (because 3:00pm is 15 hours after midnight—so $5 \cdot 3$ *is* 3, modulo 12). In Section 7.4, we'll explore exactly what division means in modular arithmetic—and some special features of division that arise when n is a prime number.

As we go, we'll see a few other applications of number theory: to error-correcting codes, secret sharing, and the apparently unrelated task of generating all 4-letter sequences (AAAA to ZZZZ). And, finally, we'll put the pieces together to explore RSA.

7.2 Modular Arithmetic

> Among those whom I like or admire, I can find no common denominator, but among those whom I love, I can: all of them make me laugh.
>
> W. H. Auden (1907–1973)

We will start with a few reminders of some basic arithmetic definitions from Chapter 2—about multiplication, division, and modular arithmetic—as these concepts are the foundations for all the work that we'll do in this chapter. We'll also introduce a few algorithms for *computing* these basic arithmetic quantities, including one of the oldest known algorithms: the *Euclidean algorithm*, from about 2300 years ago, which computes the greatest common divisor of two integers n and m (that is, the largest integer that evenly divides both n and m).

7.2.1 Remainders: A Reminder

Let's start with a few simple facts about integers. Every integer is 0 or 1 more than some even number. Every integer is 0, 1, or 2 more than a multiple of three. Every integer is at most 3 more than a multiple of four. And, in general, for any integer $k \geq 1$, every integer is r more than a multiple of k, for some $r \in \{0, 1, \ldots, k-1\}$. We'll begin with a precise statement and proof of the general version of this property:

Theorem 7.1 (Floors and Remainders: "The Division Theorem")
Let $k \geq 1$ and n be integers. Then there exist integers d and r such that (i) $0 \leq r < k$, and (ii) $kd + r = n$. Furthermore, the values of d and r satisfying (i) and (ii) are unique.

Before we prove the theorem, let's look at a few examples of what it claims:

Example 7.1 (Some examples of the Division Theorem)
For $k = 202$ and $n = 379$, the theorem states that there exist integers $r \in \{0, 1, \ldots, 201\}$ and d with $202d + r = 379$. Specifically, those values are $r = 177$ and $d = 1$, because $202 \cdot 1 + 177 = 379$.

Here are a few more examples, still with $k = 202$:

$n = 55057$	$n = 507$	$n = 177$	$n = 404$	$n = -507$	$n = -404$
$d = 272$	$d = 2$	$d = 0$	$d = 2$	$d = -3$	$d = -2$
$r = 113$	$r = 103$	$r = 177$	$r = 0$	$r = 99$	$r = 0$

You can verify that, in each of these six columns, indeed we have $202d + r = n$.

Now let's give a proof of the general result:

Proof of Theorem 7.1. Consider a fixed integer $k \geq 1$. Let $P(n)$ denote the claim

$$P(n) := \text{there exist integers } d \text{ and } r \text{ such that } 0 \leq r < k \text{ and } kd + r = n.$$

We must prove that $P(n)$ holds for all integers n. We'll first prove the result for nonnegative n (by strong induction on n), and then show the claim for $n < 0$ (making use of the result for nonnegative n).

Problem-solving tip: To prove that a property is true for all inputs, it often turns out to be easier to *first* prove a special case and then use that special case to show that the property holds in general. (Another example: it's probably easier to analyze the performance of Merge Sort on inputs whose size is an exact power of 2, and to then generalize to arbitrary input sizes.)

Case I: $n \geq 0$. We'll prove that $P(n)$ holds for all $n \geq 0$ by strong induction on n.

- For the base cases ($0 \leq n < k$), we simply select $d := 0$ and $r := n$. Indeed, these values guarantee that $0 \leq r < k$ and $kd + r = k \cdot 0 + n = 0 + n = n$.
- For the inductive case ($n \geq k$), we assume the inductive hypotheses—namely, we assume $P(n')$ for any $0 \leq n' < n$—and we must prove $P(n)$. Because $n \geq k$ and $k > 0$, it is immediate that $n' := n - k$ satisfies $0 \leq n' < n$. Thus we can apply the inductive hypothesis $P(n')$ to conclude that there exist integers d' and r' such that $0 \leq r' < k$ and $kd' + r' = n'$. Select $d := d' + 1$ and $r := r'$. Thus, indeed, $0 \leq r < k$ and

$$\begin{aligned} kd + r &= k(d'+1) + r' && \text{definition of } d \text{ and } r \\ &= kd' + k + r' && \text{distributive property} \\ &= n' + k && n' = kd' + r', \text{ by definition} \\ &= n. && \text{definition of } n' = n - k \end{aligned}$$

Case II: $n < 0$. To show that $P(n)$ holds for an arbitrary $n < 0$, we will make use of Case I. Let r' and d' be the integers guaranteed by $P(-n)$, so that $kd' + r' = -n$. We consider two cases based on whether $r' = 0$:

Case IIA: $r' \neq 0$. Then let $d := -d' - 1$ and let $r := k - r'$. (Because $k > r' > 0$, we have $0 < k - r' < k$.) Thus

$$\begin{aligned} kd + r &= k(-d' - 1) + k - r' && \text{definition of } d \text{ and } r \\ &= -kd' - k + k - r' \\ &= -(kd' + r) \\ &= -(-n) = n. && \text{definition of } d' \text{ and } r' \end{aligned}$$

Case IIB: $r' = 0$. Then let $d := -d'$ and $r := r' = 0$. Therefore

$$\begin{aligned} kd + r &= -d'k + r' && \text{definition of } d \text{ and } r \\ &= -(-n) = n. && \text{definition of } d' \text{ and } r' \end{aligned}$$

We have thus proven that $P(n)$ holds for all integers n: Case I handled $n \geq 0$, and Case II handled $n < 0$. (We have not yet proven the uniqueness of the integers r and d; this proof of uniqueness is left to you in Exercise 7.4.) □

This theorem now allows us to give a more careful definition of modular arithmetic. (In Definition 2.9, we gave the slightly less formal definition of $n \bmod k$ as the remainder when we divide n by k.)

> **Definition 7.1 (Modulus (reprise))**
> *For integers $k > 0$ and n, the quantity $n \bmod k$ is the unique integer r such that $0 \leq r < k$ and $kd + r = n$ for some integer d (whose existence is guaranteed by Theorem 7.1).*

Incidentally, the integer d whose existence is guaranteed by Theorem 7.1 is $\lfloor n/k \rfloor$: for any $k \geq 1$, we can write the integer n as

$$n = \left\lfloor \frac{n}{k} \right\rfloor \cdot k + (n \bmod k).$$

> **Taking it further:** One of the tasks that we can accomplish conveniently using modular arithmetic is *base conversion* of integers. We're used to writing numbers in decimal ("base 10"), where each digit is "worth" a factor of 10 more than the digit to its right. (For example, the number we write "31" means $1 \cdot 10^0 + 3 \cdot 10^1 = 1 + 30$.) Computers store numbers in binary ("base 2") representation, and we can convert between bases using modular arithmetic. For more, see the discussion on p. 714.

7.2.2 *Computing $n \bmod k$ and $\lfloor \frac{n}{k} \rfloor$*

So far, we've taken arithmetic operations for granted—ignoring *how* we'd figure out the numerical value of an arithmetic expression like $2^{1024} - 3^{256} \cdot 5^{202}$, which is simple to write—but not so instantaneous to calculate. (Quick! Is $2^{1024} - 3^{256} \cdot 5^{202}$ evenly divisible by 7?) Indeed, many of us spent a lot of time in elementary-school math classes learning algorithms for basic arithmetic operations like addition, multiplication, long division, and exponentiation (even if back then nobody told us that they were called algorithms).

Thinking about algorithms for some basic arithmetic operations will be useful, for multiple reasons: because they're surprisingly relevant for proving some useful facts about modular arithmetic, and because computing them efficiently turns out to be crucial in the cryptographic systems that we'll explore in Section 7.5.

```
mod-and-div(n, k):
Input: integers n ≥ 0 and k ≥ 1
Output: n mod k and ⌊n/k⌋
r := n; d := 0
while r ≥ k:
    r := r − k; d := d + 1
return r, d
```

Figure 7.1: An algorithm to compute $n \bmod k$ and $\lfloor n/k \rfloor$.

We'll start with the algorithm shown in Figure 7.1 that computes $n \bmod k$ (and simultaneously computes $\lfloor n/k \rfloor$ too). The very basic idea for this algorithm was implicit in the proof of Theorem 7.1: we repeatedly subtract k from n until we reach a number in the range $\{0, 1, \ldots, k-1\}$.

Some programming languages—Pascal, for one (admittedly dated) example—use `div` to denote integer division, so that `15 div 7` is 2.

> **Example 7.2 (An example of mod-and-div)**
> Let's compute **mod-and-div**(64, 5). We start with $r := 64$ and $d := 0$, and repeatedly decrease r by 5 and increase d by 1 until $r < 5$. Here are the values in each iteration:
>
> | r | 64 | 59 | 54 | 49 | 44 | 39 | 34 | 29 | 24 | 19 | 14 | 9 | 4 |
> |---|---|---|---|---|---|---|---|---|---|---|---|---|---|
> | d | 0 | 1 | 2 | 3 | 4 | 5 | 6 | 7 | 8 | 9 | 10 | 11 | 12 . |
>
> Thus **mod-and-div**(64, 5) returns 4 and 12—and, indeed, we can write $64 = 12 \cdot 5 + 4$, where $4 = 64 \bmod 5$ and $12 = \lfloor 64/5 \rfloor$.
>
> Similarly, **mod-and-div**(20, 17) starts with $d = 0$ and $r = 20$, and executes one (and only one) iteration of the loop, returning $d = 1$ and $r = 3$.

The **mod-and-div** algorithm is fairly intuitive, if fairly slow: it simply keeps removing multiples of k from n until there are no multiples of k left to remove. (For simplicity, the algorithm as written in Figure 7.1 only handles the case where $n \geq 0$; you'll extend the algorithm to handle negative n in Exercise 7.10. See Example 7.1 for some examples of the Division Theorem with $n < 0$.)

Lemma 7.2 (Correctness and efficiency of mod-and-div)
For any integers $n \geq 0$ and $k \geq 1$, calling **mod-and-div**(n, k) *returns* $n \bmod k$ *and* $\lfloor n/k \rfloor$*, using a total of* $\Theta(\lfloor n/k \rfloor)$ *arithmetic operations.*

Proof. We claim that, throughout the execution of the algorithm, we have $dk + r = n$ (and also $r \geq 0$). This fact is easy to see by induction on the number of iterations of the **while** loop in **mod-and-div**: it's true before the loop starts (when $r = n$ and $d = 0$), and if it's true before an iteration of the **while** loop then it's true after that iteration (when dk has increased by k, and r has decreased by k). Furthermore, when the **while** loop terminates, we also have that $r < k$. Thus the returned values satisfy $dk + r = n$ and $0 \leq r < k$—precisely as required by Definition 7.1.

The total number of iterations of the **while** loop is exactly the returned value of $d = \lfloor n/k \rfloor$, and we do three arithmetic operations per iteration: a comparison (is $r \geq k$?), a subtraction (what's $r - k$?), and an addition (what's $d + 1$?). Thus the total number of arithmetic operations is $3 \lfloor n/k \rfloor + \Theta(1) = \Theta(\lfloor n/k \rfloor)$. □

Should we consider the **mod-and-div** algorithm from Figure 7.1 fast? Let's think about how long this algorithm would take to determine whether, say, $n := 123{,}456{,}789$ is divisible by 7. (It isn't: $n = 17{,}636{,}684 \cdot 7 + 1$.) Our algorithm would take over 10 million iterations ($\lfloor n/7 \rfloor > \frac{70{,}000{,}000}{7} = 10{,}000{,}000$) iterations to compute the answer—and that seems (and is!) very slow. One way to think about the **mod-and-div**(n, k) algorithm is that it performs a *linear search* for the integer d such that $kd \leq n < (k+1)d$: we keep increasing d by one until this property holds. We could instead give a much faster algorithm based on *binary search* to find that value of d. (See Exercises 7.11–7.16.) The improved algorithm requires only logarithmically many arithmetic operations—an exponential improvement over **mod-and-div**.

Taking it further: What should count as an efficient algorithm when the inputs are numbers? In previous chapters, we've talked about the generally accepted definition of *efficient* as meaning "requiring a number of steps that is polynomial in the size of the input." That is, on an input of size n, our algorithm should run in at most $O(n^c)$ steps, for some fixed c. (See p. 628.)

So why did we say that an algorithm like **mod-and-div** isn't efficient? After all, on input n and k, the algorithm took only about n/k steps. (See Lemma 7.2.) But the key point is that an algorithm that takes a numerical input n does *not* receive an input of size n. The number 123,456,789 takes only 9 characters (the nine digits in the number!) to write down—not 123,456,789 characters. (Unless you wrote down the numbers in *unary,* using tally marks instead of digits: 卌卌卌卌卌卌卌卌卌 · · · .)

Generally, an algorithm that takes a number n as input receives that number n *written in binary.* The binary representation of n requires $\lceil \log_2 n \rceil$ bits to represent it. As usual, we consider an algorithm to be efficient if it takes time that's polynomial in the number of bits of the input—so we consider an algorithm that takes a number n as input to be efficient *if it requires a number of operations that is at most* $(\log_2 n)^c$ *for some fixed* c. (That is, the algorithm should run in time that is *polylogarithmic* in n.) Every grade-school algorithm that you learned for arithmetic—addition, subtraction, multiplication, long division, etc.—was efficient, requiring you to do a number of operations proportional to the number of digits in the numbers, and *not* to the value of the numbers themselves.

7.2.3 *Congruences, Divisors, and Common Divisors*

We argued in Lemma 7.2 that **mod-and-div**(n,k), which repeatedly subtracts k from n in a loop, correctly computes the value of n mod k. We gave a proof by induction in Lemma 7.2, but we could have instead argued for the correctness of the algorithm, perhaps more intuitively, via the following fact:

For any integers $a \geq 0$ and $k \geq 1$, we have $(a+k) \bmod k = a \bmod k$.

That is, the remainder when we divide an integer a by k isn't changed by adding an exact multiple of k to a. This property follows from the definition of mod, but it's also a special case of a useful general property of modular arithmetic, which we'll state (along with some other similar facts) in Theorem 7.3. Here are a few examples of this more general property:

Example 7.3 (The mod of a sum, and the sum of the mods)
Consider the following expressions of the form $(a+b) \bmod k$.

- $(17+43) \bmod 7 = 60 \bmod 7 = 4$. (Note $17 \bmod 7 = 3$, $43 \bmod 7 = 1$, and $3+1=4$.)
- $(18+42) \bmod 9 = 60 \bmod 9 = 6$. (Note $18 \bmod 9 = 0$, $42 \bmod 9 = 6$, and $0+6=6$.)
- $(25+25) \bmod 6 = 50 \bmod 6 = 2$. (Note $25 \bmod 6 = 1$, $25 \bmod 6 = 1$, and $1+1=2$.)

At this point it might be tempting to conjecture that $(a+b) \bmod k$ is always equal to $(a \bmod k)+(b \bmod k)$, but be careful—this claim has a bug, as this example shows:

- $(18+49) \bmod 5 = 67 \bmod 5 = 2$. (Note $18 \bmod 5 = 3$, $49 \bmod 5 = 4$, but $3+4 \neq 2$.)

Instead, it turns out that $(a+b) \bmod k = [(a \bmod k)+(b \bmod k)] \bmod k$—we had to add an "extra" mod k at the end.

Here are some of the useful general properties of modular arithmetic:

Theorem 7.3 (Properties of modular arithmetic)
For integers a and b and $k > 0$:

$$k \bmod k = 0 \tag{7.3.1}$$
$$a+b \bmod k = [(a \bmod k)+(b \bmod k)] \bmod k \tag{7.3.2}$$
$$ab \bmod k = [(a \bmod k)\cdot(b \bmod k)] \bmod k \tag{7.3.3}$$
$$a^b \bmod k = [(a \bmod k)^b] \bmod k. \tag{7.3.4}$$

We'll omit proofs of these properties, though we could give a formal proof based on the definitions of mod. (Exercise 7.17 asks you to give a formal proof for one of these properties, namely (7.3.2).) Again notice the "extra" mod k at the end of the last three of these equations—it is not the case that $ab \bmod k = (a \bmod k)\cdot(b \bmod k)$ in general. For example, $14 \bmod 6 = 2$ and $5 \bmod 6 = 5$, but $(2\cdot 5) \bmod 6 = 4 \neq 2\cdot 5$.

In the cryptographic applications that we will explore later in this chapter, it will turn out to be important to perform "modular exponentiation" efficiently—that is, we'll need to compute b^e mod n very quickly, even when e is fairly large. Fortunately, (7.3.4) will help us do this computation efficiently; see Exercises 7.23–7.25.

Congruences

We've now talked a little bit (in Theorem 7.3, for example) about two numbers a and b that have the same remainder when we divide them by k—that is, with $a \bmod k = b \bmod k$. There's useful terminology, and notation, for this kind of equivalence:

Definition 7.2 (Congruence)
Two integers a and b are congruent mod k, *written $a \equiv_k b$, if $a \bmod k = b \bmod k$.*

Typically $a \equiv_k b$ is read as "a is equivalent to b mod k" or "a is congruent to b mod k." If you're reading the statement $a \equiv_k b$ out loud, it's polite to pause slightly, as if there were a comma, before the "mod k" part.

Taking it further: Some people write $a \equiv_k b$ using the notation

$$a \equiv b \pmod{k}.$$

This notation is used to mean the same thing as our notation $a \equiv_k b$, but note the somewhat unusual precedence in this alternate notation: it says that

$$\left[a \equiv b\right] \pmod{k}$$

(and it does not, as it might appear, say that the quantity a and the quantity $[b \bmod k]$ are equivalent).

Divisors, factors, and multiples

We now return to the *divisibility* of one number by another, when the first is an exact multiple of the second. As with the previous topics in this section, we gave some preliminary definitions in Chapter 2 of divisibility (and related terminology), but we'll again repeat the definitions here, and also go into a little bit more detail.

Definition 7.3 (Divisibility, Factors, and Multiples (reprise))
For two integers $k > 0$ and n, we write $k \mid n$ to denote the proposition that $n \bmod k = 0$. If $k \mid n$, we say that k divides *n (or that k* evenly divides *n), that n is a* multiple *of k, and that k is a* factor *of n.*

(For example, we can say that $42 \mid 714$, that 6 and 17 are factors of 714, and that 714 is a multiple of 7.) Here are a few useful properties of division:

Theorem 7.4 (Properties of divisibility)
For integers a and b and c:

$$a \mid 0 \tag{7.4.1}$$

$$1 \mid a \tag{7.4.2}$$

$$a \mid a \tag{7.4.3}$$

$$a \mid b \text{ and } b \mid c \Rightarrow a \mid c \tag{7.4.4}$$

$$a \mid b \text{ and } b \mid a \Rightarrow a = b \text{ or } a = -b \tag{7.4.5}$$

$$a \mid b \text{ and } a \mid c \Rightarrow a \mid (b + c) \tag{7.4.6}$$

$$a \mid b \Rightarrow a \mid bc \tag{7.4.7}$$

$$ab \mid c \Rightarrow a \mid c \text{ and } b \mid c \tag{7.4.8}$$

These properties generally follow fairly directly from the definition of divisibility. A few are left to you in the exercises, and we'll address a few others in Chapter 8, which introduces relations. (Facts (7.4.3), (7.4.4), and a version of (7.4.5) are certain standard properties of some relations that the "divides" relation happens to have: *reflexivity, transitivity,* and so-called *antisymmetry.* See Chapter 8.) To give the flavor of these arguments, here's one of the proofs, that $ab \mid c$ implies that $a \mid c$ and $b \mid c$:

Proof of (7.4.8). Assume $ab \mid c$. Then, by definition of mod (and by Theorem 7.1), there exists an integer k such that $c = (ab) \cdot k$. Taking both sides mod a, we have

$$
\begin{aligned}
c \bmod a &= abk \bmod a && \textit{k is the integer such that } c = (ab) \cdot k \\
&= [(a \bmod a) \cdot (bk \bmod a)] \bmod a && (7.3.3) \\
&= [0 \cdot (bk \bmod a)] \bmod a && (7.3.1) \\
&= 0 \bmod a && 0 \cdot x = 0 \textit{ for any } x \\
&= 0. && 0 \bmod a = 0 \textit{ for any } a
\end{aligned}
$$

Thus $c \bmod a = 0$, so $a \mid c$. Analogously, because $b \cdot (ak) = c$, we have that $b \mid c$ too. □

Greatest common divisors and least common multiples

We now turn to our last pair of definitions involving division: for two integers, we'll be interested in two related quantities—the largest number that divides both of them, and the smallest number that they both divide.

Definition 7.4 (Greatest Common Divisor (GCD))
The greatest common divisor *of two positive integers n and m, denoted* $\gcd(n, m)$*, is the largest $d \in \mathbb{Z}^{\geq 1}$ such that $d \mid n$ and $d \mid m$.*

Definition 7.5 (Least Common Multiple (LCM))
The least common multiple *of two positive integers n and m, denoted* $\text{lcm}(n, m)$*, is the smallest $d \in \mathbb{Z}^{\geq 1}$ such that $n \mid d$ and $m \mid d$.*

Here are some examples of both GCDs and LCMs, for a few pairs of small numbers:

Example 7.4 (Examples of GCDs)
The GCD of 6 and 27 is 3, because 3 divides both 6 and 27 (and no integer $k \geq 4$ divides both). Similarly, we have $\gcd(1, 9) = 1$, $\gcd(12, 18) = 6$, $\gcd(202, 505) = 101$, and $\gcd(11, 202) = 1$.

Example 7.5 (Examples of LCMs)
The LCM of 6 and 27 is 54, because 6 and 27 both divide 54 (and no $k \leq 53$ is divided by both). Similarly, we have $\text{lcm}(1, 9) = 9$, $\text{lcm}(12, 18) = 36$, $\text{lcm}(202, 505) = 1010$, and $\text{lcm}(11, 202) = 2222$.

Both of these concepts should be (at least vaguely!) familiar from elementary school, specifically from when you learned about how to manipulate fractions:

- We can rewrite the fraction $\frac{38}{133}$ as $\frac{2}{7}$, by dividing both numerator and denominator by the common factor 19—and we can't reduce it further because 19 is the *greatest* common divisor of 38 and 133. (We have "reduced the fraction to lowest terms.")

- We can rewrite the sum $\frac{5}{12} + \frac{7}{18}$ as $\frac{15}{36} + \frac{14}{36}$ (which equals $\frac{29}{36}$) by rewriting both fractions with a denominator that's a common multiple of the denominators of the two addends—and we couldn't have chosen a smaller denominator, because 36 is the *least* common multiple of 12 and 18. (We have "put the fractions over the lowest common denominator.")

In the remainder of this section, we'll turn to the task of *efficiently computing* the greatest common divisor of two integers. (Using this algorithm, we can also find least common multiples quickly, because GCDs and LCMs turn out to be closely related quantities: for any integers a and b, we have $\text{lcm}(a, b) \cdot \gcd(a, b) = a \cdot b$.)

7.2.4 *Computing Greatest Common Divisors*

The "obvious" way to compute the greatest common divisor of two positive integers n and m is to try all candidate divisors $d \in \{1, 2, \ldots, \min(n, m)\}$ and to return the largest value of d that indeed evenly divides both n and m. This algorithm is slow—very slow!—but there is a faster way to solve the problem. Amazingly, a faster algorithm for computing GCDs has been known for approximately 2300 years: the *Euclidean algorithm*, named after the Greek geometer Euclid, who lived in the 3rd century BCE. (Euclid is also the namesake of the *Euclidean distance* between points in the plane—see Exercise 2.174—among a number of other things in mathematics.) The algorithm is shown in Figure 7.2.

```
Euclid(n, m):
Input: positive integers n and m ≥ n
Output: gcd(n, m)
if m mod n = 0 then
    return n
else
    return Euclid(m mod n, n)
```

Figure 7.2: The Euclidean algorithm for GCDs.

Taking it further: Euclid described his algorithm in his book *Elements*, from c. 300 BCE, a multivolume opus covering the fundamentals of mathematics, particularly geometry, logic, and proofs. Most people view the Euclidean algorithm as the oldest nontrivial algorithm that's still in use today; there are some older not-quite-fully-specified procedures for basic arithmetic operations like multiplication that date back close to 2000 BCE, but they're not quite laid out as algorithms.

Donald Knuth—the 1974 Turing Award winner, the inventor of TEX (the underlying system that was used to typeset virtually all scholarly materials in computer science—and this book!), and a genius of expository writing about computer science in general and algorithms in particular—describes the history of the Euclidean algorithm (among many other things!) in *The Art of Computer Programming*,[1] his own modern-day version of a multivolume opus covering the fundamentals of computer science, particularly algorithms, programming, and proofs.

Among the fascinating things that Knuth points out about the Euclidean algorithm is that Euclid's "proof" of correctness only handles the case of up to three iterations of the algorithm—because, Knuth argues, Euclid predated the idea of mathematical induction by hundreds of years. (And Euclid's version of the algorithm is quite hard to read, in part because Euclid didn't have a notion of zero, or the idea that 1 is a divisor of any positive integer n.)

[1] Donald E. Knuth. *The art of computer programming: Seminumerical algorithms (Volume 2)*. Addison-Wesley Longman, 3rd edition, 1997.

"Knuth" rhymes with "Duluth" (a city in Minnesota that Minnesotans make fun of for having harsh weather): the "K" is pronounced.

Here are three small examples of the Euclidean algorithm in action:

Example 7.6 (GCDs using the Euclidean Algorithm)
Let's compute the GCD of 17 and 42.

$$\begin{aligned}
\textbf{Euclid}(17, 42) &= \textbf{Euclid}(\underbrace{42 \bmod 17}_{=8}, 17) && \textit{42 mod 17 = 8} \neq 0\textit{, so we're in the else case.} \\
&= \textbf{Euclid}(\underbrace{17 \bmod 8}_{=1}, 8) && \textit{17 mod 8 = 1} \neq 0\textit{, so we're in the else case again.} \\
&= 1. && \textit{8 mod 1 = 0, so we're done, and we return 1.}
\end{aligned}$$

Indeed, the only positive integer that divides both 17 and 42 is 1, so $\gcd(17, 42) = 1$.
Here's another example, for 48 and 1024:

$$\begin{aligned}
\textbf{Euclid}(48, 1024) &= \textbf{Euclid}(\underbrace{1024 \bmod 48}_{=16}, 48) && \textit{1024 mod 48 = 16} \neq 0\textit{, so we're in the else case.} \\
&= 16. && \textit{48 mod 16 = 0, so we return 16.}
\end{aligned}$$

And here's one last example (written more compactly), for 91 and 287:

$$\textbf{Euclid}(91, 287) = \textbf{Euclid}(\underbrace{287 \bmod 91}_{=14}, 91) = \textbf{Euclid}(\underbrace{91 \bmod 14}_{=7}, 14) = 7.$$

Before we try to prove the correctness of the Euclidean algorithm, let's spend a few moments on the intuition behind it. The basic idea is that any common divisor of two numbers must also evenly divide their difference. For example, does 7 divide both 63 and 133? If so, then it would have to be the case that $7 \mid 63$ *and* that 7 also divides the "gap" between 133 and 63. (That's because $63 = 7 \cdot 9$, and if $7k = 133$, then $7(k - 9) = 133 - 63$.) More generally, suppose that d is a common divisor of n and $m \geq n$. Then it must be the case that d divides $m - cn$, *for any integer c where* $cn < m$. In particular, d divides $m - \lfloor \frac{m}{n} \rfloor \cdot n$; that is, d divides $m \bmod n$. (We've only argued that if d is a common divisor of n and m then d must also divide $m \bmod n$, but actually the converse holds too; we'll formalize this fact in the proof.) See Figure 7.3 for a visualization of this idea.

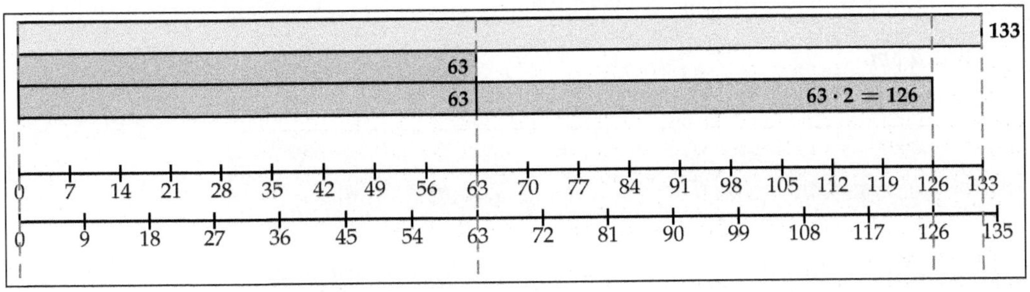

Figure 7.3: The intuition behind the Euclidean algorithm: d is a common divisor of 63 and 133 if and only if d also divides $133 - 63$ and $133 - 63 \cdot 2 = 133 - 126$. Indeed $d = 7$ is a common divisor of 63 and 133, but 9 is not (because 9 does not divide $133 - 126 = 7$).

Making the intuition formal

We will now make this intuition formal, and give a full proof of the correctness of the Euclidean algorithm: that is, we will establish that $\textbf{Euclid}(n, m) = \gcd(n, m)$ for any positive integers n and $m \geq n$, with a proof by induction. There's a crucial lemma that we'll need to prove first, based on the intuition we just described: we need to show that for any n and $m \geq n$ where $m \bmod n \neq 0$, we have $\gcd(n, m) = \gcd(n, m \bmod n)$. We will prove this fact by proving that *the common divisors of* $\{n, m\}$ are identical to *the common divisors of* $\{n, m \bmod n\}$. (Thus the *greatest* common divisor of these two pairs of integers will be identical.)

Lemma 7.5 (When $n \nmid m$, the same divisors of n divide m and $m \bmod n$)
Let n and m be positive integers such that $n \leq m$ and $n \nmid m$. Let $d \mid n$ be an arbitrary divisor of n. Then $d \mid m$ if and only if $d \mid (m \bmod n)$.

Here's a concrete example before we prove the lemma:

Example 7.7 (An example of Lemma 7.5)
Consider $n = 42$ and $m = 98$. Then $n \leq m$ and $n \nmid m$, as Lemma 7.5 requires. The divisors of 42 are $\{1, 2, 3, 6, 7, 14, 21, 42\}$. Of these divisors, the ones that also divide 98 are $\{1, 2, 7, 14\}$.

The lemma claims that the common divisors of 42 and 98 mod 42 = 14 are also precisely $\{1, 2, 7, 14\}$. And they are: because $14 \mid 42$, all divisors of 14—namely, 1, 2, 7, and 14—are common divisors of 14 and 42.

Proof of Lemma 7.5. By the assumption that $d \mid n$, we know that there's an integer a such that $n = ad$. Let $r := m \bmod n$, so that $m = cn + r$ for an integer c (as guaranteed by Theorem 7.1). We must prove that $d \mid m$ if and only if $d \mid r$.

For the forward direction, suppose that $d \mid m$. (We must prove that $d \mid r$.) By definition, there exists an integer b such that $m = bd$. But $n = ad$ and $m = bd$, so

$$m = cn + r \Leftrightarrow bd = c(ad) + r \Leftrightarrow r = (b - ac)d$$

for integers a, b, and c. Thus r is a multiple of d, and therefore $d \mid r$.

For the converse, suppose that $d \mid r$. (We must prove that $d \mid m$.) By definition, we have that $r = bd$ for some integer b. But then $n = ad$ and $r = bd$, so

$$m = cn + r \quad = \quad c(ad) + bd \quad = \quad (ac + b)d$$

for integers a, b, and c. Thus $d \mid m$. □

Corollary 7.6
Let n and $m \geq n$ be positive integers where $n \nmid m$. Then $\gcd(n, m) = \gcd(m \bmod n, n)$.

Proof. Lemma 7.5 establishes that the *set* of common divisors of $\langle n, m \rangle$ is identical to the set of common divisors of $\langle n, m \bmod n \rangle$. Therefore the *maxima* of these two sets of divisors—that is, $\gcd(n, m)$ and $\gcd(m \bmod n, n)$—are also equal. □

PUTTING IT TOGETHER: THE CORRECTNESS OF THE EUCLIDEAN ALGORITHM

Using this corollary, we can now prove the correctness of the Euclidean algorithm:

Theorem 7.7 (Correctness of the Euclidean algorithm)
For arbitrary positive integers n and m with $n \leq m$, we have **Euclid**$(n, m) = \text{gcd}(n, m)$.

Proof. We'll proceed by strong induction on n, the smaller input. Define the property

$$P(n) := \text{for any } m \geq n, \text{ we have } \textbf{Euclid}(n, m) = \text{gcd}(n, m).$$

We'll prove that $P(n)$ holds for all integers $n \geq 1$.

Base case ($n = 1$): $P(1)$ follows because both $\text{gcd}(1, m) = 1$ and **Euclid**$(1, m) = 1$: for any m, the *only* positive integer divisor of 1 is 1 itself (and indeed $1 \mid m$), and thus $\text{gcd}(1, m) = 1$. Observe that **Euclid**$(1, m) = 1$, too, because $m \bmod 1 = 0$ for any m.

Inductive case ($n \geq 2$): We assume the inductive hypotheses—that $P(n')$ holds for any $1 \leq n' < n$—and must prove $P(n)$. Let $m \geq n$ be arbitrary. There are two subcases, based on whether $n \mid m$ or $n \nmid m$:

- If $n \mid m$—that is, if $m = cn$ for an integer c—then $m \bmod n = 0$ and thus, by inspection of the algorithm, **Euclid**$(n, m) = n$. Because $n \mid n$ (and there is no $d > n$ that divides n evenly), indeed n is the GCD of n and $m = cn$.
- If $n \nmid m$—that is, if $m \bmod n \neq 0$—then

$$\begin{aligned} \textbf{Euclid}(n, m) &= \textbf{Euclid}(m \bmod n, n) && \textit{by inspection of the algorithm} \\ &= \text{gcd}(m \bmod n, n) && \textit{by the inductive hypothesis } P(m \bmod n) \\ &= \text{gcd}(n, m). && \textit{by Corollary 7.6} \end{aligned}$$

Note that $(m \bmod n) \leq n - 1$ by the definition of mod (*anything* mod n is less than n), so we can invoke the inductive hypothesis $P(m \bmod n)$ in the second step of this proof. □

Theorem 7.7 establishes the *correctness* of the Euclidean algorithm, but we introduced this algorithm because the brute-force algorithm (simply testing every candidate divisor d) was too slow. Indeed, the Euclidean algorithm *is* very efficient:

Problem-solving tip: In Theorem 7.8, it's not obvious what quantity upon which to perform induction—after all, there are two input variables, n and m. It is often useful to combine multiple inputs into a single "measure of progress" toward the base case—perhaps performing induction on the quantity $n + m$ or the quantity $n \cdot m$.

Theorem 7.8 (Efficiency of Euclidean Algorithm)
For arbitrary positive integers n and m with $n \leq m$, the recursion tree of **Euclid**(n, m) *has depth at most* $\log n + \log m$.

(The ability to efficiently compute $\text{gcd}(n, m)$ using the Euclidean algorithm—assuming we use the efficient algorithm to compute $m \bmod n$ from Exercises 7.11–7.16, at least—will be crucial in the RSA cryptographic system in Section 7.5.) You'll prove Theorem 7.8 by induction in Exercise 7.34—and you'll show that the recursion tree can be as deep as $\Omega(\log n + \log m)$, using the Fibonacci numbers, in Exercise 7.37.

Computer Science Connections

Converting Between Bases, Binary Representation, and Generating Strings

For a combination of historical and anatomical reasons—we have ten fingers and ten toes!—we generally use a *base ten*, or *decimal*, system to represent numbers. Moving from right to left, there's a 1's place, a 10's place, a 100's place, and so forth; thus 2048 denotes $8 \cdot 1 + 4 \cdot 10 + 0 \cdot 100 + 2 \cdot 1000$. This representation is an example of a *positional system*, in which each place/position has a value, and the symbol in that position tells us how many of that value the number has. Some ancient cultures used non-decimal positional systems, some of which survive to the present day: for example, the Sumarians and Babylonians used a base 60 system—and, even today, 60 seconds make a minute, and 60 minutes make an hour.

Latin: *decim* "ten." Note that *digit* is ambiguous in English between "place in a number" and "finger or toe."

In general, to represent a number n in base $b \geq 2$, we write a sequence of elements of $\{0, 1, \ldots, b-1\}$—say $[d_k d_{k-1} \cdots d_2 d_1 d_0]_b$. (We'll write the base explicitly as a subscript, for clarity.) Moving from right to left, the ith position is "worth" b^i, so this number's value is $\sum_{i=0}^{k} b^i d_i$. For example,

$$[1234]_5 = 4 \cdot 5^0 + 3 \cdot 5^1 + 2 \cdot 5^2 + 1 \cdot 5^3 = 4 + 15 + 50 + 125 = 194$$
$$[1234]_8 = 4 \cdot 8^0 + 3 \cdot 8^1 + 2 \cdot 8^2 + 1 \cdot 8^3 = 4 + 24 + 128 + 512 = 668.$$

We can use modular arithmetic to quickly convert from one base to another. For simplicity, we'll describe how to convert from base 10 into an arbitrary base b, though it's not that much harder to convert *from* an arbitrary base instead. To start, notice that $(\sum_{i=0}^{k} b^i d_i) \bmod b = d_0$. (The value $b^i d_i$ is divisible by b for any $i \geq 1$.) Therefore, to represent n in base b, we must have $d_0 := n \bmod b$. Similarly, $(\sum_{i=0}^{k} b^i d_i) \bmod b^2 = bd_1 + d_0$; thus we must choose $d_1 := \frac{n-d_0}{b} \bmod b$. (Note that $n - d_0$ must be divisible by b, because of our choice of d_0.) An algorithm following this strategy is shown in Figure 7.4. (We could also have written this algorithm without using division; see Exercise 7.5.) For example, to convert 145 to binary (base 2), we execute **baseConvert**(145, 2). Here are the values of n, i, and d_i in each iteration:

```
baseConvert(n, b):
Input: integers n and b ≥ 2
Output: n, represented in base b
i := 0
while n > 0:
    d_i := n mod b
    n := (n − d_i)/b
    i := i + 1
return [d_i d_{i−1} ··· d_1 d_0]_b
```

Figure 7.4: Base conversion algorithm, from base 10 to base b.

n	145	72	36	18	9	4	2	1	0
i	0	1	2	3	4	5	6	7	8
$d_i := n \bmod 2$	1	0	0	0	1	0	0	1	— .

Thus 145 can be written as $[10010001]_2$.

We can use the base conversion algorithm in Figure 7.4 to convert decimal numbers (base 10) into *binary* (base 2), the internal representation in computers. Or we can convert into *octal* (base 8) or *hexadecimal* (base 16), two other frequently used representations for numbers in programming. But we can also use **baseConvert** for seemingly unrelated problems. Consider the task of enumerating all 4-letter strings from the alphabet. The "easy" way to write a program to accomplish this task, with four nested loops, is painful to write—and it becomes utterly unwieldy if we needed all 10-letter strings instead. But, instead, let's count from 0 up to $26^4 - 1$—there are 26^4 different 4-letter strings—and convert each number into base 26. We can then translate each number into a sequence of letters, with the ith digit acting as an index into the alphabet that tells us which letter to put in position i. See Figure 7.5.

n in base 10	→ base 26	→ string
0	→ $[0\ 0\ 0\ 0]_{26}$	→ AAAA
1	→ $[0\ 0\ 0\ 1]_{26}$	→ AAAB
2	→ $[0\ 0\ 0\ 2]_{26}$	→ AAAC
	⋮	
25	→ $[0\ 0\ 0\ 25]_{26}$	→ AAAZ
26	→ $[0\ 0\ 1\ 0]_{26}$	→ AABA
27	→ $[0\ 0\ 1\ 1]_{26}$	→ AABB
	⋮	
1234	→ $[0\ 1\ 21\ 12]_{26}$	→ ABVM
	⋮	
456,974	→ $[25\ 25\ 25\ 24]_{26}$	→ ZZZY
456,975	→ $[25\ 25\ 25\ 25]_{26}$	→ ZZZZ

Figure 7.5: Generating all 4-letter strings. For each $n = 0, n = 1, \ldots, n = 456{,}975$, we convert n to a number in base 26; we then interpret each digit $[i]_{26} \in \{0, 1, \ldots, 25\}$ as an element of $\{\mathtt{A}, \mathtt{B}, \ldots, \mathtt{Z}\}$.

7.2.5 Exercises

Using paper and pencil only, *follow the proof of Theorem 7.1 or use* **mod-and-div** *(see Figure 7.6a) to compute integers* $r \in \{0, 1, \ldots, k-1\}$ *and* d *such that* $kd + r = n$, *for:*

7.1 $k = 17, n = 202$ **7.2** $k = 99, n = 2017$ **7.3** $k = 99, n = -2017$

7.4 When we proved Theorem 7.1, we showed that for integers $k \geq 1$ and n, there exist integers r and d such that $0 \leq r < k$ and $kd + r = n$. We stated but did not prove that r and d are unique. Prove that they are—that is, prove the following, for any integers $k \geq 1, n, r, d, r'$, and d': if $0 \leq r < k$ and $0 \leq r' < k$ and $n = dk + r = d'k + r'$, then $d' = d$ and $r' = r$.

7.5 The algorithm **baseConvert** on p. 714, which performs base conversion, is written using division. Modify the algorithm so that it uses only addition, subtraction, mod, multiplication, and comparison.

A **repdigit**$_b$ *is a number* n *that, when represented in base* b, *consists of the same symbol written over and over, repeated at least twice. (See p. 714.) For example, 666 is a repdigit*$_{10}$*: when you write* $[666]_{10}$, *it's the same digit ("6") repeated (in this case, three times). One way of understanding that 666 is a repdigit*$_{10}$ *is that* $666 = 6 + 60 + 600 = 6 \cdot 10^0 + 6 \cdot 10^1 + 6 \cdot 10^2$. *We can write* $[40]_{10}$ *as* $[130]_5$ *because* $40 = 0 + 3 \cdot 5 + 1 \cdot 5^2$, *or as* $[101000]_2$ *because* $40 = 1 \cdot 2^3 + 1 \cdot 2^5$. *So 40 is not a repdigit*$_{10}$, *repdigit*$_5$, *or repdigit*$_2$. *But 40 is a repdigit*$_3$, *because* $40 = [1111]_3$.

7.6 Prove that every number $n \geq 3$ is a repdigit$_b$ for some base $b \geq 2$, where $n = [11 \cdots 1]_b$.

7.7 Prove that every even number $n > 6$ is a repdigit$_b$ for some base $b \geq 2$, where $n = [22 \cdots 2]_b$.

7.8 Prove that *no* odd number n is a repdigit$_b$ of the form $[22 \cdots 2]_b$, for any base b.

7.9 Write $R(n)$ to denote the number of bases b, for $2 \leq b \leq n-1$, such that n is a repdigit$_b$. Conjecture a condition on n such that $R(n) = 1$, and prove your conjecture.

Recall the **mod-and-div**(n, m) *algorithm, reproduced in Figure 7.6(a), that computes* $n \bmod k$ *and* $\lfloor n/k \rfloor$ *by repeatedly subtracting* k *from* n *until the result is less than* k.

7.10 As written, the **mod-and-div** algorithm fails when given a negative value of n. Follow Case II of Theorem 7.1's proof to extend the algorithm for $n < 0$ too.

The **mod-and-div** *algorithm is slow—this algorithm computes an integer* d *such that* $nd \leq m < n(d+1)$ *by performing* linear search *for* d. *A faster version of this algorithm, called* **mod-and-div-faster**, *finds* d *using* binary search *instead; see Figure 7.6(b).*

7.11 The code for **mod-and-div-faster** as written uses division, by averaging *lo* and *hi*. Modify the algorithm so that it uses only addition, subtraction, multiplication, and comparison.

7.12 The code for **mod-and-div-faster** as written uses $hi := n + 1$ as the initial upper bound. Why is this assignment an acceptable for the correctness of the algorithm? Explain briefly.

7.13 Describe an algorithm that finds a better upper bound *hi*, by repeatedly doubling *hi* until it's large enough.

7.14 Let k be arbitrary. Describe an input n for which the doubling search from the last exercise yields a significant improvement on the running time of the algorithm for inputs k and n.

```
mod-and-div(n, k):
Input: integers n ≥ 0 and k ≥ 1
Output: n mod k and ⌊n/k⌋
  r := n; d := 0
  while r ≥ k:
      r := r − k; d := d + 1
  return r, d
```

```
mod-and-div-faster(n, k):
Input: integers n ≥ 0 and k ≥ 1
Output: n mod k and ⌊n/k⌋
  lo := 0; hi := n + 1.
  while lo < hi − 1:
      mid := ⌊(lo+hi)/2⌋
      if mid · k ≤ n then
          lo := mid
      else
          hi := mid
  return (n − k · lo), lo
```

Figure 7.6: A reminder of the algorithm to compute $n \bmod k$ and $\lfloor n/k \rfloor$, and a faster version.

7.15 *(programming required)* Implement, in a programming language of your choice, all three of these algorithms (**mod-and-div**, **mod-and-div-faster**, and the doubling-search tweaked version of **mod-and-div-faster** from the previous exercises) to compute $n \bmod k$ and $\lfloor n/k \rfloor$.

7.16 Run the three algorithms from the previous exercise to compute the following values: $2^{32} \bmod 202$, $2^{32} \bmod 2020$, and $2^{32} \bmod 3^{15}$. How do their speeds compare?

7.17 Prove (7.3.2): for integers $k > 0$, a, and b, we have $a + b \bmod k = [(a \bmod k) + (b \bmod k)] \bmod k$. Begin your proof as follows: *We can write* $a = ck + r$ *and* $b = dk + t$ *for* $r, t \in \{0, \ldots, k-1\}$ *(as guaranteed by Theorem 7.1).* Then use **mod-and-div** and Lemma 7.2.

Prove the following properties of modular arithmetic and divisibility, for any positive integers a, b, and c:

7.18 $a \bmod b = (a \bmod bc) \bmod b$

7.19 (7.4.1): $a \mid 0$

7.20 (7.4.2): $1 \mid a$

7.21 (7.4.6): if $a \mid b$ and $a \mid c$, then $a \mid (b + c)$.

7.22 (7.4.7): if $a \mid c$, then $a \mid bc$.

Consider the "repeated squaring" algorithm for modular exponentiation shown in Figure 7.7. Observe that this algorithm computes $b^e \bmod n$ *with a recursion tree of depth* $\Theta(\log e)$.

7.23 Use this algorithm to compute $3^{80} \bmod 5$ *without using a calculator.* (You should never have to keep track of a number larger than 5 except for the exponent itself when you're doing these calculations!)

7.24 Write down a recurrence relation representing the number of multiplications done by **mod-exp**(b, e, n). Prove, using this recurrence, that the number of multiplications done is between $\log e$ and $2 \log e$.

7.25 *(programming required)* Implement **mod-exp** in a programming language of your choice. Also implement a version of **mod-exp** that computes b^e and then, after that computation is complete, takes the result mod n. Compare the speeds of these two algorithms in computing $3^k \bmod 5$, for $k = 80$, $k = 800$, $k = 8000$, $\ldots$, $k = 8{,}000{,}000$. Explain.

mod-exp(b, e, n):
Input: integers $n \geq 1$, b, and $e \geq 0$
Output: $b^e \bmod n$
if $e = 0$ **then**
  return 1
else if e is even **then**
  $result$:= **mod-exp**$(b, \frac{e}{2}, n)$
  return $(result \cdot result) \bmod n$
else
  $result$:= **mod-exp**$(b, e-1, n)$
  return $(b \cdot result) \bmod n$

Figure 7.7: Modular exponentiation via repeated squaring.

There's a category of numerical tricks often called "divisibility rules" that you may have seen—quick ways of testing whether a given number is evenly divisible by some small k. The test for whether an integer n is divisible by 3 is this: add up the *digits* of n; n is divisible by 3 if and only if this sum is divisible by 3. *For example, 6,007,023 is divisible by 3 because* $6+0+0+7+0+2+3 = 18$, *and* $3 \mid 18$. *(Indeed* $3 \cdot 2{,}002{,}341 = 6{,}007{,}023$.*) This test relies on the following claim: for any sequence* $\langle x_0, x_1, \ldots, x_{n-1} \rangle \in \{0, 1, \ldots, 9\}^n$, *we have*

$$\left[\sum_{i=0}^{n-1} 10^i x_i\right] \bmod 3 \;=\; \left[\sum_{i=0}^{n-1} x_i\right] \bmod 3.$$

(For example, 6,007,023 is represented as $x_0 = 3$, $x_1 = 2$, $x_2 = 0$, $x_3 = 7$, $x_4 = 0$, $x_5 = 0$, *and* $x_6 = 6$.*)*

7.26 Prove that the test for divisibility by 3 is correct. First prove that $10^i \bmod 3 = 1$ for any integer $i \geq 0$; then prove the stated claim. Your proof should make heavy use of the properties in Theorem 7.3.

7.27 The divisibility test for 9 is to add up the digits of the given number, and test whether that sum is divisible by 9. State and prove the condition that ensures that this test is correct.

Using paper and pencil only, *use the Euclidean algorithm to compute the GCDs of the following pairs of numbers:*

7.28 $n = 111, m = 202$

7.29 $n = 333, m = 2017$

7.30 $n = 156, m = 360$

7.31 *(programming required)* Implement the Euclidean algorithm in a language of your choice.

7.32 *(programming required)* Early in Section 7.2.4, we discussed a brute-force algorithm to compute $\gcd(n, m)$: try all $d \in \{1, 2, \ldots, \min(n, m)\}$ and return the largest d such that $d \mid n$ and $d \mid m$. Implement this algorithm, and compare its performance to the Euclidean algorithm as follows: for both algorithms, find the largest n for which you can compute $\gcd(n, n-1)$ in less than 1 second on your computer.

Let's analyze the running time of the Euclidean algorithm for GCDs, to prove Theorem 7.8.

7.33 Let n and m be arbitrary positive integers where $n \leq m$. Prove that $m \bmod n \leq \frac{m}{2}$. *(Hint: what happens if* $n \leq \frac{m}{2}$*? What happens if* $\frac{m}{2} < n \leq m$*?)*

7.34 Using the previous exercise, prove that the Euclidean algorithm terminates within $O(\log n + \log m)$ recursive calls. (Actually one can prove a bound that's tighter by a constant factor, but this result is good enough for asymptotic work.)

Now let's show that, in fact, the Euclidean algorithm generates a recursion tree of depth $\Omega(\log n + \log m)$ *in the worst case—specifically, when* **Euclid**(f_n, f_{n+1}) *is run on consecutive Fibonacci numbers* f_n, f_{n+1}.

7.35 Show that, for all $n \geq 3$, we have $f_n \bmod f_{n-1} = f_{n-2}$, where f_i is the ith Fibonacci number. (Recall from Definition 6.21 that $f_1 := 1$, $f_2 := 1$ and $f_n := f_{n-1} + f_{n-2}$ for $n \geq 3$.)

7.36 Prove that, for all $n \geq 3$, **Euclid**(f_{n-1}, f_n) generates a recursion tree of depth $n - 2$.

7.37 Using the last exercise and the fact that $f_n \leq 2^n$ (Exercise 6.95), argue that the running time of the Euclidean algorithm is $\Omega(\log n + \log m)$ in the worst case.

7.3 *Primality and Relative Primality*

> Why is it that we entertain the belief that for every purpose odd numbers are the most effectual?
>
> Pliny the Elder (23–79)

Now that we've reviewed divisibility (and the related notions of factors, divisors, and multiples) in Section 7.2, we'll continue with a brief review of another concept from Chapter 2: the definition of *prime numbers.* We'll then introduce the related notion of *relatively prime* integers—pairs of numbers that share no common divisors aside from 1—and a few applications and extensions of both definitions.

7.3.1 *Primality (A Reminder) and Relative Primality (An Introduction)*

We begin with a reminder of the definitions from Chapter 2:

Definition 7.6 (Primes and composites (reprise))
An integer $p \geq 2$ is called prime *if the only positive integers that evenly divide it are 1 and p itself. An integer $n \geq 2$ that is not prime is called* composite. *(Note that 1 is neither prime nor composite.)*

For example, the integers 2, 3, 5, and 7 are all prime, but 4 (which is divisible by 2) and 6 (which is divisible by 2 and 3) are composite. It's also worth recalling two results that we saw in previous chapters:

- There are infinitely many prime numbers: Example 4.15 gave a proof by contradiction to show that there is no largest prime. (That result is attributed to Euclid—the same Euclid whose algorithm we encountered in Section 7.2.)
- Theorem 4.16 showed that any composite number $n \geq 2$ is divisible by some factor $d \leq \sqrt{n}$. (That is, $n \geq 2$ is prime if and only if $d \not\mid n$ for every $d \in \{2, 3, \ldots, \sqrt{n}\}$.)

We used the latter result to give an algorithm for the *primality testing problem*—that is, determining whether a given integer $n \geq 2$ is prime or composite—that performs $\sqrt{n}$ divisibility tests. (This algorithm simply exhaustively tests whether n is divisible by any of the candidate divisors between 2 and $\sqrt{n}$.)

> **Taking it further:** The faster divisibility algorithm that you developed in Exercises 7.11–7.16 will allow us to test primality in $\Theta(\sqrt{n} \cdot \log^k n)$ steps, for some constant k: faster than the naïve algorithm, but still not efficient. There *are* faster algorithms for primality testing that require only polylogarithmically many operations—that is, $O(\log^k n)$, for some fixed k—to test whether n is prime. See, for example, the discussion on p. 742 of a *randomized* algorithm that efficiently tests for primality, which requires only $O(\log^k n)$ steps to test whether n is prime, although it does have a small (provably small!) probability of making a mistake. There are also deterministic algorithms to solve this problem in polylogarithmic time, though they're substantially more complicated than this randomized algorithm.

Prime numbers turn out to be useful in all sorts of settings, and it will sometimes turn out to be valuable to compute a large collection of primes all at once. Of course, we can always generate more than one prime number by using a primality-testing algorithm (like the one we just suggested) more than once, until enough numbers have passed the test. But some of the work that we do in figuring out whether n is prime actually turns out to be helpful in figuring out whether $n' > n$ is prime. An algorithm called the *Sieve of Eratosthenes*, which computes a list of *all* prime numbers up to a given integer, exploits this redundancy to save some computation. The Sieve generates its list of prime numbers by successively eliminating ("sieving") all multiples of each discovered prime: for example, once we know that 2 is prime and that 4 is a multiple of 2, we will never have to test whether $4 \mid n$ in determining whether n is prime. (If n isn't prime because $4 \mid n$, then n is also divisible by 2—that is, 4 is never the smallest integer greater than 1 that evenly divides n, so we never have to bother testing 4 as a candidate divisor.) See Exercises 7.38–7.42 and Figure 7.15.

Taking it further: The Sieve of Eratosthenes is one of the earliest known algorithms, dating back to about 200 BCE. (The date isn't clear, in part because none of Eratosthenes's work survived; the algorithm was reported, and attributed to Eratosthenes, by Nicomachus about 300 years later.) The Euclidean algorithm for greatest common divisors from Section 7.2, which dates from c. 300 BCE, is one of the few older algorithms that are known.[2]

The Sieve of Eratosthenes is named after Eratosthenes, a Greek mathematician who lived in the 3rd century BCE. For more, see

[2] Donald E. Knuth. *The art of computer programming: Seminumerical algorithms (Volume 2)*. Addison-Wesley Longman, 3rd edition, 1997.

THE DISTRIBUTION OF THE PRIMES

For a positive integer n, let *primes*(n) denote the number of prime numbers less than or equal to n. Thus, for example, we have

$$\begin{aligned}
0 &= primes(1) \\
1 &= primes(2) \\
2 &= primes(3) = primes(4) \\
3 &= primes(5) = primes(6), \text{ and} \\
4 &= primes(7) = primes(8) = primes(9) = primes(10).
\end{aligned}$$

Or, to state it recursively: we have $primes(1) := 0$, and, for $n \geq 2$, we have

$$primes(n) := \begin{cases} primes(n-1) & \text{if } n \text{ is composite} \\ 1 + primes(n-1) & \text{if } n \text{ is prime.} \end{cases}$$

Figure 7.8(a) displays the value of *primes*(n) for moderately small n. An additional fact that we'll state without proof is the *Prime Number Theorem*—illustrated in Figure 7.8(b)—which describes the behavior of *primes*(n) for large n:

Theorem 7.9 (Prime Number Theorem)
As n gets large, the ratio between primes(n) *and* $\frac{n}{\ln n}$ *approaches* 1.

Formal proofs of the Prime Number Theorem are complicated beasts—far more complicated that we'll want to deal with here!—but even an intuitive understanding of the theorem is useful. Informally, this theorem says that, given an integer n, approximately a $\frac{1}{\ln n}$ fraction of the numbers "close to" n are prime. (See Exercise 7.45.)

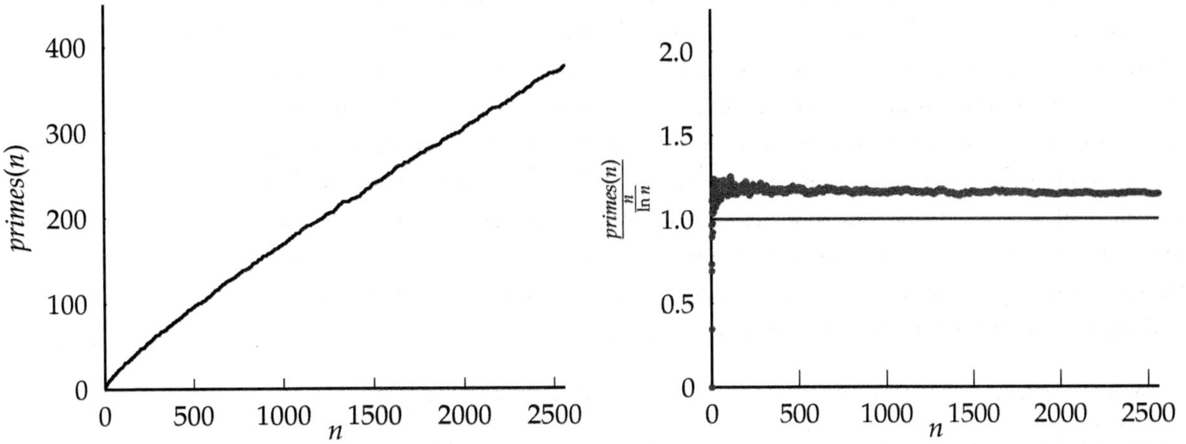

(a) A plot of n vs. $primes(n) := |\{q \leq n : q \text{ is prime}\}|$.

(b) A plot of n vs. the ratio of $\frac{n}{\ln n}$ and $primes(n)$.

Figure 7.8: The distribution of primes. The Prime Number Theorem states that the ratio $primes(n)/\frac{n}{\ln n}$, in (b), converges (slowly!) to 1.

Example 7.8 (Using the Prime Number Theorem)

Problem: Using the estimate $primes(n) \approx \frac{n}{\ln n}$, calculate (approximately) how many 10-digit integers are prime.

Solution: By definition, there are exactly *primes*(999,999,999) primes with 9 or fewer digits, and *primes*(9,999,999,999) primes with 10 or fewer digits. Thus the number of 10-digit primes is

$$\begin{aligned} primes(9{,}999{,}999{,}999) - primes(999{,}999{,}999) &\approx \frac{9{,}999{,}999{,}999}{\ln 9{,}999{,}999{,}999} - \frac{999{,}999{,}999}{\ln 999{,}999{,}999} \\ &\approx 434{,}294{,}499 - 48{,}254{,}956 \\ &= 386{,}039{,}543. \end{aligned}$$

Thus, roughly 386 million of the 9 billion 10-digit numbers (about 4.3%) are prime. (Exercise 7.46 asks you to consider how far off this estimate is.)

Problem-solving tip: Back-of-the-envelope calculations are often great as plausibility checks: although the Prime Number Theorem doesn't state a formal bound on how different *primes*(n) and $\frac{n}{\ln n}$ are, you can see whether a solution to a problem "smells right" with an approximation like this one.

The density of the primes is potentially interesting for its own sake, but there's also a practical reason that we'll care about the Prime Number Theorem. In the RSA cryptosystem (see Section 7.5), one of the first steps of the protocol involves choosing two large prime numbers p and q. The bigger p and q are, the more secure the encryption, so we would want p and q to be pretty big—say, both approximately 2^{2048}. The Prime Number Theorem tells us that, roughly, one out of every $\ln 2^{2048} \approx 1420$ integers around 2^{2048} is prime. Thus, we can find a prime in this range by repeatedly choosing a random integer n of the right size and testing n for primality, using some efficient primality testing algorithm. (More about testing algorithms soon.) Approximately one out of every 1420 integers we try will turn out to be prime, so on average we'll only need to try about 2840 values of n before we find primes to use as p and q.

Prime factorization

Recall that any integer can be *factored* into the product of primes. For example, we can write $2001 = 3 \cdot 23 \cdot 29$ and $202 = 2 \cdot 101$ and $507 = 3 \cdot 13 \cdot 13$ and $55057 = 55057$. (All of $\{2, 3, 13, 23, 29, 101, 55057\}$ are prime.) The *Fundamental Theorem of Arithmetic* (Theorem 5.5) states that any integer n can be factored into a product of primes—and that, up to reordering, there is a *unique* prime factorization of n. (In other words, any two prime factorizations of an integer n *can* differ in the ordering of the factors—for example, $202 = 101 \cdot 2$ and $202 = 2 \cdot 101$—but they can differ *only* in ordering.) We proved the "there exists" part of the theorem in Example 5.12 using induction; a bit later in this section, we'll prove uniqueness. (The proof uses some properties of prime numbers that are most easily seen using an extension of the Euclidean algorithm that we'll introduce shortly; we'll defer the proof until we've established those properties.)

Relative primality

An integer n is prime if it has no divisors except 1 and n itself. Here we will introduce a related concept for *pairs* of integers—two numbers that do not *share* any divisors except 1:

Definition 7.7 (Relative primality)
Two positive integers n and m are called relatively prime *if* $\gcd(n, m) = 1$*—that is, if* 1 *is the only positive integer that evenly divides both n and m.*

Here are a few small examples:

Example 7.9 (Some relatively prime integers)
The integers 21 and 25 are relatively prime, as $21 = 3 \cdot 7$ and $25 = 5 \cdot 5$ have no common divisor (other than 1). Similarly, 5 and 6 are relatively prime, as are 17 and 35. (But 12 and 21 are not relatively prime, because they're both divisible by 3.)

There will be a number of useful facts about relatively prime numbers that you'll prove in the exercises—for example, a prime number p and any integer n are relatively prime unless $p \mid n$; and, more generally, two numbers are relatively prime if and only if their prime factorizations do not share any factors.

Taking it further: Let $f(x)$ be a polynomial. One of the special characteristics of prime numbers is that $f(x)$ has some special properties when we evaluate $f(x)$ normally, *or if we take the result of evaluating the polynomial* $\bmod p$ *for some prime number p.* In particular, if $f(x)$ is a polynomial of degree k, then either $f(a) \equiv_p 0$ *for every* $a \in \{0, 1, \ldots, p-1\}$ or there are at most k values $a \in \{0, 1, \ldots, p-1\}$ such that $f(a) \equiv_p 0$. (We saw this property in Section 2.5.3 when we didn't take the result modulo the prime p.) As a consequence, if we have two polynomials $f(x)$ and $g(x)$ of degree k, then if f and g are not equivalent modulo p, then there are at most k values of $a \in \{0, 1, \ldots, p-1\}$ for which $f(a) \equiv_p g(a)$.

We can use the fact that polynomials of degree k "behave" in the same way modulo p (with respect to the number of roots, and the number of places that two polynomials agree) to give efficient solutions to two problems: *secret sharing,* in which n people wish to "distribute" shares of a secret so that any k of them can reconstruct the secret (but no set of $k - 1$ can); and a form of *error-correcting codes,* as we discussed in Section 4.2. The basic idea will be that by using a polynomial $f(x)$ and evaluating $f(x) \bmod p$ for a prime p, we'll be able to use *small* numbers (less than p) to accomplish everything that we'd be able to accomplish by evaluating $f(x)$ without the modulus. See the discussions of secret sharing on p. 730 and of Reed–Solomon codes on p. 731.

7.3.2 *A Structural Fact and the Extended Euclidean Algorithm*

Given an integer $n \geq 2$, quickly determining whether n is prime seems tricky: we've seen some easy algorithms for this problem, but they're pretty slow. And, though there *are* efficient but complicated algorithms for primality testing, we haven't seen (and, really, nobody knows) a genuinely simple algorithm that's also efficient. On the other hand, the analogous question about relative primality—*given integers n and m, are n and m relatively prime?*—is easy. In fact, we already know everything we need to solve this problem efficiently, just from the definition: n and m are relatively prime if and only if their GCD is 1, which occurs if and only if **Euclid**$(n, m) = 1$. So we can efficiently test whether n and m are relatively prime by testing whether **Euclid**$(n, m) = 1$.

We will start this section with a structural property about GCDs. (Right now it shouldn't be at all clear what this claim has to with anything in the last paragraph—but stick with it! The connection will come along soon.) Here's the claim:

Lemma 7.10 (There are multiples of n and m that add up to $\gcd(n, m)$)
Let n and m be any positive integers, and let $r = \gcd(n, m)$. Then there exist integers x and y such that $xn + ym = r$.

Here are a few examples of the multiples guaranteed by this lemma:

Example 7.10 (Some examples of Lemma 7.10)
In Example 7.9, we saw that $\{5, 6\}$ and $\{17, 35\}$ are both relatively prime—that is, $\gcd(5, 6) = \gcd(17, 35) = 1$—and that $\gcd(12, 21) = 3$. Also note that $\gcd(48, 1024) = 16$ (from Example 7.6), and $\gcd(16, 48) = 16$. For these pairs, we have:

$$\begin{aligned}
(-1) \cdot 5 + 1 \cdot 6 &= -5 + 6 &&= 1 = \gcd(5, 6) \\
33 \cdot 17 + (-16) \cdot 35 &= 561 - 560 &&= 1 = \gcd(17, 35) \\
2 \cdot 12 + (-1) \cdot 21 &= 24 - 21 &&= 3 = \gcd(12, 21) \\
(-21) \cdot 48 + 1 \cdot 1024 &= -1008 + 1024 &&= 16 = \gcd(48, 1024) \\
1 \cdot 16 + 0 \cdot 48 &= 16 + 0 &&= 16 = \gcd(16, 48).
\end{aligned}$$

Note that for the second example in the table, the pair $\{17, 35\}$, we could have chosen -2 and 1 instead of 33 and -16, as $-2 \cdot 17 + 1 \cdot 35 = 1 = 33 \cdot 17 + (-16) \cdot 35$.

Note that the integers x and y whose existence is guaranteed by Lemma 7.10 are not necessarily positive! (In fact, in Example 7.10 the only time that we didn't have a negative coefficient for one of the numbers was for the pair $\{16, 48\}$, where $\gcd(16, 48) = 16 = 1 \cdot 16 + 0 \cdot 48$.) Also, observe that there may be more than one pair of values for x and y that satisfy Lemma 7.10—in fact, you'll show in Exercise 7.58 that there are *always* infinitely many values of $\{x, y\}$ that satisfy the lemma.

Although, if you stare at it long enough, Example 7.10 might give a *tiny* hint about why Lemma 7.10 is true, a proof still seems distant. But, in fact, we'll be able to prove the claim based what looks like a digression: a mild extension to the Euclidean algorithm. For a little bit of a hint as to how, let's look at one more example of the Euclidean algorithm, but interpreting it as a guide to find the integers in Lemma 7.10:

Example 7.11 (An example of Lemma 7.10, using the Euclidean algorithm)
Let's find integers x and y such that $91x + 287y = \gcd(91, 287)$.

By running **Euclid**(91, 287), we make the recursive calls **Euclid**(14, 91) and **Euclid**(7, 14), which returns 7. Putting these calls into a small table—and using Definition 7.1's implied equality $m = \lfloor \frac{m}{n} \rfloor \cdot n + (m \bmod n)$, slightly rearranged—we have:

m	n	$m \bmod n$	$\lfloor \frac{m}{n} \rfloor$	$m \bmod n = m - \lfloor \frac{m}{n} \rfloor \cdot n$	
287	91	14	3	$14 = 287 - 3 \cdot 91$	(1)
91	14	7	6	$7 = 91 - 6 \cdot 14$	(2)
14	7	0			

Notice that $7 = \gcd(91, 287) =$ **Euclid**(91, 287). Using (1) and (2), we can rewrite 7 as:

$$7 = 91 - 6 \cdot 14 \qquad \textit{by (2)}$$
$$= 91 - 6 \cdot (287 - 3 \cdot 91) = -6 \cdot 287 + 19 \cdot 91. \qquad \textit{by (1) and simplification}$$

Thus $x := -6$ and $y := 19$ satisfy the requirement that $91x + 287y = \gcd(91, 287)$.

The Extended Euclidean algorithm

The *Extended Euclidean algorithm*, shown in Figure 7.9, follows the outline of Example 7.11, applying these algebraic manipulations recursively. Lemma 7.10 will follow from a proof that this extended version of the Euclidean algorithm actually *computes* three integers x, y, r such that $\gcd(n, m) = r = xn + ym$. Here are two examples:

```
extended-Euclid(n, m):
Input: positive integers n and m ≥ n.
Output: x, y, r ∈ Z where gcd(n, m) = r = xn + ym
if m mod n = 0 then
    return 1, 0, n          // 1·n + 0·m = n = gcd(n, m)
else
    x, y, r := extended-Euclid(m mod n, n)
    return y − ⌊m/n⌋ · x, x, r
```

Figure 7.9: The Extended Euclidean algorithm.

Example 7.12 (Running the Extended Euclidean Algorithm I)
Evaluating **extended-Euclid**(12, 18) recursively computes **extended-Euclid**(6, 12) = $\langle 1, 0, 6 \rangle$, and then computes its result from $\langle 1, 0, 6 \rangle$ and the values of $n = 12$ and $m = 18$:

extended-Euclid(12, 18) *(because* 18 mod 12 $\neq$ 0*, we make a recursive call).*

extended-Euclid($\underbrace{18 \bmod 12}_{=6}$, 12)

$= \boxed{1, 0, 6}$ *(because* 12 mod 6 = 0*).*

$= y - \lfloor \frac{m}{n} \rfloor \cdot x, x, r$ *where* $\boxed{x = 1, y = 0, r = 6}$ *and* $\boxed{n = 12, m = 18}$.

$= 0 - \lfloor \frac{18}{12} \rfloor \cdot 1, 1, 6$

$= -1, 1, 6.$

The recursive call returned $x = 1$, $y = 0$, and $r = 6$, and the else case of the algorithm tells us that our result is $\langle y - \lfloor \frac{m}{n} \rfloor \cdot x, x, r \rangle$ where $m = 18$ and $n = 12$. Plugging these values into the formula for the result, we see that **extended-Euclid**(12, 18) returns $\langle -1, 1, 6 \rangle$—and, indeed, $\gcd(12, 18) = 6$ and $-1 \cdot 12 + 1 \cdot 18 = 6$.

Example 7.13 (Running the Extended Euclidean Algorithm II)
For slightly more complicated example, let's compute **extended-Euclid**(18, 30):

$$
\begin{aligned}
&\textbf{extended-Euclid}(\boxed{18, 30})\\
&\quad \textbf{extended-Euclid}(\underbrace{30 \bmod 18}_{=12}, 18)\\
&\quad\quad \textbf{extended-Euclid}(\underbrace{18 \bmod 12}_{=6}, 12)\\
&\quad\quad = 1, 0, 6\\
&\quad = \boxed{-1, 1, 6} && \textit{by Example 7.12.}\\
&= y - \lfloor \tfrac{m}{n} \rfloor \cdot x, x, r && \textit{where } \boxed{x = -1, y = 1, r = 6} \textit{ and } \boxed{n = 18, m = 30}.\\
&= 1 - \lfloor \tfrac{30}{18} \rfloor \cdot (-1), -1, 6\\
&= 1 - 1 \cdot (-1), -1, 6\\
&= 2, -1, 6.
\end{aligned}
$$

Again, as required, we have $\gcd(18, 30) = 6$ and $2 \cdot 18 + -1 \cdot 30 = 36 - 30 = 6$.

We're now ready to state the correctness of the Extended Euclidean algorithm:

Theorem 7.11 (Correctness of the Extended Euclidean Algorithm)
For arbitrary positive integers n and m with $n \leq m$, **extended-Euclid**(n, m) *returns three integers x, y, r such that $r = \gcd(n, m) = xn + ym$.*

The proof, which is fairly straightforward by induction, is left to you as Exercise 7.60. And once you've proven this theorem, Lemma 7.10—which merely stated that *there exist* integers x, y, r with $r = \gcd(n, m) = xn + ym$ for any n and m—is immediate.

Note also that the Extended Euclidean algorithm is an efficient algorithm—you already proved in Exercise 7.34 that the depth of the recursion tree for **Euclid**(n, m) is upper bounded by $O(\log n + \log m)$, and the running time of **extended-Euclid**(n, m) is asymptotically the same as **Euclid**(n, m). (The only quantity that we need to use in **extended-Euclid** that we didn't need in **Euclid** is $\lfloor \frac{m}{n} \rfloor$, but we already had to find $m \bmod n$ in **Euclid**—so if we used **mod-and-div**(n, m) to compute $m \bmod n$, then we "for free" also get the value of $\lfloor \frac{m}{n} \rfloor$.)

Problem-solving tip: A nice way, particularly for computer scientists, to prove a theorem of the form "there exists x such that $P(x)$" is to *actually give algorithm that computes such an x!*

7.3.3 *The Uniqueness of Prime Factorization*

Lemma 7.10—that there are multiples of n and m that add up to $\gcd(n, m)$—and the Extended Euclidean algorithm (which computes those coefficients) will turn out to be helpful in proving some facts that are apparently unrelated to greatest common divisors. Here's a claim about divisibility related to prime numbers in that vein, which we'll be able to use to prove that prime factorizations are unique:

Lemma 7.12 (When a prime divides a product)
Let p be prime, and let a and b be integers. Then $p \mid ab$ if and only if $p \mid a$ or $p \mid b$.

Proof. We'll proceed by mutual implication.

For the backward direction, assume $p \mid a$. (The case for $p \mid b$ is strictly analogous.) Then $a = kp$ for some integer k, and thus $ab = kpb$, which is obviously divisible by p.

For the forward direction, assume that $p \mid ab$ and suppose that $p \nmid a$. We must show that $p \mid b$. Because p is prime and $p \nmid a$, we know that $\gcd(p, a) = 1$ (see Exercise 7.47), and, in particular, **extended-Euclid**(p, a) returns the GCD 1 and two integers n and m such that $1 = pm + an$. Multiplying both sides by b yields $b = pmb + anb$, and thus

$$\begin{aligned}
b \bmod p &= (pmb + anb) \bmod p \\
&= (pmb \bmod p + anb \bmod p) \bmod p && (7.3.2) \\
&= (0 + anb \bmod p) \bmod p && (7.4.7) \\
&= (0 + 0) \bmod p && p \mid ab \text{ by assumption, and (7.4.7) again} \\
&= 0.
\end{aligned}$$

That is, we've shown that if $p \nmid a$, then $p \mid b$. (And $\neg x \Rightarrow y$ is equivalent to $x \vee y$.) □

We can use this fact to prove that an integer's prime factorization is unique. (We'll prove only the uniqueness part of the theorem here; see Example 5.12 for the "there exists a prime factorization" part.)

Problem-solving tip: When you define something, you genuinely get to choose how to define it! When you can make a choice in the definition that makes your life easier, *do it!*

Taking it further: Back when we defined prime numbers, we were very careful to specify that 1 *is neither prime nor composite*. You may well have found this insistence to be silly and arbitrary and pedantic—after all, the only positive integers that evenly divide 1 are 1 and, well, 1 itself, so it sure seems like 1 ought to be prime. But there was a good reason that we chose to exclude 1 from the list of primes: *it makes the uniqueness of prime factorization true!* If we'd listed 1 as a prime number, there would be many different ways to prime factor, say, 202: for example, $202 = 2 \cdot 101$ and $202 = 1 \cdot 2 \cdot 101$ and $202 = 1 \cdot 1 \cdot 2 \cdot 101$, and so forth. So we'd have to have restated the theorem about uniqueness of prime factorization ("...is unique up to reordering *and the number of times that we multiply by* 1"), which is a much more cumbersome statement. This theorem *is* the reason that 1 is not defined as a prime number, in this book or in any other mathematical treatment.

Theorem 7.13 (Prime Factorization Theorem (Reprise))
Let $n \in \mathbb{Z}^{\geq 1}$ be any positive integer. There exist $k \geq 0$ prime numbers $p_1, p_2, \ldots, p_k$ such that $n = \prod_{i=1}^{k} p_i$. Further, up to reordering, the prime numbers $p_1, p_2, \ldots, p_k$ are unique.

Proof (of uniqueness). We'll proceed by strong induction on n.

For the base case ($n = 1$), we can write 1 as the product of zero prime numbers—recall that $\prod_{i \in \varnothing} i = 1$—and this representation is unique. (The product of one or more primes is greater than 1, as all primes are at least 2.)

For the inductive case ($n \geq 2$), we assume the inductive hypotheses, namely that any $n' < n$ has a unique prime factorization. We must prove that the prime factorization of n is also unique. We consider two subcases:

Case I: n is prime. Then the statement holds immediately: the only prime factorization is $p_1 = n$. (Suppose that there were a different way of prime factoring n, as $n = \prod_{i=1}^{\ell} q_i$ for prime numbers $\langle q_1, q_2, \ldots, q_\ell \rangle$. We'd have to have $\ell \geq 2$ for this factorization to differ from $p_1 = n$, but then each q_i satisfies $q_i > 1$ and $q_i < n$ and $q_i \mid n$—contradicting what it means for n to be prime.)

Case II: n is composite. Then suppose that $p_1, p_2, \ldots, p_k$ and $q_1, q_2, \ldots, q_\ell$ are two sequences of prime numbers such that $n = \prod_{i=1}^{k} p_i = \prod_{i=1}^{\ell} q_i$. Without loss of generality, assume that both sequences are sorted in increasing order, so that $p_1 \leq p_2 \leq \cdots \leq p_k$ and $q_1 \leq q_2 \leq \cdots \leq q_\ell$. We must prove that these two sequences are actually equal.

- *Case IIA:* $p_1 = q_1$. Define $n' := \frac{n}{p_1} = \frac{n}{q_1} = \prod_{i=2}^{k} p_i = \prod_{i=2}^{\ell} q_i$ as the product of all the other prime numbers (excluding the primes p_1 and $q_1 = p_1$). By the inductive hypothesis, n' has a unique prime factorization, and thus $p_2, p_3, \ldots, p_k$ and $q_2, q_3, \ldots, q_\ell$ are identical.
- *Case IIB:* $p_1 \neq q_1$. Without loss of generality, suppose $p_1 < q_1$. But $p_1 \mid n$, and therefore $p_1 \mid \prod_{i=1}^{\ell} q_i$. By Lemma 7.12, there exists an i such that $p_1 \mid q_i$. But $2 \leq p_1 < q_1 \leq q_i$. This contradicts the assumption that q_i was prime. □

Taking it further: How difficult is it to factor a number n? Does there exist an efficient algorithm for factoring—that is, one that computes the prime factorization of n in a number of steps that's proportional to $O(\log^k n)$ for some k? *We don't know.* But it is generally believed that the answer is *no*, that factoring large numbers cannot be done efficiently. The (believed) difficulty of factoring is a crucial pillar of widely used cryptographic systems, including the ones that we'll encounter in Section 7.5. There *are* known algorithms that factor large numbers efficiently on so-called *quantum computers* (see the discussion on p. 1016)—but nobody knows how to build quantum computers. And, while there's no known efficient algorithm for factoring large numbers on classical computers, there's also no proof of hardness for this problem. (And most modern cryptographic systems count on the difficulty of the factoring problem—which is only a conjecture!)

7.3.4 *The Chinese Remainder Theorem*

We'll close this section with another ancient result about modular arithmetic, called the *Chinese Remainder Theorem,* from around 1750 years ago. Here's the basic idea. If n is some nonnegative integer, then knowing that, say, when n is divided by 7 its remainder is 4 gives you a small clue about n's value: one seventh of integers have the right value mod 7. Knowing n mod 2 and n mod 13 gives you more clues. The Chinese Remainder Theorem says that knowing n mod k for enough values of k will (almost) let you figure out the value of n exactly—at least, if those values of k are all relatively prime. Here's a concrete example:

The name of the Chinese Remainder Theorem comes from its early discovery by the Chinese mathematician Sun Tzu, who lived around the 5th century. (This Sun Tzu is a different Sun Tzu from the one who wrote *The Art of War* about 800 years prior.)

Example 7.14 (An example of the Chinese Remainder Theorem)
Problem: What nonnegative integers n satisfy the following conditions?

$$n \bmod 2 = 0 \qquad n \bmod 3 = 2 \qquad n \bmod 5 = 1.$$

Solution: Suppose $n \in \{0, 1, \ldots, 29\}$. Then there are only six possible values for which $n \bmod 5 = 1$, namely $\{0+1, 5+1, 10+1, 15+1, 20+1, 25+1\} = \{1, 6, 11, 16, 21, 26\}$. Of these, the only even values are 6, 16, and 26. And we have 6 mod 3 = 0, 16 mod 3 = 1, and 26 mod 3 = 2. Thus $n = 26$.

Notice that, for any integer k, we have $k \equiv_b k + 30$ for all three moduli $b \in \{2, 3, 5\}$. Therefore any $n \equiv_{30} 26$ will satisfy the given conditions.

n	0	1	2	3	4	5	6	7	8	9	10	11	12	13	14	15	16	17	18	19	20	21	22	23	24	25	26	27	28	29
n mod 2	0	1	0	1	0	1	0	1	0	1	0	1	0	1	0	1	0	1	0	1	0	1	0	1	0	1	0	1	0	1
n mod 3	0	1	2	0	1	2	0	1	2	0	1	2	0	1	2	0	1	2	0	1	2	0	1	2	0	1	2	0	1	2
n mod 5	0	1	2	3	4	0	1	2	3	4	0	1	2	3	4	0	1	2	3	4	0	1	2	3	4	0	1	2	3	4

n	0	6	12	18	24	10	16	22	28	4	20	26	2	8	14	15	21	27	3	9	25	1	7	13	19	5	11	17	23	29
n mod 2	0	0	0	0	0	0	0	0	0	0	0	0	0	0	0	1	1	1	1	1	1	1	1	1	1	1	1	1	1	1
n mod 3	0	0	0	0	0	1	1	1	1	1	2	2	2	2	2	0	0	0	0	0	1	1	1	1	1	2	2	2	2	2
n mod 5	0	1	2	3	4	0	1	2	3	4	0	1	2	3	4	0	1	2	3	4	0	1	2	3	4	0	1	2	3	4

Figure 7.10: The remainders of all $n \in \{0, 1, \ldots, 29\}$, modulo 2, 3, and 5—sorted by n (above) and by the remainders (below).

The basic point of Example 7.14 is that every value of $n \in \{0, \ldots, 29\}$ has a unique "profile" of remainders mod 2, 3, and 5. (See Figure 7.10.) Crucially, every one of the 30 possible profiles of remainders occurs in Figure 7.10, and no profile appears more than once. (The fact that there are exactly 30 possible profiles follows from the Product Rule for counting; see Section 9.2.1.)

The Chinese Remainder Theorem states the general property that's illustrated in these particular tables: each "remainder profile" occurs once and only once. Here is a formal statement of the theorem. We refer to a constraint of the form $x \bmod n = a$ as a *congruence*, following Definition 7.2. We also write $\mathbb{Z}_k := \{0, 1, \ldots, k-1\}$.

Theorem 7.14 (Chinese Remainder Theorem: two congruences)
Let n and m be any two relatively prime integers. For any $a \in \mathbb{Z}_n$ and $b \in \mathbb{Z}_m$, there exists one and only one integer $x \in \mathbb{Z}_{nm}$ such that $x \bmod n = a$ and $x \bmod m = b$.

Proof. To show that there exists an integer x satisfying $x \bmod n = a$ and $x \bmod m = b$, we'll give a proof by construction—specifically, we'll *compute* the value of x given the values of $\{a, b, n, m\}$. The simple algorithm is shown in Figure 7.11. We must argue that $x \bmod n = a$ and $x \bmod m = b$. Note that $\gcd(n, m) = 1$ because n and m are relatively prime by assumption. Thus, by the correctness of the Extended Euclidean algorithm, we have

```
Input: relatively prime n, m ∈ Z; a ∈ Z_n; b ∈ Z_m.
Output: x such that x mod m = a and x mod n = b.
c, d, r := extended-Euclid(n, m)
return x := (adm + bcn) mod nm
```

Figure 7.11: An algorithm for the Chinese Remainder Theorem. (Ensure that $m \geq n$ by swapping n and m if necessary.)

$$cn + dm = 1. \qquad (*)$$

Multiplying both sides of $(*)$ by a, we know that

$$acn + adm = a. \qquad (\dagger)$$

Recall that we defined $x := (adm + bcn) \bmod nm$. Let's now show that $x \bmod n = a$:

$$\begin{aligned}
x \bmod n &= (adm + bcn) \bmod nm \bmod n && \textit{definition of } x \\
&= (adm + bcn) \bmod n && \textit{Exercise 7.18} \\
&= (adm + 0) \bmod n && bcn \bmod n = 0 \textit{ because } n \mid bcn \\
&= (adm + acn) \bmod n && acn \bmod n = 0 \textit{ because } n \mid acn \textit{ too!} \\
&= a \bmod n && (\dagger) \\
&= a. && a \in \{0, 1, \ldots, n-1\} \textit{ by assumption, so } a \bmod n = a
\end{aligned}$$

We can argue that $x = adm + bcn \equiv_m bdm + bcn \equiv_m b$ completely analogously, where the last equivalence follows by multiplying both sides of $(*)$ by b instead.

Thus we've now established that *there exists* an $x \in \mathbb{Z}_{nm}$ with $x \bmod n = a$ and $x \bmod m = b$ (because we *computed* such an x). To prove that there is a *unique* such x, suppose that $x \bmod n = x' \bmod n$ and $x \bmod m = x' \bmod m$ for two integers $x, x' \in \mathbb{Z}_{nm}$. We will prove that $x = x'$—which establishes that there's actually only one element of $\mathbb{Z}_{nm}$ with this property. By assumption, we know that $(x - x') \bmod n = 0$ and $(x - x') \bmod m = 0$, or, in other words, we know that $n \mid (x - x')$ and $m \mid (x - x')$. By Exercise 7.70 and the fact that n and m are relatively prime, then, we know that $nm \mid (x - x')$. And because both $x, x' \in \mathbb{Z}_{nm}$, we've therefore shown that $x = x'$. □

Writing tip: Now that we've done a lot of manipulations with modular arithmetic, in proofs we will start to omit some simple steps that are by now tedious—like those using (7.3.2) to say that $y + z \bmod n$ is equal to $((y \bmod n) + (z \bmod n)) \bmod n$.

Some examples

Here are two concrete examples of using the Chinese Remainder Theorem (and, specifically, of using the algorithm from Figure 7.11):

Example 7.15 (The Chinese Remainder Theorem, in action)

Let's use the algorithm from the proof of the Chinese Remainder Theorem to find the integer $x \in \mathbb{Z}_{30}$ that satisfies $x \bmod 5 = 4$ and $x \bmod 6 = 5$.

Note that 5 and 6 are relatively prime, and **extended-Euclid**(5, 6) returns $\langle -1, 1, 1 \rangle$. (Indeed, we have that $5 \cdot -1 + 6 \cdot 1 = 1 = \gcd(5, 6)$.) Thus we compute x from the values of $\langle n, m, a, b, c, d \rangle = \langle 5, 6, 4, 5, -1, 1 \rangle$ as

$$adm + bcn = 4 \cdot 1 \cdot 6 + 5 \cdot -1 \cdot 5 = 24 - 25 = -1.$$

Thus $x := -1 \bmod 30 = 29$. And, indeed, $29 \bmod 5 = 4$ and $29 \bmod 6 = 5$.

Example 7.16 (A second example of the Chinese Remainder Theorem)

<u>*Problem:*</u> We are told that $x \bmod 7 = 1$ and $x \bmod 9 = 5$. What is the value of x?

<u>*Solution:*</u> We find **extended-Euclid**(7, 9) = $\langle 4, -3, 1 \rangle$ by tracing the algorithm's execution. The algorithm in Figure 7.11 computes $x := adm + bcn \bmod nm$, where $n = 7$ and $m = 9$ are the given moduli; $a = 1$ and $b = 5$ are the given remainders; and $c = 4$ and $d = -3$ are the computed multipliers from **extended-Euclid**. Thus

$$x := (1 \cdot -3 \cdot 9) + (5 \cdot 4 \cdot 7) \bmod 7 \cdot 9 = -27 + 140 \bmod 63 = 113 \bmod 63 = 50.$$

Indeed, $50 \bmod 7 = 1$ and $50 \bmod 9 = 5$. Thus $x \equiv_{63} 50$.

Generalizing to k congruences

We've now shown the Chinese Remainder Theorem for two congruences, but Example 7.14 had *three* constraints ($x \bmod 2$, $x \bmod 3$, and $x \bmod 5$). In fact, the generalization of the Chinese Remainder Theorem to k congruences, for any $k \geq 1$, is also true—again, as long as the moduli are *pairwise relatively prime* (that is, *any* two of the moduli share no common divisors).

We can prove this generalization fairly directly, using induction and the two-congruence case. The basic idea will be to repeatedly use Theorem 7.14 to combine a pair of congruences into a single congruence, until there are no pairs left to combine. Here's a concrete example:

Example 7.17 (The Chinese Remainder Theorem, with 3 congruences)
Let's describe the values of x that satisfy the congruences

$$x \bmod 2 = 1 \qquad x \bmod 3 = 2 \qquad x \bmod 5 = 4. \tag{$*$}$$

To do so, we first identify values of y that satisfy the first two congruences, ignoring the third. Note that 2 and 3 are relatively prime, and **extended-Euclid**$(2,3) = \langle -1, 1, 1 \rangle$. Thus, $y \bmod 2 = 1$ and $y \bmod 3 = 2$ if and only if

$$y \bmod (2 \cdot 3) = (1 \cdot 1 \cdot 3 + 2 \cdot -1 \cdot 2) \bmod (2 \cdot 3) = 5.$$

In other words, $y \in \mathbb{Z}_6$ satisfies the congruences $y \bmod 2 = 1$ and $y \bmod 3 = 2$ *if and only if* y satisfies the single congruence $y \bmod 6 = 5$. Thus the values of x that satisfy $(*)$ are precisely the values of x that satisfy

$$x \bmod 6 = 5 \qquad x \bmod 5 = 4. \tag{$\dagger$}$$

And, in Example 7.15, we showed that values of x that satisfy $(\dagger)$ are precisely those with $x \bmod 30 = 29$.

Now, using the idea from this example, we'll prove the general version of the Chinese Remainder Theorem:

Theorem 7.15 (Chinese Remainder Theorem: General version)
Let $n_1, n_2, \ldots, n_k$ be a collection of pairwise relatively prime integers, for some $k \geq 1$, and let $N := \prod_{i=1}^{k} n_i$.

For any $\langle a_1, \ldots, a_k \rangle$ with each $a_i \in \mathbb{Z}_{n_i}$, there exists one and only one integer $x \in \mathbb{Z}_N$ such that $x \bmod n_i = a_i$ for all $1 \leq i \leq k$.

Proof. We proceed by induction on k.

Base case ($k = 1$): Then there's only one constraint, namely $x \bmod n_1 = a_1$, and obviously $x := a_1$ is the only element of $\mathbb{Z}_N = \mathbb{Z}_{n_1}$ that satisfies this congruence.

Inductive case ($k \geq 2$): We assume the inductive hypothesis, namely that there exists a unique $x \in \mathbb{Z}_M$ satisfying any set of $k - 1$ congruences whose moduli have product M. To make use of this assumption, we will convert the k given congruences into $k - 1$ equivalent congruences, as follows: by Theorem 7.14, there exists a (unique) value $y^* \in \mathbb{Z}_{n_1 n_2}$ such that $y^* \bmod n_1 = a_1$ and $y^* \bmod n_2 = a_2$. In Exercise 7.69 you'll prove that $n_1 n_2$ is also relatively prime to every other n_i, and, in Exercise 7.79, you will show that a value $x \in \mathbb{Z}_N$ satisfies $x \bmod n_1 = a_1$ and $x \bmod n_2 = a_2$ *if and* *only if* x satisfies $x \bmod n_1n_2 = y^*$. More formally, given the A-constraints (on the left), define the B-constraints (on the right):

$x \bmod n_1 = a_1$	(1A)	$x \bmod n_1n_2 = y^*$	(1-and-2B)
$x \bmod n_2 = a_2$	(2A)		
$x \bmod n_3 = a_3$	(3A)	$x \bmod n_3 = a_3$	(3B)
$x \bmod n_4 = a_4$	(4A)	$x \bmod n_4 = a_4$	(4B)
$\vdots$		$\vdots$	
$x \bmod n_k = a_k.$	(kA)	$x \bmod n_k = a_k.$	(kB)

Observe that the product of the moduli is the same for both the A-constraints and the B-constraints: $N := n_1 \cdot n_2 \cdot n_3 \cdots n_k$ for A, and $(n_1n_2) \cdot n_3 \cdots n_k$ for B. Thus:

- By Exercise 7.79, an integer $x \in \mathbb{Z}_N$ satisfies the A-constraints if and only if x satisfies the B-constraints.
- By the inductive hypothesis—which applies by Exercise 7.69—there's a unique $x \in \mathbb{Z}_N$ that satisfies the B-constraints.

Therefore there is a unique $x \in \mathbb{Z}_N$ that satisfies the A-constraints, as desired. □

Here we gave an inductive argument for the general version of Chinese Remainder Theorem (based on the 2-congruence version), but we could also give a version of the proof that directly echoes Theorem 7.14's proof. See Exercise 7.107.

Taking it further: One interesting implication of the Chinese Remainder Theorem is that we could choose to represent integers efficiently in a very different way from binary representation, instead using something called *modular representation.* In modular representation, we store an integer n as a sequence of values of $n \bmod b$, for a set of relatively prime values of b. To be concrete, consider the set $\{11, 13, 15, 17, 19\}$, and let $N := 11 \cdot 13 \cdot 15 \cdot 17 \cdot 19 = 692{,}835$ be their product. The Chinese Remainder Theorem tells us that we can uniquely represent any $n \in \mathbb{Z}_N$ as

$$\langle n \bmod 11, n \bmod 13, n \bmod 15, n \bmod 17, n \bmod 19 \rangle.$$

For example, $2^{17} = \langle 7, 6, 2, 2, 10 \rangle$, and $17 = \langle 6, 4, 2, 0, 17 \rangle$. Perhaps surprisingly, the representation of $2^{17} + 17$ is $\langle 2, 10, 4, 2, 8 \rangle$ and $17 \cdot 2^{17} = \langle 9, 11, 4, 0, 18 \rangle$, which are really nothing more than the result of doing component-wise addition/multiplication (modulo that component's corresponding modulus):

	mod 11	13	15	17	19
	⟨ 7,	6,	2,	2,	10 ⟩
+	⟨ 6,	4,	2,	0,	17 ⟩
=	⟨ 13,	10,	4,	2,	27 ⟩
≡	⟨ 2,	10,	4,	2,	8 ⟩

and

	mod 11	13	15	17	19
	⟨ 7,	6,	2,	2,	10 ⟩
·	⟨ 6,	4,	2,	0,	17 ⟩
=	⟨ 42,	24,	4,	0,	170 ⟩
≡	⟨ 9,	11,	4,	0,	18 ⟩.

This representation has some advantages over the normal binary representation: the numbers in each component stay small, and multiplying k pairs of 5-bit numbers is significantly faster than multiplying one pair of $5k$-bit numbers. (Also, the components can be calculated in parallel!) But there are some other operations that are slowed down by this representation. (See Exercises 7.145–7.146.)

Computer Science Connections

Secret Sharing

Although encryption/decryption is probably the most natural cryptographic problem, there are many other important problems in the same general vein. Here we'll introduce and solve a different cryptographic problem—using a solution due to Adi Shamir (the S of the RSA cryptosystem, which we'll see in Section 7.5).[3] Imagine a shared resource, collectively owned by some group, that the group wishes to keep secure—for example, the launch codes for the U.S.'s nuclear weapons. In the post-apocalyptic world in which you're imagining these codes being used, where many top officials are probably dead, we'll need to ensure that any, say, $k = 3$ of the cabinet members (out of the $n = 15$ cabinet positions) can launch the weapons. But you'd also like to guarantee that no single rogue secretary can destroy the world!

[3] Adi Shamir. How to share a secret. *Communications of the ACM*, 22(11):612–613, November 1979.

In *secret sharing*, we seek a scheme by which we distribute "shares" of the secret $s \in S$ to a group of n people such that the following properties hold:

1. If any k of these n people cooperate, then—by combining their k shares of the secret—they can compute the secret s (preferably efficiently).
2. If any $k' < k$ of these n people cooperate, then by combining their k' shares they learn *nothing* about the secret s. (Informally, to "learn nothing" about the secret means that no k' shares of the secret allow one to infer that s comes from any particular $S' \subset S$.)

(Note that just "splitting up the bits" of the secret violates condition 2.)

The basic idea will be to define a polynomial $f(x)$, and distribute the value of $f(i)$ as the the ith "share" of the secret; the secret itself will be $f(0)$. Why will this be useful? Imagine that $f(x) = ax + b$. (The secret is thus $f(0) = a \cdot 0 + b = b$.) Knowing that $f(1) = 17$ tells you that $a + b = 17$, but it doesn't tell you anything about b itself: for every possible value of the secret, there's a value of a that makes $a + b = 17$. But knowing $f(1) = 17$ and $f(2) = 42$ lets you solve for $a = 25, b = -8$. If $f(x) = ax^2 + bx + c$, then knowing $f(x_1)$ and $f(x_2)$ gives you two equations and three unknowns—but you *can* solve for c if you know the value of $f(x)$ for *three* different values of x. In general, knowing k values of a polynomial f of degree k lets you compute $f(0)$, but any $k - 1$ values of f are consistent with *any* value of $f(0)$. And this result remains true if, instead of using the value $f(x)$ as the share of the secret, we instead use $f(x) \bmod p$, for some prime p. (See p. 731.) Here's a concrete example, to distribute shares of a secret $m \in \{0,1,2,3,4\}$:

- Choose $a_1, \ldots, a_k$ uniformly and independently at random from $\{0,1,2,3,4\}$.
- Let $f(x) = m + \sum_{i=1}^{k} a_i x^i$. Distribute $\langle n, f(n) \bmod 5\rangle$ as "share" #n.

For example, let $k := 3$, and suppose you know that $f(1) \bmod 5 = 1$ and $f(2) \bmod 5 = 2$. These facts don't help you figure out $f(0)$: there are polynomials $\{f_0, f_1, \ldots, f_4\}$ with $f_b(0) = b$ that are all consistent with those observations! (See Figure 7.12.) To put this fact another way, given points $\langle x_1, y_1\rangle$ and $\langle x_2, y_2\rangle$ for $x_1, x_2 \neq 0$, *for any y-intercept b, there exists an $f(x)$ such that $f(x_1) \equiv_p y_1, f(x_2) \equiv_p y_2$, and $f(0) \equiv_p b$.* But three people *can* reconstruct the secret! There's only one quadratic that passes through three given points.

$f_0(x) = 0 + 1x + 0x^2$
$f_1(x) = 1 + 2x + 3x^2$
$f_2(x) = 2 + 3x + 1x^2$
$f_3(x) = 3 + 4x + 4x^2$
$f_4(x) = 4 + 0x + 2x^2$

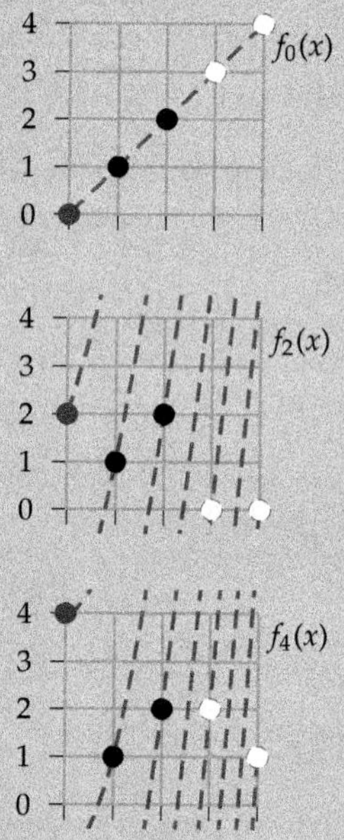

Figure 7.12: Let $f(x) := a + bx + cx^2$. Even knowing $f(1) \equiv_5 1$ and $f(2) \equiv_5 2$, we don't know $f(0) \bmod 5$; there are polynomials consistent with $f(0) \equiv_5 m$ for every $m \in \{0,1,2,3,4\}$. Here we see $f_b(x) \bmod 5$. (These polynomials can be hard to visualize, because their values "wrap around" from 5 to 0.)

Computer Science Connections

Error Correction with Reed–Solomon Codes

Earlier (see Chapter 4), we discussed *error-correcting codes:* we encode a *message* m as a *codeword* $c(m)$, so that m is (efficiently) recoverable from $c(m)$, or even from a mildly corrupted codeword $c' \approx c(m)$. (Note the difference in motivation with cryptography: in error-correcting codes, we want a codeword that makes computing the original message very easy; in cryptography, we want a ciphertext that makes computing the original message very hard.) The key property that we seek is that if $m_1 \neq m_2$, then $c(m_1)$ and $c(m_2)$ are "very different," so that decoding c' simply corresponds to finding the m that minimizes the difference between c' and $c(m)$.

There, we discussed *Reed–Solomon codes,* one of the classic schemes for error-correcting codes. Under Reed–Solomon codes, to encode a message $m \in \mathbb{Z}^k$, we define the polynomial $p_m(x) := \sum_{i=1}^{k} m_i x^i$, and encode m as $\langle p_m(1), p_m(2), \ldots, p_m(n)\rangle$. (We choose n much bigger than k, to achieve the desired error-correction properties.) For example, for the messages $m_1 = \langle 1,3,2\rangle$ and $m_2 = \langle 3,0,3\rangle$, we have $p_{m_1}(x) = x + 3x^2 + 2x^3$ and $p_{m_2}(x) = 3x + 3x^3$. For $n = 6$, we have the codewords (for m_1 and m_2, respectively)

$$\langle 6, 30, 84, 180, 330, 546\rangle \quad \text{and} \quad \langle 6, 30, 90, 204, 390, 666\rangle.$$

The key point is that *two distinct polynomials of degree k agree on at most k inputs,* which means that the codewords for m_1 and m_2 will be very different. (Here $p_{m_1}(x)$ and $p_{m_2}(x)$ agree on $x \in \{1,2\}$, but not on $x \in \{3,4,5,6\}$.) The theorem upon which this difference rests is important enough to be called the *Fundamental Theorem of Algebra*; see Figure 7.13.

Theorem 7.16
Let $f(x)$ be a polynomial of degree k. Then either $f(a) = 0$ for every $a \in \mathbb{Z}$, or the equation $f(x) = 0$ has at most k solutions for $x \in \mathbb{Z}$.

Corollary 7.17
Let f and $g \neq f$ be polynomials of degree k. Then $|\{x : f(x) = g(x)\}| \leq k$.

Figure 7.13: The Fundamental Theorem of Algebra. The corollary follows because the polynomial $h(x) = f(x) - g(x)$ also has degree at most k, and $\{x : f(x) = g(x)\}$ is precisely the set $\{x : h(x) = 0\}$.

While this fact about Reed–Solomon codes is nice, it's already evident that the numbers in the codewords get really big—546 and 666 are very big relative to the integers in the original messages! In real Reed–Solomon codes, there's another trick that's used: every value is stored *modulo a prime.* Let q be a prime. We'll actually encode our message m as

$$\langle p_m(1) \bmod q, p_m(2) \bmod q, \ldots, p_m(n) \bmod q\rangle.$$

In fact, we now encode a message $m \in \mathbb{Z}_q^k$ with a codeword in $\mathbb{Z}_q^n$. And it turns out that everything important about polynomials remains true if we take all values modulo a prime q! (See Figure 7.14.)

Theorem 7.18
Let $f(x)$ be a polynomial of degree k, and let q be a prime number. Then either $f(a) \bmod q = 0$ for every $a \in \mathbb{Z}_q$, or the equation $f(x) = 0$ has at most k solutions for $x \in \mathbb{Z}_q$.

Corollary 7.19
Let f and $g \neq f$ be polynomials of degree k. Then $|\{x : f(x) \equiv_q g(x)\}| \leq k$.

Figure 7.14: The Fundamental Theorem of Algebra, modulo a prime.

The combined message of Reed–Solomon error-correcting codes and the Shamir secret-sharing scheme (p. 730) is the following. Suppose that there is a degree-k polynomial p that is unknown to you, and suppose that you are given the evaluation of this polynomial on n distinct points.

if $n < k$: Then you know nothing about the constant term of the polynomial. (Secrets kept!)

if $n = k$: Then you can compute every coefficient of the polynomial, including the constant term. (Secrets shared!)

if $n > k$: Then you can find the degree-k polynomial consistent with the largest number of these points. (Errors corrected!)

7.3.5 Exercises

The Sieve of Eratosthenes *returns a list of all prime numbers up to a given integer n by creating a list of candidate primes* $\langle 2, 3, \ldots, n \rangle$, *and repeatedly marking the first unmarked number p as prime and striking out all entries in the list that are multiples of p. See the Sieve in action in Figure 7.15.*

2 3 4 5 6 7 8 9 10 11 12 13 14 15 16 17 18 19 20 21 22 23 24 25 26 27 28 29 30 31 32 33 34 35 36 37 38 39 40
2 3 4 5 6 7 8 9 10 11 12 13 14 15 16 17 18 19 20 21 22 23 24 25 26 27 28 29 30 31 32 33 34 35 36 37 38 39 40
2 3 4 5 6 7 8 9 10 11 12 13 14 15 16 17 18 19 20 21 22 23 24 25 26 27 28 29 30 31 32 33 34 35 36 37 38 39 40
2 3 4 5 6 7 8 9 10 11 12 13 14 15 16 17 18 19 20 21 22 23 24 25 26 27 28 29 30 31 32 33 34 35 36 37 38 39 40
2 3 4 5 6 7 8 9 10 11 12 13 14 15 16 17 18 19 20 21 22 23 24 25 26 27 28 29 30 31 32 33 34 35 36 37 38 39 40
2 3 4 5 6 7 8 9 10 11 12 13 14 15 16 17 18 19 20 21 22 23 24 25 26 27 28 29 30 31 32 33 34 35 36 37 38 39 40
⋮

Figure 7.15: A few iterations of the Sieve of Eratosthenes. Primes are underlined as they're discovered; numbers are written in light gray as they're crossed off.

7.38 Write pseudocode to describe the Sieve of Eratosthenes.

7.39 Run the algorithm, by hand, to find all primes less than 100.

7.40 *(programming required)* Implement the Sieve of Eratosthenes in a programming language of your choice. Use your program to compute all primes up to 100,000. How many are there?

7.41 *(programming required)* Earlier, we suggested another algorithm to compute all primes up to $n := 100{,}000$: for each $i = 2, 3, \ldots, n$, test whether i is divisible by any integer between 2 and $\sqrt{i}$. Implement this algorithm too, and compare their execution times. What happens for $n := 500{,}000$?

7.42 Assume that each number k is crossed off by the Sieve of Eratosthenes *every time* a divisor of it is found. (For example, 6 is crossed off when 2 is the prime in question, *and* when 3 is the prime in question.) Prove that the total number of crossings-out by **sieve**(n) is $\leq H_n \cdot n$, where H_n is the nth harmonic number. (See Definition 5.4.)

Use the Prime Number Theorem to ...

7.43 ... estimate the number of primes between $2^{127} + 1$ and 2^{128}.

7.44 ... estimate the 2^{128}th-largest prime.

7.45 ... argue that, roughly, the probability that a randomly chosen number close to n is prime is about $1/\ln n$. *(Hint: what does* primes(n) − primes($n-1$) *represent?)*

7.46 Using the same technique as in Example 7.8, estimate the number of 6-digit primes. Then, using the Sieve or some other custom-built program, determine how far off the estimate was.

Let p be an arbitrary prime number and let a be an arbitrary nonnegative integer. Prove the following facts.

7.47 If $p \nmid a$, then $\gcd(p, a) = 1$.

7.48 For any positive integer k, we have $p \mid a^k$ if and only if $p \mid a$. *(Hint: use induction and Lemma 7.12.)*

7.49 For any integers $n, m \in \{1, \ldots, p-1\}$, we have that $p \nmid nm$.

7.50 For any integer m and any prime number q distinct from p (that is, $p \neq q$), we have $m \equiv_p a$ and $m \equiv_q a$ if and only if $m \equiv_{pq} a$. *(Hint: think first about the case $a = 0$; then generalize.)*

7.51 If $0 \leq a < p$, then $a^2 \equiv_p 1$ if and only if $a \in \{1, p-1\}$. *(You may use Theorem 7.18 from p. 731.)*

Here are some pairs of integers. Using the brute force algorithm (test all candidate divisors) and paper and pencil only, *determine whether they are relatively prime.*

7.52 54321 and 12345 **7.53** 209 and 323 **7.54** 101 and 1100

Using the Extended Euclidean algorithm, compute (by hand) $\gcd(n, m)$ *and integers* x, y *such that* $xn + ym = \gcd(n, m)$ *for the following pairs of numbers:*

7.55 60 and 93 **7.56** 24 and 28 **7.57** 74 and 13

Prove the following extensions to Lemma 7.10:

7.58 There are *infinitely many pairs* of integers x, y such that $xn + ym = \gcd(n, m)$, for any nonnegative integers n and m.

7.59 The extension to $k \geq 2$ integers: if $\gcd(a_1, \ldots, a_k) = d$, then there exist integers $x_1, \ldots, x_k$ such that $\sum_{i=1}^{k} a_i x_i = d$. (Define $\gcd(x_1, x_2, \ldots, x_k) := \gcd(x_1, \gcd(x_2, \ldots, x_k))$ for $k \geq 3$.)

7.60 Prove Theorem 7.11 (the correctness of the Extended Euclidean algorithm) by induction on n: show that for arbitrary positive integers n and m with $n \leq m$, **extended-Euclid**(n, m) returns three integers x, y, r such that $r = \gcd(n, m) = xn + ym$.

7.61 *(programming required)* Write a program that implements the Extended Euclidean algorithm. (Recommended: if you did Exercises 7.11–7.16, compute $m \bmod n$ and $\lfloor \frac{m}{n} \rfloor$ with a single call to **mod-and-div-faster**(m, n).)

I have a friend named Nikki, who's from New Zealand. Nikki and I went out to eat together, and I paid for both dinners. She was going to pay me back, in cash—but she had only New Zealand dollars [NZD]. (I was happy to take NZDs.) Nikki had a giant supply of 5NZD bills; I had a giant supply of 5 U.S. dollar [USD] bills. At the time, the exchange rate was 5NZD = 3USD (or close enough to 5 : 3 for two friends to call it good).

7.62 Prove that Nikki can pay me exactly 4USD in value, through only the exchange of 5NZD and 5USD bills.

7.63 In Exercise 7.62, was there something special about the number 4? Identify for which nonnegative integers x Nikki can pay me back exactly x USD in value, through only the exchange of 5NZD and 5USD bills, and prove your answer.

7.64 In Exercises 7.62–7.63, was there something special about the number 3? Suppose that, due to geopolitical turmoil and a skyrocketing of the price of wool, the 5NZD bill is now worth b USDs, for some $b \equiv_5 3$. I still have many 5USD bills, and Nikki still has the equivalent of many b USD bills. What amounts can Nikki now pay me? Prove your answer.

7.65 In an unexpected twist, I run out of U.S. dollars and Nikki runs out of New Zealand dollars. But I discover that I have a giant supply of identical Israeli Shekel notes, each of which is worth k USD. And Nikki discovers that she has a giant supply of identical Thai Baht notes, each of which is worth ℓ USD. (Assume k and ℓ are integers.) What amounts can she pay me now? Again, prove your answer.

Prove the following facts about relative primality.

7.66 Two consecutive integers (n and $n + 1$) are always relatively prime.

7.67 Two consecutive Fibonacci numbers are always relatively prime.

7.68 Two integers a and b are relatively prime if and only if there is no prime number p such that $p \mid a$ and $p \mid b$. (Notice that this claim differs from the definition of relative primality, which required that there be no *integer* $n \geq 2$ such that $n \mid a$ and $n \mid b$.)

Let a and b be relatively prime integers. Prove the following facts:

7.69 Let $c \in \mathbb{Z}^{\geq 1}$ be relatively prime to both a and b. Then c and ab are also relatively prime.

7.70 For any integer n, we have that both $a \mid n$ and $b \mid n$ if and only if $ab \mid n$.

7.71 For every integer m, there exist integers x and y such that $ax + by = m$.

For the following constraints, describe the set of all $x \in \mathbb{Z}^{\geq 0}$ that satisfies them. Describe this set as $\{a + bk : k \in \mathbb{Z}^{\geq 0}\}$, where a is smallest x satisfying the constraints, $a + b$ is the next smallest, $a + 2b$ is the next smallest, etc.

7.72 $x \bmod 13 = 6$ and $x \bmod 19 = 2$

7.73 $x \bmod 21 = 3$ and $x \bmod 11 = 2$

7.74 $x \bmod 6 = 3$ and $x \bmod 7 = 3$

7.75 $x \bmod 5 = 4$ and $x \bmod 6 = 5$ and $x \bmod 7 = 2$

7.76 $x \bmod 5 = 4$ and $x \bmod 6 = 5$ and $x \bmod 7 = 3$

Show that relative primality was mandatory for the Chinese Remainder Theorem. Namely, show that, for two integers n and m that are not necessarily relatively prime, for some $a \in \mathbb{Z}_n$ and $b \in \mathbb{Z}_m$ …

7.77 … it may be the case that *no* $x \in \mathbb{Z}_{nm}$ satisfies $x \bmod n = a$ and $x \bmod m = b$.

7.78 … it may be the case that *more than one* $x \in \mathbb{Z}_{nm}$ satisfies $x \bmod n = a$ and $x \bmod m = b$.

7.79 Let n and m be relatively prime, and let $a \in \mathbb{Z}_n$ and $b \in \mathbb{Z}_m$. Define y^* to be the unique value in $\mathbb{Z}_{nm}$ such that $y^* \bmod n = a$ and $y^* \bmod m = b$, whose existence is guaranteed by Theorem 7.14. Prove that an integer $x \in \mathbb{Z}_{nm}$ satisfies $x \bmod n = a$ and $x \bmod m = b$ *if and only if* x satisfies $x \bmod nm = y^*$.

7.4 *Multiplicative Inverses*

> Civilization is a limitless multiplication of unnecessary necessities.
>
> Mark Twain (1835–1910)

For any integer $n \geq 2$, let $\mathbb{Z}_n$ denote the set $\{0, 1, \ldots, n-1\}$. In this section, we'll discuss *arithmetic over* $\mathbb{Z}_n$—that is, arithmetic where we think of all expressions by considering their value modulo n. For example, when $n = 9$, the expressions $4 + 6$ and $8 \cdot 7$ are equivalent to 1 and 2, respectively, because $10 \bmod 9 = 1$ and $56 \bmod 9 = 2$. When $n = 10$, the expressions $4 + 6$ and $8 \cdot 7$ are equivalent to 0 and 6, respectively.

We have already encountered addition and multiplication in the world of modular arithmetic (for example, in Theorem 7.3). But we haven't yet defined subtraction or division. (Theorem 7.3 also introduced exponentiation over $\mathbb{Z}_n$, and it turns out that, along with division, exponentiation in modular arithmetic will form the foundation of the RSA cryptographic system; see Section 7.5.) Subtraction turns out to be fairly straightforward (see Exercise 7.81), but division will be a bit trickier than $+$, $\cdot$, and $-$. In this section, we'll introduce what division over $\mathbb{Z}_n$ even means, and then discuss algorithms to perform modular division.

7.4.1 *The Basic Definitions*

Before we introduce any of the technical definitions, let's start with a tiny bit of intuition about why there's something potentially interesting going on with division in $\mathbb{Z}_n$. For concreteness, here's a small example in $\mathbb{Z}_9$:

Example 7.18 (Halving some numbers in $\mathbb{Z}_9$)

Problem: In $\mathbb{Z}_9 = \{0, 1, 2, 3, 4, 5, 6, 7, 8\}$, where every expression's value is understood mod 9, what element of $\mathbb{Z}_9$ is half of 6? Half of 8? Half of 5?

Solution: What number is half of 6? Well, easy: it's obviously 3. (Why? Because 6 is double 3, and therefore 3 is half of 6—or, in other words, 3 is half of 6 because $3 \cdot 2$ is 6.) And what number is half of 8? Easy again: it's 4 (because $4 \cdot 2$ is 8).

Okay, what number is half of 5? The first temptation is to say that it's 2.5 (or $\frac{5}{2}$, if you're more of a fan of fractions)—but that doesn't make sense as an answer: after all, which element of $\{0, 1, 2, 3, 4, 5, 6, 7, 8\}$ is 2.5?!? So the next temptation is to say that there is *no* number that's half of 5. (After all, in normal nonmodular arithmetic, there is no integer that's half of 5.) But that's not right either: there *is* an answer in $\mathbb{Z}_9$, even if it doesn't quite match our intuition. The number that's half of 5 is in fact 7(!). Why? Because $7 \cdot 2$ is 5. (Remember that we're in $\mathbb{Z}_9$, and $14 \bmod 9 = 5$.) So, in $\mathbb{Z}_9$, the number 7 is half of the number 5. (See Figure 7.16.)

$2a$	a
0	0
1	5
2	1
3	6
4	2
5	7
6	3
7	8
8	4

Figure 7.16: For each $b \in \mathbb{Z}_9$, the value of $a \in \mathbb{Z}_9$ such that $2a = b$.

Example 7.18 illustrates the basic idea of division in $\mathbb{Z}_n$: we'll define $\frac{a}{b}$ as the number k such that $k \cdot b$ is equivalent to a in $\mathbb{Z}_n$. To make this idea formal, we'll need a few definitions about modular arithmetic. But, first, we'll go back to "normal" arithmetic, for the real numbers, and introduce the two key concepts: *identity* and *inverse*.

Multiplicative inverses in $\mathbb{R}$

The number 1 is called the *multiplicative identity,* because it has the property that

$$x \cdot 1 = 1 \cdot x = x, \text{ for any } x \in \mathbb{R}.$$

(We've encountered identities in a number of contexts already. In Definition 2.41, we introduced the identity matrix I, where $MI = IM = M$ for any matrix M. And Exercises 3.13–3.16 explored the identities of logical connectives; for example, the identity of $\vee$ is False, because $p \vee \text{False} \equiv \text{False} \vee p \equiv p$ for any proposition p.)

The *multiplicative inverse* of a number x is the number by which we have to multiply x to get 1 (that is, to get the multiplicative identity) as the result. In other words, the multiplicative inverse of $x \in \mathbb{R}$ is the real number x^{-1} such that $x \cdot x^{-1} = 1$. (We generally denote the multiplicative inverse of x as x^{-1}, though it may be easier to think about the multiplicative inverse of x as $\frac{1}{x}$, because $x \cdot \frac{1}{x} = 1$. Actually the "$^{-1}$" notation is in general ambiguous between denoting inverse and denoting exponentiation with a negative exponent—though these concepts match up perfectly for the real numbers. Exercise 7.99 addresses negative exponents in modular arithmetic.) For example, the multiplicative inverse of 8 is $\frac{1}{8} = 0.125$, because $8 \cdot 0.125 = 1$.

When we think of *dividing* $y \in \mathbb{R}$ *by* $x \in \mathbb{R}$, we can instead think of this operation as *multiplying* y *by* x^{-1}. For example, we have $7/8 = 7 \cdot 8^{-1} = 7 \cdot 0.125 = 0.875$.

Not every real number has a multiplicative inverse: specifically, there is no number that yields 1 when it's multiplied by 0, so 0^{-1} doesn't exist. (And we can't divide y by 0, because 0^{-1} doesn't exist.) But for any $x \neq 0$, the multiplicative inverse of x does exist, and it's given by $x^{-1} := \frac{1}{x}$.

Problem-solving tip: When you encounter a new definition, it's often helpful to try it out in a setting that you already understand well. For example, it's easier understand Manhattan distance in $\mathbb{R}^2$ (see Example 2.40) before trying to understand it for general $\mathbb{R}^n$. In this case, you've grasped division in $\mathbb{R}$ since, what, second grade—so, before trying to make sense of the definitions for $\mathbb{Z}_n$, try to consider the analogy of each definition for $\mathbb{R}$.

Multiplicative inverses in $\mathbb{Z}_n$

Now let's turn to the analogous definitions in the world of modular arithmetic, in $\mathbb{Z}_n$. Notice that 1 is still the multiplicative identity, for any modulus n: for any $x \in \mathbb{Z}_n$, it is the case that $x \bmod n = 1 \cdot x \bmod n = x \cdot 1 \bmod n$. The definition of the multiplicative inverse in $\mathbb{Z}_n$ is identical to the definition in $\mathbb{R}$:

Definition 7.8 (Multiplicative Inverse)
Let $n \geq 2$ be any integer, and let $a \in \mathbb{Z}_n$ be arbitrary. The multiplicative inverse of a in $\mathbb{Z}_n$ *is the number $a^{-1} \in \mathbb{Z}_n$ such that $a \cdot a^{-1} \equiv_n 1$. If there is no element $x \in \mathbb{Z}_n$ such that $ax \equiv_n 1$, then a^{-1} is undefined.*

(Note that Definition 7.8 describes the multiplicative inverse as "the" a^{-1} that has the desired property. In Exercise 7.92, you'll show that there can't be two distinct values $b, c \in \mathbb{Z}_n$ where $ab \equiv_n ac \equiv_n 1$.) Here are a few examples of multiplicative inverses, and of a case where there is no multiplicative inverse:

Writing tip: Let $a \in \mathbb{Z}_n$. The notation a^{-1} doesn't explicitly indicate the modulus n anywhere, and the value of n matters! If there's *any* ambiguity about the value of n, then be sure to specify it clearly in your words surrounding the notation.

Example 7.19 (Some multiplicative inverses)
The multiplicative inverse of 2 in $\mathbb{Z}_9$ is $2^{-1} = 5$, because $2 \cdot 5 = 10 \equiv_9 1$, and the multiplicative inverse of 1 in $\mathbb{Z}_9$ is $1^{-1} = 1$, because $1 \cdot 1 \equiv_9 1$. The multiplicative

inverse of 7 in $\mathbb{Z}_{11}$ is 8 because $7 \cdot 8 = 56 \equiv_{11} 1$, and the multiplicative inverse of 7 in $\mathbb{Z}_{13}$ is 2 because $7 \cdot 2 = 14 \equiv_{13} 1$.

Example 7.20 (A nonexistent multiplicative inverse)
The number 3 has no multiplicative inverse in $\mathbb{Z}_9$, as the following table shows:

$$\begin{array}{lllllllll} 3\cdot 0 & = 0 & \equiv_9 0 & 3\cdot 3 & = 9 & \equiv_9 0 & 3\cdot 6 & = 18 & \equiv_9 0 \\ 3\cdot 1 & = 3 & \equiv_9 3 & 3\cdot 4 & = 12 & \equiv_9 3 & 3\cdot 7 & = 21 & \equiv_9 3 \\ 3\cdot 2 & = 6 & \equiv_9 6 & 3\cdot 5 & = 15 & \equiv_9 6 & 3\cdot 8 & = 24 & \equiv_9 6. \end{array}$$

All nine of these entries are not equivalent to 1 modulo 9, so there is no 3^{-1} in $\mathbb{Z}_9$.

Example 7.21 (Multiplicative inverses in $\mathbb{Z}_7$)
Problem: Find the values of $0^{-1}, 1^{-1}, 2^{-1}, 3^{-1}, 4^{-1}, 5^{-1}$, and 6^{-1} in $\mathbb{Z}_7$.

Solution: The simplest way (though not necessarily the fastest way!) to solve this problem is by building a multiplication table for $\mathbb{Z}_7$, as shown in Figure 7.17. (The entry in row a and column b of the table is the value ab mod 7—for example, $4 \cdot 5 = 20 = 2 \cdot 7 + 6$, so the entry in row 4, column 5 is the number 6.) For each row a, the value a^{-1} we seek is the column that has a 1 in it, if there is such a column in that row. (And there is a 1 in every row except $a = 0$.) Thus in $\mathbb{Z}_7$ we have $1^{-1} = 1$, $2^{-1} = 4$, $3^{-1} = 5$, $4^{-1} = 2$, $5^{-1} = 3$, and $6^{-1} = 6$—and 0^{-1} is undefined.

	0	1	2	3	4	5	6
0	0	0	0	0	0	0	0
1	0	1	2	3	4	5	6
2	0	2	4	6	1	3	5
3	0	3	6	2	5	1	4
4	0	4	1	5	2	6	3
5	0	5	3	1	6	4	2
6	0	6	5	4	3	2	1

Figure 7.17: The multiplication table for $\mathbb{Z}_7$.

Taking it further: The field of mathematics called *abstract algebra* focuses on giving and analyzing very general definitions of structures that satisfy certain properties—allowing apparently disparate objects (like Boolean logic and Rubik's cubes) to be studied at the same time. For example, a *group* is a pair $\langle G, \cdot \rangle$, where G is a set of objects and $\cdot$ is a binary operator on G, where certain properties are satisfied:

- *Closure:* for any $a, b \in G$, we have $a \cdot b \in G$.
- *Associativity:* for any $a, b, c \in G$, we have $a \cdot (b \cdot c) = (a \cdot b) \cdot c$.
- *Identity:* there is an *identity element* $e \in G$ with the property that $a \cdot e = e \cdot a = a$ for every $a \in G$.
- *Inverse:* for every $a \in G$, there exists $b \in G$ such that $a \cdot b = b \cdot a = e$ (where e is the identity element).

For example, $\langle \mathbb{Z}, + \rangle$ is a group. As we'll see, so too is $\langle \mathbb{Z}_p - \{0\}, \cdot \rangle$, where $\cdot$ denotes multiplication and p is any prime integer. Despite the very abstract nature of these definitions—and other more general or more specific algebraic structures, like *semigroups, rings,* and *fields*—they are a surprisingly useful way of understanding properties of $\mathbb{Z}_p$. See any good textbook on abstract algebra for more detail.

7.4.2 *When Multiplicative Inverses Exist (and How to Find Them)*

Examples 7.19, 7.20, and 7.21 might inspire you to ask a question that will turn out to be both useful and reasonably simple to answer: *under what circumstances does a particular number $a \in \mathbb{Z}_n$ have a multiplicative inverse?* As we saw with arithmetic over $\mathbb{R}$, there's never a multiplicative inverse for 0 in any $\mathbb{Z}_n$ (because, for any x, we have $x \cdot 0 = 0 \not\equiv_n 1$)—but what happens for nonzero a?

To take one particular case, we just found that $2^{-1} = 5$ in $\mathbb{Z}_9$ but that 3^{-1} does not exist in $\mathbb{Z}_9$. It's worth reflecting a bit on "why" 3^{-1} failed to exist in $\mathbb{Z}_9$. There are a lot of ways to think about it, but here's one convenient way to describe what went wrong: any multiple of 3 is (obviously!) divisible by 3, and numbers divisible by 3 are never one more than multiples of 9. In other words, the only possible values of $3x \bmod 9$ are $\{0, 3, 6\}$—a set that fails to include 1. (Recall from Definition 7.8 that, for 3^{-1} to exist in $\mathbb{Z}_9$, we'd have to have been able to find an x such that $3x \equiv_9 1$.) Similarly, 6^{-1} doesn't exist in $\mathbb{Z}_9$: again, the only possible values of $6x \bmod 9$ are $\{0, 3, 6\}$, which once again does not include 1.

These observations should be reminiscent of the concepts that we discussed in Section 7.3: for a number $a \in \mathbb{Z}_n$, we seem to be unable to find a multiplicative inverse a^{-1} in $\mathbb{Z}_n$ whenever a and n share a common divisor $d > 1$. In other words, when a and n are not relatively prime, then a^{-1} fails to exist in $\mathbb{Z}_n$. (That's because any multiple xa of a will also be divisible by d, and so $xa \bmod n$ will also be divisible by d, and therefore $xa \bmod n$ will not equal 1.) In fact, not being relatively prime to n is the *only* way to fail to have a multiplicative inverse in $\mathbb{Z}_n$, as we'll prove. (Note that $0 \in \mathbb{Z}_n$ is *not* relatively prime to n, because $\gcd(n, 0) \neq 1$.)

Theorem 7.20 (Existence of Multiplicative Inverses)
Let $n \geq 2$ and $a \in \mathbb{Z}_n$. Then a^{-1} exists in $\mathbb{Z}_n$ if and only if n and a are relatively prime.

Proof. By definition, a multiplicative inverse of a exists in $\mathbb{Z}_n$ precisely when there exists an integer x such that $ax \equiv_n 1$. (The definition actually requires $x \in \mathbb{Z}_n$, not just $x \in \mathbb{Z}$, but see Exercise 7.98.) But $ax \equiv_n 1$ means that ax is one more than a multiple of n—that is, there exists some integer y such that $ax + yn = 1$. In other words,

$$a^{-1} \text{ exists in } \mathbb{Z}_n \text{ if and only if there exist integers } x, y \text{ such that } ax + yn = 1. \qquad (*)$$

Observe that $(*)$ echoes the form of Lemma 7.10 (and thus also echoes the output of the Extended Euclidean algorithm), and we can use this fact to prove the theorem. We'll prove the two directions of the implication separately:

If a^{-1} exists in $\mathbb{Z}_n$, then a and n are relatively prime. We'll prove the contrapositive. Suppose that a and n are not relatively prime—that is, suppose that $\gcd(a, n) = d$ for some $d > 1$. We will show that a^{-1} does not exist in $\mathbb{Z}_n$. Because $d \mid a$ and $d \mid n$, there exist integers c and k such that $a = cd$ and $n = kd$. But then, for *any* integers x and y, we have that

$$ax + yn = cdx + ykd = d(cx + yk)$$

and thus $d \mid (ax + yn)$. Thus there are no integers x, y for which $ax + yn = 1$ and therefore, by $(*)$, a^{-1} does not exist in $\mathbb{Z}_n$.

If a and n are relatively prime, then a^{-1} exists in $\mathbb{Z}_n$. Suppose that a and n are relatively prime. Then $\gcd(a, n) = 1$ by definition. Thus, by the correctness of the Extended Euclidean algorithm (Theorem 7.11), the output of **extended-Euclid**(a, n) is $\langle x, y, 1 \rangle$ for integers x, y such that $xa + yn = \gcd(a, n) = 1$. The fact that **extended-Euclid**(a, n) outputs integers x and y such $xa + yn = 1$ means that such an x and y must exist—and so, by $(*)$, a^{-1} exists in $\mathbb{Z}_n$. □

Note that this theorem is consistent with the examples that we saw previously: we found 1^{-1} and 2^{-1} but not 3^{-1} in $\mathbb{Z}_9$ (Examples 7.19 and 7.20; 1 and 2 are relatively prime to 9, but 3 is not), and we found multiplicative inverses for all nonzero elements of $\mathbb{Z}_7$ (Example 7.21; all of $\{1, 2, \ldots, 6\}$ are relatively prime to 7).

Two implications of Theorem 7.20

There are two useful implications of this result. First, when the modulus is prime, multiplicative inverses exist for *all* nonzero elements of $\mathbb{Z}_n$, because every nonzero $a \in \mathbb{Z}_n$ and n are relatively prime for any prime number n.

Corollary 7.21
If p is prime, then every nonzero $a \in \mathbb{Z}_p$ has a multiplicative inverse in $\mathbb{Z}_p$.

(We saw an example of this corollary in Example 7.21, where we identified the multiplicative inverses of all nonzero elements in $\mathbb{Z}_7$.)

The second useful implication of Theorem 7.20 is that, whenever the multiplicative inverse of a exists in $\mathbb{Z}_n$, we can efficiently *compute* a^{-1} in $\mathbb{Z}_n$ using the Extended Euclidean algorithm—specifically, by running the (simple!) algorithm in Figure 7.18. (This problem also nicely illustrates a case in which proving a structural fact vastly improves the efficiency of a calculation—the algorithm in Figure 7.18 is *way* faster than building the entire multiplication table, as we did in Example 7.21.)

```
inverse(a, n):
Input: a ∈ Z_n and n ≥ 2
Output: a^{-1} in Z_n, if it exists
x, y, d := extended-Euclid(a, n)
if d = 1 then
    return x mod n            // xa + yn = 1, so xa ≡_n 1.
else
    return "no inverse for a exists in Z_n."
```

Figure 7.18: An algorithm for computing multiplicative inverses using the Extended Euclidean algorithm.

Corollary 7.22
For any $n \geq 2$ and $a \in \mathbb{Z}_n$, **inverse**(a, n) *returns the value of a^{-1} in $\mathbb{Z}_n$.*

Proof. We just proved that a^{-1} exists if and only if **extended-Euclid**(a, n) returns $\langle x, y, 1 \rangle$. In this case, we have $xa + yn = 1$ and therefore $xa \equiv_n 1$. Defining $a^{-1} := x \bmod n$ ensures that $a \cdot (x \bmod n) \equiv_n 1$, as required. (Again, see Exercise 7.98.) □

Here's an example, replicating the calculation of 5^{-1} in $\mathbb{Z}_7$ from Example 7.21:

Example 7.22 (5^{-1} in $\mathbb{Z}_7$, again)
To compute 5^{-1}, we run the Extended Euclidean algorithm on 5 and 7:

extended-Euclid$(5, 7)$
  extended-Euclid$(\overbrace{7 \bmod 5}^{=2}, 5)$
    extended-Euclid$(\underbrace{5 \bmod 2}_{=1}, 2)$
    $= 1, 0, 1$
  $= -2, 1, 1$
$= 3, -2, 1.$

The Extended Euclidean algorithm returns $\langle 3, -2, 1 \rangle$, implying that $3 \cdot 5 + -2 \cdot 7 = 1 = \gcd(5, 7)$. Therefore **inverse**$(5, 7)$ returns $3 \bmod 7 = 3$. And, indeed, $3 \cdot 5 \equiv_7 1$.

Example 7.23 (7^{-1} in $\mathbb{Z}_9$)
In Example 7.16, we saw that **extended-Euclid**(7,9) = $\langle 4, -3, 1 \rangle$. Thus 7 and 9 are relatively prime, and 7^{-1} in $\mathbb{Z}_9$ is 4 mod 9 = 4. And indeed $7 \cdot 4 = 28 \equiv_9 1$.

7.4.3 *Fermat's Little Theorem*

We'll now make use of the results that we've developed so far—specifically Corollary 7.21—to prove a surprising and very useful theorem, called *Fermat's Little Theorem*, which states that a^{p-1} is equivalent to 1 mod p, for any prime number p and any $a \neq 0$. (And we'll see why this result is useful for cryptography in Section 7.5.)

Fermat's Little Theorem is named after Pierre de Fermat, a 17th-century French mathematician.

Taking it further: Fermat's Little Theorem is the second-most famous theorem named after Pierre de Fermat. His more famous theorem is called *Fermat's Last Theorem*, which states the following:

For any integer $k \geq 3$, there are no positive integers x, y, z satisfying $x^k + y^k = z^k$.

There *are* integer solutions to the equation $x^k + y^k = z^k$ when $k = 2$—the so-called *Pythagorean triples*, like $\langle 3,4,5 \rangle$ (where $3^2 + 4^2 = 9 + 16 = 25 = 5^2$) and $\langle 7,24,25 \rangle$ (where $7^2 + 24^2 = 49 + 576 = 625 = 25^2$). But Fermat's Last Theorem states that there are no integer solutions when the exponent is larger than 2.

The history of Fermat's Last Theorem is convoluted and about as fascinating as the history of any mathematical statement can be. In the 17th century, Fermat conjectured his theorem, and scrawled—in the margin of one of his books on mathematics—the words "I have discovered a truly marvelous proof, which this margin is too narrow to contain . . . ." The conjecture, and Fermat's assertion, were found after Fermat's death—but the proof that Fermat claimed to have discovered was never found. And it seems almost certain that he did not have a correct proof of this claim. Some 350 years later, in 1995, the mathematician Andrew Wiles published a proof of Fermat's Last Theorem, building on work by a number of other 20th-century mathematicians.

The history of the Fermat's Last Theorem—including the history of Fermat's conjecture and the centuries-long quest for a proof—has been the subject of a number of books written for a nonspecialist audience; see, for example, the book by Simon Singh.[4]

[4] Simon Singh. *Fermat's Last Theorem: The Story of a Riddle That Confounded the World's Greatest Minds for 358 Years.* Fourth Estate Ltd., 2002.

Before we can prove Fermat's Little Theorem itself, we'll need a preliminary result. We will show that, for any prime p and any nonzero $a \in \mathbb{Z}_p$, the first $p - 1$ nonzero multiples of a—that is, $\{a, 2a, 3a, \ldots, (p-1)a\}$—are precisely the $p - 1$ nonzero elements of $\mathbb{Z}_p$. Or, to state this claim in a slightly different way, we will prove that the function $f : \mathbb{Z}_p \to \mathbb{Z}_p$ defined by $f(k) = ak \bmod p$ is both one-to-one and onto (and also satisfies $f(0) = 0$). Here is a formal statement of the result:

Lemma 7.23 ($\{1, 2, \ldots p-1\}$ and $\{1a, 2a, \ldots (p-1)a\}$ are equivalent mod p)
For prime p and any $a \in \mathbb{Z}_p$ where $a \neq 0$, we have

$$\{1 \cdot a \bmod p, 2 \cdot a \bmod p, \ldots, (p-1) \cdot a \bmod p\} = \{1, 2, \ldots, p-1\}.$$

Before we dive into a proof, let's check an example:

Example 7.24 ($\{ai \bmod 11\}$ vs. $\{i \bmod 11\}$)
Consider the prime $p = 11$ and two values of a, namely $a = 2$ and $a = 5$. Then, taking

all results modulo 11, we have

i	1	2	3	4	5	6	7	8	9	10
$2i$	2	4	6	8	10	12	14	16	18	20
$2i$ mod 11	2	4	6	8	10	1	3	5	7	9
$5i$	5	10	15	20	25	30	35	40	45	50
$5i$ mod 11	5	10	4	9	3	8	2	7	1	6

Note that every number from $\{1, 2, \ldots, p\}$ appears (once and only once) in the $\{2i \bmod 11\}$ and $\{5i \bmod 11\}$ rows of this table—exactly as desired. That is,

$$\{1, 2, 3, \ldots, 10\} \equiv_{11} \{2, 4, 6, \ldots, 20\} \equiv_{11} \{5, 10, 15, \ldots, 50\}.$$

We can also observe examples of this result in the multiplication table for $\mathbb{Z}_7$. (See Figure 7.19 for a reminder.) We can see that every (nonzero) row $\{a, 2a, 3a, 4a, 5a, 6a\}$ contains all six numbers $\{1, 2, 3, 4, 5, 6\}$, in some order, in the six nonzero columns.

	0	1	2	3	4	5	6
0	0	0	0	0	0	0	0
1	0	1	2	3	4	5	6
2	0	2	4	6	1	3	5
3	0	3	6	2	5	1	4
4	0	4	1	5	2	6	3
5	0	5	3	1	6	4	2
6	0	6	5	4	3	2	1

Figure 7.19: The multiplication table for $\mathbb{Z}_7$: a reminder.

Proof of Lemma 7.23. Consider any prime p, and any nonzero $a \in \mathbb{Z}_p$. We must prove that $\{a, 2a, \ldots, (p-1)a\} \equiv_p \{1, 2, \ldots, p-1\}$.

We will first argue that the set $\{1 \cdot a \bmod p, 2 \cdot a \bmod p, \ldots, (p-1) \cdot a \bmod p\}$ contains no duplicates—that is, the value of $i \cdot a \bmod p$ is different for every i. Let $i, j \in \{1, 2, \ldots, p-1\}$ be arbitrary. We will show that $ia \equiv_p ja$ implies that $i = j$, which establishes this first claim. Suppose that $ia \equiv_p ja$. Then, multiplying both sides by a^{-1}, we have that $iaa^{-1} \equiv_p jaa^{-1}$, which immediately yields $i \equiv_p j$ because $a \cdot a^{-1} \equiv_p 1$. (Note that, because p is prime, by Corollary 7.21, we know that a^{-1} exists in $\mathbb{Z}_p$.) Therefore, for any $i, j \in \{1, 2, \ldots, 1-p\}$, if $i \neq j$ then $ai \not\equiv_p aj$.

We now need only show that $ia \bmod p \neq 0$ for any $i > 0$. But that fact is straightforward to see: $ia \bmod p = 0$ if and only if $p \mid ia$, but p is prime and $i < p$ and $a < p$, so p cannot divide ia. (See Exercise 7.49.) □

With this preliminary result in hand, we turn to Fermat's Little Theorem itself:

Theorem 7.24 (Fermat's Little Theorem)
Let p be prime, and let $a \in \mathbb{Z}_p$ where $a \neq 0$. Then $a^{p-1} \equiv_p 1$.

As with the previous lemma, we'll start with a few examples of this claim, and then give a proof of the general result. (While this property admittedly might seem a bit mysterious, it turns out to follow fairly closely from Lemma 7.23, as we'll see.)

Example 7.25 (Some examples of Fermat's Little Theorem)
Here are a few examples, for the prime numbers 7 and 19:

$$\begin{aligned}
2^6 \bmod 7 &= 64 \bmod 7 &&= (7 \cdot 9 + 1) \bmod 7 = 1 \\
3^6 \bmod 7 &= 729 \bmod 7 &&= (104 \cdot 7 + 1) \bmod 7 = 1 \\
4^{18} \bmod 19 &= 68719476736 \bmod 19 &&= (3616814565 \cdot 19 + 1) \bmod 19 = 1.
\end{aligned}$$

THE PROOF OF FERMAT'S LITTLE THEOREM

We'll now turn to a proof of the theorem: for any prime p and any nonzero $a \in \mathbb{Z}_p$, we have that $a^{p-1} \equiv_p 1$:

Proof of Fermat's Little Theorem (Theorem 7.24). Note that, because p is prime, by Corollary 7.21, the multiplicative inverses $1^{-1}, 2^{-1}, \ldots, (p-1)^{-1}$ all exist in $\mathbb{Z}_p$.

By Lemma 7.23, we know that $\{1 \cdot a \bmod p, 2 \cdot a \bmod p, \ldots, (p-1) \cdot a \bmod p\}$ and $\{1, 2, \ldots, p\}$ are the same set, and thus have the same product:

$$\begin{aligned} & 1 \cdot 2 \cdot 3 \cdots (p-1) \\ \equiv_p \; & (1 \cdot a) \cdot (2 \cdot a) \cdot (3 \cdot a) \cdots ((p-1) \cdot a). \end{aligned} \tag{1}$$

Multiplying both sides of (1) by the product of all $p-1$ multiplicative inverses of $\{1, \ldots, p-1\}$—that is, multiplying by $1^{-1} \cdot 2^{-1} \cdot \cdots \cdot (p-1)^{-1}$—we have

$$\begin{aligned} & 1 \cdot 2 \cdot 3 \cdots (p-1) \cdot 1^{-1} \cdot 2^{-1} \cdots (p-1)^{-1} \\ \equiv_p \; & (1 \cdot a) \cdot (2 \cdot a) \cdot (3 \cdot a) \cdots ((p-1) \cdot a) \cdot 1^{-1} \cdot 2^{-1} \cdots (p-1)^{-1}. \end{aligned} \tag{2}$$

Rearranging the left-hand side of (2) and replacing $b \cdot b^{-1}$ by 1 for each $b \in \{1, \ldots, p-1\}$, we simply get 1:

$$1 \equiv_p (1 \cdot a) \cdot (2 \cdot a) \cdot (3 \cdot a) \cdots ((p-1) \cdot a) \cdot 1^{-1} \cdot 2^{-1} \cdots (p-1)^{-1}. \tag{3}$$

Rearranging the right-hand side of (3) and again replacing each $b \cdot b^{-1}$ by 1, we are left only with $p-1$ copies of a:

$$1 \equiv_p a^{p-1}.$$

□

Note that Fermat's Little Theorem is an implication, *not* an equivalence. It states that *if* p is prime, *then* for every $a \in \{1, \ldots, p-1\}$—that is, for every p relatively prime to n—we have $a^{p-1} \equiv_p 1$. The converse does not always hold: if $a^{n-1} \equiv_n 1$ for every $a \in \mathbb{Z}_n$ that's relatively prime to n, *we cannot conclude that n is prime.* For example, $a^{560} \equiv_{561} 1$ for every $a \in \{1, 2, \ldots, 560\}$ with $\gcd(a, 561) = 1$—but 561 is not prime! (See Exercise 7.110.) A number like 561, which passes the test in Fermat's Little Theorem but is not prime, is called a *Fermat pseudoprime* or a *Carmichael number.*

Carmichael numbers are named after Robert Carmichael, an American mathematician who first discovered these numbers, in the early 20th century.

Taking it further: Let $n \geq 2$ be an integer, and suppose that we need to determine whether n is prime. There's a test for primality that's implicitly suggested by Fermat's Little Theorem—for "many" different values of $a \in \mathbb{Z}_n$, test to make sure that $a^{n-1} \bmod n = 1$—but this test sometimes incorrectly identifies composite numbers as prime, because of the Carmichael numbers. (For speed, we generally test a few randomly chosen values of $a \in \mathbb{Z}_p$ instead of trying many of them—but of course testing *fewer* values of a certainly can't prevent us from incorrectly identifying Carmichael numbers as prime.) However, there are some tests for primality that have a similar spirit but that aren't fooled by certain inputs in this way. See the discussion on p. 742 for a description of a randomized algorithm called the *Miller–Rabin test* that checks primality using this approach.

Computer Science Connections

Miller–Rabin Primality Test

Fermat's Little Theorem says that $a^{n-1} \equiv_n 1$ for any prime n and any nonzero $a \in \mathbb{Z}_n$, which makes the randomized algorithm in Figure 7.20 tempting as a way to test for primality. It's clear that **bogus-isPrime?**(p) returns "prime" for any prime p—by Fermat's Little Theorem—but what's not clear is the *false negative probability*. Unfortunately, the probability can be terrible for particular values of n: for example, $n = 118{,}901{,}521$ is not prime, but the only a for which $a^{n-1} \not\equiv_n 1$ are multiples of 271, 541, or 811—less than 0.7% of $\{1, 2, \ldots, n-1\}$. (See the discussion of *Carmichael numbers*, and Exercise 7.110. And Carmichael numbers whose prime factors are all > 271 give even worse performance.)

```
bogus-isPrime?(n, k):
Input: n is a candidate prime number; k is a
    "certainty parameter" telling us how many
    tests to perform before giving up and reporting
    n as prime.
repeat
    choose a ∈ {1, 2, ..., n − 1} randomly
until a^(n−1) ≢_n 1 or we've tried k times
return "prime" if every a^(n−1) ≡_n 1; else return
    "composite"
```

Figure 7.20: A bogus primality tester based on Fermat's Little Theorem.

We can, however, give a randomized primality test using modular arithmetic that doesn't get fooled for any particular input integer. The *Miller–Rabin primality test*[5] is based on the following fact (see Exercise 7.51):

$$\text{if } p \text{ is prime, then } x^2 \equiv_p 1 \text{ if and only if } x \in \{1, p-1\}. \tag{1}$$

Or, taking the contrapositive,

$$\text{if } a^2 \equiv_n 1 \text{ for } a \notin \{1, n-1\}, \text{ then } n \text{ is not prime.} \tag{2}$$

The original version of this test, due to Miller, is a nonrandom version of this algorithm that relies on a (still!) unproven assumption in mathematics; it was subsequently modified by Rabin to remove the assumption (but at the cost of making it random instead). See

[5] Gary L. Miller. Riemann's hypothesis and tests for primality. *Journal of Computer and System Sciences*, 13(3):300–317, 1976; and Michael O. Rabin. Probabilistic algorithm for testing primality. *Journal of Number Theory*, 12(1):128–138, 1980.

The basic idea of Miller–Rabin is to look for an $a \in \mathbb{Z}_n$ with this property. (See Figure 7.21.) Consider a candidate prime number $n \geq 3$. Thus n is odd, so $n-1$ is even, and we can write $n - 1 = 2^r d$, where d is an odd number and $r \geq 1$. (For $n = 561$, for example, we can write $n - 1 = 560 = 2^4 \cdot 35$—so $r = 4$ and $d = 35$.) Let $a \in \mathbb{Z}_n$ with $a \neq 0$. Define the sequence

$$a^d,\; (a^d)^2 = a^{2d},\; (a^{2d})^2 = a^{4d},\; \ldots,\; (a^{2^{r-1}d})^2 = a^{2^r d} = a^{n-1}, \tag{3}$$

with each entry taken modulo n. For example, for $n = 561$ (so $r = 4$ and $d = 35$) and $a = 4$, this sequence (modulo n) would be

$$\langle \underbrace{166}_{a^d \equiv_n 4^{35} \equiv_n 166},\; \underbrace{67}_{a^{2d} \equiv_n 166^2 \equiv_n 67},\; \underbrace{1}_{a^{4d} \equiv_n 67^2 \equiv_n 1},\; \underbrace{1}_{a^{8d} \equiv_n 1^2 \equiv_n 1},\; \underbrace{1}_{a^{16d} \equiv_n 1^2 \equiv_n 1} \rangle.$$

By Fermat's Little Theorem, we know n is not prime if $a^{n-1} \not\equiv_n 1$. Thus if (3) ends with something $\not\equiv_n 1$, we know that n is not prime. And if there's a 1 that appears immediately after an entry x where $x \bmod n \notin \{1, n-1\}$ in (3), then we also know that n is not prime: $x^2 \equiv_n 1$ but $x \bmod n \notin \{1, n-1\}$, so by (2) we know that n is not prime. The key fact, which we won't prove here, is that *many different values of $a \in \mathbb{Z}_n$ result in one of these two violations*:[6]

```
miller-rabin-isPrime?(n, k):
Input: n is a candidate prime number; k is a
    "certainty parameter"
write n − 1 as 2^r d for an odd number d
while we've done fewer than k tests:
    choose a random a ∈ {1, ..., n − 1}
    σ := ⟨a^d, a^(2d), a^(4d), a^(8d), ..., a^(2^r d)⟩ mod n.
    if σ ≠ ⟨..., 1⟩ or if σ = ⟨..., x, 1, ...⟩ for some
        x ∉ {1, n − 1} then
        return "composite"
return "prime"
```

Figure 7.21: Miller–Rabin primality test.

> **Fact:** If n is not prime, then for at least $\frac{n-1}{2}$ different nonzero values of $a \in \mathbb{Z}_n$, the sequence (3) contains a 1 following an entry $x \notin \{1, n-1\}$ or the sequence (3) doesn't end with 1.

This fact then allows us to test for n's primality by trying k different randomly chosen values of a; the probability that every one of these tests fails when n is not prime is at most $1/2^k$.

For a proof of this fact, see

[6] Thomas H. Cormen, Charles E. Leiserson, Ronald L. Rivest, and Clifford Stein. *Introduction to Algorithms*. MIT Press, 3rd edition, 2009.

7.4.4 Exercises

7.80 Following Example 7.18, identify the numbers that are half of *every* element in $\mathbb{Z}_9$. (That is, for each $a \in \mathbb{Z}_9$, find $b \in \mathbb{Z}_9$ such that $2b = a$.)

We talked extensively in this section about multiplicative inverses, but there can be inverses for other operations, too. The next few exercises explore the additive inverse *in $\mathbb{Z}_n$. Notice that the additive identity in $\mathbb{Z}_n$ is 0: for any $a \in \mathbb{Z}_n$, we have $a + 0 \equiv_n 0 + a \equiv_n a$. The* additive inverse *of $a \in \mathbb{Z}_n$ is typically denoted $-a$.*

7.81 Give an algorithm to find the additive inverse of any $a \in \mathbb{Z}_n$. (Be careful: the additive inverse of a has to be a value from $\mathbb{Z}_n$, so you can't just say that 3's additive inverse is negative 3!)

Given your solution to the previous exercise, prove the following properties:

7.82 For any $a \in \mathbb{Z}_n$, we have $-(-a) \equiv_n a$.

7.83 For any $a, b \in \mathbb{Z}_n$, we have $a \cdot (-b) \equiv_n (-a) \cdot b$.

7.84 For any $a, b \in \mathbb{Z}_n$, we have $a \cdot b \equiv_n (-a) \cdot (-b)$.

In regular arithmetic, for a number $x \in \mathbb{R}$, a square root *of x is a number y such that $y^2 = x$. If $x = 0$, there's only one such y, namely $y = 0$. If $x < 0$, there's no such y. If $x > 0$, there are two such values y (one positive and one negative). Consider the following claim, and prove or disprove it.*

7.85 Let $n \geq 2$ be arbitrary. Then (i) there exists *one and only one* $b \in \mathbb{Z}_n$ such that $b^2 \equiv_n 0$; and (ii) for any $a \in \mathbb{Z}_n$ with $a \neq 0$, there is *not* exactly one $b \in \mathbb{Z}_n$ such that $b^2 \equiv_n a$. *(Hint: think about Exercise 7.81.)*

Using paper and pencil (and brute-force calculation), compute the following multiplicative inverses (or state that the inverse doesn't exist):

7.86 4^{-1} in $\mathbb{Z}_{11}$

7.87 7^{-1} in $\mathbb{Z}_{11}$

7.88 0^{-1} in $\mathbb{Z}_{11}$

7.89 5^{-1} in $\mathbb{Z}_{15}$

7.90 7^{-1} in $\mathbb{Z}_{15}$

7.91 9^{-1} in $\mathbb{Z}_{15}$

7.92 Prove that the multiplicative inverse is unique: that is, for arbitrary $n \geq 2$ and $a \in \mathbb{Z}_n$, suppose that $ax \equiv_n 1$ and $ay \equiv_n 1$. Prove that $x \equiv_n y$.

Write down the full multiplication table (as in Figure 7.17) for the following:

7.93 $\mathbb{Z}_5$

7.94 $\mathbb{Z}_6$

7.95 $\mathbb{Z}_8$

For arbitrary $n \geq 2$ and $a \in \mathbb{Z}_n$:

7.96 Prove or disprove the following: $(n-1)^{-1} = n - 1$ in $\mathbb{Z}_n$.

7.97 Prove that $(a^{-1})^{-1} = a$: that is, a is the multiplicative inverse of the multiplicative inverse of a.

7.98 Prove that there exists $x \in \mathbb{Z}$ with $ax \equiv_n 1$ if and only if there exists $y \in \mathbb{Z}_n$ with $ay \equiv_n 1$.

7.99 Suppose that the multiplicative inverse a^{-1} exists in $\mathbb{Z}_n$. Let $k \in \mathbb{Z}_n$ be any exponent. Prove that a^k has a multiplicative inverse in $\mathbb{Z}_n$, and, in particular, prove that the multiplicative inverse of a^k is the kth power of the multiplicative inverse of a. (That is, prove that $(a^k)^{-1} \equiv_n (a^{-1})^k$.)

Using paper and pencil *and the algorithm based on the Extended Euclidean algorithm, compute the following multiplicative inverses (or explain why they don't exist). See Figure 7.22 for a reminder.*

7.100 17^{-1} in $\mathbb{Z}_{23}$

7.101 7^{-1} in $\mathbb{Z}_{25}$

7.102 9^{-1} in $\mathbb{Z}_{33}$

7.103 *(programming required)* Implement **inverse**(a, n) from Figure 7.18 in a language of your choice.

extended-Euclid(n, m):
Input: positive integers n and $m \geq n$.
Output: $x, y, r \in \mathbb{Z}$ where $\gcd(n, m) = r = xn + ym$

if $m \bmod n = 0$ **then**
  return $1, 0, n$ // $1 \cdot n + 0 \cdot m = n = \gcd(n, m)$
else
  $x, y, r :=$ **extended-Euclid**$(m \bmod n, n)$
  return $y - \lfloor \frac{m}{n} \rfloor \cdot x, x, r$

inverse(a, n):
Input: $a \in \mathbb{Z}_n$ and $n \geq 2$
Output: a^{-1} in $\mathbb{Z}_n$, if it exists

$x, y, d :=$ **extended-Euclid**(a, n)
if $d = 1$ **then**
  return $x \bmod n$ // $xa + yn = 1$, so $xa \equiv_n 1$.
else
  return "no inverse for a exists in $\mathbb{Z}_n$."

Figure 7.22: A reminder of two algorithms.

7.104 Prove or disprove the converse of Corollary 7.21: if n is composite, then there exists $a \in \mathbb{Z}_n$ (with $a \neq 0$) that does not have a multiplicative inverse in $\mathbb{Z}_n$.

7.105 Let p be an arbitrary prime number. What value does the quantity $2^{p+1} \bmod p$ have? Be as specific as you can. Explain.

7.106 It turns out that $247^{248} \bmod 249 = 4$. From this, you can conclude at least one of following: 247 is not prime; 247 is prime; 249 is not prime; or 249 is prime. Which one(s)? Explain.

7.107 Reprove the general version of the Chinese Remainder Theorem with single constructive argument, as in the 2-congruence case, instead of using induction. Namely, assume $n_1, n_2, \ldots, n_k$ are pairwise relatively prime, and let $a_i \in \mathbb{Z}_{n_i}$. Let $N := \prod_{i=1}^{k} n_i$. Let $N_i := N/n_i$ (more precisely, let N_i be the product of all n_js *except* n_i) and let d_i be the multiplicative inverse of N_i in $\mathbb{Z}_{n_i}$. Prove that $x := \sum_{i=1}^{k} a_i N_i d_i$ satisfies the congruence $x \bmod n_i = a_i$ for all $1 \leq i \leq k$.

The totient function $\varphi : \mathbb{Z}^{\geq 1} \rightarrow \mathbb{Z}^{\geq 0}$, *sometimes called* Euler's totient function *after the 18th-century Swiss mathematician Leonhard Euler, is defined as*

$$\varphi(n) := \textit{the number of } k \textit{ such that } 1 \leq k \leq n \textit{ such that } k \textit{ and } n \textit{ have no common divisors.}$$

For example, $\varphi(6) = 2$ *because* 1 *and* 5 *have no common divisors with* 6 *(but all of* $\{2,3,4,6\}$ *do share a common divisor with* 6*). There's a generalization of Fermat's Little Theorem, sometimes called the* Fermat–Euler Theorem *or* Euler's Theorem, *that states the following: if a and n are relatively prime, then* $a^{\varphi(n)} \equiv_n 1$.

7.108 Using the Fermat–Euler theorem, argue that

(i) Fermat's Little Theorem holds.
(ii) a^{-1} in $\mathbb{Z}_n$ is $a^{\varphi(n)-1} \bmod n$, for any $a \in \mathbb{Z}_n$ that is relatively prime to n.

Verify the latter claim for the multiplicative inverses of $a \in \{7, 17, 31\}$ in $\mathbb{Z}_{60}$.

7.109 *(programming required)* Implicitly, the Fermat–Euler theorem gives a different way to compute the multiplicative inverse of a in $\mathbb{Z}_n$:

1. compute $\varphi(n)$ [say by brute force, though there are somewhat faster ways—see Exercises 9.34–9.36]; and
2. compute $a^{\varphi(n)-1} \bmod n$ [perhaps using repeated squaring; see Figure 7.7].

Implement this algorithm to compute a^{-1} in $\mathbb{Z}_n$ in a programming language of your choice.

Recall that a Carmichael number *is a composite number that passes the (bogus) primality test suggested by Fermat's Little Theorem. In other words, a Carmichael number n is an integer that is composite but such that, for any* $a \in \mathbb{Z}_n$ *that's relatively prime to n, we have* $a^{n-1} \bmod n = 1$.

7.110 *(programming required)* Write a program to verify that 561 is (a) not prime, but (b) satisfies $a^{560} \bmod 561 = 1$ for every $a \in \{1, \ldots, 560\}$ that's relatively prime to 561. (That is, verify that 561 is a Carmichael number.)

7.111 Suppose n is a *composite* integer. Argue that there exists at least one integer $a \in \{1, 2, \ldots, n-1\}$ such that $a^{n-1} \not\equiv_n 1$. (In other words, there's always *at least one* nonzero $a \in \mathbb{Z}_n$ with $a^{n-1} \not\equiv_n 1$ when n is composite. Thus, although the probability of error in **bogus-isPrime?** from p. 742 may be very high for particular composite integers n, the probability of success is nonzero, at least!)

The following theorem is due to Alwin Korselt, from 1899: an integer n is a Carmichael number if and only if n is composite, squarefree, and for all prime numbers p that divide n, we have that $p - 1 \mid n - 1$. *(An integer n is* squarefree *if there is no integer* $d \geq 2$ *such that* $d^2 \mid n$.*)*

7.112 *(programming required)* Use Korselt's theorem (and a program) to find all Carmichael numbers less than 10,000.

7.113 Use Korselt's theorem to prove that all Carmichael numbers are odd.

7.114 *(programming required)* Implement the Miller–Rabin primality test (see p. 742) in a language of your choice.

7.5 *Cryptography*

Three may keep a secret, if two of them are dead.

Benjamin Franklin (1706–1790)

In the rest of this chapter, we will make use of the number-theoretic machinery that we've now developed to explore *cryptography*. Imagine that a sender, named *Alice*, is trying to send a secret message to a receiver, named *Bob*. The goal of cryptography is to ensure that the message itself is kept secret even if an eavesdropper—named *Eve*—overhears the transmission to Bob. To achieve this goal, Alice does not directly transmit the message m that she wishes to send to Bob; instead, she *encrypts* m in some way. The resulting encrypted message c is what's transmitted to Bob. (The original message m is called *plaintext*; the encrypted message c that's sent to Bob is called the *ciphertext*.) Bob then *decrypts* c to recover the original message m. A diagram of the basic structure of a cryptographic system is shown in Figure 7.23.

Traditionally, cryptographic systems are described using an imagined crew of people whose names start with consecutive letters of the alphabet. We'll stick with these traditional names: Alice, Bob, Charlie, etc.

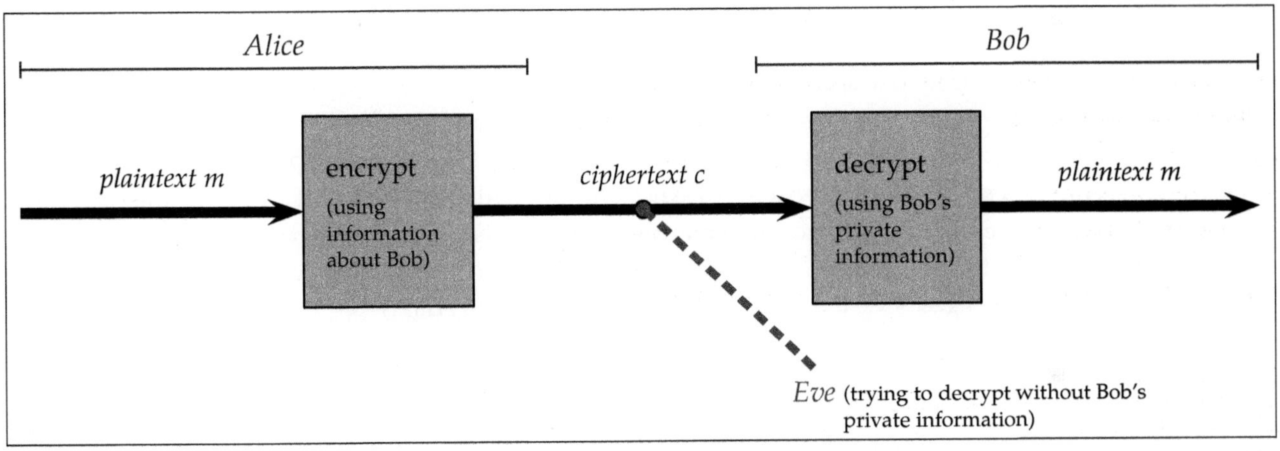

Figure 7.23: The outline of a cryptographic system.

The two obvious crucial properties of a cryptographic system are that (i) Bob can compute m from c, and (ii) Eve cannot compute m from c. (Of course, to make (i) and (ii) true simultaneously, it will have to be the case that Bob has some information that Eve doesn't have—otherwise the task would be impossible!)

One-time pads

The simplest idea for a cryptographic system is for Alice and Bob to agree on a *shared secret key* that they will use as the basis for their communication. The easiest implementation of this idea is what's called a *one-time pad*, which works as follows. Alice and Bob agree in advance on an integer n, denoting the length of the message that they would like to communicate. They also agree in advance on a secret bitstring $k \in \{0,1\}^n$, where each bit $k_i \in \{0,1\}$ is chosen independently and uniformly—so that every one of the 2^n different n-bit strings has a $\frac{1}{2^n}$ chance of being chosen as k. To encrypt a plaintext message $m \in \{0,1\}^n$, Alice computes the *bitwise exclusive or* of m and k—in other words, the ith bit of the ciphertext is $m_i \oplus k_i$. To decrypt the ciphertext $c \in \{0,1\}^n$, Bob computes the bitwise XOR of c and k.

The *pad* in the name comes from spycraft—spies might carry physical pads of paper, where each sheet has a fresh secret key written on it. The *one-time* in the name derives from the fact that this system is secure only if the same key is never reused, as we'll discuss.

Example 7.26 (A One-Time Pad)

- Alice and Bob agree (in advance) on the secret key $k = 10111000$.
- To transmit the message $m = 01101110$, Alice finds the bitwise XOR of m and k:

m	0	1	1	0	1	1	1	0
k	1	0	1	1	1	0	0	0
$c = m \oplus k$	1	1	0	1	0	1	1	0

- To decrypt the ciphertext $c = 11010110$, Bob finds the bitwise XOR of c and k:

c	1	1	0	1	0	1	1	0
k	1	0	1	1	1	0	0	0
$c \oplus k$	0	1	1	0	1	1	1	0

Observe that $c \oplus k = 01101110$ is indeed precisely $m = 01101110$, as desired.

The reason that Bob can decrypt the ciphertext to recover the original message m is simple: for any bits a and b, it's the case that $(a \oplus b) \oplus b = a$. (See Figure 7.24.) The fact that Eve *cannot* recover m from c relies on the fact that, for any message m and every ciphertext c, there is precisely one secret key k such that $m \oplus k = c$. (So Eve is just as likely to see a particular ciphertext *regardless of what the message is,* and therefore she gains no information about m by seeing c. See Exercise 7.116.) Thus the one-time pad is perfectly secure as a cryptographic system—if Alice and Bob only use it once! If Alice and Bob reuse the same key to exchange many different messages, then Eve can use frequency analysis to get a handle on the key, and therefore can begin to decode the allegedly secret messages. (See Exercises 10.72–10.76 or Exercise 7.117.)

a	b	$a \oplus b$	$(a \oplus b) \oplus b$
0	0	0	0
0	1	1	0
1	0	1	1
1	1	0	1

Figure 7.24: The truth table for $(a \oplus b) \oplus b = a$.

Taking it further: One of the earliest encryption schemes is now known as a *Caesar Cipher,* after Julius Caesar, who used it in his correspondence. It can be understood as a cryptographic system that uses a one-time pad more than once. The Caesar cipher works as follows. The sender and receiver agree on a *shift* x, an integer, as their secret key. The ith letter in the alphabet (from `A` = 0 through `Z` = 25) will be shifted forward by x positions in the alphabet. The shift "wraps around," so that we encode letter i as letter $(i + x) \bmod 26$. For example, if $x = 3$ then `A`→`D`, `L`→`O`, `Y`→`B`, etc. To send a text message m consisting of multiple letters from the alphabet, the same shift is applied to each letter. (For convenience, we'll leave nonalphabetic characters unchanged.) For example, the ciphertext `XF BSF EJTDPWFSFE; GMFF BU PODF!` was generated with the shift $x = 1$ from the message `WE ARE DISCOVERED; FLEE AT ONCE!`. Because we've reused the same shift x for each letter of the message, the Caesar Cipher is susceptible to being broken based on frequency analysis. (In the `XF BSF EJTDPWFSFE; GMFF BU PODF!` example, `F` is by far the most common letter in the ciphertext—and `E` is by far the most common letter in English text. From these two facts, you might infer that $x = 1$ is the most probable secret key. See Exercise 7.117.)

Millennia later, the Enigma machines, the encryption system used by the Germans during World War II, was—as with Caesar—a substitution cipher, but one where the shift changed with each letter. (But not in completely unpredictable ways, as in a one-time pad!) See p. 960 for more.

Public-key cryptography

In addition to being single-use-only, there's another strange thing about the one-time pad: if Alice and Bob are somehow able to communicate an n-bit string securely—as they must to share the secret key k—it doesn't seem particularly impressive that they can then communicate the n-bit string m securely.

Public-key cryptography is an idea to get around this oddity. Here is the idea, in a nutshell. Every participant will have a *public key* and a *private* (or *secret*) *key,* which will somehow be related to the public key. A user's public key is completely public—for example, posted on the web. If Alice wishes to send a message m to Bob, then Alice will (somehow!) encrypt her message to Bob using Bob's public key, producing ciphertext c. Bob, who of course knows Bob's secret key, can decrypt c to reconstruct m; Eve, not knowing Bob's secret key, cannot decrypt c.

This idea sounds a little crazy, but we will be able to make it work. Or, at least, we will make it work *on the assumption that Eve has only limited computational power*—and on the assumption that certain computational problems, like factoring large numbers, require a lot of computational power to solve. (For example, Bob's secret key cannot be easily computable from Bob's public key—otherwise Eve could easily figure out Bob's secret key and then run whatever decryption algorithm Bob uses!)

7.5.1 The RSA Cryptosystem

The basic idea of public-key cryptography was discussed in abstract terms in the 1970s—especially by Whitfield Diffie, Martin Hellman, and Ralph Merkle—and, after some significant contributions by a number of researchers, a cryptosystem successfully implementing public-key cryptography was discovered by Ron Rivest, Adi Shamir, and Leonard Adleman.[7] The *RSA cryptosystem*, named after the first initials of their three last names, is one of the most famous, and widely used, cryptographic protocols today. The previous sections of this chapter will serve as the building blocks for the RSA system, which we'll explore in the rest of this section.

[7] R. L. Rivest, A. Shamir, and L. Adleman. A method for obtaining digital signatures and public-key cryptosystems. *Communications of the ACM*, 21:120–126, February 1978.

Taking it further: The RSA cryptosystem is named after its three 1978 discoverers, and the Turing Award—the highest honor in computer science, roughly equivalent to the Nobel Prize of computer science—was conferred on Rivest, Shamir, and Adleman in 2002 for this discovery. But there is also a "shadow history" of the advances in cryptography made in the second half of the 20th century.

The British government's signal intelligence agency, called Government Communications Headquarters (GCHQ), had been working to solve precisely the same set of research questions about cryptography as academic researchers like R., S., and A. (GCHQ was perhaps best known for its success in World War II, in breaking the Enigma Code of the German military; see p. 960 for more discussion.) And it turned out that several British cryptographers at GCHQ—Clifford Cocks, James Ellis, and Malcolm Williamson—had discovered the RSA protocol several years *before* 1978. But their discovery was classified by the British government, and thus we call this protocol "RSA" instead of "CEW."

See the excellent book by Simon Singh for more on the history of cryptography, including both the published and classified advances in cryptographic systems.[8] Also see the discussion on p. 753 of the *Diffie–Hellman key exchange protocol*, one of the first (published) modern breakthroughs in cryptography, which allows Alice and Bob to solve another apparently impossible problem: exchanging secret information while communicating only over an insecure channel.

[8] Simon Singh. *The Code Book: The Secret History of Codes and Code-breaking*. Fourth Estate Ltd., 1999.

In RSA, as for any public-key cryptosystem, we must define three algorithmic components. (These three algorithms for the RSA cryptosystem are shown in Figure 7.25; an overview of the system is shown in Figure 7.26.) They are:

- *key generation:* how do Alice and Bob construct their public/private keypairs?
- *encryption:* when Alice wishes to send a message to Bob, how does she encode it?
- *decryption:* when Bob receives ciphertext from Alice, how does he decode it?

The very basic idea of RSA is the following. (The details of the protocols are in Figure 7.25.) To encrypt a numerical message m for Bob, Alice will compute $c := m^e \bmod n$, where Bob's public key is $\langle e, n \rangle$. To decrypt the ciphertext c that he receives, Bob will

Key Generation:

1. Bob chooses two large primes, p and q, and defines $n := pq$.
2. Bob chooses $e \neq 1$ such that e and $(p-1)(q-1)$ are relatively prime.
3. Bob computes $d := e^{-1}$ modulo $(p-1)(q-1)$.
4. Bob publishes $\langle e, n \rangle$ as his public key; Bob's secret key is $\langle d, n \rangle$.

Encryption: If Alice wants to send message m to Bob,

1. Alice finds Bob's public key, say $\langle e_{\text{Bob}}, n_{\text{Bob}} \rangle$, as he published it.
2. To send message $m \in \{0, \ldots, n_{\text{Bob}} - 1\}$, Alice computes $c := m^{e_{\text{Bob}}} \bmod n_{\text{Bob}}$.
3. Alice transmits c to Bob.

Decryption: When Bob receives ciphertext c,

1. Bob computes $m := c^{d_{\text{Bob}}} \bmod n_{\text{Bob}}$, where $\langle d_{\text{Bob}}, n_{\text{Bob}} \rangle$ is Bob's secret key.

Figure 7.25: The RSA cryptosystem.

compute $c^d \bmod n$, where Bob's private key is $\langle d, n \rangle$. (Of course, there's an important relationship among the quantities e, d, and n!)

AN EXAMPLE OF RSA KEY GENERATION, ENCRYPTION, AND DECRYPTION

Later we will prove that the message that Bob decrypts is always the same as the message that Alice originally sent. But we'll start with an example. First, Bob generates a public and private key, using the protocol in Figure 7.25. (All three phases can be implemented efficiently, using techniques from this chapter; see Exercises 7.129–7.132.)

Example 7.27 (Generating an RSA keypair for Bob)

For good security properties, we'd want to pick seriously large prime numbers p and q, but to make the computation easier to see we'll choose very small primes.

1. Suppose we choose the "large" primes $p = 13$ and $q = 17$. Then $n := 13 \cdot 17 = 221$.

2. We now must choose a value of $e \neq 1$ that is relatively prime to $(p-1)(q-1) = 12 \cdot 16 = 192$. Note that $\gcd(2, 192) = 2 \neq 1$, so $e = 2$ fails. Similarly $\gcd(3, 192) = 3$ and $\gcd(4, 192) = 4$. But $\gcd(5, 192) = 1$. We pick $e := 5$.

3. We now compute $d :=$ **inverse**$(e, (p-1)(q-1))$—that is, $d := e^{-1}$ in $\mathbb{Z}_{(p-1)(q-1)}$:

   extended-Euclid(5, 192)
       extended-Euclid(192 mod 5 = 2, 5)
       $= -2, 1, 1$ *exactly as in Example 7.22*
   $= y - \lfloor \frac{m}{n} \rfloor \cdot x, x, r$ *where $x = -2, y = 1, r = 1$ and $m = 192, n = 5$.*
   $= 77, -2, 1.$
   Thus **inverse**(5, 192) returns 77 mod 192 = 77.

   (Indeed, $5 \cdot 77 = 385 = 192 \cdot 2 + 1$, so $5 \cdot 77 \equiv_{192} 1$.) Thus we set $d := 77$.

Thus Bob's public key is $\langle e, n \rangle = \langle 5, 221 \rangle$, and Bob's secret key is $\langle d, n \rangle = \langle 77, 221 \rangle$.

It may seem strange that n is part of *both* Bob's secret key *and* Bob's public key—it's usually done this way for symmetry, but also to support *digital signatures*. When Alice sends Bob a message, she can encrypt it *using her own secret key*; Bob can then decrypt the message *using Alice's public key* to verify that Alice was indeed the person who sent the message.

Bob now publishes his public key somewhere, keeping his secret key to himself. If Alice now wishes to send a message to Bob, she uses his public key, as follows:

Example 7.28 (Encrypting a message with RSA)
To send message $m = 202$ to Bob, whose public key is $\langle e, n\rangle = \langle 5, 221\rangle$, Alice computes

$$m^e \bmod n = 202^5 \bmod 221 = 336{,}323{,}216{,}032 \bmod 221 = 206.$$

Thus she sends Bob the ciphertext $c := 206$.

When Bob receives an encrypted message, he uses his secret key to decrypt it:

Example 7.29 (Decrypting a message with RSA)
When Bob, whose secret key is $\langle d, n\rangle = \langle 77, 221\rangle$, receives the ciphertext $c = 206$ from Alice, he decrypts it as

$$c^d \bmod n = 206^{77} \bmod 221.$$

Computing $206^{77} \bmod 221$ by hand is a bit tedious, but we can calculate it with "repeated squaring" (using the fact that $b^{2k} \bmod n = (b^2 \bmod n)^k \bmod n$ and $b^{2k+1} \bmod n = b \cdot (b^{2k} \bmod n) \bmod n$; see Exercises 7.23–7.25):

$$\begin{aligned}
206^{77} \bmod 221 &= 206 \cdot (\underbrace{206^2 \bmod 221}_{=4})^{38} \bmod 221 \\
&= 206 \cdot (\underbrace{4^2 \bmod 221}_{=16})^{19} \bmod 221 \\
&= 206 \cdot 16 \cdot (\underbrace{16^2 \bmod 221}_{=35})^{9} \bmod 221 \\
&= 206 \cdot 16 \cdot 35 \cdot (\underbrace{35^2 \bmod 221}_{=120})^{4} \bmod 221 \\
&= 206 \cdot 16 \cdot 35 \cdot (\underbrace{120^2 \bmod 221}_{=35})^{2} \bmod 221 \\
&= 206 \cdot 16 \cdot 35 \cdot (\underbrace{35^2 \bmod 221}_{=120}) \bmod 221 \\
&= \underbrace{206 \cdot 16 \cdot 35 \cdot 120}_{=13{,}843{,}200} \bmod 221 \\
&= 202.
\end{aligned}$$

Thus Bob decrypts the ciphertext 206 as $202 = 206^{77} \bmod 221$. Indeed, then, the message that Bob receives is precisely 202—the same message that Alice sent!

We've now illustrated the full RSA protocol: generating a key, and encrypting and decrypting a message. Here's one more chance to work through the full pipeline:

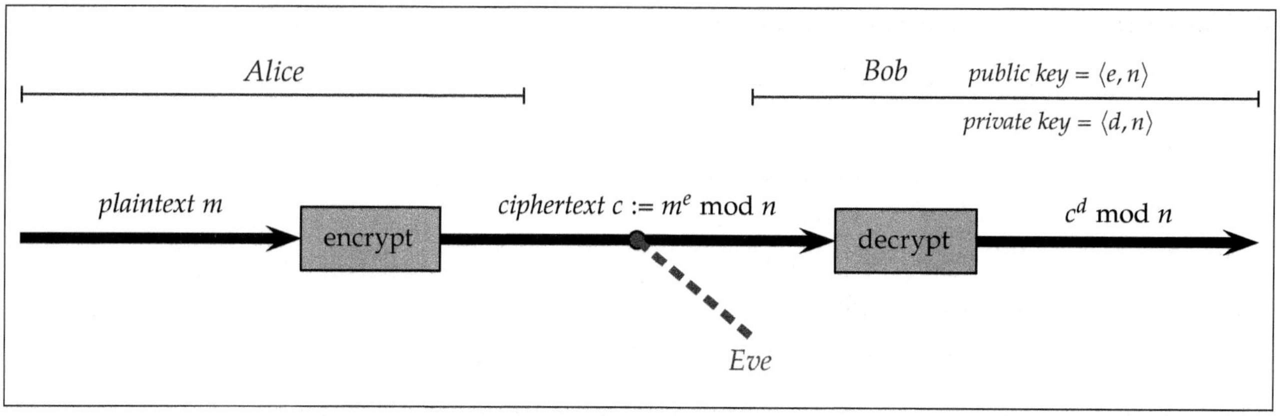

Figure 7.26: A schematic of the RSA cryptosystem, where $n = pq$ and $de \equiv_{(p-1)(q-1)} 1$, for two prime numbers p and q.

Example 7.30 (RSA, again, from end to end)

Problem: Bob generates a public/private keypair using the primes $p = 11$ and $q = 13$, choosing the smallest valid value of e. You encrypt the message 95 to send to Bob (using his generated public key). What ciphertext do you send to Bob?

Solution: For $\langle p, q\rangle = \langle 11, 13\rangle$, we have $pq = 143$ and $(p-1)(q-1) = 120$. Because 120 is divisible by 2, 3, 4, 5, and 6 but $\gcd(120, 7) = 1$, we choose $e := 7$. We find $d :=$ **inverse**$(7, 120) = 103$. Then Bob's public key is $\langle e, n\rangle = \langle 7, 143\rangle$ and Bob's private key is $\langle d, n\rangle = \langle 103, 143\rangle$.

To send Bob the message $m = 95$, we compute $m^e \bmod n = 95^7 \bmod 143$, which is 17. Thus the ciphertext is $c := 17$. (Bob would decrypt this ciphertext as $c^d \bmod n = 17^{103} \bmod 143$—which indeed is 95.)

7.5.2 *The Correctness of RSA*

Examples 7.27–7.29 gave one instance of the RSA cryptosystem working properly, in the sense that **decrypt(encrypt(**m**))** turned out to be the original message m itself—but, of course, we want this property to be true in general. Let's prove that it is. Before we give the full statement of correctness, we'll prove an intermediate lemma:

Lemma 7.25 (Correctness of RSA: decrypting the ciphertext, modulo p or q)

Suppose e, d, p, q, n are all as specified in the RSA key generation protocol—that is, $n = pq$ for primes p and q, and $ed \equiv_{(p-1)(q-1)} 1$. Let $m \in \mathbb{Z}_n$ be any message. Then

$$m' := [(m^e \bmod n)^d \bmod n] \qquad \text{(the decryption of the encryption of } m\text{)}$$

satisfies both $m' \equiv_p m$ and $m' \equiv_q m$.

Proof. We'll prove $m' \equiv_p m$; because p and q are symmetric in the definition, $m' \equiv_q m$ follows immediately. Recall that we chose d so that $ed \equiv_{(p-1)(q-1)} 1$, and thus we have

$ed = k(p-1)(q-1)+1$ for some integer k. Hence

$$
\begin{aligned}
&[(m^e \bmod n)^d \bmod n] \bmod p \\
&= (m^{ed} \bmod n) \bmod p && \text{by (7.3.4)} \\
&= (m^{k(p-1)(q-1)+1} \bmod pq) \bmod p && \text{by definition of } e, d, n, \text{ and } k \\
&= m^{k(p-1)(q-1)+1} \bmod p && \text{by Exercise 7.18} \\
&= [m \cdot m^{k(p-1)(q-1)}] \bmod p && a^{k+1} = a \cdot a^k \\
&= [(m \bmod p) \cdot (m^{k(p-1)(q-1)} \bmod p)] \bmod p && \text{by (7.3.3)} \\
&= [(m \bmod p) \cdot ((m^{k(q-1)} \bmod p)^{p-1} \bmod p)] \bmod p. && \text{by (7.3.4)}
\end{aligned}
$$

Although it's not completely obvious, we're actually almost done: we've now shown

$$
\begin{aligned}
&\left[(m^e \bmod n)^d \bmod n\right] \bmod p \\
&= \left[(m \bmod p) \cdot ((m^{k(q-1)} \bmod p)^{p-1} \bmod p)\right] \bmod p. \qquad (*)
\end{aligned}
$$

If only the highlighted portion of the right-hand side of $(*)$ were equal to 1, we'd have shown exactly the desired result, because the right-hand side would then equal $[(m \bmod p) \cdot 1] \bmod p = m \bmod p \bmod p = m \bmod p$—exactly what we had to prove! And the good news is that the highlighted portion of $(*)$ matches the form of Fermat's Little Theorem: the highlighted expression is $a^{p-1} \bmod p$, where $a := m^{k(q-1)} \bmod p$, and Fermat's Little Theorem tells us $a^{p-1} \bmod p = 1$ as long as $a \not\equiv_p 0$—that is, as long as $p \nmid a$. (We'll also have to handle the case when a *is* divisible by p, but we'll be able to do that separately.) Here are the two cases:

Problem-solving tip: If there's a proof outline that will establish a desired claim except in one or two special cases, then try to "break off" those special cases and handle them separately. Here we handled the "normal" case $a \not\equiv_p 0$ using Fermat's Little Theorem, and broke off the special $a \equiv_p 0$ case and handled it separately.

- If $a \equiv_p 0$, then notice that $m^{k(q-1)} \bmod p = 0$ and thus that $p \mid m^{k(q-1)}$. Therefore:

$$
\begin{aligned}
[(m^e \bmod n)^d \bmod n] \bmod p &= [(m \bmod p) \cdot a^{p-1} \bmod p] \bmod p && \text{by } (*) \\
&= [(m \bmod p) \cdot 0] \bmod p && \text{by the assumption that } a \equiv_p 0 \\
&= 0 \\
&= m \bmod p,
\end{aligned}
$$

  where the last equality follows because p is prime and $p \mid m^{k(q-1)}$; thus Exercise 7.48 tells us that $p \mid m$ as well.

- If $a \not\equiv_p 0$, then we can use Fermat's Little Theorem:

$$
\begin{aligned}
[(m^e \bmod n)^d \bmod n] \bmod p &= [(m \bmod p) \cdot a^{p-1} \bmod p] \bmod p && \text{by } (*) \\
&= [(m \bmod p) \cdot 1] \bmod p && \text{by Fermat's Little Theorem} \\
&= m \bmod p.
\end{aligned}
$$

We've now established that $[(m^e \bmod n)^d \bmod n] \bmod p = m \bmod p$ in both cases, and thus the lemma follows. □

Using Lemma 7.25 to do most of the work, we can now prove the main theorem:

Theorem 7.26 (Correctness of RSA)
Suppose that Bob's RSA public key is $\langle e, n \rangle$ and his corresponding private key is $\langle d, n \rangle$. Let $m \in \mathbb{Z}_n$ be any message. Then $\textbf{decrypt}_{Bob}(\textbf{encrypt}_{Bob}(m)) = m$.

Proof. Note that $\textbf{decrypt}_{\text{Bob}}(\textbf{encrypt}_{\text{Bob}}(m)) = (m^e \bmod n)^d \bmod n$. By Lemma 7.25,

$$(m^e \bmod n)^d \bmod n \equiv_p m \qquad \text{and} \qquad (m^e \bmod n)^d \bmod n \equiv_q m.$$

By Exercise 7.50, together these facts imply that $(m^e \bmod n)^d \bmod n \equiv_{pq} m$ as well. Because $n = pq$ and $m < n$, therefore $(m^e \bmod n)^d \bmod n = m \bmod n = m$. □

WHAT ABOUT EVE?

When Alice encrypts a message m for Bob and transmits the corresponding RSA-encrypted ciphertext, we've now shown in Theorem 7.26 that Bob is able to decrypt to recover the original message m. What's left to establish is that Eve *cannot* recover m from what she knows—namely, from the ciphertext $m^e \bmod n$ and from Bob's public key $\langle e, n \rangle$. (That's the desired security property of the system!)

Unfortunately, not only are we unable to *prove* this property, it's simply not true! Eve *is* able to recover m from the $m^e \bmod n$ and e and n, as follows: she factors n—that is, finds the primes p and q such that $pq = n$—and then computes d precisely as Bob did when he generated his RSA keys. (And Eve then computes the "secret" message $(m^e \bmod n)^d \bmod n$ precisely as Bob did when he decrypted.)

But the fact that Eve *has the information necessary to recover m* doesn't mean that RSA is doomed: factoring large numbers (particularly those that are the product of two large primes, perhaps) seems to be a computationally difficult problem. Even if you know that $n = 121{,}932{,}625{,}927{,}450{,}033$ it will take you quite a while to find p and q—and the best known algorithms for factoring are not fast enough for Eve.

Taking it further: The crucial property that we're using in RSA is an asymmetry in two "directions" of a problem. Taking two large prime numbers p and q and computing their product $n = pq$ is easy (that is, it can be done quickly, in polylogarithmic time). Taking a number n that happens to be the product of two primes and factoring it into $p \cdot q$ appears to be hard (that is, nobody knows how to do it quickly). Cryptographic systems have been built on a number of different problems with this kind of asymmetry; see a good textbook on cryptography for much, much more.[9]

See, for example,

[9] Jonathan Katz and Yehuda Lindell. *Introduction to Modern Cryptography*. Chapman & Hall/CRC Press, 2007.

Also see this book for a discussion of some of the ways in which "textbook RSA"—what we've described here!—is susceptible to all sorts of attacks. Industrial-level RSA implementations take all sorts of precautions that we haven't even begun to discuss.

Notice, though, that Eve could break RSA another way, too: she only needs to find m, and she knows both the ciphertext $c = m^e \bmod n$ and Bob's public key $\langle e, n \rangle$. So Eve could discover m by computing the "eth root of c"—that is, the number x such that $x^e = c$. Unfortunately for Eve, the fact that she has to compute the eth root of m^e $\underline{\bmod n}$, and not just the eth root of m^e, is crucial; this problem also seems to be computationally difficult. (See Exercise 7.139—though there's some evidence that choosing a small value of e, like $e = 3$, might weaken the security of the system.)

Note, though, that we have *not* proven that Eve is unable to efficiently break RSA encryption—for all we know, a clever student of computational number theory (you!?) will discover an efficient algorithm for factoring large numbers or computing eth roots in $\mathbb{Z}_n$. (Or, for all we know, perhaps someone already has!)

Computer Science Connections

Diffie–Hellman Key Exchange

Suppose that Alice and Bob wish to communicate to establish some shared piece of secret information—perhaps to share a key to use in a one-time pad, or to use for some other cryptographic protocol. But the only communication channel available to Alice and Bob is insecure; Eve can listen to all of their communication. This problem is called the *key exchange* problem: *two parties seek to establish a shared secret while they communicate only over an insecure channel.* Like public-key cryptography (as in RSA), this task seems completely impossible—and, also like public-key cryptography, despite its apparent impossibility, this problem was solved in the 1970s. The solution that we'll describe here is called the *Diffie–Hellman key exchange protocol.*[10]

[10] Whitfield Diffie and Martin Hellman. New directions in cryptography. *IEEE Transactions on Information Theory*, pages 644–654, November 1976.

Let p be prime. The key number-theoretic definition for Diffie–Hellman is what's called a *primitive root mod p*, which is an element $g \in \mathbb{Z}_p$ such that every nonzero element of $\mathbb{Z}_p$ is equivalent to a power of g. (In other words, $\{g^1, g^2, \ldots, g^{p-1}\} \equiv_p \{1, 2, \ldots, p-1\}$.) See Figure 7.27 for some examples. It's a theorem of number theory that every $\mathbb{Z}_p$ for prime p has at least one primitive root. Here, then, is the protocol for Diffie–Hellman key exchange:

	2^1	2^2	2^3	2^4	2^5	2^6	2^7	2^8	2^9	2^{10}
$=$	2	4	8	16	32	64	…			
$\equiv_{11}$	2	4	8	5	10	9	7	3	6	1
	5^1	5^2	5^3	5^4	5^5	5^6	5^7	5^8	5^9	5^{10}
$\equiv_{11}$	5	3	4	9	1	5	3	4	9	1
	7^1	7^2	7^3	7^4	7^5	7^6	7^7	7^8	7^9	7^{10}
$\equiv_{11}$	7	5	2	3	10	4	6	9	8	1

Figure 7.27: Primitive roots of $\mathbb{Z}_{11}$. This set has four different primitive roots $\{2, 6, 7, 8\}$, two of which are shown here: the first 10 powers of 2 and 7 (but not 5) in $\mathbb{Z}_{11}$ produce all 10 nonzero elements of $\mathbb{Z}_{11}$.

1. Alice and Bob agree on a prime p and a number g that's a primitive root mod p. They communicate p and g over the insecure channel.
2. Alice chooses a secret value $a \in \mathbb{Z}_p$ randomly, computes $A := g^a \bmod p$, and sends A to Bob. Bob chooses a secret value $b \in \mathbb{Z}_p$ randomly, computes $B := g^b \bmod p$, and sends B to Alice. (Note that A and B are sent over the channel, but the values of a and b are never transmitted.)
3. Alice, who knows a (she picked it) and $B = g^b \bmod p$ (Bob sent it to her), computes $B^a \bmod p$. Bob, who knows b (he picked it) and $A = g^a \bmod p$ (Alice sent it to him), computes $A^b \bmod p$.

Note that $A^b \equiv_p (g^a)^b = g^{ab}$ and $B^a \equiv_p (g^b)^a = g^{ab}$—so Alice and Bob now have a shared piece of information, namely $g^{ab} \bmod p$. (And they can complete their computations efficiently, as in RSA; see Exercise 7.132.)

But why is this shared piece of information a secret? Let's look at the protocol from Eve's perspective: she observes the values of p, g, $g^a \bmod p$, and $g^b \bmod p$. But it is generally believed that the problem of computing a from the values of p, g, and $g^a \bmod p$ cannot be solved efficiently. (This problem is called the *discrete logarithm problem:* it's the modular analogy of computing y from the values of x and x^y—that is, computing $\log_x(x^y)$.) Most researchers believe that the discrete log problem is difficult (as long as the prime p is of appreciable size), and thus that Eve cannot feasibly figure out the value $g^{ab} \bmod p$, shared by Alice and Bob.

It's worth pointing out that, as we've stated the protocol, Diffie–Hellman is susceptible to a so-called *man-in-the-middle attack:* a malicious party (traditionally called *Mallory)* who has control over the channel can impersonate Bob to Alice, and impersonate Alice to Bob. (There are improvements to the protocol that address this issue.) Doing so allows Mallory to intercept, decrypt, and then reencrypt subsequent communications that Alice and Bob thought were secure—and they'd never know that Mallory was involved.[11]

See any good book on cryptography, such as the following, for much more on this protocol (and the susceptibilities of Diffie–Hellman and other protocols to attacks like the man in the middle):

[11] Jonathan Katz and Yehuda Lindell. *Introduction to Modern Cryptography.* Chapman & Hall/CRC Press, 2007.

7.5.3 *Exercises*

7.115 In our encryption/decryption scheme for one-time pads, we used *exclusive or:* plaintext m was encrypted as $m \oplus k$, and ciphertext c was decrypted as $c \oplus k$. Because $(m \oplus k) \oplus k$ is logically equivalent to m, Bob always recovers the original plaintext. But there are actually three other Boolean operators that we could have used instead of $\oplus$—that is, there are three other connectives $\circ$ such that $(m \circ k) \circ k \equiv m$. (See Figure 4.31.) Identify those three other connectives. Why are these three connectives either uninteresting or actively bad choices as alternatives to $\oplus$?

7.116 *(Requires knowledge of probability; see Chapter 10.)* For a one-time pad with an n-bit key, we have

$$\Pr\left[\text{ciphertext} = c\right] = \sum_{m} \left[\Pr\left[\text{ciphertext} = c|\text{plaintext} = m\right] \cdot \Pr\left[\text{plaintext} = m\right] \right].$$

Prove that the probability that the ciphertext is a particular $c \in \{0,1\}^n$ is precisely $1/2^n$ *for any distribution over plaintext messages.*

7.117 *(programming required)* As we suggested, one-time pads are secure *if* they're used only once, but using the same key more than once compromises security. I took a (famous) document written in English, and encoded it as follows: I converted each character to an ASCII value (in binary, 00000000 to 11111111), and separated this bitstring into 40-bit chunks. (Each chunk contains 5 characters, with 8 bits each.) I generated a 40-bit key, and encoded each chunk using that key. (That is: I used a one-time pad more than one time!) The encoded document starts like this:

1110111111011100100010011010000110111101
1110100011010101111110111010011110100001
1111010110111110100010011010100010000111

You can find the full encoding at `http://cs.carleton.edu/faculty/dln/one-time-pad.txt`. Figure out (a) what 40-bit key I used, and (b) what the encoded document is.

7.118 *(programming required)* Implement one-time pads in a programming language of your choice.

7.119 Using the "large" primes $p = 19$ and $q = 23$, compute the RSA public and private keys. You may have multiple valid choices for e—if so, choose the smallest e that you can.

7.120 Repeat for $p = 31$ and $q = 37$.

7.121 Repeat for $p = 41$ and $q = 43$.

Suppose that Bob's public key is $n = 221$ and $e = 5$. (And so Bob's private key is $n = 221$ and $d = 77$.)

7.122 Compute the RSA encryption to send Bob the message $m = 42$.

7.123 Repeat for the message $m = 99$.

7.124 If Bob receives the ciphertext $c = 99$, what message was sent to Bob?

7.125 Repeat for the ciphertext $c = 17$.

7.126 *(programming required)* Suppose that Charlie's public key is $\langle e = 3, n = 1{,}331{,}191\rangle$, and the ciphertext $c = 441{,}626$. Figure out the message that was sent to Charlie by factoring n.

7.127 Repeat for the public key $\langle e = 11, n = 12{,}187{,}823\rangle$, and the ciphertext $c = 7{,}303{,}892$.

7.128 Repeat for the public key $\langle e = 5, n = 662{,}983{,}829\rangle$, and the ciphertext $c = 43{,}574{,}279$.

In both key generation and encryption/decryption, the RSA cryptosystem completes a number of steps that require some nonobvious ideas to make them efficient. Luckily, we've covered those ideas at various times in previous parts of the chapter. For each of the following, explain how to compute the desired quantity efficiently (that is, with a number of primitive arithmetic operations that's $O(\log^k n)$ for the value of n in the RSA protocol, for some constant k).

(For some of these problems, you'll simply be able to cite a previously developed algorithm in a few words; in others, you'll need to combine more than one algorithm or use an algorithm in a nontrivial way.)

7.129 Find a large prime number: say, find the smallest prime number greater than a given number x.

7.130 Given primes p and q, find a number $e \neq 1$ such that e and $(p-1)(q-1)$ are relatively prime.

7.131 Given primes p, q and e relatively prime to $(p-1)(q-1)$, compute e^{-1} modulo $(p-1)(q-1)$.

7.132 Given n, e, and $m \in \mathbb{Z}_n$, compute $m^e \bmod n$. (Similarly, given n, d, and c, compute $c^d \bmod n$.)

7.133 Prove that, in the RSA key-generation protocol, the number e that we choose is always odd.

7.134 Prove that, in the RSA key-generation protocol, the number d that we choose is also always odd.

Imagine the following modifications to the RSA key generation protocol. What goes wrong if we use change the algorithm as described? Be precise. Is there a step of the protocol that can no longer be executed? Does Bob no longer have the information necessary to decrypt the ciphertext? Does Eve now have the power to decrypt the ciphertext?

7.135 The protocol tells us to choose two large primes p, q. But, instead, we choose *one* prime p, and set $q := p$.

7.136 The protocol tells us to choose two large primes p and q. But, instead, we choose two large numbers p and q that aren't actually prime.

7.137 The protocol tells us to choose $e \neq 1$ that's relatively prime to $(p-1)(q-1)$. But, instead, we choose $e = 1$.

7.138 The protocol tells us to choose $e \neq 1$ that's relatively prime to $(p-1)(q-1)$. But, instead, we choose an e that is not relatively prime to $(p-1)(q-1)$.

7.139 Explain precisely how to use binary search to find the eth root of m^e efficiently. Then explain precisely why this binary-search approach doesn't work to find the eth root of $m^e \bmod n$ in general.

Implement the RSA cryptosystem in a programming language of your choice. Use the results from Exercises 7.129–7.132 to make your solutions efficient. Your code should implement the following components:

7.140 *(programming required)* Key generation. Given two prime numbers p and q as input, produce a public and private RSA keypair $\langle e, n\rangle$ and $\langle d, n\rangle$. *(Hint: Exercises 7.31 and 7.103 will be helpful. To pick e, you may wish to simply try all odd numbers and use Exercise 7.31—you could make this step faster, but generally speaking this slightly slower approach will still be fast enough.)*

7.141 *(programming required)* Encryption and decryption. For encryption, given a public key $\langle e, n\rangle$ and a message $m \in \mathbb{Z}_n$, compute the corresponding ciphertext $c := m^e \bmod n$. Similarly, for decryption: given a private key $\langle d, n\rangle$ and a ciphertext $c \in \mathbb{Z}_n$, compute $m := c^d \bmod n$. *(Hint: Exercise 7.25 will be helpful.)*

Generally, a user of a cryptographic system will want to send text *rather than a* number, *so you'll need to add a capacity for converting text into an integer. And RSA will only support encrypting elements of $\mathbb{Z}_n$, not $\mathbb{Z}$, so you'll actually need to convert the text into a* sequence of elements of $\mathbb{Z}_n$.

7.142 *(programming required)* Write a pair of functions `string->intlist`(s, n) and `intlist->string`(L, n) that convert between strings of characters and a list of elements from $\mathbb{Z}_n$. You may do this conversion in many ways, but it must be the case that these operations are inverses of each other: if `string->intlist`$(s^*, n) = L^*$, then `intlist->string`$(L^*, n) = s^*$. *(Hint: the easiest way to do this conversion is to view text encoded as a sequence of ASCII symbols, each of which is an element of $\{0, 1, \ldots, 255\}$. Thus you can view your input text as a number written in base 256. Your output is a number written in base n. Use* **baseConvert** *from p. 714.)*

7.143 *(programming required)* Demonstrate that your implementations from Exercises 7.140, 7.141, and 7.142 are working properly by generating keys, encrypting, and decrypting using the primes p = 5,277,019,477,592,911 and q = 7,502,904,222,052,693, and the message `"THE SECRET OF BEING BORING IS TO SAY EVERYTHING."` (Voltaire (1694–1778)).

Complete the last missing piece of your RSA implementation:

7.144 *(programming required)* Prime generation. The key generation implementation from Exercise 7.140 relies on being given two prime numbers. Write a function that, given a (sufficiently large) range of possible numbers between n_{min} and n_{max}, repeatedly does the following: choose a random integer between n_{min} and n_{max}, and test whether it's prime using the Miller–Rabin test (see Exercise 7.114).

The Chinese Remainder Theorem tells us that $m \in \mathbb{Z}_{pq}$ is uniquely described by its value modulo p and q—that is, $m \bmod p$ and $m \bmod q$ fully describe m. Here's one way to improve the efficiency of RSA using this observation: instead of computing $m := c^d \bmod pq$ directly, instead compute $a := c^d \bmod p$ and $b := c^d \bmod q$. Then use the algorithm implicit in Theorem 7.14 to compute the value m with $m \bmod p = a$ and $m \bmod q = b$.

7.145 *(programming required)* Modify your implementation of RSA to use the above idea.

7.146 Actually, instead of computing $a := c^d \bmod p$ and $b := c^d \bmod q$, we could have computed $a := c^{d \bmod p-1} \bmod p$ and $b := c^{d \bmod q-1} \bmod q$. Explain why this modification is valid. (This change can improve the efficiency of RSA, because now both the base and the exponent may be substantially smaller than they were in the regular RSA implementation.)

7.6 *Chapter at a Glance*

Modular Arithmetic

Given integers $k \geq 1$ and n, there exist unique integers d and r such that $0 \leq r < k$ and $kd + r = n$. The value of d is $\lfloor \frac{n}{k} \rfloor$, the (whole) number of times k goes into n; the value of r is $n \bmod k$, the remainder when we divide n by k.

Two integers a and b are *equivalent* or *congruent mod n*, written $a \equiv_n b$, if a and b have the same remainder when divided by n—that is, when $a \bmod n = b \bmod n$. For expressions taken mod n, we can always freely "reduce" mod n (subtracting multiples of n) before performing addition or multiplication. (See Theorem 7.3.)

We write $k \mid n$ to denote the proposition that $n \bmod k = 0$. If $k \mid n$, we say that k *(evenly) divides* n, that k is a *factor* of n, and that n is a *multiple* of k. See Theorem 7.4 for some useful properties of divisibility: for example, if $a \mid b$ then, for any integer c, it's also the case that a divides bc as well. The *greatest common divisor* $\gcd(n, m)$ of two positive integers n and m is the largest d that evenly divides both n and m; the *least common multiple* is the smallest $d \in \mathbb{Z}^{\geq 1}$ that n and m both evenly divide. GCDs can be computed efficiently using the *Euclidean algorithm*. (See Figure 7.28.)

Euclid(n, m):
Input: positive integers n and $m \geq n$
Output: $\gcd(n, m)$

```
if m mod n = 0 then
    return n
else
    return Euclid(m mod n, n)
```

Figure 7.28: The Euclidean algorithm for GCDs.

Primality and Relative Primality

An integer $p \geq 2$ is *prime* if the only positive integers that evenly divide it are 1 and p itself; an integer $n \geq 2$ that is not prime is called *composite*. (Note that 1 is neither prime nor composite.) Let *primes*(n) denote the number of prime numbers less or equal than n. The *Prime Number Theorem* states that, as n gets large, the ratio between *primes*(n) and $\frac{n}{\log n}$ converges (slowly!) to 1. Every positive integer can be factored into a product of zero or more prime numbers, and that factorization is unique up to the ordering of the factors.

Two positive integers n and m are called *relatively prime* if they have no common factors aside from 1—that is, if $\gcd(n, m) = 1$. A tweak to the Euclidean algorithm, called the *Extended Euclidean algorithm*, takes arbitrary positive integers n and m as input, and (efficiently) computes three integers x, y, r such that $r = \gcd(n, m) = xn + ym$. (See Figure 7.29.) We can determine whether n and m are relatively prime using the (Extended) Euclidean algorithm.

extended-Euclid(n, m):
Input: positive integers n and $m \geq n$.
Output: $x, y, r \in \mathbb{Z}$ where $\gcd(n, m) = r = xn + ym$

```
if m mod n = 0 then
    return 1, 0, n          // 1·n + 0·m = n = gcd(n, m)
else
    x, y, r := extended-Euclid(m mod n, n)
    return y − ⌊m/n⌋ · x, x, r
```

Figure 7.29: The Extended Euclidean algorithm.

Let $n_1, n_2, \ldots, n_k$ be a collection of integers, any pair of which is relatively prime. Let $N := \prod_{i=1}^{k} n_i$. Writing $\mathbb{Z}_m := \{0, 1, \ldots, m-1\}$, the *Chinese Remainder Theorem* states that, for any sequence of values $\langle a_1, \ldots, a_k \rangle$ with each $a_i \in \mathbb{Z}_{n_i}$, there exists one and only one integer $x \in \mathbb{Z}_N$ such that $x \bmod n_i = a_i$ for all $1 \leq i \leq k$.

Multiplicative Inverses

For any integer $n \geq 2$, let $\mathbb{Z}_n$ denote the set $\{0, 1, \ldots, n-1\}$. Let $a \in \mathbb{Z}_n$ be arbitrary. The *multiplicative inverse of a in $\mathbb{Z}_n$* is the number $a^{-1} \in \mathbb{Z}_n$ such that $a \cdot a^{-1} \equiv_n 1$ if any such number exists. (If no such number exists, then a^{-1} is undefined.) For example, the multiplicative inverse of 2 in $\mathbb{Z}_9$ is $2^{-1} = 5$ because $2 \cdot 5 = 10 \equiv_9 1$; the multiplicative inverse of 1 in $\mathbb{Z}_9$ is $1^{-1} = 1$ because $1 \cdot 1 \equiv_9 1$; and the multiplicative inverse of 3 in $\mathbb{Z}_9$ is undefined (because $3a \not\equiv_9 1$ for any $a \in \mathbb{Z}_9$).

Let $n \geq 2$ and $a \in \mathbb{Z}_n$. The multiplicative inverse a^{-1} exists in $\mathbb{Z}_n$ if and only if n and a are relatively prime. Furthermore, when a^{-1} exists, we can find it using the Extended Euclidean algorithm. We compute $\langle x, y, r \rangle :=$ **extended-Euclid**(a, n); when $\gcd(a, n) = 1$ (as it is when a and n are relatively prime), the returned values satisfy $xa + yn = 1$, and thus $a^{-1} := x \bmod n$ is the multiplicative inverse of a in $\mathbb{Z}_n$. For a prime number p, every nonzero $a \in \mathbb{Z}_p$ has a multiplicative inverse in $\mathbb{Z}_p$.

Fermat's Little Theorem states that, for any prime p and any integer a with $p \nmid a$, the $(p-1)$st power of a must equal 1 modulo p. (That is: for prime p and nonzero $a \in \mathbb{Z}_p$, we have $a^{p-1} \equiv_p 1$. For example, because 17 is prime, Fermat's Little Theorem—or arithmetic!—tells us that $5^{16} \bmod 17 = 1$.)

Cryptography

A sender ("Alice") wants to send a private message to a receiver ("Bob"), but they can only communicate using a channel that can be overheard by an eavesdropper ("Eve"). In *cryptography,* Alice *encrypts* the message m (the "plaintext") and transmits the encrypted version c (the "ciphertext"); Bob then *decrypts* it to recover the original message m. The simplest way to achieve this goal is with a *one-time pad:* Alice and Bob agree on a shared secret bitstring k; the ciphertext is the bitwise XOR of m and k, and Bob decrypts by computing the bitwise XOR of c and k.

A more useful infrastructure is *public-key cryptography,* in which Alice and Bob do not have to communicate a secret in advance. Every user has a *public key* and a (mathematically related) *private key*; to communicate with Bob, Alice uses Bob's public key for encryption (and Bob uses his private key for decryption). The *RSA cryptosystem* is a widely used protocol for public-key cryptography; it works as follows:

- **Key generation:** Bob finds large primes p and q; he chooses an $e \neq 1$ that's relatively prime to $(p-1)(q-1)$; and he computes $d := e^{-1}$ modulo $(p-1)(q-1)$. Bob's public key is $\langle e, n \rangle$ and his private key is $\langle d, n \rangle$, where $n := pq$.
- **Encryption:** When Alice wants to send m to Bob, she encrypts m as $c := m^e \bmod n$.
- **Decryption:** Bob decrypts c as $c^d \bmod n$.

By our choices of n, p, q, d, and e, Fermat's Little Theorem allows us to prove that Bob's decryption of the encryption of message m is always the original message m itself. And, under commonly held beliefs about the difficulty of factoring large numbers (and computing "eth roots mod n"), Eve cannot compute m without spending an implausibly large amount of computation time.

Key Terms and Results

Key Terms

Modular Arithmetic

- modulus; $n \bmod k$ and $\lfloor \frac{n}{k} \rfloor$
- congruence/equivalence ($\equiv_n$)
- (evenly) divides, factor, multiple
- greatest common divisor
- least common multiple
- Euclidean algorithm

Primality and Relative Primality

- prime vs. composite numbers
- Prime Number Theorem
- prime factorization
- relative primality
- Extended Euclidean algorithm
- Chinese Remainder Theorem

Multiplicative Inverses

- $\mathbb{Z}_n$
- multiplicative inverse (a^{-1} in $\mathbb{Z}_n$)
- Fermat's Little Theorem
- Carmichael number

Cryptography

- Alice, Bob, Eve
- plaintext, ciphertext
- one-time pad
- public-key cryptography
- public key; private key
- key generation; encryption/decryption
- RSA

Key Results

Modular Arithmetic

1. For any integers $k \geq 1$ and n, there exist unique integers d and r such that $0 \leq r < k$ and $kd + r = n$. (And $r = n \bmod k$ and $d = \lfloor \frac{n}{k} \rfloor$.)
2. For arbitrary positive integers n and $m \geq n$, the Euclidean algorithm efficiently computes $\gcd(n, m)$.

Primality and Relative Primality

1. The *Prime Number Theorem:* as n gets large, the ratio between $\frac{n}{\log n}$ and the number of primes less than or equal to n approaches 1.
2. Every positive integer has a prime factorization (which is unique up to reordering).
3. Given positive integers n and m, the Extended Euclidean algorithm efficiently computes three integers x, y, r such that $r = \gcd(n, m) = xn + ym$.
4. The *Chinese Remainder Theorem:* Suppose $n_1, n_2, \ldots, n_k$ are all relatively prime, and let $N := \prod_{i=1}^{k} n_i$. Then, for any $\langle a_1, \ldots, a_k \rangle$ with each $a_i \in \mathbb{Z}_{n_i}$, there exists a unique $x \in \mathbb{Z}_N$ such that $x \bmod n_i = a_i$ for all $1 \leq i \leq k$.

Multiplicative Inverses

1. In $\mathbb{Z}_n$, the multiplicative inverse a^{-1} of a exists if and only if n and a are relatively prime. When it does exist, we can find a^{-1} using the Extended Euclidean algorithm.
2. *Fermat's Little Theorem:* for any prime number p and any nonzero $a \in \mathbb{Z}_p$, we have $a^{p-1} \equiv_p 1$.

Cryptography

1. In the RSA cryptosystem, Alice can use Bob's public key to encrypt a message m so that Bob can decrypt it efficiently. (And, under reasonable assumptions about certain numerical problems' hardness, Eve *can't* recover m without an exorbitant amount of computation.)

8
Relations

In which our heroes navigate a sea of many related perils, some of which turn out to be precisely equivalent to each other.

8.1 *Why You Might Care*

> Reality must take precedence over public relations, for Nature cannot be fooled.
>
> Richard Feynman (1918–1988)

In Chapter 2, we encountered *functions*, a basic data type that maps each element of some input set A to an element of an output set B. Here, we'll explore a generalization of functions, called *relations*, that represent arbitrary subsets of $A \times B$. For example, a large retailer might be interested in the relation *purchased*, a subset of $Customers \times Products$. (A function is a special kind of relation where each input element is related to one and only one element of the output set; notice that the same customer may have purchased many different products—or one, or none at all—so *purchased* is not a function.) Or a college might be interested in the relation *prerequisiteOf*, a subset of $Courses \times Courses$, where a student can only register for a course c if, for every course c' where $\langle c', c \rangle \in prerequisiteOf$, she's already passed c'. (And so the college would also want to compute the relation $passed \subseteq Students \times Courses$.)

Relations are the critical foundation of *relational databases*, an utterly widespread modern area of CS, underlying many of the web sites we all use regularly. (One classical special-purpose programming language for relational databases is called SQL, for "structured query language"; there are other platforms, too.) A relational database stores a (generally quite large!) collection of structured data. Logically, a database is organized as a collection of *tables*, each of which represents a relation, where each *row* of a table represents an element contained in that relation. Fundamental manipulations of these relations can then be used to answer more sophisticated questions about the underlying data. For example, using standard operations in relational databases (and the relations *prerequisiteOf* and *passed* above), we could compute things like (i) a list of every class c for which you have satisfied all prerequisites of c but have not yet passed c; or (ii) a list of people with whom you've taken at least one class; or (iii) a list of people p with whom you've taken at least one class and where p has also taken at least one class that meets condition (i). (Those are the friends you could ask for help when you take that class.) Or that large retailer might want, for a particular user u, to find the 10 products not purchased by u that were most frequently purchased by other users who share, say, at least half of their purchases with u. All of these queries—though sometimes rather brutally complicated to state in English—can be expressed fairly naturally in the language of relations.

Relational databases are probably the most prominent (and, given the name, the most obvious!) practical application of relations, but there are many others, too. In a sense, Chapter 11, on trees and graphs, is filled with a long list of other applications of relations; a *directed graph* is really nothing more than a relation on a set of nodes. And in this chapter, we'll also encounter some other applications in asymptotics, in computer graphics (figuring out an order in which to draw shapes so that the right piece ends up "on top" on the screen), and in regular expressions (a widely used formalism for specifying patterns for which we might search in text).

8.2 *Formal Introduction*

> A man is a bundle of relations, a knot of roots, whose flower and fruitage is the world.
>
> Ralph Waldo Emerson (1803–1882)

⟨blue, green⟩
⟨blue, purple⟩
⟨red, orange⟩
⟨red, purple⟩
⟨yellow, green⟩
⟨yellow, orange⟩

Figure 8.1: A relation between the primary colors and secondary colors.

Informally, a *(binary) relation* describes a pairwise relationship that holds for certain pairs of elements from two sets A and B. One particular relation is shown in Figure 8.1, expressing the "is a component of" relationship between primary and secondary colors: that is, Figure 8.1 denotes a particular relation on the sets $A = \{\text{red}, \text{yellow}, \text{blue}\}$ and $B = \{\text{green}, \text{purple}, \text{orange}\}$. This description of a relation—a pairwise relationship between some of the elements of two sets A and B—is obviously very general. But let's start by considering a few specific examples, which together will begin to show the range of the kinds of properties that relations can represent:

Example 8.1 (Satisfaction)
Let $A := \{f : \text{truth assignments for } p \text{ and } q\}$ and $B := \{\varphi : \text{propositions over } p \text{ and } q\}$. One interesting relation between elements of A and B denotes whether a particular truth assignment makes a particular proposition true. (This relation is usually called *satisfies*.) For a proposition φ, a truth assignment f either satisfies φ or it doesn't satisfy φ. For example:

- the truth assignment $[{}^{p=T}_{q=F}]$ satisfies $p \vee q$ (as do all truth assignments except $[{}^{p=F}_{q=F}]$);
- the truth assignment $[{}^{p=T}_{q=F}]$ satisfies $p \wedge \neg q$ (and no other truth assignment does);
- every truth assignment in A satisfies $p \vee \neg p$; and
- no truth assignment in A satisfies $q \wedge \neg q$.

(Thus an element of B might be satisfied by zero, one, or more elements of A. Similarly, an element of A might satisfy many different elements of B.)

Example 8.2 (Numbers that are not too different)
Consider the following relationship between two elements of $\mathbb{R}$: we'll say that two real numbers $x, y \in \mathbb{R}$ are *withinHalf* of each other if $|x - y| \leq 0.5$. For example, we have $\mathit{withinHalf}(2.781828, 3.0)$ and $\mathit{withinHalf}(3.14159, 3.0)$ and $\mathit{withinHalf}(2.5, 3.0)$ and $\mathit{withinHalf}(2.5, 2.0)$. Note that $\mathit{withinHalf}(x, x)$ holds for any real number x.

Example 8.3 (Being related to)
In keeping with the word "relation," we actually use the phrase "is related to" in English to express one specific binary relation on pairs of people—"being in the same family as" (or "being a (blood) relative of"). For example, we can make the true claim that *Rosemary Clooney is related to George Clooney.* (And a related statement is also true: *George Clooney is related to Rosemary Clooney.* The fact that these two statements convey the same information follows from the fact that the *is related to* relation has a

property called *symmetry:* for any x and y, it's the case that x is related to y if and only if y is related to x. Not all relations are symmetric, as we'll see in Section 8.3.)

Some qualitatively different types of relations are already peeking out in these few examples (and more properties of relations will reveal themselves as we go further). Sometimes the relation contains a finite number of pairs, as in Figure 8.1 (primary/secondary colors); sometimes the relation contains an infinite number of pairs, as in *withinHalf*. Sometimes a relation connects elements from two different sets, as in Example 8.1 (satisfaction, which connected truth assignments to propositions); sometimes it connects two elements from the same set, as in Example 8.3 ("is a (blood) relative of," which connects people to people). Sometimes a particular element x is related to every candidate element, sometimes to none. And sometimes the relation has some special properties like *reflexivity,* in which every x is related to x itself (as in *withinHalf*), or *symmetry* (as in "is a (blood) relative of").

8.2.1 *The Definition of a Relation, Formalized*

Later in the chapter, we'll return to the types of properties that we just introduced, but before we can look at properties of relations, we first need to define them. Technically, a binary relation is simply a subset of the Cartesian product of two sets:

Definition 8.1 ((Binary) relation)
A (binary) relation on $A \times B$ *is a subset of* $A \times B$.

Often we'll be interested in a relation on $A \times A$*, where the two sets are the same. If there is no danger of confusion, we may refer to a subset of* $A \times A$ *as simply a* relation on A.

Here are a few formal examples of relations:

Example 8.4 (A few relations, formally)
The following sets are all relations:

- $\{\langle 12,1\rangle, \langle 1,2\rangle, \langle 2,3\rangle, \langle 3,4\rangle, \langle 4,5\rangle, \langle 5,6\rangle, \langle 6,7\rangle, \langle 7,8\rangle, \langle 8,9\rangle, \langle 9,10\rangle, \langle 10,11\rangle, \langle 11,12\rangle\}$ is a relation on $\{1, \ldots, 12\}$. (Informally, this relation expresses "is one hour before.")
- $|$ ("divides") is a relation on $\mathbb{Z}$, where $|$ denotes the set $\{\langle d,n\rangle : n \bmod d = 0\}$.
- $\leq$ is a relation on $\mathbb{R}$, where $\leq$ denotes the set $\{\langle x,y\rangle : x \text{ is no bigger than } y\}$.
- As a reminder, the *power set* of a set S, denoted $\mathscr{P}(S)$, is the set of all subsets of S. For any set S, then, we can define $\subseteq$ as a relation on $\mathscr{P}(S)$, where $\subseteq$ denotes the set

$$\subseteq = \{\langle A,B\rangle \in \mathscr{P}(S) \times \mathscr{P}(S) : [\forall x \in S : x \in A \Rightarrow x \in B]\}.$$

For the set $S = \{1,2\}$, for example, the relation $\subseteq$ is

$$\subseteq = \left\{ \begin{array}{lllll} \langle\varnothing,\varnothing\rangle, & \langle\varnothing,\{1\}\rangle, & \langle\varnothing,\{2\}\rangle, & \langle\varnothing,\{1,2\}\rangle, & \\ \langle\{1\},\{1\}\rangle, & \langle\{1\},\{1,2\}\rangle, & \langle\{2\},\{2\}\rangle, & \langle\{2\},\{1,2\}\rangle, & \langle\{1,2\},\{1,2\}\rangle \end{array} \right\}.$$

- $\{\langle \text{Ron Rivest}, 2002\rangle, \langle \text{Adi Shamir}, 2002\rangle, \langle \text{Len Adleman}, 2002\rangle, \langle \text{Alan Kay}, 2003\rangle, \langle \text{Vint Cerf}, 2004\rangle, \langle \text{Robert Kahn}, 2004\rangle, \langle \text{Peter Naur}, 2005\rangle, \langle \text{Frances Allen}, 2006\rangle\}$ is a relation on the set *People* $\times$ $\{2002, 2003, 2004, 2005, 2006\}$, representing the relationship between people and any year in which they won a Turing Award.

Rivest, Shamir, and Adleman won Turing Awards for their work in cryptography; see Section 7.5. Kay was an inventor of the paradigm of object-oriented programming. Cerf and Kahn invented the communication protocols that undergird the Internet. Naur made crucial contributions to the design of programming languages, compilers, and software engineering. Allen made foundational contributions to optimizing compilers and parallel computing.

For some relations—for example, $|$ and $\leq$ and $\subseteq$ from Example 8.4—it's traditional to write the symbol for the relation *between* the elements that are being related, using so-called *infix notation.* (So we write $3 \leq 3.5$, rather than $\langle 3, 3.5\rangle \in \leq$.) In general, for a relation R, we may write either $\langle x, y\rangle \in R$ or $x\,R\,y$, depending on context.

Taking it further: Most programming languages use infix notation in their expressions: that is, they place their operators between their operands, as in `(5 + 3) / 2` in Java or Python or C to denote the value $\frac{5+3}{2}$. But some programming languages, like Postscript (the language commonly used by printers) or the language of Hewlett–Packard calculators, use *postfix* notation, where the operator follows the operands. Other languages, like Scheme, use *prefix* notation, in which the operator comes before the operands. (In Postscript, we would write `5 3 add 2 div`; in Scheme, we'd write `(/ (+ 5 3) 2)`.) While we're all much more accustomed to infix notation, one of the advantages of pre- or postfix notation is that the order of operations is unambiguous: compare the ambiguous `5 + 3 / 2` to its two postfix alternatives, namely `5 3 2 div add` and `5 3 add 2 div`.

Example 8.5 (Bitstring prefixes)

Problem: Let *isPrefix* denote the following relation: for two bitstrings x and y, we have $\langle x, y\rangle \in$ *isPrefix* if and only if the bitstring y starts with precisely the symbols contained in x. (After the bits of x, the bitstring y may contain zero or more additional bits.) For example, 001 is a prefix of <u>001</u>110 and <u>001</u>, but 001 is not a prefix of 1001. Write down the relation *isPrefix* on bitstrings of length ≤ 2 explicitly, using set notation.

Solution: Denoting the empty string by ε, the relation is

$$\textit{isPrefix} = \left\{\begin{array}{lllllll} \langle\varepsilon,\varepsilon\rangle, & \langle\varepsilon,0\rangle, & \langle\varepsilon,1\rangle, & \langle\varepsilon,00\rangle, & \langle\varepsilon,01\rangle, & \langle\varepsilon,10\rangle, & \langle\varepsilon,11\rangle, \\ \langle 0,0\rangle, & \langle 0,00\rangle, & \langle 0,01\rangle, & \langle 1,1\rangle, & \langle 1,10\rangle, & \langle 1,11\rangle, & \\ \langle 00,00\rangle, & \langle 01,01\rangle, & \langle 10,10\rangle, & \langle 11,11\rangle & & & \end{array}\right\}.$$

Visualizing binary relations

For a relation R on $A \times B$ where both A and B are finite sets, instead of viewing R as a list of pairs, it can be easier to think of R as a two-column table, where each row corresponds to an element $\langle a, b\rangle \in R$. Alternatively, we can visualize relations in a way similar to the way that we visualized functions in Chapter 2: we place the elements of A in one column, the elements of B in a second column, and draw a line connecting $a \in A$ to $b \in B$ whenever $\langle a, b\rangle \in R$. Note that when we drew functions using these two-column pictures, every element in the left-hand column had exactly one arrow leaving it. That's not necessarily true for a relation; elements in the left-hand column could have none, one, or two or more arrows leaving them.

Figure 8.2 shows a relation represented in these two ways. (For a relation that's a subset of $A \times A$, the graphical version of this two-column representation is less appropriate because there's really only one kind of element; see Section 8.3 for a different way of visualizing these relations, and see Figure 8.13(a) for *isPrefix* as an example.)

Month	Days
Jan	31
Feb	28
Feb	29
Mar	31
Apr	30
May	31
Jun	30
Jul	31
Aug	31
Sep	30
Oct	31
Nov	30
Dec	31

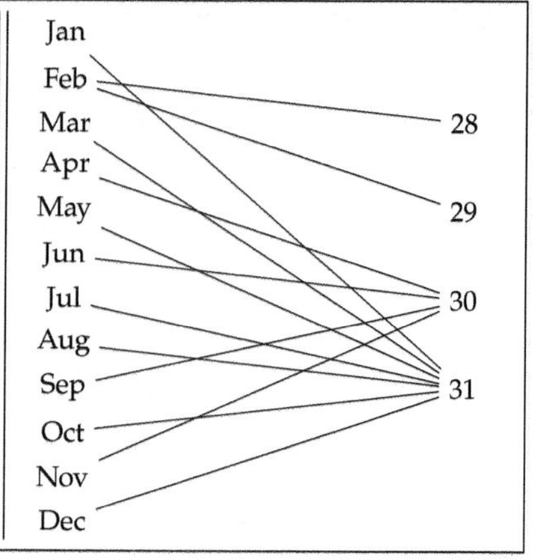

Figure 8.2: The relation indicating the number of days per month. (Note that Feb is related to *both* 28 and 29.)

Taking it further: Recall from Chapter 3 that we defined a *predicate* as a Boolean-valued function—that is, P is a function $P : U \to \{\text{True}, \text{False}\}$ for a set U, called the *universe*. (See Definition 3.18.) For example, we considered the predicate $P_{\text{alphabetical}}(x, y)$ = "string x is alphabetically before string y."

Binary predicates—when the universe is a set of pairs $U = A \times B$—are very closely related to binary relations. The main difference is that in Chapter 3 we thought of a binary predicate P as a *function* $P : A \times B \to \{\text{True}, \text{False}\}$, whereas here we're thinking of a relation R on $A \times B$ as a *subset* $R \subseteq A \times B$. For example, the relation $R_{\text{alphabetical}}$ is the set $\{\langle \text{AA}, \text{AAH}\rangle, \langle \text{AA}, \text{AARDVARK}\rangle, \ldots, \langle \text{ZYZZYVA}, \text{ZYZZYVAS}\rangle\}$. And $P_{\text{alphabetical}}(\text{AA}, \text{AAH}) = \text{True}$, $P_{\text{alphabetical}}(\text{AA}, \text{ZYZZYVA}) = \text{True}$, and $P_{\text{alphabetical}}(\text{BEFORE}, \text{AFTER}) = \text{False}$.

But there's a direct translation between these two worldviews. Given a relation $R \subseteq A \times B$, we can define the predicate P_R such that

$$P_R(a,b) = \begin{cases} \text{True} & \text{if } \langle a,b\rangle \in R \\ \text{False} & \text{if } \langle a,b\rangle \notin R. \end{cases}$$

The function P_R is known as the *characteristic function* of the set R: that is, it's the function such that $P_R(x) = \text{True}$ if and only if $x \in R$. ($P_{\text{alphabetical}}$ is the characteristic function of $R_{\text{alphabetical}}$.)

We can also go the other direction, and translate a Boolean-valued binary function into a relation. Given a predicate $P : A \times B \to \{\text{True}, \text{False}\}$, define the relation $R_P := \{\langle a,b\rangle : P(a,b)\}$—that is, define R_P as the set of pairs for which the function P is true. In either case, we have a direct correspondence between (i) the elements of the relation, and (ii) the inputs to the function that make the output true.

8.2.2 *Inverse and Composition of Binary Relations*

Because a relation on $A \times B$ is simply a subset of $A \times B$, we can combine relations on $A \times B$ using all the normal set-theoretic operations: if R and S are both relations on $A \times B$, then $R \cup S$, $R \cap S$, and $R - S$ are also relations on $A \times B$, as is the set $\sim R := \{\langle a,b\rangle \in A \times B : \langle a,b\rangle \notin R\}$.

But we can also generate new relations in ways that are specific to relations, rather than being generic set operations. Two of the most common are the *inverse* of a relation (which turns a relation on $A \times B$ into a relation on $B \times A$ by "flipping around" every pair in the relation) and the *composition* of two relations (which turns two relations on $A \times B$ and $B \times C$ into a single relation on $A \times C$, where a and c are related if there's a "two-hop" connection from a to c via some element $b \in B$).

Inverting a relation

Here is the formal definition of the inverse of a relation:

Definition 8.2 (Inverse of a Relation)
Let R be a relation on $A \times B$. The inverse R^{-1} *of R is a relation on $B \times A$ defined by $R^{-1} := \{\langle b,a\rangle \in B \times A : \langle a,b\rangle \in R\}$.*

Here are a few examples of the inverses of some simple relations:

Example 8.6 (Some inverses)

- The inverse of the relation $\leq$ is the relation $\geq$.
- The inverse of the relation $=$ is the relation $=$ itself. (That is, $=$ is its own inverse.)
- The inverse of the months–days relation from Figure 8.2 is shown in Figure 8.3.
- Define the relation

$$R := \left\{ \begin{array}{l} \langle 1,2\rangle, \langle 1,3\rangle, \langle 1,4\rangle, \langle 1,5\rangle, \langle 1,6\rangle, \\ \langle 2,2\rangle, \langle 2,4\rangle, \langle 2,6\rangle, \langle 3,3\rangle, \langle 3,6\rangle, \langle 4,4\rangle, \langle 5,5\rangle, \langle 6,6\rangle \end{array} \right\}.$$

The inverse of R is the relation

$$R^{-1} = \left\{ \begin{array}{l} \langle 2,1\rangle, \langle 3,1\rangle, \langle 4,1\rangle, \langle 5,1\rangle, \langle 6,1\rangle, \\ \langle 2,2\rangle, \langle 4,2\rangle, \langle 6,2\rangle, \langle 3,3\rangle, \langle 6,3\rangle, \langle 4,4\rangle, \langle 5,5\rangle, \langle 6,6\rangle \end{array} \right\}.$$

(Note that R is $\{\langle d,n\rangle : d \text{ divides } n\}$, and R^{-1} is $\{\langle n,d\rangle : n \text{ is a multiple of } d\}$.)

Month	Days
Jan	31
Feb	28
Feb	29
Mar	31
Apr	30
May	31
Jun	30
Jul	31
Aug	31
Sep	30
Oct	31
Nov	30
Dec	31

Days	Month
31	Jan
28	Feb
29	Feb
31	Mar
30	Apr
31	May
30	Jun
31	Jul
31	Aug
30	Sep
31	Oct
30	Nov
31	Dec

Figure 8.3: The relation from Figure 8.2, and its inverse.

Note that, as in the month–day example, the inverse of any relation shown in table form is simply the relation resulting from swapping the two columns of the table.

Composing two relations

The second way of creating a new relation from existing relations is *composition,* which, informally, represents the successive "application" of two relations. Two elements x and y are related under the relation $S \circ R$, denoting the composition of two relations R and S, if there's some intermediate element b that connects x and y under R and S, respectively. (We already saw how to compose *functions,* in Section 2.5, by applying one function immediately after the other. Functions are a special type of relation—see Section 8.2.3—and the composition of functions will similarly be a special case of the composition of relations.) Let's start with an informal example to build some intuition:

Warning! The composition of R and S is, as with functions, denoted $S \circ R$: the function $g \circ f$ *first* applies f and *then* applies g, so $(g \circ f)(x)$ gives the result $g(f(x))$. The order in which the relations are written may initially be confusing.

Example 8.7 (Relation composition, informally)

Consider a relation *allergicTo* on *People* $\times$ *Ingredients* and a relation *containedIn* on *Ingredients* $\times$ *Entrees*. Then the composition of *allergicTo* and *containedIn* is a relation on *People* $\times$ *Entrees* identifying pairs $\langle p,e\rangle$ for which *entree e contains at least one ingredient to which person p is allergic.*

Here's the formal definition:

Definition 8.3 (Composition of two relations)

Let R be a relation on $A \times B$ and let S be a relation on $B \times C$. Then the composition *of R*

and S is a relation on $A \times C$, denoted $S \circ R$, where $\langle a, c \rangle \in S \circ R$ if and only if there exists an element $b \in B$ such that $\langle a, b \rangle \in R$ and $\langle b, c \rangle \in S$.

Perhaps the easiest way to understand the composition of relations is through the picture-based view that we introduced in Figure 8.2: the relation $S \circ R$ contains pairs of elements that are joined by "two-hop" connections, where the first hop is defined by R and the second hop is defined by S. (See Figure 8.4.)

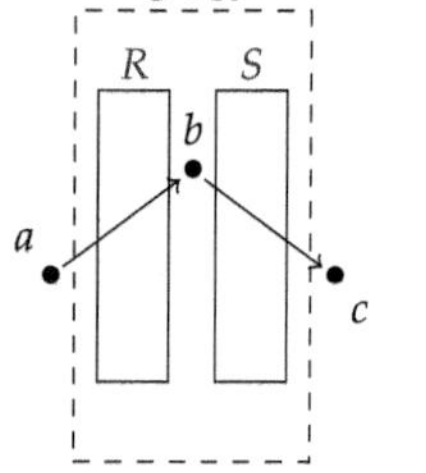

Figure 8.4: The composition of R and S. A pair $\langle a, c \rangle$ is in $S \circ R$ when, for some b, both $\langle a, b \rangle \in R$ and $\langle b, c \rangle \in S$.

Some examples of composing relations

Here are a few examples of the composition of some relations:

Example 8.8 (The composition of two small relations)
Consider the following two relations:

- Let $R := \{\langle 0, a \rangle, \langle 0, b \rangle, \langle 0, c \rangle, \langle 1, c \rangle, \langle 1, d \rangle\}$ be a relation on $\{0, 1\} \times \{a, b, c, d\}$.
- Let $S := \{\langle b, \pi \rangle, \langle b, \sqrt{3} \rangle, \langle c, \sqrt{2} \rangle, \langle d, \sqrt{2} \rangle\}$ be a relation on $\{a, b, c, d\} \times \mathbb{R}$.

Then $S \circ R \subseteq \{0, 1\} \times \mathbb{R}$ is the relation that consists of all pairs $\langle x, z \rangle$ such that there exists an element $y \in \{a, b, c, d\}$ where $\langle x, y \rangle \in R$ and $\langle y, z \rangle \in S$. That is,

$$S \circ R = \{ \underbrace{\langle 0, \pi \rangle, \langle 0, \sqrt{3} \rangle,}_{\text{because of } b} \quad \underbrace{\langle 0, \sqrt{2} \rangle}_{\text{because of } c}, \quad \underbrace{\langle 1, \sqrt{2} \rangle}_{\text{because of } c \text{ and } d} \}.$$

See Figure 8.5 for the visual representation of the relation composition from Example 8.8: because there are "two-hop" paths from 0 to $\{\pi, \sqrt{3}, \sqrt{2}\}$ and from 1 to $\{\sqrt{2}\}$, the relation $S \circ R$ is as described. (Again: the relation $S \circ R$ consists of pairs related by a two-step chain, with the first step under R and the second under S.)

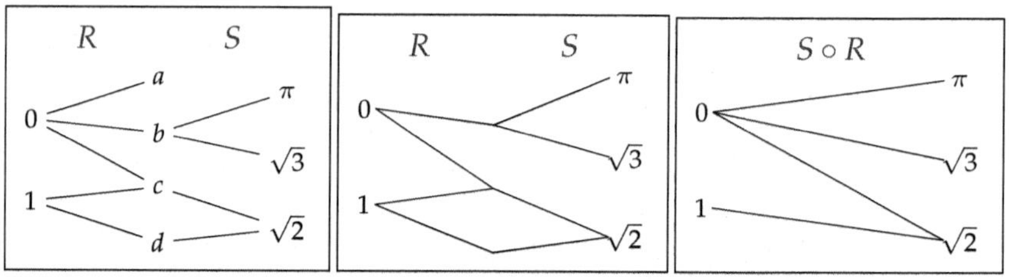

Figure 8.5: The composition of two relations, visualized.

Here's a second example of composing relations, this time where the relations being composed are more meaningful:

Example 8.9 (Relations in the U.S. Senate)
The United States Senate has two senators from each state, each of whom is affiliated with zero or one political parties. See Figure 8.6 for two relations: the relation S, between all U.S. states whose names start with the letter "I" and the senators who represented them in the year 2016; and the relation T, between senators and their political party.

Figure 8.6(c) shows the composition of these relations, which is a relation between *IStates* and *Parties*. Notice that $\langle state, party \rangle \in T \circ S$ if and only if there exists a senator s such that *state* is represented by s *and* s is affiliated with party *party*.

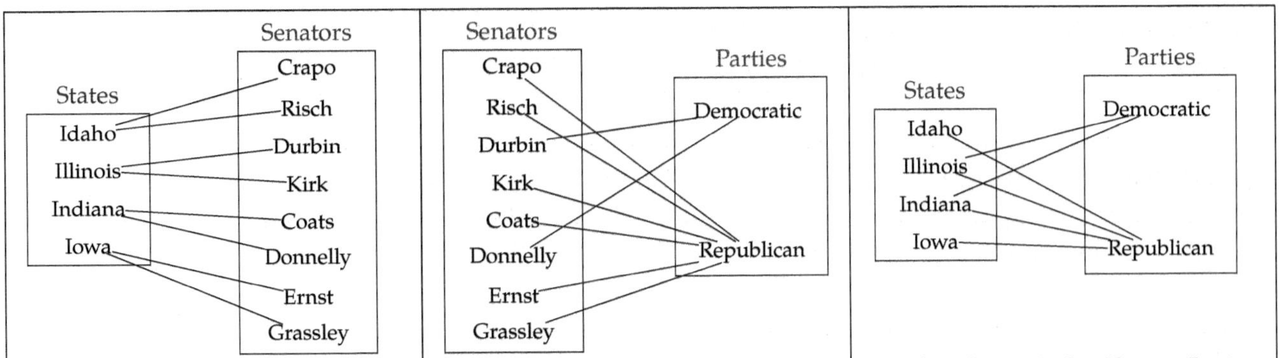

(a) The relation $S \subseteq IStates \times Senators$ of each state's senators.

(b) The relation $T \subseteq Senators \times Parties$ of each senator's party affiliation.

(c) The relation $T \circ S \subseteq IStates \times Parties$.

Figure 8.6: Two relations S and T, and their composition $T \circ S$.

So far we've considered composing relations on $A \times B$ and $B \times C$ for three distinct sets A, B, and C. But we can also consider a relation $R \subseteq A \times A$, and in this case we can also compose R *with itself*. Here are some brief examples:

Example 8.10 (Composing a relation with itself)
<u>*Problem*</u>: For each of the following relations R on $\mathbb{Z}^{\geq 1}$, describe the relation $R \circ R$:

1. *successor*, namely the set $\{\langle n, n+1 \rangle : n \in \mathbb{Z}^{\geq 1}\}$.
2. $=$, namely the set $\{\langle n, n \rangle : n \in \mathbb{Z}^{\geq 1}\}$.
3. *relativelyPrime*, defined as the set of pairs of relatively prime (positive) integers, so that $relativelyPrime := \{\langle n, m \rangle : \gcd(n, m) = 1\}$.

<u>*Solution*</u>: 1. By definition, $\langle x, z \rangle \in successor \circ successor$ if and only if there exists an integer y such that *both* $\langle x, y \rangle \in successor$ *and* $\langle y, z \rangle \in successor$. Thus the only possible y is $y = x + 1$, and the only possible z is $z = y + 1 = x + 2$. Thus

$$successor \circ successor = \{\langle n, n+2 \rangle : n \in \mathbb{Z}^{\geq 1}\}.$$

2. (We'll write *equals* instead of $=$; otherwise the notation becomes indecipherable.) By definition, the pair $\langle x, z \rangle$ is in the relation *equals* $\circ$ *equals* if and only if there exists an integer y such that $x = y$ and $y = z$. But that's true if and only if $x = z$. That is, $\langle x, z \rangle \in equals \circ equals$ if and only if $\langle x, z \rangle \in equals$. Thus composing *equals* with itself doesn't change anything: *equals* $\circ$ *equals* is identical to *equals*.

3. We must identify all pairs $\langle x, z \rangle \in \mathbb{Z}^{\geq 1} \times \mathbb{Z}^{\geq 1}$ such that there exists an integer y where $\langle x, y \rangle \in relativelyPrime$ *and* $\langle y, z \rangle \in relativelyPrime$. But notice that $y = 1$ is relatively prime to *every* positive integer. Thus, for any $\langle x, z \rangle \in \mathbb{Z}^{\geq 1} \times \mathbb{Z}^{\geq 1}$, we have that $\langle x, 1 \rangle \in relativelyPrime$ and $\langle 1, z \rangle \in relativelyPrime$. Thus

$$relativelyPrime \circ relativelyPrime = \mathbb{Z}^{\geq 1} \times \mathbb{Z}^{\geq 1}.$$

Problem-solving tip: Just as you do with a program, always make sure that your mathematical expressions "type check." (For example, just as the Python expression `0.33 * "atomic"` doesn't make sense, the composition $R \circ R$ for the relation $R = \{\langle 1, \texttt{A} \rangle, \langle 2, \texttt{B} \rangle\}$ doesn't denote anything useful.)

AN EXAMPLE OF COMPOSING A RELATION WITH ITS OWN INVERSE

We'll close with one last example of composing relations, this time by taking the composition of a relation R and its inverse R^{-1}:

Month	Days
Jan	31
Feb	28
Feb	29
Mar	31
Apr	30
May	31
Jun	30
Jul	31
Aug	31
Sep	30
Oct	31
Nov	30
Dec	31

Days	Month
31	Jan
28	Feb
29	Feb
31	Mar
30	Apr
31	May
30	Jun
31	Jul
31	Aug
30	Sep
31	Oct
30	Nov
31	Dec

Figure 8.7: The relations R and R^{-1}, from Figure 8.3.

Example 8.11 (Composing a relation and its inverse)

Problem: Let $R \subseteq M \times D$ be the relation between the months and the numbers of days in that month, and let $R^{-1} \subseteq D \times M$ be its inverse. (See Figure 8.7 for a reminder.) What is $R^{-1} \circ R$?

Solution: First, because $R \subseteq M \times D$ and $R^{-1} \subseteq D \times M$, we know that $R^{-1} \circ R \subseteq M \times M$. We have to identify

$$\langle x, y\rangle \in M \times M \text{ such that } \exists z \in D : \langle x, z\rangle \in R \text{ and } \langle z, y\rangle \in R^{-1}$$

$$\Leftrightarrow \exists z \in D : \langle x, z\rangle \in R \text{ and } \langle y, z\rangle \in R. \qquad \textit{definition of inverse}$$

In other words, we seek pairs of months that are related by R to at least one of the same values. The exhaustive list of pairs in $R^{-1} \circ R$ is

$$\left\{\begin{array}{l}
\langle \text{Jan},\text{Jan}\rangle, \langle \text{Jan},\text{Mar}\rangle, \langle \text{Jan},\text{May}\rangle, \langle \text{Jan},\text{Jul}\rangle, \langle \text{Jan},\text{Aug}\rangle, \langle \text{Jan},\text{Oct}\rangle, \langle \text{Jan},\text{Dec}\rangle,\\
\langle \text{Mar},\text{Jan}\rangle, \langle \text{Mar},\text{Mar}\rangle, \langle \text{Mar},\text{May}\rangle, \langle \text{Mar},\text{Jul}\rangle, \langle \text{Mar},\text{Aug}\rangle, \langle \text{Mar},\text{Oct}\rangle, \langle \text{Mar},\text{Dec}\rangle,\\
\langle \text{May},\text{Jan}\rangle, \langle \text{May},\text{Mar}\rangle, \langle \text{May},\text{May}\rangle, \langle \text{May},\text{Jul}\rangle, \langle \text{May},\text{Aug}\rangle, \langle \text{May},\text{Oct}\rangle, \langle \text{May},\text{Dec}\rangle,\\
\langle \text{Jul},\text{Jan}\rangle, \langle \text{Jul},\text{Mar}\rangle, \langle \text{Jul},\text{May}\rangle, \langle \text{Jul},\text{Jul}\rangle, \langle \text{Jul},\text{Aug}\rangle, \langle \text{Jul},\text{Oct}\rangle, \langle \text{Jul},\text{Dec}\rangle,\\
\langle \text{Oct},\text{Jan}\rangle, \langle \text{Oct},\text{Mar}\rangle, \langle \text{Oct},\text{May}\rangle, \langle \text{Oct},\text{Jul}\rangle, \langle \text{Oct},\text{Aug}\rangle, \langle \text{Oct},\text{Oct}\rangle, \langle \text{Oct},\text{Dec}\rangle,\\
\langle \text{Dec},\text{Jan}\rangle, \langle \text{Dec},\text{Mar}\rangle, \langle \text{Dec},\text{May}\rangle, \langle \text{Dec},\text{Jul}\rangle, \langle \text{Dec},\text{Aug}\rangle, \langle \text{Dec},\text{Oct}\rangle, \langle \text{Dec},\text{Dec}\rangle,\\
\\
\langle \text{Apr},\text{Apr}\rangle, \langle \text{Apr},\text{Jun}\rangle, \langle \text{Apr},\text{Sep}\rangle, \langle \text{Apr},\text{Nov}\rangle,\\
\langle \text{Jun},\text{Apr}\rangle, \langle \text{Jun},\text{Jun}\rangle, \langle \text{Jun},\text{Sep}\rangle, \langle \text{Jun},\text{Nov}\rangle,\\
\langle \text{Sep},\text{Apr}\rangle, \langle \text{Sep},\text{Jun}\rangle, \langle \text{Sep},\text{Sep}\rangle, \langle \text{Sep},\text{Nov}\rangle,\\
\langle \text{Nov},\text{Apr}\rangle, \langle \text{Nov},\text{Jun}\rangle, \langle \text{Nov},\text{Sep}\rangle, \langle \text{Nov},\text{Nov}\rangle,\\
\\
\langle \text{Feb},\text{Feb}\rangle
\end{array}\right\}.$$

Note that $R^{-1} \circ R$ in Example 8.11 is different from the relation $R \circ R^{-1}$: the latter is the set of numbers that are related by R^{-1} to at least one of the same months, while the former is the set of months that are related by R to at least one of the same numbers. Thus $R \circ R^{-1} = \{\langle 31,31\rangle, \langle 30,30\rangle, \langle 29,29\rangle, \langle 28,28\rangle, \langle 28,29\rangle, \langle 29,28\rangle\}$. (The only distinct numbers related by $R \circ R^{-1}$ are 28 and 29, because of February.)

Also note that the relation $R^{-1} \circ R$ from Example 8.11 has a special form: this relation "partitions" the twelve months into three clusters—the 31-day months, the 30-day months, and February—so that any two months in the same cluster are related by $R^{-1} \circ R$, and no two months in different clusters are related by $R^{-1} \circ R$. (See Figure 8.13(b) for a visualization.) A relation with this structure, where elements are partitioned into clusters (and two elements are related if and only if they're in the same cluster) is called an *equivalence relation*; see Section 8.4.1 for much more.

8.2.3 *Functions as Relations*

Back in Chapter 2, we defined a *function* as something that maps each element of the set of legal inputs (the *domain*) to an element of the set of legal outputs (the *range*):

Definition 2.44 (functions): Let A and B be sets. A *function* f *from* A *to* B, written $f : A \to B$, assigns to each input value $a \in A$ a unique output value $b \in B$; the unique value b assigned to a is denoted by $f(a)$. We sometimes say that f *maps* a *to* $f(a)$.

While we've begun this chapter defining relations as a completely different kind of thing from functions, we can actually view functions as simply a special type of relation. For example, the "one hour later than" relation $\{\langle 12,1\rangle, \langle 1,2\rangle, \ldots, \langle 10,11\rangle, \langle 11,12\rangle\}$ from Example 8.4 really *is* a function $f : \{1,\ldots,12\} \to \{1,\ldots,12\}$, where we could write f more compactly as $f(x) := (x \bmod 12) + 1$.

In general, to think of a function $f : A \to B$ as a relation, we will view f as defining the set of ordered pairs $\langle x, f(x)\rangle$ for each $x \in A$, rather than as a mapping:

Definition 8.4 (Functions, viewed as relations)
Let A and B be sets. A function f from A to B, *written $f : A \to B$, is a relation on $A \times B$ with the additional property that, for every $a \in A$, there exists one and only one element $b \in B$ such that $\langle a, b\rangle \in f$.*

That is, we view the function $f : A \to B$ as the set $F := \{\langle x, f(x)\rangle : x \in A\}$, which is a subset of $A \times B$. The restriction of the definition requires that F has a unique output defined for every input: there cannot be two distinct pairs $\langle x, y\rangle$ and $\langle x, y'\rangle$ in F, and furthermore there cannot be any x for which there's no $\langle x, \bullet\rangle$ in F.

Example 8.12 (A function as a relation)
(Write $\mathbb{Z}_{11}$ to denote $\{0,1,2,\ldots,10\}$, as in Chapter 7.) The function $f : \mathbb{Z}_{11} \to \mathbb{Z}_{11}$ defined as $f(x) = x^2 \bmod 11$ can be written as

$$\{\langle 0,0\rangle, \langle 1,1\rangle, \langle 2,4\rangle, \langle 3,9\rangle, \langle 4,5\rangle, \langle 5,3\rangle, \langle 6,3\rangle, \langle 7,5\rangle, \langle 8,9\rangle, \langle 9,4\rangle, \langle 10,1\rangle\}.$$

Observe that f^{-1}, the inverse of f, is *not* a function—for example, the pairs $\langle 5,4\rangle$ and $\langle 5,7\rangle$ are both in f^{-1}, and there is no element $\langle 2, \bullet\rangle \in f^{-1}$. But f^{-1} *is* still a relation.

Example 8.13 (Composing functions)
Problem: Suppose that $f \subseteq A \times B$ and $g \subseteq B \times C$ are functions (in the sense of Definition 8.4). Prove that the relation $g \circ f$ is a function from A to C.

Solution: By definition, the composition of the relations f and g is

$$g \circ f := \{\langle x, z\rangle : \text{there exists } y \text{ such that } \langle x, y\rangle \in f \text{ and } \langle y, z\rangle \in g\}.$$

Because f is a function, there exists one and only one y^* such that $\langle x, y^*\rangle \in f$. Furthermore, because g is a function, for this particular y^* there exists a unique z such that $\langle y^*, z\rangle \in g$. Thus there exists one and only one z such that $\langle x, z\rangle \in g \circ f$. By definition, then, the relation $g \circ f$ is a function.

Under this functions-as-relations view, the definitions of the inverse and composition of functions—Definitions 2.48 and 2.52—precisely line up with the definitions of the inverse and composition of relations from this section. Furthermore, if a function is just a special type of relation, then the special types of functions that we defined in Chapter 2—one-to-one and onto functions—are just further restrictions on relations. Under the relation-based view of functions, the function $f \subseteq A \times B$ is called *one-to-one* if, for every $b \in B$, there exists at most one element $a \in A$ such that $\langle a, b \rangle \in f$. The function $f \subseteq A \times B$ is called *onto* if, for every $b \in B$, there exists at least one element $a \in A$ such that $\langle a, b \rangle \in f$.

Observe that, if $f \subseteq A \times B$ is a function, then the inverse f^{-1} of f—that is, the set $f^{-1} = \{\langle b, a \rangle : \langle a, b \rangle \in f\}$—is guaranteed to be a relation on $B \times A$, but it is a function from B to A if and only if f is both one-to-one and onto. In Exercises 8.38–8.43, you'll explore some other properties of the composition of functions/relations.

8.2.4 *n-ary Relations*

The relations that we've explored so far have all expressed relationships between *two* elements. But some interesting properties might involve more than two entities; for example, you might assemble all of your friends' birthdays as a collection of *triples* of the form $\langle name, birthdate, birthyear \rangle$. Or we might consider a relation on integers of the form $\langle a, b, k \rangle$ where $a \equiv_k b$. A relation involving tuples with n components, called an *n-ary relation*, is a natural generalization of a (binary) relation:

Definition 8.5 (n-ary relation)
An n-ary relation on the set $A_1 \times A_2 \times \cdots \times A_n$ is a subset of $A_1 \times A_2 \times \cdots \times A_n$. If there is no danger of confusion, we may refer to a subset of A^n as an n-ary relation on A.

(We generally refer to 2-ary relations as *binary relations* and 3-ary relations as *ternary relations*.) Here are a few examples:

Example 8.14 (Summing to 8)
Define *sumsTo8* as a ternary relation on the set $\{0,1,2,3,4\}$, where

$$sumsTo8 = \{\langle a,b,c \rangle \in \{0,1,2,3,4\}^3 : a+b+c = 8\}.$$

Then the elements in *sumsTo8* are:

$$\left\{ \begin{array}{l} \langle 0,4,4\rangle, \langle 1,3,4\rangle, \langle 1,4,3\rangle, \langle 2,2,4\rangle, \langle 2,3,3\rangle, \langle 2,4,2\rangle, \langle 3,1,4\rangle, \langle 3,2,3\rangle, \\ \langle 3,3,2\rangle, \langle 3,4,1\rangle, \langle 4,0,4\rangle, \langle 4,1,3\rangle, \langle 4,2,2\rangle, \langle 4,3,1\rangle, \langle 4,4,0\rangle \end{array} \right\}.$$

Example 8.15 (Betweenness)
The set $B := \{\langle x,y,z \rangle \in \mathbb{R}^3 : x \leq y \leq z \text{ or } x \geq y \geq z\}$ is a ternary relation on $\mathbb{R}$ that expresses "betweenness"—that is, the triple $\langle x,y,z \rangle \in B$ if x, y, and z are in a consistent order (either ascending or descending).

For example, we have $\langle -1,0,1 \rangle \in B$ and $\langle 6,5,4 \rangle \in B$, because $-1 \leq 0 \leq 1$ and $6 \geq 5 \geq 4$. But $\langle -7,8,-9 \rangle \notin B$, because these three numbers are neither in ascending order (because $8 \not\leq -9$) nor descending order (because $-7 \not\geq 8$).

Example 8.16 (RGB colors)
A 4-ary relation on *Names* $\times$ $\{0,1,\ldots,255\}$ $\times$ $\{0,1,\ldots,255\}$ $\times$ $\{0,1,\ldots,255\}$ is shown below: a collection of colors, each with its official name in HTML/CSS and its red/green/blue components (all three of which are elements of the set $\{0,1,\ldots,255\}$).

HTML (hypertext markup language) and *CSS (cascading style sheet)* are languages used to express the format, style, and layout of web pages.

White	255	255	255
Red	255	0	0
Lime	0	255	0
Blue	0	0	255
Cyan	0	255	255
Magenta	255	0	255
Yellow	255	255	0
Black	0	0	0
Gray	128	128	128
Maroon	128	0	0
Green	0	128	0
Navy	0	0	128
Teal	0	128	128
Purple	128	0	128
Olive	128	128	0

(This relation contains the full set of RGB colors with component values all drawn from either $\{0,128\}$ or $\{0,255\}$.)

Taking it further: *Databases*—systems for storing and accessing collections of structured data—are a widespread modern application of computer science. Databases store student records for registrars, account information for financial institutions, and even records of who liked whose posts on Facebook; in short, virtually every industrial system that has complex data with nontrivial relationships among data elements is stored in a database. More specifically, a *relational database* stores information about a collection of entities and relationships among those entities: fundamentally, a relational database is a collection of n-ary relations, which can then be manipulated and queried in various ways. Designing databases well affects both how easy it is for a user to pose the questions that he or she wishes to ask about the data, *and* how efficiently answers to those questions can be computed. See p. 815 for more on relational databases and how they connect with the types of relations that we've discussed so far.

Expressing n-ary relations as a collection of binary relations

Non-binary relations, like those in the last few examples, represent complex interactions among more than two entities. For example, the "betweenness" relation

$$B := \{\langle x,y,z\rangle \in \mathbb{R}^3 : x \leq y \leq z \text{ or } x \geq y \geq z\}$$

from Example 8.15 fundamentally expresses a relationship regarding triples of numbers: for any three real numbers x, y, and z, there are triples $\langle x,y,\bullet\rangle \in B$ and $\langle \bullet,y,z\rangle \in B$ and $\langle x,\bullet,z\rangle \in B$—but whether $\langle x,y,z\rangle$ itself is in the relation B genuinely depends on how all three numbers relate to each other. Similarly, the *sumsTo8* relation from Example 8.14 is a genuinely three-way relationship among elements—not something that can be directly reduced to a pair of pairwise relationships. But we *can* represent an n-ary relation R by a collection of binary relations, if we're a little creative in defining the sets that are being related. (Decomposing n-ary relations into multiple binary relations may be helpful if we store this type of data in a database; there may be advantages of clarity and efficiency in this view of an n-ary relation.)

This idea is perhaps easiest to see for the colors from Example 8.16: because each color name appears once and only once in the table, we can treat the name as unique "key" that allows us to treat the 4-ary relation as three separate binary relations, corresponding to the red, green, and blue components of the colors. (See Figure 8.8.) But how would we represent an n-ary relation like the ternary *sumsTo8* using multiple binary relations? (Recall the relation

R	
White	255
Red	255
Lime	0
Blue	0
Cyan	0
⋮	⋮

G	
White	255
Red	0
Lime	255
Blue	0
Cyan	255
⋮	⋮

B	
White	255
Red	0
Lime	0
Blue	255
Cyan	255
⋮	⋮

Figure 8.8: The colors from the 4-ary relation in Example 8.16, represented as three binary relations.

$$sumsTo8 = \left\{ \begin{array}{l} \langle 0,4,4\rangle, \langle 1,3,4\rangle, \langle 1,4,3\rangle, \langle 2,2,4\rangle, \langle 2,3,3\rangle, \langle 2,4,2\rangle, \langle 3,1,4\rangle, \langle 3,2,3\rangle, \\ \langle 3,3,2\rangle, \langle 3,4,1\rangle, \langle 4,0,4\rangle, \langle 4,1,3\rangle, \langle 4,2,2\rangle, \langle 4,3,1\rangle, \langle 4,4,0\rangle \end{array} \right\}$$

from Example 8.14.) One idea is to introduce a new set of fake "entities" that correspond to each of the tuples in *sumsTo8*, and then build binary relations between each component and this set of entities. For example, define the set

$$E := \{044, 134, 143, 224, 233, 242, 314, 323, 332, 341, 404, 413, 422, 431, 440\},$$

and then define the three binary relations *first*, *second*, and *third* shown in Figure 8.9. Now $\langle a,b,c\rangle \in sumsTo8$ if and only if there exists an $e \in E$ such that $\langle e,a\rangle \in first$, $\langle e,b\rangle \in second$, and $\langle e,c\rangle \in third$. (See Exercise 8.44 for a similar way to think of betweenness using binary relations.)

first	
044	0
134	1
143	1
224	2
233	2
⋮	⋮

second	
044	4
134	3
143	4
224	2
233	3
⋮	⋮

third	
044	4
134	4
143	3
224	4
233	3
⋮	⋮

Figure 8.9: The relation *sumsTo8*, as three binary relations.

COMPUTER SCIENCE CONNECTIONS

RELATIONAL DATABASES

A *database* is a (generally large!) collection of structured data. A user can both "query" the database (asking questions about existing entries, like "which states are the homes of at least two students who have GPAs above 3.0 in CS classes?") and edit it (adding or updating existing entries). The bulk of modern attention to databases focuses on *relational databases,* based explicitly on the types of relations explored in this chapter.[1] (Previous database systems were generally based on rigid top-down organization of the data.) One of the most common ways to interact with this sort of database is with a special-purpose programming language, the most common of which is *SQL.*

The era of relational databases is generally seen as starting with a massively influential paper by Edgar Codd:

[1] Edgar F. Codd. A relational model of data for large shared data banks. *Communications of the ACM*, 13(6):377–387, 1970.

"SQL" is short for *Structured Query Language;* it's pronounced either like "sequel" or by spelling out the letters (to rhyme with "Bless you, Mel!").

In a relational database, the fundamental unit of storage is the *table,* which represents an n-ary relation $R \subseteq A_1 \times A_2 \times \cdots \times A_n$. A table consists of a collection of *columns,* each of which represents a component of R; the columns are labeled with the name of the corresponding component so that it's possible to refer to columns by name rather than solely by their index. The *rows* of the table correspond to elements of the relation: that is, each row is a value $\langle a_1, a_2, \ldots, a_n \rangle$ that's in R. An example of a table of this form, echoing Example 8.16 but with labeled columns, is shown in Figure 8.10.

name	red	green	blue
Green	0	128	0
Lime	0	255	0
Magenta	255	0	255
Maroon	128	0	0
Navy	0	0	128
Olive	128	128	0
Purple	128	0	128
Red	255	0	0
Teal	0	128	128
White	255	255	255
Yellow	255	255	0

Figure 8.10: Some RGB colors.

Thus a relational database is at its essence a collection of n-ary relations. (There are other very interesting aspects of databases; for example, how should the database organize its data on the hard disk to support its operations as efficiently as possible?)[2] Operations on relational databases are based on three fundamental operations on n-ary relations. The first two basic operations either choose some of the rows or some of the columns from a relation:

- *select:* for a function $\varphi : A_1 \times \cdots \times A_n \to \{\text{True}, \text{False}\}$ and an n-ary relation $R \subseteq A_1 \times \cdots \times A_n$, we can *select* those elements of R that satisfy φ.
- *project:* for an n-ary relation $R \subseteq A_1 \times \cdots \times A_n$, we can *project* R into a smaller set of columns by deleting some A_is.

We will only just brush the surface of relational databases here—there's a full course's worth of material on databases (and then some!) that we've left out. For more, see a good book on databases, like

[2] Avi Silberschatz, Henry F. Korth, and S. Sudarshan. *Database System Concepts.* McGraw-Hill, 6th edition, 2010.

For example, we might select those colors with blue component equal to zero, or we might project the colors relation down to just red and blue values. (In SQL, these operations are done with unified syntax; we can write

```
SELECT name, red FROM colors WHERE green > blue;
```

to get the first result shown in Figure 8.11.) The third key operation in relational databases, called *join,* corresponds closely to the composition of relations. In a join, we combine two relations by insisting that an identified shared column of the two relations matches. Unlike with the composition of relations, we *continue to include that matching column* in the resulting table:

- *join:* for two binary relations $X \subseteq S \times T$ and $Y \subseteq T \times U$, the *join* of X and Y, denoted $X \bowtie Y$, is a *ternary* relation on $S \times T \times U$, defined as $X \bowtie Y := \{\langle a, c, b \rangle \in S \times T \times U : \langle a, c \rangle \in X \text{ and } \langle c, b \rangle \in Y\}$.

In SQL syntax, this operation is denoted by `INNER JOIN`; for example, with S and T as in Figure 8.6, we can generate the second table in Figure 8.11 with

```
SELECT * FROM T INNER JOIN S ON T.senator = S.senator;
```

name	red
Lime	0
Yellow	255
Green	0
Olive	128

senator	party	state
Crapo	R	ID
Risch	R	ID
Durbin	D	IL
⋮		

Figure 8.11: Selecting colors with *green* > *blue* and projecting to *name, red;* and joining S and T from Figure 8.6.

8.2.5 Exercises

Here are a few English-language descriptions of relations on a particular set. For each, write out (by exhaustive enumeration) the full set of pairs in the relation, as we did in Example 8.5.

8.1 *divides*, written $|$, on $\{1, 2, \ldots, 8\}$ (so $\langle d, n \rangle \in |$ if and only if $n \bmod d = 0$, as in Example 8.4).

8.2 *subset*, written $\subset$, on $\mathscr{P}(\{1,2,3\})$ (so $\langle S, T \rangle \in \subset$ if and only if $S \neq T$ and $\forall x : x \in S \Rightarrow x \in T$).

8.3 *isProperPrefix* on bitstrings of length ≤ 3. See Example 8.5, but here we are considering *proper* prefixes only. A string x is prefix, but not a proper prefix, of itself: more formally, x is a proper prefix of y if x starts with precisely the symbols of y, followed by one or more other symbols.

Problem-solving tip: It's easy to miss an element of these relations if you solve these problems by hand. Consider writing a small program to enumerate all the pairs meeting the descriptions in Exercises 8.1–8.5.

For two strings x and y, we say that x is a substring *of y if the symbols of x appear consecutively somewhere in y. We say that x is a* subsequence *of y if the symbols of x appear in order, but not necessarily consecutively, in y. (For example, 001 is a substring of 1001 but not of 0101. But 001 is a subsequence of 1001 and also of 0101.) A string x is called a* proper *substring/subsequence of y if x is a substring/subsequence of y but $x \neq y$. Again, write out (by exhaustive enumeration) the full set of pairs in these relations:*

8.4 *isProperSubstring* on bitstrings of length ≤ 3

8.5 *isProperSubsequence* on bitstrings of length ≤ 3

Let $\subseteq$ and $\subset$ denote the subset and proper subset relations on $\mathscr{P}(\mathbb{Z})$. (That is, we have $\langle A, B \rangle \in \subset$ if $A \subseteq B$ but $A \neq B$.) What relation is represented by each of the following?

8.6 $\subseteq \cup \subset$

8.7 $\subseteq - \subset$

8.8 $\subset - \subseteq$

8.9 $\subset \cap \subseteq$

8.10 $\sim\subset$

Consider the following two relations on $\{1, 2, 3, 4, 5, 6\}$: $R = \{\langle 2,2 \rangle, \langle 5,1 \rangle, \langle 2,3 \rangle, \langle 5,2 \rangle, \langle 2,1 \rangle\}$ and $S = \{\langle 3,4 \rangle, \langle 5,3 \rangle, \langle 6,6 \rangle, \langle 1,4 \rangle, \langle 4,3 \rangle\}$. What pairs are in the following relations?

8.11 R^{-1}

8.12 S^{-1}

8.13 $R \circ R$

8.14 $R \circ S$

8.15 $S \circ R$

8.16 $R \circ S^{-1}$

8.17 $S \circ R^{-1}$

8.18 $S^{-1} \circ S$

Five so-called mother sauces *of French cooking were codified by the chef Auguste Escoffier in the early 20th century. (Many other sauces—"daughter" or "secondary" sauces—used in French cooking are derived from these basic recipes.) They are:*

- Sauce Béchamel *is made of milk, butter, and flour.*
- Sauce Espagnole *is made of stock, butter, and flour.*
- Sauce Hollandaise *is made of egg, butter, and lemon juice.*
- Sauce Velouté *is made of stock, butter, and flour.*
- Sauce Tomate *is made of tomatoes, butter, and flour.*

8.19 Write down the "is an ingredient of" relation on *Ingredients* $\times$ *Sauces* using the tabular representation of relations introduced in Figure 8.2.

8.20 Writing R to denote the relation that you enumerated in Exercise 8.19, what is $R \circ R^{-1}$? Give both a list of elements and an English-language description of what $R \circ R^{-1}$ represents.

8.21 Again for the R from Exercise 8.19, what is $R^{-1} \circ R$? Again, give both a list of elements and a description of the meaning.

Suppose that a Registrar's office has computed the following relations:

taughtIn $\subseteq$ Classes $\times$ Rooms   taking $\subseteq$ Students $\times$ Classes   at $\subseteq$ Classes $\times$ Times.

For the following exercises, express the given additional relation using taughtIn, taking, *and* at, *plus relation composition and/or inversion (and no other tools).*

8.22 $R \subseteq$ *Students* $\times$ *Times*, where $\langle s, t \rangle \in R$ indicates that student s is taking a class at time t.

8.23 $R \subseteq$ *Rooms* $\times$ *Times*, where $\langle r, t \rangle \in R$ indicates that there is a class in room r at time t.

8.24 $R \subseteq$ *Students* $\times$ *Students*, where $\langle s, s' \rangle \in R$ indicates that students s and s' are taking at least one class in common.

8.25 $R \subseteq$ *Students* $\times$ *Students*, where $\langle s, s' \rangle \in R$ indicates that there's at least one time when s and s' are both taking a class (but not necessarily the same class).

Let parent $\subseteq$ People $\times$ People *denote the relation* $\{\langle p,c\rangle : p$ *is a parent of* $c\}$. *What familial relationships are represented by the following relations?*

For the sake of simplicity in the following questions, assume that there are no divorces, remarriages, widows, widowers, adoptions, single parents, etc. That is, you should assume that each child has exactly two parents, and any two children who share one parent share both parents.

8.26 $parent \circ parent$

8.27 $(parent^{-1}) \circ (parent^{-1})$

8.28 $parent \circ (parent^{-1})$

8.29 $(parent^{-1}) \circ parent$

8.30 $parent \circ parent \circ (parent^{-1}) \circ (parent^{-1})$

8.31 $parent \circ (parent^{-1}) \circ parent \circ (parent^{-1})$

8.32 Suppose that the relations $R \subseteq \mathbb{Z} \times \mathbb{Z}$ and $S \subseteq \mathbb{Z} \times \mathbb{Z}$ contain, respectively, n pairs and m pairs of elements. In terms of n and m, what's the largest possible size of $R \circ S$? The smallest?

Consider the following claims about the composition of relations.

8.33 For arbitrary relations R, S, and T, prove that $R \circ (S \circ T) = (R \circ S) \circ T$.

8.34 For arbitrary relations R and S, prove that $(R \circ S)^{-1} = (S^{-1} \circ R^{-1})$.

8.35 Let R be any relation on $A \times B$. Prove or disprove: $\langle x,x\rangle \in R \circ R^{-1}$ for every $x \in A$.

8.36 What set is represented by the relation $\leq \circ \geq$, where $\leq$ and $\geq$ are relations on $\mathbb{R}$?

8.37 What set is represented by the relation *successor* $\circ$ *predecessor*, where *successor* $= \{\langle n,n+1\rangle : n \in \mathbb{Z}\}$ and *predecessor* $= \{\langle n,n-1\rangle : n \in \mathbb{Z}\}$?

Suppose that $R \subseteq A \times B$ *and* $T \subseteq B \times C$ *are relations. Prove the following:*

8.38 If R and T are both functions, then $T \circ R$ is a function too.

8.39 If R and T are both one-to-one functions, then $T \circ R$ is one-to-one too.

8.40 If R and T are both onto functions, then $T \circ R$ is onto too.

The next few exercises ask you to address the converse of the last few. Supposing that $T \circ R$ *has the listed property, can you infer that both relations* R *and* T *have the same property? Only* R*? Only* T*? Neither? Prove your answers.*

8.41 $T \circ R$ is a function. Must T be a function? R? Both?

8.42 $T \circ R$ is a one-to-one function and R and T are both functions. Must T be one-to-one? R? Both?

8.43 $T \circ R$ is an onto function and R and T are both functions. Must T be onto? R? Both?

On p. 815, we introduced three operations on relations that are used frequently in relational databases:

- select, *which chooses a subset of the elements of an n-ary relation. For* $R \subseteq A_1 \times \cdots \times A_n$ *and a function* $\varphi : A_1 \times \cdots \times A_n \to \{\text{True}, \text{False}\}$, *we can select only those elements of* R *that satisfy* φ.
- project, *which turns an n-ary relation into an* n'*-ary relation for some* $n' \leq n$ *by eliminating components. For* $R \subseteq A_1 \times \cdots \times A_n$ *and* $S \subseteq \{1,2,\ldots,n\}$, *we can project* R *into a smaller set of columns by removing the ith component of each pair in* R *for any* $i \notin S$.
- join, *which combines two binary relations* $R \subseteq A \times B$ *and* $S \subseteq B \times C$ *into a single ternary relation containing triples* $\langle a,b,c\rangle$ *such that* $\langle a,b\rangle \in R$ *and* $\langle b,c\rangle \in S$.

For example, let $R = \{\langle 1,2,3\rangle, \langle 4,5,6\rangle\}$, *let* $S = \{\langle 6,7\rangle, \langle 6,8\rangle\}$, *and let* $T = \{\langle 7,9\rangle, \langle 7,10\rangle\}$. *Then*

- select(R, xzEven) $= \{\langle 4,5,6\rangle\}$ *for* $\text{xzEven}(x,y,z) = (2 \mid x) \wedge (2 \mid z)$.
- project(R, $\{1,2\}$) $= \{\langle 1,2\rangle, \langle 4,5\rangle\}$ *and* project(R, $\{1,3\}$) $= \{\langle 1,3\rangle, \langle 4,6\rangle\}$.
- join(S,T) $= \{\langle 6,7,9\rangle, \langle 6,7,10\rangle\}$.

Solve the following using the relation operators $^{-1}$ *(inverse),* $\circ$ *(composition), select, project, and join:*

8.44 Recall from Example 8.15 the "betweenness" relation, defined as the ternary relation $B := \{\langle x,y,z\rangle \in \mathbb{R}^3 : x \leq y \leq z \text{ or } x \geq y \geq z\}$. Show how to construct B using only $\leq$, the relation operators ($^{-1}$, $\circ$, join, select, project), and standard set-theoretic operations ($\cup, \cap, \sim, -$).

Using the relation C *defined in Figure 8.12, and select/project/join, write a set that corresponds to the following:*

8.45 the names of all colors that have red component 0.

8.46 the names of all pairs of colors whose amount of blue is the same.

8.47 the names of all colors that are more blue than red.

Color	R	G	B
White	255	255	255
Red	255	0	0
Lime	0	255	0
Blue	0	0	255
Cyan	0	255	255
Magenta	255	0	255
Yellow	255	255	0
Black	0	0	0
Gray	128	128	128
Maroon	128	0	0
Green	0	128	0
Navy	0	0	128
Teal	0	128	128
Purple	128	0	128
Olive	128	128	0

Figure 8.12: A 4-ary relation C (see Example 8.16).

Let X *denote the set of color names from Example 8.16. Define three relations* Red, Green, *and* Blue *on* $X \times \{0,1,\ldots,255\}$ *such that* $\langle x,r,g,b\rangle \in C$ *if and only if* $\langle x,r\rangle \in$ Red, $\langle x,g\rangle \in$ Green, *and* $\langle x,b\rangle \in$ Blue.

8.48 Repeat Exercise 8.46 using only $^{-1}$, $\circ$, and the relations *Red*, *Green*, *Blue*, $\leq$, and $=$.

8.49 Do the same for Exercise 8.47—or, at least, compute the set of $\langle x,x\rangle$ such that x is the name of a color that's more blue than red. (You may construct a relation R on colors, and then take $R \cap =$.)

8.3 *Properties of Relations: Reflexivity, Symmetry, and Transitivity*

> Pride destroys all symmetry and grace, and affectation is a more terrible enemy to fine faces than the small-pox.
>
> Sir Richard Steele (1672–1729)

Let $R \subseteq A \times A$ be a relation on a single set A (as in the *successor* or $\leq$ relations on $\mathbb{Z}$, or the *is a (blood) relative of* relation on people). We've seen a two-column approach to visualizing a relation $R \subseteq A \times B$, but this layout is misleading when the sets A and B are identical. (Weirdly, we'd have to draw each element twice, in both the A column and the B column.) Instead, it will be more convenient to visualize a relation $R \subseteq A \times A$ without differentiated columns, using a *directed graph*: we simply write down each element of A, and draw an arrow from a_1 to a_2 for every pair $\langle a_1, a_2 \rangle \in R$. (See Chapter 11 for much more on directed graphs.) A few small examples are shown in Figure 8.13.

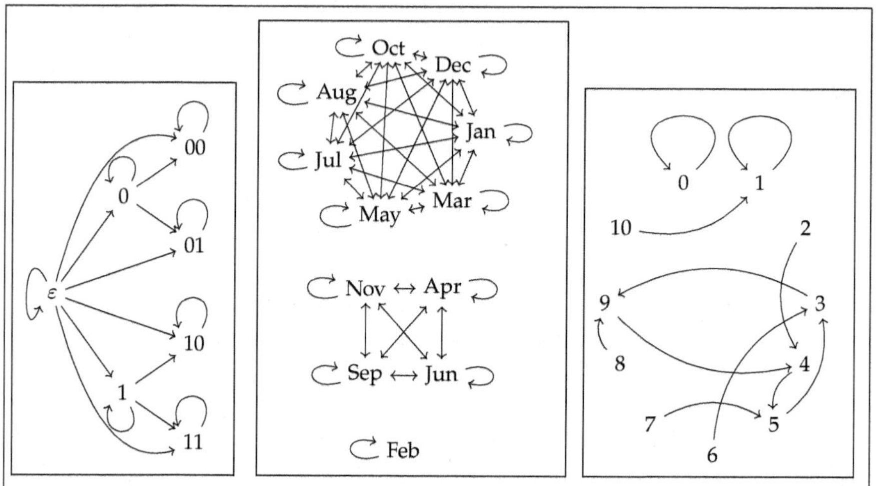

(a) *isPrefix* (b) Months of the same length (c) $\langle x, x^2 \bmod 11 \rangle$ for $x \in \mathbb{Z}_{11}$

Figure 8.13: Visualizations of three relations, from Example 8.5 (prefixes of bitstrings), Example 8.11 (months), and Example 8.12 ($\langle x, x^2 \rangle \bmod 11$).

This directed-graph visualization of relations will provide a useful way of thinking intuitively about relations in general—and about some specific types of relations in particular. There are several important structural properties that some relations on A have (and that some relations do not), and we'll explore these properties throughout this section. We'll consider three basic categories of properties:

reflexivity: whether elements are related to themselves. That is, is an element x necessarily related to x itself?

symmetry: whether order matters in the relation. That is, if x and y are related, are y and x necessarily related too?

transitivity: whether chains of related pairs are themselves related. That is, if x and y are related and y and z are related, are x and z necessarily related too?

These properties turn out to characterize several important types of relations—for example, some relations divide A into clusters of "equivalent" elements (as in Figure 8.13(b)), while other relations "order" A in some consistent way (as in Figure 8.13(a))—and we'll see these special types of relations in Section 8.4. But first we'll examine these three categories of properties in turn, and then we'll define *closures* of relations, which expand any relation R as little as possible while ensuring that the expansion of R has any particular desired subset of these properties.

8.3.1 *Reflexivity*

The *reflexivity* of a relation $R \subseteq A \times A$ is based on whether elements of A are related to themselves. That is, are pairs $\langle a, a \rangle$ in R? The relation R is *reflexive* if $\langle a, a \rangle$ is always in R (for every $a \in A$), and it's *irreflexive* if $\langle a, a \rangle$ is never in R (for any $a \in A$):

Latin: *re* "back" + *flect* "bend."

Definition 8.6 (Reflexive and Irreflexive Relations)
A relation R on A is reflexive *if, for every $x \in A$, we have that $\langle x, x \rangle \in R$.*
A relation R on A is irreflexive *if, for every $x \in A$, we have that $\langle x, x \rangle \notin R$.*

Using the visualization style from Figure 8.13, a relation is reflexive if every element $a \in A$ has a "loop" from a back to itself—and it's irreflexive if no $a \in A$ has a loop back to itself. (See Figure 8.14.)

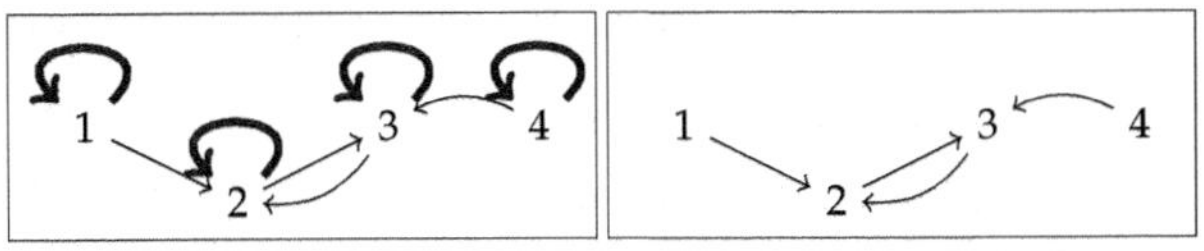

Figure 8.14: A relation on A is reflexive if every $a \in A$ has a self-loop (the dark arrows in the left panel), and it is irreflexive if no $a \in A$ does (as in the right panel).

Example 8.17 (Reflexivity of $=$, $\equiv_{17}$, and $\langle x, x^2 \rangle$ mod 11)
The relations $=$ and $\equiv_{17}$ on $\mathbb{Z}$—that is, the relations $\{\langle x, y \rangle : x = y\}$ and $\{\langle x, y \rangle : x \bmod 17 = y \bmod 17\}$—are both reflexive, because $x = x$ and $x \bmod 17 = x \bmod 17$ for any $x \in \mathbb{Z}$. But the relation $R := \{\langle x, x^2 \bmod 11 \rangle : x \in \mathbb{Z}_{11}\}$ from Figure 8.13(c) is not reflexive, because (among other examples) we have $\langle 7, 7 \rangle \notin R$.

Note that there are relations that are neither reflexive *nor* irreflexive. For example, the relation $S = \{\langle 0, 1 \rangle, \langle 1, 1 \rangle\}$ on $\{0, 1\}$ isn't reflexive (because $\langle 0, 0 \rangle \notin S$), but it's also not irreflexive (because $\langle 1, 1 \rangle \in S$).

Example 8.18 (A few arithmetic relations)
Problem: Which of the following relations on $\mathbb{Z}^{\geq 1}$ are reflexive? Irreflexive?

1. divides: $R_1 = \{\langle n, m \rangle : m \bmod n = 0\}$
2. greater than: $R_2 = \{\langle n, m \rangle : n > m\}$
3. less than or equal to: $R_3 = \{\langle n, m \rangle : n \leq m\}$
4. square: $R_4 = \{\langle n, m \rangle : n^2 = m\}$
5. equivalent mod 5: $R_5 = \{\langle n, m \rangle : n \bmod 5 = m \bmod 5\}$

Solution:
1. **reflexive.** For any positive integer n, we have that $n \bmod n = 0$. Thus $\langle n, n \rangle \in R_1$ for any n.
2. **irreflexive.** For any $n \in \mathbb{Z}^{\geq 1}$, we have that $n \not> n$. Thus $\langle n, n \rangle \notin R_2$ for any n.
3. **reflexive.** For any positive integer n, we have $n \leq n$, so every $\langle n, n \rangle \in R_3$.
4. **neither.** The square relation is not reflexive because $\langle 9, 9 \rangle \notin R_4$ and it is also not irreflexive because $\langle 1, 1 \rangle \in R_4$, for example. (That's because $9 \neq 9^2$, but $1 = 1^2$.)
5. **reflexive.** For any $n \in \mathbb{Z}^{\geq 1}$, we have $n \bmod 5 = n \bmod 5$, so $\langle n, n \rangle \in R_5$.

Note again that, as with *square*, it is possible to be *neither* reflexive *nor* irreflexive. (But it's not possible to be *both* reflexive *and* irreflexive, as long as $A \neq \emptyset$: for any $a \in A$, if $\langle a, a \rangle \in R$, then R is not irreflexive; if $\langle a, a \rangle \notin R$, then R is not reflexive.)

8.3.2 *Symmetry*

The *symmetry* of a relation $R \subseteq A \times A$ is based on whether the order of the elements in a pair matters. That is, if the pair $\langle a, b\rangle$ is in R, is the pair $\langle b, a\rangle$ always also in R? (Or is it never in R? Or sometimes but not always?) The relation R is *symmetric* if, for every a and b, the pairs $\langle a, b\rangle$ and $\langle b, a\rangle$ are both in R or both not in R.

Greek: *syn* "same" + *metron* "measure."

There are two accompanying notions: a relation R is *antisymmetric* if the only time $\langle a, b\rangle$ and $\langle b, a\rangle$ are both in R is when $a = b$, and R is *asymmetric* if $\langle a, b\rangle$ and $\langle b, a\rangle$ are never both in R (whether $a = b$ or $a \neq b$). Here are the formal definitions:

An important etymological note: *anti-* means "against" rather than "not." *Asymmetric* (no $\langle a,b\rangle, \langle b,a\rangle \in R$) is different from *antisymmetric* (if $\langle a,b\rangle, \langle b,a\rangle \in R$ then $a = b$) is different from *not symmetric* (there is some $\langle a,b\rangle \in R$ but $\langle b,a\rangle \notin R$).

Definition 8.7 (Symmetric, Antisymmetric, and Asymmetric Relations)
A relation R on A is symmetric *if, for every $a, b \in A$, if $\langle a, b\rangle \in R$ then $\langle b, a\rangle \in R$.*

A relation R on A is antisymmetric *if, for every $a, b \in A$ such that $\langle a, b\rangle \in R$ and $\langle b, a\rangle \in R$, we have $a = b$.*

A relation R on A is asymmetric *if, for every $a, b \in A$, if $\langle a, b\rangle \in R$ then $\langle b, a\rangle \notin R$.*

Again thinking about the visualization from Figure 8.13: a relation is symmetric if every arrow $a \to b$ is matched by an arrow $b \to a$ in the opposite direction. It's antisymmetric if there are no matched bidirectional pairs of arrows between two distinct elements a and b; and it's asymmetric if there also aren't even any self-loops. (An a-to-a self-loop is, in a weird way, a "pair" of arrows $a \to b$ and $b \to a$, just with $a = b$.) See Figure 8.15.

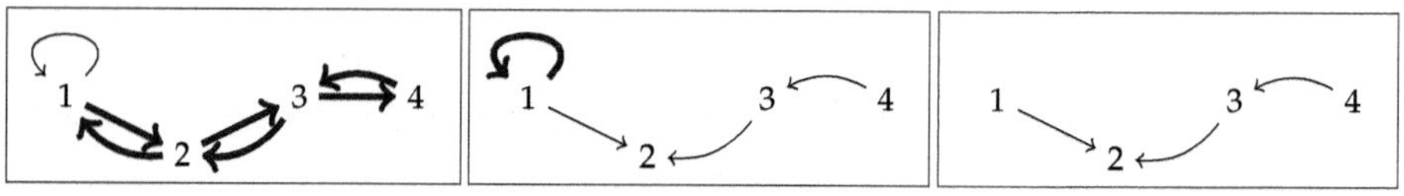

Figure 8.15: R is symmetric if every $a \to b$ is matched by $b \to a$ (as in the left panel). R is antisymmetric if no $a \leftrightarrow b$ exists for $a \neq b$ (as in the middle or right panel), and asymmetric if it also has no self-loops (as in the right panel).

Example 8.19 (Some symmetric relations)
The relations

$$\{\langle w, w'\rangle : w \text{ and } w' \text{ have the same length}\} \qquad \textit{(on the set of English words)}$$
$$\{\langle s, s'\rangle : s \text{ and } s' \text{ sat next to each other in class today}\} \qquad \textit{(on the set of students)}$$

are both symmetric. If w contains the same number of letters as w', then w' also contains the same number of letters as w. And if I sat next to you, then you sat next to me! (The first relation is also reflexive—`ZEUGMA` contains the same number of letters as `ZEUGMA`—but the latter is irreflexive, as no student sits beside herself in class.)

zeugma, n.: grammatical device in which words are used in parallel construction syntactically, but not semantically, as in

Yesterday, Alice caught a rainbow trout and hell from Bob for fishing all day.

Example 8.20 (A few arithmetic relations, again)
Problem: Which of these relations from Example 8.18 (see below for a reminder) are symmetric? Antisymmetric? Asymmetric?

$$\begin{aligned}
R_1 &= \{\langle n, m\rangle : m \bmod n = 0\}\\
R_2 &= \{\langle n, m\rangle : n > m\}\\
R_3 &= \{\langle n, m\rangle : n \leq m\}\\
R_4 &= \{\langle n, m\rangle : n^2 = m\}\\
R_5 &= \{\langle n, m\rangle : n \bmod 5 = m \bmod 5\}.
\end{aligned}$$

Solution: 1. **antisymmetric.** Because $n \bmod m = m \bmod n = 0$ if and only if $n = m$, if $\langle n, m\rangle \in R_1$ and $\langle m, n\rangle \in R_1$ then $n = m$. But the relation is neither symmetric (for example, $3 \mid 6$ but $6 \nmid 3$) nor asymmetric (for example, $3 \mid 3$).

2. **asymmetric.** If $x < y$ then $y \not< x$, even if $x = y$. So R_2 is asymmetric.

3. **antisymmetric.** Similar to (1), R_3 is antisymmetric: if $x \leq y$ and $y \leq x$, then $x = y$. (But $3 \leq 6$ and $6 \not\leq 3$, and $3 \leq 3$, so R_3 is neither symmetric nor asymmetric.)

4. **antisymmetric.** The square relation is neither symmetric nor asymmetric because $\langle 3, 9\rangle \in R_4$ but $\langle 9, 3\rangle \notin R_4$, and $\langle 1, 1\rangle \in R_4$. (That's because $3^2 = 9$ but $9^2 \neq 3$, and $1^2 = 1$.) But it is antisymmetric, because the only way that $x^2 = y$ and $y^2 = x$ is if $x = y$ (specifically $x = y = 0$ or $x = y = 1$).

5. **symmetric.** The "equivalent mod 5" relation is symmetric because equality is: for any n and m, we have $n \bmod 5 = m \bmod 5$ if and only if $m \bmod 5 = n \bmod 5$. But it's not antisymmetric: $\langle 17, 202\rangle \in R_5$ and $\langle 202, 17\rangle \in R_5$.

Note that it is possible for a relation to be *both* symmetric and antisymmetric; see Exercise 8.69. And it's also possible for a relation R not to be symmetric, but also for R to fail to be either antisymmetric or asymmetric:

Example 8.21 (A non-symmetric, non-asymmetric, non-antisymmetric relation)
The relation $R := \{\langle 0,1\rangle, \langle 0,2\rangle, \langle 1,0\rangle\}$ on $\{0,1,2\}$ isn't symmetric ($0 \rightarrow 2$ but $2 \not\rightarrow 0$), and it isn't asymmetric or antisymmetric ($0 \rightarrow 1$ and $1 \rightarrow 0$ but $0 \neq 1$).

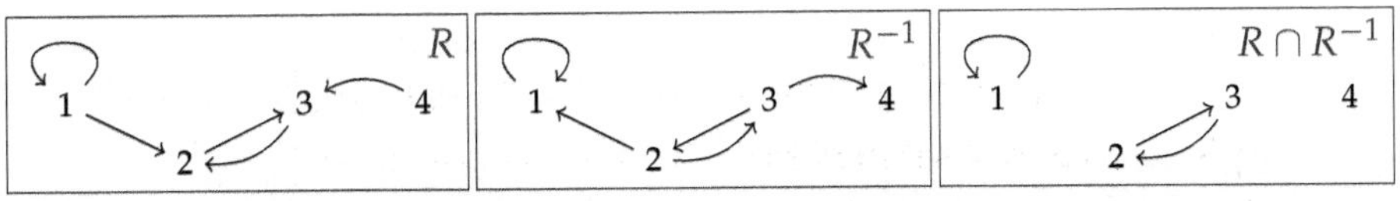

Figure 8.16: A relation R on A, its inverse R^{-1}, and $R \cap R^{-1}$. From the right panel, we see that R isn't symmetric ($1 \rightarrow 2$ and $4 \rightarrow 3$ are missing), asymmetric (1 has a self-loop) or antisymmetric ($2 \leftrightarrow 3$ is present). (But $R - \{\langle 2,3\rangle\}$ is antisymmetric.)

One other useful way to think about the symmetry (or antisymmetry/asymmetry) of a relation R is by considering the inverse R^{-1} of R. Recall that R^{-1} reverses the direction of all of the arrows of R, so $\langle a, b\rangle \in R$ if and only if $\langle b, a\rangle \in R^{-1}$. A symmetric relation is one in which every $a \rightarrow b$ arrow is matched by a $b \rightarrow a$ arrow, so reversing the arrows doesn't change the relation. For an antisymmetric relation R, the inverse R^{-1} has only self-loops in common with R. And an asymmetric relation has no arrows in common with its inverse. (See Figure 8.16.) Specifically:

Theorem 8.1 (Symmetry in terms of inverses)
Let $R \subseteq A \times A$ be a relation and let R^{-1} be its inverse. Then:

- *R is symmetric if and only if $R \cap R^{-1} = R = R^{-1}$.*
- *R is antisymmetric if and only if $R \cap R^{-1} \subseteq \{\langle a, a\rangle : a \in A\}$.*
- *R is asymmetric if and only if $R \cap R^{-1} = \varnothing$.*

You'll prove this theorem formally in Exercises 8.66–8.68.

8.3.3 *Transitivity*

The *transitivity* of a relation $R \subseteq A \times A$ is based on whether the relation always contains a "short circuit" from a to c whenever two pairs $\langle a, b \rangle$ and $\langle b, c \rangle$ are in R. An alternative view is that a transitive relation R is one in which "applying R twice" doesn't yield any new connections. For example, consider the relation "lives in the same town as": if a person x lives in the same town as a person y you live in same town as, then in fact x directly (without reference to the intermediary y) lives in the same town as you. Here is the formal definition:

Latin: *trans* "across/through."

Definition 8.8 (Transitive Relation)
A relation R on A is transitive *if, for every $a, b, c \in A$, if $\langle a, b \rangle \in R$ and $\langle b, c \rangle \in R$, then $\langle a, c \rangle \in R$ too.*

Or, using the visualization from Figure 8.13, a relation is transitive if there are no "open triangles": if $a \to b$ and $b \to c$, then $a \to c$. (In any "chain" of connected elements in a transitive relation, every element is also connected to all elements that are "downstream" of it.) See Figure 8.17.

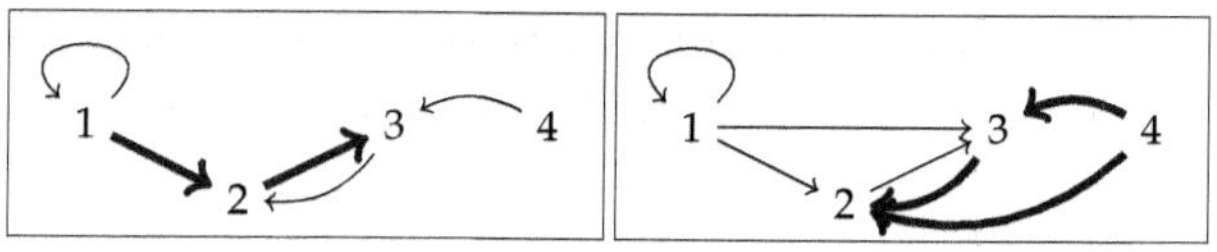

Figure 8.17: A relation on A is transitive if every triangle is closed. The left panel shows a relation that is not transitive (the dark arrows form an open triangle). The right panel shows a transitive relation, with a highlighted closed triangle.

Example 8.22 (Some transitive relations)
The relations

$$\{\langle w, w' \rangle : w \text{ and } w' \text{ have the same length}\} \qquad \textit{(on the set of English words)}$$
$$\{\langle s, s' \rangle : s \text{ arrived in class before } s' \text{ today}\} \qquad \textit{(on the set of students)}$$

are both transitive. If w contains the same number of letters as w', and w' contains the same number of letters as w'', then w certainly contains the same number of letters as w'' too. And if Alice got to class before Bob, and Bob got to class before Charlie, then Alice got to class before Charlie.

Example 8.23 (A few arithmetic relations, one more time)
Problem: Which of the relations from Examples 8.18 and 8.20 are transitive?

$$\begin{aligned} R_1 &= \{\langle n, m \rangle : m \bmod n = 0\} \\ R_2 &= \{\langle n, m \rangle : n > m\} \\ R_3 &= \{\langle n, m \rangle : n \leq m\} \\ R_4 &= \{\langle n, m \rangle : n^2 = m\} \\ R_5 &= \{\langle n, m \rangle : n \bmod 5 = m \bmod 5\} \end{aligned}$$

Solution: 1. **transitive.** Suppose that $a \mid b$ and $b \mid c$. We need to show that $a \mid c$. But that's easy: by definition $a \mid b$ and $b \mid c$ mean that $b = ak$ and $c = b\ell$ for integers k and ℓ. Therefore $c = a \cdot (k\ell)$—and thus $a \mid c$. (This fact was Theorem 7.4.4.)

2. **transitive.** If $x > y$ and $y > z$, then we know $x > z$.

3. **transitive.** Just as in (2), R_3 is transitive: if $x \leq y$ and $y \leq z$, then $x \leq z$.
4. **not transitive.** The square relation isn't transitive, because, for example, we have $\langle 2,4 \rangle \in R_4$ and $\langle 4,16 \rangle \in R_4$—but $\langle 2,16 \rangle \notin R_4$. (That's because $2^2 = 4$ and $4^2 = 16$ but $2^2 \neq 16$.)
5. **transitive.** The "equivalent mod 5" relation is transitive because equality is: if $n \bmod 5 = m \bmod 5$ and $m \bmod 5 = p \bmod 5$, then $n \bmod 5 = p \bmod 5$.

While we can understand the transitivity of a relation R directly from Definition 8.8, we can also think about the transitivity of R by considering the relationship between R and $R \circ R$—that is, R and the composition of R with itself. (Earlier we saw how to view the symmetry of R by connecting R and its inverse R^{-1}.)

Theorem 8.2 (Transitivity in terms of self-composition)
Let $R \subseteq A \times A$ be a relation. Then R is transitive if and only if $R \circ R \subseteq R$.

Again, you'll prove this theorem in the exercises (Exercise 8.85).

Taking it further: Imagine a collection of n people who have individual preferences over k candidates. That is, we have n relations $R_1, R_2, \ldots, R_n$, each of which is a relation on the set $\{1, 2, \ldots, k\}$. We wish to aggregate these individual preferences into a single preference relation for the collection of people. Although this description is much more technical than our everyday usage, the problem that we've described here is well known: it's otherwise known as *voting*. (Economists also call this topic the theory of *social choice*.) Some interesting and troubling paradoxes arise in voting problems, related to transitivity—or, more precisely, to the absence of transitivity.

Suppose that we have three candidates: Alice, Bob, and Charlie. For simplicity, let's suppose that we also have exactly three voters: #1, #2, and #3. (This paradox also arises when there are many more voters.) Consider the situation in which Voter #1 thinks Alice > Bob > Charlie; Voter #2 thinks Charlie > Alice > Bob; and Voter #3 thinks Bob > Charlie > Alice. Then, in head-to-head runoffs between pairs of candidates, the results would be:

- Alice beats Bob: 2 votes (namely #1 and #2) for Alice, to 1 vote (just #3) for Bob.
- Bob beats Charlie: 2 votes (namely #1 and #3) for Bob, to 1 vote (just #2) for Charlie.
- Charlie beats Alice: 2 votes (namely #2 and #3) for Charlie, to 1 vote (just #1) for Alice.

That's pretty weird: we have taken strict preferences (each of which is certainly transitive!) from each of the voters, and aggregated them into a nontransitive set of societal preferences. This phenomenon—no candidate would win a head-to-head vote against every other candidate—is called the *Condorcet paradox*. (The *Condorcet criterion* declares the winner of a vote to be the candidate who would win a runoff election against any other individual candidate.)

The Condorcet paradox is troubling, but an even more troubling result says that, more or less, there's *no* good way of designing a voting system! *Arrow's Theorem*, proven around 1950, states that there's no way to aggregate individual preferences to society-level preferences in a way that's consistent with three "obviously desirable" properties of a voting system: (1) if every voter prefers candidate A to candidate B, then A beats B; (2) there's no "dictator" (a single voter whose preferences of the candidates directly determines the outcome of the vote); and (3) "independence of irrelevant alternatives" (if candidate A beats B when candidate C is in the race, then A still beats B if C were to drop out of the race).[3]

The *Condorcet paradox* is named after the 18th-century French philosopher/mathematician Marquis de Condorcet (rhymes with *gone for hay*). *Arrow's Theorem* is named after Kenneth Arrow, a 20th-century American economist (who won the 1972 Nobel Prize in Economics, largely for this theorem). See

[3] Kenneth Arrow. *Social Choice and Individual Values*. Wiley, 1951.

8.3.4 *Properties of Asymptotic Relationships*

Now that we've introduced the three categories of properties of relations (reflexivity, symmetry, and transitivity), let's consider one more set of relations in light of these properties: the *asymptotics* of functions. Recall from Chapter 6 that, for two functions

$f : \mathbb{R}^{\geq 0} \to \mathbb{R}^{\geq 0}$ and $g : \mathbb{R}^{\geq 0} \to \mathbb{R}^{\geq 0}$, we say that

$$\begin{aligned}
&f(n) \text{ is } O(g(n)) && \text{if and only if} && \exists n_0 \geq 0, c > 0 : \left(\forall n \geq n_0 : f(n) \leq c \cdot g(n)\right). \\
&f(n) \text{ is } \Theta(g(n)) && \text{if and only if} && f(n) \text{ is } O(g(n)) \text{ and } g(n) \text{ is } O(f(n)). \\
&f(n) \text{ is } o(g(n)) && \text{if and only if} && f(n) \text{ is } O(g(n)) \text{ and } g(n) \text{ is not } O(f(n)).
\end{aligned}$$

(Actually we previously phrased the definitions of $\Theta(\cdot)$ and $o(\cdot)$ in terms of $\Omega(\cdot)$, but the definition we've given here is completely equivalent, as proven in Exercise 6.30.) We can view these asymptotic properties as relations on the set $F := \{f : \mathbb{R}^{\geq 0} \to \mathbb{R}^{\geq 0}\}$ of functions.

The standard asymptotic notation doesn't match the standard notation for relations—we write $f = \Theta(g)$ rather than $f \,\Theta\, g$ or $\langle f, g \rangle \in \Theta$—but Θ genuinely is a relation on F, in the sense that some pairs of functions are related by Θ and some pairs are not. And O and o are relations on F in the same way.

Example 8.24 (O and Θ and o: reflexivity)

O is reflexive: For any function f, we can easily show that $f = O(f)$ by choosing the constants $n_0 := 1$ and $c := 1$, because it is immediate that $\forall n \geq 1 : f(n) \leq 1 \cdot f(n)$. Therefore O is reflexive, because every function f satisfies $f = O(f)$.

Θ is reflexive: This fact follows immediately from the fact that O is reflexive:

$$\begin{aligned}
\Theta \text{ is reflexive} &\Leftrightarrow \forall f \in F : f = \Theta(f) && \textit{definition of reflexivity} \\
&\Leftrightarrow \forall f \in F : f = O(f) \text{ and } f = O(f) && \textit{definition of } \Theta \\
&\Leftrightarrow \forall f \in F : f = O(f) && p \wedge p \equiv p \\
&\Leftrightarrow O \text{ is reflexive.} && \textit{definition of reflexivity}
\end{aligned}$$

o is irreflexive: This fact follows by similar logic: for any function $f \in F$,

$$f = o(f) \Leftrightarrow f = O(f) \text{ and } f \neq O(f). \qquad \textit{definition of } o(\cdot)$$

But $p \wedge \neg p \equiv$ False (including when p is "$f = O(f)$"), so o is irreflexive.

Example 8.25 (O and Θ and o: symmetry)

O is not symmetric, antisymmetric, or asymmetric: Define the functions $t_1(n) = n$ and $t_2(n) = n^2$ and $t_3(n) = 2n^2$. O is not symmetric because, for example, $t_1 = O(t_2)$ but $t_2 \neq O(t_1)$. O is not asymmetric because, for example, $t_1 = O(t_1)$. And O is not antisymmetric because, for example, $t_2 = O(t_3)$ and $t_3 = O(t_2)$ but $t_2 \neq t_3$.

Θ is symmetric: This fact follows immediately from the definition: for arbitrary f and g,

$$\begin{aligned}
f = \Theta(g) &\Leftrightarrow f = O(g) \text{ and } g = O(f) && \textit{definition of } \Theta \\
&\Leftrightarrow g = O(f) \text{ and } f = O(g) && p \wedge q \equiv q \wedge p \\
&\Leftrightarrow g = \Theta(f). && \textit{definition of } \Theta
\end{aligned}$$

(Θ is not anti/asymmetric, because $t_2 = \Theta(t_3)$ for $t_2(n)$ and $t_3(n)$ as defined above.)

o is asymmetric: This fact follows immediately, by similar logic: for arbitrary f and g, we have $f = o(g)$ and $g = o(f)$ if and only if $f = O(g)$ and $g \neq O(f)$ *and* $g = O(f)$ and $f \neq O(g)$—a contradiction! So if $f = o(g)$ then $g \neq o(f)$. Therefore o is asymmetric.

You proved in Exercises 6.18, 6.46, and 6.47 that O, Θ, and o are all transitive, so we won't repeat the proofs here.

In sum, then, we've argued that O is reflexive and transitive (but not symmetric, asymmetric, or antisymmetric); o is irreflexive, asymmetric, and transitive; and Θ is reflexive, symmetric, and transitive.

Taking it further: Among the computer scientists, philosophers, and mathematicians who study formal logic, there's a special kind of logic called *modal logic* that's of significant interest. Modal logic extends the type of logic we introduced in Chapter 3 to also include logical statements about whether a true proposition is *necessarily* true or *accidentally* true. For example, the proposition *Canada won the 2014 Olympic gold medal in curling* is true—but the gold-medal game *could* have turned out differently and, if it had, that proposition would have been false. But *Either it rained yesterday or it didn't rain yesterday* is true, and there's no possible scenario in which this proposition would have turned out to be false. We say that the former statement is "accidentally" true (it was an "accident" of fate that the game turned out the way it did), but the latter is "necessarily" true.

In modal logic, we evaluate the truth value of a particular logical statement multiple times, once in each of a set W of so-called *possible worlds.* Each possible world assigns truth values to every atomic proposition. Thus every logical proposition φ of the form we saw in Chapter 3 has a truth value in each possible world $w \in W$. But there's another layer to modal logic. In addition to the set W, we are also given a relation $R \subseteq W \times W$, where $\langle w, w' \rangle \in R$ indicates that w' *is possible relative to* w. In addition to the basic logical connectives from normal logic, we can also write two more types of propositions:

$\Diamond\varphi$	"possibly φ"	$\Diamond\varphi$ is true in w if $\exists w' \in W$ such that $\langle w, w' \rangle \in R$ and φ is true in w'.
$\Box\varphi$	"necessarily φ"	$\Box\varphi$ is true in w if $\forall w' \in W$ such that $\langle w, w' \rangle \in R$, φ is true in w'.

Of course, these operators can be nested, so we might have a proposition like $\Box(\Diamond p \Rightarrow \Box p)$.

Different assumptions about the relation R will allow us to use modal logic to model different types of interesting phenomena. For example, we might want to insist that $\Box\varphi \Rightarrow \varphi$ ("if φ is necessarily true, then φ is true": that is, if φ is true in every world $w' \in W$ possible relative to w, then φ is true in w). This axiom corresponds to the relation R being reflexive: w is always possible relative to w. Symmetry and transitivity correspond to the axioms $\varphi \Rightarrow \Box\Diamond\varphi$ and $\Box\varphi \Rightarrow \Box\Box\varphi$.

The general framework of modal logic (with different assumptions about R) has been used to represent logics of knowledge (where $\Box\varphi$ corresponds to "I know φ"); logics of provability (where $\Box\varphi$ corresponds to "we can prove φ"); and logics of possibility and necessity (where $\Box\varphi$ corresponds to "necessarily φ" and $\Diamond\varphi$ to "possibly φ"). Others have also studied *temporal logics* (where $\Box\varphi$ corresponds to "always φ" and $\Diamond\varphi$ to "eventually φ"); these logical formalisms have proven to be very useful in formally analyzing the correctness of programs.[4]

For a good introduction to modal logic, see

[4] G. E. Hughes and M. J. Cresswell. *A New Introduction to Modal Logic.* Routledge, 1996.

8.3.5 *Closures of Relations*

Until now, in this section we've discussed some important properties that certain relations $R \subseteq A \times A$ may or may not happen to have. We'll close this section by looking at how to "force" the relation R to have one or more of these properties. Specifically, we will introduce the *closure* of a relation with respect to a property like symmetry: we'll take a relation R and expand it into a relation R' that has the desired property, while adding as few pairs to R as possible. That is, the *symmetric closure* of R is the smallest set $R' \supseteq R$ such that the relation R' is symmetric.

Taking it further: In general, a set S is said to be *closed under the operation* f if, whenever we apply f to an arbitrary element of S (or to an arbitrary k-tuple of elements from S, if f takes k arguments), then the result is also an element of S. For example, the integers are closed under $+$ and $\cdot$, because the sum of two integers is always an integer, as is their product. But the integers are *not* closed under $/$: for example, $2/3$ is not an integer even though $2, 3 \in \mathbb{Z}$. The *closure* of S under f is the smallest superset of S that is closed under f.

Here are the formal definitions:

Definition 8.9 (Reflexive, symmetric, and transitive closures)
Let $R \subseteq A \times A$ be a relation.

- *The* reflexive closure *of R is the smallest relation $R' \supseteq R$ such that R' is reflexive.*
- *The* symmetric closure *of R is the smallest relation $R'' \supseteq R$ such that R'' is symmetric.*
- *The* transitive closure *of R is the smallest relation $R^+ \supseteq R$ such that R^+ is transitive.*

We'll illustrate these definitions with an example of the symmetric, reflexive, and transitive closures of a small relation, and then return to a few of our running examples of arithmetic relations.

Example 8.26 (Closures of a small relation)
Consider the relation $R := \{\langle 1,5\rangle, \langle 2,2\rangle, \langle 2,4\rangle, \langle 4,1\rangle, \langle 4,2\rangle\}$ on $\{1,2,3,4,5\}$. Then we have the following closures of R. (See Figure 8.18 for visualizations.)

$$\text{reflexive closure} = R \cup \{\langle 1,1\rangle, \langle 3,3\rangle, \langle 4,4\rangle, \langle 5,5\rangle\}.$$

$$\text{symmetric closure} = R \cup \Big\{ \underbrace{\langle 5,1\rangle}_{\text{because of } \langle 1,5\rangle}, \underbrace{\langle 1,4\rangle}_{\text{because of } \langle 4,1\rangle} \Big\}.$$

$$\text{transitive closure} = R \cup \Big\{ \underbrace{\langle 2,1\rangle}_{\text{because of } \langle 2,4\rangle \text{ and } \langle 4,1\rangle}, \underbrace{\langle 4,4\rangle}_{\text{because of } \langle 4,2\rangle \text{ and } \langle 2,4\rangle}, \underbrace{\langle 4,5\rangle}_{\text{because of } \langle 4,1\rangle \text{ and } \langle 1,5\rangle}, \underbrace{\langle 2,5\rangle}_{\text{because of } \langle 2,4\rangle \text{ and } \langle 4,5\rangle} \Big\}.$$

It's worth noting that $\langle 2,5\rangle$ had to be in the transitive closure R^+ of R, even though there was no x such that $\langle 2,x\rangle \in R$ and $\langle x,5\rangle \in R$. There's one more intermediate step in the chain of reasoning: the pair $\langle 4,5\rangle$ had to be in R^+ because $\langle 4,1\rangle, \langle 1,5\rangle \in R$, and therefore both $\langle 2,4\rangle$ and $\langle 4,5\rangle$ had to be in R^+—so $\langle 2,5\rangle$ had to be in R^+ as well.

Example 8.27 (Closures of divides)
Recall the "divides" relation $R = \{\langle n,m\rangle : m \bmod n = 0\}$. Because R is both reflexive and transitive, the reflexive closure and transitive closure of R are both just R itself. The symmetric closure of R is the set of pairs $\langle n,m\rangle$ where one of n and m is a divisor of the other (in either order): $\{\langle n,m\rangle : n \bmod m = 0 \text{ or } m \bmod n = 0\}$.

Example 8.28 (Closures of $>$)
Recall the "greater than" relation $\{\langle n,m\rangle : n > m\}$. The reflexive closure of $>$ is $\geq$—that is, the set $\{\langle n,m\rangle : n \geq m\}$. The symmetric closure of $>$ is the relation $\neq$—that is, the set $\{\langle n,m\rangle : n > m \text{ or } m > n\} = \{\langle n,m\rangle : n \neq m\}$. The relation $>$ is already transitive, so the transitive closure of $>$ is $>$ itself.

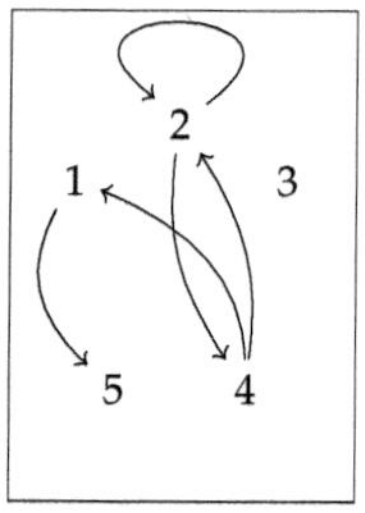

(a) The relation R.

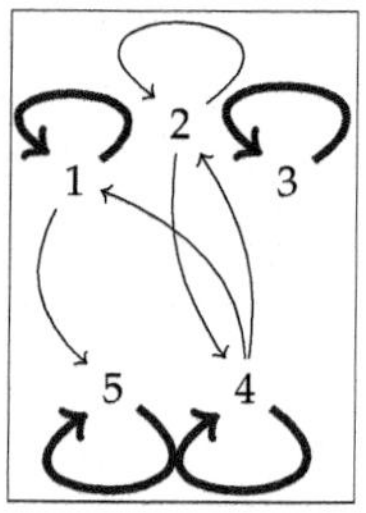

(b) The reflexive closure of R.

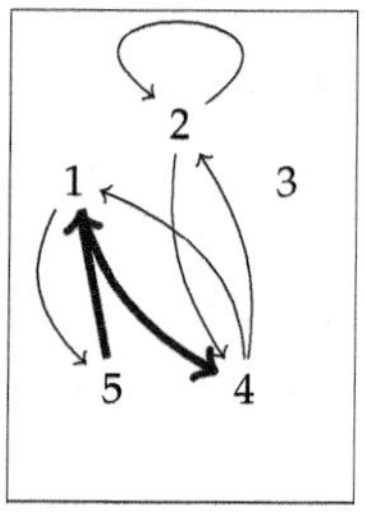

(c) The symmetric closure of R.

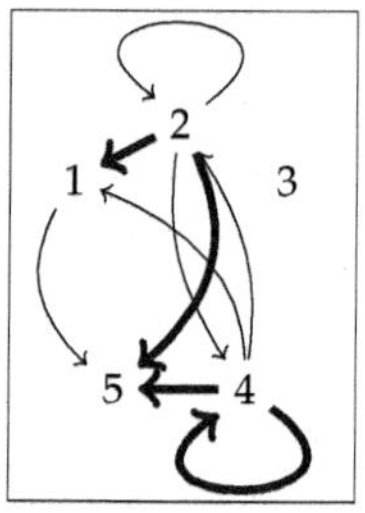

(d) The transitive closure of R.

Figure 8.18: A relation R, and several closures. In each, the dark arrows had to be added to R to achieve the desired property.

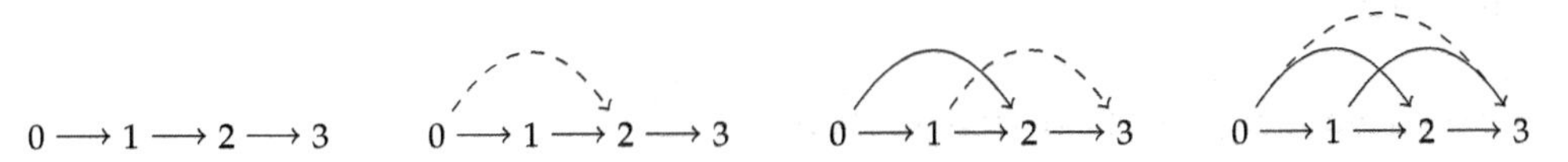

(a) The relation R.

(b) $\langle 0,1 \rangle$ and $\langle 1,2 \rangle$ mean that we must add $\langle 0,2 \rangle$.

(c) $\langle 1,2 \rangle$ and $\langle 2,3 \rangle$ mean that we must add $\langle 0,2 \rangle$.

(d) $\langle 0,2 \rangle$, which we added in (b), and $\langle 2,3 \rangle$ mean that we must now add $\langle 0,3 \rangle$ too.

Figure 8.19: Computing the transitive closure of the relation $\{\langle 0,1 \rangle, \langle 1,2 \rangle, \langle 2,3 \rangle\}$. Note that in panel (d), we could have instead argued that we had to add $\langle 0,3 \rangle$ because of $\langle 0,1 \rangle$ and $\langle 1,3 \rangle$ (from panel (c)), rather than because of $\langle 0,2 \rangle$ (from panel (b)) and $\langle 2,3 \rangle$.

Computing the closures of a relation

How did we compute the closures in the last few examples? The approach itself is simple: starting with $R' = R$, we repeatedly look for a violation of the desired property in R' (an element of R' required by the property but missing from R'), and repair that violation by adding the necessary element to R'. For the reflexive and symmetric closures, this idea is straightforward: the violations of reflexivity are precisely those elements of $\{\langle a,a \rangle : a \in A\}$ not already in R, and the violations of symmetry are precisely those elements of R^{-1} that are not already in R.

For the transitive closure, things are slightly trickier: as we resolve existing violations by adding missing pairs to the relation, new violations of transitivity can crop up. (See Figure 8.19.) Thus, to compute the transitive closure, we can simply iterate as described above: starting with $R' := R$, repeatedly add to R' any missing $\langle a,c \rangle$ with $\langle a,b \rangle, \langle b,c \rangle \in R'$, until there are no more violations of transitivity. (While we won't prove it here, it's an important fact that the order in which we add elements to the transitive closure turns out not to affect the final result.) See Figure 8.20 for algorithms to compute these closures for $R \subseteq A \times A$ for a finite set A. (Note that these algorithms are *not* guaranteed to terminate if A is infinite! Also, there are faster ways to find the transitive closure based on graph algorithms—see Chapter 11—but the basic idea is captured here.)

reflexive-closure(R):
Input: a relation $R \subseteq A \times A$
Output: the smallest reflexive $R' \supseteq R$
return $R \cup \{\langle a,a \rangle : a \in A\}$

symmetric-closure(R):
Input: a relation $R \subseteq A \times A$
Output: the smallest symmetric $R' \supseteq R$
return $R \cup R^{-1}$

transitive-closure(R):
Input: a relation $R \subseteq A \times A$
Output: the smallest transitive $R' \supseteq R$
$R' := R$
while there exist $a, b, c \in A$ such that $\langle a,b \rangle \in R$ and $\langle b,c \rangle \in R$ and $\langle a,c \rangle \notin R'$:
  $R' := R' \cup \{\langle a,c \rangle\}$
return R'

Figure 8.20: Algorithms to compute reflexive, symmetric, and transitive closures of a relation $R \subseteq A \times A$, when A is finite.

Alternatively, here's another way to view the transitive closure of $R \subseteq A \times A$. The relation $R \circ R$ denotes precisely those pairs $\langle a,c \rangle$ where $\langle a,b \rangle, \langle b,c \rangle \in R$ for some $b \in A$. Thus the "direct" violations of transitivity are pairs that are in $R \circ R$ but not R. But, as we saw in Figure 8.19, the relation $R \cup (R \circ R)$ might have violations of transitivity, too: that is, a pair $\langle a,d \rangle \notin R \cup (R \circ R)$ but where $\langle a,b \rangle \in R$ and $\langle b,d \rangle \in R \circ R$ for some $b \in A$. So we have to add $R \circ R \circ R$ as well. And so on! In other words, the transitive closure R^+ of R is given by $R^+ = R \cup R^2 \cup R^3 \cup \cdots$, where $R^k := R \circ R \circ \cdots \circ R$ is the result of composing R with itself k times. Thus:

- the reflexive closure of R is $R \cup \{\langle a,a \rangle : a \in A\}$.
- the symmetric closure of R is $R \cup R^{-1}$.
- the transitive closure of R is $R \cup R^2 \cup R^3 \cup \cdots$.

(Exercise 8.104 asks you to prove correctness, and Exercise 8.105 asks you to show that the transitive closure can be much bigger than the relation itself.)

Closures with respect to multiple properties at once

In addition to defining the closure of a relation R with respect to one of the three properties (reflexivity, symmetry, or transitivity), we can also define the closure with respect to two or more of these properties simultaneously. Any subset of these properties makes sense in this context, but the two most common combinations require reflexivity and transitivity, with or without requiring symmetry:

Definition 8.10 (Reflexive (symmetric) transitive closure)
Let $R \subseteq A \times A$ be a relation.

- *The* reflexive transitive closure *of R is the smallest relation $R^* \supseteq R$ such that R^* is both reflexive and transitive.*
- *The* reflexive symmetric transitive closure *of R is the smallest relation $R^{\equiv} \supseteq R$ such that $R^{\equiv}$ is reflexive, symmetric, and transitive.*

Example 8.29 (Parent)
Consider the relation $parent := \{\langle p, c\rangle : p \text{ is a parent of } c\}$ over a set S. (This example makes sense if we think of S as a set of people where "parent" has biological meaning, or if we think of S as a set of nodes in a tree.) Then:

- The transitive closure of *parent* is

$$parent \cup grandparent \cup greatgrandparent \cup greatgreatgrandparent \cdots.$$

- The reflexive transitive closure of *parent* is *ancestor*. That is, $\langle x, y\rangle$ is in the reflexive transitive closure of *parent* if and only if x is a direct ancestor of y, counting x as a direct ancestor of x herself. (Compared to the transitive closure, the reflexive transitive closure also includes the relation $yourself := \{\langle x, x\rangle : x \in S\}$.)

Example 8.30 (Adjacent seating at a concert)
Consider a set S of people attending a concert held in a theater with rows of seats. Let R denote the relation of "sat immediately to the right of," so that $\langle x, y\rangle \in R$ if and only if x sat one seat to y's right in the same row. (See Figure 8.21.)

The transitive closure of R is "sat (not necessarily immediately) to the right of." The symmetric closure of R is "sat immediately next to." The symmetric transitive closure of R is "sat in the same row as." The reflexive symmetric transitive closure of R is also "sat in the same row as." (You sit in the same row as yourself.)

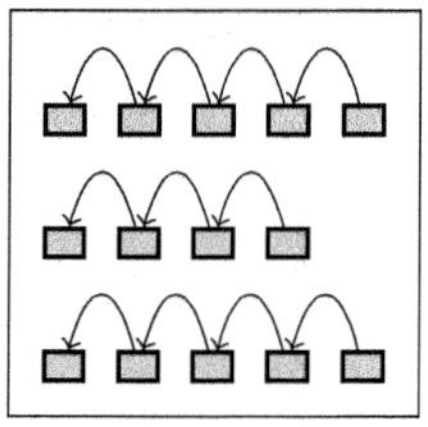

Figure 8.21: The sat-immediately-to-the-right-of relation.

As we discussed previously, we can think of the transitive closure R^+ of the relation R as the result of repeating R one or more times: in other words, we have that

$R^+ := R \cup R^2 \cup R^3 \cup \cdots$. The *reflexive* transitive closure of R also adds $\{\langle a,a \rangle : a \in A\}$ to the closure, which we can view as the result of repeating R *zero* or more times. In other words, we have that the reflexive transitive closure R^* is $R^* = R^0 \cup R^+$, where $R^0 := \{\langle a,a \rangle : a \in A\}$ represents the "zero-hop" application of R.

Taking it further: The basic idea underlying the (reflexive) transitive closure of a relation R—allowing (zero or) one or more repetitions of a relation R—also comes up in a widely useful tool for pattern matching in text, called *regular expressions*. Using regular expressions, you can search a text file for lines that match certain kinds of patterns (like: find all violations in the dictionary of the "I before E except after C" rule), or apply some operation to all files with a certain name (like: remove all `.txt` files). For more discussion of regular expressions more generally, and a little more on the connection between (reflexive) transitive closure and regular expressions, see p. 830.

We'll end with one last example of closures of an arithmetic relation:

Example 8.31 (Closures of the successor relation)

Problem: The *successor* relation on the integers is $\{\langle n, n+1 \rangle : n \in \mathbb{Z}\}$. What are the reflexive, symmetric, transitive, reflexive transitive, and reflexive symmetric transitive closures of this relation?

Solution:

- The reflexive closure of *successor* is the relation $\{\langle n,m \rangle : m = n \text{ or } m = n+1\}$—that is, pairs of integers where the second component is equal to or one greater than the first component.
- The symmetric closure of *successor* is $\{\langle n,m \rangle : m = n-1 \text{ or } m = n+1\}$—that is, pairs of integers where the second component is exactly one less or one greater than the first component.
- The transitive closure of *successor* is the relation $<$—that is, the relation $\{\langle n,m \rangle : n < m\}$. In fact, the infinite version of Figure 8.20 illustrates why: for any n, we have $\langle n, n+1 \rangle$ and $\langle n+1, n+2 \rangle$ in *successor*, so the transitive closure includes $\langle n, n+2 \rangle$. But $\langle n+2, n+3 \rangle$ is in *successor*, so the transitive closure also includes $\langle n, n+3 \rangle$. But $\langle n+3, n+4 \rangle$ is in *successor*, so the transitive closure also includes $\langle n, n+4 \rangle$. And so forth! (See Exercise 8.106 for a formal proof.)
- The reflexive transitive closure of the successor relation $\{\langle x, x+1 \rangle : x \in \mathbb{Z}\}$ is $\leq$.
- Finally, the reflexive symmetric transitive closure of *successor* is actually $\mathbb{Z} \times \mathbb{Z}$: that is, *every* pair of integers is in this relation.

Incidentally, we can view $\leq$ (the reflexive transitive closure of *successor*) as either *the reflexive closure of* $<$ (the transitive closure of *successor*), or we can view $\leq$ as *the transitive closure of* $\{\langle n,m \rangle : m = n \text{ or } m = n+1\}$ (the reflexive closure of *successor*). It's true in general that the reflexive closure of the transitive closure equals the transitive closure of the reflexive closure.

Computer Science Connections

Regular Expressions

Regular expressions (sometimes called *regexps* or *regexes* for short) are a mechanism to express pattern-matching searches in strings. (Their name is also a bit funny; more on that below.) Regular expressions are used by a number of useful utilities on Unix-based systems, like `grep` (which prints all lines of a file that match a given pattern) and `sed` (which can perform search-and-replace operations for particular patterns). And many programming languages have a capability for regular-expression processing—they're a tremendously handy tool for text processing.

Let Σ denote an *alphabet* of symbols. (For convenience, think of $\Sigma = \{\texttt{A}, \texttt{B}, \ldots, \texttt{Z}\}$, but generally it's the set of all ASCII characters.) Let Σ^* denote the set of all finite-length strings of symbols from Σ. (Note that the * notation echoes the notation for the reflexive transitive closure: Σ^* is the set of elements resulting from "repeating" Σ zero or more times.)

The basics of regular expressions are shown in Figure 8.22. Essentially the syntax of regular expressions (recursively) defines a relation $\mathit{Matches} \subseteq \mathit{Regexps} \times \Sigma^*$, where certain strings match a given pattern α. Figure 8.22 says that, for example, $\{s : \langle \alpha\beta, s \rangle \in \mathit{Matches}\}$ is precisely the set of strings that can be written xy where $\langle \alpha, x \rangle$ and $\langle \beta, y \rangle$ are in *Matches*. There's some other shorthand for common constructions, too: for example, a list of characters in square brackets matches any of those characters (for example, `[AEIOU]` is shorthand for `(A|E|I|O|U)`). (Other syntax allows a range of characters or everything *but* a list of characters: for example, `[A-Z]` for all letters, and `[^AEIOU]` for consonants.) A few other regexp operators correspond to the types of closures that we introduced in this section. (See Figure 8.23.)

A	matches the single character A
B	matches the single character B
⋮	⋮
Z	matches the single character Z
.	matches any single character in Σ
$\alpha\beta$	matches any string xy where x matches α and y matches β
$\alpha\vert\beta$	matches any string x where x matches α *or* x matches β

Figure 8.22: The basics of regexps.

α?	matches any string that matches α *or* the empty string
α+	matches any string $x_1x_2\ldots x_k$, with $k \geq 1$, where each x_i matches α
α*	matches any string $x_1x_2\ldots x_k$, with $k \geq 0$, where each x_i matches α

Figure 8.23: Some more regexp operators. The + operator is roughly analogous to transitive closure—α+ matches any string that consists of one or more repetitions of α—while ? is roughly analogous to the reflexive closure and * to the reflexive transitive closure. The only difference is that here we're combining repetitions by *concatenation* rather than by *composition*.

For example, the following regular expressions match words in a dictionary that have some vaguely interesting properties:

1. `.*(CIE|[^C]EI).*`
2. `.*[^AEIOU][^AEIOU][^AEIOU][^AEIOU][^AEIOU].*`
3. `[^AEIOU]*A[^AEIOU]*E[^AEIOU]*I[^AEIOU]*O[^AEIOU]*U[^AEIOU]*`

Respectively, these regexps match (1) words that violate the "`I` before `E` except after `C`" rule (like `WEIRD` or `GLACIER`); (2) words with five consecutive consonants (like `LENGTHS` or `WITCHCRAFT`); and (3) words with all five vowels, once each, in alphabetical order (like `FACETIOUS` and `ABSTEMIOUS`).

The odd-sounding name "regular expression" derives from a related notion, called a "regular language." A *language* $L \subseteq \Sigma^*$ is a subset of all strings; in the subfield of theoretical computer science called *formal language theory*, we're interested in how easy it is to determine whether a given string $x \in \Sigma^*$ is in L or not, for a particular language L. (Some example languages: the set of words containing only type of vowel, or the set of binary strings with the same number of `1`s and `0`s.) A *regular language* is one for which it's possible to determine whether $x \in L$ by reading the string from left to right and, at each step, remembering only a constant amount of information about what you've seen so far. (The set of univocalic words is regular; the set of "balanced" bitstrings is not.)[5]

We have only hinted at the depth of regular languages, regular expressions, and formal language theory here. There's a whole courseload of material about these languages: for a bit more, see p. 846; for a lot more, see a good textbook on computational complexity and formal languages, like

[5] Michael Sipser. *Introduction to the Theory of Computation*. Course Technology, 3rd edition, 2012; and Dexter Kozen. *Automata and Computability*. Springer, 1997.

8.3.6 *Exercises*

8.50 Draw a directed graph representing the relation $\{\langle x, x^2 \bmod 13\rangle : x \in \mathbb{Z}_{13}\}$.
8.51 Repeat for $\{\langle x, 3x \bmod 13\rangle : x \in \mathbb{Z}_{15}\}$.
8.52 Repeat for $\{\langle x, 3x \bmod 15\rangle : x \in \mathbb{Z}_{15}\}$.

Which of the following relations on $\{0,1,2,3,4\}$ are reflexive? Irreflexive? Neither?
8.53 $\{\langle x,x\rangle : x^5 \equiv_5 x\}$
8.54 $\{\langle x,y\rangle : x+y \equiv_5 0\}$
8.55 $\{\langle x,y\rangle$: there exists z such that $x \cdot z \equiv_5 y\}$
8.56 $\{\langle x,y\rangle$: there exists z such that $x^2 \cdot z^2 \equiv_5 y\}$

Let $R \subseteq A \times A$ and $T \subseteq A \times A$ be relations. Prove or disprove the following:
8.57 R is reflexive if and only if R^{-1} is reflexive.
8.58 if R and T are both reflexive, then $R \circ T$ is reflexive.
8.59 if $R \circ T$ is reflexive, then R and T are both reflexive.
8.60 R is irreflexive if and only if R^{-1} is irreflexive.
8.61 if R and T are both irreflexive, then $R \circ T$ is irreflexive.

Which relations from Exercises 8.53–8.56 on $\{0,1,2,3,4\}$ are symmetric? Antisymmetric? Asymmetric? Explain.
8.62 $\{\langle x,x\rangle : x^5 \equiv_5 x\}$
8.63 $\{\langle x,y\rangle : x+y \equiv_5 0\}$
8.64 $\{\langle x,y\rangle$: there exists z such that $x \cdot z \equiv_5 y\}$
8.65 $\{\langle x,y\rangle$: there exists z such that $x^2 \cdot z^2 \equiv_5 y\}$

Prove Theorem 8.1, connecting the symmetry/asymmetry/antisymmetry of a relation R to the inverse R^{-1} of R.
8.66 Prove that R is symmetric if and only if $R \cap R^{-1} = R = R^{-1}$.
8.67 Prove that R is antisymmetric if and only if $R \cap R^{-1} \subseteq \{\langle a,a\rangle : a \in A\}$.
8.68 Prove that R is asymmetric if and only if $R \cap R^{-1} = \varnothing$.

8.69 Be careful: it's possible for a relation $R \subseteq A \times A$ to be both symmetric and antisymmetric! Describe, as precisely as possible, the set of relations on A that are both.
8.70 Prove or disprove: if R is asymmetric, then R is antisymmetric.

Fill in each cell in Figure 8.24 with a relation on $\{0,1\}$ that satisfies the given criteria. Or, if the criteria are inconsistent, explain why there is no such a relation.
8.71 a reflexive, symmetric relation on $\{0,1\}$.
8.72 a reflexive, antisymmetric relation on $\{0,1\}$.
8.73 a reflexive, asymmetric relation on $\{0,1\}$.
8.74 an irreflexive, symmetric relation on $\{0,1\}$.
8.75 an irreflexive, antisymmetric relation on $\{0,1\}$.
8.76 an irreflexive, asymmetric relation on $\{0,1\}$.
8.77 a symmetric relation on $\{0,1\}$ that's neither reflexive nor irreflexive.
8.78 an antisymmetric relation on $\{0,1\}$ that's neither reflexive nor irreflexive.
8.79 an asymmetric relation on $\{0,1\}$ that's neither reflexive nor irreflexive.

	symmetric	antisymmetric	asymmetric
reflexive	Exer. 8.71	Exer. 8.72	Exer. 8.73
irreflexive	Exer. 8.74	Exer. 8.75	Exer. 8.76
neither	Exer. 8.77	Exer. 8.78	Exer. 8.79

Figure 8.24: Some fill-in-the-blank relations.

Which relations from Exercises 8.53–8.56 on $\{0,1,2,3,4\}$ are transitive? Explain.
8.80 $\{\langle x,x\rangle : x^5 \equiv_5 x\}$.
8.81 $\{\langle x,y\rangle : x+y \equiv_5 0\}$.
8.82 $\{\langle x,y\rangle$: there exists z such that $x \cdot z \equiv_5 y\}$.
8.83 $\{\langle x,y\rangle$: there exists z such that $x^2 \cdot z^2 \equiv_5 y\}$.

Formally prove the following statements about a relation $R \subseteq A \times A$, using the definitions of the given properties.
8.84 Prove that, if R is irreflexive and transitive, then R is asymmetric.
8.85 Prove Theorem 8.2: show that R is transitive if and only if $R \circ R \subseteq R$.
8.86 Theorem 8.2 cannot be stated with an = instead of $\subseteq$ (although I actually made this mistake in a previous draft!). Give an example of a transitive relation R where $R \circ R \subset R$ (that is, where $R \circ R \neq R$).

The following exercises describe a relation with certain properties. For each, say whether it is possible for a relation $R \subseteq A \times A$ to simultaneously have all of the stated properties. If so, describe as precisely as possible what structure the relation R must have. If not, prove that it is impossible.

8.87 Is it possible for R to be simultaneously symmetric, transitive, and irreflexive?

8.88 Is it possible for R to be simultaneously transitive and a function?

8.89 Identify *all* relations R on $\{0,1\}$ that are transitive.

8.90 Of the transitive relations on $\{0,1\}$ from Exercise 8.89, which are also reflexive and symmetric?

Consider the relation $R := \{\langle 2,4\rangle, \langle 4,3\rangle, \langle 4,4\rangle\}$ *on the set* $\{1,2,3,4\}$.

8.91 What is the reflexive closure of R?

8.92 What is the symmetric closure of R?

8.93 What is the transitive closure of R?

8.94 What is the reflexive transitive closure of R?

8.95 What is the reflexive symmetric transitive closure of R?

Now consider the relation $T := \{\langle 1,2\rangle, \langle 1,3\rangle, \langle 2,1\rangle, \langle 2,3\rangle, \langle 3,1\rangle, \langle 3,2\rangle, \langle 3,4\rangle, \langle 4,5\rangle\}$ *on* $\{1,2,3,4,5\}$.

8.96 What is the reflexive closure of T?

8.97 What is the symmetric closure of T?

8.98 What is the transitive closure of T?

8.99 What is the symmetric closure of $\geq$?

The next few exercises ask you to implement relations (and the standard relation operations) in a programming language of your choice. Don't worry too much about efficiency in your implementation; it's okay to run in time $\Theta(n^3)$, $\Theta(n^4)$ or even $\Theta(n^5)$ when relation R is on a set of size n.

8.100 *(programming required)* Develop a basic implementation of relations on a set A. Also implement inverse (R^{-1}) and composition ($R \circ T$, where both R and T are subsets of $A \times A$).

8.101 *(programming required)* Write functions **reflexive?**, **irreflexive?**, **symmetric?**, **antisymmetric?**, **asymmetric?**, and **transitive?** to test whether a given relation R has the specified property.

8.102 *(programming required)* Implement the closure algorithms (reproduced in Figure 8.25) for relations.

8.103 *(programming required)* Using your implementations from the last few exercises, verify your answers to Exercises 8.71–8.79 (see Figure 8.24).

8.104 Prove that the transitive closure of R is indeed $R^+ := R \cup R^2 \cup R^3 \cup \cdots$, as follows: show that if $S \supseteq R$ is any transitive relation, then $R^k \subseteq S$. (We'd also need to prove that R^+ is transitive, but you can omit this part of the proof. You may find a recursive definition of R^k most helpful: $R^1 = R$ and $R^k = R \circ R^{k-1}$.)

reflexive-closure(R):
Input: a relation $R \subseteq A \times A$
Output: the smallest reflexive $R' \supseteq R$
return $R \cup \{\langle a,a\rangle : a \in A\}$

symmetric-closure(R):
Input: a relation $R \subseteq A \times A$
Output: the smallest symmetric $R' \supseteq R$
return $R \cup R^{-1}$

transitive-closure(R):
Input: a relation $R \subseteq A \times A$
Output: the smallest transitive $R' \supseteq R$
$R' := R$
while there exist $a,b,c \in A$ such that $\langle a,b\rangle \in R$ and $\langle b,c\rangle \in R$ and $\langle a,c\rangle \notin R'$:
    $R' := R' \cup \{\langle a,c\rangle\}$
return R'

Figure 8.25: A reminder of algorithms to compute the reflexive, symmetric, and transitive closures of a relation on a finite set.

8.105 Give an example of a relation $R \subseteq A \times A$, for a finite set A, such that the transitive closure of R contains at least $c \cdot |R|^2$ pairs, for some constant $c > 0$. Make c as big as you can.

8.106 Recall the relation *successor* $:= \{\langle x, x+1\rangle : x \in \mathbb{Z}^{\geq 0}\}$. Prove by induction on k that, for any integer x and any positive integer k, we have that $\langle x, x+k\rangle$ is in the transitive closure of successor. (In other words, you're showing that the transitive closure of *successor* is $\geq$. Note that you cannot rely on the algorithm in Figure 8.25 because $\mathbb{Z}^{\geq 0}$ is not finite!)

8.107 We talked about the X closure of a relation R, for X being any nonempty subset of the properties of reflexivity, symmetry, and transitivity. But we didn't define the "antisymmetric closure" of a relation R—with good reason! Why doesn't the antisymmetric closure make sense?

8.4 *Special Relations: Equivalence Relations and Partial/Total Orders*

Talking with you is sort of the conversational equivalent of an out of body experience.

Bill Watterson (b. 1958), *Calvin & Hobbes*

In Section 8.3, we introduced three key categories of properties that a particular relation $R \subseteq A \times A$ might have: (ir)reflexivity, (a/anti)symmetry, and transitivity. Here we'll consider relations R that have one of two particular combinations of those three categories of properties. Two very different "flavors" of relations emerge from these two particular constellations of properties:

- *equivalence relations* (reflexive, symmetric, and transitive), which divide the elements of A into one or more groups of equivalent elements, so that all elements in the same group are "the same" under R; and
- *order* relations (reflexive or irreflexive, antisymmetric, and transitive), which "rank" the elements of A, so that some elements of A are "more R" than others.

In this section, we'll give formal definitions of these two types of relations, and look at a few applications.

8.4.1 *Equivalence Relations*

An *equivalence relation* $R \subseteq A \times A$ separates the elements of A into one or more groups, where any two elements in the same group are *equivalent* according to R:

Definition 8.11 (Equivalence relation)
An equivalence relation *is a relation that is reflexive, symmetric, and transitive.*

The most important equivalence relation that you've seen is equality (=): certainly, for any objects a, b, and c, we have that (i) $a = a$; (ii) $a = b$ if and only if $b = a$; and (iii) if $a = b$ and $b = c$, then $a = c$.

The relation *sat in the same row as* (see Example 8.30) is also an equivalence relation: it's reflexive (you sat in the same row as you yourself), symmetric (anyone you sat in the same row as also sat in the same row as you), and transitive (you sat in the same row as anyone who sat in the same row as someone who sat in the same row as you). And we already saw another example in Example 8.11: the relation

$$\{\langle m_1, m_2\rangle : \text{months } m_1 \text{ and } m_2 \text{ have the same number of days (in some years)}\}$$

(see Figure 8.26 for a reminder) is also an equivalence relation. It's tedious but simple to verify by checking all pairs that the relation in Figure 8.26 is reflexive, symmetric, and transitive. (See also Exercises 8.115–8.117.)

Here are a few more examples of equivalence relations:

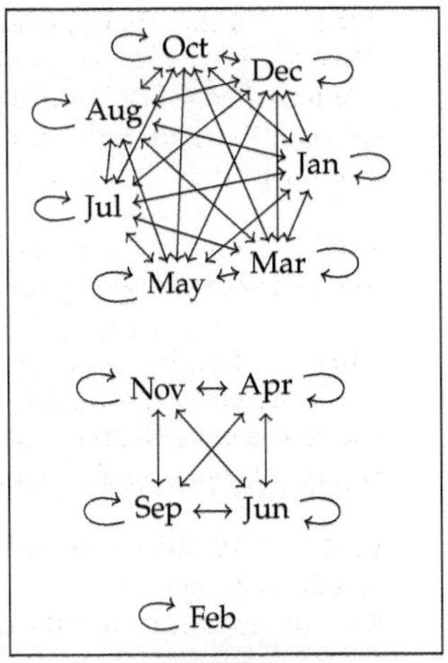

Figure 8.26: The months-of-the-same length relation (a reminder).

Example 8.32 (Some equivalence relations)
All of the following are equivalence relations:

1. The set of pairs from $\{0, 1, \ldots, 23\}$ with the same representation on a 12-hour clock:

$$\left\{ \begin{array}{c} \langle 0,0\rangle, \langle 0,12\rangle, \langle 12,0\rangle, \langle 12,12\rangle, \\ \langle 1,1\rangle, \langle 1,13\rangle, \langle 13,1\rangle, \langle 13,13\rangle, \\ \vdots \\ \langle 11,11\rangle, \langle 11,23\rangle, \langle 23,11\rangle, \langle 23,23\rangle \end{array} \right\}.$$

2. The asymptotic relation Θ (that is, for two functions f and g, we have $\langle f, g\rangle \in \Theta$ if and only if f is $\Theta(g)$). We argued in Examples 8.24–8.25 and Exercise 6.46 that Θ is reflexive, symmetric, and transitive.

3. The relation $\equiv$ on logical propositions, where $P \equiv Q$ if and only if P and Q are true under precisely the same set of truth assignments. (We even used the word "equivalent" in defining $\equiv$, which we called *logical equivalence* back in Chapter 3.)

Example 8.33 (All equivalence relations on a small set)
Problem: List all equivalence relations on the set $\{a, b, c\}$.

Solution: There are five different equivalence relations on this set:

$\{\langle a,a\rangle, \langle b,b\rangle, \langle c,c\rangle\}$	*"no element is equivalent to any other"*
$\{\langle a,a\rangle, \langle a,b\rangle, \langle b,a\rangle, \langle b,b\rangle, \langle c,c\rangle\}$	*"a and b are equivalent, but they're different from c"*
$\{\langle a,a\rangle, \langle a,c\rangle, \langle b,b\rangle, \langle c,a\rangle, \langle c,c\rangle\}$	*"a and c are equivalent, but they're different from b"*
$\{\langle a,a\rangle, \langle b,b\rangle, \langle b,c\rangle, \langle c,b\rangle, \langle c,c\rangle\}$	*"b and c are equivalent, but they're different from a"*
$\{\langle a,a\rangle, \langle a,b\rangle, \langle a,c\rangle, \langle b,a\rangle, \langle b,b\rangle, \langle b,c\rangle, \langle c,a\rangle, \langle c,b\rangle, \langle c,c\rangle\}$.	*"all elements are equivalent"*

Equivalence classes

The descriptions of the quintet of equivalence relations on the set $\{a, b, c\}$ from Example 8.33 makes more explicit the other way that we've talked about an equivalence relation R on A: as a relation that carves up A into one or more *equivalence classes*, where any two elements of the same equivalence class are related by R (and no two elements of different classes are). Here's the formal definition:

Definition 8.12 (Equivalence class)
Let $R \subseteq A \times A$ be an equivalence relation. The equivalence class *of $a \in A$ is defined as the set $\{b \in A : \langle a, b\rangle \in R\}$ of elements related to A under R. The equivalence class of $a \in A$ under R is denoted by $[a]_R$—or, when R is clear from context, just as $[a]$.*

The equivalence classes of an equivalence relation on A form a *partition* of the set A—that is, every element of A is in one and only one equivalence class. (See Definition 2.30 for a reminder of the definition of "partition.")

Example 8.34 (Equivalent mod 5)
Define the relation $\equiv_5$ on $\mathbb{Z}$, so that $\langle x, y \rangle \in \equiv_5$ if and only if $x \bmod 5 = y \bmod 5$. It's easy to check that all three requirements (reflexivity, symmetry, and transitivity) are met; see Examples 8.18, 8.20, and 8.23. There are five equivalence classes under $\equiv_5$:

$$\{0,5,10,\ldots\}, \{1,6,11,\ldots\}, \{2,7,12,\ldots\}, \{3,8,13,\ldots\}, \text{ and } \{4,9,14,\ldots\},$$

corresponding to the five possible values mod 5.

Example 8.35 (Some equivalence classes)
The five different equivalence relations on $\{\texttt{a}, \texttt{b}, \texttt{c}\}$ in Example 8.33 correspond to five different sets of equivalence classes:

$\Big\{ \{\texttt{a}\}, \{\texttt{b}\}, \{\texttt{c}\} \Big\}$ *"no element is equivalent to any other"*

$\Big\{ \{\texttt{a}, \texttt{b}\}, \{\texttt{c}\} \Big\}$ *"a and b are equivalent, but they're different from c"*

$\Big\{ \{\texttt{a}, \texttt{c}\}, \{\texttt{b}\} \Big\}$ *"a and c are equivalent, but they're different from b"*

$\Big\{ \{\texttt{a}\}, \{\texttt{b}, \texttt{c}\} \Big\}$ *"b and c are equivalent, but they're different from a"*

$\Big\{ \{\texttt{a}, \texttt{b}, \texttt{c}\} \Big\}.$ *"all elements are equivalent"*

AN EXAMPLE: EQUIVALENCE OF RATIONAL NUMBERS

Back in Chapter 2, we defined the rational numbers (that is, fractions) as the set $\mathbb{Q} := \mathbb{Z} \times \mathbb{Z}^{\neq 0}$—that is, as two-element sequences of integers, respectively called the numerator and the denominator, where the denominator must be nonzero. (See Example 2.39.) Here you will give a formal treatment of two rational numbers like $\langle 17, 34 \rangle$ and $\langle 101, 202 \rangle$ being equivalent, in the sense that $\frac{17}{34} = \frac{101}{202} = \frac{1}{2}$:

Example 8.36 (Equivalence of rationals by reducing to lowest terms)
Problem: Formally define a relation $\equiv$ on $\mathbb{Q}$ that captures the notion of equality for fractions, and prove that $\equiv$ is an equivalence relation.

Solution: We define two rationals $\langle a, b \rangle$ and $\langle c, d \rangle$ as equivalent if and only if $ad = bc$—that is, we define the relation $\equiv$ as the set

$$\Big\{ \Big\langle \langle a, b \rangle, \langle c, d \rangle \Big\rangle : ad = bc \Big\}.$$

To show that $\equiv$ is an equivalence relation, we must prove that $\equiv$ is reflexive, symmetric, and transitive. These three properties follow fairly straightforwardly from the fact that the relation = on integers is an equivalence relation. We'll prove symmetry (reflexivity and transitivity can be proven analogously): for arbitrary $\langle a,b \rangle, \langle c,d \rangle \in \mathbb{Q}$ we have

$$\begin{aligned} \langle a,b \rangle \equiv \langle c,d \rangle &\Rightarrow ad = bc && \textit{definition of } \equiv \\ &\Rightarrow bc = ad && \textit{symmetry of } = \\ &\Rightarrow \langle c,d \rangle \equiv \langle a,b \rangle. && \textit{definition of } \equiv \end{aligned}$$

Taking it further: Recall that the equivalence class of a rational $\langle a,b \rangle \in \mathbb{Q}$ under $\equiv$, denoted $[\langle a,b \rangle]_\equiv$, represents the set of all rationals equivalent to $\langle a,b \rangle$. For example,

$$[\langle 17,34 \rangle]_\equiv = \{\langle 1,2 \rangle, \langle -1,-2 \rangle, \langle 2,4 \rangle, \langle -2,-4 \rangle, \ldots, \langle 17,34 \rangle, \ldots\}.$$

For equivalence relations like $\equiv$ for $\mathbb{Q}$, we may agree to associate an equivalence class with a *canonical element* of that class—here, the representative that's "in lowest terms." So we might agree to write $\langle 1,2 \rangle$ to denote the equivalence class $[\langle 1,2 \rangle]$, for example. This idea doesn't matter too much for the rationals, but it plays an important (albeit rather technical) role in figuring out how to define the real numbers in a mathematically coherent way. One standard way of defining the real numbers is as *the equivalence classes of converging infinite sequences of rational numbers*, called *Cauchy sequences* after the 19th-century French mathematician Augustin Louis Cauchy. (Two converging infinite sequences of rational numbers are defined to be equivalent if they converge to the same limit—that is, if the two sequences eventually differ by less than ε, for all $\varepsilon > 0$.) Thus when we write π, we're actually secretly denoting an infinitely large set of equivalent converging infinite sequences of rational numbers—but we're representing that equivalence class using a particular canonical form. Actually producing a coherent definition of the real numbers is a surprisingly recent development in mathematics, dating back less than 150 years. For more, see a good textbook on the subfield of math called *analysis*.[6]

For example, this book is a classic:

[6] Walter Rudin. *Principles of mathematical analysis*. McGraw–Hill, third edition, 1976.

Coarsening and refining equivalence relations

An equivalence relation $\equiv$ on A slices up the elements of A into equivalence classes—that is, disjoint subsets of A such that any two elements of the same class are related by $\equiv$. For example, you might consider two restaurants equivalent if they serve food from the same cuisine (Thai, Indian, Ethiopian, Chinese, British, Minnesotan, ...). But, given $\equiv$, we can imagine further subdividing the equivalence classes under $\equiv$ by making finer-grained distinctions (that is, *refining* $\equiv$)—perhaps dividing Indian into North Indian and South Indian, and Chinese into Americanized Chinese and Authentic Chinese. Or we could make $\equiv$ less specific (that is, *coarsening* $\equiv$) by combining some of the equivalence classes—perhaps having only two equivalence classes, Delicious (Thai, Indian, Ethiopian, Chinese) and Okay (British, Minnesotan). See Figure 8.27.

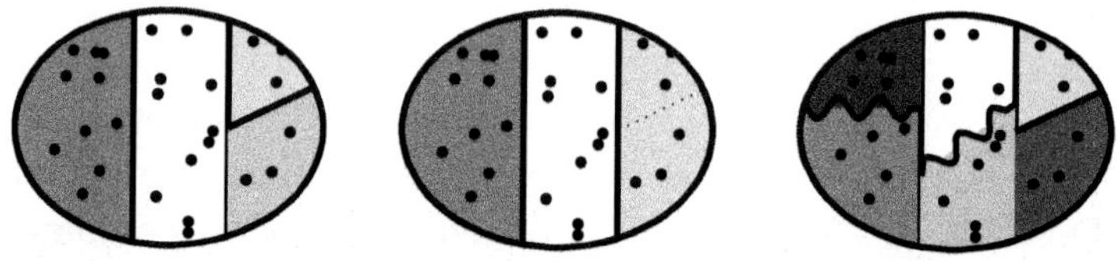

(a) An equivalence relation $\equiv$. (b) A coarsening of $\equiv$. (c) A refinement of $\equiv$.

Figure 8.27: Refining/coarsening an equivalence relation. In (a), dots represent elements; each colored region denotes an equivalence class under $\equiv$. Panel (b) shows a new equivalence relation formed by merging classes from $\equiv$; (c) shows a new equivalence relation formed by subdividing classes from $\equiv$.

Definition 8.13 (Coarsening/refining equivalence relations)
Consider two equivalence relations $\equiv_c$ and $\equiv_r$ on the same set A. We say that $\equiv_r$ is a refinement *of $\equiv_c$, or that $\equiv_c$ is a* coarsening *of $\equiv_r$, if $(a \equiv_r b) \Rightarrow (a \equiv_c b)$ for any $\langle a,b \rangle \in A \times A$. We can also refer to $\equiv_c$ as* coarser *than $\equiv_r$, and $\equiv_r$ as* finer *than $\equiv_c$.*

For example, equivalence mod 10 is a refinement of equivalence mod 5: whenever $n \equiv_{10} m$—that is, when $n \bmod 10 = m \bmod 10$—we know for certain that $n \bmod 5 = m \bmod 5$ too. (In other words, we have $(n \equiv_{10} m) \Rightarrow (n \equiv_5 m)$.) An equivalence class of the coarser relation is formed from the union of one or more equivalence classes of the finer relation. Here $\equiv_{10}$ is a refinement of $\equiv_5$, and, for example, the equivalence class $[3]_{\equiv_5}$ is the union of two equivalence classes from $\equiv_{10}$, namely $[3]_{\equiv_{10}} \cup [8]_{\equiv_{10}}$.

Taking it further: A *deterministic finite automaton (DFA)* is a simple model of a so-called "machine" that has a finite amount of memory, and processes an input string by moving from state to state according to a fixed set of rules. DFAs can be used for a variety of applications (for example, in computer architecture, compilers, or in modeling simple behavior in computer games). And they can also be understood in terms of equivalence relations. See p. 846 for more.

Example 8.37 (Refining/coarsening equivalence relations on {a, b, c})

In Example 8.35, we considered five different equivalence relations on {a, b, c}:

{{a}, {b}, {c}}

{{a, b}, {c}}   {{a, c}, {b}}   {{a}, {b, c}}

{{a, b, c}}

Of these, all three equivalence relations in the middle row *refine* the one-class equivalence relation {{a, b, c}} and *coarsen* the three-class equivalence relation {{a}, {b}, {c}}. (And the three-class equivalence relation {{a}, {b}, {c}} also refines the one-class equivalence relation {{a, b, c}}.)

Taking it further: This is a very meta comment, but we can think of "is a refinement of" as a relation *on equivalence relations on a set A*. In fact, the relation "is a refinement of" is reflexive, antisymmetric, and transitive: $\equiv$ refines $\equiv$; if $\equiv_1$ refines $\equiv_2$ and $\equiv_2$ refines $\equiv_1$ then $\equiv_1$ and $\equiv_2$ are precisely the same relation on A; and if $\equiv_1$ refines $\equiv_2$ and $\equiv_2$ refines $\equiv_3$ then $\equiv_1$ refines $\equiv_3$. Thus "is a refinement of" is, as per the definition to follow in the next section, a partial order on equivalence relations on the set A. Thus, for example, there is a *minimal element* according to the "is a refinement of" relation on the set of equivalence relations on any finite set A—that is, an equivalence relation $\equiv_{\min}$ such that $\equiv_{\min}$ is refined by no relation aside from $\equiv_{\min}$ itself. (Similarly, there's a maximal relation $\equiv_{\max}$ that refines no relation except itself.) See Exercises 8.118 and 8.119.

8.4.2 *Partial and Total Orders*

An equivalence relation $\equiv$ on a set A has properties that "feel like" a form of equality—differing from = only in that there might be multiple elements that are unequal but nonetheless cannot be distinguished by $\equiv$. Here we'll introduce a different special type of relation, more akin to $\leq$ than =, that instead describes a consistent *order* among the elements of A:

Definition 8.14 (Partial Order)
Let A be a set. A relation $\preceq$ on A that is reflexive, antisymmetric, and transitive is called a partial order. *(A relation $\prec$ on A that is* irreflexive, *antisymmetric, and transitive is called a* strict partial order.*)*

(Actually, the requirement of antisymmetry in a strict partial order is redundant; see Exercise 8.84.) Here are a few examples, from arithmetic and sets:

Example 8.38 (Some (strict) partial orders on $\mathbb{Z}$: $|$, $>$, and $\leq$)
In Examples 8.18, 8.20, and 8.23, we showed that the following relations are all antisymmetric, transitive, and either reflexive or irreflexive:

1. divides (reflexive): $R_1 = \{\langle n,m \rangle : m \bmod n = 0\}$ is a partial order.
2. greater than (irreflexive): $R_2 = \{\langle n,m \rangle : n > m\}$ is a strict partial order.
3. less than or equal to (reflexive): $R_3 = \{\langle n,m \rangle : n \leq m\}$ is a partial order.

Example 8.39 (The subset relation)
Consider the relation $\subseteq$ on the set $\mathscr{P}(\{0,1\})$, which consists of the following pairs of sets:

- $\{\} \subseteq \{0\}$, $\{\} \subseteq \{1\}$, and $\{\} \subseteq \{0,1\}$.
- $\{0\} \subseteq \{0\}$ and $\{0\} \subseteq \{0,1\}$.
- $\{1\} \subseteq \{1\}$ and $\{1\} \subseteq \{0,1\}$.
- $\{0,1\} \subseteq \{0,1\}$.

It's easy to verify that $\subseteq$ is reflexive, antisymmetric, and transitive. (One easy way to see this fact is via Figure 8.28, which abbreviates the visualizations in Figure 8.13 by leaving out an a-to-c arrow if their relationship is implied by transitivity because of a-to-b and b-to-c arrows. We'll see more of this type of abbreviated diagram in a moment.)

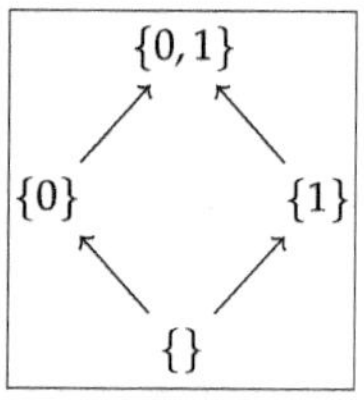

Figure 8.28: The $\subseteq$ relation on $\mathscr{P}(\{0,1\})$: $A \subseteq B$ if we can get from A to B by following arrows in this diagram.

COMPARABILITY AND TOTAL ORDERS

Note that, in a partial order $\preceq$, there can be two elements $a, b \in A$ such that *neither* $a \preceq b$ *nor* $b \preceq a$. For example, for the subset relation from Example 8.39 we have $\{0\} \not\subseteq \{1\}$ and $\{1\} \not\subseteq \{0\}$, and for the divides relation we have $17 \nmid 21$ and $21 \nmid 17$. In this case, the relation $\preceq$ does not say which of these elements is "smaller." This phenomenon is the reason that $\preceq$ is called a *partial* order, because it only specifies how *some* pairs compare.

There's a very misleading common-language use of "incomparable" (or "beyond compare") to mean "unequaled"—as in *Cheese from France is incomparable to cheese from Wisconsin.* Be careful! "Incomparable" means "cannot be compared" and *not* "cannot be matched."

Definition 8.15 (Comparability)
Let $\preceq$ be a partial order on A. We say that two elements $a \in A$ and $b \in A$ are comparable *under $\preceq$ if either $a \preceq b$ or $b \preceq a$. Otherwise we say that a and b are* incomparable.

When there are no incomparable pairs under $\preceq$, then we call $\preceq$ a *total* order:

Definition 8.16 (Total Order)
A relation $\preceq$ on A is a total order *if it's a partial order and every pair of elements in A is comparable. (A relation $\prec$ is a* strict total order *if $\prec$ is a strict partial order and every pair of* distinct *elements in A is comparable.)*

A FEW EXAMPLES OF PARTIAL AND TOTAL ORDERS

Here are a few examples of orders, related to strings and to asymptotics:

Example 8.40 (Ordering strings)

Problem: Let Σ^* denote the set of all (finite-length) strings of letters. Which of the following relations on Σ^* are partial orders? Total orders? Which are strict?

1. $\langle x, y \rangle \in R$ if $|x| \geq |y|$. (The length of a string x—the number of letters in x—is denoted $|x|$.)
2. $\langle x, y \rangle \in S$ if x comes alphabetically no later than y. (See Example 3.46.)
3. $\langle x, y \rangle \in T$ if the number of As in x is less than the number of As in y.

Solution: 1. The relation $\{\langle x, y \rangle : |x| \geq |y|\}$ is reflexive and transitive, but it is not antisymmetric: for example, both $\langle$PASCAL, RASCAL$\rangle$ and $\langle$RASCAL, PASCAL$\rangle$ are in the relation, but RASCAL $\neq$ PASCAL. So this relation isn't a partial order.

2. The relation "comes alphabetically no later than" is reflexive (every word w comes alphabetically no later than w), antisymmetric (the only word that comes alphabetically no later than w *and* no earlier than w is w itself), and transitive (if w_1 is alphabetically no later than w_2 and w_2 is no later than w_3, then indeed w_1 is no later than w_3). Thus S is a partial order.

   In fact, any two words are comparable under S: either w is a prefix of w' (and $\langle w, w' \rangle \in S$) or there's a smallest index i in which $w_i \neq w'_i$ (and either $\langle w, w' \rangle \in S$ or $\langle w', w \rangle \in S$, depending on whether w_i is earlier or later in the alphabet than w'_i). Thus S is actually a total order.

3. The relation "contains fewer As than" is irreflexive (any word w contains exactly the same number of As as it contains, not *fewer* than that!) and transitive (if we have $a_w < a_{w'}$ and $a_{w'} < a_{w''}$, then we also have $a_w < a_{w''}$). Therefore the relation is antisymmetric (by Exercise 8.84), and thus T is a strict partial order.

   But neither $\langle$PASCAL, RASCAL$\rangle$ nor $\langle$RASCAL, PASCAL$\rangle$ are in T—both words contain 2 As, so neither has fewer than the other—and thus RASCAL and PASCAL are incomparable, and T is not a (strict) total order.

Example 8.41 (O and o as orders?)

We've argued that o is irreflexive (Example 8.24), transitive (Exercise 6.47), and asymmetric (Example 8.25). Thus o is a strict partial order. But o is *not* a (strict) total order:

we saw a function $f(n)$ in Example 6.6 such that $f(n) \neq o(n^2)$ *and* $n^2 \neq o(f(n))$, so these two functions are incomparable.

And, though we showed that O is reflexive and transitive (Exercise 6.18), we showed that O is *not* antisymmetric (Example 8.25), because, for example, the functions $f(n) = n^2$ and $g(n) = 2n^2$ are O of each other. Thus O is not a partial order.

Taking it further: A relation like O that is both reflexive and transitive (but not necessarily antisymmetric) is sometimes called a *preorder*. Although O is not a partial order, it very much has an "ordering-like" feel to it: it *does* rank functions by their growth rate, but there are clusters of functions that are all equivalent under O. We can think of O as defining *a partial order on the equivalence classes under* Θ. We saw another preorder in Example 8.40, with the relation R ("x and y have the same length"): although there are many pairs of nonidentical strings x and y where $\langle x, y \rangle, \langle y, x \rangle \in R$, it is only because of ties in lengths that R fails to be a partial order—indeed, a total order.

Hasse diagrams

Let R be any relation on A. For $k \geq 1$, we will call a sequence $\langle a_1, a_2, \ldots, a_k \rangle \in A^k$ a *cycle* if $\langle a_1, a_2 \rangle, \langle a_2, a_3 \rangle, \cdots, \langle a_{k-1}, a_k \rangle \in R$ and $\langle a_k, a_1 \rangle \in R$. A cycle is a sequence of elements, each of which is related by R to the next element in the sequence (where the last element is related to the first). For a partial order $\preceq$, there are cycles with $k = 1$ (because a partial order is reflexive, $a_1 \preceq a_1$ for any a_1), but there are no longer cycles. (You'll prove this fact in Exercise 8.130.)

Recall the "directed graph" visualization of a relation $R \subseteq A \times A$ that we introduced earlier (see Figure 8.13): we write down every element of A, and then, for every pair $\langle a_1, a_2 \rangle \in R$, we draw an arrow from a_1 to a_2. For a relation R that's a partial order, we'll introduce a simplified visualization, called a *Hasse diagram*, that allows us to figure out the full relation R but makes the diagram dramatically cleaner.

Hasse diagrams are named after Helmut Hasse, a 20th-century German mathematician.

Let $\preceq$ be a partial order. Consider three elements a, b, and c such that $a \preceq b$ and $b \preceq c$ and $a \preceq c$. Then *the very fact that* $\preceq$ *is a partial order* means that $a \preceq c$ can be inferred from the fact that $a \preceq b$ and $b \preceq c$. (That's just transitivity.) Thus we will omit from the diagram any arrows that can be inferred via transitivity. Similarly, we will leave out self-loops, which can be inferred from reflexivity. Finally, as we discussed above, there are no nontrivial cycles (that is, there are no cycles other than self-loops) in a partial order. Thus we will arrange the elements so that when $a \preceq b$ we will draw *a physically below b* in the diagram; all arrows will implicitly point upward in the diagram. Here are two examples:

Example 8.42 (A small Hasse diagram)
A Hasse diagram for the partial order

$$\{\langle 0,0\rangle, \langle 0,1\rangle, \langle 0,2\rangle, \langle 0,3\rangle, \langle 0,4\rangle, \langle 1,1\rangle, \langle 2,2\rangle, \langle 2,3\rangle, \langle 2,4\rangle, \langle 3,3\rangle, \langle 3,4\rangle, \langle 4,4\rangle\}$$

is shown in Figure 8.29. Note that we've omitted all arrow directions (they all point up), all five self-loops (they can be inferred from reflexivity), and the pairs $\langle 0,3\rangle$, $\langle 0,4\rangle$, and $\langle 2,4\rangle$ (they can be inferred from transitivity).

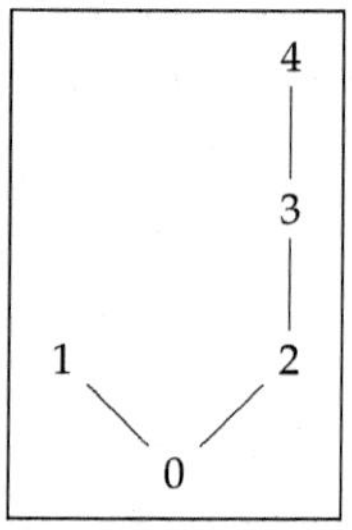

Figure 8.29: A small Hasse diagram.

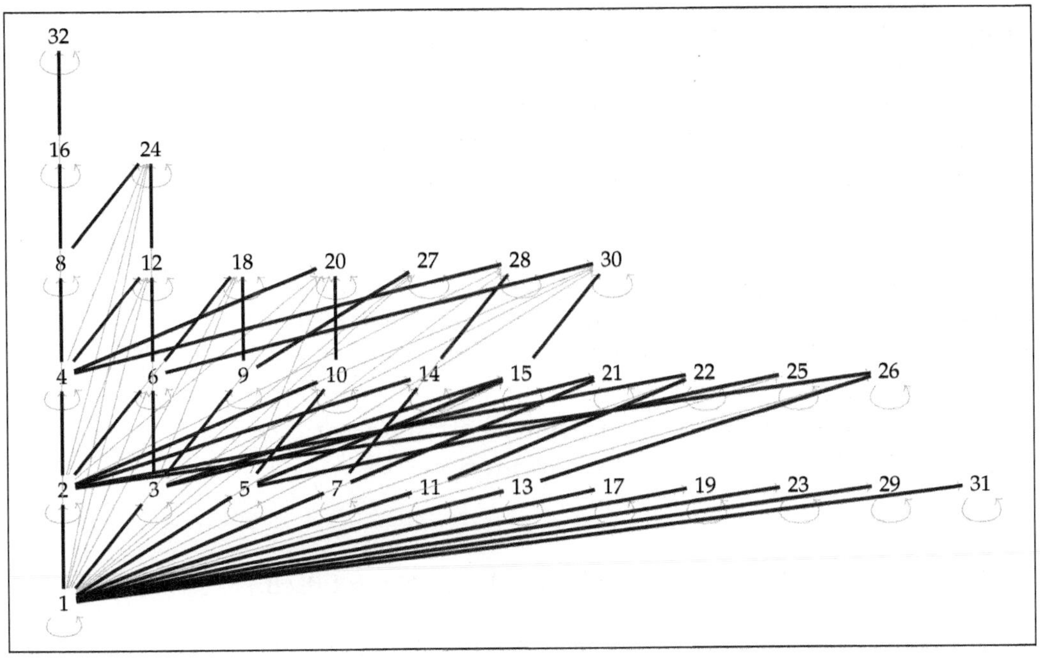

Figure 8.30: A Hasse diagram for "divides" on $\{1,2,\ldots,32\}$. The darker lines represent the Hasse diagram; the lighter arrows give the full picture of the relation, including all of the relationships that can be inferred from the fact that the relation is a partial order.

Example 8.43 (Hasse diagram for divides)
A Hasse diagram for the relation $\mid$ (divides) on the set $\{1,2,\ldots,32\}$ is shown in Figure 8.30. Again, the diagram omits arrow directions, self-loops, and "indirect" connections that can be inferred by transitivity. For example, the fact that $2 \mid 20$ is implicitly represented by the arrows $2 \to 4 \to 20$ (or $2 \to 10 \to 20$).

Which arrows must be shown in a Hasse diagram? Those arrows that cannot be inferred by the definition of a partial order—so we must draw a direct connections for all those relationships that are not "short circuits" of pairs of other relationships. In other words, we must draw lines for all those pairs $\langle a, c\rangle$ where $a \preceq c$ *and there is no* $b \notin \{a, c\}$ *such that* $a \preceq b$ *and* $b \preceq c$. Such a c is called an *immediate successor* of a.

Warning! When $a \preceq b$ holds for a partial order $\preceq$, we think of a as "smaller" than b under $\preceq$—a view that can be a little misleading if, for example, the partial order in question is $\geq$ instead of $\leq$. One example of this oddity: for $\geq$, the immediate successor of 42 is 41.

Minimal/maximal elements in a partial order

Consider the partial order $\preceq := \{\langle 1,1\rangle, \langle 1,2\rangle, \langle 1,3\rangle, \langle 1,4\rangle, \langle 2,2\rangle, \langle 2,4\rangle, \langle 3,3\rangle, \langle 4,4\rangle\}$—that is, the divides relation on the set $\{1,2,3,4\}$. There's a strong sense in which 1 is the "smallest" element under $\preceq$: *every* element a satisfies $1 \preceq a$. And there's a slightly weaker sense in which 3 and 4 are both "largest" elements under $\preceq$: *no* element a satisfies $3 \preceq a$ or $4 \preceq a$. These ideas inspire two related pairs of definitions:

Definition 8.17 (Minimum/maximum element)
For a partial order $\preceq$ *on* A:

- a minimum element *is* $x \in A$ *such that, for every* $y \in A$, *we have* $x \preceq y$.
- a maximum element *is* $x \in A$ *such that, for every* $y \in A$, *we have* $y \preceq x$.

Definition 8.18 (Minimal/maximal element)
For a partial order $\preceq$ on A:

- *a* minimal element *is $x \in A$ such that, for every $y \in A$ with $y \neq x$, we have $y \npreceq x$.*
- *a* maximal element *is $x \in A$ such that, for every $y \in A$ with $y \neq x$, we have $x \npreceq y$.*

A maxim*al* whatzit is any whatzit that loses its whatzitness if we add anything to it. A maxim*um* whatzit is the largest possible whatzit. If you've studied calculus, you've seen a similar distinction under a different name: *maximal* corresponds to a local maximum; *maximum* corresponds to a global maximum.

Note that x being a minimal element does *not* demand that every other element be larger than x—only that no element is smaller! (Again, we're talking about a *partial* order—so $x \npreceq y$ doesn't imply that $y \preceq x$.) In other words, a minimal element is one for which every other element y either satisfies $x \preceq y$ *or* is incomparable to x.

Example 8.44 (Minimal/maximal/maximum/minimum elements in "divides")
For the divides relation on $\{1, 2, \ldots, 32\}$ (Example 8.43 and Figure 8.30):

- 1 is a minimum element. (Every $n \in \{1, 2, \ldots, 32\}$ satisfies $1 \mid n$.)
- 1 is also a minimal element. (No $n \in \{1, 2, \ldots, 32\}$ satisfies $n \mid 1$, except $n = 1$ itself.)
- There is no maximum element. (No $n \in \{1, 2, \ldots, 32\}$ aside from 32 satisfies $n \mid 32$, so 32 is the only candidate—but $31 \nmid 32$.)
- There are a slew of maximal elements: each of $\{17, 18, \ldots, 32\}$ is a maximal element. (None of these elements divides any $n \in \{1, 2, \ldots, 32\}$ other than itself.)

(You'll prove that any minimum element is also minimal, and that there can be at most one minimum element in a partial order, in Exercises 8.143 and 8.144.)

We've already seen partial orders that don't have minimum or maximum elements, but every partial order must have at least one minimal element and at least one maximal element—at least, as long as the partial order is over a set A that's finite:

Theorem 8.3 (Every (finite) partial order has a minimal/maximal element)
Let $\preceq \subseteq A \times A$ be a partial order on a finite set A. Then $\preceq$ has at least one minimal element and at least one maximal element.

Proof. We'll prove that there's a minimal element; the proof for the maximal element is completely analogous. Our proof is constructive; we'll give an algorithm to *find* a minimal element. (See Figure 8.31.)

```
Input: a partial order ⪯ on a finite set A
Output: a ∈ A that's minimal under ⪯
i := 1
x_1 := an arbitrarily chosen element in A
while there exists any y ≠ x_i with y ⪯ x_i:
    x_{i+1} := any such y (with y ≠ x_i and y ⪯ x_i)
    i := i + 1
return x_i
```

Figure 8.31: An algorithm to find a minimal element.

It's easy to see that *if this algorithm terminates, then it returns a minimal element.* After all, the **while** loop only terminates if we've found an $x_i \in A$ such that there's no $y \neq x_i$ with $y \preceq x_i$—which is precisely the definition of x_i being a minimal element. Thus the real work is in proving that this algorithm actually terminates.

We claim that after $|A|$ iterations of the **while** loop—that is, after we've defined $x_1, x_2, \ldots, x_{|A|+1}$—we must have found a minimal element. Suppose not. Then we have found elements $x_1 \succeq x_2 \succeq \cdots \succeq x_{|A|+1}$, where $x_{i+1} \neq x_i$ for each i. Because there are only $|A|$ different elements in A, in a sequence of $|A| + 1$ elements we must have encountered the same element more than once. (This argument implicitly makes use of the *pigeonhole principle*, which we'll see in much greater detail in Chapter 9.) But that's a cycle containing two or more elements! And Exercise 8.130 asks you to show that there are no such cycles in a partial order. □

Problem-solving tip: A good visualization of data often makes an apparently complicated statement much simpler. Another way of stating Theorem 8.3 and its proof: start anywhere, and follow lines downward in the Hasse diagram; eventually, you must run out of elements below you, and you can't go any lower. Thus there's at least one bottommost element in any (finite) Hasse diagram.

Note that Theorem 8.3 only claimed that a minimal element must exist in a partial order *on a finite set A*. The claim would be false without that assumption! If A is an infinite set, then there may be no minimal element in A under a partial order. (See Exercise 8.141.)

We can identify minimal and maximal elements of a partial order very easily from the Hasse diagram: they're simply the elements that aren't connected to anything above them (the maximal elements), and those that aren't connected to anything below them (the minimal elements). And, indeed, there are always topmost element(s) and bottommost element(s) in a Hasse diagram, and thus there are always maximal/minimal elements in any partial order—if the set of elements is finite, at least!

8.4.3 *Topological Ordering*

Partial orders can be used to specify constraints on the order in which certain tasks must be completed. For example, the printer must be loaded with paper before the document can be printed; the document must be written before the document can be printed; the paper must be purchased before the printer can be loaded with paper. Or, as another example: a computer science major at a certain college in the midwest must take courses following the prerequisite structure specified in Figure 8.32.

But, while these types of constraints impose on a *partial* order on elements, the jobs must actually be completed in some sequence. (Likewise, the courses must be taken in some sequence—for a major who avoids "doubling up" on CS courses in the same term, at least.) The task we face here is to *extend* a partial order into a total order—that is, to create a total order that obeys all of the constraints of the partial order, while making comparable all previously incomparable pairs.

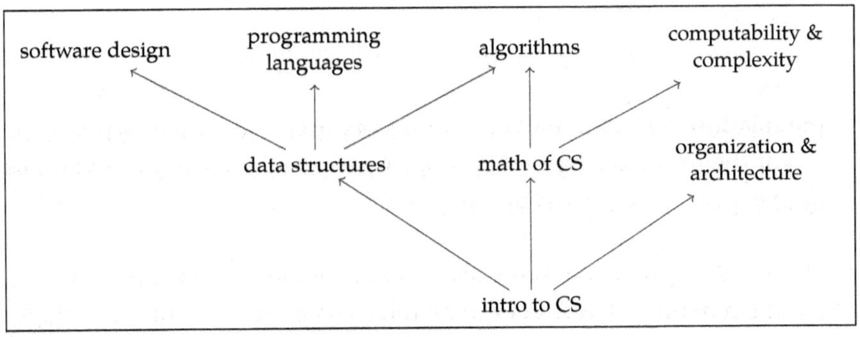

Figure 8.32: The CS major at a certain college in the midwest.

Definition 8.19 (Consistency of a total order with a partial order)
A total order $\preceq_{total}$ is consistent with the partial order $\preceq$ *if* $a \preceq b$ *implies that* $a \preceq_{total} b$.

In general, there are many total orders that are consistent with a given partial order. Here's an example:

Example 8.45 (Ordering CS classes)
The following course orderings are consistent with the prerequisites in Figure 8.32. (There are many other valid orderings, too.)

- intro to CS → data structures → math of CS → organization & architecture → software design → programming languages → algorithms → computability & complexity.
- intro to CS → data structures → software design → programming languages → math of CS → algorithms → computability & complexity → organization & architecture.

The first of these orderings corresponds to reading the elements of the Hasse diagram from the bottom-to-top (and left-to-right within a "row"); the second corresponds to completing the top row left-to-right (first recursively completing the requirements to make the next element of the top row valid).

As in these examples, we can construct a total order that's consistent with any given partial order on the set A. Such an ordering of A is called a *topological ordering* of A. (Some people will refer to a topological ordering as a *topological sort* of A.) We'll prove this result inductively, by repeatedly identifying a minimal element a from the set of unprocessed elements, and then adding constraints to make a be a minim*um* element (and not just a minim*al* element).

Theorem 8.4 (Extending any partial order to a total order)
Let A be any finite set with a partial order $\preceq$. Then there is a total order $\preceq_{total}$ on A that's consistent with $\preceq$.

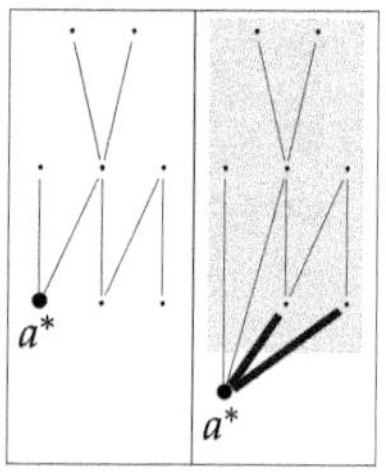

Figure 8.33: A sketch of the proof of Theorem 8.4. First, we identify some minim*al* element a^* in $\preceq$ (left panel). Then we turn a^* into a minim*um* element by adding constraints (thick lines in the right panel), and then we inductively find a total ordering of the remaining partial order (the shaded box at right).

Proof. We'll proceed by induction on $|A|$.

For the base case ($|A| = 1$), the task is trivial: there's simply nothing to do! The relation $\preceq$ must be $\{\langle a,a\rangle\}$, where $A = \{a\}$, because partial orders are reflexive. And the relation $\{\langle a,a\rangle\}$ *is* a total order on $\{a\}$ that's consistent with $\preceq$.

For the inductive case ($|A| \geq 2$), we assume the inductive hypothesis (for any set A' of size $|A'| = |A| - 1$ and any partial order on A', there's a total order on A' consistent with that partial order). We must show how to extend $\preceq$ to be a total order on all of A. Here's the idea: we'll remove some element of A that can go first in the total order, inductively find a total order of all the remaining elements, and then add the removed element to the beginning of the order.

More specifically, let $a^* \in A$ be an arbitrary minimal element under $\preceq$ on A—in other words, let a^* be any element such that no $b \in A - \{a^*\}$ satisfies $b \preceq a^*$. Such an element is guaranteed to exist by Theorem 8.3. Add any missing pair $\langle a^*, b\rangle$ to $\preceq$. It's easy to see that $\preceq$ is still a partial order on A: by the definition of a minimal element, we haven't introduced any violations of transitivity or antisymmetry. Now, inductively, we extend the partial order $\preceq$ on $A - \{a^*\}$ to a total order; the result is a total order on A that's consistent with $\preceq$. (See Figure 8.33.)

(Slightly more formally: note that $\preceq' := \{\langle x,y\rangle \in (A - \{a^*\}) \times (A - \{a^*\}) : x \preceq y\}$ is a partial order on $A - \{a^*\}$; by the inductive hypothesis, there exists a total order $\preceq'_{\text{total}}$

on $A - \{a^*\}$ consistent with $\preceq'$. Define

$$\preceq_{\text{total}} = \left\{ \langle x, y \rangle \in A \times A : \langle x, y \rangle \in \preceq'_{\text{total}} \text{ or } x = a^* \right\}.$$

It's easy to verify that $\preceq_{\text{total}}$ is a total order on A that's consistent with $\preceq$.) □

Taking it further: Deciding the order in which to compute the cells of a spreadsheet (where a cell might depend on a list of other cells' contents) is solved using a topological ordering. In this setting, let C denote the set of cells in the spreadsheet, and define a relation $R \subseteq C \times C$ where $\langle c, c' \rangle \in R$ if we need to know the value in cell c before we can compute the value for c'. (For example, if cell `C4`'s value is determined by the formula `A1 + B1 + C1`, then the three pairs $\langle$`A1, C4`$\rangle$, $\langle$`B1, C4`$\rangle$, and $\langle$`C1, C4`$\rangle$ are all in R. Note that it's not possible to compute all the values in a spreadsheet if there's a cell x whose value depends on cell y, which depends on $\cdots$, which depends on cell x—in other words, the "depends on" relationship cannot have a cycle! Furthermore, we're in trouble if there's a cell x whose value depends on x itself. In other words, we can compute the values in a spreadsheet if and only if R is irreflexive and transitive—that is, if R is a strict partial order.

Another problem that can be solved using the idea of topological ordering is that of *hidden-surface removal* in computer graphics: we have a 3-dimensional "scene" of objects that we'd like to display on a 2-dimensional screen, as if it were being viewed from a camera. We need to figure out which of the objects are invisible from the camera (and therefore need not be drawn) because they're "behind" other objects. One classic algorithm, called the *painter's algorithm*, solves this problem using ideas from relations and topological ordering. See the discussion on p. 847.

Computer Science Connections

Deterministic Finite Automata (DFAs)

As we hinted at previously (see the discussion of regular expressions on p. 830), there are some interesting computational applications of *finite-state machines*, a formal model for a computational device that uses a fixed (finite) amount of memory to respond to input. Variations on these machines can be used in building very simple characters in a video game, in computer architecture, in software systems to do automatic speech recognition, and other tasks. They can also identify which strings match a given regular expression—in fact, for a set of strings L, it's a theorem that there exists a finite-state machine M that recognizes precisely the strings in L if and only if there's a regular expression α that matches precisely the strings in L.

Formally, a *deterministic finite automaton (DFA)*—the simplest version of a finite-state machine—is a quintuple $M = \langle \Sigma, Q, \delta, s, F \rangle$, where:

- Σ is a finite *alphabet*, the set of input symbols the machine can handle;
- Q is a finite set of *states*; the machine is always in one of these states. (The fact that Q is finite corresponds to M having only finite memory.)
- $\delta : Q \times \Sigma \to Q$ is a *transition function:* when the machine is in state $q \in Q$ and sees an input symbol $a \in \Sigma$, the machine moves into state $\delta(q, a)$.
- $s \in Q$ is the *start state*, where M begins before having seen any input.
- $F \subseteq Q$ is the set of *final states*. If, after processing a string x, M ends up in a state $q \in F$, then M *accepts* x; if M ends in a state $q \notin F$, then M *rejects* x.

An example of a DFA that accepts all bitstrings whose first two symbols are the same is shown in Figure 8.34.

- $\Sigma = \{0, 1\}$
- $Q = \{a, b, c, win, lose\}$
- δ is defined by the following table:

	0	1
a	b	c
b	win	$lose$
c	$lose$	win
win	win	win
$lose$	$lose$	$lose$

- the start state is a.
- the only final state is win.

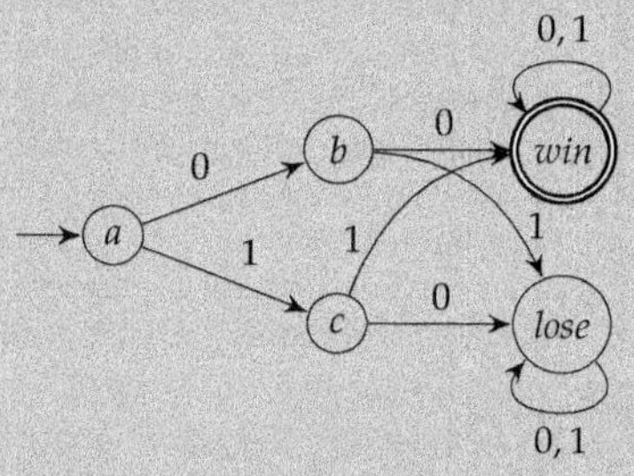

Figure 8.34: A DFA accepting all bitstrings whose first two symbols are the same—both by defining all five components, and by a picture. The *start state* is marked with an unattached incoming arrow; from state q on input symbol a, the arrow leaving q with label a points to $\delta(q, a)$. Final states are circled.

We can also understand DFAs—and the sorts of sets of strings that they can recognize—by thinking about equivalence relations. To see this connection, suppose that we wish to identify binary strings representing integers that are evenly divisible by 3. (So 11 and 1001 and 1111 are all "yes" because $3 \mid 3$ and $3 \mid 9$ and $3 \mid 15$, but 10001 is "no" because $3 \nmid 17$.)

Here's one way to solve this problem. Let's define an equivalence relation on binary strings, where $x \equiv y$ if and only if, for any bitstring z, we have that (xz is divisible by 3) $\Leftrightarrow$ (yz is divisible by 3). In other words, two bitstrings x and y are equivalent if, no matter what additional bitstring suffix we add to both of them, the two resulting bitstrings are either both divisible by three or both not divisible by three. For example, it turns out that $11 \equiv 1001$ (11 and 1001 are both 'yes'; 110 and 10010 are both 'yes'; 111 and 10011 are both 'no'; 1110 and 100010 are both 'no'; etc.). Similarly, we have $1000 \equiv 10$. It's not hard to prove that $\equiv$ is an equivalence relation. It's also true, though a bit harder to prove, that there are only three equivalence classes for $\equiv$. (Those equivalence classes are: bitstrings that are 0 mod 3, those that are 1 mod 3, and those that are 2 mod 3.) Thus we can actually figure out whether a bitstring is evenly divisible by 3 with the simple DFA in Figure 8.35. The three states of this machine, going from left to right, correspond to the three equivalence classes for $\equiv$—namely [0], [1], and [10]. (For a set of strings that cannot be recognized by a DFA—for example, bitstrings with an equal number of 0s and 1s—there are an infinite number of equivalence classes for $\equiv$.)[7]

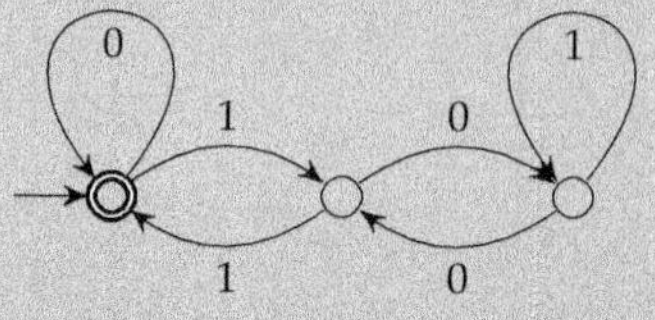

Figure 8.35: A DFA for bitstrings representing numbers divisible by 3. The input is divisible by three if and only if we end up in the leftmost state.

These particular DFAs merely hint at the kind of problem that can be solved with this kind of machine—for much more, see a good textbook in formal languages, such as

[7] Dexter Kozen. *Automata and Computability*. Springer, 1997; and Michael Sipser. *Introduction to the Theory of Computation*. Course Technology, 3rd edition, 2012.

Computer Science Connections

The Painter's Algorithm and Hidden-Surface Removal

At a high level, the goal in computer graphics is to take a 3-dimensional scene—a set of objects in $\mathbb{R}^3$ (with differing shapes, colors, surface reflectivities, textures, etc.)—as seen from a particular vantage point (a point and a direction, also in $\mathbb{R}^3$). The task is then to *project* the scene into a 2-dimensional image. There are a lot of components to this task, and we've already talked a bit about some of them: typically we'll approximate the shapes of the objects using a large collection of triangles (see p. 528), and then compute where each triangle shows up in the camera's view, in $\mathbb{R}^2$, via rotation (see p. 249).

Figure 8.36: A house in a golden wood.

Even after triangulation and rotation, we are still left with another important step: when two triangles overlap in the 2-dimensional image, we have to figure out which to draw—that is, which one is obscured by the other. This task is also known *hidden-surface removal:* we want to omit whatever pieces of the image aren't visible. For example, when we wish to render the humble forest scene in Figure 8.36, we have to draw trees in front of and behind the house, and one particular tree in front of another. One approach to hidden-surface removal is called the *Painter's Algorithm,* named after a hypothetical artist at an easel: we can "paint" the shapes in the image "from back to front," simply painting over faraway shapes with the closer ones as we go:

How might we implement this approach? Let S be the set of shapes that we have to draw. We can compute a relation $obscures \subseteq S \times S$, where a pair $\langle s_1, s_2 \rangle \in obscures$ tells us that we have to draw s_2 before we draw s_1. We seek a total order on S that is consistent with the *obscures* relation; we'll draw the shapes in this order.

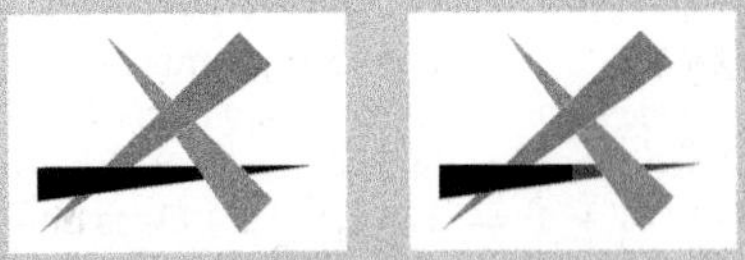

Figure 8.37: A cycle of obscurity, and splitting one of the cycle's pieces to break the cycle.

Unfortunately *obscures* isn't a total order—or even a partial order! The biggest problem with *obscures* is that we can have "cycles of obscurity"—s_1 obscures s_2 which obscures s_3 which, eventually, obscures a shape s_k that obscures s_1. (See Figure 8.37; although it may look like an M. C. Escher drawing, there's nothing strange going on—just three triangles that overlap a bit like a pretzel.) This issue can be resolved using some geometric algorithms specific to the particular task: we'll *split up* shapes in each cycle of obscurity—splitting the black triangle into a left-half and a right-half object, for example—so that we no longer have any cycles. (Again see Figure 8.37.)

We now have an expanded set S' of shapes, and a cycle-free relation *obscures* on S'. We can use this relation to compute the order in which to draw the shapes, as follows:

- compute the reflexive, transitive closure of *obscures* on S'. The resulting relation is a partial order on S'.
- extend this partial order to a total order on S', using Theorem 8.4.

We now have a total ordering on the shapes that respect the *obscures* relation, so we can draw the shapes in precisely this order.[8]

While the Painter's Algorithm does correctly accomplish hidden-surface removal, it's pretty slow (particularly as we've described it here). For example, when there are many layers to a scene, we actually have to "paint" each pixel in the resulting image many many times. Every computation of a pixel's color before the last is a waste of time. You can learn about cleverer approaches to hidden-surface removal, like the "z-buffer," in a good textbook on computer graphics, such as

[8] John F. Hughes, Andries van Dam, Morgan McGuire, David F. Sklar, James D. Foley, Steven K. Feiner, and Kurt Akeley. *Computer Graphics: Principles and Practice.* Addison-Wesley, 3rd edition, 2013.

8.4.4 Exercises

List all equivalence relations . . .

8.108 . . . on $\{0, 1\}$.

8.109 . . . on $\{0, 1, 2, 3\}$.

Are the following relations on $\mathscr{P}(\{0, 1, 2, 3\})$ equivalence relations? If so, list the equivalence classes under the relation; if not, explain why not.

8.110 $\langle A, B\rangle \in R_1$ if and only if (i) A and B are nonempty and the largest element in A equals the largest element in B, or (ii) if $A = B = \varnothing$.

8.111 $\langle A, B\rangle \in R_2$ if and only if the sum of the elements in A equals the sum of the elements in B.

8.112 $\langle A, B\rangle \in R_3$ if and only if the sum of the elements in A equals the sum of the elements in B and the largest element in A equals the largest element in B. (That is, $R_3 = R_1 \cap R_2$.)

8.113 $\langle A, B\rangle \in R_4$ if and only $A \cap B \neq \varnothing$.

8.114 $\langle A, B\rangle \in R_5$ if and only $|A| = |B|$.

In Example 8.11, we considered the relation $M := \{\langle m, d\rangle : \text{in some years, month } m \text{ has } d \text{ days}\}$, and computed the pairs in the relation $M^{-1} \circ M$. By checking all the requirements (or by visual inspection of Figure 8.13(b)), we see that $M^{-1} \circ M$ is an equivalence relation. But it turns out that the fact that $M^{-1} \circ M$ is an equivalence relation says something particular about M, and is not true in general. Let $R \subseteq A \times B$ be an arbitrary relation. Prove or disprove whether $R^{-1} \circ R$ must have the three required properties of an equivalence relation (at least one of these is false!):

8.115 Prove or disprove: $R^{-1} \circ R$ must be reflexive.

8.116 Prove or disprove: $R^{-1} \circ R$ must be symmetric.

8.117 Prove or disprove: $R^{-1} \circ R$ must be transitive.

Let A be any set. There exist two equivalence relations $\equiv_{coarsest}$ and $\equiv_{finest}$ with the following property: if $\equiv$ is an equivalence relation on A, then (i) $\equiv$ refines $\equiv_{coarsest}$, and (ii) $\equiv_{finest}$ refines $\equiv$.

8.118 Identify $\equiv_{\text{coarsest}}$, prove that it's an equivalence relation, and prove property (i) above.

8.119 Identify $\equiv_{\text{finest}}$, prove that it's an equivalence relation, and prove property (ii) above.

8.120 In many programming languages, there are two distinct but related notions of "equality": *has the same value as* and *is the same object as*. In Python, these are denoted as `==` and `is`, respectively; in Java, they are `.equals()` and `==`, respectively. (Confusingly!) (For example, in Python, `1776 + 1 is 1777` is false, but `1776 + 1 == 1777` is true.) Does one of these equality relations refine the other? Explain.

8.121 List all partial orders on $\{0, 1\}$.

8.122 List all partial orders on $\{0, 1, 2\}$.

Are the following relations on $\mathscr{P}(\{0, 1, 2, 3\})$ partial orders, strict partial orders, or neither? Explain.

8.123 $\langle A, B\rangle \in R_1 \Leftrightarrow \sum_{a \in A} a \leq \sum_{b \in B} b$

8.124 $\langle A, B\rangle \in R_2 \Leftrightarrow \prod_{a \in A} a \leq \prod_{b \in B} b$

8.125 $\langle A, B\rangle \in R_3 \Leftrightarrow A \subseteq B$

8.126 $\langle A, B\rangle \in R_4 \Leftrightarrow A \supseteq B$

8.127 $\langle A, B\rangle \in R_5 \Leftrightarrow |A| < |B|$

8.128 Prove that $\preceq$ is a partial order if and only if $\preceq^{-1}$ is a partial order.

8.129 Prove that if $\preceq$ is a partial order, then $\{\langle a, b\rangle : a \preceq b \text{ and } a \neq b\}$ is a strict partial order.

8.130 A *cycle* in a relation R is a sequence of k distinct elements $a_0, a_1, \ldots, a_{k-1} \in A$ where $\langle a_i, a_{i+1 \bmod k}\rangle \in R$ for each $i \in \{0, 1, \ldots, k-1\}$. A cycle is *nontrivial* if $k \geq 2$. Prove that there are no nontrivial cycles in any transitive, antisymmetric relation R. (Hint: use induction on the length k of the cycle.)

Let $S \in \mathbb{Z}^{\geq 1} \times \mathbb{Z}^{\geq 1}$ be a collection of points. Define the relation $R \subseteq S \times S$ as follows: $\langle\langle a, b\rangle, \langle x, y\rangle\rangle \in R$ if and only if $a \leq x$ and $b \leq y$. (You can think of $\langle a, b\rangle \in S$ as an a-by-b picture frame, and $\langle f, f'\rangle \in R$ if and only if f fits inside f'. Or you can think of $\langle a, b\rangle \in S$ as a job that you'd get a "happiness points" from doing and that pays you b dollars, and $\langle j, j'\rangle \in R$ if and only if j generates no more happiness and pays no more than j'.

8.131 Show that R might not be a total order by identifying two incomparable elements of $\mathbb{Z}^{\geq 1} \times \mathbb{Z}^{\geq 1}$.

8.132 Prove that R must be a partial order.

8.133 Write out all pairs in the relation represented by the Hasse diagram in Figure 8.38(a).

8.134 Repeat for Figure 8.38(b).

8.135 Draw the Hasse diagram for the partial order $\subseteq$ on the set $\mathscr{P}(1, 2, 3)$.

8.136 Draw the Hasse diagram for the partial order $\preceq$ on the set $S := \{0, 1\} \cup \{0, 1\}^2 \cup \{0, 1\}^3$, where, for two bitstrings $x, y \in S$, we have $x \preceq y$ if and only if x is a prefix of y.

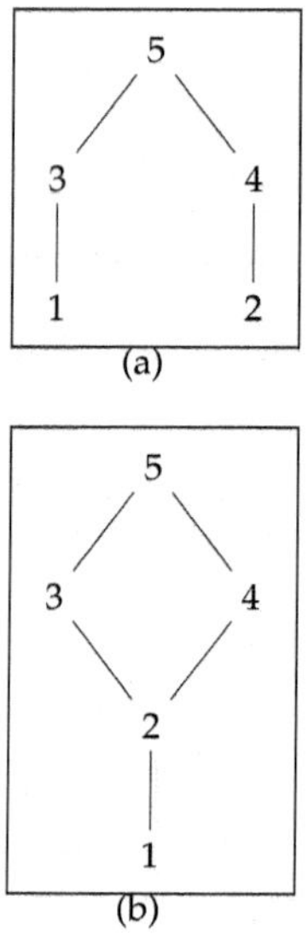

Figure 8.38: Some Hasse diagrams.

Let $\preceq$ be a partial order on A. Recall that an immediate successor *of $a \in A$ is an element c such that (i) $a \preceq c$, and (ii) there is no $b \notin \{a, c\}$ such that $a \preceq b$ and $b \preceq c$. In this case a is said to be an* immediate predecessor *of c.*

8.137 For the partial order $\geq$ on $\mathbb{Z}^{\geq 1}$, identify all the immediate predecessor(s) and immediate successor(s) of 202.

8.138 For the partial order $|$ (divides) on $\mathbb{Z}^{\geq 1}$, identify all the immediate predecessor(s) and immediate successor(s) of 202.

8.139 Give an example of a strict partial order on $\mathbb{Z}^{\geq 1}$ such that *every* integer has exactly two different immediate successors.

8.140 Prove that for a partial order $\preceq$ on A *when A is finite* there must be an $a \in A$ that has fewer than two immediate successors.

8.141 Consider the partial order $\geq$ on the set $\mathbb{Z}^{\geq 0}$. Argue that there is *no* maximal element in $\mathbb{Z}$.

8.142 Note that there *is* a minimal element under the partial order $\geq$ on $\mathbb{Z}^{\geq 0}$—namely 0, which is also the minimum element. Give an example of a partial order on an infinite set that has *neither* a minimal *nor* a maximal element.

8.143 Let $\preceq$ be a partial order on a set A. Prove that there is at most one minimum element in A under $\preceq$. (That is, prove that if $a \in A$ and $b \in A$ are both minimum elements, then $a = b$.)

8.144 Let $\preceq$ be a partial order on a set A, and let $a \in A$ be a minimum element under $\preceq$. Prove that a is also a minim*al* element.

Here's a (surprisingly addictive) word game that can be played with a set of Scrabble tiles. Each player has a set of words that she "owns"; there is also a set of individual tiles in the middle of the table. At any moment, a player can form a new word by taking both (1) one or more tiles from the middle, and (2) zero or more words owned by any of the players; and reordering those letters to form a new word, which the player now owns. For example, from the word `GRAMPS` *and the letters* `R` *and* `O`*, a player could make the word* `PROGRAMS`*.*

Define a relation $\preceq$ on the set W of English words (of three or more letters), as follows: $w \preceq w'$ if w' can be formed from word w plus one or more individual letters. For example, we showed above that `GRAMPS` $\preceq$ `PROGRAMS`*.*

8.145 Give a description (in English) of what it means for a word w to be a minimal element under $\preceq$, and what it means for a word w' to be a maximal element under $\preceq$.

8.146 *(programming required)* Write a program that, given a word w, finds all immediate successors of w. (You can find a dictionary of English words on the web, or `/usr/share/dict/words` on Unix-based operating systems.) Report all immediate successors of `GRAMPS` using your dictionary.

8.147 *(programming required)* Write a program to find the English word that is the *longest* minimal element under $\preceq$ (that is, out of all minimal elements, find the one that contains the most letters).

(If you're bored and decide to waste time playing this game: it's more fun if you forbid stealing words with "trivial" changes, like changing `COMPUTER` *into* `COMPUTERS`*. Each player should also get a fair share of the tiles, originally face down; anyone can flip a new tile into the middle of the table at any time.)*

8.148 Consider a spreadsheet containing a set of cells C. A cell c can contain a *formula* that depends on zero or more other cells. Write $\preceq$ to denote the relation $\{\langle p, s \rangle : \text{cell } s \text{ depends on cell } p\}$. For example, the value of cell `C2` might be the result of the formula `A2 * B1`; here `A2` $\preceq$ `C2` and `B1` $\preceq$ `C2`. A spreadsheet is only meaningful if $\preceq$ is a strict partial order. Give a description (in English) of what it means for a cell c to be a minimal element under $\preceq$, and what it means for a cell c' to be a maximal element under $\preceq$.

8.149 List all total orders consistent with the partial order reproduced in Figure 8.39(a).

8.150 Repeat for the partial order reproduced in Figure 8.39(b).

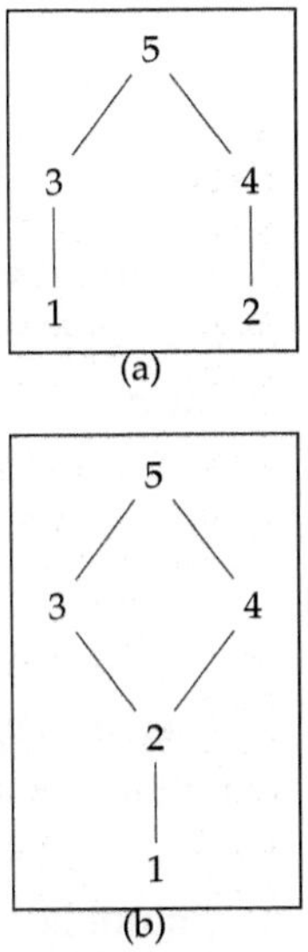

Figure 8.39: Reproductions of the Hasse diagrams from Figure 8.38.

A chain *in a partial order $\preceq$ on A is a set $C \subseteq A$ such that $\preceq$ imposes a total order on C—that is, writing the elements of C as $C = \{c_1, c_2, \ldots, c_k\}$ [in an appropriate order], we have $c_1 \preceq c_2 \preceq \cdots \preceq c_k$.*

8.151 Identify all chains of $k \geq 2$ elements in the partial order in Figure 8.39(a).

8.152 Repeat for the partial order reproduced in Figure 8.39(b).

An antichain *in a partial order $\preceq$ on A is a set $S \subseteq A$ such that no two distinct elements in S are comparable under $\preceq$—that is, for any distinct $a, b \in S$ we have $a \not\preceq b$.*

8.153 Identify all antichains S with $|S| \geq 2$ in the partial order in Figure 8.39(a).

8.154 Repeat for the partial order reproduced in Figure 8.39(b).

8.155 Consider the set $A := \{1, 2, \ldots, n\}$. Consider the following claim: *there exists a relation $\preceq$ on the set A that is* both *an equivalence relation* and *a partial order.* Either prove that the claim is true (and describe, as precisely as possible, the structure of any such relation $\preceq$) or disprove the claim.

8.5 *Chapter at a Glance*

Formal Introduction

A *(binary) relation on $A \times B$* is a subset of $A \times B$. For a relation R on $A \times B$, we can write $\langle a,b \rangle \in R$ or $a\,R\,b$. When A and B are both finite, we can describe R using a two-column table, where a row containing a and b corresponds to $\langle a,b \rangle \in R$. Or we can view R graphically: draw all elements of A in one column, all elements of B in a second column, and draw a line connecting $a \in A$ to $b \in B$ whenever $\langle a,b \rangle \in R$.

We'll frequently be interested in a relation that's a subset of $A \times A$, where the two sets are the same. In this case, we may refer to a subset of $A \times A$ as simply a *relation on A*. For a relation $R \subseteq A \times A$, it's more convenient to visualize R using a *directed graph*, without separated columns: we simply draw each element of A, with an arrow from a_1 to a_2 whenever $\langle a_1, a_2 \rangle \in R$.

The *inverse* of a relation $R \subseteq A \times B$ is a new relation, denoted R^{-1}, that "flips around" every pair in R: the relation $R^{-1} := \{\langle b,a \rangle : \langle a,b \rangle \in R\}$ is a subset of $B \times A$. The *composition* of two relations $R \subseteq A \times B$ and $S \subseteq B \times C$ is a new relation, denoted $S \circ R$, that, informally, represents the successive "application" of R and S. A pair $\langle a,c \rangle$ is related under $S \circ R \subseteq A \times C$ if and only if there exists an element $b \in B$ such that $\langle a,b \rangle \in R$ and $\langle b,c \rangle \in S$.

For sets A and B, a *function f from A to B*, written $f : A \to B$, is a special kind of relation on $A \times B$ where, for every $a \in A$, there exists one and only one element $b \in B$ such that $\langle a,b \rangle \in f$.

An *n-ary relation* is a generalization of a binary relation ($n = 2$) to describe a relationship among n-tuples, rather than just pairs. An n-ary relation on the set $A_1 \times A_2 \times \cdots \times A_n$ is just a subset of $A_1 \times A_2 \times \cdots \times A_n$; an n-ary relation on a set A is a subset of A^n.

Properties of Relations: Reflexivity, Symmetry, and Transitivity

A relation R on A is *reflexive* if, for every $a \in A$, we have that $\langle a,a \rangle \in R$. It's *irreflexive* if $\langle a,a \rangle \notin R$ for every $a \in A$. (In the visualization described above, where we draw an arrow $a_1 \to a_2$ whenever $\langle a_1, a_2 \rangle \in R$, reflexivity corresponds to every element having a "self-loop" and irreflexivity corresponds to no self-loops.) Note that a relation might be *neither* reflexive nor irreflexive.

A relation R on A is *symmetric* if, for every $a, b \in A$, we have $\langle a,b \rangle \in R$ if and only if $\langle b,a \rangle \in R$. The relation is *antisymmetric* if the only time both $\langle a,b \rangle \in R$ and $\langle b,a \rangle \in R$ is when $a = b$, and it's *asymmetric* if it's never the case that $\langle a,b \rangle \in R$ and $\langle b,a \rangle \in R$ whether $a \neq b$ or $a = b$. Note that, while asymmetry implies antisymmetry, they are different properties—and they're both different from "not symmetric"; a relation might not be symmetric, antisymmetric, *or* asymmetric. (In the visualization, a relation is symmetric if every arrow $a \to b$ is matched by an arrow $b \to a$; it's antisymmetric if there are no matched bidirectional pairs of arrows between a and $b \neq a$; and it's asymmetric if it's antisymmetric and furthermore there aren't even any self-loops.) An alternative view is that a relation R is symmetric if and only if $R \cap R^{-1} = R = R^{-1}$; it's antisymmetric if and only if $R \cap R^{-1} \subseteq \{\langle a,a\rangle : a \in A\}$; and it's asymmetric if and only if $R \cap R^{-1} = \varnothing$.

A relation R on A is *transitive* if, for every $a, b, c \in A$, if $\langle a,b\rangle \in R$ and $\langle b,c\rangle \in R$, then $\langle a,c\rangle \in R$ too. In the visualization, R is transitive if there are no "open triangles": in a chain of connected elements, every element is also connected to all "downstream" connections. The relation R is transitive if and only if $R \circ R \subseteq R$.

For a relation $R \subseteq A \times A$, the *closure* of R with respect to some property is the smallest relation $R' \supseteq R$ that has the named property. For example, the *symmetric closure of* R is the smallest relation $R'' \supseteq R$ such that R'' is symmetric. We also define the *reflexive closure* R'; the *transitive closure* R^+; the *reflexive transitive closure* R^*; and the *reflexive symmetric transitive closure* $R^{\equiv}$. When A is finite, we can compute any of these closures by repeatedly adding any missing elements to the set. The reflexive closure of R is given by $R \cup \{\langle a,a\rangle : a \in A\}$; the symmetric closure of R is $R \cup R^{-1}$; and the transitive closure of R is $R \cup R^2 \cup R^3 \cup \cdots$.

Special Relations: Equivalence Relations and Partial/Total Orders

There are two special kinds of relations that emerge from particular combinations of these properties: *equivalence relations* and *partial/total orders*.

Equivalence relations: An *equivalence relation* is a relation $\equiv$ that's reflexive, symmetric, and transitive. Such a relation partitions the elements of A into one or more categories, called *equivalence classes;* any two elements in the same equivalence class are related by $\equiv$, and no two elements in different equivalence classes are related.

A *refinement* of $\equiv$ is another equivalence relation $\equiv_r$ on the same set A where $a \equiv b$ whenever $a \equiv_r b$. Each equivalence class of $\equiv$ is partitioned into one or more equivalence classes by $\equiv_r$, but no equivalence class of $\equiv_r$ intersects with more than one equivalence class of $\equiv$. We also call $\equiv$ a *coarsening* of $\equiv_r$.

Partial and total orders: A *partial order* is a reflexive, antisymmetric, and transitive relation $\preceq$. (A *strict partial order* $\prec$ is *irreflexive*, antisymmetric, and transitive.) Elements a and b are *comparable under* $\preceq$ if either $a \preceq b$ or $b \preceq a$; otherwise they're *incomparable*. A *Hasse diagram* is a simplified visual representation of a partial order where we draw a physically below c whenever $a \preceq c$, and we omit the $a \to c$ arrow if there's some other element b such that $a \preceq b \preceq c$. (We also omit self-loops.)

For a partial order $\preceq$ on A, a *minimum element* is an element $a \in A$ such that, for every $b \in A$, we have $a \preceq b$; a *minimal element* is an $a \in A$ such that, for every $b \in A$ with $b \neq a$, we have $b \not\preceq a$. (*Maximum* and *maximal elements* are defined analogously.) Every minimum element is also minimal, but a minimal element a isn't minimum unless a is comparable with every other element. There's at least one minimal element in any partial order on a finite set.

A *total order* is a partial order under which all pairs of elements are comparable. A total order $\preceq_{\text{total}}$ is *consistent with the partial order* $\preceq$ if $a \preceq b$ implies that $a \preceq_{\text{total}} b$. For any partial order $\preceq$ on a finite set A, there is a total order $\preceq_{\text{total}}$ on A that's consistent with $\preceq$. Such an ordering of A is called a *topological ordering* of A.

Key Terms and Results

Key Terms

FORMAL INTRODUCTION

- (binary) relation
- inverse (of a relation)
- composition (of two relations)
- functions (as relations)
- n-ary relation

PROPERTIES OF RELATIONS

- reflexivity
- irreflexivity
- symmetry
- asymmetry
- antisymmetry
- transitivity
- closures (of a relation)

SPECIAL RELATIONS

- equivalence relation
- equivalence class
- coarsening, refinement
- partial order
- comparability
- total order
- Hasse diagram
- minimal/maximal element
- minimum/maximum element
- consistency (of a total order with a partial order)
- topological ordering

Key Results

FORMAL INTRODUCTION

1. For relations $R \subseteq A \times B$ and $S \subseteq B \times C$, the relations $R^{-1} \subseteq B \times A$ and $S \circ R \subseteq A \times C$—the inverse of R and the composition of R and S—are defined as

$$R^{-1} := \{\langle b,a\rangle : \langle a,b\rangle \in R\}$$
$$S \circ R := \{\langle a,c\rangle : \exists b \in B \text{ such that } \langle a,b\rangle \in R \text{ and } \langle b,c\rangle \in S\}.$$

2. A function $f : A \to B$ is a special case of a relation on $A \times B$, where, for every $a \in A$, there exists one and only one element $b \in B$ such that $\langle a,b\rangle \in f$.

PROPERTIES OF RELATIONS

1. A relation R is symmetric if and only if $R \cap R^{-1} = R = R^{-1}$; it's antisymmetric if and only if $R \cap R^{-1} \subseteq \{\langle a,a\rangle : a \in A\}$; and it's asymmetric if and only if $R \cap R^{-1} = \varnothing$.
2. A relation R is transitive if and only if $R \circ R \subseteq R$.
3. The reflexive closure of R is $R \cup \{\langle a,a\rangle : a \in A\}$; the symmetric closure of R is $R \cup R^{-1}$; and the transitive closure of R is $R \cup R^2 \cup R^3 \cup \cdots$.

SPECIAL RELATIONS

1. For a partial order $\preceq \subseteq A \times A$ on a *finite* set A, there is at least one minimal element and at least one maximal element under $\preceq$.
2. Let A be any finite set with a partial order $\preceq$. Then there is a total order $\preceq_{\text{total}}$ (a *topological ordering* of A) on A that's consistent with $\preceq$.

9
Counting

In which our heroes encounter many choices, some of which may lead them to live more happily than others, and a precise count of their number of options is calculated.

9.1 *Why You Might Care*

> How do I love thee? Let me count the ways.
>
> Elizabeth Barrett Browning (1806–1861)

This chapter is devoted to the apparently trivial task of *counting*. By "counting," we mean the following problem: given a potentially convoluted description of a set S, compute the cardinality of S—that is, compute the number of elements in S. It may seem bizarre that counting could somehow be harder than at the preschool level (just count! *one, two, three*), but it will turn out that we can solve surprisingly subtle problems with some useful and general (and subtle) techniques.

We'll start in Section 9.2 by introducing basic counting techniques—how to compute the cardinality of a union $A \cup B$ of two sets, or sequences from the Cartesian product $A \times B$ of two sets. We then turn in Section 9.3 to one of the best counting strategies: *being lazy!* If we can show that $|A| = |B|$ and we already know the value of $|B|$, then figuring out $|A|$ is easy; we'll often use functions to relate two sets so that we can then lazily compute the size of the apparently harder-to-count set. Finally, in Section 9.4, we will explore *combinations* ("how many ways are there to choose an unordered collection of k items out of a set of n possibilities?") and *permutations* ("how many ways are there to put a set of n items into some order?").

Why does counting matter in computer science? There are, again, surprisingly many applications. Here are a few examples. One common (though very basic) style of algorithm is a *brute-force algorithm*, which finds the best *whatzit* by trying every possible *whatzit* and seeing which one is best. Determining whether a brute-force algorithm is fast enough depends on counting how many possible *whatzit*s there are. A more advanced algorithmic design technique, called *dynamic programming*, can be used to design efficient recursive solutions to problems—as long as there aren't too many *distinct* subproblems. Counting techniques are even powerful enough to establish a mind-bending result about *computability*: we will be able to prove that *there are more problems than computer programs*—which means that there are some problems that cannot be solved by any program!

Probability (see Chapter 10) has a plethora of applications in computer science, ranging from randomized algorithms in sorting (algorithms that process their input by making random decisions about how to act) to models of random noise in speech recognition or random errors in typing (if I'm trying to type the letter `p`, what is the chance that I accidentally type `o` instead?). We can think of the probability of some event X happening, roughly, as two counting problems: the numerator and denominator of the ratio

$$\frac{\text{the number of ways } X \text{ can happen}}{\text{the number of ways } X \text{ can either happen or not happen}}.$$

There are many other applications of counting scattered throughout computer science, and we will discuss a few more along the way: breaking cryptographic systems, compressing audio/image/video files, and changing the addressing scheme on the internet because we've run out of smaller addresses, to name a few.

9.2 *Counting Unions and Sequences*

> If a man who cannot count finds a four-leaf clover, is he entitled to happiness?
>
> Stanislaw J. Lec (1909–1966)

Suppose that we have two sets A and B from which we must choose an element. There are two different natural scenarios that meet this one-sentence description: we must choose a total of one element from *either* A or B, or we must choose one element from *each* of A and B. For example, consider a restaurant that offers soups $A = \{\text{chicken noodle, beer cheese, minestrone}, \ldots\}$ and salads $B = \{\text{caesar, house, arugula}, \ldots\}$. A lunch special that includes soup *or* salad involves choosing an $x \in A \cup B$. A dinner special including soup *and* salad involves choosing an $x \in A$ and also choosing a $y \in B$—that is, choosing an element $\langle x, y \rangle \in A \times B$. In Section 9.2.1, we'll start with two basic rules for computing these cardinalities:

- *Sum Rule:* If A and B are disjoint, then $|A \cup B| = |A| + |B|$.
- *Product Rule:* The number of pairs $\langle x, y \rangle$ with $x \in A$ and $y \in B$ is $|A \times B| = |A| \cdot |B|$.

These rules will handle the simple restaurant scenarios above, but there are a pair of extensions that we'll introduce to handle slightly more complex situations. The first (Section 9.2.2) extends the Sum Rule to allow us to calculate the cardinality of a union of two sets *even if* those sets may contain elements in common:

- *Inclusion–Exclusion:* $|A \cup B| = |A| + |B| - |A \cap B|$.

The second extension (Section 9.2.3) generalizes the Product Rule to allow us to calculate the cardinality of a set of pairs $\langle x, y \rangle$ even if the choice of x changes the list (but not the number) of possible choices for y:

- *Generalized Product Rule:* Consider pairs $\langle x, y \rangle$ of the following form: we can choose any $x \in A$, and, for each such x, there are precisely n different choices for y. Then the total number of pairs meeting this description is $|A| \cdot n$.

The remainder of this section will give the details of these four rules, and how to use these rules individually and in combination.

9.2.1 *The Basics: The Sum and Product Rules*

Sum Rule: counting unions

Our first rule addresses the *union* of two sets: if two sets A and B are disjoint, then the cardinality of their union is simply the sum of their sizes:

Theorem 9.1 (Sum Rule)
Let A and B be sets. If $A \cap B = \varnothing$, then $|A \cup B| = |A| + |B|$.

More generally, consider a collection of $k \geq 1$ sets $A_1, A_2, \ldots, A_k$. If these sets are all disjoint—that is, if $A_i \cap A_j = \varnothing$ whenever $i \neq j$—then the cardinality of their union is the sum of their cardinalities: $|A_1 \cup A_2 \cup \cdots \cup A_k| = |A_1| + |A_2| + \cdots + |A_k|$.

The Sum Rule captures an intuitive fact: if a box contains some red things and some blue things, then the total number of things in the box is the number of red things plus the number of blue things. Here are a few examples that use this rule:

Example 9.1 (Counting disjoint unions)

- Let $A := \{1, 2\}$ and $B := \{3, 4, 5, 6\}$. Thus $|A| = 2$ and $|B| = 4$. Observe that the sets A and B are disjoint. By the sum rule, $|A \cup B| = |A| + |B| = 2 + 4 = 6$. Indeed, we have $A \cup B = \{1, 2, 3, 4, 5, 6\}$, which contains 6 elements.
- There are 11 starters on your school's women's soccer team. Suppose there are 8 nonstarters on the team. The total number of people on the team is $19 = 11 + 8$.
- At a certain school in the midwest, there are currently 30 computer science majors who are studying abroad. There are 89 computer science majors who are studying on campus. Then the total number of computer science majors is $119 = 89 + 30$.
- Consider a computer lab that contains 32 Macs and 14 PCs and 1 PDP-8 (a 1960s-era machine, one of the first computers that was sold commercially). Then the total number of computers in the lab is $47 = 32 + 14 + 1$.

Example 9.2 (Students in classes)

Problem: During this term, there are 19 students taking Data Structures, and 39 students taking Mathematics of Computer Science. Let S denote the set of students taking Data Structures *or* Mathematics of Computer Science this term. What is $|S|$?

Solution: There isn't enough information to answer the question!

- *If* there are no students who are taking both classes (that is, if $DS \cap MOCS = \varnothing$), then $|S| = |DS| + |MOCS| = 19 + 39 = 58$.
- But, for all we know from the problem statement, every student in Data Structures is also taking Mathematics of Computer Science. In this case, we have $DS \subset MOCS$ and thus $S = DS \cup MOCS = MOCS$; therefore $|S| = |MOCS| = 39$.

(The *Inclusion–Exclusion Rule*, in Section 9.2.2, formalizes the calculation of $|A \cup B|$ in terms of $|A|$, $|B|$, and $|A \cap B|$, in the manner that we just considered.)

Taking it further: The logic that we used in Example 9.2 to conclude that there were at most 58 students in the two classes combined is an application of the general fact that $|A \cup B| \leq |A| + |B|$. While this fact is pretty simple, it turns out to be remarkably useful in proving facts about probability. The *Union Bound* states that the probability that *any* of $A_1, A_2, \ldots, A_k$ occurs is at most $p_1 + p_2 + \cdots + p_k$, where p_i denotes the probability that A_i occurs. The Union Bound turns out to be useful when each A_i is a "bad event" that we're worried might happen, and these bad events may have complicated probabilistic dependencies—but if we can show that the probability that every particular one of these bad events is some very small ε, then we can use the Union Bound to conclude that the probability of experiencing *any* bad event is at most $k \cdot \varepsilon$. (See Exercise 10.141, for example.)

Using the Sum Rule in less obvious settings

As a general strategy for solving counting problems, we can try to find a way to apply the Sum Rule—even if it does not superficially seem to be applicable. If we can find a way to partition an apparently complicated set S into simple disjoint sets $S_1, S_2, \ldots, S_k$ such that $\bigcup_{i=1}^{k} S_i = S$, then we can use the Sum Rule to find $|S|$.

In this spirit, here's a somewhat more complex example of using the Sum Rule, where we have to figure out the subsets ourselves: let's determine how many 8-bit strings contain precisely two ones. (The full list of the bitstrings meeting this condition appears in Figure 9.1.)

11000000	01100000	00110000	00011000	00001100	00000110	00000011
	10100000	01010000	00101000	00010100	00001010	00000101
		10010000	01001000	00100100	00010010	00001001
			10001000	01000100	00100010	00010001
				10000100	01000010	00100001
					10000010	01000001
						10000001

Figure 9.1: All bitstrings in $\{0,1\}^8$ that contain exactly two ones.

Example 9.3 (8-bit strings with exactly 2 ones)

Problem: How many elements of $\{0,1\}^8$ have precisely two 1s?

Solution: Obviously, we can just count the number of bitstrings in Figure 9.1, which yields the answer: there are 28 such bitstrings. But let's use the Sum Rule instead.

What does a bitstring $x \in \{0,1\}^8$ with two ones look like? There must be two indices i and j—say with $i > j$—such that $x_i = x_j = 1$, and all other components of x must be 0:

$$x = \underbrace{00\cdots0}_{j-1 \text{ zeros}} \overbrace{1}^{\text{one in position } j} \underbrace{00\cdots0}_{i-j-1 \text{ zeros}} \overbrace{1}^{\text{one in position } i} \underbrace{00\cdots0}_{8-i \text{ zeros}}.$$

(For example, the bitstring 01001000 has ones in positions $j = 2$ and $i = 5$, interspersed with an initial block of $j - 1 = 1$ zero, a block of $i - j - 1 = 2$ between-the-ones zeros, and a block of $8 - i = 3$ final zeros.)

Problem-solving tip: When you're trying to find the cardinality of a complicated set S, try to find a way to split S into a collection of simpler disjoint sets, and then apply the Sum Rule.

We are going to divide the set of 8-bit strings with two 1s *based on the index i*. That is, suppose that $x \in \{0,1\}^8$ contains two ones, and the *second* 1 in x appears in bit position #i. Then there are $i - 1$ positions in which the *first* one could appear—any of the slots $j \in \{1, 2, \ldots, i-1\}$ that come before i. (See Figure 9.1, where the $(i-1)$st column contains all $i - 1$ bitstrings whose second 1 appears in position #i. For example, column #3 contains the 3 bitstrings with $x_{4,5,6,7,8}$ = 10000: that is, 10010000, 01010000, and 00110000.) Because every x with exactly two ones has an index i of its second 1, we can use the Sum Rule to say that the answer to the given question is

$$\begin{aligned} \sum_{i=1}^{8} \left[\text{number of bitstrings with the second 1 in position } i\right] &= \sum_{i=1}^{8} (i-1) \\ &= 0 + 1 + \cdots + 7 \\ &= 28. \end{aligned}$$

(We'll also see another way to solve this example later, in Example 9.39.)

Let's also generalize this example to bitstrings of arbitrary length:

Example 9.4 (k-bit strings with exactly 2 ones)
Consider the set $S := \{x \in \{0,1\}^k : x \text{ has precisely two 1s}\}$. As in Example 9.3, every bitstring $x \in S$ has an index i of its second 1; we'll use the value of i to partition S into sets that can be easily counted, and then use the Sum Rule to find $|S|$. Specifically, for each index i with $1 \leq i \leq k$, define the set

$$\begin{aligned} S_i &= \{x \in S : x_i = 1 \text{ and } x_{i+1} = x_{i+2} = \cdots = x_k = 0\}. \\ &= \{x \in \{0,1\}^k : [\exists j \leq i-1 : x_i = x_j = 1 \text{ and } x \text{ has no other 1s}]\}. \end{aligned}$$

Observe that $|S_i| = i - 1$: there are $i-1$ different possible values of j. Also, observe that $S = \bigcup_{i=1}^{k} S_i$ and that, for any $i \neq i'$, the sets S_i and $S_{i'}$ are disjoint. Thus

$$|S| = \left|\bigcup_{i=1}^{k} S_i\right| = \sum_{i=1}^{k} |S_i| = \sum_{i=1}^{k} (i-1) = \frac{k(k-1)}{2} \qquad (*)$$

by the Sum Rule and the formula for the sum of the first n integers (Example 5.4).

As a check of our formula, let's verify our solution for some small values of k:

- For $k = 2$, $(*)$ says there are $\frac{2(2-1)}{2} = 1$ strings with two 1s. Indeed, there's just one: `11`.
- For $k = 3$, indeed there are $\frac{3(3-1)}{2} = 3$ strings with two 1s: `011`, `101`, and `110`.
- For $k = 4$, there are $\frac{4 \cdot 3}{2} = 6$ such strings: `1100`, `1010`, `0110`, `1001`, `0101`, and `0011`.

Note that $(*)$ matches Example 9.3: for $k = 8$, we have $28 = \frac{8 \cdot 7}{2}$ strings with two 1s.

Problem-solving tip: Check to make sure your formulas are reasonable by testing them for small inputs (as we did in Example 9.4).

Product Rule: counting sequences

Our second basic counting rule addresses the *Cartesian product* of sets. Recall that, for sets A and B, the Cartesian product $A \times B$ consists of all pairs $\langle a, b\rangle$ with $a \in A$ and $b \in B$. (For example, $\{1,2,3\} \times \{\texttt{x},\texttt{y}\} = \{\langle 1,\texttt{x}\rangle, \langle 1,\texttt{y}\rangle, \langle 2,\texttt{x}\rangle, \langle 2,\texttt{y}\rangle, \langle 3,\texttt{x}\rangle, \langle 3,\texttt{y}\rangle\}$.) The cardinality of $A \times B$ is the product of the cardinalities of A and B:

Theorem 9.2 (Product Rule)
Let A and B be sets. Then $|A \times B| = |A| \cdot |B|$.

More generally, consider a collection of k arbitrary sets $A_1, A_2, \ldots, A_k$, and consider the set of k-element sequences where, for each i, the ith component is an element of A_i. The number of such sequences is given by the product of the sets' cardinalities:

$$|A_1 \times A_2 \times \cdots \times A_k| = |A_1| \cdot |A_2| \cdot \cdots \cdot |A_k|.$$

Here are a few examples of counting using the Product Rule:

Example 9.5 (Counting sequences)

- Let $A := \{1,2\}$ and $B := \{3,4,5,6\}$. By the product rule, $|A \times B| = |A| \cdot |B| = 2 \cdot 4 = 8$. Indeed, $A \times B = \{\langle 1,3\rangle, \langle 1,4\rangle, \langle 1,5\rangle, \langle 1,6\rangle, \langle 2,3\rangle, \langle 2,4\rangle, \langle 2,5\rangle, \langle 2,6\rangle\}$, which contains 8 elements.

- At a certain school in the midwest, there are currently 56 senior computer science majors and 63 junior computer science majors. Then the number of ways to choose a pair of class representatives, one senior and one junior, is $56 \cdot 63 = 3528$.

- Consider a tablet computer that is sold with three different options: a choice of protective cover, a choice of stylus, and a color. If there are 7 different styles of protective cover, 5 different styles of stylus, and 3 different colors, then there are $7 \cdot 5 \cdot 3 = 105$ different configurations of the computer.

Like the Sum Rule, the Product Rule should be reasonably intuitive: if we are choosing a pair $\langle a,b\rangle$ from $A \times B$, then we have $|A|$ different choices of the first component a—and, for each of those $|A|$ choices, we have $|B|$ choices for the second component b. (Thinking of A as $A = \{a_1, a_2, \ldots, a_{|A|}\}$, we can even view $\{\langle a,b\rangle : a \in A, b \in B\}$ as

$$\{\langle a_1,b\rangle : b \in B\} \cup \{\langle a_2,b\rangle : b \in B\} \cup \quad \cdots \quad \cup \{\langle a_{|A|},b\rangle : b \in B\}.$$

By the Sum Rule, this set has cardinality $|B| + |B| + \cdots + |B|$, with one term for each element of A—in other words, it has cardinality $|A| \cdot |B|$.) Here are a few more examples:

Example 9.6 (32-bit strings)

Problem: How many different 32-bit strings are there?

Solution: The set of 32-bit strings is $\{0,1\}^{32}$—that is, elements of

$$\underbrace{\{0,1\} \times \{0,1\} \times \{0,1\} \times \cdots \times \{0,1\}}_{\text{32 times}}.$$

Because $|\{0,1\}| = 2$, the Product Rule lets us conclude that $|\{0,1\}^{32}|$ is

$$\underbrace{2 \cdot 2 \cdot 2 \cdot \quad \cdots \quad \cdot 2}_{\text{32 times}} = 2^{32}.$$

(We can use the same type of analysis to show that there are $2^4 = 16$ strings of 4 bits; for concreteness, they're all listed in Figure 9.2.)

```
0000
0001
0010
0011
0100
0101
0110
0111
1000
1001
1010
1011
1100
1101
1110
1111
```

Figure 9.2: The set of all 4-bit strings.

Example 9.7 (Number of possible shortened URLs)

A *URL-shortening service* like `bit.ly` or `snipurl.com` allows a user to compress a long URL into a much shorter sequence of characters. (The shorter URL can then be used in emails or tweets or other contexts in which a long URL is unwieldy.) For example, by entering the URL of Alan Turing's Wikipedia page into `bit.ly`, I got the URL `http://bit.ly/1o6HPM` as a shortened form of `http://en.wikipedia.org/wiki/Alan_Turing`.

If a shortened URL consists of 6 characters, each of which is a digit, lowercase letter, or uppercase letter, the number of possible shortened URLs is, using the Product Rule,

$$|C \times C \times C \times C \times C \times C| = |C| \cdot |C| \cdot |C| \cdot |C| \cdot |C| \cdot |C| = |C|^6,$$

where $C = \{0,\ldots,9\} \cup \{a,\ldots,z\} \cup \{A,\ldots,Z\}$ is the set of possible characters. Because $|C| = 10 + 26 + 26 = 62$ via the Sum Rule, we know that there are $62^6 = 56{,}800{,}235{,}584$ possible shortened 6-character URLs.

Taking it further: The point of a URL-shortening service is to translate long URLs into short ones, but it's theoretically impossible for *every* URL to be shortened by this service: there are more possible URLs of length k than there are URLs of length strictly less than k. A similar issue arises with *file compression* algorithms, like ZIP, that try to reduce the space required to store a file. See the discussion on p. 938.

Product Rule: counting sequences from a fixed set

This use of the Product Rule—to count the number of sequences of length k with elements all drawn from a fixed set S, rather than having a different set of options for each component—is common enough that we'll note it as a separate rule:

Theorem 9.3 (Product Rule: sequences of elements from a single set S)
For any set S and any $k \in \mathbb{Z}^{\geq 1}$, the number of k-tuples from the set $S^k = \underbrace{S \times S \times \cdots \times S}_{k \text{ times}}$ is $|S^k| = |S|^k$.

A notational reminder regarding Theorem 9.3: S^k is the set

$$S \times S \times \cdots \times S,$$

that is, the set of k-tuples where each component is an element of S. On the other hand, $|S|^k$ is the number $|S|$ raised to the kth power.

Here's another example using this special case of the Product Rule:

Example 9.8 (MAC addresses)

Problem: A *media access control address*, or *MAC address*, is a unique identifier for a network adapter, like an ethernet card or wireless card. A MAC address consists of a sequence of six groups of pairs of hexadecimal digits. (A *hexadecimal digit* is one of 0123456789ABCDEF.) For example, F7:DE:F1:B6:A4:38 is a MAC address. (The pairs of digits are traditionally separated by colons when written down.) How many different MAC addresses are there?

Solution: There are 16 different hexadecimal digits. Thus, using the Product Rule, there are $16 \cdot 16 = 256$ different pairs of hexadecimal digits, ranging from 00 to *FF*. Using the Product Rule again, as in Example 9.7, we see that there are 256^6 different sequences of six pairs of hexadecimal digits. Thus there are $256^6 = [16^2]^6 = [(2^4)^2]^6 = 2^{48} = 281{,}474{,}976{,}710{,}656$ total different MAC addresses.

Taking it further: In addition to the numerical addresses assigned to particular hardware devices—the MAC addresses from Example 9.8—each device that's connected to the internet is also assigned an address, akin to a mailing address, that's used to identify the destination of a packet of information. But we've had to make a major change to the way that information is transmitted across the internet because of a counting problem: we've run out of addresses! See the discussion on p. 919.

9.2.2 *Inclusion–Exclusion: Unions of Nondisjoint Sets*

The counting techniques that we've introduced so far have some important restrictions. We can only use the Sum Rule to calculate $|A \cup B|$ when A and B are *disjoint*. And we are only able to use the Product Rule to calculate the number of sequences when the set of options for the second component does not depend on the choice that we made in the first component. In the remainder of this section, we will extend our techniques to remove these restrictions so that we can handle more general problems. Let's start with a specific example of the cardinality of the union of nondisjoint sets:

Example 9.9 (Primes and odds)
Consider the set $O = \{1, 3, 5, 7, 9\}$ of odd numbers less than 10 and the set $P = \{2, 3, 5, 7\}$ of prime numbers less than 10. What is $|O \cup P|$?

It might be tempting to use the Sum Rule to conclude that $|O \cup P| = |O| + |P| = 5 + 4 = 9$. But this conclusion is incorrect, because $P \cap O = \{3, 5, 7\} \neq \varnothing$, so the Sum Rule doesn't apply. In particular, $O \cup P = \{1, 2, 3, 5, 7, 9\}$, so $|O \cup P| = 6$.

The issue with the naïve application of the Sum Rule in Example 9.9 is called *double counting*: in the expression $|O| + |P|$, we counted the elements in the intersection $O \cap P$ twice, which gave us the incorrect total count. The idea underlying the Inclusion–Exclusion Rule is to correct for this error: to compute the size of the union of two sets A and B, we extend the Sum Rule to correct for the double counting by subtracting $|A \cap B|$ from the final result. (See Figure 9.3.) This counting rule is called inclusion–exclusion because we *include* (add) the cardinalities of the two individual sets, and then *exclude* (subtract) the cardinality of the intersection of the pairs:

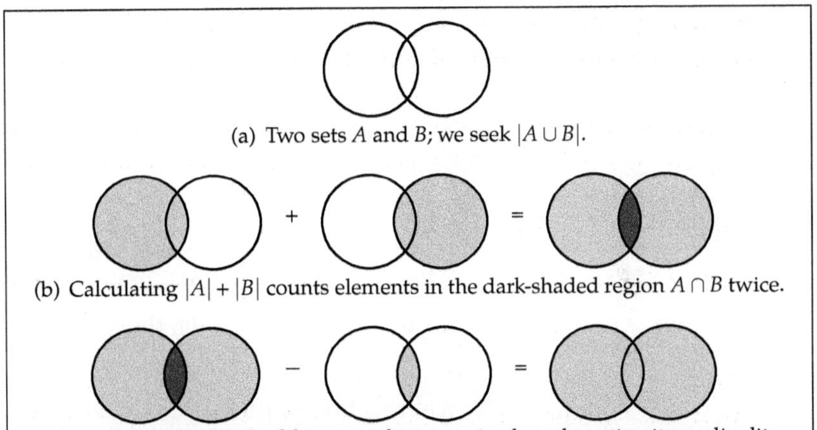

(a) Two sets A and B; we seek $|A \cup B|$.

(b) Calculating $|A| + |B|$ counts elements in the dark-shaded region $A \cap B$ twice.

(c) We correct for the double-counted intersection by subtracting its cardinality.

Figure 9.3: The Inclusion–Exclusion Rule.

Theorem 9.4 (Inclusion–Exclusion)
Let A and B be sets. Then $|A \cup B| = |A| + |B| - |A \cap B|$.

Here are a few small examples:

Example 9.10 (Counting not necessarily disjoint unions)

- Let $A := \{1, 2, 3\}$ and $B := \{3, 4, 5, 6\}$. Thus $A \cap B = \{3\}$, and so $|A| = 3$ and $|B| = 4$ and $|A \cap B| = 1$. By the inclusion–exclusion rule, $|A \cup B| = |A| + |B| - |A \cap B| = 3 + 4 - 1 = 6$. Indeed, we have $A \cup B = \{1, 2, 3, 4, 5, 6\}$, which contains 6 elements.

Problem-solving tip: Sometimes the easiest way to solve a problem—in CS or in life!—is to find an imperfect approximation to the solution, and then correct for whatever inaccuracies result. Inclusion–Exclusion is a good example of this estimate-and-fix strategy.

- At a certain school in the midwest, there are 119 computer science majors and 65 math majors. There are 7 students double majoring in CS and math. Thus a total of $119 + 65 - 7 = 177$ different students are majoring in either of the two fields.
- There are 21 consonants (`BCDFGHJKLMNPQRSTVWXYZ`) in English. There are 6 vowels in English (`AEIOUY`). There is one letter that's both a vowel and a consonant (`Y`). Thus there are $21 + 6 - 1 = 26$ total letters.
- Let E be the set of even integers between 1 and 100. Let O be the set of odd integers between 1 and 100. Note that $|E| = 50$, $|O| = 50$, and $|E \cap O| = 0$. Thus $|E \cup O| = 50 + 50 - 0 = 100$.

Here's an example that uses Inclusion–Exclusion to compute the cardinality of a slightly more complicated set:

Example 9.11 (ATM machine PIN numbers)

Problem: A certain bank's customers can select a 4-digit number (called a *PIN*) to access their accounts, but the bank insists that the PIN may not start with the same digit repeated three times (for example, 7770) or end with the same digit repeated three times (for example, 0111). How many *invalid* PINs are there?

Solution: Let S denote the set of PINs that start with three repeated digits. Let E denote the set of PINs that end with three repeated digits. Then the set of invalid PINs is $S \cup E$.

- Note that $|S| = 100$: we can view a PIN in S as a sequence of two digits $\langle x, y \rangle \in \{0, 1, \ldots, 9\}^2$, with x repeated three times in the PIN. (So $\langle 3, 1 \rangle$ corresponds to the PIN 3331.) By the Product Rule, there are $10^2 = 100$ such codes.
- Similarly, $|E| = 100$: we can think of an element of E as a sequence of two digits $\langle x, y \rangle \in \{0, 1, \ldots, 9\}^2$, where y is repeated three times in the PIN.

If $S \cap E$ were empty, then we could apply the Sum Rule to compute $|S \cup E|$. But there *are* PINs that are in both S and E:

- A 4-digit number $\langle x, y, z, w \rangle$ is in $S \cap E$ if and only if $x = y = z$ (because $\langle x, y, z, w \rangle \in S$) and $y = z = w$ (because $\langle x, y, z, w \rangle \in E$). That is, any 4-digit number that consists of the same digit repeated four times is in $S \cap E$. Thus

$$S \cap E = \{0000, 1111, 2222, 3333, 4444, 5555, 6666, 7777, 8888, 9999\},$$

and $|S \cap E| = 10$.

(See Figure 9.4 for S, E, and $S \cap E$.) Applying the Inclusion–Exclusion rule, we see that the set $S \cup E$ of invalid PINs has cardinality $|S| + |E| - |S \cap E| = 100 + 100 - 10 = 190$. (So $10{,}000 - 190 = 9810$ PINs are valid.)

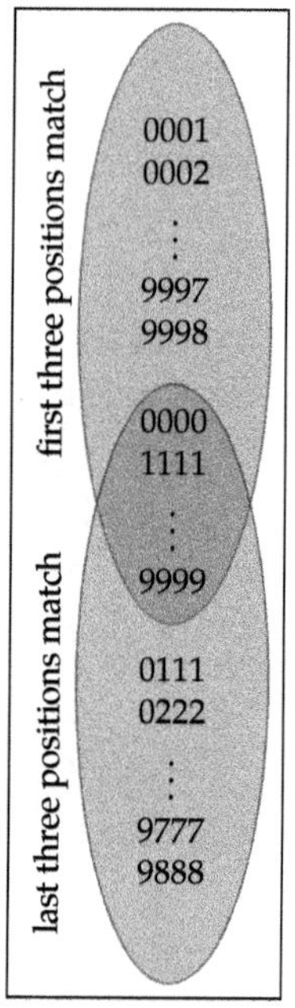

Figure 9.4: Invalid PINs, starting or ending with the same digit repeated three times.

The basic Sum Rule is actually a special case of the Inclusion–Exclusion Rule: if A and B are disjoint, then $|A \cap B| = \varnothing$, so $|A \cup B| = |A| + |B| - |A \cap B| = |A| + |B| - 0 = |A| + |B|$.

Inclusion–Exclusion for three sets

Theorem 9.4 describes how to calculate the cardinality of the union of *two* sets, but this idea can be generalized. The basic idea is simple: we will try counting in the easiest way possible, and then we'll correct for any overcounting or undercounting.

For example, we can compute the cardinality of the union of *three* sets $A \cup B \cup C$ using a more complicated version of Inclusion–Exclusion:

- We add (include) the three singleton sets ($|A| + |B| + |C|$), but this sum counts any element contained in more than one of the three sets more than once.
- So we subtract (exclude) the three pairwise intersections ($|A \cap B| + |A \cap C| + |B \cap C|$) from the sum. But we're not done: imagine an element contained in *all three* of A, B, and C; such an element was included three times and then excluded three times, so it hasn't been counted at all.
- So we add (include) the three-way intersection $|A \cap B \cap C|$.

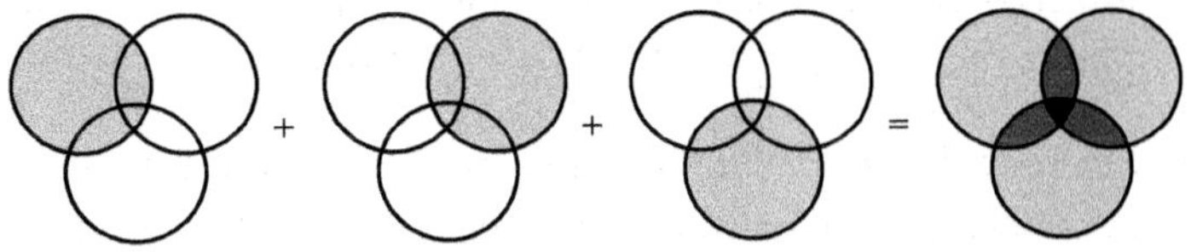

(a) If we start to compute $|A \cup B \cup C|$ as $|A| + |B| + |C|$, we correctly count the light-shaded regions, but we count elements in the medium-shaded regions twice, and elements in the dark-shaded region three times.

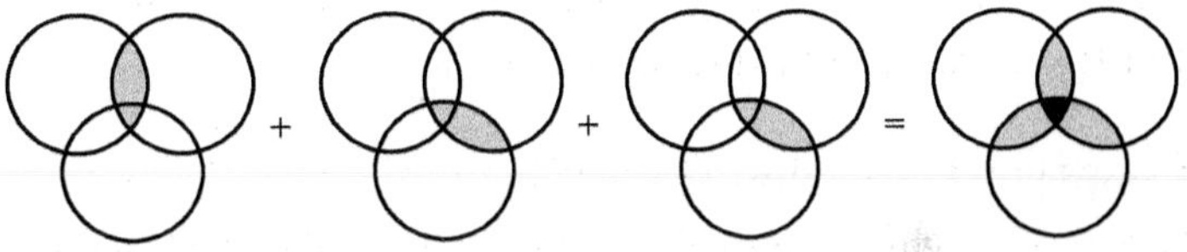

(b) Subtracting the sum of the sizes of the pairwise intersections $|A \cap B| + |B \cap C| + |A \cap C|$ almost corrects for the double counting from (a), but it also triple counts the elements of $A \cap B \cap C$.

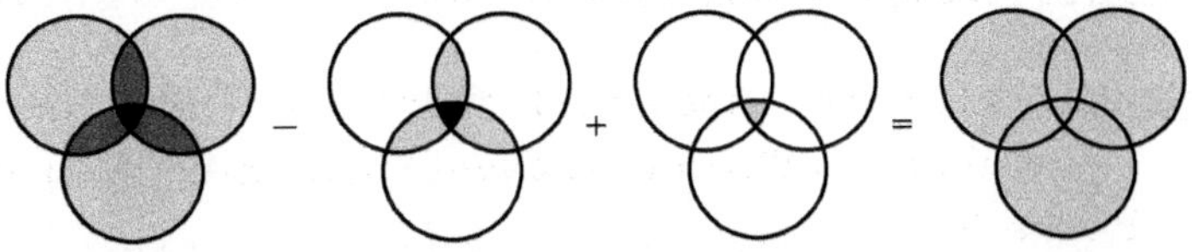

(c) The result of (a) minus (b) hasn't counted the elements of $A \cap B \cap C$ at all, so we can achieve the final count by adding $|A \cap B \cap C|$.

Figure 9.5: The Inclusion–Exclusion Rule for three sets A, B, and C. See Theorem 9.5.

This calculation yields the following three-set rule for inclusion–exclusion. (Or see Figure 9.5 for a visual illustration of why this calculation is correct.)

Theorem 9.5 (Inclusion–Exclusion for three sets)
Let A, B, and C be sets. Then $|A \cup B \cup C|$ is given by

$$|A| + |B| + |C| - |A \cap B| - |A \cap C| - |B \cap C| + |A \cap B \cap C|.$$

Here are a couple of small examples of the three-set version of inclusion–exclusion:

Example 9.12 (Counting three-set unions)

- Let $A := \{0, 1, 2, 3, 4\}$ and $B := \{0, 2, 4, 6\}$ and $C := \{0, 3, 6\}$. Then

$$\begin{aligned}
&|A \cup B \cup C| \\
&= \underbrace{5}_{|A|} + \underbrace{4}_{|B|} + \underbrace{3}_{|C|} - \underbrace{3}_{|A\cap B| = |\{0,2,4\}|} - \underbrace{2}_{|A \cap C| = |\{0,3\}|} - \underbrace{2}_{|B \cap C| = |\{0,6\}|} + \underbrace{1}_{|A\cap B\cap C| = |\{0\}|} \\
&= 12 - 7 + 1 = 6,
\end{aligned}$$

by Inclusion–Exclusion. Indeed, $A \cup B \cup C = \{0, 1, 2, 3, 4, 6\}$. (See Figure 9.6.)

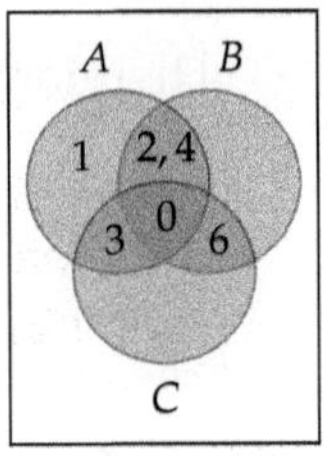

Figure 9.6: Some small sets.

- Consider the words `ONE`, `TWO`, `THREE`, `FOUR`, `FIVE`, `SIX`, `SEVEN`, and `EIGHT`. Let E be the set of these words containing at least one `E`, let T be the words containing a `T`, and let R be the words containing an `R`. Then

$$\begin{aligned}&|E \cup T \cup R|\\ &= \underbrace{5}_{E=\{\texttt{ONE, THREE, FIVE, SEVEN, EIGHT}\}} + \underbrace{3}_{T=\{\texttt{TWO, THREE, EIGHT}\}} + \underbrace{2}_{R=\{\texttt{THREE, FOUR}\}} - \underbrace{2}_{E\cap T=\{\texttt{THREE, EIGHT}\}} - \underbrace{1}_{E\cap R=\{\texttt{THREE}\}} - \underbrace{1}_{T\cap R=\{\texttt{THREE}\}} + \underbrace{1}_{E\cap T\cap R=\{\texttt{THREE}\}}\\ &= 7,\end{aligned}$$

and, indeed, seven of the eight words are in $E \cup T \cup R$ (the only one missing is `SIX`).

We'll close with a slightly bigger example, about integers divisible by 2, 3, or 5:

Example 9.13 (Divisibility)

Problem: How many integers between 1 and 1000, inclusive, are evenly divisible by any of 2, 3, or 5?

Solution: Define the following sets:

$$\begin{aligned}A &= \{n \in \{1,\ldots,1000\} : 2 \mid n\}\\ B &= \{n \in \{1,\ldots,1000\} : 3 \mid n\}\\ C &= \{n \in \{1,\ldots,1000\} : 5 \mid n\}.\end{aligned}$$

We must compute $|A \cup B \cup C|$.

- It's fairly easy to see that $|A| = 500$, $|B| = 333$, and $|C| = 200$, because $A = \{2n : 1 \leq n \leq 500\}$, $B = \{3n : 1 \leq n \leq 333\}$, and $C = \{5n : 1 \leq n \leq 200\}$.
- Observe that $A \cap B$ is the set of integers between 1 and 1000 that are divisible by both 2 and 3—that is, the set of integers divisible by 6. By the same logic that we used to compute $|A|$, $|B|$, and $|C|$, we see
  - $|A \cap B| = |\{6n : 1 \leq n \leq 166\}| = 166$,
  - $|A \cap C| = |\{10n : 1 \leq n \leq 100\}| = 100$, and
  - $|B \cap C| = |\{15n : 1 \leq n \leq 66\}| = 66$.
- And, using the same approach, we can conclude that $A \cap B \cap C = \{n : 30 \mid n\} = \{30n : 1 \leq n \leq 33\}$, so $|A \cap B \cap C| = 33$.

Therefore, using the Inclusion–Exclusion Rule, $|A \cup B \cup C|$ is

$$\underbrace{500}_{|A|} + \underbrace{333}_{|B|} + \underbrace{200}_{|C|} - \underbrace{166}_{|A\cap B|} - \underbrace{100}_{|A\cap C|} - \underbrace{66}_{|B\cap C|} + \underbrace{33}_{|A\cap B\cap C|} = 734.$$

Problem-solving tip: To verify a calculation like this one, it's a good idea (and very easy!) to write a short program.

We can further generalize the inclusion–exclusion principle to calculate the cardinality of the union of an arbitrary number of sets. (See Exercises 9.30 and 9.181.)

9.2.3 *The Generalized Product Rule*

The Product Rule (Theorem 9.2) tells us how to compute the number of 2-element sequences where the first element is drawn from the set A and the second from the set B—specifically, it says that $|A \times B|$ is $|A| \cdot |B|$. But there are many types of sequences that do not precisely fit this setting: the Product Rule only describes the set of sequences where each component is selected from a *fixed* set of options. If the set of options for choice #2 depends on choice #1, then we cannot directly apply the Product Rule. However, the basic principle of the Product Rule still applies if the *number* of different choices for the second component is the same regardless of the choice of the first component, even if the *particular set of choices* can differ:

Theorem 9.6 (Generalized Product Rule)
Let S denote a set of sequences, each of length k, where for each index $i \in \{1, \ldots, k\}$ the following condition holds: for each choice of the first $i-1$ components of the sequence, there are exactly n_i choices for the ith component. Then $|S| = \prod_{i=1}^{k} n_i$.

Here are a few examples using the Generalized Product Rule:

Example 9.14 (Gold, silver, and bronze)
Problem: A set S of eight sprinters qualify for the finals of the 100-meter dash in the Olympics. One will win the gold medal, another the silver, and a third the bronze. How many different trios of medalists are possible?

Solution: It "feels" like we can solve this problem using the Product Rule, by choosing a sequence of three elements from S, where we forbid duplication in our choices. But our choice of gold, silver, and bronze medalists would be from

$$S \times (S - \{\text{the gold medalist}\}) \times (S - \{\text{the gold and silver medalists}\})$$

and the Product Rule doesn't permit the set of choices for the second component to depend on the first choice, or the options for the third choice to depend on the first two choices.

Instead, observe that there are 8 choices for the gold medalist. For each of those choices, there are 7 choices for the silver medalist. For each of these pairs of gold and silver medalists, there are 6 choices for the bronze medalist. Thus, by the Generalized Product Rule, the total number of trios of medalists is $8 \cdot 7 \cdot 6 = 336$.

Example 9.15 (Opening moves in a chess game)
In White's very first move in a chess game, there are $n_1 = 10$ pieces that can move: any of White's 8 pawns or 2 knights. Each of these pieces has $n_2 = 2$ legal moves: the pawns can move forward either 1 or 2 squares, and the knights can move either ↰ or ↱. (See Figure 9.7.) Thus there are $n_1 \cdot n_2 = 10 \cdot 2 = 20$ legal first moves.

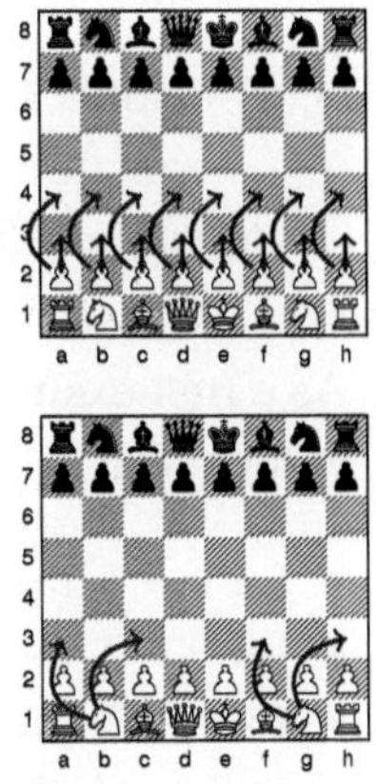

Figure 9.7: The valid first moves in a chess game.

Example 9.16 (Students in classes)
At a certain school in the midwest, each of 2023 students enrolls in exactly 3 classes per term. The set

$$Enrollments := \{\langle s,c\rangle : s \text{ is a student enrolled in class } c \text{ during the current term}\}$$

has cardinality $2023 \cdot 3 = 6069$, by the Generalized Product Rule: for each of the $n_1 = 2023$ choices of student, there are $n_2 = 3$ choices of classes. (Note that the original Product Rule does not apply, because the set *Enrollments* is not a Cartesian product: in general, two students are not enrolled in the same *classes*—just the same *number* of classes.)

Although we didn't say we were doing so, we actually used the underlying idea of the Generalized Product Rule in Example 9.11. Let's make its use explicit here:

Example 9.17 (4-digit PINs starting with a triplicated digit)
Let $S \subseteq \{0,1,\ldots,9\}^4$ denote the set of 4-digit PINs that start with three repeated digits. We claim that $|S| = 100$, as follows:

- There are $n_1 = 10$ choices for the first digit.
- There is only $n_2 = 1$ choice for the second digit: it must match the first digit.
- There's also only $n_3 = 1$ choice for the third digit: it must match the first two.
- There are $n_4 = 10$ choices for the fourth digit.

Thus there are $n_1 \cdot n_2 \cdot n_3 \cdot n_4 = 10 \cdot 1 \cdot 1 \cdot 10 = 100$ elements of S.

Permutations

The Generalized Product Rule sheds some light on a concept that arises in a wide range of contexts: a *permutation* of a set S, which is any ordering of the elements of S.

Definition 9.1 (Permutation)
A permutation *of a set S is a sequence of elements from S that is of length $|S|$ and contains no repetitions. In other words, a permutation of S is an ordering of the elements of S.*

As a first example, let's list all the permutations of the set $\{1,2,\ldots,n\}$ for a few small values of n:

- for $n = 1$, there's just one ordering: $\langle 1\rangle$.
- for $n = 2$, there are two orderings: $\langle 1,2\rangle$ and $\langle 2,1\rangle$.
- for $n = 3$, there are six: $\langle 1,2,3\rangle$, $\langle 1,3,2\rangle$, $\langle 2,1,3\rangle$, $\langle 2,3,1\rangle$, $\langle 3,1,2\rangle$, and $\langle 3,2,1\rangle$.
- for $n = 4$, there are twenty-four: six with 1 as the first element (which can then be followed by any of the six permutations of $\langle 2,3,4\rangle$), six with 2 as the first element, six with 3 first, and six with 4 first, yielding a total of $4 \cdot 6 = 24$ orderings.

How many permutations of an n-element set are there? There are several ways to see the general pattern, including recursively, but it may be easiest to use the Generalized Product Rule to count the number of permutations:

Theorem 9.7 (Number of permutations)
Let S be any set, and write $n := |S|$. The number of different permutations of S is $n!$.

Proof. There are n choices for the first element of a permutation of S. For the second element, there are $n - 1$ choices (all but the element chosen first). There are $n - 2$ choices for the third slot (all but the elements chosen first and second). In general, for the ith element, there are $n - i + 1$ choices. Thus the number of permutations of S is

$$\prod_{i=1}^{n}(n - i + 1) = \prod_{j=1}^{n} j = n!$$

by the Generalized Product Rule. □

Here's a small example for a concrete set S:

Example 9.18 (10-digit numbers)
Problem: What fraction of integers between 0 and 9,999,999,999 (all written as 10-digit numbers, including any leading zeros) have no repeated digits?

Solution: We seek a 10-digit sequence with no repetitions—that is, a permutation of $\{0, 1, \ldots, 9\}$. There are $10! = 3{,}628{,}800$ such permutations, by Theorem 9.7. There are a total of 10^{10} integers between 0 and 9,999,999,999, by the Product Rule. Thus the fraction of these integers with no repeated digits is $\frac{10!}{10^{10}} \approx 0.00036\cdots$, about one out of every 2750 integers in this range.

Taking it further: A permutation of a set S is an ordering of that set S—so thinking about permutations is closely related to thinking about *sorting algorithms* that put an out-of-order array into a specified order. By using the counting techniques of this section, we can prove that algorithms *must* take a certain amount of time to sort; see the discussion on p. 920.

We will also return to permutations frequently later in the chapter. For example, in Section 9.4, we will address counting questions like the following: *how many different 13-card hands can be drawn from a standard 52-card deck of playing cards?* (Here's one way to think about it: we can lay out the 52 cards in any order—any permutation of the cards—and then pick the first 13 of them as a hand. We'll have to correct for the fact that any ordering of the first 13 cards—and, for that matter, any ordering of the last 39—will count as the same hand. But permutations will also help us to think about this correction!)

9.2.4 *Combining Products and Sums*

Suppose that we select a pair $\langle a, b\rangle$ from a set of possible choices. The Product Rule tells us how many ways to make these choices if the particular choice of a does not affect the set of options from which b is chosen. The Generalized Product Rule tells us how many ways to make these choices if the particular choice of a does not affect *the size of* the set of options from which b is chosen. But if the *number* of options for the choice of b differs based on the choice of a, even the Generalized Product Rule does not apply. In this case, we can use a combination of the Sum Rule and the Generalized Product Rule to calculate the number of results. We'll close this section with a few examples of these somewhat more complex counting questions.

Example 9.19 (Ordering coffee)
A certain coffeeshop sells the following espresso-based drinks:

americano*, cappuccino, espresso*, latte, macchiato, mocha.

The drinks marked with an asterisk do not contain milk; the others do. All drinks can be made with either decaf or regular espresso. All milk-containing drinks can be made with any of $\{\text{soy}, \text{skim}, 2\%, \text{whole}\}$ milk. How many different drinks are sold by this coffeeshop?

We can think of a chosen drink as a sequence of the form

$$\langle \text{drink type}, \text{milk type (or "none")}, \text{espresso type} \rangle.$$

There are $4 \cdot 4 \cdot 2 = 32$ choices of milk-based drinks (4 drink types, 4 milk types, and 2 espresso types). There are $2 \cdot 1 \cdot 2 = 4$ choices of non-milk-based drinks (2 drink types, 1 "milk" type ["none"], and 2 espresso types). Thus the total number of different drinks sold by this coffeeshop is $32 + 4 = 36$.

Example 9.20 (Text numbers)
Problem: In the United States, a text message can be sent either to a regular 10-digit phone number, or to a so-called *short code* which is a 5- or 6-digit number. Neither a phone number nor a short code can start with a 0 or a 1. How many different textable numbers are there in the United States?

Solution: Let $D = \{2, 3, \ldots, 9\}$. Note $|D| = 8$. The set of valid textable numbers is:

$$\underbrace{D \times (D \cup \{0,1\})^9}_{\text{phone numbers}} \cup \underbrace{D \times (D \cup \{0,1\})^4}_{\text{5-digit short codes}} \cup \underbrace{D \times (D \cup \{0,1\})^5}_{\text{6-digit short codes}}.$$

The Product Rule tells us that $|D \times (D \cup \{0,1\})^i| = |D| \cdot |D \cup \{0,1\}|^i = 8 \cdot 10^i$ for any i. (To be totally pedantic: we're using the Sum Rule to conclude that $|D \cup \{0,1\}| = |D| + |\{0,1\}| = 10$, because D and $\{0,1\}$ are disjoint.) Therefore:

$$\begin{aligned} &\left|D \times (D \cup \{0,1\})^9 \cup D \times (D \cup \{0,1\})^4 \cup D \times (D \cup \{0,1\})^5\right| \\ &= \left|D \times (D \cup \{0,1\})^9\right| + \left|D \times (D \cup \{0,1\})^4\right| + \left|D \times (D \cup \{0,1\})^5\right| && \text{Sum Rule: the three types of numbers are disjoint because they have different lengths} \\ &= 8 \cdot 10^9 + 8 \cdot 10^4 + 8 \cdot 10^5 && \text{Product Rule, as described in the previous paragraph} \\ &= 8{,}000{,}880{,}000. \end{aligned}$$

Problem-solving tip: When you're confronted with a counting problem that appears complicated, try to find a nice way of splitting the problem into several disjoint options. Often a difficult counting problem is actually the sum of two simple counting problems.

Combining sums and products: prefix-free codes

We'll end the section with two somewhat more complicated counting problems, where we're asked to calculate the number of objects meeting some particular condition: sets of bitstrings such that no string is a prefix of another, and results of a best-of-five series of games. In both cases, we can give a solution based entirely on a brute-force approach by simply enumerating all possible sequences, eliminating any that don't meet the stated condition, and counting the uneliminated sequences one by one. But there are also ways to break down the set of objects of interest into subsets that we can count using the Sum and (Generalized) Product Rules.

A *prefix-free code* is a set C of bitstrings with the property that no $x \in C$ is a prefix of any other $y \in C$. (For example, if $010 \in C$, then we must have $0101 \notin C$, because 010 is a prefix of 0101.) Let's compute the number of prefix-free codes where all of the codewords are only 1 or 2 bits long:

0	1	00	01	10	11	ok?
						✓
					✓	✓
				✓		✓
				✓	✓	✓
			✓			✓
			✓		✓	✓
			✓	✓		✓
			✓	✓	✓	✓
		✓				✓
		✓			✓	✓
		✓		✓		✓
		✓		✓	✓	✓
		✓	✓			✓
		✓	✓		✓	✓
		✓	✓	✓		✓
		✓	✓	✓	✓	✓

0	1	00	01	10	11	ok?
	✓					✓
	✓				✓	11
	✓			✓		10
	✓			✓	✓	10
	✓		✓			✓
	✓		✓		✓	11
	✓		✓	✓		10
	✓		✓	✓	✓	10
	✓	✓				✓
	✓	✓			✓	11
	✓	✓		✓		10
	✓	✓		✓	✓	10
	✓	✓	✓			✓
	✓	✓	✓		✓	11
	✓	✓	✓	✓		10
	✓	✓	✓	✓	✓	10

0	1	00	01	10	11	ok?
✓						✓
✓					✓	✓
✓				✓		✓
✓				✓	✓	✓
✓			✓			01
✓			✓		✓	01
✓			✓	✓		01
✓			✓	✓	✓	01
✓		✓				00
✓		✓			✓	00
✓		✓		✓		00
✓		✓		✓	✓	00
✓		✓	✓			00
✓		✓	✓		✓	00
✓		✓	✓	✓		00
✓		✓	✓	✓	✓	00

0	1	00	01	10	11	ok?
✓	✓					✓
✓	✓				✓	11
✓	✓			✓		10
✓	✓			✓	✓	10
✓	✓		✓			01
✓	✓		✓		✓	01
✓	✓		✓	✓		01
✓	✓		✓	✓	✓	01
✓	✓	✓				00
✓	✓	✓			✓	00
✓	✓	✓		✓		00
✓	✓	✓		✓	✓	00
✓	✓	✓	✓			00
✓	✓	✓	✓		✓	00
✓	✓	✓	✓	✓		00
✓	✓	✓	✓	✓	✓	00

Figure 9.8: All 64 subsets of $\{0, 1, 00, 01, 10, 11\}$, with indication of whether the subset is prefix-free or not. In each row (a subset), if the set is not prefix-free, then one violation found in the set is listed.

Example 9.21 (Prefix-free codes)

One simple way to find the number of prefix-free codes $C \subseteq \{0,1\}^1 \cup \{0,1\}^2$ is to write down all subsets of $S := \{0,1\}^1 \cup \{0,1\}^2$, and then check each subset to eliminate any set that violates the prefix rule. (See Figure 9.8, which was generated by a computer program; there are 25 codes in the table that pass the prefix test.) There are $2^{|S|} = 2^6 = 64$ subsets of S: we can describe each subset of S as an element of $\{\text{yes}, \text{no}\}^{|S|}$ where the ith component tells us whether the ith element of S is in the set. The Product Rule tells that $|\{\text{yes}, \text{no}\}^{|S|}| = 2^6 = 64$. (See Lemma 9.10.)

Here's a different approach, involving more thinking and less brute-force calculation. Let's partition the set of valid codes into four classes based on whether $0 \in C$ and $1 \in C$:

- If $0 \notin C$ and $1 \notin C$, then any subset of $\{00, 01, 10, 11\}$ can be in C.
- If $0 \notin C$ and $1 \in C$, then any subset of $\{00, 01\}$ can also be in C.
- If $0 \in C$ and $1 \notin C$, then any subset of $\{10, 11\}$ can also be in C.
- If $0 \in C$ and $1 \in C$, then no 2-bit strings can be included.

By the Product Rule, there are, respectively, 2^4 and 2^2 and 2^2 and 2^0 choices corresponding to these classes. (The four classes correspond to the four columns of Figure 9.8.) By the Sum Rule, the total number of prefix-free codes using 1- and 2-bit strings is $16 + 4 + 4 + 1 = 25$.

Taking it further: Prefix-free codes are useful in that they can be transmitted unambiguously, without a special marker that separates codewords. For example, consider the prefix-free code $\{0, 10, 11\}$. Then a sequence 0101111100 can only be interpreted as $0\|10\|11\|11\|10\|0$. If a code is not prefix-free—like the English language!—then a sequence of codewords cannot be unambiguously decoded: for example, THEME might be one word (*theme*) or it might be two (*the me*).

Huffman coding—named after David Huffman, a 20th-century American computer scientist—is an algorithm for computing a prefix-free code that can be used for data compression for English (for example), by allowing us to translate each letter into a corresponding code word. Huffman coding carefully assigns shorter codewords to more commonly used letters, and thus has a special property: among all prefix-free codes, its codewords have the smallest length, on average. A Huffman code can be constructed using a simple greedy approach; for more, see a good textbook on algorithms.

Combining sums and products: a best-of-five series

Here's one more example of using our counting rules in combination:

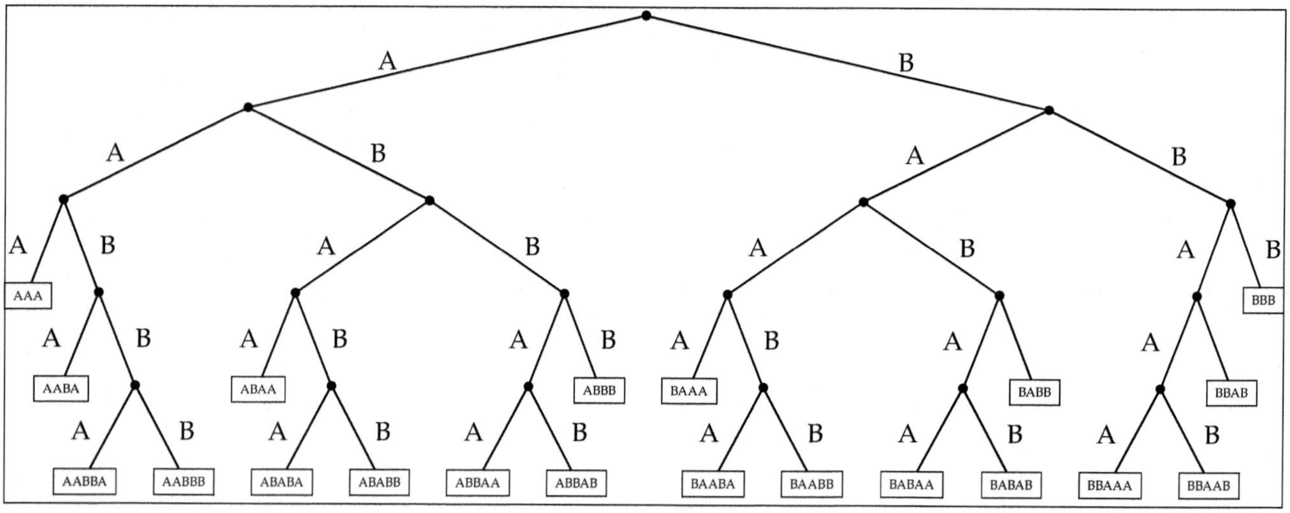

Figure 9.9: A tree representing each best-of-five series of games between two teams, A and B. The branch points correspond to the games, and are labeled by the winner of the game. The 20 different sequences of outcomes are shown at the bottom of the tree.

Example 9.22 (A best-of-five series)

Problem: Suppose that two teams A and B play a best-of-five series of games: the teams play until one team has won three games, at which point the match is over, and that team is the winner. How many different sequences of outcomes are there?

Solution: The simplest approach is to use brute force: simply write out all possible sequences of outcomes, and count them up. This approach is shown in Figure 9.9. However, there's another way to count. Suppose that team A wins the series:

- There's 1 outcome in which A never loses: A wins games 1, 2, and 3.
- There are 3 outcomes in which A loses once: A loses immediately before its first win ($BAAA$), before its second win ($ABAA$), or before its third win ($AABA$).
- If A loses twice, then A must have won the fifth game, and exactly two of the first four. Thinking of the outcomes of the first four games as 4-bit strings with 1s denoting A's wins, Example 9.4 says there are precisely 6 such outcomes.

In sum, there are $1 + 3 + 6 = 10$ ways for A to win the series. There are 10 analogous ways for B to win, so there are 20 outcomes in total.

COMPUTER SCIENCE CONNECTIONS

RUNNING OUT OF IP ADDRESSES, AND IPv6

A crucial component of the internet is the assignment of an *address* to every machine connected to the network. This address is called an *IP address*, where "IP" stands for *Internet Protocol*—the algorithm by which packets of information are handled while they're being transmitted across the internet. Each packet of information to be transmitted stores a variety of pieces of information, including (1) some basic header information; (2) a *source address* (the sender of the information); (3) a *destination address* (the intended recipient of the information); and (4) the data to be transmitted (the "payload").

The subfield of computer science called *computer networking* is devoted to everything about how the internet (or some smaller network) works: design of the network, physical systems, protocols for routing, and more.[1] Here we are going to concentrate on the *IP address itself*, and a particular issue related to how many—or how few!—addresses there are.

For more, see a good textbook on computer networks, like

[1] James F. Kurose and Keith W. Ross. *Computer Networking: A Top-Down Approach*. Addison–Wesley, 6th edition, 2013.

Each device on the internet that can send or receive information needs an address by which to do so. For almost the entire history of the internet, an IP address has simply been a 32-bit string. These IP addresses are typically represented as an element of $\{0, \ldots, 255\}^4$ instead of as an element of $\{0,1\}^{32}$, by converting 8 bits at a time into base-10 numbers, and then writing each 8-bit chunk separated by periods. For example, the site `cs.carleton.edu` is associated with the IP address

$$\underbrace{10001001}_{137} . \underbrace{00010110}_{22} . \underbrace{00000100}_{4} . \underbrace{00010111}_{23}.$$

You can find the IP address of your favorite site using a tool called `nslookup` on most machines, which checks a so-called *name server* to translate a site's name (like `whitehouse.gov`) into an IP address (like `173.223.132.110`).

As an easy counting problem, we can check that there only $2^{32} = 4{,}294{,}967{,}296$ different possible 32-bit IP addresses—about 4.3 billion addresses. Every machine connected to the internet needs to be addressable to receive data, so that means that we can only support about 4.3 billion connected devices. In the 1990s and 2000s, more and more people began to have machines connected to the internet, and each person also began to have more and more devices that they wanted to connect. It became clear that we were facing a dire shortage of IP addresses! As such, a new version of the Internet Protocol (version six, hence called *IPv6*) has been introduced.

There are some strategies from computer networking for conserving addresses by "translation," so that several computers $c_1, c_2, \ldots$ can be connected via an access point p—where p is the only machine that has a public, visible IP address. All of those computers' traffic is handled by p, but p must be able to reroute the traffic it receives to the correct one of the c_i computers. For more information, see the Kurose–Ross textbook cited previously.

In IPv6, instead of using 32-bit addresses, we now use 128-bit addresses. There are some tricky elements to the transition from 32-bit to 128-bit addresses—your computer better keep working!—but there are now 2^{128} different addresses available. That's $340{,}282{,}366{,}920{,}938{,}463{,}463{,}374{,}607{,}431{,}768{,}211{,}456 \approx 3.4 \times 10^{38}$, which should hold us for a few millennia. For example, `whitehouse.gov` is associated with a 32-bit address `173.223.132.110`, *and* a 128-bit address `2600:1408:0010:019a:0fc4`, represented by 5 blocks of 4 hexadecimal numbers—that is, as an element of

$$\left[\{\texttt{0,1,2,3,4,5,6,7,8,9,a,b,c,d,e,f}\}^4\right]^5.$$

Computer Science Connections

A Lower Bound for Comparison-Based Sorting

Most people who encounter the *sorting* problem—given an array $A[1 \ldots n]$, rearrange A so that it's in ascending order—initially devise a quadratic-time algorithm. (For simplicity, suppose that we're sorting *distinct* elements.) The most common examples of $\Theta(n^2)$-time algorithms are Selection Sort, Insertion Sort, and Bubble Sort. Then, after a lot of thought (and, usually, some help), those people often are able to devise a $O(n \log n)$-time sorting algorithm, like Merge Sort, Quick Sort, or Heap Sort. (See Section 6.3.)

But suppose that you were extra impatient with the speed of your sorting algorithm, and you were extra, extra clever. Could you do asymptotically better than $O(n \log n)$ in the worst case? The answer, we'll show, is no—with a footnote: any "comparison-based" sorting algorithm requires $\Omega(n \log n)$ time. (The footnote is that it depends on what we mean by "sort," as we'll see.)

```
selectionSort(A[1...n]):
  for i := 1 to n:
      minIndex := i
      for j := i+1 to n:
          if A[j] < A[minIndex] then
              minIndex := j
      swap A[i] and A[minIndex]
```

Figure 9.10: Selection Sort.

A Warm-up: Selection Sort

First, recall Selection Sort, shown in Figure 9.10. One way to analyze its running time is as we did in Example 6.7: there are n iterations, and in the $(n-i)$th iteration we require i steps. In other words, the running time of Selection Sort is $\sum_{i=1}^{n} i$. We could repeat the straightforward inductive proof that $\sum_{i=1}^{n} i = n(n+1)/2$, but instead Figure 9.11 gives a more visual way of seeing this result. Figure 9.11(a) shows a shaded triangle that represents the running time of selection sort: $\sum_{i=1}^{n} i$, where row i of the triangle has i steps in it. Figure 9.11(b) shows that this triangle is *contained within an n-by-n square* and also *contains an $\frac{n}{2}$-by-$\frac{n}{2}$ square*. Thus the area of the triangle is upper bounded by $n \cdot n = n^2$ and lower bounded by $\frac{n}{2} \cdot \frac{n}{2} = \frac{n^2}{4}$, and therefore is $\Theta(n^2)$. This picture is a visual representation of a more algebraic proof:

$$\sum_{i=1}^{n} i \le \sum_{i=1}^{n} n = n^2, \qquad \text{and} \qquad \sum_{i=1}^{n} i \ge \sum_{i=\frac{n}{2}+1}^{n} i \ge \sum_{i=\frac{n}{2}+1}^{n} \frac{n}{2} = \frac{n^2}{4}.$$

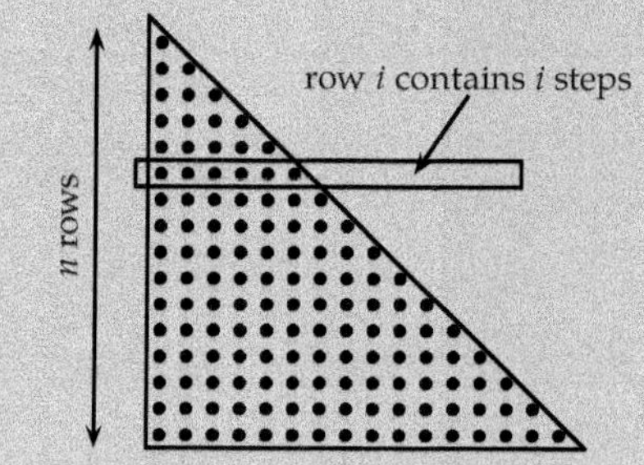

(a) Selection Sort's running time.

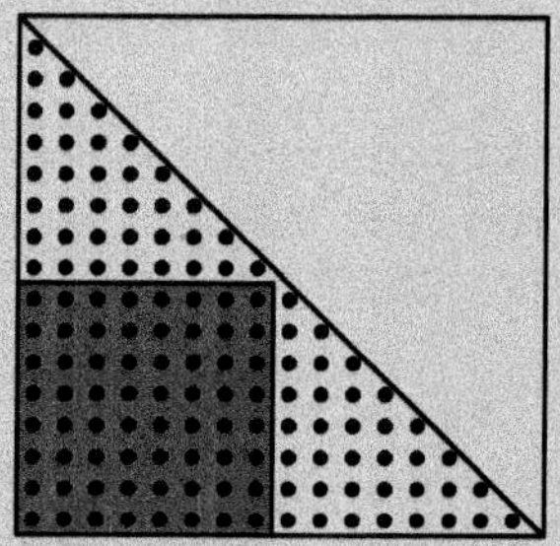

(b) The analysis of the running time.

Figure 9.11: A visual representation of the proof that Selection Sort runs in $\Theta(n^2)$ time.

While the analysis of Selection Sort isn't necessary for our main proof, the style of analysis from Figure 9.11 will be useful in a moment.

There Are No $O(n)$ Comparison-Based Sorting Algorithms

All of the sorting algorithms that we've encountered in the book are *comparison-based sorting algorithms*: they proceed by repeatedly *comparing* the values of two elements x_i and x_j from the input array without considering *the values themselves*. Depending on the result of the comparison, the algorithm may then swap some elements of the array. (Comparison-based sorting algorithms probably include every sorting algorithm that you've ever seen, except counting, radix, and bucket sorts.)

One way to view a comparison-based sorting algorithm is through a *decision tree*, like the one shown in Figure 9.12 for Selection Sort on a 3-element array. The internal nodes encode the comparisons made by the algorithm. The leaves correspond to sorted orders—the output of the sorting algorithm.

COMPUTER SCIENCE CONNECTIONS

SORTING LOWER BOUNDS, CONTINUED

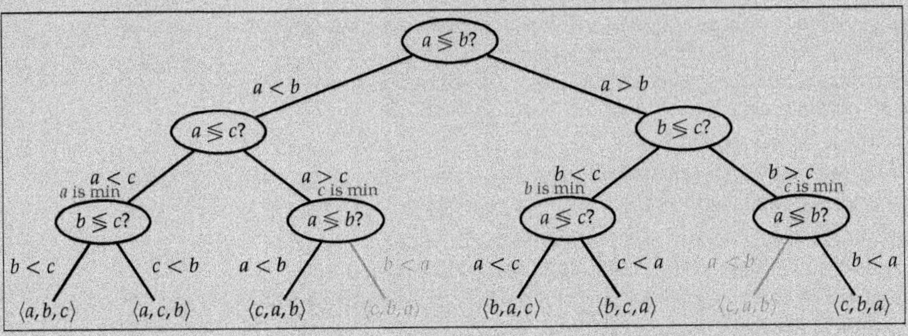

Figure 9.12: The decision tree for Selection Sort on the input array $\langle a, b, c \rangle$. Selection Sort first does two comparisons to find the minimum value of $\{a, b, c\}$, and subsequently compares the remaining two elements to decide on the final order. The two lighter-shaded branches of the tree are logically inconsistent, but Selection Sort reexecutes the a-versus-b comparison in those cases.

The running time of the sorting algorithm whose input corresponds to a particular leaf is Ω(number of comparisons on that root-to-leaf path) because, although the algorithm might do more than compare—in fact, it must (for example, it has to perform swaps)—it must do at least these comparisons.

We will use the decision tree to establish a lower bound on the running time of comparison-based sorting algorithms:

Theorem 9.8
Any comparison-based sorting algorithm requires $\Omega(n \log n)$ time.

Proof. Consider the decision tree T of the sorting algorithm. First, observe that T *must have at least* $n!$ *leaves.* There are $n!$ different permutations of the input, and a correct algorithm must be capable of producing any of these permutations as output. Second, observe that T *has at most* 2^d *nodes at depth* d. (It's a binary tree!) Thus the height h of T satisfies $2^h \geq n!$. Taking logarithms of both sides, we have

$$\begin{aligned} h \geq \log_2(n!) &= \log_2 [n \cdot (n-1) \cdot (n-2) \cdot \ \cdots \ \cdot (\tfrac{n}{2}+1) \cdot (\tfrac{n}{2}) \cdot \ \cdots \ \cdot 1] \\ &\geq \log_2 [n \cdot (n-1) \cdot (n-2) \cdot \ \cdots \ \cdot (\tfrac{n}{2}+1)] \\ &\geq \log_2 \left[(\tfrac{n}{2})^{(n/2)} \right] \\ &= (\tfrac{n}{2}) \cdot \log_2 (n/2) \\ &= \Omega(n \log n). \end{aligned}$$

□

The crucial fact here is precisely analogous to the one in Figure 9.11:

$$\prod_{i=1}^{n} i \geq \prod_{i=\frac{n}{2}+1}^{n} i \geq \prod_{i=\frac{n}{2}+1}^{n} \tfrac{n}{2} = (\tfrac{n}{2})^{n/2}.$$

The only difference is that here we're using products instead of summations.

A LINEAR-TIME SORTING ALGORITHM

While we've now shown that every *comparison-based* sorting algorithm takes $\Omega(n \log n)$ time, there are faster algorithms for special cases. Figure 9.13 shows one, called *counting sort*, which allows us to sort without comparing elements to each other. As long as the elements of the array are *integers from a small range*, then this algorithm is fast: the running time is $\Theta(c + n)$ (the last nested loop requires $\sum_v count[v] = n$ time); as long as c is small, this algorithm runs in linear time.

```
countingSort(A[1...n]):
Input: array (A[1...n]) where each
    A[i] ∈ {1,2,...,c}.
for v := 1 to c:
    count[v] := 0
for i := 1 to n:
    count[A[i]] := count[A[i]] + 1
i := 1
for v := 1 to c:
    for t := 1 to count[v]:
        A[i] := v
        i := i + 1
```

Figure 9.13: Counting Sort.

9.2.5 *Exercises*

9.1 A *tweet* (a message posted on Twitter) is a sequence of at most 140 characters. Assuming there are 256 valid different characters that can appear in each position, how many distinct tweets are possible?

Cars in the United States display license plates containing an alphanumeric code, whose format varies from state to state. Each of the following questions describes the format of currently issued license plates by some state. For each, determine the number of different license plate codes (often misleadingly called license plate numbers *despite the presence of letters) of the listed form. All letters in all codes are upper case.*

9.2 Minnesota: digit-digit-digit-letter-letter-letter (as in `400GPA`).

9.3 Pennsylvania: letter-letter-letter-digit-digit-digit-digit (as in `EEE2718`).

9.4 Connecticut: digit-letter-letter-letter-letter-digit (as in `4FIVE6`).

9.5 You have been named Secretary of Transportation for the State of [Your Favorite State]. Congratulations! You're considering replacing the current license plate format `ABCD-1234` (4 letters followed by 4 digits) with a sequence of any k symbols, each of which can be either a letter or a digit. How large must k be so that your new format has at least as many options as the old format did?

9.6 Until recently, France used license plates that contain codes of any of the following forms:

- digit-digit-digit-letter-digit-digit.
- digit-digit-digit-letter-letter-digit-digit, where the first letter is alphabetically $\leq$ `P`.
- digit-digit-digit-digit-letter-letter-digit-digit, where the first letter is alphabetically $\geq$ `Q`.
- digit-digit-digit-letter-letter-letter-digit-digit.

How many license plates, in total, met the French requirements?

9.7 A particular voicemail system allows numerical passwords of length 3, 4, or 5 digits. How many passwords are possible in this system?

9.8 What about numerical passwords of length 4, 5, or 6?

A contact lens is built with the following parameters: a (spherical) power *(for correcting near- or farsightedness); and, possibly, a* cylindrical power *and an* axis *(for correcting astigmatism). For a particular brand of contacts, the possible parameters for a lens that corrects near- or farsightedness only are*

- *a power between* -6.00 *and* $+6.00$ *inclusive in* 0.25 *steps (excluding* 0.00*); between* 6.50 *and* 8.00 *inclusive in* 0.50 *steps; and between* -6.50 *and* -10.00 *inclusive in* 0.50 *steps,*

and the parameters for a lens that corrects astigmatism are

- *one of the powers listed previously;*
- *a cylindrical power in* $\{-0.75, -1.25, -1.75, -2.25\}$*; and*
- *an axis between* $10°$ *and* $180°$ *in steps of* $10°$.

9.9 How many different contact lenses are there?

9.10 A patient needing vision correction in both eyes may get different contact lenses for each eye. A *prescription* assigns a lens for the left eye and for the right eye. How many contact prescriptions are there?

9.11 During the West African Ebola crisis that started in 2014, geneticists were working to trace the spread of the disease. To do so, they acquired DNA samples of the viruses from a number of patients, and affixed a unique "tag" to each patient's sample.[2] A *tag* is a sequence of 8 nucleotides—each an element of {`A`, `C`, `G`, `T`}—attached to the end of a virus sample from each patient, so that subsequently it will be easy to identify the patient associated with a particular sample. How many different such tags are there?

[2] Richard Preston. The Ebola wars. *The New Yorker*, 27 October 2014.

9.12 In a computer science class, there are 14 students who have previously written a program in Java, and 12 students who have previously written a program in Python. How many students have previously written a program in at least one of the two languages? (If you can't give a single number as a definitive answer, give as narrow a range of possible values as you can.)

9.13 True story: a relative was given a piece of paper with the password to a wireless access point that was written as follows: a154bc0401011. But she couldn't tell from this handwriting whether each "1" was `1` (one), `l` (ell), or `I` (eye); or whether "0" was `0` (zero) or `O` (oh). How many possible passwords would she have to try before having exhausted all of the possibilities?

A Rubik's cube—*named after the 20th-century Hungarian architect Ernő Rubik—is a 3-by-3-by-3 grid of cells, where any of the six nine-cell faces (top, bottom, left, right, front, back) can be rotated 90° clockwise or counterclockwise in a single move. (See Figure 9.14.) Each face of each cell is colored with one of six colors (blue, red, green, yellow, white, and orange); initially, all nine cell-faces on each cube-face have the same color, but the cube can then be scrambled. The challenge is to use rotations to configure a scrambled cube such that each face of the cube contains nine cells of the same color.*

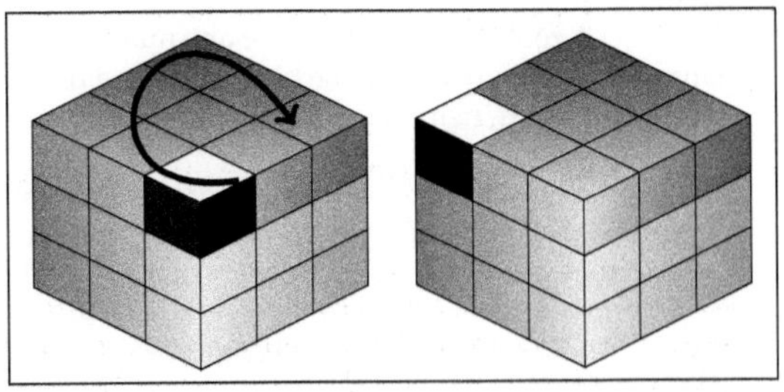

Figure 9.14: A Rubik's cube with the top face (and one cell in particular) highlighted, and the result of a single move—rotating the top face clockwise.

9.14 How many Rubik's cube moves are there?

9.15 It is known that, from any configuration, 26 moves suffice to solve the cube. (Note that we're counting every 90° rotation as a move; if you rotate the same face 180° by using two consecutive 90° moves, it counts as two moves.) How many sequences of 26 moves are possible?

9.16 It's useless to rotate a face clockwise in one move, and rotate the same face counterclockwise in the next move. (You've just undone the previous move.) A counterclockwise move followed by a clockwise move is analogous. How many sequences are there of 26 moves that never undo the previous move?

Emacs *is a widely used software program for—among other things—editing text documents (including this book!). Here's a mildly simplified description of Emacs (to make this problem more manageable). In Emacs, a* command character *is produced by pressing a letter key while holding down either the* Control *key, the* Meta *key, or both. (For example, Control+*`Y` *or Meta+*`B` *or Control+Meta+*`U` *are command characters.)*

9.17 How many command characters are there in Emacs?

Emacs is complicated enough that it needs more commands than Exercise 9.17 allows. To allow for more commands, Emacs has been extended, as follows. Meta+`X` *and Control+*`X`*—as in eXtended—are command* prefixes, *so that neither Meta+*`X` *nor Control+*`X` *is a valid command, but, for example, "Control+*`X` *Control+*`U`*" is (and it's different from Control+*`U`*). A valid command can be formed by Control+*`X` *or Meta+*`X` *followed by any letter or any command character (including Control+*`X` *or Meta+*`X`*). All other command characters from Exercise 9.17 are still valid.*

9.18 How many command characters are there now?

9.19 Argue that, for any sets A and B, $|A \cup B| = |A - B| + |B - A| + |A \cap B|$. (Use the Sum Rule.)

9.20 How many 100-bit strings have *at most* 2 ones? (Use Example 9.4.)

9.21 Determine how many k-bit strings have exactly three 1s using the approach in Example 9.4—that is, by dividing the set of bitstrings based on the position of the third one.

9.22 *(programming required)* Write a program, in a language of your choice, to enumerate all bitstrings in $\{0,1\}^{16}$ and count the number that have 0, 1, 2, and 3 ones. Use this program to verify your answer to the last exercise and your approach to Exercise 9.20.

9.23 The following is a simpler "solution" to Example 9.4, where we computed the number of elements of $\{0,1\}^k$ that have precisely two 1s. What, exactly, is wrong with this argument?

We wish to determine $|S|$, where S is the set of k-bit strings with exactly 2 ones. Define $S_i := \{x \in S : x_i = 1\}$, for each $i \in 1, 2, \ldots, k$. Observe that $S = \bigcup_{i=1}^{k} S_i$ and that $|S_i| = k - 1$. Therefore, by the Sum Rule, $|S| = \sum_{i=1}^{k} |S_i| = \sum_{i=1}^{k} (k-1) = k(k-1)$.

Unicode *is a character set frequently used on the web; it supports hundreds of thousands of characters from many languages—English, Greek, Chinese, Arabic, and all other scripts in current use. A very common encoding scheme for Unicode, called* UTF-8, *uses a variable number of bits to represent different characters (with more commonly used characters using fewer bits). Valid UTF-8 characters can be of any of the following forms, using 1, 2, 3, or 4 bytes, and have one of the following forms (where* `x` *represents an arbitrary bit):*

- `0xxxxxxx`
- `110xxxxx 10xxxxxx`
- `1110xxxx 10xxxxxx 10xxxxxx`
- `11110yyy 10yyxxxx 10xxxxxx 10xxxxxx`, *with a further restriction: the first five bits (marked* `yyyyy`*) must be either of the form* `0xxxx` *or* `10000`.

The ith character in the Unicode character set is encoded by the ith legal UTF-8 representation, resulting from converting i into binary and filling in the `x` *(and* `y`*) bits from the templates.*

9.24 How many characters can be encoded using UTF-8?

9.25 There's a rule for Unicode that doesn't allow excess zero padding: if a character can be encoded using one byte, then the two-byte encoding is illegal. For example, `01010101` encodes the same character as `11000001 10010101`; thus the latter is illegal. How many of the characters from the last exercise can be encoded without violating this rule?

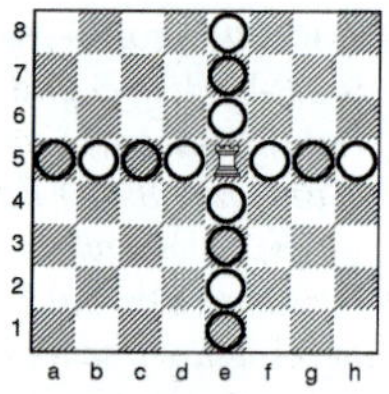

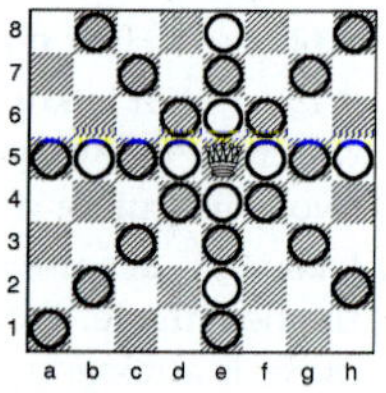

Figure 9.15: Two chess boards, showing the legal moves for a rook (above) and queen (below).

9.26 A rook in chess can move any number of spaces horizontally or vertically. (See Figure 9.15.) How many ways are there to put one black rook and one white rook on an 8-by-8 chessboard so they can't capture each other (that is, neither can move onto the other's square)?

9.27 A queen in chess can move any number of spaces horizontally, vertically, or diagonally. (Again, see Figure 9.15.) How many ways are there to put one black queen and one white queen on an 8-by-8 chessboard so they can't capture each other (that is, neither can move onto the other's square)? *(Hint: think about how far the black queen is from the edge of the board.)*

9.28 *(programming required)* Write a program to verify your solution to the previous exercise.

9.29 You have a wireless-enabled laptop, phone, and tablet. Each device needs to be assigned a unique "send" frequency and a unique "receive" frequency to communicate with a base station. Let $S := \{1, \ldots, 8\}$ denote send frequencies and $R := \{a, \ldots, h\}$ receive frequencies. A *frequency assignment* is an element of $S \times R$. A set of frequency assignments is *noninterfering* if no elements of S or R appears twice. How many noninterfering frequency assignments are there for your three devices?

9.30 Write down an inclusion–exclusion formula for $|A \cup B \cup C \cup D|$.

9.31 How many integers between 1 and 1000, inclusive, are divisible by one or more of 3, 5, and 7?
9.32 How many integers between 1 and 1000, inclusive, are divisible by one or more of 6, 7, and 8?
9.33 How many integers between 1 and 10000, inclusive, are divisible by at least one of 2, 3, 5, or 7?

In Chapter 7, we encountered the totient function $\varphi : \mathbb{Z}^{\geq 1} \to \mathbb{Z}^{\geq 0}$, defined as

$$\varphi(n) := \text{the number of } k \text{ with } 1 \leq k \leq n \text{ such that } k \text{ and } n \text{ have no common divisors.}$$

We can always compute the totient of n by brute force (just test all $k \in \{1, \ldots, n\}$ for common divisors using the Euclidean algorithm, for example). But the next few exercises will give a hint at another way to do this computation more efficiently. For a fixed integer n:

9.34 Suppose $m \in \mathbb{Z}^{\geq 1}$ evenly divides n. Define $M := \{k \in \{1, \ldots, n\} : m \mid k\}$. Argue that $|M| = \frac{n}{m}$.

9.35 *(A number-theoretic interlude.)* Let the prime factorization of n be $n = p_1^{e_1} \cdot p_2^{e_2} \cdots p_\ell^{e_\ell}$, for distinct prime numbers $\{p_1, \ldots, p_\ell\}$ and integers $e_1, \ldots, e_\ell \geq 1$. Let $k \leq n$ be arbitrary. Argue that k and n have no common divisors greater than 1 if and only if, for all i, we have $p_i \nmid k$.

9.36 Let n be an integer such that $n = p^i q^j$ for two distinct prime numbers p and q, and integers $i \geq 1$ and $j \geq 1$. (For example, we can write $544 = 17^1 \cdot 2^5$; here $p = 17, q = 2, i = 1$, and $j = 5$.) Let $P := \{k \in \{1, \ldots, n\} : p \mid k\}$ and $Q := \{k \in \{1, \ldots, n\} : q \mid k\}$. Argue that $\varphi(n) = n(1 - \frac{1}{p})(1 - \frac{1}{q})$ by using Inclusion–Exclusion to compute $|P \cup Q|$. (You should find the last two exercises helpful.)

In the sport of cricket, a team consists of 11 players who come up to bat in pairs. Initially, players #1 and #2 bat. When one of those two players gets out, then player #3 replaces the one who got out. When one of the two batting players—player #3 and whichever player of {#1, #2} didn't get out—gets out, then player #4 joins the one who isn't out. This process continues until the 10th player gets out, leaving the last player not out (but stranded without a partner).

Thus, in total, there are 11 players who bat together in 10 partnerships. As an example, consider the lineup Anil, Brendan, Curtly, Don, Eoin, Freddie, Glenn, Hansie, Inzamam, Jacques, Kumar. We could have the following batting partnerships: Anil & Brendan; Anil & Curtly; Anil & Don; Don & Eoin; Don & Freddie; …; Don & Kumar.

9.37 How many different partnerships (pairs of players) are possible?

9.38 How many different sequences of partnerships (like the example list of partnerships given previously) are possible? (It doesn't matter which of the last two players gets out.)

9.39 A team's batting lineup may be truncated (by winning the game or by choosing not to bat any longer) at any point after the first pair starts batting. Now how many different sequences of partnerships are possible? Here, it *does* matter whether one of the last two players gets out or not (but not which of the two was the one who got out).

9.40 Suppose that, as in Example 9.11, a bank allows 4-digit PINs, but doesn't permit a PIN that starts with the same digit repeated twice (for example, 7730) or ends with the same digit repeated twice (for example, 0122). Now how many invalid PINs are there?

9.41 Let S_k denote the set of PINs that are k digits long, where the PIN may not start with three repeated digits or end with three repeated digits. In terms of k, what is $|S_k|$? (Example 9.11 computed $|S_4|$.)

Checkers *is a game, like chess, played on an 8-by-8 grid. Chinook, a recently built checkers-playing program that never loses a game,*[3] *computes all possible board positions with up to k tokens, for small k. Over the next few exercises, you'll compute the scope of that task for very small k—namely,* $k \in \{1, 2\}$. *Figuring out how many board positions have two tokens—note that two tokens can't occupy the same square!—will take a little more work.*

Briefly, the rules of checkers are as follows. Two players, Red and Black, move tokens *diagonally on an 8-by-8 grid; tokens can only occupy shaded squares. There are two types of tokens:* pieces *and* kings. *Any piece that has reached the opposite side of the board from its starting side (row 8 or row 1) becomes a king. (So Black cannot have a piece in row 8, because that piece would have become a king.) Note that Black occupying square C3 is different from Red occupying C3. (See Figure 9.16.)*

[3] Jonathan Schaeffer, Neil Burch, Yngvi Bjornsson, Akihiro Kishimoto, Martin Muller, Rob Lake, Paul Lu, and Steve Sutphen. Checkers is solved. *Science*, 317(5844):1518–1522, 14 September 2007.

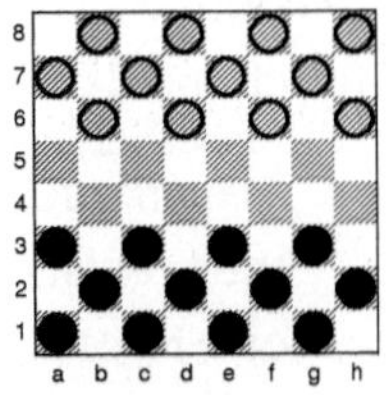

Figure 9.16: A checker board. Pieces can occupy any shaded square; a black piece that reaches row 8 or a red piece that reaches row 1 becomes a king.

9.42 How many board positions have exactly one token (of either color)?

9.43 How many board positions have two kings, one of each color?

9.44 How many board positions have two Red kings? (Notice that two Red kings cannot be distinguished, so it doesn't matter "which" one comes first.)

9.45 How many board positions have two Black pieces?

9.46 How many board positions have two pieces, one of each color?

9.47 How many board positions have one Red king and one Red piece?

9.48 How many board positions have one Black king and one Red piece?

9.49 Use the last six exercises to determine how many total board positions have two tokens.

9.50 *(programming required)* Write a program, in a language of your choice, to verify your answer to the last few exercises (particularly the total count, in the last exercise).

9.51 How many subsets of $\{0, 1\}^1 \cup \{0, 1\}^2 \cup \{0, 1\}^3$ are prefix free? (See Example 9.21.) You will probably find it easiest to solve this problem by writing a program.

A text-to-speech system takes written language (text) and reads it aloud as audio (speech). One of the simplest ways to build a text-to-speech system is to prerecord each syllable, and then paste together those sounds. (Pasting separate recordings is difficult, and this system as described will produce very robotic-sounding speech. But it's a start.) A syllable consists of a consonant or cluster of consonants called the onset, *then a vowel called the* nucleus, *and finally the consonant(s) called the* coda. *In many languages, only some combinations of choices for these parts are allowed—there are fascinating linguistic constraints based on ordering or place of articulation (for example, English allows stay but not tsay, and allows clay and play but not tlay) that we're almost entirely omitting here.*

9.52 A consonant can be described by a *place of articulation* (one of 11 choices: the lips, the palate, etc.); a *manner of articulation* (one of 8 choices: stopping the airflow, stopping the oral airflow with the nasal passage open, etc.); and a *voicing* (the vocal cords are either vibrating, or not). According to this description, how many consonants are there?

9.53 A vowel can be described as either *lax* or *tense*; as either *high* or *mid* or *low*; and as either *front* or *central* or *back*. According to this description, how many vowels are there?

9.54 As a (very!) rough approximation, Japanese syllables consist of one of 25 consonants followed by one of 5 vowels, with one consonant that can appear as a coda (or the coda can be left off). How many Japanese syllables are there?

9.55 As a rough (even rougher!) approximation, English syllables consist of an onset that is either one of 25 consonants or a *cluster* of any two of these consonants, followed by one of 16 vowels, followed optionally by one of 25 consonants. How many English syllables are there?

9.56 To cut down on the large number of syllables that you found in the last exercise, some systems are instead based on *demisyllables*—the first half or the second half of a syllable. (We glue the sounds together in the middle of the vowel.) That is, a demisyllable is either a legal onset followed by a vowel, or a vowel followed by a legal coda. How many demisyllables are there in English (making the same very rough assumptions as the last question)?

9.3 *Using Functions to Count*

The sun's shining bright
Everything seems all right
When we're poisoning pigeons in the park.

Tom Lehrer (b. 1928), "Poisoning Pigeons In The Park"

Our focus in Section 9.2 was on counting sequences of choices (the Generalized Product Rule) and choices of choices (the Sum Rule). But what about counting other kinds of sets? Our basic plan is simple: *be lazy!* In this section, we'll introduce ways of counting the cardinality of a given set A *in terms of* $|B|$ *for some other set* B, by using functions that translate between the elements of A and the elements of B:

- *Mapping Rule:* There exists a bijection $f : A \to B$ if and only if $|A| = |B|$. Similarly, there exists an onto function $f : A \to B$ if and only if $|A| \geq |B|$, and there exists a one-to-one function $f : A \to B$ if and only if $|A| \leq |B|$.
- *Division Rule:* Suppose there exists a function $f : A \to B$ such that, for every $b \in B$, we have $|\{a \in A : f(a) = b\}| = k$. Then $|A| = k \cdot |B|$.

In particular, we'll hope to "translate" a choice from an arbitrary set into a sequence of choices from very simple sets—which, using the tools from Section 9.2, we know how to count. Here's a first example to illustrate the basic idea:

Example 9.23 (Number of valid Hamming codewords)

Problem: In Section 4.2, we introduced the Hamming code, an error-correcting code that encodes any 4-bit *message* $m \in \{0,1\}^4$ as a 7-bit *codeword* $x \in \{0,1\}^7$. Specifically, the encoding function *encode* : $\{0,1\}^4 \to \{0,1\}^7$ maps $\langle a,b,c,d \rangle$ to $\langle a,b,c,d,b \oplus c \oplus d, a \oplus b \oplus d, a \oplus b \oplus d \rangle$, where $\oplus$ is exclusive or. That is, a valid Hamming codeword x is an element of $\{0,1\}^7$ satisfying three conditions:

$$x_2 + x_3 + x_4 \equiv_2 x_5 \qquad x_1 + x_3 + x_4 \equiv_2 x_6 \qquad x_1 + x_2 + x_4 \equiv_2 x_7.$$

How many different valid codewords does the Hamming code have?

Solution: We can count the number of valid codewords by looking at all $2^7 = 128$ elements of $\{0,1\}^7$ and testing these three conditions (✓ = pass; ✗ = fail):

codeword	codeword	codeword	codeword	codeword	codeword	codeword	codeword
0000000 ✓✓✓	0010000 ✗✗✓	0100000 ✗✓✗	0110000 ✓✗✗	1000000 ✓✗✗	1010000 ✗✓✗	1100000 ✗✗✓	**1110000 ✓✓✓**
0000001 ✓✓✗	0010001 ✗✗✗	0100001 ✗✓✓	0110001 ✓✗✓	1000001 ✓✗✓	1010001 ✗✓✓	1100001 ✗✗✗	1110001 ✓✓✗
0000010 ✓✗✓	0010010 ✗✓✓	0100010 ✗✗✗	0110010 ✓✓✗	1000010 ✓✓✗	1010010 ✗✗✗	1100010 ✗✓✓	1110010 ✓✗✓
0000011 ✓✗✗	0010011 ✗✓✗	0100011 ✗✗✓	**0110011 ✓✓✓**	**1000011 ✓✓✓**	1010011 ✗✗✓	1100011 ✗✓✗	1110011 ✓✗✗
0000100 ✗✓✓	0010100 ✓✗✓	0100100 ✓✓✗	0110100 ✗✗✗	1000100 ✗✗✗	1010100 ✓✓✗	1100100 ✓✗✓	1110100 ✗✓✓
0000101 ✗✓✗	0010101 ✓✗✗	**0100101 ✓✓✓**	0110101 ✗✗✓	1000101 ✗✗✓	**1010101 ✓✓✓**	1100101 ✓✗✗	1110101 ✗✓✗
0000110 ✗✗✓	**0010110 ✓✓✓**	0100110 ✓✗✗	0110110 ✗✓✗	1000110 ✗✓✗	1010110 ✓✗✗	**1100110 ✓✓✓**	1110110 ✗✗✓
0000111 ✗✗✗	0010111 ✓✓✗	0100111 ✓✗✓	0110111 ✗✓✓	1000111 ✗✓✓	1010111 ✓✗✓	1100111 ✓✓✗	1110111 ✗✗✗
0001000 ✗✗✗	0011000 ✓✓✗	0101000 ✓✗✓	0111000 ✗✓✓	1001000 ✗✓✓	1011000 ✓✗✓	1101000 ✓✓✗	1111000 ✗✗✗
0001001 ✗✗✓	**0011001 ✓✓✓**	0101001 ✓✗✗	0111001 ✗✓✗	1001001 ✗✓✗	1011001 ✓✗✗	**1101001 ✓✓✓**	1111001 ✗✗✓
0001010 ✗✓✗	0011010 ✓✗✗	**0101010 ✓✓✓**	0111010 ✗✗✓	1001010 ✗✗✓	**1011010 ✓✓✓**	1101010 ✓✗✗	1111010 ✗✓✗
0001011 ✗✓✓	0011011 ✓✗✓	0101011 ✓✓✗	0111011 ✗✗✗	1001011 ✗✗✗	1011011 ✓✓✗	1101011 ✓✗✓	1111011 ✗✓✓
0001100 ✓✗✗	0011100 ✗✓✗	0101100 ✗✗✓	**0111100 ✓✓✓**	**1001100 ✓✓✓**	1011100 ✗✗✓	1101100 ✗✓✗	1111100 ✓✗✗
0001101 ✓✗✓	0011101 ✗✓✓	0101101 ✗✗✗	0111101 ✓✓✗	1001101 ✓✓✗	1011101 ✗✗✗	1101101 ✗✓✓	1111101 ✓✗✓
0001110 ✓✓✗	0011110 ✗✗✗	0101110 ✗✓✓	0111110 ✓✗✓	1001110 ✓✗✓	1011110 ✗✓✓	1101110 ✗✗✗	1111110 ✓✓✗
0001111 ✓✓✓	0011111 ✗✗✓	0101111 ✗✓✗	0111111 ✓✗✗	1001111 ✓✗✗	1011111 ✗✓✗	1101111 ✗✗✓	**1111111 ✓✓✓**

By checking every entry in the table, we see that there are 16 valid codewords.

Problem-solving tip: Use programming to help you! If you're going to use the simple-but-tedious way to count legal Hamming code codewords, via enumeration, write a program rather than doing it by hand. (For example, the table in Example 9.23 was generated with a Python program!)

This table-based approach is fine, but here's a less tedious way to count. By the definition of the encoding function, every possible message in $\{0,1\}^4$ is encoded as a different codeword in $\{0,1\}^7$. Furthermore, every valid codeword is the encoding of a message in $\{0,1\}^4$. Thus the number of valid codewords equals the number of messages, and there are $|\{0,1\}^4| = 16$ valid codewords.

9.3.1 *The Mapping Rule*

The approach that we used in Example 9.23 is based on *functions* that translate from one set to another. In the remainder of this section, we will formalize this style of reasoning as a general technique for counting problems. To build intuition about how to use functions to count, let's start with some small, informal examples:

Example 9.24 (Some mappings, informally)

- Let S be a collection of documents, where each document is labeled with one of 5 genres: *poem*, *essay*, *memoir*, *drama*, or *novel*.
  - Suppose every genre appears as the label for at least one document. Then $|S| \geq 5$. (We see 5 different kinds of labels on documents, and every document has only one label. Thus there must be at least 5 different documents.)
  - Suppose there's no genre that appears as the label for two distinct documents. Then $|S| \leq 5$. (No label is reused—that is, no label appears on more than one document—so we can only possibly observe 5 total labels. Every document is labeled, so we can't have more than 5 documents.)
- You're taking a class in which no two students' last names start with the same letter. Then there are at most 26 students in the class.
- You're in a club on campus that has at least one member from every state in the U.S. Then the club has at least 50 members.
- You're out to dinner with friends, and you and each of your friends order one of 8 desserts on the menu. Suppose that each dessert is ordered at least once, and no two of you order the same dessert. Then your group has exactly 8 people.

Taking it further: The document/genre scenario in Example 9.24 is an example of a *classification problem*, where we must *label* some given input data ("instances") as belonging to exactly one of k different *classes*. Classification problems are one of the major types of tasks encountered in the subfield of CS called *machine learning*. In machine learning, we try to build software systems that can "learn" how to better perform a task on the basis of some training data. Other problems in machine learning include *anomaly detection*, where we try to identify which instances from a set "aren't like" the others; or *clustering problems* (see p. 234), where we try to separate a collection of instances into coherent subgroups—for example, separating a collection of documents into "topics." Classification problems are very common in machine learning: for example, we might want to classify a written symbol as one of the 26 letters of the alphabet (*optical character recognition*); or classify a portion of an audio speech stream as one of 40,000 common English words (*speech recognition*); or classify an email message as either "spam" or "not spam" (*spam detection*).

FORMALIZING THE RULE

How can we generalize the intuition of Example 9.24 into a rule for counting? Think about the first scenario, the documents and the genres: we can view the labels on the documents in S as being given by a function

$$label : S \to \{poem, essay, memoir, drama, novel\}.$$

If there exists any function that behaves in the way that *label* did in Example 9.24—that is, either "covering" all of the possible outputs at least once each, or covering all of the possible outputs *at most* once each—then we can infer whether the set of possible inputs or the set of possible outputs is bigger.

The formal statements of the counting rules based on this intuition rely on the definition of three special types of functions that we defined in Chapter 2: onto functions, one-to-one functions, and bijections. (See Figure 9.17 for a reminder of the definitions.) Formally, the existence of a function $f : A \to B$ with one of these properties will let us relate $|A|$ and $|B|$:

Definition reminder: onto, one-to-one, and bijective functions.
Let A and B be two sets, and let $f : A \to B$ be a function. Then:

- f is *onto* if, for all $b \in B$, there exists an $a \in A$ such that $f(a) = b$.
- f is *one-to-one* if, for all $a \in A$ and $a' \in A$, if $f(a) = f(a')$ then $a = a'$.
- f is a *bijection* if it is both one-to-one and onto.

Slightly less formally: the function f is onto if "every possible output is hit"; f is one-to-one if "no output is hit more than once"; and f is a bijection if "every output is hit exactly once."

Figure 9.17: A reminder of Definitions 2.49, 2.50, and 2.51 (onto, one-to-one, and bijective functions).

Theorem 9.9 (Mapping Rule)
Let A and B be arbitrary sets. Then:

- *An onto function $f : A \to B$ exists if and only if $|A| \geq |B|$.*
- *A one-to-one function $f : A \to B$ exists if and only if $|A| \leq |B|$.*
- *A bijection $f : A \to B$ exists if and only if $|A| = |B|$.*

See Figure 9.18 for a visual representation of the Mapping Rule, and for the intuition as why it's correct: the number of arrows leaving A is precisely $|A|$; if $|A|$ arrows are enough to "cover" all elements of B, then $|B| \leq |A|$; and if $|A|$ arrows can be directed into $|B|$ elements without any duplication, then $|B| \geq |A|$. (And, actually, the third part of the Mapping Rule is implied by the first two parts: if there's a bijection $f : A \to B$ then f is both onto and one-to-one, so the first two parts of the Mapping Rule imply that $|A| \geq |B|$ *and* $|A| \leq |B|$, and thus that $|A| = |B|$.)

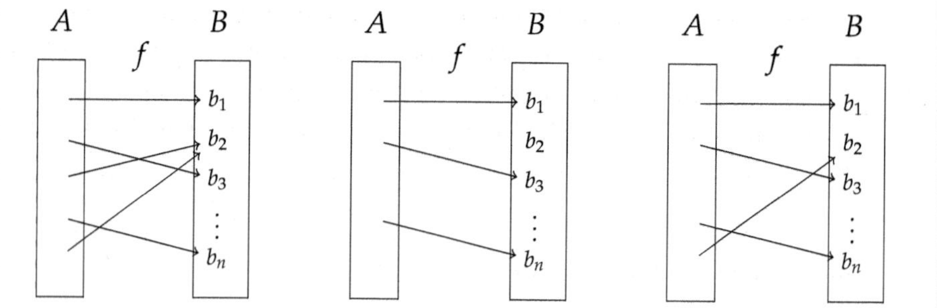

(a) f is onto: every element of B has an incoming arrow, so $|A| \geq |B|$.

(b) f is one-to-one: no element of B has more than one incoming arrow, so $|A| \leq |B|$.

(c) f is a bijection: every element of B has exactly one incoming arrow, so $|A| = |B|$.

Figure 9.18: The Mapping Rule. The number of arrows equals $|A|$.

A FEW EXAMPLES

We'll start with another example—like those in Example 9.24—of the logic underlying the Mapping Rule, but this time using function terminology:

Example 9.25 (Students and assignments)
Let S be a set of 128 students in a computer science class, let A be a set of programming assignments, and suppose that $mine : S \to A$ is a function so that $mine(s)$ is the assignment that has the name of student s written on it. (Because *mine* is a function, each student's name is by definition on one and only one submitted assignment.)

- Suppose the function *mine* is onto. Then every assignment in A has at least one student's name on it—and therefore there are at least as many students as assignments: each name is written only once, and every assignment has a name on it. So $|A| \leq 128$. (There could be fewer than 128 if, for example, assignments were allowed to be submitted by pairs of students.)

- Suppose the function *mine* is one-to-one. Then no assignment has more than one name on it—and therefore there are at least as many students as assignments: each assignment has at most one name, so there can't be more names than assignments. So $|A| \geq 128$. (There could be more than 128 if, for example, there are assignments in the pile that were submitted by students in a different section of the course.)

- Suppose the function *mine* is both onto and one-to-one. Then each assignment has exactly one name written on it, and thus $|A| = |S| = 128$.

Let's also rewrite two of the informal scenarios from Example 9.24 to explicitly use functions and the Mapping Rule:

Example 9.26 (Classes, names, and states, formalized)

- Let S be the set of students taking a particular class. Define the function $f : S \to \{\texttt{A}, \texttt{B}, \ldots, \texttt{Z}\}$, where $f(s)$ is the first letter of the last name of student s. If no two students' last names start with the same letter, then $f(s) = f(s')$ only when $s = s'$—in other words, the function f is one-to-one. Then, by the Mapping Rule, $|S| \leq |\{\texttt{A}, \texttt{B}, \ldots, \texttt{Z}\}|$: there are at most 26 students in the class.

- Let T be the set of people in a particular club. Let $T' \subseteq T$ be those people in T who are from one of the 50 states. Because $T' \subseteq T$, we have $|T| \geq |T'|$.

  Define the function $g : T' \to \{Alabama, Alaska, \ldots, Wyoming\}$, where $g(x)$ is the home state of person x. If there is at least one student from every state, then for all $s \in \{Alabama, Alaska, \ldots, Wyoming\}$ there's an $x \in T'$ such that $g(x) = s$—in other words, the function g is onto. Then, by the Mapping Rule, $|T'| \geq |\{Alabama, Alaska, \ldots, Wyoming\}|$: there are at least 50 people in the club.

We'll close this section with an example of using the Mapping Rule to count the cardinality of a set that we have not yet been able to calculate. We'll do so by giving a bijection between this new set (with previously unknown cardinality) and a set whose cardinality we *do* know.

The set that we'll analyze here is the *power set* of a set X—the set of all subsets of X, defined as $\mathscr{P}(X) := \{Y : Y \subseteq X\}$. (See Definition 2.31.) For example, $\mathscr{P}(\{0,1\})$ is $\{\{\}, \{0\}, \{1\}, \{0,1\}\}$. Let's look at the power set of $\{1,2,\ldots,8\}$:

Example 9.27 (Power set of $\{1,2,\ldots,8\}$)

Problem: What is $|\mathscr{P}(\{1,2,\ldots,8\})|$?

Solution: We'll give a bijection between $\{0,1\}^8$ and $\mathscr{P}(\{1,2,\ldots,8\})$—that is, we'll define a function $b : \{0,1\}^8 \to \mathscr{P}(\{1,2,\ldots,8\})$ that's a bijection. Here is the correspondence: for every 8-bit string $y \in \{0,1\}^8$, define $b(y)$ to be the subset $Y \subseteq \{1,2,\ldots,8\}$ such that $i \in Y$ if and only if the ith bit of y is 1. For example:

$y = 11101010$	$\to$	$Y = \{1,2,3,5,7\}$	*that is,* $b(11101010) = \{1,2,3,5,7\}$,
$y = 00001000$	$\to$	$Y = \{5\}$	*and* $b(00001000) = \{5\}$,
$y = 00000000$	$\to$	$Y = \{\}$	*and* $b(00000000) = \{\}$.

Because every subset corresponds to some bitstring, and no subset corresponds to more than one bitstring, the function $b : \{0,1\}^8 \to \mathscr{P}(\{1,2,\ldots,8\})$ is a bijection between $\{0,1\}^8$ and $\mathscr{P}(\{1,2,\ldots,8\})$.

Because a bijection from $\{0,1\}^8$ to $\mathscr{P}(\{1,2,\ldots,8\})$ exists, the Mapping Rule says that $|\mathscr{P}(\{1,2,\ldots,8\})| = |\{0,1\}^8| = 2^8 = 256$.

The idea of the mapping from Example 9.27 applies for an arbitrary finite set X. Here is the general result:

Lemma 9.10 (Cardinality of the Power Set)

Let X be any finite set. Then $|\mathscr{P}(X)| = 2^{|X|}$.

Lemma 9.10 is the reason for the power set's name: the cardinality of $\mathscr{P}(X)$ is 2 to the *power* of $|X|$.

Proof. Let $n = |X|$. Let $X = \{x_1, x_2, \ldots, x_n\}$ be an arbitrary ordering of the elements of X. Define a function $f : \{0,1\}^n \to \mathscr{P}(X)$ as follows:

$$f(y) = \{x_i : \text{the } i\text{th bit of } y \text{ is } 1\}.$$

It is easy to see that f is onto: for any subset Y of X, there exists a $y \in \{0,1\}^n$ such that $f(y) = Y$. It is also easy to see that f is one-to-one: if $y \neq y'$ then there exists an i such that $y_i \neq y'_i$, so $[x_i \in f(y)] \neq [x_i \in f(y')]$. Therefore f is a bijection, and by the Mapping Rule we can conclude $|\mathscr{P}(X)| = |\{0,1\}^{|X|}| = 2^{|X|}$. □

Taking it further: Although our focus in this chapter is on finding the cardinality of *finite* sets, we can also apply the Mapping Rule to think about *infinite cardinalities.* Infinite sets are generally more the focus of mathematicians than of computer scientists, but there are some fascinating (and completely mind-bending) results that are relevant for computer scientists, too. For example, we can prove that the number of even integers *is the same* as the number of integers (even though the former is a proper subset of the latter!). But we can also prove that $|\mathbb{R}| > |\mathbb{Z}|$. More relevantly for computer science, we can prove that there are strictly more *problems* than there are *computer programs,* and therefore that *there are problems that cannot be solved by a computer.* See the discussion on p. 937.

9.3.2 *The Division Rule*

When we introduced the Inclusion–Exclusion Rule, we used an approach to counting that we might call *count first, apologize later:* to compute the cardinality of a set $A \cup B$, we found $|A| + |B|$ and then "fixed" our count by subtracting the number of elements that we'd counted twice—namely, subtracting $|A \cap B|$. Here we'll consider an analogous count-and-correct rule, called the *Division Rule*, that applies when we count every element of a set multiple times (and where each element is recounted the same number of times); we'll then correct our total by dividing by this "redundancy factor." Let's start with some informal examples:

Example 9.28 (Some redundant counting, informally)

- Suppose that the Juggling Club on campus sells 99 juggling torches to its members, in sets of three. Then there are 33 people who purchased torches.
- There are 42 people at a party. Suppose that every person shakes hands with every other person. How many handshakes have occurred? There are many ways to solve this problem, but here's an approach that uses division: each person shakes hands with all 41 other people, for a total of (42 people) $\cdot$ (41 shakes/person) = 1722 shakes. But each handshake involves *two* people, so we've counted every shake exactly twice; thus there are actually a total of $861 = \frac{1722}{2} = \frac{42 \cdot 41}{2}$ handshakes.
- In Game 5 of the 1997 NBA Finals, the Chicago Bulls had 10 players who were on the court for some portion of the game. The number of minutes played by these ten were $\langle 45, 44, 26, 24, 24, 24, 23, 23, 4, 3 \rangle$. The total number of minutes played was $45 + 44 + 26 + 24 + 24 + 24 + 23 + 23 + 4 + 3 = 240$. In basketball, five players are on the court at a time. Thus the game lasted $\frac{240}{5} = 48$ minutes.

We'll phrase the Division Rule using the same general structure as the Mapping Rule, in terms of a function that maps from one set to another. Specifically, if we have a function $f : A \to B$ that always maps exactly the same number of elements of A to each element of B—for instance, exactly three torches are mapped to any particular juggler in Example 9.28—then $|A|$ and $|B|$ differ exactly by that factor:

Theorem 9.11 (Division Rule)
Let A and B be arbitrary sets. Suppose that there exists a function $f : A \to B$ such that, for every $b \in B$, there are exactly k elements $a_1, \ldots, a_k \in A$ such that $f(a_i) = b$. (That is, $|\{a \in A : f(a) = b\}| = k$ for all $b \in B$.) Then $|A| = k \cdot |B|$.

(The Division Rule with $k = 1$ simply *is* the bijection case of the Mapping Rule: what it means for $f : A \to B$ to be a bijection is precisely that $|\{a \in A : f(a) = b\}| = 1$ for every $b \in B$. If such a function f exists, then both the Mapping Rule and the Division Rule say that $|A| = 1 \cdot |B|$.)

Here are two simple examples to illustrate the formal version of the Division Rule:

Example 9.29 (Redundant counting, formally)

- Let M be the set of members of the Juggling Club, and let T be the set of torches bought by the members of the club. Consider the function $boughtBy : T \to M$. Assuming that each member bought precisely three torches—that is, assuming that $|\{t \in T : boughtBy(t) = m\}| = 3$ for every $m \in M$—then $|T| = 3 \cdot |M|$.

- Consider the sets $A = \{0, 1, \ldots, 31\}$ and $B = \{0, 1, \ldots, 15\}$. Define the function $f : A \to B$ as $f(n) = \lfloor n/2 \rfloor$. For each $b \in B$, there are exactly two input values whose output under f is b, namely $2b$ and $2b + 1$. Thus by the Division Rule $|A| = 2 \cdot |B|$.

This basic idea—if we've counted each thing k times, then dividing our total count by k gives us the number of things—is pretty obvious, and it'll also turn out to be surprisingly useful. Here's a sequence of examples, starting with a warm-up exercise and continuing with two (slightly less obvious) applications of the Division Rule:

Example 9.30 (Rearranging PERL, PEER, and SMALLTALK)

Problem: How many different ways can you arrange the letters of ...

1. ... the name of the programming language PERL?
2. ... the word PEER?
3. ... the name of the programming language SMALLTALK?

Solution: PERL: There are 4 different letters, and any permutation of them is a different ordering. Thus there are $4! = 4 \cdot 3 \cdot 2 \cdot 1 = 24$ orderings. (See Theorem 9.7.)

PEER: We'll answer this question using the solution for PERL. Define the function L->E as follows: given a 4-character input string, it produces a 4-character output string in which every L has been replaced by an E. For example, L->E(PERL) = PERE. Let S denote the orderings of the word PERL, and let T denote the orderings of PEER. Note that the function L->E : $S \to T$ has the property that, for every $t \in T$, there are exactly two strings $x \in S$ such that L->E(x) = t. (For example, L->E(PERL) = PERE and L->E(PLRE) = PERE.) See Figure 9.19. Thus, by the Division Rule, there are $\frac{4!}{2} = \frac{24}{2} = 12$ ways to order the letters of PEER.

SMALLTALK: There are 9! different orderings of the nine "letters" in the word S M A_1 L_1 L_2 T A_2 L_3 K. (We are writing L_1 and L_2 and L_3 to denote three different "letters," and similarly for A_1 and A_2.) We will use the Division Rule repeatedly to "erase" subscripts:

- The function that erases subscripts on the As maps two inputs to each output: one with A_1 before A_2, and one with A_2 before A_1. Thus there are $\frac{9!}{2}$ different orderings of the "letters" in the word S M A L_1 L_2 T A L_3 K.
- The function that takes an ordering of S M A L_1 L_2 T A L_3 K and erases the subscripts on the Ls maps precisely *six* inputs to each output: one for each of the 3! possible orderings of the Ls.

Thus there are $\frac{9!}{2 \cdot 3!} = \frac{362{,}880}{12} = 30{,}240$ different orderings of the letters in the word S M A L L T A L K.

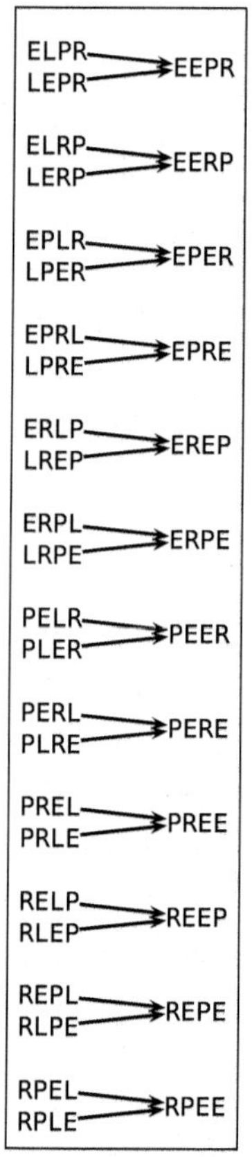

Figure 9.19: The 24 different orderings of PERL and the 12 different orderings of PEER. The function that replaces L by E is displayed by the arrows.

Counting orderings when some elements are indistinguishable

Although we phrased Example 9.30 in terms of the number of ways to rearrange the letters of some particular words, there's a very general idea that underlies the PEER and SMALLTALK examples. We'll state the underlying idea as a theorem:

Theorem 9.12 (Rearranging with duplicates)
The number of ways to rearrange a sequence containing k different distinct elements $\{x_1, \ldots, x_k\}$*, where element* x_i *appears* n_i *times, is*

$$\frac{(n_1 + n_2 + \cdots + n_k)!}{(n_1!) \cdot (n_2!) \cdot \cdots \cdot (n_k!)}.$$

For example, PERL has $k = 4$ distinct elements, which appear $n_P = n_E = n_R = n_L = 1$ time each; the theorem says that there are $\frac{(1+1+1+1)!}{1! \cdot 1! \cdot 1! \cdot 1!} = 4!$ ways to arrange the letters. On the other hand, SMALLTALK has $k = 6$ distinct elements, which appear $n_A = 2$, $n_L = 3$, and $n_S = n_M = n_T = n_K = 1$ times each; the theorem says that there are $\frac{(2+3+1+1+1+1)!}{2! \cdot 3! \cdot 1! \cdot 1! \cdot 1! \cdot 1!} = \frac{9!}{2! \cdot 3!}$ ways to arrange the letters. Let's prove the theorem:

Proof of Theorem 9.12. Let's handle a simpler case first: suppose that we have n different elements that we can put into any order, and precisely k of these n elements are indistinguishable. Then there are exactly $\frac{n!}{k!}$ different orderings of those n elements. To see this fact, imagine "decorating" each of those k items with some kind of artificial distinguishing mark, like the numerical subscripts of the letters of SMALLTALK from Example 9.30. Then there are $n!$ different orderings of the n elements. The *erase* function that eliminates our artificial distinguishing marks has $k!$ inputs that yield the same output—namely, one *for each ordering* of the k artificially marked elements. Therefore, by the Division Rule, there are $\frac{n!}{k!}$ different orderings of the elements, without the distinguishing markers.

The full theorem is just a mild generalization of this argument, to allow us to consider more than one set of indistinguishable elements. (In particular, we could give a formal proof by induction on the number of elements with $n_i \geq 2$.) In total, there are $(n_1 + n_2 + \cdots + n_k)!$ different orderings of the elements themselves, but there are $n_1!$ equivalent orderings of the first element, $n_2!$ of the second, and so forth. The function that "erases subscripts" as in Example 9.30 has $(n_1!) \cdot (n_2!) \cdot \cdots \cdot (n_k!)$ different equivalent orderings, and thus the total number of orderings is, by the Division Rule,

$$\frac{(n_1 + n_2 + \cdots + n_k)!}{(n_1!) \cdot (n_2!) \cdot \cdots \cdot (n_k!)}.$$ □

Here's another simple example that we can solve using this theorem:

Example 9.31 (Writing 232,848 as a sequence of prime factors)
Problem: How many ways can we write 232,848 as a product $p_1 p_2 \cdots p_k$, where each p_i is prime? (The *set* of prime factors, and *the number of occurrences of each factor*, are the same in every product, because the prime factorization of any positive integer is unique. But the *order* may change: for example, we can write $6 = 3 \cdot 2$ or $6 = 2 \cdot 3$.)

Solution: The prime factorization of 232,848 is $232{,}848 = 2^4 \cdot 3^3 \cdot 7^2 \cdot 11$. Thus a product of primes that equals 232,848 consists of 4 copies of two, 3 copies of three, 2 copies of seven, and one copy of eleven—in some order. (For example, $2 \cdot 2 \cdot 7 \cdot 3 \cdot 3 \cdot 7 \cdot 2 \cdot 11 \cdot 3 \cdot 2$.) By Theorem 9.12, the number of orderings of these elements is

$$\frac{(4+3+2+1)!}{4! \cdot 3! \cdot 2! \cdot 1!} = \frac{10!}{4! \cdot 3! \cdot 2!} = \frac{3{,}628{,}800}{24 \cdot 6 \cdot 2} = 12{,}600.$$

Problem-solving tip: There are often many different ways to solve a given problem—and you can use whatever approach makes the most sense *to you!* For example, Exercise 9.106 explores a completely different way to solve Example 9.32, based on the Generalized Product Rule instead of the Division Rule.

A SLIGHTLY MORE COMPLICATED EXAMPLE

Here is one final example of the Division Rule, in which we'll use this approach on a slightly more complicated problem:

Example 9.32 (Assigning partners)

Problem: The professor divides the n students in a CS class into $\frac{n}{2}$ partnerships, with two students per partnership. (Assume that n is even.) The order of partners within a pair doesn't matter, nor does the order of the partnerships. (That is, the listings

Paul and George John and Ringo	and	Ringo and John George and Paul

represent exactly the same set of partnerships.) How many ways are there to divide the class into partnerships?

Solution: Let's line up the students in some order, and then pair the first two students, then pair the third and fourth, and so on. There are $n!$ different orderings of the students, but there are fewer than $n!$ possible partnerships, because we've double counted each set of pairs in two different ways:

- there are two equivalent orderings of the first pair of students, and two equivalent orderings of the second pair, and so on.
- the ordering of the pairs doesn't matter, so the partnerships themselves can be listed in any order at all (without changing who's paired with whom).

Each of the $\frac{n}{2}$ pairs can be listed in 2 orders, so—by the Product Rule—there are $2^{n/2}$ different possible within-pair orderings. And there are $(n/2)!$ different orderings of the pairs. Applying the Division Rule, then, we see that there are

$$\frac{n!}{(n/2)! \cdot 2^{n/2}} \qquad (*)$$

total possible ways to assign partners.

Let's make sure that $(*)$ checks out for some small values of n. For $n = 2$, there's just one pairing, and indeed $(*)$ is $\frac{2!}{1! \cdot 2^1} = \frac{2}{2} = 1$. For $n = 4$, the formula $(*)$ yields $\frac{4!}{2^3} = \frac{4 \cdot 3 \cdot 2}{8} = 3$ pairings; indeed, for the quartet Paul, John, George, and Ringo, there are three possible partners for Paul (and once Paul is assigned a partner there are no further choices to be made). See Figure 9.20 for an illustration: we try all $4! = 24$ orderings of the four people, then we reorder the names within each pair, and finally we reorder the pairs.

ordering	*reordered within pairs*	
AB CD	AB CD	AB + CD
AB DC	AB CD	
BA CD	AB CD	
BA DC	AB CD	
CD AB	CD AB	
CD BA	CD AB	
DC AB	CD AB	
DC BA	CD AB	
AC BD	AC BD	AC + BD
AC DB	AC BD	
BD AC	BD AC	
BD CA	BD AC	
CA BD	AC BD	
CA DB	AC BD	
DB AC	BD AC	
DB CA	BD AC	
AD BC	AD BC	AD + BC
AD CB	AD BC	
BC AD	BC AD	
BC DA	BC AD	
CB AD	BC AD	
CB DA	BC AD	
DA BC	AD BC	
DA CB	AD BC	

Figure 9.20: Partnerships for $n = 4$ students: the 4! orderings, then the orderings sorted within pairs, and then with the pairs sorted.

9.3.3 *The Pigeonhole Principle*

We'll close this section with a very simple—but also surprisingly useful—theorem based on the Mapping Rule, called the *pigeonhole principle.* Here are a few informal examples to introduce the underlying idea:

A *pigeonhole* refers to one of the "cells" in a grid of compartments that are open in the front, and which can house either snail mail or, back in the day, roosting pigeons. (There's also a related verb: to *pigeonhole* someone/something is to categorize that person/thing into one of a small number of—misleadingly simple—groups.)

Example 9.33 (What happens when there are more things than kinds of things)

- If there are more socks in your drawer than there are colors of socks in your drawer, then you must have two socks of the same color.
- If there are only 5 possible letter grades and there are 6 or more students in a class, then there must be two students who receive the same letter grade.
- If you take 9 or more CS courses during the 8 semesters that you're in college, then there must be at least one semester in which you doubled up on CS courses.
- In the antiquated language in which this result is generally stated: if there are n pigeonholes, and $n+1$ pigeons that are placed into those pigeonholes, then there must be at least one pigeonhole that contains more than one pigeon.

Here is the general statement of the theorem, along with its proof:

Theorem 9.13 (Pigeonhole Principle)
Let A and B be sets with $|A| > |B|$, and let $f : A \to B$ be any function. Then there exist distinct elements $a \in A$ and $a' \in A$ such that $f(a) = f(a')$.

Proof. We can prove the Pigeonhole Principle using the Mapping Rule. Given the sets A and B, and the function $f : A \to B$, the Mapping Rule tells us that

$$\text{if } f : A \to B \text{ is one-to-one, then } |A| \leq |B|. \tag{1}$$

Taking the contrapositive of (1), we have

$$\text{if } |A| > |B|, \text{ then } f : A \to B \text{ is not one-to-one.} \tag{2}$$

By assumption, we have that $|A| > |B|$, so $f : A \to B$ is not one-to-one. The theorem follows by the definition of a one-to-one function: the fact that $f : A \to B$ is not one-to-one means precisely that there is some $b \in B$ that's "hit" twice by f. In other words, there exist distinct $a \in A$ and $a' \in A$ such that $a \neq a'$ and $f(a) = f(a')$. □

A slight generalization of this idea is also sometimes useful: if there are n total objects, each of which has one of k types, then there must be a type that has at least $\lceil n/k \rceil$ objects. (We'll omit the proof, but the idea is very similar to Theorem 9.13.)

Theorem 9.14 (Pigeonhole Principle: Extended Version)
Let A and B be sets, and let $f : A \to B$ be any function. Then there exists some $b \in B$ such that the set $\{a \in A : f(a) = b\}$ contains at least $\lceil |A|/|B| \rceil$ elements.

(Another less formal way of stating this fact is "the maximum must exceed the average": the number of elements in A that "hit" a particular $b \in B$ is $|A|/|B|$ on average, and there must be some element of B that's hit at least this many times.)

We'll start with two simpler examples of the pigeonhole principle, and close with a slightly more complicated application. (In the last example, the slightly tricky part of applying the pigeonhole principle is figuring out what corresponds to the "holes.")

Example 9.34 (Congressional voting)
Suppose that there were 5 different bills upon which the House of Representatives voted yesterday. (There are 435 representatives in the U.S. House.) The pigeonhole principle implies that there are two representatives who voted identically on yesterday's bills. A representative's vote can be expressed as an element of $\{aye, nay, abstain\}^5$, which has cardinality $3^5 = 243$. Because $243 < 435$, the pigeonhole principle says that there are two representatives with the same voting record.

Example 9.35 (Logical equivalence)
Let S be a set of 17 different logical propositions over the Boolean variables p and q.

A truth table for a proposition $\varphi \in S$ is an element of $\{\text{True}, \text{False}\}^4$ (the rows of the truth table correspond to each of the four truth assignments for p and q), and there are only $|\{\text{True}, \text{False}\}^4| = 2^4 = 16$ different such values. Therefore, our 17 different propositions have only 16 different possible truth tables—so, by the pigeonhole principle, there must be two different propositions that have the same truth table.

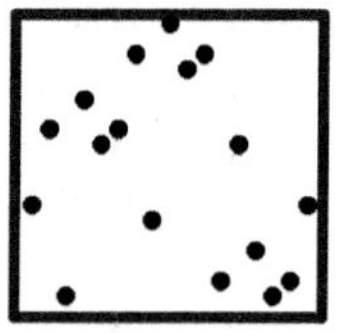

(a) 17 points in a 1-by-1 square.

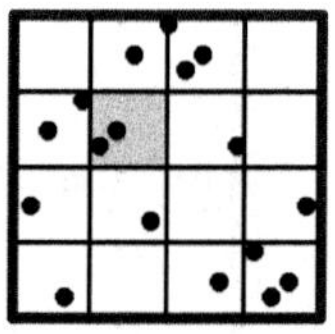

(b) The square divided into 16 subsquares, and one of the several doubly occupied subsquares.

Figure 9.21: Putting $n^2 + 1$ points in the unit square.

Example 9.36 (Points in a square)
Problem: Suppose that there are $n^2 + 1$ points in a 1-by-1 square, as in Figure 9.21(a). Show that there must be two points within distance $\frac{\sqrt{2}}{n}$ of each other.

Solution: We will use the pigeonhole principle. Divide the unit square into n^2 equal-sized disjoint subsquares—each with dimension $\frac{1}{n}$-by-$\frac{1}{n}$. (To prevent overlap, we'll say that every shared boundary line is included in the square to the left or below the shared line.) There are n^2 subsquares, and $n^2 + 1$ points. By the pigeonhole principle, at least one subsquare contains two or more points. (See Figure 9.21(b).)

Notice that the farthest apart that two points in a subsquare can be is when they are at opposite corners of the subsquare. In this case, they are $\frac{1}{n}$ apart in x-coordinate and $\frac{1}{n}$ apart in y-coordinate—in other words, they are separated by a distance of

$$\sqrt{(\tfrac{1}{n})^2 + (\tfrac{1}{n})^2} = \sqrt{\tfrac{2}{n^2}} = \tfrac{\sqrt{2}}{n}.$$

Taking it further: The pigeonhole principle can be used to show that *compression* of data files (for example, ZIP files or compressed image formats like GIF) must either lose information about the original data (so-called *lossy compression*) or must, for some input files, actually cause the "compressed" version to be larger than the original file. See the discussion on p. 938.

Computer Science Connections

Infinite Cardinalities (and Problems that Can't Be Solved by Any Program)

Recall the Mapping Rule: *for any two sets A and B, a bijection $f : A \to B$ exists if and only if $|A| = |B|$.* Although we were thinking about finite sets when we stated this rule, the statement holds even for infinite sets A and B; we can even think of this rule as *defining* what it means for two sets to have the same cardinality. Those sets S such that $|S| = |\mathbb{Z}|$, called *countable* sets, will turn out to be particularly important. Surprisingly, some sets that "seem" much bigger or much smaller than the integers have the same cardinality as $\mathbb{Z}$. For example, the set of nonnegative integers has the same cardinality as the set of all integers! (See Figure 9.22 for a bijection between these sets.) This fact is very strange—after all, we're looking at sets A and B *where A is a proper subset of B* and we've now established that $|A| = |B|$! But, indeed, because we have a bijection between A and B, they really are the same size.

Define the function $f : \mathbb{Z}^{\geq 0} \to \mathbb{Z}$ as $f(n) = \lceil \frac{n}{2} \rceil \cdot (-1)^n$. Then:

$$\begin{aligned}
f(0) &= \lceil \tfrac{0}{2} \rceil \cdot (-1)^0 = 0 \cdot 1 = 0 \\
f(1) &= \lceil \tfrac{1}{2} \rceil \cdot (-1)^1 = 1 \cdot -1 = -1 \\
f(2) &= \lceil \tfrac{2}{2} \rceil \cdot (-1)^2 = 1 \cdot 1 = 1 \\
f(3) &= \lceil \tfrac{3}{2} \rceil \cdot (-1)^3 = 2 \cdot -1 = -2 \\
f(4) &= \lceil \tfrac{4}{2} \rceil \cdot (-1)^4 = 2 \cdot 1 = 2 \\
&\vdots
\end{aligned}$$

Figure 9.22: A bijection between $\mathbb{Z}^{\geq 0}$ and $\mathbb{Z}$. Thus $|\mathbb{Z}^{\geq 0}| = |\mathbb{Z}|$.

p	r	i	n	t		"	h	e	l	l	o		w	o
112	114	105	110	116	32	34	104	101	108	108	111	32	119	111
1110000	1110010	1101001	1101110	1110100	100000	100010	1101000	1100101	1101100	1101100	1101111	100000	1110111	1101111

Figure 9.23: Converting a Python program into an integer. This program corresponds to the integer whose binary representation is 1110000 1110010 1101001 1101110 $\cdots$.

Or consider a Python program p. Think of the source code of p as a file—which thus represents p as a sequence of characters, each of which is represented as a sequence of bits, which can therefore be interpreted as an integer written in binary. (See Figure 9.23.) Therefore there is a bijection f between the integers and the set of Python programs, where $f(i)$ is the ith-largest Python program (sorted numerically by its binary representation).

With all of these sets that have the same cardinality, it might be tempting to think that *all* infinite sets have the same cardinality as $\mathbb{Z}$. But they don't!

Theorem 9.15
The set of all subsets of $\mathbb{Z}^{\geq 0}$—that is, $\mathscr{P}(\mathbb{Z}^{\geq 0})$—is strictly bigger than $\mathbb{Z}^{\geq 0}$.

Proof. Suppose for a contradiction that $f : \mathbb{Z}^{\geq 0} \to \mathscr{P}(\mathbb{Z}^{\geq 0})$ is an onto function. We'll show that there's a set $S \in \mathscr{P}(\mathbb{Z}^{\geq 0})$ such that *for every $n \in \mathbb{Z}^{\geq 0}$ we have $f(n) \neq S$.* Define the set S as follows:

$$S := \{i \in \mathbb{Z}^{\geq 0} : i \notin f(i)\} \qquad \textit{(So } i \in S \Leftrightarrow \textit{the set } f(i) \textit{ does } \text{not} \textit{ contain } i.)$$

Observe that the set S *differs from $f(i)$ for every i*: specifically, for every i we have $i \in S \Leftrightarrow i \notin f(i)$. Thus S is never "hit" by f—contradicting the assumption that f was onto. Therefore there is no onto function $f : \mathbb{Z}^{\geq 0} \to \mathscr{P}(\mathbb{Z}^{\geq 0})$, and, by the Mapping Rule, $|\mathbb{Z}^{\geq 0}| < |\mathscr{P}(\mathbb{Z}^{\geq 0})|$. (This argument is called a proof by *diagonalization*; see Figure 9.24.) □

	0	1	2	3	4	
$f(0)$	**1**	0	1	0	1	$\cdots$
$f(1)$	0	**0**	0	1	1	$\cdots$
$f(2)$	0	1	**1**	0	1	$\cdots$
$f(3)$	1	1	0	**1**	1	$\cdots$
$f(4)$	1	0	1	0	**0**	$\cdots$
	$\vdots$	$\vdots$	$\vdots$	$\vdots$	$\vdots$	$\ddots$

Figure 9.24: Diagonalization. Suppose that $f : \mathbb{Z}^{\geq 0} \to \mathscr{P}(\mathbb{Z}^{\geq 0})$. In a table, write row n corresponding to $f(n)$—so that $f(n)$ has a "1" in column j when $j \in f(n)$. Define $S := \{i : i \notin f(i)\}$—that is, the opposite of the diagonal element. For this table we have $0 \notin S$ (because $0 \in f(0)$), $1 \in S$ (because $1 \notin f(1)$), etc.

We can think of any subset of $\mathbb{Z}$ as defining a *problem* that we might want to write a Python program to solve. For example, the set $\{0, 2, 4, 6, \ldots\}$ is the problem of identifying even numbers. The set $\{1, 2, 4, 8, 16, \ldots\}$ is exact powers of 2. The set $\{2, 3, 5, 7, 11, \ldots\}$ is prime numbers. What does all of this say? *There are more problems than there are Python programs!* And thus there are problems that cannot be solved by any program![4]

Problems that can't be solved by any computer program are called *uncomputable*. Section 4.4.4 identifies some particular uncomputable problems, or see a good book on computability, like

[4] Dexter Kozen. *Automata and Computability*. Springer, 1997; and Michael Sipser. *Introduction to the Theory of Computation*. Course Technology, 3rd edition, 2012.

Computer Science Connections

Lossy and Lossless Compression

The task in *compression* is to take a large (potentially massively large!) piece of data and to represent it, somehow, using a smaller amount of space. Compression techniques are tremendously common, for a wide variety of data: text, images, audio, and video, for example. There are two fundamentally different approaches to compression of an original data file d into a compressed form d': *lossy* and *lossless* compression.

Lossy Compression. In *lossy compression*, d' does not represent exactly all of the information in d—that is, we've "lost" some information through compression. (That's why the compression is called "lossy.") In fact, many of the standard file formats for images, audio, and video are just standard methods for lossy compression. For example, JPEG is a lossy image compression format, and MP3 is a lossy audio compression format. The general goal with a lossy compression technique is to maintain, to the extent possible, "perceptual indistinguishability." For example, a digital audio stream can be represented precisely as a sequence of *intensities at each time t* ("how loud is the sound at time t?"). A lossy compression technique for sound might round the intensities: instead of representing an intensity as one of 2^{16} values ("a 16-bit sound," which is CD quality), we could round to the nearest of 2^8 values. (This idea is called *quantization*; see Example 2.56.) As long as the lost precision is smaller than the level of human perception, the new audio file would "sound the same" as the original.

Lossless Compression. In *lossless compression*, the precise contents of the original data file d can be reconstructed when the compressed data file d' is uncompressed. This approach is the one commonly used, for example, when compressing text using a program like ZIP.

The typical idea of lossless compression is to exploit redundancy in the stored data and to avoid wasting space storing the "same" information twice. For example, take the complete works of Shakespeare. By replacing every occurrence of `the` with `QQ` (two letters that don't occur consecutively in Shakespeare) the resulting file takes "only" about 99.2% of the original size. We can then set up a "translation table" telling us that `QQ` → `the` when we're decompressing. One interesting fact about lossless compression, though, is that it is *impossible* to actually compress every input file into a smaller size:

The word `the` appears over 20,000 times in the complete works of Shakespeare. The words `thee`, `them`, `their`, `they`, `there`, and `these` also appear over 1000 times each.

Theorem 9.16
Let C be any lossless compression function. Then there exists an input file d such that C(d) takes up at least as much space as d.

Proof. Suppose that C compresses all n-bit inputs into $n-1$ or fewer bits. That is, $C : \{0,1\}^n \to \bigcup_{i=0}^{n-1} \{0,1\}^i$. Observe that the domain has size 2^n and the range has size $\sum_{i=0}^{n-1} 2^i = 2^n - 1$. By the pigeonhole principle, there must be two distinct input files d_1 and d_2 such that $C(d_1) = C(d_2)$. But this C cannot be a lossless compression technique: if the compressed versions of the files are identical, the decompressed versions must be identical too! □

Here's an example of a lossless "compression" function making a file bigger: I downloaded the complete works of Shakespeare from Project Gutenberg, `http://www.gutenberg.org`. It took 5,590,193 bytes uncompressed, and 2,035,948 bytes when run through `gzip`. But `shakespeare.zip.zip.zip` (2,035,779 bytes), run through `gzip` three times, is actually bigger than `shakespeare.zip.zip` (2,035,417 bytes).

9.3.4 *Exercises*

9.57 Use the idea of Example 9.23 to determine how many bitstrings $x \in \{0,1\}^7$ *fail all three* Hamming code tests—those marked "✗✗✗" in the table in Example 9.23, or satisfying these three conditions:

$$x_2 + x_3 + x_4 \not\equiv_2 x_5 \qquad x_1 + x_3 + x_4 \not\equiv_2 x_6 \qquad x_1 + x_2 + x_4 \not\equiv_2 x_7.$$

9.58 Prove that the set P of legal positions in a chess game satisfies $|P| \leq 13^{64}$. *(Hint: Define a one-to-one function from* $\{1,2,\ldots,13\}^{64}$ *to P.)*

Let Σ *be a nonempty set. A* string *over* Σ *is a sequence of elements of* Σ*—that is,* $x \in \Sigma^n$ *for some* $n \geq 0$.

9.59 How many strings of length n over the alphabet $\{\texttt{A},\texttt{B},\ldots,\texttt{Z},␣\}$ are there? How many contain exactly 2 "words" (that is, contain exactly one space ␣ that is not in the first or last position)?

9.60 Let $n \geq 3$. How many n-symbol strings over this alphabet contain exactly 3 "words"? *(Hint: use Example 9.4 to account for n-symbol strings with exactly two* ␣*s; then use Inclusion–Exclusion to prevent initial/final/consecutive spaces, as in* ␣ABC$\cdots$, $\cdots$XYZ␣, *and* $\cdots$JKL␣␣MNO$\cdots$.*)*

A string over the alphabet $\{\texttt{[},\texttt{]}\}$ *is called a string of* balanced parentheses *if two conditions hold: (i) every* [*is later closed by a*]*; and (ii) every*] *closes a previous* [. *(You must close everything, and you never close something you didn't open.) Let* $B_n \subseteq \{\texttt{[},\texttt{]}\}^n$ *denote the set of strings of balanced parentheses that contain n symbols.*

9.61 Show that $|B_n| \leq 2^n$: define a one-to-one function $f : B_n \to \{0,1\}^n$ and use the Mapping Rule.

9.62 Show that $|B_n| \geq 2^{n/4}$ by defining a one-to-one function $g : \{0,1\}^{n/4} \to B_n$ and using the Mapping Rule. *(Hint: consider* `[][]` *and* `[[]]`.*)*

A certain college in the midwest requires its users' passwords to be 15 characters long. Inspired by an XKCD comic (see `http://xkcd.com/936/`*), a certain faculty member at this college now creates his passwords by choosing three 5-letter English words from the dictionary, without spaces. (An example password is* `ADOBESCORNADORN`*, from the words* `ADOBE` *and* `SCORN` *and* `ADORN`*.) There are 8636 five-letter words in the dictionary that he found.*

9.63 How many passwords can be made from any 15 (uppercase-only) letters? How many passwords can be made by pasting together three 5-letter words from this dictionary?

9.64 How many passwords can be made by pasting together three *distinct* 5-letter words from this dictionary? (For example, the password `ADOBESCUBAADOBE` is forbidden because `ADOBE` is repeated.)

The faculty member in question has a hard time remembering the order of the words in his password, so he's decided to ensure that the three words he chooses from this dictionary are different and appear in alphabetical order in his password. *(For example, the password* `ADOBESCUBAFOXES` *is forbidden because* `SCUBA` *is alphabetically after* `FOXES`.*)*

9.65 How many passwords fit this criterion? Solve this problem as follows. Let P denote the set of three-distinct-word passwords (the set from Exercise 9.64). Let A denote the set of three-distinct-alphabetical-word passwords. Define a function $f : P \to A$ that sorts. Then use the Division Rule.

9.66 After play-in games, the NCAA basketball tournament involves 64 teams, arranged in a *bracket* that specifies who plays whom in each round. (The winner of each game goes on to the next round; the loser is eliminated. See Figure 9.25.) How many different outcomes (that is, lists of winners of all games) of the tournament are there?

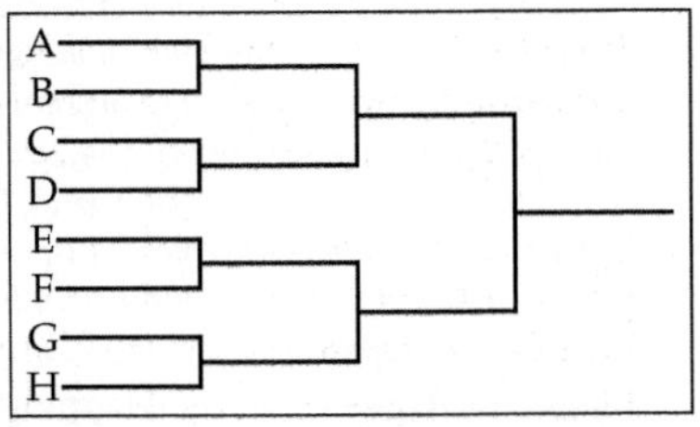

Figure 9.25: An 8-team tournament bracket. In the first round, A plays B, C plays D, etc. The A/B winner plays the C/D winner in the second round, and so forth.

A palindrome *over* Σ *is a string* $x \in \Sigma^n$ *that reads the same backward and forward—like* `0110`, `TESTSET`, *or (ignoring spaces and punctuation)* `SIT ON A POTATO PAN, OTIS!`.

9.67 How many 6-letter palindromes (elements of $\{\texttt{A},\texttt{B},\ldots,\texttt{Z}\}^6$) are there?

9.68 How many 7-letter palindromes (elements of $\{\texttt{A},\texttt{B},\ldots,\texttt{Z}\}^7$) are there?

9.69 Let $n \geq 1$ be an integer, and let P_n denote the set of palindromes over Σ of length n. Define a bijection $f : P_n \to \Sigma^k$ (for some $k \geq 0$ that you choose). Prove that f is a bijection, and use this bijection to write a formula for $|P_n|$ for arbitrary $n \in \mathbb{Z}^{\geq 1}$.

Let n be a positive integer. Recall an integer $k \geq 1$ *is a* factor *of n if* $k \mid n$. *The integer n is called* squarefree *if there's no integer* $m \geq 2$ *such that* $m^2 \mid n$.

9.70 How many positive integer factors does 100 have? How many are squarefree?

9.71 How many positive integer factors does 12! have? *(Hint: calculate the prime factorization of 12!.)*

9.72 How many squarefree factors does 12! have? Explain your answer.

9.73 *(programming required)* Write a program that, given $n \in \mathbb{Z}^{\geq 1}$, finds all squarefree factors of n.

9.74 Consider two sets A and B. Consider the following claim: if there is a function $f : A \to B$ that is not onto, then $|A| < |B|$. Why does this claim not follow directly from the Mapping Rule?

The genre-counting problem (Example 9.24) considered a function $f : \{1, 2, \ldots, n\} \to \{1, 2, 3, 4, 5\}$. *When* $n = 5 \ldots$

9.75 How many different functions $f : \{1, 2, \ldots, 5\} \to \{1, 2, \ldots, 5\}$ are there?

9.76 How many one-to-one functions $f : \{1, 2, \ldots, 5\} \to \{1, 2, \ldots, 5\}$ are there?

9.77 How many bijections $f : \{1, 2, \ldots, 5\} \to \{1, 2, \ldots, 5\}$ are there?

9.78 Let $n \geq 1$ and $m \geq n$ be integers. Consider the set G of functions $g : \{1, 2, \ldots n\} \to \{1, 2, \ldots, m\}$. How many functions are in G? How many one-to-one functions are there in G? How many bijections?

9.79 Show that the number of bijections $f : A \to B$ is equal to the number of bijections $g : B \to A$. *(Hint: define a bijection between* $\{$*bijections* $f : A \to B\}$ *and* $\{$*bijections* $g : B \to A\}$*, and use the bijection case of the mapping rule!)*

9.80 A *Universal Product Code (UPC)* is a numerical representation of the bar codes used in stores, with an error-detecting feature to handle misscanned codes. A UPC is a 12-digit number $\langle x_1, x_2, \ldots, x_{12} \rangle$ where $[\sum_{i=1}^{6} 3x_{2i-1} + x_{2i}] \bmod 10 = 0$. (That is, the even-indexed digits plus three times the odd-indexed digits should be divisible by 10.) Prove that there exists a bijection between the set of 11-digit numbers and the set of valid 12-digit UPC codes. Use this fact to determine the number of valid UPC codes.

9.81 A *strictly increasing sequence* of integers is $\langle i_1, i_2, \ldots, i_k \rangle$ where $i_1 < i_2 < \cdots < i_k$. How many strictly increasing sequences start with 1 and end with 1024? (That is, we have $i_1 = 1$ and $i_k = 1024$. The value of k can be anything you want; you should count both $\langle 1, 1024 \rangle$ and $\langle 1, 2, 3, 4, \ldots, 1023, 1024 \rangle$.)

A subsequence *of a sequence* $x = \langle x_1, x_2, \ldots, x_n \rangle$ *is a sequence* $\langle x_{i_1}, x_{i_2}, \ldots, x_{i_k} \rangle$ *of* $k \geq 0$ *elements of* x*, where* $\langle i_1, i_2, \ldots, i_k \rangle$ *is a strictly increasing sequence. For example,* PYTHON *is a subsequence of* PYTHAGOREAN *and* BASIC *is a subsequence of* BRAINSICKNESS.

9.82 Suppose the components of $x = \langle x_1, x_2, \ldots, x_n \rangle$ are all different (as in PYTHON but not PYTHAGOREAN). Use the Mapping Rule to figure out how many subsequences of x there are.

9.83 Suppose the components of $x = \langle x_1, x_2, \ldots, x_n \rangle$ are all different, *except for a single pair of identical elements that are separated by k other elements.* For example, PYTHAGOREAN has $n = 11$ and $k = 4$, because there are four entries (GORE) between the As (at index 5 and 10), which are the only repeated entries. In terms of n and k, how many subsequences of x are there?

As Example 9.23 describes, the Hamming Code adds 3 different parity bits to a 4-bit message m, where each added bit corresponds to the parity of a carefully chosen subset of the message bits, creating a 7-bit codeword c. Let k and n, respectively, denote the number of bits in the message and the codeword. (For the Hamming Code, we have $k = 4$ *and* $n = 7$.*)*

A decoding *algorithm takes a received (and possibly corrupted) codeword* c' *and determines which message has a corresponding codeword* c *that is most similar to* c'. *(See Section 4.2, or Figure 9.26 for a brief reminder. See also Exercises 4.25–4.28.) We can view the decoding algorithm as a function* decode : $\mathscr{P}(1, 2, \ldots, n-k) \to \{0, 1, 2, \ldots, n\}$*— where* decode(S) *tells us which bit (if any) to flip in the received codeword when S is the set of mismatched parity bits. (If* decode(S) $= 0$*, then no bits should be flipped.)*

9.84 Argue using the Mapping Rule (that is, without reference to the precise function in Figure 9.26) that for the Hamming Code's parameters ($n = 7$ and $k = 4$) that there exists a bijection decode : $\mathscr{P}(\{1, 2, \ldots, n-k\}) \to \{0, 1, 2, \ldots, n\}$.

9.85 Suppose that we choose $n = 9$ and $k = 4$. Does there exist a bijection from $\mathscr{P}(\{1, 2, \ldots, n-k\})$ to $\{0, 1, 2, \ldots, n\}$? Why or why not?

9.86 Suppose that we choose $n = 31$. For what value(s) of k does there exist a bijection from $\mathscr{P}(\{1, 2, \ldots, n-k\})$ to $\{0, 1, 2, \ldots, n\}$? Prove your answer.

9.87 Prove that, for any n that is not one less than a power of 2, there does *not* exist a bijection from $\mathscr{P}(\{1, 2, \ldots, n-k\})$ to $\{0, 1, 2, \ldots, n\}$.

The Hamming code

For the message $m = \langle a, b, c, d \rangle$, we compute three parity bits:

- *parity bit #1:* $b \oplus c \oplus d$
- *parity bit #2:* $a \oplus c \oplus d$
- *parity bit #3:* $a \oplus b \oplus d$

and send $c := \langle a, b, c, d, \text{parity \#1}, \text{parity \#2}, \text{parity \#3} \rangle$.

Having received a (possibly corrupted) codeword c', we compute what the parity bits would have been for the received message bits, and check for mismatches between the computed and received parity bits:

parity bit mismatches	error (which bit to flip)
$\{\}$	no error!
$\{1\}$	parity #1
$\{2\}$	parity #2
$\{3\}$	parity #3
$\{1, 2\}$	bit c
$\{1, 3\}$	bit b
$\{2, 3\}$	bit a
$\{1, 2, 3\}$	bit d

Figure 9.26: Decoding the Hamming Code. Every single-bit error is corrected.

In the corporate and political worlds, there's a dubious technique called URL squatting, *where someone creates a website whose name is very similar to a popular site and uses it to skim the traffic generated by poor-typing internet users. For example, Google owns the addresses* `gogle.com` *and* `googl.com`, *which redirect to* `google.com`. *(But, as of this writing, someone else owns* `oogle.com`, `goole.com`, *and* `googe.com`.*) Consider an n-letter company name. How many single-typo manglings of the name are there if we consider the following kinds of errors? Consider only uppercase* letters *throughout. (If your answers depend on the particular n-letter company name, then say* how *they depend on that name. Note that no transposition errors are possible for the company name* `MMM`, *for example.)*

9.88 one-letter substitutions

9.89 one-letter insertions

9.90 one-pair transpositions (two adjacent letters written in the wrong order)

9.91 one-letter deletions

How many different ways can you arrange the letters of the following words?

9.92 `PASCAL`
9.93 `GRACEHOPPER`
9.94 `ALANTURING`
9.95 `CHARLESBABBAGE`
9.96 `ADALOVELACE`
9.97 `PEERTOPEERSYSTEM`

9.98 *(programming required)* Write a function that, given an input string, computes the number of ways to rearrange the string's letters. Use your program to verify your answers to the last few exercises.

9.99 *(programming required)* In Example 9.31, we analyzed the number of ways to write a particular integer n as the product of primes. (Because the prime factorization of n is unique, the only difference between these products is the order in which the primes appear.) Write a program, in a language of your choice, to compute the number x_n of ways we can write a given number n as $p_1 \cdot p_2 \cdots p_k$, where each p_i is prime. For what number $n \leq 10{,}000$ is x_n the greatest?

In Chapter 3, we discussed the application of Boolean logic to AI-based approaches to playing games like Tic-Tac-Toe. (See p. 344, or Figure 9.27 for a 2-by-2 version of the game [Tic-Tac; the 3-by-3 version is Tic-Tac-Toe].)

Specifically, recall the Tic-Tac-Toe game tree: the root of the tree is the empty board, and the children of any node in the tree are the boards that result from any move made in any of the empty squares. We talked briefly about why chess is hard to solve using an approach like this. (In brief: it's huge.) The next few problems will explore why a little bit of cleverness helps a lot in solving even something as simple as Tic-Tac-Toe.

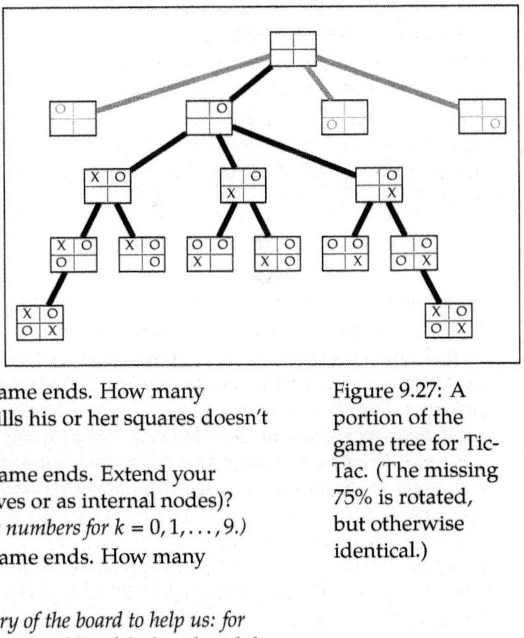

Figure 9.27: A portion of the game tree for Tic-Tac. (The missing 75% is rotated, but otherwise identical.)

9.100 Tic-Tac-Toe ends when either player completes a row, column, or diagonal. But for this question, assume that even after somebody wins the game, the board is completely filled in before the game ends. (That is, every leaf of the game tree has a completely filled board.) How many leaves are in the game tree?

9.101 Continue to assume that the board is completely filled in before the game ends. How many *distinct* leaves are there in the tree? (That is, suppose that the order in which O fills his or her squares doesn't matter; if the same squares are filled, the boards count as the same.)

9.102 Continue to assume that the board is completely filled in before the game ends. Extend your answer to Exercise 9.100: how many total boards appear in the game tree (as leaves or as internal nodes)? *(Hint: it may be easiest to compute the number of boards after k moves, and add up your numbers for k = 0, 1, . . . , 9.)*

9.103 Continue to assume that the board is completely filled in before the game ends. How many *distinct* total boards—internal nodes or leaves—are there in the tree?

There are still two optimizations left that we haven't tried. The first is using the symmetry of the board to help us: for example, there are really only three first moves that can be made in Tic-Tac-Toe: a corner, the middle of the board, and the middle of a side. The second optimization is to truncate the tree when there's a winner. These are both a bit tedious to track by hand, but it's manageable with a small program.

9.104 *(programming required)* We can cut the size of the game tree down to less than a third of the original size—actually substantially more!—by exploiting symmetry in plays. (We're down to a third of the original size just within the first move.) Write a program to compute the entire Tic-Tac-Toe game tree, and use it to determine the number of unique boards (counting as equivalent two boards that match with respect to rotational or reflectional symmetry) in the game tree. How many boards are now in the tree?

9.105 *(programming required)* We can reduce the size of the game tree just a bit further by not expanding the portions of the game tree where one of the players has already won. Extend your implementation from the last exercise so that no moves are made in any board in which O or X has already won. How many boards are in the tree now?

Recall Example 9.32: we must put n students (where n is even) into $\frac{n}{2}$ partnerships. (We don't care about the order of the partnerships, nor about the order of partners within a pair.) Here is an alternative way of solving this problem:

9.106 Consider sorting the n people alphabetically by name. Repeat the following $\frac{n}{2}$ times: for the unmatched person p whose name is alphabetically first, choose a partner for p from the set of all other unmatched people. How many choices are there in iteration i? How many choices are there, in total?

9.107 Algebraically prove the following identity. *(Hint: what does $(n/2)! \cdot 2^{n/2}$ represent?)*

$$\prod_{i=1}^{n/2}(n-2i+1) = \frac{n!}{(n/2)! \cdot 2^{n/2}}$$

Think of an n-gene chromosome as a permutation of the numbers $\{1, 2, \ldots, n\}$, representing the order in which these n genes appear. The following questions ask you to determine how many chromosome-level rearrangement events of a particular form there are. (See, for example, Figure 3.38.)

9.108 A *prefix reversal* inverts the order of the first j genes, for some $j > 1$ and $j \leq n$. For example, for the chromosome $\langle 5,9,6,2,1,4,7,3,8\rangle$ we could get the result $\langle \underline{6,9,5},2,1,4,7,3,8\rangle$ or $\langle \underline{1,2,6,9,5},4,7,3,8\rangle$ from a prefix reversal. How many different prefix reversals are there for a 1000-gene chromosome?

9.109 A *reversal* inverts the order of the genes between index i and index j, for some i and $j > i$. For example, for the chromosome $\langle 5,9,6,2,1,4,7,3,8\rangle$ we could get the result $\langle \underline{6,9,5},2,1,4,7,3,8\rangle$ or $\langle 5,9,6,\underline{4,1,2},7,3,8\rangle$ from a reversal. How many different reversals are there for a 1000-gene chromosome?

9.110 A *transposition* takes the genes between indices i and j and places them between indices k and $k+1$, for some i and $j > i$ and $k \notin \{i, i+1, \ldots, j\}$. For example, for the chromosome $\langle 5,9,6,2,1,4,7,3,8\rangle$ we could get the result $\langle 5,\underline{1,4,7,3},9,6,2,\sqcup 8\rangle$ or $\langle \sqcup 1,4,\underline{5,9,6,2},7,3,8\rangle$ from a transposition. How many different transpositions are there for a 1000-gene chromosome?

A cellular automaton *is a formalism that's sometimes used to model complex systems—like the spatial distribution of populations, for example. Here is the model, in its simplest form. We start from an n-by-n toroidal lattice of cells: a two-dimensional grid, that "wraps around" so that that there's no edge. (Think of a donut.) Each cell is connected to its eight immediate neighbors.*

Cellular automata are a model of evolution over time: our model will proceed in a sequence of time steps. *At every time step, each cell u is in one of two states:* active *or* inactive. *A cell's state may change from time t to time t + 1. More precisely, each cell u has an* update rule *that describes u's state at time t + 1 given the state of u and each of u's neighbors at time t. (For example, see Figure 9.28.)*

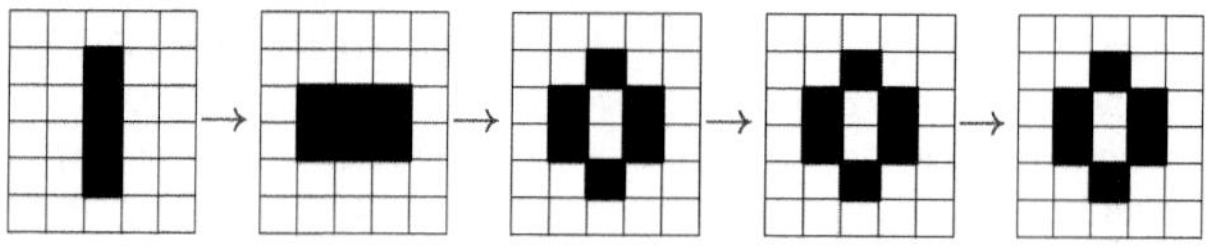

Figure 9.28: In the *Game of Life*, each cell has an identical update rule: an active cell with ≤ 1 live neighbors dies (from "loneliness"), a live cell with ≥ 4 live neighbors dies (from "overcrowding"), and a dead cell with exactly three living neighbors becomes alive.

9.111 An *update rule* is a function that takes the state of a cell and the state of its eight neighbors as input, and produces the new state of the cell as output. How many different update rules are there?

9.112 Let's call an update rule a *strictly cardinal update rule* if—as in the Game of Life—the state of a cell u at time $t+1$ depends only the following: (i) the state of cell u at time t, and (ii) the *number* of active neighbors of cell u at time t. How many different strictly cardinal update rules are there?

Suppose that we have an 10-by-10 lattice of 100 cells, and we have an update rule f_u for every cell u. (These update rules might be the same or differ from cell to cell.) Suppose the system begins in an initial configuration M_0. Suppose we start the system at time $t = 0$ in configuration M_0, and derive the configuration M_t at time $t \geq 1$ by computing

$$M_t(u) = f_u(\text{the states of } u\text{'s neighbors in } M_{t-1}).$$

Let's consider the possible outcomes of the sequence $M_0, M_1, M_2, \ldots$. Say that this sequence exhibits eventual convergence *if the following holds: there exists a time $t \geq 0$ such that, for all times $t' \geq t$, we have $M_{t'} = M_t$. (So the Life example in Figure 9.28 exhibits eventual convergence.) Otherwise, we'll say that this sequence* oscillates.

9.113 Given M_0 and the f_u's, we'd like to know what the long-run behavior of this system is: does it eventually converge or does it oscillate? Prove that, for a sufficiently large value of K, we have eventual convergence if and only if the following algorithm returns True. Also compute the smallest value of K for which this algorithm is guaranteed to be correct.

- Start with $M := M_0$ and $t := 0$.
- Repeat the following K times: update M to the next time step (that is, for each u compute the updated $M'(u)$ by evaluating f_u on u's neighbor cells in M).
- If M would be unchanged by one additional round of updates, return True. Else return False.

9.114 Suppose that we place 1234 items into 17 buckets. (For example, consider hashing 1234 items into a 17-cell hash table.) Call the number of items in a bucket its *occupancy*, and the *maximum occupancy* the number of items in the most-occupied bucket. What's the smallest possible maximum occupancy?

9.115 Consider a function $f : A \to B$. Fill in the blank with a statement relating $|A|$ and $|B|$, and then prove the resulting claim: if ____, then, for some $b \in B$, we have $|\{a \in A : f(a) = b\}| \geq 202$.

9.116 Suppose that we quantize a set of values from $S = \{1, 2, \ldots, n\}$ into $\{k_1, k_2, \ldots, k_5\} \subset S$. (See Example 2.56.) Namely, we choose these 5 values and then define a function $q : S \to \{k_1, k_2, \ldots, k_5\}$. The *maximum error* of this quantization is $\max_{x \in S} |x - q(x)|$. Use the Pigeonhole Principle (or the "the maximum must exceed the average" generalization) to determine the smallest possible maximum error.

Imagine a round-robin chess tournament for 150 players, each of whom plays 7 games. (In other words, each player is guaranteed to participate in precisely 7 games with 7 different opponents. Remember that each game has two players.)

9.117 There are 20 possible first moves for White in a chess game, and 20 possible first moves for Black in response. (See Example 9.15.) Prove that there must be two different games in the tournament that began with the same first two moves (one by White and one by Black).

9.118 Suppose that would-be draws in this tournament are resolved by a coin flip, so that every game has a winner and a loser. Prove that there must be two participants in such a tournament who have precisely the same sequence of wins and losses (for example, WWWLLLW).

A win–loss record reports a number of wins and a number of losses (for example, 6 wins and 1 loss, or 3 wins and 4 losses), without reference to the order of these results.

9.119 Continuing to suppose that there are no draws in this tournament, identify as large a value of k as you can for which the following claim is true, and prove that it's true for your value of k: there is some win–loss record that is shared by at least k competitors.

9.120 Now suppose that draws are allowed, so that competitors have a win–loss–draw record (for example, 2 wins, 1 loss, and 4 draws). Identify the largest k for which there is some win–loss–draw record that is shared by at least k competitors, and prove that this claim holds for the k you've identified.

9.4 *Combinations and Permutations*

> Not everything that can be counted counts, and not everything that counts can be counted.
>
> William Bruce Cameron (1921–2002)

So far in this chapter, we've been working to develop a toolbox of general techniques for counting problems: the Sum Rule and Inclusion–Exclusion, the (Generalized) Product Rule, the Mapping Rule, and the Division Rule. This section will be different; instead of a new technique, here we will devote our attention to a particularly common kind of counting problem: the number of ways to *choose* a subset from a given set of candidate elements. Let's start with an illustrative example:

Example 9.37 (Printing t-shirts)

Problem: Suppose you run a t-shirt shop. There is a collection of *jobs* that you're asked to run, but there's limited time so you must choose which ones to actually print. There are 17 requested jobs $\{\texttt{a}, \texttt{b}, \ldots, \texttt{q}\}$, but there is only time to print 4 different jobs. How many ways are there to select 4 of these 17 candidate jobs?

Solution: There are two answers, depending on how we interpret the problem: does the *order* of the printed jobs matter, or does it only matter *whether* a job was printed? (Are we choosing an ordered 4-tuple? Or an unordered subset of size 4?)

Order matters: Then the Generalized Product Rule immediately gives us the answer: there are 17 choices for the first job, 16 for the second job, 15 for the third, and 14 for the fourth; thus there are $17 \cdot 16 \cdot 15 \cdot 14$ total choices.

Another way to write $17 \cdot 16 \cdot 15 \cdot 14$ is $\frac{17!}{13!}$: every multiplicand between 1 and 13 appears in both the numerator and denominator, leaving only $\{17, 16, 15, 14\}$ uncancelled. We can justify the $\frac{17!}{13!}$ version of the answer using the Division Rule: we choose one of the 17! orderings of all 17 jobs, and then print the first 4 jobs in this order—but we've counted each 4-job ordering 13! times (once for each ordering of the 13 unprinted jobs), so we must divide by 13!.

Order doesn't matter: As in the previous case, there are $\frac{17!}{13!}$ ways of choosing an ordered sequence of 4 jobs. Because order doesn't matter, we have counted each set of four chosen jobs 4! times, once for each ordering of them. By the Division Rule, then, there are $\frac{17!}{13! \cdot 4!}$ ways of selecting 4 unordered jobs from a set of 17.

Two different fundamental notions of choice are illustrated by Example 9.37: *permutations*, in which the order of the chosen elements matters, and *combinations*, in which the order doesn't matter. These two notions will be our focus in this section. Here's another example to further illustrate combinations:

Example 9.38 (Arranging letters of a bitstring)

Problem: How many different ways can you arrange the symbols in the "word" `000111`? What about the "word" `00...011...1` containing k zeros and $n - k$ ones?

Solution: This problem is just another application of the techniques we used for `PERL` and `PEER` and `SMALLTALK` in Example 9.30. (We can think of the word `000111` just like a word like `DEEDED`: two different letters, appearing three times each.) There are 6 total characters in the word, each appearing 3 times, so the total number of arrangements is $\frac{6!}{3! \cdot 3!}$. (See Theorem 9.12.)

For the general version of the problem—the word `00...011...1`, with k zeros and $n-k$ ones—we have a total of n characters, so there are $n!$ ways of writing them down. But $k!$ orderings of the zeros, and $(n-k)!$ orderings of the ones, are identical. Hence, by the Division Rule, the total number of orderings is $\frac{n!}{k! \cdot (n-k)!}$.

Combinations

The quantity that we computed in Example 9.38 is called the number of *combinations* of k elements chosen from a set of n candidates:

The quantity $\binom{n}{k}$ is also sometimes called a *binomial coefficient*, for reasons that we'll see in Section 9.4.3. It's also sometimes denoted $C(n,k)$ ("C" as in "Combination").

Definition 9.2 (Combinations)
Consider nonnegative integers n and k with $k \leq n$. The quantity $\binom{n}{k}$ is defined as

$$\binom{n}{k} := \frac{n!}{k! \cdot (n-k)!},$$

and is read as "n choose k."

As we just argued in Example 9.38, the quantity $\binom{n}{k}$ denotes the number of ways to choose a k-element subset of a set of n elements. For convenience, define $\binom{n}{k} := 0$ whenever $n < 0$ or $k < 0$ or $k > n$: there are *zero* ways to choose a k-element subset of a set of n elements under these circumstances.

Taking it further: When there are annoying complications (or divide-by-zero errors or the like) in the boundary cases of a definition, it's often easiest to tweak the definition to make those cases less special. (Here, for example, instead of having $\binom{7}{8}$ be undefined, we treat $\binom{7}{8}$ as 0.)

A similar idea in programming can make life much simpler when you encounter data structures with complicated edge conditions—for example, a node in a linked list that might not have a successor. A *sentinel* is a "fake" element that you might add to the boundary of a data structure that makes the edge elements of the data structure less special. For example, in image processing, we might augment an n-by-m image with an extra 0th and $(m+1)$st column, and an extra 0th and $(n+1)$st row, of blank pixels. Once these "border pixels" are added, *every pixel in the image has a neighbor in each cardinal direction.* Thus there's no special code required for edge pixels in code to, for example, apply a blur filter to the image.

Here are a few small examples of counting problems that use combinations:

Example 9.39 (8-bit strings with 2 ones)
How many different 8-bit strings have exactly 2 ones?

We solved this precise problem in Example 9.3 using the Sum Rule, but combinations give us an easier way to answer this question. We must choose 2 out of 8 indices to make equal to one. There are $\binom{8}{2} = \frac{8!}{2! \cdot (8-2)!} = \frac{8!}{2! \cdot 6!} = \frac{8 \cdot 7}{2} = 28$ such choices of indices, and thus $\binom{8}{2}$ different 8-bit bitstrings with exactly 2 ones. These 28 strings are shown in Figure 9.29.

```
11000000
10100000
10010000
10001000
10000100
10000010
10000001
01100000
01010000
01001000
01000100
01000010
01000001
00110000
00101000
00100100
00100010
00100001
00011000
00010100
00010010
00010001
00001100
00001010
00001001
00000110
00000101
00000011
```

Figure 9.29: All 8-bit bitstrings with exactly 2 ones.

Example 9.40 (32-bit strings with < 3 ones)
How many different 32-bit strings have fewer than 3 ones?

We will use the Sum Rule, plus the formula for combinations. (We can partition the set of 32-bit strings that have fewer than 3 ones into those with 0, 1, or 2 ones.) Thus there are $\binom{32}{0} + \binom{32}{1} + \binom{32}{2} = 1 + 32 + \frac{32 \cdot 31}{2} = 1 + 32 + 496 = 529$ total such strings.

(Recall that $0! = 1$, so $\binom{32}{0} = \frac{32!}{0! \cdot (32-0)!} = \frac{32!}{0! \cdot 32!} = \frac{32!}{1 \cdot 32!} = \frac{32!}{32!} = 1$.)

Finally, here's an example of counting using combinations that relates counting to probability. (There's much more about probability in Chapter 10.) If we flip an *unbiased coin* (in other words, a coin that comes up heads with probability $\frac{1}{2}$ and tails with probability $\frac{1}{2}$ each time we flip it), then every sequence of coin flips is equally likely. The probability that an "event" E happens when we flip an unbiased coin is the fraction of possible flip sequences for which E actually occurs.

Example 9.41 (Exactly 50% heads)
Suppose we flip an unbiased coin 10 times. What is the probability that precisely 5 flips come up heads?

There are $2^{10} = 1024$ total sequences, of which $\binom{10}{5} = \frac{10!}{5! \cdot 5!} = 252$ have precisely 5 heads. Thus there's a $\frac{252}{1024} \approx 0.2461$ chance of exactly half of the flips being heads.

9.4.1 *Four Different Ways to Select k out of n Options*

In Example 9.37, we saw two different ways in which we can imagine choosing a subset of k distinct elements from a set S of n candidates, depending on whether the *order* in which we choose those k elements matters.

There is another dichotomy that can arise in counting problems: we can imagine circumstances in which we choose k elements from a set S, but where *repetition* is allowed (that is, we can choose the same element more than once). In other scenarios, repetition might not make sense. Here are some examples of all four situations (see also Figure 9.30):

order matters *repetition allowed* (9 ways)	*order matters* *repetition not allowed* (6 ways)	*order irrelevant* *repetition allowed* (6 ways)	*order irrelevant* *repetition not allowed* (3 ways)
A, then A		A and A	
A, then B	A, then B	A and B	A and B
B, then A	B, then A		
A, then C	A, then C	A and C	A and C
C, then A	C, then A		
B, then B		B and B	
B, then C	B, then C	B and C	B and C
C, then B	C, then B		
C, then C		C and C	

Figure 9.30: Four ways of choosing 2 elements from the candidates A, B, and C—depending on whether we can choose the same element more than once, and whether the order of choices matters.

• You order a two-scoop ice cream cone from a list of flavors. Order matters: a chocolate scoop on top of a mint scoop $\neq$ mint on top of chocolate. Repetition is allowed: you can choose vanilla for both scoops.

• Your soccer game is tied, and you must choose 5 of your 11 players to take penalty kicks to break the tie. Order matters: the kicks are taken in sequence, so Pelé then Maradona $\neq$ Maradona then Pelé. Repetition is forbidden: each player is allowed to take only one kick.

• You order a three-salad salad sampler from a list of salads. Order doesn't matter: salads are served on a round plate, so it doesn't matter which one is "first." Repetition is allowed: you can choose the Caesar as two or all three of your salads.

• You select a starting lineup of 5 basketball players from your 13-person team. Order doesn't matter: all 5 chosen players are equivalent in starting the game. Repetition is forbidden: you must choose five different players.

Here we will consider all four types of counting problems—ordered/unordered choice with/without repetition—and do a few examples. See Figure 9.31 for a summary of the number of ways to make these different types of choices.

	order matters	*order doesn't matter*
repetition forbidden	$\frac{n!}{(n-k)!}$	$\binom{n}{k}$
repetition allowed	n^k	$\binom{n+k-1}{k}$

Figure 9.31: Four ways of selecting k of n items, and the number of ways to make that selection.

WHEN ORDER MATTERS AND REPETITION IS FORBIDDEN

Suppose that we choose a *sequence* of k *distinct* elements from a set S: that is, the *order of the selected elements matters* and *repetition is not allowed.* (For example, in a player draft for a sports league, no player can be chosen more than once—"repetition is forbidden"—and the outcome of the draft depends not just on whether Babe Ruth was chosen, but also whether it was the Eagles or the Wildcats that selected him.)

In other words, we make k successive selections from S, but no candidate can be chosen more than once. Such a sequence is sometimes called a *k-permutation* of S—an ordered sequence of k distinct elements of S. (Recall from Definition 9.1 that a *permutation* of a set S is an ordering of S's elements.)

Some people denote the number of ways of choosing an ordered sequence of k distinct selections from a set of n options by $P(n,k)$, because "permutation" starts with "P."

There are $\frac{n!}{(n-k)!}$ different k-permutations of an n-element set S, by the Generalized Product Rule. (Specifically, there are

$$\underbrace{(n)}_{\text{choices of first element}} \cdot \underbrace{(n-1)}_{\text{choices of second element}} \cdot \cdots \cdot \underbrace{(n-k+1)}_{\text{choices of } k\text{th element}}$$

total choices, and $\frac{n!}{(n-k)!} = n \cdot (n-1) \cdot (n-2) \cdot \cdots \cdot (n-k+1)$.)

Example 9.42 (4 of 10)
Suppose that you are asked to place four of the cards $\{A\heartsuit, 2\heartsuit, \cdots, 10\heartsuit\}$ on the table, arranged from left to right in an order of your choosing. There are $10 \cdot 9 \cdot 8 \cdot 7 = \frac{10!}{(10-4)!}$ such arrangements: order matters (A234♡ ≠ 432A♡) and repetition is not allowed (4444♡ isn't a valid arrangement, because you only have one 4♡ card).

WHEN ORDER MATTERS AND REPETITION IS ALLOWED

Suppose that we simply choose a sequence of k (not necessarily distinct) elements: that is, *order matters* and *repetition is allowed.* In other words, we make k successive selections from S, and we're allowed to make the same choice multiple times. (For example, suppose you and $k-1$ friends go to a Chinese restaurant with n items on the menu, and each of you orders something for dinner. You're allowed to order the same dish as your friends—"repetition is allowed"—but you getting the Tofu with Black Bean Sauce and your vegan friend getting Twice-Cooked Pork is definitely different from the other way around.)

Then there are n^k different ways to make this choice, by the Product Rule: at every stage, there are n possible choices, and there are k stages.

Example 9.43 (4 of 10, a second way)
Suppose that you are asked to create a 4-digit integer. There are 10^4 such integers: order matters ($1234 \neq 4321$) and repetition is allowed (4444 is a valid 4-digit number).

WHEN ORDER DOESN'T MATTER AND REPETITION IS FORBIDDEN

Suppose that we choose an *unordered* set of k *distinct* elements: that is, *order does not matter* and *repetition is not allowed.* (For example, suppose you and $n - 1$ friends enter a raffle in which k identical new cell phones will be given away. Each of you puts your name on one of n cards that are placed in a hat, and k cards are drawn to choose the winners. Cards for winners are not put back into the hat after they're drawn, so nobody can win twice—"repetition is forbidden"—but Alice and Bob winning is the same as Bob and Alice winning.)

When we choose an unordered set of k distinct elements from a set of n options, there are $\binom{n}{k}$ different ways to make this choice, by the definition of combination. Such a subset is sometimes called a *k-combination* of S—an unordered set of k distinct elements of S. (Recall from Definition 9.2 that a *combination* of elements from a set S is precisely an unordered subset of elements from S.)

Example 9.44 (4 of 10, another way)
Suppose that you're asked to create a 10-bit number with exactly 4 ones. You do so by starting with `0000000000` and choosing 4 indices to change from `0` to `1`. There are $\binom{10}{4}$ such bitstrings: the order in which you choose a bit to make a 1 doesn't matter (changing bit #2 and then bit #7 to 1 yields the same bitstring as changing bit #7 and then bit #2 to 1) and repetition is not allowed (you have to change 4 *different* bits to 1).

Problem-solving tip: When you encounter a problem that seems completely novel, run through the techniques you know about and try them on for size, even if they're not an obvious fit. The type of counting in Example 9.45 doesn't seem like it has a lot to do with combinations, but by changing the way you view this problem it can be transformed into a problem you've seen before.

WHEN ORDER DOESN'T MATTER AND REPETITION IS ALLOWED

While these three types of selecting k out of n elements are the most frequent, the fourth possibility can sometimes arise, too: *order doesn't matter* but *repetition is allowed.* Let's build some intuition for this case with a longer example:

Example 9.45 (Taking notes on six sheets of paper in three classes)
Problem: You discover that your school notebook has only $k = 6$ sheets of paper left in it. You are attending $n = 3$ different classes today: Archaeology (A), Buddhism (B), and Computer Science (C). How many ways are there to allocate your six sheets of paper across your three classes? (No paper splitting or hoarding: each sheet must be allocated to one and only one class!)

(Here's another way to phrase the question: you must choose how many pages to assign to A, how many to B, and how many to C. That is, you must choose three nonnegative integers a, b, and c with $a + b + c = 6$. How many ways can you do it?)

Solution: The 28 ways of allocating your paper are shown in the following tables, sorted by the number of pages allocated to Archaeology (and breaking ties by the number of pages allocated to Buddhism). The allocations are shown in three ways:

- Pages are represented by the class name.
- Pages are represented by □, with | marking divisions between classes: we allocate the number of pages before the first divider to A, the number between the dividers to B, and the number after the second divider to C.
- Pages are represented by 0, with 1 marking divisions between classes: as in the □-and-| representation, we allocate pages before the first 1 to A, those between the 1s to B, and those after the second 1 to C.

Here are the 28 different allocations:

AAAAAA		
AAAAA	B	
AAAAA		C
AAAA	BB	
AAAA	B	C
AAAA		CC
AAA	BBB	
AAA	BB	C
AAA	B	CC
AAA		CCC
AA	BBBB	
AA	BBB	C
AA	BB	CC
AA	B	CCC
AA		CCCC
A	BBBBB	
A	BBBB	C
A	BBB	CC
A	BB	CCC
A	B	CCCC
A		CCCCC
	BBBBBB	
	BBBBB	C
	BBBB	CC
	BBB	CCC
	BB	CCCC
	B	CCCCC
		CCCCCC

```
A      |B      |C
□□□□□□ |       |
□□□□□  |□      |
□□□□□  |       |□
□□□□   |□□     |
□□□□   |□      |□
□□□□   |       |□□
□□□    |□□□    |
□□□    |□□     |□
□□□    |□      |□□
□□□    |       |□□□
□□     |□□□□   |
□□     |□□□    |□
□□     |□□     |□□
□□     |□      |□□□
□□     |       |□□□□
□      |□□□□□  |
□      |□□□□   |□
□      |□□□    |□□
□      |□□     |□□□
□      |□      |□□□□
□      |       |□□□□□
       |□□□□□□ |
       |□□□□□  |□
       |□□□□   |□□
       |□□□    |□□□
       |□□     |□□□□
       |□      |□□□□□
       |       |□□□□□□
```

00000011
00000101
00000110
00001001
00001010
00001100
00010001
00010010
00010100
00011000
00100001
00100010
00100100
00101000
00110000
01000001
01000010
01000100
01001000
01010000
01100000
10000001
10000010
10000100
10001000
10010000
10100000
11000000

All three versions of this table accurately represent the full set of 28 allocations, but let's concentrate on the representation in the second and third columns—particularly the third. The 0-and-1 representation in the third column contains *exactly* the same strings as Figure 9.29, which listed all $28 = \binom{8}{2}$ of the 8-bit strings that contain exactly 2 ones.

In a moment, we'll state a theorem that generalizes this example into a formula for the number of ways to select k out of n elements when order doesn't matter but repetition is allowed. But, first, here's a slightly different way of thinking about the result in Example 9.45 that may be more intuitive.

Suppose that we're trying to allocate a total of k pages among n classes. Imagine placing the k pages into a three-ring binder along with $n-1$ "divider tabs" (the kind that separate sections of a binder), as in Figure 9.32. There are now $n+k-1$ things in your binder. (In Example 9.45, there were 6 pages and 2 dividers, so 8 total things are in the binder.) The ways of allocating the pages precisely correspond to the ways of ordering the things in the binder—that is, choosing which of the $n+k-1$ things in the binder should be blank sheets of paper, and which should be dividers. So there are $\binom{n+k-1}{k}$ ways of doing so. In Example 9.45, we had $n=3$ and $k=6$, so there were $\binom{8}{6}=28$ ways of doing this allocation.

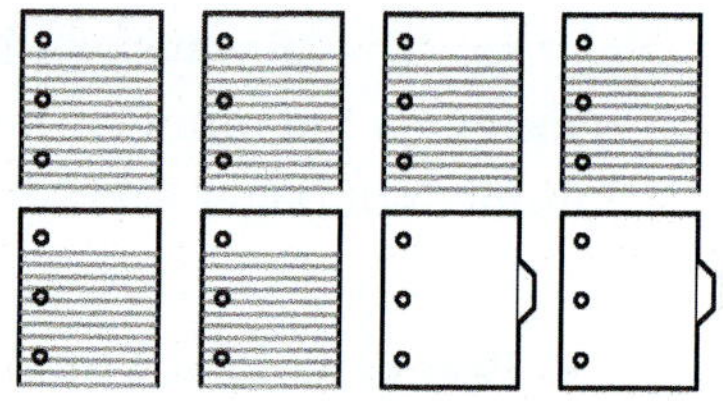

Figure 9.32: Any ordering of 6 pieces of paper and 2 divider tabs defines three sections (before, between, and after the dividers).

While the description in Example 9.45 wasn't stated in precisely these terms, our paper-allocation task was really a task about choosing with repetition: six times (once for each piece of paper), we select one of the elements of the set $\{A, B, C\}$ of classes. We may select the same class as many times as we wish ("repetition is allowed"), and the pieces of paper are indistinguishable ("order doesn't matter"). Here is the general statement of the number of ways to select k out of n elements for this scenario:

Theorem 9.17 (Choosing with repetition when order doesn't matter)
The number of ways to select k out of n elements when order doesn't matter but repetition is allowed is $\binom{n+k-1}{k}$.

Proof. We'll give a proof based on the Mapping Rule. We can represent a particular choice of k elements from the set of n candidates as a sequence $x \in (\mathbb{Z}^{\geq 0})^n$ such that $\sum_{i=1}^{n} x_i = k$. (Specifically, x_i tells us how many times we chose element i.) Define

$$X := \{x \in (\mathbb{Z}^{\geq 0})^n : \textstyle\sum_{i=1}^{n} x_i = k\}$$
$$\text{and } S := \{x \in \{0,1\}^{n+k-1} : x \text{ contains exactly } n-1 \text{ ones and } k \text{ zeros}\}.$$

We claim that there is a bijection between X and S. Specifically, define $f : X \to S$ as

$$f(x_1, x_2, \ldots, x_n) = \underbrace{00\cdots0}_{x_1 \text{ times}}\ 1\ \underbrace{00\cdots0}_{x_2 \text{ times}}\ 1\ \cdots\ 1\ \underbrace{00\cdots0}_{x_n \text{ times}}$$

(This representation is precisely the one in Example 9.45.) It's easy to see that f is a bijection: every element of S corresponds to one and only one element of X. As we argued in Example 9.38, the cardinality of S is $\binom{n+k-1}{k}$. □

Here's another example of this type of choice:

Example 9.46 (4 of 10, one last way)
Suppose that you have decided to buy 4 total drinks for a group of 10 of your friends. (You may buy multiple drinks for the same friend.) You can think of lining your friends up and performing a total of 13 successive actions, each of which is either (a) buying a drink for the friend immediately in front of you, or (b) shouting "next!". Of your 13 actions, 4 must be drink purchases. (The other 9 must be shouts of "next!") There are $\binom{13}{4}$ ways to choose these actions.

CHOOSING k OF n ELEMENTS, SUMMARIZED

We've now discussed four notions of choosing k elements from a set of n candidates, depending on whether we could choose the same option more than once and whether the order of our choices mattered:

- order matters and repetition is allowed: n^k ways.
- order matters and repetition is forbidden: $\frac{n!}{(n-k)!}$ ways.
- order doesn't matter and repetition is allowed: $\binom{n+k-1}{k}$ ways.
- order doesn't matter and repetition is forbidden: $\binom{n}{k}$ ways.

(Or see Figure 9.31 for a summary.) We've also considered the same example—choosing 4 of 10 options—in each setting, and the number of ways to do so was different in each of the four different scenarios:

- order matters and repetition is allowed: $10{,}000 = 10^4$ ways.
- order matters and repetition is forbidden: $5040 = 10 \cdot 9 \cdot 8 \cdot 7$ ways.
- order doesn't matter and repetition is allowed: $715 = \binom{13}{4}$ ways.
- order doesn't matter and repetition is forbidden: $210 = \binom{10}{4}$ ways.

Taking it further: In CS, we frequently encounter tasks where we must identify the best solution from a set of possibilities. For example, we might want to find the *longest increasing subsequence (LIS)* of a sequence of n integers. A *brute-force algorithm* is one that solves the problem by literally trying every possible solution and selecting the best. (For LIS, there are 2^n subsequences, so this algorithm is very slow.) But if there's a certain kind of structure and enough repetition in the subproblems that arise in a naïve recursive solution, a more advanced algorithmic design technique called *dynamic programming* can yield a much faster algorithm. And counting the number of subproblems—and the number of distinct subproblems!—is what establishes when algorithms using brute force or dynamic programming are good enough. See the discussion on p. 959.

9.4.2 *Some Properties of $\binom{n}{k}$, and Combinatorial Proofs*

Of the four ways of choosing k elements from n candidates that we explored in Section 9.4.1, perhaps the most common is the setting when order doesn't matter and repetition is forbidden. In this section, we'll explore some of the remarkable mathematical properties of the numbers—the values of $\binom{n}{k}$—that arise in this scenario.

The properties that we'll prove here (and those that you'll establish in the exercises) will be equalities of the form $x = y$ for two expressions x and y. We'll generally be able to give two very different styles of proof that $x = y$. One type of proof uses algebra, typically using the definition of $\binom{n}{k}$ and algebraic manipulations to show that x and y are equal. The other type of proof will be a more story-based approach, called a *combinatorial proof,* where we argue that $x = y$ by explaining how x and y are really just two ways of looking at the same set:

Definition 9.3 (Combinatorial Proof)
A combinatorial proof *establishes that two quantities x and y are equal by defining a set S and proving that $|S| = x$ and $|S| = y$ by counting $|S|$ in two different ways.*

The algebraic approach is perhaps apparently more straightforward, but combinatorial proofs can be more fun. Here's a first example:

Theorem 9.18 (A symmetry in choosing)
For any positive integer n and any integer $k \in \{0, 1, \ldots, n\}$, we have $\binom{n}{k} = \binom{n}{n-k}$.

Proof #1 of $\binom{n}{k} = \binom{n}{n-k}$, via algebra. We simply follow our noses through the definition:

$$\begin{aligned}
\binom{n}{k} &= \frac{n!}{k! \cdot (n-k)!} && \textit{definition of combinations} \\
&= \frac{n!}{(n-k)! \cdot k!} && \textit{commutativity of multiplication} \\
&= \frac{n!}{(n-k)! \cdot (n-(n-k))!} && \textit{antisimplification: } k = n-(n-k) \\
&= \binom{n}{n-k}. \quad \square && \textit{definition of combinations}
\end{aligned}$$

Here is a second proof of Theorem 9.18—this time a combinatorial proof. The basic idea is that we will construct a set S such that we can prove that $|S| = \binom{n}{k}$ and we can prove that $|S| = \binom{n}{n-k}$. (Thus we can conclude $\binom{n}{k} = \binom{n}{n-k}$.)

Proof #2 of $\binom{n}{k} = \binom{n}{n-k}$, via a combinatorial proof: Suppose that n students submit implementations of Bubble Sort in a computer science class. The instructor has k gold stars, and he will affix a gold star to each of k different implementations. Let S be the set of ways to affix gold stars. Here are two ways of computing $|S|$:

- First, we claim that $|S| = \binom{n}{k}$. Specifically, the instructor will choose k of the n submissions and affix gold stars to the k chosen elements. There are $\binom{n}{k}$ ways of doing so.
- Second, we claim that $|S| = \binom{n}{n-k}$. Specifically, the instructor will choose $n - k$ of the n submissions that he will *not* adorn with gold stars. The remaining unchosen submissions will be adorned. There are $\binom{n}{n-k}$ ways of choosing the unadorned submissions.

But $|S|$ is the same regardless of how we count it! So $\binom{n}{k} = |S| = \binom{n}{n-k}$ and the theorem follows. $\square$

Problem-solving tip: The hard part in a combinatorial proof is coming up with a story that explains both sides of the equation. Understanding what the more complicated side of the equation means is often a good place to start.

(Another way to think about the combinatorial proof: an n-bit string with k ones *is* an n-bit string with $n - k$ zeros; the number of choices for where the ones go is identical to the number of choices for where the zeros go.)

A combinatorial proof requires creativity—*what set S should we consider?*—but the argument that the proof is correct is generally comparatively straightforward. Thus the challenge in proving an identity with a combinatorial proof is a challenge of narrative: we must find a story in which the two sides of the equation both capture the set described by that story.

Pascal's Identity

Here's another example claim with both algebraic and combinatorial proofs:

Theorem 9.19 (Pascal's Identity)
For any integer $n \geq 1$ and any $k \in \{0, 1, \ldots, n\}$:

$$\binom{n-1}{k} + \binom{n-1}{k-1} = \binom{n}{k}.$$

Pascal's identity is named after Blaise Pascal, a 17th-century French mathematician. The programming language Pascal was also named in his honor.

Proof #1 of Pascal's Identity (algebra). Observe that if $k = 0$ or $k = n$, the identity follows immediately: by definition, we have $\binom{n}{0} = 1 = 1 + 0 = \binom{n-1}{0} + \binom{n-1}{-1}$ and similarly $\binom{n}{n} = 1 = 0 + 1 = \binom{n-1}{n} + \binom{n-1}{n-1}$. For the non-boundary cases, we'll manipulate the left-hand side until it's equal to the right-hand side:

$$\binom{n-1}{k} + \binom{n-1}{k-1}$$

$$= \frac{(n-1)!}{k! \cdot (n-1-k)!} + \frac{(n-1)!}{(k-1)! \cdot (n-k)!} \qquad \textit{definition of combinations}$$

$$= \frac{(n-1)!}{k! \cdot (n-1-k)!} \cdot \frac{n-k}{n-k} + \frac{(n-1)!}{(k-1)! \cdot (n-k)!} \cdot \frac{k}{k} \qquad \textit{multiplying by } 1 = \tfrac{x}{x}$$

$$= \frac{(n-1)! \cdot (n-k)}{k! \cdot (n-k)!} + \frac{(n-1)! \cdot k}{k! \cdot (n-k)!} \qquad (k-1)! \cdot k = k! \textit{ and } (n-1-k)! \cdot (n-k) = (n-k)!$$

$$= \frac{(n-1)! \cdot [(n-k)+k]}{k! \cdot (n-k)!} \qquad \textit{factoring}$$

$$= \frac{n!}{k! \cdot (n-k)!} \qquad n-k+k = n, \textit{ and } (n-1)! \cdot n = n!$$

$$= \binom{n}{k}. \quad \Box \qquad \textit{definition of combinations}$$

Proof #2 of Pascal's Identity (combinatorial proof). For the case of $k = 0$ or $k = n$, the argument is the same as in Proof #1. Otherwise, consider a set of $n \geq 1$ employees, one of whom is named Babbage. How many ways can we select a subset of k different employees? Here are two different ways of counting the number of these subsets:

- We choose k of the n employees. There are $\binom{n}{k}$ ways to do so.
- We decide whether to include Babbage, and then fill in the rest of the team:
  - If we pick Babbage, we need to pick $k-1$ further employees from the $n-1$ other (non-Babbage) employees; thus there are $\binom{n-1}{k-1}$ ways to select a team that includes Babbage.
  - If we don't pick Babbage, we pick all k employees from the $n-1$ others; thus there are $\binom{n-1}{k}$ ways to select a team that does not include Babbage.

  By the Sum Rule, there are therefore $\binom{n-1}{k-1} + \binom{n-1}{k-1}$ ways to choose a team.

Because we've counted the cardinality of one set in two different ways, the two sizes must be equal. Therefore $\binom{n}{k} = \binom{n-1}{k-1} + \binom{n-1}{k}$ and the theorem follows. □

Taking it further: World War II was perhaps the first major historical moment in which computer science—and, by the end of the war, the computer—was central to the story. The German military used a complex cryptographic device called the *Enigma machine* for encryption of military communication during the war. The Enigma machine, which was partially mechanical and partially electrical, had a large (though not unfathomably large) set of possible physical configurations, each corresponding to a different cryptographic "key." Among the first applications of an electronic computer—and the reason that one of the first computers was designed and built in the first place—was in breaking these codes, in part by exhaustively exploring the set of possible keys. As such, understanding the number of different keys in the system (a counting problem!) was crucial to the Allies' success in breaking the Enigma code. For more, see the discussion on p. 960.

9.4.3 *The Binomial Theorem*

The quantity $\binom{n}{k}$ is sometimes called a *binomial coefficient*, for reasons that we'll see in this section. First, a reminder: the product of two binomials $(x + y)$ and $(a + b)$ is $xa + xb + ya + yb$. (You may have once learned the "FOIL" mnemonic for the terms of the product: first = xa; outer = xb; inner = ya; and last = yb.) Thus when we square $x + y$—that is, multiply it by itself—we get

$$(x+y)\cdot(x+y) = xx + xy + yx + yy = 1\cdot x^2 + 2\cdot xy + 1\cdot y^2.$$

A *binomial* (Latin *bi* "two" + *nom* "name") is a special kind of polynomial—*poly* "many" + *nom* "name"—that has precisely two terms.

Observe that the three coefficients of these terms, in order, are $\langle 1,2,1\rangle = \langle \binom{2}{0}, \binom{2}{1}, \binom{2}{2}\rangle$. The *binomial theorem* is a general statement of this pattern: when we multiply out the expression $(x + y)^n$, the coefficient of the $x^k y^{n-k}$ term is $\binom{n}{k}$:

Theorem 9.20 (The Binomial Theorem)
For any $a \in \mathbb{R}$, any $b \in \mathbb{R}$, and any $n \in \mathbb{Z}^{\geq 0}$, we have

$$(a+b)^n = \sum_{i=0}^{n} \binom{n}{i} a^i b^{n-i}.$$

Before we prove the binomial theorem, let's start with some intuition about *why* these coefficients arise. For example, let's compute $(x + y)^4 = (x + y)\cdot(x + y)\cdot(x + y)\cdot(x + y)$, without doing any simplification by combining like terms:

$$\begin{aligned}
&(x+y)\cdot(x+y)\cdot(x+y)\cdot(x+y)\\
&= (xx+xy+yx+yy)\cdot(x+y)\cdot(x+y)\\
&= (xxx+xyx+yxx+yyx+xxy+xyy+yxy+yyy)\cdot(x+y)\\
&= xxxx+xyxx+yxxx+yyxx+xxyx+xyyx+yxyx+yyyx\\
&\quad +xxxy+xyxy+yxxy+yyxy+xxyy+xyyy+yxyy+yyyy.
\end{aligned}$$

Every term of the resulting expression consists of 4 multiplicands, one from each of the 4 copies of $(x + y)$. How many of these 16 terms contain, say, 2 copies of x and 2 copies of y? There are 6—$yyxx$, $xyyx$, $yxyx$, $xyxy$, $yxxy$, and $xxyy$—which is just the number of elements of $\{x, y\}^4$ that contain precisely two copies of x. While the symbols are different, it's easy to see that this quantity is precisely the number of elements of $\{0, 1\}^4$ that contain precisely two ones—which is just $\binom{4}{2}$.

We will prove the Binomial Theorem in generality in a moment, but to build a little bit of intuition for the proof, let's look at a special case first:

Problem-solving tip: When you're asked to solve a problem for a general value of n, one good way to get started is to try to solve it for a specific small value of n—and then try to generalize your solution to an arbitrary n. It's often easier to generalize from a particular n to a general n than to give a fully generally answer "from scratch."

Example 9.47 (The coefficients of $(x + y)^3$)

We're going to show that $(x+y)^3 = x^3 + 3x^2y + 3xy^2 + y^3$ *in the same style that we'll use in the full proof of the Binomial Theorem.* We'll start with the observation, made previously, that $(x + y)^2 = x^2 + 2xy + y^2 = \binom{2}{0}x^2 + \binom{2}{1}xy + \binom{2}{2}y^2$. A key step will make use of Theorem 9.19 to move from the coefficients of $(x + y)^2$ to the coefficients of $(x + y)^3$.

$$\begin{aligned}(x+y)^3 &= (x+y)\cdot(x+y)^2 \\ &= (x+y)\cdot\left[\binom{2}{0}x^2+\binom{2}{1}xy+\binom{2}{2}y^2\right] \\ &= \underbrace{\binom{2}{0}x^3+\binom{2}{1}x^2y+\binom{2}{2}xy^2}_{x\cdot\left(\binom{2}{0}x^2+\binom{2}{1}xy+\binom{2}{2}y^2\right)} + \underbrace{\binom{2}{0}x^2y+\binom{2}{1}xy^2+\binom{2}{2}y^3}_{y\cdot\left(\binom{2}{0}x^2+\binom{2}{1}xy+\binom{2}{2}y^2\right)}\end{aligned}$$

which, collecting like terms, simplifies to

$$(x+y)^3 = \binom{2}{0}x^3 + \left[\binom{2}{1}+\binom{2}{0}\right]x^2y + \left[\binom{2}{2}+\binom{2}{1}\right]xy^2 + \binom{2}{2}y^3.$$

By Theorem 9.19, we have that $\binom{2}{1} + \binom{2}{0} = \binom{3}{1}$ and $\binom{2}{2} + \binom{2}{1} = \binom{3}{2}$, so

$$(x+y)^3 = \binom{2}{0}x^3 + \binom{3}{1}x^2y + \binom{3}{2}xy^2 + \binom{2}{2}y^3$$

Because $\binom{n}{n} = 1$ and $\binom{n}{0} = 1$ for any n, we have that $\binom{2}{0} = \binom{3}{0}$ and $\binom{2}{2} = \binom{3}{3}$, and thus

$$\begin{aligned}(x+y)^3 &= \binom{3}{0}x^3 + \binom{3}{1}x^2y + \binom{3}{2}xy^2 + \binom{3}{3}y^3 \\ &= x^3 + 3x^2y + 3xy^2 + y^3.\end{aligned}$$ □

The combination notation can sometimes obscure the structure of the proof; for further intuition, here is what this proof looks like, without the notational overhead:

$$\begin{aligned}(x+y)^3 &= (x+y)\cdot(x+y)^2 \\ &= (x+y)\cdot(x^2+2xy+y^2) \\ &= (x^3+2x^2y+xy^2)+(x^2y+2xy^2+y^3) \\ &= x^3+(2+1)x^2y+(1+2)xy^2+y^3 \\ &= x^3+3x^2y+3xy^2+y^3.\end{aligned}$$

Proof of the Binomial Theorem

We're now ready to give a proof of the general form of the Binomial Theorem. Our proof will use mathematical induction on the exponent, and the structure of the inductive case of the proof will precisely mimic that of Example 9.47.

Proof of Binomial Theorem. Let a and b be arbitrary real numbers. We wish to prove that, for any integer $n \geq 0$,

$$(a+b)^n = \sum_{i=0}^{n} \binom{n}{i} a^i b^{n-i}.$$

We proceed by induction on n.

The base case ($n = 0$) is straightforward: anything to the 0th power is 1, so in particular $(a+b)^0 = 1$. And $\sum_{i=0}^{0} \binom{0}{i} a^i b^{0-i} = \binom{0}{0} \cdot 1 \cdot 1 = 1$.

For the inductive case ($n \geq 1$), we assume the inductive hypothesis $(a+b)^{n-1} = \sum_{i=0}^{n-1} \binom{n-1}{i} a^i b^{n-1-i}$. We must prove that $(a+b)^n = \sum_{i=0}^{n} \binom{n}{i} a^i b^{n-i}$. Our proof echoes the structure of Example 9.47:

$$\begin{aligned}
(a+b)^n &= (a+b)\cdot(a+b)^{n-1} && \textit{definition of exponentiation}\\
&= (a+b)\cdot \sum_{i=0}^{n-1} \binom{n-1}{i} a^i b^{n-1-i} && \textit{inductive hypothesis}\\
&= a\cdot\left[\sum_{i=0}^{n-1} \binom{n-1}{i} a^i b^{n-1-i}\right] + b\cdot\left[\sum_{i=0}^{n-1} \binom{n-1}{i} a^i b^{n-1-i}\right] && \textit{distributing the multiplication}\\
&= \left[\sum_{i=0}^{n-1} \binom{n-1}{i} a^{i+1} b^{n-1-i}\right] + \left[\sum_{i=0}^{n-1} \binom{n-1}{i} a^i b^{n-i}\right] && \textit{distributing the multiplication, again}\\
&= \left[\sum_{j=1}^{n} \binom{n-1}{j-1} a^j b^{n-j}\right] + \left[\sum_{i=0}^{n-1} \binom{n-1}{i} a^i b^{n-i}\right]. && \textit{reindexing the first summation } (j := i+1)
\end{aligned}$$

By separating out the $i = 0$ and $j = n$ terms from the two summations, and then combining like terms, we have

$$\begin{aligned}
(a+b)^n &= \left[\sum_{j=1}^{n-1} \binom{n-1}{j-1} a^j b^{n-j}\right] + \left[\sum_{i=1}^{n-1} \binom{n-1}{i} a^i b^{n-i}\right] + \binom{n-1}{n-1} a^n b^{n-n} + \binom{n-1}{0} a^0 b^{n-0}\\
&= \left[\sum_{j=1}^{n-1} \left(\binom{n-1}{j-1} + \binom{n-1}{j}\right) a^j b^{n-j}\right] + \binom{n-1}{n-1} a^n b^{n-n} + \binom{n-1}{0} a^0 b^{n-0}.
\end{aligned}$$

Applying Theorem 9.19 to substitute $\binom{n}{j}$ for $\binom{n-1}{j-1} + \binom{n-1}{j}$ and using the fact that $\binom{n-1}{n-1} = 1 = \binom{n}{n}$ and $\binom{n-1}{0} = 1 = \binom{n}{0}$, we have

$$\begin{aligned}
(a+b)^n &= \left[\sum_{j=1}^{n-1} \binom{n}{j} a^j b^{n-j}\right] + \binom{n-1}{n-1} a^n b^{n-n} + \binom{n-1}{0} a^0 b^{n-0} && \binom{n}{j} = \binom{n-1}{j-1} + \binom{n-1}{j}\\
&= \left[\sum_{j=1}^{n-1} \binom{n}{j} a^j b^{n-j}\right] + \binom{n}{n} a^n b^{n-n} + \binom{n}{0} a^0 b^{n-0} && \binom{n-1}{n-1} = 1 = \binom{n}{n} \textit{ and } \binom{n-1}{0} = 1 = \binom{n}{0}\\
&= \left[\sum_{j=0}^{n} \binom{n}{j} a^j b^{n-j}\right], && \textit{incorporating the } j = 0 \textit{ and } j = n \textit{ terms back into the summation}
\end{aligned}$$

which proves the theorem. □

9.4.4 *Pascal's Triangle*

Much of this section has been devoted to understanding the binomial coefficients, through the Binomial Theorem and through combinatorial proofs of a number of their other properties. We'll close our discussion of binomial coefficients with a visual representation of these quantities, called *Pascal's triangle.* Pascal's triangle arranges the binomial coefficients in a classical and very useful way: the nth row of Pascal's triangle consists of all of the $n+1$ binomial coefficients $\binom{n}{0}, \binom{n}{1}, \cdots, \binom{n}{n}$, in order. Figure 9.33 shows the first nine rows of Pascal's triangle:

Like Pascal's identity, Pascal's triangle is named after the 17th-century French mathematician Blaise Pascal.

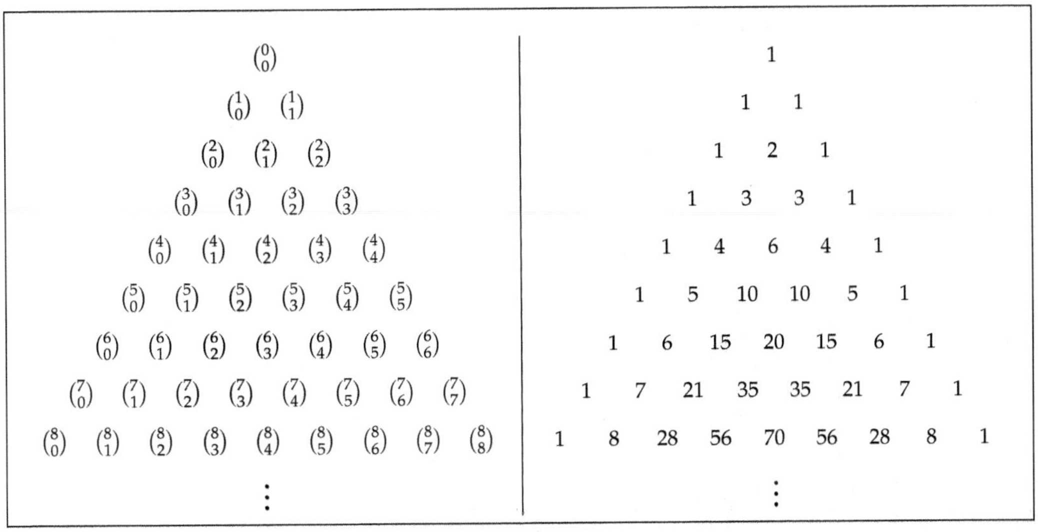

Figure 9.33: The first several rows of Pascal's triangle, in both "choose" notation and in numerical form.

Many of the properties of the binomial coefficients that we've established previously can be seen by looking at patterns visible in Pascal's triangle—as can some others that we'll prove here, or that you'll prove in the exercises.

For example, Figure 9.34 gives visualizations of two properties that we've already proven. Theorem 9.18 states that $\binom{n}{k} = \binom{n}{n-k}$; this theorem is reflected by the fact that the numerical values of Pascal's triangle are symmetric around a vertical line drawn down through the middle of the triangle. And Theorem 9.19 ("Pascal's Identity"), which states that $\binom{n-1}{k} + \binom{n-1}{k-1} = \binom{n}{k}$, is illustrated by the fact that each entry in Pascal's triangle is the sum of the two elements immediately above it (up-and-left and up-and-right).

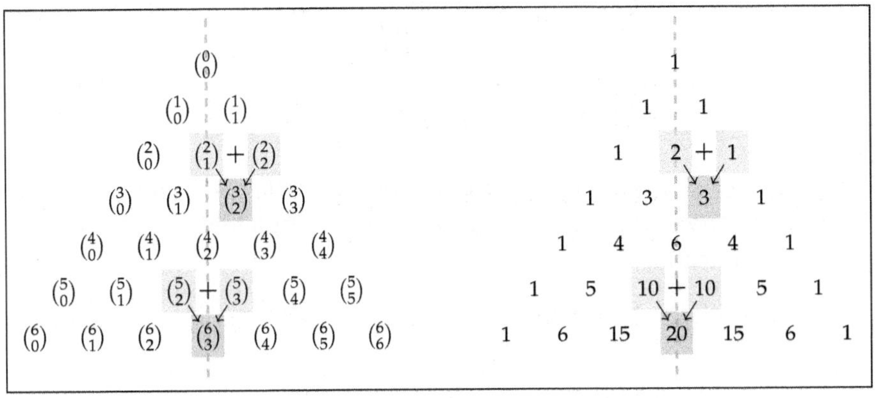

Figure 9.34: Theorems 9.18 and 9.19 reflected in Pascal's triangle.

There are many other notable properties of the binomial coefficients, many of which we can see more easily by looking at Pascal's triangle. Here's one example; a number of other properties are left to you in the exercises. Let's look at the *row sums* of Pascal's triangle—that is, computing $\binom{n}{0} + \binom{n}{1} + \cdots + \binom{n}{n}$ for different values of n. (See Figure 9.35.)

From calculating the row sum for a few small values of n, we see that the nth row appears to have value equal to 2^n. (Incidentally, the sum of the *squares* of the numbers in any particular row in Pascal's triangle also has a special form, as you'll see in Exercise 9.170.) Indeed, the power-of-two pattern for the row sums of Pascal's triangle that we observe in Figure 9.35 holds for arbitrary n—and we'll prove this theorem here, in several different ways.

1	$= 1$
$1+1$	$= 2$
$1+2+1$	$= 4$
$1+3+3+1$	$= 8$
$1+4+6+4+1$	$= 16$
$1+5+10+10+5+1$	$= 32$
$1+6+15+20+15+6+1$	$= 64$
$\vdots$	

Figure 9.35: The row sums of Pascal's triangle.

Theorem 9.21 (Sum of a row of Pascal's triangle)
$\sum_{i=0}^{n} \binom{n}{i} = 2^n$.

Proof #1 (algebraic/inductive) [sketch]. We can gain a bit of intuition for this claim from Theorem 9.19 (Pascal's Identity): each entry $\binom{n}{k}$ in the nth row is added into *exactly two* entries in the $(n+1)$st row, namely $\binom{n+1}{k}$ and $\binom{n+1}{k+1}$. Therefore the values in row #n of Pascal's triangle each contribute *twice* to the values in row #$(n+1)$, and therefore the $(n+1)$st row's sum is twice the sum of the nth row. This intuition can be turned into an inductive proof, which you'll give in Exercise 9.169. □

Proof #2 (combinatorial). Let $S := \{1, 2, \ldots, n\}$ be a set with n elements. Let's count the number of subsets of S in two different ways.

On one hand, there are 2^n such subsets: there is a bijection between subsets of S and $|S|$-bit strings. (See Lemma 9.10.)

On the other hand, let's account for the subsets of S by *first* choosing a size k of the subset, and then counting the number of subsets of that size. By the Sum Rule, the total number of subsets of S is exactly

$$\sum_{k=0}^{n} (\text{the number of subsets of } S \text{ of size } k).$$

By definition, there are exactly $\binom{n}{k}$ subsets of size k. Therefore the total number of subsets is $\sum_{k=0}^{n} \binom{n}{k}$. Thus $2^n = \sum_{k=0}^{n} \binom{n}{k}$. □

Proof #3 (making clever use of the Binomial Theorem). We'll start from the right-hand side of the theorem statement, and begin with a completely unexpected, but obviously true, antisimplification:

$$\begin{aligned} 2^n &= (1+1)^n && \textit{obviously } 2 = 1+1\textit{; therefore } 2^n = (1+1)^n \\ &= \sum_{i=0}^{n} \binom{n}{i} 1^i 1^{n-i} && \textit{binomial theorem} \\ &= \sum_{i=0}^{n} \binom{n}{i}. \quad \square && 1^k = 1 \textit{ for any value of } k \end{aligned}$$

You'll explore some of the many other interesting and useful properties of Pascal's triangle, and of the binomial coefficients in general, in the exercises.

Computer Science Connections

Brute Force Algorithms and Dynamic Programming

In an *optimization problem*, we're given a set S of valid solutions and some measure of quality $f : S \to \mathbb{R}$, and asked to compute the element $x \in S$ that's the best according to f. (That is, we want to find the $x \in S$ that optimizes $f(x)$.) Two examples are shown in Figure 9.36: the *traveling salesman problem (TSP)*—the problem solved every day by delivery drivers, who have to visit a given list of addresses and return to the depot—and the *cheapest vertical seam (CVS)* problem, which arises in a remarkable computer graphics application.[5] (For an example of the latter problem, see Figure 9.37.)

For both TSP and CVS, there are very simple, but very slow, *brute-force algorithms* that solve the problem by computing the list of all possible solutions (all orderings of the cities; all top-to-bottom paths) and identifying the best of these possible solutions. It's by now a reasonably straightforward counting exercise to show that there are $n!$ orderings and between $2^n \cdot n$ and $3^n \cdot n$ paths (it takes some work to avoid counting paths that fall off the left/right edges of the grid). These running times are unimpressive—even n around 100 would require decades of computing time—and this is, more or less, the best known algorithm for TSP! (See p. 326.)

But we can do better for CVS, with another view of the problem. Given a grid G, define $best(i,j)$ as the *cost of the cheapest path from grid cell $\langle i,j \rangle$ to the bottom of the grid.* Then we can solve the CVS problem using a recursive algorithm that computes $best(i,j)$ for every cell $\langle i,j \rangle$, as in Figure 9.38. Unfortunately, this algorithm is just as slow as the brute-force approach: to compute **best**(i,j), we make three recursive calls, at least two of which remain inside the grid. Thus the running time $T(i)$ to find **best**$(n-i,j)$ with i rows beneath cell $\langle i,j \rangle$ is given by the recurrence $T(1) = 1$ and $T(i) \geq 2T(i-1)+1$—which satisfies $T(n) \geq 2^n$, just as slow as before.

But a key algorithmic observation is that the number of *different* cells in the grid is much smaller—only n^2 different cells! So, while the algorithm in Figure 9.38 does take $\Omega(2^n)$ time, it actually "should" require only $\Theta(n^2)$ time—*as long as we avoid recomputing* **best**(i,j) *multiple times for the same value of $\langle i,j \rangle$!* Once we've figured out **best**$(3,7)$ (because we needed that value to figure out **best**$(4,6)$), we don't bother recomputing **best**$(3,7)$ when we need it again (while we're computing **best**$(4,7)$ and **best**$(4,8)$); instead, we just remember the value and reuse it without doing any further computation.

The most straightforward way to implement this basic idea is called *memoization*: we build a data structure in which we check to see whether we've already stored the value of **best**(i,j) before computing the value via the three recursive calls, and we always add all values we compute to the data structure before returning them. A slightly more efficient way of implementing this idea is called *dynamic programming*, where we transform this recursive solution into one using loops—and build up the values of **best**(i,j) from the bottom up. (See Figure 9.39).

In general, dynamic programming is an algorithmic design technique that can save us a massive amount of computation—as long as the number of *different* problems encountered in the recursive solution is small.

Traveling Salesman Problem (TSP):
Input: A set C of n cities, and distance function d giving the driving time between any two cities.
Output: An ordering π of C such that the sum of the driving times $\sum_i d(\pi_i, \pi_{i+1})$ is minimized.

Cheapest Vertical Seam (CVS):
Input: An n-by-n grid of integers.
Output: A path from the top row to the bottom row, moving in direction $\{\swarrow, \downarrow, \searrow\}$ at each step, such that the sum of the integers along the path is minimized.

Figure 9.36: Two problems.

[5] Shai Avidan and Ariel Shamir. Seam carving for content-aware image resizing. In *ACM SIGGRAPH*, 2007.

9	8	7	1	9
7	3	2	9	1
2	8	5	6	9
4	7	5	3	4
3	8	2	8	1

Figure 9.37: A small example of CVS.

best(i,j): // *Assume $G_{1...n,1...n}$ is given.*
 if $i = n$ **then**
  return $G_{i,j}$ *(in the last row)*
 else if $j \leq 0$ or $j \geq n$ **then**
  return $+\infty$ *(outside the grid)*
 else
  return the minimum of:

$$\begin{cases} G_{i,j} + \mathbf{best}(i+1, j-1), \\ G_{i,j} + \mathbf{best}(i+1, j), \\ G_{i,j} + \mathbf{best}(i+1, j+1). \end{cases}$$

Figure 9.38: A recursive algorithm for CVS. (To solve CVS itself, return the smallest **best**$(1,j)$ for every $1 \leq j \leq n$.)

CVS$(G_{1...n,1...n})$:
 for $j := 1, \ldots, n$:
  $T[n,j] := G_{i,j}$
 for $i := n-1, \ldots, 1$:
  for $j := 1, \ldots, n$:
   $T[i,j]$:= the minimum of:
   (Treat $T[\cdot,j] = \infty$ if j out of range.)

$$\begin{cases} G_{i,j} + T[i+1, j-1], \\ G_{i,j} + T[i+1, j], \\ G_{i,j} + T[i+1, j+1]. \end{cases}$$

 return $\min_j T[1,j]$.

Figure 9.39: A dynamic programming algorithm for CVS.

COMPUTER SCIENCE CONNECTIONS

THE ENIGMA MACHINE AND THE FIRST COMPUTER

The Enigma machine was a physical cryptographic device used by the Germans during World War II to communicate between German high command and their military units in the field. The basic structure of the machine involved *rotors* and *cables*. A *rotor* was a 26-slot physical wheel that encoded a permutation π: when the wire corresponding to input i is active, the output wire corresponding to π_i is active. A *plugboard* allowed an arbitrary matching of keys on the keyboard to the inputs to the rotors—a *cable* was what actually connected a key to the first rotor. (The machine did not require any cables in the plugboard; if there was no cable, then the key pressed was what went into the rotor in the first place.) The basic encryption in the Enigma machine proceeded as follows:

1. The user pressed a key, say A, on the keyboard. If there was a cable from the A key, then the key would be remapped to the other end of the cable; otherwise the procedure proceeded using the A. (See Figure 9.40.)
2. The pressed key was permuted by rotor #1; the output of rotor #1 was permuted by rotor #2; the output of rotor #2 was permuted by rotor #3. (See Figure 9.41.) The output of rotor #3 was "reflected" by a fixed permutation, and then the reflector's output pass through the three rotors, in reverse order and backward: the output of the reflector was permuted by rotor #3, then by #2, and then by #1. (See Figure 9.42.)
3. A light corresponding to the output of rotor #1, passed through the plugboard cable if present, lights up; the illuminated letter is the encoding.

The tricky part is that the rotors rotate by one notch when the key is pressed, so that the encoding changes with every keypress.

The "secret key" that the two communicating entities needed to agree upon was which rotors to use in which order ($5 \cdot 4 \cdot 3 = 60$; there were 5 standard rotors in an Enigma), what the initial position of the rotors should be ($26^3 = 17{,}576$), and what plugboard matching to use ($\frac{26!}{13! \cdot 2^{13}} \approx 8 \times 10^{12}$ choices if all 26 letters were matched; see Example 9.32). Interestingly, almost all of the complexity came from the plugboards.

Perhaps surprisingly, the fact that there were so many possible settings for the Enigma led to the invention of one of the first programmable computers, by Alan Turing at Bletchley Park, in England, during the war. Turing built a machine that could test many of these configurations, by brute force. (If there were fewer possibilities, it could have been cracked by hand; if there were many more, it couldn't have been cracked by brute force.) Turing and his team developed a device called the Bombe to exhaustively try to compute the shared German secret key—each day!

Many other cryptographic tricks related to the way the Enigma was being used were also part of the analysis. For example, the construction of the device meant that no letter could encrypt to itself; this fact was exploited in the analysis. Another crucial part of the code breaking was a *known plaintext attack* on the Enigma: the British also used knowledge of what the Germans tended to communicate (like weather reports) to narrow their search.

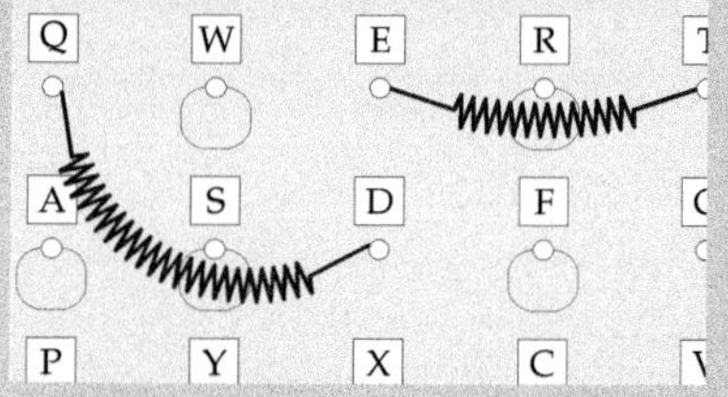

Figure 9.40: The effect of the plugboard. Each of the 26 keys is either mapped to itself (like W here), or is matched with another key (like Q ↔ D here). Pressing an unmatched key x yields x itself; pressing a matched key x yields whatever letter is matched to x.

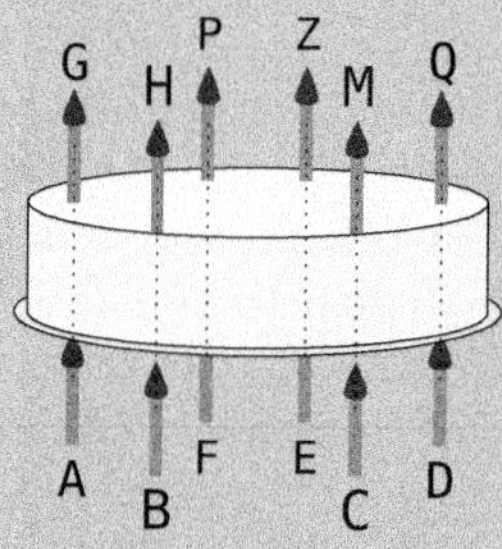

Figure 9.41: The effect of a rotor. Each rotor encodes a permutation of the letters; when the input letter i comes into the rotor, the output π_i comes out. (Here, for example, an input B turns into an output of H.) After each keypress, the top portion of the rotor would rotate by one notch, so that B would now turn into G.

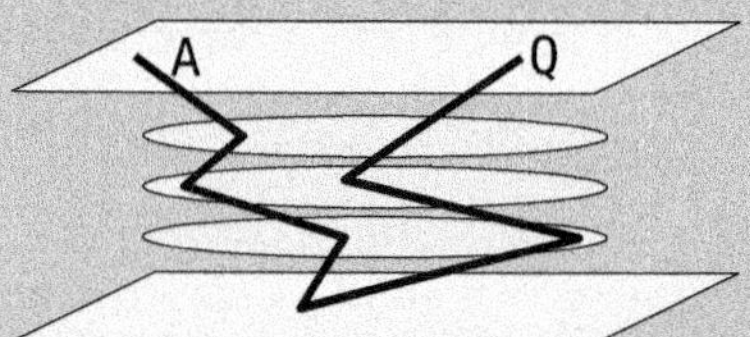

Figure 9.42: The Enigma machine's operation. The operator types an A, which (after going through the plugboard) is permuted by rotor #1, rotor #2, rotor #3, the fixed permutation of the machine, rotor #3, rotor #2, and rotor #1. It then (after passing through the plugboard) lights up the output, Q. The rotors advance by one notch, and encoding continues with the next letter.

9.4.5 Exercises

For two strings x and y, let's call a shuffle *of x and y any interleaving of the letters of the two strings (that maintains the order of the letters within each string, but may repeatedly alternate between blocks of x letters and blocks of y letters). For example, the words* ALE *and* LID *can be shuffled into* ALLIED *or* ALLIDE *or* ALLIDE *or* LIDALE. *How many* different *strings can be produced as shuffles of the following pairs of words?*

9.121 BACK and FORTH
9.122 DAY and NIGHT
9.123 SUPPLY and DEMAND
9.124 LIFE and DEATH
9.125 ON and ON
9.126 OUT and OUT

9.127 *(programming required)* Write a program, in a language of your choice, that computes all shuffles of two given words x and y. A recursive approach works well: a shuffle consists either of the first character of x followed by a shuffle of $x_{2\ldots|x|}$ and y, or the first character of y followed by a shuffle of x and $y_{2\ldots|y|}$. (Be sure to eliminate any duplicates from your resulting list.)

The next few questions ask you to think about shuffles of generic strings, instead of particular words. (Assume that the alphabet is an arbitrarily large set—you are not restricted to the 26 letters in English.) Consider two strings x and y, and let $n := |x| + |y|$ be the total number of characters between them. Note that the number of distinct shuffles of x and y may depend both *on the lengths of x and y* and *on the particular strings themselves; for example, if some letters are shared between or within the two strings, there may be fewer possible shuffles.*

9.128 In terms of n, what is the *maximum* possible number of different shuffles of x and y?
9.129 In terms of n, what's the *minimum* possible number of distinct shuffles of x and y?
9.130 What is the largest possible number of different shuffles of *three* strings of length a, b, and c?

9.131 How many 42-bit strings have exactly 16 ones?
9.132 How many 23-bit strings have at most 3 ones? (The coincidental arithmetic structure of the answer actually turns out to be helpful for error-correcting codes; see Exercise 4.30.)
9.133 How many 32-bit strings have a number of ones within ± 2 of the number of zeros?
9.134 The set of 64-bit strings with k ones is largest for $k = 32$. What's the smallest m for which

$$|\{\text{the number of 64-bit strings with} \leq m \text{ ones}\}| \geq |\{\text{the number of 64-bit strings with 32 ones}\}|?$$

9.135 What is the smallest even integer n for which the following statement is true? If we flip an unbiased coin n times, as in Example 9.41, the probability that we get exactly $\frac{n}{2}$ heads is less than 10%.

A bridge hand consists of 13 cards from a standard 52-card deck, with 13 ranks *(2 through ace) and 4* suits *($\clubsuit$, $\diamondsuit$, $\heartsuit$, and $\spadesuit$). (That is, the cards in the deck are $\{2, 3, \ldots, 10, J, Q, K, A\} \times \{\clubsuit, \diamondsuit, \heartsuit, \spadesuit\}$.) How many different bridge hands are there that meet the following conditions?*

9.136 A *void in spades:* a 13-card hand that contains only cards from the suits $\clubsuit$, $\diamondsuit$, and $\heartsuit$.
9.137 A *singleton in hearts:* exactly one of the 13 cards comes from the suit $\heartsuit$.
9.138 All four kings.
9.139 No queens at all.
9.140 Exactly two jacks.
9.141 Exactly two jacks and exactly two queens.
9.142 A bridge hand has *high honors* if it contains the five highest-ranked cards $\{10, J, Q, K, A\}$ in the same suit. How many bridge hands have high honors? *(Warning: be careful about double counting!)*

Many bridge players evaluate their hands by the following system of points. *First, give yourself one* high-card point *for a jack, two for a queen, three for a king, and four for an ace. Furthermore, give yourself three* distribution points *for each void (a suit in which you have zero cards), two points for a singleton (a suit with one card), and one point for a doubleton (a suit with two cards).*

9.143 How many bridge hands have a high-card point count of zero?
9.144 How many bridge hands have a high-card point count of zero *and* a distribution point count of zero? What fraction of all bridge hands is this?

How many ways are there to choose 32 out of 202 options if . . .

9.145 . . . repetition is allowed and order matters?
9.146 . . . repetition is forbidden and order matters?
9.147 . . . repetition is allowed and order doesn't matter?
9.148 . . . repetition is forbidden and order doesn't matter?

The first 10 prime numbers are $\{2,3,5,7,11,13,17,19,23,29\}$. How many different integers have exactly ...

9.149 ... 5 prime factors (all from this set), where all of these factors are different?

9.150 ... 5 prime factors (all from this set)? (Note that $32 = 2 \cdot 2 \cdot 2 \cdot 2 \cdot 2$ is an example.)

How many different integers have exactly 10 prime factors ...

9.151 ... all of which come from the set of the first 20 prime numbers?

9.152 ... all of which come from the set of the first 20 prime numbers, and where all 10 of these factors are different from each other?

Suppose that we have two sequences $\langle x_1, x_2, \ldots, x_n \rangle$ and $\langle y_1, y_2, \ldots, y_{2n} \rangle$ of data points—perhaps representing a sequence of intensities from two streams of speech. We wish to align x to y by matching up elements of x to elements of y. (For example, y might represent a reference stream, where we're trying to match x up to it.) We insist that each element of x is assigned to one and only one element of y. (See Figure 9.43.)

9.153 How many ways are there to assign each of the n elements of x to one of the $2n$ elements of y?

9.154 How many ways are there to assign each of the n elements of x to one of the $2n$ elements of y so that no element of y is matched to more than one element of x?

In many applications, we can only consider alignments of the elements of x and y that "maintain order": that is, we can't have x_5 assigned to an element of y that comes after the element assigned to x_6. (If $f : \{1, \ldots, n\} \to \{1, \ldots, 2n\}$ represents the alignment, then we require that $i \leq j$ implies that $f(i) \leq f(j)$.)

9.155 How many ways are there to assign each of the n elements of x to one of the $2n$ elements of y in a way that maintains order?

9.156 How many ways are there to assign each of the n elements of x to one of the $2n$ elements of y in a way that maintains order so that no element of y is matched to more than one element of x?

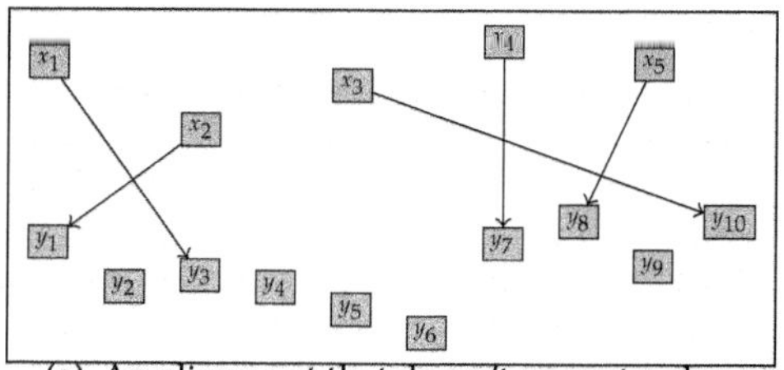

(a) An alignment that doesn't respect order.

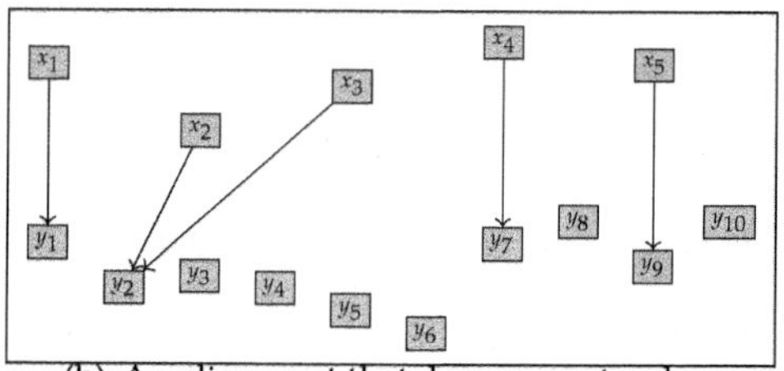

(b) An alignment that does respect order.

Figure 9.43: An alignment between two sequences, for Exercises 9.153–9.156. (Thanks to Roni Khardon, from whom I learned a version of the exercises.)

9.157 Consider the equation $a + b + c = 202$. How many solutions are there where a, b, and c are all nonnegative integers?

9.158 How many different solutions are there to the equation $a+b+c+d+e = 8$, where all of $\{a,b,c,d,e\}$ have to be nonnegative integers?

9.159 What about for $a+b+c+d+e = 88$, again where all variables must be nonnegative integers?

9.160 What about for $a + 2b + c = 128$, again where a, b, and c must be nonnegative integers? *(Hint: sum over the possible values of b and use Theorem 9.17.)*

The Association for Computing Machinery (the ACM)—a major professional society for computer scientists—puts on student programming competitions regularly. Teams of students spend a few hours working on some programming problems (of various levels of difficulty).

9.161 Suppose that, at a certain college in the midwest, there are 141 computer science majors. A programming contest team consists of 3 students. How many ways are there to choose a team?

9.162 Suppose that, at a certain programming contest, teams are given 10 problems to try to solve. When the contest begins, each of the 3 members of the team has to choose a problem to think about first. (More than one team member can think about the same problem.) How many ways are there for the 3 team members to choose a problem to think about first?

9.163 In most programming contests, teams are scored by the number of problems they correctly solve. (There are tiebreakers based on time and certain penalties.) A team can submit multiple solutions to the same problem. Suppose that a particular team has calculated that they have time to code up and submit 20 different attempted answers to the 10 questions in the contest. How many different ways can they allocate their 20 submissions across the 10 problems? (The order of their submissions doesn't matter.)

See the discussion on p. 730, or

9.164 Solve the following problem, posed by Adi Shamir in his original paper on secret sharing:[6]

> Eleven scientists are working on a secret project. They wish to lock up the documents in a cabinet so that the cabinet can be opened if and only if six or more of the scientists are present. What is the smallest number of locks needed? What is the smallest number of keys to the locks each scientist must carry?

[6] Adi Shamir. How to share a secret. *Communications of the ACM*, 22(11):612–613, November 1979.

9.165 In machine learning, we try to use a collection of *training data*—for example, a large collection of ⟨image, letter⟩ pairs of images of handwritten letters and the English letter that they represent—to compute a predictor that will do well on predicting answers on a set of novel *test data*. One danger in such a system is *overfitting:* we might build a predictor that's overly affected by idiosyncrasies of the training data. One way to address the risk of overfitting is a technique called *cross-validation:* we divide the training data into several subsets, and then, for each subset S, train our predictor based on $\sim S$ and test it on S. We might then average the parameters of our predictor across the subsets S. In *ten-fold cross-validation* on a n-element training set, we would split our n training examples into disjoint sets $S_1, S_2, \ldots, S_{10}$ where $|S_i| = \frac{n}{10}$.

How many ways are there to split an n-element set into disjoint subsets $S_1, S_2, \ldots, S_{10}$ of size $\frac{n}{10}$ each? (Note the order of the subsets themselves doesn't matter, nor does the order of the elements within a subset.)

9.166 Consider the set of bitstrings $x \in \{0,1\}^{n+k}$ with n zeros and k ones with the additional condition that *no ones are adjacent.* (For $n = 3$ and $k = 2$, for example, the legal bitstrings are `00101`, `01001`, `01010`, `10001`, `10010`, and `10100`.) Prove by induction on n that the number of such bitstrings is $\binom{n+1}{k}$.

9.167 Consider the set of bitstrings $x \in \{0,1\}^{n+k}$ with n zeros and k ones with the additional condition that *every block of ones has even length.* (For $n = 3$ and $k = 2$, for example, the legal bitstrings are `00011`, `00110`, `01100`, `11000`.) Prove that, for any even k, the number of such bitstrings is $\binom{n+(k/2)}{n}$.

9.168 Prove that $k \cdot \binom{n}{k} = n \cdot \binom{n-1}{k-1}$ twice, using both an algebraic and a combinatorial proof.

9.169 Using induction on n, prove Theorem 9.21—that is, prove that

$$\sum_{i=0}^{n} \binom{n}{i} = 2^n.$$

9.170 Prove the following identity about the squares of the binomial coefficients. (For example, for $n = 4$, this identity states that $\binom{4}{0}^2 + \binom{4}{1}^2 + \binom{4}{2}^2 + \binom{4}{3}^2 + \binom{4}{4}^2 = 1^2 + 4^2 + 6^2 + 4^2 + 1^2 = 70$ is equal to $\binom{8}{4}$. And, indeed, $\binom{8}{4} = \frac{8!}{4! \cdot 4!} = 70$.) Use a combinatorial proof.

$$\sum_{k=0}^{n} \binom{n}{k}^2 = \binom{2n}{n}.$$

9.171 Prove the following identity by algebraic manipulation:

$$\binom{n}{m}\binom{m}{k} = \binom{n}{k}\binom{n-k}{m-k}.$$

9.172 Now prove the identity from Exercise 9.171 with a combinatorial proof. *(Hint: think about choosing a team of m people from a pool of n candidates, and picking k managers from the team that you've chosen.)*

9.173 Prove the following identity, using an algebraic, inductive, or combinatorial proof:

$$\sum_{k=0}^{n} \binom{k}{m} = \binom{n+1}{m+1}.$$

Recall that $\binom{a}{b} = 0$ for any $b < 0$ or $b > a$, so many of the terms of the summation are zero. For example, for $m = 3$ and $n = 5$, the claim states that $\binom{6}{4} = \binom{0}{3} + \binom{1}{3} + \binom{2}{3} + \binom{3}{3} + \binom{4}{3} + \binom{5}{3} = 0 + 0 + 0 + \binom{3}{3} + \binom{4}{3} + \binom{5}{3}$.

9.174 Prove the following identity about the binomial coefficients and the Fibonacci numbers (where f_i is the ith Fibonacci number), by induction on n:

$$\sum_{k=0}^{\lfloor n/2 \rfloor} \binom{n-k}{k} = f_{n+1}.$$

9.175 Prove *van der Monde's identity*:

$$\binom{n+m}{k} = \sum_{r=0}^{k} \binom{m}{k-r} \cdot \binom{n}{r}.$$

(Hint: suppose you have a deck of n red cards and m black cards, from which you choose a hand of k total cards.)

A common subsequence *of two strings x and y is a string z that's a subsequence of both. A subsequence of an n-character string corresponds to a subset of $\{1, 2, \ldots, n\}$, indicating which indices are included (and which aren't). (See Exercise 9.82.) For example,* `BASIC` *is a common subsequence of* `BRAINSICKNESS` *and* `BIOACOUSTICS`.

9.176 Suppose that you have been asked to find the *number of common subsequences* of two n-character strings $x, y \in \Sigma^n$, by brute force. An algorithm to do so is shown in Figure 9.44(a). How many times do we execute Line 3 (testing whether $a = b$)?

9.177 Using the fact that common subsequences must have the same length, we can modify the algorithm as shown in Figure 9.44(b). Now how many times do we execute Line 4 (testing whether $a = b$)?

9.178 Using *Stirling's approximation* of the factorial function, which states that $n! \approx \sqrt{2\pi n}(n/e)^n$ (where $\pi = 3.1415\cdots$ and $e = 2.7182\cdots$), argue that Figure 9.44(b) is an improvement on Figure 9.44(a).

```
1: for each subsequence a of x:
2:     for each subsequence b of y:
3:         check if a = b
```

(a) A brute-force algorithm.

```
1: for k = 0...n:
2:     for each subsequence a of x of length k:
3:         for each subsequence b of y of length k:
4:             check if a = b
```

(b) A length-aware brute-force algorithm.

Figure 9.44: Two algorithms for common subsequences.

9.179 Use the Binomial Theorem to prove the following identity:

$$\sum_{k=0}^{n} (-1)^k \cdot \binom{n}{k} = 0.$$

9.180 Use the Binomial Theorem to prove the following identity:

$$\sum_{k=0}^{n} \frac{\binom{n}{k}}{2^k} = \left(\frac{3}{2}\right)^n.$$

9.181 In Section 9.2.2, we introduced the *Inclusion–Exclusion* rule for counting the union of 2 or 3 sets:

$$|A \cup B| = |A| + |B| - |A \cap B|$$
$$|A \cup B \cup C| = |A| + |B| + |C| - |A \cap B| - |A \cap C| - |B \cap C| + |A \cap B \cap C|$$

Exercise 9.30 asked you to give a formula for a 4-set intersection, but here's a completely general solution:

$$\left|\bigcup_{i=1}^{k} A_i\right| = \sum_{i=1}^{k}\left[(-1)^{i+1} \cdot \sum_{j_1 < j_2 < \cdots < j_i} |A_{j_1} \cap A_{j_2} \cap \cdots \cap A_{j_i}|\right].$$

(Recall that $\bigcup_{i=1}^{k} A_i = A_1 \cup A_2 \cup \cdots \cup A_k$.) Argue that this formula correctly expresses the Inclusion–Exclusion Rule for any number of sets. *(Hint: figure out how many ℓ-set intersections each element x appears in. Then use the Binomial Theorem—specifically, Exercise 9.179.)*

9.182 In Example 8.4, we looked at the subset relation for a set S: that is, we defined the set of pairs

$$subset := \{\langle A, B\rangle \in \mathscr{P}(S) \times \mathscr{P}(S) : [\forall x \in S : x \in A \Rightarrow x \in B]\}.$$

For any particular set $B \in \mathscr{P}(S)$, the *number of sets A such that $\langle A, B\rangle \in subset$* is precisely $2^{|B|}$. The total number of pairs in the *subset* relation on S is thus 2^k times *the number of subsets of S of size k*, summed over all k. We've already seen that the number of subsets of S of size k is $\binom{|S|}{k}$. Thus the total number of pairs in the *subset* relation on S is

$$\sum_{k=0}^{|S|} (\text{number of subsets of } S \text{ of size } k) \cdot 2^k = \sum_{k=0}^{|S|} \binom{|S|}{k} \cdot 2^k.$$

Use the Binomial Theorem to compute a simple formula for this summation.

9.5 *Chapter at a Glance*

Counting is the problem of, given a potentially convoluted description of a set S, computing the cardinality of S. Our general strategy for counting is to develop techniques for counting simple sets like unions and sequences, and then to handle more complicated counting problems by "translating" them into these simple problems.

Counting Unions and Sequences

The *Sum Rule* describes how to compute the cardinality of the union of sets: if A and B are disjoint sets, then $|A \cup B| = |A| + |B|$. More generally, if the sets $A_1, A_2, \ldots, A_k$ are all disjoint, then $\left|\bigcup_{i=1}^{k} A_i\right| = \sum_{i=1}^{k} |A_i|$. If the sets A and B are not disjoint, then the Sum Rule doesn't apply. Instead, we can use *Inclusion–Exclusion* to count $|A \cup B|$. This rule states that $|A \cup B| = |A| + |B| - |A \cap B|$ for any sets A and B. For three sets,

$$|A \cup B \cup C| = |A| + |B| + |C| - |A \cap B| - |A \cap C| - |B \cap C| + |A \cap B \cap C|.$$

To compute the cardinality of the Cartesian product of sets, we can use the *Product Rule*: for sets A and B, we have $|A \times B| = |A| \cdot |B|$. More generally, for arbitrary sets $A_1, A_2, \ldots, A_k$, we have $|A_1 \times A_2 \times \cdots \times A_k| = \prod_{i=1}^{k} |A_i|$. Applying the Product Rule to a set $S \times S \times \cdots \times S$, we see that, for any set S and any $k \in \mathbb{Z}^{\geq 1}$, we have $|S^k| = |S|^k$. If the set of options for one choice depends on previous choices, then we cannot directly apply the Product Rule. However, the basic idea still applies: the *Generalized Product Rule* says that $|S| = \prod_{i=1}^{k} n_i$ if S denotes a set of sequences of length k, where, for each choice of the first $i - 1$ components of the sequence, there are exactly n_i choices for the ith component.

A *permutation* of a set S is sequence of elements from S that contains no repetitions and has length $|S|$. In other words, a permutation of S is an ordering of the elements of S. By the Generalized Product Rule, there are precisely $n! = n \cdot (n-1) \cdot (n-2) \cdot \cdots \cdot 1$ permutations of an n-element set.

Using Functions to Count

Let A and B be arbitrary sets. We can use a function $f : A \to B$ to relate $|A|$ and $|B|$. The *Mapping Rule* says that:

- There exists a function $f : A \to B$ that's onto if and only if $|A| \geq |B|$.
- There exists a function $f : A \to B$ that's one-to-one if and only if $|A| \leq |B|$.
- There exists a function $f : A \to B$ that's a bijection if and only if $|A| = |B|$.

The Mapping Rule implies, among other things, that the power set $\mathscr{P}(S)$ of a set S has cardinality $|\mathscr{P}(S)| = 2^{|S|}$.

The *Division Rule* says the following: suppose that there exists a function $f : A \to B$ such that, for every $b \in B$, there are exactly k elements $a_1, \ldots, a_k \in A$ such that $f(a_i) = b$. Then $|A| = k \cdot |B|$. The Division Rule implies, among other things, that the number of ways to rearrange a sequence containing k different distinct elements $\{x_1, \ldots, x_k\}$, where element x_i appears n_i times, is

$$\frac{(n_1 + n_2 + \cdots + n_k)!}{(n_1!) \cdot (n_2!) \cdot \cdots \cdot (n_k!)}.$$

The *pigeonhole principle* says that if A and B are sets with $|A| > |B|$, and $f : A \to B$, then there exist distinct a and $a' \in A$ such that $f(a) = f(a')$. That is, if there are more pigeons than holes, and we place the pigeons into the holes, then there must be (at least) one hole containing more than one pigeon.

Combinations and Permutations

Consider nonnegative integers n and k with $k \leq n$. The quantity $\binom{n}{k}$ is defined as

$$\binom{n}{k} := \frac{n!}{k! \cdot (n-k)!},$$

and is read as "n choose k." The quantity $\binom{n}{k}$ denotes the number of ways to choose a k-element subset of a set of n elements, called a *combination*, when each element can only be selected at most once and the order of the selected elements doesn't matter. The quantity $\binom{n}{k}$ is also sometimes called a *binomial coefficient*.

Depending on whether we allow the same candidate to be chosen more than once and whether we care about the order in which the candidates are chosen, there are many versions of selecting k out of a set of n candidates:

- If the order of the selected elements doesn't matter and repetition of the chosen elements is not allowed, then there are $\binom{n}{k}$ ways to choose.
- If order matters and repetition is not allowed, there are $\frac{n!}{(n-k)!}$ ways.
- If order matters and repetition is allowed, there are n^k ways.
- If order doesn't matter and repetition is allowed, there are $\binom{n+k-1}{k}$ ways.

A *combinatorial proof* establishes that two quantities x and y are equal by defining a set S and proving that $|S| = x$ and $|S| = y$ by counting $|S|$ in two different ways. We can give combinatorial proofs of the following facts about the binomial coefficients, among others:

$$\binom{n}{k} = \binom{n}{n-k} \qquad \binom{n}{k} = \binom{n-1}{k} + \binom{n-1}{k-1} \qquad \textstyle\sum_{i=0}^{n} \binom{n}{i} = 2^n.$$

The *binomial theorem* states that, for any $a, b \in \mathbb{R}$ and any $n \in \mathbb{Z}^{\geq 0}$,

$$(a+b)^n = \sum_{i=0}^{n} \binom{n}{i} a^i b^{n-i}.$$

We can prove the binomial theorem by induction on the exponent n.

Many of the interesting properties of the binomial coefficients can be seen by looking at patterns visible in *Pascal's triangle*, which arranges the binomial coefficients so that the nth row contains the $n+1$ binomial coefficients $\binom{n}{0}, \binom{n}{1}, \cdots, \binom{n}{n}$. See Figure 9.45 for the first few rows of Pascal's triangle.

$$\binom{0}{0}$$
$$\binom{1}{0} \; \binom{1}{1}$$
$$\binom{2}{0} \; \binom{2}{1} \; \binom{2}{2}$$
$$\binom{3}{0} \; \binom{3}{1} \; \binom{3}{2} \; \binom{3}{3}$$
$$\binom{4}{0} \; \binom{4}{1} \; \binom{4}{2} \; \binom{4}{3} \; \binom{4}{4}$$
$$\binom{5}{0} \; \binom{5}{1} \; \binom{5}{2} \; \binom{5}{3} \; \binom{5}{4} \; \binom{5}{5}$$
$$\binom{6}{0} \; \binom{6}{1} \; \binom{6}{2} \; \binom{6}{3} \; \binom{6}{4} \; \binom{6}{5} \; \binom{6}{6}$$
$$\vdots$$

Figure 9.45: The first several rows of Pascal's triangle.

Key Terms and Results

Key Terms

COUNTING UNIONS AND SEQUENCES

- Sum Rule
- Product Rule
- double counting
- Inclusion–Exclusion
- Generalized Product Rule
- permutation

USING FUNCTIONS TO COUNT

- Mapping Rule
- Division Rule
- pigeonhole principle

COMBINATIONS AND PERMUTATIONS

- combinations
- permutations
- $\binom{n}{k}$ / binomial coefficient
- binomial theorem
- combinatorial proof
- Pascal's triangle

Key Results

COUNTING UNIONS AND SEQUENCES

1. The Sum Rule: if the sets $A_1, A_2, \ldots, A_k$ are all disjoint, then $\left|\bigcup_{i=1}^{k} A_i\right| = \sum_{i=1}^{k} |A_i|$. The Inclusion–Exclusion Rule allows us to handle nondisjoint sets; for example, for any sets A, B we have $|A \cup B| = |A| + |B| - |A \cap B|$.
2. The Product Rule: $|A_1 \times A_2 \times \cdots \times A_k| = \prod_{i=1}^{k} |A_i|$. For any set S and any $k \in \mathbb{Z}^{\geq 1}$, we have $|S^k| = |S|^k$.
3. The Generalized Product Rule: if S is a set of sequences of length k, where, for each choice of the first $i-1$ components of the sequence, there are exactly n_i choices for the ith component, then $|S| = \prod_{i=1}^{k} n_i$.

USING FUNCTIONS TO COUNT

1. The Mapping Rule: an onto function $f : A \to B$ means $|A| \geq |B|$; a one-to-one function $f : A \to B$ means $|A| \leq |B|$; and a bijection $f : A \to B$ means $|A| = |B|$.
2. For any set S, $|\mathscr{P}(S)| = 2^{|S|}$.
3. The Division Rule: if $f : A \to B$ satisfies $|\{a \in A : f(a) = b\}| = k$ for all $b \in B$, then $|A| = k \cdot |B|$.
4. The number of ways to arrange a sequence containing elements $\{x_1, \ldots, x_k\}$, where x_i appears n_i times, is $\frac{(n_1+n_2+\cdots+n_k)!}{(n_1!)\cdot(n_2!)\cdot\cdots\cdot(n_k!)}$.
5. Pigeonhole principle: if $f : A \to B$ and $|A| > |B|$, then there exist $a, a' \neq a \in A$ such that $f(a) = f(a')$.

COMBINATIONS AND PERMUTATIONS

1. There are four versions of selecting k out of n candidates, depending on whether the order of the chosen elements matters and whether we can choose the same element twice. (See Figure 9.31.) The binomial coefficient $\binom{n}{k}$ denotes the number of ways to choose when repetition is forbidden and order doesn't matter (called *combinations*).
2. Some useful properties: $\binom{n}{k} = \binom{n}{n-k}$ and $\binom{n-1}{k} + \binom{n-1}{k-1} = \binom{n}{k}$ and $\sum_{i=0}^{n} \binom{n}{i} = 2^n$.
3. The binomial theorem: $(a+b)^n = \sum_{i=0}^{n} \binom{n}{i} a^i b^{n-i}$.

10
Probability

In which our heroes evade threats and conquer their fears by flipping coins, rolling dice, and spinning the wheels of chance.

10.1 Why You Might Care

> Fortune can, for her pleasure, fools advance,
> And toss them on the wheels of Chance.
>
> Juvenal (c. 55—c. 127)

This chapter introduces *probability*, the study of randomness. Our focus, as will be no surprise by this point of the book, is on building a formal mathematical framework for analyzing random processes. We'll begin with a definition of the basics of probability: defining a random process that chooses one particular *outcome* from a set of possibilities (any one of which occurs some fraction of the time). We'll then analyze the likelihood that a particular *event* occurs—in other words, asking whether the chosen outcome has some particular property that we care about. We then consider *independence* and *dependence* of events, and *conditional probability*: how, if at all, does knowing that the randomly chosen outcome has one particular property change our calculation of the probability that it has a different property? (For example, perhaps 90% of all email is spam. Does knowing that a particular email contains the word `ENLARGE` make that email more than 90% likely to be spam?) Finally, we'll turn to *random variables* and *expectation*, which give quantitative measurements of random processes: for example, if we flip a coin 1000 times, how many heads would we see (on average)? How many runs of 10 or more consecutive heads? Probabilistic questions are surprisingly difficult to have good intuition about; the focus of the chapter will be on the tools required to rigorously settle these questions.

Probability is relevant almost everywhere in computer science. One broad application is in *randomized algorithms* to solve computational problems. In the same way that the best strategy to use in a game of rock–paper–scissors involves randomness (throw rock $\frac{1}{3}$ of the time, throw paper $\frac{1}{3}$ of the time, throw scissors $\frac{1}{3}$ of the time), there are some problems—for example, finding the median element of an unsorted array, or testing whether a given large integer is a prime number—for which the best known algorithm (the fastest, the simplest, the easiest to understand, ...) proceeds *by making random choices.* The same idea occurs in data structures: a *hash table* is an excellent data structure for many applications, and it's best when it assigns elements to (approximately) random cells of a table. (See Section 10.1.1.) Randomization can also be used for *symmetry breaking*: we can ensure that 1000 identical drones do not clog the airwaves by all trying to communicate simultaneously: each drone will choose to try to communicate at a random time. And we can generate more realistic computer graphics of flame or hair or, say, a field of grass by, for each blade, randomly perturbing the shape and configuration of an idealized piece of grass.

As a rough approximation, we can divide probabilistic applications in CS into two broad categories: those uses in which the randomness is internally generated by our algorithms or data structures, and those cases in which the randomness comes "from the outside." The first type we discussed above. In the latter category, consider circumstances in which we wish to build some sort of computational model that addresses some real-world phenomenon. For example, we might wish to model social behavior (a social network of friendships), or traffic on a road network or on the internet, or to build a speech recognition system. Because these applications interact with extremely complex real-world behaviors, we will typically think of them as being generated according to some deterministic (nonrandom) underlying rule, but with hard-to-model variation that is valuably thought of as generated by a random process. In systems for speech recognition, it works well to treat a particular "frame" of the speech stream (perhaps tens of milliseconds in duration) as a noisy version of the sound that the speaker intended to produce, where the noise is essentially a random perturbation of the intended sound.

Finally, you should care about probability because *any* well-educated person must understand something about probability. You need probability to understand political polls, weather forecasting, news reports about medical studies, wagers that you might place (either with real money or by choosing which of two alternatives is a better option), and many other subjects. Probability is everywhere!

10.1.1 *Hashing: A Running Example*

Throughout this chapter, we will consider a running sequence of examples that are about *hash tables,* a highly useful data structure that also conveniently illustrates a wide variety of probabilistic concepts. So we'll start here with a short primer on hash tables. (See also p. 267, or a good textbook on data structures.)

A *hash table* is a data structure that stores a set of elements in a table $T[1 \ldots m]$—that is, an array of size m. (Remember that, throughout this book, arrays are indexed starting at 1, not 0.) The set of possible elements is called the *universe* or the *keyspace.* We will be asked to store in this table a particular small subset of the keyspace. (For example, the keyspace might be the set of all 8-letter strings; we might be asked to store the user IDs of all students on campus.) We use a *hash function* h to determine in which cell of the table $T[1 \ldots m]$ each element will be stored. The hash function h takes elements of the keyspace as input, and produces as output an index identifying a cell in T. To store an element x in T using hash function h, we compute $h(x)$ and place x into the cell $T[h(x)]$. (We say that the element x *hashes to* the cell $T[h(x)]$.)

We must somehow handle *collisions,* when we're asked to store two different elements that hash to the same cell of T. We will usually consider the simplest solution, where we use a strategy called *chaining* to resolve collisions. To implement chaining, we store all elements that hash to a cell *in that cell,* in an unsorted list. Thus, to find whether an element y is stored in the hash table T, we look one-by-one through the list of elements stored in $T[h(y)]$.

Example 10.1 (A small hash table)
Let the keyspace be $\{1, 2, 3, 4\}$, and consider a 2-cell hash table with the hash function h given by $h(x) = (x \bmod 2) + 1$. (Thus $h(1) = h(3) = 2$ and $h(2) = h(4) = 1$.)

- If we store the elements $\{1, 4\}$, then the table would be

$T[1]$	$T[2]$
[4]	[1]

.

- If we store the elements $\{2, 4\}$, then the table would be

$T[1]$	$T[2]$
[2, 4]	[]

.

More formally, we are given a finite set K called the *keyspace*, and we are also given a positive integer m representing the table size. We will base the data structure on a hash function $h : K \to \{1, \ldots, m\}$. For the purposes of this chapter, we choose h *randomly*, specifically choosing the hash function so that *each function from K to $\{1, \ldots, m\}$ is equally likely to be chosen as h.*

Let's continue our above example with a randomly chosen hash function. For the moment, we'll treat the process of randomly choosing a hash function informally. (The precise definitions of what it means to choose randomly, and what it means for certain "events" to occur, will be defined in the following sections.)

Example 10.2 (A small hash table)

As before, let $K = \{1,2,3,4\}$ and $m = 2$. There are $m^{|K|} = 2^4 = 16$ different functions $h : K \to \{1,2\}$, and each of these functions is equally likely to be chosen. (The functions are listed in Figure 10.1.) Each of these functions is chosen a $\frac{1}{16}$ fraction of the time. Thus:

- a $\frac{8}{16} = \frac{1}{2}$ fraction of the time, we have $h(4) = h(1)$.
  (These functions are marked with an 'A' in Figure 10.1.)
- a $\frac{6}{16} = \frac{3}{8}$ fraction of the time, the hash function is "perfectly balanced"—that is, hashes an equal share of the keys to each cell.
  (These functions are marked with a 'B' in Figure 10.1.)
- a $\frac{1}{16}$ fraction of the time, the hash function hashes every element of K into cell #2.
  (This one function is marked with a 'C' in Figure 10.1.)

$h(1)$	$h(2)$	$h(3)$	$h(4)$	
1	1	1	1	A
1	1	1	2	
1	1	2	1	A
1	1	2	2	B
1	2	1	1	A
1	2	1	2	B
1	2	2	1	AB
1	2	2	2	
2	1	1	1	
2	1	1	2	AB
2	1	2	1	B
2	1	2	2	A
2	2	1	1	B
2	2	1	2	A
2	2	2	1	
2	2	2	2	A C

Figure 10.1: All functions from $\{1,2,3,4\}$ to $\{1,2\}$. Each row is a different function h; the ith column records the value of $h(i)$. The letters mark some functions as described in Example 10.2.

Taking it further: In practice, the function h will not be chosen completely at random, for a variety of practical reasons (for example, we'd have to write down the whole function to remember it!), but throughout this chapter we will model hash tables as if h is chosen completely randomly. The assumption that the hash function is chosen randomly, with every function $K \to \{1,2,\ldots,m\}$ equally likely to be chosen, is called the *simple uniform hashing assumption.* It is very common to make this assumption when analyzing hash tables.

It may be easier to think of choosing a random hash function using an iterative process instead: for every key $x \in K$, we choose a number i_x uniformly at random and independently from $\{1,2,\ldots,m\}$. (The definitions of "uniformly" and "independently" are coming in the next few sections. Informally, this description means that each number in $\{1,2,\ldots,m\}$ is equally likely to be chosen as i_x, regardless of what choices were made for previous numbers.) Now define the function h as follows: on input x, output i_x. One can prove that this process is completely identical to the process illustrated in Example 10.2: write down every function from K to $\{1,2,\ldots,m\}$ (there are $m^{|K|}$ of them), and pick one of these functions at random.

After we've chosen the hash function h, a set of actual keys $\{x_1, \ldots, x_n\} \subseteq K$ will be given to us, and we will store the element x_i in the table slot $T[h(x_i)]$. Notice that the *only* randomly determined quantity is the hash function h. Everything else—the keyspace K, the table size m, and the set of to-be-stored elements—is fixed.

10.2 *Probability, Outcomes, and Events*

> Anyone who does not know how to make the most of his luck has no right to complain if it passes by him.
>
> Miguel de Cervantes (1547–1616)

This section will give formal definitions of the fundamental concepts in probability, giving us a framework to use in thinking about the many computational applications that involve chance. These definitions are somewhat technical, but they'll allow us reason about some fairly sophisticated probabilistic settings fairly quickly.

Warning! It is very rare to have good intuition or instincts about probability questions. Try to hold yourself back from jumping to conclusions too quickly, and instead use the systematic approaches to probabilistic questions that are introduced in this chapter.

10.2.1 *Outcomes and Probability*

Here's the very rough outline of the relevant definitions; we'll give more details in a moment. Imagine a scenario in which some quantity is determined in some random way. We will consider a set S of possible *outcomes*. Each outcome has an associated *probability*, which is a number between 0 and 1. The set S is called the *sample space*. In any particular result of this scenario, one outcome from S is selected randomly (by "nature"); the frequency with which a particular outcome is chosen is given by that outcome's associated probability. (Sometimes we might talk about the *process* by which a sequence of random quantities is selected, and the *realization* as the actual choice made according to this process.) For example, for flipping an unweighted coin we would have $S = \{\text{Heads}, \text{Tails}\}$, where Heads has probability 0.5 and Tails has probability 0.5. Our particular outcome might be Heads.

Here are the formal definitions:

Definition 10.1 (Outcomes and sample space)
An outcome *of a probabilistic process is the sequence of results for all randomly determined quantities. (An outcome can also be called a* realization *of the probabilistic process.) The* sample space *S is the set of all outcomes.*

Definition 10.2 (Probability function)
Let S be a sample space. A probability function $\texttt{Pr} : S \to \mathbb{R}$ *describes, for each outcome $s \in S$, the fraction of the time that s occurs. (We denote probabilities using square brackets, so the probability of $s \in S$ is written $\texttt{Pr}[s]$.) We insist that the following two conditions hold of the probability function* $\texttt{Pr}$:

$$\sum_{s \in S} \texttt{Pr}[s] = 1 \tag{10.1}$$

$$\texttt{Pr}[s] \geq 0 \text{ for all } s \in S. \tag{10.2}$$

Intuitively, condition (10.1) says that *something has to happen*: when we flip a coin, then either it comes up heads or it comes up tails. (And so $\texttt{Pr}[\text{Heads}] + \texttt{Pr}[\text{Tails}] = 1$.) The other condition, (10.2), formalizes the idea that $\texttt{Pr}[s]$ denotes the fraction of the time that the outcome s occurs: *the least frequently that an outcome can occur is never.*

The probability function $\texttt{Pr}$ is also sometimes called a *probability distribution over S*. (This function "distributes" one unit of probability across the set S of all possible outcomes, as in (10.1).)

Taking it further: Bizarrely, in *quantum computation*—an as-yet-theoretical type of computation based on quantum mechanics—we can have outcomes whose probabilities are not restricted to be real numbers between 0 and 1. This model is (very!) difficult to wrap one's mind around, but a computer based on this idea turns out to let us solve interesting problems, and faster than on "normal" computers. For example, we can factor large numbers efficiently on a quantum computer. (Though we don't know how to build quantum computers of any nontrivial size.) See p. 1016 for some discussion.

A few examples: cards, coins, and words

Here are a few examples of sample spaces with probabilities naturally associated with each outcome:

Example 10.3 (One card from the deck)
We draw one card from a perfectly shuffled deck of 52 cards. Then we can denote the sample space as $S = \{2, 3, \ldots, 10, J, Q, K, A\} \times \{\clubsuit, \diamondsuit, \heartsuit, \spadesuit\}$. Each card $c \in S$ has $\texttt{Pr}[c] = \frac{1}{52}$. Note that condition (10.1) is satisfied because

$$\textstyle\sum_{c \in S} \texttt{Pr}[c] = \sum_{c \in S} \frac{1}{52} = 52 \cdot \frac{1}{52} = 1,$$

and (10.2) is obviously satisfied because $\texttt{Pr}[c] = \frac{1}{52} \geq 0$ for each c.

Example 10.4 (Coin flips)
You flip a quarter and Bill Gates flips a platinum trillion-dollar coin. Assume that both coins are fair (equally likely to come up Heads and Tails) and that flips of the quarter and the platinum coin do not affect each other in any way. Then the four outcomes are—writing the quarter's result first—$\langle$Heads, Heads$\rangle$, $\langle$Heads, Tails$\rangle$, $\langle$Tails, Heads$\rangle$, and $\langle$Tails, Tails$\rangle$. Each of these four outcomes has probability 0.25.

Example 10.5 (A word on the page)
Consider the following sentence, which—excluding spaces—contains a total of 29 different symbols (namely `N`, `o`, `w`, `i`, `s`, `t`, ..., `t`):

```
Now is the winter of our discontent.
```

Now is the winter of our discontent/ Made glorious summer by this sun of York;/And all the clouds that lour'd upon our house/In the deep bosom of the ocean buried.
— William Shakespeare (1564–1616) *King Richard III*

We are going to select a word from this sentence, according to the following process: choose one of the 29 non-space symbols from the sentence with equal likelihood; the selected word is the one in which the selected symbol appears. (Thus longer words will be chosen more frequently than shorter words, because longer words contain more symbols—and are therefore more likely to be selected.)

The sample space is $S = \{\texttt{Now}, \texttt{is}, \texttt{the}, \texttt{winter}, \texttt{of}, \texttt{our}, \texttt{discontent}\}$. There are $3+2+3+6+2+3+10 = 29$ total symbols, and thus $\texttt{Pr}[\texttt{Now}] = \frac{3}{29}$, $\texttt{Pr}[\texttt{is}] = \frac{2}{29}$, and so on, through $\texttt{Pr}[\texttt{discontent}] = \frac{10}{29}$. Again, the conditions for being a probability are satisfied: each outcome's probability is nonnegative, and $\sum_{w \in S} \texttt{Pr}[w] = 1$.

Examples 10.3 and 10.4 are scenarios of *uniform probability*, in which each outcome in the sample space is chosen with equal likelihood. (Specifically, each $s \in S$ has probability $\Pr[s] = \frac{1}{|S|}$.) Example 10.5 illustrates *nonuniform probability*, in which some outcomes occur more frequently than others.

Note that for a single sample space S, we can have many different distinct processes by which we choose an outcome from S. For example:

Example 10.6 (Two ways of choosing from $S = \{0, 1, 2, \ldots, 7\}$)
One process for selecting an element of S is to flip three fair coins and treat their results as a binary number (HHH = 111 $\to$ 7, HHT = 110 $\to$ 6, ..., TTT = 000 $\to$ 0). This process gives a uniform distribution over S: each sequence of coin flips occurs with the same probability. For example, $\Pr[4] = \frac{1}{8} = 0.125$ and $\Pr[7] = \frac{1}{8} = 0.125$.

A second process for selecting an element of S is to flip 7 fair coins and to let the outcome be the number of heads that we see in those 7 flips (HHHHHHH $\to$ 7, HHHHHHT $\to$ 6, HHHHHTH $\to$ 6, ..., TTTTTTT $\to$ 0). This process gives a *nonuniform* distribution over S, because the number of sequences that have k heads is different for different values of k. For example:

$$\Pr[4] = \frac{\binom{7}{4}}{2^7} = \frac{35}{128} \approx 0.2734, \qquad \text{but} \qquad \Pr[7] = \frac{\binom{7}{7}}{2^7} = \frac{1}{128} \approx 0.0078.$$

As a word of warning, notice that probabilistic statements *about a particular realization* don't make sense; the only kind of probabilistic statement that makes sense is a statement *about a probabilistic process*. If you happen to be one of the $\approx 10\%$ of the population that's red–green colorblind, and a friend says "what are the odds that you're colorblind!?", the correct answer is: the probability is 1 (because it happened!).

10.2.2 Events

Many of the probabilistic questions that we'll ask are about whether the realization has some particular property, rather than whether a single particular outcome occurs. For example, we might ask for the probability of getting more heads than tails in 1000 flips of a fair coin. Or we might ask for the probability that a hand of seven cards (dealt from a perfectly shuffled deck) contains at least two pairs. There may be many different outcomes in the sample space that have the property in question. Thus, often we will be interested in the probability of a *set* of outcomes, rather than the probability of a *single* outcome. Such a set of outcomes is called an *event*:

Definition 10.3 (Event)
Let S be a sample space with probability function Pr. *An* event *is a subset of S. The* probability of an event *E is the sum of the probabilities of the outcomes in E, and it is written* $\Pr[E] = \sum_{s \in E} \Pr[s]$.

The probability of an event $E \subseteq S$ follows by a probabilistic version of the Sum Rule, from counting: because one (and only one) outcome is chosen in a particular realization, the probability of either outcome x or y occurring is $\Pr[x] + \Pr[y]$.

Note that the notation in Definition 10.3 generalizes the function Pr by allowing us to write *either* elements of S *or* subsets of S as inputs to Pr. That is, previously we considered a function $\Pr : S \to [0,1]$; we have now "extended" our notation so that it's a function $\Pr : \mathscr{P}(S) \to [0,1]$. (To be more precise, we're actually extending the notation to be a function $\Pr : (S \cup \mathscr{P}(S)) \to [0,1]$, because we're still letting ourselves write outcomes as arguments too.)

Our mixture of Pr [outcome] and Pr [event] is an abuse of notation; we're mixing the type of input willy nilly. But, because $\Pr[x]$ for an outcome x and $\Pr[\{x\}]$ for the singleton event $\{x\}$ are identical, we can write probabilities this way without risk of confusion.

A few examples

Here are a few examples of events and their probabilities:

Example 10.7 (At least one head)
You and Bill Gates each flip fair coins, as in Example 10.4. Define the event $H = \{\langle \text{Heads}, \text{Heads}\rangle, \langle \text{Heads}, \text{Tails}\rangle, \langle \text{Tails}, \text{Heads}\rangle\}$ as "at least one coin comes up heads." Then $\Pr[H] = 0.25 + 0.25 + 0.25 = 0.75$.

Example 10.8 (Aces up)
Problem: Suppose that you draw one card from a perfectly shuffled deck, as in Example 10.3. What is the probability that you draw an ace?

Solution: The event in question is $E = \{A♣, A♢, A♡, A♠\}$. Each of these four outcomes has a probability of $\frac{1}{52}$, so $\Pr[E] = \frac{1}{52} + \frac{1}{52} + \frac{1}{52} + \frac{1}{52} = \frac{4}{52} = \frac{1}{13}$.

Example 10.9 (Full house)
Problem: You're dealt 5 cards from a shuffled deck, so that each set of 5 cards is equally likely to be your hand. A hand is a *full house* if 3 cards share one rank, and the other 2 cards share a second rank. (For example, the hand 3♡, 3♠, 9♡, 9♣, 3♣ is a full house.) What's the probability of being dealt a full house?

Solution: There are $\binom{52}{5}$ possible hands, each of which is dealt with probability $1/\binom{52}{5}$. Thus the key question is a counting question: *how many full houses are there?* We can compute this number using the Generalized Product Rule; specifically, we can view a full house as the result of the following sequence of selections:

- we choose the rank of which to have three of a kind;
- we choose which 3 of the 4 cards of that rank are in the hand;
- we choose the rank of the pair (any of the 12 remaining ranks); and
- we choose which 2 of the 4 cards of that rank are in the hand.

Thus there are $\binom{13}{1} \cdot \binom{4}{3} \cdot \binom{12}{1} \cdot \binom{4}{2}$ full houses, and the probability of a full house is

$$\frac{\binom{13}{1} \cdot \binom{4}{3} \cdot \binom{12}{1} \cdot \binom{4}{2}}{\binom{52}{5}} = \frac{3744}{2598960} \approx 0.00144.$$

Here's a slightly more complex example, with multiple events of interest:

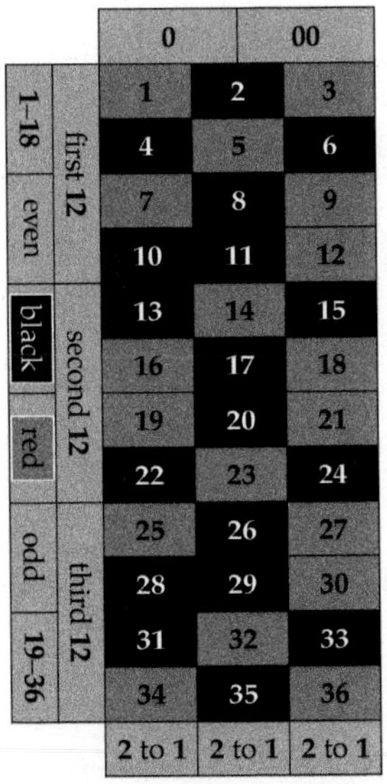

event name	outcomes	probability
1–18	$\{1,2,3,4,5,6,7,8,9,10,11,12,13,14,15,16,17,18\}$	$\frac{18}{38}$
even	$\{2,4,6,8,10,12,14,16,18,20,22,24,26,28,30,32,34,36\}$	$\frac{18}{38}$
1st 12	$\{1,2,3,4,5,6,7,8,9,10,11,12\}$	$\frac{12}{38}$
black	$\{2,4,6,8,10,11,13,15,17,20,22,24,26,28,29,31,33,35\}$	$\frac{18}{38}$
red	$\{1,3,5,7,9,12,14,16,18,19,21,23,25,27,30,32,34,36\}$	$\frac{18}{38}$
2nd 12	$\{13,14,15,16,17,18,19,20,21,22,23,24\}$	$\frac{12}{38}$
odd	$\{1,3,5,7,9,11,13,15,17,19,21,23,25,27,29,31,33,35\}$	$\frac{18}{38}$
19–36	$\{19,20,21,22,23,24,25,26,27,28,29,30,31,32,33,34,35,36\}$	$\frac{18}{38}$
3rd 12	$\{25,26,27,28,29,30,31,32,33,34,35,36\}$	$\frac{12}{38}$
"2 to 1" A	$\{1,4,7,10,13,16,19,22,25,28,31,34\}$	$\frac{12}{38}$
"2 to 1" B	$\{2,5,8,11,14,17,20,23,26,29,32,35\}$	$\frac{12}{38}$
"2 to 1" C	$\{3,6,9,12,15,18,21,24,27,39,33,36\}$	$\frac{12}{38}$

Figure 10.2: The Roulette board, and the corresponding events.

Example 10.10 (Roulette)

In *roulette*, a wheel is spun, and a metal ball comes to rest in one of the wheel's 38 segments. The segments are numbered 1–36 (each colored red or black, as shown in Figure 10.2), and there are two more segments labeled 0 and 00 (both colored green). Assume that the ball is equally likely to land in each segment, and that the sample space consists of $\{00, 0, 1, 2, \ldots, 36\}$. There are 38 outcomes in the sample space.

In addition to particular outcomes or pairs/triples/quadruples of outcomes whose numbered squares are adjacent in the board, a roulette player can bet on a number of different events defined by the twelve panels along the left-hand and bottom sides of the grid. These events, and their probabilities, are shown in Figure 10.2. (For roulette purposes, the numbers 0 and 00 count as *neither* even nor odd—for reasons related only to casinos' business models, and not to the value of 0 mod 2.)

This version of roulette is the American wheel; the European wheel has only one green zero segment, and there are only 37 outcomes. The player in the European version does better on average.

The details of the particular roulette events in Example 10.10 aren't particularly important, but the distinction between outcomes and events—which this example should make starkly (it's the difference between "the ball stops on number 17" and "the ball stops on an odd number")—is crucial in probability.

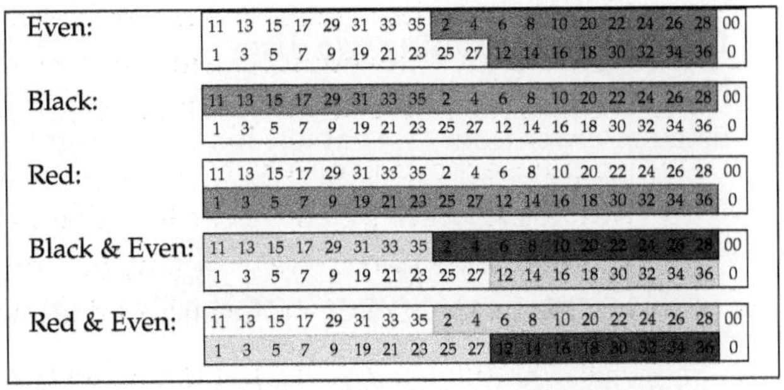

Figure 10.3: A visualization of the sample space and a few events from roulette.

It is often useful to visualize a sample space, and the events of interest, using a Venn diagram–like representation. It can be particularly helpful to draw the subsets/events in such a way that their area corresponds to their probability. A small example of this visualization, for some of the roulette events from Example 10.10, is shown in Figure 10.3. This figure also shows a few intersections of pairs of events: because an event is just a subset of the sample space, the intersection of two events is still a subset of the sample space, and therefore is also an event.

Here are a few useful general properties of the probability of events:

Theorem 10.1 (Some properties of event probabilities)
Let S be a sample space, and let $A \subseteq S$ and $B \subseteq S$ be events. Then, writing $\overline{A} := S - A$ to denote the complement of the event A, we have:

$$\Pr[S] = 1 \tag{10.1.1}$$
$$\Pr[\varnothing] = 0 \tag{10.1.2}$$
$$\Pr\left[\overline{A}\right] = 1 - \Pr[A] \tag{10.1.3}$$
$$\Pr[A \cup B] = \Pr[A] + \Pr[B] - \Pr[A \cap B]. \tag{10.1.4}$$

These properties all follow directly from the definition of the probability of an event.

10.2.3 *Tree Diagrams in Probability*

Many probabilistic processes involve a *sequence* of randomly determined quantities, rather than just a single random choice. Much like in counting, we can use a *tree diagram* to represent the sequence of random choices—and then we can look for the probability of a particular outcome as reflected in the sequence of choices in the tree. In a tree diagram for a probabilistic sequence of choices:

Problem-solving tip: Tree diagrams are generally a very good way to solve probability questions; they force you to systematically think about all of the steps of a probabilistic process (and also about all of the steps of solving probability problems!).

- Every internal node in the tree corresponds to a random decision; every edge leaving that internal node is labeled with the probability of a particular decision. The probability labels of all edges leaving any particular internal node u must add up to 1. (The interpretation is: if the probabilistic process reaches node u, then each branch leaving u is chosen with frequency in proportion to its label.)
- Every leaf corresponds to an outcome. The probability of reaching a particular leaf is precisely equal to the product of the labels on the edges leading from the root to that leaf. As usual, the probability of an event is the sum of the probabilities of the outcomes contained in that event.

Here is a first small example:

Example 10.11 (Rolling two dice)
Here's the probability tree for rolling two fair dice, one after the other. Outcomes are listed in order from $\langle 1,1 \rangle$ at left to $\langle 6,6 \rangle$ at right; every edge has probability $\frac{1}{6}$.

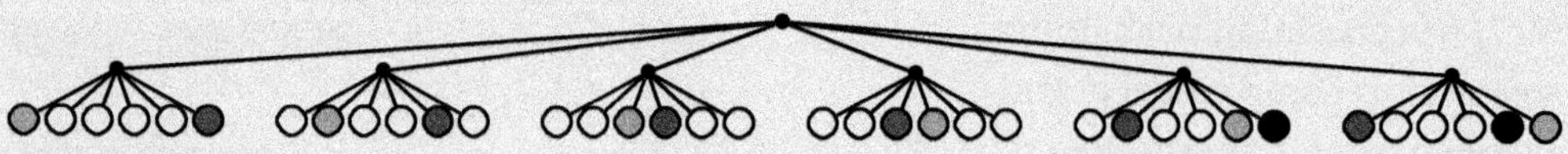

- All edges have probability $\frac{1}{6}$; thus each outcome's probability is $\frac{1}{6} \cdot \frac{1}{6} = \frac{1}{36}$.
- The event "doubles are rolled" (light-shaded outcomes) has probability $6 \cdot \frac{1}{36} = \frac{1}{6}$.
- The event "a 7 is rolled" (medium-shaded outcomes) has probability $6 \cdot \frac{1}{36} = \frac{1}{6}$.
- The event "an 11 is rolled" (black-shaded outcomes) has probability $2 \cdot \frac{1}{36} = \frac{1}{18}$.

Incidentally, one of the calculations from Example 10.11 can also be rephrased to address a question about hashing. Suppose that we hash two elements into a hash table with 6 slots using a uniform random hash function. (See Section 10.1.1.) What is the probability that we have a collision? In fact, this question is precisely the same as asking for the probability of rolling doubles with two fair dice—that is, $\frac{1}{6}$, by Example 10.11.

When we introduced hash tables in Section 10.1.1, we described resolving collisions by *chaining:* an element x is stored in cell $T[h(x)]$; if that cell is already occupied, then we simply add x to a *list* of elements in cell $T[h(x)]$. But there are several other strategies for resolving collisions in a hash table, the simplest of which is called *linear probing.* In linear probing, when we insert an element x into the table, we put x in *the first unoccupied cell,* moving from left to right, starting at cell $h(x)$. (That is, we try to put x in cell $T[h(x)]$, but if that cell already has an element, then we try to put x into cell $T[h(x)+1]$, and then cell $T[h(x)+2]$, and so on. We wrap around to $T[1]$ after we reach the right edge of T.) See Figure 10.4 for an example.

Taking it further: One of the downsides of resolving collisions in a hash table using linear probing is a phenomenon called "clustering": contiguous blocks of filled cells develop, and these filled blocks tend to get longer and longer as more and more elements are added to the table. (This problem is beginning to occur in Figure 10.4.) Other collision-resolution schemes can mitigate this problem; see Exercises 10.45–10.50.

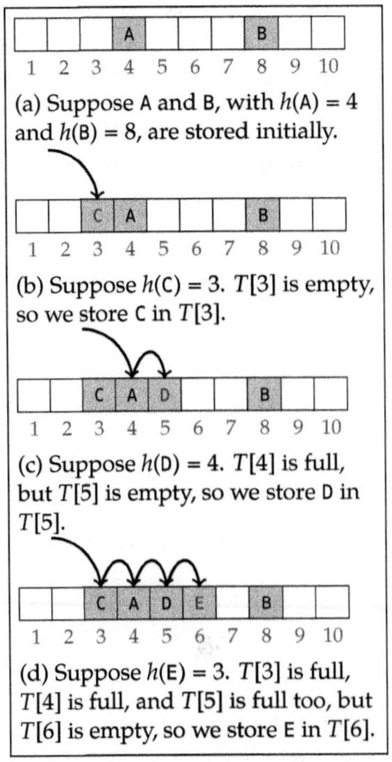

(a) Suppose A and B, with $h(\text{A}) = 4$ and $h(\text{B}) = 8$, are stored initially.

(b) Suppose $h(\text{C}) = 3$. $T[3]$ is empty, so we store C in $T[3]$.

(c) Suppose $h(\text{D}) = 4$. $T[4]$ is full, but $T[5]$ is empty, so we store D in $T[5]$.

(d) Suppose $h(\text{E}) = 3$. $T[3]$ is full, $T[4]$ is full, and $T[5]$ is full too, but $T[6]$ is empty, so we store E in $T[6]$.

Figure 10.4: Linear probing.

Example 10.12 (Hashing with linear probing)

Problem: Suppose that we hash 2 elements into a hash table with 6 slots using a uniform random hash function h, where we resolve collisions by linear probing. What is the probability that we end up with 2 consecutive slots of the hash table filled?

Solution: The sample space is $S = \{1,2,\ldots,6\} \times \{1,2,\ldots,6\}$: we first randomly choose a value for $h(\text{A})$, and then randomly choose a value for $h(\text{B})$. We'll build a tree diagram to represent these choices, as shown (in part) here:

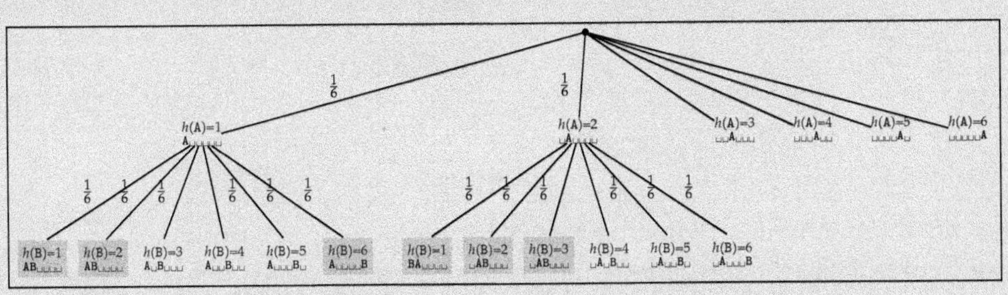

The highlighted outcomes have A and B hashed to adjacent cells. (The remainder of the tree is analogous; it's good practice to try drawing the other branches.)

Each branch of the tree is equally likely, so each outcome occurs with probability $\frac{1}{36}$. How many different outcomes result in A and B being stored in adjacent cells? For each of the 6 possible hash values for A, there are 3 hash values for B that cause A and B to be adjacent, when $h(\text{B})$ is one of $h(\text{A}) - 1$, $h(\text{A})$, and $h(\text{A}) + 1$. So the final probability of a cluster forming is $(6 \cdot 3) \cdot \frac{1}{36} = \frac{18}{36} = \frac{1}{2}$.

Here's another (by now famous) example, called the *Monty Hall Problem*, in which using a probability tree helps resolve a potentially confusing probability question:

The Monty Hall Problem is named after Monty Hall, the host of the television game show *Let's Make A Deal* in the 1960s–1980s. The problem became famous after a kerfuffle involving Marilyn vos Savant, in *Parade* magazine.

Example 10.13 (Monty Hall Problem)

Problem: Here is the problem (based on the *Let's Make a Deal* setup):

> You are given the choice of three doors, behind which are a car, a goat, and another goat. You choose a door. Monty Hall opens one of the doors that you didn't choose to reveal a goat. He then offers you the chance to switch to the other (unopened) door that you didn't initially choose. Should you switch?

(To make this question concrete, assume that the car is initially placed randomly; you choose an initial door randomly; the host always opens one of the two doors you didn't choose to reveal a goat, choosing a goat at random if there are two unchosen goats; and the host will always give you an opportunity to switch.)

Solution: There are three randomly chosen quantities: where the car is placed, which door you choose, and which goat is revealed (if there are two possibilities). We can express the process using the following probability tree:

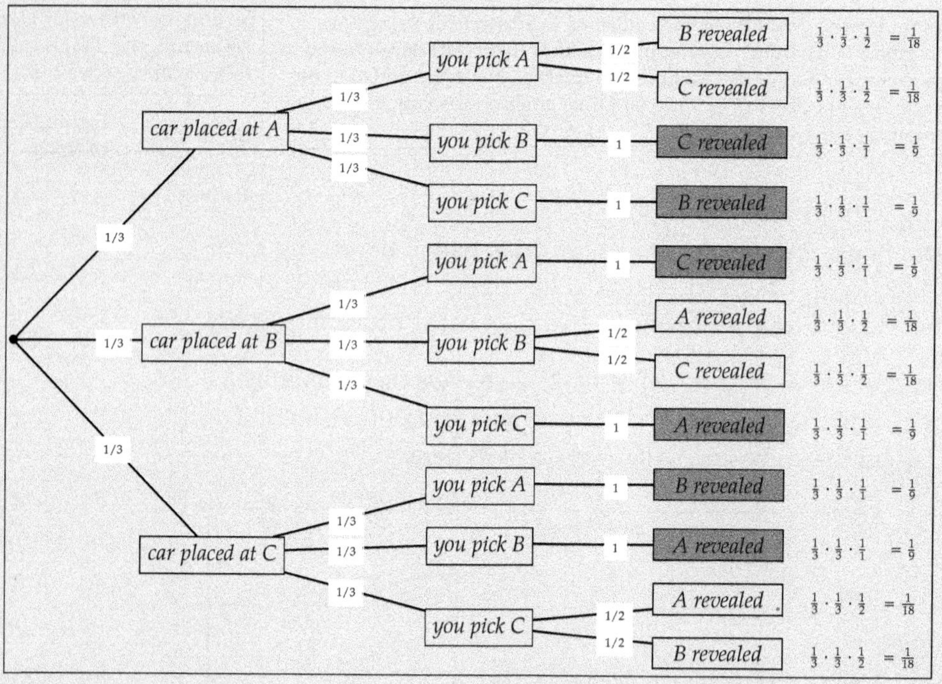

car	*your choice*	*revealed goat*	probability
A	A	B	1/18
A	A	C	1/18
A	B	C	1/9
A	C	B	1/9
B	A	C	1/9
B	B	A	1/18
B	B	C	1/18
B	C	A	1/9
C	A	B	1/9
C	B	A	1/9
C	A	B	1/18
C	B	A	1/18

Figure 10.5: The 12 outcomes in the Monty Hall sample space, and their associated probabilities. The shaded outcomes are those where you win by switching.

The shaded outcomes are those in which switching from your initially chosen door causes your new door to hide a car; the unshaded outcomes are those in which not switching causes you to win. These outcomes and their associated probabilities are also shown in Figure 10.5; again, in the shaded outcomes you win by switching.

There are six outcomes in which switching causes you to win the car. Each of these outcomes has probability $\frac{1}{9}$, so the probability of winning a car by switching is $6 \cdot \frac{1}{9} = \frac{2}{3}$. The other six outcomes are those in which *not* switching causes you to get a car (and switching gets you a goat); these outcomes each have probability $\frac{1}{18}$, and so the probability of winning by not switching is $6 \cdot \frac{1}{18} = \frac{1}{3}$. *You should switch.*

Problem-solving tip: It is usually worth the time to make the probabilistic process concrete, and to make explicit any hidden assumptions about the process, before solving the problem. (That's how we began Example 10.13.)

Taking it further: Section 10.2.3 has been devoted to tree diagrams—a systematic way of analyzing probabilistic settings in which a sequence of random choices is made. Typically we think of—or at least model—these random choices as being made "by nature": if you flip a coin, you act as though the universe "chooses" (via microdrafts of wind, the precise topology of the ground where the coin bounces, etc.) whether the coin will come up Heads or Tails.

But, in many scenarios in computer science, we want to generate the randomness *ourselves*, perhaps in a program: choose a random element of the set A; go left with probability $\frac{1}{2}$ and go right with probability $\frac{1}{2}$; generate a random 8-symbol password. The process of *actually generating* a sequence of "random" numbers on a computer is difficult, and (perhaps surprisingly) very closely tied to notions of cryptographic security. A *pseudorandom generator* is an algorithm that produces a sequence of bits that seem to be random, at least to someone examining the sequence of generated bits with limited computational power. It turns out that building a difficult-to-break encryption system is in a sense equivalent to building a difficult-to-distinguish-from-random pseudorandom generator.[1]

For more, see:
[1] Oded Goldreich. *Foundations of Cryptography*. Cambridge University Press, 2006.

10.2.4 *Some Common Probability Distributions*

We'll end this section by spending a few words on some of the common probabilistic processes (and therefore some common probability distributions) that arise in computer science applications.

UNIFORM DISTRIBUTION

Under the *uniform distribution*, every outcome is equally likely. We can define a uniform distribution for any finite sample space S:

Definition 10.4 (Uniform distribution)
Let S be a finite sample space. Under the uniform distribution, the probability of any particular outcome $s \in S$ is given by $\Pr[s] = \frac{1}{|S|}$.

Some familiar examples of the uniform distribution include:

- flipping a fair coin ($\Pr[\text{Heads}] = \Pr[\text{Tails}] = \frac{1}{2}$).
- rolling a fair 6-sided die ($\Pr[1] = \Pr[2] = \Pr[3] = \Pr[4] = \Pr[5] = \Pr[6] = \frac{1}{6}$).
- choosing one card from a shuffled deck ($\Pr[c] = \frac{1}{52}$ for any card c).

Note that, if outcomes are chosen uniformly at random, then the probability of an event is simply its fraction of the sample space. That is, for any event $E \subseteq S$, we have

$$\Pr[E] = \frac{|E|}{|S|}.$$

Taking it further: We often make use of a uniform distribution in randomized algorithms. For example, in randomized quicksort or randomized select applied to an array $A[1 \ldots n]$, a key step is to choose a "pivot" value uniformly at random from A, and then use the chosen value to guide subsequent operation of the algorithm. (See Exercises 10.24–10.27.)

BERNOULLI DISTRIBUTION

The next several distributions are related to "flipping coins" in various ways. "Coin flipping" is a common informal way of referring to any probabilistic process is which we have one or more *trials*, where each trial has the same "success probability," also known as "getting heads." We will refer to flipping an actual coin as a coin flip, but we will also refer to other probabilistic processes that succeed with some fixed probability as a coin flip. We will consider a (possibly) *biased coin*—that is, a coin that comes up heads with probability p, and comes up tails with probability $1 - p$. The coin is called *fair* if $p = \frac{1}{2}$; that is, if the probability distribution is uniform. We can call the coin *p-biased* when $\Pr[\text{heads}] = p$. It's important that the result of one trial has no effect on the success probability of any subsequent trial. (That is, these flips are *independent*; see Section 10.3.)

The first coin-related distribution is simply the one associated with a single trial:

Definition 10.5 (Bernoulli distribution)
The Bernoulli distribution with parameter p *is the probability distribution that results from flipping one p-biased coin. Thus the sample space is* $\{H, T\}$, *where* $\Pr[H] = p$ *and* $\Pr[T] = 1 - p$.

The Bernoulli distribution is named after Jacob Bernoulli, a 17th-century Swiss mathematician.

Taking it further: Imagine a sequence of Bernoulli trials performed with $p = 0.01$, and another sequence of Bernoulli trials performed with $p = 0.48$. The former sequence will consist almost entirely of zeros; the latter will be about half zeros and about half ones. There's a precise technical sense in which the second sequence *contains more information* than the first, measured in terms of the *entropy* of the sequence. See p. 1017 for some discussion.

Binomial distribution

A somewhat more interesting distribution results from considering a *sequence* of flips of a biased coin. Consider the following probabilistic process: perform n flips of a p-biased coin, and then count the number of heads in those flips. The *binomial distribution with parameters n and p* is a distribution over the sample space $\{0, 1, \ldots, n\}$, where $\Pr[k]$ denotes the probability of getting precisely k heads in those flips. Figure 10.6 shows several examples of binomial distributions, for different settings of the parameters n and p. Each panel of Figure 10.6 shows the probability $P[k]$ of getting precisely k heads in n flips of a p-biased coin, for each k in the sample space.

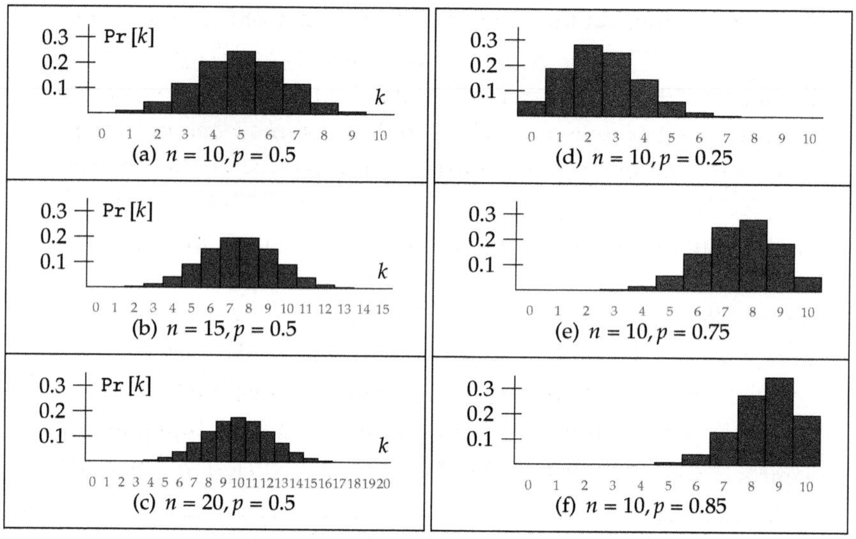

Figure 10.6: Several binomial distributions, for different values of n and p.

If we flip a p-biased coin n times, what is the probability of the event of getting exactly k heads? For example, consider the outcome

$$\underbrace{\text{HH}\cdots\text{H}}_{k \text{ times}}\ \underbrace{\text{TT}\cdots\text{T}}_{n-k \text{ times}}.$$

The probability of this outcome is $p^k \cdot (1-p)^{n-k}$: the first k flips must come up heads, and the next $n - k$ flips must come up tails. In fact, *any* ordering of k heads and $n - k$ tails has probability $p^k \cdot (1-p)^{n-k}$. One way to see this fact is by imagining the probability tree, which is a binary tree with left branches (heads) having probability p and right branches (tails) having probability $1 - p$. The outcomes in question have k left branches and $n - k$ right branches, and thus have probability $p^k \cdot (1-p)^{n-k}$. There are $\binom{n}{k}$ different outcomes with k heads—a sequence of n flips, out of which we choose which k come up heads. Therefore:

Definition 10.6 (Binomial distribution)
The binomial distribution with parameters n and p *is a distribution over the sample space* $\{0, 1, \ldots, n\}$, *where for each* $k \in \{0, 1, \ldots, n\}$ *we have*

$$\Pr[k] = \binom{n}{k} \cdot p^k \cdot (1-p)^{n-k}.$$

For an unbiased coin, when $p = \frac{1}{2}$, the expression for $\Pr[k]$ from Definition 10.6 simplifies to $\Pr[k] = \binom{n}{k}/2^n$, because $(\frac{1}{2})^k \cdot (1 - \frac{1}{2})^{n-k} = (\frac{1}{2})^k \cdot (\frac{1}{2})^{n-k} = (\frac{1}{2})^n$.

Geometric distribution

Another interesting coin-derived distribution comes from the "waiting time" before we see heads for the first time. Consider a p-biased coin, and continue to flip it until we get a heads. The output of this probabilistic process is the number of flips that were required, and the *geometric distribution with parameter* p is defined by this process. (The name "geometric" comes from the fact that the probability of needing k flips looks a lot like a geometric series, from Chapter 5.) See Figure 10.7 for a few such distributions.

What is the probability of needing precisely k flips to get heads for the first time? We would have to have $k - 1$ initial flips come up tails, and then one flip come up heads. As with the binomial distribution, one nice way to think about the probability of this outcome uses the probability tree. This tree has left branches (heads) having probability p and right branches (tails) having probability $1 - p$; the outcome k follows $k - 1$ right branches and one left branch, and thus has probability $(1-p)^{k-1} \cdot p$. Therefore:

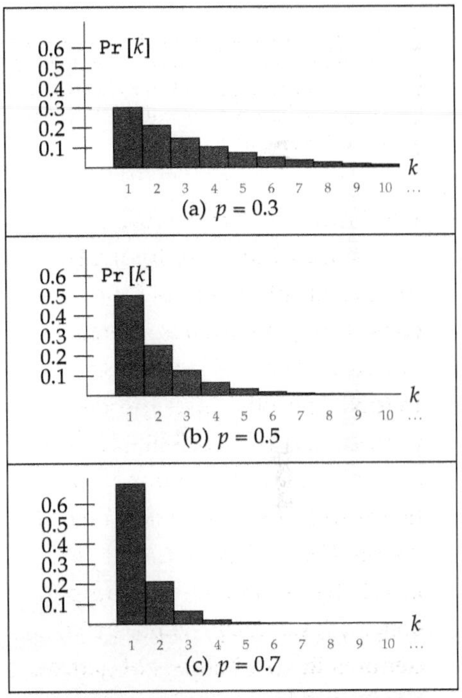

Figure 10.7: Several geometric distributions, for different values of p. Although these plots are truncated at $k = 10$, the distribution continues infinitely: $\Pr[k] > 0$ for all positive integers k.

Definition 10.7 (Geometric distribution)
Let p *be a real number satisfying* $0 < p \leq 1$. *The* geometric distribution with parameter p *is a distribution over the sample space* $\mathbb{Z}^{\geq 1} = \{1, 2, 3, \ldots\}$, *where for each* k *we have*

$$\Pr[k] = (1-p)^{k-1} \cdot p.$$

Notice that the geometric distribution is our first example of an *infinite* sample space: every positive integer is a possible result.

Computer Science Connections

Quantum Computing

As the 20th-century revolution in physics brought about by the discovery of quantum mechanics unfolded, some researchers working at the boundary of physics and computer science developed a new model of computation based on these quantum ideas. This model of *quantum computation* relies deeply on some very deep physics, far too deep for one page, but here is a brief summary—without any of the details of the physics.

"Anyone who is not shocked by quantum theory has not understood it."
— attributed to Niels Bohr (1885–1962)

The most basic element of data in a quantum computer is a *quantum bit*, or *qubit*. Like a bit (the basic element of data on a *classical* computer), a qubit can be in one of two basic states. These two states are written as $|0\rangle$ and $|1\rangle$. (A classical bit is in state 0 or 1). The quantum magic is that a qubit can *be in both states simultaneously*, in what's called a *superposition* of these basic states. A qubit will be in a state $\alpha|0\rangle + \beta|1\rangle$, where α and β are "weights" where $|\alpha|^2 + |\beta|^2 = 1$. (Actually, the weights α and β are *complex* numbers, but the basic idea will come across if we think of them as real numbers—possibly negative!—instead.) Thus, while there are only two states of a bit, there are infinitely many states that a qubit can be in. So a qubit's state contains a huge amount of information. *But*, by the laws of quantum physics, we are limited in how we can extract that information from a qubit. Specifically, we can *measure* a qubit, but we only see 0 or 1 as the output. When we measure a qubit $\alpha|0\rangle + \beta|1\rangle$, the probability that we see 0 is $|\alpha|^2$; the probability that we see 1 is $|\beta|^2$. For example, we might have a qubit in the state

$$\tfrac{1}{2}|0\rangle + \tfrac{\sqrt{3}}{2}|1\rangle. \qquad \textit{(Note } \left(\tfrac{1}{2}\right)^2 + \left(\tfrac{\sqrt{3}}{2}\right)^2 = \tfrac{1}{4} + \tfrac{3}{4} = 1.\textit{)}$$

When we measure this qubit, 25% of the time we'd see a 0, and 75% of the time we'd see a 1.

There are two more crucial points. First, when there are multiple qubits—say n of them—the qubits' state is a superposition of 2^n basic states. (For example, two qubits are in a state $\alpha_{00}|00\rangle + \alpha_{01}|01\rangle + \alpha_{10}|10\rangle + \alpha_{11}|11\rangle$.) Second, even though we only see one value when we measure qubits, there can be "cancellation" (or *interference*) among coefficients. There are notable restrictions on how we can operate on qubits, based on constraints of physics, but at a very rough level, we can run an operation on an n-qubit quantum computer in parallel in each of the 2^n basic states and, if the process is designed properly, still read something useful from our single measurement.[2]

This cursory description of qubits and quantum computation is nowhere close to a full accounting of how qubits work, or what a quantum computer might do. For much more, see the wonderful text

[2] Michael A. Nielsen and Isaac L. Chuang. *Quantum Computation and Quantum Information*. Cambridge University Press, 2000.

Why does anyone care about any of this? The main interest in quantum computation stems from a major breakthrough, *Shor's algorithm* (named after its discoverer, Peter Shor): an algorithm that solves the factoring problem—given a large integer n, determine n's prime factorization—efficiently on a quantum computer. An efficient factoring problem is deeply problematic for most currently deployed cryptographic systems (see Chapter 7), so a functional quantum computer would be a big deal. *But*, at least as of this writing, no one has been able to build a quantum computer of any appreciable size. So at the moment, at least, it's a theoretical device—but there's active research both on the physics side (can we actually build one?) and on the algorithmic side (what *else* could we do if we did build one?).

Computer Science Connections

Information, Charles Dickens, and the Entropy of English

Consider the following two (identical-length) sequences of letters and spaces—one from Charles Dickens's *A Tale of Two Cities* and one generated by uniformly randomly choosing a sequence of elements of $\{A, \ldots, Z, ␣\}$:

```
IT WAS THE BEST OF TIMES, IT WAS THE WORST OF TIMES, IT WAS THE AGE OF
WISDOM, IT WAS THE AGE OF FOOLISHNESS, IT WAS THE EPOCH OF BELIEF, IT
WAS THE EPOCH OF INCREDULITY.
```

```
TUYSSUWWYVOZULF XZQBSFS AFNBMAOOGWZPAHGREAYC SUSCMBOWDCNCYEJBHPVCRO
MLVTGVHTVCZXHSCQFULCMBO CDIWTXOCUPKTFZVNBHRGDWAKZSZPFTZKEWKWIH O
QFIUWTCDKUBTQSPLXSYXGQZA DLXBHKFILFPZ.
```

Which sequence contains more information? It is very tempting to choose the first (information about contrast, and irony, and the opposition of ideas!)—but, in a precise technical sense, Random contains far more information than Dickens. The basic reason is that, in Dickens, certain letters occur far more frequently than others—E occurs 17 times and there are six letters that don't appear at all. (In Random, all 26 letters appear.) With such a lopsided distribution, you already know a lot about what letter is (probably) going to come next, and so there's less new information conveyed by a typical letter.

Figure 10.8: A sequence of bits, produced independently at random with probability $p = 0.25$ (top), $p = 0.5$ (middle), and $p = 0.9$ (bottom) of a one. Their entropies are, respectively, 0.8113, 1.0000, and 0.4690.

Formally, the *entropy* of a sequence of letters (or bits, or whatever) is a measure of "how surprising" each element of the sequence is, averaged over the sequence. We'll convert the "unit of surprise" into a real number between zero and one, where zero corresponds to *the next letter is 100% predictable* and one corresponds to *we have absolutely no idea what the next letter will be.* Formally, the entropy H of a probability distribution over S is given by

$$-\sum_{x \in S} \Pr[x] \cdot \log(\Pr[x]).$$

For example, if we produce a sequence of coin flips where each flip comes up heads with probability p (see Figure 10.8), then the entropy of the sequence will be $-\left(p \log p + (1-p)\log(1-p)\right)$, as shown in Figure 10.9.

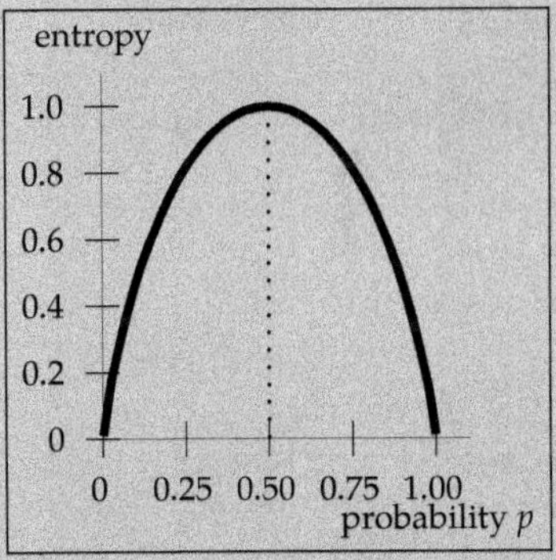

Figure 10.9: The entropy of a biased coin whose heads probability is p.

This definition of entropy comes from the 1940s, in a paper by Claude Shannon,[3] and has found all sorts of useful applications since. Here is one example: the entropy of a sequence of bits expresses a theoretical limit on the *compressibility* of that sequence. (And that theoretical limit is, in fact, achievable.) That is, if the entropy of a string of n bits is very low—say around 0.25—then with some clever algorithms we can represent that string (without any error) using only about $\frac{n}{4}$ bits. But we can't represent it in fewer bits with perfect fidelity ("lossless" compression; see p. 938).

There is significant redundancy in English text, as we've already mentioned, based on the nonuniformity in the probability distribution of individual letters. But there's even more redundancy based on the fact that the probability that the ith character of an English document is an H is affected by whether the $(i-1)$st character was a T. (In the language of Section 10.3, these events are not independent.) If you've seen the letters ␣TH in succession, you can make a very good bet that E is coming next. Compression schemes for English make use of this phenomenon.[4]

[3] Claude E. Shannon. A mathematical theory of communication. *Bell System Technical Journal*, 27:379–423, 1948.

For more about entropy, compressibility, and information generally, see a textbook about information theory. A great classic reference is:

[4] Thomas M. Cover and Joy A. Thomas. *Elements of Information Theory*. Wiley, 1991.

10.2.5 *Exercises*

Philippe flips a fair coin 100 times. Let the outcome be the number of heads that he sees.

10.1 What is the sample space?

10.2 What is $\Pr[0]$?

10.3 What is $\Pr[50]$?

10.4 What is $\Pr[64]$?

Philippe now flips his fair coin n times. He is interested in the event "there are (strictly) more heads than tails." What's the probability of this event for the following values of n?

10.5 $n = 2$

10.6 $n = 3$

10.7 $n = 1001$ *(Hint:* $\Pr[k] = \Pr[1001 - k]$*.)*

10.8 an arbitrary positive integer n

Bridget plays Bridge. Bridge is a card game played with a standard 52-card deck. Each player is initially dealt a hand of 13 cards; assume a fair deal in which each of the $\binom{52}{13}$ hands is equally likely.

10.9 What is the probability of being dealt both A♣ and A♢?

10.10 Suppose Bridget receives a uniformly drawn hand of 13 cards, in a uniformly random order. Because your ex-friend Peter was trying to cheat at poker with this deck, the A♣ card is marked. You observe that the card the fourth-from-the-right position in Bridget's hand is A♣. What is the probability that Bridget also has the A♢ in her hand?

Most casual bridge players sort their hands by suit (♠, ♡, ♣, ♢ from left to right), and decreasing from left to right by rank within each suit. (So one might have a hand like ♠AK4 ♡983 ♣AKQ ♢AJ98, reading from left to right.) Professional players are taught not *to sort their hands, because doing so causes which card they play to leak information about the rest of their hand to the other players. Suppose Bridget receives a uniformly drawn hand of 13 cards, and* sorts the cards in her hand. *Peter's card marking is still present, and you observe the A♣ in a particular position in Bridget's hand. In the following scenarios, what is the probability that Bridget also has the A♢ in her hand?* (That is: out of all hands for which A♣ is highest/lowest/etc. card, what fraction also have the A♢?)

10.11 A♣ is the fourth-from-the-right (that is, fourth-from-the-lowest) card

10.12 A♣ is the *rightmost* (that is, lowest) card

10.13 A♣ is the *leftmost* (that is, highest) card

Chrissie plays Cribbage. Cribbage is a card game played with a standard 52-card deck. For the purposes of these questions, assume that a player is dealt one of the $\binom{52}{4}$ different 4-card hands, chosen uniformly at random. Cribbage hands are awarded points for having a variety of special configurations:

- *A* flush *is a hand with all four cards from the same suit.*
- *A* run *is a set of at least 3 cards with consecutive rank. (For example, the hand 3♡, 9♣, 10♢, J♣ contains a run.)*
- *A* pair *is a set of two cards with identical rank.*

Aces are low in Cribbage, so A, 2, 3 is a valid run, but Q, K, A is not.

10.14 What's the probability that Chrissie is dealt a flush?

10.15 What's the probability that Chrissie is dealt a run of length 4?

10.16 What's the probability of getting *two* runs of length 3 that is not a run of 4? (For example, the hand 9♡, 9♣, 10♢, J♣ contains two runs of length 3: the first is 9♡, 10♢, J♣ and the second is 9♣, 10♢, J♣.)

10.17 What's the probability of getting *one* (and only one) run of length 3 (and not a run of length 4)?

10.18 What's the probability of getting at least one pair? *(Hint:* $\Pr[\text{getting a pair}] = 1 - \Pr[\text{getting no pair}]$*.)*

10.19 What's the probability of getting two or more pairs? (In cribbage, any two cards with the same rank count as a pair; for example, the hand 2♡2♢2♠8♣ has *three* pairs: 2♡2♢ and 2♡2♠ and 2♢2♠.)

10.20 *(programming required)* Write a program to approximately verify your calculations from these Cribbage exercises, as follows: generate 1,000,000 random hands from a standard deck, and count the number of those samples in which there's a flush, run (of the three flavors), pair, or multiple pairs.

10.21 *(programming required)* Modify your program to exactly verify your calculations: exhaustively generate *all* 4-card hands, and count the number of hands with the various features (flushes, runs, pairs).

10.22 A *fifteen* is a subset of cards whose ranks sum to 15, where an A counts as 1 and each of $\{10, J, Q, K\}$ counts as 10. (For example, the hand 3♡, 2♣, 5♢, J♣ contains two fifteens: 3♡ + 2♣ + J♣ = 15 and 5♢ + J♣ = 15.) What's the probability a 4-card hand contains at least one fifteen? *(Hint: use a program.)*

10.23 A bitstring $x \in \{0,1\}^5$ is stored in vulnerable memory, subject to corruption—for example, on a spacecraft. An α-ray strikes the memory and resets one bit to a random value (both the new value and which bit is affected are chosen uniformly at random). A second α-ray strikes the memory and resets one bit (again chosen uniformly at random). What's the probability that the resulting bitstring is identical to x?

```
quickSort(A[1...n]):
1: if n ≤ 1 then
2:     return A
3: else
4:     choose pivot ∈ {1,...,n}, somehow.
5:     L := list of all A[i] where A[i] < A[pivot].
6:     R := list of all A[i] where A[i] > A[pivot].
7:     return quickSort(L) + ⟨A[pivot]⟩ + quickSort(R)
```

Figure 10.10: Quick Sort, briefly. (See Figure 5.20(a) for more detail.) Assume that the elements of A are all distinct.

Recall the quick sort *algorithm for sorting an array A: we choose a "pivot" value x; we partition A into those elements less than x and those greater than x; and we return x and those two sublists, recursively sorted, in the correct order. (See Figure 10.10.) This algorithm is efficient if the two sublists are close to equal in size. There are many ways to choose the pivot value, but one common (and good!) strategy is to choose x* randomly *from A.*

Assume that the elements of A are all distinct. If we select pivot *in Line 4 by choosing* uniformly at random *from the set* $\{1,\ldots,n\}$*:*

10.24 As a function of n, what is the probability that $|L| \leq 3n/4$ *and* $|R| \leq 3n/4$? (You may assume that n is divisible by 4.)

10.25 As a function of n and $\alpha \in [0,1]$, what is the probability $|L| \leq \alpha n$ *and* $|R| \leq \alpha n$? (You may neglect issues of integrality: assume αn is an integer.)

Suppose that we choose pivot *in Line 4 by choosing* three elements p_1, p_2, p_3 uniformly at random *from the set* $\{1,\ldots,n\}$*, and taking as* pivot *the* p_i *whose corresponding element of A is the median of the three. (Assume that the same index can be chosen as both* p_1 *and* p_3*, for example.) For example, for the array* $A = \langle 94, 32, 29, 85, 64, 8, 12, 99\rangle$*, we might randomly choose* $p_1 = 1$*,* $p_2 = 7$*, and* $p_3 = 2$*. Then the pivot will be* p_3 *because* $A[p_3] = 32$ *is between* $A[p_2] = 12$ *and* $A[p_1] = 94$*. Under this "median of three" strategy:*

10.26 What is the probability that $|L| \leq 3n/4$ *and* $|R| \leq 3n/4$? Assume n is large; for ease, you may neglect issues of integrality in your answer.

10.27 As a function of $\alpha \in [0,1]$, what is the probability $|L| \leq \alpha n$ *and* $|R| \leq \alpha n$? Again, you may assume that n is large, and you may neglect issues of integrality in your answer.

Suppose that Team Emacs and Team VI play a best-of-five series of softball games. Emacs, being better than VI, wins each game with probability 60%.

"Emacs" rhymes with "ski wax"; "VI" rhymes with "knee-high." The teams are named after two text editors frequently used by computer scientists to write programs or emails or textbooks.

10.28 Use a tree diagram to compute the probability that Team Emacs wins the series.

10.29 What is the probability that the series goes five games? (That is, what is the probability that neither team wins 3 of the first 4 games?)

10.30 Update your last two answers if Team Emacs wins each game with probability 70%.

(Calculus required.) *Now assume that Team Emacs wins each game with probability p, for an arbitrary value* $p \in [0,1]$*. For the following questions, write down a formula expressing the probability of the listed event. Also find the value of p that maximizes the probability, and the probability of the specified event for this maximizing p.*

10.31 There is a fifth game in the series.

10.32 There is a fourth game of the series.

10.33 There is a fourth game of the series *and* Team Emacs wins that fourth game.

Let S be a sample space, and let $\text{Pr} : S \to [0,1]$ *be an arbitrary function satisfying the requirements of being a probability function (Definition 10.2). That is, we have*

$$\sum_{s \in S} \text{Pr}[s] = 1 \qquad \text{and} \qquad \text{Pr}[s] \geq 0 \text{ for all } s \in S.$$

Argue briefly that the following properties hold.

10.34 For any outcome $s \in S$, we have $\text{Pr}[s] \leq 1$.

10.35 For any event $A \subseteq S$, we have $\text{Pr}\left[\overline{A}\right] = 1 - \text{Pr}[A]$. (Recall that $\overline{A} = S - A$.)

10.36 For any events $A, B \subseteq S$, we have $\text{Pr}[A \cup B] = \text{Pr}[A] + \text{Pr}[B] - \text{Pr}[A \cap B]$.

10.37 The *Union Bound:* for any events $A_1, A_2, \ldots, A_n$, we have $\text{Pr}[\bigcup_i A_i] \leq \sum_i \text{Pr}[A_i]$.

Imagine n identical computers that share a single radio frequency for use as a network connection. Each of the n computers would like to send a packet of information out across the network, but if two or more different computers simultaneously try to send a message, no message gets through. Here you'll explore another use of randomization: using randomness for symmetry breaking.

10.38 Suppose that each computer flips a coin that comes up heads with probability p. What is the probability that *exactly* one of the n machines' coins comes up heads (and thus that machine can send its message)? Your answer should be a formula that's in terms of n and p.

(The next two exercises require calculus.)

10.39 Given the formula you found in Exercise 10.38, what p should you choose to maximize the probability of a message being successfully sent?

10.40 What is the probability of success if you choose p as in Exercise 10.39? What is the limit of this quantity as n grows large? *(You may use the following fact:* $(1 - \frac{1}{m})^m \to e^{-1}$ *as* $m \to \infty$*.)*

We hash items into a 10-slot hash table using a hash function h that uniformly assigns elements to $\{1, \ldots, 10\}$. Compute the probability of the following events if we hash 3 elements into the 10-slot table:

10.41 no collisions occur

10.42 all 3 elements have the same hash value

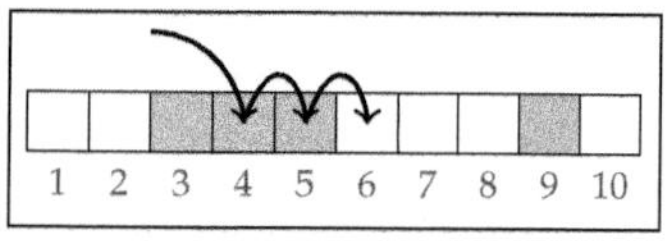

Figure 10.11: A reminder of linear probing. If $h(x) = 4$, then we try to store x in slot 4, then 5, then 6. Because slot 6 is empty, x is placed into that slot.

Suppose that we resolve collisions by linear probing, *wherein an element x that hashes to an occupied cell $h(x)$ is placed in the first unoccupied cell after $h(x)$. (That is, we try to put x into $h(x)$, then $h(x) + 1$, then $h(x) + 2$, and so forth—wrapping back around to the beginning of the table after the 10th slot. See Figure 10.11.) If we hash 3 elements into the 10-slot table, what is the probability that ...*

10.43 at least 2 adjacent slots are filled. (Count slot #10 as adjacent to #1.)

10.44 3 adjacent slots are filled.

One issue with resolving collisions by linear probing is called clustering: *if there's a large block of occupied slots in the hash table, then there's a relatively high chance that the next element placed into the table extends that block.*

10.45 Suppose that we currently have a single block of k adjacent slots full in an n-slot hash table, and all other slots are empty. What's the probability that the next element inserted into the hash table extends that block (that is, leaves $k + 1$ adjacent slots full).

10.46 *(programming required)* Write a program to hash 5000 elements into a 10,007-slot hash table using linear probing. Record which cell x_{5000} ends up occupying—that is, how many hops from $h(x_{5000})$ is x_{5000}? Run your program 2048 times, and report how far, on average, x_{5000} moved from $h(x_{5000})$. Also report the *maximum* distance that x_{5000} moved.

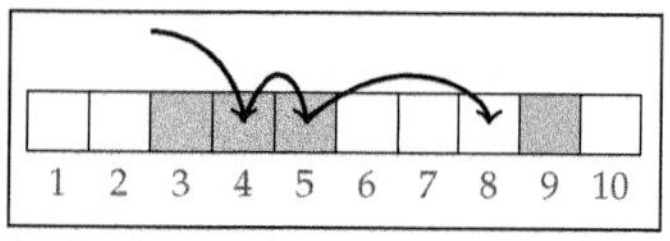

Figure 10.12: Quadratic probing. We try to store x in slot $h(x)$, then $h(x) + 1^2$, then $h(x) + 2^2$, etc.

Because linear probing suffers from this clustering issue, other mechanisms for resolving collisions are sometimes used. Another choice is called quadratic probing: *we change the cell number we try by an increasing step size at every stage, instead of by one every time. Specifically, to hash x into an n-slot table, first try to store x in $h(x)$; if that cell is full, try putting x into $h(x) + i^2$, wrapping back around to the beginning of the table as usual, for $i = 1, 2, \ldots$. (Linear probing tried slot $h(x) + i$ instead.)*

10.47 *(programming required)* Modify your program from Exercise 10.46 to use quadratic probing instead, and report the same statistics: the mean and maximum number of cells probed for x_{5000}.

10.48 In about one paragraph, explain the differences that you observed between linear and quadratic probing. A concern called *secondary clustering* arises in quadratic probing: if $h(x) = h(y)$ for two elements x and y, then the sequence of cells probed for x and y is identical. These sequences were also identical for linear probing. In your answer, explain why secondary clustering from quadratic probing is less of a concern than the clustering from linear probing.

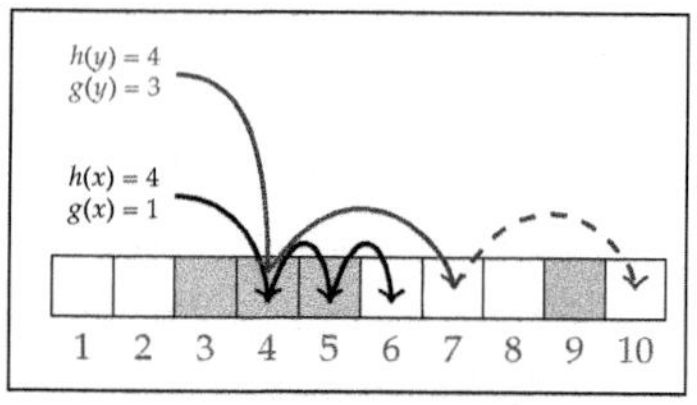

Figure 10.13: Double hashing. We try to store x in slot $h(x)$, then $h(x) + g(x)$, then $h(x) + 2g(x)$, etc. (wrapping around the table as necessary).

A fourth way of handling collisions in hash tables (after chaining, linear probing, and quadratic probing) is what's called double hashing: *we move forward by the same number of slots at every stage, but that number is randomly chosen, as the output of a different hash function. Specifically, to hash x into an n-slot table, first try to store x in $h(x)$; if that cell is full, try putting x into $h(x) + i \cdot g(x)$, wrapping back around to the beginning of the table as usual, for $i = 1, 2, \ldots$. (Here g is a* different *hash function, crucially one whose output is never zero.) See Figure 10.13.*

10.49 *(programming required)* Modify your program from Exercises 10.46 and 10.47 to use double hashing. Again report the mean and maximum number of cells probed for x_{5000}.

10.50 In about one paragraph, explain the differences you observe between chaining, linear probing, quadratic probing, and double hashing. Is there any reason you wouldn't always use double hashing?

Consider a randomized algorithm that solves a problem on a particular input correctly with probability p, and it's wrong with probability $1 - p$. Assume that each run of the algorithm is independent of every other run, so that we can think of each run as being an (independent) coin flip of a p-biased coin (where heads means "correct answer").

10.51 *(Requires calculus.)* Suppose that the probability p is unknown to you. You observe that exactly k out of n trials gave the correct answer. Then the number k of correct answers follows a binomial distribution with parameters n and p: that is, the probability that exactly k runs give the correct answer is

$$\binom{n}{k} \cdot p^k \cdot (1-p)^{n-k}. \qquad (*)$$

Prove that the *maximum likelihood estimate* of p is $p = \frac{k}{n}$—that is, prove that $(*)$ is maximized by $p = \frac{k}{n}$.

10.52 *(Requires calculus.)* Suppose that the probability p is unknown to you. You observe that it takes n trials before the first time you get a correct answer. Then n follows a geometric distribution with parameter p: that is, the probability that n runs were required is given by

$$(1-p)^{n-1}p. \qquad (\dagger)$$

Prove that the maximum likelihood estimate of p is $p = \frac{1}{n}$—that is, prove that $(\dagger)$ is maximized by $p = \frac{1}{n}$.

10.3 *Independence and Conditional Probability*

> If your parents never had children, chances are you won't, either.
>
> Dick Cavett (b. 1936)

Imagine that you're interviewing to be a consultant for Premier Passenger Pigeon Purveyors, a company that pitches its products to prospective pigeon purchasers using online advertising—specifically, by displaying ads to users of a particular search engine on the web. PPPP makes \$50 profit from each sale, and, from historical data, they have determined that 0.02% of searchers who see an ad buy a pigeon. The interviewer asks you how much PPPP should be willing to pay to advertise to a searcher. A good answer is \$0.01: on average, PPPP earns $\$50 \cdot 0.0002 = \$0.01$ per ad, so paying anything up to a penny per ad yields a profit, on average. But you realize that there's a better answer (and, by giving it, you get the job): *it depends on what the user is searching for!* A user who searches for `BIRD` or `PIGEON` or `BUYING A PET TO COMBAT LONELINESS` is far more likely to respond to a PPPP ad than an average user, while a user who searches for `ORNITHOPHOBIA` is much less likely to respond to an ad.

It is a general phenomenon in probability that *knowing that event A has occurred* may tell you that *an event B is much more likely (or much less likely) to occur* than you'd previously known. In this section, we'll discuss when knowing that an event A has occurred does or does not affect the probability that B occurs (that is, whether A and B are *dependent* or *independent*, respectively). We'll then introduce *conditional probability*, which allows us to state and manipulate quantities like "the probability that B happens *given that A happens.*"

10.3.1 *Independence and Dependence of Events*

We'll start with *independence* and *dependence* of events. Intuitively, two events A and B are dependent if A's occurrence/nonoccurrence gives us some information about whether B occurs; in contrast, A and B are independent when A occurs with the same probability when B occurs as it does when B does not occur. More formally:

Definition 10.8 (Independent and dependent events)
Two events A and B are independent *if and only if* $\text{Pr}[A \cap B] = \text{Pr}[A] \cdot \text{Pr}[B]$. *The events A and B are called* dependent *if they are not independent.*

If A and B are dependent events, then we can also say that A and B are *correlated*; independent events are said to be *uncorrelated.*

This definition is phrased a bit differently from the intuition above, but a little manipulation of the equation from Definition 10.8 may help to make the connection clearer. Assume for the moment that $\text{Pr}[B] \neq 0$. (Exercise 10.70 addresses the case of $\text{Pr}[B] = 0$.) Dividing both sides of the equality $\text{Pr}[A] \cdot \text{Pr}[B] = \text{Pr}[A \cap B]$ by $\text{Pr}[B]$, we see that the events A and B are independent if and only if

$$\text{Pr}[A] = \frac{\text{Pr}[A \cap B]}{\text{Pr}[B]}.$$

The left-hand side ($\Pr[A]$) denotes the fraction of the time that A occurs. The right-hand side ($\Pr[A \cap B] / \Pr[B]$) denotes the fraction of the time *when B occurs* that A occurs too. If these two fractions are equal, then A occurs with the same probability when B occurs as it does when B does not occur. (And if these two fractions are equal, then *both* when B occurs and when B does not occur, A occurs with probability $\Pr[A]$—that is, the probability of A without reference to B.)

EXAMPLES OF INDEPENDENT AND DEPENDENT EVENTS

To establish that two events A and B are independent, we can simply compute $\Pr[A]$, $\Pr[B]$, and $\Pr[A \cap B]$, and show that the product of the first two quantities is equal to the third. Here are a few examples:

Example 10.14 (Some independent events)

The following pairs of events are independent:

1. I flip a fair penny and a fair nickel. Define the following events:

   - *Event A:* The penny is heads.
   - *Event B:* The nickel is heads.

   Then $\Pr[A] = 0.5$ and $\Pr[B] = 0.5$ and $\Pr[A \cap B] = 0.25 = 0.5 \cdot 0.5$.

2. I draw a card from a randomly shuffled deck. Define the following events:

   - *Event A:* I draw an ace.
   - *Event B:* I draw a heart.

   For these events, we have
   $$\begin{aligned} \Pr[A] &= \Pr\left[\{A\clubsuit, A\diamondsuit, A\heartsuit, A\spadesuit\}\right] = \tfrac{1}{13} \\ \Pr[B] &= \Pr\left[\{A\heartsuit, 2\heartsuit, \ldots, K\heartsuit\}\right] = \tfrac{1}{4} \\ \Pr[A \cap B] &= \Pr\left[\{A\heartsuit\}\right] = \tfrac{1}{52} = \tfrac{1}{4} \cdot \tfrac{1}{13}. \end{aligned}$$

3. I roll a fair red die and a fair blue die. Define the following events:

   - *Event A:* The red die is odd.
   - *Event B:* The sum of the rolled numbers is odd.

   Then, writing outcomes as $\langle$the red roll, the blue roll$\rangle$, we have
   $$\begin{aligned} \Pr[A] &= \Pr\left[\{1,3,5\} \times \{1,2,3,4,5,6\}\right] &= \tfrac{18}{36} = 0.5 \\ \Pr[B] &= \Pr[\underbrace{\{1,3,5\} \times \{2,4,6\}}_{\text{red odd, blue even}} \cup \underbrace{\{2,4,6\} \times \{1,3,5\}}_{\text{red even, blue odd}}] &= \tfrac{18}{36} = 0.5 \end{aligned}$$

   Observe that $A \cap B = \{1,3,5\} \times \{2,4,6\}$, and so $\Pr[A \cap B] = \frac{9}{36} = (0.5) \cdot (0.5)$.

Any time the processes by which A and B come to happen are completely unrelated, it's certainly true that A and B are independent. But events can also be independent in other circumstances, as we saw in Example 10.14.3: both events in this example in some way incorporated the result of the value rolled on the red die, but the stated events themselves are independent anyway.

A visual representation of independent and dependent events is shown in Figure 10.14.

In Example 10.14, we showed that a few pairs of events are independent by showing that $\Pr[A \cap B] = \Pr[A] \cdot \Pr[B]$. By contrast, we can establish that two events are not independent—that is, they are *dependent*—directly from the definition by showing that $\Pr[A \cap B] \neq \Pr[A] \cdot \Pr[B]$. Here are a few examples:

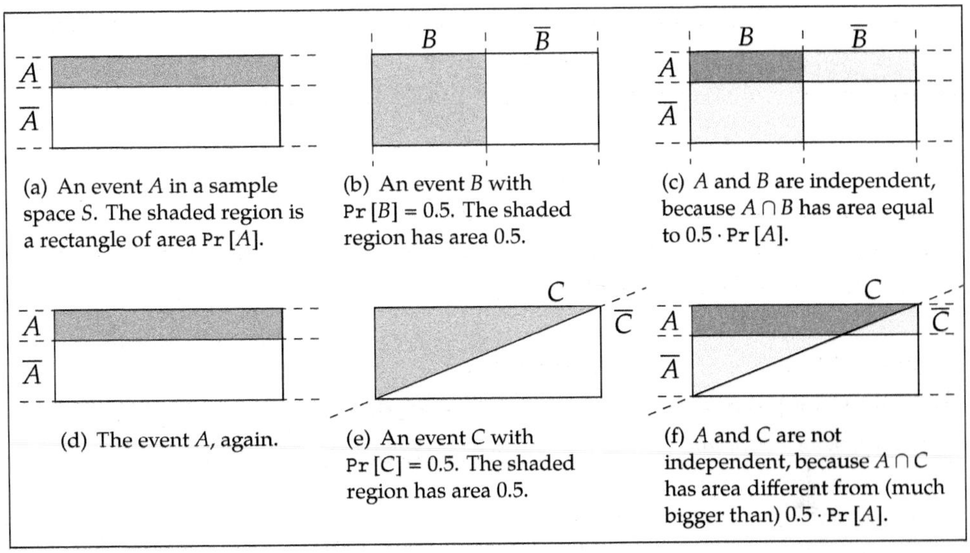

(a) An event A in a sample space S. The shaded region is a rectangle of area $\Pr[A]$.
(b) An event B with $\Pr[B] = 0.5$. The shaded region has area 0.5.
(c) A and B are independent, because $A \cap B$ has area equal to $0.5 \cdot \Pr[A]$.
(d) The event A, again.
(e) An event C with $\Pr[C] = 0.5$. The shaded region has area 0.5.
(f) A and C are not independent, because $A \cap C$ has area different from (much bigger than) $0.5 \cdot \Pr[A]$.

Figure 10.14: Pairs of independent and dependent events, represented visually. Think of the area of a region as its probability; the area of the sample space (the enclosing rectangle) is 1.

Example 10.15 (Some dependent events)
The following pairs of events are *not* independent:

1. I roll a fair die. Define the following events:
   - *Event A:* I roll an odd number.
   - *Event B:* I roll a prime number.

$$\begin{aligned} \Pr[A] &= \Pr[\{1,3,5\}] &= \tfrac{1}{2} \\ \Pr[B] &= \Pr[\{2,3,5\}] &= \tfrac{1}{2} \\ \Pr[A \cap B] &= \Pr[\{3,5\}] &= \tfrac{2}{6}. \end{aligned}$$

   Because the probability of $A \cap B$ is $\frac{2}{6} = \frac{1}{3}$, but $\Pr[A] \cdot \Pr[B] = \frac{1}{2} \cdot \frac{1}{2} = \frac{1}{4} \neq \frac{1}{3}$, the events A and B are dependent.

   Similarly, define *Event C* as "I roll an even number." Because $\Pr[B] = \Pr[C] = \frac{1}{2}$ and $\Pr[C \cap B] = \Pr[\{2\}] = \frac{1}{6} \neq \frac{1}{2} \cdot \frac{1}{2}$, the events B and C are dependent too.

2. I draw a card from a randomly shuffled deck. Define the following events:
   - *Event A:* I draw a heart.
   - *Event B:* I draw a spade.

$$\begin{aligned} \Pr[A] &= \Pr[\{A\heartsuit, 2\heartsuit, \ldots, K\heartsuit\}] &= \tfrac{1}{4} \\ \Pr[B] &= \Pr[\{A\spadesuit, 2\spadesuit, \ldots, K\spadesuit\}] &= \tfrac{1}{4} \\ \Pr[A \cap B] &= \Pr[\varnothing] &= 0, \end{aligned}$$

   where $A \cap B = \varnothing$ because no cards are both a heart *and* a spade. Because $0 \neq \frac{1}{16} = \frac{1}{4} \cdot \frac{1}{4}$, we have $\Pr[A \cap B] \neq \Pr[A] \cdot \Pr[B]$. These events are dependent.

3. I flip a fair penny and a fair nickel. Define the following events:

   - *Event A:* The penny is heads.
   - *Event B:* Both coins are heads.

   Then $\Pr[A] = 0.5$ and $\Pr[B] = 0.25$ and $\Pr[A \cap B] = 0.25 = \Pr[B] \neq \Pr[A] \cdot \Pr[B]$.

CORRELATION OF EVENTS

The pairs of dependent events from Example 10.15 are of two different qualitative types. Knowing that the first event occurred can make the second event more likely to occur ("rolling an odd number" and "rolling a prime number" for the dice) or less likely to occur ("rolling an even number" and "rolling a prime number"):

Definition 10.9 (Positive and negative correlation)
When two events A and B satisfy $\Pr[A \cap B] > \Pr[A] \cdot \Pr[B]$, *we say that A and B are* positively correlated. *When* $\Pr[A \cap B] < \Pr[A] \cdot \Pr[B]$, *we say that A and B are* negatively correlated. *(If* $\Pr[A \cap B] = \Pr[A] \cdot \Pr[B]$, *then A and B are* uncorrelated.*)*

At the extreme, knowing that the first event occurred can ensure that the second event definitely does not occur ("drawing a heart" and "drawing a spade" from Example 10.15) or can ensure that the second event definitely does occur ("both coins are heads" and "the first coin is heads" from Example 10.15).

Here are some further examples in which you're asked to figure out whether certain pairs of events are independent or dependent:

Example 10.16 (Encryption by random substitution)

Problem: One simple form of encryption for text is a *substitution cipher*, in which (in the simplest version) we choose a permutation of the alphabet, and then replace each letter with its permuted variant. (For example, we might permute the letters as `ABCDE···` $\rightarrow$ `XENBG···`; thus `DECADE` would be written as `BGNXBG`.) Suppose we choose a random permutation for this mapping, so that each of the 26! orderings of the alphabet is equally likely. Are the following events Q and Z independent or dependent?

- Q = "the letter `Q` is mapped to itself (that is, `Q` is 'rewritten' as `Q`)."
- Z = "the letter `Z` is mapped to itself."

Solution: We must compute $\Pr[Q]$, $\Pr[Z]$, and $\Pr[Q \cap Z]$. Because each permutation is equally likely to be chosen, we have

$$\Pr[Q] = \frac{\#\text{ permutations } \pi_{1,2,\ldots,26} \text{ where } \pi_{17} = 17}{\#\text{ permutations } \pi_{1,2,\ldots,26}} = \frac{25!}{26!} = \frac{1}{26}$$

because we can choose any of 25! orderings of all non-`Q` letters. Similarly,

$$\Pr[Z] = \frac{\#\text{ permutations } \pi_{1,2,\ldots,26} \text{ where } \pi_{26} = 26}{\#\text{ permutations } \pi_{1,2,\ldots,26}} = \frac{25!}{26!} = \frac{1}{26}.$$

To compute $\text{Pr}[Q \cap Z]$, we need to count the number of permutations $\pi_{1...26}$ with both $\pi_{17} = 17$ and $\pi_{26} = 26$. Any of the 24 other letters can go into any of the remaining 24 slots of the permutation, so there are 24! such permutations. Thus

$$\text{Pr}[Q \cap Z] = \frac{\#\text{ permutations } \pi_{1,2,...,26} \text{ where } \pi_{17} = 17 \text{ and } \pi_{26} = 26}{\#\text{ permutations } \pi_{1,2,...,26}} = \frac{24!}{26!} = \frac{1}{25 \cdot 26}.$$

Thus we have

$$\text{Pr}[Q \cap Z] = \tfrac{1}{25 \cdot 26} \qquad \text{and} \qquad \text{Pr}[Q] \cdot \text{Pr}[Z] = \tfrac{1}{26} \cdot \tfrac{1}{26} = \tfrac{1}{26 \cdot 26}.$$

There's only a small difference between $\frac{1}{26 \cdot 26} \approx 0.00148$ and $\frac{1}{25 \cdot 26} \approx 0.00154$, but they're indubitably different, and thus Q and Z are *not independent.*

(Incidentally, substitution ciphers are susceptible to *frequency analysis*: the most common letters in English-language texts are `ETAOIN`—almost universally in texts of reasonable length—and the frequencies of various letters is surprisingly consistent. See Exercises 10.72–10.76.)

Example 10.17 (Matched flips of two fair coins)

Problem: I flip two fair coins (independently). Consider the following events:

- *Event A:* the first flip comes up heads.
- *Event B:* the second flip comes up heads.
- *Event C:* the two flips match (are both heads or are both tails).

Which pairs of these events are independent, if any?

Solution: The sample space is $\{\text{HH}, \text{HT}, \text{TH}, \text{TT}\}$, and the events from the problem statement are given by $A = \{\text{HH}, \text{HT}\}$, $B = \{\text{HH}, \text{TH}\}$, and $C = \{\text{HH}, \text{TT}\}$. Thus $A \cap B = A \cap C = B \cap C = \{\text{HH}\}$—that is, HH is the only outcome that results in more than one of these events being true. (See Figure 10.15.)

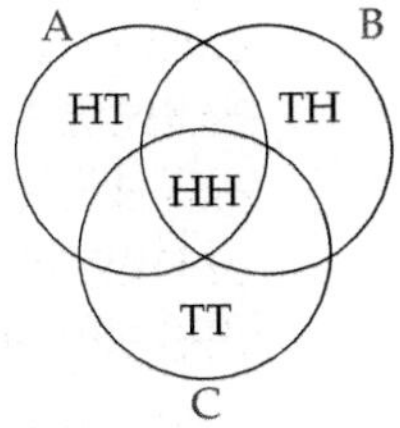

Figure 10.15: Two coin flips and three events.

Because the coins are fair, every outcome in this sample space has probability $\frac{1}{4}$. Focusing on the events A and B, we have

$$\begin{aligned} \text{Pr}[A] &= \text{Pr}\left[\{\text{HH}, \text{HT}\}\right] = \tfrac{1}{2} \\ \text{Pr}[B] &= \text{Pr}\left[\{\text{HH}, \text{TH}\}\right] = \tfrac{1}{2} \\ \text{Pr}[A \cap B] &= \text{Pr}\left[\{\text{HH}\}\right] = \tfrac{1}{4}. \end{aligned}$$

Thus $\text{Pr}[A] \cdot \text{Pr}[B] = \frac{1}{2} \cdot \frac{1}{2} = \frac{1}{4}$, and $\text{Pr}[A \cap B] = \frac{1}{4}$. Because $\text{Pr}[A] \cdot \text{Pr}[B] = \text{Pr}[A \cap B]$, the two events are independent.

The calculation is identical for the other two pairs of events, and so A and B are independent; A and C are independent; and B and C are independent.

Example 10.18 (Matched flips of two biased coins)

Problem: How would your answers to Example 10.17 change if the coins are p-biased instead of fair?

Solution: The sample space and events remain as in Example 10.17 (see Figure 10.16), but the outcomes now have different probabilities:

outcome	HH	HT	TH	TT
probability	$p \cdot p$	$p \cdot (1-p)$	$(1-p) \cdot p$	$(1-p) \cdot (1-p)$

Using these outcome probabilities, we compute the event probabilities as follows:

$$\Pr[A] = \Pr[\{HH, HT\}] = p \cdot p + p \cdot (1-p) = p \quad (1)$$
$$\Pr[B] = \Pr[\{HH, TH\}] = p \cdot p + (1-p) \cdot p = p \quad (2)$$
$$\Pr[C] = \Pr[\{HH, TT\}] = p \cdot p + (1-p) \cdot (1-p) = p^2 + (1-p)^2. \quad (3)$$

Because $A \cap B = B \cap C = A \cap C = \{HH\}$, we also have

$$\Pr[A \cap B] = \Pr[B \cap C] = \Pr[A \cap C] = \Pr[HH] = p^2. \quad (4)$$

Thus A and B are still independent, because $\Pr[A] \cdot \Pr[B] = p \cdot p = p^2 = \Pr[A \cap B]$ by (1), (2), and (4). But what about the events A and C? By (1), (3), and (4), we have

$$\Pr[A] \cdot \Pr[C] = p \cdot \left[p^2 + (1-p)^2\right] \quad \text{and} \quad \Pr[A \cap C] = p^2.$$

By a bit of algebra, we see that $\Pr[A \cap C] = \Pr[A] \cdot \Pr[C]$ if and only if

$$\begin{aligned} p^2 = p(p^2 + (1-p)^2) &\Leftrightarrow 0 = p(p^2 + (1-p)^2) - p^2 \\ &\Leftrightarrow 0 = 2p^3 - 3p^2 + p \\ &\Leftrightarrow 0 = p(2p-1)(p-1). \end{aligned}$$

So the events A and C are independent—that is, $\Pr[A \cap C] = \Pr[A] \cdot \Pr[C]$—if and only if $p \in \{0, \frac{1}{2}, 1\}$.

Thus events A and B are independent for any value of p, while events A and C (and similarly B and C) are independent if and only if $p \in \{0, \frac{1}{2}, 1\}$.

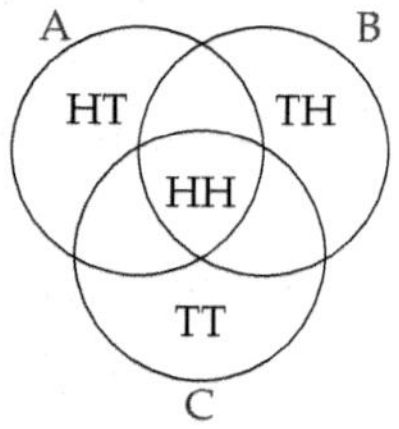

Figure 10.16: The flips and events, again. Recall the events:
A: 1st flip heads.
B: 2nd flip heads.
C: flips match.

Taking it further: While *any two of the events* from Example 10.17 (or Example 10.18 with $p = \frac{1}{2}$) are independent, *the third event is not independent of the other two.* Another way to describe this situation is that the events A and $B \cap C$ are *not* independent: in particular, $\Pr[A \cap (B \cap C)] / \Pr[B \cap C] = 1 \neq \Pr[A]$. A set of events $A_1, A_2, \ldots, A_n$ is said to be *pairwise independent* if, for any two indices i and $j \neq i$, the events A_i and A_j are independent. More generally, these events are said to be *k-wise independent* if, for any subset S of up to k of these events, the events in S are all independent. (And we say that the set of events is *fully independent* if every subset of any size satisfies this property.)

Sometimes it will turn out that we "really" care only about pairwise independence. For example, if we think about a hash table that uses a "random" hash function, we're usually only concerned with the question "do elements x and y collide?"—which is a question about just one pair of events. Generally, we can create a pairwise-independent random hash function more cheaply than creating a fully independent random hash function. If we view random bits as a scarce resource (like time and space, in the style of Chapter 6), then this savings is valuable.

10.3.2 *Conditional Probability*

In Section 10.3.1, we discussed the black-and-white distinction between pairs of independent events and dependent events: if A and B are independent, then knowing whether or not B happened gives you no information about whether A happened; if A and B are dependent, then the probability that A happens if B happened is different from the probability that A happens if B did not happen. But *how* does knowing that B occurred change your estimate of the probability of A? Think about events like "the sky is clear" and "it is very windy" and "it will rain today": sometimes B means that A is less likely or even impossible; sometimes B means that A is more likely or even certain. Here we will discuss *quantitatively* how one event's probability is affected by the knowledge of another event.

The *conditional probability of A given B* represents the probability of A occurring *if we know that B occurred*:

Definition 10.10 (Conditional probability)
The conditional probability of A given B, *written* $\Pr\left[A|B\right]$, *is given by*

$$\Pr\left[A|B\right] = \frac{\Pr\left[A \cap B\right]}{\Pr\left[B\right]}.$$

(The quantity $\Pr\left[A|B\right]$ *is also sometimes called the* probability of A conditioned on B.*)*
We will treat $\Pr\left[A|B\right]$ *as undefined when* $\Pr\left[B\right] = 0$.

Here are a few simple examples:

Example 10.19 (Odds and primes)
I choose a number uniformly at random from $\{1, 2, \ldots, 10\}$. Define these two events:

- *Event A:* The chosen number is odd.
- *Event B:* The chosen number is prime.

For these events, we have $\Pr\left[A|B\right] = \dfrac{\Pr\left[A \cap B\right]}{\Pr\left[B\right]} = \dfrac{\Pr\left[\{3,5,7\}\right]}{\Pr\left[\{2,3,5,7\}\right]} = \dfrac{3}{4}$

and $\Pr\left[B|A\right] = \dfrac{\Pr\left[A \cap B\right]}{\Pr\left[A\right]} = \dfrac{\Pr\left[\{3,5,7\}\right]}{\Pr\left[\{1,3,5,7,9\}\right]} = \dfrac{3}{5}.$

Example 10.20 (Dominoes)
Problem: Shuffle the dominoes in Figure 10.17, and draw one uniformly at random.

1. What is the probability that you drew a domino with a 2 () on it?
2. You make a draw and see the domino . (Imagine the shaded side of the domino is covered by your hand.) What's the probability your domino has a 2?
3. You make a draw and see that the domino is . What is the probability that you drew a domino with a 2?

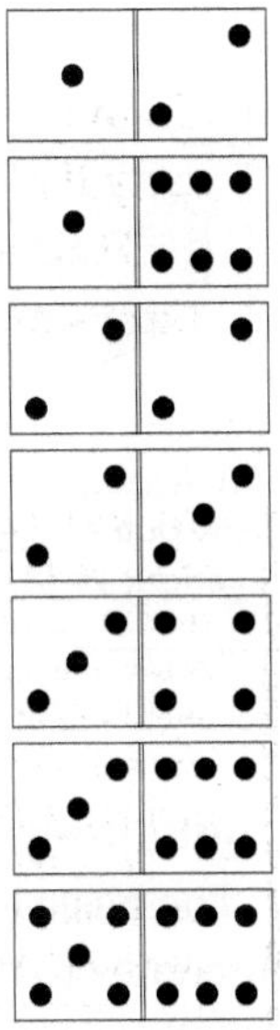

Figure 10.17: Some dominoes.

<u>Solution:</u> 1. We are asked for the probability of drawing a domino with a 2:

contain a 2 do not contain a 2

Thus 3 of the 7 dominoes have a 2, so Pr [⚁] = $\frac{3}{7}$.

2. We observe ⚀ on our drawn domino. We're asked for the probability of a 2:

contains a 2 does not contain a 2 impossible (do not contain a 1)

We know that the domino you drew must have been either ⚀⚁ or ⚀⚅. These two dominoes were equally likely to be drawn, and 1 of these 2 has a ⚁, so there's a $\frac{1}{2}$ probability that you drew a ⚁. Using conditional probability notation, we can write this quantity as Pr [⚁|⚀] = $\frac{1}{2}$.

3. We are computing Pr [⚁|⚄], the probability of a ⚁ *given that* we observed a ⚄. By the definition of conditional probability, we have

$$\text{Pr}[⚁|⚄] = \frac{\text{Pr}[⚁ \cap ⚄]}{\text{Pr}[⚄]} = \frac{0}{\frac{1}{7}} = 0.$$

(A less notationally heavy way of writing this argument: because we see a ⚄, we know that the domino you drew must have been ⚄⚅. This domino doesn't have a ⚁ and so there's zero chance that we observe a ⚁.)

Conditional probability as "zooming in" (and another example)

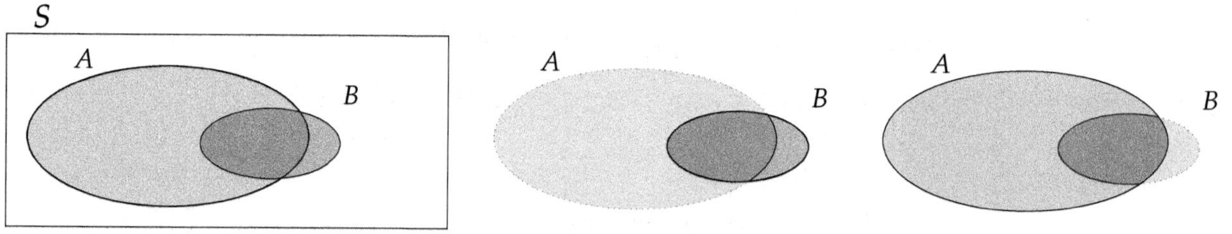

(a) A sample space S and two events A and B. Any outcome in S can be chosen, and so in this example Pr $[A] \approx 0.4$ and Pr $[B] \approx 0.1$.

(b) Conditioning on the event B. Any outcome in B can be chosen, and so Pr $[A|B]$ is the fraction of those outcomes for which A occurs, so here Pr $[A|B] \approx 0.8$.

(c) Conditioning on the event A. Any outcome in A can be chosen, and so Pr $[B|A]$ is the fraction of those outcomes for which B occurs, so here Pr $[B|A] \approx 0.2$.

Figure 10.18: A view of conditional probability.

Intuitively, we can think of Pr $[A|B]$ as "zooming" the universe down to the set B. The basic idea that we used in Example 10.20 was to narrow the set of possible outcomes to those consistent with the observed partial data about the drawn domino, and then compute the fraction of the narrowed sample space for which A occurs. This view of conditional probability is illustrated in Figure 10.18.

Here's one more example, where we condition on slightly more complex events.

Example 10.21 (Coin flips)

Problem: Flip a fair coin 10 times (with all flips independent: the ith flip has no effect on the jth flip for $j \neq i$). Write H to denote the event of getting at least 9 heads.

1. What is $\Pr[H]$?
2. Let A be the event "the first flip comes up heads." What is $\Pr[H|A]$?
3. Let B be the event "the first flip comes up tails." What is $\Pr[H|B]$?
4. Let C be the event "the first three flips come up heads." What is $\Pr[H|C]$?
5. Let D be the event "we get at least 8 heads." What is $\Pr[H|D]$?

Solution: 1. Observe that every outcome—every element of $\{H, T\}^{10}$—is equally likely, each with probability $1/2^{10}$. The number of sequences of 10 flips with 9 or 10 heads is $\binom{10}{9} + \binom{10}{10} = 10 + 1 = 11$, so $\Pr[H] = 11/2^{10} \approx 0.0107$.

For the conditional probabilities, we will compute $\Pr[H \cap X]$ and $\Pr[X]$ for each of the stated events X. The final answer is their ratio. Because each outcome is equally likely, we only have to compute the cardinality of the given events (and the cardinality of their intersection with H) to answer the questions.

2. For A (the first flip comes up H), we have $|A \cap H| = 10$: there are 9 outcomes with one Tails that start with a Heads (HTHHHHHHHH, HHTHHHHHHH, ..., HHHHHHHHHT) and 1 outcome with zero Tails (HHHHHHHHHH). Thus $\Pr[A \cap H] = 10/2^{10}$. Obviously $\Pr[A] = \frac{1}{2}$. Thus

$$\Pr[H|A] = \frac{\Pr[A \cap H]}{\Pr[A]} = \frac{10/2^{10}}{1/2} = \frac{10}{2^9} \approx 0.01953.$$

3. For B (the first flip comes up T), we've already "used up" the single permitted non-heads in the first flip, so there's only one outcome in $B \cap H$, namely THHHHHHHHH. And, again, obviously $\Pr[B] = \frac{1}{2}$. Therefore we have

$$\Pr[H|B] = \frac{\Pr[B \cap H]}{\Pr[B]} = \frac{1/2^{10}}{1/2} = \frac{1}{2^9} \approx 0.00195.$$

4. For C (the first three flips come up H), we have $\Pr[C] = \frac{1}{8}$. The outcomes in $C \cap H$ are exactly those that start with HHH followed by 6+ heads in the last 7 flips. There are $\binom{7}{7} + \binom{7}{6} = 8$ such outcomes. Thus

$$\Pr[H|C] = \frac{\Pr[C \cap H]}{\Pr[C]} = \frac{8/2^{10}}{1/8} = \frac{64}{2^{10}} \approx 0.0625.$$

5. For D (there are at least 8 heads), we have $\Pr[H \cap D] = \Pr[H] = 11/2^{10}$. (There are no outcomes in which we get 9+ heads but fail to get 8+ heads!) The probability of getting 8+ heads in 10 fair flips is

$$\Pr[D] = \frac{\binom{10}{8} + \binom{10}{9} + \binom{10}{10}}{2^{10}} = \frac{45 + 10 + 1}{2^{10}} = \frac{56}{2^{10}}.$$

And therefore

$$\Pr[H|D] = \frac{\Pr[D \cap H]}{\Pr[D]} = \frac{11/2^{10}}{56/2^{10}} = \frac{11}{56} \approx 0.1964.$$

To repeat the word of warning from early in this chapter: *it can be very difficult to have good intuition about probability questions.* For example, the last problem in Example 10.21 asked for the probability of getting 9+ heads in 10 flips *conditioned on getting* 8+ *heads.* It may be easy to talk yourself into believing that, of the times that we get 8+ heads, there's a $\approx 50\%$ chance of getting 9 or more heads. ("Put aside the first 8 heads, and look at one of the other flips—it's heads with probability $\frac{1}{2}$, so we get a 9th heads with probability $\frac{1}{2}$.") But this intuition is blatantly wrong. Another way of thinking about the calculation in the last part of Example 10.21 is to observe that there are 56 outcomes with 8, 9, or 10 heads. Only 11 of these outcomes have 9 or 10 heads. Each outcome is equally likely. So if we're promised that one of the 56 outcomes occurred, then there's an $\frac{11}{56}$ chance that one of the 11 occurred.

Taking it further: So far, we have considered only random processes in which each outcome that can occur does so with probability $\varepsilon > 0$—that is, there have been no infinitesimal probabilities. But we can imagine scenarios in which infinitesimal probabilities make sense.

For example, imagine a probabilistic process that chooses a real number x between 0 and 1, where each element of the sample space $S = \{x : 0 \leq x \leq 1\}$ is equally likely to be chosen. We can make probabilistic statements like $\Pr[x \leq 0.5] = \frac{1}{2}$—half the time, we end up with $x \leq 0.5$, half the time we end up with $x \geq 0.5$—but for *any* particular value c, *the probability that $x = c$ is zero!* (Perhaps bizarrely, $\Pr[x \leq 0.5] = \Pr[x < 0.5]$. Indeed, $\Pr[x = 0.5]$ *cannot be* $\varepsilon > 0$, for any ε. Every possible outcome has to have that same probability ε of occurring, and for any $\varepsilon > 0$ there are more than $\frac{1}{\varepsilon}$ real numbers between 0 and 1. So we'd violate (10.1) if we had $\Pr[x = 0.5] > 0$.)

To handle infinitesimal probabilities, we need calculus. We can describe the above circumstance with a *probability density function* $p : S \to [0, 1]$, so that, in place of (10.1), we require

$$\int_{x \in S} p(x)dx = 1.$$

(For a uniformly chosen $x \in [0, 1]$, we have $p(x) = 1$; for a uniformly chosen $x \in [0, 100]$, we have $p(x) = \frac{1}{100}$.) Some of the statements that we've made in this chapter don't apply in the infinitesimal case. For example, the "zooming in" view of conditional probability from Figure 10.18 doesn't quite work in the infinitesimal case. In fact, we can consider questions about $\Pr[A|B]$ even when $\Pr[B] = 0$, like *what is the probability that a uniformly chosen $x \in [0, 100]$ is an integer, conditioned on x being a rational number?*. (And Exercise 10.70—if $\Pr[B] = 0$, then A and B are independent—isn't true with infinitesimal probabilities.) But details of this infinitesimal version of probability theory are generally outside of our concern here, and are best left to a calculus-based/analysis-based textbook on probability.

The restriction to non-infinitesimal probabilities is generally a reasonable one to make for CS applications, but it *is* a genuine restriction. (It's worth noting that we *have* encountered an infinite sample space before—just one that didn't have any infinitesimal probabilities. In a geometric distribution with parameter $\frac{1}{2}$, for example, any positive integer k is a possible outcome, with $\Pr[k] = 1/2^k$, which is a finite, albeit very small, probability for any positive integer k.)

10.3.3 *Bayes' Rule and Calculating with Conditional Probability*

Here, we'll briefly introduce a few simple but useful ways of thinking about conditional probability: the connection between independence of events and conditional probability; a few ways of thinking about plain (unconditional) probability using conditional probability; and, finally, *Bayes' Rule*, a tremendously useful formula that relates $\Pr[A|B]$ and $\Pr[B|A]$.

Relating independence of events and conditional probability

Consider two events A and B for which $\Pr[B] \neq 0$. Observe that A and B are independent if and only if $\Pr[A|B] = \Pr[A]$:

$$\begin{aligned} A \text{ and } B \text{ are independent} &\Leftrightarrow \Pr[A] \cdot \Pr[B] = \Pr[A \cap B] && \textit{definition of independence} \\ &\Leftrightarrow \Pr[A] = \frac{\Pr[A \cap B]}{\Pr[B]} && \textit{dividing by } \Pr[B] \\ &\Leftrightarrow \Pr[A] = \Pr[A|B]. && \textit{definition of } \Pr[A|B] \end{aligned}$$

(Note that this calculation doesn't work when $\Pr[B] = 0$—we can't divide by 0, and $\Pr[A|B]$ is undefined—but see Exercise 10.70.) Notice again that this relationship is an if-and-only-if relationship: when A and B are not independent, then $\Pr[A]$ and $\Pr[A|B]$ *must* be different. Here is a small example:

Example 10.22 (Self-mapped letters in substitution ciphers)
In Example 10.16, we showed that, for a random permutation π of the alphabet, the events Q (Q is mapped to itself by π) and Z (Z is mapped to itself by π) were not independent: specifically, $\Pr[Q] = \frac{1}{26}$, $\Pr[Z] = \frac{1}{26}$, and $\Pr[Q \cap Z] = \frac{1}{25 \cdot 26}$. Thus

$$\Pr[Q|Z] = \frac{\Pr[Q \cap Z]}{\Pr[Z]} = \frac{1/(25 \cdot 26)}{1/26} = \frac{1}{25}.$$

Compare $\Pr[Q|Z] = \frac{1}{25}$ to $\Pr[Q] = \frac{1}{26}$: thus, knowing that Z is mapped to itself makes it *slightly more likely* that Q is also mapped to itself. The reason that Z makes Q slightly more probable is that, when Z occurs, Z cannot be mapped to Q, so there are only 25 letters "competing" to be mapped to Q instead of 26.

Problem-solving tip: Often it is easier to get intuition about a probabilistic statement by imagining an absurdly small variant of the problem. Here, for example, imagine a 2-letter alphabet Q,Z. Then if Z is mapped to itself *then* Q *must also be mapped to itself.* So $\Pr[Q] = \frac{1}{2}$, but $\Pr[Q|Z] = 1$.

INTERSECTIONS AND CONDITIONAL PROBABILITY

The definition of conditional probability (Definition 10.10) states that

$$\Pr[A|B] = \frac{\Pr[A \cap B]}{\Pr[B]}.$$

Multiplying both sides of this equality by $\Pr[B]$ yields a useful way of thinking about the probability of intersections:

Theorem 10.2 (The Chain Rule)
Let A and B be arbitrary events. Then

$$\Pr[A \cap B] = \Pr[B] \cdot \Pr[A|B].$$

And, more generally, for events $A_1, A_2, \ldots, A_k$, we have

$$\begin{aligned} &\Pr[A_1 \cap A_2 \cap A_3 \cap \cdots \cap A_k] \\ &= \Pr[A_1] \cdot \Pr[A_2|A_1] \cdot \Pr[A_3|A_1 \cap A_2] \cdot \cdots \cdot \Pr[A_k|A_1 \cap \cdots \cap A_{k-1}]. \end{aligned}$$

If we're interested in the probability that A and B occur, then we need it to be the case that A occurs—and, *conditioned on A occurring*, B occurs too.

Example 10.23 (Drawing a heart flush in poker)

Problem: A *flush* in poker is a 5-card hand, all of which are the same suit. What is the probability of drawing a heart flush from a randomly shuffled deck?

Solution: We can draw any heart first. We have to keep drawing hearts to get a flush, so for $2 \leq k \leq 5$, the kth card we draw must be one of the remaining $14 - k$ hearts from the $53 - k$ cards left in the deck. That is, writing H_i to denote the event that the ith card drawn is a heart:

$$\begin{aligned}
&\Pr[H_1 \cap H_2 \cap H_3 \cap H_4 \cap H_5] \\
&= \Pr[H_1] \cdot \Pr[H_2 | H_1] \cdot \Pr[H_3 | H_{1,2}] \cdot \Pr[H_4 | H_{1,2,3}] \cdot \Pr[H_5 | H_{1,2,3,4}] \\
&= \frac{13}{52} \cdot \frac{12}{51} \cdot \frac{11}{50} \cdot \frac{10}{49} \cdot \frac{9}{48} \\
&= \frac{154440}{311875200} \approx 0.00049519807.
\end{aligned}$$

(We could also have directly computed this quantity via counting: there are $\binom{13}{5}$ hands of 5 hearts, and $\binom{52}{5}$ total hands. Thus the fraction of all hands that are heart flushes is

$$\frac{\binom{13}{5}}{\binom{52}{5}} = \frac{\frac{13!}{5! \cdot 8!}}{\frac{52!}{5! \cdot 47!}} = \frac{13! \cdot 47!}{8! \cdot 52!} = \frac{13 \cdot 12 \cdot 11 \cdot 10 \cdot 9}{52 \cdot 51 \cdot 50 \cdot 49 \cdot 48},$$

which is the same quantity that we found above.)

We can use the chain rule to compute the probability of an event A by making the (obvious!) observation that another event B either occurs or doesn't occur:

Theorem 10.3 (The Law of Total Probability)

Let A and B be arbitrary events. Then

$$\Pr[A] = \Pr[A | B] \cdot \Pr[B] + \Pr[A | \overline{B}] \cdot \Pr[\overline{B}].$$

Proof. We'll proceed by splitting A into two disjoint subsets, $A \cap B$ and $A - B$ (which is otherwise known as $A \cap \overline{B}$):

$$\begin{aligned}
\Pr[A] &= \Pr[(A \cap B) \cup (A \cap \overline{B})] && A = (A \cap B) \cup (A \cap \overline{B}) \\
&= \Pr[A \cap B] + \Pr[A \cap \overline{B}] && A \cap B \text{ and } A \cap \overline{B} \text{ are disjoint} \\
&= \Pr[A | B] \cdot \Pr[B] + \Pr[A | \overline{B}] \cdot \Pr[\overline{B}]. && \text{the chain rule}
\end{aligned}$$

Thus the theorem follows. □

Here's a simple example of using the law of total probability:

Example 10.24 (Binary Symmetric Channel)

We wish to transmit a 1-bit message from a sender to a receiver. The sender's message is 0 with probability 0.3, and it's 1 with probability 0.7. The sender sends this data using a communication channel that corrupts (that is, flips) every transmitted bit with probability 0.25. Then the probability that the receiver receives a "1" message is

$$\begin{aligned}\Pr[\text{receive } 1] &= \Pr\left[\text{receive } 1|\text{send } 1\right] \cdot \Pr[\text{send } 1] + \Pr\left[\text{receive } 1|\text{send } 0\right] \cdot \Pr[\text{send } 0]\\ &= (0.75 \cdot 0.7) + (0.25 \cdot 0.3)\\ &= 0.525 + 0.075 = 0.6.\end{aligned}$$

Taking it further: The *binary symmetric channel* is given this name because it transmits a bit (it's *binary*) and it corrupts a 0 with the same probability as it corrupts a 1 (it's *symmetric*). (See Figure 10.19; view each arrow in the channel as transforming a particular input bit to a particular output bit, with the indicated probability.)

The binary symmetric channel is one of the most basic forms of a noisy communication channel (that is, a channel that does not perfectly transmit its input without any chance of corruption). The subfield of *information theory* is devoted to analyzing topics like the (theoretical) efficiency of communication channels, including the binary symmetric channel. For much more, see a textbook on information theory.[5]

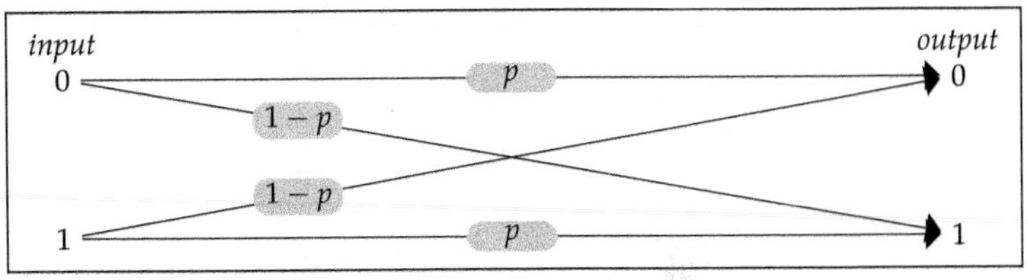

Figure 10.19: The binary symmetric channel.

[5] Thomas M. Cover and Joy A. Thomas. *Elements of Information Theory*. Wiley, 1991.

BAYES' RULE

Bayes' Rule is a simple—but tremendously useful—rule for "flipping around" a conditional probability statement. It allows us to express the conditional probability of A given B in terms of the conditional probability of B given A:

Theorem 10.4 (Bayes' Rule)

For any two events A and B:

$$\Pr\left[A|B\right] = \frac{\Pr\left[B|A\right] \cdot \Pr[A]}{\Pr[B]}.$$

Bayes' Rule is named after Thomas Bayes, an 18th-century English mathematician.

Proof. Applying the chain rule to break apart $\Pr[A \cap B]$ "in both orders," we have

$$\begin{aligned}\Pr[A \cap B] &= \Pr\left[A|B\right] \cdot \Pr[B]\\ \Pr[B \cap A] &= \Pr\left[B|A\right] \cdot \Pr[A].\end{aligned}$$

The left-hand sides of these equations are identical because $A \cap B = B \cap A$ (and therefore $\Pr[A \cap B] = \Pr[B \cap A]$), so their right-hand sides are equal, too:

$$\Pr\left[A|B\right] \cdot \Pr[B] = \Pr\left[B|A\right] \cdot \Pr[A].$$

Dividing both sides of this equality by $\Pr[B]$ yields the desired equation:

$$\Pr\left[A|B\right] = \frac{\Pr\left[B|A\right] \cdot \Pr[A]}{\Pr[B]}.$$

□

Here are a couple of simple examples of using Bayes' Rule:

Example 10.25 (Binary Symmetric Channel, again)
As in Example 10.24, assume a sender transmits a 0 with probability 0.3 and a 1 with probability 0.7 across a channel that corrupts every bit with probability 0.25. We showed in Example 10.24 that $\Pr[\text{receive } 1] = 0.6$ and thus $\Pr[\text{receive } 0] = 0.4$. Then the probability that the receiver receiving a "1" message was indeed sent a 1 is

$$\begin{aligned}\Pr\left[\text{message sent was } 1 \middle| \text{receive } 1\right] &= \frac{\Pr\left[\text{receive } 1 \middle| \text{send } 1\right] \cdot \Pr[\text{send } 1]}{\Pr[\text{receive } 1]} && \textit{by Bayes' Rule} \\ &= \frac{0.75 \cdot 0.7}{0.6} = 0.875.\end{aligned}$$

And the probability that the receiver receiving a "0" message was indeed sent a 0 is

$$\begin{aligned}\Pr\left[\text{message sent was } 0 \middle| \text{receive } 0\right] &= \frac{\Pr\left[\text{receive } 0 \middle| \text{send } 0\right] \cdot \Pr[\text{send } 0]}{\Pr[\text{receive } 0]} && \textit{by Bayes' Rule} \\ &= \frac{0.75 \cdot 0.3}{0.4} = 0.5625.\end{aligned}$$

(Qualitatively, these numbers tell us that most of received ones were actually sent as ones, but barely more than half of the received zeros were actually sent as zeros.)

Example 10.26 (9+ heads, again)
We flip a fair coin 10 times. As in Example 10.21, let A denote the event that the first flip comes up heads and let H denote the event that there are 9 or more heads in the 10 flips. (There we showed $\Pr[H] = 11/2^{10}$, $\Pr[A] = \frac{1}{2}$, and $\Pr\left[H \middle| A\right] = 10/2^9$.) Then

$$\Pr\left[A \middle| H\right] = \frac{\Pr\left[H \middle| A\right] \cdot \Pr[A]}{\Pr[H]} = \frac{(10/2^9) \cdot \frac{1}{2}}{11/2^{10}} = \frac{10}{11}.$$

Taking it further: A *speech recognition system* is supposed to "listen" to speech in a language like English, and recognize the words that are being spoken. Bayes' Rule allows us to think about two different types of evidence that such a system uses in deciding what words it "thinks" are being said; see p. 1036.

A particularly important application of Bayes' Rule is in "updating" one's beliefs about the world by observing new information. (Here "beliefs" take the form of a probability distribution.) One starts with a *prior distribution* which one then updates based on *evidence* to produce a *posterior distribution*. Here are two examples:

The prior (*pre* = before) is your best guess of the probability of the event prior to seeing the produced evidence; the posterior (*post* = after) is your best guess after seeing the evidence.

Example 10.27 (Alice the CS major)
We are interested in whether a student (let's call her Alice) is a computer science major. Our prior for Alice might be $\Pr\left[\text{CS major}\right] = 0.05$ because 5% of students are CS majors. We learn that Alice took Ancient Philosophy. If we know that 10% of students as a whole take Ancient Philosophy, and 50% of CS majors do, then

$$\text{Pr}\left[\text{CS major}|\text{phil}\right] = \frac{\text{Pr}\left[\text{phil}|\text{CS major}\right] \cdot \text{Pr}\left[\text{CS major}\right]}{\text{Pr}\left[\text{phil}\right]} = \frac{0.5 \cdot 0.05}{0.10} = 0.25.$$

Our posterior distribution (that is, the updated guess) is that there is now a 25% chance that Alice is a CS major.

Example 10.28 (Flipping a coin to decide which coin to flip)
I have two coins in an opaque bag. The coins are visually indistinguishable, but one coin is fair ($\text{Pr}[H] = 0.5$); the other coin is 0.75-biased ($\text{Pr}[H] = 0.75$). I pull one of the two coins out at random.

- My *prior distribution* is that there is a 50% chance I'm holding the fair coin, and a 50% chance I'm holding the biased coin. (That is, $\text{Pr}[\text{biased}] = \text{Pr}[\text{fair}] = 0.5$.)

I flip the coin that I'm holding. It comes up heads.

- The *evidence* is the Heads flip.

Because the biased coin is more likely to produce Heads flips than the fair coin is (and we saw Heads), this evidence should make us view it as more likely that the coin that I'm holding is the biased coin. Let's compute my *posterior probability:*

- The posterior probability of an event is the probability of that event *conditioned on the observed evidence.* So we wish to compute $\text{Pr}\left[\text{biased}|\text{H}\right]$:

$$\begin{aligned}
\text{Pr}\left[\text{biased}|\text{H}\right] &= \frac{\text{Pr}\left[\text{H}|\text{biased}\right] \cdot \text{Pr}[\text{biased}]}{\text{Pr}[\text{H}]} && \textit{Bayes' Rule} \\
&= \frac{\text{Pr}\left[\text{H}|\text{biased}\right] \cdot \text{Pr}[\text{biased}]}{\text{Pr}\left[\text{H}|\text{biased}\right] \cdot \text{Pr}[\text{biased}] + \text{Pr}\left[\text{H}|\text{fair}\right] \cdot \text{Pr}[\text{fair}]} && \textit{Law of Total Probability} \\
&= \frac{0.75 \cdot \text{Pr}[\text{biased}]}{(0.75 \cdot \text{Pr}[\text{biased}]) + (0.5 \cdot \text{Pr}[\text{fair}])} && \textit{the given biases of the coins: 0.75 for biased, 0.5 for fair} \\
&= \frac{0.75 \cdot 0.5}{(0.75 \cdot 0.5) + (0.5 \cdot 0.5)} && \text{Pr}[\textit{biased}] = \text{Pr}\left[\textit{fair}\right] = 0.5\textit{, as defined by the prior} \\
&= \frac{0.375}{0.375 + 0.25} = 0.6.
\end{aligned}$$

So the posterior probability is $\text{Pr}\left[\text{biased}|\text{H}\right] = 0.6$ and $\text{Pr}\left[\text{fair}|\text{H}\right] = 0.4$.

Taking it further: The idea of Bayesian reasoning is used frequently in many applications of computer science—any time a computational system weighs various pieces of evidence in deciding what kind of action to take in a particular situation. One of the most noticeable examples of this type of reasoning occurs in *Bayesian spam filters*; see p. 1037 for more.

COMPUTER SCIENCE CONNECTIONS

SPEECH RECOGNITION, BAYES' RULE, AND LANGUAGE MODELS

A software system for *speech recognition* must solve the following problem: given an audio stream $\mathcal{S}$ of spoken English as input, produce as output a transcript $\mathcal{W}$ of the words in $\mathcal{S}$. There will be many candidate transcripts of $\mathcal{S}$, and generally the task of the system is to produce the *most likely sequence of words given the audio stream*—that is, to find the $\mathcal{W}^*$ maximizing $\Pr[\mathcal{W}^*|\mathcal{S}]$.

Using Bayes' Rule, we can rephrase $\Pr[\mathcal{W}^*|\mathcal{S}]$ into an expression that's easier to understand:

$$\begin{aligned} &\text{the } \mathcal{W}^* \text{ maximizing } \Pr[\mathcal{W}^*|\mathcal{S}] \\ &= \text{the } \mathcal{W}^* \text{ maximizing } \frac{\Pr[\mathcal{S}|\mathcal{W}^*] \cdot \Pr[\mathcal{W}^*]}{\Pr[\mathcal{S}]} && \textit{Bayes' Rule} \\ &= \text{the } \mathcal{W}^* \text{ maximizing } \Pr[\mathcal{S}|\mathcal{W}^*] \cdot \Pr[\mathcal{W}^*]. && \Pr[\mathcal{S}] \textit{ is the same for each } \mathcal{W}^* \end{aligned}$$

Thus there are two valuable sources of data for evaluating a candidate $\mathcal{W}$:

- $\Pr[\mathcal{S}|\mathcal{W}]$, the *likelihood of the observation*: the probability that this sound stream would have been produced if $\mathcal{W}$ were the sequence of words; and
- $\Pr[\mathcal{W}]$, the *probability of this output*: the probability of this sequence of words being uttered at all.

For example, *even if* the audio stream is a better acoustic match for the phrase *whirled Siri string*, you'd want your system to prefer the phrase *World Series ring*—because an English speaker is far more likely to say the latter phrase than the former. (That is, $\Pr[\textit{World Series ring}]$ is much higher than $\Pr[\textit{whirled Siri string}]$.) Of course, we still must take into account the audio stream $\mathcal{S}$—otherwise, *regardless of the audio*, we'd end up with a system that produced precisely the same output sentence (the most common sentence in English: *I'm sorry!*, or whatever it is) for any input sound stream.

Generally speaking, the quantity $\Pr[\mathcal{S}|\mathcal{W}]$ would be estimated by an acoustic model of the vocal tract: if I'm trying to say *Camp Utah seance*, what is the probability that I produce a particular stream $\mathcal{S}$ of sounds?

The quantity $\Pr[\mathcal{W}]$ is estimated by what's called a *language model*. We would acquire a large collection of English text, and then try to use that data to estimate how likely a particular sequence is. The simplest language model is the *unigram* model:

- from a giant data set with N total words, for each word w we count up the number of times $n(w)$ that w appears.
- if $\mathcal{W} = w_1, w_2, \ldots, w_k$, we estimate $\Pr[\mathcal{W}]$ as $\frac{n(w_1)}{N} \cdot \frac{n(w_2)}{N} \cdot \; \cdots \; \cdot \frac{n(w_k)}{N}$.

A more complex language model might use *bigrams*—two-word sequences—instead; we count the number of occurrences of w_i, w_{i+1} consecutively in the giant data set, and estimate $\Pr[\mathcal{W}]$ based on these counts. Other more complex language models are used in real systems.[6] There's also a great deal of complication with avoiding *overfitting* of the language model to the training data. (In addition to speech recognition, a variety of other natural language processing problems are generally solved with the same general approach.)

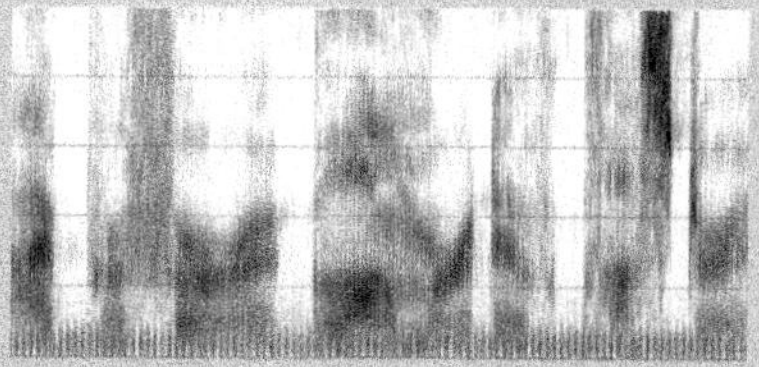

Figure 10.20: A *spectrogram* representation of an audio stream: the x-axis represents time, the y-axis represents frequency, and the darkness of the shading denotes the intensity of sound at that particular frequency at that particular time. (See p. 234 for more discussion.) The task is to turn this representation into its most probable sequence of words—in this case, the sentence "I prefer agglomerative clustering."

For much more, see

[6] Daniel Jurafsky and James H. Martin. *Speech and Language Processing: An Introduction to Natural Language Processing, Computational Linguistics, and Speech Recognition*. Pearson Prentice Hall, 2nd edition, 2008.

COMPUTER SCIENCE CONNECTIONS

BAYESIAN MODELING AND SPAM FILTERING

There are, it's estimated, a few hundred billion email messages sent on earth per day. Of those, a significant fraction of those messages are unsolicited, unwanted bulk messages—that is, what's commonly known as *spam*. Somewhere between 50% and 95% of emails are currently spam. (It's hard to be precise; statistics and definitions of spam vary, and there's change over time as certain spammers are shut down, or not.)

See statistics on email and spam produced by the Radicati Group, for example: `www.radicati.com`.

The basic idea of a *spam filter* is to estimate the probability that a particular message m is spam. The email client, or possibly the individual user, can choose a threshold p; a message m for which $\text{Pr}\left[m \text{ is spam}\right] \geq p$ is placed into a spam folder. The choice of p depends on the user's relative concern about false positives (nonspam messages that end up being incorrectly treated as spam) versus false negatives (spam messages that end up being incorrectly left in the inbox). So, how might a spam filter actually make its decisions? Here's one approach, based fundamentally on Bayes' Rule. Consider a message consisting of words $w_1, w_2, \ldots, w_n$; we must compute $\text{Pr}\left[\text{spam}|w_1, w_2, \ldots w_n\right]$. Using Bayes' Rule, we turn around this probability:

$$\text{Pr}\left[\text{spam}|w_1, w_2, \ldots w_n\right] = \frac{\text{Pr}\left[w_1, w_2, \ldots w_n|\text{spam}\right] \cdot \text{Pr}\left[\text{spam}\right]}{\text{Pr}\left[w_1, w_2, \ldots w_n\right]}$$

And, by the law of total probability (every message is either spam or not spam), we can further rewrite this probability as

$$\frac{\text{Pr}\left[w_1, w_2, \ldots w_n|\text{spam}\right] \cdot \text{Pr}\left[\text{spam}\right]}{\text{Pr}\left[w_1, w_2, \ldots w_n|\text{spam}\right] \text{Pr}\left[\text{spam}\right] + \text{Pr}\left[w_1, w_2, \ldots w_n|\text{not spam}\right] \text{Pr}\left[\text{not spam}\right]}.$$

That is, we want to know: what is the probability that the sequence of words $w_1, \ldots, w_n$ would have been generated in a spam message, relative to the probability that $w_1, \ldots, w_n$ would have been generated in a spam or nonspam message? (These "relative probabilities" are weighted by the background probability of spam-vs.-nonspam messages.)

It's a good test of your probabilistic intuition to ask: supposing that we have a spam filter that correctly classifies 90% of email messages as spam/nonspam, and 95% of email messages are spam, what fraction of email in your inbox is nonspam? The answer, by Bayes' Rule:

$$\text{Pr}\left[\text{nonspam}|\text{inbox}\right]$$
$$= \frac{\text{Pr}\left[\text{inbox}|\text{nonspam}\right] \text{Pr}\left[\text{nonspam}\right]}{\left(\begin{array}{l}\text{Pr}\left[\text{inbox}|\text{nonspam}\right] \text{Pr}\left[\text{nonspam}\right] \\ + \text{Pr}\left[\text{inbox}|\text{spam}\right] \text{Pr}\left[\text{spam}\right]\end{array}\right)}$$
$$= \frac{0.9 \cdot 0.05}{0.9 \cdot 0.05 + 0.1 \cdot 0.95}$$
$$= \frac{0.045}{0.045 + .095}$$
$$= 0.3214 \cdots.$$

In other words, a full two thirds of messages in your inbox would be spam!

A *naïve Bayes classifier* uses an additional assumption: that the appearance of every word in an email is an independent event. That is, we're going to estimate $\text{Pr}\left[w_1, w_2, \ldots w_n\right]$ as if the probability of each w_i appearing does not depend on any other word appearing. (Obviously that assumption isn't right: the probability of the word `MORTGAGE` appearing is *not* independent of the probability of the word `RATE` appearing, in either spam or nonspam.)

$$\text{Pr}\left[w_1, w_2, \ldots w_n|\text{spam}\right] \approx \text{Pr}\left[w_1|\text{spam}\right] \cdot \text{Pr}\left[w_2|\text{spam}\right] \cdot \cdots \cdot \text{Pr}\left[w_n|\text{spam}\right].$$

Thus a naïve Bayes classifier estimates the probability of a message being generated as spam by multiplying a measure of "how spammy" each word is. A spam filter would still need to have two numbers associated with each word w_i—namely $\text{Pr}\left[w_i|\text{spam}\right]$ and $\text{Pr}\left[w_i|\text{nonspam}\right]$. We can estimate these numbers from a *training set* of spam/nonspam emails, with some sort of "smoothing" mechanism to improve our estimate of the spamminess of a word that doesn't appear in any of the training emails.[7]

For more about the training of these estimates, and about *text classification*—the broader version of the problem that we're trying to solve in spam filtering—again see:

[7] Daniel Jurafsky and James H. Martin. *Speech and Language Processing: An Introduction to Natural Language Processing, Computational Linguistics, and Speech Recognition*. Pearson Prentice Hall, 2nd edition, 2008.

10.3.4 Exercises

Choose one of the 12 months of the year uniformly at random. (That is, choose a number uniformly from the set $\{1,2,\ldots,12\}$.) Indicate whether the following pairs of events are independent or dependent. Justify your answers.

10.53 "The month number is even" and "the month number is divisible by 3."

10.54 "The month number is even" and "the month number is divisible by 5."

10.55 "The month number is even" and "the month number is divisible by 6."

10.56 "The month number is even" and "the month number is divisible by 7."

We flip a fair coin 6 times. Which of these events are independent or dependent? Justify your answers.

10.57 "The number of heads is even" and "the number of heads is divisible by 3."

10.58 "The number of heads is even" and "the number of heads is divisible by 4."

10.59 "The number of heads is even" and "the number of heads is divisible by 5."

10.60 We flip three fair coins, called a, b, and c. Are the events "The number of heads in $\{a,b\}$ is odd" and "The number of heads in $\{b,c\}$ is odd" independent or dependent?

10.61 How (if at all) would your answer to the previous exercise change if the three coins are p-biased? (That is, assume $\Pr[a = H]$, $\Pr[b = H]$, and $\Pr[c = H]$ are all equal to p.)

Consider the list of words and the events in Figure 10.21. Choose a word at random from this list. Which of these pairs of events are independent? For the pairs that are dependent, indicate whether the events are positively or negatively correlated. Justify your answers.

10.62 A and B

10.63 A and C

10.64 B and C

10.65 A and D

10.66 A and E

10.67 $A \cap B$ and E

10.68 $A \cap C$ and E

10.69 $A \cap D$ and E

Let A and B be arbitrary events in a finite sample space.

10.70 Prove that if $\Pr[B] = 0$, then A and B are independent.

10.71 Prove that A and B are independent if and only if A and $\overline{B}$ are independent.

A substitution cipher *(see Example 10.16) is a simple cryptographic scheme in which we choose a permutation π of the alphabet, and replace each letter i with π_i. (Decryption is the same process, but backward: replace π_i by i.) However, substitution ciphers are susceptible to* frequency analysis, *in which an eavesdropper who observes the encrypted message (the* ciphertext*) infers that the most common letter in the ciphertext probably corresponds to the most common letter in English text (the letter* `E`*), the second-most common to the second-most common (*`T`*), and so on.*

10.72 *(programming required)* Write a program that generates a random permutation π of the alphabet, and encrypts a given input text using π. (Leave all non-alphabetic characters unchanged.)

10.73 *(programming required)* Write a program that takes a text as input, converts it to upper case, and produces as output a vector $\langle f_A, f_B, \ldots, f_Z \rangle$, where $f_\bullet$ is the fraction of letters in the input text that are the letter $\bullet$. (So f will be a probability distribution over the alphabet.)

10.74 *(programming required)* Write a program that, given a reference text and a text encrypted with an unknown substitution cipher, attempts to decrypt by mapping the most common encrypted letters, in order, to the most common reference letters. You can find useful reference files—for example, the complete works of Shakespeare—from Project Gutenberg, `http://www.gutenberg.org/`.

A Caesar cipher *is a special kind of substitution cipher in which the permutation π is generated by choosing a numerical* shift *s and moving all letters s steps forward in the alphabet, wrapping back to the beginning of the alphabet as necessary. (For example, with a shift of 5,* `A` $\to$ `F` *and* `W` $\to$ `B`*.)*

10.75 *(programming required)* Write a Caesar cipher encryption program that encrypts a given input text file with a randomly chosen shift in $\{0,1,\ldots,25\}$.

10.76 *(programming required)* If you run your decryption program from Exercise 10.74 on Caesar-ciphered text, you'll find that your program generally doesn't work perfectly. Write a Caesar-cipher-decrypting program that takes advantage of the fact that every letter is shifted by the same amount. Find the most probable s—the s that minimizes the difference in the probabilities of each letter from the reference text and the deciphered text. That is, minimize $\sum_i |f'_i - f_{i+s}|$, where f comes from the ciphertext and f' comes from the reference text.

`ABIDES`
`BASES`
`CAJOLED`
`DATIVE`
`EXUDE`
`FEDORA`
`GASOLINES`
`HABANERO`

(a) Some words.

A : "the first letter of the word is a consonant."
B : "the second letter of the word is a consonant."
C : "the second letter of the word is a vowel."
D : "the last letter of the word is a consonant."
E : "the word has even length."

(b) Some events.

Figure 10.21: A word list from which we choose a random word, and some events.

Flip n fair coins. For any two distinct indices i and j with $1 \leq i < j \leq n$, define the event $A_{i,j}$ as

$$A_{i,j} := \text{(the ith coin flip came up heads) XOR (the jth coin flip came up heads).}$$

For example, for $n = 4$ and the outcome $\langle T, T, H, H \rangle$, the events $A_{1,3}$, $A_{1,4}$, $A_{2,3}$, and $A_{2,4}$ all occur; $A_{1,2}$ and $A_{3,4}$ do not. Thus, from n independent coin flips, we've defined $\Omega(n^2)$ different events—$\binom{n}{2}$, to be specific. In the next few exercises, you'll show that these $\binom{n}{2}$ events are pairwise independent, *but not* fully independent.

10.77 Let i and $j > i$ be arbitrary. Show that $\Pr[A_{i,j}] = \frac{1}{2}$.

10.78 Let i and $j > i$ be arbitrary, and let i' and $j' > i'$ be arbitrary. Show that any two distinct events $A_{i,j}$ and $A_{i',j'}$ are independent. That is, show that $\Pr\left[A_{i,j} | A_{i',j'}\right] = \Pr\left[A_{i,j} | \overline{A_{i',j'}}\right] = \frac{1}{2}$ if $\{i,j\} \neq \{i',j'\}$.

10.79 Show that there is a set of three distinct A events that are *not* mutually independent. That is, identify three events $A_{i,j}$, $A_{i',j'}$, and $A_{i'',j''}$ where the sets $\{i,j\}$, $\{i',j'\}$, and $\{i'',j''\}$ are all different (though not necessarily disjoint). Then show that if you know the value of $A_{i,j}$ and $A_{i',j'}$, the probability of $A_{i'',j''} \neq \frac{1}{2}$.

Suppose that you have the dominoes in Figure 10.22, and you shuffle them and draw one domino uniformly at random. (More specifically, you choose any particular domino with probability $\frac{1}{12}$. After you've chosen the domino, you choose an orientation, with a 50–50 chance of either side pointing to the left.) What are the following conditional probabilities? ("Even total" means that the sum of the two halves of the domino is even. "Doubles" means that the two halves are the same.)

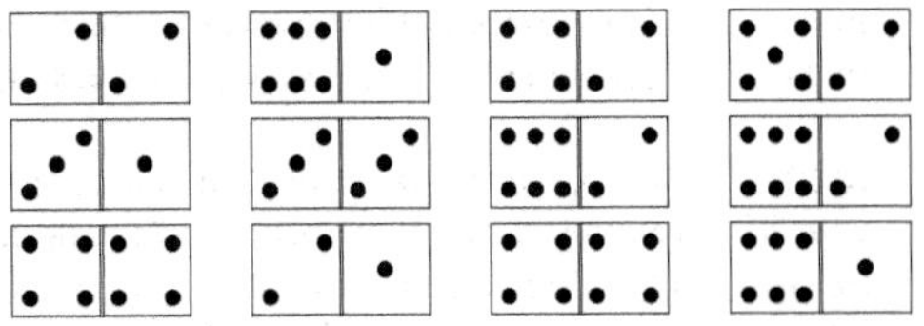

Figure 10.22: Some dominoes.

10.80 $\Pr$ [even total|doubles]

10.81 $\Pr$ [doubles|even total]

10.82 $\Pr$ [doubles|at least one ⚃]

10.83 $\Pr$ [at least one ⚃|doubles]

10.84 $\Pr$ [total ≥ 7|doubles]

10.85 $\Pr$ [doubles|total ≥ 7]

10.86 $\Pr$ [even total|total ≥ 7]

10.87 $\Pr$ [doubles|*left half* of drawn domino is ⚂]

10.88 Suppose A and B are mutually exclusive events—that is, $A \cap B = \varnothing$. Prove or disprove the following claim: A and B cannot be independent.

10.89 Let A and B be two events such that $\Pr[A|B] = \Pr[B|A]$. Which of the following is true? (a) A and B must be independent; (b) A and B must *not* be independent; or (c) A and B may or may not be independent (there's not enough information to tell). Justify your answer briefly.

Suppose, as we have done throughout the chapter, that $h : K \to \{1, \ldots, n\}$ is a random hash function.

10.90 Suppose that there are currently k cells in the array that are occupied. Consider a key $x \in K$ not currently stored in the hash table. What is the probability that the cell $h(x)$ into which x hashes is empty?

10.91 Suppose that you insert n distinct values $x_1, x_2, \ldots, x_n$ into an initially empty n-slot hash table. What is the probability that there are no collisions? *(Hint: if the first i elements have had no collisions, what is the probability that the $(i+1)$st hashed element does not cause a collision? Use Theorem 10.2 and Exercise 10.90.)*

There's a disease BCF ("base-case failure") that afflicts a small but very unfortunate fraction of the population. One in a thousand people in the population have BCF. Explain your answers to the following questions:

10.92 Doctor Genius has invented a BCF-detection test. Her test, though, isn't perfect:

- it has *false negatives*: if you do have BCF, then her test says that you're not sick with probability 0.01.
- it has *false positives*: if you don't have BCF, then her test says that you're sick with probability 0.03.

What is the probability p that Dr. Genius gives a random person x an erroneous diagnosis?

10.93 "Doctor" Quack has invented a BCF-detection test, too. He was a little confused by the statement "one in a thousand people in the population have BCF," so his test is this: no matter who the patient is, with probability $\frac{1}{1000}$ report "sick" and with probability $\frac{999}{1000}$ report "not sick." What is p now?

Alice wishes to send a 3-bit message 011 *to Bob, over a noisy channel that corrupts (flips) each transmitted bit independently with some probability. To combat the possibility of her transmitted message differing from the received message, she adds a parity bit to the end of her message (so that the transmitted message is* 0110*). [Bob checks that he receives a message with an even number of* 1*s, and if so interprets the first three received bits as the message.]*

10.94 Assume that each bit is flipped with probability 1%. Conditioned on receiving a message with an even number of 1s, what is the probability that the message Bob received is the message that Alice sent?

10.95 What if the probability of error is 10% per bit?

Suppose, as in Example 10.28, I have two coins—one fair and one p-biased. I pull one uniformly at random from an opaque bag, and flip it. What is $\Pr$ *[I pulled the biased coin|the following observed flips]? Justify your answers.*

10.96 $p = \frac{2}{3}$, and I observe a single Heads flip.

10.97 $p = \frac{3}{4}$, and I observe the flip sequence HHHT.

10.98 $p = \frac{3}{4}$, and I observe the flip sequence HTTTHT.

A Bloom filter *is a probabilistic data structure designed to store a set of elements from a universe U, allowing very quick query operations to determine whether a particular element has been stored.*[8] *Specifically, it supports the operations* **Insert**(x), *which adds x to the stored set, and* **Lookup**(x), *which reports whether x was previously stored. But, unlike most data structures for this problem, we will allow ourselves to (occasionally) make mistakes in lookups, in exchange for making these operations fast.*

Here's how a Bloom filter works. We will choose k different hash functions $h_1, \ldots, h_k : U \rightarrow \{1, \ldots, m\}$, and we will maintain an array of m bits, all initially set to zero. The operations are implemented as follows:

- *To insert x into the data structure, we set the k slots $h_1(x), h_2(x), \ldots, h_k(x)$ of the array to 1. (If any of these slots was already set to 1, we leave it as a 1.)*
- *To look up x in the data structure, we check that the k slots $h_1(x), h_2(x), \ldots, h_k(x)$ of the array are all set to 1. If they're all 1s, we report "yes"; if any one of them is a 0, we report "no."*

For an example, see Figure 10.98. Note that there can be a false positive *in a lookup: if all k slots corresponding to a query element x happen to have been set to 1 because of other insertions, then x will incorrectly be reported to be present.*

As usual, we treat each of the k hash functions as independently assigning each element of U to a uniformly chosen slot of the array. Suppose that we have an m-slot Bloom filter, with k independent hash functions, and we insert n elements into the data structure.

10.99 Suppose we have $k = 1$ hash functions, and we've inserted $n = 1$ element into the Bloom filter. Consider any particular slot of the m-slot table. What is the probability that this particular slot is still set to 0? (That is, what is the probability that this slot is *not* the slot set to 1 when the single element was inserted?)

10.100 Let the number k of hash functions be an arbitrary number $k \geq 1$, but continue to suppose that we've inserted only $n = 1$ element in the Bloom filter. What is the probability a particular slot is still set to 0 after this insertion?

10.101 Let the number k of hash functions be an arbitrary number $k \geq 1$, and suppose that we've inserted an arbitrary number $n \geq 1$ of elements into the Bloom filter. What is the probability a particular slot is still set to 0 after these insertions?

1	2	3	4	5	6	7	8	9	10	11	12	13
0	0	0	0	0	0	0	0	0	0	0	0	0
0	0	0	1	0	0	0	0	0	1	0	0	0
0	0	0	1	0	0	0	1	0	1	1	0	0

(a) The table initially; after inserting 3; and after inserting 7. Note $h_1(3) = 4$, $h_2(3) = 10$, $h_1(7) = 8$, and $h_2(7) = 11$.

1	2	3	4	5	6	7	8	9	10	11	12	13
0	0	0	1	0	0	0	1	0	1	1	0	0
0	0	0	1	0	0	0	1	0	1	1	0	0
0	0	0	1	0	0	0	1	0	1	1	0	0

(b) Testing for 3 (yes!), 15 (no!), and 10 (yes!?!). Note $h_1(15) = 3$, $h_2(15) = 5$, $h_1(10) = 11$, and $h_2(10) = 10$—so 10 is a *false positive.*

Figure 10.23: An example of a Bloom filter with $k = 2$ hash functions: $h_1(x) = x \bmod 13 + 1$ and $h_2(x) = x^2 \bmod 13 + 1$.

Define the false-positive rate *of a Bloom filter (with m slots, k hash functions, and n inserted elements) to be the probability that we incorrectly report that y is in the table when we query for an uninserted element y.*

For many years (starting with Bloom's original paper about Bloom filters), people in computer science believed that the false positive rate was precisely p^k, where $p = (1 -$ [your answer to Exercise 10.101]). *The justification was the following. Let B_i denote the event "slot $h_i(y)$ is occupied." We have a false positive if and only if $B_1, B_2, \ldots, B_k$ are all true. Thus*

$$\text{the false positive rate} = \Pr[B_1 \text{ and } B_2 \text{ and } \cdots \text{ and } B_k].$$

You showed in the previous exercise that $\Pr[B_i] = p$. **Everything up until here is correct; the next step in the argument, however, was not!** *Therefore, because the B_i events are independent,*

$$\text{the false positive rate} = \Pr[B_1 \text{ and } B_2 \text{ and } \cdots \text{ and } B_k] = \Pr[B_1] \cdot \Pr[B_2] \cdots \Pr[B_k] = p^k.$$

But it turns out that B_i and B_j are not *independent!*[9] *(This error is a prime example of how hard it is to have perfect intuition about probability!)*

10.102 Let $m = 2$, $k = 2$, and $n = 1$. Compute *by hand* the false-positive rate. *(Hint: there are "only" 16 different outcomes, each of which is equally likely: the random hash functions assign values in $\{1, 2\}$ to $h_1(x)$, $h_2(x)$, $h_1(y)$, and $h_2(y)$. In each of these 16 cases, determine whether a false positive occurred.)*

10.103 Compute p^2—the answer you would have gotten by using

$$\text{false-positive rate} = (1 - [\text{your answer to Exercise 10.101}])^2.$$

Which is bigger—p^2 or [your answer to Exercise 10.102]? In approximately one paragraph, explain the difference, including an explanation of *why* the events B_1 and B_2 are not independent.

10.104 While the actual false-positive rate is not exactly p^k, it turns out that p^k is a very good approximation to the false-positive rate as long as m is sufficiently big and k is sufficiently small. Write a program that creates a Bloom filter with $m = 1{,}000{,}000$ slots and $k = 20$ hash functions. Insert $n = 100{,}000$ elements, and estimate the false positive probability by querying for n additional uninserted elements $y \notin X$. What is the false-positive rate *that you observe* in your experiment? How does it compare to p^k?

[8] Burton H. Bloom. Space/time trade-offs in hash coding with allowable errors. *Communications of the ACM*, 13(7):422–426, July 1970.

[9] Prosenjit Bose, Hua Guo, Evangelos Kranakis, Anil Maheshwari, Pat Morin, Jason Morrison, Michiel Smid, and Yihui Tang. On the false-positive rate of Bloom filters. *Information Processing Letters*, 108(4):210–213, 2008; and Ken Christensen, Allen Roginsky, and Miguel Jimeno. A new analysis of the false positive rate of a Bloom filter. *Information Processing Letters*, 110:944–949, 2010.

10.4 *Random Variables and Expectation*

> Acts of sacrifice, charity and penance are not to be given up but should be performed. . . . All these activities should be performed without any expectation of result.
>
> *Bhagavad Gita* 18:5–6

Thus far, we have been considering *whether or not* something occurs—that is, using the language of probability, we have been interested in *events*. But often we will also be interested in *how many?* questions and not just *did it or did it not?* questions. How many heads came up in 1000 coin flips? How many times do we have to flip a coin before it comes up heads for the 1000th time? For a randomly ordered array $A[1 \ldots n]$ of the integers $\{1, \ldots, n\}$, for how many indices i is $A[i] < A[i+1]$? To address these types of questions, we will introduce the concept of a *random variable*, which measures some numerical quantity that varies from outcome to outcome. We will also consider the *expectation* of a random variable, which is the value of that variable averaged over all of the outcomes in the sample space.

Warning! A "random variable" is one of the worst-named concepts in this entire book. A random variable is not a variable—rather, it's a *function* that maps each outcome to a numerical value. But everyone calls it a random variable, so that's what we'll call it, too.

10.4.1 *Random Variables*

We begin with the definition of a random variable itself:

Definition 10.11 (Random variable)
A random variable X assigns a numerical value to every outcome in the sample space S. (In other words, a random variable is a function $X : S \to \mathbb{R}$.)

Here are a few simple examples:

Example 10.29 (Counting heads in 3 flips)
Suppose that we flip a fair coin independently, three times. (Then the sample space is $S = \{H, T\}^3$, and $\Pr[x] = \frac{1}{8}$ for any $x \in S$.) Define the random variables

$$X = \text{the number of heads}$$
$$Y = \text{the number of initial consecutive tails.}$$

These random variables take on the following values:

$$\begin{aligned}
X(HHH) &= 3 & Y(HHH) &= 0\\
X(HHT) &= 2 & Y(HHT) &= 0\\
X(HTH) &= 2 & Y(HTH) &= 0\\
X(HTT) &= 1 & Y(HTT) &= 0\\
X(THH) &= 2 & Y(THH) &= 1\\
X(THT) &= 1 & Y(THT) &= 1\\
X(TTH) &= 1 & Y(TTH) &= 2\\
X(TTT) &= 0 & Y(TTT) &= 3.
\end{aligned}$$

Example 10.30 (Word length, and number of vowels)
Select a word from the sample space {`Now`, `is`, `the`, `winter`, `of`, `our`, `discontent`} by choosing word w with probability proportional to the number of letters in w, as in Example 10.5. Define a random variable L to denote the number of letters in the word chosen. Thus L(`discontent`) = 10 and L(`winter`) = 6, for example. We can also define a random variable V to denote the number of *vowels* in the word chosen. Thus V(`discontent`) = 3 and V(`winter`) = 2, for example. Here are the values for these two random variables for each outcome in the sample space:

w	Pr $[w]$	$L(w)$	$V(w)$
`Now`	3/29	3	1
`is`	2/29	2	1
`the`	3/29	3	1
`winter`	6/29	6	2
`of`	2/29	2	1
`our`	3/29	3	2
`discontent`	10/29	10	3

Although it's an abuse of notation, often we just write X to denote the value of a random variable X *for a realization chosen according to the probability distribution* `Pr`. (So we might write "$X = 3$ with probability $\frac{1}{8}$" or "there are L letters in the chosen word.")

We can state probability questions about events based on random variables, as the following example illustrates:

Example 10.31 (More word length and vowel counts)
Choose a word as in Example 10.30. Define L as the number of letters in the word, and define V as the number of vowels in the word. Then Pr $[L = 3]$ denotes the probability that we choose an outcome w for which $L(w) = 3$. (In other words, $L = 3$ denotes the event $\{w : L(w) = 3\}$.) Thus (see the table in Example 10.30)

$$\begin{aligned} \Pr[L = 3] &= \Pr[\{\texttt{Now}, \texttt{the}, \texttt{our}\}] = \tfrac{9}{29} \\ \Pr[V = 3] &= \Pr[\{\texttt{discontent}\}] = \tfrac{10}{29}. \end{aligned}$$

We will also abuse notation by performing arithmetic on random variables (remember, these are functions!): for two random variables X and Y, we write $X + Y$ as a new random variable that, for any outcome x, denotes the sum of $X(x)$ and $Y(x)$. We will interpret similarly any other arithmetic expression that involves random variables. (The notational analogue here is writing "sin + cos" to denote the function $f(x) = \sin(x) + \cos(x)$.) Here's a small example:

Example 10.32 (Number of consonants)
We can express the number of consonants in the randomly chosen word from our running example (see Example 10.30) as $L - V$. For example, $L - V = 1$ when the chosen word is `our`, and $L - V = 4$ when the chosen word is `winter`.

Indicator random variables

One special type of random variable that will come up frequently is an *indicator random variable,* which only takes on the values 0 and 1. (Such a random variable "indicates" whether a particular event has occurred.) Here's a simple example:

Example 10.33 (Indicator random variables in coin flips)
Suppose that we flip three fair coins independently. Let X_1 be an indicator random variable that reports whether the first flip came up heads. Similarly, let X_2 and X_3 be indicator random variables for the second and third flips. Then:

outcome	X_1	X_2	X_3
HHH	1	1	1
HHT	1	1	0
HTH	1	0	1
HTT	1	0	0
THH	0	1	1
THT	0	1	0
TTH	0	0	1
TTT	0	0	0

Note that the *total number of heads* is given by the random variable $X_1 + X_2 + X_3$.

Independence of random variables

Just as with independence for events, we will often be concerned with whether knowing the value of one random variable tells us something about the value of another. Two random variables X and Y are *independent* if every two events of the form "$X = x$" and "$Y = y$" are independent: for every value x and y, it must be the case that $\Pr[X = x \text{ and } Y = y] = \Pr[X = x] \cdot \Pr[Y = y]$. For example:

Example 10.34 (Some independent/dependent random variables)
The random variables X_2 and X_3 from Example 10.33—we flip 3 fair coins independently; X_2 and X_3 indicate whether the 2nd and 3rd flips are heads—are independent. You can check all four possibilities; for example,

$$\Pr[X_2 = 1 \text{ and } X_3 = 1] = \tfrac{1}{4} = \tfrac{1}{2} \cdot \tfrac{1}{2} = \Pr[X_2 = 1] \cdot \Pr[X_3 = 1] \text{ and}$$
$$\Pr[X_2 = 1 \text{ and } X_3 = 0] = \tfrac{1}{4} = \tfrac{1}{2} \cdot \tfrac{1}{2} = \Pr[X_2 = 1] \cdot \Pr[X_3 = 0].$$

On the other hand, the random variables X and Y from Example 10.29—we flip 3 fair coins independently; X is the number of heads and Y is the number of consecutive initial tails—are not independent; for example,

$$\Pr[X = 3] \cdot \Pr[Y = 3] = \tfrac{1}{8} \cdot \tfrac{1}{8} \qquad \text{but} \qquad \Pr[X = 3 \text{ and } Y = 3] = 0.$$

10.4.2 *Expectation*

A random variable X measures a numerical quantity that varies from realization to realization. We will often be interested in the "average" value of X, which is otherwise known as the random variable's *expectation:*

Definition 10.12 (Expectation)
The expectation *of a random variable X, denoted $\mathrm{E}[X]$, is the average value of X, defined as*

$$\mathrm{E}[X] = \sum_{x \in S} X(x) \cdot \Pr[x].$$

The expectation of X is also sometimes called the mean *of X.*
We can equivalently write $\mathrm{E}[X] = \sum_y (y \cdot \Pr[X = y])$ by summing over each possible value y that X can take on, rather than by summing over outcomes.

The alternate version of the summation for expectation in Definition 10.12 follows by collecting together each outcome x that has the same value of the random variable $X(x)$:

$$\begin{aligned}&\sum_{x \in S} X(x) \cdot \Pr[x]\\ &= \sum_{y \in \mathbb{R}} \sum_{\substack{x \in S:\\ X(x)=y}} y \cdot \Pr[x]\\ &= \sum_{y \in \mathbb{R}} y \cdot \sum_{\substack{x \in S:\\ X(x)=y}} \Pr[x]\\ &= \sum_{y \in \mathbb{R}} y \cdot \Pr[X = y].\end{aligned}$$

In other words, $\mathrm{E}[X]$ is the average value of X over all outcomes (where the average is weighted, with weights defined by the probability function). For example:

Example 10.35 (Expectation of a Bernoulli random variable)
Let X be an indicator random variable for a Bernoulli trial with parameter p—that is, $X = 1$ with probability p and $X = 0$ with probability $1 - p$. Then $\mathrm{E}[X]$ is precisely

$$\begin{aligned}\mathrm{E}[X] &= 1 \cdot \Pr[X = 1] + 0 \cdot \Pr[X = 0] && \textit{definition of expectation (alternative version)}\\ &= 1 \cdot p + 0 \cdot (1 - p) && \textit{definition of a Bernoulli trial with parameter } p\\ &= p.\end{aligned}$$

Example 10.36 (Counting heads in 3 flips, again)
Problem: Recall Example 10.29, where the random variable X denotes the number of heads in three independent flips of a fair coin. (The sample space was $S = \{H, T\}^3$, and $\Pr[x] = \frac{1}{8}$ for any $x \in S$.) What is $\mathrm{E}[X]$?

Solution: The expectation of X is

$$\begin{aligned}\mathrm{E}[X] &= \sum_{x \in \{H,T\}^3} \Pr[x] \cdot X(x)\\ &= \tfrac{1}{8}X(HHH) + \tfrac{1}{8}X(HHT) + \tfrac{1}{8}X(HTH) + \tfrac{1}{8}X(HTT)\\ &\quad + \tfrac{1}{8}X(THH) + \tfrac{1}{8}X(THT) + \tfrac{1}{8}X(TTH) + \tfrac{1}{8}X(TTT)\\ &= \tfrac{1}{8} \cdot [3 + 2 + 2 + 1 + 2 + 1 + 1 + 0]\\ &= \tfrac{12}{8} = 1.5.\end{aligned}$$

In other words, in three flips of a fair coin, we expect 1.5 flips to come up Heads.

Warning! Just because $\mathrm{E}[X] = 1.5$ doesn't mean that $\Pr[X = 1.5]$ is big! (If you ever flip three fair coins and see exactly 1.5 heads, it might be a sign that the world is ending.) Remember that "average" and "typical" aren't the same thing!

Example 10.37 (Counting letters and vowels, again)

Recall the probabilistic process of choosing a word from the sentence `Now is the winter of our discontent` in proportion to word length. Recall also the random variables from Example 10.30: L denotes the chosen word's length, and V the number of vowels in the chosen word. (See Figure 10.24 for a reminder.) Then we have

$$\begin{aligned} \mathrm{E}[L] &= 3 \cdot \tfrac{3}{29} + 2 \cdot \tfrac{2}{29} + 3 \cdot \tfrac{3}{29} + 6 \cdot \tfrac{6}{29} + 2 \cdot \tfrac{2}{29} + 3 \cdot \tfrac{3}{29} + 10 \cdot \tfrac{10}{29} \\ &= \tfrac{171}{29} \\ &\approx 5.8966. \end{aligned}$$

$$\begin{aligned} \mathrm{E}[V] &= 1 \cdot \tfrac{3}{29} + 1 \cdot \tfrac{2}{29} + 1 \cdot \tfrac{3}{29} + 2 \cdot \tfrac{6}{29} + 1 \cdot \tfrac{2}{29} + 2 \cdot \tfrac{3}{29} + 3 \cdot \tfrac{10}{29} \\ &= \tfrac{57}{29} \\ &\approx 1.9656. \end{aligned}$$

outcome	Pr	L	V
`Now`	$\frac{3}{29}$	3	1
`is`	$\frac{2}{29}$	2	1
`the`	$\frac{3}{29}$	3	1
`winter`	$\frac{6}{29}$	6	2
`of`	$\frac{2}{29}$	2	1
`our`	$\frac{3}{29}$	3	2
`discontent`	$\frac{10}{29}$	10	3

Figure 10.24: A reminder of the sample space, probabilities, and random variables for Example 10.37.

Taking it further: If we think about it without a great deal of care, there's something apparently curious about the result from Example 10.37. We've plopped down our thumb on a random letter in the sentence `Now is the winter of our discontent`, and we've computed that the word that our thumb lands on has an average length of about 5.9 letters. That seems a little puzzling, because there are 7 words in the sentence, with an average word length of $\frac{29}{7} = 4.1428$ letters. But there's a good reason for this discrepancy: *longer words are more likely to be chosen* because they have more letters, and therefore the average word that's chosen has more letters than average. An analogous phenomenon occurs in many other settings, too. When you're driving, you spend most of your time on longer-than-average trips. Most people in Canada live in a larger-than-average-sized Canadian city. Most 3rd-grade students in California are in a larger-than-average-size 3rd-grade class. (In fact, this broader phenomenon is sometimes called the *class-size paradox*.) Perhaps even more jarringly, a random person x knows fewer people than the average number of people known by someone x knows—that is, on average, your friends are more popular than you are.[10] (Why? A very popular person—call her Oprah—is, by definition, the friend of many people, and therefore Oprah's astronomical popularity is averaged into the popularity of many people x. In computing the popularity of a randomly chosen person x, Oprah only contributes her popularity once for x = Oprah—but she contributes it many times to the popularity of x's friends.)

This phenomenon may illustrate an example of a *sampling bias*, in which we try to draw a uniform sample from a population but we end up with some kind of bias that overweights some members of the population at the expense of others. Sampling biases are a widespread concern in any statistical approach to understanding a population. For example, consider a telephone-based political poll that collects voters' preferences for candidates one evening by randomly dialing phone numbers until somebody answers, and records the answerer's preference. This poll will overweight those people who are sitting around at home during the evening—which correlates with the voter's age, which correlates with the voter's political affiliation.

[10] Scott L. Feld. Why your friends have more friends than you do. *American Journal of Sociology*, 96(6):1464–1477, May 1991.

Example 10.38 (Number of aces in a bridge hand)

Problem: Suppose that we are dealt a 13-card hand from a standard 52-card deck. What is the expected number of aces in our hand?

Solution: Later we'll solve this problem more easily (see Example 10.41), but here we'll do it the hard way. We'll compute the probability of getting $0, 1, \ldots, 4$ aces:

- There are $\binom{52}{13}$ different hands.
- There are $\binom{4}{k} \cdot \binom{48}{13-k}$ hands with exactly k aces. (We have to pick k ace cards from the 4 aces in the deck, and $13 - k$ non-ace cards from the 48 non-aces.)

Because each hand is equally likely to be chosen, therefore

$$\text{Pr}\left[\text{drawing exactly } k \text{ aces}\right] = \frac{\binom{4}{k} \cdot \binom{48}{13-k}}{\binom{52}{13}}.$$

And thus, letting A be a random variable denoting the number of aces, we have

$$\begin{aligned}
\text{E}[A] &= \sum_{h} \text{Pr}\left[\text{being dealt hand } h\right] \cdot (\text{number of aces in } h) \\
&= \sum_{i=0}^{4} i \cdot \text{Pr}[A = i] \qquad \textit{(reordering sum by collecting all hands with the same number of aces)} \\
&= \frac{\overbrace{0 \cdot \binom{4}{0} \cdot \binom{48}{13}}^{0 \cdot \text{Pr}[A=0]} + \overbrace{1 \cdot \binom{4}{1} \cdot \binom{48}{12}}^{1 \cdot \text{Pr}[A=1]} + \overbrace{2 \cdot \binom{4}{2} \cdot \binom{48}{11}}^{2 \cdot \text{Pr}[A=2]} + \overbrace{3 \cdot \binom{4}{3} \cdot \binom{48}{10}}^{3 \cdot \text{Pr}[A=3]} + \overbrace{4 \cdot \binom{4}{4} \cdot \binom{48}{9}}^{4 \cdot \text{Pr}[A=4]}}{\binom{52}{13}} \\
&= \frac{0 \cdot 1 \cdot \binom{48}{13} + 1 \cdot 4 \cdot \binom{48}{12} + 2 \cdot 6 \cdot \binom{48}{11} + 3 \cdot 4 \cdot \binom{48}{10} + 4 \cdot 1 \cdot \binom{48}{9}}{\binom{52}{13}} \\
&= \frac{0 + 278{,}674{,}137{,}872 + 271{,}142{,}404{,}416 + 78{,}488{,}590{,}752 + 6{,}708{,}426{,}560}{635{,}013{,}559{,}600} \\
&= \frac{635{,}013{,}559{,}600}{635{,}013{,}559{,}600} \\
&= 1.
\end{aligned}$$

That is, the expected number of aces in a 13-card hand is precisely 1.

A USEFUL PROPERTY OF EXPECTATION

We've now seen several examples of computing the expectation of random variables by directly following the definition of expectation. Here we'll introduce a transformation that can often make expectation calculations easier, at least for positive integer–valued random variables:

Theorem 10.5 (A new formula for expectation, for nonnegative integers)
Let $X : S \to \mathbb{Z}^{\geq 0}$ be a random variable. Then $\text{E}[X] = \sum_{i=1}^{\infty} \text{Pr}[X \geq i]$.

(Note that by definition $\text{E}[X] = \sum_{i=0}^{\infty} i \cdot \text{Pr}[X = i]$, so we're trading the multiplication of i for the replacement of $=$ by $\geq$.)

The proof will follow by changing the order of summation in the expectation formula. We'll give an algebraic proof in a moment, but it may be easier to follow the idea by looking at a visualization first. See Figure 10.25.

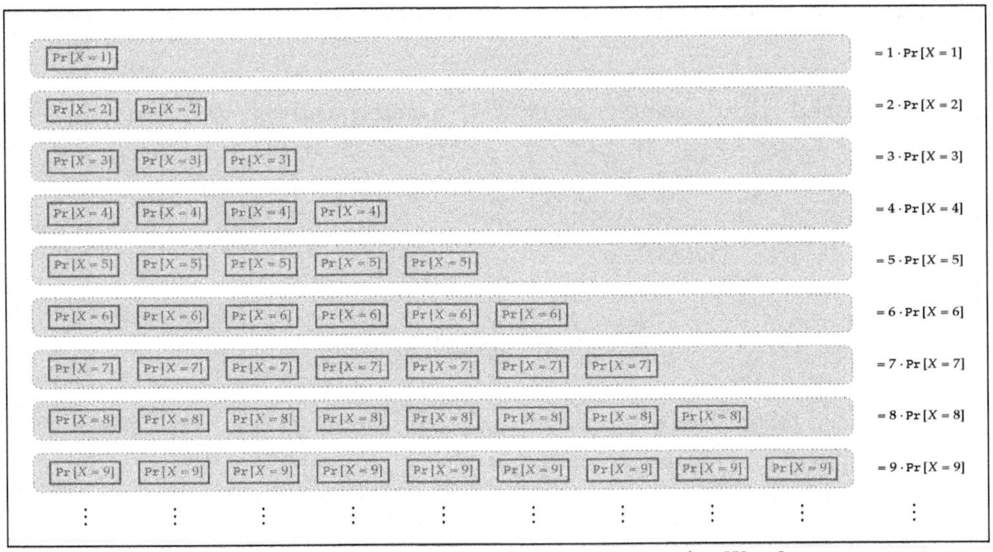

(a) The sum $\sum_{i=0}^{\infty} i \cdot \Pr[X = i]$ can be visualized as adding up i copies of $\Pr[X = i]$ one row at a time ...

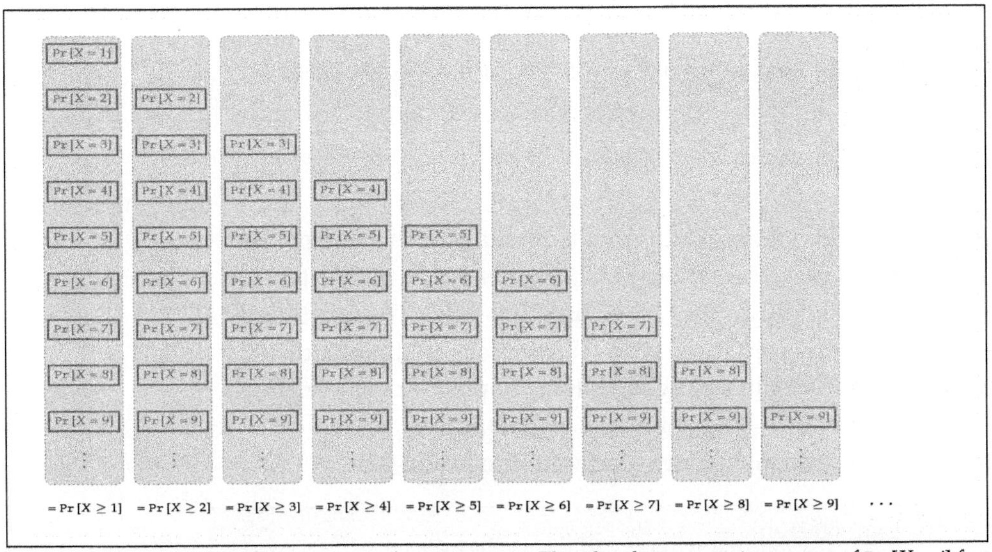

(b) ... or by visualizing adding up one *column* at a time. The jth column contains a copy of $\Pr[X = i]$ for every i greater than or equal to j, so the value of the sum of the jth column is $\Pr[X \geq j]$.

Figure 10.25: A change of summation. View $E[X] = \sum_{i=0}^{\infty} i \cdot \Pr[X = i]$ as the sum of the entries of an infinite table, where the ith row of the table contains i copies of $\Pr[X = i]$. By computing column sums instead of row sums, we see $\sum_{i=0}^{\infty} i \cdot \Pr[X = i] = \sum_{j=1}^{\infty} \Pr[X \geq j]$.

Proof of Theorem 10.5. We proceed using the manipulation from Figure 10.25:

$$
\begin{aligned}
E[X] &= \sum_{i=0}^{\infty} i \cdot \Pr[X = i] && \textit{definition of expectation} \\
&= \sum_{i=0}^{\infty} \sum_{j=1}^{i} \Pr[X = i] && i = \textstyle\sum_{j=1}^{i} 1 \\
&= \sum_{j=1}^{\infty} \sum_{i=j}^{\infty} \Pr[X = i] && \textit{changing the order of summation (see Figure 10.25)} \\
&= \sum_{j=1}^{\infty} \Pr[X \geq j]. \qquad \square && \Pr[X \geq j] = \textstyle\sum_{i=j}^{\infty} \Pr[X = i]
\end{aligned}
$$

We can use this theorem to find the expected value of a geometric random variable:

Example 10.39 (Expectation of a geometric random variable)
Let X be a geometric random variable with parameter p. (That is, X measures the number of flips of a p-biased coin before we get Heads for the first time.) Then $\mathrm{E}[X]$ is precisely $\frac{1}{p}$:

$$\begin{aligned}
\mathrm{E}[X] &= \sum_{i=1}^{\infty} \Pr[X \geq i] && \textit{Theorem 10.5 } (\mathrm{E}[X] = \textstyle\sum_{i=1}^{\infty} \Pr[X \geq i]) \\
&= \sum_{i=1}^{\infty} \Pr\left[\text{fail to get heads in } i-1 \text{ flips}\right] && \textit{definition of geometric random variable} \\
&= \sum_{i=1}^{\infty} (1-p)^{i-1} && \textit{need } i-1 \textit{ consecutive tails flips} \\
&= \sum_{i=0}^{\infty} (1-p)^{i} && \textit{changing index of summation} \\
&= \frac{1}{1-(1-p)} = \frac{1}{p}. && \textit{formula for geometric summations}
\end{aligned}$$

For example, we expect to flip a fair coin (with $p = \frac{1}{2}$) *twice* before we get heads.

10.4.3 *Linearity of Expectation*

Here's a very useful general property of expectation, called *linearity of expectation*: the expectation of a sum is the sum of the expectations. (A *linear function* is a function f that satisfies $f(a+b) = f(a) + f(b)$—for example, $f(x) = 3x$ or $f(x) = 0$.) The usefulness of Linearity of Expectation will come from the way in which it lets us "break down" a complicated random variable into the sum of a collection of simple random variables. (We can then compute $\mathrm{E}\left[\text{Complicated}\right] = \mathrm{E}\left[\sum_i \text{Simple}_i\right] = \sum_i \mathrm{E}\left[\text{Simple}_i\right]$.)

We'll see several useful examples soon, but let's start with the proof:

Theorem 10.6 (Linearity of Expectation)
Consider a sample space S, and let $X : S \to \mathbb{R}$ and $Y : S \to \mathbb{R}$ be any two random variables. Then $\mathrm{E}[X+Y] = \mathrm{E}[X] + \mathrm{E}[Y]$.

Proof. We'll be able to prove this theorem by just invoking the definition of expectation and following our algebraic noses:

$$\begin{aligned}
\mathrm{E}[X+Y] &= \sum_{s \in S} (X+Y)(s) \cdot \Pr[s] && \textit{definition of expectation} \\
&= \sum_{s \in S} \left[X(s) + Y(s)\right] \cdot \Pr[s] && \textit{definition of the random variable } X+Y \\
&= \left[\sum_{s \in S} X(s) \cdot \Pr[s]\right] + \left[\sum_{s \in S} Y(s) \cdot \Pr[s]\right] && \textit{distributing the multiplication; rearranging} \\
&= E[X] + E[Y]. && \textit{definition of expectation}
\end{aligned}$$

Therefore $\mathrm{E}[X+Y] = \mathrm{E}[X] + \mathrm{E}[Y]$, as desired. □

Notice that Theorem 10.6 does *not* impose any requirement of independence on the random variables X and Y: even if X and Y are highly correlated (positively or negatively), we *still* can use linearity of expectation to conclude that $\mathrm{E}[X+Y] = \mathrm{E}[X] + \mathrm{E}[Y]$. There are many apparently complicated problems in which using linearity of expectation makes a solution totally straightforward. Here are a few examples:

Example 10.40 (Expectation of a binomial random variable)

Problem: We have a p-biased coin (that is, $\Pr[\text{heads}] = p$) that we flip 1000 times. What is the expected number of heads that come up in these 1000 flips?

Solution: The intuition is fairly straightforward: a p-fraction of flips are heads, so we should expect $1000p$ heads in 1000 flips. But doing the math requires a bit of work.

An abandoned first attempt: Let's compute the probability that there are exactly k heads in a sequence of 1000 flips, and then apply the definition of expectation directly. There are $\binom{1000}{k}$ sequences of 1000 flips that have exactly k heads, and the probability of any one of these sequences is $p^k(1-p)^{1000-k}$, so

$$\begin{aligned}
&\mathrm{E}[\text{number of heads}] \\
&= \sum_{k=0}^{1000} k \cdot \Pr[\text{number of heads} = k] && \textit{definition of expectation} \\
&= \sum_{k=0}^{1000} k \cdot \binom{1000}{k} \cdot p^k \cdot (1-p)^{1000-k}. && \textit{above analysis of } \Pr[\textit{number of heads} = k]
\end{aligned}$$

We could try to simplify this expression (but it turns out to be pretty hard!). Instead, let's start over with a different approach.

A second try: Here's a strategy that ends up being much easier. Define 1000 random variables $X_1, X_2, \ldots, X_{1000}$, where X_i is the indicator random variable

$$X_i = \begin{cases} 1 & \text{if the } i\text{th flip of the coin comes up Heads} \\ 0 & \text{if the } i\text{th flip of the coin comes up Tails.} \end{cases}$$

The total number of heads in the 1000 coin flips is given by the random variable

$$X = X_1 + X_2 + \cdots + X_{1000}.$$

We can use this definition of X and linearity of expectation to compute the expected number of heads much more easily:

$$\begin{aligned}
\mathrm{E}[\text{number of heads}] = \mathrm{E}[X] = \mathrm{E}\left[\sum_{i=1}^{1000} X_i\right] && \textit{definition of } X \\
= \sum_{i=1}^{1000} \mathrm{E}[X_i] && \textit{linearity of expectation} \\
= \sum_{i=1}^{1000} p && \textit{Example 10.35 (expectation of a Bernoulli variable)} \\
= 1000p.
\end{aligned}$$

Problem-solving tip: Often, the easiest way to compute an expectation is by finding a way to express the quantity of interest in terms of a sum of indicator random variables.

Example 10.41 (Number of aces in a bridge hand, better)
Recall Example 10.38, where we showed that the number A of aces in a randomly chosen 13-card hand from a standard 52-card deck has $\mathrm{E}[A] = 1$. Here is a *much* easier way of solving that problem:

Number your cards from 1 to 13. Let A_i be an indicator random variable that reports whether the ith card in your hand is an ace. Then $A = A_1 + A_2 + \ldots + A_{13}$. Note that $\Pr[A_i = 1] = \frac{1}{13}$ (there are $\frac{4}{52} = \frac{1}{13}$ aces in the deck), so

$$\begin{aligned} \mathrm{E}[A] &= \mathrm{E}[A_1 + A_2 + \cdots + A_{13}] \\ &= \mathrm{E}[A_1] + \mathrm{E}[A_2] + \cdots + \mathrm{E}[A_{13}] && \textit{linearity of expectation} \\ &= 13 \cdot \tfrac{1}{13} && \Pr[A_i = 1] = \tfrac{1}{13} \textit{ as above, and so } \mathrm{E}[A_i] = \tfrac{1}{13} \textit{ (Example 10.35)} \\ &= 1. \end{aligned}$$

(The random variables A_i and A_j are correlated—but, again, linearity of expectation doesn't care! We can still use it to conclude that $\mathrm{E}[A_i + A_j] = \mathrm{E}[A_i] + \mathrm{E}[A_j]$.)

Some examples about hashing

Here are two more problems about expectation, both involving hashing:

Example 10.42 (Hashing)
Problem: Suppose that we hash 1000 elements into a 1000-slot hash table, using a completely random hash function, resolving collisions by chaining. (See Section 10.1.1.) How many empty slots do we expect?

Solution: Let's compute the probability that some particular slot is empty:

$$\begin{aligned} &\Pr[\text{slot } i \text{ is empty}] \\ &= \Pr[\text{none of the 1000 elements hash to slot } i] \\ &= \Pr[\text{every element } j \in \{1, 2, \ldots, 1000\} \text{ hashes to a slot other than } i] \\ &= \prod_{j=1}^{1000} \Pr[\text{element } j \text{ hashes to a slot other than } i] && \textit{elements are hashed independently} \\ &= \prod_{j=1}^{1000} \tfrac{999}{1000} && \textit{elements are hashed uniformly, and there are 999 other slots} \\ &= \left(\tfrac{999}{1000}\right)^{1000} = 0.3677\cdots. \end{aligned}$$

We'll finish with the by-now-familiar calculation that also concluded the last two examples: we define a collection of indicator random variables and use linearity of expectation. Let X_i be an indicator random variable that's 1 if slot i is empty and 0 if slot i is full. Then the expected number of empty slots is

$$\mathrm{E}\left[\sum_{i=1}^{1000} X_i\right] = \sum_{i=1}^{1000} \mathrm{E}[X_i] = 1000 \cdot \left(\tfrac{999}{1000}\right)^{1000} \approx 367.7.$$

Taking it further: If we stated the question from Example 10.42 in full generality, we would ask: *if we hash n elements into n slots, how many empty slots are there in expectation?* Using the same approach as in Example 10.42, we'd find that the fraction of empty slots is, in expectation, $(1 - 1/n)^n$. Using calculus, it's possible to show that $(1 - 1/n)^n$ approaches $1/e \approx 0.367879$ as $n \to \infty$. So, for large n, we'd expect to have $\frac{n}{e}$ empty slots when we hash n elements into n slots.

We can also turn this hashing problem on its head: we've been asking "if we hash n elements into n slots, how many slots do we expect to find empty?" Instead we can ask "how many elements do we expect have to hash into n slots before all n slots are full?" This problem is called the *coupon-collector problem*; see Exercises 10.136–10.137 for more.

Let's also consider a second example about hashing—this time counting the (expected) number of collisions, rather than the (expected) number of empty slots:

Example 10.43 (Expected collisions in a hash table)

Problem: Hash n elements $A = \{x_1, \ldots, x_n\}$ into an m-slot hash table. Recall that a *collision* between two elements x_i and x_j (for $i \neq j$) occurs when $h(x_i) = h(x_j)$.

1. Consider two elements $x_i \neq x_j$. What's $\Pr$ [there's a collision between x_i and x_j]?
2. What is the expected number of collisions among the elements of A?

Solution: 1. A collision between x_i and x_j occurs precisely when, for some index k, we have $h(x_i) = k$ *and* $h(x_j) = k$. Thus:

$$\begin{aligned}
&\Pr\left[\text{collision between } x_i \text{ and } x_j\right] \\
&= \Pr\left[\left[h(x_i) = h(x_j) = 1\right] \text{ or } \left[h(x_i) = h(x_j) = 2\right] \text{ or } \cdots \text{ or } \left[h(x_i) = h(x_j) = m\right]\right] \\
&= \sum_{k=1}^{m} \Pr\left[h(x_i) = k \text{ and } h(x_j) = k\right] && \text{by the sum rule; these events are disjoint} \\
&= \sum_{k=1}^{m} \Pr\left[h(x_i) = k\right] \cdot \Pr\left[h(x_j) = k\right] && \text{hashing assumption: hash values are independent} \\
&= \sum_{k=1}^{m} \frac{1}{m} \cdot \frac{1}{m} && \text{hashing assumption: hash values are uniform} \\
&= \frac{m}{m^2} = \frac{1}{m}.
\end{aligned}$$

So the probability that a particular pair of elements collides is precisely $\frac{1}{m}$.

2. Given (1), we can again compute the expected number of collisions using indicator random variables and linearity of expectation. The number of collisions between elements of A is precisely the number of unordered pairs $\{x_i, x_j\}$ that collide. For indices i and $j > i$, then, define $X_{i,j}$ as the indicator random variable

$$X_{i,j} = \begin{cases} 1 & \text{if } x_i \text{ and } x_j \text{ collide} \\ 0 & \text{if they do not.} \end{cases}$$

Thus the expected number of collisions among the elements of A is given by

$$\begin{aligned} & \mathrm{E}\Big[\sum_{1 \le i < j \le n} X_{i,j}\Big] && \textit{summing over all unordered pairs of elements} \\ &= \sum_{1 \le i < j \le n} \mathrm{E}\left[X_{i,j}\right] && \textit{linearity of expectation} \\ &= \sum_{1 \le i < j \le n} \frac{1}{m} && \textit{part 1 of this example: we showed } \mathrm{E}\left[X_{i,j}\right] = \Pr\left[X_{i,j} = 1\right] = \tfrac{1}{m} \\ &= \frac{\binom{n}{2}}{m} = \frac{n(n-1)}{2m}. && \textit{there are } \tbinom{n}{2} = \tfrac{n(n-1)}{2} \textit{ unordered pairs of elements} \end{aligned}$$

One consequence of this analysis is that we'd expect the first collision in an m-slot hash table to occur when the number n of hashed elements reaches approximately $\sqrt{2m}$: for $n = \sqrt{2m} + 1$, the expected number of collisions would be

$$\frac{n(n-1)}{2m} = \frac{(\sqrt{2m}+1) \cdot \sqrt{2m}}{2m} \approx \frac{2m}{2m} = 1.$$

Taking it further: Example 10.43 also explains the so-called *birthday paradox*. Assume that a person's birthday is uniformly and independently chosen from the $m = 365$ days in the year. (Close, but not quite true; certain parts of the year are nine months before days whose probabilities are notably more than $\frac{1}{365}$.) Under this assumption, you can think of "birthday" as a random hash function mapping people to $\{1, 2, \ldots, 365\}$. By Example 10.43, if you're in a room with more than $\sqrt{2 \cdot 365} = 27.018$ people, you'd expect to find a pair that shares a birthday. (It's called a "paradox" because most people's intuition is that you'd need way more than 28 people in a room before you'd find a shared birthday.)

Two more examples of expectation: breaking PINs and Insertion Sort

Here's another example of expectation, in a simple security context:

Example 10.44 (Brute-force breaking of PINs)

Problem: I steal a debit card from a (former) friend. The card has a 4-digit PIN, between 0000 and 9999, that I need to know to get all my friend's money. Here are two strategies:

1. every day, I try a random PIN.
2. every day, I try a random PIN *that I haven't tried before.*

How many days would I expect to wait before I get into my friend's account?

Solution: 1. Observe that the probability of getting the correct PIN on a particular day is $\frac{1}{10000}$. Thus we have a geometric random variable with parameter $\frac{1}{10000}$, so by Example 10.39 we expect to need 10000 days to break the PIN.

2. As usual, there are multiple ways to solve this problem—and, for illustrative purposes, we'll describe two of them, using fairly different approaches.

Solution A: $\Pr\left[\textbf{winning on day \#}i\right]$**.** The key will be to find the probability of breaking the code for the first time on day i. (For the purposes of this analysis, imagine that we keep guessing new PINs even after we find the correct answer.)

Because we make $i-1$ guesses on the $i-1$ days before day i, we know

$$\Pr\left[\text{getting the PIN } \textit{before} \text{ day } i\right] = \tfrac{i-1}{10000} \qquad (1)$$

$$\Pr\left[\underline{\text{not}} \text{ getting the PIN } \textit{before} \text{ day } i\right] = 1 - \tfrac{i-1}{10000} = \tfrac{10001-i}{10000}. \qquad (2)$$

Furthermore, on day i there are $10000-(i-1)$ untried guesses, and so

$$\Pr\left[\text{getting the PIN } \textit{on} \text{ day } i \mid \underline{\text{not}} \text{ getting it } \textit{before} \text{ day } i\right] = \tfrac{1}{10000-(i-1)} = \tfrac{1}{10001-i}. \qquad (3)$$

Thus the expected number of days that we have to keep guessing is:

$$\begin{aligned}
&\sum_{i=1}^{10000} i \cdot \Pr\left[\text{we first break the code on day \#i}\right] && \textit{definition of expectation}\\
&= \sum_{i=1}^{10000} i \cdot \Pr\left[\text{wrong on days } 1,\ldots,i-1\right] \cdot \Pr\left[\text{right on day } i \mid \text{wrong on days } 1,\ldots,i-1\right] && \textit{Chain Rule}\\
&= \sum_{i=1}^{10000} i \cdot \tfrac{10001-i}{10000} \cdot \tfrac{1}{10001-i} && \textit{(2) and (3), as argued above}\\
&= \tfrac{1}{10000} \cdot \sum_{i=1}^{10000} i && \textit{algebra}\\
&= \tfrac{1}{10000} \cdot \tfrac{10000 \cdot 10001}{2} = 5000.5. && \textit{arithmetic summation (Example 5.4)}
\end{aligned}$$

(Another way to view this solution: our PIN-guessing strategy corresponds to choosing a permutation of $\{0000,\ldots,9999\}$ uniformly at random, and guessing in the chosen order. The correct PIN is equally likely to be at any position in the permutation so, for any i, we require exactly i days with probability precisely $\frac{1}{10000}$.)

Solution B: $\Pr\left[\textbf{have to guess on day \#}i\right]$**.** Define an indicator random variable X_i, where $X_i = 1$ if we have to make a guess on day $\#i$, and $X_i = 0$ if we do not. Thus the number of days that we have to guess is precisely $X := \sum_{i=1}^{10000} X_i$. Observe that

$$\mathrm{E}\left[X_i\right] = \Pr\left[X_i\right] = \Pr\left[\text{lose on days } 1,\ldots,i-1\right] = \tfrac{10001-i}{10000}$$

by the same reasoning as in Solution A. Thus

$$\begin{aligned}
\mathrm{E}[X] &= \sum_{i=1}^{10000} \mathrm{E}[X_i] && \textit{linearity of expectation}\\
&= \sum_{i=1}^{10000} \frac{10001-i}{10000} && \textit{the above argument}\\
&= \sum_{j=1}^{10000} \frac{j}{10000} && \textit{change of variables } j = 10001 - i\\
&= 5000.5. && \textit{just as in Solution A}
\end{aligned}$$

So avoiding duplication saves, in expectation, just less than half of the days: we expect to use 10000 days if we allow duplication, and 5000.5 days if we avoid it.

(Incidentally, the argument in Solution B is just another way of viewing the transformation from Theorem 10.5: instead of calculating the value of $\sum_i i \cdot \Pr\left[\text{exactly } i \text{ days}\right]$, we calculated $\sum_i \Pr\left[\text{at least } i \text{ days}\right]$.)

Let's conclude with one last example of another type: analyzing the expected performance of an algorithm on a randomly chosen input. In Example 6.13, we gave a brief intuition for the average-case (expected) performance of Insertion Sort. (See Figure 10.26 for a reminder of the algorithm.) Here is a somewhat different version of the analysis, which comes out with the same result:

```
insertionSort(A[1 ... n]):
  for i := 2 to n:
     j := i
     while j > 1 and A[j] < A[j − 1]:
        swap A[j] and A[j − 1]
        j := j − 1
```

Figure 10.26: A reminder of Insertion Sort.

Example 10.45 (Expected performance of Insertion Sort)

Problem: Let the array A be a permutation of $\{1, \ldots, n\}$ chosen uniformly at random. What is the expected number of swaps performed by **insertionSort**($A[1 \ldots n]$)?

Solution: Define an indicator random variable $X_{j,i}$ for indices $j < i$:

$$X_{j,i} = \begin{cases} 1 & \text{if the (original) elements } A[j] \text{ and } A[i] \text{ are swapped by } \textbf{insertionSort} \\ 0 & \text{if not.} \end{cases}$$

Note that $\mathrm{E}[X_{j,i}] = \Pr[X_{j,i} = 1] = \frac{1}{2}$: precisely half of permutations have their ith element larger than their jth element. (There's a bijection between the set of permutations with their ith element larger than their jth element and the set of permutations with their ith element smaller than their jth element. Because these sets have the same size, the probability of choosing one of the former is $\frac{1}{2}$.)

Because **insertionSort** correctly sorts its input and only swaps out-of-order pairs once per pair, the total number of swaps done is precisely

$$X = \sum_{i=2}^{n} \sum_{j=1}^{i-1} X_{i,j}.$$

Note that the number of indicator random variables in this sum is

$$\sum_{i=2}^{n}\sum_{j=1}^{i-1} 1 = \sum_{i=2}^{n}(i-1) = \sum_{i=1}^{n-1} i = \frac{(n-1)\cdot n}{2} = \binom{n}{2}.$$

Thus by linearity of expectation we have

$$\mathrm{E}[X] = \binom{n}{2} \cdot \mathrm{E}[X_{i,j}] = \binom{n}{2} \cdot \frac{1}{2}.$$

10.4.4 Conditional Expectation

Just as we did with conditional probability in Section 10.3, we can define a notion of *conditional expectation:* that is, the average value of a random variable X *when a particular event occurs.*

Definition 10.13 (Conditional expectation)
The conditional expectation *of a random variable X given an event E, denoted* $\mathrm{E}\left[X|E\right]$, *is the average value of X over all outcomes where E occurs:*

$$\mathrm{E}\left[X|E\right] = \sum_{x \in E} X(x) \cdot \Pr\left[x|E\right].$$

In the original definition of expectation, we summed over all x in the whole sample space; here we sum only over the outcomes in the event E. Furthermore, here we weight the value of X by $\Pr\left[x|E\right]$ rather than by $\Pr[x]$. We'll omit the details, but conditional expectation has analogous properties to those of the original (nonconditional) version of expectation, including linearity of expectation.

Here's a brief example of computing some conditional expectations:

Example 10.46 (Hearts in Poker)
Problem: In Texas Hold 'Em, a particular variant of poker, after a standard deck of cards is randomly shuffled, you are dealt two "personal" cards, and then five "community" cards are dealt. Let P denote the number of your personal cards that are hearts, and let C denote the number of community cards that are hearts. What are the following?

1. $\mathrm{E}[P]$
2. $\mathrm{E}[C]$
3. $\mathrm{E}\left[C|P=0\right]$
4. $\mathrm{E}\left[C|P=2\right]$

Solution: 1 & 2. Each card that's dealt has a $\frac{13}{52} = \frac{1}{4}$ chance of being a heart. By linearity of expectation, then, $\mathrm{E}[P] = \frac{2}{4} = 0.5$ and $\mathrm{E}[C] = \frac{5}{4} = 1.25$. (Implicitly, we're defining indicator random variables for "the ith card is a heart," so $P = P_1 + P_2$ and $C = C_1 + \cdots + C_5$.)

3. Given that 2 of the 39 non-heart cards were dealt as your personal cards, there are still 13 undealt hearts among the remaining 50 undealt cards. Thus there is a $\frac{13}{50} = 0.26$ chance that any particular undealt card is a heart. Thus, again by linearity of expectation, we have that $\text{E}\left[C|P=0\right] = 5 \cdot \frac{13}{50} = 1.30$.
4. Similarly, there are 11 undealt hearts among the remaining 50 undealt cards. Thus there is an $\frac{11}{50} = 0.22$ chance that any particular undealt card is a heart, and $\text{E}\left[C|P=2\right] = 5 \cdot \frac{11}{50} = 1.10$.

We'll omit the proof, but it's worth noting a useful property that connects expectation to conditional expectation, an analogy to the law of total probability:

Theorem 10.7 (Law of Total Expectation)
For any random variable X and any event E:

$$\text{E}[X] = \text{E}\left[X|E\right] \cdot \Pr[E] + \text{E}\left[X|\,\overline{E}\,\right] \cdot (1 - \Pr[E]).$$

That is, the expectation of X is the (weighted) average of the expectation of X when E occurs and when E does not occur.

Taking it further: One tremendously valuable use of probability is in *randomized algorithms*, which flip some coins as part of solving some problem. There is a massive variety in the ways that randomization is used in these algorithms, but one example—the computation of the *median* element of an unsorted array of numbers—is discussed on p. 1060. (We'll make use of Theorem 10.7.) Median finding is a nice example of problem for which there is a very simple, efficient algorithm that makes random choices in its solution. (There *are* deterministic algorithms that solve this problem just as efficiently, but they are *much* more complicated than this randomized algorithm.)

10.4.5 Deviation from Expectation

Let X be a random variable. By definition, the value of $\text{E}[X]$ is the average value that X takes on, where we're averaging over many different realizations. But how far away from $\text{E}[X]$ is X, on average? That is, what is the average difference between (a) X, and (b) the average value of X? We might care about this quantity in applications like political polling or scientific experimentation, for example. Suppose X is a random variable defined as follows:

$$X = \begin{cases} -1 & \text{the voter will vote for the Democratic candidate} \\ 0 & \text{the voter will vote for neither the Democratic nor Republican candidates} \\ +1 & \text{the voter will vote for the Republican candidate} \end{cases}$$

for a voter chosen uniformly at random from the population. If $\text{E}[X] < 0$, then the Democrat will beat the Republican in the election; if $\text{E}[X] > 0$, then the Republican will beat the Democrat. We might estimate $\text{E}[X]$ by calling, say, 500 uniformly chosen voters from the population and averaging their responses. We'd like to know whether our estimate is accurate (that is, if our estimate is close to $\text{E}[X]$). This kind of question is the core of statistical reasoning. We'll only begin to touch on these questions, but here are a few of the most important concepts.

Definition 10.14 (Variance)
Let X be a random variable. The variance *of X is*

$$\text{var}(X) = \mathrm{E}\left[(X - \mathrm{E}[X])^2\right].$$

The standard deviation *is* $\text{std}(X) = \sqrt{\text{var}(X)}$.

(Exercise: why didn't we just define $\text{std}(X) = \mathrm{E}[X - \mathrm{E}[X]]$?)

Here's a simple example:

Example 10.47 (Variance/standard deviation of a Bernoulli random variable)
Let X be the outcome of a flipping a p-biased coin. (That is, X is a Bernoulli random variable.) We previously showed that $\mathrm{E}[X] = p$, so the variance of X is

$$\begin{aligned}
\text{var}(X) &= \mathrm{E}[\,(X - \mathrm{E}[X])^2\,] && \textit{definition of expectation}\\
&= \mathrm{E}[\,(X - p)^2\,] && \textit{expectation of a Bernoulli random variable (Example 10.35)}\\
&= \Pr[X = 0] \cdot (0 - p)^2 + \Pr[X = 1] \cdot (1 - p)^2 && \textit{definition of expectation}\\
&= (1 - p) \cdot (0 - p)^2 + p \cdot (1 - p)^2 && \textit{definition of Bernoulli random variable}\\
&= (1 - p)p^2 + p(1 - p)^2\\
&= (1 - p)p \cdot (p + 1 - p)\\
&= (1 - p)p.
\end{aligned}$$

Thus the standard deviation is $\text{std}(X) = \sqrt{\text{var}(X)} = \sqrt{(1 - p)p}$.

(For example, for a fair coin, the standard deviation is $\sqrt{(1 - 0.5)0.5} = \sqrt{0.25} = 0.5$: an average coin flip is 0.5 units away from the mean 0.5. In fact, every coin flip is that far away from the mean!)

Here's another simple example, illustrating the fact that two random variables can have the same mean but wildly different variances:

Example 10.48 (Roulette bets)
Here are two bets available to a player in roulette (see Figure 10.27 for a reminder):

- Bet \$1 on "red": If the spin lands on one of the 18 red numbers, you get \$2 back; otherwise you get nothing.
- Bet \$1 on "17": If the spin lands on the number 17, you get \$36 back; otherwise you get nothing.

Let X denote the payoff from playing the first bet, so $X = 0$ with probability $\frac{20}{38}$ and $X = 2$ with probability $\frac{18}{38}$. Let Y denote the payoff from playing the second bet, so $Y = 0$ with probability $\frac{37}{38}$ and $X = 36$ with probability $\frac{1}{38}$. The expectations match:

$$\begin{aligned}
\mathrm{E}[X] &= \tfrac{20}{38} \cdot 0 + \tfrac{18}{38} \cdot 2 = \tfrac{36}{38}\\
\mathrm{E}[Y] &= \tfrac{37}{38} \cdot 0 + \tfrac{1}{38} \cdot 36 = \tfrac{36}{38}.
\end{aligned}$$

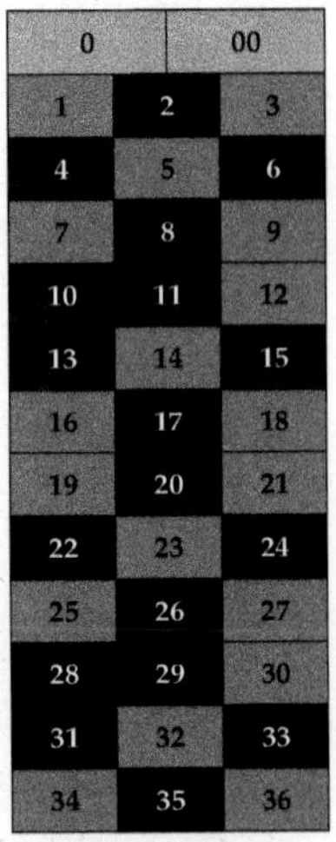

Figure 10.27: A reminder of the roulette outcomes. A number in the set $\{0, 00, 1, 2, \ldots, 36\}$ is chosen uniformly at random by a spinning wheel; there are 18 *red* numbers and 18 *black* numbers; 0 and 00 are neither red nor black.

But the variances are very different:

$$\begin{aligned}\text{var}(X) &= \tfrac{20}{38} \cdot (0 - \tfrac{36}{38})^2 + \tfrac{18}{38} \cdot (2 - \tfrac{36}{38})^2 &&= 0.9972\cdots \\ \text{var}(Y) &= \tfrac{37}{38} \cdot (0 - \tfrac{36}{38})^2 + \tfrac{1}{38} \cdot (36 - \tfrac{36}{38})^2 &&= 33.2077\cdots.\end{aligned}$$

Generally speaking, the expectation of a random variable measures "how good it is" (on average), while the variance measures "how risky it is."

VARIANCE, THE SQUARED EXPECTATION, AND THE EXPECTATION OF THE SQUARE

Here's a useful property of variance, which sometimes helps us avoid tedium in calculations. We can write $\text{var}(X)$ as $\text{var}(X) = \text{E}[X^2] - (\text{E}[X])^2$, that is, the difference between the *expectation of the square of* X and the *square of the expectation of* X:

Theorem 10.8 (Variance = expectation of the square minus the expectation^2)
For any random variable X*, we have*

$$\text{var}(X) = \text{E}\left[X^2\right] - (\text{E}[X])^2.$$

Proof. Writing $\mu := \text{E}[X]$, we have

$$\begin{aligned}
&\text{var}(X) \\
&= \text{E}\left[(X-\mu)^2\right] && \textit{definition of expectation} \\
&= \text{E}\left[X^2 - 2X\mu + \mu^2\right] && \textit{multiplying out} \\
&= \text{E}\left[X^2\right] + \text{E}[-2X\mu] + \text{E}\left[\mu^2\right] && \textit{linearity of expectation} \\
&= \text{E}\left[X^2\right] - 2\mu \cdot \text{E}[X] + \mu^2 && \textit{Exercise 10.151} \\
&= \text{E}\left[X^2\right] - 2\mu \cdot \mu + \mu^2 && \textit{definition of } \mu = \text{E}[X] \\
&= \text{E}\left[X^2\right] - \mu^2 \\
&= \text{E}\left[X^2\right] - (\text{E}[X])^2. && \square
\end{aligned}$$

Here is a simple example in which Theorem 10.8 eases the computation:

Example 10.49 (Variance/standard deviation of a uniform random variable)
Problem: Let X be the result of a roll of a fair die. What is $\text{var}(X)$?

Solution: Because $\Pr[X = k] = \frac{1}{6}$ for all $k \in \{1, \ldots, 6\}$, we have that

$$\begin{aligned}\text{E}[X] &= \tfrac{1}{6} \cdot (1+2+3+4+5+6) \\ &= \tfrac{1}{6} \cdot 21 \\ &= 3.5.\end{aligned}$$

Similarly, we can compute $\mathrm{E}\left[X^2\right]$ as follows:

$$\begin{aligned}\mathrm{E}\left[X^2\right] &= \tfrac{1}{6} \cdot (1^2 + 2^2 + 3^2 + 4^2 + 5^2 + 6^2) \\ &= \tfrac{1}{6} \cdot 91 \\ &\approx 15.1666\cdots.\end{aligned}$$

Therefore, by Theorem 10.8,

$$\mathrm{var}\,(X) = \mathrm{E}\,[X^2] - (\mathrm{E}\,[X])^2 = \tfrac{91}{6} - \tfrac{49}{4} = \tfrac{35}{12} \approx 2.9116\cdots,$$

and $\mathrm{std}\,(X) = \sqrt{35/12} \approx 1.7078\cdots$.

(In Exercise 10.150, you'll show that the standard deviation of the average result of two independent dice rolls is much smaller.)

Taking it further: Suppose that we need to estimate the fraction of [very complicated objects] that have [easy-to-verify property]: would I win a higher fraction of chess games with Opening Move A or B? Roughly how many different truth assignments satisfy Boolean formula φ? Roughly how many integers in $\{2, 3, \ldots, n-1\}$ evenly divide n? Is the array A "mostly" sorted?

One nice way to approximate the answer to these questions is the *Monte Carlo method,* one of the simplest ways to use randomization in computation. The basic idea is to compute many *random* candidate elements—chess games, truth assignments, possible divisors, etc.—and test each one; we can then estimate the answer to the question of interest by calculating the fraction of those random candidates that have the property in question. See p. 1062 for more discussion.

Computer Science Connections

A Randomized Algorithm for Finding Medians

The *median* element of an array $A[1 \ldots n]$ is the item that would appear in the $\lceil n/2 \rceil$th slot of the sorted order if we sorted A. For example, the median of $[1,3,5,7,9]$ is 5, and the median of $[4,3,2,1]$ is 2. (We arbitrarily chose to find the $\lceil n/2 \rceil$th element instead of the $\lfloor n/2 \rfloor$th.) This description already suggests a solution to the median problem: sort A, and then return $A[\lceil n/2 \rceil]$. But we can do better than the sorting-based approach: we'll give a faster algorithm for finding the median element of an unsorted array. Our algorithm will be randomized, and the *expected* running time of the algorithm will be linear. It will turn out to be easier to solve a generalization of the median problem, called Select. See Figure 10.28.

> Select:
>
> *Given:* an array $A[1 \ldots n]$ and an index $k \in \{1, \ldots, n\}$.
>
> *Output:* the element x in A such that, if you were to sort A, x would appear in the kth slot of the sorted array.

Figure 10.28: The Select problem.

A recursive solution to Select is given in Figure 10.29; we can solve the median problem by calling **randSelect**($A[1 \ldots n]$, $\lceil n/2 \rceil$). A proof of correctness of the algorithm—that is, a proof that **randSelect** actually solves the Select problem—is reasonably straightforward by induction. (In fact, correctness is guaranteed *regardless of how we choose x in Line 3* of the algorithm.) But we still have to analyze the running time.

```
randSelect(A[1...n], i):
// Find the ith-largest element of A.
// If i ∉ {1,2,...,n}, then error.
 1: if n = 1 then
 2:     return A[1]. (If i ≠ 1, then error.)
 3: choose x ∈ {1,...,n} randomly
 4: Losers[1...ℓ] := {y ∈ A : y < A[x]}
 5: Winners[1...w] := {y ∈ A : y > A[x]}.
 6: if i < ℓ+1 then
 7:     return randSelect(Losers, i)
 8: else if i = ℓ+1 then
 9:     return A[x]
10: else if i > ℓ+1 then
11:     return randSelect(Winners, i − ℓ − 1)
```

Figure 10.29: Randomized Median finding. (We build *Losers* and *Winners* by going through A element-by-element.)

Running Time: The Big Picture

Think about an invocation of **randSelect**(A), and imagine the array A in sorted order and divided into quartiles:

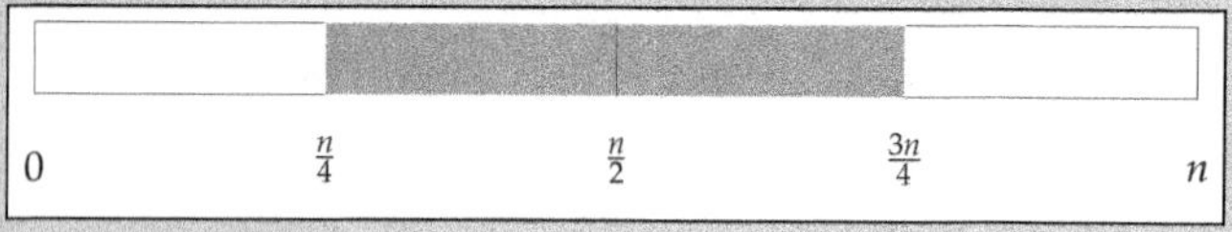

Here are two crucial observations:

1. Suppose that the element $A[x]$ chosen in step 3—call $A[x]$ the *pivot*—falls within the shaded region of the quartile picture above. Then we know that $|Losers| \leq \frac{3n}{4}$ *and* $|Winners| \leq \frac{3n}{4}$.
2. The shaded region contains half of the elements of A.

(Why? To put it briefly: because half of the elements of A are in the middle half of the array A.) So what? Let's think intuitively for a moment, and defer the formal analysis. Whenever we choose an element from the middle half of the sorted order, the next recursive call is on an array of size at most $\frac{3}{4}$ the size of the original input. Also observe that the running time of any particular call (aside from the recursive call) is linear in the input size. Thus, if we got lucky every time and picked an element from the middle half of the array, we'd have a recurrence like the following:

$$T(1) = 1 \qquad\qquad T(n) \leq n + T(3n/4)$$

That's a classic Master Method recurrence with a solution of $T(n) = \Theta(n)$. (Actually the master method only says that $T(n) = O(n)$, because we have an inequality in the recurrence. But it's trivial that the running time is $\Omega(n)$ as well, because just building *Losers* and *Winners* at the root takes $\Omega(n)$ time.)

Computer Science Connections

A Randomized Algorithm for Finding Medians, Continued

Running Time: Making it Formal

We engaged in wishful thinking in the last paragraph: it's obviously not true that we get a pivot in the middle half of the array every time. In fact, it's only half the time! But this isn't so bad: *even if we imagine that picking a pivot outside the middle half yields zero progress at all toward the base case,* we'd only double the estimate of the running time! Let's make this formal. Define

C_n := the number of comparisons performed by **randSelect** on an input of size n.

Notice that C_n is a random variable: the number of comparisons that are performed depends on which pivots are chosen! But we can analyze $\mathrm{E}[C_n]$.

Before we start, let's make one quick observation: the expected running time of this algorithm is monotonic in its input size. That is, $\mathrm{E}[C_n] \le \mathrm{E}[C_{n'}]$ if $n \le n'$. (This fact is tedious to prove rigorously, but is still pretty obvious.)

Theorem: $\mathrm{E}[C_n] \le 8n$.

Proof (by strong induction on n). Base case (n = 1): In fact, when $n = 1$, the algorithm performs zero comparisons, and indeed $0 \le 8$.

Inductive case ($n \ge 2$): We assume the inductive hypothesis, namely that for any $n' < n$, we have that $\mathrm{E}[C'_n] \le 8n'$. We must prove that $\mathrm{E}[C_n] \le 8n$.

Let's consider the comparisons that are made on an input array of size n. First, there are n comparisons performed in Lines 4–5, to compute *Losers* and *Winners*. Then there are whatever comparisons are made in the recursive call. Because we're trying to compute a worst-case bound, we'll make do with the following observation: $C_n \le n + C_{\max(|Losers|,|Winners|)}$.

Let $\mathcal{M}$ denote the event that our pivot is in the middle half of A (that is, falls in the shaded region of the diagram on the previous page). Thus:

$$\mathrm{E}[C_n] \le \mathrm{E}\left[n + C_{\max(|Losers|,|Winners|)}\right] \quad \text{the above accounting of the comparisons}$$

$$= n + \mathrm{E}\left[C_{\max(|Losers|,|Winners|)}\right] \quad \text{linearity of expectation}$$

$$= n + \mathrm{E}\left[C_{\max(|Losers|,|Winners|)} \middle| \mathcal{M}\right] \cdot \Pr[\mathcal{M}] + \mathrm{E}\left[C_{\max(|Losers|,|Winners|)} \middle| \overline{\mathcal{M}}\right] \cdot \Pr\left[\overline{\mathcal{M}}\right] \quad \text{Law of Total Expectation (Theorem 10.7)}$$

$$= n + \tfrac{1}{2} \cdot \left[\mathrm{E}\left[C_{\max(|Losers|,|Winners|)} \middle| \mathcal{M}\right] + \mathrm{E}\left[C_{\max(|Losers|,|Winners|)} \middle| \overline{\mathcal{M}}\right]\right] \quad \text{Crucial observation \#2: } \Pr[\mathcal{M}] = \Pr\left[\overline{\mathcal{M}}\right] = \tfrac{1}{2}$$

$$\le n + \tfrac{1}{2} \cdot \left[\mathrm{E}\left[C_{3n/4}\right] + \mathrm{E}[C_n]\right]. \quad \text{Crucial observation \#1: if } \mathcal{M} \text{ occurs, we recurse on} \le \tfrac{3n}{4} \text{ elements; else it's certainly on} \le n \text{ elements.}$$

Thus we have argued that

$$\mathrm{E}[C_n] \le n + \tfrac{1}{2} \cdot \mathrm{E}\left[C_{3n/4}\right] + \tfrac{1}{2} \cdot \mathrm{E}[C_n] \text{ and therefore}$$

$$\mathrm{E}[C_n] \le 2n + \mathrm{E}\left[C_{3n/4}\right]. \quad \text{starting with the previous inequality and subtracting } \tfrac{1}{2} \cdot \mathrm{E}[C_n] \text{ from both sides, and then multiplying both sides by 2}$$

The inductive hypothesis says that $\mathrm{E}\left[C_{3n/4}\right] \le 8 \cdot \frac{3n}{4} = 6n$, so we therefore have

$$\mathrm{E}[C_n] \le 2n + 6n = 8n.$$

□

COMPUTER SCIENCE CONNECTIONS

THE MONTE CARLO METHOD

If we need to compute some (potentially very complicated) quantity, one way to do so is the *Monte Carlo method*. Let's take a computation of area of a potentially complicated shape as an example. If we identify a bounding box (a rectangle surrounding the shape) and then generate a sequence of random points in the bounding box, we can count how many of those points fall into the shape in question.

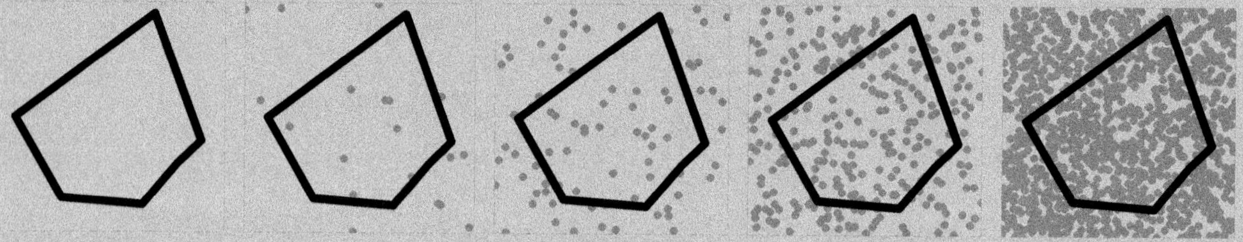

Figure 10.30: A shape, and an estimate of its area with random points: we simply estimate the area using the fraction of the chosen points that fall within the shape. The more points, the more accurate the estimate.

For example, to find the area of the shape in Figure 10.30, we can throw a random point into the bounding box. The probability that the randomly chosen point is inside the polygon is precisely the ratio of the area of the polygon to the area of the bounding box—and thus the expected fraction of points that land inside the shape precisely yields the area of the shape. Of course, the more points we throw at the bounding box, the more accurate our estimate of the area will be: the fraction of heads in n flips of a p-biased coin has a much lower variance (but the same expectation) as n gets bigger and bigger. (See Exercise 10.155.)

There are a few issues complicating this approach. First, we must find a bounding box for which the shape in question covers a "large" fraction of the bounding box. (If the probability p of a random point falling into the shape is tiny, then a little bad luck in sampling—2 points land inside instead of 3?—causes huge relative [multiplicative] error in our area estimate.) Second, we've described this process as choosing a uniform point from the bounding box—which requires infinitesimal probabilities associated with each of the infinitely many points inside the bounding box. The handle this, typically we would define a "mesh" of points: we specify a "resolution" ε and choose a coordinate of a random point as k/ε for a random $k \in \{0, 1, \ldots, 1/\varepsilon\}$.

The example in Figure 10.30 is a nice way of being lazy—we *could* have calculated the area of the polygon with some tedious algebra—but there are some other examples in which this technique is even more useful. Some of the simplest methods for estimating the value of π in the last century were based on Monte Carlo methods. One option is to throw a point $\langle x, y \rangle$ into the unit square $[0,1] \times [0,1]$ and test what fraction have $x^2 + y^2 \leq 1$. Another is an algorithm called *Buffon's needle*—named after an 18th-century French mathematician—in which we throw unit-length "needles" onto a surface with parallel lines one unit apart; one can show that $\Pr[\text{a needle crosses a line}] = \frac{2}{\pi}$. See Figure 10.31.

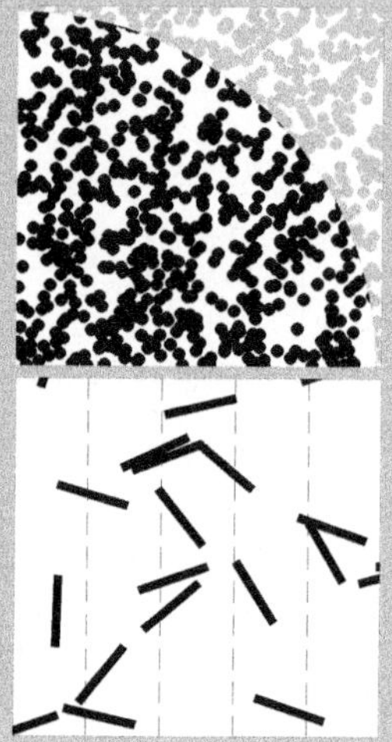

Figure 10.31: Estimating π with a point in the unit square, or with Buffon's needle.

10.4.6 Exercises

Choose a word in $S = \{\texttt{Computers}, \texttt{are}, \texttt{useless}, \texttt{They}, \texttt{can}, \texttt{only}, \texttt{give}, \texttt{you}, \texttt{answers}\}$ (a quote attributed to Pablo Picasso) by choosing a word w with probability proportional to the number of letters in w. Let L be a random variable denoting the number of letters in the chosen word, and let V be a random variable denoting the number of vowels.

10.105 Give a table of outcomes and their probabilities, together with the values of L and V.

10.106 What is $\Pr[L = 4]$? What is $E[V|L = 4]$?

10.107 Are L and V independent?

10.108 What are $E[L]$ and $E[V]$?

10.109 What is $\text{var}(L)$?

10.110 What is $\text{var}(V)$?

Flip a fair coin 16 times. Define the following two random variables:

- *let H be an indicator random variable that's 1 if at least one of the 16 flips comes up heads, and 0 otherwise.*
- *let R be a random variable equal to the length of the longest "run" in the flips. (A* run *of length k is a sequence of k consecutive flips that all come up Heads, or k consecutive flips that all come up Tails.)*

10.111 What's $E[H]$?

10.112 What's $E[R]$? *(Hint: write a program—*not *by simulating many sequences of 16 coin flips, but rather by listing exhaustively all outcomes.)*

10.113 Are H and R independent?

In 1975, a physicist named Michael Winkelmann invented a dice-based game with the following three (fair) dice:

Blue die: *sides 1, 2, 5, 6, 7, 9* **Red die:** *sides 1, 3, 4, 5, 8, 9* **Black die:** *sides 2, 3, 4, 6, 7, 8*

There are some weird properties of these dice, as you'll see.

10.114 Choose one of the three dice at random, roll it, and call the result X. Show that $\Pr[X = k] = \frac{1}{9}$ for any $k \in \{1, \ldots, 9\}$.

10.115 Choose one of the three dice at random, roll it, and call the result X. Put that die back in the pile and again (independently) choose one of the three dice at random, roll it, and call the result Y. Show that $\Pr[9X - Y = k] = \frac{1}{81}$ for any $k \in \{0, \ldots, 80\}$.

10.116 Roll each die. Call the results B (blue), R (red), and K (black). Compute $E[B]$, $E[R]$, and $E[K]$.

10.117 Define B, R, and K as in the last exercise. Compute $\Pr[B > R | B \neq R]$, $\Pr[R > K | R \neq K]$, and $\Pr[K > B | K \neq B]$—in particular, show that all three of these probabilities (strictly) exceed $\frac{1}{2}$.

The last exercise demonstrates that the red, blue, and black dice are nontransitive, *using the language of relations (Chapter 8): you'd bet on Blue beating Red and you'd bet on Red beating Black, but (surprisingly) you'd want to bet on Black beating Blue. Here's another, even weirder, example of nontransitive dice. (And if you're clever and mildly unscrupulous, you can win some serious money in bets with your friends using these dice.)*

Kelly die: *sides 3, 3, 3, 3, 3, 6* **Lime die:** *sides 2, 2, 2, 5, 5, 5* **Mint die:** *sides 1, 4, 4, 4, 4, 4*

These dice are fair; each side comes up with probability $\frac{1}{6}$. Roll each die, and call the resulting values K, L, and M.

10.118 Show that the expectation of each of these three random variables is identical.

10.119 Show that $\Pr[K > L]$, $\Pr[L > M]$, and $\Pr[M > K]$ are all strictly greater than $\frac{1}{2}$.

You can think of the last exercise as showing that, if you had to bet on which of K or L would roll a higher number, you should bet on K. (And likewise for L over M, and for M over K.) Now let's think about rolling each die twice *and adding the two rolled values together. Roll each die twice, and call the resulting values K_1, K_2, L_1, L_2, M_1, and M_2, respectively.*

10.120 Show that the expectation of the three values $K_1 + K_2$, $L_1 + L_2$, and $M_1 + M_2$ are identical.

10.121 *(programming required)* Show that the following probabilities are all strictly *less* than $\frac{1}{2}$:

$$\Pr[K_1 + K_2 > L_1 + L_2], \Pr[L_1 + L_2 > M_1 + M_2], \text{ and } \Pr[M_1 + M_2 > K_1 + K_2].$$

(Notice that which die won switched directions—and all we did was go from rolling the dice once to rolling them twice!) To show this result, write a program to check how many of the 6^4 outcomes cause $K_1 + K_2 > L_1 + L_2$, etc.

Suppose that you are dealt a 5-card hand from a standard deck. For the purposes of the next two questions, a pair *consists of any two cards with the same rank—so ♣A♡A◇A23 contains* three *pairs (♡A◇A and ♣A◇A and ♣A♡A). Let P denote the number of pairs in your hand.*

10.122 Compute $E[P]$ "the hard way," by computing $\Pr[P = 0]$, $\Pr[P = 1]$, $\Pr[P = 2]$, and so forth. (There can be as many as 6 pairs in your hand, if you have four-of-a-kind.)

10.123 Compute $E[P]$ "the easy way," by defining an indicator random variable $R_{i,j}$ that's 1 if and only if cards #i and #j are a pair, computing $E[R_{i,j}]$, and using linearity of expectation.

In bridge, you are dealt a 13-card hand from a standard deck. A hand's high-card points *are awarded for face cards: 4 for an ace, 3 for a king, 2 for a queen, and 1 for a jack. A hand's* distribution points *are awarded for having a* small *number of cards in a particular suit: 1 point for a "doubleton" (only two cards in a suit), 2 points for a "singleton" (only one card in a suit), and 3 points for a "void" (no cards in a suit).*

10.124 What is the expected number of high-card points in a bridge hand? *(Hint: define some simple random variables, and use linearity of expectation.)*

10.125 What is the expected number of distribution points *for hearts* in a bridge hand? *(Hint: calculate the probability of having exactly 2 hearts, exactly 1 heart, or no hearts in a hand.)*

10.126 Using the results of the last two exercises and linearity of expectation, find the expected number of points (including both high-card and distribution points) in a bridge hand.

We've shown linearity of expectation—the expectation of a sum equals the sum of the expectations—even when the random variables in question aren't independent. It turns out that the expectation of a product equals the product of the expectations when the random variables are independent, but not in general when they're dependent.

10.127 Let X and Y be independent random variables. Prove that $\mathrm{E}[X \cdot Y] = \mathrm{E}[X] \cdot \mathrm{E}[Y]$.

On the other hand, suppose that X and Y are dependent *random variables. Prove that ...*

10.128 ... $\mathrm{E}[X \cdot Y]$ is not necessarily equal to $\mathrm{E}[X] \cdot \mathrm{E}[Y]$.

10.129 ... $\mathrm{E}[X \cdot Y]$ is also not necessarily *unequal* to $\mathrm{E}[X] \cdot \mathrm{E}[Y]$.

We showed in Example 10.39 that the expected number of flips of a p-biased coin before we get Heads is precisely $\frac{1}{p}$.

10.130 How many flips would you expect to have to make before you see 1000 heads *in total* (not necessarily consecutive)? *(Hint: define a random variable X_i denoting the number of coin flips* after the $(i-1)$st Heads *before you get another Heads. Then use linearity of expectation.)*

10.131 How many flips would you expect to make before you see two *consecutive* heads?

In Insertion Sort, we showed in Example 10.45 that the expected number of swaps is $\binom{n}{2}/2$ for a randomly sorted input. With respect to comparisons, *it's fairly easy to see that each element participates in one more comparison than it does swap—with one exception: those elements that are swapped all the way back to the beginning of the array. Here you'll precisely analyze the expected number of comparisons.*

```
insertionSort(A[1...n]):
for i := 2 to n:
    j := i
    while j > 1 and A[j] < A[j - 1]:
        swap A[j] and A[j - 1]
        j := j - 1
```

Figure 10.32: A reminder of Insertion Sort.

10.132 What is the probability that the ith element of the array is swapped all the way back to the beginning of the array?

10.133 What's the expected number of comparisons done by Insertion Sort on a randomly sorted n-element input?

Suppose we hash n elements into an 100,000-slot hash table, resolving collisions by chaining.

10.134 Use Example 10.43 to identify the smallest n for which the expected number of collisions first reaches 1. What the smallest n for which the expected number of collisions exceeds 100,000?

10.135 *(programming required)* Write a program to empirically test your answers from the last exercise, by doing $k = 1000$ trials of loading *[your first answer from Exercise 10.134]* elements into a 100,000-slot hash table. Also do $k = 100$ trials of loading *[your second answer from Exercise 10.134]* elements. On average, how many collisions did you see?

Consider an m-slot hash table that resolves collisions by chaining. In the next few problems, we'll figure out the expected number of elements that must be hashed into this table before every *slot is "hit"—that is, until every cell of the hash table is full.*

10.136 Suppose that the hash table currently has $i-1$ filled slots, for some number $i \in \{1, \ldots, m\}$. What is the probability that the next element that's hashed falls into an *unoccupied* slot? Let the random variable X_i denote the number of elements that are hashed *until one more cell is filled.* What is $\mathrm{E}[X_i]$?

10.137 Argue that the total number X of elements hashed before the entire hash table is full is given by $X = \sum_{i=1}^{m} X_i$. Using Exercise 10.136 and linearity of expectation, prove that $\mathrm{E}[X] = m \cdot H_m$.

(Recall that H_m denotes the mth harmonic number, *where $H_m := \sum_{i=1}^{m} \frac{1}{i}$. See Definition 5.4.)*

The problem you've addressed in the last two exercises is called the coupon collector problem *among computer scientists: imagine, say, a cereal company that puts one of n coupons into each box of cereal that it sells, choosing which coupon type goes into each box randomly. How many boxes of cereal must a serial cereal eater buy before he collects a complete set of the n coupons?*

True story: some nostalgic friends and I were trying to remember all of the possible responses on a Magic 8 Ball, a pseudopsychic toy that reveals one of 20 answers uniformly at random when it's shaken—things like

$$\{\text{ask again later, signs point to yes, don't count on it}, \ldots\}.$$

We found a toy shop with a Magic 8 Ball in stock and started asking it questions. We hoped to have learned all 20 different answers before we got kicked out of the store.

10.138 What is the probability that we'd get 20 different answers in our first 20 trials?

10.139 In expectation, how many trials would we need before we found all 20 answers? (Use the result on coupon collecting from Exercise 10.137.)

In Exercise 10.139, you determined the number of trials that, on average, *are necessary to get all 20 answers. But how* likely *are we to succeed with a certain number of trials?*

10.140 Suppose we perform 200 trials. What is the probability that a *particular* answer (for example, "ask again later") was never revealed in any of those 200 trials?

10.141 Use the Union Bound (Exercise 10.37) and the previous exercise to argue that the probability that we need more than 200 trials to see all 20 answers is less than 0.1%.

10.142 Suppose that one random bit in a 32-bit number is corrupted (that is, flipped from 0 to 1 or from 1 to 0). What is the expected size of the error (thinking of the change of the value in binary)? What about for a random bit in an n-bit number?

10.143 Suppose that the numbers $\{1, \ldots, n\}$ are randomly ordered—that is, we choose a random permutation π of $\{1, \ldots, n\}$. For a particular index i, what is the probability that $\pi_i = i$—that is, the ith biggest element is in the ith position?

10.144 Let X be a random variable denoting the number of indices i for which $\pi_i = i$. What is $\mathrm{E}[X]$? *(Hint: define indicator random variables and use linearity of expectation.)*

Markov's inequality *states that, for a random variable X that is always nonnegative (that is, for any x in the sample space, we have $X(x) \geq 0$), the following statement is true, for any $\alpha \geq 1$:*

$$\Pr[X \geq \alpha] \leq \frac{\mathrm{E}[X]}{\alpha}.$$

Markov's inequality is named after Andrey Markov, a 19th-to-20th-century Russian mathematician. A number of other important ideas in probability are also named after him, like Markov processes, Hidden Markov models, and more.

10.145 Prove Markov's inequality. *(Hint: use conditional expectation.)*

10.146 The *median* of a random variable X is a value x such that

$$\Pr[X \leq x] \geq \tfrac{1}{2} \quad \text{and} \quad \Pr[X \geq x] \geq \tfrac{1}{2}.$$

Using Markov's inequality, prove that the median of a nonnegative random variable X is at most $2 \cdot \mathrm{E}[X]$.

Take a fair coin, and repeatedly flip it until it comes up heads. Let K be a random variable indicating the number of flips performed. (We've already shown that $\mathrm{E}[K] = 2$, in Example 10.39.) You are offered a chance to play a gambling game, for the low low price of y dollars to enter. A fair coin will be flipped until it comes up heads, and you will be paid $(3/2)^K$ dollars if K flips were required. (So there's a $\frac{1}{2}$ chance that you'll be paid \$1.50 because the first flip comes up heads; a $\frac{1}{4}$ chance that you'll be paid \$2.25 = $(1.50)^2$ because the first flip comes up tails and the second comes up heads, and so forth.)

10.147 Assuming that you care *only* about expected value—that is, you're willing to play if and only if $\mathrm{E}[(3/2)^K] \geq y$—then what value of y is the break-even point? (In other words, what is $\mathrm{E}[(3/2)^K]$?)

10.148 Let's sweeten the deal slightly: you'll be paid 2^K dollars if K flips are required. Assuming that you still care *only* about expected value, then what value of y is the break-even point? *(Be careful!)*

10.149 Let X be the number of heads flipped in 4 independent flips of a fair coin. What is $\mathrm{var}(X)$?

10.150 Let Y be the average of two independent rolls of a fair die. What is $\mathrm{var}(Y)$?

10.151 Let $a \in \mathbb{R}$, and let X be a random variable. Prove that $\mathrm{E}[a \cdot X] = a \cdot \mathrm{E}[X]$.

10.152 Let $a \in \mathbb{R}$, and let X be a random variable. Prove that $\mathrm{var}(a \cdot X) = a^2 \cdot \mathrm{var}(X)$.

10.153 Prove that $\mathrm{var}(X + Y) = \mathrm{var}(X) + \mathrm{var}(Y)$ for two independent random variables X and Y. *(Hint: use Exercise 10.127.)*

10.154 Let X be a random variable following a binomial distribution with parameters n and p. (That is, X is the number of heads found in n flips of a p-biased coin.) Using Exercise 10.153 and the logic as in Example 10.40, show that $\mathrm{E}[X] = np$ and $\mathrm{var}(X) = np(1-p)$.

10.155 Flip a p-biased coin n times, and let Y be a random variable denoting the *fraction* of those n flips that came up heads. What are $\mathrm{E}[Y]$ and $\mathrm{var}(Y)$?

In the next few exercises, you'll find the variance of a geometric random variable. This derivation will require a little more work than the result from Exercise 10.154 (about the variance of a binomial random variable); in particular, we'll need a preliminary result about summations first:

10.156 *(Calculus required.)* Prove the following two formulas, for any real number r with $0 \leq r < 1$:

$$\sum_{i=0}^{\infty} ir^i = \frac{r}{(1-r)^2} \qquad \sum_{i=0}^{\infty} i^2 r^i = \frac{r(1+r)}{(1-r)^3}.$$

(Hint: use the geometric series formula $\sum_{i=0}^{n} r^i = \frac{r^{n+1}-1}{r-1}$ from Theorem 5.2, differentiate, and take the limit as n grows. Repeat for the second derivative.)

10.157 Let X be a geometric random variable with parameter p. (That is, X denotes the number of flips of a p-biased coin we need before we see heads for the first time.) What is $\mathrm{var}(X)$? *(Hint: compute both $\mathrm{E}[X]^2$ and $\mathrm{E}[X^2]$. The previous exercise will help with at least one of those computations.)*

Recall from Chapter 3 that a proposition is in 3-conjunctive normal form (3CNF) *if it is the conjunction of* clauses, *where each clause is the disjunction of three different variables/negated variables. For example,*

$$(\neg p \vee q \vee r) \wedge (\neg q \vee \neg r \vee x)$$

is in 3CNF. Recall further that a proposition φ is satisfiable *if it's possible to give a truth assignment for the variables of φ to true/false so that φ itself turns out to be true. We've previously discussed that it is believed to be computationally very difficult to determine whether a proposition φ is satisfiable (see p. 326)—and it's believed to be very hard to determine whether φ is satisfiable* even if φ is in 3CNF. *But you'll show here an easy way to satisfy "most" clauses of a proposition φ in 3CNF, using randomization.*

10.158 Let φ be a proposition in 3CNF. Consider a *random truth assignment* for φ—that is, each variable is set independently to True with probability $\frac{1}{2}$. Prove that a particular clause of φ is true under this truth assignment with probability $\geq \frac{7}{8}$.

10.159 Suppose that φ has m clauses and n variables. Prove that the *expected* number of satisfied clauses under a random truth assignment is at least $\frac{7m}{8}$.

10.160 Prove the following general statement about any random variable: $\Pr[X \geq \mathrm{E}[X]] > 0$. *(Hint: use conditional expectation.)* Then, using this general fact and Exercise 10.159, argue that, for any 3CNF proposition φ, *there exists a truth assignment that satisfies at least $\frac{7}{8}$ of φ's clauses.*

Taking it further: One can also show that there's a very good chance—at least $8/m$—that a random truth assignment satisfies at least $7m/8$ clauses, and therefore we expect to find such a truth assignment within $m/8$ random trials. This algorithm is called *Johnson's algorithm*, named after the researcher David Johnson; for details of this and other randomized algorithms for satisfiability, see a good book on randomized algorithms.[11]

[11] Michael Mitzenmacher and Eli Upfal. *Probability and computing: randomized algorithms and probabilistic analysis*. Cambridge University Press, 2005; Rajeev Motwani and Prabhakar Raghavan. *Randomized Algorithms*. Cambridge University Press, 1995; and Jon Kleinberg and Éva Tardos. *Algorithm Design*. Addison–Wesley, 2006.

10.5 Chapter at a Glance

Probability, Outcomes, and Events

Imagine a process by which some quantities of interest are determined in some random way. An *outcome*, or *realization*, of this probabilistic process is the sequence of results for all randomly determined quantities. The *sample space* S is the set of all possible outcomes. A *probability function* $\text{Pr} : S \to \mathbb{R}$ describes, for each outcome $s \in S$, the fraction of the time that s occurs. The probability function Pr must satisfy two conditions: (i) $\sum_{s \in S} \text{Pr}[s] = 1$, and (ii) $\text{Pr}[s] \geq 0$ for every $s \in S$.

An *event* is a subset of S, and the *probability of an event* E, written $\text{Pr}[E]$, is the sum of the probabilities of the outcomes contained in E. We have that $\text{Pr}[S] = 1$ and $\text{Pr}[\varnothing] = 0$. For events A and B, writing $\overline{A}$ ("not A") to denote the event $\overline{A} = S - A$, we have that $\text{Pr}[\,\overline{A}\,] = 1 - \text{Pr}[A]$, and $\text{Pr}[A \cup B] = \text{Pr}[A] + \text{Pr}[B] - \text{Pr}[A \cap B]$.

We can use a *tree diagram* to represent a sequence of random choices, where internal nodes of the tree correspond to random decisions made by the probabilistic process; leaves correspond to the outcomes in the sample space. Every edge leaving an internal node is labeled with the probability of the corresponding random decision; the probability of a particular outcome is precisely equal to the product of the labels on the edges leading from the root to its corresponding leaf.

The *uniform distribution* is the probability distribution in which all outcomes in the sample space S are equally likely—that is, when $\text{Pr}[s] = \frac{1}{|S|}$ for each $s \in S$. (*Nonuniform probability* is when this equality does not hold.)

The *Bernoulli distribution with parameter* p is the probability distribution that results from flipping one coin, where the sample space is $\{\text{H}, \text{T}\}$ and $\text{Pr}[\text{H}] = p$ (and thus $\text{Pr}[\text{T}] = 1 - p$). Such a coin is called *p-biased.* Each coin flip is called a *trial*; the flip is called *fair* if $p = \frac{1}{2}$.

The *binomial distribution with parameters* n *and* p is a distribution over the sample space $\{0, 1, \ldots, n\}$ determined by flipping a p-biased coin n times and counting the number of times the coin comes up heads. Here $\text{Pr}[k] = \binom{n}{k} \cdot p^k \cdot (1-p)^{n-k}$ denotes the probability that there are precisely k heads in the n flips.

The *geometric distribution with parameter* p is a distribution over the positive integers, where the output is determined by the number of flips of a p-biased coin required before we first see a heads; thus $\text{Pr}[k] = (1-p)^{k-1} \cdot p$ for any integer $k \geq 1$.

Independence and Conditional Probability

When there are multiple events of interest, then one useful way understanding the relationship between two events is to understand whether one event's occurrence changes the likelihood of the other event also occurring. When there's no change, the events are called *independent*; when there is a change in the probability, the events are called *dependent*. More formally, two events A and B are *independent* (or *uncorrelated*) if and only if $\text{Pr}[A \cap B] = \text{Pr}[A] \cdot \text{Pr}[B]$. Otherwise the events A and B are called *dependent* (or *correlated*). Intuitively, A and B are dependent if A's occurrence/nonoccurrence tells us something about whether B occurs. When knowing that A occurred makes B

more likely to occur, we say that A and B are *positively correlated*; when A makes B less likely to occur, we say that A and B are *negatively correlated*.

The *conditional probability of A given B* is

$$\Pr\left[A|B\right] = \frac{\Pr\left[A \cap B\right]}{\Pr\left[B\right]}.$$

(Treat $\Pr\left[A|B\right]$ as undefined when $\Pr[B] = 0$.) Intuitively, we can think of $\Pr\left[A|B\right]$ as "zooming" the universe down to the set B. Two events A and B for which $\Pr[B] \neq 0$ are independent if and only if $\Pr\left[A|B\right] = \Pr[A]$.

There are a few useful equivalences based on conditional probability. For any events A and B, the *chain rule* says that $\Pr[A \cap B] = \Pr[B] \cdot \Pr\left[A|B\right]$; more generally,

$$\begin{aligned}&\Pr[A_1 \cap A_2 \cap A_3 \cap \cdots \cap A_k]\\&= \Pr[A_1] \cdot \Pr\left[A_2|A_1\right] \cdot \Pr\left[A_3|A_1 \cap A_2\right] \cdot \;\cdots\; \cdot \Pr\left[A_k|A_1 \cap \cdots \cap A_{k-1}\right].\end{aligned}$$

The *law of total probability* says that $\Pr[A] = \Pr\left[A|B\right] \cdot \Pr[B] + \Pr\left[A|\,\overline{B}\,\right] \cdot \Pr[\,\overline{B}\,]$.

Bayes' Rule is a particularly useful rule that allows us to "flip around" a conditional probability statement: for any two events A and B, we have

$$\Pr\left[A|B\right] = \frac{\Pr\left[B|A\right] \cdot \Pr[A]}{\Pr[B]}.$$

Random Variables and Expectation

The probabilistic statements that we've considered so far are about events ("whether or not" questions); we can also consider probabilistic questions about "how much" or "how often." A *random variable* X assigns a numerical value to every outcome in the sample space S—that is, a random variable is a function $X : S \to \mathbb{R}$. (Often we write X to denote the value of a random variable X for a realization chosen according to Pr, or perform arithmetic on random variables.) An *indicator random variable* is a $\{0, 1\}$-valued random variable. Two random variables X and Y are *independent* if every two events of the form "$X = x$" and "$Y = y$" are independent.

The *expectation* of a random variable X, denoted $\mathrm{E}[X]$, is the average value of X, defined as $\mathrm{E}[X] = \sum_{x \in S} X(x) \cdot \Pr[x]$. A Bernoulli random variable with parameter p has expectation p. A binomial random variable with parameters p and n has expectation pn. A geometric random variable with parameter p has expectation $\frac{1}{p}$.

Linearity of expectation is the very useful fact that the expectation of a sum is the sum of the expectations. That is, for random variables $X : S \to \mathbb{R}$ and $Y : S \to \mathbb{R}$, we have $\mathrm{E}[X + Y] = \mathrm{E}[X] + \mathrm{E}[Y]$. (Note that there is no requirement of independence on X and Y!) Another useful fact is that, for a positive integer–valued random variable $X : S \to \mathbb{Z}^{\geq 0}$, we have $\mathrm{E}[X] = \sum_{i=1}^{\infty} \Pr[X \geq i]$.

The *conditional expectation* of a random variable X given an event E is the average value of X over outcomes where E occurs, defined as $\mathrm{E}\left[X|E\right] = \sum_{x \in E} X(x) \cdot \Pr\left[x|E\right]$.

The *variance* of a random variable X is

$$\mathrm{var}(X) = \mathrm{E}\left[(X - \mathrm{E}[X])^2\right] = \mathrm{E}\left[X^2\right] - (\mathrm{E}[X])^2.$$

The *standard deviation* is $\mathrm{std}(X) = \sqrt{\mathrm{var}(X)}$.

Key Terms and Results

Key Terms

PROBABILITY, OUTCOMES, AND EVENTS

- outcome/realization
- sample space
- probability function/distribution
- event
- tree diagram
- uniform vs. nonuniform probability
- fair vs. biased coin flips
- uniform distribution
- Bernoulli distribution
- binomial distribution
- geometric distribution

INDEPENDENCE AND CONDITIONAL PROBABILITY

- independent/uncorrelated events
- dependent/correlated events
- positive/negative correlation
- conditional probability
- chain rule
- law of total probability
- Bayes' Rule

RANDOM VARIABLES AND EXPECTATION

- random variable
- indicator random variable
- independent random variables
- expectation
- linearity of expectation
- conditional expectation
- variance
- standard deviation

Key Results

PROBABILITY, OUTCOMES, AND EVENTS

1. For a sample space S and events A and B, writing $\overline{A}$ ("not A") to denote the event $S - A$, we have that $\Pr[S] = 1$, $\Pr[\varnothing] = 0$, $\Pr[\,\overline{A}\,] = 1 - \Pr[A]$, and $\Pr[A \cup B] = \Pr[A] + \Pr[B] - \Pr[A \cap B]$.
2. Under the uniform distribution, $\Pr[s] = \frac{1}{|S|}$ for every $s \in S$. Consider parameters p and n. Under a Bernoulli distribution, $\Pr[\text{H}] = p$ and $\Pr[\text{T}] = 1 - p$. Under a binomial distribution, $\Pr[k] = \binom{n}{k} p^k (1-p)^{n-k}$. Under a geometric distribution, $\Pr[k] = (1-p)^{k-1} p$.

INDEPENDENCE AND CONDITIONAL PROBABILITY

1. Events A and B are independent if and only if $\Pr[A \cap B] = \Pr[A] \cdot \Pr[B]$, or, equivalently, if $\Pr\left[A|B\right] = \Pr[A]$.
2. The chain rule: $\Pr[A \cap B] = \Pr[B] \cdot \Pr\left[A|B\right]$.
3. The law of total probability: $\Pr[A] = \Pr\left[A|B\right] \cdot \Pr[B] + \Pr\left[A|\,\overline{B}\,\right] \cdot \Pr[\,\overline{B}\,]$.
4. Bayes' Rule: $\Pr\left[A|B\right] = \frac{\Pr[B|A] \cdot \Pr[A]}{\Pr[B]}$.

RANDOM VARIABLES AND EXPECTATION

1. The *expectation* of a random variable X is the average value of X, defined as $\mathrm{E}[X] = \sum_{x \in S} X(x) \cdot \Pr[x]$.
2. A Bernoulli random variable with parameter p has expectation p. A binomial random variable with parameters p and n has expectation pn. A geometric random variable with parameter p has expectation $\frac{1}{p}$.
3. Linearity of expectation: for any two random variables X and Y, we have $\mathrm{E}[X+Y] = \mathrm{E}[X] + \mathrm{E}[Y]$. (Note that there is no requirement of independence on X and Y!)
4. For a random variable $X : S \to \mathbb{Z}^{\geq 0}$, we have that $\mathrm{E}[X] = \sum_{i=1}^{\infty} \Pr[X \geq i]$.
5. For a random variable X, we have $\mathrm{var}(X) = \mathrm{E}\left[(X - \mathrm{E}[X])^2\right] = \mathrm{E}\left[X^2\right] - (\mathrm{E}[X])^2$.

11
Graphs and Trees

In which our heroes explore the many twisting paths through the gnarled forest, emerging in the happy and peaceful land in which their computational adventures will continue.

11.1 *Why You Might Care*

Oh what a tangled web we weave,
When first we practise to deceive!

Sir Walter Scott (1771–1832), *Marmion* (1808)

It's possible to make graphs sound hopelessly abstract and utterly uninteresting: *a* graph *is a pair* $\langle V, E\rangle$, *where* V *is a nonempty collection of entities called* nodes *and* E *is a collection of* edges *that join pairs of nodes.* But graphs are fascinating—at least, when the entities and the relationship represented by the edges are themselves interesting! Here are a few of the many examples of types of graphs:

- *social networks* like Facebook (or LinkedIn or Pinterest or ...): the nodes are people, and an edge between two people represents a friendship (or at least a "friendship").
- the *world-wide web*: the nodes are web pages, and an edge represents a hyperlink from one page to another. These hyperlinks between pages form the basis for the ranking of web pages by search engines like Google.[1]
- *dating networks*: nodes represent people; an edge connects two people who have been involved in a romantic relationship. These networks have implications for the spread of certain communicable diseases, particularly sexually transmitted infections.
- *road networks* and other transportation networks: edges represent roads; nodes represent intersections. For example, United Parcel Service (UPS) saves gas (and money!) by using a route-finding algorithm through this network that avoid turns across traffic.[2]
- *food webs*: nodes represent species within a particular ecosystem, and an edge from one species to another indicates that the first species preys on the latter.
- *co-purchase networks*: nodes are products that are sold by a retailer like Walmart or Amazon; an edge between two products indicates the number of customers who bought both products. These networks have implications for *recommender systems*, the "people who bought x also bought y" feature of Amazon.
- *the internet*: nodes are computers (personal computers, servers, and other networking hardware like routers), and edges represent physical wires connecting two machines together. When you request a video from `youtube.com`, the computers involved in the network must collectively construct a path along which YouTube's bits can flow so that they reach your computer.

Graphs are ubiquitous. Indeed, any pairwise relationship among entities is really underlyingly a graph: web pages and links, computers and fiber optic cables, kidney patients/donors and compatibility for transplants. The applications are innumerable, and this chapter will barely scratch the surface. Graphs and graph-theoretic reasoning will arise again and again well beyond the end of this book.

[1] Sergei Brin and Larry Page. The anatomy of a large-scale hypertextual web search engine. In *7th International World-Wide Web Conference*, 1998.

[2] Joel Lovell. Left-hand-turn elimination. *The New York Times*, 9 December 2007.

11.2 *Formal Introduction*

> The Bible tells us to love our neighbors, and also to love our enemies; probably because they are generally the same people.
>
> G. K. Chesterton (1874–1936)

We begin by defining the terminology for the two different basic types of graphs. In both, we have a set of entities called *nodes*, some pairs of which are joined by a relationship called an *edge*. (A node can also be called a *vertex*.) The two types of graph differ in whether the relationship represented by an edge is "between two nodes" or "from one node to another." In an *undirected graph*, the relationship denoted by the edges is symmetric (for example, "u and v are genetically related"):

Definition 11.1 (Undirected Graph)
A undirected graph *is a pair* $G = \langle V, E \rangle$ *where* V *is a nonempty set of* vertices *or* nodes, *and* $E \subseteq \Big\{ \{u, v\} : u, v \in V \Big\}$ *is a set of* edges *joining pairs of vertices.*

vertex, n.: a node. plural: *vertices*.

We will use the terms *node/nodes* and *vertex/vertices* interchangeably throughout this chapter. (Both terms are used commonly in CS.) A graph can also be called a *network*; edges are also sometimes called *links*, or occasionally *arcs* in directed graphs.

The second basic kind of graph is a *directed graph*, in which the relationship denoted by the edges need not be reciprocated (for example, "u has texted v"):

Definition 11.2 (Directed Graph)
A directed graph *is a pair* $G = \langle V, E \rangle$ *where* V *is a nonempty set of nodes, and* $E \subseteq V \times V$ *is a set of edges joining (ordered) pairs of vertices.*

In other words, in a directed graph an edge is an *ordered* pair of vertices ("an edge from u to v") and in an undirected graph an edge is an *unordered* pair of vertices ("an edge between u and v"). Think about the difference between Twitter followers (directed) and Facebook friendships (undirected): Alice can follow Bob without Bob following Alice, but they're either friends or they're not friends.

Graphs are generally drawn with nodes represented as circles, and edges represented by lines. Each edge in directed graphs is drawn with an arrow indicating its *orientation* ("which way it goes"). Here is an example of each:

Example 11.1 (A sample undirected graph)
Here is an undirected graph:

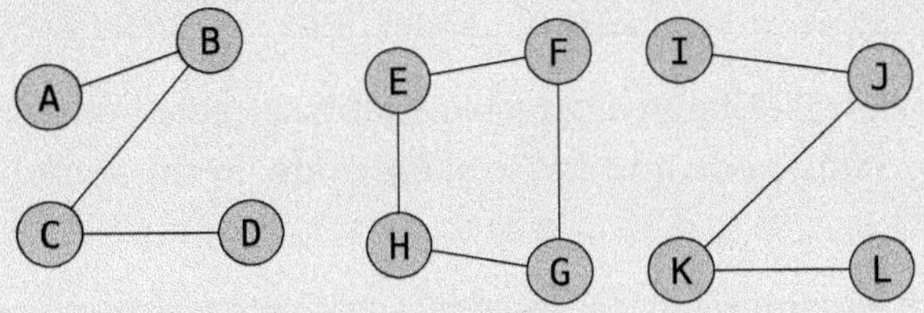

This graph contains:

- 12 nodes: $\{$A, B, C, D, E, F, G, H, I, J, K, L$\}$.
- 10 edges: $\Big\{ \{$A, B$\}, \{$B, C$\}, \{$C, D$\}, \{$E, F$\}, \{$E, H$\}, \{$F, G$\}, \{$G, H$\}, \{$I, J$\}, \{$J, K$\}, \{$K, L$\} \Big\}$.

Example 11.2 (Streets of Manhattan: a sample directed graph)
The following directed graph contains 9 nodes, each corresponding to an intersection of a "street" running east–west and an "avenue" running north–south in Manhattan:

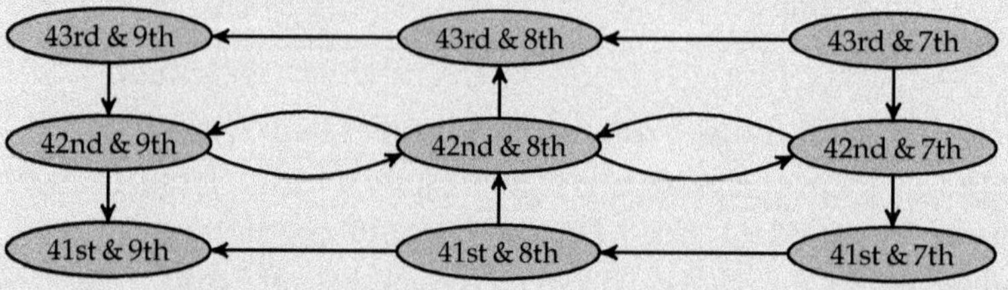

There are 14 edges in this graph. There's something potentially tricky in counting to 14: edges in a directed graph are *ordered* pairs, so there are *two* edges between *42nd & 9th* and *42nd & 8th*, one in each direction—$\langle$*42nd & 9th*, *42nd & 8th*$\rangle$ and $\langle$*42nd & 8th*, *42nd & 9th*$\rangle$. The pair of nodes *42nd & 8th* and *42nd & 7th* is similar.

For many of the concepts that we'll explore in this chapter, it will turn out that there are no substantive differences between the ideas for directed and undirected graphs. To avoid being tedious and unhelpfully repetitive, whenever it's possible we'll state definitions and results about both undirected and directed graphs simultaneously. But doing so will require a little abuse of notation: we'll allow ourselves to write an edge as an ordered pair $\langle u, v \rangle$ *even for an undirected graph.* In an undirected graph, we will agree to understand both $\langle u, v \rangle$ and $\langle v, u \rangle$ as meaning $\{u, v\}$.

Simple graphs

For many of the real-world phenomena that we will be interested in modeling, it will make sense to make a simplifying assumption about the edges in our graphs. Specifically, we will typically restrict our attention to so-called *simple* graphs, which forbid two different kinds of edges: edges that connect nodes to themselves, and edges that are precise duplicates of other existing edges. (See Figure 11.1.)

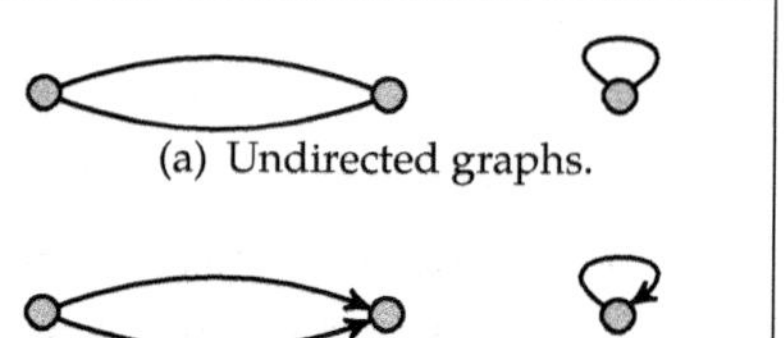

(a) Undirected graphs.

(b) Directed graphs.

Figure 11.1: Parallel edges and self-loops.

Definition 11.3 (Self-loops and parallel edges)
A self-loop *is an edge from a node u to itself. Two edges are called* parallel *if they both go from same node u and both go to the same node v.*

Note that the edges $\langle u, v \rangle$ and $\langle v, u \rangle$ are not parallel in a directed graph: directed edges are parallel only if they both go *from* the same node and *to* the same node, in the same orientation.

Definition 11.4 (Simple graph)
A graph is simple *if it contains no parallel edges and no self-loops.*

In general, the particular real-world phenomenon that we seek to model will dictate whether self-loops, parallel edges, or both will make sense. Here are a few examples:

Example 11.3 (Self-loops and parallel edges)

Problem: Suppose that we construct a graph to model each of the following phenomena. In which settings do self-loops or parallel edges make sense?

1. A social network: nodes correspond to people; (undirected) edges represent friendships.
2. The web: nodes correspond to web pages; (directed) edges represent links.
3. The flight network for a commercial airline: nodes correspond to airports; (directed) edges denote flights scheduled by the airline in the next month.
4. The email network at a college: nodes correspond to students; there is a (directed) edge $\langle u, v \rangle$ if u has sent at least one email to v within the last year.

Solution:
1. Neither self-loops nor parallel edges make sense. A self-loop would correspond to a person being a friend of himself, and parallel edges between two people would correspond to them being friends "twice." (But two people are either friends or not friends.)
2. Both self-loops and parallel edges are reasonable. It is easy to imagine a web page p that contains a hyperlink to p itself. It is also easy to imagine a web page p that contains two separate links to another web page q. (For example, as of this writing, the "CNN" logo on `www.cnn.com` links to `www.cnn.com`. And, as of the end of this sentence, this page has three distinct references to `www.cnn.com`.)
3. In the flight network, many parallel edges will exist: there are generally many scheduled commercial flights from one airport to another—for example, there are dozens of flights every week from BOS (Boston, MA) to SFO (San Francisco, CA) on most major airlines. However, there are no self-loops: a commercial flight from an airport back to the same airport doesn't go anywhere!
4. Self-loops are reasonable but parallel edges are not. A student u has either sent email to v in the last year or she has not, so parallel edges don't make sense in this network. However, self-loops exist if any student has sent an email to herself (as many people do to remind themselves to do something later).

Throughout, we assume that all graphs are simple unless otherwise noted.

Taking it further: Actually, the way that we phrased our definitions of graphs in Definitions 11.1 and 11.2 doesn't even *allow* us to consider parallel edges. (Our definitions do allow self-loops, though.) That's because we defined the edges as a subset E of $V \times V$ or $\{\, \{u, v\} : u, v \in V \}$, and sets don't allow duplication—which means that we can't have $\langle u, v \rangle$ in E "twice." There are alternate ways to formalize graphs that do permit parallel edges, but they're needlessly complicated for the applications that we'll focus on in this chapter.

11.2.1 *Neighborhoods and Degree*

Imagine a social network in which two people, Ursula and Victor, are friends—or, more generally, imagine an undirected graph in which nodes u and v are joined by an edge. Here's the vocabulary for referring to these nodes and the edge between them:

Definition 11.5 (Adjacency, neighbors, endpoints, incidence)
For an edge $e = \{u, v\}$ in an undirected graph (see Figure 11.2), we say that:

- *the nodes u and v are* adjacent;
- *the node v is a* neighbor *of the node u (and vice versa);*
- *the nodes u and v are the* endpoints *of the edge e; and*
- *the nodes u and v are both* incident *to the edge e.*

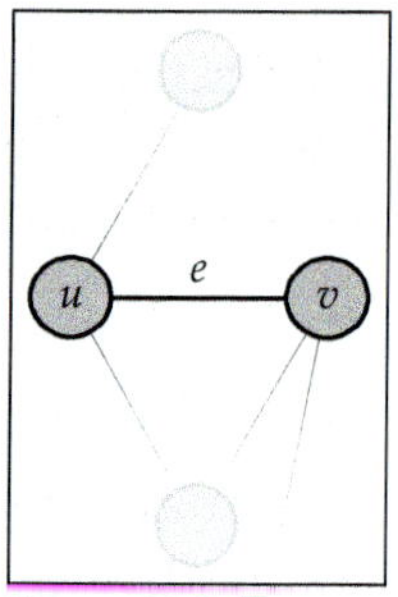

Figure 11.2: Two nodes joined by an edge.

It's important to distinguish between two distinct concepts:

- the *direct* connection between two nodes u and v that are adjacent—that is, a single edge that joins u and v directly; and
- an *indirect* connection between two nodes that follows a sequence of edges.

At the moment, we're talking *only* about the first kind of connection, a direct connection via a single edge. (A multihop connection is called a *path*; we'll talk about paths in Section 11.3.) Here's an example of the vocabulary from Definition 11.5:

Example 11.4 (Disney World to Disney Land)
Here is a small portion of the U.S. Interstate system between Orlando, FL and Los Angeles, CA. Each of the roads is labeled by its name.

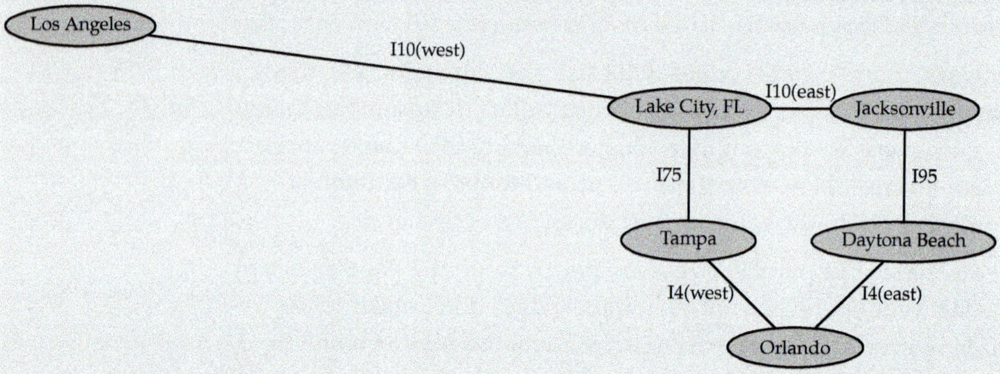

In this graph:

- Orlando is adjacent to Tampa and Daytona Beach.
- None of the other nodes (Lake City, Jacksonville, Los Angeles) is a neighbor of Orlando. Orlando is also not a neighbor of itself.
- The endpoints of edge I75 are Tampa and Lake City.
- Jacksonville is incident to I95, as is Daytona Beach.

The *neighborhood* of a node is the set of all nodes adjacent to it:

Definition 11.6 (Neighborhood)
Let $G = \langle V, E\rangle$ be an undirected graph, and let $u \in V$ be a node. The neighborhood *of u is the set $\big\{v \in V : \{u, v\} \in E\big\}$—that is, the set of all neighbors of u.*

For example, in the graph from Example 11.4 (reproduced in abbreviated form in Figure 11.3), the neighborhood of Lake City (LC) is {Los Angeles (LA), Tampa (TA), Jacksonville (JA)}. Or, for a graph G that represents a social network, the neighborhood of a node u is the set of people who are u's friends.

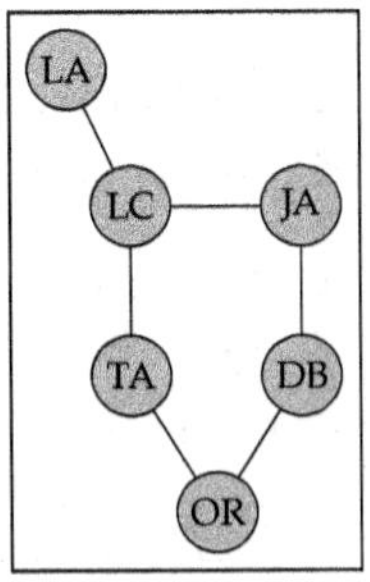

Figure 11.3: The road network from Example 11.4, abbreviated.

DEGREE

It's also common to refer the *number* of neighbors that a node has (without reference to which particular nodes happen to be that node's neighbors):

> **Definition 11.7 (Degree)**
> *The* degree *of a node u in an undirected graph G is the size of the neighborhood of u in G—that is, the number of nodes adjacent to u.*

For example, in the graph in Figure 11.3, Lake City (LC) has degree 3 and Los Angeles (LA) has degree 1. Or, in a social network, the degree of a node u is the popularity of u—the number of friends that u has. Here are a few practice questions:

Example 11.5 (Neighborhood and degree)
Problem: Consider the following graph:

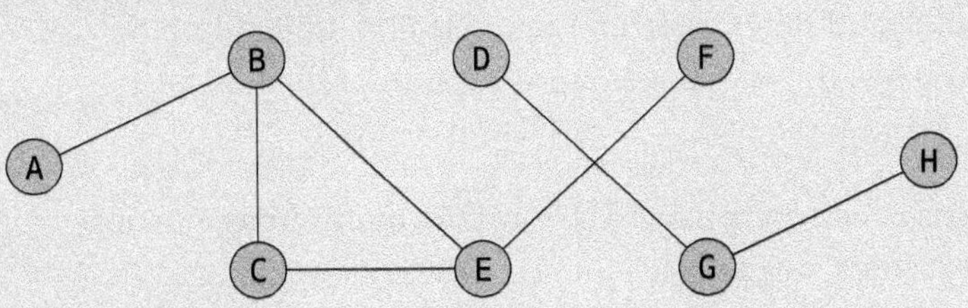

1. What are the neighbors of node C?
2. What nodes, if any, have degree equal to one?
3. What node has the highest degree in this graph?
4. What nodes, if any, are in the neighborhoods of both nodes B and E?

Solution: 1. Node C has two neighbors, namely the nodes B and E.

2. The nodes with degree one are those with precisely one neighbor. These nodes are: A, D, F, and H. (Their solitary neighbors are, respectively: B, G, E, and G.)
3. We simply count neighbors for each node, and we find that nodes B and E both have degree three, and are tied as the nodes with the highest degree.
4. The neighborhood of node B is {A, C, E}, and the neighborhood of node E is {B, C, F}. Taking the intersection of those sets yields the one node in the neighborhood of both B and E, namely node C.

Taking it further: Consider a population of people—say, the current residents of Canada—represented as a social network, in an undirected graph whose edges represent friendship. For a node in the social network (also known as a person), we can calculate many numbers that may be interesting: height, age, income, number of cigarettes smoked per day, self-reported happiness, etc. Then, for any one of these numerical properties, we can consider the *distribution* over the population: for example, the distribution of heights, or the distribution of ages. (The height distribution will follow a roughly bell-shaped curve; the age distribution is more complicated, both because of death and because of variation in the birth rate over time.) Another interesting numerical property of a person u is the *degree* of u: that is, the number of friends that u has. The *degree distribution* of a graph describes how popularity varies across the nodes of the network. The degree distribution has some interesting properties—very different from the distribution of heights or ages. See p. 1123 for some discussion.

The Handshaking Lemma

Before we move on from degree, we'll prove a basic but valuable fact, colloquially called the "handshaking lemma." (We can represent a group of people, some pairs of whom shake hands, using an undirected graph: an edge joins u and v if and only if u and v shook hands; the theorem describes the number of shakes.) The handshaking lemma relates the sum of nodes' degrees to the number of edges in the graph:

Theorem 11.1 ("Handshaking Lemma")
Let $G = \langle V, E \rangle$ be an undirected graph. Then

$$\sum_{u \in V} \text{degree}(u) = 2|E|.$$

For example, Figure 11.4 shows our road network from Example 11.4, with all nodes labeled by their degree. This graph has $|E| = 6$ edges, and the sum of the nodes' degrees is $1 + 3 + 2 + 2 + 2 + 2 = 12$, and indeed $12 = 2 \cdot 6$. Here is a proof:

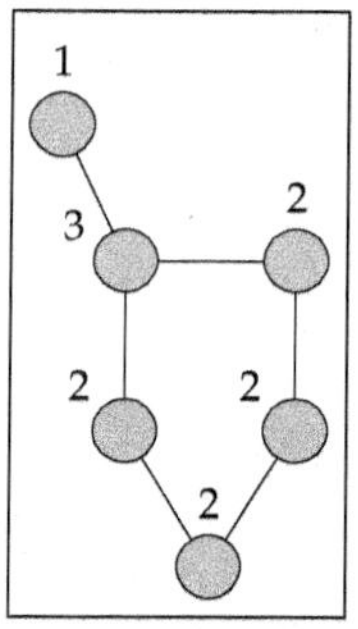

Figure 11.4: The road network from Figure 11.3, with nodes labeled by their degree.

Proof of Theorem 11.1. Every edge has two endpoints! Or, more formally, imagine looping over each edge to compute all nodes' degrees:

```
initialize d_u to 0 for each node u
for each edge {u, v} ∈ E:
    d_u := d_u + 1
    d_v := d_v + 1
```

"Look on every exit as being an entrance somewhere else."
— Tom Stoppard (b. 1937), *Rosencrantz and Guildenstern are Dead* (1966)

In each iteration of the **for** loop, we increment two different $d_\bullet$ values; thus, after i iterations, we have that $\sum_u d_u = 2i$. (We could give a fully rigorous proof of this fact by induction.) We complete $|E|$ iterations of the **for** loop, one for each edge, and thus at the end of the algorithm we have that $\sum_{u \in V} d_u = 2|E|$. Furthermore, after the loop, it's clear that $d_u = degree(u)$ for every node u. Thus

$$\sum_{u \in V} d_u = \sum_{u \in V} degree(u) = 2|E|. \qquad \square$$

Here's a useful corollary of Theorem 11.1 (the proof is left to you as Exercise 11.17):

Corollary 11.2
Let n_{odd} denote the number of nodes whose degree is odd. Then n_{odd} is even.

(For example, for the graph in Figure 11.4, we have $n_{\text{odd}} = 2$: the two nodes with odd degree are those with degree 1 and 3. And 2 is an even number.)

NEIGHBORHOODS AND DEGREE: DIRECTED GRAPHS

The definitions of adjacency, neighbors, and degree from Definitions 11.5–11.7 were all for *undirected* graphs. Here we'll introduce the analogous notions for directed graphs, all of which are slightly more complicated because they must account for the orientation of each edge. We start with the directed version of "neighbors":

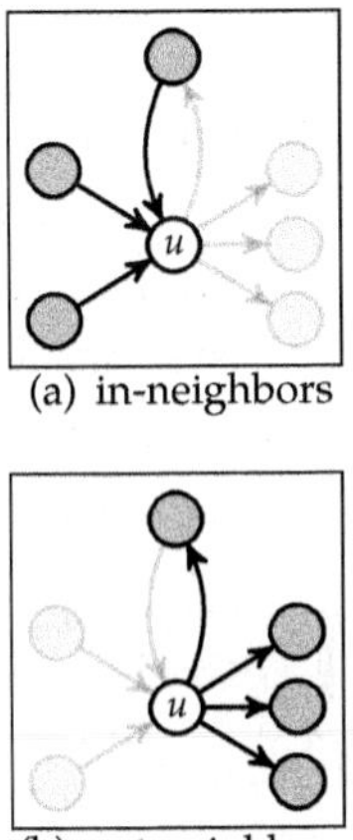

(a) in-neighbors
(b) out-neighbors

Figure 11.5: The in- and out-neighbors of a node u.

Definition 11.8 (Neighbors in directed graphs)
For an edge $\langle u, v \rangle$ from node u to node v in a directed graph, we say that:

- *the node v is an* out-neighbor *of the node u; and*
- *the node u is an* in-neighbor *of the node v.*

For example, if G represents a flight network (with nodes as airports and directed edges corresponding to flights), then the out-neighbors of node u are those airports that have direct flights from u, and the in-neighbors of u are those airports that have direct flights to u. (See Figure 11.5.) Now, using these definitions, we can define the analogues of neighborhoods and degree in directed graphs:

Definition 11.9 (Neighborhoods and degrees in directed graphs)
For a node u in an directed graph, we say that:

- *the* in-neighborhood *of u is $\{v : \langle v, u \rangle \in E\}$, the set of in-neighbors of v;*
- *the* in-degree *of u is its number of in-neighbors (its in-neighborhood's cardinality);*
- *the* out-neighborhood *of u is $\{v : \langle u, v \rangle \in E\}$, the set of out-neighbors of u; and*
- *the* out-degree *of u is its number of out-neighbors (its out-neighborhood's cardinality).*

Here are a few practice questions about in- and out-neighborhoods:

Example 11.6 (Neighborhood and degree in a directed graph)
Problem: Consider the following directed graph:

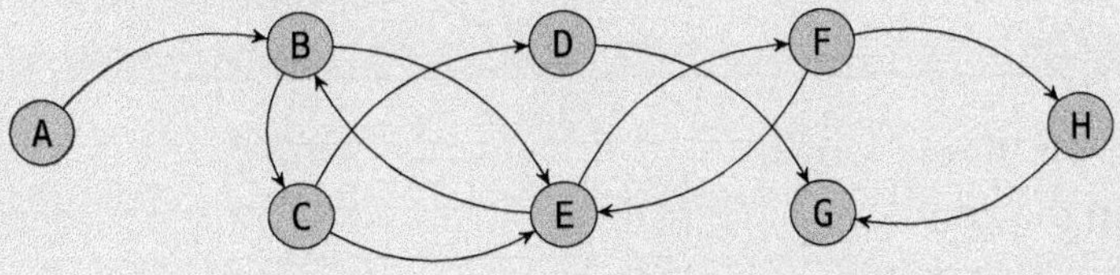

1. What are the in-neighbors of node C? The out-neighbors of C?
2. What nodes, if any, are in both the in-neighborhood and out-neighborhood of node E?
3. What nodes, if any, have in-degree zero? Out-degree zero?

Solution: 1. Node C has one in-neighbor, namely B, and two out-neighbors, namely D and E.

2. Node E has three in-neighbors (B, C, and F) and two out-neighbors (B and F). So nodes B and F are in both E's in-neighborhood and E's out-neighborhood.

3. Node A has no in-neighbors, so A's in-degree is zero. Node G has no out-neighbors, so G's out-degree is zero.

11.2.2 *Representing Graphs: Data Structures*

The graphs that we've considered so far have been presented visually: as a picture, with nodes drawn as circles and edges drawn as lines or arrows. But, of course, when we represent a graph on a computer, we'll need to use some data structure to store a network, not just some image file. Here we will give a brief summary of the two major data structures used to represent graphs. If you've had a course on data structures, then this material may be a review; if not, it will be a preview.

Taking it further: A visual representation is great for some smaller networks, and a well-designed layout can sometimes make even large networks easy to understand at a glance. *Graph drawing* is the problem of algorithmically laying out the nodes of a graph well—in an aesthetic and informative manner. There's a physics analogy that's often used in laying out graphs, in which we imagine nodes "attracting" and "repelling" each other depending on the presence or absence of edges. See p. 1124 for some discussion, including an application of this graph-drawing idea to the 9/11 Memorial in New York City. Some other gorgeous visualizations of network (and other!) data can be found online at sites like Flowing Data (`http://flowingdata.com/`), Information Is Beautiful (`http://informationisbeautiful.net`), or some of the beautiful books on data visualization like the *Atlas of Science*.[3]

[3] Katy Börner. *Atlas of Science: Visualizing What We Know*. MIT Press, 2010.

The most straightforward data structure for a graph is just a list of nodes and a list of edges. But this straightforward representation suffers for some standard, natural questions that are typically asked about graphs. Many of the natural questions that we will find ourselves asking are things like: *What are all of the neighbors of* `A`*?* or *Are* `B` *and* `C` *joined by an edge?* There are two standard data structures for graphs, each of which is tailored to make it possible to answer one of these two questions quickly.

ADJACENCY LISTS

The first standard data structure for graphs is an *adjacency list*, which—as the name implies—stores, for each node u, a list of the nodes adjacent to u:

Definition 11.10 (Adjacency list)
In an adjacency list *of a graph $G = \langle V, E \rangle$, for each node $u \in V$, we store an unsorted list of all of u's neighbors in the graph.*

The schematic for an adjacency list is illustrated in Figure 11.6: each node in the graph corresponds to a row of the table, which points to an unsorted list of that node's neighbors. (These lists are unsorted so that it's faster to add a new edge to the data structure.)

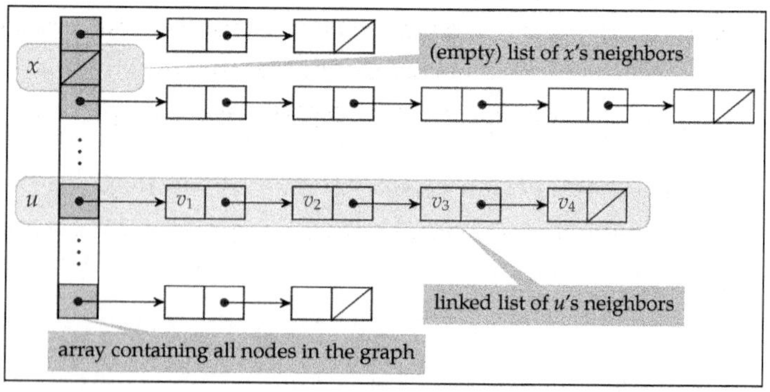

Figure 11.6: A schematic of an adjacency list.

There's no significant difference between adjacency lists for undirected graphs and for directed graphs: for an undirected graph, we list the *neighbors* for each node u; for a directed graph, we list the *out-neighbors* of each node. (Every edge $\langle u, v \rangle$ in a directed graph appears only once in the data structure, in u's list. Every edge $\{u, v\}$ in an undirected graph is represented twice: v appears in u's list, and u appears in v's list. This observation is another way of thinking of the proof of Theorem 11.1.)

Here are example adjacency lists for two graphs, one undirected and one directed:

Example 11.7 (Two sample adjacency lists)
Consider the following two graphs:

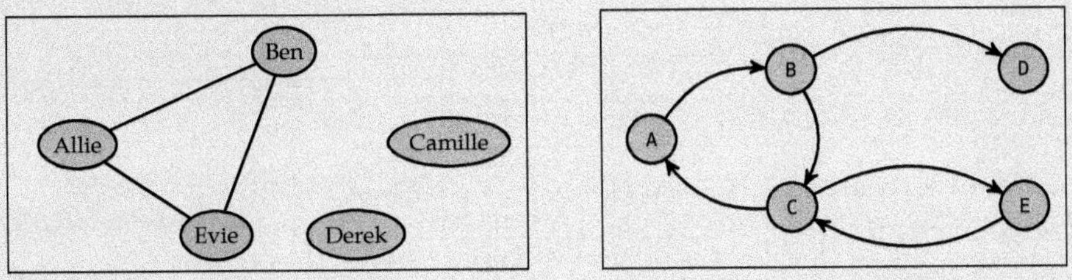

The adjacency lists for these two graphs are as follows.

Allie:	Evie, Ben
Ben:	Allie, Evie
Camille:	--
Derek:	--
Evie:	Allie, Ben

A:	B
B:	C, D
C:	E, A
D:	--
E:	C

Note that the order of the (out-)neighbors of any particular node isn't specified: for example, we could just as well said that `Evie`'s neighbors were `[Ben, Allie]` as `[Allie, Ben]`.

Adjacency matrices

The second standard data structure for representing graphs is an *adjacency matrix:*

Definition 11.11 (Adjacency matrix)
In an adjacency matrix *of a graph $G = \langle V, E \rangle$, we store the graph using an $|V|$-by-$|V|$ table. The ith row of the table corresponds to the neighbors of node i. A True (or* 1*) in column j indicates that the edge $\langle i, j \rangle$ is in E; a False (or* 0*) indicates that $\langle i, j \rangle \notin E$.*

In a directed graph, the ith row corresponds to the *out-neighbors* of node i, so that the $\langle i, j \rangle$th entry of the matrix corresponds to the presence/absence of an edge *from i to j*. The ith column corresponds to the in-neighbors of i. Here are two examples of adjacency matrices, for the graphs from Example 11.7:

Example 11.8 (Two sample adjacency matrices)
The following adjacency matrices represent the graphs from Example 11.7:

	Allie	Ben	Camille	Derek	Evie
Allie	0	1	0	0	1
Ben	1	0	0	0	1
Camille	0	0	0	0	0
Derek	0	0	0	0	0
Evie	1	1	0	0	0

	A	B	C	D	E
A	0	1	0	0	0
B	0	0	1	1	0
C	1	0	0	0	1
D	0	0	0	0	0
E	0	0	1	0	0

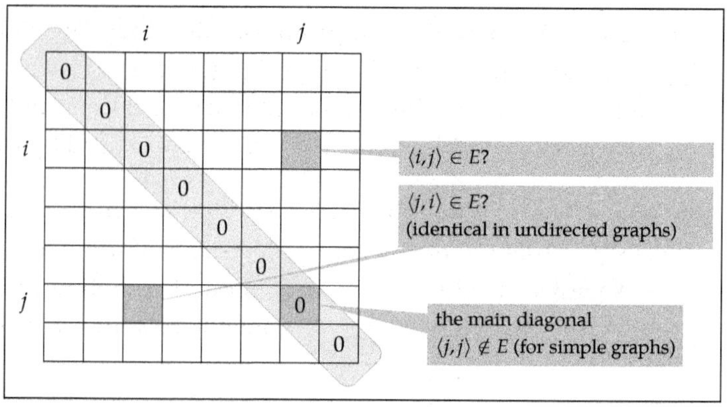

Figure 11.7: A schematic of an adjacency matrix.

The adjacency matrix has two properties that are worth a note. (See Figure 11.7.)

- *The main diagonal contains all zeros:* a 1 in the $\langle i,i \rangle$th position of the matrix would correspond to an edge between node i and node i—that is, a self-loop, which is forbidden in a simple graph.
- *For an undirected graph, the matrix is symmetric:* the $\langle i,j \rangle$th position of the matrix records the presence or absence of an edge from i to j, which is identical to the presence or absence of an edge from j to i in an undirected graph. Adjacency matrices are not necessarily symmetric in directed graphs: there may be an edge from u to v without an edge from v to u.

Choosing between adjacency lists and matrices

Which of the two data structures that we've seen for graphs should we choose? Are adjacency lists better than adjacency matrices, or the other way around? Recall the two basic questions about graphs that we wish to answer quickly:

(A) is v a neighbor of u?
(B) what are all of u's neighbors?

Figuring the details of how efficiently we can answer these questions with an adjacency list or an adjacency matrix is better suited to a data-structures textbook than this one, but here's a brief summary of the reasoning.

Meta–problem-solving tip: The answer to "which is better?" in a class or textbook is almost always *It depends!* After all, why would we waste time/pages on a solution that's always worse!? (The only plausible answer is that it warms us up conceptually for a better but more complex solution.) The real question here *what does it depend on?*

Adjacency Lists: An adjacency list is perfectly tailored to answering Question (B): we've stored precisely the list of u's neighbors for each node u, so we simply iterate through that list to output u's neighborhood. To answer Question (A), we need to search through that same unsorted list to see if v is present. In both cases, we have to spend constant time finding u's list in the table, and then we examine a list of length *degree*(u) to answer the question.

Adjacency Matrices: An adjacency matrix is perfect for answering Question (A): we just look at the appropriate spot in the table. If the $\langle u,v \rangle$th entry is True, then the edge $\langle u,v \rangle$ exists. This lookup takes constant time. Answering Question (B) requires looking at one entire row of the table, entry by entry. There are $|V|$ entries in the row, so this loop requires $|V|$ operations.

Thus adjacency matrices solve Question (A) faster, while adjacency lists are faster at solving Question (B). In addition to the time to answer these questions, we'd also want the *space*—the amount of memory—consumed by the data structure to be as small as possible. (You can think of "the amount of memory" as the total number of boxes that appear in the diagrams in Figures 11.6 and 11.7.)

Example 11.9 (Space consumption for adjacency lists and matrices)

Problem: Consider a graph $G = \langle V, E \rangle$ stored using an adjacency list or an adjacency matrix. In terms of the number of nodes and the number of edges in G—that is, in terms of $|V|$ and $|E|$—how much memory is used by these data structures?

Solution: An adjacency matrix is a $|V|$-by-$|V|$ table, and thus contains exactly $|V|^2$ cells. (Of them, the $|V|$ cells on the diagonal are always 0, but they're still there!)

An adjacency list is a $|V|$-element table pointing to $|V|$ lists; the length of the list for node u is exactly $degree(u)$. Thus the total number of cells in the data structure is

$$|V| + \sum_{u \in V} degree(u).$$

In an undirected graph we have $\sum_u degree(u) = 2|E|$, by Theorem 11.1; in a directed graph we have $\sum_u out\text{-}degree(u) = |E|$ by Exercise 11.18. Thus the total amount of memory used is

$$\begin{cases} |V| + 2|E| & \text{for an undirected graph} \\ |V| + |E| & \text{for a directed graph.} \end{cases}$$

Here's the summary of the efficiency differences between these data structures (using asymptotic notation from Chapter 6):

	adjacency list	adjacency matrix
is v a neighbor of u?	$1 + \Theta(degree(u))$	$\Theta(1)$
what are all of u's neighbors?	$1 + \Theta(degree(u))$	$\Theta(\|V\|)$
space	$\Theta(\|V\| + \|E\|)$	$\Theta(\|V\|^2)$

The better data structure in each row is highlighted. (Note that, in a simple graph, we have that $degree(u) \leq |V|$ and $|E| \leq |V|^2$.) So, is an adjacency list or an adjacency matrix better? *It depends!*

First, it depends on what kind of questions—Question (A) or Question (B) listed previously, for example—we want to answer: if we will ask few "is v a neighbor of u?" questions, then adjacency lists will be faster. If we will ask many of those questions, then we probably prefer adjacency matrices. Similarly, it might depend on how much, if at all, the graph changes over time: adjacency lists are harder to update than adjacency matrices.

Second, it depends on how many edges are present in the graph. If the total number of edges in the graph is relatively small—and thus most nodes have only a few neighbors—then $degree(u)$ will generally be small, and the adjacency list will win. If the total number of edges in the graph is relatively large, then $degree(u)$ will generally be larger, and the adjacency matrix will perform better. (Many of the most interesting real-world graphs are sparse: for example, the typical degree of a person in a social network like Facebook is perhaps a few hundred or at most a few thousand—very small in relation to the hundreds of millions of Facebook users.)

11.2.3 *Relationships between Graphs: Isomorphism and Subgraphs*

Now that we have the general definitions, we'll turn to a few more specific properties that certain graphs have. We'll start in this section with two different relationships between pairs of graphs—when two graphs are "the same" and when one is "part" of another; in Section 11.2.4, we'll look at single graphs with a particular structure.

GRAPH ISOMORPHISM

When two graphs G and H are identical except for how we happen to have arranged the nodes when we drew them on the page (and except for the names that we happen to have assigned to the nodes), then we call the graphs *isomorphic*. Informally, G and H are isomorphic if there's a way to relabel (and rearrange) the nodes of G so that G and H are exactly identical. More formally:

Greek: *iso* "same"; *morph* "form."

Definition 11.12 (Graph isomorphism)
Consider two graphs $G = \langle V, E\rangle$ and $H = \langle U, F\rangle$. We say that G and H are isomorphic *if there exists a bijection $f : V \to U$ such that*

$$\text{for all } a \in V \text{ and } b \in V, \qquad \langle a,b\rangle \in E \Leftrightarrow \langle f(a), f(b)\rangle \in F.$$

(By abusing notation as we described earlier, this definition works for either undirected or directed graphs G and H.) Here are some small examples:

Example 11.10 (Two isomorphic graphs)
Let's show that the following two directed graphs are isomorphic. (The first graph's edges could also have be written as $\{\langle a,b\rangle : a < b \text{ and } a \text{ evenly divides } b\}$.)

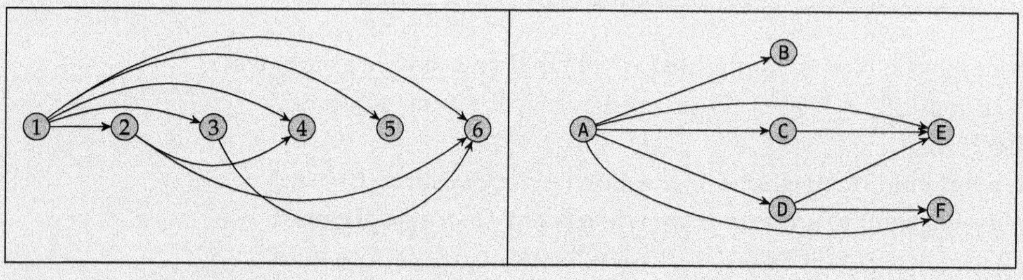

To do so, define the following bijection $f : \{1,2,\ldots,6\} \to \{\mathsf{A},\mathsf{B},\ldots,\mathsf{F}\}$:

x	1	2	3	4	5	6
$f(x)$	A	D	C	F	B	E

The tables of edges in the graphs now match exactly, so they are isomorphic:

	1	2	3	4	5	6
1		✓	✓	✓	✓	✓
2				✓		✓
3						✓
4						
5						
6						

	A	D	C	F	B	E
$f(1) = \mathsf{A}$		✓	✓	✓	✓	✓
$f(2) = \mathsf{D}$				✓		✓
$f(3) = \mathsf{C}$						✓
$f(4) = \mathsf{F}$						
$f(5) = \mathsf{B}$						
$f(6) = \mathsf{E}$						

Example 11.11 (Isomorphic graphs)

Problem: Which pairs, if any, of the following graphs are isomorphic?

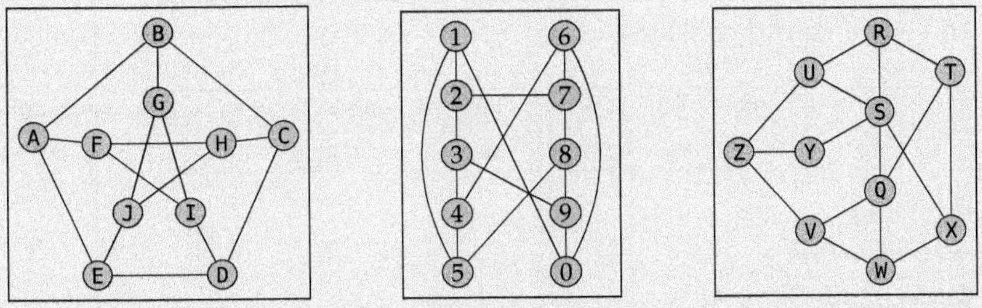

Solution: The first two graphs are isomorphic. The easiest way to see this fact is to show the mapping between the nodes of the two graphs:

A	B	C	D	E	F	G	H	I	J
1	2	3	4	5	0	7	9	6	8

It's easy to verify that all 15 edges now match up between the first two graphs. But the third graph is not isomorphic to either of the others. The easiest justification is that node S in the third graph has degree 5, and no node in either of the first two graphs has degree 5. No matter how we reshuffle the nodes of graph #3, there will still be a node of degree 5—so the third graph can never match the others.

Problem-solving tip: When you're trying to prove or disprove a claim about graphs, you may find it useful to test out the claim against the following four "trivial" graphs:

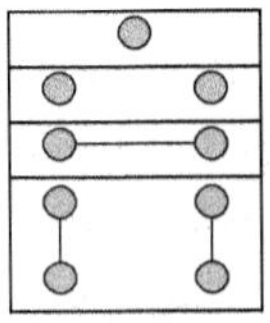

A lot of bogus claims about graphs turn out to be false on one of these four examples—or, unexpectedly, the so-called *Petersen graph*, the first graph in Example 11.11. (The Petersen graph is named after Julius Petersen, a 19th-century Danish mathematician.) It's a good idea to try out any conjecture on all five of these graphs before you let yourself start to believe it!

Taking it further: In general, it's easy to test whether two graphs are isomorphic by brute force (try all permutations!), but no substantially better algorithms are known. The computational complexity of the graph isomorphism problem has been studied extensively over the last few decades, and there has been substantial progress—but no complete resolution.

It's easy to convince someone that two graphs G and H are isomorphic: we can simply describe the relabeling of the nodes of G so that the resulting graphs are identical. (The "convincee" then just needs to verify that the edges really do match up.) When G and H are not isomorphic, it *might* be easy to demonstrate their nonisomorphism: for example, if they have a different number of nodes or edges, or if the degrees in G aren't identical to the degrees in H. But the graphs may have identical degree distributions and yet *not* be isomorphic; see Exercise 11.49.

Subgraphs

When a graph H is isomorphic to a graph G, we can think of having created H by moving around some of the nodes and edges of G. When H is a *subgraph* of G, we can think of having created H by deleting some of the nodes and edges of G. (Of course, it doesn't make sense to delete either endpoint of an edge e without also deleting the edge e.) Here's the definition, for either undirected or directed graphs:

Definition 11.13 (Subgraph)

Let $G = \langle V, E \rangle$ be a graph. A subgraph of G *is a graph $G' = \langle V', E' \rangle$ where $V' \subseteq V$ and $E' \subseteq E$ such that every edge $\langle u, v \rangle \in E'$ satisfies $u \in V'$ and $v \in V'$.*

Note that Definition 11.13 uses the abuse of notation that we mentioned earlier: we "ought" to have written $\{u, v\} \in E'$ for the case that G is undirected.

For example, consider the graph $G = \langle V, E \rangle$ with nodes $V = \{\texttt{A}, \texttt{B}, \texttt{C}, \texttt{D}\}$ and edges $E = \{\{\texttt{A}, \texttt{B}\}, \{\texttt{A}, \texttt{C}\}, \{\texttt{B}, \texttt{C}\}, \{\texttt{C}, \texttt{D}\}\}$. Then the graph G' with nodes $\{\texttt{B}, \texttt{C}, \texttt{D}\}$ and edges $\{\{\texttt{B}, \texttt{C}\}, \{\texttt{C}, \texttt{D}\}\}$ is a subgraph of G. In fact, G has *many* different subgraphs:

Example 11.12 (All 3-node subgraphs of G)

Here are all of the 3-node subgraphs of the graph G with nodes $V = \{\texttt{A}, \texttt{B}, \texttt{C}, \texttt{D}\}$ and edges $E = \{\{\texttt{A}, \texttt{B}\}, \{\texttt{A}, \texttt{C}\}, \{\texttt{B}, \texttt{C}\}, \{\texttt{C}, \texttt{D}\}\}$. (There are many other subgraphs—about 50 total—when we consider subgraphs with 1, 2, 3, or 4 nodes.)

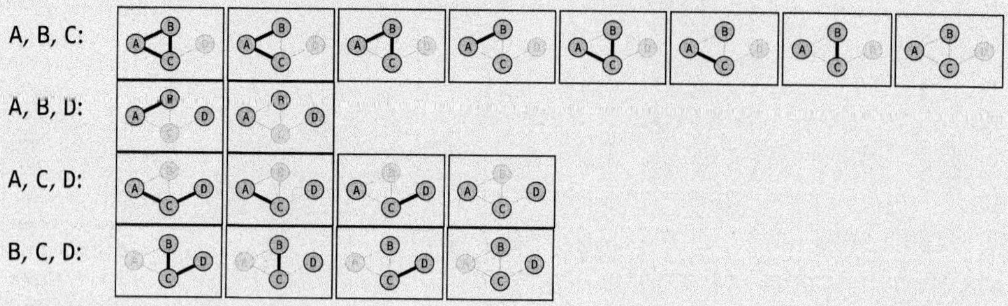

Taking it further: One of the earliest applications of a formal, mathematical perspective to networks—a collaboration between a psychologist and mathematician, in the 1950s—was based on subgraphs. Consider a *signed social network*, an undirected graph where each edge is labeled with '+' to indicate friends, or '−' to indicate enemies. (See Figure 11.8(a).) The adages "the enemy of my enemy of my friend" and "the friend of my friend is my friend" correspond to the claim that the subgraphs in Figure 11.8(b) would not appear. Dorwin Cartwright (the psychologist) and Frank Harary (the mathematician) proved some very interesting structural properties of any signed social network G that does not have either triangle in Figure 11.8(b) as a subgraph—a property that they called "structural balance"—and in the process helped launch much of the mathematical and computational work on graphs that's followed.[4]

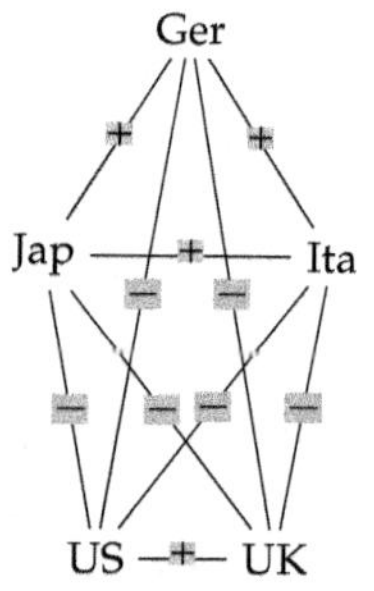

(a) A signed network from 1941

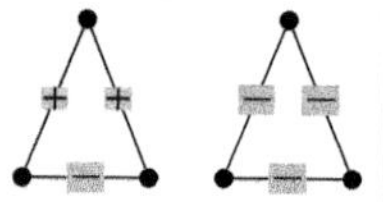

(b) Two triangles

Figure 11.8: Signed social networks. For more about signed networks and these results, see

[4] Dorwin Cartwright and Frank Harary. Structural balance: a generalization of Heider's theory. *Psychological Review*, 63(5):277–293, 1956.

We sometimes refer to a special kind of subgraph: the subgraph of $G = \langle V, E \rangle$ *induced* by a set $V' \subseteq V$ of nodes is the subgraph of G where every edge between nodes in V' is retained. The first subgraph in each row of Example 11.12 is the induced subgraph for its nodes. Here's a brief description of one application of (induced) subgraphs:

Example 11.13 (Motifs in biological networks)

At any particular moment in any particular cell, some of the genes in the organism's DNA are being *expressed*—that is, some genes are "turned on" and the proteins that they code for are being produced by the cell. Furthermore, one gene g can *regulate* another gene g': when g is being expressed, gene g can cause the expression of gene g' to increase or decrease over the baseline level. A great deal of recent biological research has allowed us to construct *gene-regulation networks* for different such settings: that is, a directed graph G whose nodes are genes, and whose edges represent the regulation of one gene by another.

Consider the induced subgraph of a particular set of genes in such a graph G—that is, the interactions among the particular genes in that set. Certain patterns of these subgraphs, called *motifs*, occur significantly more frequently in gene-regulation networks than would be expected by chance. Biologists generally believe that these repeated patterns indicate something important in the way that our genes work, so computational biologists have been working hard to build efficient algorithms to identify induced subgraphs that are overrepresented in a network.

11.2.4 *Special Types of Graphs: Complete, Bipartite, Regular, and Planar Graphs*

In Section 11.2.3, we looked at two ways in which a pair of graphs might be related. Here, we'll consider special characteristics that a single graph might have—that is, subcategories of graphs with some particular structural properties. These special types of graphs arise frequently in various applications.

COMPLETE GRAPHS

Our first special type of graph is a *complete graph* (also called a *clique*), which is an undirected graph in which every possible edge exists:

In CS, the word *clique* usually rhymes with *bleak* or *sleek*. In common-language usage, the word usually rhymes with *slick* or *flick*.

Definition 11.14 (Complete graph/clique)
A complete graph *or* clique *is an undirected graph* $G = \langle V, E\rangle$ *such that* $\{u, v\} \in E$ *for any two distinct nodes* $u \in V$ *and* $v \in V$.

See Figure 11.9 for examples of complete graphs of varying sizes. (In everyday usage, a *clique* is a small, tight-knit, and exclusionary group of friends that doesn't mingle with outsiders. If you think about a graph as a social network, the common-language meaning is similar to Definition 11.14.)

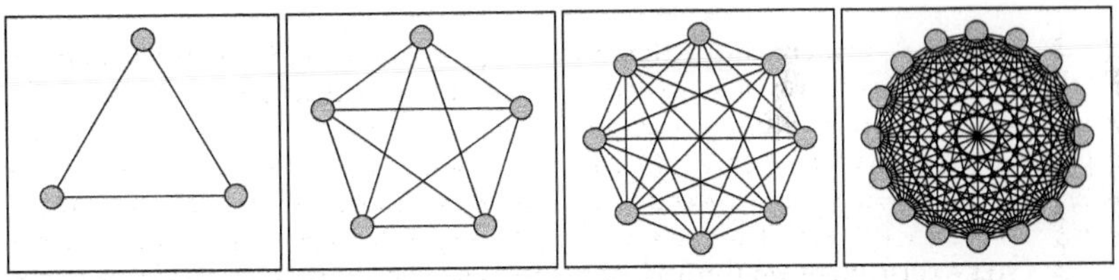

Figure 11.9: Complete graphs with 3, 5, 8, and 16 nodes.

Observe that an undirected graph with n nodes has $\binom{n}{2}$ unordered pairs of nodes, and therefore an n-node complete graph has $\binom{n}{2} = n(n-1)/2$ edges.

A complete graph with n nodes is sometimes denoted by $\mathcal{K}_n$.

The word *clique* can also refer to a *subgraph* that's complete—that is, in which every possible edge actually exists. For example, the graph $G = \langle V, E\rangle$ with $V = \{\mathtt{A}, \mathtt{B}, \mathtt{C}, \mathtt{D}\}$ and $E = \big\{ \{\mathtt{A}, \mathtt{B}\}, \{\mathtt{A}, \mathtt{C}\}, \{\mathtt{B}, \mathtt{C}\}, \{\mathtt{C}, \mathtt{D}\} \big\}$ contains a 3-node clique $\{\mathtt{A}, \mathtt{B}, \mathtt{C}\}$. Here's one small example of an interesting application in which cliques arise:

There are two different prevailing explanations for the $\mathcal{K}_n$ notation:
- the K is as in complete—or, rather, as in *komplett*; the notation was invented by a German speaker.
- the K is in honor of Kazimierz Kuratowski, a 20th-century Polish mathematician who made major contributions to the study of graphs (among other mathematical topics).

Example 11.14 (Collaboration networks and cliques)
Imagine a setting in which different groups of people can work together in different teams, with each person allowed to participate in multiple teams. For example:

- actors in movies. (A "team" is the cast of a single movie.)
- scientific researchers. (A "team" is the set of coauthors of a published paper.)
- employees of a company. (A "team" is a group that worked on a specific project.)

A *collaboration network* is a graph G that represents a setting like these: the nodes of G are the people involved; there is an edge between any two people who have worked together on at least one team. (You may have heard of a challenge in the collaboration network: in the *Kevin Bacon Game*, you're given the name of some actor A; your job is to find a sequence of edges that connects A to the "Kevin Bacon" node in the movie collaboration network. There's a similar game that computer scientists play in the scientific collaboration network, trying to connect themselves to the Hungarian polymath Paul Erdős. See p. 438.)

For example, for the teams listed below, we get the collaboration network at right:

- Tigers: Deborah, George, Hicham, Josh, Lauren
- Unicorns: Anita, Bev, Eva, Fernan
- Vultures: Cathy, Eva, Kelly

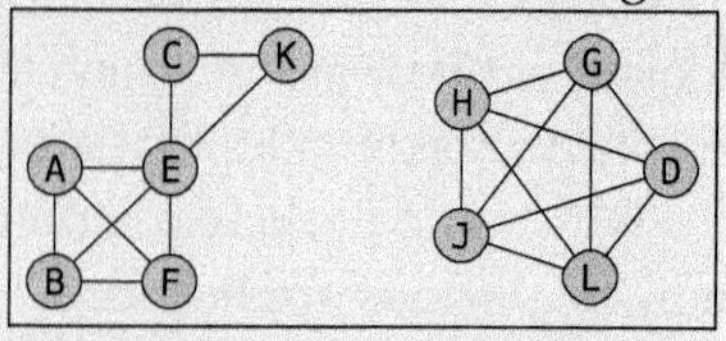

Notice that each team results in a clique inside the collaboration graph—every pair of members of that team is joined by an edge—in this case, creating a K_5, K_4, and K_3 in the graph:

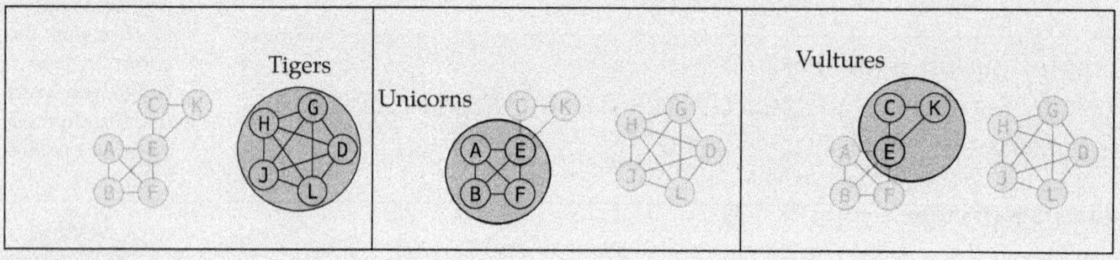

BIPARTITE GRAPHS

Our second special kind of graph is a *bipartite* graph. In a bipartite graph, the nodes can be divided into two groups such that no edges join two nodes that are in the same group: that is, there are two "kinds" of nodes, and all edges join a node of Type A to a node of Type B. Formally:

Latin: *bi* "two"; *part* "part."

Definition 11.15 (Bipartite graph)
A bipartite graph *is an undirected graph* $G = \langle V, E \rangle$ *such that* V *can be partitioned into two disjoint sets* L *and* R *where, for every edge* $e \in E$*, one endpoint of* e *is in* L *and the other endpoint of* e *is in* R.

For example, consider the graph $G = \langle V, E \rangle$ whose nodes are $V = \{\mathtt{A}, \mathtt{B}, \mathtt{C}, \mathtt{D}, \mathtt{E}, \mathtt{F}\}$ and whose edges are $E = \Big\{ \{\mathtt{A}, \mathtt{B}\}, \{\mathtt{A}, \mathtt{C}\}, \{\mathtt{C}, \mathtt{E}\}, \{\mathtt{D}, \mathtt{E}\} \Big\}$. The graph G is bipartite: for example, we can split the nodes into two groups—the vowels $\{\mathtt{A}, \mathtt{E}\}$ and the consonants $\{\mathtt{B}, \mathtt{C}, \mathtt{D}, \mathtt{F}\}$—such that every edge joins a vowel and a consonant. (There's another split that would also have worked: $\{\mathtt{A}, \mathtt{E}, \mathtt{F}\}$ and $\{\mathtt{B}, \mathtt{C}, \mathtt{D}\}$.) See Figure 11.10 for a visualization of the vowel–consonant split.

Bipartite graphs are traditionally drawn with the nodes arranged in two columns, one for each part: *left* ("L") and *right* ("R"). But notice that the definition only requires that it be *possible* to divide the nodes into two groups, with no within-group edges.

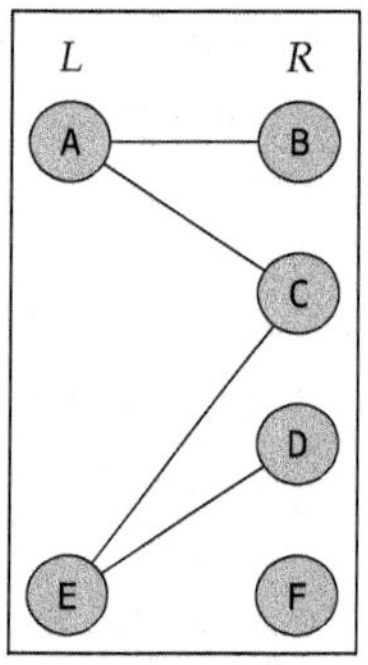

Figure 11.10: A bipartite graph.

Example 11.15 (Bipartite or nonbipartite?)
<u>*Problem:*</u> Which of the following graphs are bipartite?

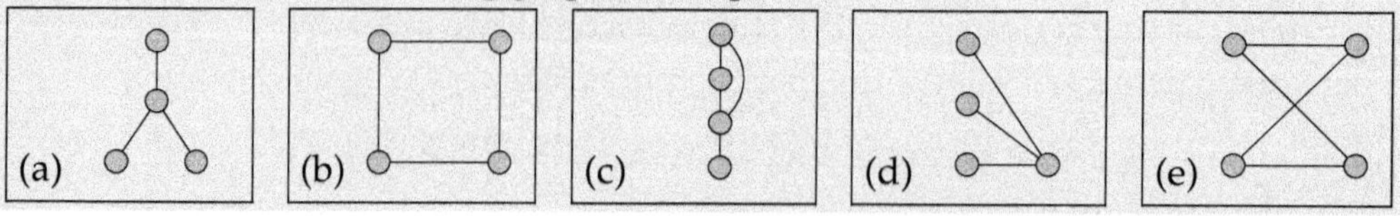

Solution: All of them except (c)! Although (d) and (e) are the only graphs drawn in the "two-column" format, both (a) and (b) can be rearranged into two columns. In fact, aside from node positioning, graphs (a) and (d) are identical. And, similarly, graphs (b) and (e) are isomorphic!

Only (c) is not bipartite: if we attempt to put the topmost node in one group, then both of the next higher two nodes must both be in the other group—but they're joined by an edge themselves, and so we're stuck.

Many interesting real-world phenomena can be modeled using bipartite graphs:

Example 11.16 (Bipartite graphs as models)
Here are just a few of the scenarios that are naturally modeled using bipartite graphs:

- dating relationships in a strictly heterosexual community: the nodes are the boys B and the girls G; every edge connects some boy to some girl.
- nodes are courses and students; an edge joins a student to each class she's taken.
- *affiliation networks*: people and organizations are the nodes; an edge connects person p and organization o if p is a member of o.

There's one further refinement of bipartite graphs that we'll mention: a *complete bipartite graph* is a bipartite graph in which every possible edge exists. In other words, a complete bipartite graph has the form $G = \langle L \cup R, E \rangle$ where $\{\ell, r\} \in E$ for every node $\ell \in L$ and $r \in R$. A complete bipartite graph with ℓ nodes in the left group and r nodes in the right group is sometimes denoted by $\mathcal{K}_{\ell,r}$. See Figure 11.11 for a few examples. (Note again that, as with the $\mathcal{K}_{2,4}$ in Figure 11.11, we don't have to draw a bipartite graph in two-column format—if it's bipartite, then it's still bipartite no matter how we draw it!)

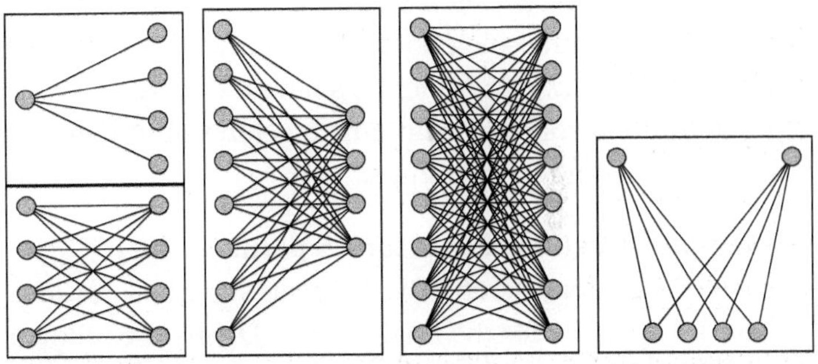

Figure 11.11: Complete bipartite graphs of varying sizes: $\mathcal{K}_{1,4}$, $\mathcal{K}_{4,4}$, $\mathcal{K}_{8,4}$, $\mathcal{K}_{8,8}$, and $\mathcal{K}_{2,4}$.

REGULAR GRAPHS

Our next type of graph is defined in terms of the degree of its nodes: a *regular graph* is one in which all of the nodes have an identical number of neighbors.

Definition 11.16 (Regular graph)
Let $d \geq 0$ be an integer. A d-regular graph *is a graph G such that every node has degree precisely equal to d. If G is d-regular for any d, then we say that G is a* regular graph.

(Most of the time one talks about regular graphs that are undirected, but we can speak of regular directed graphs, too; we'd generally require that all in-degrees match each other *and* all out-degrees match each other.)

For example, consider the graph $G = \langle V, E \rangle$ whose nodes are $V = \{\texttt{A}, \texttt{B}, \texttt{C}, \texttt{D}, \texttt{E}, \texttt{F}\}$ and whose edges are $E = \Big\{ \{\texttt{A},\texttt{B}\}, \{\texttt{A},\texttt{E}\}, \{\texttt{B},\texttt{C}\}, \{\texttt{C},\texttt{F}\}, \{\texttt{D},\texttt{E}\}, \{\texttt{D},\texttt{F}\} \Big\}$. The graph G is 2-regular: you can check that each node has exactly two neighbors. As another example, note that the complete graph $\mathcal{K}_n$ is $(n-1)$-regular, as each node has all $n-1$ other nodes as neighbors. Or see Figure 11.12 for another example of a regular graph.

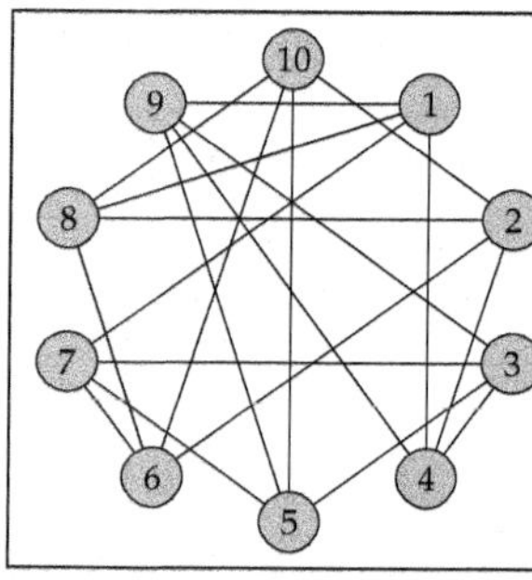

Figure 11.12: A 4-regular 10-node graph.

There are many real-world examples in which regular graphs are useful: for example, imagine constructing a physical network of computers in which each machine only has the capacity for a fixed number of connections. Here are two other useful applications of regular graphs:

Example 11.17 (Scheduling sports with a regular graph)
You are the League Commissioner for an intramural ultimate frisbee league. There are 10 teams in the league, each of whom should play four games. No two teams should play each other twice. Suppose that you construct an undirected graph $G = \langle V, E \rangle$, where $V = \{1, 2, \ldots, 10\}$ is the set of teams, and E is the set of games to be played. If G is an 4-regular graph, then all of the listed requirements are met. Figure 11.12 is a randomly generated example of such a graph; you could use that graph to set the league schedule.

A 1-regular graph is called a *perfect matching*, because each node is "matched" with one—and only one—neighbor. (If every node has degree *at most* 1, then the graph is just called a *matching*.) Matchings have a variety of applications—for example, see p. 960 for their role in the Enigma machine—but here's another specific use of matchings, in assigning partnerships:

Example 11.18 (Matchings for CS partnerships)
Each of n students in an Intro CS class submits a list of people whom they'd like to have as a partner for the final project. Define the following undirected graph G:

- the set V of nodes is $\{1, 2, \ldots, n\}$, one per student.
- the set E of edges includes $\{u, v\}$ if *both* of the following are true: student u wants to work with student v, *and* student v wants to work with student u.

The instructor can assign partnerships by finding a 1-regular graph $G' = \langle V, E' \rangle$ with $E' \subseteq E$—that is, a subgraph of G that includes all of the nodes of G. For example:

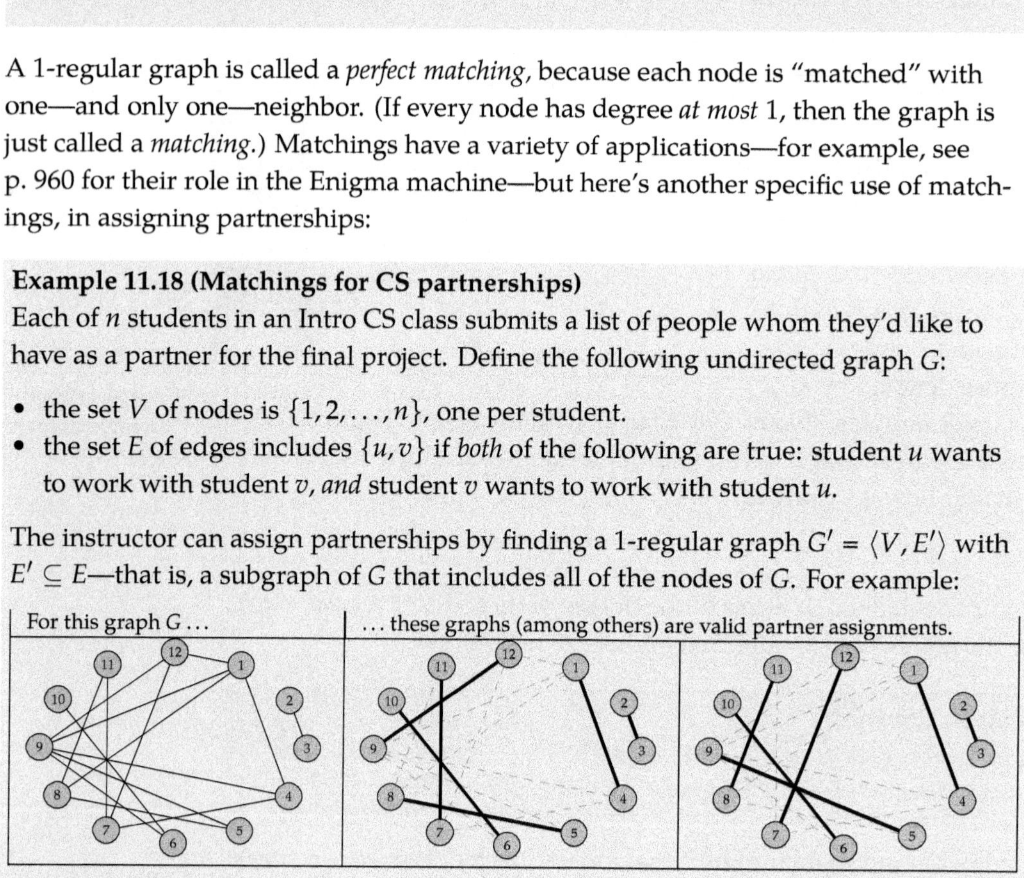

(Incidentally, Example 9.32 asked: how many perfect matchings are there in $\mathcal{K}_n$?)

Planar graphs

Our last special type of graph is a *planar graph*, which is one that can be drawn on a sheet of paper without any lines crossing:

Definition 11.17 (Planar graph)
A planar graph *is a graph G such that it is possible to draw G on a plane (that is, on a piece of paper) such that no edges cross.*

It's important to note that a graph is planar if it is *possible* to draw it with no crossing edges; just because a graph is drawn with edges crossing does not mean that it isn't planar. Here is an example of a planar graph:

Example 11.19 (New England, in a plane)
Here are two copies of the same graph—one drawn with edge crossings, and another with the nodes rearranged to avoid edge crossing:

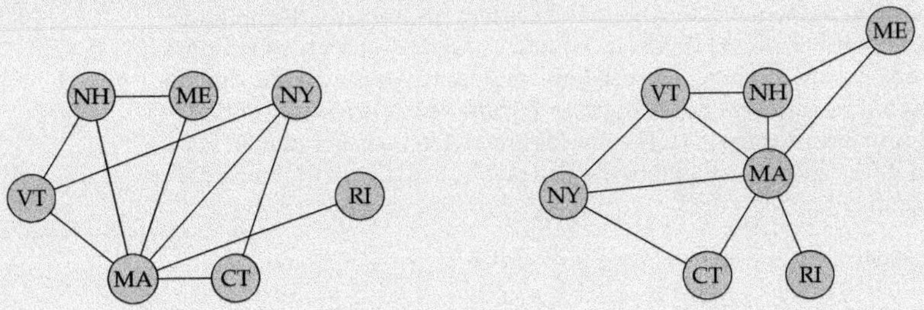

Example 11.19 shows one of the most famous types of planar graph, one derived from a map: we can think of the countries on a map as nodes, and we draw an edge between two country–nodes if those two countries share a border. (See p. 437 for a discussion of the *four-color theorem* for maps, which we could have phrased as a result about planar graphs instead.)

There are other applications of planar graphs in computer science, too. For example, we can view a *circuit* (see Section 3.3.3) as a graph, where the logic gates correspond to nodes and the wires correspond to edges. Most modern circuits are now *printed* on a board (where the "ink" is the conducting material that serves as the wire), and the question of whether a particular circuit can be printed on a single layer is precisely the question of whether its corresponding graph is planar. (If it's not planar, we'd like to minimize the number of edges that cross, or more specifically the number of layers we'd need in the circuit.)

Here's one more set of planarity challenges for you to try:

Example 11.20 (Two planar challenges)
Problem: Are these graphs planar? 1.

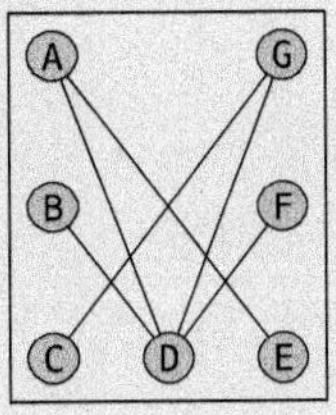

2.

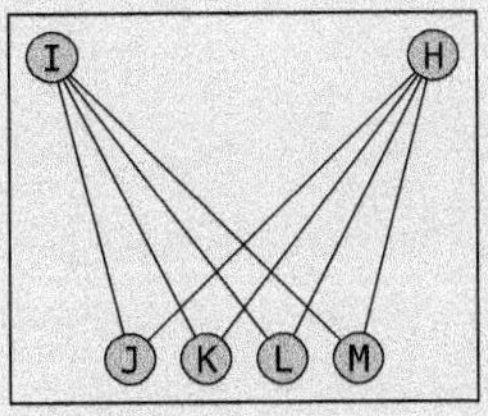

Solution: Yes, both: we can rearrange the nodes so that there are no edges that cross.

1.
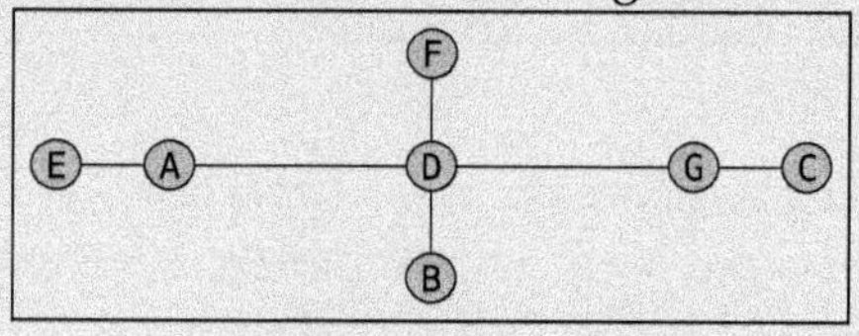

2.
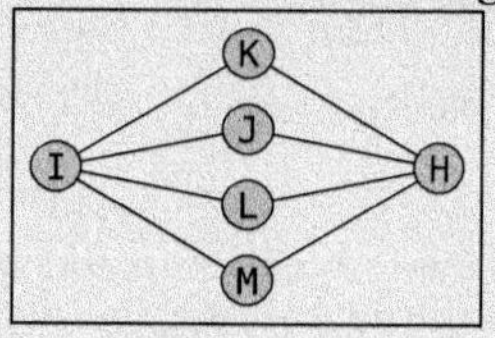

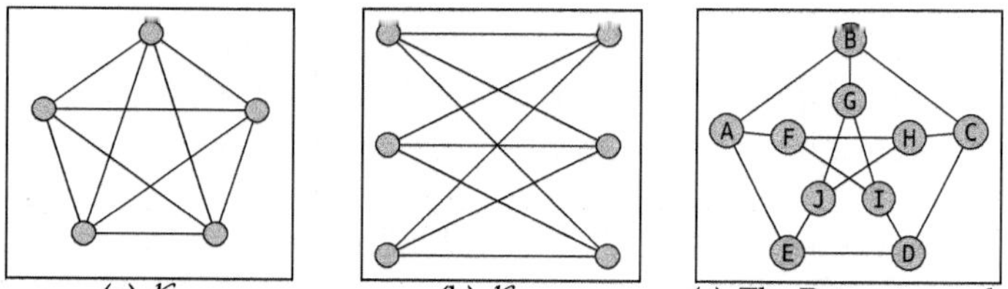

(a) $\mathcal{K}_5$ (b) $\mathcal{K}_{3,3}$ (c) The Petersen graph

Figure 11.13: Nonplanar graphs.

Taking it further: Determining how to lay out a planar graph without edge crossings can be an interesting amusement—see `www.planarity.net` for a surprisingly fun game based on planar graphs. So far we haven't seen any examples of graphs that *can't* be rearranged so that no edges cross. But, if you play around long enough, you should be able to convince yourself that neither $\mathcal{K}_5$ and $\mathcal{K}_{3,3}$ are planar; see Figure 11.13. And, while this shouldn't be at all obvious, it turns out that $\mathcal{K}_5$ and $\mathcal{K}_{3,3}$ are in a sense the only "reasons" that a graph can be nonplanar. A theorem known as *Kuratowski's Theorem*—after the Polish mathematician who may have lent his initial to the notation for complete graphs—says that every graph is planar unless it "contains" $\mathcal{K}_5$ or $\mathcal{K}_{3,3}$ for a subgraph-like notion of "containment." (It's not exactly the subgraph relation, because there are graphs that do not contain $\mathcal{K}_5$ or $\mathcal{K}_{3,3}$ as subgraphs but nonetheless are nonplanar in some sense "because" of one of them. For example, the Petersen Graph from Example 11.11—see Figure 11.13(c)—is nonplanar, but it doesn't have $\mathcal{K}_5$ as a subgraph. But if we "collapse" together the nodes `A/F`, `B/G`, `C/H`, `D/I`, and `E/J` into "supernodes" then the resulting graph *is* $\mathcal{K}_5$.)

Computer Science Connections

Degree Distributions and the Heavy Tail

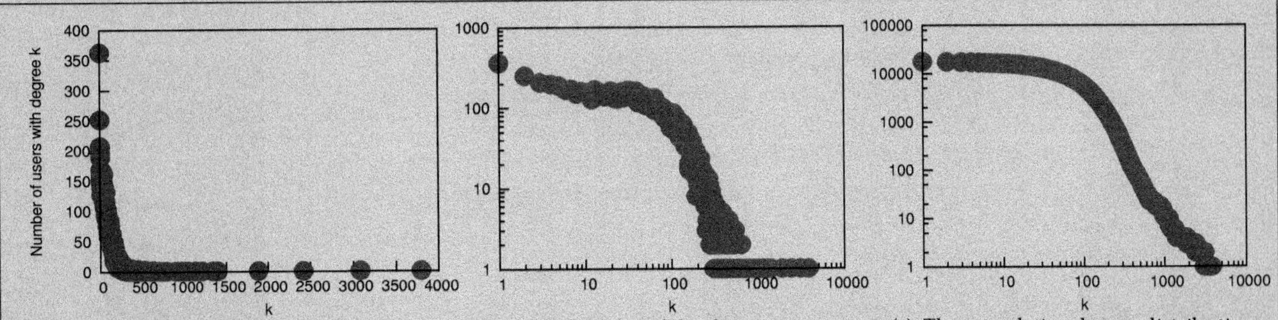

(a) The degree distribution

(b) A log–log plot of the degree distribution

(c) The cumulative degree distribution

Figure 11.14: The degree distribution of ≈ 18,000 Facebook users at the University of North Carolina. Figure 11.14(b) shows a log–log plot of the same data as the linear plot in Figure 11.14(a). Figure 11.14(c) shows a log–log plot of the *cumulative degree distribution*: the number of people with degree $\geq k$, whereas Figures 11.14(a) and 11.14(b) showed the number with degree $= k$.

When we think about massive graphs like the World-Wide Web (with nodes representing web pages and edges representing hyperlinks from one page to another) or an online social network (with nodes representing people and edges representing "friendships"), it is interesting to look at how properties of individual nodes are distributed across the population. We can look at the distribution of any node-by-node property—the physical height of Twitter users, or the number of words of text per web page, for example. But in addition to demographic properties like height and length, we can also look at the distribution of network-type properties.

The *degree distribution* of a graph G shows, for each possible degree d, the number of nodes in G whose degree is d. While one might initially expect degree distributions to look similar to the distribution of heights, it turns out that the degree distribution of an online social network has very different properties. Figure 11.14 shows the degree distribution (in linear, log–log, and cumulative form) for members of the University of North Carolina.[5]

From the Facebook5 dataset, from Mason Porter via the International Network for Social Network Analysis:

[5] Amanda L. Traud, Peter J. Mucha, and Mason A. Porter. Social structure of Facebook networks. *CoRR*, abs/1102.2166, 2011.

Figure 11.14 shows, for each value of k, the number of people who have precisely k Facebook friends. About 350 people have only 1 friend, which is the most common number of friends to have. There are about 750,000 friendships represented in this dataset; the *average degree* is ≈ 84. But, looking at the far-right end of Figure 11.14(a) and 11.14(b), we see a handful of people with very high degrees: 2000, 2500, 3000, and even ≈ 3800. One of the interesting facts about degree distributions in real social networks (or the web) is that there are people whose popularity is massively larger than average: the highest-degree person in this dataset is about $3800/84 \approx 45$ times more popular than average. (Imagine the tallest person at the University of North Carolina being 45 times taller than average!)

Significant research by computer scientists (and many others!) interested in the structure of social networks and the world-wide web has focused on this so-called *heavy-tailed degree distribution*.[6] Some of the literature debates the particular form of this distribution; for example, whether the distribution has the particular form of a *power law*, where the number of people with degree k is roughly k^α for some small constant α, usually around 2.

You can read more about power laws and heavy-tailed degree distributions:

[6] David A. Easley and Jon M. Kleinberg. *Networks, Crowds, and Markets: Reasoning About a Highly Connected World.* Cambridge University Press, 2010.

Computer Science Connections

Graph Drawing, Graph Layouts, and the 9/11 Memorial

Visual representations of most large graphs are too cluttered for a human viewer to process: there are just too many nodes and edges crammed into a small space to see much of anything. Visually presenting a graph like Facebook (billions of nodes, tens of billions of edges) without it looking like a grade-school scribble is daunting. But there is an entire subfield of computer science called *graph drawing*, which is devoted to taking networks and producing good—clear, aesthetic, informative—images of the networks.

In some large graphs, each node has a "natural location" and thus it is clear where on the page it should be placed. For example, graphs may represent data in which the nodes have a precise location situated in the physical world. When we have that kind of layout information for each node, presenting the graph well is easier. (See Figure 11.15.) But many large graphs do not have obvious coordinates associated with each node: while you and your college classmates do have geographic locations (dorm rooms), it's not clear that your dorm really best describes "where" you fit in the social scene of your institution.

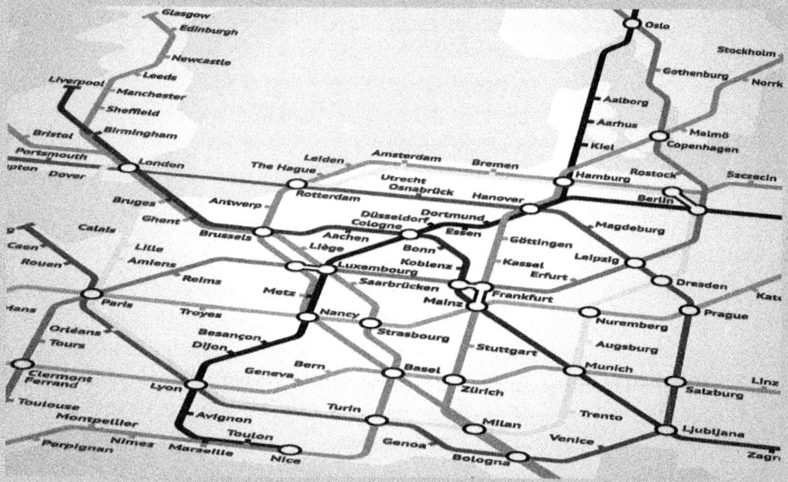

Figure 11.15: A visualization of selected European train routes, where each node's position corresponds to the city's spatial location. Image reproduced with permission from RGBAlpha/Getty Images, Inc.

For graphs whose nodes don't have obvious coordinates, we have to do something else. One approach that's often used in graph drawing is to arrange the nodes based on a physics analogy, as follows. Imagine each node as a charged particle: any two nodes that are joined by an edge are pulled together by an attractive force, and any two nodes that are not joined by an edge are pushed apart by a repulsive force. Then figuring out how to place nodes on the page can be done by starting them in a random configuration and letting the attractive/repulsive forces move the nodes around until they're "happy" in their current positions.

An idea like this one was actually used in designing the 9/11 memorial at the site of the World Trade Center. The memorial was designed with bronze panels inscribed with the 2982 names of victims. A team of computer scientists, architects, and visual artists collaborated to organize the names in a meaningful way. Families were invited to submit "meaningful adjacencies" between victims—which would cause two names to be as close together in the bronze panels as possible. (One of the other algorithmic issues regarding the layout of this memorial was that the designers wanted the names to be placed at evenly spaced intervals on the bronze panels; this constraint added to the computational complexity of the process.) The team used an algorithm to organize the names in an arrangement that respected these requests, which was then used in the final design of the memorial.[7]

In addition to the broader news reports on the wrenching emotional and historical aspects of 9/11 Memorial, the algorithmic aspects of the memorial were also covered in the popular press. You can read more about it here:

[7] Nick Paumgarden. The names. *The New Yorker*, 16 May 2011.

11.2.5 Exercises

For each of the following, draw a graph $G = \langle V, E\rangle$ for the following sets of nodes and edges. Does it make sense to use a directed or undirected graph? Is the graph you've drawn simple?

11.1 nodes $V = \{1, 2, \ldots, 10\}$; an edge connects x and y if $gcd(x, y) = 1$.

11.2 nodes $V = \{1, 2, \ldots, 10\}$; an edge connects x and y if x divides y.

11.3 nodes $V = \{1, 2, \ldots, 10\}$; an edge connects x and y if $x < y$.

For the following undirected graphs, list the edges of the graph, and identify the node(s) with the highest degree. For the directed graphs, identify the node(s) with the highest in-degree, and the node(s) with the highest out-degree.

11.4

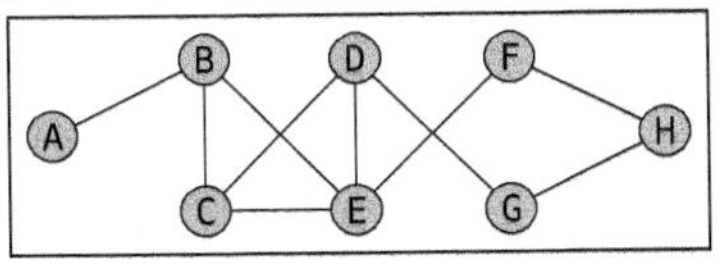

11.5

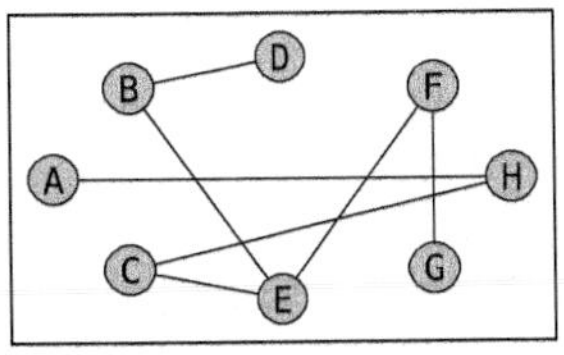

11.6

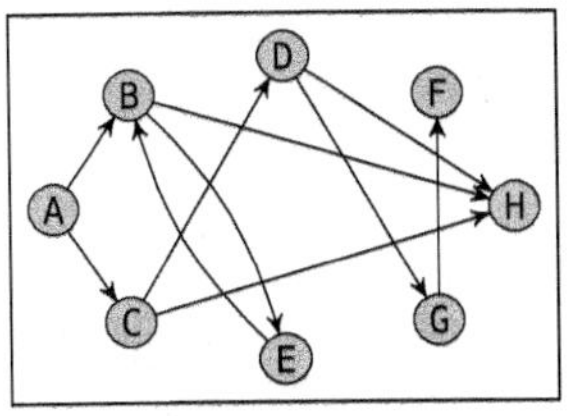

11.7

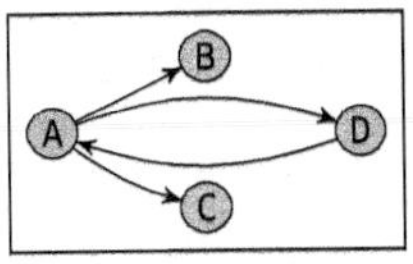

Consider a graph $G = \langle V, E\rangle$ with $n := |V|$ nodes. State your answers in terms of n. Justify.

11.8 If G is an *undirected*, simple graph, what's the largest that $|E|$ can be? The smallest?

11.9 If G is a *directed*, simple graph, what's the largest that $|E|$ can be? The smallest?

11.10 How do your answers to Exercise 11.9 change if self-loops are allowed?

11.11 How do your answers to Exercise 11.9 change if self-loops and parallel edges are allowed?

The anthropologist Robin Dunbar has argued that humans have a mental capacity for only ≈ 150 friends.[8] *(This argument is based in part on the physical size of the human brain, and cross-species comparisons; 150 is now occasionally known as* Dunbar's Number.*)*

[8] Robin Dunbar. *How Many Friends Does One Person Need?: Dunbar's Number and Other Evolutionary Quirks.* Harvard University Press, 2010.

Thanks to Michael Kearns, from whom I learned a somewhat related version of these exercises.

Suppose that Alice has exactly 150 friends, and each of her friends has exactly 150 friends—that is, a friend of Alice knows Alice and 149 other people. (Note that Alice's friends' sets of friends can overlap.) Let S denote the set of people that Alice knows directly or with whom Alice has a mutual friend.

11.12 What's the largest possible value of $|S|$?

11.13 What's the smallest possible value of $|S|$?

Continue to assume that everyone has precisely 150 friends. Let S_k denote the set of all people that Bob knows via a chain of k or fewer intermediate friends:

- *Bob's friends are in S_0;*
- *the people in S_0 and the friends of people in S_0 are in S_1;*
- *the people in S_1 and the friends of people in S_1 are in S_2; and so forth.*

11.14 Let $k \geq 0$ be arbitrary. What's the largest possible value of $|S_k|$?

11.15 Let $k \geq 0$ be arbitrary. What's the smallest possible $|S_k|$?

Prove the following properties of graphs, related to Theorem 11.1 or degree more generally:

11.16 Let u be a node in an undirected graph G. Prove that u's degree is at most the sum of the degrees of u's neighbors.

11.17 Prove Corollary 11.2: in an undirected graph $G = \langle V, E\rangle$, let n_{odd} denote the number of nodes whose degree is odd. Prove that n_{odd} is an even number. That is: prove that

$$|\{u \in V : degree(u) \bmod 2 = 1\}| \bmod 2 = 0.$$

11.18 Prove the analogy of Theorem 11.1 for directed graphs: for a directed graph $G = \langle V, E\rangle$,

$$\sum_{u \in V} \text{in-degree}(v) = \sum_{u \in V} \text{out-degree}(v) = |E|.$$

A linked list *is a data structure consisting of a collection of* nodes, *each of which contains two fields: a* data *field (whatever the node stores) and a* `next` *field that is either* `null` *or points to a node in the linked list. A particular node is designated as the* head node. *Note that a* circular linked list *in which a node points back to a previously encountered node meets this definition. See Figure 11.16.*

Define a not-necessarily-simple directed graph $G = \langle V, E \rangle$*, where* V *is the set of all nodes reachable by following any number of* `next` *pointers starting at the head node, and* $\langle u, v \rangle \in E$ *if* u*'s* `next` *field points to* u*. Observe that each node* u *in* G *has out-degree* $d \in \{0, 1\}$*.*

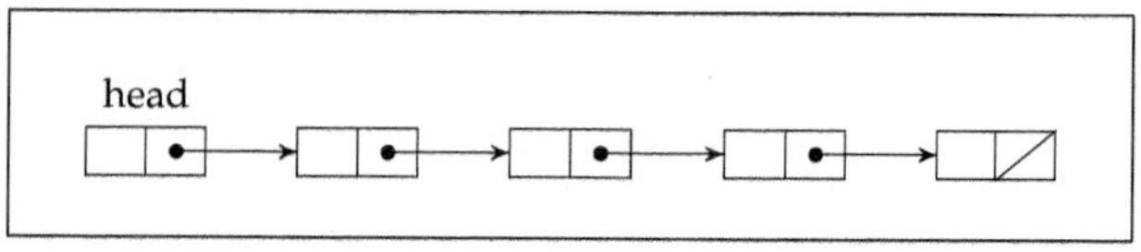

Figure 11.16: A linked list. Each rectangle is a node, and shows two fields: `data` on the left and `next` on the right.

Describe a 5-node linked list in which ...

11.19 ... every node has in-degree $d = 1$.

11.20 ... some node has in-degree $d = 2$.

11.21 ... the resulting graph G is not simple.

11.22 *(This exercise is a tougher algorithmic challenge.)* You are given access to the head node h of an n-node linked list. The value of n is unknown to you. The only operations permitted are (a) to save a node; (b) test whether two saved nodes are the same or different; and (c) given a node u, fetch the node pointed to by u.`next`. Give an algorithm to determine whether the given list is circular *using only a constant amount of memory*—that is, remembering only a constant number of nodes at a time.

A doubly linked list *has* n *nodes with data and* two pointers, `previous` *and* `next`*, to other nodes (or* `null`*). (See Figure 11.17 for an example.) Let* C_n *denote an* n*-node doubly linked list with nodes* $\{1, 2, \ldots, n\}$*, where, for each node* u*,*

- u*'s* `next` *node is* $v = (u \bmod n) + 1$
- v*'s* `previous` *node is* u*.*

Define a directed graph $G_n = \langle V, E \rangle$*, where* V *is the set* $\{1, 2, \ldots, n\}$ *of nodes, and every node has two edges leaving it: one edge* $\langle u, u.\mathtt{next} \rangle$*, and one edge* $\langle u, u.\mathtt{previous} \rangle$*.*

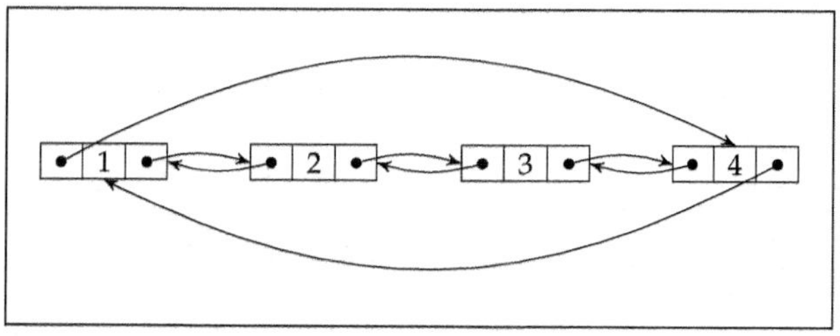

Figure 11.17: A doubly linked list. Each rectangle is a node, and shows three fields: `previous` on the left, `data` in the middle, and `next` on the right.

11.23 Draw G_5.

11.24 Give an example of a G_n that contains a self-loop.

11.25 Give an example of a G_n that contains parallel edges.

Write down an adjacency list representing each of the following graphs.

11.26

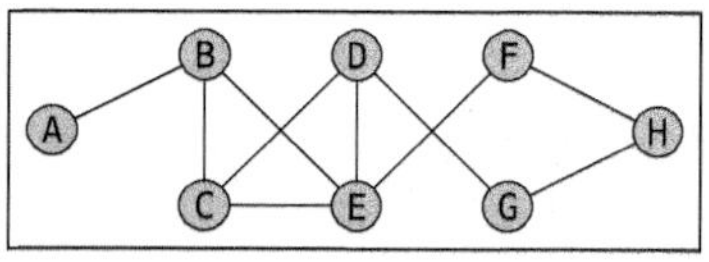

11.27

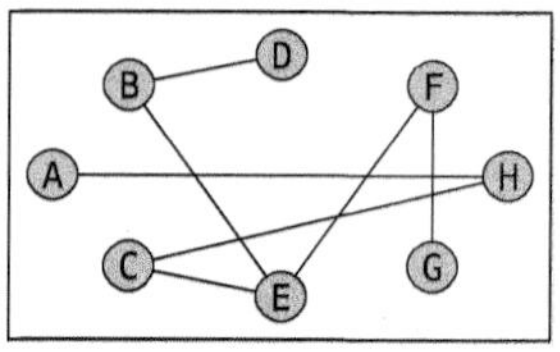

11.28

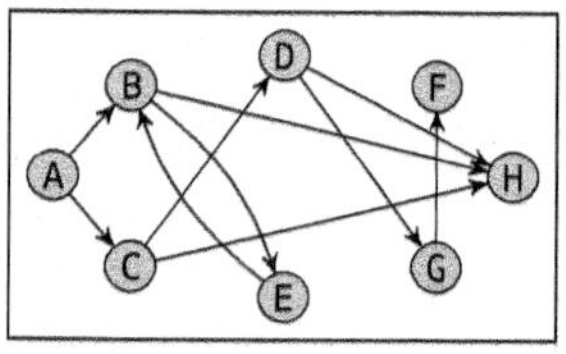

11.29

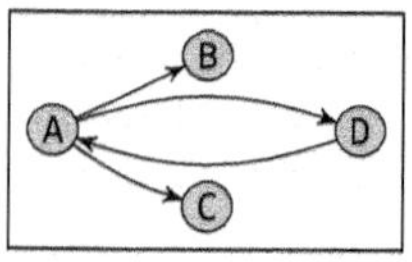

Now give an adjacency matrix for the graphs shown in the above exercises:

11.30 Exercise 11.26

11.31 Exercise 11.27

11.32 Exercise 11.28

11.33 Exercise 11.29

11.34 Suppose that a (possibly directed or undirected) simple graph G is represented by an adjacency list. Suppose further that, for every node u in G, the list of (out-)neighbors of u has a different length. *True or False: G must be a directed graph.* Justify your answer.

11.35 Describe a directed graph G meeting the specifications of Exercise 11.34.

The density *of a graph $G = \langle V, E\rangle$ is the fraction of all possible edges that actually exist: that is,*

$$\text{density} = \frac{|E|}{\text{[your answer to the first part of Exercise 11.8/Exercise 11.9]}}.$$

> **Taking it further:** *Informally, a* dense *graph is one for which most pairs of nodes are joined by an edge, and a* sparse *graph is one in which few pairs of nodes are joined by an edge. We will use these terms informally; a graph is dense if its density is close to 1, and sparse if its density is close to 0. Some people define graphs as* dense *if $|E| = \Theta(|V|^2)$ and as* sparse *if $|E| = O(|V|)$. (These asymptotic definitions only make sense for a* family *of graphs—one for each size n.) There are (families of) graphs that are neither sparse nor dense according to this definition; see Exercise 6.37.*

As a function of n, what are the densities of the following undirected graphs, with nodes $V = \{1, 2, \ldots n\}$? (See Figure 11.18 for small versions of each of these graphs.)

11.36 an n-node path: $E = \{\{1,2\}, \{2,3\}, \ldots, \{n-1, n\}\}$.

11.37 an n-node cycle: $E = \{\{1,2\}, \{2,3\}, \ldots, \{n-1, n\}, \{n,1\}\}$.

11.38 $\frac{n}{3}$ disconnected triangles (assume that $n \bmod 3 = 3$):

$$E = \{\underbrace{\{1,2\},\{2,3\},\{3,1\}}_{\text{triangle on }1,2,3}, \underbrace{\{4,5\},\{5,6\},\{6,4\}}_{\text{triangle on }4,5,6}, \ldots \underbrace{\{n-2,n-1\},\{n-1,n\},\{n,n-2\}}_{\text{triangle on }n-2,n-1,n}\}.$$

11.39 3 separate $\frac{n}{3}$-node cliques (assume that $n \bmod 3 = 3$): $E = \{\{x, y\} : x \bmod 3 = y \bmod 3\}$.

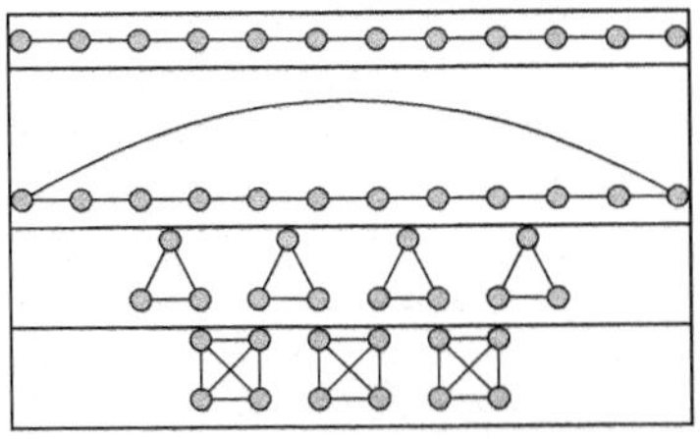

Figure 11.18: A 12-node path, cycle, collection of $\frac{n}{3}$ triangles, and collection of three $\frac{n}{3}$-node cliques.

A hypercube *H_n is a graph in which the 2^n different nodes are all elements of $\{0,1\}^n$. There is an edge between x and y if they differ in only one bit position. (Using the language of Chapter 4.2, there's an edge between any two nodes whose Hamming distance is 1.)*

11.40 Draw H_3.

11.41 Write down an adjacency list for H_4.

11.42 Write down an adjacency matrix for H_4.

11.43 In terms of n, how many edges does H_n have? What is its density?

Decide whether the following pairs of graphs are isomorphic, and prove your answers.

11.44

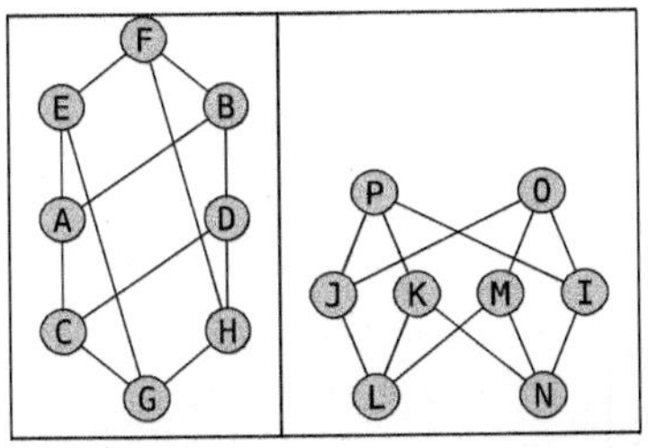

11.45

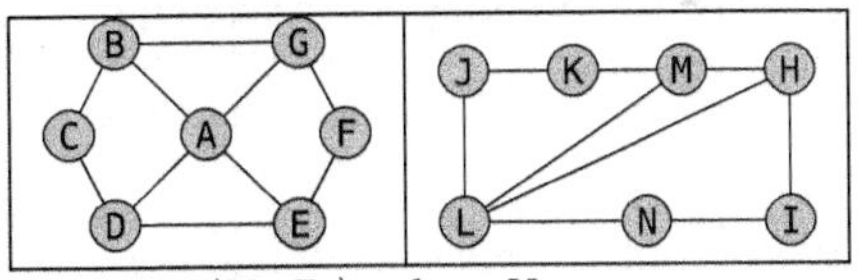

11.46 $G_1 = \langle V_1, E_1\rangle$, where $V_1 = \{10, 11, 12, 13, 14, 15\}$ and $\langle x, y\rangle \in E_1$ if and only if x and y are not relatively prime.

$G_2 = \langle V_2, E_2\rangle$, where $V_2 = \{20, 21, 22, 23, 24, 25\}$ and $\langle x, y\rangle \in E_2$ if and only if x and y are not relatively prime.

Prove or disprove the following claims about isomorphism:

11.47 All 5-node graphs with degrees 1, 1, 1, 1, and 0 are isomorphic.

11.48 All 5-node graphs with degrees 4, 4, 4, 3, and 3 are isomorphic.

11.49 All 5-node graphs with degrees 3, 3, 2, 2, and 2 are isomorphic.

11.50 All n-node, 3-regular graphs are isomorphic.

The computational problem of finding the largest clique (complete graph) that's a subgraph of a given graph G is believed to be very difficult. But for small graphs it's possible to do, even by brute force. For each of the following graphs, identify the size of the largest clique that's a subgraph of the given graph.

11.51

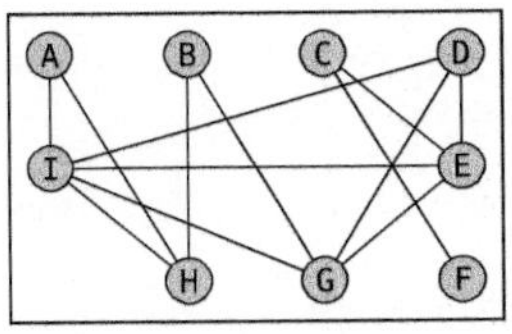

11.52

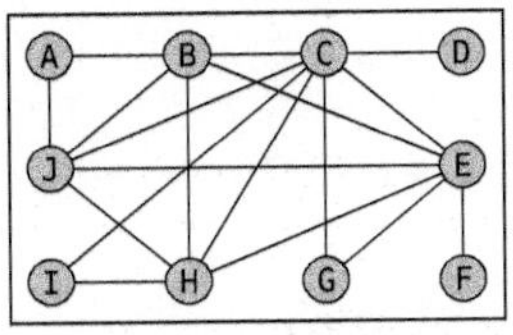

11.53

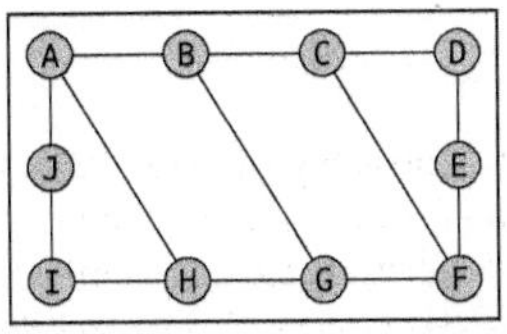

11.54 Consider the collaboration network (see Example 11.14) in Figure 11.19. Assuming that the nodes correspond to actors in movies, what is the *smallest number* of movies that could possibly have generated this collaboration network?

11.55 Are you certain that there weren't more movies than *[your answer to the previous exercise]* that generated this graph? Explain.

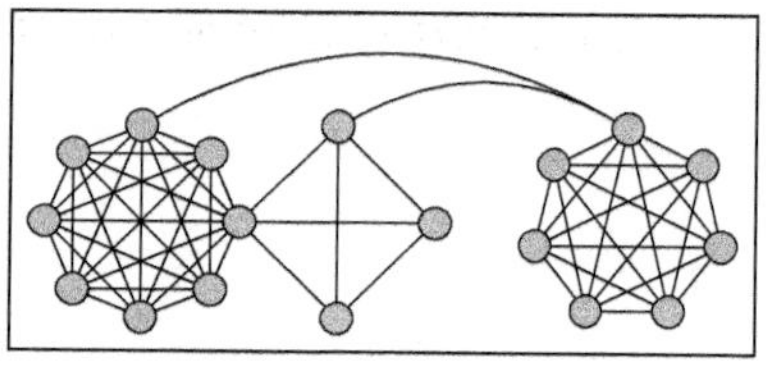

Figure 11.19: A collaboration network.

For which integers n are the following graphs bipartite? Prove your answers.

11.56 $V = \{1, 2, \ldots, n\}$; $E = \{\langle i, i-1 \rangle : i \geq 2\}$.

11.57 $V = \{0, 1, \ldots, n-1\}$; $E = \{\langle i, i+1 \bmod n \rangle : i \geq 1\}$.

11.58 $\mathcal{K}_n$. That is, a complete graph of n nodes: $V = \{1, 2, \ldots, n\}$; $E = \{\{u, v\} : u \in V \text{ and } v \in V\}$.

11.59 $V = \{0, 1, \ldots, 2n-1\}$; $E = \{\langle i, (i+n) \bmod 2n \rangle : i \in V\}$.

Are either of the following graphs bipartite? Explain.

11.60

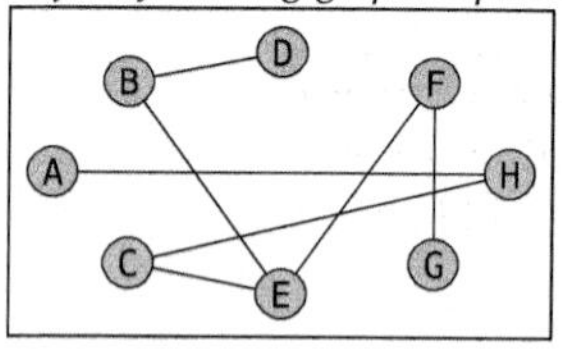

11.61

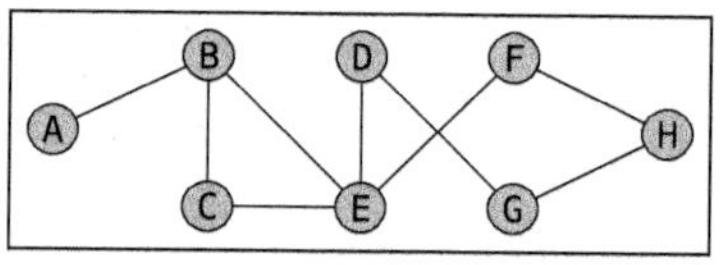

Consider a bipartite graph with a set L of nodes in the left column and a set of nodes R on the right column, where $|L| = |R|$. Prove or disprove the following claims:

11.62 The sum of the degrees of the nodes in L must equal the sum of the degrees of the nodes in R.

11.63 The sum of the degrees of the nodes in L must be even.

11.64 The sum of the degrees of all nodes (that is, all nodes in $L \cup R$) must be an even number.

Suppose that G is a complete bipartite graph with n nodes—that is, $G = \mathcal{K}_{|L|,|R|}$ for $|L| + |R| = n$.

11.65 What's the largest number of edges that can appear in G?

11.66 What's the smallest number of edges that can appear in G? *(Careful!)*

11.67 Prove or disprove: any graph that does not contain a triangle (that is, three nodes a, b, and c with the edges $\{a, b\}$ and $\{b, c\}$ and $\{c, a\}$ in the graph) as a subgraph is bipartite.

11.68 Definition 11.16 describes a regular undirected graph. In a *directed* regular graph, we require that there be two integers d_{in} and d_{out} such that every node's in-degree is d_{in} and every node's out-degree is d_{out}. Prove that we must have $d_{in} = d_{out}$.

Show that both of the following graphs are planar.

11.69

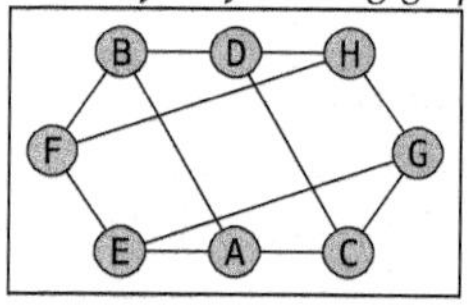

11.70

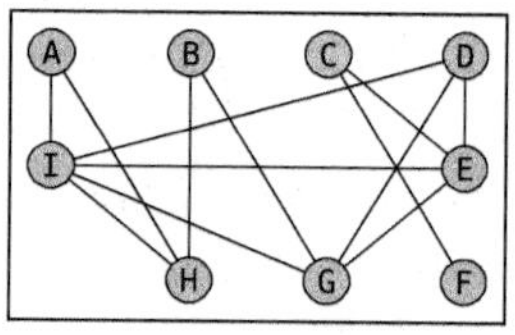

11.71 Prove that any 2-regular graph is planar.

11.3 *Paths, Connectivity, and Distances*

> Well, you can go west to the next intersection, get onto the turnpike, go north through the toll gate at Augusta, 'til you come to that intersection . . . well, no. You keep right on this tar road; it changes to dirt now and again. Just keep the river on your left. You'll come to a crossroads and . . . let me see. Then again, you can take that scenic coastal route that the tourists use. And after you get to Bucksport . . . well, let me see now. Millinocket. Come to think of it, you can't get there from here.
>
> Marshall Dodge (1935–1982) and Robert Bryan (b. 1931), "Which Way to Millinocket?" *Bert and I* (1958)

One of the most basic questions that one can ask about a graph is whether it is possible to get from some given node s to some given node t by following a sequence of edges. Is there some chain of friends that connects Barack Obama to Phil Collins? Can you get from Missoula to Madison by car? (And, if there is a way to get from s to t, what is the *shortest* way to get there?) These basic questions concern the existence of *paths* in the graph:

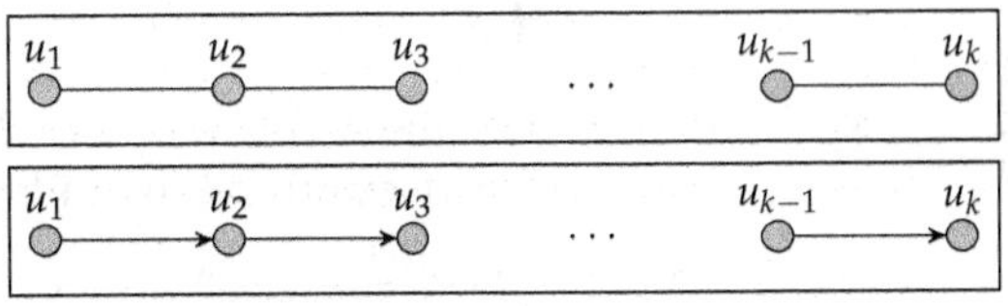

Figure 11.20: Paths in undirected and directed graphs.

Definition 11.18 (Path)
Consider a (directed or undirected) graph $G = \langle V, E \rangle$. A path *in G is a sequence $\langle u_1, u_2, \ldots, u_k \rangle$ of $k \geq 1$ nodes such that:*

- *$u_i \in V$ for every $i \in \{1, \ldots, k\}$, and*
- *$\langle u_i, u_{i+1} \rangle \in E$ for every $i \in \{1, \ldots, k-1\}$.*

(See Figure 11.20.) We say that such a sequence of nodes is a path from u_1 to u_k*, and that this path has* length $k - 1$*. We also say that this path* traverses *the edges $\langle u_i, u_{i+1} \rangle$.*

(Note that this definition includes both directed and undirected graphs: if the edges are directed, we have to follow them "in the right direction.") For example, in both of the graphs shown in Figure 11.21, there is no path from A to X. But, in both, the sequence $\langle$A, C, E, Z$\rangle$ is a path of length 3 from A to Z. In both cases, the edges traversed by the path are $\{\langle$A, C$\rangle, \langle$C, E$\rangle, \langle$E, Z$\rangle\}$. Notice that the length of a path is the number of *edges* that it traverses, which is one fewer than the number of nodes in the path.

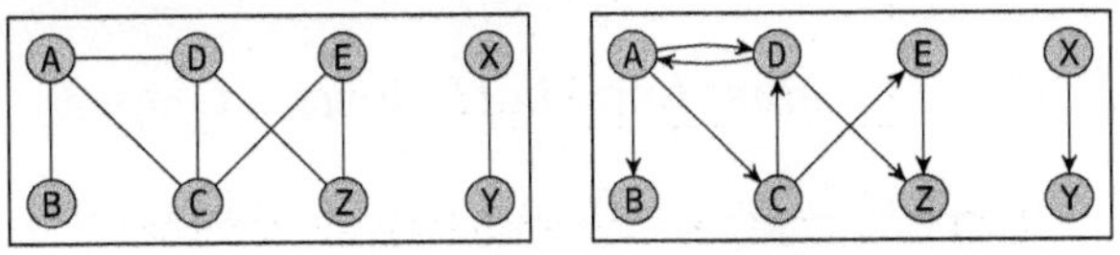

Figure 11.21: Two graphs with paths from A to Z.

Taking it further: A common mistake made by novice (and not-so-novice) programmers is an *off-by-one error* in specifying the bounds on a loop, by iterating either one time too many or one time too few. These errors are also sometimes called *fencepost errors*: if you build a 10-yard fence with posts placed every yard, then there are *eleven* fenceposts (at yard 0, yard 1, . . ., yard 10). Be careful! A path $\langle$A, C, E, Z$\rangle$ contains four nodes, but it traverses three edges (A $\rightarrow$ C, C $\rightarrow$ E, and E $\rightarrow$ Z) and has length 3.

Here's an example of finding paths in a small graph:

Example 11.21 (Finding paths)
Problem: Consider the following undirected graph:

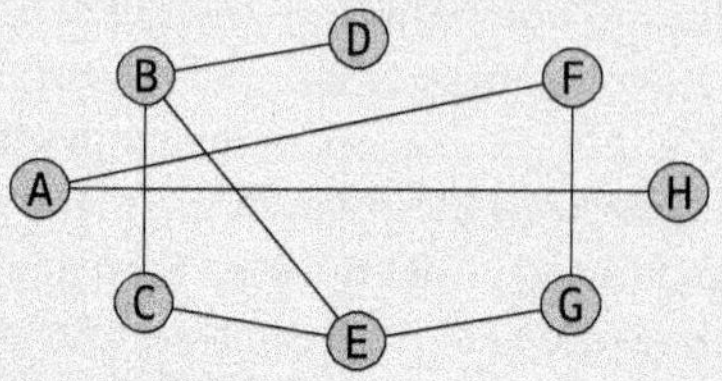

1. Is there a path from node H to node E?
2. Name three different paths from node D to node F. What is the length of each path?

Solution: 1. Yes; ⟨H, A, F, G, E⟩ is a path from node H to E.

2. The following sequences are paths from D to F:
   - ⟨D, B, E, G, F⟩, which has length 4.
   - ⟨D, B, C, E, G, F⟩, which has length 5.

   Finding a third path might seem harder, but Definition 11.18 did not require that the nodes in a path be distinct from each other. (In other words, nothing forbade the repetition of nodes in a path.) So a third path from D to F is:

   - ⟨D, B, C, E, B, C, E, G, F⟩, which has length 8.

We will often restrict our attention to paths that never go back to a vertex that they've already visited, which are called *simple paths*:

Definition 11.19 (Simple Path)
A path $\langle u_1, u_2, \ldots, u_k \rangle$ *is* simple *if all of the nodes* $u_1, \ldots, u_k$ *are distinct.*

Of the three paths identified in Example 11.21, the first two are simple paths, but the third path is not simple because it repeated nodes {B, C, E}.

11.3.1 Connectivity in Undirected Graphs

The most basic question about two nodes in a graph is whether it's possible to get from one to another—that is, are these two nodes *connected?* We start with a formal definition of connectivity for undirected graphs, because the relevant notions are simpler in the undirected setting.

Definition 11.20 (Connected nodes and connected graphs)
Let $G = \langle V, E \rangle$ *be an undirected graph.*

- *Two nodes* $u \in V$ *and* $v \in V$ *are* connected *if there exists a path from* u *to* v.
- *The graph G is* connected *if* u *and* v *are connected for* any *two nodes* $u \in V$ *and* $v \in V$.
- *The graph G is called* disconnected *if it is not connected.*

For example, Figure 11.22 shows one disconnected graph—there's no path from A to H, for example—and one connected graph. You can check that the second graph is connected by testing all pairs of nodes. (Exercise 11.87 asks you to show that connectivity is symmetric in an undirected graph: if there exists a path from u to v, then there exists a path from v to u.)

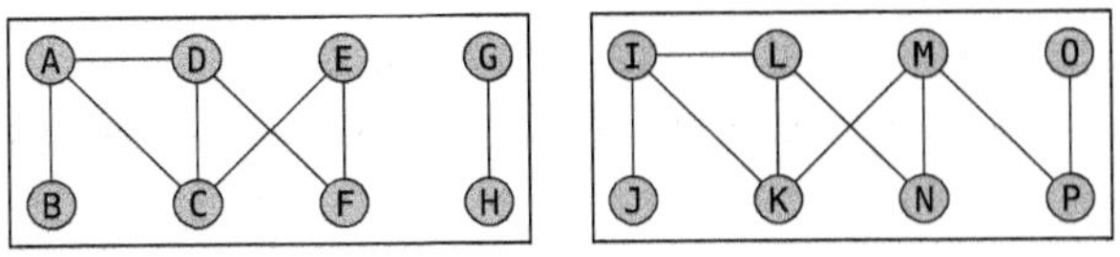

Figure 11.22: A disconnected and connected undirected graph.

Example 11.22 (Connectivity of an undirected graph)
Problem: Is the following graph connected?

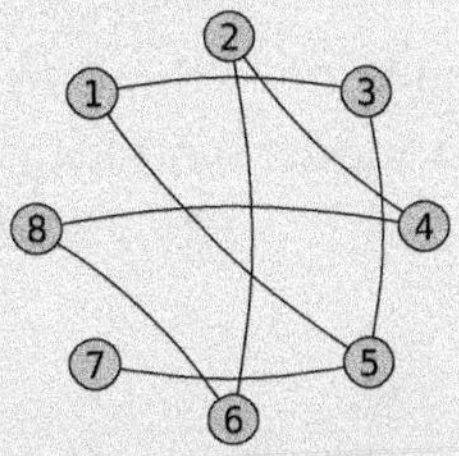

Solution: No: odd-numbered nodes have edges only to other odd-numbered nodes, and even-numbered nodes have edges only to other even-numbered nodes. So there is no path from, for example, node 1 to node 2; this graph is disconnected.

Problem-solving tip: Sometimes it's very helpful to redraw a graph that you're given, with nodes placed more meaningfully. For example, the graph from Example 11.22 can be redrawn as

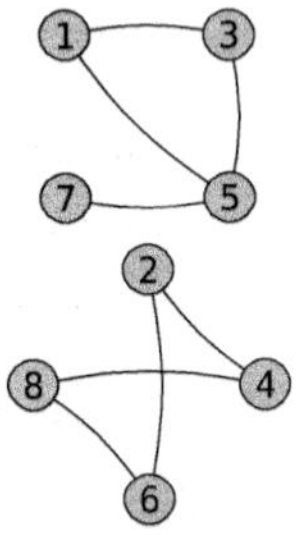

just by sliding the even-numbered nodes down. This visualization makes it clear that the graph is disconnected.

Connected components

More generally, we will talk about the *connected components* of an undirected graph $G = \langle V, E \rangle$—"subsections" of the graph in which all pairs of nodes are connected.

Definition 11.21 (Connected component)
In an undirected graph $G = \langle V, E \rangle$, a connected component *is a set $C \subseteq V$ such that:*

(i) any two nodes $s \in C$ and $t \in C$ are connected.
(ii) for any node $x \in V - C$, adding x to C would make (i) false.

A subset $C \subseteq V$ of nodes is a connected component of an undirected graph $G = \langle V, E \rangle$ if, intuitively, it forms its own "section" of the graph: any two nodes in C are connected, and no node in C is connected to any node not in C. For example, Figure 11.23 shows a graph with three connected components—one with 4 nodes, one with 3 nodes, and one with just a single node.

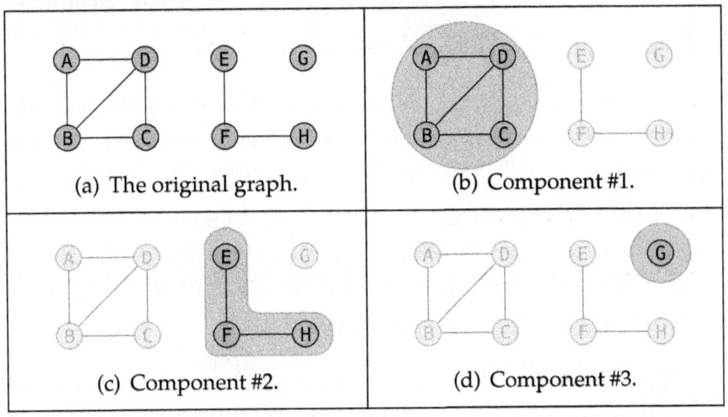

Figure 11.23: A graph's connected components.

Note that we could have defined a "connected graph" in terms of the definition of connected components (instead of Definition 11.20): an undirected graph $G = \langle V, E \rangle$ is *connected* if it contains only one connected component, namely the entire node set V.

Example 11.23 (Connected components of an undirected graph)
Problem: What are the connected components of the following graph?

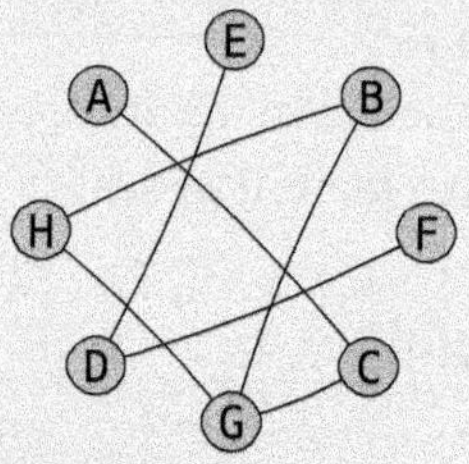

Solution: The set $S = \{\text{A}, \text{B}, \text{C}, \text{G}, \text{H}\}$ is a connected component; there are paths from every node $u \in S$ to every node $v \in S$, and furthermore no node in S is connected to any node not in S. To be thorough, here are paths connecting each pair of nodes from S:

	A	B	C	G	H
A	⟨A⟩	⟨A, C, G, B⟩	⟨A, C⟩	⟨A, C, G⟩	⟨A, C, G, H⟩
B		⟨B⟩	⟨B, G, C⟩	⟨B, G⟩	⟨B, H⟩
C			⟨C⟩	⟨C, G⟩	⟨C, G, H⟩
G				⟨G⟩	⟨G, H⟩
H					⟨H⟩

Note that we haven't bothered to write down a path from u to v when we'd already recorded a path from v to u, because the graph is undirected and paths are symmetric. We also had many choices of paths for many of these entries: for example, other paths from B to H included ⟨B, G, H⟩ or ⟨B, G, H, B, G, H⟩.

There's a second connected component in the graph: the nodes {D, E, F}. It's easy to check that both clauses of Definition 11.21 are also satisfied for this set.

Observe that, in *any* undirected graph $G = \langle V, E \rangle$, there is a path from each node $u \in V$ to itself. Namely, the path is $\langle u \rangle$, and it has length 0. Check Definition 11.18!

Taking it further: There are many computational settings in which undirected paths are relevant; here's one example, in brief. In *computer vision,* we try to build algorithms to process—"understand," even—images. For example, before it can decide how to react to them, a self-driving car must partition the image of the world from a front-facing camera into separate objects: painted lines on the road, trees, other cars, pedestrians, etc. Here's a crude way to get started (real systems use far more sophisticated techniques): define a graph whose nodes are the image's pixels; there is an edge between pixels p and p' if (i) the two pixels are adjacent in the image, and (ii) the colors of p and p' are within a threshold of acceptable difference. The connected components of this graph are a (very rough!) approximation to the "objects" in the image.

This description misses all sorts of crucial features of good algorithms for the image-segmentation problem, but even as stated it may be familiar from a different context: the "region fill" tool in image-manipulation software uses something very much like what we've just described.

11.3.2 *Connectivity in Directed Graphs*

Recall that we have to follow edges "in the right direction" in a directed graph G: as in Definition 11.18, a path from u_1 to u_k in G is a sequence $\langle u_1, u_2, \ldots, u_k \rangle$ where every pair $\langle u_i, u_{i+1} \rangle$ is an edge in G. Thus notions of connectivity in directed graphs are more complicated: the existence of a path from u to v does not imply the existence of a path from v to u. We will speak of a node t as being *reachable* from a node s if it's possible to go from s to t, and of pairs of nodes as being *strongly connected* when it's possible to "go in both directions" between them:

Definition 11.22 (Reachability and strongly connected nodes/graphs)
Let $G = \langle V, E \rangle$ be a directed graph.

- *A node $u \in V$ is* reachable *from a node $v \in V$ if there is a directed path from u to v.*
- *Two nodes $u \in V$ and $v \in V$ are* strongly connected *if u is reachable from v, and v is reachable from u.*
- *The graph G is* strongly connected *if every pair of nodes in V is strongly connected.*

For example, you can check that the first graph in Figure 11.24 is strongly connected by testing for directed paths between all pairs of nodes, in both directions. But the second graph in Figure 11.24 is not strongly connected: there's no path from any node in the right-hand side (nodes $\{\texttt{M}, \texttt{N}, \texttt{O}, \texttt{P}\}$) to any node in the left-hand side (nodes $\{\texttt{I}, \texttt{J}, \texttt{K}, \texttt{L}\}$).

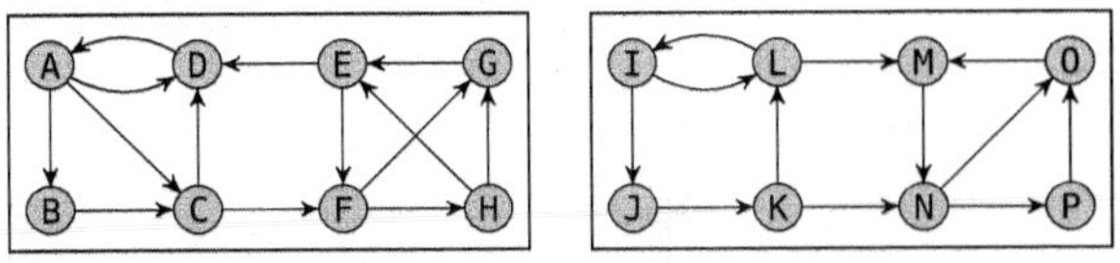

Figure 11.24: Two directed graphs, one that's strongly connected and one that's not.

STRONGLY CONNECTED COMPONENTS

As with undirected graphs, for a directed graph we will divide the graph into "sections"—subsets of the nodes—each of which is strongly connected. These sections are called *strongly connected components* of the graph:

Definition 11.23 (Strongly connected component)
In a directed graph $G = \langle V, E \rangle$, a strongly connected component (SCC) *is a set $C \subseteq V$ such that:*

(i) any two nodes $s \in C$ and $t \in C$ are strongly connected.
(ii) for any node $x \in V - C$, adding x to C would make (i) false.

Figure 11.25 shows an example of a directed graph G and the three strongly connected components in G. The easiest strongly connected component to identify is $\{\texttt{A}, \texttt{B}, \texttt{C}, \texttt{D}\}$: we can go counterclockwise around the loop $\texttt{A} \to \texttt{B} \to \texttt{C} \to \texttt{D} \to \texttt{A}$, so we can go from any one of these four nodes to any other, and we can't get from any of these four nodes to any of the other nodes. The other two strongly connected components are $\{\texttt{E}, \texttt{F}, \texttt{H}\}$ and, separately, $\{\texttt{G}\}$ on its own. The reason is that G is not strongly connected to any other node: we can't get *from* G *to* any other node. (We can go around the $\texttt{E} \to \texttt{F} \to \texttt{H} \to \texttt{E}$ loop, so these three nodes are together in the other strongly connected component.)

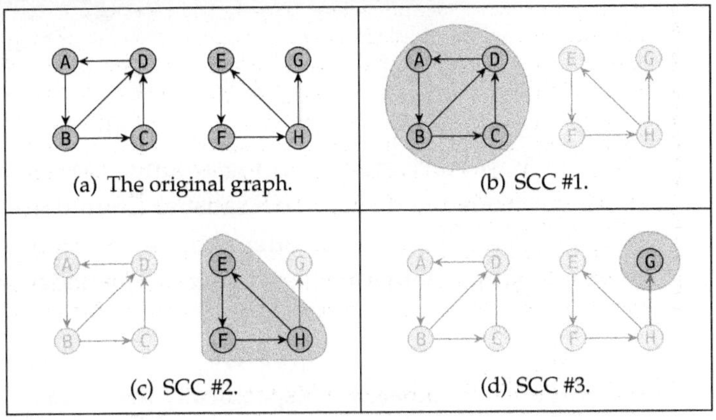

Figure 11.25: A graph and its connected components.

Here's another example of finding strongly connected components:

Example 11.24 (Finding strongly connected components)

Problem: What are the strongly connected components of the following graph?

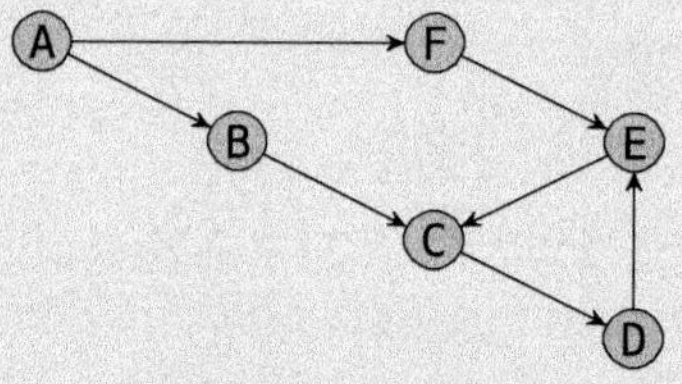

Solution: The three nodes {C, D, E} form a strongly connected component: there is a path from any one of them to any other of them (C → D → E → C → D → E · · ·), and furthermore there is no path from any {C, D, E} to any other node in the graph.

In fact, every other node in the graph is alone in a strongly connected component by itself. For example, while there is a path from A to every node in the graph, there is no path *from* any other node to A. (There is a path from A to A, so the set {A} is a strongly connected component.) Thus the four strongly connected components of the graph are {A}, {B}, {F}, and {C, D, E}.

Here's an example that shows why the second clause of Definition 11.23 is crucial:

Example 11.25 (A non-SCC)

Problem: In the following graph, the set $S :=$ {A, B, C, E, F} is *not* a strongly connected component. Why not?

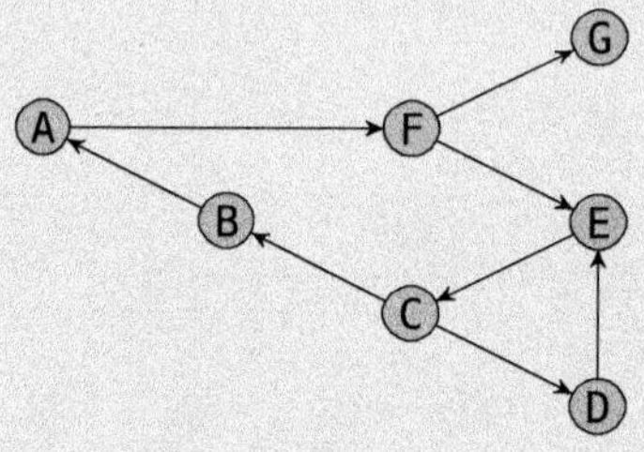

Solution: It is indeed the case that there is a path in both directions between any two nodes in S: we can just keep "going around" clockwise in S and we eventually reach every other node in S. So S satisfies Definition 11.23(i). But it fails to satisfy Definition 11.23(ii): if we considered the set $S^+ := S \cup$ {D}, it is still the case that there is a path in both directions between any nodes in S^+. Thus S is not a strongly connected component!

On the other hand, $S^+ =$ {A, B, C, D, E, F} *is* a strongly connected component: we can't add any other node (specifically G; it's the only other node) to S^+ without falsifying this property—because there's no path from G to A, for example. Thus the two strongly connected components are {A, B, C, D, E, F} and {G}.

Taking it further: There are many computational settings in which directed paths, reachability, and strongly connected components are relevant. For example, for a spreadsheet, consider a directed graph whose nodes are the spreadsheet's cells, and an edge $\langle u, v \rangle$ indicates that u's contents affect the contents of cell v; when a user changes the content of cell c, we must update all cells that are reachable from node c. For a chess-playing program, consider a directed graph whose nodes are board configurations, and there's an edge $\langle u, v \rangle$ if a legal move in u can result in v; any configuration u that's unreachable from the starting board configuration can never occur in chess, and thus your program doesn't have to bother evaluating what move to make in position u.

See p. 1142 for a discussion of another application of reachability and strongly connected components: the structure of the world-wide web, understood with respect to the directed paths in the graph defined by the pages and the hyperlinks of the web.

11.3.3 *Shortest Paths and Distance*

So far we have concentrated on the basic question of connectivity: for a given pair of nodes, does any path exist from one node to the other? Here we address a more refined question: what is the *shortest* path that goes from one node to the next?

Definition 11.24 (Shortest Paths)
Let $G = \langle V, E \rangle$ be a graph (undirected or directed), and let $s \in V$ and $t \in V$ be two nodes. A path from s to t is a shortest path *if its length is the smallest out of all s-to-t paths.*

(Recall that the *length* of a path $\langle u_1, u_2, \ldots, u_k \rangle$ is $k - 1$, the number of edges that it traverses.) Observe that there may be more than one shortest path from a node s to a node t, if there are multiple paths that are tied in length.

Definition 11.25 (Distance)
The distance *from s to t is the length of a shortest path from s to t. If there is no path from s to t, then we say that the distance from s to t is infinite (written as "∞").*

For example, consider the undirected graph in Figure 11.26. We have the following distances from node A in this graph:

A	B	C	D	E	F
0	1	2	2	1	1

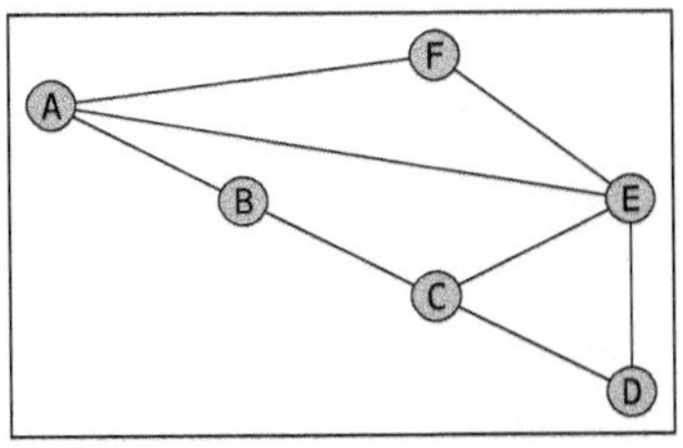

Figure 11.26: An undirected graph.

The distance from A to A is 0 because $\langle A \rangle$ is a path from A to A. This graph also has an example of a pair of nodes connected by two different shortest paths, going from A to C (via either B or E).

For the directed graph in Figure 11.27, we have the following distances from node G:

G	H	I	J	K	L
0	1	2	3	1	∞

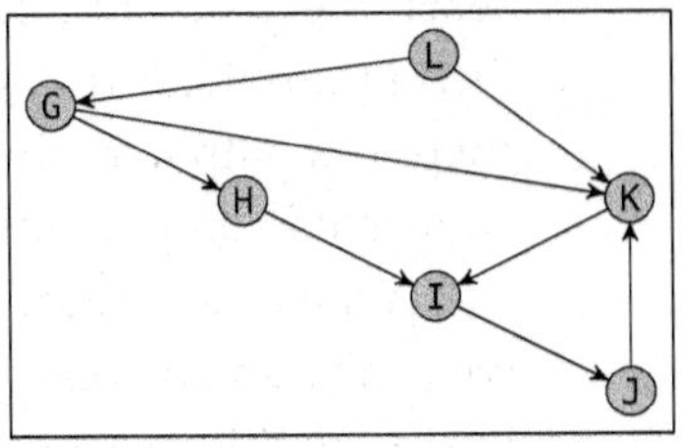

Figure 11.27: A directed graph.

Again, there's a path from G to G of length zero, so the distance from G to G is 0. Note that there's no G-to-J path of length two (because the edge from J to K goes in the wrong direction), so the distance from G to J is 3 (via K and I, or via H and I). Similarly, there is no directed path from G to L, so the distance is infinite.

Here's another example of finding shortest paths in a small graph:

Example 11.26 (Shortest paths in directed graphs)

Problem: Find the shortest path from A to L in the graph with this adjacency list:

```
A:  B, D, E, F, G
B:  C, D, I
C:  B, D, I
D:  E
E:  A, F
F:
G:  F
H:  E, F
I:  B, H, K
J:  C, K
K:  L
L:  F
```

Solution:

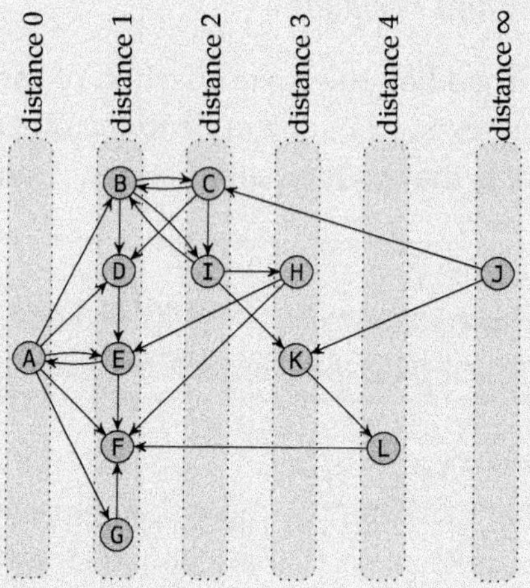

The nodes at distance 1 from A are B, D, E, F, and G. There's no edge from any of those nodes to L—or indeed to K, which is L's only in-neighbor. Thus the distance from A to L cannot be any smaller than 4. But there is an edge from I to K, and one from B to I. We can assemble these edges into the path ⟨A, B, I, K, L⟩. This path has length 4. So the distance from A to L is 4. (Drawing the graph, as on the right, with nodes arranged by their distance from A, can make these facts easier to see.)

Problem-solving tip: In solving any graph problem with a small graph, a good first move is to draw the graph.

11.3.4 *Finding Paths: Breadth-First Search (BFS)*

There are many aspects of graphs that are valuable for interesting computational applications, but perhaps the single most important graph algorithm is *breadth-first search (BFS)*. BFS is a path-finding algorithm: it explores outward from a given source node s in a given graph G until it finds every node reachable from s in G. BFS can be used to solve all sorts of graph-related problems, as we'll see.

Here's the intuition of the algorithm. (See Figure 11.28.) We maintain a set L of nodes that are reachable from the given node s (the shaded nodes in Figure 11.28). To start, we set $L := \{s\}$. Now we find all as-yet-undiscovered neighbors of nodes in L, and add those nodes (the dark-shaded nodes in Figure 11.28) to L: if $\langle u, v \rangle \in E$ and you can reach the node u from s, then you can also reach v from s, via u. But now we've found some more nodes that can be reached from s, which means that we can also reach any nodes that are directly connected to *them* from s. So we'll repeat that process with the updated list L. And we'll do it again, and again, and again, until we stop finding new nodes.

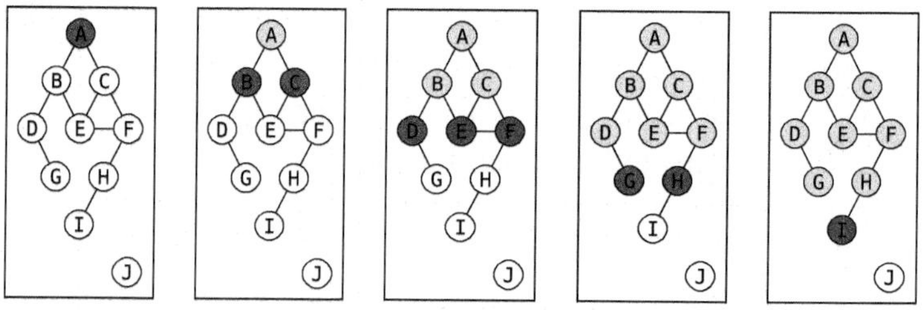

Figure 11.28: The intuition of breadth-first search: the steps of BFS on a small graph, starting at node A.

Observe that BFS discovers nodes in order of their *distance* from the source node. Every expansion of L takes the full breadth of the frontier and expands it out by one more "layer" in the graph. (That's why the algorithm is called breadth-first search.) You can think of BFS as throwing a pebble onto the graph at the node s, and then watching the ripples expanding out from s.

Breadth-first search is presented more formally in Figure 11.29. (While we've described BFS in terms of undirected graphs for simplicity, it works equally well for directed graphs. The only change is that Line 6 should say "for every *out*-neighbor" for a directed graph.)

Breadth-First Search (BFS):
Input: a graph $G = \langle V, E \rangle$ and a source node $s \in V$
Output: the set of nodes reachable from s in G

```
1: Frontier := ⟨s⟩
        // Frontier will be a list of nodes to process, in order.
2: Known := ∅
        // Known will be the set of already-processed nodes.
3: while Frontier is nonempty:
4:     u := the first node in Frontier
5:     remove u from Frontier
6:     for every neighbor v of u:
7:         if v is in neither Frontier nor Known then
8:             add v to the end of Frontier
9:     add u to Known
10: return Known
```

Figure 11.29: The pseudocode for breadth-first search.

Here's another example of breadth-first search in action, running the algorithm in full detail (precisely as specified in Figure 11.29):

Example 11.27 (Sample run of BFS, in detail)
We'll trace BFS starting at node A in the following graph (shown here in the form of a picture and as an adjacency list):

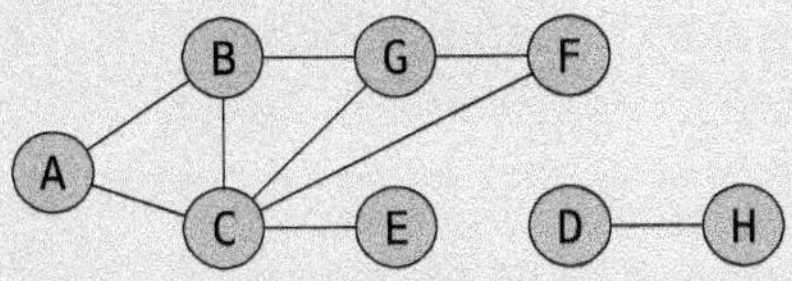

```
A:  B, C
B:  A, C, G
C:  A, B, E, F, G
D:  H
E:  C
F:  C, G
G:  B, C, F
H:  D
```

● = Frontier ● = just moved from Frontier to Known ● = Known ○ = neither Known nor Frontier	Known	Frontier	Explanation
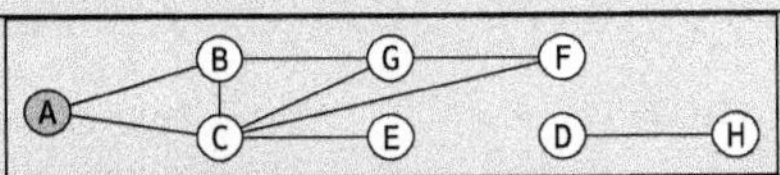	{}	⟨A⟩	initialization (Lines 1–2)
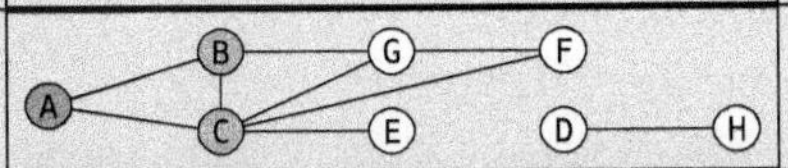	{A}	⟨B, C⟩	processing A (Lines 4–9)
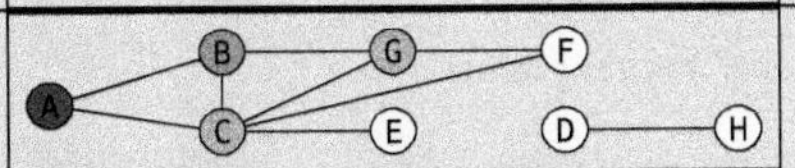	{A, B}	⟨C, G⟩	processing B (Lines 4–9)
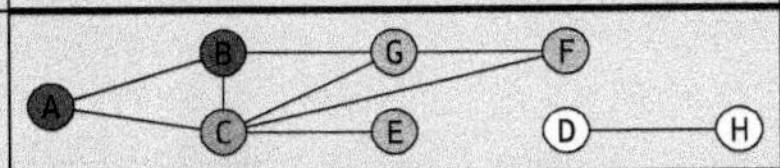	{A, B, C}	⟨G, E, F⟩	processing C (Lines 4–9)
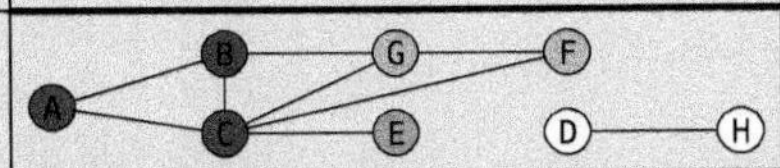	{A, B, C, G}	⟨E, F⟩	processing G (Lines 4–9)
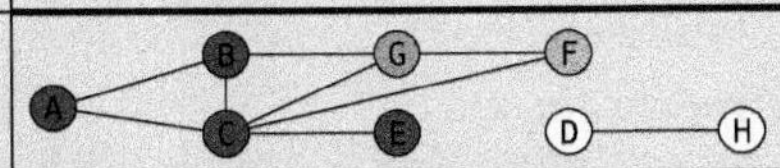	{A, B, C, G, E}	⟨F⟩	processing E (Lines 4–9)
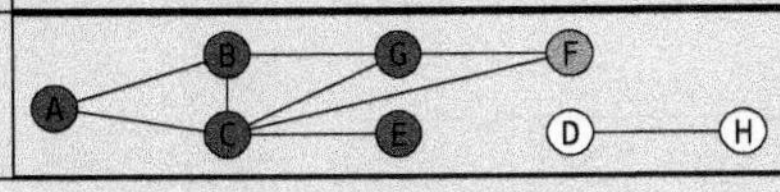	{A, B, C, G, E, F}	⟨⟩	processing F (Lines 4–9)

Because *Frontier* is now empty, the **while** loop in BFS terminates. The algorithm returns the set *Known*, {A, B, C, G, E, F}.

Correctness of BFS

We'll prove two important properties of BFS. The first is *correctness:* the set that BFS returns is precisely those nodes that are reachable from the starting node. The second is *efficiency:* BFS finds this set quickly. The first claim might seem obvious—and thus proving it may feel annoyingly pedantic—but there's a bit of subtlety to the argument, and it's good practice at using induction in proofs besides.

> **Theorem 11.3 (Correctness of BFS)**
> *Let $G = \langle V, E \rangle$ be any graph, and let $s \in V$ be an arbitrary node. Then the set of nodes discovered by* **BFS**(G, s) *is exactly* $\{t \in V : t \text{ is reachable from } s \text{ in } G\}$.

Proof. We'll prove the result by showing two set inclusions: the discovered nodes form a subset of the reachable nodes, and the reachable nodes form a subset of the discovered nodes. Both proofs will use induction, though on different quantities.

Claim #1: **BFS**$(G, s) \subseteq \{t \in V : t \text{ is reachable from } s \text{ in } G\}$. By inspection, we see that (i) BFS returns the set of nodes that end up in the *Known* set, and (ii) the only way that a node ends up in *Known* is having previously been in *Frontier*. Thus it will suffice to prove the following property for all $k \geq 0$, by strong induction on k:

$Q(k)$:= if a node $t \in V$ is added to the list *Frontier* during the kth iteration of the **while** loop of BFS, then there is a path from s to t.

Base case ($k = 0$): If the node t was added to *Frontier* during the 0th iteration of the **while** loop—that is, before the **while** loop begins—then t was added in Line 1 of **BFS**. Therefore t is actually the node s itself. There is a path from s to s itself in any graph, and thus $Q(0)$ holds.

Inductive case ($k \geq 0$): We assume the inductive hypotheses $Q(0), \ldots, Q(k-1)$, and we must prove $Q(k)$. Consider a node t that was added to *Frontier* during the kth iteration of the **while** loop—in other words, t was added in the **for** loop (Lines 6–8) because t is a neighbor of some node u that was already in *Frontier*. That is, we know that $\langle u, t \rangle \in E$ and that u was added to *Frontier* in the (k')th iteration, for some $k' < k$. By the inductive hypothesis $Q(k')$, there is a path P from s to u. Therefore there is a path from s to t, too:

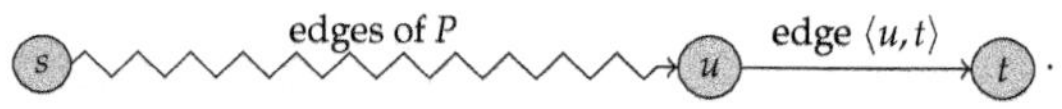

Problem-solving tip: The hard part here is figuring *on what quantity* to do induction. One way to approach this question is to figure out a recursive way of stating the correctness claim.
Q: why is there a path to every node added to *Frontier*? (A: there was a path to every previous node in *Frontier*, and there's an edge from some previously added node to this one!)
Q: why is every node u reachable from s eventually added to *Frontier*? (A: because a neighbor of u that's closer to s is eventually added to *Frontier*, and every neighbor of a node in *Frontier* is eventually added to *Frontier*!)

Claim #2: **BFS**$(G, s) \supseteq \{t \in V : t \text{ is reachable from } s \text{ in } G\}$. If a node t is reachable from s in G, then by definition the distance from s to t is some integer $d \geq 0$. Furthermore, by inspection of the algorithm, we see that any node that's added to *Frontier* is eventually moved to *Known*. Thus it will suffice to prove the following property for all $d \geq 0$, by (weak) induction on d:

$R(d)$:= if a node $t \in V$ at distance d from s, then t is eventually added to *Frontier*.

Base case ($d = 0$): We must prove $R(0)$: any node t at distance 0 is eventually added to *Frontier*. But the only node at distance 0 from s is s itself, and **BFS** adds s itself to *Frontier* in Line 1 of the algorithm.

Inductive case ($d \geq 1$): We assume the inductive hypothesis $R(d-1)$, and we must prove $R(d)$. Let t be a node at distance d from s. Then by definition of distance there is a shortest path P of length d from s to t. Let u be the node immediately before t in P. Then the distance from s to u must be $d-1$, and therefore by the inductive hypothesis $R(d-1)$ the node u is added to *Frontier* in some iteration of the **while** loop. There are at most $|V|$ iterations of the loop, and thus eventually u is the first node in *Frontier*. In that iteration, the node t is added to *Frontier* (if it had not already been added). Thus $R(d)$ follows. □

(In the exercises, you'll show how to modify BFS so that it actually computes *distances* from s, using an idea very similar to the proof of Claim #2 of Theorem 11.3.)

Running time of BFS

Theorem 11.4 (Efficiency of BFS)
For a graph $G = \langle V, E \rangle$ represented using an adjacency list, BFS takes $\Theta(|V| + |E|)$ time.

Proof. See Figure 11.30 for a reminder of the algorithm. Lines 1, 2, and 10 take $\Theta(1)$ time, so the only question is how long the **while** loop takes. In the worst case, every node in the graph is reachable from the node from which BFS is run. In this case, there is one iteration of the **while** loop for every node $u \in V$. How long does the body of the **while** loop (Lines 4–9) take for a particular node u?

- Lines 4, 5, and 9 take $\Theta(1)$ time.
- The **for** loop in Lines 6–8 has one iteration *for each neighbor of u*. (In an adjacency list, the loop simply steps through the list of neighbors, one by one.) Each **for**-loop iteration takes $\Theta(1)$ time, and there are *degree*(u) iterations for node u.

```
Breadth-First Search (BFS):
Input: a graph G = ⟨V, E⟩ and a source node s ∈ V
Output: the set of nodes reachable from s in G

1: Frontier := ⟨s⟩
2: Known := ∅
3: while Frontier is nonempty:
4:     u := the first node in Frontier
5:     remove u from Frontier
6:     for every neighbor v of u:
7:         if v is in neither Frontier nor Known then
8:             add v to the end of Frontier
9:     add u to Known
10: return Known
```

Figure 11.30: A reminder of BFS.

Therefore, ignoring multiplicative constants, the worst-case running time of BFS is

$$1 + \sum_{u \in V} \left[1 + degree(u)\right]$$

$$= 1 + \left[\sum_{u \in V} 1\right] + \left[\sum_{u \in V} degree(u)\right] \qquad \text{rearranging the summation}$$

$$= 1 + |V| + 2|E| \qquad \text{or } 1 + |V| + |E| \text{ for a directed graph} \qquad \text{Theorem 11.1/Exercise 11.18}$$

$$= \Theta(|V| + |E|).$$ □

Taking it further: BFS arises in applications throughout computer science, from network routing to artificial intelligence. Another application of BFS occurs (hidden from your view) as you use programming languages like Python and Java, through a language feature called *garbage collection*. In garbage-collected languages, when you as a programmer are done using whatever data you've stored in some chunk of memory, you just "drop it on the floor"; the "garbage collector" comes along to reclaim that memory for other use in the future of your program. The garbage collector runs BFS-like algorithms to determine whether a particular piece of memory is actually trash. See p. 1143 for more.

11.3.5 *Finding Paths: Depth-First Search (DFS)*

Another important algorithm for exploring graphs is called *depth-first search (DFS)*, which can be described informally as follows. Instead of exploring outward from the source node s in "layers" as in BFS, we will try to explore a new node at every stage of the search. We start at s, and at every stage we move to an unvisited neighbor of our current node. If at any stage we're stuck at a node u that has no unvisited neighbors, we go back from u to the node from which we first reached u and continue exploring from there. Here is an example of DFS in a small graph, informally:

Example 11.28 (Sample run of depth-first search)

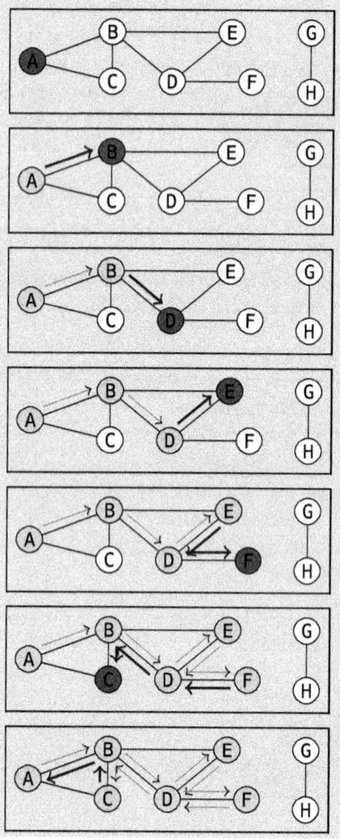

We start exploring node A; in each frame, the dark-shaded node is the current node.

Previously discovered nodes are lightly shaded. Arrows indicate the steps of the exploration.

In each of the first four frames, we move from the current node to a neighbor that is unexplored. (We pick the alphabetically first node if there's a choice.)

The current node E has no unvisited neighbors, so we backtrack from E to D to find D's unvisited neighbor F.

We backtrack from F to D to B to discover the new node C.

We backtrack from C to B to A; there are no further unexplored nodes from any of these nodes, and thus the algorithm terminates.

Intuitively, depth-first search is a close match for the way that you would explore a maze: you start at the entrance, follow a passageway to a location you've never visited before; using breadcrumbs or a pencil, you remember where you've been and backtrack if you get stuck. You may have heard of another algorithm for mazes:

> Place your right hand on the wall as you go in the entrance. Continue to walk forward, always keeping your right hand on the wall. Eventually, you will get out of the maze.

In fact, this right-hand-on-the-wall algorithm is identical in spirit to DFS: whenever you encounter a choice, you always choose the first (right-most) unexplored passageway, and if you ever get stuck at a dead end you turn around and go back from whence you came.

We can implement DFS with only a small change to BFS, as shown in Figure 11.31: instead of putting a newly discovered node u at the *end* of the list *Frontier* of nodes from which to explore (as in BFS), we put a newly discovered node u at the *beginning* of *Frontier*. (In other words, *BFS treats the list* Frontier *as a* queue—*first in, first out*—*while DFS treats the list* Frontier *as a* stack—*last in, first out*.) Another small change is necessary, to allow a node already in *Frontier* to be "moved" earlier in the list of nodes to explore.

Depth-First Search (DFS):
Input: a graph $G = \langle V, E \rangle$ and a source node $s \in V$
Output: the set of nodes reachable from s in G

```
Frontier := ⟨s⟩
Known := ∅
while Frontier is nonempty:
    u := the first node in Frontier
    remove u from Frontier
    if u is not in Known then
        for every neighbor v of u:
            if v is not in Known then
                add v to the start of Frontier
        add u to Known
return Known
```

Figure 11.31: The pseudocode for depth-first search. The only changes from BFS are underlined.

Because this alteration of BFS changes only the order in which the nodes in *Frontier* are explored, DFS does precisely the same work as BFS, and is correct for the same reasons: DFS returns precisely the set of nodes reachable from the given source node s. (With a little more cleverness in moving nodes to the front of *Frontier*, DFS can also be implemented in $\Theta(|V| + |E|)$ time.) Here's a fully detailed example of DFS:

Example 11.29 (Sample run of DFS, in detail)
We'll trace DFS starting at node A in this graph:

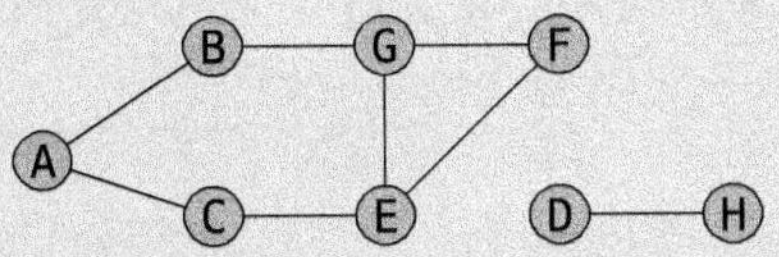

A:	B, C
B:	A, G
C:	A, E
D:	H
E:	C, F, G
F:	E, G
G:	B, E, F
H:	D

○ = Frontier ○ = just moved from Frontier to Known ● = Known ○ = neither Known nor Frontier	Known	Frontier u: just added.	Explanation
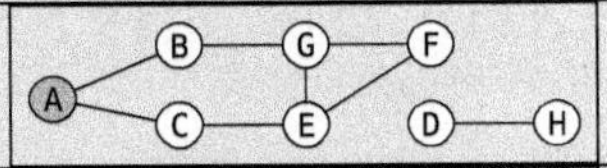	{}	⟨A⟩	initialization
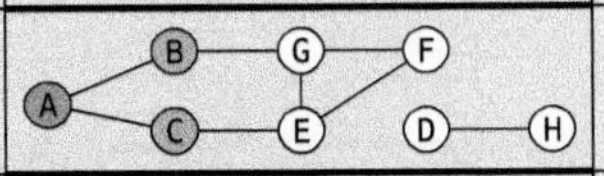	{A}	⟨B, C⟩	processing A
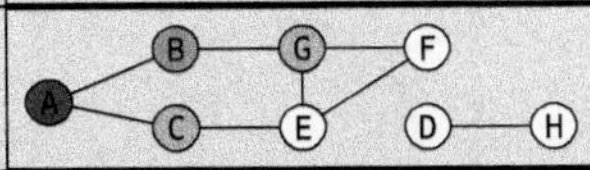	{A, B}	⟨G, C⟩	processing B (A known ⇒ not re-added)
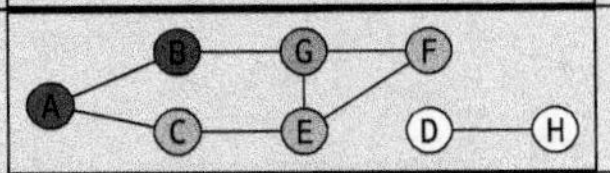	{A, B, G}	⟨E, F, C⟩	processing G (B known ⇒ not re-added)
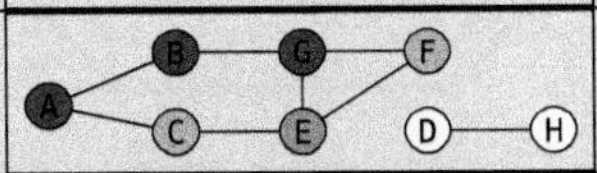	{A, B, G, E}	⟨C, F, F, C⟩	processing E (G known ⇒ not re-added)
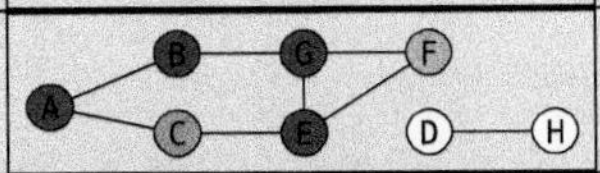	{A, B, G, E, C}	⟨_F, F, C⟩	processing C (A,E known ⇒ not re-added)
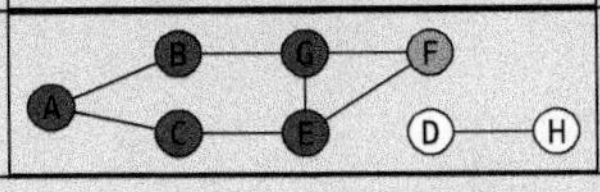	{A, B, G, E, C, F}	⟨_F, C⟩	processing F

There are two more iterations that remove the last two entries in *Frontier* (making no changes to *Known* and adding nothing further to *Frontier*), because both F and C are already in *Known*. The **while** loop then terminates, and DFS returns {A, B, G, E, C, F}.

Computer Science Connections

The Bowtie Structure of the Web

As the web has grown more and more central in the daily lives of us all, it has garnered increasing attention from researchers in computer science. A great deal of work has been performed to characterize the web in terms of its degree distribution (see p. 1123) or in terms of the "small-world phenomenon" (see p. 438). But one foundational and influential paper sought to characterize the web's structure in terms of its strongly connected components.[9] In the early days of the web, eight researchers from AltaVista, IBM, and Compaq downloaded around 200 million web pages, comprising about 1.5 billion links. They then analyzed the structure of the resulting graph, by categorizing the pages:

1. Let CORE denote those web pages contained in the largest SCC of the web graph. Like many other networks (for example, social networks and collaboration networks), the web graph has a *giant component* that contains many more nodes than the second-largest SCC. Denote by CORE those nodes in the largest SCC in the web graph.
2. Let IN denote those web pages p such that (i) $p \notin$ CORE, and (ii) there is a path from p to some node in CORE. That is, there is a path from p to every page in CORE, but there's no path from any node in CORE to p.
3. Let OUT denote those web pages p such that (i) $p \notin$ CORE, and (ii) there is a path from some node in CORE to p. That is, there is a path from every page in CORE to page p, but there's no path from p to any node in CORE.

When displayed graphically, as in Figure 11.32, these categories of web pages look like a bowtie, and so the paper by Broder et al. came to be known as "the bowtie paper."

[9] Andrei Broder, Ravi Kumar, Farzin Maghoul, Prabhakar Raghavan, Sridhar Rajagopalan, Raymie Stata, Andrew Tomkins, and Janet Wiener. Graph structure in the web. *Computer Networks*, 33(1–6):309–320, 2000.

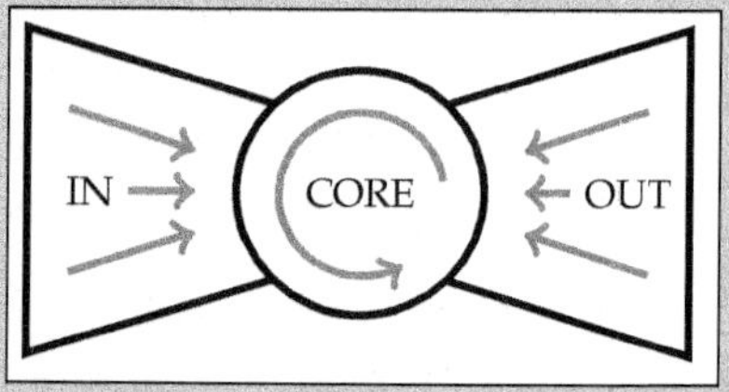

Figure 11.32: The "bowtie structure" of the web graph, in its basic form. Broder et al. found that roughly 25% of web pages fell into each of these categories: 56M pages (of 200M) in CORE, 43M pages in IN, and 43M pages in OUT.

To complete the picture of the bowtie structure of the web, we must note that not all web pages are included in Figure 11.32. There are three further categories of nodes:

4. Let TUBES denote those pages p that (i) are reachable from a node of IN (that is, there's a page $q \in$ IN that has a path to p), *and* (ii) can reach a node of OUT (that is, there's a page $q \in$ OUT to which p has a path), and (iii) $p \notin$ CORE.
5. Let TENDRILS denote those pages p that are *either* reachable from a node of IN, or can reach a node of OUT, but not both.
6. Let DISCONNECTED denote those pages p that are not in CORE, IN, OUT, TUBES, or TENDRILS—that is, those pages p that can neither reach nor be reached by any node in those sets.

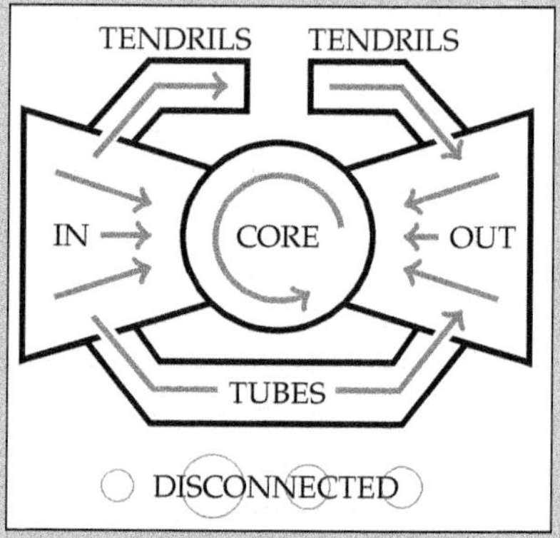

Figure 11.33: The remainder of the "bowtie structure" of the web graph. There were about 44M pages in TENDRILS and TUBES, and about 17M pages in DISCONNECTED.

One of the unexpected facts found by Broder et al. was the extent to which the web is actually *not* particularly well connected. In particular, if we were to choose web pages p and q uniformly at random from the web graph, there was only a roughly 24% chance of that a directed path from p to q exists—far lower than the "small world" phenomenon would suggest.

COMPUTER SCIENCE CONNECTIONS

GARBAGE COLLECTION

In many modern programming languages, including Python and Java, the burden of managing memory is lifted from the shoulders of the programmer. When a new object is needed, the programmer just creates it. After a program has been running for a while, there may be objects that were stored in memory but are now *inaccessible* because the programmer has no way to refer to them ever again. This stored but inaccessible data is called *garbage*. Figure 11.34 shows an example of garbage being created. In Python- and Java-like languages, the system provides a *garbage collector* that periodically runs to clean up the garbage, which allows that memory to be reused for future allocations. (In contrast, in languages like C or C++, when you as a programmer are done using a chunk of memory, it's your responsibility to declare to the system that you're done using that memory by explicitly "deallocating" or "freeing" it.)

There are many sophisticated garbage-collection algorithms that are employed in real systems, but fundamentally the algorithmic idea is based on finding reachable nodes in a graph. There is a *root set* of memory locations that are reachable—essentially every variable that's defined in any currently active function call on the stack. Furthermore, if a memory location ℓ is pointed to by a reachable memory location, then ℓ too is reachable. Two simpler algorithms that are sometimes used in garbage collection are based on some corresponding simple graph-theoretic approaches. Here's a brief description of these two garbage-collection algorithms:[10]

Reference counting: For each block b of memory, we maintain a *reference count* of the number of other blocks of memory (or root set variables) that refer to b. When the garbage collector runs, any block b that has a reference count equal to 0 is marked as garbage and reclaimed for future use.

Mark-and-sweep: When the garbage collector runs, we iteratively *mark* each block b that is accessible. Specifically, for every variable v in the root set, we mark the block to which v refers. Then, for any block b that is marked, we also mark any block to which b refers. Once the marking process is completed, we *sweep* through memory, and reclaim all unmarked blocks.

In graph-theoretic terms, we view memory as a directed graph, with an edge from each block b to the block(s) to which b refers. Reference counting declares as garbage any node with in-degree 0; mark-and-sweep declares as garbage any node that is not reached by BFS starting from the root set.

Reference counting is a simpler algorithm, but it has a problem with cyclical structures. If two inaccessible blocks of memory refer to each other, they both have nonzero reference count, and therefore won't be marked as garbage. An example is shown in Figure 11.35. There are issues of efficiency with mark-and-sweep (the entire system has to pause while the garbage collector runs), and so other, more sophisticated algorithms are generally used in real systems.

Suppose that `Node(data,next)` creates a new node for a singly linked list, with data `data` and with a pointer `next` to the next node in the list. Imagine executing the following code:

```
1  L = Node(7,NULL)
2  L = Node(5,L)
3  L = Node(3,L)
4  L = Node(2,L)
5  L.next = L.next.next
```

Then the state of memory after executing lines 1–4 is

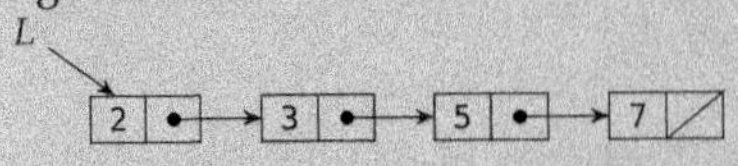

But when we execute line 5, the state of memory becomes

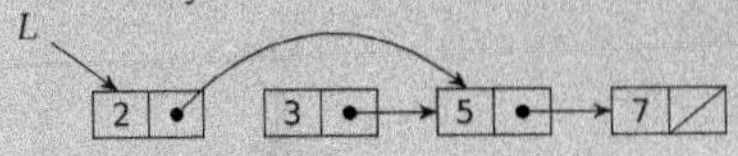

The node with data = 3 is *garbage* now: there is no way to access that memory again, because there is no way for the programmer to refer to it.

Figure 11.34: Garbage being created.

You can learn more about garbage collection in any good textbook on programming languages. A few of these are:

[10] Michael L. Scott. *Programming Language Pragmatics*. Morgan Kaufmann Publishers, 3rd edition, 2009; and Kenneth C. Louden and Kenneth A. Lambert. *Programming Languages: Principles and Practices*. Course Technology, 3rd edition, 2011.

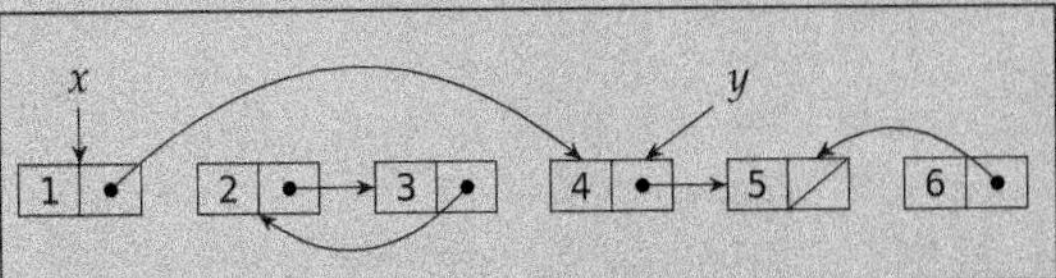

Figure 11.35: A memory diagram with six blocks of memory, and two root set variables x and y. Reference counting would show block #6 with a reference count of zero, and therefore it would be reclaimed. Mark-and-sweep would mark blocks #1, #4, and #5; thus it would reclaim blocks #2, #3, and #6.

11.3.6 Exercises

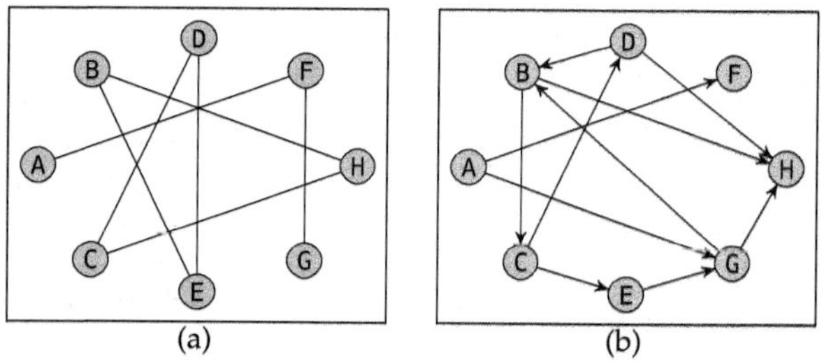


```
A:  B, E, F
B:  A
C:  D
D:  C, F
E:  A
F:  A, D
```
(c)

```
A:  B, C
B:  A, C
C:  A, B, F
D:  E
E:  B, D
F:  C
```
(d)

	A	B	C	D	E	F	G	H
A				1				1
B			1		1		1	
C		1		1				
D	1		1			1		
E		1						1
F				1				1
G		1						
H	1				1	1		

(e)

Figure 11.36: Several graphs.

For the graphs defined in Figure 11.36, identify the following specified objects (or indicate why no such thing exists):

11.72 a path from D to B in Figure 11.36(a)
11.73 two different paths from C to H in Figure 11.36(a)
11.74 a path from C to B in Figure 11.36(b)
11.75 two different paths from A to H in Figure 11.36(b)
11.76 a path from D to H in Figure 11.36(b) that is *not* simple.
11.77 a path from B to C in the graph defined by the adjacency list in Figure 11.36(c)
11.78 a shortest path from B to F in Figure 11.36(d)
11.79 a *non–shortest* path from B to C in the graph defined by the adjacency matrix in Figure 11.36(e)
11.80 all nodes reachable from A in Figure 11.36(d)
11.81 all nodes reachable from A in Figure 11.36(e)

Which of these graphs are (strongly) connected? Explain your answers. Identify all of the connected components for the undirected graphs, and all of the strongly connected components for the directed graphs.

11.82 Figure 11.36(a)
11.83 Figure 11.36(b) (strong connectivity)
11.84 Figure 11.36(c)
11.85 Figure 11.36(d) (strong connectivity)
11.86 Figure 11.36(e)

Let $G = \langle V, E \rangle$ be an undirected graph, and let $s \in V$ and $t \in V$ be any two nodes in G. Prove the following:

11.87 If there's a path of length k from s to t, then there's a path of length k from t to s.
11.88 Every shortest path between s and t is a simple path.

For a directed graph $G = \langle V, E \rangle$, the diameter *of G is the largest node-to-node distance in the graph. That is,*

$$\text{diameter}(G) = \max_{s \in V, t \in V} d(s,t),$$

where $d(s,t)$ denotes the length of the shortest path from node s to node t in G. Prove your answers:

Although the context is different, our version of "diameter" matches the idea from geometry: the diameter of a circle is the distance between the two points in the circle that are farthest apart. That's still true for a graph.

11.89 In terms of n, what is the *smallest* diameter that an n-node undirected graph can have?
11.90 In terms of n, what is the *largest* diameter that a connected n-node undirected graph can have? Give an example of a graph where the diameter is this large. *(In other words, assuming that G is connected, what's the largest possible distance between two nodes in G? Note that, without the restriction that the graph be connected, the answer would be ∞.)*

Consider an n-node 3-regular undirected graph G. (That is, we're considering a graph $G = \langle V, E \rangle$ with $|V| = n$, where each node $u \in V$ has degree exactly equal to 3.) In terms of n:

11.91 What is the largest possible number of connected components in a 3-regular graph?
11.92 What is the smallest possible number of connected components in a 3-regular graph?
11.93 Describe a connected 3-regular graph with n nodes with a diameter that's at least $\frac{n}{8}$.
11.94 Describe a connected 3-regular graph with n nodes with a diameter that's at most $8 \log n$.

11.95 Prove or disprove: let $G = \langle L \cup R, E \rangle$ be a bipartite graph with $|L| = |R|$. Suppose that every node in the graph (that is, all nodes in L and R) has at least one neighbor. Then the graph is connected.

Consider an undirected graph G. Recall that a simple path from s to t in G is a path that does not go through any node more than once. A Hamiltonian path *from s to t in G is a path from s to t that goes through each node of G* precisely *once. In general, finding Hamiltonian paths in a graph is believed to be computationally very difficult. But there are some specific graphs in which it's easy to find one.*

Hamiltonian paths are named after William Rowan Hamilton, a 19th-century Irish mathematician/physicist.

11.96 Find a Hamiltonian path in the Petersen graph:

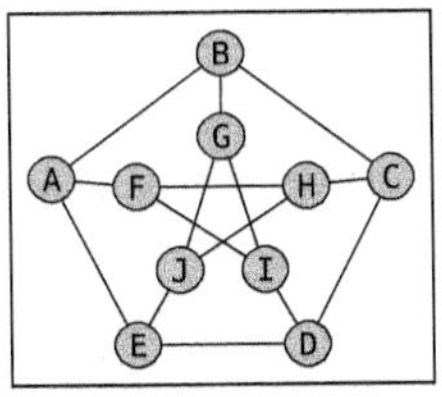

11.97 Let K_n be a complete graph, and let s and t be two distinct nodes in the graph. How many different Hamiltonian paths are there from s to t?

11.98 Let $K_{n,m}$ be a complete bipartite graph with $n + m$ nodes, and let s and t be two distinct nodes in the graph. How many different Hamiltonian paths are there from s to t? *(Careful; your answer may depend on s and t.)*

The diameter *of an undirected graph $G = \langle V, E \rangle$ is defined as the* maximum *distance between any two nodes $s \in V$ and $t \in V$. (See Exercises 11.89 and 11.90.) The maximum distance is one measure of how far a graph "sprawls," but another way of measuring this idea is by looking at the* average *distance instead. That is, for a pair of distinct nodes $\langle s, t \rangle$ chosen uniformly from the set V, what's the distance from s to t? That is, the average distance of a graph $G = \langle V, E \rangle$ is defined as*

$$\text{the average distance of } G = \frac{\sum_{s \in V} \sum_{t \in V: t \neq s} \text{distance}(s, t)}{n(n-1)}.$$

(There are $n(n-1)$ ordered pairs of distinct nodes.) Often the average distance is a bit harder to calculate than the maximum distance, but in the next few exercises you'll look at the average distance for a pair of simple graphs.

11.99 Consider an n-node cycle, where n is odd. (We'll see a formal definition of a cycle in Section 11.4, but for now just look at the 15-node example in Figure 11.37(a).) Compute the average distance in this n-node graph. *(Hint: every node is positioned symmetrically, so you can just figure out the average distance from some particular node u.)*

11.100 What is the average distance for an n-node cycle where n is even? (See the 16-node example in Figure 11.37(b).)

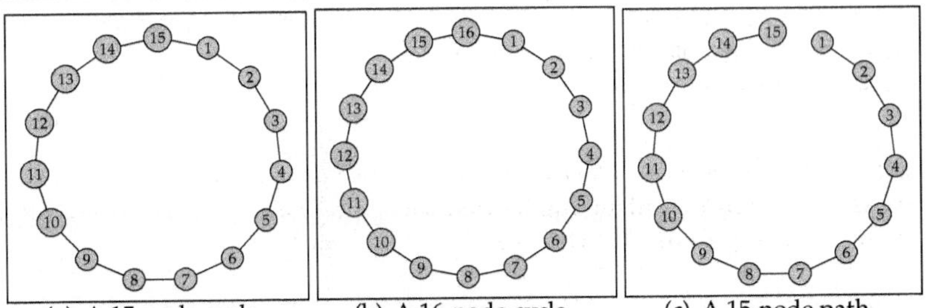

(a) A 15-node cycle. (b) A 16-node cycle. (c) A 15-node path.

Figure 11.37: Three graphs.

11.101 What is the average distance for an n-node path? (See the 15-node example in Figure 11.37(c).) *(Hint: for any particular integer k, how many pairs of nodes have distance k? Then simplify the summation.)*

11.102 *(programming required)* Write a program, in a language of your choice, to verify your answers to the last three exercises: build a graph of the appropriate size and structure, sum all of the node-to-node distances, and compute their average.

Suppose that G is an undirected graph with n nodes. Answer the following questions in terms of n:

11.103 If G is disconnected, what is the largest possible number of edges that G can contain?

11.104 If G is connected, what is the smallest possible number of edges that G can contain?

Suppose that G is a directed graph with n nodes. Answer the following questions in terms of n:

11.105 If G is strongly connected, what is the smallest number of edges that G can contain?

11.106 If every node of G is in its own strongly connected component (that is, there are n different SCCs, one per node), what is the largest number of edges that G can contain?

A metric *on a set V is a function $d : V \times V \to \mathbb{R}^{\geq 0}$ that obeys the following conditions (see Exercise 4.6 for more):*

- reflexivity: *for any $u \in V$ and $v \in V$, we have $d(u,u) = 0$ and $d(u,v) \neq 0$ whenever $u \neq v$.*
- symmetry: *for any $u \in V$ and $v \in V$, we have $d(u,v) = d(v,u)$.*
- triangle inequality: *for any $u \in V$ and $v \in V$ and $z \in V$, we have $d(u,v) \leq d(u,z) + d(z,v)$.*

Let $d_G(u,v)$ denote the distance (shortest path length) between nodes $u \in V$ and $v \in V$ for a graph $G = \langle V, E \rangle$.

11.107 Prove that d_G is a metric if G is any connected undirected graph.

11.108 Prove that d_G is not necessarily a metric for a directed graph G, even if G is strongly connected.

11.109 Definition 11.23 defined a strong connected component in a graph $G = \langle V, E \rangle$ as *a set $C \subseteq V$ such that: (i) any two nodes $s \in C$ and $t \in C$ are strongly connected; and (ii) for any node $x \in V - C$, adding x to C would make (i) false.* Suppose that we'd instead defined clause (i) as *for any two nodes $s \in C$ and $t \in C$, the node t is reachable from node s.* (But we don't require that s be reachable from t.) This alternate definition is equivalent to the original. Why?

11.110 Prove that the strongly connected components (SCCs) of a directed graph partition the nodes of the graph: that is, prove that the relation $R(u,v)$ denoting mutual reachability (u is reachable from v, and v is reachable from u) is an equivalence relation (reflexive, symmetric, and transitive).

Consider the directed graphs represented in Figure 11.38, one by picture and one by adjacency list. Identify the strongly connected components . . .

11.111 . . . in Figure 11.38(a).

11.112 . . . in Figure 11.38(b).

Suppose that we run breadth-first search from the following nodes. What is the last *node that BFS discovers? (If there's a tie, then list all the tied nodes.)*

11.113 BFS from node `A` in Figure 11.38(a).

11.114 BFS from node `B` in Figure 11.38(a).

11.115 BFS from node `0` in Figure 11.38(b).

11.116 BFS from node `12` in Figure 11.38(b).

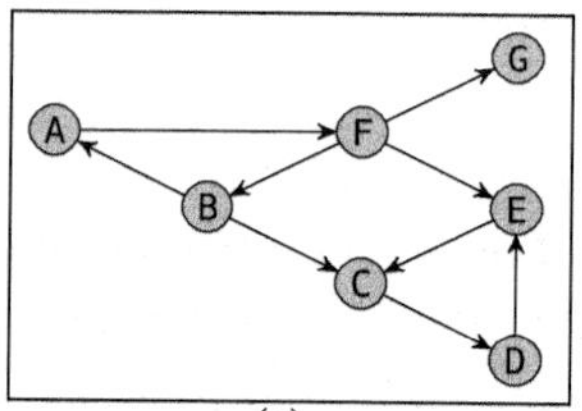

(a)

0:	3, 7
1:	9, 2, 5
2:	1, 10, 9
3:	0, 7, 1
4:	10, 7
5:	1
6:	7, 11
7:	0, 4, 6, 8
8:	11, 12
9:	1
10:	2, 4
11:	6, 8
12:	8

(b)

Figure 11.38: Two graphs.

Breadth-first search as described in Figure 11.29 finds all nodes reachable from a given source node in a given graph, and, in fact, it discovers nodes in increasing order of their distance from s. But we didn't actually record distances during the computation.

11.117 Modify the pseudocode for BFS to compute distances instead of just whether a path exists, by annotating every node added to *Frontier* with its distance from the source node s.

11.118 Argue that in your modified version of BFS, there are never more than two different distances stored in the Frontier.

11.119 Argue that the claim from the previous exercise may be false for *depth*-first search.

11.120 Consider a graph G represented by an adjacency matrix M. What does the $\langle i,j \rangle$th entry of MM (the matrix that results from squaring the matrix M) represent?

A word chain *is a sequence $\langle w_1, w_2, \ldots, w_k \rangle$ of words, where each w_i is a word in English, and w_{i+1} is one letter different from w_i. For example, a word chain from* `FROWN` *to* `SMILE` *for my dictionary is*

`FROWN` → `FLOWN` → `FLOWS` → `SLOWS` → `SLOTS` → `SLITS` → `SKITS` → `SKITE` → `SMITE` → `SMILE`.

(`SKITE` is a word of Scottish origin, meaning "an oblique blow.")

11.121 *(programming required)* Write a program that uses a BFS-like algorithm to find a shortest word chain between two given words w_1 and w_2 of the same length. (You can find a dictionary of English words on the web, or `/usr/share/dict/words` on Unix-based operating systems. You'll want to cull your dictionary to only words of the right length before you start.) There are faster solutions that involve searching "in both directions" out from w_1 and into w_2 until you find a match, but BFS from w_1 will work.

11.4 *Trees*

> I think that I shall never see
> A poem lovely as a tree.
>
> Joyce Kilmer (1886–1918), "Trees"
> *Trees and Other Poems* (1914)

Informally, a *tree* is a graph that grows from a *root*, branching outward and eventually leading to the *leaves*. (We computer scientists are always upside down compared to botanists: unlike an oak or maple or tamarack, the root of a tree in CS is at the top, and it grows downward toward the leaves.) See Figure 11.39.

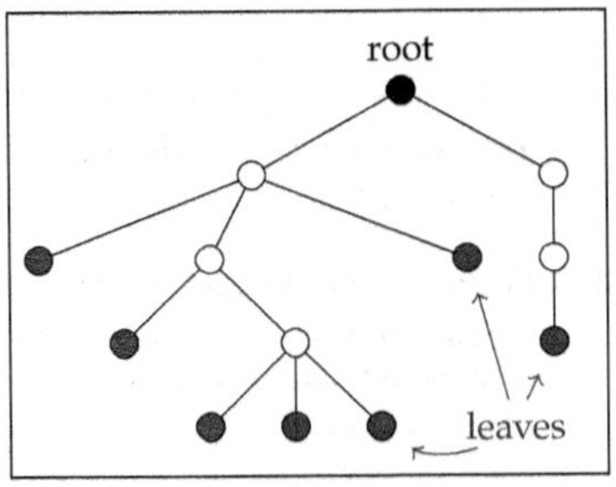

Figure 11.39: A small tree.

Trees arise very frequently in computer science: to name just a few examples, they're the class hierarchies of object-oriented programming, the binary search trees of data structures (see p. 1160), the game trees describing the progression of Tic-Tac-Toe or chess (p. 344), the parse trees that describe formal or natural languages (p. 543), the recursion trees that describe the execution of recursive algorithms (Section 6.4). Trees are also frequently used in computational models of important phenomena from outside of CS: for example, in reconstructing evolutionary phylogenies (in computational biology), or in reconstructing the paths by which rumors spread from the originator of the information (in social network analysis). In this section, we'll introduce trees formally—including definitions, properties, algorithms, and applications—as a special type of graph.

11.4.1 *Cycles*

Before we can define trees properly, we must first define another notion about graphs in general—a *cycle*, which is way to get from a node back to itself:

Definition 11.26 (Cycle)
A cycle $\langle u_1, u_2, \ldots, u_k, u_1\rangle$ *is a path of length* ≥ 2 *from a node* u_1 *back to node* u_1 *that does not traverse the same edge twice. Just as for any other path, the* length of the cycle $\langle u_1, u_2, \ldots, u_k, u_1\rangle$ *is the number of edges it traverses—that is, k.*

Figure 11.40 shows examples of an undirected and directed graph with a cycle $\langle$A, B, C, A$\rangle$. Note that the edges $\langle s,t\rangle$ and $\langle t,s\rangle$ in a directed graph are different; in an undirected graph, the edges $\{s,t\}$ and $\{t,s\}$ are the same. Thus a cycle in a directed graph *can* use both $\langle s,t\rangle$ and $\langle t,s\rangle$, but a cycle in an undirected graph cannot use both $\langle s,t\rangle$ and $\langle t,s\rangle$. In Figure 11.40, the path $\langle$C, E, C$\rangle$ *is* a cycle in the directed graph, but *is not* a cycle in the undirected graph because it reuses an edge.

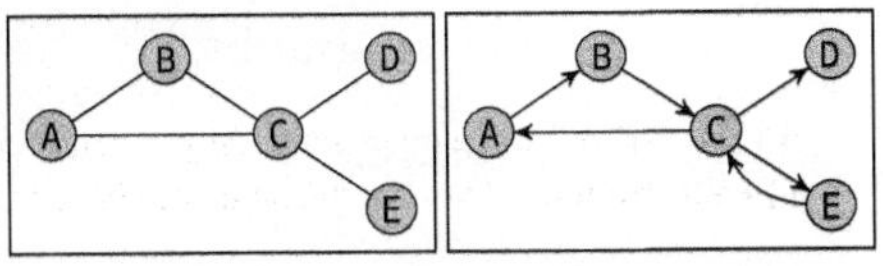

Figure 11.40: Two graphs with cycles $\langle$A, B, C, A$\rangle$.

Technically speaking, the definition of a cycle in Definition 11.26 says that the undirected graph in Figure 11.40 has six different cycles:

- $\langle$A, B, C, A$\rangle$, $\langle$C, A, B, C$\rangle$, and $\langle$B, C, A, B$\rangle$ (going clockwise), and
- $\langle$A, C, B, A$\rangle$, $\langle$C, B, A, C$\rangle$, and $\langle$B, A, C, B$\rangle$ (going counterclockwise).

However, we will adopt the convention that there is one and only one cycle in this graph. Because we can "start anywhere" in a cycle, we consider a cycle to be defined only by the relative ordering of the nodes involved, regardless of where we start. In an undirected graph, we can "go either direction" (clockwise or counterclockwise), so we also ignore the direction of travel in distinguishing cycles. In a directed graph, the direction of travel *does* matter; we may be able to go in one direction around a cycle without being able to go in the other. In other words, we say that Figure 11.41(a) and Figure 11.41(b) have one cycle each, while Figure 11.41(c) has two.

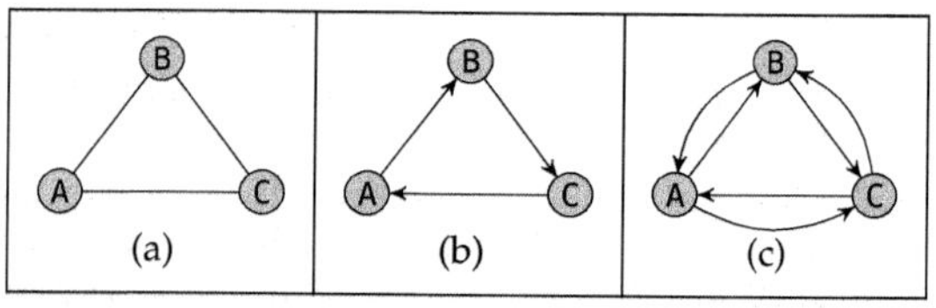

Figure 11.41: Some cycles.

A cycle is by definition forbidden from traversing the same edge twice. A *simple* cycle also does not visit any *node* more than once:

Definition 11.27 (Simple cycle)
A cycle $\langle u_1, u_2, \ldots, u_k, u_1 \rangle$ *is* simple *if each* u_i *is distinct—that is, no nodes in the cycle are duplicated aside from the last node (which equals the first node).*

(We've now used the word "simple" in three different contexts: simple graphs have no parallel edges or self-loops, and simple paths and cycles have no repeated vertices. Intuitively, all three definitions correspond to an entity that's not unnecessarily complicated.) For one example, see Figure 11.42; here are two more:

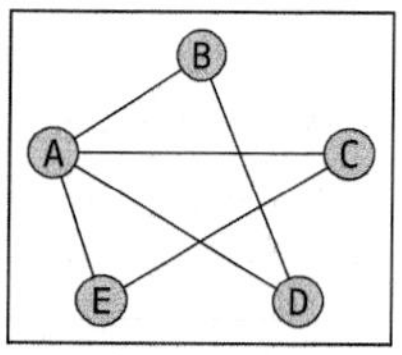

Figure 11.42: In this graph, ⟨D, B, A, C, E, A, D⟩ is a non-simple cycle. This graph also has two simple cycles: ⟨D, B, A, D⟩ and ⟨C, E, A, C⟩.

Example 11.30 (Finding cycles)
Problem: Identify all simple cycles in the following graphs:

1.

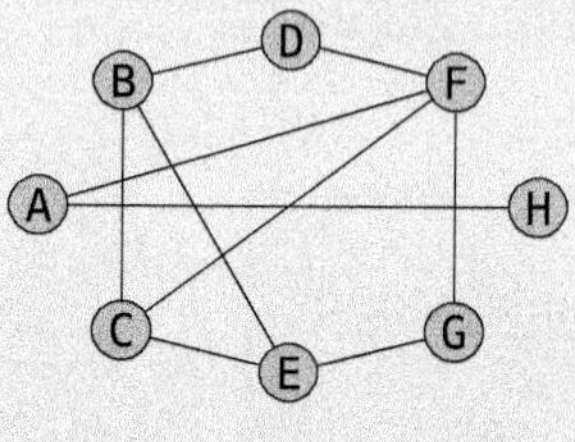

2.

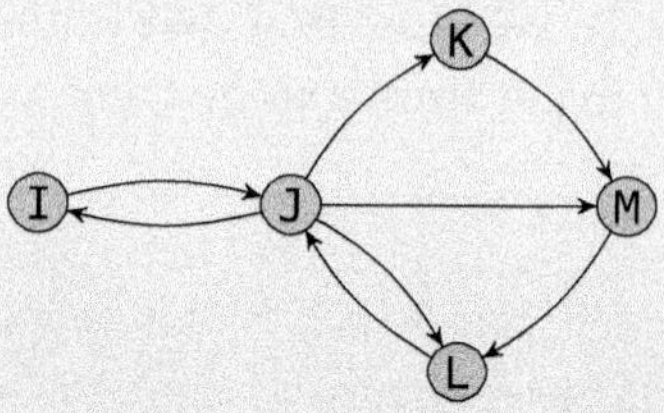

Solution: A nice way to identify cycles systematically is to look for cycles of all possible lengths: 2-node cycles, 3-node cycles, etc. (Actually 2-node cycles are possible only in directed graphs. *Exercise: why?*) Here are the simple cycles in these graphs:

1.	⟨B, E, C, B⟩	2.	⟨I, J, I⟩
	⟨B, D, F, C, B⟩		⟨J, L, J⟩
	⟨C, F, G, E, C⟩		⟨J, M, L, J⟩
	⟨B, D, F, G, E, B⟩		⟨J, K, M, L, J⟩
	⟨B, D, F, G, E, C, B⟩		

Note that (to name one of several examples) the sequence ⟨I, J, L, J, I⟩ is also a cycle in the second graph—it traverses four distinct directed edges and goes from node I to I—but this cycle is not simple, because node J is repeated.

We can use a modification of breadth-first search to identify cycles algorithmically. Specifically, suppose that we wish to find out whether a node u is involved in a cycle in a directed graph. We run BFS starting at node u, and if we ever encounter a node v that has u as a neighbor, then we have found a cycle involving node u. (An extra modification is necessary for undirected graphs; see Exercise 11.129.)

Taking it further: Kidneys are the most frequently transplanted organ today, in part because—unlike for other organs—humans generally have a "spare": we're born with two kidneys, but only need one functioning kidney to live a healthy life. Thus patients suffering from kidney failure may be able to get a transplant from friends or family members who are willing to donate one of their kidneys. But this potential transplant relies on the donor and the patient being compatible in dimensions like blood type and the physical size of the organs. Recently a computational solution to the problem of incompatibility has emerged, using algorithms based on finding (short) cycles in a particular graph: there is now national exchange for matching up two (or a few) patients with willing-but-incompatible donors, and doing a multiway transplant. See p. 1159 for more discussion.

Acyclic graphs

While cycles are important on their own, their relevance for trees is actually when they *don't* exist:

Definition 11.28 (Acyclic Graphs)
A graph is acyclic *if it contains no cycles.*

Let's prove a useful structural fact about acyclic graphs. (Recall that we are considering *finite* graphs, where the set of nodes in the graph is finite. The following claim would be false if graphs could have an infinite number of nodes!)

Lemma 11.5 (Every acyclic graph has a node with degree 0 or 1)
Let $G = \langle V, E \rangle$ be an acyclic undirected graph. Then there exists a node in V whose degree is zero or one.

Proof. We'll give a constructive proof of the claim—specifically, we'll give an algorithm that finds a node with the stated property:

```
let u_0 be an arbitrary node in the graph, and let i := 0
while the current node u_i has no unvisited neighbors:
    let u_{i+1} be a neighbor of u_i that has not previously been visited.
    increment i
```

Observe that this process must terminate in at most $|V|$ iterations, because we must visit a new node in each step. Suppose that this algorithm goes through k iterations of the **while** loop, and let t be the last node visited by the algorithm. (So $t = u_k$.)

- If $k = 0$, then $t = u_0$ has degree zero, so the claim follows immediately.
- If $k \geq 1$, then we'll argue that t has degree one. Because the algorithm terminated, there cannot be an edge between t and any unvisited node. Furthermore, if there were an edge from t to any previously visited node u_j for $j < k - 1$, then there would be a cycle in the graph, namely $\langle u_j, u_{j+1}, \ldots, u_{k-1}, u_k, u_j \rangle$. Therefore t's only neighbor is u_{k-1}, and the degree of t is one. $\square$

For directed graphs, the claim analogous to Lemma 11.5 is *every directed acyclic graph contains a node with outdegree zero.* (You'll prove it in Exercise 11.130.)

Taking it further: A *directed acyclic graph* (often just called a *DAG*) is, perhaps obviously, a directed graph that contains no cycles. A DAG G corresponds to a (strict) partial order (see Chapter 8); a cycle in G corresponds to a violation of transitivity. In fact, we can think of *any* directed graph $G = \langle V, E \rangle$ as a relation—specifically, the edge set E is a subset of $V \times V$. Like transitivity and acyclicity, many of the concepts that we explored in Chapter 8 have analogues in the world of graphs.

11.4.2 Trees

With the definition of cycles in hand, we can now define trees themselves:

Definition 11.29 (Tree)
A tree is an undirected graph that is connected and acyclic.

An irrelevant note about Chinese: the character for *tree* is 木; the character for *forest* is 森 (a disconnected collection of trees!).

We will also sometimes talk about graphs that satisfy only the latter requirement: a *forest* is an undirected graph that is acyclic (but not necessarily connected). Every connected component of a forest is a tree, and note that a tree is itself a forest.

Several examples of trees are shown in Figure 11.43: all six graphs have a single connected component and contain no cycles. Therefore all six are trees.

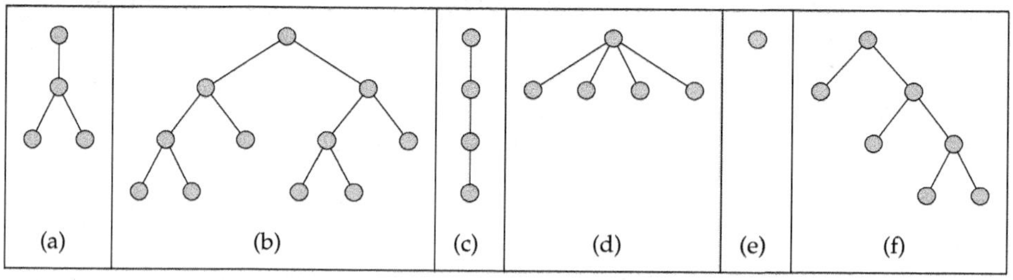

Figure 11.43: Some sample trees.

We'll prove several structural facts about trees in this section, beginning with one concerning the number of edges in a tree. To start, let's look at the number of nodes and edges in each of the trees in Figure 11.43:

	(a)	(b)	(c)	(d)	(e)	(f)
number of nodes	4	11	4	5	1	7
number of edges	3	10	3	4	0	6

In each of these trees, the number of nodes is one more than the number of edges, and that's no coincidence; here's the statement and proof of the general fact:

Theorem 11.6 (Number of edges in a tree)
Let $T = \langle V, E \rangle$ be a tree. Then $|E| = |V| - 1$.

Proof. Let $P(n)$ denote the property that any n-node tree has precisely $n - 1$ edges. We will prove that $P(n)$ holds for all $n \geq 1$ by induction on n.

Base case ($n = 1$): We must prove $P(1)$: any 1-node tree has $1 - 1 = 0$ edges. But the only 1-node (simple) graph is the one shown in Figure 11.43(e), which has zero edges, and so we're done immediately.

Inductive case ($n \geq 2$): We assume the inductive hypothesis $P(n-1)$—that is, every $(n-1)$-node tree has $n-2$ edges. We must prove $P(n)$.

Consider an arbitrary tree $T = \langle V, E \rangle$ with $|V| = n$. By definition, T is acyclic and connected. By Lemma 11.5, then, there exists a node $u \in V$ with degree 0 or 1 in T. Furthermore, because T is connected, the degree of u cannot be 0. Thus u is a node with $degree(u) = 1$. Let $v \in V$ be the unique neighbor of u in T. Let T' be T with node u and the edge $\{u,v\}$ between u and v deleted. (See Figure 11.44.)

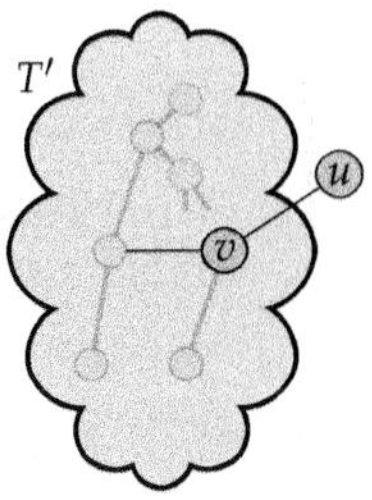

Figure 11.44: A tree T, with a node u of degree = 1 and its neighbor v. The tree T' is T without the node u and the edge $\{u,v\}$.

We claim that the graph $T' = \langle V - \{u\}, E - \{\{u,v\}\} \rangle$ is a tree, too. The acyclicity and connectivity of T' both follow from the fact that T was acyclic and connected, and the fact that the eliminated node u was of degree 1.

The tree T' contains $n-1$ nodes, and thus, by the inductive hypothesis $P(n-1)$, contains $n-2$ edges. Therefore T, whose edges are precisely the edges of T' plus the eliminated edge $\{u,v\}$, contains precisely $(n-2)+1 = n-1$ edges. □

An immediate consequence of Theorem 11.6 is that every tree is teetering on the edge of being disconnected and of having a cycle (see Figure 11.45):

Corollary 11.7 (A tree with an edge added or removed is not a tree)
Let $T = \langle V, E \rangle$ be any tree. Then:

1. *adding any edge $e \notin E$ to T creates a cycle; and*
2. *removing any edge $e \in E$ from T disconnects the graph.*

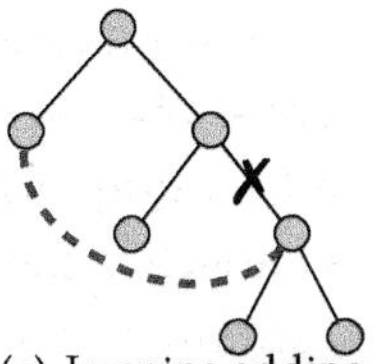

(a) Imagine adding the dashed edge, or removing the edge marked with ✗.

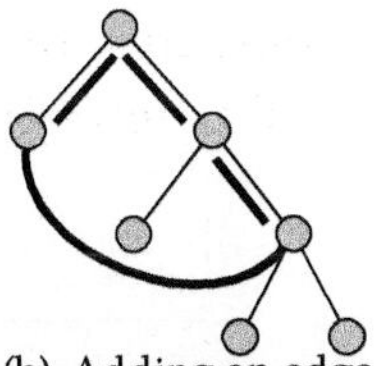

(b) Adding an edge creates a cycle.

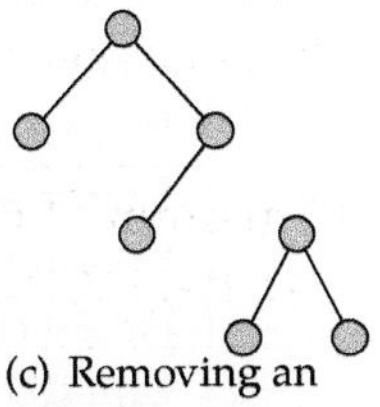

(c) Removing an edge disconnects the graph.

Figure 11.45: Adding/removing an edge from a tree.

Proof. 1. Define the graph $G = \langle V, E \cup \{e\} \rangle$ as the result of adding the new edge e to the tree T. Because adding an edge to a graph can never disrupt connectivity and T was already connected, we know that G must be connected too. Thus if G were acyclic, then G would be a tree. But G has one more edge than T—specifically, G has $(|V|-1)+1 = |V|$ edges—and therefore isn't a tree by Theorem 11.6.

2. The proof is similar: let G' be T with e removed. Removing an edge cannot create a cycle, so G' is acyclic. But G' has too few edges to be a tree by Theorem 11.6, so G' must be disconnected. □

(Here's an alternative proof of Corollary 11.7.1. Let $\langle u,v \rangle$ be an edge not in the tree T. Because T is connected, there is already a (simple) path P from u to v in T. If we add $\langle u,v \rangle$ to T, then there is a cycle: follow P from u to v and then follow the new edge from v back to u. Therefore G contains a cycle.)

Rooted trees

We often designate a particular node of a tree T as the *root*, which is traditionally drawn as the topmost node. (Note that we could designate any node as the root and—just like that mobile of zoo animals from your crib from infancy—"hang" the tree by that node.) We will adopt the standard convention that, whenever we draw trees, the vertically highest node is the root.

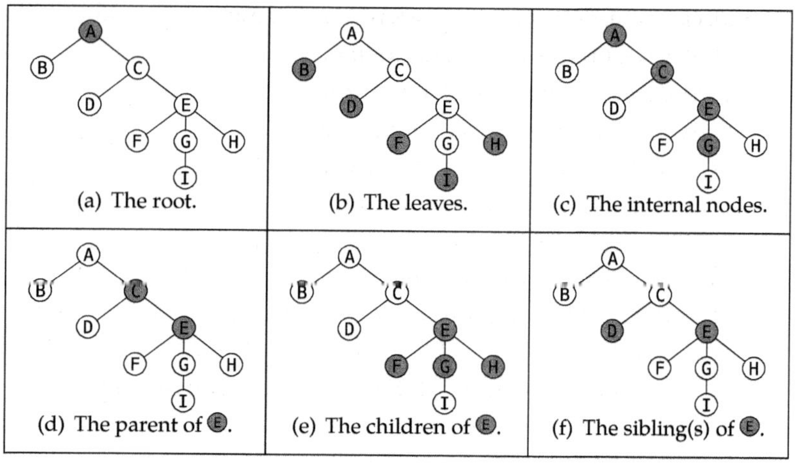

(a) The root. (b) The leaves. (c) The internal nodes. (d) The parent of E. (e) The children of E. (f) The sibling(s) of E.

Figure 11.46: The root, leaves, and internal nodes of the tree; the parent, children, and siblings of a particular node.

There's a lot of terminology about trees in computer science that's borrowed from the world of family trees:

- For a node u in a tree with root $r \neq u$, the *parent* of u is the unique neighbor of u that is closer to r than u is. (The root is the only node that has no parent.)
- A node v is one of the *children* of a node u if v's parent is u.
- A node v is a *sibling* of a node $u \neq v$ if v and u have the same parent.

A node with zero children is called a *leaf*. A node with one or more children is called an *internal node*. (Note that the root is an internal node unless the tree is the trivial one-node graph.) See Figure 11.46 for an illustration of all of these definitions. Note that Figure 11.46 is correct only when the root is the topmost node in the image; with a different root, all of the panels could change. Here's a concrete example:

Example 11.31 (A sample tree)

Here are two trees. (The second tree is just the first, rerooted to make E the new root.)

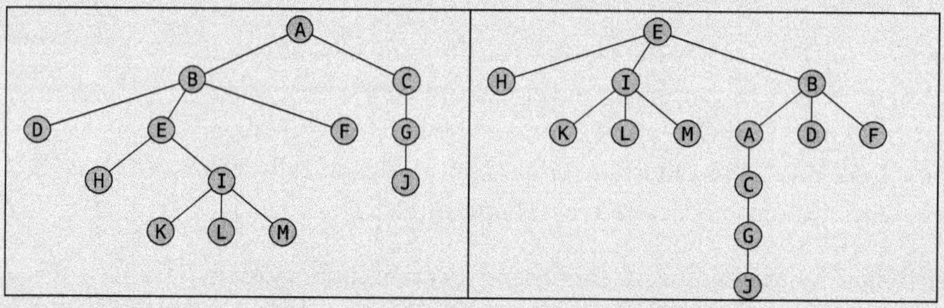

Then we have:

Root: A Leaves: {D, F, H, J, K, L, M} Internal nodes: {A, B, C, E, G, I}	Root: E Leaves: {D, F, H, J, K, L, M} Internal nodes: {A, B, C, E, G, I}
Parent of B: A Children of B: {D, E, F}	Parent of B: E Children of B: {A, D, F}
Parent of A: none Children of A: {B, C}	Parent of A: B Children of A: {C}

While the leaves and internal nodes are identical in these two trees, note that if we'd rerooted the tree at any of the erstwhile leaves instead, the new root would become an internal node instead of a leaf. For example, if we reroot this tree at H, then the leaves would be {D, F, J, K, L, M} and the internal nodes would be {A, B, C, E, G, H, I}.

SUBTREES, DESCENDANTS, AND ANCESTORS

Let T be a rooted tree, and let u be any node in T. The *subtree rooted at u* consists of u and all those nodes and edges "below" u in T. (In other words, a node v is in the subtree rooted at u if and only if v is no closer to the root of T than u is; the subtree is the induced subgraph of these nodes.) Such a node v in the subtree rooted at u is called a *descendant of u* if $v \neq u$. The node u is called an *ancestor of v*. See Figure 11.47 for illustrations of these three definitions. Here's an example:

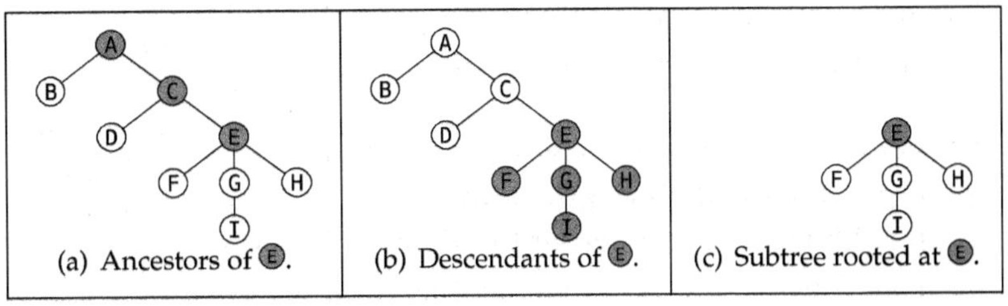

Figure 11.47: Ancestors, descendants, and subtrees.

Example 11.32 (Descendants and ancestors)
Recall the trees from Example 11.31:

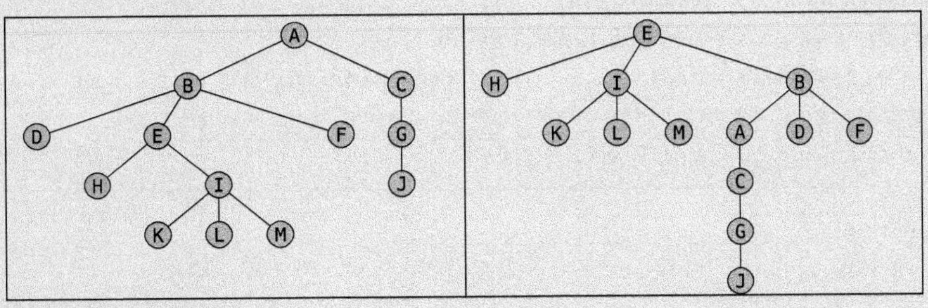

Then we have:

Descendants of B: {D, E, F, H, I, K, L, M} Ancestors of B: {A}	Descendants of B: {A, C, D, F, G, J} Ancestors of B: {E}
Descendants of H: none Ancestors of H: {A, B, E}	Descendants of H: none Ancestors of H: {E}
Subtree rooted at B: 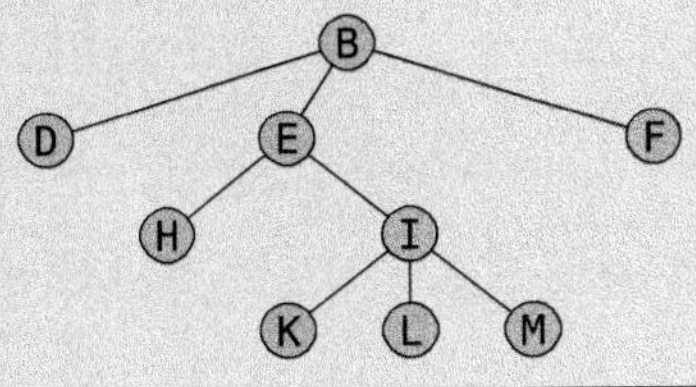	Subtree rooted at B: 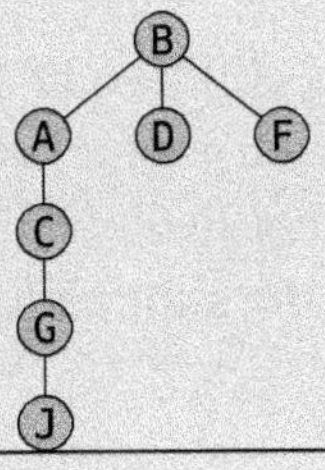

We have one final pair of definitions to (at last!) conclude our parade of terminology about rooted trees, related to how "tall" a tree is. Consider a rooted tree T with root node r. The *depth* of a node u is the distance from u to r. The *height* of a tree is the maximum, over all nodes u in the tree, of the depth of node u.

For example, every node in the tree in Figure 11.48 is labeled by its depth: the root has depth 0, its children have depth 1, their children (the "grandchildren" of the root) have depth 2, and so forth. The height of the tree is the largest depth of any of its nodes—in this case, the height is 4.

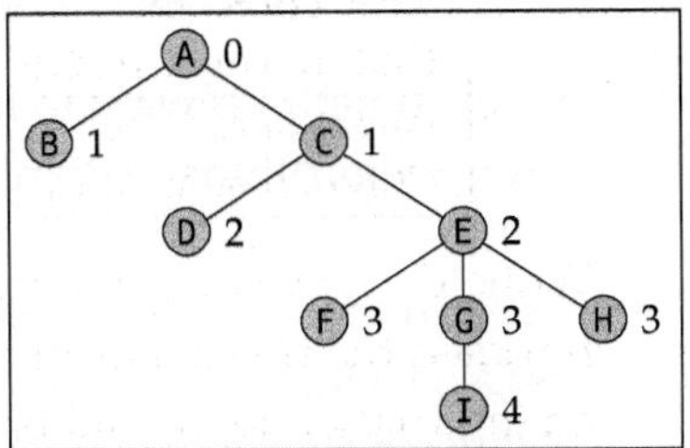

Figure 11.48: A rooted tree's nodes, labeled by depth.

Taking it further: Alternatively, we could give several of the definitions about rooted trees recursively. For example, we could define ancestors and descendants of a node u be in a rooted tree T as follows:

- A node v is an *ancestor* of u if (i) v is the parent of u; or (ii) v is the parent of any ancestor of u.
- A node v is a *descendant* of u if (i) v is a child of u; or (ii) v is a child of any descendant of u.

We can also think of the depth of a node, or the height of a tree, recursively. The depth of the root is zero; the depth of a node with a parent p is $1 +$ (the depth of p). For height:

- the height of a one-node tree T is zero; and
- the height of a tree T with root r with children $\{c_1, c_2, \ldots, c_k\}$ is

$$1 + \max_{i \in \{1,\ldots,k\}} \text{the height of the subtree rooted at } c_i.$$

BINARY TREES

We'll often encounter a special type of tree in which nodes have a limited number of children:

Definition 11.30 (Binary trees and k-ary trees)
A binary tree *is a rooted tree in which each node has* 0, 1, *or* 2 *children. More generally, if every node in a rooted tree T has k or fewer children, then T is called a* k-ary tree. *(In other words, a binary tree is* 2-ary.*)*

For example, consider the trees in Figure 11.49. Of them, only the tree in Figure 11.49(d) is not a binary tree, because its root has four children. (This tree is a 4-ary tree.) But the other five trees are all binary: in each, every internal node has either 1 child or 2 children.

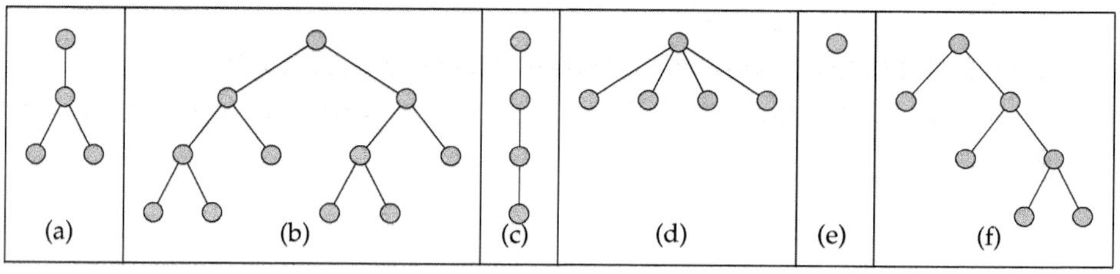

Figure 11.49: The trees from Figure 11.43, repeated. All but (d) are binary trees.

In a binary tree, the possible children of a node are called its *left child* and *right child*. (Even for a node u in a binary tree that has only one child, we'll insist that the lone child be designated as either the left child of u or the right child of u.) For a node u, we say that u's *left subtree* is the subtree rooted at u's left child; the *right subtree* is analogous.

11.4.3 *Tree Traversal*

We will sometimes want to list all of the nodes contained in a tree T. There are three standard algorithms that are used for this purpose, called *pre-order*, *in-order*, and *post-order traversal*. While these algorithms can be generalized to non-binary trees, they're easier to understand for binary trees (and they're most frequently deployed for binary trees), so we'll consider them that way.

All three algorithms are recursive, and all three algorithms execute precisely the same steps—just in a different order. On an empty tree T, we do nothing; on a non-empty tree T, all three algorithms perform the following steps:

- we "visit" the root of the tree T. (You can think of "visiting" the root as printing out the contents of the root node, or as adding it to the end of an accumulating list of the nodes that we've encountered in the tree.)
- we recursively traverse the left subtree of T, finding all nodes there.
- we recursively traverse the right subtree of T, finding all nodes there.

But the three traversal algorithms execute the three steps in different orders, either visiting the root *before both recursive calls* ("pre-order"); *between the recursive calls* ("in-order"); or *after both recursive calls* ("post-order"). We always recurse on the left subtree before we recurse on the right subtree. Here are the details:

pre-order-traverse(T):	**in-order-traverse**(T):	**post-order-traverse**(T):
if T is empty **then**	**if** T is empty **then**	**if** T is empty **then**
do nothing.	do nothing.	do nothing.
else	**else**	**else**
visit the root of T	**in-order-traverse**(T's left subtree)	**post-order-traverse**(T's left subtree)
pre-order-traverse(T's left subtree)	visit the root of T	**post-order-traverse**(T's right subtree)
pre-order-traverse(T's right subtree)	**in-order-traverse**(T's right subtree)	visit the root of T

Figure 11.50: Three different algorithms to traverse a binary tree.

Let's take a look at an example of traversing a small tree using these algorithms. First we'll look at the *pre-order* traversal, in which the first node visited in any subtree is the root of that subtree:

Example 11.33 (Traversing a small tree: pre-order traversal)
Let's determine the order of nodes' visits by a pre-order traversal of the following tree:

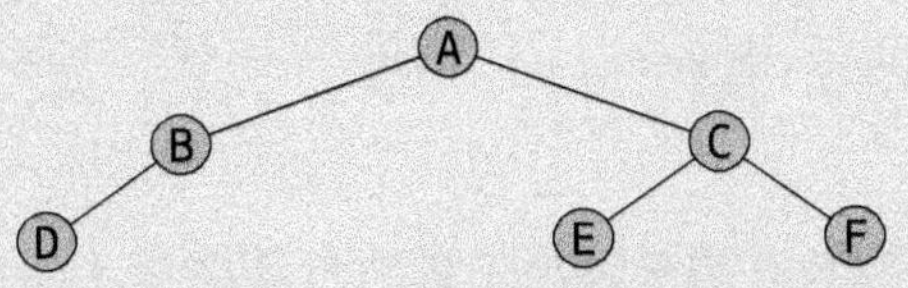

In a pre-order traversal, we first visit the root, then pre-order-traverse the left subtree, then pre-order-traverse the right subtree. In other words, we first visit the root A, then pre-order-traverse 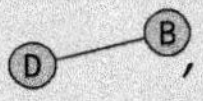, then pre-order-traverse 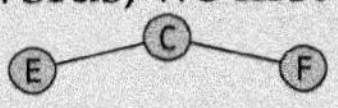:

Step #1: visit the root. We visit the root A.

Step #2: pre-order-traverse the left subtree. To pre-order-traverse 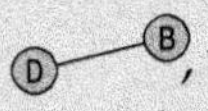, we first visit the root B, then pre-order-traverse the left subtree D, then pre-order-traverse the (empty) right-subtree. In order, these steps visit B and D.

Step #3: pre-order-traverse the right subtree. To pre-order-traverse 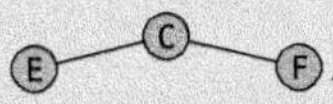, we first visit C, then pre-order-traverse the left subtree E, and then pre-order-traverse the right subtree F. Pre-order-traversing E just results in visiting E, and pre-order-traversing F just visits F. In order, these steps visit C, E, and F.

Putting this all together, the pre-order traversal of the tree visits the nodes in this order:

$$\underbrace{\text{A}}_{\text{step \#1}},\ \underbrace{\text{B, D}}_{\text{step \#2}},\ \underbrace{\text{C, E, F}}_{\text{step \#3}}.$$

Here are examples of the other two traversal algorithms, on the same tree:

Example 11.34 (Traversing a small tree: in-order and post-order traversals)
Problem: Recall the tree from Example 11.33:

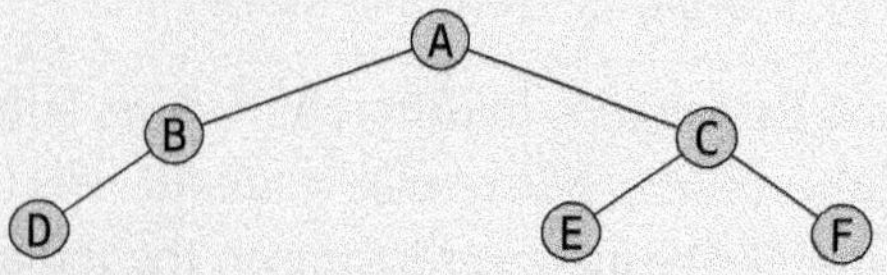

1. In what order are the nodes visited by an *in-order* traversal of this tree?
2. What about a *post-order* traversal?

Solution: 1. We first traverse 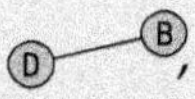, then visit A, then traverse 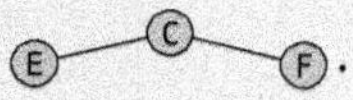.
Traversing 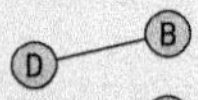 visits D and B: first the left subtree, then the root.
Traversing 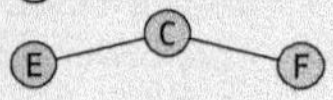 visits E, then C, then F.
Thus an in-order traversal visits the nodes in the order D, B, A, E, C, F.

2. For a post-order traversal, the root of each subtree is the *last* node traversed in that subtree: we first traverse 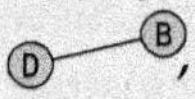, then traverse 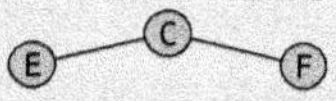, then visit A.
Traversing 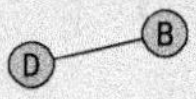 visits D and B: first the left subtree, then the nonexistent right subtree, then the root.
Traversing 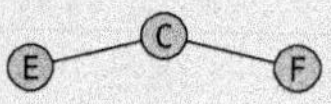 visits E, then F, then C.
Thus a post-order traversal visits the tree's nodes in the order D, B, E, F, C, A.

Here's another example, of using traversals to reconstruct a binary tree:

Example 11.35 (Trees from traversals)
Problem: Here is the output of all three traversals on a binary tree T. What's T?

pre-order traversal	in-order traversal	post-order traversal
9, 2, 7, 4, 5, 3	2, 9, 5, 4, 3, 7	2, 5, 3, 4, 7, 9

Solution: We'll reassemble T from the root down. The root is first in the pre-order traversal (and last in the post-order), so 9 is the root. The root separates the left subtree from the right subtree in the in-order traversal; thus the left subtree contains just 2 and the right contains $\{3, 4, 5, 7\}$. So the tree has the following form:

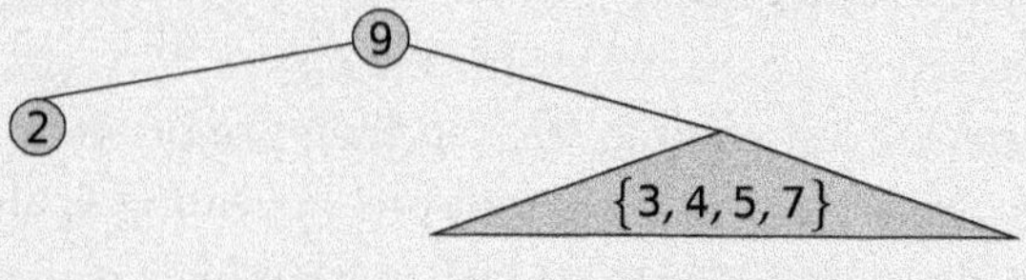

The post-order 5, 3, 4, 7 and in-order 5, 4, 3, 7 show that 7 is the root of the unknown portion of the tree and that 7's right subtree is empty. The last three nodes are pre-ordered 4, 5, 3; in-ordered 5, 4, 3; and post-ordered 5, 3, 4. In sum, that says that 4 is the root, 5 is the left subtree, and 3 is the right subtree. Assembling these pieces yields the final tree:

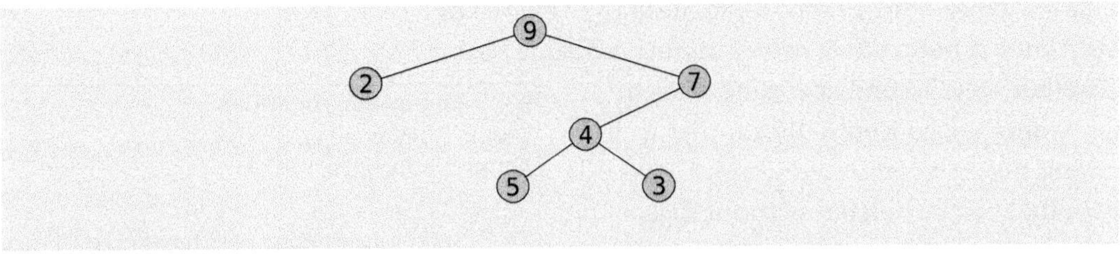

Taking it further: One particularly important type of binary tree is the *binary search tree (BST)*, a widely used data structure—one that's probably very familiar if you've taken a course on data structures. A BST is a binary tree in which each node has some associated "key" (a piece of data), and the nodes of the tree are stored in a particular sorted order: all nodes in the left subtree have a key smaller than the root, and all nodes in the right subtree have a key larger than the root. Thus an in-order traversal of a binary search tree yields the tree's keys in sorted order. For more, see p. 1160. An even more specific form of binary search tree, called a *balanced binary search tree*, adds an additional structural property related to the depth of nodes in the tree. See p. 643 for a discussion of one scheme for balanced binary search trees, called *AVL trees*.

11.4.4 *Spanning Trees*

Let $G = \langle V, E \rangle$ be an undirected graph. For example, imagine that each node in V represents a dorm room on your campus, and each edge in E denotes a possible fiber optic cable that can be laid to build an ethernet connection throughout the residence halls. A reasonable goal is to actually place only some of those possible cables, a subset $E' \subseteq E$, while ensuring that network traffic can be sent between any two dorm rooms—that is, ensuring that the resulting network is connected. In other words, one seeks a *spanning tree* of the graph G:

Definition 11.31 (Spanning tree)
Let $G = \langle V, E \rangle$ be a connected undirected graph. A spanning tree *of G is a tree $T = \langle V, E' \rangle$ with the same nodes as G and with edges $E' \subseteq E$ that are a subset of G's edges.*

A spanning tree of G is called "spanning" because it connects (that is, spans) all nodes in G. Figure 11.51 shows a small example: the first panel shows a small graph G; the remaining panels show the 8 different spanning trees of G.

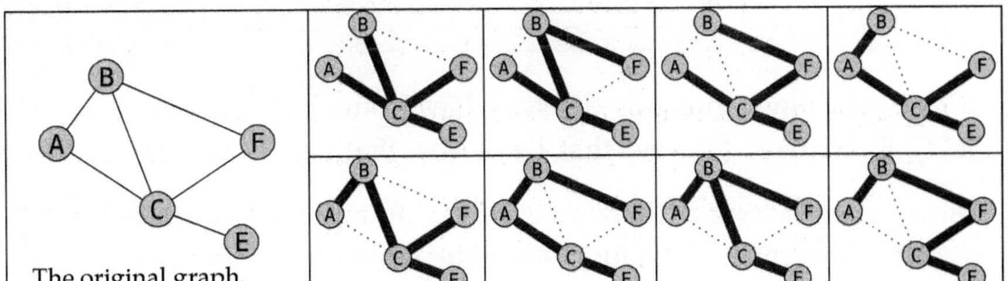

Figure 11.51: All 8 spanning trees of the graph shown in the first panel.

A graph G has a spanning tree if and only if G is connected: we can be sure to only remove "redundant" edges that aren't required for connectivity, and removing edges from G can never cause a disconnected graph to become connected. (For disconnected graphs, people sometimes talk about a *spanning forest:* a forest $F = \langle V, E' \rangle$ with $E' \subseteq E$, where the connected components of the original graph G and the connected components of the forest F are identical.)

Although we didn't talk about it this way when we introduced breadth- and depth-first search (see Figures 11.29 and 11.31), these algorithms can find spanning trees, with a small change: as we explore the graph, we include in E' every edge $\langle u, v \rangle$ that leads from a previously known node u to a newly discovered node v.

We'll also see some other ways to find spanning trees in Section 11.5.2, but here's another, conceptually simpler technique. To find a spanning tree in a connected graph G, we repeatedly find an edge that can be deleted without disconnecting G—that is, an edge that's in a cycle—and delete it. See Figure 11.52 for the algorithm. Here's an example:

Cycle Elimination Algorithm:
Input: a connected graph $G = \langle V, E \rangle$
Output: a spanning tree of G

```
while there exists a cycle C in G:
    let e be an arbitrary edge traversed by C
    remove e from E
return the resulting graph ⟨V, E⟩.
```

Figure 11.52: The pseudocode for an algorithm to find a spanning tree.

Example 11.36 (Finding a spanning tree via cycle elimination)
Here are the iterations of the Cycle Elimination algorithm in computing a spanning tree of a given connected graph. In each iteration, we've selected an arbitrary cycle (lightly shaded) and then selected an arbitrary edge from that cycle (heavily shaded) and removed it. After three iterations, the resulting graph has no cycles, and remains connected; the resulting graph is a spanning tree of the original graph.

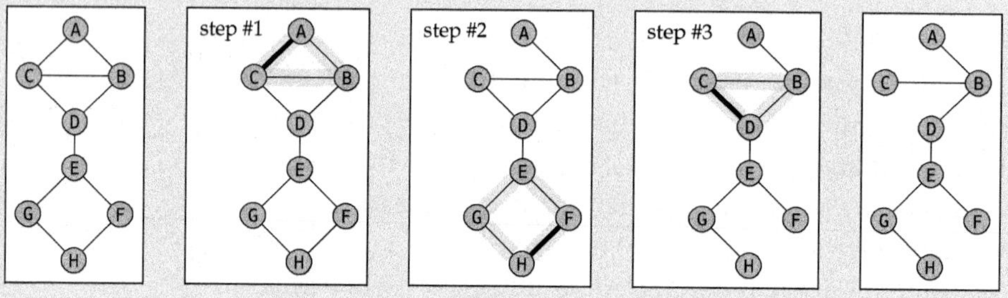

We can prove that the Cycle Elimination algorithm correctly finds spanning trees, given an arbitrary connected graph as input:

Theorem 11.8 (Correctness of the Cycle Elimination algorithm)
Given any connected graph $G = \langle V, E \rangle$, the Cycle Elimination algorithm returns a spanning tree T of G.

Proof. The algorithm only deletes edges from G, so certainly $T = \langle V, E' \rangle$ satisfies $E' \subseteq E$. We need to prove that T is a tree: that is, T is acyclic and T is connected.

Acyclicity: As long as there's a cycle remaining, the algorithm stays in the **while** loop. Thus we only exit the loop when the remaining graph is acyclic. (And the loop terminates in at most $|E|$ iterations, because an edge is deleted in every iteration.)

Connectivity: We claim that the graph is connected throughout the algorithm. It's true at the beginning of the algorithm, by assumption. Now consider an iteration in which we delete the edge $\{u, v\}$ from a cycle C. Let s and t be arbitrary nodes; we will argue that there is still a path from s to t. Before we deleted $\{u, v\}$, there was a path P from s to t. If P didn't traverse the edge $\{u, v\}$, then P is still a path from s to t. Otherwise, we can still get from s to t by going "the long way around" the cycle C instead of following the single edge $\{u, v\}$. (See Figure 11.53.) Thus there is still a path from any node s to any node t, and so the graph stays connected. □

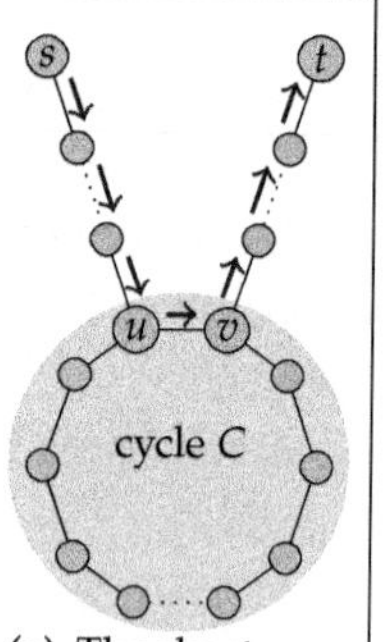

(a) The short way from s to t, via $\{u, v\}$.

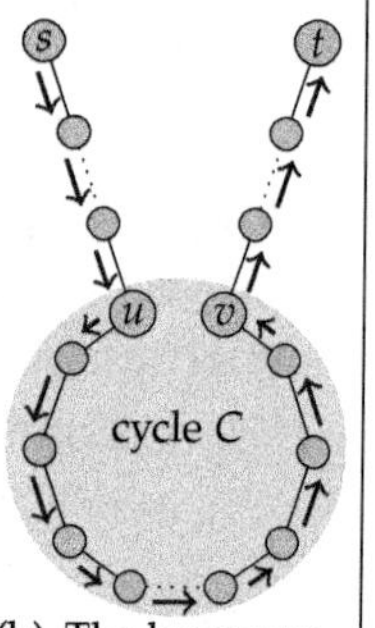

(b) The long way from s to t.

Figure 11.53: Maintaining connectivity in the Cycle Elimination Algorithm.

Computer Science Connections

Directed Graphs, Cycles, and Kidney Transplants

Kidneys are essential to human life; they play an essential filtering role in the body without which we would all die. Although we are born with two kidneys, humans need only one functioning kidney to live healthy lives. Because we're all naturally equipped with a "spare," kidney transplants are the most common form of transplant surgery performed today. Thousands of lives are saved annually through kidney transplants.

Typically a patient in need of a kidney identifies a friend or relative who is willing to donate. If the patient and donor are compatible—for example, blood type and physical size of the donor's kidney must be appropriate—then medical teams perform two simultaneous operations: one to remove the "spare" kidney from the donor, and one to implant it in the patient. (Some patients instead receive kidneys from strangers who chose to donate their organs in case of an untimely death.) Unfortunately, many patients who need kidneys have a friend or relative willing to donate to them—but they are incompatible with their prospective donor's kidney. These patients may spend years on a waiting list for a transplant, undergoing painful, expensive, and only partially effective dialysis while they wait and hope.

In recent years, medical personnel have begun a program of *kidney exchanges*. Suppose that a patient p_1 is incompatible with her prospective donor d_1, another patient p_2 is incompatible with his prospective donor d_2, but pairs $\langle p_1, d_2 \rangle$ and $\langle p_2, d_1 \rangle$ are both compatible with each other. *Four* teams of doctors can then do a "paired exchange" with four surgeries, in which d_1 donates to p_2 and d_2 donates to p_1. (To ensure that everybody follows through, the surgeries must be simultaneous: if d_1 donates to p_2 *before* d_2 undergoes surgery, then d_2 has no incentive to go through the surgery, as d_2's friend p_2 has already received his kidney.) We can even consider larger exchanges (three or more simultaneous donations)—though as the number of surgeries increases, the logistical difficulty increases as well.

Deciding which transplants to complete is done using a graph-based algorithm. Each patient p_i comes to the system with a donor d_i who is willing to donate to p_i. Define a directed graph G as follows. There is a node for each patient p_i and a node for each donor d_i. Add a directed edge $\langle p_i, d_i \rangle$ for every i. Also add a directed edge $\langle d_i, p_j \rangle$ if donor d_j is compatible with patient p_j. A cycle in G then corresponds to a set of surgeries that can be completed: every donor in the cycle donates a kidney, and every patient in the cycle receives a compatible kidney. See Figure 11.54 for an example.

The algorithm that's actually used in the real kidney exchange network in the United States computes a *set* of node-disjoint cycles that will be performed.[11] To limit the number of simultaneous surgeries that are required, the algorithm seeks a set of *cycles of length* 4 *or length* 6—that is, 2 or 3 transplants—in G that maximizes the total number of nodes included. (The constraint on cycle length makes the computational problem much more difficult, so the algorithm requires significant computational power to compute the surgeries to complete.)

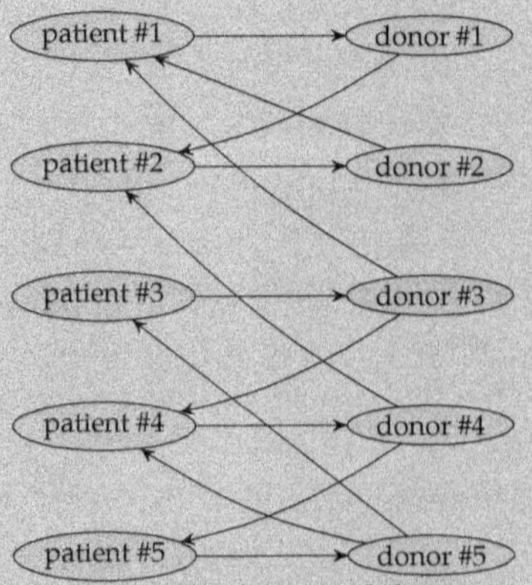

(a) The graph of compatibilities. A directed edge goes from every patient to her corresponding donor. There is a directed edge from a donor to a patient if that patient can receive a kidney from that donor.

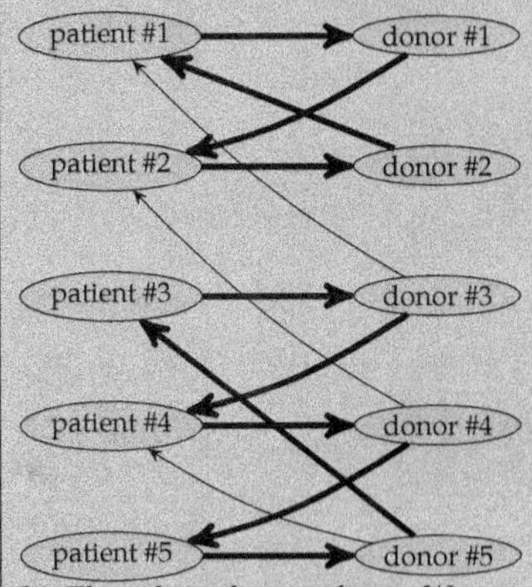

(b) The selected transplants. We "cover" this graph with two cycles; if we perform the transplants highlighted (the darker donor-to-patient edges), then every patient receives a compatible kidney.

Figure 11.54: An example of a kidney exchange network, and the cycle-based algorithm to select transplants.

[11] David Abraham, Avrim Blum, and Tuomas Sandholm. Clearing algorithms for barter exchange markets: Enabling nationwide kidney exchanges. In *Proceedings of the ACM Conference on Electronic Commerce (EC)*, 2007.

Computer Science Connections

Binary Search Trees

Trees are the basis of many important data structures, of which *binary search trees* are perhaps most frequently used. Binary search trees are data structures that implement the abstract data type called a *dictionary*: we have a set of *keys*, each of which has a corresponding *value*. (For example, the keys might be words and the values definitions, or they might be student names and GPAs, or usernames and encrypted passwords.) The data structure must support operations like *insert*(k, v) (add a new key/value pair) and *lookup*(k) (report the value associated with key k, if any).

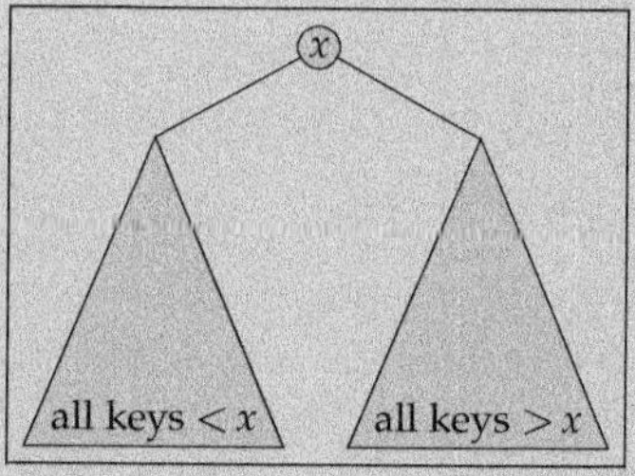

Figure 11.55: The binary search tree condition. For every node with key x: all keys in the left subtree of the node have a key $< x$; and all keys in the right subtree of the node have a key $> x$.

A *binary search tree (BST)* is a binary tree for which every node u satisfies the *BST condition* illustrated in Figure 11.55: every node v in u's left subtree has a key that is less than u's key, and every node v in u's right subtree has a key that is greater than u's key. (For simplicity, assume that all keys are distinct.)

An example of a binary search tree is shown in Figure 11.56. Incidentally, the BST condition implies the following claim: *an in-order traversal of a binary search tree visits the keys in sorted order.* This claim can be proven formally by induction, but the intuition is straightforward: an in-order traversal of a node with key x first visits nodes $< x$ (while traversing the left subtree), then x itself, and then nodes $> x$ (while traversing the right subtree). Because, recursively, the nodes of the left and right subtrees are themselves visited in sorted order, the entire tree's keys are visited in sorted order.

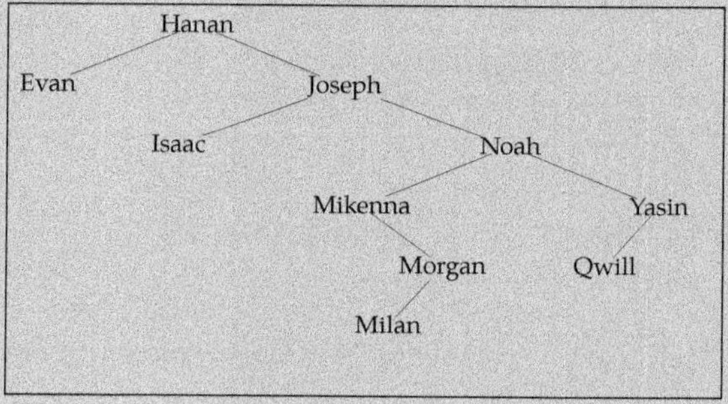

Figure 11.56: A binary search tree storing a set of 10 keys. The key is shown in each node; the accompanying value isn't drawn.

Binary search trees are good data structures for dictionaries because *insert* and *lookup* can be implemented simply and efficiently. If we perform a lookup for a key k in an empty BST T, we return "not found." (For simplicity, we allow a BST to be empty—that is, to contain zero nodes.) Otherwise, compare k to the key r stored in the root node of T:

- if $k = r$, then return the value stored at the root.
- if $k < r$, then perform a lookup for k in the left subtree.
- if $k > r$, then perform a lookup for k in the right subtree.

The BST condition guarantees that we find the node with key k if it's in the tree. (You can prove this fact by induction.) The *insert* operation can be implemented similarly, by adding a new node exactly where a lookup for the key k would have found k.

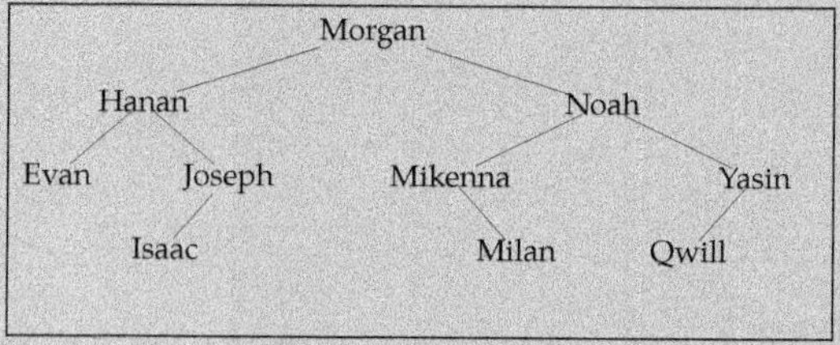

Figure 11.57: Another binary search tree with the same set of keys.

The worst-case running time of *lookup* and *insert* is proportional to the height of the binary search tree. More "balanced" BSTs—in which every internal node has a left subtree with roughly the same height as its right subtree—have better performance. (There are many different BSTs with the same set of keys; for example, another BST that has the same keys as the BST in Figure 11.56 is shown in Figure 11.57.)

Most software therefore uses *balanced binary search trees* instead—for example, *AVL trees* or *red–black trees*.[12] (See p. 643 for further discussion of AVL trees, and a proof of their efficiency.)

See the details in any good textbook on data structures, or in

[12] Thomas H. Cormen, Charles E. Leisersen, Ronald L. Rivest, and Clifford Stein. *Introduction to Algorithms*. MIT Press, 3rd edition, 2009.

11.4.5 *Exercises*

Identify all of the simple cycles in the following graphs:

11.122

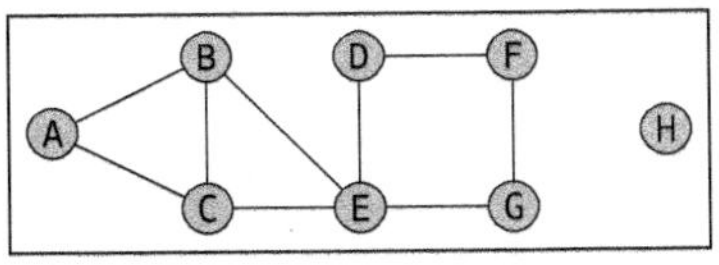

11.123

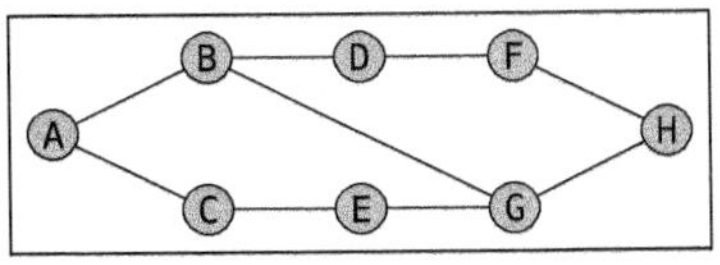

11.124

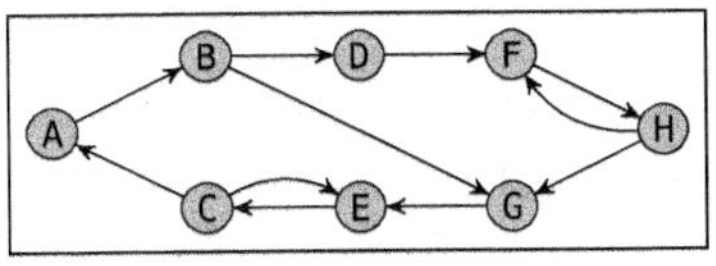

Consider an undirected graph G with n nodes. In terms of n . . .

11.125 . . . what is the longest simple cycle that G can contain? Explain.

11.126 . . . what is the longest cycle (not necessarily simple) that G can contain? Explain.

Prove your answers to the following questions, and simplify your answer as n gets large. (For handling large n, a useful fact from calculus: $\sum_{i=0}^{n} \frac{1}{i!}$ approaches $e = 2.71828\cdots$ as n grows.)

11.127 In the n-node complete graph $\mathcal{K}_n$, how many simple cycles is a particular node u involved in?

11.128 Let u be a node in a n-node complete *directed* graph: all edges except for self-loops are present. How many simple cycles is node u involved in?

11.129 A small modification to BFS can detect cycles involving a node s a directed graph, as shown in Figure 11.58. However, this modification doesn't quite work for undirected graphs. Give an example of an acyclic graph in which the algorithm Figure 11.58 falsely claims that there is a cycle. Then describe briefly how to modify this algorithm to correctly detect cycles involving node s in undirected graphs.

```
Input: a graph G = ⟨V, E⟩ and a source node s ∈ V
Output: is s involved in a cycle in G?

Frontier := ⟨s⟩
Known := ∅
while Frontier is nonempty:
    u := the first node in Frontier
    remove u from Frontier
    if s is a neighbor of u then
        return "s is involved in a cycle."
    for every neighbor v of u:
        if v is in neither Frontier nor Known then
            add v to the end of Frontier
    add u to Known
return "s is not involved in a cycle."
```

Figure 11.58: BFS modified (slightly buggily) to detect cycles involving the node s.

Recall Lemma 11.5: in any acyclic undirected graph, there exists a node whose degree is zero or one. Prove the following two extensions/variations of this lemma:

11.130 Prove that every directed acyclic graph contains a node with out-degree zero.

11.131 Prove that there are *two* nodes of degree 1 in any acyclic undirected graph that contains at least one edge.

Recall Definition 11.26: a *cycle* $\langle u_0, u_1, \ldots, u_k, u_0 \rangle$ is a path of length ≥ 2 from a node u_0 back to node u_0 that does not traverse the same edge twice. *At various times in class, I've tried to define cycles in all of the following ways—and they're all bogus definitions, in the sense that they describe something different from Definition 11.26. For each of the following broken definitions, explain why I was wrong:*

11.132 A *cycle* is a simple path from s to s.

11.133 A *cycle* is a path of length ≥ 2 from s to s.

11.134 A *cycle* is a path from s to s that doesn't traverse any edge more than once.

11.135 A *cycle* is a path from s to s that includes at least 3 distinct nodes.

11.136 A *cycle* is a path of length ≥ 2 from s to s that doesn't traverse any edge twice consecutively.

11.137 Definition 11.28 defines an acyclic graph as one containing no cycles, but it would have been equivalent to define acyclic graphs as those containing no *simple* cycles. Prove that G has a cycle if and only if G has a simple cycle.

Recall that $G = \langle V, E \rangle$ is a regular graph *if every $u \in V$ has* degree(u) $= d$, *for some fixed constant d.*

11.138 Identify two different regular graphs that are trees.

11.139 It turns out that there are two and only two different trees T that are regular graphs. Prove that there are no other regular graphs that are trees.

A triangle *is a simple cycle containing exactly three nodes. A* square *is a simple cycle containing exactly four nodes.*

11.140 What is the largest number of triangles possible in an undirected graph of n nodes?

11.141 What is the largest number of squares possible in an undirected graph of n nodes?

Let's analyze the largest number of edges that are possible in an n-node undirected graph that contains no triangles.

11.142 Consider a triangle-free graph $G = \langle V, E \rangle$. For nodes $u \in V$ and $v \in V$, argue that if $\{u, v\} \in E$, then we have $degree(u) + degree(v) \leq |V|$.

11.143 Prove the following claim by induction on the number of nodes in the graph: if $G = \langle V, E \rangle$ is triangle-free, then $|E| \leq |V|^2/4$. *(Hint: use the previous exercise.)*

11.144 Give an example of an n-node triangle-free graph that contains $\frac{n^2}{4}$ edges.

Consider the following adjacency lists. Is the graph that each represents a tree? Justify your answers.

11.145

A:	B, E
B:	A
C:	D
D:	C, F
E:	A
F:	D

11.146

A:	C
B:	C, E
C:	A, B, F
D:	E
E:	B, D
F:	C

11.147

A:	D
B:	E, F
C:	D, F
D:	A, C
E:	B
F:	B, C

11.148

A:	C, D, F
B:	F
C:	A, E, F
D:	A
E:	C
F:	A, B, C

Prove or disprove the following claims about trees:

11.149 There is a node of degree equal to 2 in any tree with ≥ 3 nodes.

11.150 In any rooted binary tree (all nodes have 0, 1, or 2 children), there are an even number of leaves.

11.151 If a graph $G = \langle V, E \rangle$ has $|V| - 1$ edges, then G must be a forest.

11.152 The following pair of definitions is subtly broken: the *root* of a tree is a node that is not a child, and a *leaf* is a node that is a child but not a parent. What's broken?

For the tree in Figure 11.59, with node A *as the root ...*

11.153 ... what are the leaves?

11.154 ... which nodes are internal nodes?

11.155 ... what the are parent, children, and siblings of node D?

11.156 ... what are the descendants of node D?

11.157 ... what are the ancestors of node F?

11.158 ... what is the height of the tree?

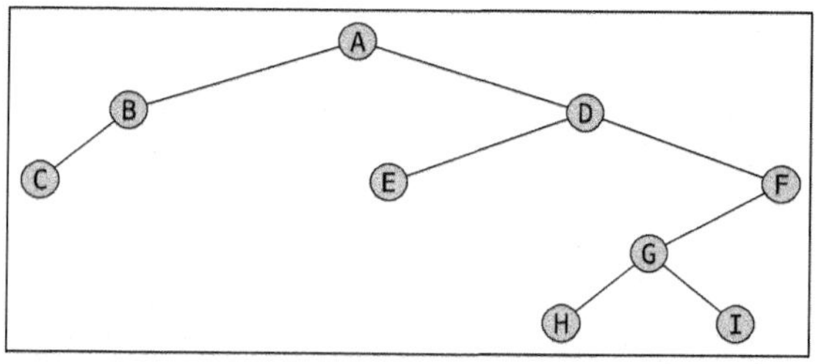

Figure 11.59: A rooted tree.

11.159 Let T be an arbitrary n-node rooted tree, with root r and with ℓ different leaves. Prove or disprove: if we reroot T at a new node $r' \neq r$, then the number of leaves remains exactly ℓ.

A complete binary tree of height h *has "no holes": reading from top-to-bottom and left-to-right, every node exists. Complete binary trees form a subset of nearly complete binary trees: a* nearly complete binary tree *has every node until the last row, which is allowed to stop early. (See Figure 11.60, and see also p. 529 for a discussion of* heaps, *which are a data structure represented as a nearly complete binary tree.)*

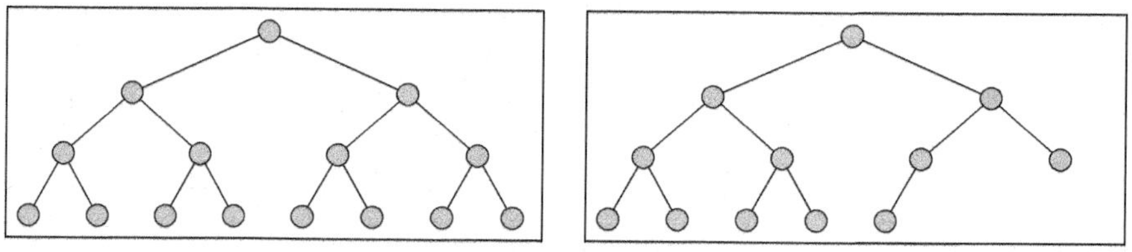

Figure 11.60: A complete and nearly complete binary tree of height 3.

11.160 Prove by induction that a complete binary tree of height h contains precisely $2^{h+1} - 1$ nodes.

11.161 How many leaves does a nearly complete binary tree of height h have? Give the smallest and largest possible values, and explain.

11.162 What is the diameter of a nearly complete binary tree of height h? Again, give the smallest and largest possible values, and explain your answer. (Recall that the *diameter* of a graph $G = \langle V, E \rangle$ is $\max_{s,t \in V} d(s,t)$, where $d(s,t)$ denotes the length of the shortest path from u to v in G.)

Suppose that we "rerooted" a complete binary tree of height h by instead designating one of the erstwhile leaves as the root. In the rerooted tree, what are the following quantities?

11.163 the height

11.164 the diameter

11.165 the number of leaves

Justify your answers to the following questions: describe an 1000*-node binary tree with ...*

11.166 ... height as large as possible.

11.167 ... height as small as possible.

11.168 ... as many leaves as possible.

11.169 ... as few leaves as possible.

11.170 What is the largest possible height for an n-node binary tree in which *every node has precisely zero or two children*? Justify your answer.

In what order are nodes of the tree in Figure 11.61 traversed ...

11.171 ... by a pre-order traversal?

11.172 ... by an in-order traversal?

11.173 ... by a post-order traversal?

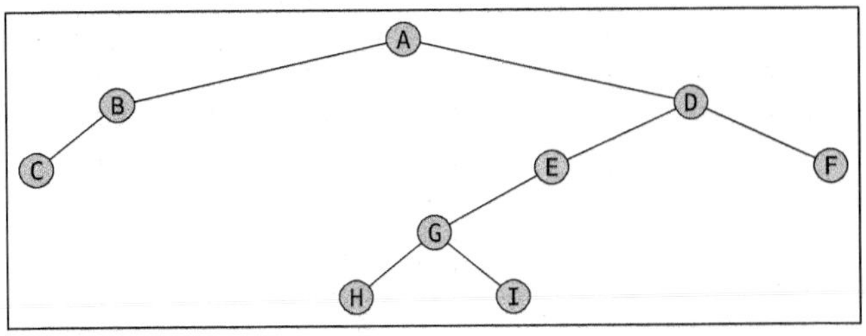

Figure 11.61: A rooted tree.

11.174 Draw the binary tree with in-order traversal $4, 1, 2, 3, 5$; pre-order traversal $1, 4, 3, 2, 5$; and post-order traversal $4, 2, 5, 3, 1$.

11.175 Do the same for the tree with in-order traversal $1, 3, 5, 4, 2$; pre-order traversal $1, 3, 5, 2, 4$; and post-order traversal $4, 2, 5, 3, 1$.

11.176 Describe (that is, fully explain the structure of) an n-node binary tree T for which the *pre-order* and *in-order* traversals of T result in precisely the same ordering of T's nodes. (That is, **pre-order-traverse**(T) = **in-order-traverse**(T).)

11.177 Describe a binary tree T for which the *pre-order* and *post-order* traversals result in precisely the same ordering of T's nodes. (That is, **pre-order-traverse**(T) = **post-order-traverse**(T).)

11.178 Prove that there are two distinct binary trees T and T' such that pre-order and post-order traversals are both identical on the trees T and T'. (That is, **pre-order-traverse**(T) = **pre-order-traverse**(T') and **post-order-traverse**(T) = **post-order-traverse**(T') but $T \neq T'$.)

11.179 Give a recursive algorithm to reconstruct a tree from the in-order and post-order traversals.

11.180 Argue that we didn't leave out any spanning trees of G in Figure 11.51, reproduced here for your convenience:

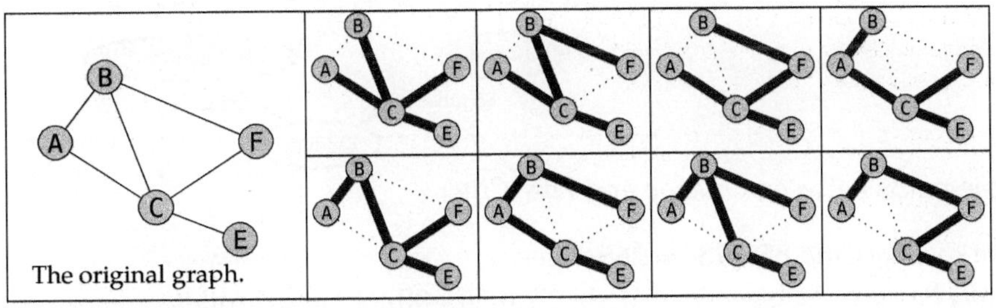

How many spanning trees do the following graphs have? Explain.

11.181

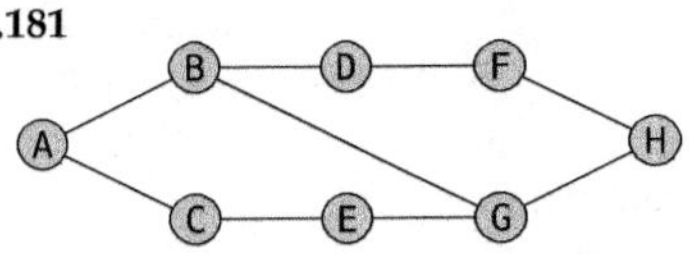

11.182

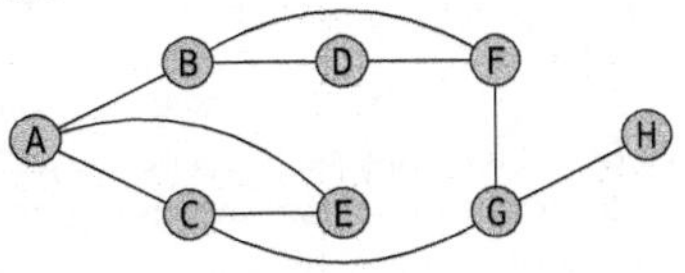

11.5 Weighted Graphs

Force without wisdom falls of its own weight.

Horace (65–8 BCE), *Odes* (23 BCE)

Many real-world situations are naturally modeled by different edges having different "weights": the price of an airplane flight, the closeness of a friendship, the physical length of a road, the time required to transmit data across an internet connection. These graphs are called *weighted graphs:*

Definition 11.32 considers only nonnegative weights—every $w_e \geq 0$—which is a genuine restriction. (For example, the "signed" social networks from Figure 11.8(a) have positive and negative weights signifying friendship and enmity.) Some, but not all, of the results that we'll discuss in this section carry over to the setting of negative weights.

Definition 11.32 (Weighted graph)
A weighted graph *is a graph* $G = \langle V, E \rangle$ *and a weight function* $w : E \to \mathbb{R}^{\geq 0}$*, so that each edge* $e \in E$ *has a* weight $w(e) \geq 0$*. For simplicity of notation, we'll often write* w_e *instead of* $w(e)$*; we'll also sometimes refer to* w_e *as the* length *of the edge e.*

In a weighted graph, the length of a path *in a weighted graph is the sum of the lengths of the edges traversed by the path. (A* shortest path *is, as before, one with the smallest length.)*

Either undirected or directed graphs can be weighted. Aside from the length of a path, all of the other notions and terminology from unweighted graphs carry over: neighbors and degree, paths and connectivity, and so forth. Weighted graphs can be represented just as unweighted graphs were: we typically store the weight of edge $\langle u, v \rangle$ directly in the $\langle u, v \rangle$th entry of the adjacency matrix, or attach the edge weight as an additional slot in the adjacency list entries. Here's an example:

Example 11.37 (A weighted graph)
Here's the highway system from Example 11.4, where each road is labeled with its length:

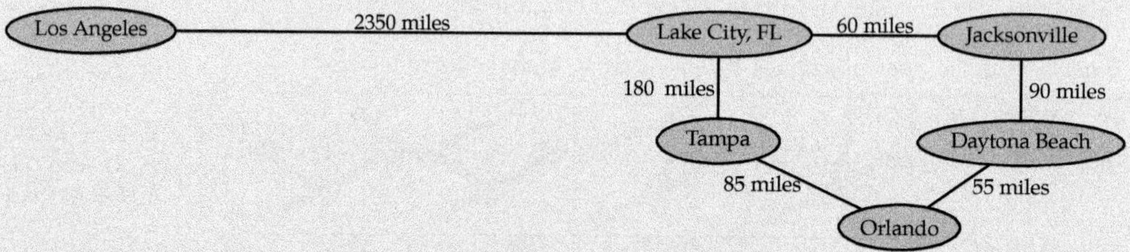

There are two simple paths between *Orlando* and *Lake City*:

- *Orlando* ↔ *Tampa* ↔ *Lake City*: $85 + 180 = 265$ miles.
- *Orlando* ↔ *Daytona Beach* ↔ *Jacksonville* ↔ *Lake City*: $55 + 90 + 60 = 205$ miles.

The second path is shorter, even though it traverses more edges, as $265 > 205$.

Taking it further: The primary job of a web search engine is to respond to a user's search query ("give me web pages about Horace") with a list of relevant pages. There's a complex question of data structures, parallel computing, and networking infrastructure in solving even the simplest part of this task: identifying the set R of web pages (out of many billions) that contain the search term. A subtler challenge—and at least as important—is figuring out how to *rank* the set R. What pages in R are the "most important," the ones that we should display on the first page of results? See p. 1174 for some discussion of how Google uses a weighted graph (and probability) to do this ranking.

11.5.1 *Shortest Paths in Weighted Graphs: Dijkstra's Algorithm*

A *shortest path* from s to t in a weighted graph is the path connecting s and t that has shortest total length. In many natural applications where shortest paths are useful, we have weights on edges: you want the *shortest* walking route from the bar back to your apartment, for example, not necessarily the one with the fewest turns. In Example 11.37, we already saw a case in which the shortest path used more edges than necessary. Thus we cannot directly use breadth-first search to compute distances in weighted graphs.

But we *can* compute distances using an algorithm that's very similar in spirit to BFS. The basic idea of breadth-first search is to "expand outward" from the source node s in layers, accumulating a set of nodes u for which we know the distance from s to u. We add nodes in increasing order of their distance from s, and eventually we've computed distances from s to all nodes in the graph. (See Figure 11.62.) The trouble for weighted graphs is that the order in which BFS builds up its knowledge about shortest paths doesn't always work (as in Example 11.37). But we can use a cleverer way of building up knowledge about the network to find shortest paths in weighted graphs, too.

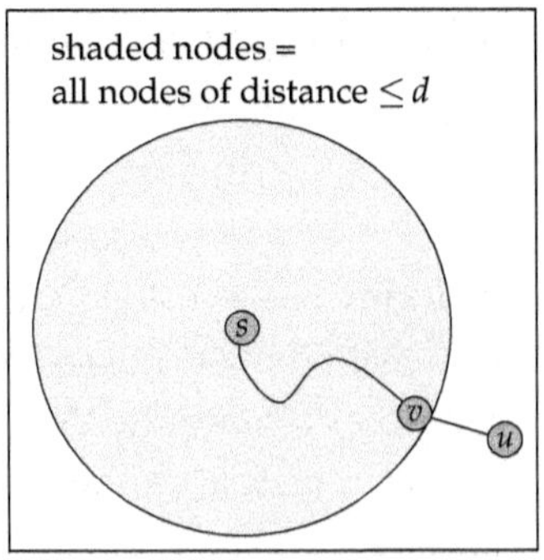

Figure 11.62: The intuition of BFS. Assume the shaded set S contains every node within distance d of s, and that $u \notin S$ is a neighbor of $v \in S$. The distance from s to u must be $d + 1$.

The algorithm that we'll describe is due to Edsger Dijkstra, and hence it is known as *Dijkstra's algorithm*. The key idea of Dijkstra's algorithm has parallels with BFS:

> Suppose that we know the distance from a source node s to every node in some set S of nodes. (Assume that $s \in S$.) We will find some node *not* in S for which we can determine the shortest path from s.

Edsger Dijkstra was a 20th-century Dutch computer scientist—one of the founders of theoretical computer science, and the 1972 Turing Award winner.

For now, let's not worry about where this set S came from; the key point is just that we are assuming that we know distances to certain nodes (those in S), and we seek to leverage that existing knowledge to learn the distance to some other node (not previously in S). We'll then add that new node to S and iterate.

Before we state the formal result, let's look at an example:

Irrelevant quotation: "Computer science is no more about computers than astronomy is about telescopes." — attributed to Edsger W. Dijkstra (1930–2002)

Irrelevant challenge: Name a common English word that, like `DIJKSTRA`, has at least five (or 6 or even 7, which is technically possible) consecutive consonants. (Not `SYZYGY` or `RHYTHMS`; `Y` is a vowel if it's used as a vowel!)

Example 11.38 (An example of distances)
Consider the following weighted, undirected graph (with edge weights marked on the edges):

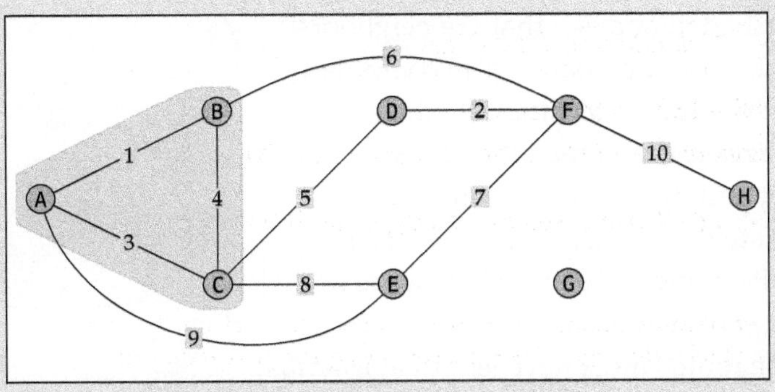

Suppose we know the distances from `A` to every node in the shaded set $S = \{\texttt{A}, \texttt{B}, \texttt{C}\}$:

$$d(\texttt{A}, \texttt{A}) = 0 \qquad d(\texttt{A}, \texttt{B}) = 1 \qquad d(\texttt{A}, \texttt{C}) = 3.$$

We wish to expand our set of known nodes by adding a neighbor of an already shaded node. The candidate nodes that are neighbors of nodes with known distances are $\{D, E, F\}$. In particular, their candidate distances are:

node	distance
F (via B)	$d(A, B) + w_{B,F} = 1 + 6 = 7$
E (via A)	$d(A, A) + w_{A,E} = 0 + 9 = 9$
E (via C)	$d(A, C) + w_{C,E} = 3 + 8 = 11$
D (via C)	$d(A, C) + w_{C,D} = 3 + 5 = 8$

Let's argue that we can now conclude that $d(A, F) = 7$.

The key reason is that, to get from A to F, we have to "escape" the set of shaded nodes—and every "escape route" (path to F) must reach its last shaded node v (that's $d(A, v)$) and then follow an edge to its first unshaded node u (that's $w_{v,u}$). Because this table tells us that every path out of the shaded region has length at least 7, and we've found a path from A to F with exactly that length, we conclude that $d(A, F) = 7$.

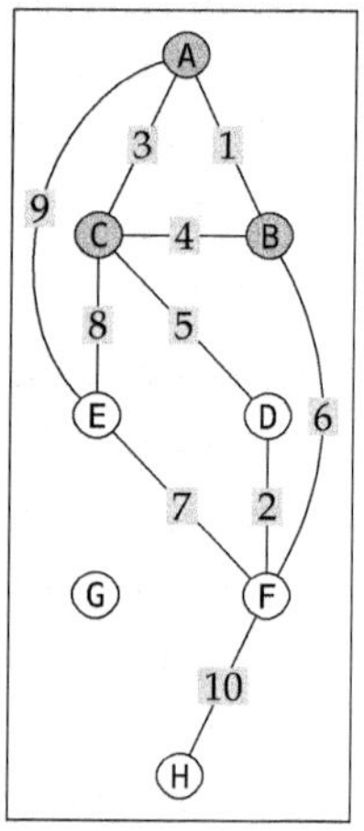

Figure 11.63: The graph for Example 11.38, repeated and rotated. We've computed that $d(A, A) = 0$ and $d(A, B) = 1$ and $d(A, C) = 3$.

Computing the distance to a new node

The same basic reasoning that we used in Example 11.38 will allow us to prove a general observation that's the foundation of Dijkstra's algorithm:

Lemma 11.9 (Foundation of Dijkstra's Algorithm)
Let $G = \langle V, E\rangle$ be a graph with edge weights w, let $S \subset V$ be a set of nodes, and let $s \in S$ be a source node. Let $d(s, v)$ denote the distance from s to v for every node v in S. For a node $u \notin S$, define

$$d_u := \min_{v \in S \,:\, u \text{ is a neighbor of } v} d(s, v) + w_{v,u}.$$

Let u^* be the node $u \notin S$ for which d_u is minimized. Then the distance from s to u^* is d_{u^*}.

Before we prove the lemma, let's restate the claim in slightly less notation-heavy English. (See Figure 11.64.) We have a set S of nodes—the shaded nodes in the picture—for which we know the distance from s. We examine all unshaded nodes u that are neighbors of shaded nodes v. For each shaded/unshaded pair, we've computed the sum of the distance $d(s, v)$ and the edge weight $w_{v,u}$. And we've chosen the pair $\langle v^*, u^*\rangle$ that minimizes this quantity.

The lemma says that the shortest path from s to this particular u^* must have length precisely equal to $d_{u^*} := d(s, v^*) + w_{v^*,u^*}$. The intuition matches the argument in Example 11.38: to get from s to u^*, we have to somehow "escape" the set of shaded nodes—and, by the way that we chose u^*, every "escape route" must have length at least d_{u^*}.

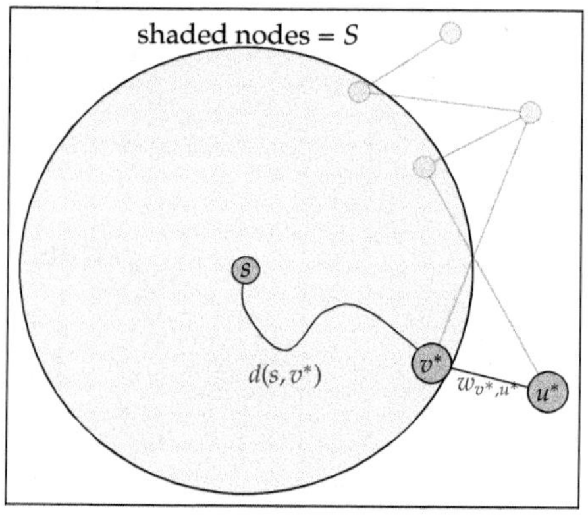

Figure 11.64: The intuition for Lemma 11.9.

Proof of Lemma 11.9. We must show that the distance from s to u^* is d_{u^*}, and we'll do it in two steps: by showing that the distance is no more than d_{u^*}, and by showing that the distance is no less than d_{u^*}.

The distance from s to u^* is $\leq d_{u^*}$. We must argue that there *is* a path of length $d(s, v^*) + w_{v^*,u^*}$ from s to u^*. By assumption and the fact that $v^* \in S$, we know that $d(s, v^*)$ is the distance from s to v^*, so there must exist a path P of length $d(s, v^*)$ from s to v^*. (It's the curved line in Figure 11.64.) By tacking u^* onto the end of P, we've constructed a path from s to u^* via v^* with length $d(s, v^*) + w_{v^*,u^*}$.

The distance from s to u^* is $\geq d_{u^*}$. Consider an arbitrary path P from s to u^*. We must show that P has length at least $d(s, v^*) + w_{v^*,u^*}$.

What does P look like? The node s is in the set S, so P starts out at $s \in S$, then wanders around for a while inside S, then crosses outside of S for the first time, wanders around outside S for a while, and eventually ends up at $u^* \notin S$. Nothing prevents P from re-entering (and later re-exiting) S after its first departure—indeed, it can go in and out of S several times—but it definitely has to leave S at least once. Thus P has to look like the following:

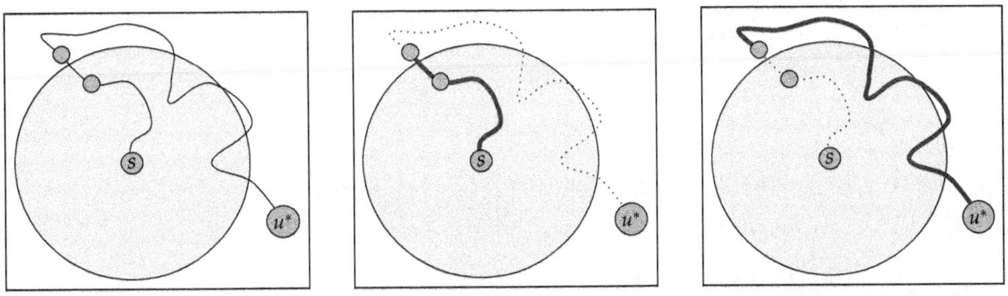

(a) the entire path P (b) the portion of P up to the first exit from S (c) the portion of P after the first exit from S

Therefore we know that

the length of P

= (the length of P up to the first exit) + (the length of P after the first exit)

$\geq$ (the length of the shortest path exiting S) + (the length of P after the first exit)

P up to the first exit is a path exiting S, so its length is at least the length of the shortest such path

$\geq d(s, v^*) + w_{v^*,u^*}$ + (the length of P after the first exit)

we chose u^* and v^* so that $d(s, v^*) + w_{v^*,u^*}$ is exactly the length of the shortest path exiting S

$\geq d(s, v^*) + w_{v^*,u^*} + 0$ *all edge weights are nonnegative, so all path lengths are ≥ 0 too*

$= d_{u^*}$. *definition of d_{u^*}*

Thus the length of P is at least d_{u^*}.

We've therefore argued that the distance from s to u^* is both $\leq d_{u^*}$ and $\geq d_{u^*}$. Thus the distance is precisely d_{u^*}, and the lemma follows. □

Problem-solving tip: When we want to prove that $x = y$, it's sometimes easier to prove $x \geq y$ and $x \leq y$ separately.

Dijkstra's Algorithm

With Lemma 11.9 proven, we can now put together the pieces of the entire algorithm. The lemma describes a way to take a set S of nodes with known distance from the source node s, and correctly calculate the distance from s to a new node $u \notin S$.

In Dijkstra's algorithm, the idea is to apply the calculation from Lemma 11.9 repeatedly to find all distances from the given source node s. We'll need a base case to get started, but that's straightforward: we start with the set of nodes with known distance from s as $S = \{s\}$, where the distance from s to s is zero. The full algorithm is shown in Figure 11.65.

Before we prove the algorithm's correctness, let's run through an example:

Dijkstra's Algorithm:
Input: a weighted graph $G = \langle V, E \rangle$, nonnegative edge weights $w_e \geq 0$, and a source node $s \in V$.
Output: the distance from s to every node in G

Let $S := \{s\}$ and let $d(s,s) := 0$. *// S is the set of nodes with known distances.*
while there exists a node in S with a neighbor not in S:
  for every node $u \notin S$, define

$$d_u := \min_{v \in S \,:\, u \text{ is a neighbor of } v} d(s,v) + w_{v,u}.$$

  $u^* :=$ the node with the smallest d_u.
  Add u^* to S and set $d(s,u^*) := d_{u^*}$.
for every node $u \in V - S$:
  $d(s,u) := \infty$
return the recorded values $d(s,u)$.

Figure 11.65: The pseudocode for Dijkstra's algorithm.

Example 11.39 (Dijkstra's algorithm in action)
Let's run Dijkstra's algorithm on the network from Example 11.37, with the graph rotated for compactness. We'll start from the Orlando (OR) node. Here is the execution:

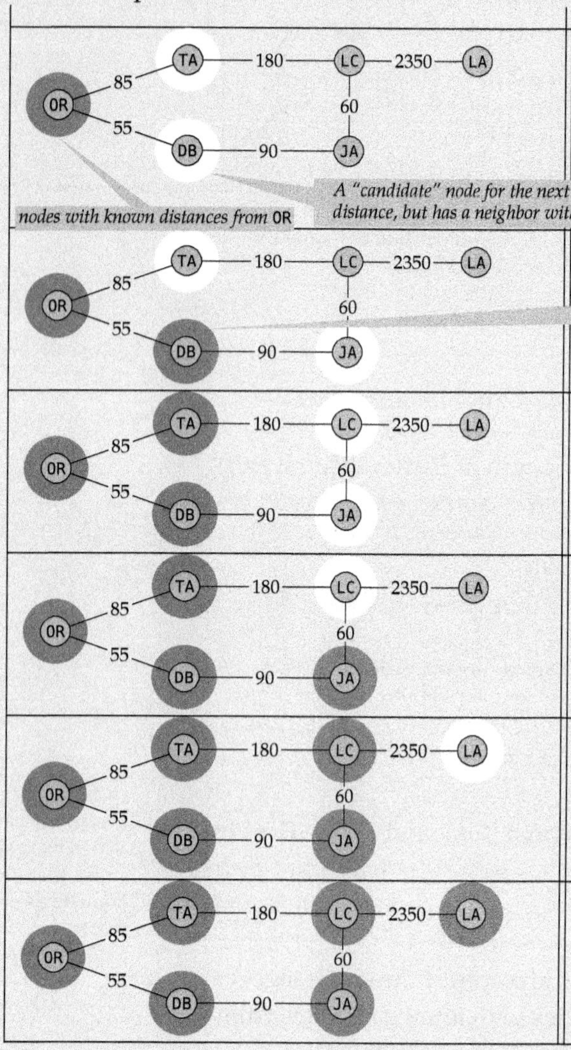

DB	JA	LA	LC	OR	TA
				0	
55				0	
55				0	85
55	145			0	85
55	145		205	0	85
55	145	2555	205	0	85

THE CORRECTNESS OF DIJKSTRA'S ALGORITHM

We'll now prove the correctness of the algorithm, using Lemma 11.9 and induction:

Theorem 11.10 (Correctness of Dijkstra's Algorithm)
Let $G = \langle V, E \rangle$ be a graph with nonnegative edge weights w_e for each edge. Let $s \in V$ be a source node, and let $d(s, \bullet) :=$ **Dijkstra**(G, w, s) *be the values computed by Dijkstra's Algorithm. Then, for every node u, we have that $d(s, u)$ is the length of the shortest path from s to u in G under w.*

Proof. Looking at the algorithm, we see that Dijkstra's Algorithm records finite distances from s in Line 1 (for s itself) and Line 5 (for other nodes reachable from s). Suppose that Dijkstra's algorithm executes n iterations of the loop in Line 2, thus recording $n + 1$ total distances in Lines 1 and 5—say in the order $u_0, u_1, \ldots, u_n$. Let $P(i)$ denote the claim that $d(s, u_i)$ is the length of the shortest s-to-u_i path. We claim by strong induction on i that $P(i)$ holds for all $i \in \{0, 1, \ldots, n\}$.

Base case ($i = 0$): We must prove that $d(s, u_0)$ is recorded correctly. The 0th node u_0 is recorded in Line 1, so u_0 is the source node s itself. And the shortest path from s to s in any graph with nonnegative edge weights is the 0-hop path $\langle s \rangle$, of length 0.

Inductive case ($i \geq 1$): We assume the inductive hypothesis $P(0), P(1), \ldots, P(i-1)$: that is, all recorded distances $d(s, u_0), d(s, u_1), \ldots, d(s, u_{i-1})$ are correct. We must prove $P(i)$: that is, that the recorded distance $d(s, u_i)$ is correct. But this follows immediately from Lemma 11.9: the algorithm chooses u_i as the $u \notin S$ minimizing

$$d_u := \min_{v \in S \,:\, u \text{ is a neighbor of } v} d(s, v) + w_{v,u},$$

where $S = \{u_0, u_1, \ldots, u_{i-1}\}$. Lemma 11.9 states precisely that this value d_u is the length of the shortest path from s to u.

Finally, observe that any node u that's only discovered in Line 6 is not reachable from s, and so indeed $d(s, u) = \infty$. (A fully detailed argument that the ∞ values are correct can follow the structure in Theorem 11.3, which proved the correctness of BFS.) □

Taking it further: Dijkstra's algorithm as written in Figure 11.65 can be straightforwardly implemented to run in $O(|V| \cdot |E|)$ time: each iteration of the **while** loop (Line 2) can look at each edge to compute the smallest d_u. But with cleverer data structures, Dijkstra's algorithm can be made to run in $O(|E| \log |V|)$ time. This improved running-time analysis, as well as other shortest-path algorithms—for example, handling the case in which edge weights can be negative (it's worth thinking about where the proof of Lemma 11.9 fails if an edge e can have $w_e < 0$), or computing distances between *all pairs* of nodes instead of just every distance from a *single source*—is a standard topic in a course on algorithms. Any good algorithms text should cover these algorithms and their analysis.

Before we leave Dijkstra's algorithm, it's worth reflecting on its similarities with BFS. In both cases, we start from a seed set S of nodes for which we know the distance from s—namely $S = \{s\}$. Then we build up the set of nodes for which we know the distance from s by finding the unknown nodes that are closest to s, and adding them to S. Of course, BFS is conceptually simpler, but Dijkstra's algorithm solves a more complicated problem. It's a worthwhile exercise to think about what happens if Dijkstra's algorithm is run on an unweighted graph. (How does it relate to BFS?)

11.5.2 *Spanning Trees in Weighted Graphs: Minimum Spanning Trees*

Recall from Definition 11.31 that a *spanning tree* of a connected graph $G = \langle V, E \rangle$ is a tree $T = \langle V, E' \rangle$ where $E' \subseteq E$. As with shortest paths, in many of the applications in which spanning trees are interesting, we actually want to find a spanning tree whose edges have minimum possible total cost. For example, when a college wants to put down networking cable in a new dorm building, they wish to ensure that the resulting network is connected, while minimizing the cost of construction.

Formally, in a weighted graph, the *cost* of a spanning tree T is the sum of the weights of its edges: $\sum_{e \in E'} w_e$. A *minimum spanning tree (MST)* is a spanning tree whose cost is as small as possible. Here are two small examples:

Example 11.40 (Some minimum spanning trees)
Consider the following two graphs (the road network from Example 11.37 and the larger connected component from Example 11.38):

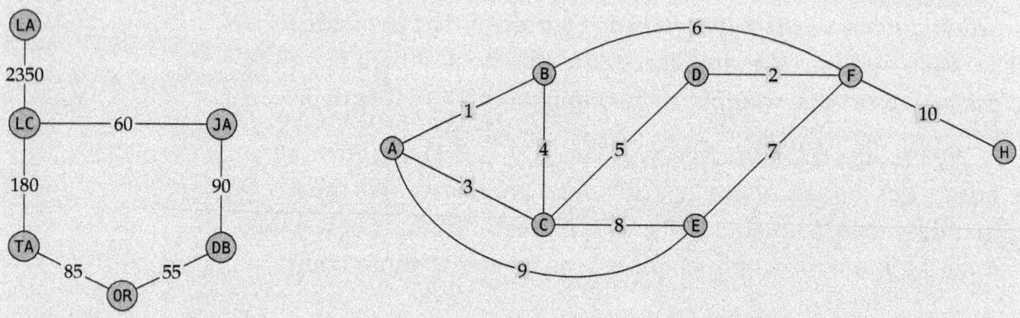

Here are the minimum spanning trees. (For the first, every spanning tree omits exactly one edge from the lone cycle; the cheapest tree omits the most expensive edge.)

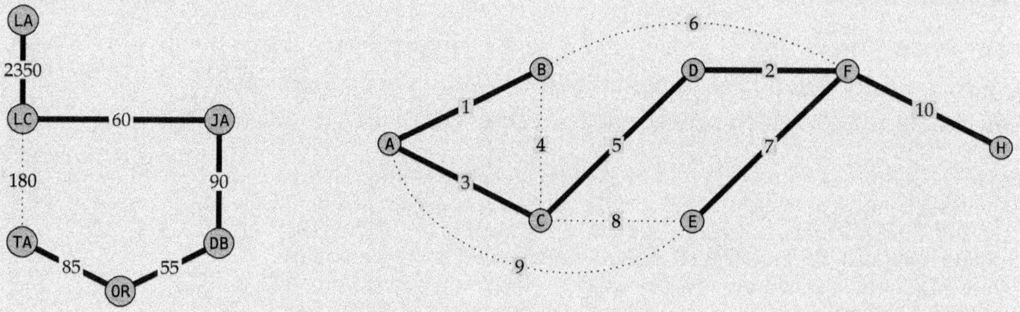

As with shortest paths in weighted graphs, the question of how to find a minimum spanning tree most efficiently is more appropriate to an algorithms text than this book. But, between the Cycle Elimination Algorithm (Figure 11.52) and Example 11.40, we've already done almost all the work to develop a first algorithm.

Assume throughout that all edge weights are distinct. (This assumption lets us refer to "*the* most expensive edge" in a set of edges. Removing this assumption complicates the language that we have to use, but it doesn't fundamentally change anything about the MST problem or its solution.)

> **Lemma 11.11 (The "cycle rule")**
> *Let C be a cycle in a connected undirected graph $G = \langle V, E \rangle$, and let e be the heaviest edge in C. Then e is not in any minimum spanning tree of G.*

Proof. Consider a spanning tree T of G, and suppose that $e = \{u, v\}$ is included in T. We'll show that T is not a *minimum* spanning tree. (Thus the only minimum spanning trees of G do not include e.)

By definition, the spanning tree T is connected. If we delete $\{u, v\}$ from T, the resulting graph will have two connected components, one containing u and the other containing v. (This fact follows by Corollary 11.7.) Call those connected components U and V, respectively. See Figure 11.66(a).

Imagine following the cycle C from u to v the "long way" around C. This part of C starts at u, wanders around U for a while, and eventually crosses over into V, before finally arriving at v. Let $a \in U$ be the last node in U and b the first node in V as we go around C. (Note that C might go back and forth between U and V multiple times, but define a and b based on the *first* time C leaves U.) See Figure 11.66(b).

Now define the graph T' as T with the edge $\{u, v\}$ removed and with the edge $\{a, b\}$ inserted instead. Crucially, T' is a spanning tree of G; because we've only swapped *which* edge connected the connected sets U and V. Thus T' remains connected and acyclic.

Now observe that the cost of T' is less than the cost of T, because the edge $\{u, v\}$ is heavier than the edge $\{a, b\}$. (Both $\{u, v\}$ and $\{a, b\}$ are in the cycle C, and by assumption $\{u, v\}$ is the heaviest edge in C.) But therefore T' is a cheaper spanning tree than T, and thus T isn't a minimum spanning tree. □

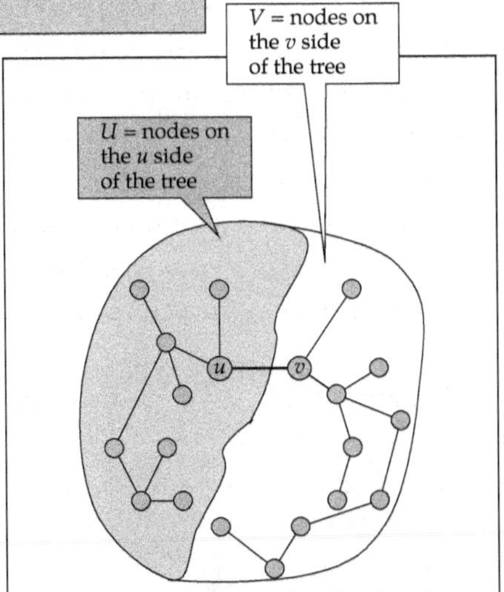

(a) Removing the edge $\{u, v\}$ splits the tree into two connected components.

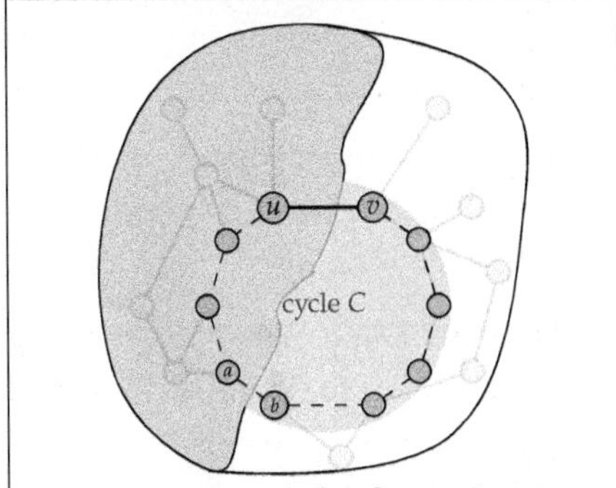

(b) C is a cycle with $\{u, v\}$ as its heaviest edge. Some other edge $\{a, b\}$ from the cycle has $a \in U$ and $b \in V$.

Figure 11.66: The cycle rule for MSTs.

Finding MSTs by removing cycles

Lemma 11.11 immediately suggests that we can find minimum spanning trees using a modification of the Cycle Elimination Algorithm:

Weighted Cycle Elimination Algorithm
Input: a weighted connected graph $G = \langle V, E \rangle$ with edge weights w_e
Output: a minimum spanning tree of G

```
while there exists a cycle C in G:
    let e be the heaviest edge traversed by C
    remove e from E
return the resulting graph ⟨V, E⟩.
```

While the Weighted Cycle Elimination Algorithm is correct and reasonably efficient, there are more efficient algorithms based on Lemma 11.11. One such algorithm is called *Kruskal's Algorithm*, named after its discoverer Joseph Kruskal. The key idea of Kruskal's Algorithm is that by *sorting* the edges in increasing order, we can be more efficient: we add edges in increasing order of their weight, as long as doing so doesn't create a cycle.

Joseph Kruskal was a 20th-century American computer scientist/mathematician/statistician. He published his MST algorithm in 1956.

The insight of this algorithm is that, by considering edges in increasing order of weight, if including an edge e creates a cycle, then we know that e must be the heaviest edge in that cycle. See Figure 11.67. Kruskal's algorithm is reasonably efficient: the sorting step takes $O(|E| \log |E|)$ time, and each of the $|E|$ iterations of the **for** loop can be implemented using one call to BFS to test for a cycle. (And, in fact, there are some cleverer ways to implement Line 4 so that the entire algorithm runs in $O(|E| \log |E|)$ time.) Here's an example:

Kruskal's Algorithm
Input: a weighted connected graph $G = \langle V, E\rangle$ with distinct edge weights w_e
Output: a minimum spanning tree of G

1: Sort the edges e in increasing order of weight.
2: $S := \varnothing$
3: **for** each edge e (taken in increasing order of weight):
4:     **if** the graph $\langle V, S \cup \{e\}\rangle$ doesn't contain a cycle **then**
5:         add e to S
6: **return** the resulting graph $\langle V, S\rangle$

Figure 11.67: Kruskal's Algorithm.

Example 11.41 (Sample run of Kruskal's algorithm)
In each panel, the highlighted edge is being considered for inclusion in the tree. Black edges have already been included; light edges have not yet been considered.

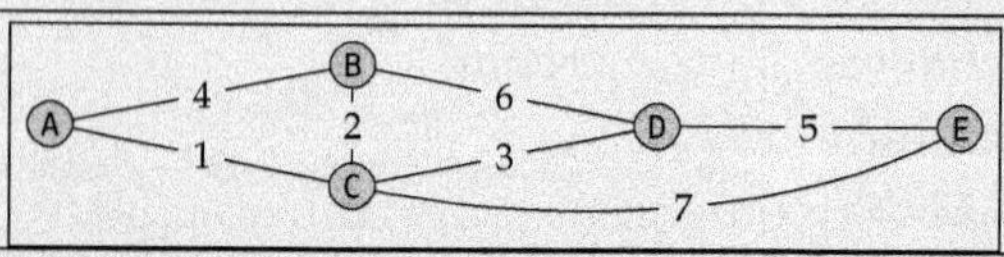

The original graph.

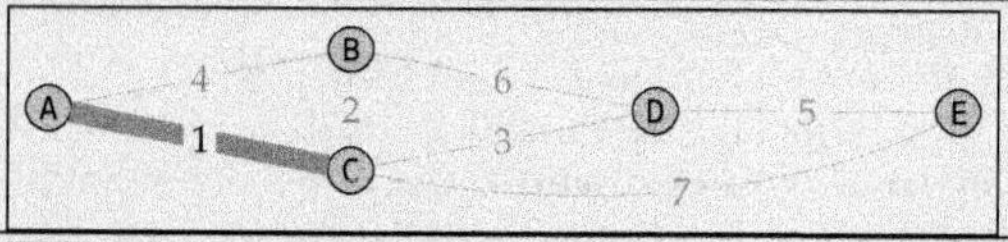

We examine the cheapest edge {A, C}. It doesn't create a cycle, so we keep it.

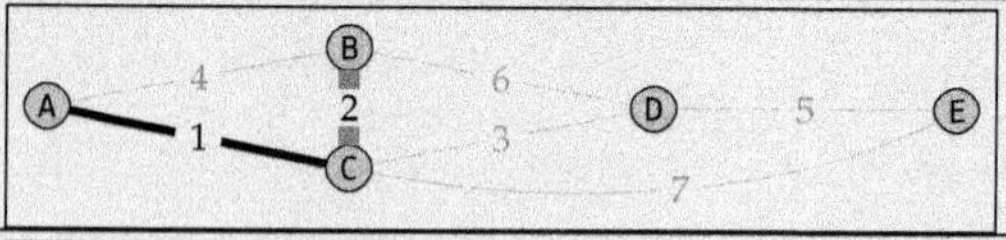

We examine the next cheapest edge {B, C}. It doesn't create a cycle, so we keep it.

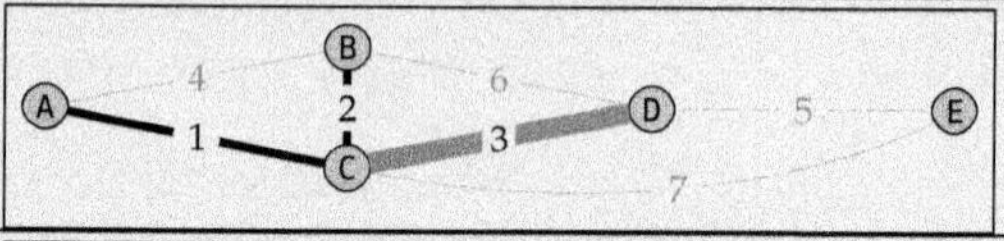

We examine the next cheapest edge {C, D}. It doesn't create a cycle, so we keep it.

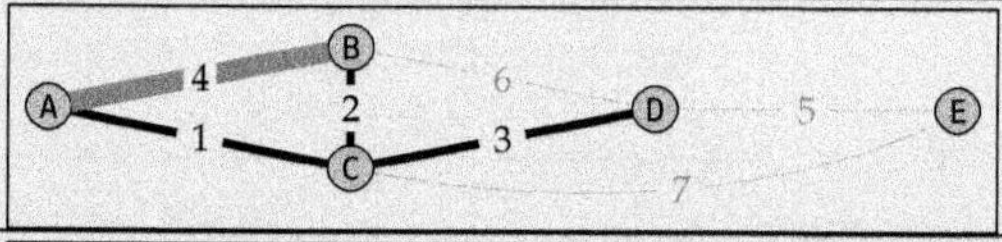

We examine the next cheapest edge {A, B}. It creates a cycle ⟨A, B, C, A⟩, so we discard it.

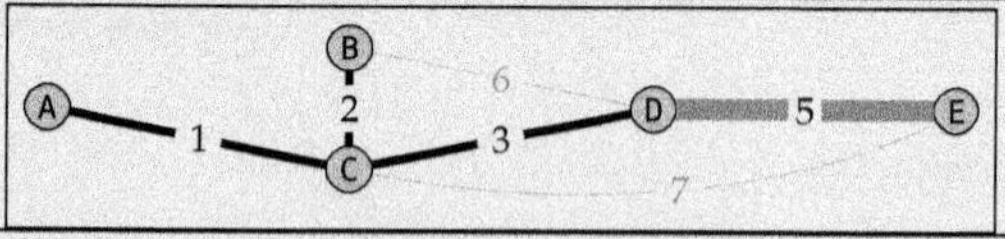

The next edge is {D, E}; we keep it.

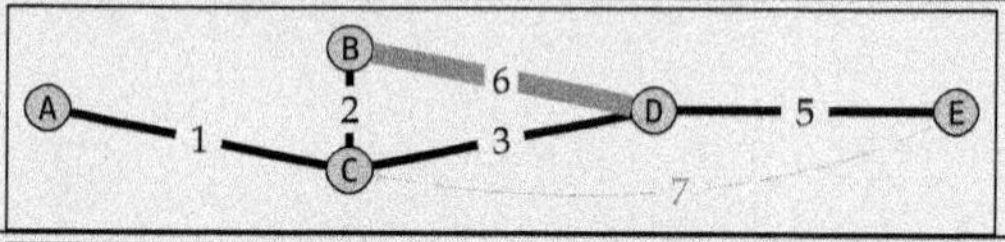

The next edge is {B, D}; it creates a cycle, so we discard it.

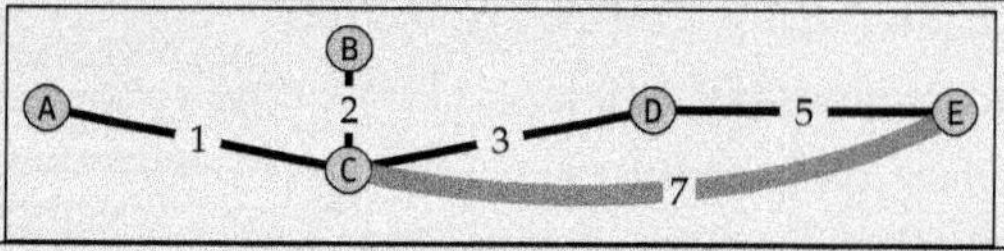

We last edge is {C, E}; it creates a cycle, so we discard it too.

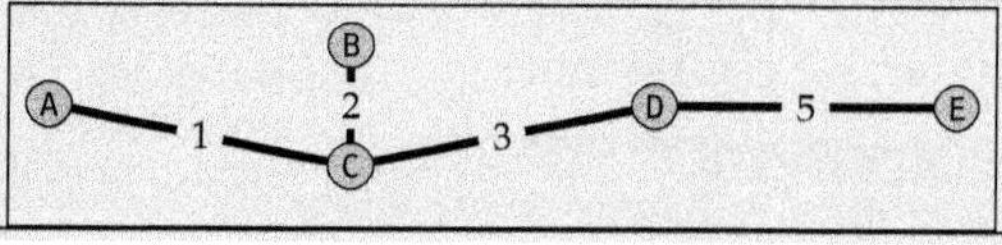

The final spanning tree.

Here is the general statement of correctness for both algorithms:

Theorem 11.12 (Correctness of minimum spanning tree algorithms)
The Weighted Cycle Elimination Algorithm and Kruskal's Algorithm both return a minimum spanning tree for any weighted connected undirected graph.

Proof. The correctness of the Weighted Cycle Elimination Algorithm follows immediately from Lemma 11.11 (the cycle rule) and from Theorem 11.8 (the correctness of the Cycle Elimination Algorithm): the heaviest edge in any cycle does not appear in any MST, and we terminate with a spanning tree when we repeatedly eliminate any edge from an arbitrarily chosen cycle.

For Kruskal's algorithm, consider an edge e that is *not* retained—that is, when e is considered, it is not included in the set S. The only reason that e wasn't included is that adding it would create a cycle C involving e and previously included edges—but because the edges are considered in increasing order of weight, that means that e is the heaviest edge in C. Thus by Lemma 11.11, Kruskal's algorithm removes only edges not contained in any minimum spanning tree. Because it only excludes edges that create cycles, the resulting graph is also connected—and thus a minimum spanning tree. □

Taking it further: There are several other commonly used algorithms for minimum spanning trees, using different structural properties than the Cycle Rule. For much more on these other algorithms, and for the clever data structures that allow Kruskal's Algorithm to be implemented in $O(|E| \log |E|)$ time, see any good textbook on algorithms.

Computer Science Connections

Random Walks and Ranking Web Pages

When Google launched as a web search engine, one of its major innovations over its competition was in how it ranked the pages returned in response to a user's query. Here are two key ideas in Google's ranking system, called *PageRank* (named after Larry Page, one of Google's founders):

- view a link from page u to page v as implicit "endorsement" of v by u.
- not all endorsements are equal: if a page u is endorsed by many other pages, then being endorsed by u is a bigger deal.

These point can be restated more glibly as: *a page is important if it is pointed to by many important pages.* The idea of PageRank is to break this apparent circularity using the *Random Surfer Model.* Imagine a hypothetical web user who starts at a random web page, and, at every time step, clicks on a randomly chosen link from the page she's currently visiting. The more frequently that this hypothetical user visits page u, the more important we'll say u is.

The Random Surfer explores the web using what's called a *random walk* on the web graph. In its simplest form, a random walk on a directed graph $G = \langle V, E \rangle$ visits a sequence $u_0, u_1, u_2, \ldots$ of nodes in G as follows:

1. choose a node $u_0 \in V$, uniformly at random.
2. in step $t = 1, 2, \ldots$, the next node u_t is chosen by picking a node uniformly at random from the out-neighborhood of the previous node u_{t-1}.

(See Figure 11.68(a) for an example.)

As you'll explore in Exercises 11.204–11.208, under mild assumptions about G, there's a special probability distribution p over the nodes of the graph, called the *stationary distribution* of the graph, that has the following property: if we choose an initial node u with probability $p(u)$, and we then take one step of the random walk from u, the resulting probability distribution over the nodes is still p. And, it turns out, we can approximate p by the probability distribution observed simply by running the random walk for many steps, as in Figure 11.68(b). We'll use p as our measure of importance.

We've already made a lot of progress toward the stated goals: $p(u)$ is higher the more in-neighbors u has, but $p(u)$ will be increased even more when the in-neighbors of u have a high probability themselves. In Figure 11.68(c), for example, we see that $p(\texttt{D}) > p(\texttt{B})$ and $p(\texttt{D}) > p(\texttt{C})$, despite B and C having higher in-degree than D.

But there are a few complications that we still have to address to get to the full PageRank model.[13] One is that the Random Surfer has nowhere to go if she ends up at a page u that has no out-neighbors. (The random walk's next step isn't even defined.) In this case, we'll have the Random Surfer jump to a completely random page (each of the $|V|$ nodes is chosen with probability $\frac{1}{|V|}$). Second, this model allows the Random Surfer to get stuck in a "dead end" if there's a group of nodes that has no edges leaving it. Thus—and this change probably makes the Random Surfer more realistic anyway—we'll add a *restart probability* of 15% to every stage of the random walk: with probability 85%, we behave as previously described; with probability 15%, we jump to a randomly chosen node. (See Figure 11.68(d) for the updated probabilities.)

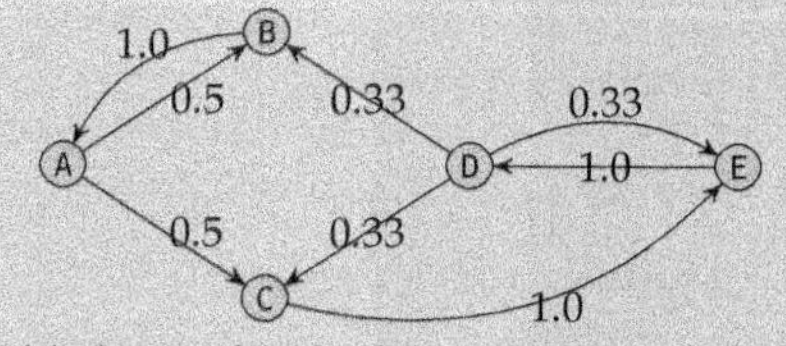

(a) A sample 5-node graph. Edges are annotated with their probabilities in a random walk; we can view the resulting weighted graph as defining the process.

node	steps
A	166,653
B	166,652
C	166,155
D	250,270
E	250,271

(b) The number of steps spent at each node in a computer-generated 1,000,000-step random walk starting at A. This particular walk began ABABABABABABACEDCEDEDBABAC.

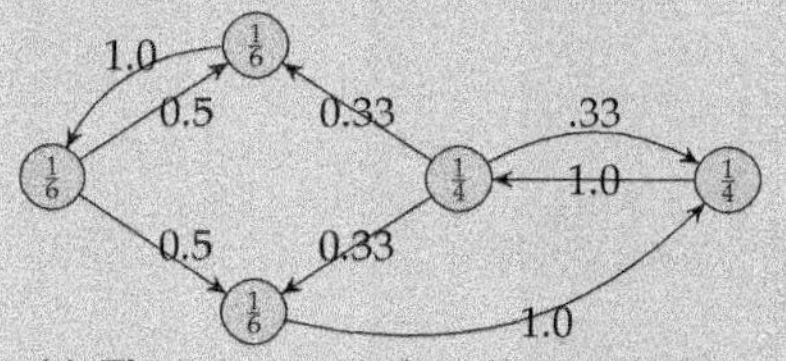

(c) The stationary distribution for G.

	A	B	C	D	E
A	0.03	0.45	0.45	0.03	0.03
B	0.88	0.03	0.03	0.03	0.03
C	0.03	0.03	0.03	0.03	0.88
D	0.03	0.31	0.31	0.03	0.31
E	0.03	0.03	0.03	0.88	0.03

(d) The updated link probabilities, with random restarts.

Figure 11.68: A random walk.

You can find more about the Random Surfer model and PageRank (including interesting questions about how to calculate it on a graph with nodes numbering in the billions) in a good textbook on data mining, like

[13] Jure Leskovec, Anand Rajaraman, and Jeff Ullman. *Mining of Massive Datasets.* Cambridge University Press, 2nd edition, 2014.

There are also many other ingredients in Google's ranking recipe beyond PageRank, though PageRank was an early and important one.

11.5.3 *Exercises*

For the following graphs, find all shortest paths between the given nodes. Give both the path length and the path itself.

11.183 From A to E:

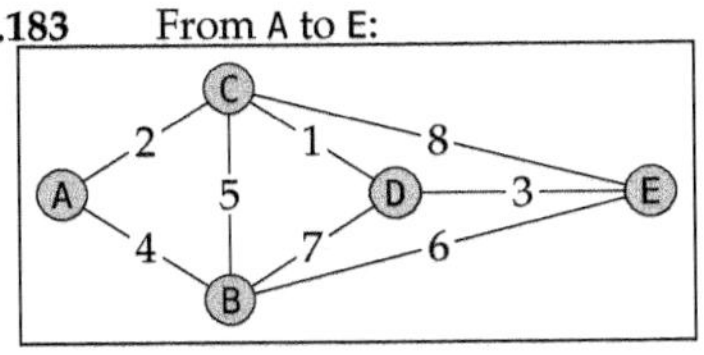

11.184 From A to E:

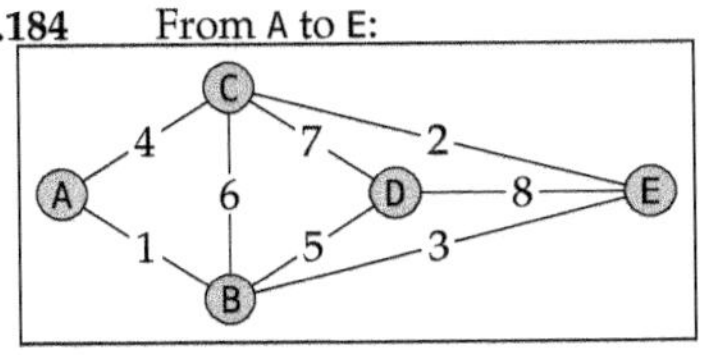

11.185 From A to E:

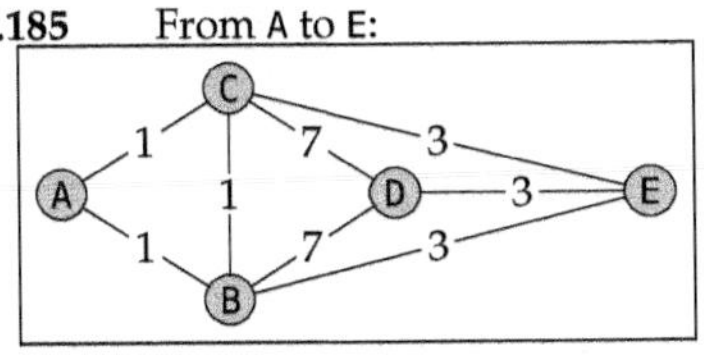

11.186 From A to H:

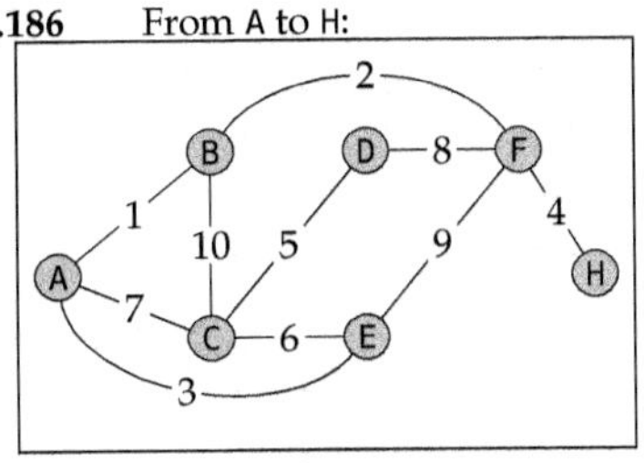

11.187 From A to H:

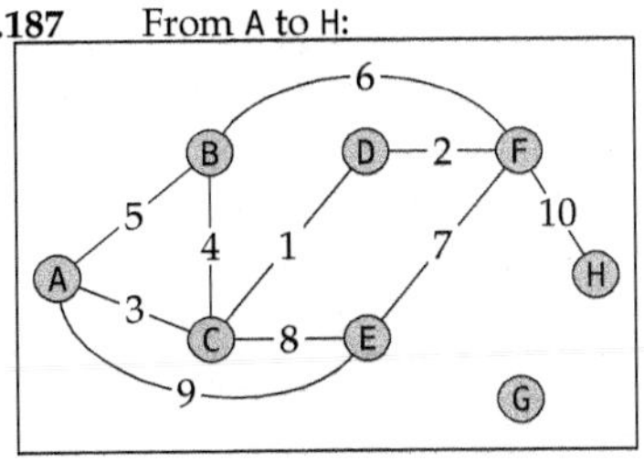

11.188 Let n be arbitrary. Give an example of an n-node weighted graph $G = \langle V, E \rangle$ with designated nodes $s \in V$ and $t \in V$ in which both of the following conditions hold:

(i) all edge weights are distinct (for any $e \in E$ and $e' \in E$, we have $w(e) \neq w(e')$ if $e \neq e'$), and
(ii) for some $\alpha > 1$ and $c > 0$, there are at least $c \cdot \alpha^n$ different shortest paths between s and t.

Suppose that we are running Dijkstra's Algorithm on the graph shown in Figure 11.69 to compute distances from the node A*. So far Dijkstra's Algorithm has computed four distances:*

$d(\text{A}, \text{A}) = 0 \qquad d(\text{A}, \text{B}) = 1 \qquad d(\text{A}, \text{C}) = 3 \qquad d(\text{A}, \text{F}) = 7$

If we continue Dijkstra's algorithm for further iterations, it records the distance for a new node in each iteration.

11.189 What is the next node recorded, and what is its distance?

11.190 What is the next node (after the one from Exercise 11.189) for which Dijkstra's algorithm records a distance, and what is its distance? List all subsequently discovered nodes, and their distances.

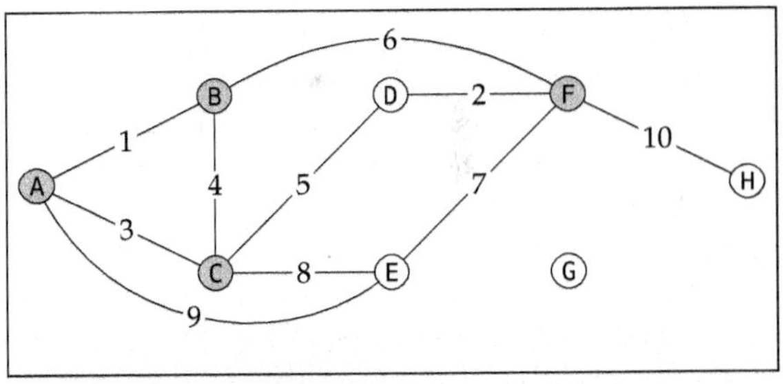

Figure 11.69: A weighted graph.

11.191 Trace Dijkstra's algorithm on the graph shown in Figure 11.69 to compute distances from the node H. List all discovered nodes and their distances, in the order in which they're discovered.

11.192 Identify *exactly* where the proof of correctness for Dijkstra's algorithm (specifically, in the proof of Lemma 11.9) the argument fails if edge weights can be negative. Then give an example of a graph with negative edge weights in which Dijkstra's algorithm fails.

Suppose that $G = \langle V, E \rangle$ is a weighted, directed graph in which nodes represent physical states of a system, and an edge $\langle u, v \rangle$ indicates that one can move from state u to state v. The weight $w_{\langle u,v \rangle}$ of edge $\langle u, v \rangle$ denotes the multiplicative cost of the exchange: one can trade $w_{u,v}$ units of u for 1 unit of v. For example, if there's an edge ⟨A, B⟩ *with weight 1.04, then I can trade 2.08 units of energy in state* A *for 2 units of energy in state* B.

Suppose that we wish to find a shortest multiplicative path (SMP) from a given node s to a given node t in G, where the cost of the path is the product of the edge weights along it. For example, in Figure 11.70, the SMP from A *to* D *is* A → B → C → D *at cost* $1.1 \cdot 1.5 \cdot 1.4 = 2.31$, *which is better than* A → B → D *at cost* $1.1 \cdot 2.5 = 2.75$.

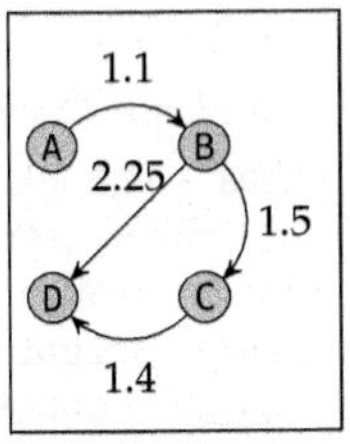

Figure 11.70: A weighted graph.

11.193 Describe how to modify Dijkstra's algorithm to find the shortest SMP in a given weighted graph G. Alternatively, describe how to modify a given weighted graph G into a graph G' so that Dijkstra's algorithm run on G' finds an SMP in G.

11.194 As you argued in Exercise 11.192, Dijkstra's algorithm may fail if edge weights are negative. State the condition that guarantees that your algorithm from Exercise 11.193 properly computes SMPs.

List all minimum spanning trees of the following graphs. (Note that some have edges with nondistinct weights.)

11.195

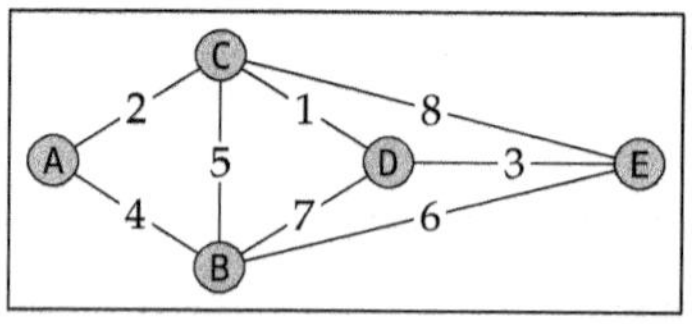

11.196

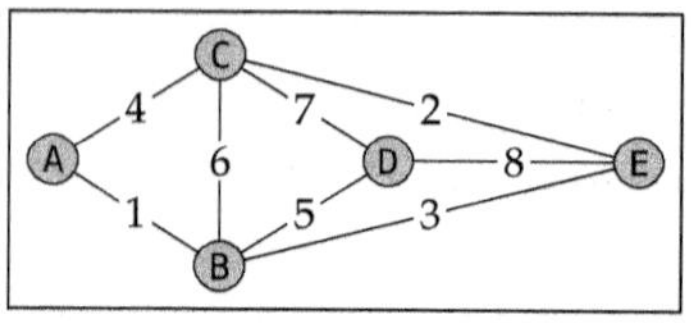

11.197

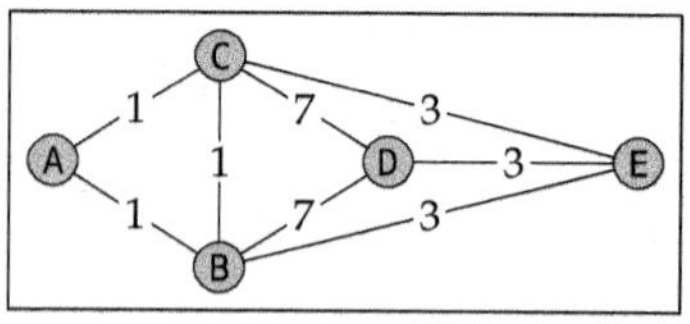

11.198

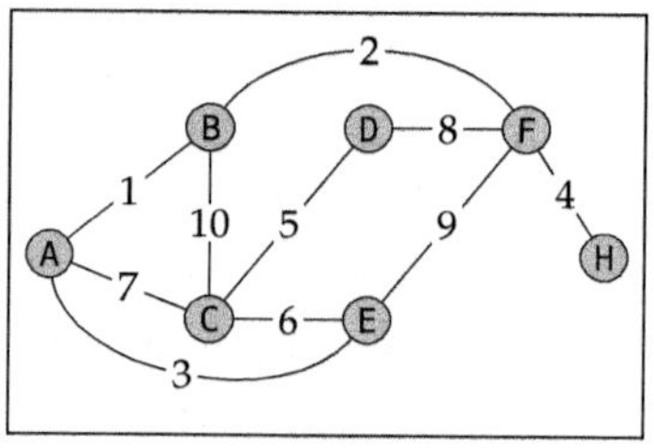

11.199

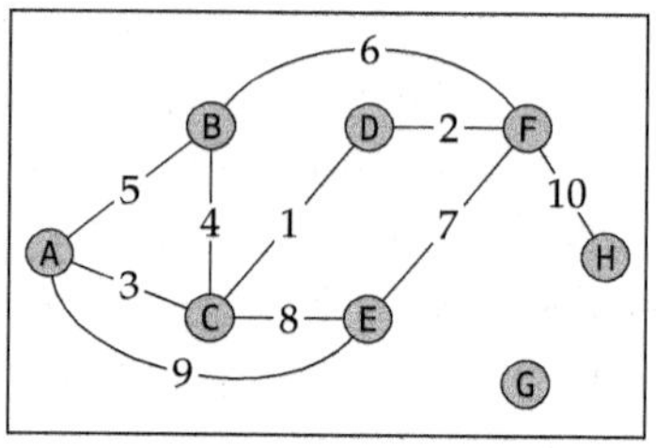

Consider the undirected 9-node complete graph $\mathcal{K}_9$. There are $\binom{9}{2} = \frac{9 \cdot 8}{2} = 36$ unordered pairs of nodes in this graph, so there are 36 different edges in the graph. Suppose that you're asked to assign each of these 36 edges a distinct weight from the set $\{1, 2, \ldots, 36\}$. (You get to choose which edges have which weights.)

11.200 What's the cheapest possible minimum spanning tree of $\mathcal{K}_9$?

11.201 What's the most expensive edge that can appear in a minimum spanning tree of $\mathcal{K}_9$?

11.202 What's the costliest possible minimum spanning tree of $\mathcal{K}_9$?

11.203 Generalize Exercise 11.200 and 11.202: what are the cheapest and most expensive possible MSTs for the graph $\mathcal{K}_n$ if all edges have distinct weights chosen from $\{1, 2, \ldots, \binom{n}{2}\}$? *(Hint: see Exercise 9.173.)*

Recall from p. 1174 that a random walk *in a graph $G = \langle V, E \rangle$ proceeds as follows: we start at a node $u_0 \in V$, and, at every time step, we select as the next node u_{i+1} a uniformly chosen (out-)neighbor of u_i.*

Suppose we choose an initial node u_0 according to a probability distribution p, and we then take one step of the random walk from u_0 to get a new node u_1. The probability distribution p is a stationary distribution *if it satisfies the following condition: for every node $s \in V$, we have that $\Pr[u_0 = s] = \Pr[u_1 = s] = p(s)$. Such a distribution is called "stationary" because, if p is the probability distribution before a step of the walk, then p is still the probability distribution after a step of the walk (and thus the distribution "hasn't moved"—that is, is stationary).*

11.204 Argue that $p(\mathsf{A}) = p(\mathsf{B}) = p(\mathsf{C}) = \frac{1}{3}$ is a stationary distribution for the graph in Figure 11.71(a).

11.205 Argue that the graph in Figure 11.71(b) has at least two distinct stationary distributions.

Suppose that we start a random walk at node A *in the graph in Figure 11.71(a). The following chart shows the probability of being at any particular node after each step of the random walk:*

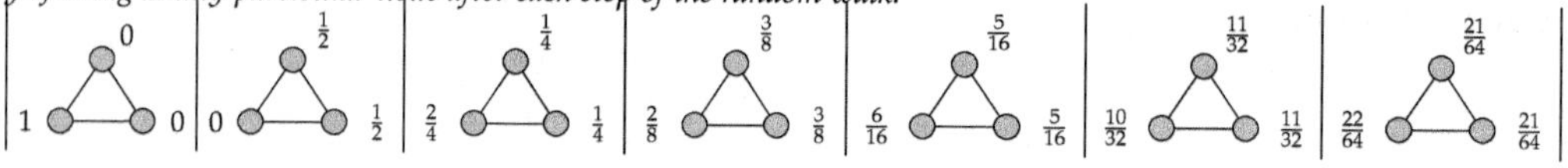

Let $p_k(u)$ denote the probability of the kth step of this random walk being at node u. Although we'll skip the proof, the following theorem turns out to be true of random walks on undirected graphs G:

> *If G is connected and nonbipartite, then a unique stationary distribution p exists for this random walk on G (regardless of which node we choose as the initial node for the walk). Furthermore, the stationary distribution is the limit of the probability distributions p_k of where the random walk is in the kth step.*

11.206 *(programming required)* Write a random-walk simulator: take an undirected graph G as input, and simulate 2000 steps of a random walk starting at an arbitrary node. Repeat 2000 times, and report the fraction of walks that are at each node. What are your results on the graph from Figure 11.71(a)?

11.207 Argue that the above process doesn't converge to a unique stationary distribution in a bipartite graph. (For example, what's p_{1000} if a random walk starts at node J in the graph in Figure 11.71(c)? Node K?)

11.208 Let $G = \langle V, E \rangle$ be an arbitrary connected undirected graph. For any $u \in V$, define

$$p(u) := \frac{\mathit{degree}(u)}{2 \cdot |E|}.$$

Prove that the probability distribution p is a stationary distribution for the random walk on G.

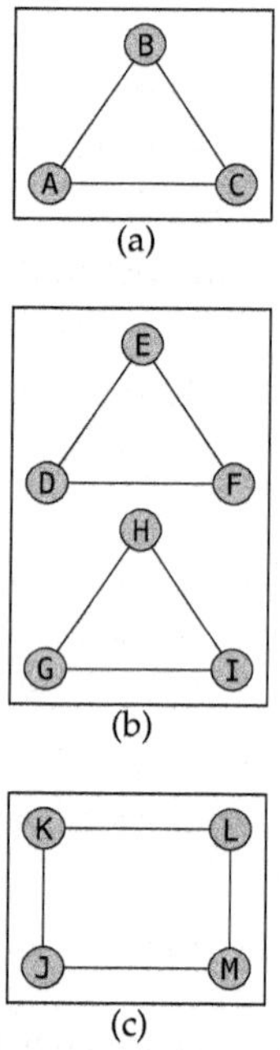

Figure 11.71: Some undirected graphs upon which a random walk can be performed.

11.6 *Chapter at a Glance*

Formal Introduction

A *graph* is a pair $G = \langle V, E \rangle$ where V is a set of *vertices* or *nodes*, and E is a set of *edges*. In a *directed graph*, the edges $E \subseteq V \times V$ are ordered pairs of vertices; in an *undirected graph*, the edges $E \subseteq \{\{u,v\} : u,v \in V\}$ are unordered pairs. A directed edge $\langle u,v \rangle$ goes from u to v; an undirected edge $\langle u,v \rangle$ goes between u and v. We sometimes write $\langle u,v \rangle$ even for an undirected graphs. A *simple graph* has no *parallel edges* joining the same two nodes and also has no *self loops* joining a node to itself.

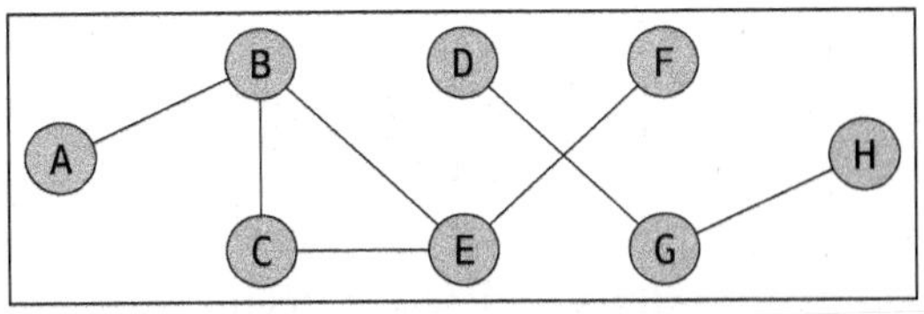

For an edge $e = \langle u,v \rangle$, we say that u and v are *adjacent*; v is a *neighbor* of u; u and v are the *endpoints* of e; and u and v are both *incident* to e. The *neighborhood* of a node u is $\{v : \langle u,v \rangle \in E\}$, its set of neighbors. The *degree* of u is the cardinality of u's neighborhood. In a directed graph, the *in-neighbors* of u are the nodes that have an edge pointing to u; the *out-neighbors* are the nodes to which u has an edge pointing; and the *in-degree* and *out-degree* of u are the number of in- and out-neighbors, respectively.

An *adjacency list* stores a graph using an array with $|V|$ entries; the slot for node u is a linked list of u's neighbors. An *adjacency matrix* stores the graph using a two-dimensional Boolean array of size $|V| \times |V|$; the value in $\langle \text{row } u, \text{column } v \rangle$ indicates whether the edge $\langle u,v \rangle$ exists.

Two graphs are *isomorphic* if they are identical except for the naming of the nodes. A *subgraph* of G contains a subset V' of G's nodes and a subset E' of G's edges joining elements of V'. An *induced subgraph* is a subgraph in which every edge that joins elements of V' is included in E'. A *complete graph* or *clique* is a graph $\mathcal{K}_n$ in which every possible edge exists. A *bipartite graph* is one in which nodes can be partitioned into sets L and R such that every edge joins a node in L to a node in R. A *regular graph* is one in which every node has identical degree. A *planar graph* is one that can be drawn on paper without any edges crossing.

Paths, Connectivity, and Distances

A *path* is a sequence of $k \geq 1$ nodes $\langle v_1, v_2, \ldots, v_k \rangle$, where $\langle v_{i-1}, v_i \rangle \in E$ for every index $i \in \{1, 2, \ldots, k-1\}$. The path is *simple* if all the v_is are distinct. This path has *length* $k-1$—the number of edges that it traverses—and is a *path from v_1 to v_k*.

In an undirected graph, nodes u and v are *connected* if there exists a path from u to v. A *connected component* of $G = \langle V, E \rangle$ is a set $S \subseteq V$ such that (i) every $u \in S$ and $v \in S$ are connected; and (ii) for every $w \notin S$, the set $S \cup \{w\}$ does not satisfy condition (i). The entire graph is *connected* if it has only one connected component, namely V.

In a directed graph, node u is *reachable from node v* if there exists a path from v to u; u and v are *strongly connected* if each is reachable from the other. A *strongly connected component* is a set S of nodes such that any two nodes in S are strongly connected and no node $x \notin S$ is strongly connected to any node $s \in S$.

Connectivity can be tested in time $\Theta(|V| + |E|)$ time using *breadth-first search (BFS;* see Figure 11.72) or *depth-first search (DFS).* The *distance* from node s to node t is the length of a shortest path from s to t. BFS can also be used to compute distances.

Breadth-First Search (BFS):
Input: a graph $G = \langle V, E \rangle$ and a source node $s \in V$
Output: the set of nodes reachable from s in G

```
Frontier := ⟨s⟩
        // Frontier will be a list of nodes to process, in order.
Known := ∅
        // Known will be the set of already-processed nodes.
while Frontier is nonempty:
    u := the first node in Frontier
    remove u from Frontier
    for every neighbor v of u:
        if v is in neither Frontier nor Known then
            add v to the end of Frontier
    add u to Known
return Known
```

Figure 11.72: Breadth-first search.

Trees

A *cycle* $\langle v_1, v_2, \ldots, v_k, v_1 \rangle$ is a path of length ≥ 2 from a node v_1 back to itself that does not traverse the same edge twice. The *length* of the cycle is k. The cycle is *simple* if each v_i is distinct. Cycles can be identified using BFS.

A graph is *acyclic* if it contains no cycles. Every acyclic graph has a node of degree 0 or 1. A *tree* is a connected, acyclic graph. (A *forest* is any acyclic graph.) A tree has one more node than it has vertex. A tree becomes disconnected if any edge is deleted; it becomes cyclic if any edge is added.

One node in a tree can be designated as the *root*. Every node other than the root has a *parent* (its neighbor that's closer to the root). If p is v's parent, then v is one of p's *children*. Two nodes with the same parent are *siblings*. A *leaf* is a node with no children; an *internal node* is a node with children. The *depth* of a node is its distance from the root; the *height* of the entire tree is the depth of deepest node. The *descendants* of u are those nodes that go through u to get the root; the *ancestors* are those nodes through which u's path to the root goes. The *subtree* rooted at u is the induced subgraph consisting of u and all descendants of u.

All nodes in *binary trees* have at most two children, called *left* and *right*. A *traversal* of a binary tree visits every node of the tree. An *in-order* traversal recursively traverses the root's left subtree, visits the root, and recursively traverses the root's right subtree. A *pre-order* traversal visits the root and recursively traverses the root's left and right subtrees; a *post-order* traversal recursively traverses the root's left and right subtrees and then visits the root.

A *spanning tree* of a connected graph $G = \langle V, E \rangle$ is a graph $T = \langle V, E' \subseteq E \rangle$ that's a tree. A spanning tree can by found by repeatedly identifying a cycle in G and deleting any edge in that cycle.

Weighted Graphs

In a *weighted graph*, each edge e has a weight $w_e \in \mathbb{R}^{\geq 0}$. (Although graphs with negative edge weights are possible, we haven't addressed them in any detail.) The length of a path in a weighted graph is the sum of the weights of the edges that it traverses. Shortest paths in weighted graphs can be found with Dijkstra's Algorithm (Figure 11.65), which expands a set of nodes of known distance one by one. Minimum spanning trees—spanning trees of the smallest possible total weight—in weighted graphs can be found with Kruskal's Algorithm (Figure 11.67) or by repeatedly identifying a cycle in G and deleting the heaviest edge in that cycle.

Key Terms and Results

Key Terms

Formal Introduction

- undirected and directed graphs
- nodes/vertices, edges
- parallel edges, self loops
- simple graphs
- adjacent node, incident edge
- (in/out-)neighbors, neighborhood
- (in/out-)degree
- adjacency list, adjacency matrix
- isomorphic graphs
- subgraphs
- complete, bipartite, regular, planar graphs

Paths and Connectivity

- path
- connected (nodes), connected (graph)
- connected component
- reachability
- strongly connected component
- shortest path/distance
- breadth-first search (BFS)
- depth-first search (DFS)

Trees

- cycle
- tree, forest
- root, leaf, internal node, child, parent, sibling, ancestor, descendant, depth, height, subtree
- spanning tree

Weighted graphs

- Dijkstra's algorithm
- minimum spanning trees
- Kruskal's algorithm

Key Results

Formal Introduction

1. The "handshaking lemma": for any undirected graph $G = \langle V, E \rangle$, we have $\sum_{u \in V} degree(u) = 2|E|$.
2. Representing G with an adjacency matrix requires $\Theta(|V|^2)$ space; we can answer "what are all of u's neighbors?" in $\Theta(|V|)$ time and "is there an edge between u and v?" in $\Theta(1)$ time. Representing $G = \langle V, E \rangle$ with an adjacency list requires $\Theta(|V| + |E|)$ space; both questions take $1 + \Theta(degree(u))$ time.

Paths, Connectivity, and Distances

1. Connectivity can be tested using *breadth-first search (BFS)* (Figure 11.29) or *depth-first search (DFS)* (Figure 11.31). BFS can also be used to compute the distance between nodes in a graph, and it runs in $\Theta(|V| + |E|)$ time.

Trees

1. Any tree with n nodes has exactly $n - 1$ edges. Adding any edge to a tree creates a cycle; deleting any edge disconnects the graph.
2. A spanning tree of a graph G can by found by repeatedly identifying a cycle in G and deleting an arbitrary edge in that cycle.

Weighted Graphs

1. Shortest paths in weighted graphs can be found with Dijkstra's Algorithm (Figure 11.65) if all edges have nonnegative weights.
2. Minimum spanning trees in weighted graphs can be found with Kruskal's Algorithm (Figure 11.67) or by repeatedly identifying a cycle in G and deleting the heaviest edge in that cycle.

12
Index

Printed in the USA
K084397SCI032718 01S29053000000003008